W9-ASY-379

Essentials of
BIOLOGY

Essentials of

BIOLOGY

Leland G. Johnson
Augustana College
Rebecca L. Johnson

Wm. C. Brown Publishers
Dubuque, Iowa

Book Team

John Stout
Executive Editor

Mary J. Porter
Assistant Editor

Teresa E. Webb
Designer

Julie Avery Kennedy
Senior Production Editor

Shirley M. Charley
Photo Research Editor

Mavis M. Oeth
Permissions Manager

wcb

Wm. C. Brown Publishers, College Division

Lawrence E. Cremer
President

James L. Romig
Vice-President, Product Development

David A. Corona
Vice-President, Production and Design

E. F. Jogerst
Vice-President, Cost Analyst

Bob McLaughlin
National Sales Manager

Marcia H. Stout
Marketing Manager

Craig S. Marty
Director of Marketing Research

Marilyn A. Phelps
Manager of Design

Eugenia M. Collins
Production Editorial Manager

Mary M. Heller
Photo Research Manager

wcb group

Wm. C. Brown
Chairman of the Board

Mark C. Falb
President and Chief Executive Officer

Cover photograph © John F. O'Connor, M.D./ PHOTO/NATS

Library of Congress Catalog Card Number: 84-073394

ISBN 0-697-00149-0

Printed in the United States of America
10 9 8 7 6 5 4 3 2 1

Cover A snowy egret. This marsh-dwelling bird, with its black legs and yellow feet, is often called the egret with "golden slippers." Snowy egrets were once hunted for their long silky plumes that were prized as ornaments for women's hats.

To Eleanor and Leon Johnson, for all of their encouragement and support

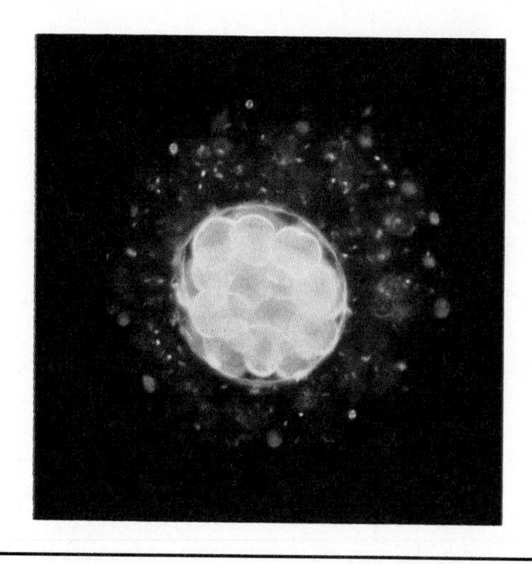

Brief Contents

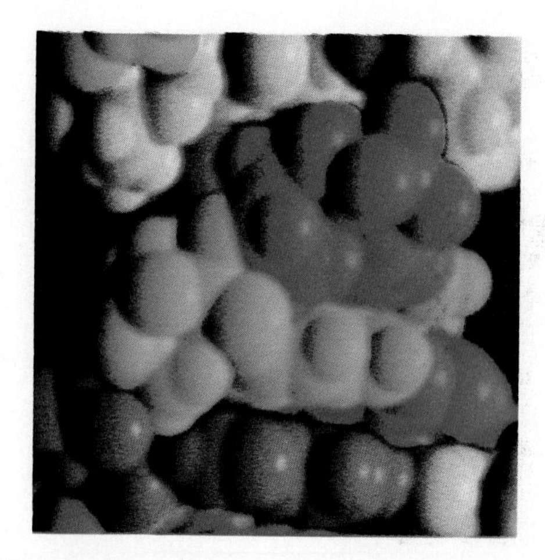

Expanded Contents

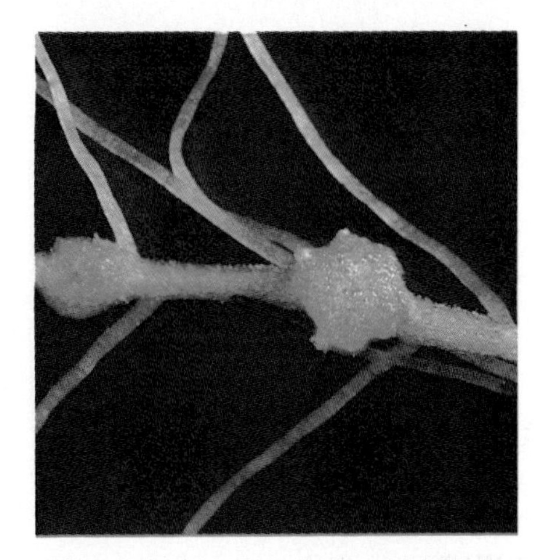

Chapter **3**

Cells and Cellular Functions 55

Chapter **4**

Energy, Reactions, and Life 81

Chapter **5**

Photosynthesis 93

Chapter **6**

Cellular Metabolism 111

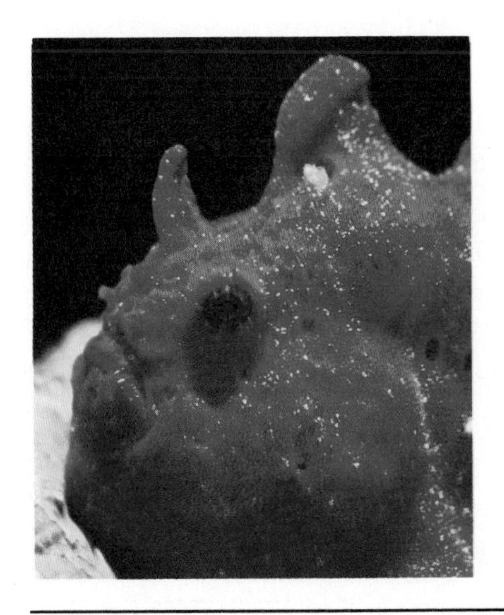

Part **3**
Regulation in Organisms 231

Chapter **11**
Animal Hormones 233

Chapter **12**
Nerve Cells and Nervous Systems 255

Chapter **13**
Reception and Response 277

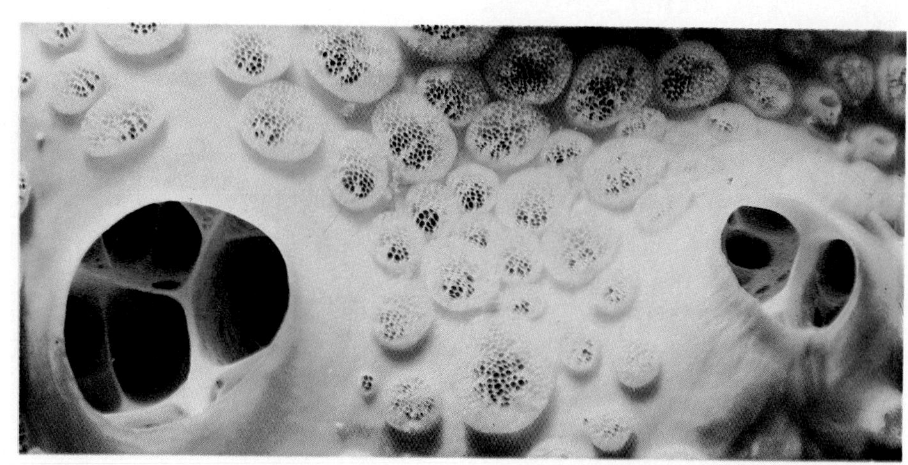

Chapter 14
Regulation in Plants 297

Part 4
The Continuity of Life 319

Chapter 15
Cell Division 321

Chapter 16
Patterns of Inheritance 337

Chapter **17**

Chromosomes and Genes 353

Chapter **18**

Molecular Aspects of Genetics 371

Chapter **19**

Reproduction and Development 395

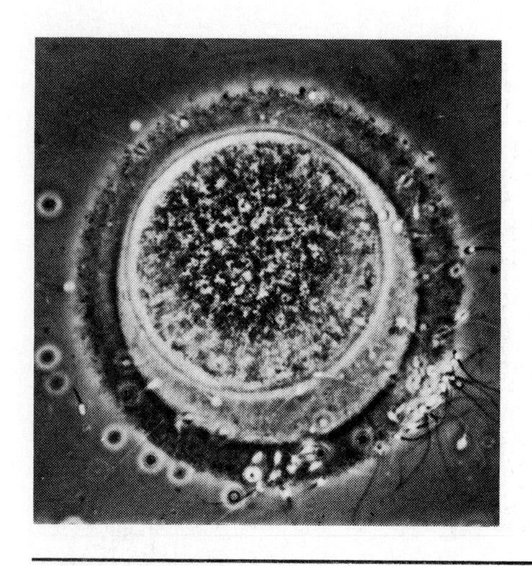

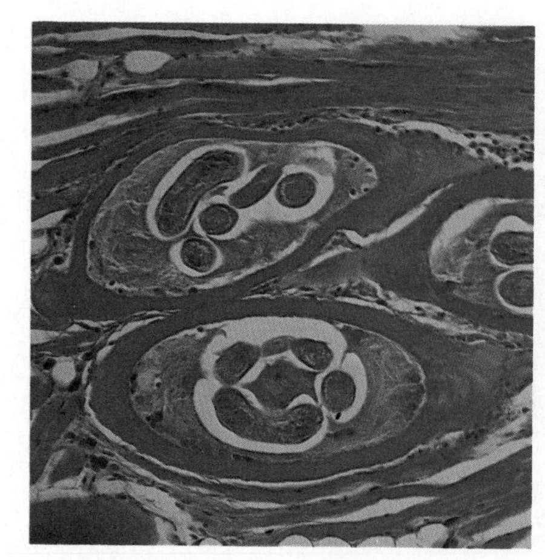

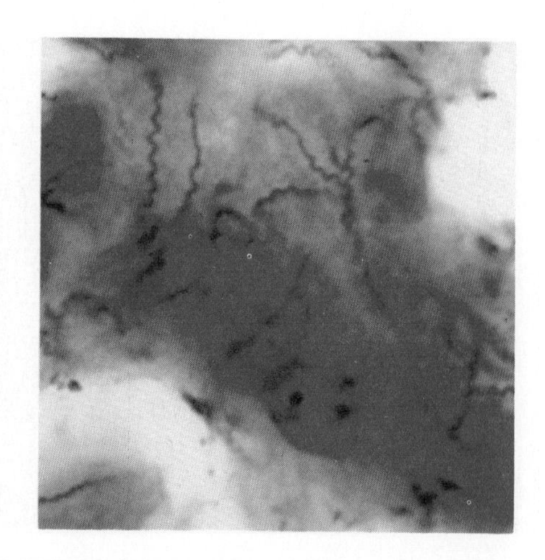

Chapter **20**
Human Reproduction 423

Chapter **21**
Lifelong Developmental Change 453

Part **5**

Evolution, Organisms, and the Environment 477

Chapter **25**
Ecology: Organisms and Environment 543

Chapter **26**
Ecology: Ecological Systems and Environmental Problems 561

Part **6**
The Diversity of Life 585

Chapter **27**
Viruses, Monera, and Protists 587

Chapter **28**
Fungi and Plants 613

Chapter **29**
Animals 637

Preface

There are many levels of understanding of the living world. For most people, a certain general understanding arises from casual and chance observation of living things around them. A more satisfying and deeper understanding of life comes from the study of biology. The goal of this book is to help students effectively plan and organize their observation, analysis and search for understanding and appreciation of the biological world, and to provide all students with an introduction to biology that is interesting, readable and accessible.

Essentials of Biology is a concise treatment of biology. It is derived from the larger text, *Biology* by Leland G. Johnson, and it is intended for use in any course where *Biology* may be viewed as more rigorous or more detailed than is required for the course.

Organization and Coverage

Essentials of Biology develops many of the same themes presented in *Biology,* but in order to better meet the needs of students in shorter courses, we have organized several sets of topics in ways that are significantly different from the organization of *Biology.* For example, fundamentals of chemistry and biologically important organic molecules are examined in a single chapter (chapter 2). These concepts are presented early in the text as the foundation for analyzing more complex organismic functions and processes. After these fundamental processes have been presented, the broader topics of development, evolution and ecology are discussed.

Treatment of the organization of cell membranes is included in our chapter on cell structure and function (chapter 3), where it provides important background for discussion of the plasma membrane and various membranous organelles.

Because of the close relationships between gas exchange and transport mechanisms, we combined topics of gas exchange and circulation in chapter 9.

Chapter 10, Blood and Body Fluid Homeostasis, examines homeostatic mechanisms, such as kidney function, that contribute to the maintenance of organismic integrity.

Principles and problems of cell differentiation are presented within the context of our descriptions of plant and animal reproduction and development.

In the ecology section of *Essentials of Biology,* temperature effects and problems associated with water balance are incorporated into a chapter on organisms and the environment (chapter 25). This focus on interactions between individual organisms and the environment sets the stage for the population, community and ecosystem analysis that follows.

As in *Biology,* we have placed the survey of organisms at the end of the book for purposes of flexibility. We feel that instructors should be free to insert the survey wherever needed. We also designed all parts and chapters to be sufficiently autonomous to make the text functional in virtually any course organization.

Finally, significant advances in knowledge, and changes in understanding have developed since *Biology* was written. *Essentials of Biology* incorporates many new, biological topics of current interest. For example, Acquired Immune Deficiency Syndrome (AIDS) is discussed in the context of immune system problems, and prions and viroids are considered along with viruses.

Principle Themes

Essentials of Biology examines all the basic principles and concepts of biology addressed in introductory biology courses. However, several key themes receive repeated and continuing attention. In developing these major themes we emphasize the organismic aspects of biology. Whether an organism is unicellular, a loosely integrated multicellular form, or a complex, highly organized plant or animal, it is a living, functioning unit that must maintain the integrity of its internal environment and interact successfully with its external environment.

One key theme is evolution. All organisms are evolutionary units of populations in that they are subjected to natural selection. Present-day organisms are the products of evolutionary processes, and some of them reproduce to become parents of future generations. This evolutionary theme puts the life of the individual organism into perspective in relation to the continuing life of the species.

A second key theme is developed at various points where the book focuses directly on the human organism. These points include physiology of organ systems, nervous and hormonal regulation, reproduction and development, and evolution. However, we use a comparative approach and human biology is always treated as part of biology in general because we humans are clearly part of the world of life and have many problems and processes in common with other organisms.

Preface *Continued*

A third key theme is homeostasis, the state of internal stability that is vital to life. All organisms are treated with a view toward understanding the mechanisms by which they maintain homeostasis. Every living thing must maintain the integrity of its internal environment through integration of many internal functions and successful regulation of interactions and exchanges with the external environment.

A fourth key theme is diversity. While there is underlying evolutionary and cellular unity, living things are fantastically diverse. We have sampled the fascinating variety of organisms and have examined and compared many of the different individual adaptations that make possible successful interactions with widely different kinds of environments.

A fifth key theme is scientific inquiry. Most beginning textbooks discuss the scientific method. Unfortunately, these presentations often are philosophical discussions not suited to the experience of students or are merely a list of steps in *the* scientific method. In reality, biologists work in a variety of ways. We have chosen to introduce the scientific method in chapter 1. Then at key points in the text, we present steps in the discovery of certain biological phenomena that illustrate the nature of scientific progress and the application of the scientific method in biology. Some of these points are the determination of the structure of DNA (chapter 2), plant nutrition and transport mechanisms (chapter 7), animal hormones (chapter 11), regulation in plants (chapter 14), molecular aspects of genetic expression (chapter 18), genetic expression in development (chapters 18 and 19), discovery of categories of immune responses (chapter 21), discovery and interpretation of human fossil ancestors (chapter 23), and analysis of behavioral responses (chapter 24).

Approach

Essentials of Biology reflects a thorough and complete integration of plant and animal biology. Fundamental biological unity is recognized where it is scientifically and educationally sound to do so. Thus, we consider plants and animals together when discussing basic cell biology, genetics, reproduction, development, evolution, and ecology.

But plants and animals also differ in basic ways. For example, while water and mineral ion absorption by roots of vascular plants and the digestive and absorptive processes by which animals obtain nutrients are processes that might be broadly categorized as nutrient procurement, they both are worthy of careful, separate examination. Likewise, plant transport processes and animal circulation are fundamentally different sets of processes. Separate consideration also is given to plant hormonal regulation, which is primarily involved in growth control, and animal hormonal regulation, which is involved much more in short-term modulation of physiological processes.

We have taken a special organizational approach in discussing chemical reaction mechanisms, photosynthesis, and cell respiration. Chapter 4 deals with energy relationships and chemical reaction mechanisms, and it includes information relevant to both chapter 5 (photosynthesis) and chapter 6 (cellular energy conversions). But chapters 5 and 6 are specifically designed to be interchangeable in sequence, so that instructors are free to treat either photosynthesis or cell respiration first.

We also took a special approach to one of the most rapidly expanding fields in modern biology: the study of resistance and immune mechanisms. Because expressions of immune responses involve developmental events that occur throughout the lives of organisms, these topics have been included in a chapter on lifelong development (chapter 21). Chapter 21 also deals with wound healing and regeneration, cancer and other abnormal growths, and aging.

In chapter 2 we discuss DNA structure, including the history of its discovery, as part of the general discussion of biologically important molecules. This strategy is logical from a biochemical viewpoint and avoids a prolonged digression into DNA structure that might interrupt the discussion of genetics, where only those aspects of DNA structure that are most directly relevant to genetic mechanisms are discussed.

Pedagogy

Student learning aids, outlined in detail in Aids to the Reader, include the text introduction, part introductions, chapter concepts, chapter outlines, chapter introductions, key terms, boxed essays, chapter summaries, chapter questions, suggested readings, the glossary, appendixes, and the index. The use of technical terminology has been limited only to essential terms in the discussions and development of biological principles. The International System of Units (SI units) has been used for all physical measurements, except for heat energy, where we have used the more familiar calories that have a strong reference framework in everyday life. All learning aids are designed to facilitate the understanding and retention of biology for the student, and to make the learning process an enjoyable and interesting one.

To the Student

We are pleased that you are using *Essentials of Biology* and we think that this book can play an important part in your study of biology. A number of aids have been planned and prepared for you and we invite you to read about them in "Aids to the Reader."

Biology is a very dynamic science. Reports of exciting discoveries in biology, medicine, agriculture, and related areas are found not only in scientific journals, but appear almost daily in newspapers and magazines. We anticipate that your biology course will prepare you to better understand and evaluate these discoveries. Furthermore, we feel that you will find *Essentials of Biology* a useful reference book and source of background information that will aid you in interpreting the biological advances of the coming years.

Aids to the Reader

Text Introduction
Chapter 1 sets the tone of the book and introduces the text's themes, goals, and principal ideas. This overview gives students an idea of the scope of the text and of the field of biology.

Part Introductions
Part introductions are brief overviews that provide continuing perspective for students and preview each part in terms of organization and purpose. They also help to relate each individual part to the other parts of the text.

Chapter Outlines
Chapter outlines for each chapter provide students with an organizational guide useful for reading, reviewing, and note-taking.

Chapter Concepts
Each chapter of this book begins with a set of fundamental concepts that are stressed throughout the chapter. Students can refer to these concepts before reading the chapter and return to them for review.

Chapter Introductions
We have introduced each chapter with a familiar or interest-catching reference that is related to concepts to be developed in the chapter. Chapter introductions preview the chapter contents and reinforce the integration of the textual material.

Key Terms
Key terms appear in bold type throughout each chapter. The terms will help students to recall the essential elements of each basic concept.

Boxed Essays
Biology is an exciting subject. In each of forty-one boxed essays in this text, we have told an intriguing story or considered a recent discovery so that students can take a stimulating, closer look at these processes or concepts. We have not used the essays, as some texts do, to segregate "difficult or advanced material" to be assigned only to selected students or omitted entirely. Throughout preparation of this book, our goal was to provide clear and straightforward explanations of all concepts that are meaningful to all student readers in the book's intended audience. We used the same approach in preparation of the boxed essays.

Chapter Summaries
Summaries appear at the end of each chapter. These reinforce concepts, terms, and interrelationships for students.

Chapter Questions
Questions at the end of each chapter review major concepts and suggest avenues for thought and further inquiry on the part of students.

Suggested Readings
The readings suggested at the end of each chapter amplify concepts, processes, or discoveries covered in the chapter. We have cited readable secondary sources that we think will provide students with further information and a realistic route to primary scientific literature.

Appendices
The appendices provide valuable reference information. Appendix 1 is a classification of organisms. Appendix 2 provides useful prefixes, and temperature and metric unit conversion tables.

Glossary
The glossary provides concise definitions of the more important recurring scientific terms used in the text. It is a valuable tool for refreshing and reinforcing key information.

Illustrations, Photographs, and Tables
The illustration and photograph program in this text is extensive and thorough. Organisms, processes, and interrelationships are clearly represented in the visual program both in artwork and in photos. Tables appear throughout the text to organize information and to provide background data for many important discussions.

Index
The index is complete, accurate and comprehensive.

Supplementary Materials

Essentials of Biology is accompanied by a comprehensive set of supplementary materials that are designed to work with the text to facilitate learning by students and to aid instructors in presenting essential concepts of biology.

Instructor's Manual

The Instructor's Manual, written by Douglas Fratianne, Ohio State University, has been carefully coordinated with the text. Each chapter includes a chapter outline, a list of learning objectives, and a list of visual aids. Approximately fifty objective questions per chapter are designed to test chapter concepts and students' mastery of the material. All objective questions are keyed to the learning objectives in the student study guide. (These are the same objectives listed in the Instructor's Manual.) Each question also has a mate test question, providing the instructor with two different sets of test questions. This feature will save the instructor time in test preparation, provide a wider variety of test questions, and help to maintain test security. True/false, multiple choice, and essay questions are provided. A helpful appendix is also included, containing the mailing addresses of the film sources given in the visual aids list.

Student Study Guide

The Student Study Guide, also written by Douglas Fratianne, provides carefully planned exercises to accompany each chapter of the text. The exercises are designed to motivate and reinforce student learning. Each study guide chapter includes a chapter outline, learning objectives, and a variety of exercises designed to help students master key terms and names, important chapter topics and concepts. Answers follow each section to provide immediate feedback to the student and to encourage independent student learning.

Laboratory Manual

The Laboratory Manual, written by Warren Dolphin and Nina Pearlmutter, Iowa State University, provides twenty-one laboratory exercises designed to allow students to explore basic scientific principles, while teaching them the rudiments of experimentation, data analysis, and scientific method. Each exercise contains the following: purposes, introduction, directions, in-text questions, list of supplies, and lab summary reports. All exercises are written in a narrative style to allow students to work independently.

Additional Materials

A separate set of forty-eight helpful acetate transparencies is also available.

wcb TestPak rounds out the supplementary materials. It is a computerized testing service for adopters of this text. It provides either a call-in/mail-in testing service or, if you have access to an Apple® II or IIe, or to an IBM® PC, you can receive the complete test item file on a microcomputer diskette.

Acknowledgments

We would like to thank all the people who have helped to make *Essentials of Biology* a reality. In particular, we wish to thank our executive editor, John Stout, and our very helpful and conscientious production editor, Julie Kennedy. We are grateful to our designer, Teresa Webb, for her creative work and careful attention to detail. Shirley Charley's diligent efforts in photo research added much to the visual appeal of this book.

Although *Essentials of Biology* is a distinctive book written to its own plan, it is related to the larger text, *Biology* by Leland G. Johnson. Thus, we would like to acknowledge again those who contributed to *Biology* in one way or another.

We are grateful to colleagues at Augustana College for their general interest and support, but we want to offer a very special thanks to Gilbert Blankespoor. He helped us to organize the ecology chapters, keeping in mind the needs of the students for whom *Essentials of Biology* is intended. He also provided much appreciated moral support at several points during the project.

Finally, we wish to thank our excellent reviewers. They provided help, advice, and constructive criticism throughout this project. Their individual contributions varied, but all were important for the development of the book. The reviewers were:

Lester Bazinet
Community College of Philadelphia

John D. Cunningham
Keene State College

David A. Francko
Oklahoma State University

Sally K. Frost
University of Kansas

D. Marvin Glick
Northern Virginia Community College

H. Bernard Hartman
Texas Tech University

Robert J. Huskey
University of Virginia

William H. Leonard
Louisiana State University

Harris J. Linder
University of Maryland

Janie C. Park
Florida Institute of Technology

Wayne C. Rosing
Middle Tennessee State University

John L. Zimmerman
Kansas State University

Essentials of

BIOLOGY

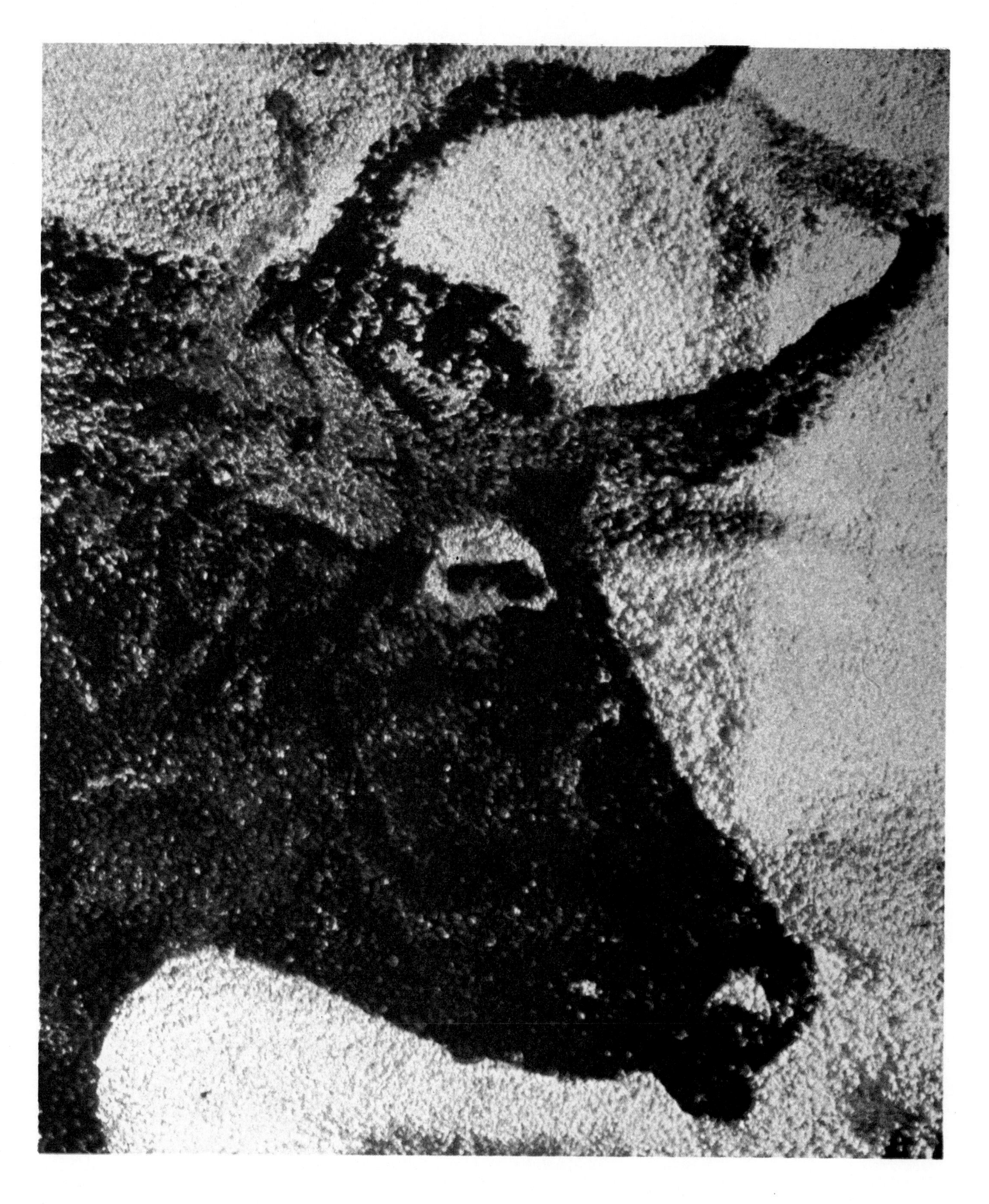

1

Studying Biology

Chapter Outline

Chapter Concepts

1. Biology is the study of life. Biology has been part of human culture since its beginnings.
2. All living things are products of change over the course of time. We call this change evolution.
3. Both Charles Darwin and Alfred Russel Wallace proposed the idea of natural selection, which provides an explanation for the evolutionary process.
4. Biology includes many specialized areas of study.
5. Although living things are very diverse, there is a fundamental unity that underlies all life on earth.
6. We can divide all living things into five kingdoms. Organizing and classifying organisms in this way is called taxonomy.
7. Biologists use a variety of approaches and methods to study life.

Facing page A cave painting of the head and chest of a bull from the "picture gallery" in a cave at Lascaux, France. Such cave paintings were produced 10 to 30 thousand years ago by Stone Age artists, possibly as part of magic rituals intended to ensure hunting success. Though that period is a relatively recent part of the total span of human existence, these cave paintings are among our earliest records of human impressions of other living things.

We human beings have always been keenly aware of the living world around us. From the dawn of the human race, knowledge of other living things has been vital in the search for food, clothing, and shelter, and in avoiding danger. As a result, shared knowledge of the biological world became part of early human culture.

Our prehistoric ancestors left records of their impressions of other living things in the form of cave paintings, rock carvings, and stone sculpture. Of those that survive, the Ice Age cave paintings are some of the most strikingly detailed and artistically beautiful (see page 2). Such early observations of other living things could be considered the beginning of **biology,** the study of life.

As human culture developed, interest in the living world moved beyond the simple observation of nature. Practical biological knowledge grew and was applied in the domestication and breeding of plants and animals (figure 1.1). Thus, in one way or another, all early human interest in biology was associated with meeting everyday, practical human needs, such as acquiring food and shelter.

As the centuries passed, a different kind of biological observation and study emerged in some ancient civilizations. The aim of this form of biology was to meet a less practical, but nevertheless very important need: the need to satisfy human curiosity. What are the differences between plants and animals? How do fish breathe under water? If a lizard loses its tail, how does it manage to grow a new one? The world in which we live was, and still is, filled with mysteries and unanswered questions. And it was because of the human urge to answer questions, the need to know, that biological knowledge began to grow.

Early Biological Studies

Some of the most impressive early studies in biology were the work of the Greek philosopher Aristotle (384–322 B.C.). Aristotle made many thorough observations of individual living things, especially marine animals, and interpreted his observations with remarkably keen insight.

In the years following Aristotle's time, many curious-minded people added to the growing body of knowledge of living things. Although the majority of these studies were simply direct biological observations, a tradition of experimental biology also developed. For example, during the seventeenth century William Harvey did brilliant research on the circulation of blood. Harvey showed by careful observation and simple experimentation that blood does not pass back and forth through the same vessels, but rather passes through vessels in only one direction as it circulates throughout the body (figure 1.2).

Figure 1.1 Ancient Egyptian record showing how the development of agriculture changed relationships between humans and other living things. Humans were no longer dependent on hunting and gathering but could systematically apply practical biological knowledge to help supply food and other needed materials.

A Change in World View: Darwin and Wallace

Have all the varieties of organisms always existed as they are, or are they products of change over the course of time? This deceptively simple question has intrigued biologists and other thoughtful people at least as far back as the time of Aristotle, but thinking on this subject took a decisive turn in the second half of the nineteenth century.

Early Evolutionary Proposals

Before the mid 1880s, some biologists had suggested that living things are subject to gradual change, or **evolution.** The French biologist Jean Baptiste Lamarck (1744–1829) proposed the interesting but erroneous concept of evolution by inheritance of acquired characteristics. Lamarck suggested that gradual change took place because organisms were able to pass on to their offspring certain traits that they themselves had acquired during their lifetimes. For example, giraffes were thought to have long necks because generation after generation of giraffes have stretched their necks to reach for food near the tops of trees, causing members of each subsequent generation to inherit longer necks. We now know that such acquired characteristics are not inherited by offspring. But Lamarck's basic idea that *living things do change over the course of time* was correct.

Figure 1.2 William Harvey demonstrating experiments that he used to prove that human blood circulates and does not flow back and forth through vessels.

However, most people continued to believe that life on earth had remained unchanged since a specific moment of creation. Many nineteenth-century biologists found it difficult to accept the notion of evolutionary change, even when faced with the accumulating fossil record. Part of this difficulty can be traced to their lack of understanding of how old the earth really is and to the absence of a reasonable explanation of the evolutionary process. Furthermore, for some biologists and especially for most nonbiologists, there was an entirely different kind of barrier to the acceptance of evolutionary change: they were comfortable with the idea that the natural world was constant and unchanging. This view lent stability to their understanding of nature, their ideas about various social relationships, and to some of their religious beliefs.

But a change was coming that would cause a revolution not only in biology, but also in much of human thought.

Background for Discovery

The two nineteenth-century naturalists who were responsible for bringing about this revolution in thought were Charles Darwin (1809–1892) and Alfred Russel Wallace (1823–1913). Each independently proposed nearly identical explanations for the process of evolution at about the same time (see box 1.1).

Darwin and Wallace had a great deal in common. Both were familiar with the work of the geologists James Hutton and Charles Lyell, who proposed that there is continuity of geological processes from the past to the present and that the earth is many times older than was previously thought. They were also aware of evolutionary ideas proposed by Lamarck and others. But possibly the most important thing that Darwin and Wallace had in common was that each had travelled extensively and made observations of living things around the world. Both were especially impressed by the diversity of plants and animals found on oceanic islands (figure 1.3).

Darwin and Wallace, working independently, developed a set of ideas known as the **theory of natural selection.** A scientific **theory** is a well-documented set of ideas that best explains a large set of scientific observations. Their theory provides an explanation for the process of evolution. Darwin went on to publish a more thorough and completely documented statement in his book *Origin of Species,* a book that has had a dramatic effect on biology and the entire intellectual world.

Box 1.1

Darwin and Wallace: Two Scientists with the Same Idea at the Same Time

After his around-the-world voyage on the exploring ship H.M.S. *Beagle,* Darwin occupied himself for years systematically organizing the data that he collected during and after the voyage. He lived an isolated life in a country house in Downe, Kent, near London and spent his time writing about his observations and gradually putting all of the pieces of the theory of natural selection together in his mind. Darwin actually wrote a preliminary outline of the theory as early as 1842, but delayed publishing it so that he could add more supporting details.

In contrast to Darwin's long, systematic progression to a statement of natural selection, Wallace focused his thinking on the idea in a flash of inspiration that occurred one night in 1858 during a spell of tropical fever! He wrote that he was so excited by the idea that he lay impatiently waiting for the fever to lift so that he would have the strength to get up and write down an outline of his ideas.

Since Wallace knew of Darwin's interest in evolution, he sent Darwin an outline of his ideas. Darwin was so shocked to see ideas so similar to his own that he felt rather discouraged for a time. Then Darwin sought the advice of friends and they decided that the best solution would be that he and Wallace should have their ideas presented at a joint public reading. Afterwards, Darwin would proceed with publication of his own book on the subject.

Because Darwin was hurried by this series of events, his *Origin of Species,* published in 1859, was much shorter than he had originally intended, although it still comprised a very substantial book. This "abstract," as Darwin called it, proposed a theory that rocked the biological and intellectual worlds.

Figure 1.3 A saddleback tortoise from the Galápagos Islands. Darwin visited the Galápagos Islands on his voyage and was very impressed that tortoises, and other organisms, from each of the islands were clearly distinguishable from those on the other islands in the group.

The Theory of Natural Selection

The theory of natural selection, which provides a plausible explanation of the mechanisms of the evolutionary process, is based on certain observations and a set of conclusions drawn from those observations. In developing this theory, Darwin and Wallace pointed out that organisms can reproduce at rates that could lead to enormous increases in the population sizes of most species. Yet, despite these vast reproductive potentials, population sizes tend to remain fairly constant.

Darwin and Wallace then noted that there is considerable variability among individuals within a species and that some variations appear to be advantageous in terms of survival. The two scientists proposed that individuals possessing these beneficial characteristics would be more likely to survive and succeed in reproducing themselves. Gradually, the organisms possessing the favorable characteristics would make up a greater proportion of the population. Over a long

(a)

(b)

(c)

Box figure 1.1 (*a*) Charles Darwin. He thought out the principles of natural selection over a period of twenty or thirty years of quiet contemplation at his country house in Downe, Kent. (*b*) Darwin's study. He did most of his writing using the lapboard on the chair in the corner. (*c*) Alfred Russel Wallace. He thought out the principles of natural selection during a one-night bout of tropical fever in the South Seas.

period of time the entire population would be modified. However, if environmental conditions changed, new sets of characteristics might be favored and the whole process could move off in another direction. Darwin and Wallace suggested that these mechanisms have collectively guided the evolution of life on earth.

Reactions to Evolutionary Theory

While Darwin and Wallace were not the first evolutionists, their statement of the principles of natural selection caused great changes in modern thinking because it provided a reasonable explanation for the evolutionary process. The time was right, the work was scientifically powerful, and Darwin's book received enough publicity to make it known to the general reading public. Darwin and Wallace's statement of the principles of natural selection was indeed a milestone in the history of biology and a turning point in the history of human thought.

The notion of evolution, or continuous change, went far beyond biology. It shook the foundations of social systems and religious beliefs and even raised questions about the stability and orderliness of Victorian English society. No doubt Darwin must have sometimes wished that he had been able to follow his earlier plan to have his work published only after his death!

The Road to Modern Biology

Although the theory of natural selection had been published for only a short time, several other lines of nineteenth-century biological research were answering key questions about the evolutionary process itself.

Mendel: The Genetic Theory

In Darwin's day, most explanations of inheritance emphasized the mixing and blending of characteristics from both parents. This idea posed a problem for Darwin and his colleagues, because the theory of natural selection requires that characteristics be passed to offspring without blending for evolution to occur. In other words, evolution by natural selection depends on the transmission of specific variations from one generation to the next.

An Austrian monk named Gregor Mendel (1822–1884) was already working on a solution to this problem at the time Darwin published his theory. However, Darwin was apparently unaware of Mendel's research, as were most other biologists of that time (figure 1.4).

Figure 1.4 Gregor Mendel, the founder of the science of genetics.

In the solitude of a monastery garden, Mendel bred common pea plants. He showed through his research that characteristics pass from one generation to the next in discrete units. These units, which we now call **genes,** express themselves in different ways. Mendel demonstrated that genes are not lost or in any way blended or changed as they are passed from generation to generation.

The existence of distinct hereditary units was critical to Darwin's theory of natural selection, but the task of unifying the work of Darwin and Mendel fell to later biologists. Mendel's results did not receive the immediate international attention they deserved. As is often the case with an innovative concept, the biologists of Mendel's time may not have been ready to accept and appreciate his conclusions. However, even before Mendel's time, certain biologists were developing another powerful set of ideas that would eventually provide the frame of reference needed for Mendel's principles. This set of concepts would become known as the **cell theory.**

The Cell Theory

Are organisms homogeneously constructed, or are they made up of recognizable, fundamental structural units? If structural units exist, what are they, and how do they function? Nineteenth-century scientists answered these questions with the cell theory. But in contrast with the rather abrupt events that changed evolutionary theory and genetics, the development of the cell theory was a gradual process involving the work of many biologists over many years. In fact, work leading to the cell theory actually began well before the time of Darwin and Mendel.

During the seventeenth and eighteenth centuries, scientists developed skills in making microscopes and began to explore the world of the small in biology. For example, the Dutch merchant, wine taster, and amateur scientist Anton van Leeuwenhoek (1632–1723) observed and described many single-celled microorganisms, which he called "animalcules" ("little animals"). He also reported that larger organisms were constructed of separate structural components.

Robert Hooke (1635–1703), a contemporary of Leeuwenhoek, described the microscopic structure of cork, which, he discovered, contained pores or spaces. Hooke named these compartments "cells" because they reminded him of small monastery rooms or cells. We have used this term ever since to describe the units that make up living organisms (figure 1.5).

With the description of the cell, its nucleus, and chromosomes (the small bodies you can see in dividing cells stained with certain dyes) in the early 1880s, it became possible for biologists to make a connection between cell structure and genetic function. This connection added a new dimension to the work of Gregor Mendel.

Figure 1.5 (*a*) The microscope that Robert Hooke used to make his observations of the cellular nature of cork. (*b*) The "cells" of cork as Robert Hooke saw them. The two parts of the picture represent thin slices of cork cut at different angles.

(a)

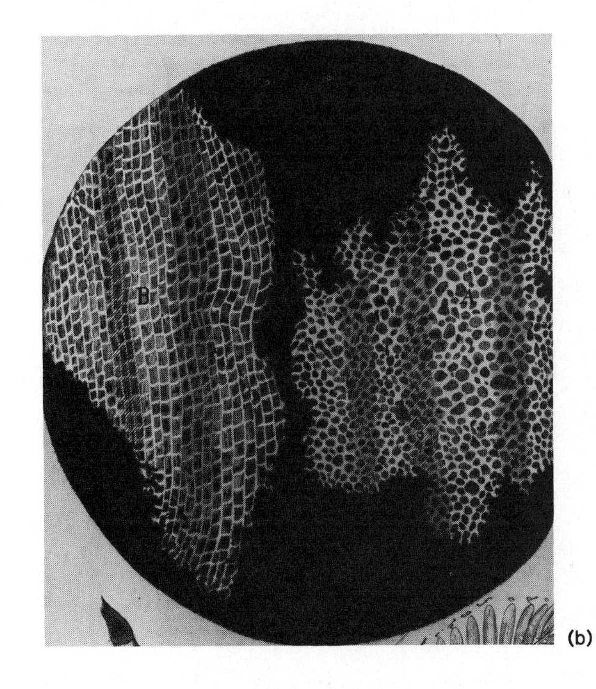

(b)

Genes and Chromosomes

Where are the hereditary factors (genes) that Mendel described actually located within living organisms? Scientists hypothesized that genes are physically linked with chromosomes. Eventually, Thomas Hunt Morgan and his colleagues proved that genes are distributed along chromosomes in linear fashion (figure 1.6).

Figure 1.6 Thomas Hunt Morgan. Morgan and his co-workers used experiments on the fruitfly, *Drosophila*, to clarify the physical relationships between genes and chromosomes. The "fly room" at Columbia was the center of the world of genetics research for a number of years.

The Development of Molecular Biology

In the early 1950s, James Watson and Francis Crick proposed the double-helix model of the DNA molecule, the carrier of genetic information. This model not only described the structure of the DNA molecule, it also gave rise to testable predictions concerning the mechanisms of transmission of genetic information during cell division.

By the 1960s, **molecular biology** was a major branch of biological research, which it remains today. Molecular biologists have subsequently studied the mechanisms of genetic expression and the control of the flow of genetic information. Today, new techniques make it possible to

Figure 1.7 The genetically modified mouse (left) grew several times faster and became nearly twice as large as the untreated control mouse (right). DNA containing the rat gene for growth hormone production was injected into the egg from which this large mouse developed. Human growth hormone genes have also been injected into mouse eggs and expressed in those mice with similarly spectacular results. These techniques were pioneered by Ralph L. Brinster, Richard D. Palmiter, and their colleagues.

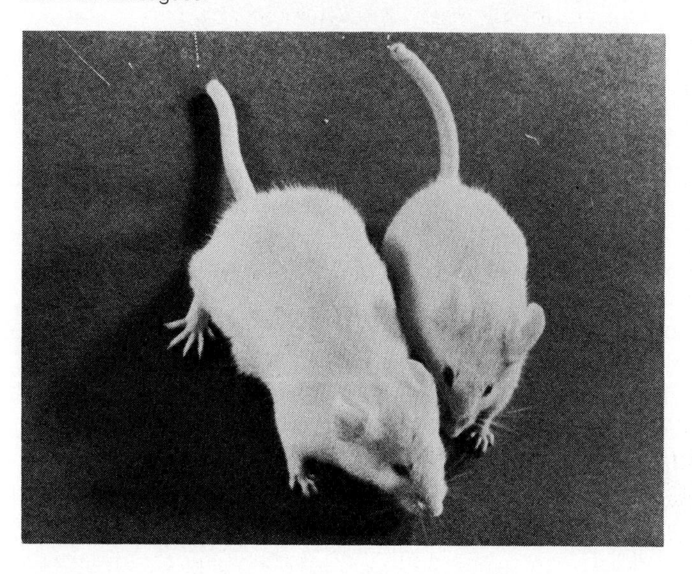

Figure 1.8 Is there danger that the world may become unfit for human habitation? The growing use of energy can stress and threaten the environment. This facility, Three Mile Island nuclear power plant near Harrisburg, Pennsylvania, had a malfunction in 1979 that threatened the area with serious nuclear contamination, and since then there have been several problems with other nuclear power plants.

Table 1.1
A Few of the Specialized Areas of Biological Study

Botany	The study of plants
Ecology	The study of relationships between organisms and environments
Entomology	The study of insects
Ethology	The study of behavior
Herpetology	The study of reptiles
Ichthyology	The study of fish
Mammalogy	The study of mammals
Microbiology	The study of microorganisms
Morphology	The study of structure
Mycology	The study of fungi
Ornithology	The study of birds
Paleontology	The study of fossils
Phycology	The study of algae
Physiology	The study of functions
Virology	The study of virus
Zoology	The study of animals

modify the genetic makeup of microorganisms so that they will produce large quantities of medically important substances. We can look ahead to the possibility of genetically modifying plants to make more efficient and productive crops, and the possibility of specific genetic therapy for individuals suffering from genetic diseases (figure 1.7).

Other Areas of Modern Biology

Molecular biology is just one of many areas of study in modern biology. As you can see in table 1.1, there is a field of study for almost every interest. Some areas of biology are named on the basis of the organism being studied. For example, **botany** is the study of plants, while **zoology** is the study of animals. Other areas are named for their particular emphasis. **Physiology** involves the study of biological functions. **Ecology** is the study of interactions of organisms with each other and with the environment.

In the 1960s and 1970s, biologists were confronted by a set of urgent practical problems that remain with us today. The interrelated problems of population growth, famine, pollution, and the depletion of natural resources are major threats to the stability of the modern world (figure 1.8).

Because the world's environmental problems are far from being solved, it is the responsibility of everyone, not just professional biologists, to understand the problems and to contribute to the solutions. Environmental concerns will probably continue to play a central role in the study of biology for many years to come (figure 1.9).

Figure 1.9 Humans have almost lost sight of the illuminating vision of oneness with other creatures in the natural world. How ironic that even the magnificent bald eagle, a national symbol of the United States of America, is endangered by environmental contamination, habitat destruction, and illegal hunting.

Figure 1.10 In unicellular organisms (*a*), the cell surface is also the boundary between organism and environment (cell = organism). All exchanges are directly between the cell and the external environment. In multicellular organisms (*b*), many cells are exposed only to the "internal environment" of the organism, and they exchange materials with that environment.

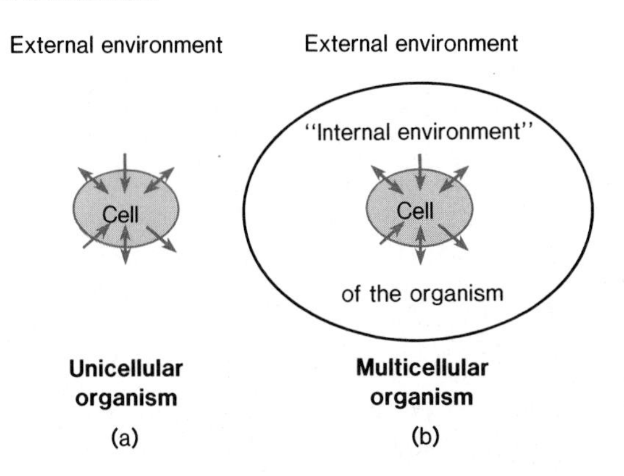

Unity and Diversity of Life

All biologists, no matter what their specialty, recognize that there is a fundamental unity of life that underlies the apparent diversity of living things. This unity is apparent when we seek an answer to the fundamental question, What is life?

All living things are products of evolution and participants in the continuing evolutionary process. They possess structural and functional features that adapt them to live successfully in their environment. All living things either are cells, in the case of **unicellular** (one-celled) organisms, or are composed of cells, in the case of **multicellular** organisms. Living things use energy gained from their environment to maintain their many functions. The chemical reactions that convert this energy to a usable form are known collectively as an organism's **metabolism.**

Furthermore, living things are **irritable;** that is, they respond to changes in their environments, and some (but certainly not all) organisms are capable of moving. Living things grow and develop. But **reproduction,** the ability to make "copies" of oneself, is the most distinctive general characteristic of living things.

Thus, a definition of life might be based on these characteristics of living things: adaptation, complex organization, metabolism, irritability, growth and development, and reproduction.

We will examine each of these characteristics of life later in this book and we would like you to keep them in mind as we go along. Some other ideas about living things that we will refer to again and again have to do with the relationships of organisms and their cells with the environment that surrounds them.

Cells, Organisms, and Environments

Every living cell has certain requirements that must be met if it is to live and interact successfully with its environment. Each cell must obtain nutrients, some of which are used as raw materials for building structures within the cell. Other nutrients are broken down to provide energy to maintain vital cell processes. All cells produce waste, which must be released to prevent buildup and damage to the inside of the cells. Cells must maintain water balance with their environments; that is, they must function so that their internal water content remains relatively stable. Cells must also exchange gases and a variety of other chemical substances with their environments. Meeting all of these requirements requires a continuous exchange of materials between the inside of each cell and its immediate environment.

In unicellular organisms (a single cell *is* the whole organism), each cell makes its required exchanges directly with the external environment. But a very different relationship exists in individual cells in a specialized, multicellular organism. Within multicellular organisms, each individual cell exchanges materials with the internal environment of the organism. The internal environment of a specialized, multicellular organism provides a relatively stable, controlled habitat for cells (figure 1.10).

Figure 1.11 (*a*) Generalized drawing of a complex plant body showing some of the exchanges that occur at specialized boundaries.
(*b*) Generalized drawing of an "average" complex animal showing some of the exchanges that occur at specialized boundaries: (1) food absorption from the gut; (2) gas exchange in the lung; (3) metabolic waste excretion, ionic regulation, and maintenance of body water balance in the kidney; and (4) sensory reception of information about environmental conditions.

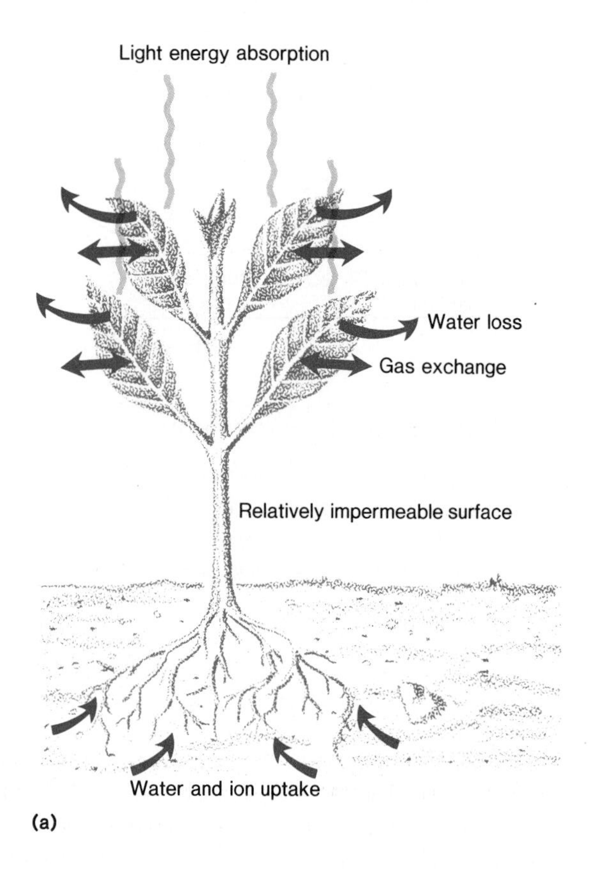

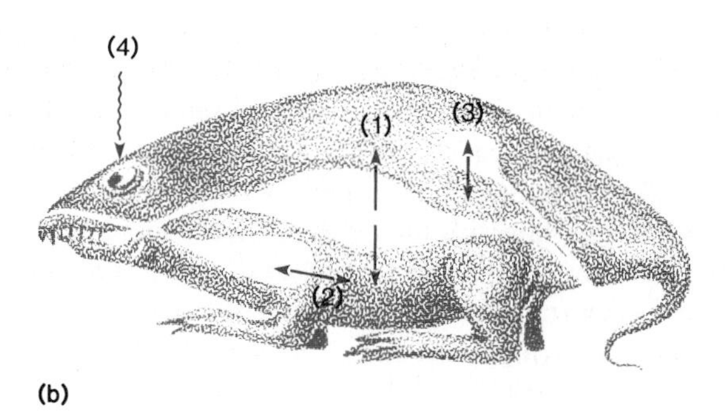

In multicellular organisms, there is a "division of labor" in which specific functions are localized in limited body areas. Some of these specialized regions function as exchange boundaries at which materials are exchanged with the external environment. Parts of the body surface not involved in these exchanges enclose and protect the internal body environment in which the majority of the cells live (figure 1.11).

Homeostasis: A Dynamic Steady State

The idea that a relatively stable internal cellular environment is maintained within multicellular organisms has intrigued biologists for a long time. French physiologist Claude Bernard recognized the importance of what he called the *milieu interieur* ("internal environment") and said that constancy of the internal environment is necessary for life in varying external environments.

Since the time of Bernard, however, we have learned that the internal environment of a large complex organism is not *absolutely* constant. The condition of the internal environment fluctuates, and the rates of various metabolic processes, concentrations of many substances, and other factors change as the organism's nutritional state, activity levels, and environmental conditions change.

Under normal conditions, each of the factors that fluctuate in the internal environment is held within an acceptable range of values. This is the range within which body cells can thrive and make necessary exchanges with their environments. Only when the organism is injured, ill, or subject to excessive environmental stress is this internal stability upset.

Normally, then, the condition of the internal environment is dynamic; that is, it changes from one moment to the next. Despite all of this change, a "steady state" is maintained because all variations are kept within acceptable limits (figure 1.12).

This dynamic steady state is called **homeostasis.** The idea that much of body functioning is aimed at maintenance of homeostasis is one of the most important concepts of modern biology. Homeostasis is important because it permits organisms to exploit more variable external environments. Environmental changes may swirl about an organism, but its body cells live and function in a sheltered, stable internal environment. Environmental fluctuation is especially prevalent in terrestrial (dry land) environments. Not surprisingly, therefore, complex multicellular organisms are the most successful occupants of terrestrial environments. The ability to maintain homeostasis in the face of environmental change is essential to life on land.

Figure 1.12 A generalized representation of some of the patterns of variation that can occur as part of homeostatic regulation: (1) short-term changes resulting from changes in activity or nutritional state; (2) cyclical regular changes with time of day or night (biological rhythm); (3) a change resulting from illness, injury or excessive environmental stress.

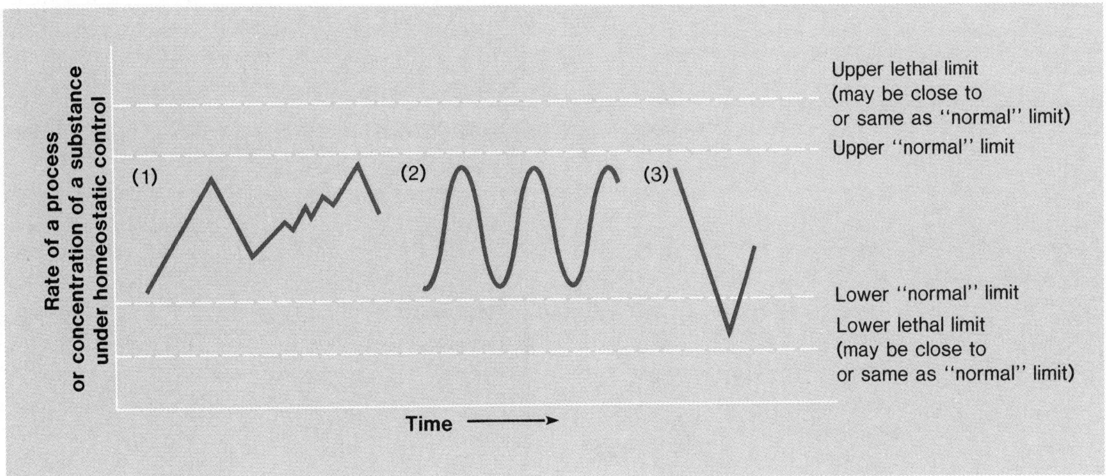

Diversity

Having recognized the unity of life on earth, we must also recognize that living things are very diverse. They come in a great variety of shapes and sizes, and they are adapted to life in a broad range of habitats (places where organisms live). There are organisms that live in fresh water, in seawater, in freezing cold, in blistering heat, and in all environments in between.

There are literally millions of kinds of living things. And, you may wonder, how do biologists make sense of this bewildering array of living things?

Biologists describe and name organisms, and arrange them in groups on the basis of similarities and differences. We need to use a system of **taxonomy** (naming and grouping of living things) in order to investigate, describe, and discuss various organisms. In this text, we will use a taxonomic scheme known as the **five-kingdom system.** The five kingdoms are **Monerans, Protists, Fungi, Plants,** and **Animals.** Let us now take a brief look at these five kingdoms and see generally what sorts of organisms are included in each.

Monerans

This kingdom includes bacteria and cyanobacteria (blue-green algae). The key characteristic of monerans is that they are **prokaryotic,** which means that their cells do not have a membrane-enclosed nucleus nor several other membrane-enclosed cellular structures found in the cells of members of the other four kingdoms.

Bacteria are microscopic single-celled organisms, which often occur in colonies containing thousands of cells (figure 1.13). Many bacteria cause infection and disease in humans, other animals, and plants, but not all bacteria are harmful or dangerous. Bacteria play vital roles in natural decay processes, which release important nutrients from dead organisms and make them available for use by other organisms. Bacteria also are critically important in sewage treatment and in many industrial processes including manufacture of cheeses and other foods.

Cyanobacteria are prokaryotic organisms that carry on **photosynthesis.** Photosynthesis is the use of light energy to synthesize nutrients from carbon dioxide and water. Cyanobacteria are found in a variety of environments including very harsh ones such as hot springs. Usually they play a positive role as a nutrient source for organisms that eat them, but occasionally they grow so numerous in a lake or pond that they make the water unpleasant or even dangerous for other living things.

Figure 1.13 Monerans. (*a*) Testis tissue of a rabbit infected with the sexually-transmitted disease syphilis. Note the spiral-shaped bacterial cells. (*b*) The cyanobacterium (blue-green alga) *Oscillatoria* that grows in many freshwater environments. Note that *Oscillatoria* grows in filaments, end-to-end chains of cells. Cyanobacteria are important food sources for many animals.

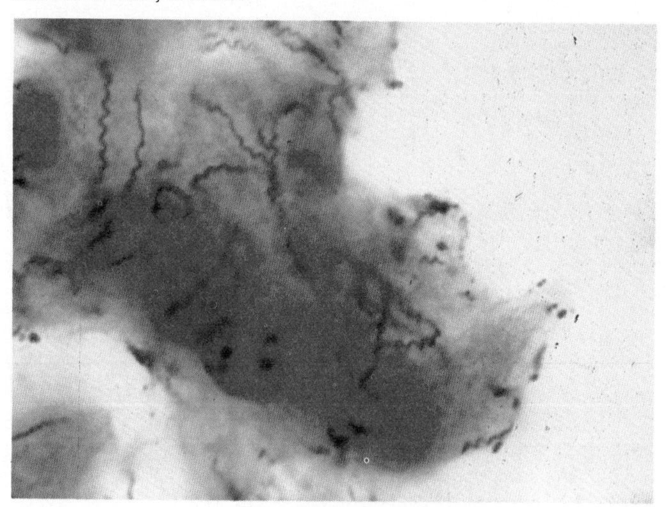

(a)

(b)

Protists

All protists are single **eukaryotic** cells or colonies of eukaryotic cells. Eukaryotic cells possess a nucleus that is enclosed by a membrane, and they have several other kinds of membrane-enclosed structures in their cells. This kingdom includes protozoa and unicellular algae.

In a single drop of pond water, you can usually see hundreds of protozoans. Some whip through the water with amazing speed, while others slowly glide along (figure 1.14). Most protozoa live independently, but some are disease-causing parasites. For example, malaria is caused by a protozoan.

Many kinds of single-celled algae are included in the protists. They carry out photosynthesis, and small as they are, they are very important as major producers of organic nutrients, especially in the oceans of the world. The nutrients that they produce support all life in the oceans ranging from tiny organisms to the largest whales.

Fungi

The fungi include many familiar organisms such as molds, mildews, mushrooms, and puffballs (figure 1.15). Although some mushrooms are delicious, there is nothing appetizing about the fuzzy green mold covering the "forgotten" orange found in the back of the refrigerator! Fungi obtain nutrients by absorbing chemical substances from their environment and are very important participants in natural decay. A few fungi cause diseases.

Plants

Plants are multicellular organisms with eukaryotic cells enclosed by rigid cell walls. Plants possess the green pigment chlorophyll and carry out photosynthesis. Members of the plant kingdom range from microscopic multicellular algae, which form delicate spheres or chains of cells, to the ivy on your campus buildings to trees, some of which are the largest of all living things (figure 1.16).

Animals

Animals are distinguished from the other kingdoms by several key characteristics. They are multicellular and their cells do not have cell walls. Animals cannot manufacture their own organic nutrients by photosynthesis. Unlike fungi, which absorb organic nutrients produced by other organisms, animals ingest (eat) other organisms or material produced by other organisms.

Animals range in diversity from simply-organized animals such as sponges and jellyfish to animals such as butterflies, frogs, and cheetahs, who can withstand the rigorous challenges of life on land (figure 1.17).

Figure 1.14 Protists. (*a*) *Paramecium,* a single-celled organism that is common in many freshwater environments. These protists obtain nutrients by engulfing small particles and smaller organisms in the water around them. (*b*) A diatom. Diatoms are single-celled algae that have beautiful, glassy cell walls. Diatoms carry out photosynthesis and they are so numerous in all of the world's waters that they provide food for many animals and supply much of the atmospheric oxygen used by all living things.

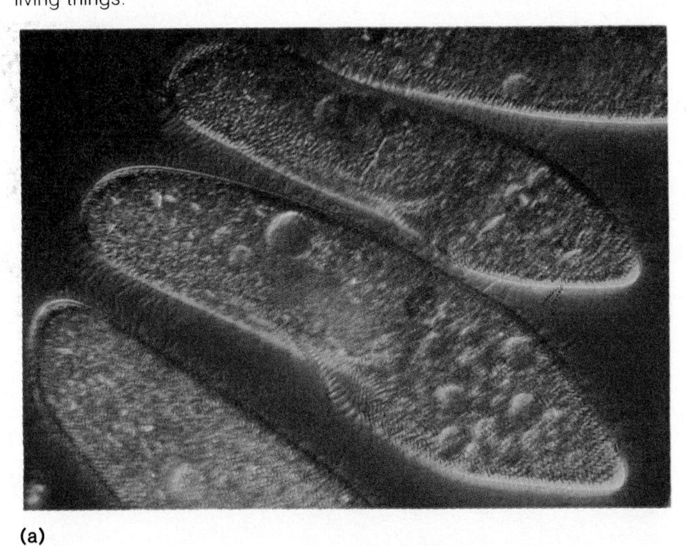

(a)

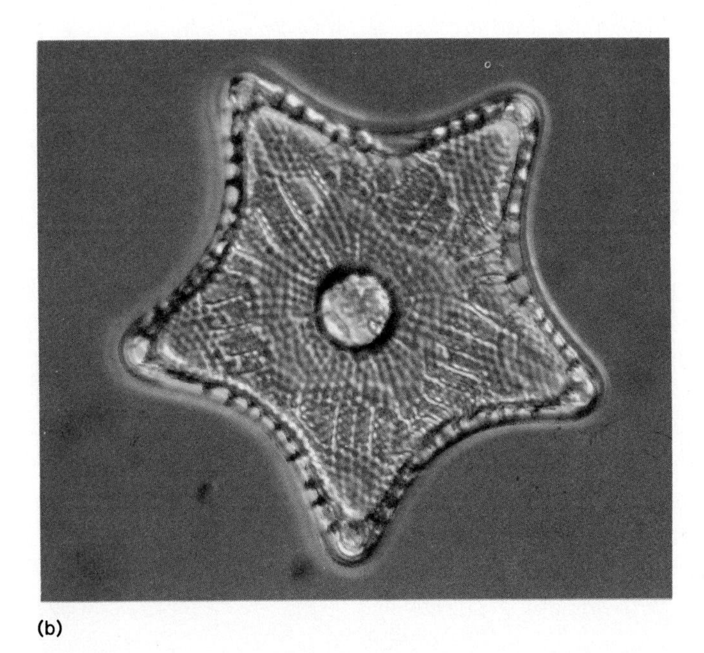

(b)

Figure 1.15 Mushrooms, members of the kingdom Fungi. Most of the complex tubular network that makes up this fungus is underground. The familiar mushroom is actually a specialized reproductive structure that is produced only occasionally.

Figure 1.16 The plant kingdom. (*a*) *Volvox,* a multicellular green alga that is a colony of cells arranged in a sphere. Note the daughter colonies developing inside the sphere. They will be released later and become new independent colonies. (*b*) Sea lettuce (*Ulva*), a green alga that grows in thin flat sheets in the ocean. (*c*) Kelp (brown algae) exposed on a seashore at low tide. (*d*) *Marchantia,* a liverwort. These small plants grow in moist environments. At times, upright stalks with egg-producing structures (shown here) develop on some plants and sperm-producing structures develop on others. (*e*) Ferns. Most ferns grow in moist, sheltered environments, but a few hardy ferns grow in open meadows. (*f*) Flowers and buds of a marigold. There are a great many flowering plants, ranging from grasses to large trees. (*g*) A giant sequoia tree, which is one of the largest living things and an example of a coniferous (cone-bearing) plant. Many common trees such as pines and spruces are also conifers.

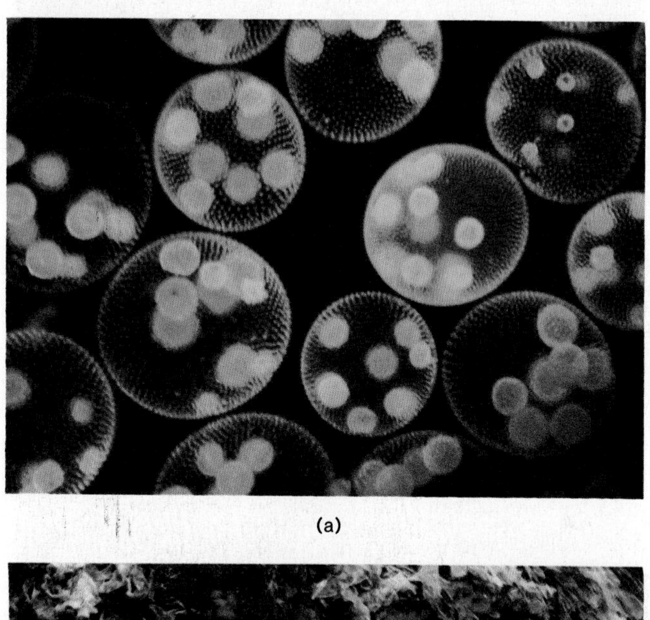

(a)

(b)

(c)

(d)

(e)

(f)

(g)

Figure 1.17 The animal kingdom. (*a*) Sponges on the ocean floor. (*b*) The Portuguese man-of-war. This is a colony of individuals living together. One is specialized as a gas-filled float. Others are specialized for defense and can deliver painful stings when touched. (*c*) Muscle tissue with three cysts containing *Trichinella,* a parasitic roundworm. (*d*) Snails belong to a group of animals called molluscs, most of whom have shells. Other molluscs include clams, oysters, octopuses, squids, and slugs. (*e*) A reef lobster. Lobsters, crabs, and crayfish are grouped together as crustaceans, one of the subdivisions of the arthropods ("joint-footed animals"). Other major groups of arthropods are the arachnids (scorpions, spiders, ticks, and mites) and the insects. (*f*) The Viceroy butterfly, an insect. Insects are the largest group of arthropods. (*g*) A variety of kinds of fish in a coral reef off the coast of Florida. Fish are the most numerous vertebrate animals (animals with backbones). The other major groups of vertebrate animals are amphibians, reptiles, birds, and mammals. (*h*) A poison arrow frog, an amphibian. Toads and salamanders are also amphibians. (*i*) A female lion with her cubs. Lions and other cats are mammals, as are humans and other familiar vertebrate animals.

(a)

(b)

(c)

(d)

(e) (f) (g) (h) (i)

Biologists' Methods

Now that we have considered the nature of life and introduced some of the major groups of living things, we need to consider one final question. How, then, do modern biologists study life? Biologists actually use a variety of approaches and methods to study life processes.

Some biologists devote themselves to making observations of nature and to recording and organizing the data that they gather. Usually, such biologists try to interfere as little as possible with the systems they study.

Other biologists use what is commonly described as the **scientific method.** They gather and study facts about a biological process. They consider possible explanations of observed facts and develop **hypotheses.** A hypothesis is sometimes called an "educated guess," but a good hypothesis is a suggested explanation that can be tested. Testing a good hypothesis sometimes yields a definite positive or negative answer, but often the test results suggest a new hypothesis, which in turn must be tested.

A hypothesis is tested either by making more observations or, more commonly, by performing an **experiment,** a rigorous scientific test involving carefully controlled manipulations. When biologists have completed their tests, they draw conclusions from the results. They often report their conclusions to other biologists by writing scientific papers. A good paper should discuss the hypothesis, the techniques used to test the hypothesis, the results of the test, and conclusions drawn from the results.

Whatever the outcome of their research, biologists must recognize that what is thought and known today will be modified in the future, just as the viewpoints of earlier biologists have been modified. Present knowledge is only a set of ideas to work with, not unalterable truth. Realizing that scientific knowledge itself is undergoing continuing evolution helps to promote objectivity.

Sometimes, however, a body of scientific knowledge becomes very widely recognized as a correct and complete description or explanation of part of biology. As we said earlier, such bodies of knowledge are called theories. We cannot emphasize too strongly that the word theory has a very different meaning in science than it does in everyday life. A scientific theory is *not* a hunch or a vague guess. It is a reliable and well-documented set of ideas that best explains a large set of scientific observations. Some examples of important biological theories are the cell theory and the theory of natural selection.

Figure 1.18 Albert Szent-Györgyi. "Discovery consists in seeing what everybody else has seen and thinking what nobody else has thought."

If we were to try to summarize the nature of science in a few words, we might say that science is a process in which open and inquiring minds explore the world of the unknown in nature. A certain freshness of vision marks the successful scientist and, for that matter, the successful person in any walk of life. As the biologist Albert Szent-Györgyi put it: "Discovery consists in seeing what everybody else has seen and thinking what nobody else has thought" (figures 1.18 and 1.19).

Figure 1.19 The lifespans of some biologists mentioned in this chapter (top) and some of their contemporaries (bottom).

Adapted from *Genetics Human Aspects* by Arthur P. Mange and Elaine Johansen Mange. Copyright © 1980 by Saunders College/Holt, Rinehart and Winston. Reprinted by permission of CBS College Publishing.

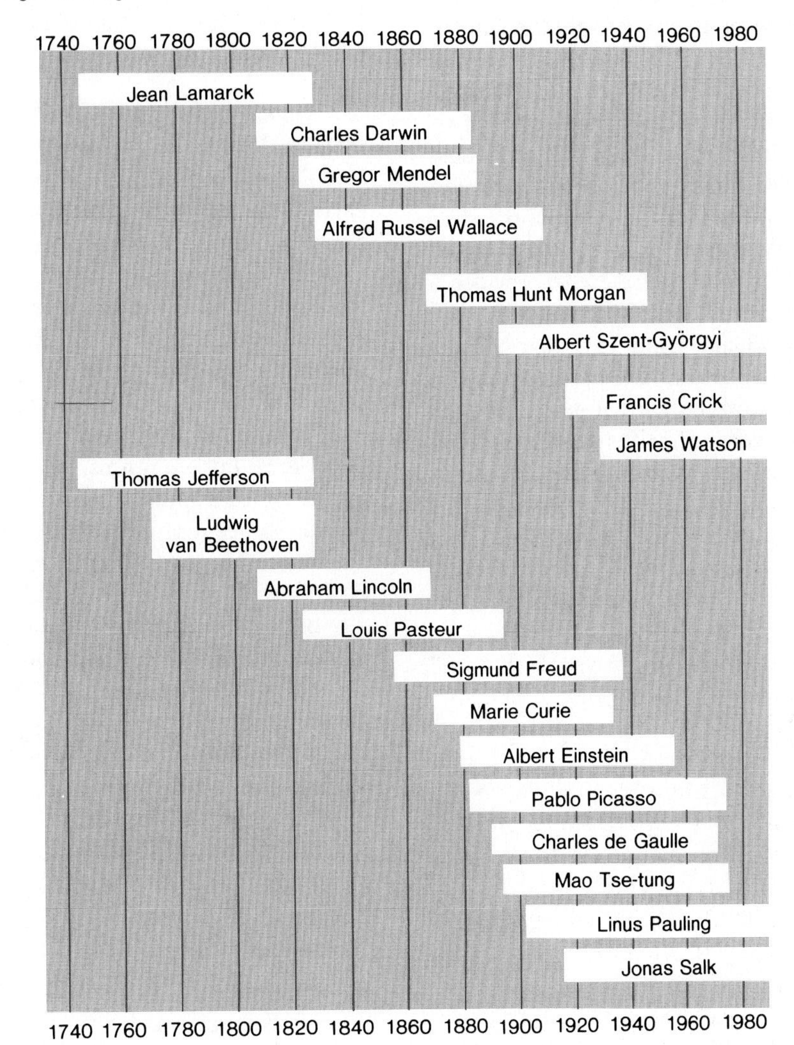

Summary

Humans have always been aware of the world around them. Motivated by curiosity, people throughout human history have added to biological knowledge by direct observation. Others tried simple experiments to answer questions about the world of life.

Before the 1880s, most people thought that life on earth had remained unchanged since a specific moment of creation. However, some biologists, such as Lamarck, suggested that living things change over many generations. A revolution in biology and society came about as a result of the ideas of Charles Darwin and Alfred Russel Wallace, who proposed the theory of natural selection. Natural selection provided a plausible explanation for the evolution of life on earth.

A contemporary of Darwin and Wallace, Gregor Mendel experimented on how characteristics were passed from one generation to the next. His work on garden peas, although not appreciated until many years after his death, provided genetic information necessary for understanding the evolutionary process.

All organisms are made up of recognizable fundamental units called cells. With the development of the cell theory, it became possible to make a connection between cell structure and genetic function.

In the 1950s James Watson and Francis Crick proposed a model for the structure of the DNA molecule. This gave rise to predictions of how genes are passed during cell division.

Modern biology is divided into many areas, ranging from molecular biology to ecology.

There is a fundamental unity among all forms of life on earth. The shared characteristics of all living things are adaptation, complex organization, metabolism, irritability, growth and development, and reproduction.

Unicellular organisms meet all of their needs by making exchanges directly with the external environment. In multicellular organisms, there is a division of labor that contributes to maintenance of an internal environment. Homeostasis is a dynamic steady state maintained in the internal body environment of organisms. The idea that much of body function is aimed at maintaining homeostasis in all living things is a key concept of modern biology.

We can organize the diverse life on our planet into five kingdoms: Monerans, Protists, Fungi, Plants, and Animals.

Biologists study life using a variety of approaches. The most common is the scientific method. A biologist investigates a question by forming a hypothesis. The hypothesis is tested by using experiments. Results of experimental tests are used to develop conclusions. A large body of sound conclusions forms a theory.

Whatever the outcome of research, biologists must recognize that present scientific knowledge is only a set of ideas to work with, not unalterable truth.

Questions for Review

1. What is biology?
2. Explain Lamarck's concept of evolution by inheritance of acquired characteristics. Can you think of an example that shows why this concept is not correct?
3. What is natural selection? How does natural selection provide an explanation for the evolutionary process?
4. Briefly describe the cell theory.
5. What are the fundamental characteristics shared by all living things?
6. Define homeostasis.
7. What is the five-kingdom system of classification?
8. Describe the steps in the scientific method.

Questions for Analysis and Discussion

1. Why do biologists find it necessary to use a system of taxonomy?
2. Discuss the difference between the way the word "theory" is used in everyday conversation and the way it is used in science.

Suggested Readings

Books

Bronowski, J. 1973. *The ascent of man*. Boston: Little, Brown, and Co.

Gardner, E. J. 1972. *History of biology*. 3d ed. Minneapolis: Burgess Publishing.

Hardy, R. M. 1976. *Homeostasis*. Studies in Biology no. 63. Baltimore: University Park Press.

Sayre, A. 1975. *Rosalind Franklin and DNA*. New York: Norton.

Watson, J. D. 1968. *The double helix*. New York: Signet Books.

Articles

Eiseley, L. C. February 1956. Charles Darwin. *Scientific American* (offprint 108).

———. February 1959. Alfred Russel Wallace. *Scientific American*.

Keele, K. D. 1978. The life and work of William Harvey. *Endeavour* 2:104.

Marshack, A. 1975. Exploring the mind of ice age man. *National Geographic* 147:64.

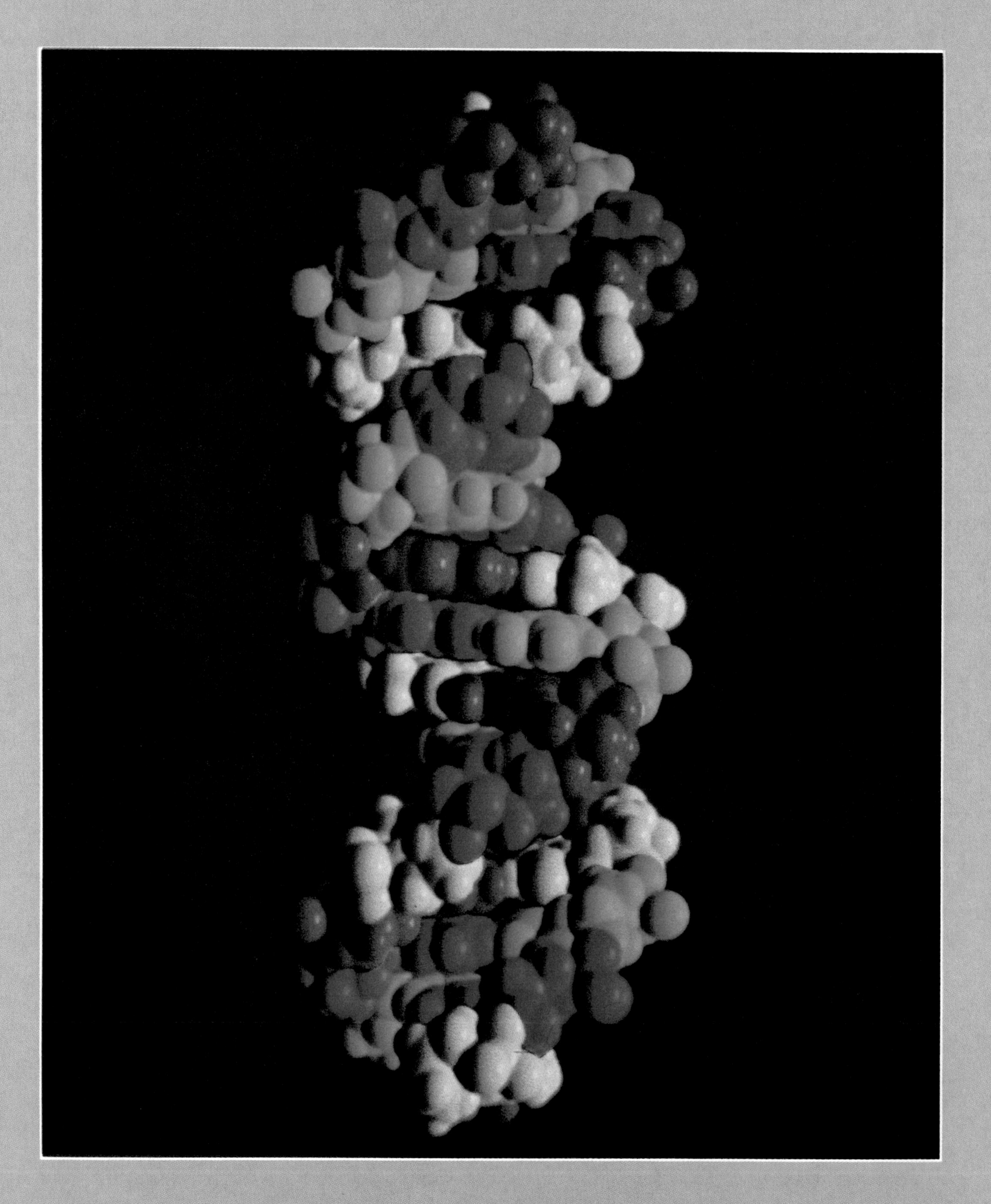

Cells, Molecules, and Life

Part 1

One of the fundamental principles of modern biology is that processes occurring in living things obey the laws of chemistry and physics, just as processes in nonliving systems do. The special properties of life arise because of the complex organization of living things and the chemical processes that occur in them. Biologists must understand certain principles of chemistry in order to better analyze the organization and functions of life on earth.

Another key to understanding life on our planet is to recognize the role of cells. The cell is a fundamental organizational unit of life. As you learn more about cell function and structure, you will find that many cellular components and chemical processes are identical or very similar in cells from a variety of organisms. Recognition of these similarities at the cellular level provides an important unifying framework within which biologists approach the diversity of living things.

All living things require energy for maintenance, growth, and reproduction. Some organisms obtain energy by harnessing light energy from the sun and then producing the materials they need (photosynthesis). Other organisms, however, are not capable of using light energy in this way. They must use photosynthesizing organisms as sources of energy and materials.

In chapters 2 through 6 we will examine some biologically important concepts of chemistry, some basic characteristics of the cells that constitute living things, and some of the energy relationships upon which all of life depends.

Facing page A computer-generated model of part of a DNA molecule.

Atoms, Molecules, and Life

Chapter Outline

Chapter Concepts

1. Chemical processes in living things obey the laws of chemistry and physics.
2. Biologists must understand some characteristics of atoms and the ways in which atoms combine in molecules to understand and analyze the characteristics of organisms.
3. Water is the most common substance in living things and the properties of water determine many properties of life.
4. Most types of molecules in organisms are organic compounds, those compounds containing carbon.
5. Many biologically important organic molecules are polymers, chains of individual molecular units linked together.

One of the oldest arguments in the history of biology concerned the basic nature of living things. Many biologists thought that the substances composing living things were fundamentally different from those in nonliving things. They thought that there was a special life chemistry, with its own set of laws differing from the laws governing nonliving matter. They also proposed that there was a special force or energy, called the "vital force," that was unique to life. Despite all efforts to detect it, no such vital force has ever been found, and chemical laws are the same whether the chemical processes occur in living things or nonliving things.

Today, biologists recognize that all matter, living and nonliving, consists of atoms and molecules. Many of the molecules in living things are larger and more complex than those commonly found in nonliving systems. As we will see shortly, these relatively large and complex molecules are arranged into highly organized systems such as membranes and cells. Because of the high degree of organization among their molecules, living things have properties that are not found in nonliving systems.

In order to be able to explore the properties of living things, we need to understand some things about the atoms and molecules that constitute them.

Elements and Compounds

All matter is composed of basic substances called **elements.** An element cannot be broken down into simpler units by chemical reactions; it contains only one kind of atom. An **atom** is the smallest characteristic unit of an element.

Atoms of most elements do not occur in isolation. Usually they are combined with other atoms to form **molecules.** A molecule is a combination of like or different atoms. For example, we do not find individual oxygen atoms by themselves in the atmosphere. Rather, oxygen in air is in the form of oxygen molecules, each of which contains two oxygen atoms.

There are ninety-two naturally occurring elements, as well as a number of artificial elements that can be made in laboratories. We can symbolize each element by one or two different letters that abbreviate the element's English or Latin name. For example, the symbols for hydrogen (H) and oxygen (O) come from English names, while the symbols for sodium (Na, for *Natrium*) and potassium (K, for *Kalium*) are derived from Latin names.

A **compound** is a substance that can be split into two or more elements; in other words, a compound is a combination of two or more kinds of atoms. Water is a compound because we can split it into its components, hydrogen and oxygen.

Table 2.1
Some of the Elements Important in Biology

Element	Symbol	Atomic Number	Atomic Weight
Hydrogen	H	1	1.01
Carbon	C	6	12.01
Nitrogen	N	7	14.01
Oxygen	O	8	16.00
Phosphorus	P	15	30.97
Sulfur	S	16	32.06
Sodium	Na	11	23.00
Magnesium	Mg	12	24.31
Chlorine	Cl	17	35.45
Potassium	K	19	39.10
Calcium	Ca	20	40.08
Iron	Fe	26	55.85
Copper	Cu	29	63.54
Zinc	Zn	30	65.37

The **formula** of a compound gives information about the kinds and numbers of atoms that make up each molecule of a compound. A formula contains the abbreviations for the atoms that make up the molecule and subscripts that indicate the number of each kind of atom. For example, the formula for water, H_2O, indicates that a water molecule contains two hydrogen atoms and one oxygen atom. By the same token, the sugar called glucose, $C_6H_{12}O_6$, contains six carbon atoms, twelve hydrogen atoms, and six oxygen atoms.

Atomic Structure

While atoms are the smallest units that have all the properties of elements, each atom is actually composed of still smaller particles.

Protons, Neutrons, and Electrons

There are several kinds of subatomic particles, but only three are crucial to the properties of elements we will consider here: protons, neutrons, and electrons. **Protons** and **neutrons** are located in a central area in the atom called the **nucleus. Electrons** are located at various distances from the nucleus. Protons have a positive electric charge, and electrons have a negative charge; thus, protons attract electrons. Neutrons have no electric charge.

Although protons and neutrons have nearly the same mass, their mass is much greater than the mass of electrons (more than 1,800 times as great). Chemists customarily use the **atomic mass unit** (AMU) to describe the mass of subatomic particles. Both the proton and the neutron have masses of approximately one.

Figure 2.1 A ^{12}Carbon atom. Carbon's mass number is twelve because its nucleus contains six protons and six neutrons. The shaded area around the nucleus represents the area in which the six electrons of the atom are located.

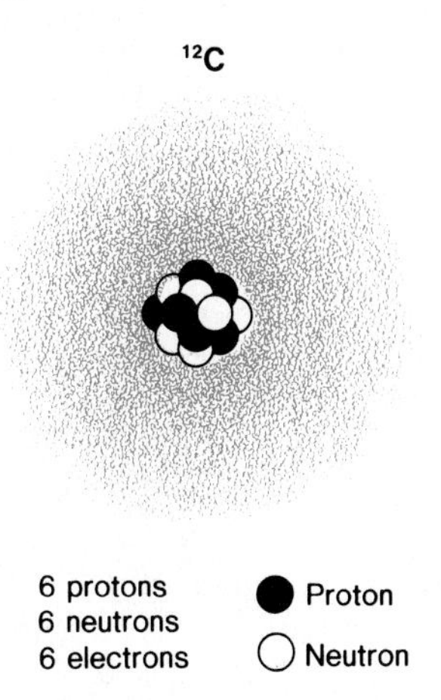

Each atom of an element has a characteristic **number** of protons in its nucleus, and this number is known as the element's **atomic number.** For example, because a carbon atom has six protons in its nucleus, carbon's atomic number is six. The atomic numbers of some biologically important elements are given in table 2.1.

Normally, the number of electrons outside the nucleus of the atom is equal to the number of protons in the nucleus. Because the positive charges of the protons equal the negative charges of the electrons, the charges cancel one another and the atom is neutral.

The **mass number** of an element indicates how many protons *and* neutrons are in the nucleus of one of its atoms. For example, most carbon atoms contain six protons and six neutrons. Carbon's mass number, 12, is symbolized with a superscript numeral preceding the element's symbol: ^{12}C (figure 2.1). The **atomic weight** is the *actual measured weight* of an element and is usually very close to, but not exactly the same as the mass number (see table 2.1).

Electrons and Orbitals

Although electrons have a much smaller mass than protons and neutrons, they are by no means less important in determining the chemical properties of atoms. The location of electrons in relation to the nucleus has a great influence on the way atoms and molecules react with each other.

Because electrons are constantly in motion around the nucleus in every atom, we cannot say exactly where a particular electron is located at any given moment. However, we can say where electrons are most likely to be located most of the time. We call the space in which an electron is located 90 percent of the time the electron's **orbital.**

Figure 2.2 Electron orbitals. (*a*) The three dumbbell-shaped orbitals of the second shell are at right angles to one another. For clarity, the spherical orbital of the first shell is not included in this sketch. Densities of dots represent relative probabilities of electrons being located in various parts of the orbitals. (*b*) All four orbitals of the second shell on a flat plane. In this example, all orbitals are complete as each contains two electrons. The nucleus is in the center. (*c*) A simpler way of representing the orbital shells with the number of electrons contained in each. This sketch represents both the first and second orbital shells.

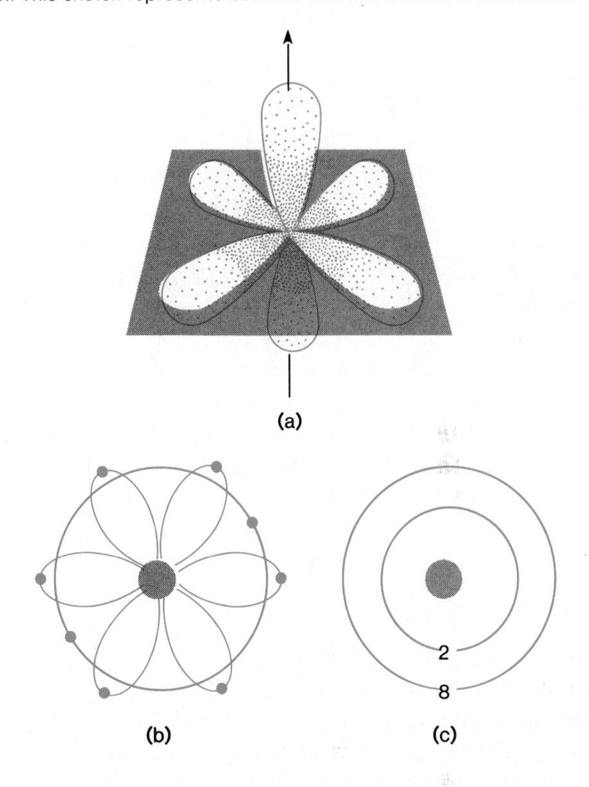

Each orbital can contain a maximum of two electrons. An atom can have several or many orbitals, but the orbitals are arranged in a specific way relative to the nucleus of the atom. Orbitals are grouped into **shells.** The first shell, the one closest to the nucleus, contains only one orbital (a spherical one). Thus, the first shell can contain only two electrons. The second shell contains four orbitals, one spherical and the other three shaped like dumbbells. This second shell can contain up to eight electrons. Although the third shell and those beyond can hold more than eight electrons, these shells are in a particularly stable condition when they hold exactly eight electrons (figure 2.2).

Under normal circumstances, electrons will be in the innermost shells in which orbital space is available. Each inner shell, therefore, is filled before the next shell (figure 2.3). When an atom's outermost shell contains eight electrons, the atom is very stable and does not react readily with other atoms. This octet (eight) rule, as it is called, is very important in the associations of atoms to form molecules.

Figure 2.3 Electron orbital shells of a chlorine atom. The chlorine atom illustrates the rule that if an atom has an incompletely filled orbital shell, it normally will be the outermost electron-containing shell that is not completely filled. The two inner shells are filled, having two and eight electrons respectively. The third shell contains only seven electrons.

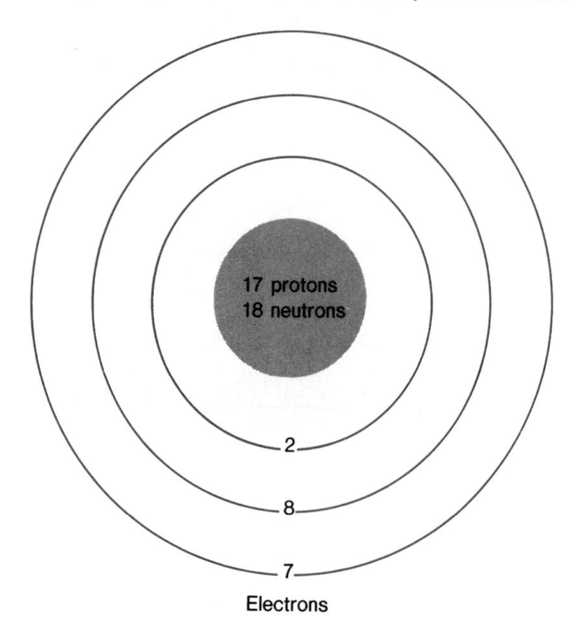

Figure 2.4 Ionic bonding in sodium chloride. Sodium donates an electron to chlorine. Thus, sodium has a stable configuration with eight electrons in its second shell, which is now its outer shell, and chlorine also has eight electrons in its outer shell. Thus, both atoms have satisfied the octet rule (octet for eight electrons in the outer shell). This electron transfer leaves both atoms with net charges; they are ions. Na^+ and Cl^- are strongly attracted to one another and form NaCl.

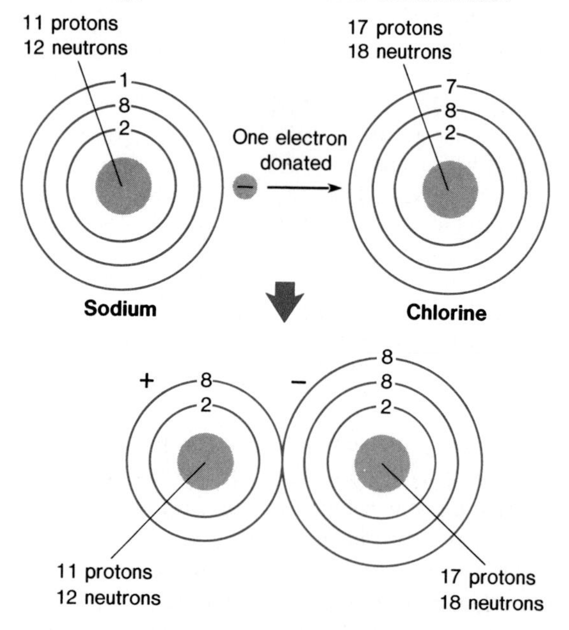

Chemical Bonding

Free atoms are quite rare in nature—at least in the earth's crust and in living things. Instead, atoms usually bind together in molecules such as O_2 or H_2O, or in larger aggregates such as crystals, which contain enormous numbers of atoms and molecules.

How do atoms bind together to become molecules? Atoms are held together in molecules by **chemical bonds.** By forming chemical bonds, atoms tend to achieve a more stable electron configuration than they have by themselves. Atoms combine into molecules by rearranging their electrons in different ways. In **ionic bonding,** electrons are *transferred* from one atom to another. In **covalent bonding,** pairs of electrons are *shared* between atoms.

Ionic Bonding

To understand ionic bonding, let us examine the reaction between sodium (Na) and chlorine (Cl) in which sodium chloride (NaCl), or common table salt is formed. Sodium's atomic number is eleven; thus, a sodium atom has eleven protons in its nucleus and eleven electrons orbiting around the nucleus. The inner shell, containing two electrons, is complete, as is the second shell with eight electrons; but the third shell of a sodium atom contains only one electron. An atom with only one electron in its outer shell tends to be an electron donor; that is, it tends to get rid of its lone electron.

Chlorine has an atomic number of seventeen, which means that a chlorine atom has two electrons in its first shell, eight in its second shell, and seven in its outer shell. Because it needs only one electron to complete its outer shell (remember the octet rule), a chlorine atom tends to be an electron acceptor.

When a sodium atom and a chlorine atom come together, an electron is transferred from the sodium atom to the chlorine atom. This electron transfer changes the balance between the protons and electrons within each of the two atoms. The sodium atom is left with one more proton than it has electrons and the chlorine atom with one more electron than it has protons. The sodium atom now has a net charge of +1 (symbolized by Na^+), while the chlorine atom's net charge is −1 (symbolized by Cl^-). Charged atoms are called **ions** and are attracted to one another by their opposite charges. Thus, an ionic compound such as sodium chloride (NaCl) is held together by the attraction between its positive and negative ions, that is, by an **ionic bond** (figure 2.4).

The ionic bonds of sodium chloride are quite strong when salt exists as a dry solid, but when an ionic compound such as NaCl is dissolved in water, it separates into Na^+ and Cl^- ions. Ionic compounds are most commonly found in this **dissociated (ionized)** form in the watery environment inside living things.

Box 2.1
What Are Isotopes?

The word **isotope** has become a common part of our language. We hear almost daily about the use of isotopes, particularly radioactive isotopes (or simply "radioisotopes") in medical diagnosis and in cancer treatment. We also hear these terms in connection with nuclear power plants and nuclear waste disposal. What do we mean by the term "isotope," and what are radioactive isotopes?

All atoms of an element have the same number of protons, but different atoms of the same element can have different numbers of neutrons. Recall that the mass number of an atom is equal to the number of protons plus the number of neutrons present in its nucleus. Thus, the different forms, or isotopes, of atoms of a given element have different mass numbers.

For example, the most common form of carbon atom, with six protons and six neutrons in its nucleus, has a mass number of 12. It is called ^{12}Carbon (^{12}C) (read "carbon-12"). A carbon isotope with six protons and seven neutrons is ^{13}Carbon, and a carbon isotope with six protons and eight neutrons is ^{14}Carbon (box figure 2.1*A*).

Some isotopes are stable, while others are unstable and tend to decay (break down) by spontaneously emitting small particles. These unstable isotopes are **radioactive.** ^{12}Carbon and ^{13}Carbon are stable isotopes, while ^{14}Carbon is radioactive.

The isotopes of some elements emit particles with a great deal of energy. Although some isotopes can be used in medicine to kill tumor cells, some of the radioisotopes produced in nuclear reactors emit very high energy radiation and are extremely dangerous to handle and difficult to store safely.

Certain isotopes that emit lower energy radiation are very useful in biological research and medical diagnosis. Radioactive isotopes can be used as chemical "labels" that mark certain compounds in cells or tissues. Because the different isotopes of an element react chemically in the same way, radioactive isotopes may be substituted for stable isotopes to make chemical substances under observation easier to detect.

Scientists can detect the location of radioactive isotopes by using photographic film. During a period of storage in the dark, the film is "exposed" by the particles emitted by radioactive atoms. When the film is developed, an image emerges that shows precisely where the radioactive atoms are located (box figure 2.1*B*). This localization technique, called autoradiography, is an extremely valuable tool in biological and medical research.

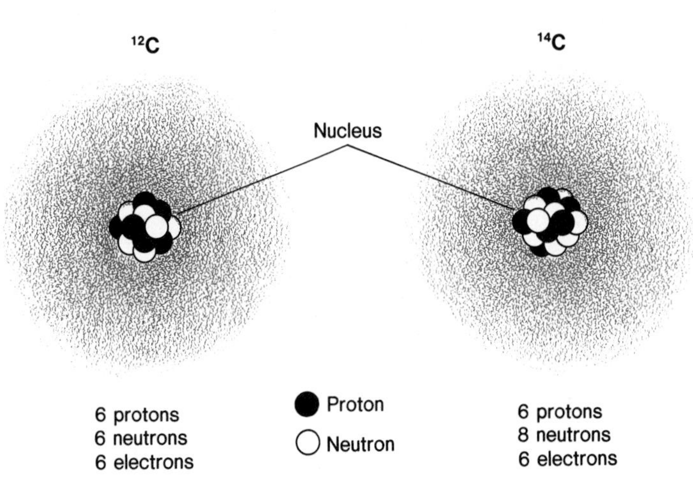

Box figure 2.1*A* Nuclei of the isotopes of an element differ only in the number of neutrons present in the nucleus. ^{12}Carbon, the most common form of carbon atom, and ^{14}Carbon, a radioactive isotope, are shown here. Note that each has six protons, but that ^{12}Carbon has six neutrons while ^{14}Carbon has eight neutrons. The shaded area around the nuclei represents the areas in which six electrons are located in each case.

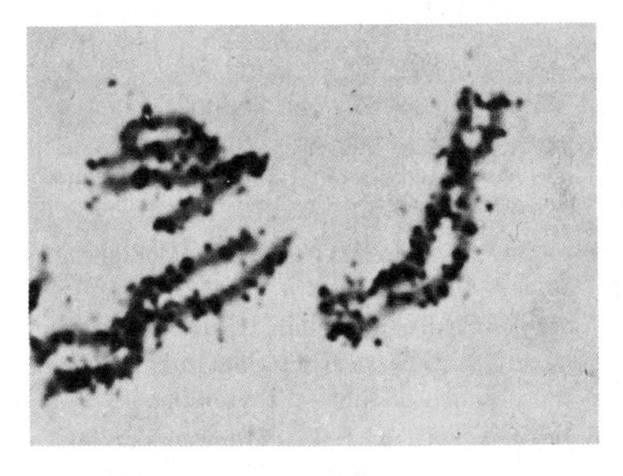

Box figure 2.1*B* An autoradiograph showing chromosomes that have their DNA labeled with a radioisotope. The black spots are places where radioactive decay has "exposed" the photographic emulsion spread over the chromosomes.

Figure 2.5 Two covalently bonded compounds with single bonds, water (H_2O) and methane (CH_4). The small circles in the diagrams represent the electrons in the outer shell of each atom. Of course, all electrons are the same, but they are represented as open and solid circles to demonstrate the electron sharing in the bond more clearly. A dash (—) represents a shared pair of electrons in the alternative representation shown to the right in each case.

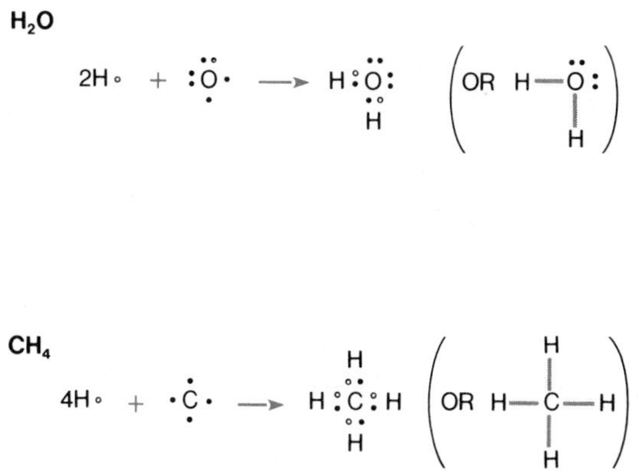

H_2O

CH_4

Figure 2.6 Some compounds with multiple covalent bonds. Each dash in the molecular structures represents a pair of shared electrons. (*a*) Carbon dioxide. (*b*) Ethylene. (*c*) Acetylene (ethyne).

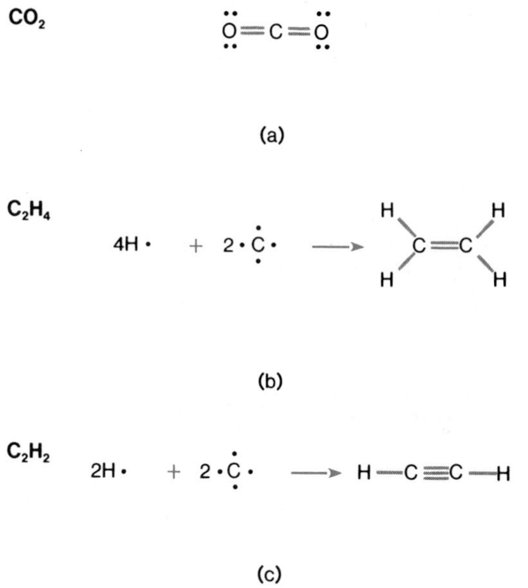

CO_2 (a)

C_2H_4 (b)

C_2H_2 (c)

Covalent Bonding

In **covalent bonds** electrons are not donated from one atom to another. In covalent bonding, electrons are shared between atoms. Thus, a covalent bond consists of a pair of electrons that are shared between two atoms and that occupy two stable orbitals, one from each atom. You can see the electron sharing in covalent bonds in figure 2.5. If a single pair of electrons is shared between a pair of atoms, it is called a **single bond.** The bonds shown in figure 2.5 are single bonds. Sometimes two atoms share two or three pairs of electrons to form **double** or **triple bonds** (figure 2.6).

Weaker Interactions

In addition to ionic and covalent bonds, there are also several types of weaker interactions among atoms and molecules. These weaker interactions are largely responsible for the three-dimensional shapes and some functional properties of many large biological molecules, such as proteins.

In some molecules that are formed by covalent bonds, the electrons tend to be attracted more towards one atom than the other. This unequal distribution of the electrons is called **polarity.** Polarity results in molecules that have a small positive charge at one end and a small negative charge at the other (figure 2.7). Because of these slight charges at their ends or poles, polar molecules are attracted to each other.

An example of this attraction between polar molecules is **hydrogen bonding,** which is very important in the chemical structure of living things. Hydrogen bonding occurs between a hydrogen atom that is already bonded covalently to nitrogen or oxygen and a *different* nitrogen or oxygen atom (see figure 2.8). Hydrogen bonding between different molecules is crucial to the role of water in living systems.

Water in Living Systems

Water is the most abundant material in living things (60 to 95 percent by weight). Understanding some properties of water is essential to understanding virtually all processes that occur in living things.

Some Properties of Water

The structure of the water molecule is nonlinear (bent) with a bond angle of 104.5° between the HOH (hydrogen-oxygen-hydrogen) atoms (figure 2.9). Water is a polar molecule, with the oxygen atom forming the negative end and the hydrogen atoms the positive end. Therefore, hydrogen bonding occurs

Figure 2.7 Examples of polar molecules. δ^+ and δ^- symbolize small charge differences at different ends of polar molecules.

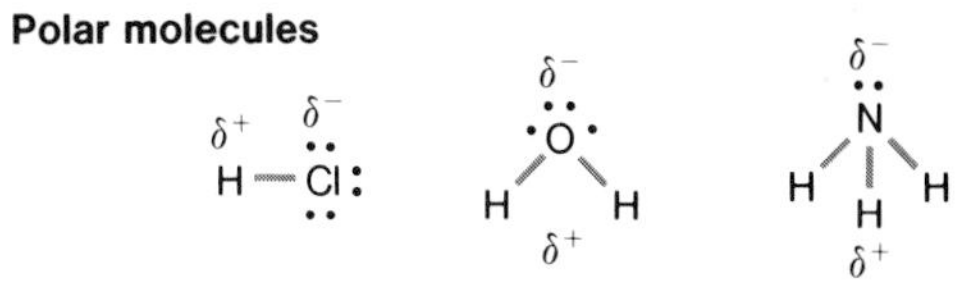

Polar molecules

δ^+ H—Cl: δ^- δ^- :O: with H H δ^+ δ^- :N with H H H δ^+

Figure 2.8 Hydrogen bonding (indicated by colored dashed lines).

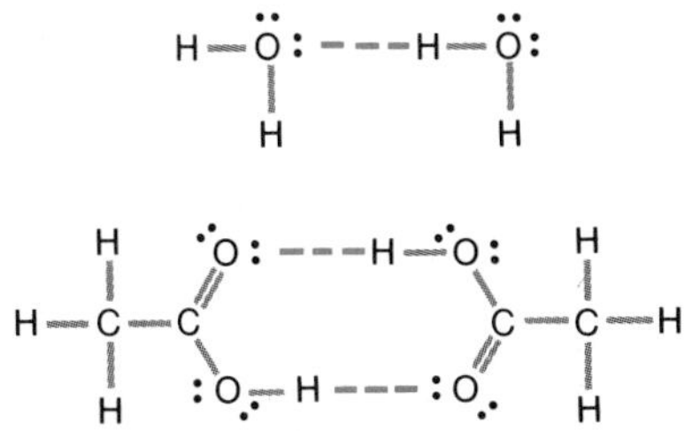

H—O:---H—O: (each O bonded to H)

H_3C—C(=O)—O—H---:O=C—CH_3 ; O:---H—O (acetic acid dimer)

between water molecules. We can see the importance of hydrogen bonding in water most strikingly by examining the structure of ice. In ice, each oxygen atom is surrounded on four sides (tetrahedrally) by four hydrogen atoms, two close (covalent bonds) and two distant (hydrogen bonds). As figure 2.9*c* shows, hydrogen bonding in ice produces a very open structure, resulting in a low density (a large volume per molecule). The density of ice is, in fact, less than the density of water in its liquid state. Water reaches its maximum density at 4°C. This explains why ice floats on cold water and why lakes freeze at the top first, leaving water below.

When a body of water does freeze on the surface, the ice acts as an insulator to prevent much of the water below it from freezing. If ice were denser than water, lakes and ponds would freeze solid, from the bottom up, and most aquatic organisms would die during the winter.

Several other physical properties of liquid water are important for water's role in living things. Because of the hydrogen bonds holding its molecules together, water has much higher melting and boiling points than we would otherwise expect for a molecule of its size. It also has a distinctively high **heat capacity** (the amount of heat energy that must be added or subtracted to raise or lower its temperature). The high heat capacity of water helps stabilize and protect organisms from rapid fluctuations in temperature. On a larger scale, the heat capacity of water stabilizes the temperature of lakes and oceans and makes the climate of land areas near large bodies of water relatively moderate.

Figure 2.9 (*a*) Structure of a water molecule. (*b*) Tetrahedral hydrogen bonding of water molecules in ice. (*c*) The open structure of ice. (*d*) An iceberg, a huge mass of floating ice. Because of its relatively open structure, ice has a lower density than liquid water and will float in water.

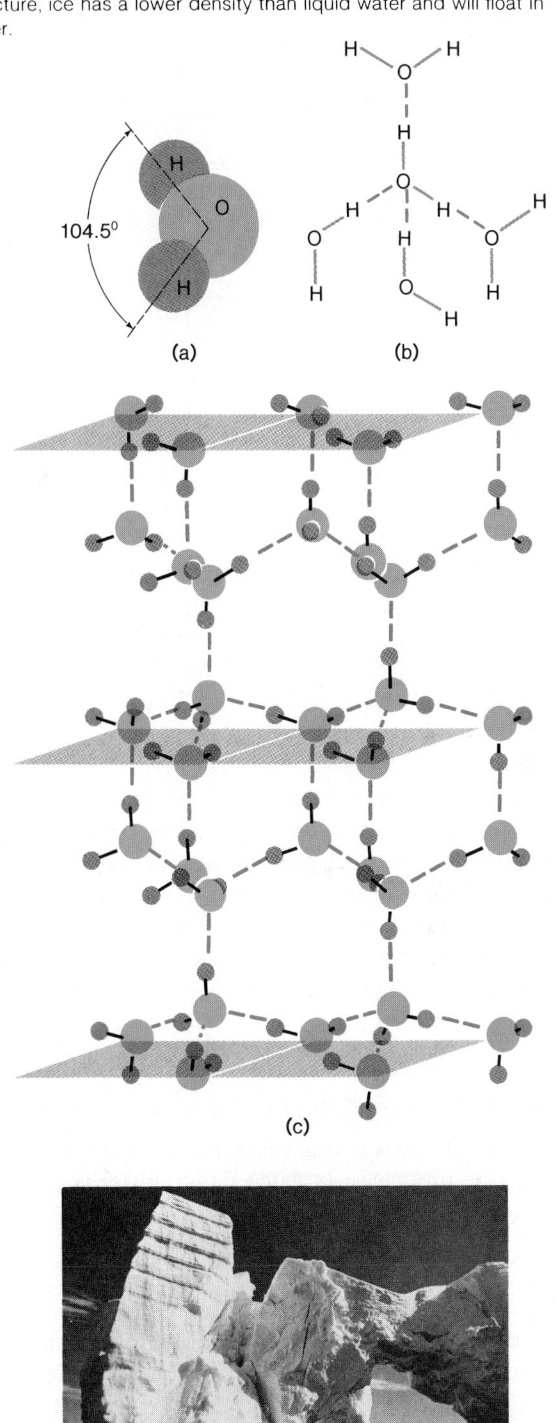

(d)

Figure 2.10 A water strider supported by surface tension.

Water also has a high **heat of vaporization.** Heat of vaporization is the amount of heat energy required to convert water from a liquid to a gaseous state. Water's high heat of vaporization makes cooling by evaporation an efficient process. For example, the evaporation of a single gram of water cools 540 grams of water by 1°C. This property makes it possible for perspiration to help to dissipate excess body heat. This cooling effect of evaporation is demonstrated dramatically when you climb out of a swimming pool on a breezy day. You can become chilled even though the air temperature is quite warm.

The hydrogen bonding that takes place among water molecules also contributes to the large amount of **surface tension** in water. In other words, the molecules making up the surface film of water resist being pulled apart. Water's surface tension is directly exploited by insects such as the water strider, which can "skate" on a water surface, its small weight supported by the surface tension of the water (figure 2.10).

Water as a Solvent

Major parts of all living things consist of watery solutions. In a **solution,** there is a uniform mixture of the molecules of two or more substances. In living things, water is by far the most important **solvent,** the dissolving substance that is present in the greatest amount. Thus, molecules of various **solutes** (dissolved substances) are dispersed among molecules of water in solutions in living things.

How does water function as a solvent and what determines which substances will or will not dissolve in water? Water's properties as a solvent relate to the polarity of water molecules. Recall that one end (pole) of each water molecule has a slight positive charge and the other a slight negative charge, and that this polarity explains hydrogen bonding among water molecules. Many molecules that are important

Figure 2.11 The formation of a micelle by molecules that have a polar "head" and a nonpolar "tail."

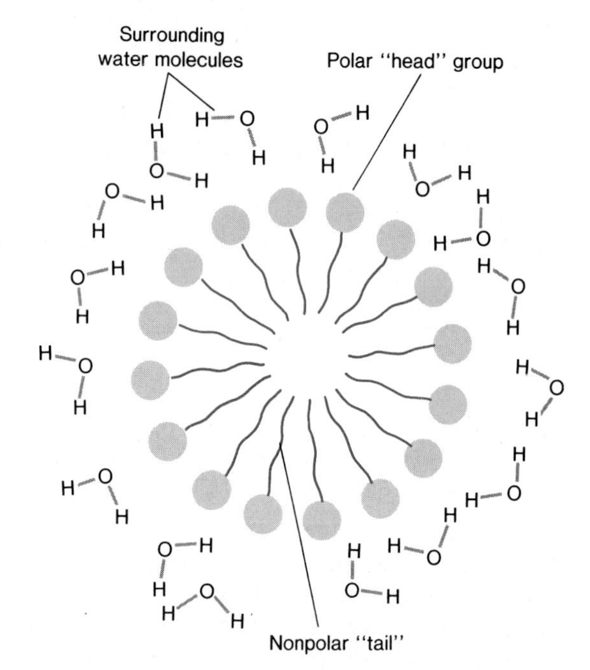

in living things are polar or have polar regions. The positive regions of such molecules are attracted to the negative ends of water molecules and vice versa. Water molecules are similarly attracted to them. The attractive forces involved are called **hydrophilic** ("water-loving") interactions. Because of these hydrophilic interactions with water molecules, such polar molecules become interspersed among water molecules. Sugars, salts, some alcohols, and ammonia are a few examples of water-soluble substances.

Other molecules, however, are nonpolar or have large nonpolar parts, and such molecules do not dissolve in water. Because the forces attracting water molecules to one another are stronger than those acting between water and such nonpolar molecules, the nonpolar molecules are "squeezed out" from between water molecules. Because nonpolar molecules do not interact with water molecules, they cannot become dispersed among the water molecules to form a solution. Fats and oils are good examples of such substances. No matter how vigorously you stir oil in water, the oil remains in droplets and does not go into solution. The interactions between water molecules and nonpolar molecules or nonpolar parts of molecules are called **hydrophobic** ("water-hating") interactions.

What happens when a molecule that has both a small, polar (hydrophilic) portion and a large, nonpolar (hydrophobic) portion is added to water? When molecules with polar "heads" and long, nonpolar "tails" are mixed with water, they form structures known as **micelles** (figure 2.11). In a micelle, the nonpolar tails turn toward each other and away from the H_2O molecules. However, hydrophilic attraction does occur between the polar head and adjacent H_2O molecules. Detergents provide a familiar, everyday example of micelle formation. A detergent disperses grease by incorporating the grease molecules into the interiors of micelles.

The phenomenon of micelle formation reveals much about the structure of biological membranes because, as we shall see in the next chapter, biological membranes are largely made up of molecules that have both hydrophilic and hydrophobic ends.

Acid-Base Phenomena

In biological systems, one of the most important ions is H^+, the hydrogen ion. The concentration of hydrogen ions strongly affects the rates of many reactions, the shape and function of enzymes and other protein molecules, and even the integrity of the cell itself.

A hydrogen atom is the simplest of all atoms. It contains one proton in the nucleus and one electron in an orbital outside the nucleus. When the electron is lost (forming the H^+ ion), all that is left is the proton in the nucleus. In fact, the H^+ ion is often simply called a proton.

In the internal environment of organisms, the most abundant molecule is water, which carries two pairs of electrons that are not involved in bonding:

$$\text{H}-\ddot{\text{O}}:-\text{H}$$

A proton (H^+) is strongly attracted to one of the pairs of electrons in a water molecule.

$$H^+ + :\ddot{O}(H)-H \rightleftharpoons [H-\ddot{O}(H)-H]^+$$

In fact, this proton attraction to water is so strong that no *free* hydrogen ions (H^+) exist in solution in water. The H^+ ion is always attached to a water molecule. But for convenience in our discussion, we will represent the hydrogen ion simply as H^+.

H^+ Concentration and pH

At any given time, a small proportion of the water molecules in pure water are **ionized;** that is, they are split into hydrogen ions (H^+) and hydroxide ions (OH^-) by this reaction:

$$H_2O \rightleftharpoons H^+ + OH^-$$

This reaction is reversible, as indicated by the arrows pointing in both directions. We say that this reaction is at **equilibrium** because in pure water, the rate of ionization of water molecules

$$H_2O \longrightarrow H^+ + OH^-$$

is equal to the rate of reaction among the ions.

$$H_2O \longleftarrow H^+ + OH^-$$

Thus, there is no further *net* change in concentration, even though both reactions continually occur.

Of course, the concentration of H^+ ions in pure water has to equal the concentration of OH^- ions because neither can be formed without the other. This is an important point because the situation that exists in pure water stands at the midpoint of a scale known as the **pH scale,** which is a measure of the concentration of H^+ ions. Pure water, and solutions in which there is a similar equality between the concentrations of H^+ ions and OH^- ions, are **neutral.** Neutral solutions have a value of 7.0 and stand at the midpoint of the pH scale, which runs from 0 to 14.

In some solutions, there is an excess of H^+ ions; that is, the concentration of H^+ ions is greater than the concentration of OH^- ions. Such solutions are **acidic,** and they have pH values of less than 7.0. In other solutions there is an excess of OH^- ions. Such solutions are **basic,** and they have pH values greater than 7.0.

The pH scale encompasses a broad range of H^+ ion concentrations (figure 2.12). It is constructed so that a difference of one pH unit represents a tenfold difference in H^+ ion concentration. For example, a solution with a pH of 6.0 has a H^+ ion concentration that is ten times as large as the H^+ ion concentration of a solution that has a pH of 7.0.

Acids and Bases

Some substances change the concentration of H^+ ions or OH^- ions in a solution when they dissolve in water. Such substances are known as **acids** and **bases.**

An acid causes an increase of H^+ ion concentration when it ionizes in solution. For example, hydrochloric acid (HCl) is ionized quite completely by the reaction:

$$HCl \longrightarrow H^+ + Cl^-$$

Adding hydrochloric acid to a solution thus increases the relative number of H^+ ions and, therefore, lowers the pH of the solution. The solution of gastric juices in the human stomach,

Figure 2.12 The pH scale showing the pH of various common fluids.

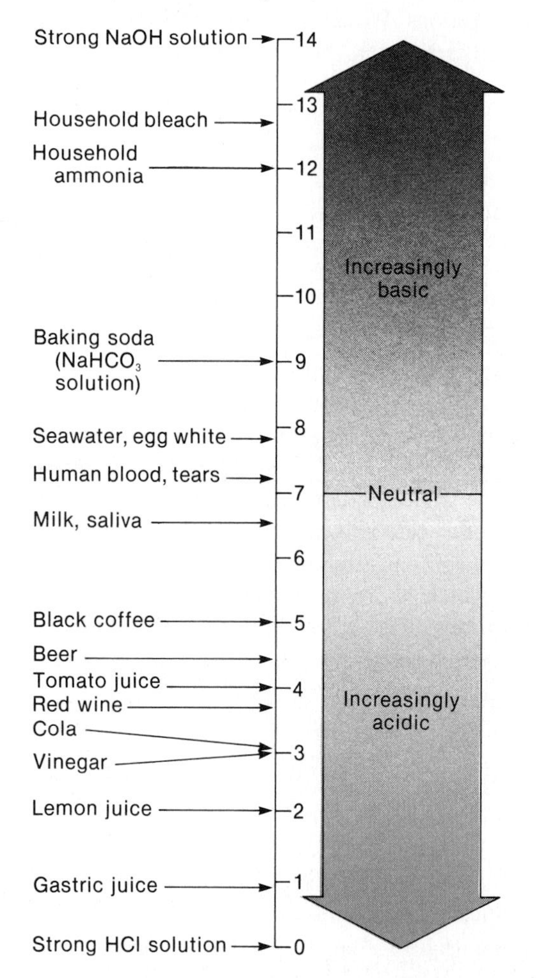

for example, contains a great deal of hydrochloric acid and sometimes has a pH as low as 1.0. This is a very acidic solution indeed. In fact, stomach contents poured on skin could cause a serious acid burn.

Correspondingly, a base causes an increase in the relative concentration of OH^- ions when it ionizes in solution. For example, sodium hydroxide (NaOH) ionizes almost completely and strongly increases the relative concentration of OH^- ions. Hydroxide ions (OH^-) combine with H^+ ions to form water, and thereby decrease the H^+ ion concentration.

Buffers

Even small pH changes may have great effects on biological processes. Because many processes are sensitive to pH changes, it is important that the pH of body fluids remains stable. Cell interiors have an average pH of 7.0 to 7.3. In humans, for example, blood and tissue fluids have a pH of 7.4 to 7.5. These values remain very stable despite the fact that many biochemical reactions either release or incorporate H^+ ions.

Why does pH remain so constant in living things? Such pH stability is possible because organisms have built-in mechanisms to prevent pH change. The most important of these mechanisms are **buffers.** A buffer resists pH changes by removing H^+ ions when the H^+ ion concentration rises and by releasing H^+ ions when the H^+ ion concentration falls. In this way, the H^+ ion concentration is kept relatively constant.

An example of a buffer is the carbonic acid-bicarbonate ion buffer system, which is the most important buffer system in human blood.

$$\underset{\text{(Carbonic acid)}}{H_2CO_3} \rightleftharpoons H^+ + \underset{\text{(Bicarbonate ion)}}{HCO_3^-}$$

In this example, if H^+ ions are added to the system, they combine with HCO_3^- to form H_2CO_3. This reaction removes extra H^+ ions and keeps the pH from changing. If H^+ ions are removed from the system (for example, if OH^- ions are added and H_2O is formed), more H_2CO_3 will ionize and replace the H^+ ions that were used. Again, pH stability is maintained. This and other buffer systems prevent potentially harmful pH changes in living things.

Organic Molecules

Most of the thousands of kinds of molecules found in living things are **organic compounds,** compounds based on the element carbon. The characteristics of the living cells that make up organisms depend on the properties and arrangements of various organic molecules. And the most important characteristics of organic compounds depend on the properties of carbon.

The carbon atom has an atomic number of six, with six protons in its nucleus and six electrons around its nucleus. Two of these electrons are in the orbital in its first shell and four electrons are in the second shell (page 29). This means that a carbon atom can form covalent bonds with as many as four other atoms. A carbon atom can bond to a variety of elements, but it most commonly bonds to hydrogen, oxygen, nitrogen, and carbon.

Many organic compounds consist of just carbon and hydrogen, but others contain **functional groups** as well. Functional groups are characteristic patterns of atoms other than those involved in carbon-carbon and carbon-hydrogen bonds.

Figure 2.13 Some examples of functional groups contained in organic molecules.

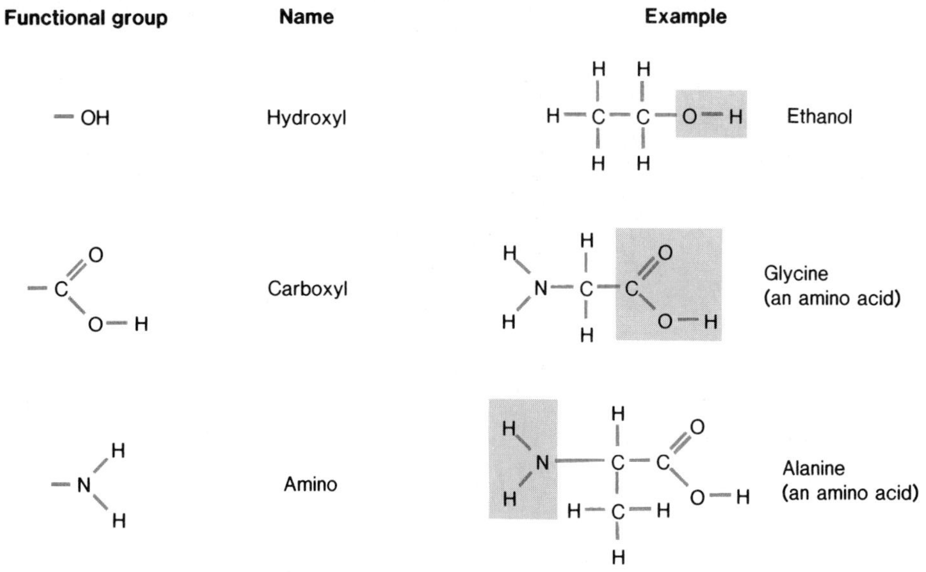

Functional group	Name	Example
$-OH$	Hydroxyl	$H-C(H)(H)-C(H)(H)-O-H$ Ethanol
$-C(=O)-O-H$	Carboxyl	$H_2N-C(H)(H)-C(=O)-O-H$ Glycine (an amino acid)
$-N(H)(H)$	Amino	$H_2N-C(H)(CH_3)-C(=O)-O-H$ Alanine (an amino acid)

These functional groups are the parts of organic molecules that are most often involved in chemical reactions. Some important functional groups are shown in figure 2.13.

Another factor contributing to the diversity of organic compounds is the presence of numerous **isomers.** Isomers are compounds having the same molecular formulas but different atom arrangements. *Structural isomers* are completely different chemical compounds; they share the same molecular formula but have different pairs of atoms connected by bonds (figure 2.14*a*). *Geometric isomers* differ only in the orientations of bonds between particular pairs of atoms (figure 2.14*b*).

Monomers and Polymers

Several prominent types of organic compounds in cells are very large molecules called **macromolecules** (macro = large). Macromolecules are **polymers** (poly = many; mer = unit); that is, they are large molecular chains of smaller organic molecules linked together. The smaller organic molecules that form the individual units of polymer chains are called **monomers.**

Monomers are joined together into polymers by **condensation reactions** in which the components of water molecules are removed (figure 2.15). The reverse of a condensation reaction is **hydrolysis** (the addition of water molecules), which separates the monomers from one another. This pattern of condensation and hydrolysis is essential to many synthesis processes in living things, as well as to the breakdown of large molecules in processes such as digestion.

Carbohydrates

Polysaccharides ("many sugars") function as energy-storage and structural compounds in organisms. The monomer units that make up polysaccharides are molecules of single sugars (**monosaccharides**) such as glucose. Both polysaccharides and the sugar units from which they are constructed belong to the class of organic compounds called **carbohydrates** because they contain carbon and the components of water—hydrogen and oxygen.

Although there are sugars that contain three, four, five, six, or more carbons, the common polysaccharides are polymers of six-carbon sugars. The polysaccharides described here are all polymers of glucose ($C_6H_{12}O_6$). Glucose is the monomer that makes up several important types of polysaccharides in living things. Glucose occurs in several forms, including a straight chain form and two different ring forms (figure 2.16). Pay special attention to these two different ring structures, the α (alpha) and β (beta) forms, and the position of the hydroxyl (OH) group attached to the number 1 carbon in each structure. It is important for you to understand this apparently slight difference between these two forms of glucose in order to grasp some very significant differences among the structures of various glucose polymers.

Figure 2.14 Some examples of isomers. (*a*) Structural isomers are molecules with the same molecular formula, but different pairs of atoms connected by bonds. The sugars glucose and fructose both have the molecular formula ($C_6H_{12}O_6$), but their atoms are arranged differently. (*b*) Geometric isomers are molecules that have different orientations of bonds in space but that are not mirror images. Each line indicates a C-C bond in retinal. Hydrogens are not shown. In the retina of the eye, *cis*-retinal is present (combined with a protein) as rhodopsin. A photon of light is absorbed by *cis*-retinal, converting it to *trans*-retinal. Thus, geometric isomers are important in the initial step in vision.

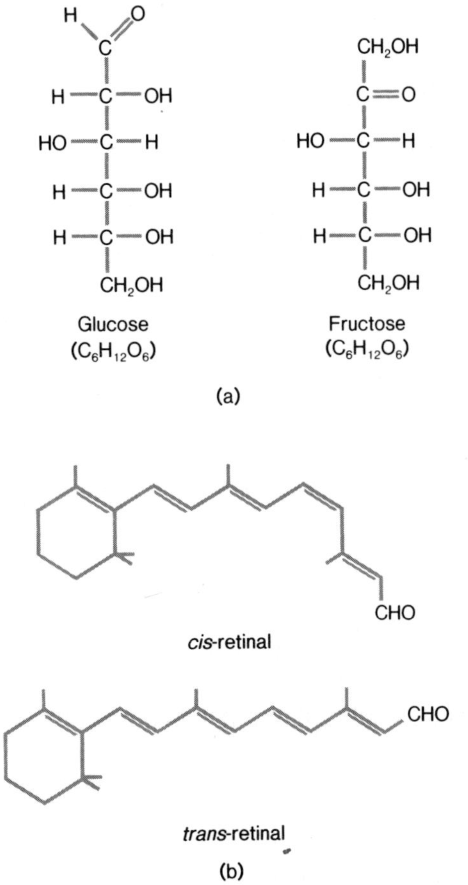

Glucose
($C_6H_{12}O_6$)

Fructose
($C_6H_{12}O_6$)

(a)

cis-retinal

trans-retinal

(b)

Linking Glucose Molecules

Polymers of glucose, like other biologically important polymers, are formed by condensation reactions. We can see a simple example of such a reaction in the formation of **disaccharides** (double sugars) from monosaccharides. For example, two glucose molecules may combine to produce a molecule of maltose (malt sugar), a disaccharide. The molecular formulas involved in this condensation reaction are:

$$C_6H_{12}O_6 + C_6H_{12}O_6 \longrightarrow C_{12}H_{22}O_{11} + H_2O$$

(Glucose) (Glucose) (Maltose)

Figure 2.15 The reactions that link and separate the monomers in an organic polymer. (*a*) Joining of molecules by condensation (H_2O is removed). Condensations can connect a series of monomers to produce a long-chain polymer. (*b*) Hydrolysis (separation with addition of H_2O) is the reverse of condensation.

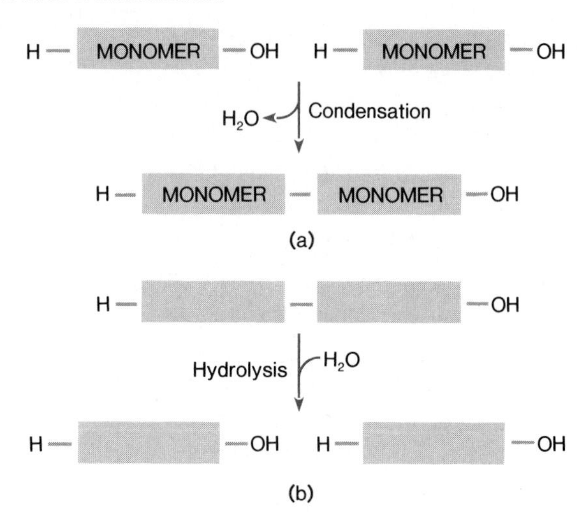

The structural formulas in figure 2.17*a* show how the actual linkage of the glucose units in maltose is formed between the number 1 carbon of one glucose molecule and the number 4 carbon of the other. Because these are α-glucose molecules (see figure 2.16), this is called an **$\alpha(1 \longrightarrow 4)$ linkage.**

Sucrose is a disaccharide, familiar to you as common table sugar. Sucrose is a very common sugar in nature because when sugar is transported from one part of a plant to another, it is usually in the form of sucrose. Sucrose consists of one molecule of glucose and one molecule of fructose, another common six-carbon sugar (figure 2.17*b*).

Starch and Glycogen

Starch and glycogen are two glucose polymers that are used in carbohydrate storage in living things. Plant cells store extra carbohydrates as starch.

Starch is a polymer of glucose that occurs in two forms. One form consists of unbranched, coiled chains of α-glucose units. The other common form of starch consists of branched chains of glucose units (figure 2.18).

Glycogen or "animal starch" is another glucose polymer. Glycogen functions in carbohydrate storage in animals. For example, human liver cells contain considerable quantities of glycogen.

Figure 2.16 Representations of glucose molecular structure. Cyclic (ring) structures of β- and α-glucose with the straight-chain form between them. Note that in the ring structure, "corners" represent carbons and short, unlabeled lines show where hydrogens are attached. Colored numbers represent the conventional numbering scheme that permits identification of each carbon atom.

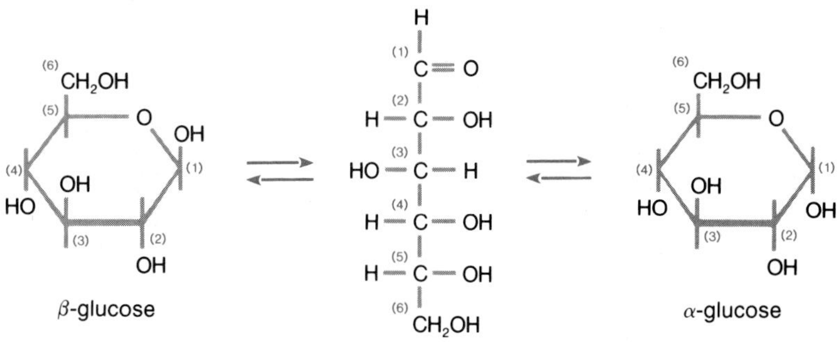

β-glucose

α-glucose

Figure 2.17 Disaccharides. (*a*) Maltose is formed from two molecules of glucose in a condensation reaction. Components of a water molecule (colored box) are removed as the bond is formed between two molecules of α-glucose. Note that the α(1→4) bond is between the number 1 carbon of one glucose and the number 4 carbon of the other. (*b*) Sucrose, a disaccharide composed of one glucose molecule and one fructose molecule.

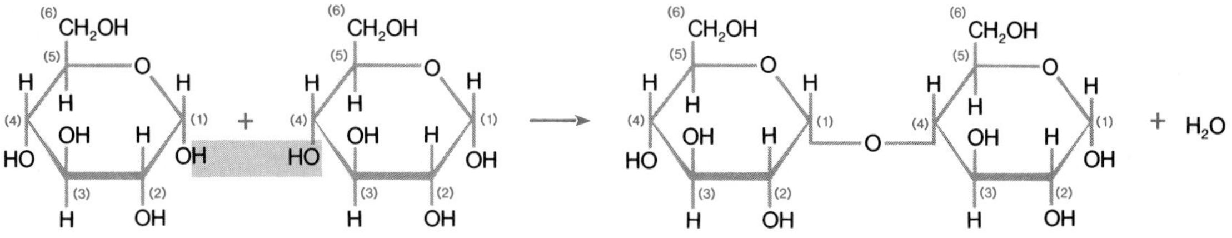

Maltose

(a)

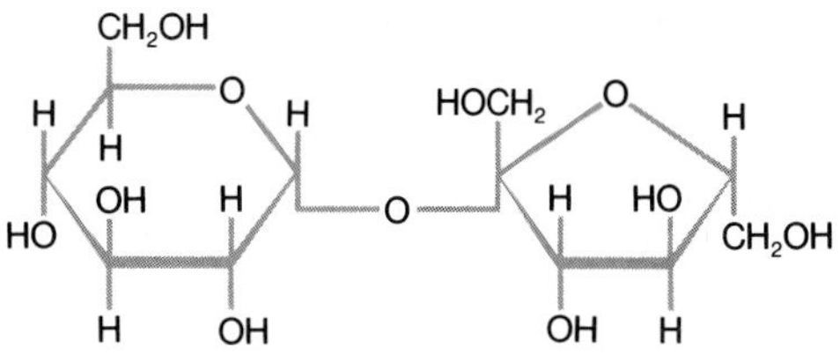

Sucrose

(b)

Cellulose

Cellulose is an important structural polysaccharide because it forms a major part of the rigid **cell walls** surrounding plant cells. Because of its structural role in all plant cells, cellulose is the most abundant organic compound in the world.

Cellulose differs from starch because cellulose is a polymer of β-glucose (figures 2.16 and 2.19). The **β(1 → 4) linkages** in cellulose are linear, so cellulose molecules tend to form in straight chains. When these chains lie side by side, they interact through extensive hydrogen bonding. The bonds hold cellulose molecules firmly together to form the strong cellulose fibers, called **microfibrils,** found in cell walls. Mature plant cells are enclosed in rigid boxes made of layers of cellulose fibers.

Although the human digestive system cannot digest cellulose, it nevertheless is economically important. Cellulose is used to produce the paper this book is written on, and it forms the linen and cotton in your clothing. Cellulose can even be modified for use as an explosive (nitrocellulose).

Figure 2.18 Starch. (*a*) A portion of a branched starch molecule showing a branch point. Note that the branch point is an $\alpha(1\rightarrow6)$ linkage. The number 1 carbon of one glucose is linked to the number 6 carbon of another glucose. (*b*) An unbranched starch chain that coils to form a helix because of characteristics of the $\alpha(1\rightarrow4)$ linkages. Individual glucose units have been further simplified in this diagram. (*c*) Plant cells store carbohydrate as starch. This electron micrograph shows many starch grains inside chloroplasts in photosynthesizing cells in a corn (*Zea mays*) leaf. Plants also store considerable quantities of starch in specialized storage areas in roots or, in some cases, stems (magnification X 10,688).

Branch

Branch point
$\alpha(1 \rightarrow 6)$ linkage

Main $\alpha(1 \rightarrow 4)$ chain

(a)

(b)

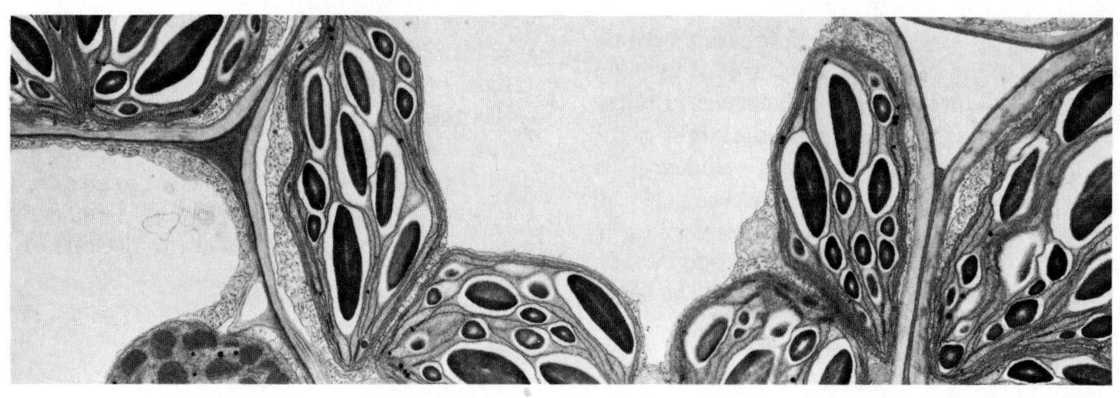

(c)

Figure 2.19 Cellulose. (*a*) A highly diagrammatic comparison of the $\alpha(1 \rightarrow 4)$ linkages in starch and the $\beta(1 \rightarrow 4)$ linkages in cellulose. Refer again to figure 2.16 to see the difference between α- and β-glucose. (*b*) A surface view of part of the cell wall of the alga *Valonia,* showing the arrangement of cellulose microfibrils (magnification $\times$ 27,200).

Branched starch

Cellulose

(a)

(b)

Box 2.2
Cellulose as Food and Fuel

Although the variations between the kinds of linkages that join glucose molecules in various polysaccharides may seem insignificant, they actually make a great difference for animals that eat plants. Polysaccharides must be broken down into glucose molecules, or at least into disaccharides, before they can be absorbed in an animal's digestive system. The digestion of polysaccharides is accomplished by enzymes that break the bonds linking glucose molecules. These enzymes, however, are very specific in the types of bonds they will break. For example, enzymes that break the bonds linking glucose units in starch will not break the bonds of cellulose.

Most animal digestive systems produce **amylases,** enzymes that split the $\alpha(1 \rightarrow 4)$ linkages in starch molecules. But **cellulases,** enzymes that break the $\beta(1 \rightarrow 4)$ linkages of cellulose molecules, are extremely rare in the animal kingdom. Because of the absence of cellulases, foods such as lettuce have practically no caloric value for humans and many other animals because they pass through the digestive tract with their cellulose virtually intact. But cellulose molecules would provide a great deal of energy if animals could use the glucose that cellulose contains.

Some animals do derive energy from the cellulose they eat. Certain snails and a very few other animals actually seem to produce cellulases and can digest cellulose. Other animals, such as termites, derive energy from cellulose through the microorganisms that live in their digestive tracts. These organisms produce cellulases, breaking down cellulose and making glucose available for themselves and their hosts. Thus, when termites do their familiar damage to wood, which is mostly cellulose, they depend on these microorganisms to produce cellulase. Termites produce no cellulase of their own, and in fact, a young termite raised in sterile isolation from other termites will starve to death even when it is provided with all the cellulose it can eat. Apparently, termites must be "infected" with these gut-dwelling microorganisms early in life.

A ruminant mammal (such as a cow, sheep, or bison) has a special stomach chamber, the **rumen,** where a culture of microorganisms digests cellulose, making absorbable material available for the host. By this means, ruminant mammals are able to utilize the cellulose in plant material for energy and growth.

When people insist on eating animals that are fed grain—grain that could be used to nourish humans—they may aggravate the world's food supply problem. Therefore, some biologists propose that humans adjust their tastes and learn to eat meat from basically grass-fed ruminants rather than grain-fed animals. In this way, meat would be produced using only plant material not suitable for direct use as human food.

Microorganisms that produce cellulases may also eventually be used to help solve general energy supply problems. Scientists are trying to develop efficient processes for using inedible plant material to produce alcohol for fuel. There are a number of technical problems involved in conducting these reactions on a large scale, but finding the solutions seems worthwhile. The development of practical methods for alcohol production from cellulose would provide a large new source of liquid fuel and perhaps stop the use of edible grain in alcohol-production processes.

Box figure 2.2 Animals that can derive nutritional value from cellulose. Some snails' digestive tracts secrete cellulases, enzymes that hydrolyze the $\beta(1 \rightarrow 4)$ linkages between glucose units in cellulose. Flagellated protozoans in the termite gut digest cellulose, thus permitting termites to thrive on a wood diet. Cows and other ruminant mammals (see chapter 6) have a permanent culture of cellulose-digesting microorganisms in the saclike rumen, a part of the stomach.

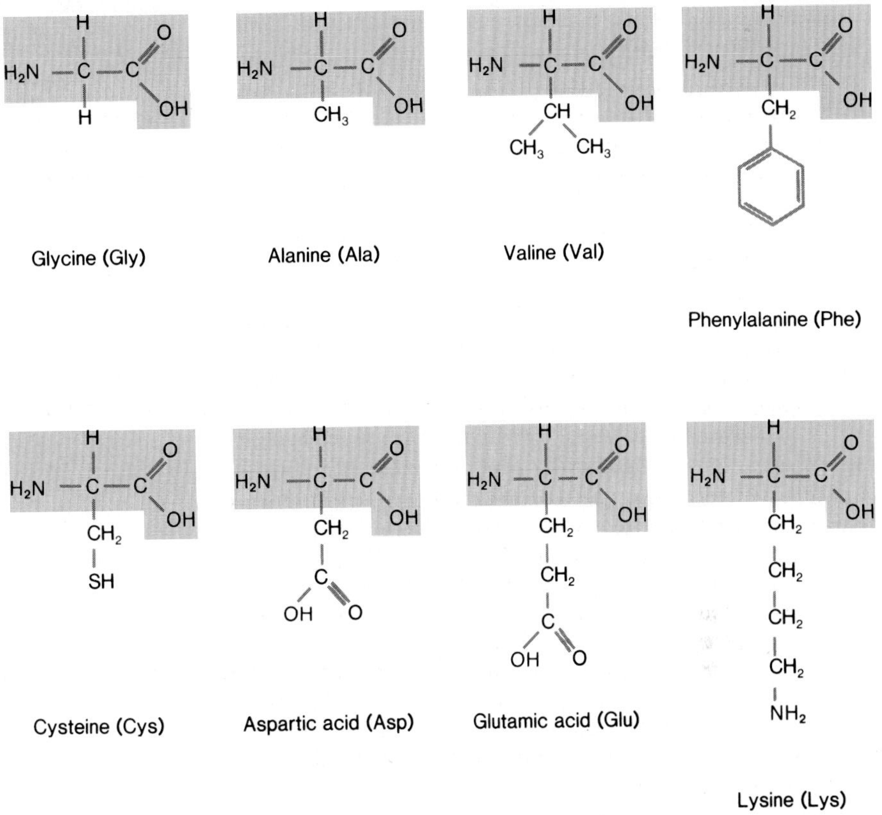

Figure 2.20 Eight of the amino acids that occur in proteins. Three-letter abbreviations commonly used for the different amino acids are included. The portion of the molecule common to all amino acids is enclosed in the color-shaded box in each case. Amino acids differ in their various side chains. The other amino acids that occur in proteins are: leucine (Leu), isoleucine (Ile), methionine (Met), tryptophan (Trp), proline (Pro), serine (Ser), threonine (Thr), asparagine (Asn), glutamine (Gln), tyrosine (Tyr), arginine (Arg), and histidine (His).

Glycine (Gly)

Alanine (Ala)

Valine (Val)

Phenylalanine (Phe)

Cysteine (Cys)

Aspartic acid (Asp)

Glutamic acid (Glu)

Lysine (Lys)

Proteins

It would be difficult to overemphasize the importance of proteins in living systems. Certain types of proteins form important parts of structural components in all cells and in spaces between cells. Enzymes, the contracting elements in muscle cells, and antibodies, which are involved in resistance to infection, are also proteins.

Proteins consist of precisely defined linear sequences of amino acids; that is, protein molecules are polymers made up of amino acids linked together by condensation reactions. An **amino acid** is an organic compound with an amino group (NH_2) and a carboxyl group (COOH) bonded to the same carbon atom (figure 2.20). Twenty different types of amino acids occur in proteins, allowing for an enormous variety of protein molecules, all with different structures and properties. In this respect proteins contrast sharply with polysaccharides such as starch and cellulose, which are polymers constructed of only one type of monomer.

Amino acids are linked together to form proteins by means of **peptide bonds.** These bonds are formed by condensation reactions between carboxyl and amino groups (figure 2.21). Because proteins are chains of amino acids connected by these bonds, proteins often are called **polypeptide chains** or simply **polypeptides.**

Figure 2.21 Formation of the peptide bond from the amino and carboxyl groups (colored boxes) of two amino acids.

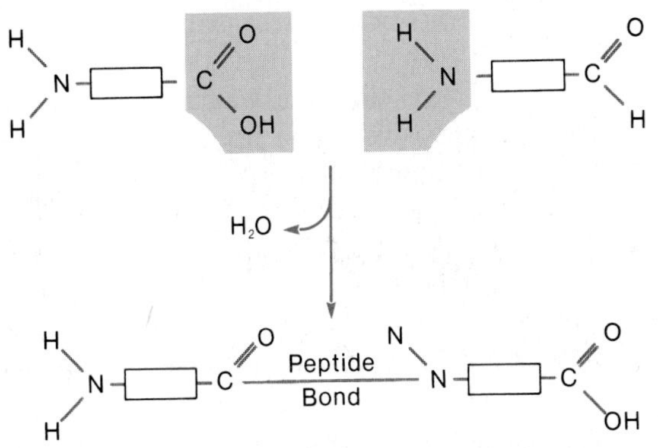

Any one particular polypeptide always has exactly the same number of amino acids arranged in exactly the same sequence. This remarkable constancy in individual polypeptide structures is an essential characteristic of protein structure and function.

Levels of Protein Structure

Proteins are not simple, straight chains of amino acids; their molecules are coiled, folded, and sometimes grouped together with other protein molecules in complex arrangements. This complexity has been cataloged into four structural levels, called (quite logically) the primary, secondary, tertiary, and quaternary structures of proteins.

The **primary structure** of a protein is the sequence of the amino acids joined together by peptide bonds in the protein molecule. Determining the primary structure of a protein is almost like solving a chemical mystery story (figure 2.22). To do so, it is necessary to remove just one amino acid (or a small group of amino acids) at a time from the end of the polypeptide chain for identification.

A polypeptide has a specific three-dimensional shape (conformation) that depends on several levels of organization beyond the primary structure. The **secondary structure** of a protein consists of specific three-dimensional folding arrangements within a polypeptide. A common example of such a pattern is the alpha-helix (α-helix) found in many proteins (figure 2.23). This α-helix is a coil running in a single direction. A large number of weak bonds (including hydrogen bonds) between amino acids in the polypeptide chain stabilize the α-helix.

The **tertiary structure** of a protein, which also is stabilized by a number of weak bonds, is formed when the polypeptide chain, with its orderly secondary structure, is bent and folded into an even more complex three-dimensional arrangement. Thus, a protein chain, which may already be arranged in an α-helix, folds upon itself to form its tertiary structure, a more complex, often globular shape (figure 2.24).

Many proteins are constructed of two or more polypeptide chains, each with its own primary, secondary, and tertiary structures. These individual chains, called **subunits,** are associated with one another in specific ways. We designate the spatial relationships among them as the **quaternary structure,** the fourth level of protein structure. **Hemoglobin,** the oxygen-transporting substance in blood, is an example of a protein made up of several (four in this case) subunits (see figure 2.25).

Figure 2.22 Primary structure of bovine (cow) insulin. Insulin has two polypeptide chains held together by sulfur to sulfur (—S—S—) bonds between specific amino acids. Discovery of the structure of insulin was the first successful determination of the complete sequence of amino acids in a protein molecule. Frederick Sanger and his colleagues completed this job in 1953, after ten years of difficult and painstaking work. Today, modern analytical equipment makes amino acid sequence determination much easier and faster.

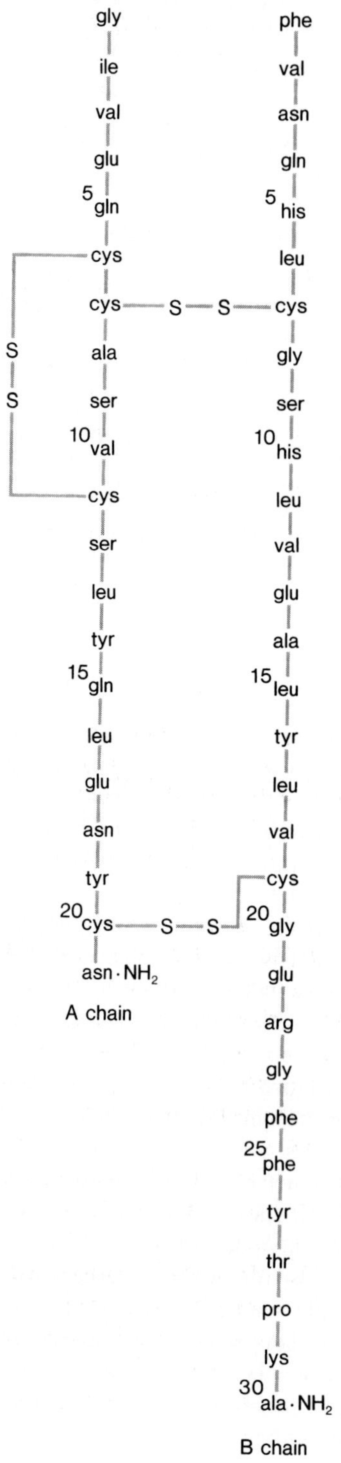

A chain: gly – ile – val – glu – 5 gln – cys – cys – ala – ser – 10 val – cys – ser – leu – tyr – 15 gln – leu – glu – asn – tyr – 20 cys – asn · NH_2
(S—S bond between A chain cys 6 and cys 11)

B chain: phe – val – asn – gln – 5 his – leu – cys – gly – ser – 10 his – leu – val – glu – ala – 15 leu – tyr – leu – val – cys – 20 gly – glu – arg – gly – phe – 25 phe – tyr – thr – pro – lys – 30 ala · NH_2
(cys—S—S—cys bonds between A chain cys 7 and B chain cys 7, and between A chain cys 20 and B chain cys 19)

A chain

B chain

Figure 2.23 The alpha-helix (α-helix). (*a*) Carbon and nitrogen atoms are the "backbone" of the protein α-helix. The structure of the α-helix was discovered by Linus Pauling and Robert Corey. (*b*) More detailed model of the α-helix showing how it is stabilized by C-O———H-N hydrogen bonds (colored dashed lines). A few of the hydrogen atoms (small circles) are shown, including those that are involved in hydrogen bonding.

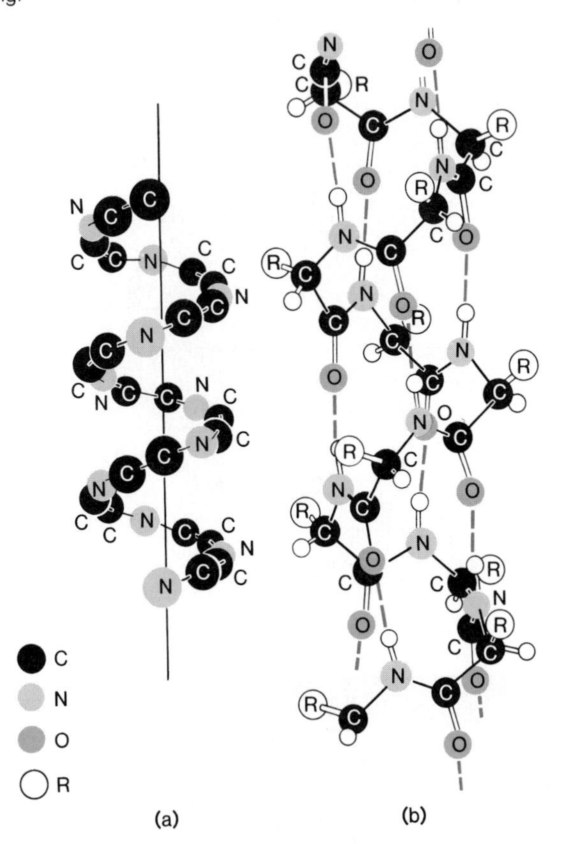

Figure 2.24 The relationship of primary, secondary, and tertiary protein structure. Expanded diagram of a small part of the polypeptide chain shows peptide bonds connecting several amino acids.

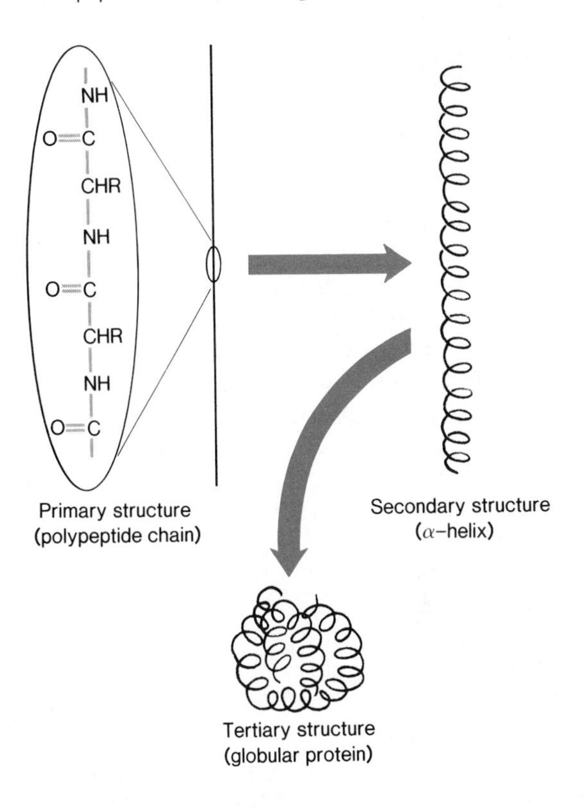

Fibrous Proteins

On the bases of structural organization and solubility, proteins can be divided into two major groups: **fibrous proteins** and **globular proteins.** Fibrous proteins are mainly structural proteins, characterized by parallel polypeptide chains lined up along an axis to form fibers or sheets. Fibrous proteins are major components of the connective tissues of many living things. Globular proteins, on the other hand, generally have more dynamic functions involving interactions with other molecules. For example, enzymes are globular proteins.

Collagen is a good example of a fibrous protein. It is the most abundant protein in vertebrate animals (animals with backbones, including humans), comprising as much as a quarter to a third of the total amount of protein in their bodies. Collagen is ideally suited for its connecting and supporting functions in the body because of its flexibility and great tensile strength (resistance to stretching).

Collagen fibers are found in all types of connective tissue, including ligaments, cartilage, bone, tendons, and the cornea of the eye. In these tissues, collagen is arranged in stable extracellular fibers to make the tissues resistant to tearing or other damage. In fact, it is possible to skin vertebrate animals primarily because of the large amount of collagen in the lower layer of the skin. The collagen makes the skin so tough that it can be pulled away from underlying tissues without tearing (figure 2.26).

Globular Proteins

In contrast to fibrous proteins, the polypeptide chains in globular proteins are coiled into compact, spherical shapes. Globular proteins are flexible and fulfill dynamic functions in organisms. Enzymes, antibodies, and transport proteins are all globular proteins. The oxygen-transporting protein hemoglobin is a good example of a globular protein (see figure 2.25). Even though the amino acid sequences of hemoglobin

Figure 2.25 Hemoglobin structure. (*a*) The heme group of hemoglobin. This large, flat molecule has an iron atom complexed with its central nitrogens. The iron atom can reversibly bind oxygen molecules. It is this binding that makes it possible for hemoglobin to function as an oxygen-transporting molecule. (*b*) The structure of oxygenated hemoglobin. The heme groups are pictured as rectangles with spherical iron atoms at their centers. (*c*) Red blood cells of a person suffering from sickle-cell anemia (left) compared with normal red blood cells (right). One amino acid substitution in the hemoglobin β-chains changes the properties of sickle-cell hemoglobin and causes the cell distortion that can result in severe circulatory problems.

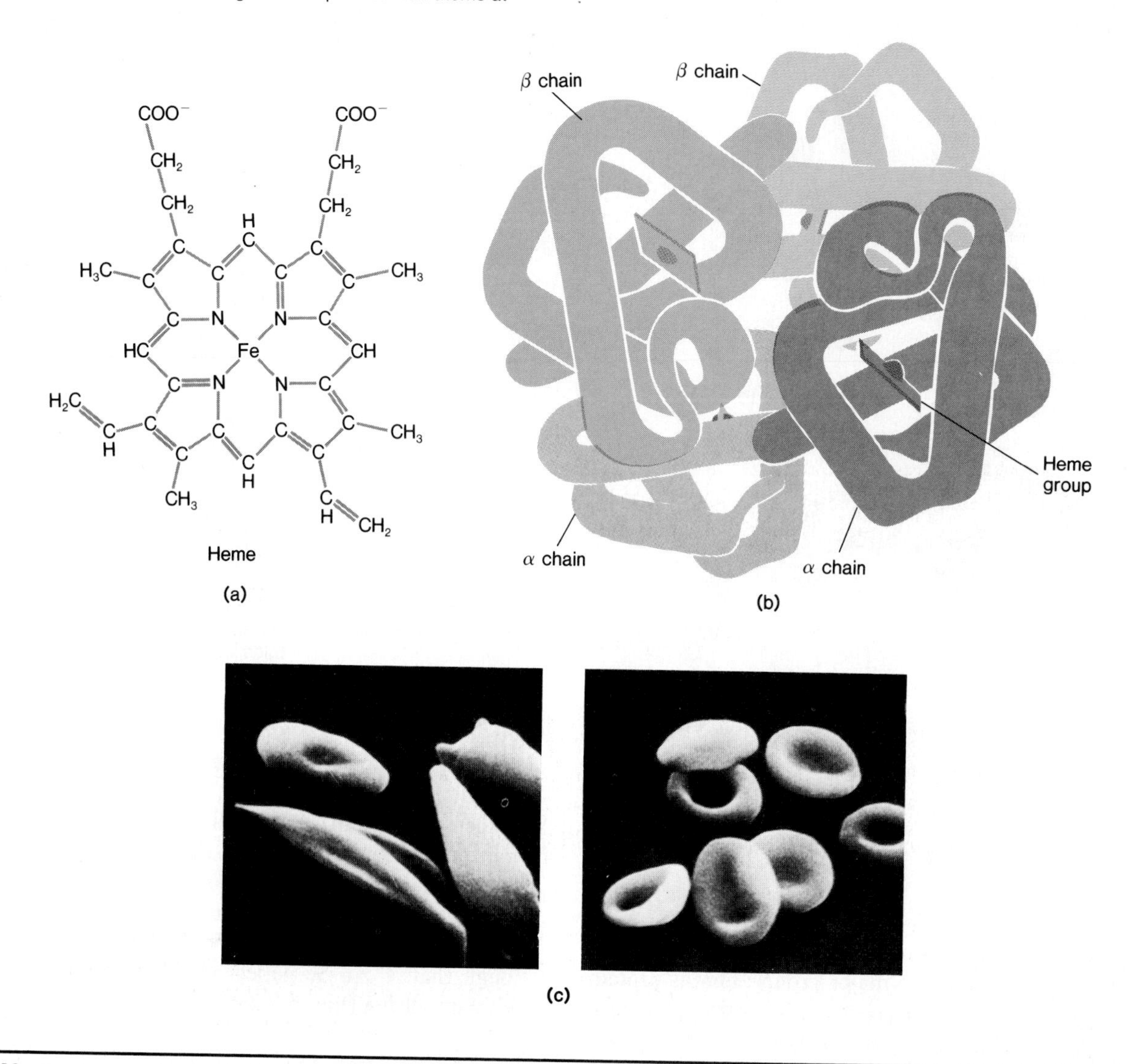

Figure 2.26 An electron micrograph showing skin collagen fibers. Collagen fibers self-assemble from polypeptide chains outside the cells that produce the polypeptides (magnification × 25,194).

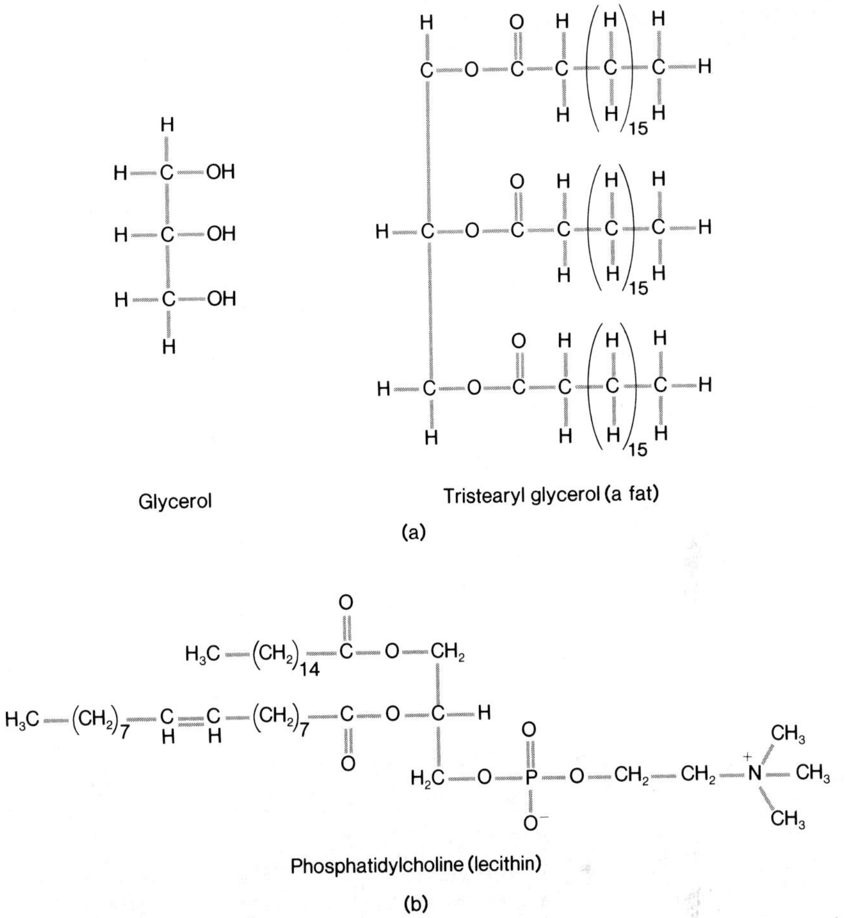

Figure 2.27 Lipids. (*a*) Glycerol and a fat. Biochemists call fats **triglycerides** because a fat molecule is formed by joining three fatty acid molecules to a molecule of glycerol by condensation reactions (see page 37). (*b*) An example of a phospholipid, phosphatidylcholine (lecithin). Such phospholipids are important constituents of biological membranes.

Glycerol

Tristearyl glycerol (a fat)

(a)

Phosphatidylcholine (lecithin)

(b)

polypeptide chains vary from species to species, the overall conformations of all hemoglobin molecules are very similar. However, the substitution of certain amino acids in the polypeptide chains of hemoglobin can have serious consequences. A striking example of the effect of amino acid substitution in hemoglobin is **sickle cell anemia.** In this human disease, a single amino acid substitution has disastrous effects. When oxygen molecules leave the hemoglobin of a person suffering from sickle cell anemia, the defective hemoglobin molecules change shape, actually altering the shape of whole red blood cells. The normally spherical cells elongate into "sickles" that have trouble passing through small vessels (see figure 2.25*c*). Circulation to various body areas is blocked and tissues are damaged. Thus, the dramatic consequence of the substitution of a single amino acid shows clearly that higher levels of protein structure arise from the primary structures, the specific amino acid sequences, of polypeptides.

Lipids

Lipids, including fats, oils, phospholipids, and steroids, are structurally much more diverse than polysaccharides or proteins. In fact, the term lipid is simply a convenient name for any compound in the cell that is insoluble in water but soluble in hydrophobic (see page 34) solvents. Some lipids are large molecules, but unlike polysaccharides and proteins, they are not polymers (chains of repeated smaller units).

Lipids resemble polysaccharides in fulfilling both structural and storage-compound roles. Fats, which are formed from glycerol and long-chain fatty acids, are used as storage compounds (figure 2.27). Because of variations in the number of carbon atoms, in the number of double bonds (C=C bonds), and in the positions of these bonds in the fatty acids, there are many different kinds of fat molecules.

Because the components of fatty acid chains in fats are generally hydrophobic, fat molecules do not dissolve in water. But if one of the fatty acids is replaced by a group that is

Figure 2.28 Steroids. (*a*) Cholesterol. This diagram shows the molecule in skeleton form. Each "corner" represents a carbon atom with hydrogens attached to it, as does each projecting "stick." Cholesterol is an important constituent of membranes in cells. (*b*) Progesterone, a hormone. Progesterone and a number of other hormones are built around the basic cholesterol structure.

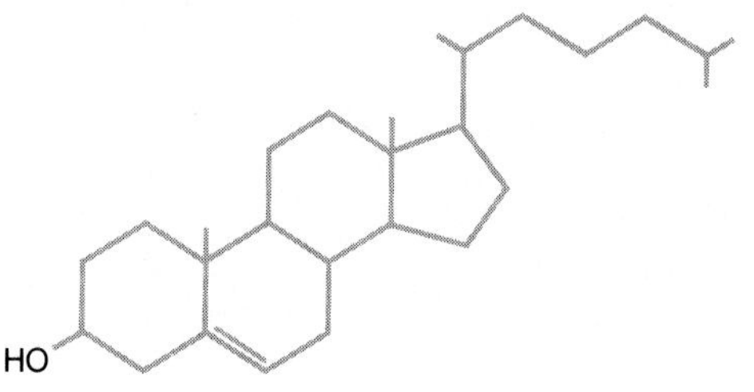

Cholesterol

(a)

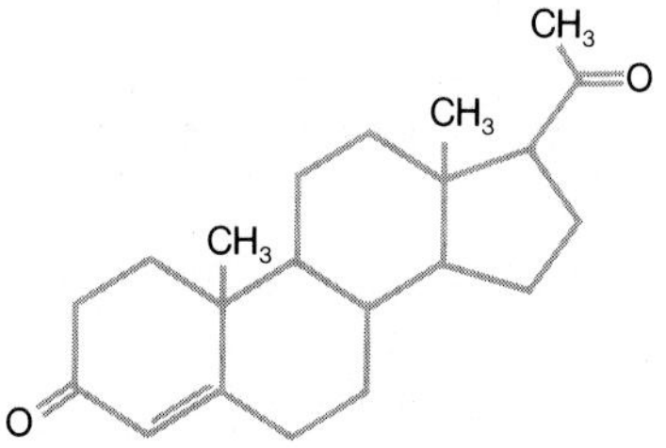

Progesterone

(b)

hydrophilic, that is, one with an affinity for water, the situation becomes more complex. For example, phospholipids are partly hydrophobic and partly hydrophilic (figure 2.27*b*). The presence of both hydrophobic and hydrophilic tendencies in such compounds is important in the formation of biological membranes. In the next chapter we will learn how such molecules form vital components of the membranes in living cells.

Biological membranes also contain **steroids.** One of the most important steroids is cholesterol (figure 2.28). Steroids contain four fused rings of carbon atoms in a characteristic arrangement. The basic steroid structure is also the foundation for many other biologically important molecules, such as the steroid hormones of vertebrate animals (figure 2.28*b*).

Nucleic Acids

Nucleic acids are macromolecules that carry genetic information. There are two major types of nucleic acids, **deoxyribonucleic acid (DNA)** and **ribonucleic acid (RNA).** Structurally, nucleic acids are linear polymers of **nucleotides.** Each nucleotide is composed of three substances: a nitrogen-containing organic base (either a **purine** or a **pyrimidine**), a five-carbon sugar, and phosphoric acid (figure 2.29). The sequence in which these purine- or pyrimidine-containing nucleotides are arranged determines the genetic coding contained in nucleic acids.

There are several differences between DNA and RNA. DNA contains the sugar deoxyribose, and the bases adenine, guanine, cytosine, and thymine. The composition of RNA differs from that of DNA in two major ways: the five-carbon sugar found in RNA is ribose rather than deoxyribose, and RNA contains uracil rather than thymine. As figure 2.29 indicates, adenine and guanine are purines, while cytosine, thymine, and uracil are pyrimidines. Bases and sugars are linked together by phosphates to form polynucleotide chains in both DNA and RNA (figure 2.30).

The Search for Nucleic Acid Structure

In the 1920s, most scientists thought that DNA was a very small molecule, probably only four nucleotides long, with its four bases occurring in equal proportions and arranged in a fixed, unchangeable sequence. This early view may partly explain why biologists were confident for years that nucleic acids were too simple in structure to convey complex genetic information. They concluded that genetic information must be coded in proteins rather than in nucleic acids.

But by the late 1940s, chemist Erwin Chargaff had begun to analyze the base composition of DNA from a number of species. He soon found that the early work on DNA structure was in error. Chargaff found that the base composition of DNA varies from species to species, just as one would expect from a substance carrying genetic material. Furthermore, Chargaff demonstrated that there are two consistencies in DNA composition, regardless of source: the total amount of purines always equals the total amount of pyrimidines, and both the adenine/thymine and guanine/cytosine ratios are 1.0 (table 2.2). In other words, adenine content equals thymine content ([A]=[T]) and guanine content equals cytosine content ([G]=[C]). These findings came to be known as *Chargaff's rules* and were the key to a great deal of later research.

In 1951, Rosalind Franklin joined Maurice Wilkins in efforts to prepare DNA fibers for study by **X-ray crystallography.** In this technique an X-ray beam is passed through a crystal of the substance being studied. Part of the X-ray beam is scattered (diffracted) as it passes through the crystal. The way that it scatters depends on the molecule's structure. A photographic plate on the other side of the crystal records a pattern of spots representing the intensity of the emergent X rays (figure 2.31*a*). This pattern reveals information about

Figure 2.29 Components of nucleic acid molecules. (*a*) Two purines and three pyrimidines found in DNA and RNA. Thymine is found in DNA, but not RNA. Uracil is a component only of RNA. (*b*) Nucleic acids are polymers of nucleotides that are composed of purine or pyrimidine bases, pentose sugars (ribose or 2-deoxyribose), and phosphoric acid. The diagram shows components of RNA. In DNA, the OH in the colored box is replaced by H.

Purines found in nucleic acids

Adenine

Guanine

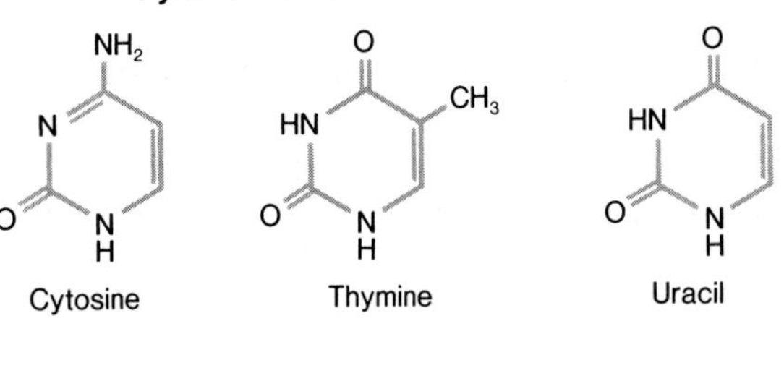

(a)

(1) Base: The purine and pyrimidine components of DNA or RNA

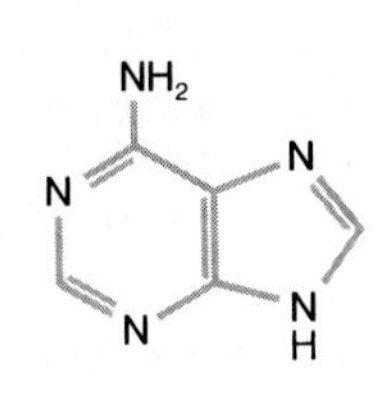

Adenine

(2) Nucleoside: The base plus the pentose sugar

Adenosine

(3) Nucleotide: The base plus the pentose sugar plus phosphoric acid

Adenosine–5′–phosphate

(b)

Table 2.2
DNA Base Composition

Organism	Molar Ratio of Bases: Adenine	Guanine	Cytosine	Thymine	$\frac{A+G}{C+T}$
Escherichia coli (bacterium)	24.6	25.5	25.6	24.3	1.00
Saccharomyces cerevisiae (yeast)	31.3	18.7	17.1	32.9	1.00
Carrot	26.7	23.1	23.2	26.9	0.99
Rana pipiens (frog)	26.3	23.5	23.8	26.4	0.99
Human (liver)	30.3	19.5	19.9	30.3	0.99

Figure 2.30 A short stretch of a single strand of a DNA molecule. Each nucleoside in the polymer is linked to neighboring nucleosides by phosphate groups. These phosphates connect the 3′ carbon of one sugar with the 5′ carbon of the adjacent nucleoside sugar. The structure of an RNA strand is similar except that ribose replaces the deoxyribose.

DNA polynucleotide strand structure

the locations of the various atoms in the crystal. Finally, the information could be used to determine the three-dimensional shapes of molecules. This work culminated with the production of Franklin's X-ray diffraction photograph of DNA (figure 2.31*b*).

Also in 1951, American biologist James Watson went to Cambridge University and began to work with Francis Crick on a model of DNA structure. Watson and Crick gathered information on DNA from diverse sources. They combined Chargaff's discoveries on base composition and information from Franklin's X-ray photographs with various predictions of how genetic material might logically be expected to behave. Chargaff's rule that [A]=[T] and [G]=[C] would hold true if there were two strands of nucleotides in a DNA molecule and if each purine (adenine or guanine) in one chain was hydrogen bonded with its corresponding pyrimidine (thymine or cytosine) in the other chain. Watson and Crick found that only the A-T and G-C combinations actually would hydrogen bond properly. Finally, when a complete model was constructed, it had to be compatible with the data obtained from Franklin's X-ray photographs. After they put together and analyzed all of the information available to them, Watson and Crick designed a model of the structure of the DNA molecule.

The structure that Watson and Crick proposed is known as the **double-helix** model of DNA. Their model looks somewhat like a twisted ladder, with two polynucleotide strands running in opposite directions and winding about each other. The A-T and G-C base pairs form the "rungs" in the interior of the helix (figure 2.32). A purine (for example, adenine) on one strand is hydrogen bonded with the appropriate pyrimidine (in this case, thymine) on the opposite strand.

Figure 2.31 X-ray crystallography of the DNA molecule. (*a*) Diagram of the technique. (*b*) X-ray diffraction photograph of DNA taken by Rosalind Franklin. The crossing pattern of dark spots in the center of the picture indicates that DNA is helical. Other parts of the picture provided additional clues for Watson and Crick in their search for the structure of DNA.

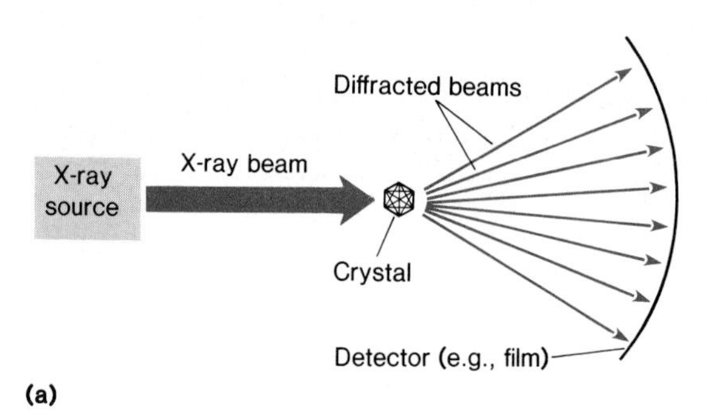

(a)

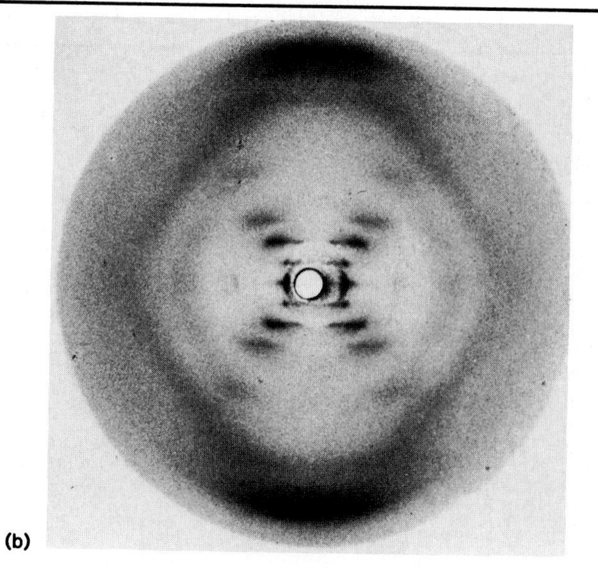

(b)

Figure 2.32 DNA structure. All distances are in nanometers (nm). One nanometer equals 10^{-9} meter or 10^{-6} millimeter; that is, one nanometer equals 0.000001 mm. (*a*) Association of bases by hydrogen bonding. Hydrogen bonds are shown as colored dashes. (*b*) Diagrammatic structure of the DNA double helix. The DNA double helix can be visualized as a twisted ladder, where the sugar-phosphate "backbones" of the strands form the rails of the ladder and the paired bases are the rungs. (*c*) James Watson (left) and Francis Crick with their model of DNA.

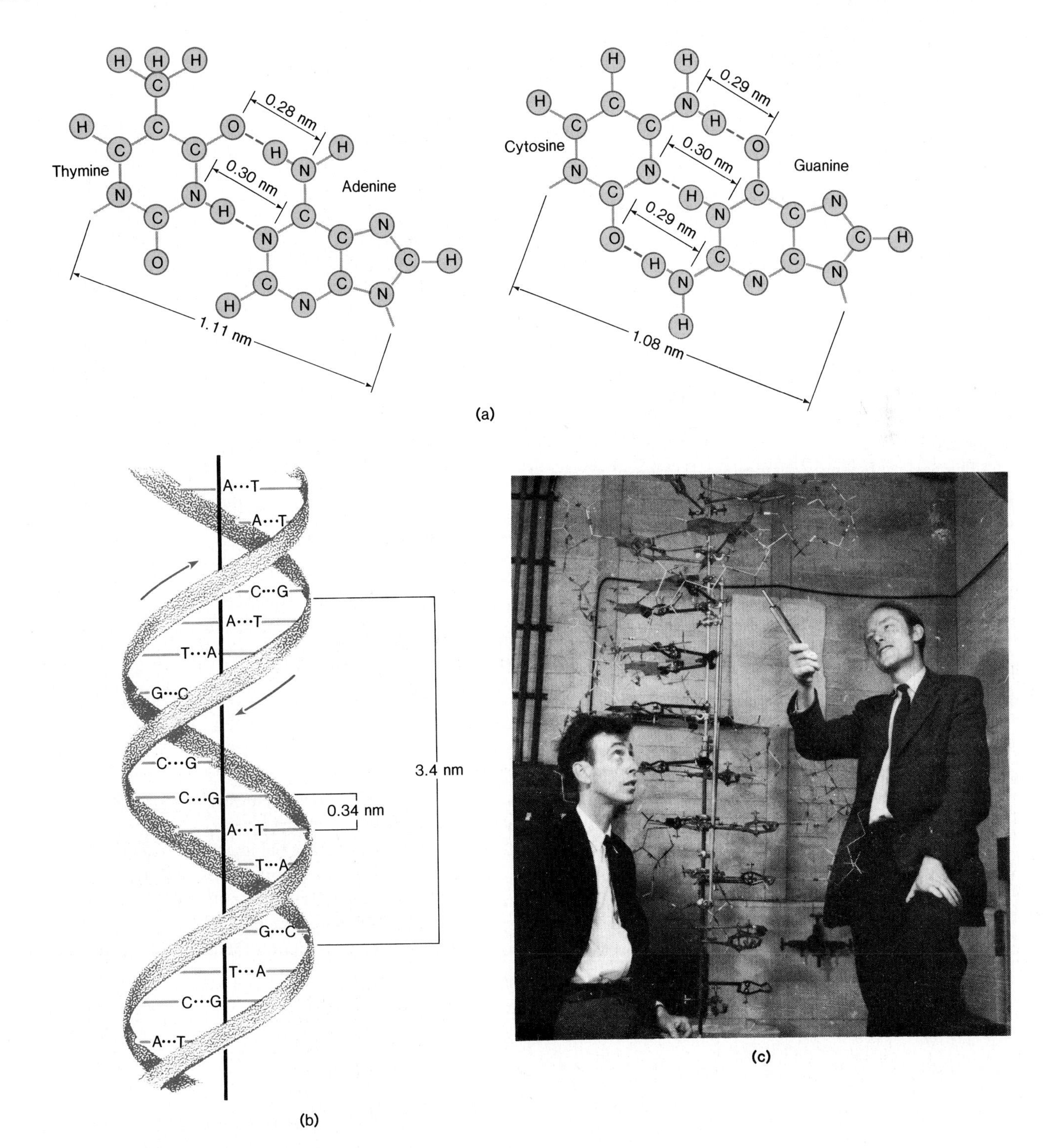

The overall structure of DNA with its hydrogen-bonded base pairs has important genetic implications. It provides a mechanism by which genetic material can be copied (replicated). The two strands of the DNA molecule can separate and a new strand, complementary to each of the lone strands, can be made so that two new and identical DNA molecules exist where there was one. Thus, during cell reproduction, each one of two new cells can receive a complete and accurate copy of the DNA in the original cell.

The work of five scientists—Watson, Crick, Franklin, Wilkins, and Chargaff—transformed the study of biology by opening the door to new research on genetic mechanisms and many functions of living cells that had not been accessible before that time.

Summary

Because chemical processes in living organisms obey the laws of chemistry and physics, it is not necessary to propose a special life chemistry. Life's special properties arise instead from the complex and diverse organizations of molecules, cells, and organisms.

The basic substances composing all forms of matter are called elements. Atoms are the smallest units of elements. Atoms are made up of smaller units called subatomic particles. Protons (positively charged particles) and neutrons are located in the nucleus of an atom and electrons (negatively charged particles) circle the nucleus in orbitals in a series of shells.

An element's atomic number is equal to the number of protons in its nucleus. Normally, the number of electrons in shells around the nucleus is the same as the number of protons. Different isotopes of an element have different numbers of neutrons in their nuclei.

The numbers and locations of electrons in the orbital shells of different atoms determine the chemical properties of those atoms and their bonding characteristics.

Ionic bonding involves the transfer of electrons between atoms. Covalent bonds form when atoms share pairs of electrons in their outer orbital shells. Weaker bonds, such as hydrogen bonding, occur both between molecules and within molecules.

Water is the most abundant material in living things. Water is made up of individual polar molecules, each with a negative charge at the oxygen end and positive charges at the hydrogen ends. This polarity explains water's solvent properties. Polar substances are soluble in water and nonpolar substances are not. Water's physical properties, such as high heat capacity, high heat of vaporization, and high surface tension also serve useful functions in living systems.

Hydrogen ion concentration can be expressed through the concept of pH. Acidic solutions have high H^+ concentrations and low pH numbers, while basic solutions have low H^+ concentrations and high pH numbers. Buffer systems resist pH changes in cells and in body fluids and are very important for the maintenance of normal biochemical functions.

Almost all of the chemical compounds in living organisms are organic compounds, compounds containing carbon. A carbon atom can form four covalent bonds and may bond to a variety of elements, especially hydrogen, oxygen, nitrogen, and carbon.

Macromolecules are polymers (chains of individual molecular units called monomers).

Monosaccharides (individual sugar molecules), disaccharides, and polysaccharides are all carbohydrates. The major polysaccharides are all polymers of the monosaccharide glucose. Various glucose polymers have different properties.

Proteins (polypeptides) are polymers of amino acids, linked together by peptide bonds. The primary structure of a polypeptide is the sequence of amino acids in the protein. Secondary structure involves the orderly folding of the amino acid chain, in forms such as the α-helix. The folded polypeptide is then folded further into the tertiary structure of the protein. Many proteins are constructed of more than one polypeptide chain. The spatial relationships among these polypeptide subunits constitute the quaternary structure of proteins.

Lipids are insoluble in water. One of the most vital roles of lipids is in cell membranes.

Nucleic acids are polymers of nucleotides. A nucleotide is composed of three substances: a purine or pyrimidine, a five-carbon sugar (ribose or deoxyribose), and phosphoric acid. The Watson-Crick model of DNA structure describes the molecule as a double helix, which resembles a spiral ladder.

Questions for Review

1. Oxygen (O) has an atomic number of 8, and phosphorus (P) has an atomic number of 15. Determine the number of neutrons in the nucleus of each of the following isotopes. The mass number is given in each case.
   a. ^{16}O
   b. ^{18}O
   c. ^{31}P
   d. ^{32}P
   e. ^{33}P
2. How many electrons are there in the outer shell of each of the following atoms? (The atomic number of nitrogen (N) is 7, sodium (Na) 11, and magnesium (Mg) 12.)
   a. N
   b. Mg
   c. Na
   d. Na^+
3. Explain what hydrophilic interactions are and why they are important in the role of water as a solvent.
4. A change from pH 5 to pH 6 in a solution means that the H^+ ion concentration has decreased by a factor of:
   a. 6
   b. 10
   c. 5
   d. 100
5. Explain how a carbon atom can form covalent bonds with four other atoms as it does, for example, in methane (CH_4).
6. Although the molecular formula of monosaccharides such as glucose and fructose is $C_6H_{12}O_6$, the molecular formula of sucrose and several other disaccharides, which consist of two monosaccharides linked together, is $C_{12}H_{22}O_{11}$ and not $C_{12}H_{24}O_{12}$. Explain this in terms of condensation reactions.
7. Which particular atoms of two amino acids become linked when a peptide bond forms?
8. Explain the relationship of Chargaff's rules to the structure of DNA.

Questions for Analysis and Discussion

1. Explain what would happen to many aquatic organisms if ice had a greater density than liquid water.
2. What biochemical evidence would convince a person concerned about world food shortages that eating grass-fed cattle does not directly deprive the hungry world of food in the same way that eating grain-fed animals does?

Suggested Readings

Books

Baker, J. J. W., and Allen, G. E. 1981. *Matter, energy, and life.* 4th ed. Reading, Mass.: Addison-Wesley.

Lehninger, A. L. 1982. *Principles of biochemistry.* New York: Worth Publishers.

Pauling, L., and Pauling, P. 1975. *Chemistry.* San Francisco: W. H. Freeman.

Stryer, L. 1981. *Biochemistry.* 2d ed. San Francisco: W. H. Freeman.

Articles

Frieden, E. July 1972. The chemical elements of life. *Scientific American.*

Koshland, D. E., Jr. October 1973. Protein shape and biological control. *Scientific American* (offprint 1280).

Milne, L. J., and Milne, M. April 1978. Insects on the water surface. *Scientific American* (offprint 1387).

Phillips, D. C., and North, A. C. T. 1979. Protein structure. *Carolina Biology Readers* no. 34. Burlington, N.C.: Carolina Biological Supply Co.

Cells and Cellular Functions

Chapter Outline

Chapter Concepts

1. The cell is the fundamental unit of life. All living things are cells or aggregates of cells and cell products.
2. The development of modern techniques of observation and analysis has greatly advanced our understanding of cells.
3. Membranes, which are fundamental to cell structure and function, are orderly arrays of lipid and protein molecules.
4. Eukaryotic cells contain a number of organelles. Each type of organelle is associated with specific cellular functions.
5. Although most plant and animal cell organelles are similar, there are some functionally significant differences between plant and animal cell structures.
6. Prokaryotic cells lack a true nucleus and membrane-enclosed organelles.
7. Materials penetrate the plasma membrane that encloses the cell by a variety of mechanisms.

Figure 3.1 (*a*) A human zygote divides to produce two cells that look very much alike. As more cells are produced by subsequent divisions, cells will specialize. (*b*) A newborn human infant made up of billions of cells specialized in thousands of different ways to produce various body parts.

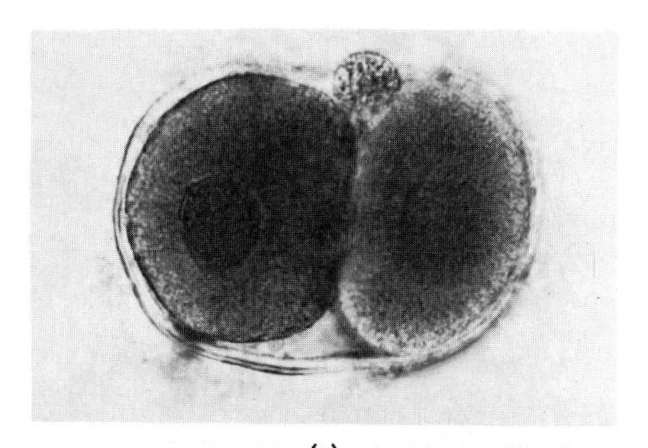

(a)

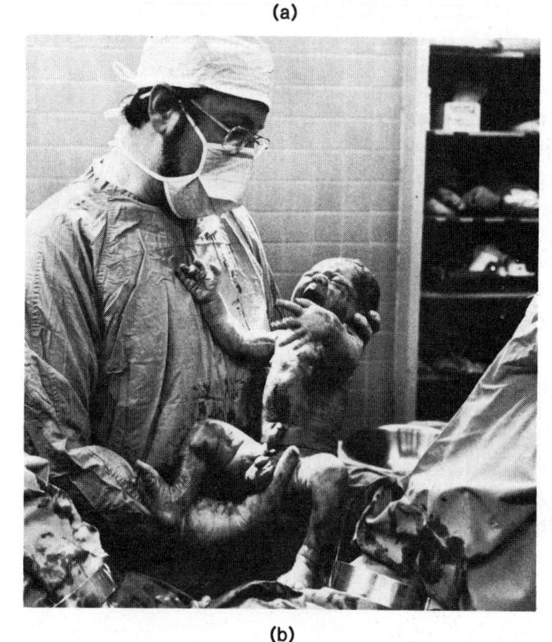

(b)

Virtually every living thing is a single cell at some stage of its life. In animals, for example, a swimming sperm cell reaches an egg cell and unites with it to form a single cell, the zygote (fertilized egg), that will give rise to all the cells that make up the body. This single cell divides to produce two cells, and each of those cells divides to produce more. Before long, there is a large population of cells. At first, these cells look and act very much alike, but before long, cells begin to specialize as various body parts form. Some cells become muscle cells, others become nerve cells, and so forth. Together all of these cells make up a body (figure 3.1).

The cell is the basic building block of all living things, from barely visible water plants and tiny worms to gigantic trees and enormous whales. It is small wonder, therefore, that

Figure 3.2 Drawings of a back view (left) and a side view (right) of one of Leeuwenhoek's microscopes. This simple microscope consisted of (1) a plate with a single lens, (2) a mounting pin, the tip of which held the specimen to be observed, (3) a focusing screw, and (4) a specimen-centering screw.

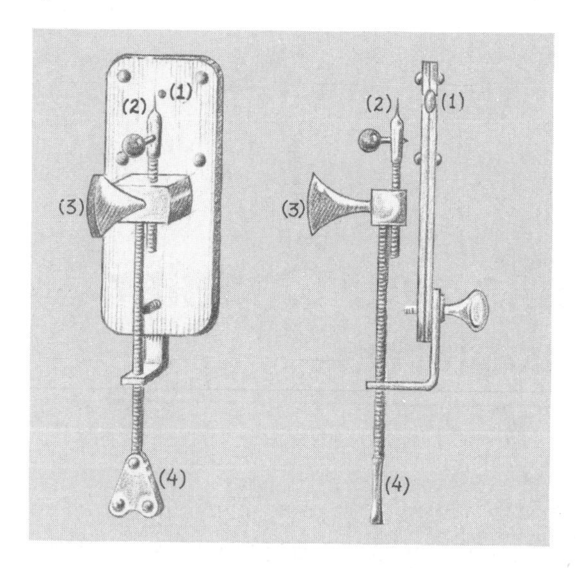

biologists have worked long and hard to understand the structure and function of living cells of many kinds.

The study of cells began with the development of the simple microscope in the seventeenth century, and the appearance of cells was first described in 1666 by Robert Hooke in his book *Micrographia.* However, Anton van Leeuwenhoek (1632–1723) of Delft, Holland, contributed the most to early cell research. Leeuwenhoek made his living as a draper and haberdasher and held minor political posts in Delft—including that of official wine taster. Despite his many activities, Leeuwenhoek found time to construct a large number of simple microscopes with such high-quality lenses that some could magnify approximately 200 times (figure 3.2). Over a period of fifty years, he observed literally hundreds of different cells and described them in amazing detail.

By the mid–1800s, this early work eventually led to a statement of one of the fundamental principles of biology, the **cell theory.** This theory was published by M. J. Schleiden, Theodor Schwann, and Rudolf Virchow. The theory states simply that the bodies of all organisms are made up of cells and the products of cells, and that every cell comes from a preexisting cell. Thus, by the middle of the nineteenth century, biologists clearly recognized the cell as the fundamental unit of life.

Unfortunately, the amount that biologists were able to learn about cells was limited for a long time by the research tools and methods available to them. The primary tool at their disposal was the **light microscope.** The problem is that light

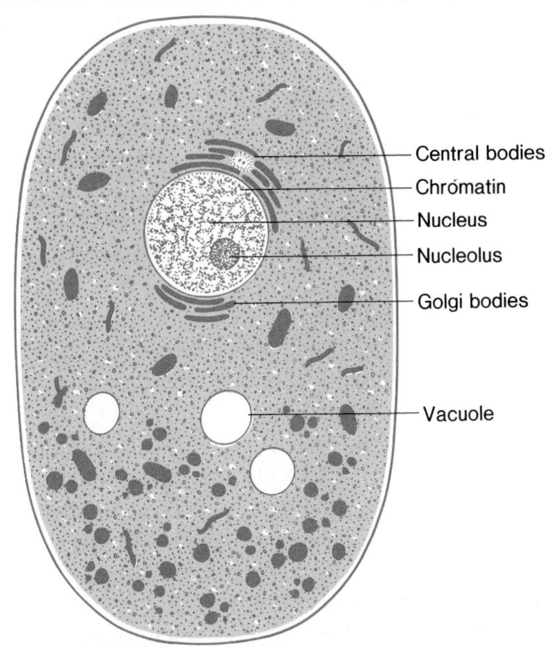

Figure 3.3 Cell structure as understood in 1925 and drawn by E. B. Wilson. It is now known that the cytoplasm lying between the nucleus and the outer membrane of the cell contains many more structures that cannot be seen with the ordinary light microscope. What Wilson called "central bodies" are now known as centrioles (page 70).

microscopes are somewhat limited for viewing very small objects, and cells and their internal components are very small indeed. Thus, when E. B. Wilson, a pioneer of cell biology in the United States, published a diagram in 1925 of the details of cell structure that were then known, his illustration was relatively simple (figure 3.3). "Wilson's cell" had a nucleus, cytoplasm around the nucleus, a few other internal structures present in the cytoplasm, and a bounding membrane (which was inferred, since it could not actually be seen with the light microscope).

We now know that the cell actually contains a complex array of membranous structures that are both structurally and functionally specialized for various vital processes in the cell. This "division of labor" within cells is similar to the division of labor in whole organisms, where various individual organs carry out specific functions for the whole body. Thus, the name **organelle** ("little organ") has been applied to these specialized structures inside cells.

Cell Size

Cells come in a variety of sizes and shapes, but almost all of them are small in terms of everyday frame of reference. A few gigantic cells, such as the yolk of a chicken's egg, are probably very familiar to you, but almost all other cells are too small to be seen with the unaided eye (figure 3.4).

Figure 3.4 Variations in the size of cells. The cells have been drawn to scale, but because of the great range in sizes, the magnification of the four lower cells is ten times that of the upper cells. The actual size of the liver cell in the lower diagram is only one-fifth the size of the human egg shown in the upper diagram.

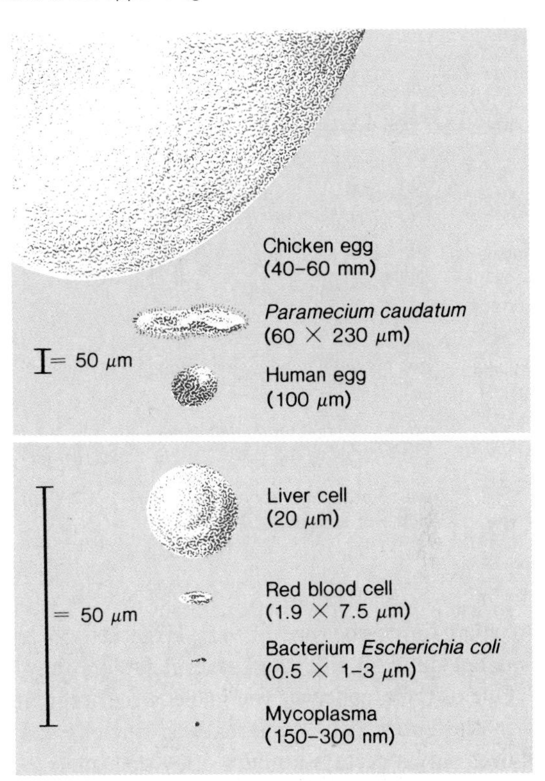

Table 3.1
Units of Measurement

Unit	Symbol	Value
Meter	m	39.37 inches
Centimeter	cm	0.01 meter
Millimeter	mm	0.1 cm (10^{-3} meter)
Micrometer (micron)	μm (μ)	0.001 mm (10^{-6} meter)
Nanometer	nm	0.001 μm (10^{-9} meter)
Angstrom	Å	0.1 nm (10^{-10} meter)

The smallest free-living bacteria are around 0.2 to 0.3 μm in diameter, while bird eggs may reach 10 cm or more in diameter. Despite this enormous range in size, the great majority of cells fall into the size range between 0.5 μm and 40 or 50 μm in diameter (table 3.1).

Even the smallest cell must be large enough to possess certain components necessary for life. Thus, there is probably a minimum size limit for functional living cells.

Figure 3.5 Comparison of surface areas and volumes of spheres of different sizes showing how cell volumes increase faster than cell surface areas as cells get larger. Note that the ratio of surface area to volume gets smaller as size grows larger. Formulas used are shown below the calculated values. Of course, most living cells are much smaller than either the spheres pictured here or the sizes used for the calculations.

Radius (r)	1 cm	2 cm	4 cm
Surface area (A)	12.57 cm^2	50.26 cm^2	201.06 cm^2
Volume (V)	4.19 cm^3	33.51 cm^3	268.08 cm^3
A/V	3.0	1.50	0.75

Surface area of a sphere $= 4 \pi r^2$
Volume of sphere $= 4/3 \pi r^3$

But what limits the size of very large cells? Here the answer is not quite so clear and several factors may be involved. One of them concerns the nucleus that regulates activities in the entire cell. Apparently a nucleus can exert control over only a certain amount of cytoplasm.

Another aspect of cell volume also works to limit cell size. As the diameter of a cell becomes greater, its volume increases more rapidly than its surface area (figure 3.5). The demand for nutrients and the rate of waste production are proportional to the volume of the cell, but the cell takes up nutrients and eliminates wastes through its surface. If the cell becomes too large, its surface area may not be great enough to carry on an adequate exchange of nutrients and wastes.

Enormous cells, such as certain bird egg cells, are not really exceptions to this rule. Giant egg cells contain huge quantities of stored nutrients in the form of yolk, which is not metabolically active and does not place demands on the nutrient-obtaining or waste-removing functions of the cell.

Methods Used in Studying Cells

Despite the very small size of most cells and the difficulties inherent in studying the details of such small, delicate objects, knowledge of cell structure and function has been expanding rapidly in recent years. This growth in knowledge is, to a great extent, the result of the development of new techniques for studying cells.

The Light Microscope

One of biology's oldest and most familiar tools is the light microscope. However, although light microscopes are useful for certain kinds of studies, they are of somewhat limited use in cell research. You might assume that it is the magnifying power of light microscopes that limits their usefulness and that we could learn much more about cells if only we could build a light microscope with greater magnifying power. Unfortunately, this is not true. In fact, it *is* possible to build light microscopes that do magnify objects much more than microscopes currently in use, but light microscopes with greater magnifying power do not provide more information about cells.

The greatest limitation of the light microscope is not magnification but resolution. **Resolution** is what makes it possible for you to discern separate, small objects that are very close together. Resolution, in a microscope with good lenses, depends upon the wavelength of the light illuminating the object under observation. Of course, we must use light that we can see, and thus the nature of visible light sets limits to the resolving power of the microscope. At best, a good microscope using visible light can resolve two points that are approximately 0.2 μm apart. With such a microscope, you can study overall cell structure and some of the larger structures inside cells at magnifications up to about 1,000 to 1,500 times their actual size. But the resolving power of a light microscope is not great enough to permit detailed study of the many smaller structures in cells.

The Electron Microscope

The **electron microscope** has a much greater resolving power than the light microscope—about 0.5 nm or less, as opposed to 0.2 μm. Hence, the electron microscope can be used to effectively magnify cells 200,000 times or more. This much greater resolution is possible because an electron microscope uses a beam of electrons instead of a beam of visible light. There are two basic types of electron microscopes, the **transmission electron microscope (TEM)** and the **scanning electron microscope (SEM).**

In a transmission electron microscope, an electron beam is passed through a very thin section or slice of the cell (figure 3.6). After leaving the cell, the electron beam is magnified and projected either on a fluorescent screen (resembling a television screen) or on photographic film.

The scanning electron microscope is used to study surfaces rather than thin sections of cells. SEM pictures, with their three-dimensional quality, reveal some remarkable details of the surfaces of cells and other objects. Surfaces being

Figure 3.6 A highly diagrammatic and simplified comparison of principles of transmission (left) and scanning (right) electron microscopy. In transmission electron microscopy the focused electron beam passes through a very thin section of the specimen. A final image of the object is viewed on a fluorescent screen or photographic plate at the bottom of the column. In scanning electron microscopy, the scanning generator makes the electron beam scan over the surface of the specimen. Secondary electrons come off atoms on the specimen's surface. The pattern of secondary electrons striking the secondary electron detector is amplified and displayed on a cathode-ray tube.

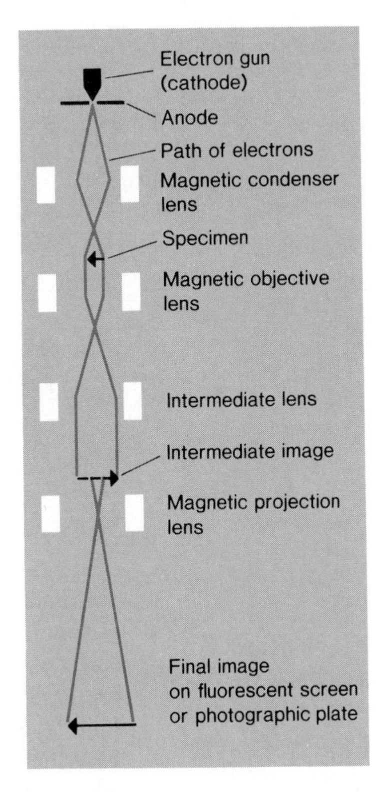

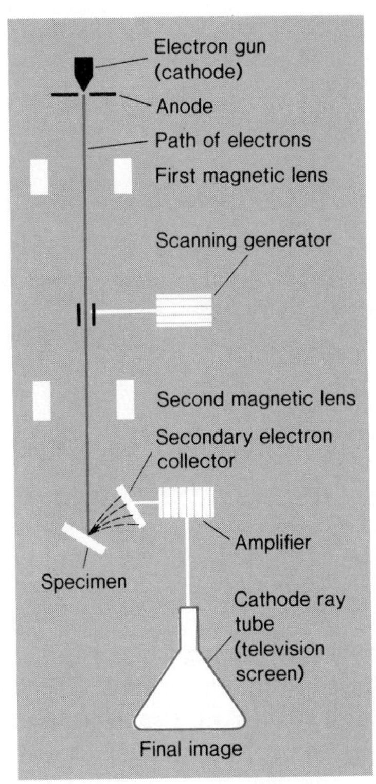

studied are usually coated with a very thin layer of metal, often gold. In the SEM, electron beams scan over the surface of the specimen, and the striking electrons drive off electrons (called secondary electrons) from the surface atoms. The pattern of these secondary electrons landing on a detector is amplified, and the image is displayed on a cathode ray tube much like that in a television set.

Taking Living Cells Apart

The cell-research techniques we have described so far—even such useful and informative techniques as electron microscopy—share one significant shortcoming: they all produce "stop-action" pictures of cell components. At some point, the cells are killed and prepared for observation. Thus, what is observed is the condition of cells and cell organelles at the moment the cells were killed. Since cells and organelles are structurally and functionally dynamic, there is obviously a need for techniques that allow observers to study cells, and especially organelles taken directly from living cells. This need has led to the development of **cell fractionation.** In cell-fractionation techniques, cells are broken open and their internal components are separated for study, under conditions where the organelles remain able to perform their specialized functions.

Cell fractionation makes possible dynamic functional studies of cell organelles—studies that complement the information obtained through techniques that give "stop-action" pictures of cells and organelles at single moments in time. All these techniques, as well as others, have been used in developing our present knowledge of cells and their organelles.

The Structure of Eukaryotic Cells

All cells can be placed in one of two categories. Bacteria and cyanobacteria (blue-green algae) are **prokaryotic** (*pro* means *before; karyon* means *nucleus*) cells. Prokaryotic cells lack a distinct nucleus bound by a membrane and do not have membrane-enclosed organelles. All other cells—the cells of plants, animals, protists, and fungi—are **eukaryotic** ("true nucleus") cells. Eukaryotic cells possess nuclei with membranes and are much more structurally complex. The cytoplasm also contains membranous networks and organelles that are enclosed in membranes.

Figure 3.7 A diagram showing the more deeply imbedded proteins in the lipid bilayer of a membrane. The proteins that are more loosely connected to the surfaces of the lipid layers are omitted to keep the diagram simple.

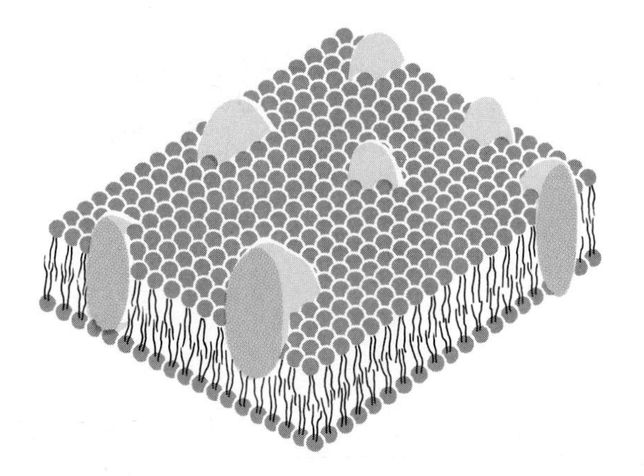

Figure 3.8 An electron micrograph showing part of a plasma membrane of a red blood cell. The dark area below the membrane is part of the inside of the cell. The lighter area above the membrane is outside the cell. The two dark lines are assumed to be the outer portions of the membrane while the lighter middle represents the interior of the lipid bilayer where the hydrophobic "tails" of lipid molecules are located (magnification × 250,000).

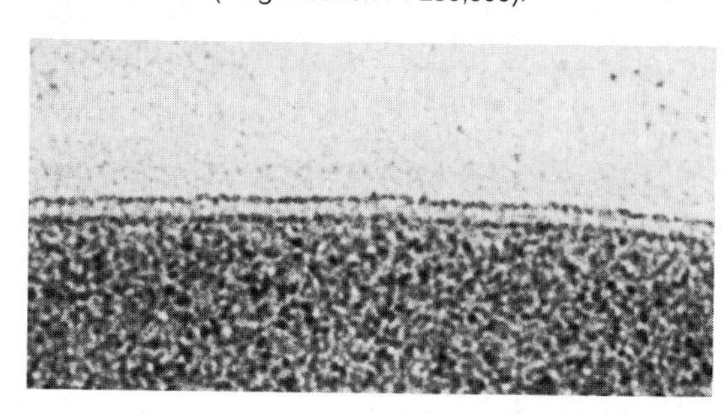

Membranes

Modern biology strongly emphasizes the importance of membranes in cell organization. The old-fashioned view of the cell as a sac of watery soup enclosed in a membrane has been replaced by the concept of an extremely complex array of membranes not only enclosing the cell, but arranged throughout the interior of the cell. In the most general terms, a biological **membrane** is a thin, hydrophobic (page 34) layer separating two areas with materials dissolved and suspended in water. Membranes separate the cell into a number of individual compartments since water-soluble molecules in one area cannot readily pass through a hydrophobic membrane into another area. But membranes are not just passive barriers; they can control the passage of molecules, and the membranes themselves are the site of some biochemical reactions.

Biological membranes are so thin that they cannot be seen with a light microscope. Therefore, before development of the electron microscope, biologists could study membranes only by chemical analysis. These chemical studies revealed that membranes contain both lipids and proteins. The lipid molecules in membranes have two parts. Each lipid molecule has a hydrophilic (page 34) head that is attractive to water molecules and a hydrophobic tail that repels water molecules. All evidence indicates that membranes have a bimolecular (two molecules thick) lipid layer. The lipids are arranged with their hydrophilic ends at the surface and their hydrophobic ends in the interior of the membrane (figure 3.7).

There seem to be two types of membrane proteins. Some of them are relatively loosely connected to the membrane surface and can be easily removed. The majority of membrane proteins, however, are more deeply imbedded in the lipid layers of the membrane. These imbedded proteins can be removed only if the membrane is harshly disrupted, as when membranes are treated with detergent.

The Plasma Membrane

One of the most important parts of a cell is the **plasma membrane,** which encloses the entire cell. The plasma membrane forms the cell's point of contact with its environment and determines what materials enter and leave the cell.

The plasma membrane, like all cellular membrane structures, is an ordered but dynamic array of lipids and proteins. It consists of a lipid bilayer with proteins attached to its surfaces and imbedded in its lipid layers. In electron microscope pictures, the plasma membrane at the surface of a cell appears as two dark lines separated by a lighter space (figure 3.8).

The plasma membrane is a **selectively permeable** barrier between the outside and the inside of the cell; that is, it permits some atoms and molecules to pass through, but prevents others from passing. Some substances are actively moved ("pumped") in or out of the cell. Many of the proteins imbedded in the plasma membrane are very much involved in the various processes by which materials enter and leave cells. We will discuss some of these processes later.

Box 3.1

Fused Hybrid Cells and Protein Icebergs

At least some of the protein molecules imbedded in the lipids of membranes seem to move around from one location to another. In fact, some biologists have stated that membrane proteins are like "icebergs floating in a sea of lipids." This is explained by the **fluid mosaic model** of membrane structure.

The movement of membrane lipids has been strikingly demonstrated in **cell fusion** experiments. In these experiments, a human cell and a mouse cell are made to fuse with one another to form "hybrid" cells. Before the cells are fused, they are treated so that membrane proteins on each are distinctively marked. The markers used are fluorescent antibodies, molecules that combine with membrane proteins and that glow in a particular color when properly illuminated under a microscope.

Membrane proteins of cultured mouse cells are labeled with green fluorescent antibodies, while those of cultured human cells are labeled red. This treatment makes it possible to trace the movements of membrane proteins and to keep track of which proteins were originally in the human cell membrane and which were in the mouse cell membrane.

Then the cells are brought together in the presence of a virus, inactivated Sendai virus, which will attach to the cells, but not proceed further in infecting them. Virus particles adhere to the cell surfaces and "glue" the two cells together. The result is a fusion of the cell membranes to form a single cell hybrid. At first the two types of color-labeled proteins lie in separate halves of the single fused cell. However, within about forty minutes the red-and-green-labeled proteins are distributed evenly over the outer cell surface. This experiment clearly demonstrates that membrane structure is dynamic. Membrane proteins are not fixed in one position; they can move to new positions in the membrane of a hybrid cell.

Cell-fusion techniques have found other uses besides demonstration of the mobility of membrane proteins in the lipid bilayer. Cell fusion has also been used in many kinds of biological and medical research. Recently, cell-fusion techniques have been used to produce cultures of hybrid cells that can make large quantities of molecules useful for medical research, diagnosis, and treatment. Surprisingly, some of the most useful kinds of hybrid cells are produced when normal cells and tumor cells are fused. In fact, there is some evidence that such hybrid cells might be utilized to produce molecules that can be used to treat tumors in the body.

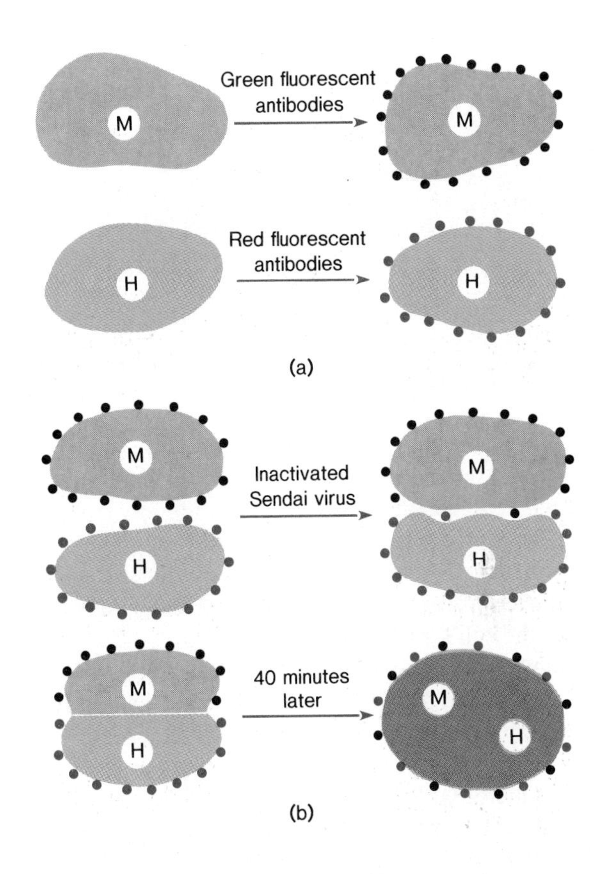

Box figure 3.1 Cell fusion experiment. (*a*) M is the nucleus of the mouse cell; H is the nucleus of the human cell. Proteins on the mouse cell surface are labeled with green fluorescent antibodies; those on the human cell surface are labeled with red fluorescent antibodies. (*b*) After fusion induced by inactivated Sendai virus, labeled proteins originally on the two separate cells become interspersed on the fused cell membrane. This demonstrates mobility of membrane proteins.

Figure 3.9 A diagram showing the carbohydrates that are bonded to some of the proteins and lipids at the outer surface of a plasma membrane.

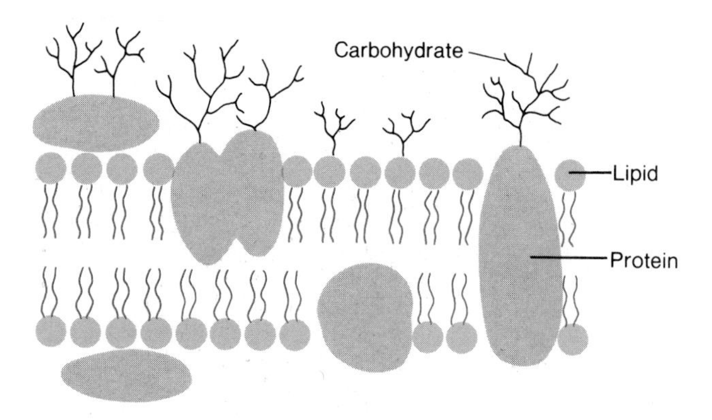

Finally, the outer surface of the plasma membrane has some special features that make it very different from the inner surface. Many of the protein molecules and some of the lipid molecules at the plasma membrane's outer surface are bonded to carbohydrates. These attached carbohydrates protrude from the cell's surface (figure 3.9).

An interesting thing about these cell-surface carbohydrates is that they differ from one cell to another and from one individual to another. For example, differences in carbohydrates bonded to proteins on the surfaces of red blood cell plasma membranes determine the human ABO blood types. People with type A blood have a particular set of carbohydrate molecules that differ from the set present on the cells of people with type B blood. People with type O blood do not have either of the molecules that characterize type A or type B.

The carbohydrates that determine the ABO blood types make up only a portion of the complex molecules on plasma membrane surfaces. It now appears that every individual may have a unique set of complex cell-surface molecules. In fact, the most individual thing about you may well be the surfaces of the plasma membranes of your cells.

The Cytoplasmic Matrix

Inside the plasma membrane is the **cytoplasm** of the cell. In the cytoplasm, the various organelles of the cell reside in the **cytoplasmic matrix,** a watery environment containing a complex network of tiny filaments. When observed with the electron microscope, the cytoplasmic matrix appears to be homogeneous without obvious special features. This does not mean that it is unimportant. It is the "environment" of the organelles and the site of a number of processes. Approximately 85 to 90 percent of the cytoplasmic matrix is water, but dissolved in that water are a host of substances vital to the functions of the cell's various organelles and the overall life of the cell. There is also a great deal of protein in the cytoplasmic matrix. Much of it is in the network of filaments, but other protein molecules are enzymes involved in cell metabolism.

Figure 3.10 The nucleus shown in this electron micrograph has a distinct nucleolus (nu) and irregular patches of chromatin. The chromatin contains DNA and condenses to form chromosomes during cell division.

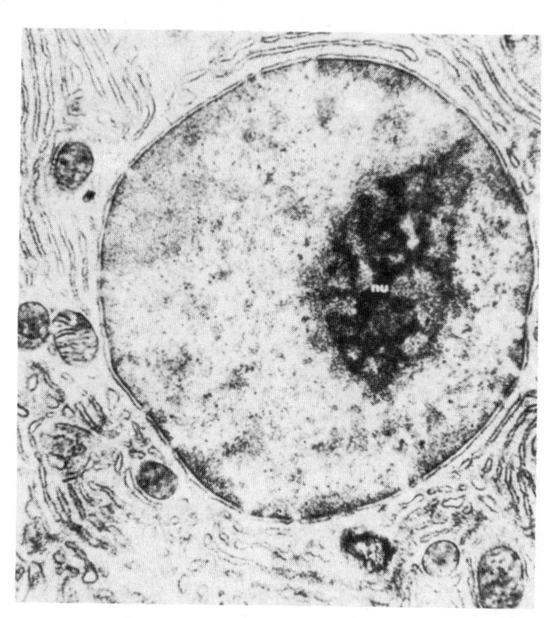

The Nucleus and Nucleolus

Because it is visible under the light microscope, the nucleus was discovered early in the study of cell structure. Nuclei are membrane-enclosed spherical bodies around 5 to 7 μm in diameter. A nucleus contains relatively dense, fibrous patches of **chromatin,** which are the DNA-containing regions (figure 3.10). In nondividing cells, this chromatin exists in a dispersed condition, but during cell division, it condenses into **chromosomes.**

The **nucleolus** is a distinctive, dense portion of the nucleus that is visible in nondividing cells, but is usually dispersed and not visible during cell division. After cell division is complete, the nucleolus reforms. The nuclei of many types of cells have more than one nucleolus, with two being a common number.

The nucleolus plays a major role in the formation of specialized bodies, the ribosomes (page 63), which are involved in protein synthesis in cells. It is not surprising, therefore, that nucleoli in cells that are actively synthesizing proteins are larger than nucleoli in cells that are not doing so.

Figure 3.11 An electron micrograph showing a close-up view of part of a nuclear envelope and some of its many pores. Patches of chromatin are visible inside the nucleus (magnification × 19,845).

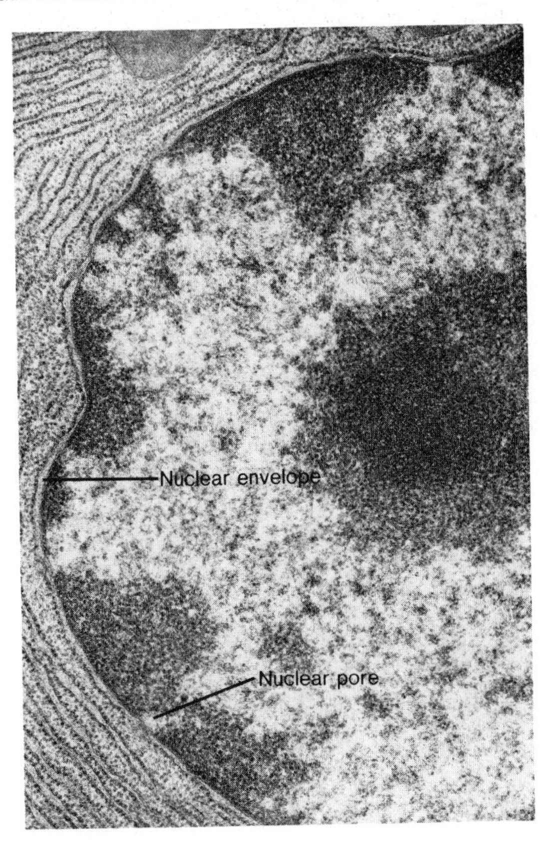

A complex, membranous **nuclear envelope** encloses the nucleus (figure 3.11). A large number of **nuclear pores** penetrate the nuclear envelope, and they appear to function as routes of transport between the nucleus and the surrounding cytoplasm. For example, RNA produced in the nucleus passes out through nuclear pores to the cytoplasm.

Ribosomes

Ribosomes are small particles that are the sites of protein synthesis in cells. Each ribosome is about 20 nm in diameter and consists of two subunits (figure 3.12). Each subunit is made up of a variety of protein and **ribosomal RNA (rRNA)** molecules. The subunits of ribosomes are free in the cytoplasmic matrix, but they come together to form whole ribosomes when they are actively involved in manufacturing proteins. During periods of active protein synthesis, ribosomes are attached to a network of membranes in the cytoplasm known as the endoplasmic reticulum.

Figure 3.12 A diagram showing that a whole ribosome consists of a large subunit and a small subunit. The subunits come together when the ribosome is involved in protein synthesis and separate when synthesis is complete. One nanometer (nm) is equal to 10^{-9} meter or 10^{-6} millimeter (0.000001 mm).

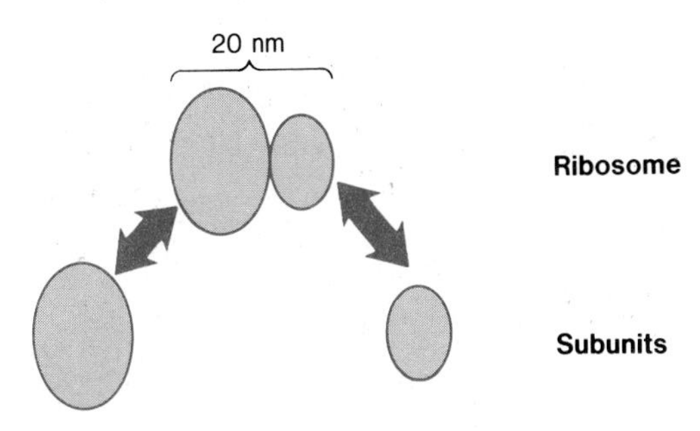

The Endoplasmic Reticulum

One fact that has become clear as a result of modern studies of cells with the transmission electron microscope is that eukaryotic cells contain complex membranous structures, the most extensive of which is the **endoplasmic reticulum (ER).** The endoplasmic reticulum, however, is not a single structure, but an irregular network of branching and fusing membranous tubules and a large number of flattened sacs (figure 3.13*a*). This network of tubules and sacs permeates much of the cytoplasmic matrix.

In cells that are synthesizing a great deal of protein, a large part of the outside of the ER is coated with ribosomes and is called **rough ER.** In other cells—for example, those producing lipids—the ER is virtually devoid of ribosomes. This is called **smooth ER** (figure 3.13*b*).

The Golgi Apparatus

The **Golgi apparatus** (also called **Golgi body**) is an organelle that was discovered near the turn of the century, and is named for one of the early cell biologists who studied it, Camillo Golgi.

A Golgi apparatus consists of stacked saclike structures made up of membranes (figure 3.14). The Golgi apparatus is involved in the packaging and secretion (release) of cell products that are produced in other parts of the cytoplasm. For example, proteins that are synthesized on ribosomes of the rough endoplasmic reticulum (ER) are transported to the Golgi apparatus for packaging.

Figure 3.13 (*a*) Electron micrograph of the rough ER in a cell. The cell's cytoplasm is packed with flattened sacs that are coated with ribosomes. (*b*) An area of cytoplasm full of smooth ER tubules. These tubules lack ribosomes and range between 40 nm and 70 nm in diameter.

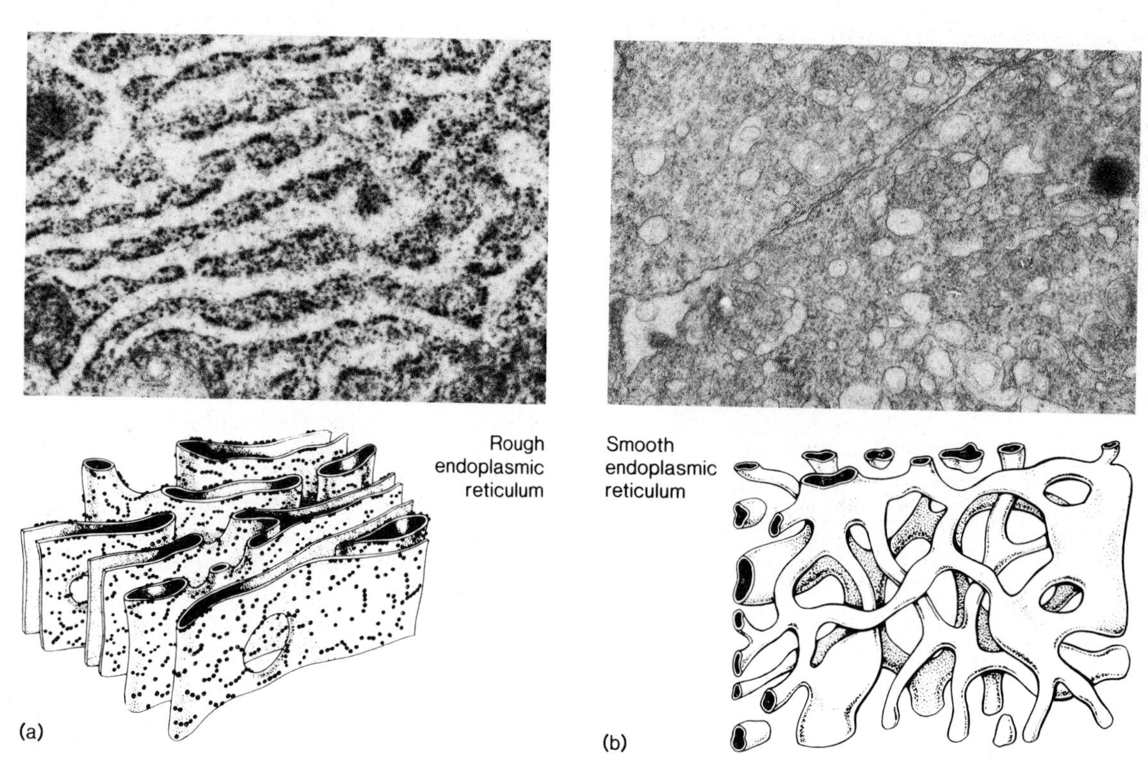

Figure 3.14 (*a*) Electron micrograph of the Golgi apparatus of a cell showing a stack of membranous sacs (magnification × 36,000). (*b*) The drawing represents the structure of the Golgi apparatus as a whole.

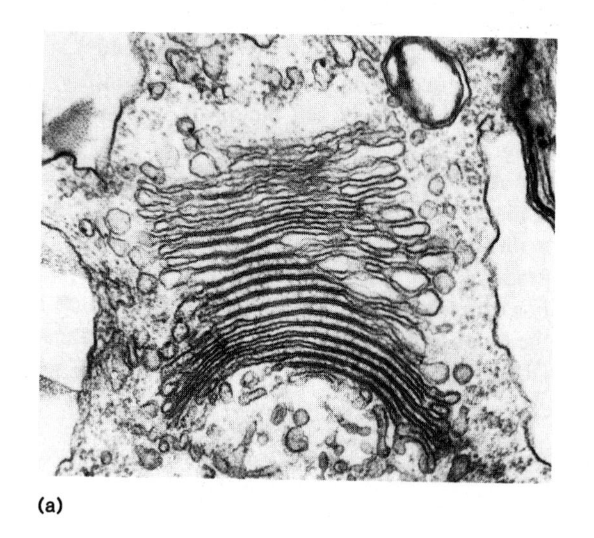

Figure 3.15 The function of the Golgi apparatus in packaging and secretion (release) of cell products. Secretory vesicles move to the cell surface and fuse with the plasma membrane, and their contents are released to the outside.

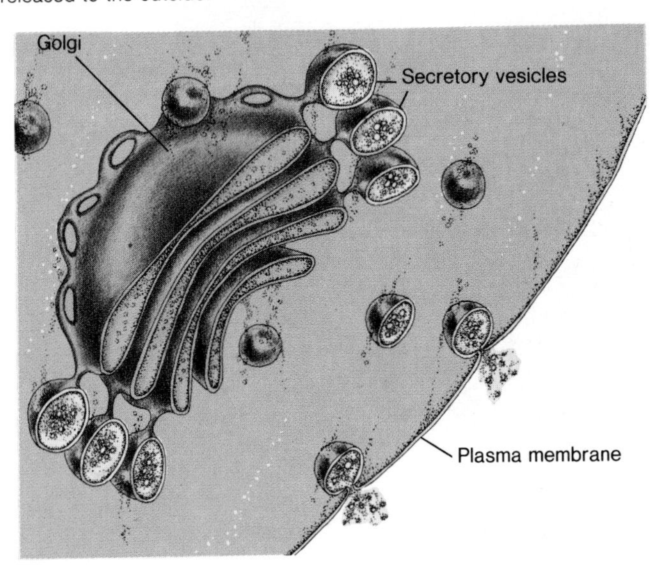

What do we mean by "packaging" of cell products? At the edges of the Golgi apparatus, little hollow spheres called **secretory vesicles** are pinched off (figure 3.15). These membrane-bound spheres contain the products being "packaged" by the Golgi apparatus. Such packaging is a step toward secreting cell products from the cell because these vesicles can move to the plasma membrane and empty their contents through it to the outside of the cell.

Lysosomes

The Golgi apparatus and endoplasmic reticulum are responsible for much more than the construction of secretory vesicles of various types. They also produce another important cell organelle, the **lysosome** (figure 3.16).

Lysosomes contain the enzymes necessary to digest all types of macromolecules. In fact, the enzymes contained in lysosomes could digest the contents of the cell itself. This explains why digestive enzymes are enclosed in a membranous sphere. The lysosome membrane protects the rest of the cell from the contents of the lysosome.

But why do cells need these membrane-bound packages of potentially dangerous enzymes? They are used selectively to digest materials that need to be broken down for the welfare of the cell. For example, material from outside the cell is sometimes taken inside via a membrane sphere (vesicle) that pinches off the plasma membrane. This is how white blood cells engulf bacteria in the body's fight against infection. Newly formed lysosomes, called primary lysosomes, fuse

Figure 3.16 Lysosomes in macrophages in the lung. Macrophages are cells that protect the lung by engulfing bits of foreign material and bacteria breathed in with the air. Secondary lysosomes contain partially digested material and are formed by fusion of primary lysosomes and vesicles containing engulfed foreign material (magnification × 14,137).

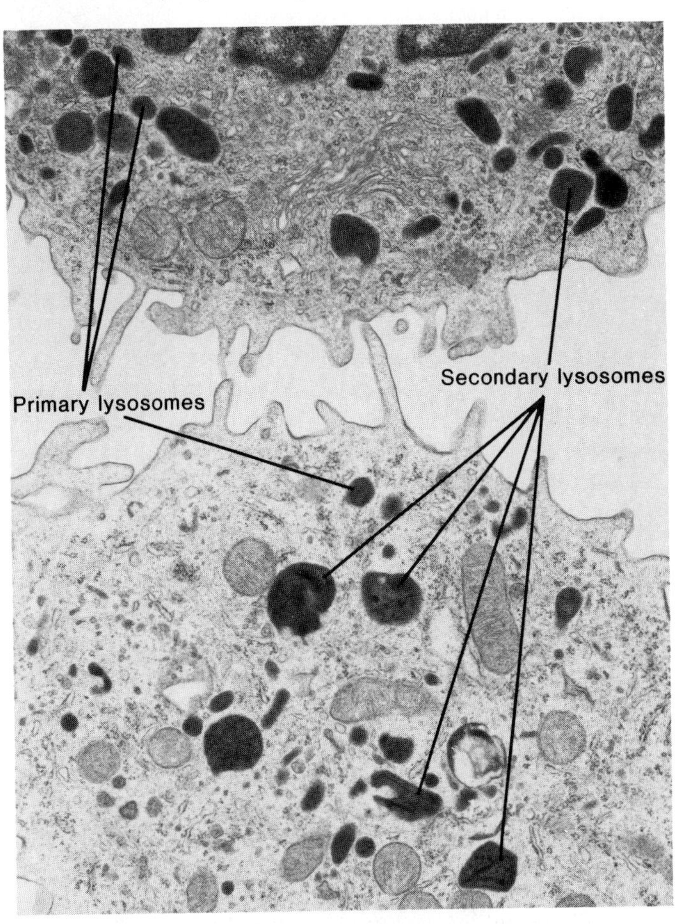

with these vesicles so that the foreign material is then exposed to the digestive enzymes contained in the lysosome. The foreign material is digested inside this fused secondary lysosome. Some products of the digestion may be absorbed through the membrane of the vesicle and serve as nutrients for the cell. But wastes often remain in residual bodies; these can then go to the plasma membrane and dump their contents out of the cell (figure 3.17).

Sometimes worn out parts of the cell itself are enclosed in vesicles, which fuse with primary lysosomes. This self-digestion of used parts is a way by which nutrients can be recycled in the cell. It also can help prevent a cell from starving when it is not receiving adequate nutrients, because a cell can selectively digest portions of itself to remain alive.

Figure 3.17 Lysosome formation and function. Secondary lysosomes form in several different ways. Primary lysosomes, which contain digestive enzymes, may fuse either with vesicles that contain small amounts of material taken in from outside the cell or with vesicles that contain part of the cell's cytoplasm enclosed in a membranous vesicle pinched off the ER. Within secondary lysosomes, the original vesicle contents are digested. Following digestion, the material remaining in residual bodies is expelled from the cell.

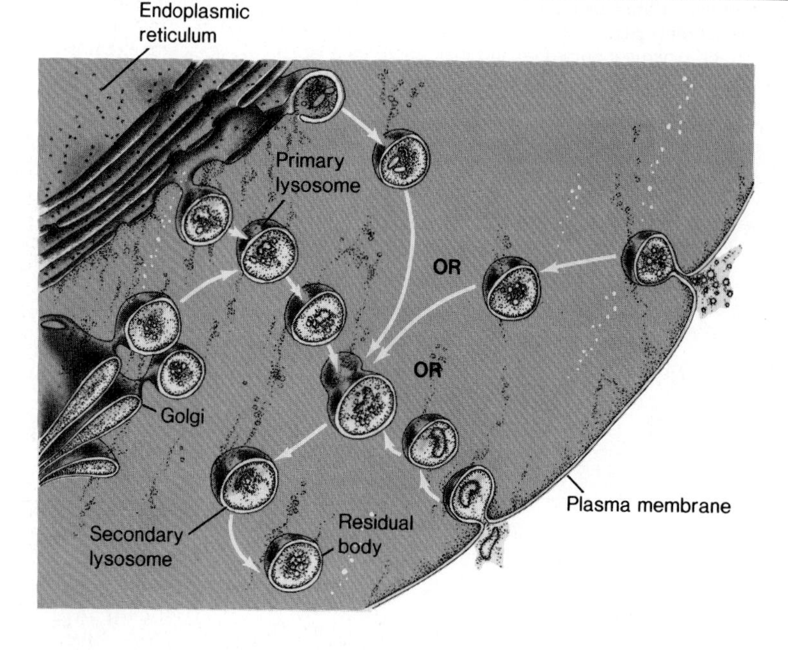

Figure 3.18 Electron micrograph showing the structure of a typical mitochondrion. Note the outer and inner mitochondrial membranes, the cristae, the mitochondrial matrix, and dense mitochondrial granules lying in the matrix (magnification × 40,000).

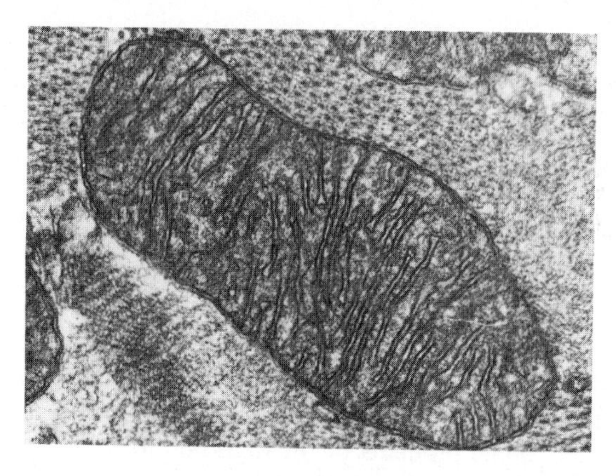

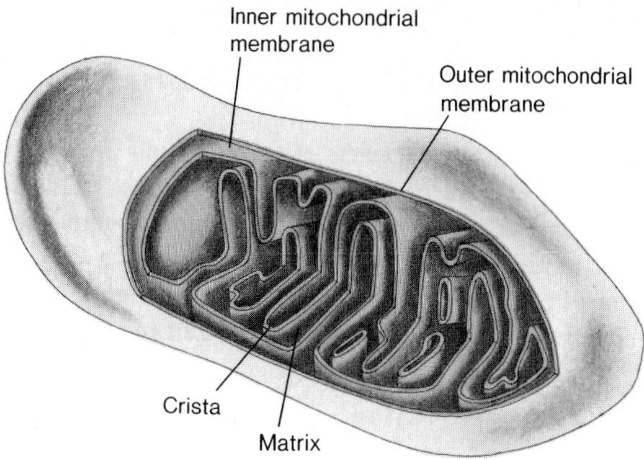

Mitochondria

Mitochondria (singular: **mitochondrion**) are frequently called the "powerhouses" of the cell. They are the sites of certain reactions that provide energy for many cell processes. Most eukaryotic cells have many mitochondria—in fact, some cells may possess as many as a thousand or more.

Most mitochondria are oblong structures measuring about 0.3 to 1.0 μm by 5 to 10 μm. The electron microscope reveals that mitochondria are complex membranous structures. Each mitochondrion is enclosed by a double membrane: the **outer mitochondrial membrane** and the **inner mitochondrial membrane,** which are separated by a tiny space (figure 3.18). The inner membrane's surface area is greatly increased by foldings called **cristae.** The cristae vary in shape among different kinds of cells. The inner membrane encloses a **mitochondrial matrix.** This matrix, which is fairly dense, contains ribosomes, DNA, and often various large granules.

We will discuss the role of mitochondria as cell "powerhouses" and describe how they provide energy for cell processes in chapter 6.

Plastids

Plastids are plant cell organelles in which the synthesis and storage of food reserves takes place. Some plastids manufacture and store starch, while others store fats, oils, and proteins. Generally, these storage plastids are colorless.

However, there are also colored plastids that contain yellow, orange, or red pigments. Such plastids give flowers, seeds, fruits, and some autumn leaves their many colors.

Figure 3.19 The structure of a chloroplast from a tobacco leaf. A complex network of grana lies within the chloroplast matrix or stroma. The organization of this membranous network is diagrammed in the insert (magnification × 19,200).

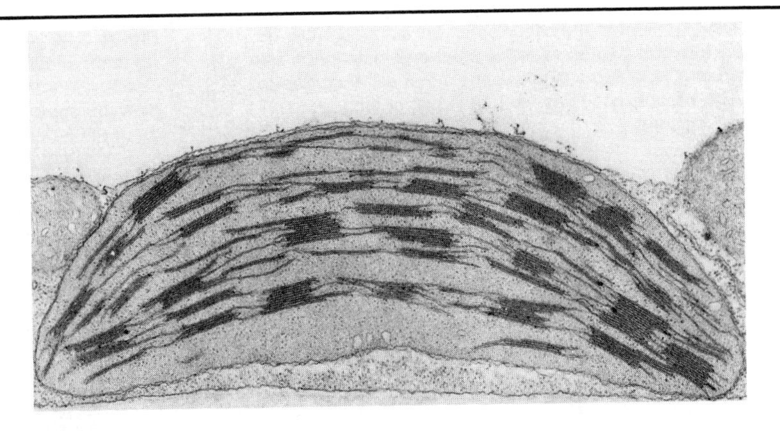

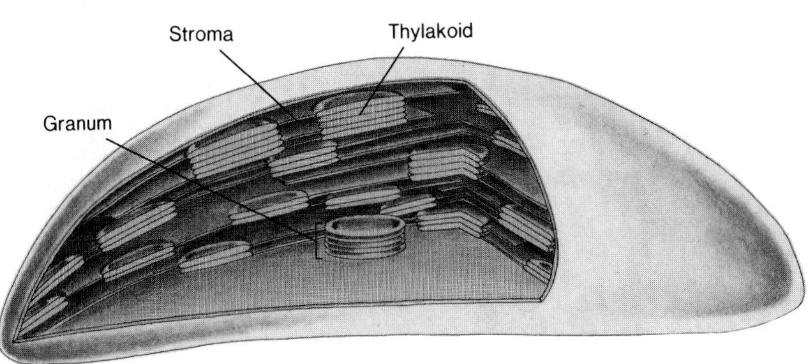

Chloroplasts are colored plastids that contain large quantities of the green pigment chlorophyll, which is crucial to photosynthesis. Photosynthesis is the absorption of light energy to convert carbon dioxide and water to carbohydrates and oxygen.

In most plants, chloroplasts are usually oval and around 2 to 4 μm by 5 to 10 μm. Like mitochondria, chloroplasts are encompassed by a double membrane (figure 3.19). Chloroplasts also have a matrix, the **stroma,** that lies within the inner membrane and contains DNA, ribosomes, and starch granules.

Also like mitochondria, a chloroplast has a complex internal membrane system. The most prominent components of this system are the numerous **grana** (singular: **granum**). These grana are stacks of between five and thirty flattened, hollow sacs or vesicles called **thylakoids.** In an electron micrograph, a granum looks much like a stack of coins.

We will further explore the structure and function of chloroplasts in the section on photosynthesis in chapter 5.

Cell Vacuoles

Most mature plant cells, and some animal and protist cells, contain **vacuoles.** A vacuole is a fluid-filled space enclosed by a membrane. The fluid in a vacuole contains relatively high concentrations of stored materials such as sugars, organic acids, proteins, and salts. Vacuoles in many plant cells contain bright-colored pigments that are responsible for the blue, violet, and scarlet colors of parts of plants such as radishes, beets, plums, cherries, and a number of flowers.

Vacuoles are much more prominent in plant cells because they occupy a greater proportion of the space in mature plant cells than the vacuoles in animal cells do. In fact, well over half of the volume of a mature plant cell may be occupied by its vacuole (figure 3.20).

Cell Movement Organelles

Movement is part of the life of cells. Various organelles move around inside all cells, and all cells exhibit movement during cell division. But some cells may move quite dramatically. For example, amoebas change shape as they move about seeking food particles, which they wrap around and engulf. Some animal cells also can change shape and move to engulf particles and even other cells (figure 3.21).

There are two types of organelles that are important in various cell movements. These are **microfilaments** and **microtubules.**

Microfilaments are minute threads of protein present in many cells (figure 3.22*a*). These tiny filaments are responsible for cell shape changes, for movements of small granules

Figure 3.20 Development of the vacuole in a growing plant cell. A young plant cell (top) has several vacuoles scattered in its cytoplasm. These small vacuoles fuse into a single vacuole that expands due to water uptake. The vacuole occupies more than half of the volume of a mature plant cell (bottom).

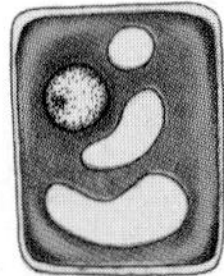

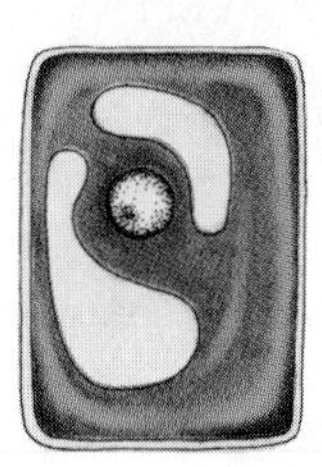

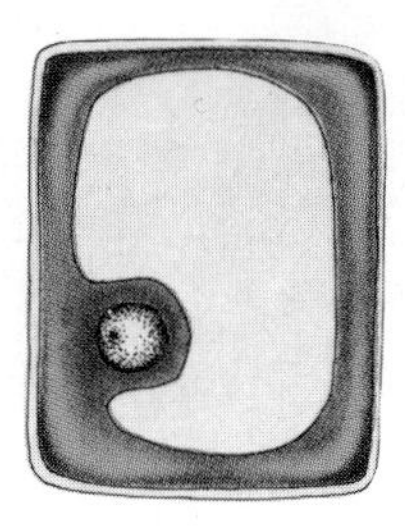

Figure 3.21 Cell movement. A scanning electron microscope picture of a large cell, known as a macrophage (M), from the liver in the process of engulfing two old, misshapen red blood cells (Er). The macrophage moves to wrap its membrane around the erythrocytes (arrows) and then takes them in as vesicles. The vesicles fuse with lysosomes (page 65) and the old cells are digested. Useful substances such as iron recovered from the digested cells can be used again by the body. Other macrophages engulf and destroy invading bacteria in much the same way (magnification × 5,180).

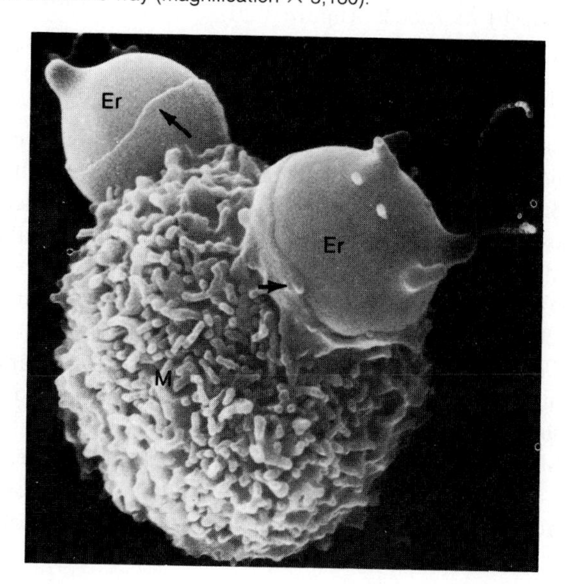

Figure 3.22 Microfilaments and microtubules. (*a*) Microfilaments in a rat embryo cell made visible by a treatment that makes them glow (fluoresce) with appropriate illumination while the rest of the cell remains dark (magnification × 632). (*b*) Microtubule structure. A microtubule is a thin cylinder composed of two types of spherical protein (tubulin) subunits, which are shown as colored and gray spheres. These subunits are stacked in a helix, and in most microtubules, the tubule wall is thirteen subunits in circumference.

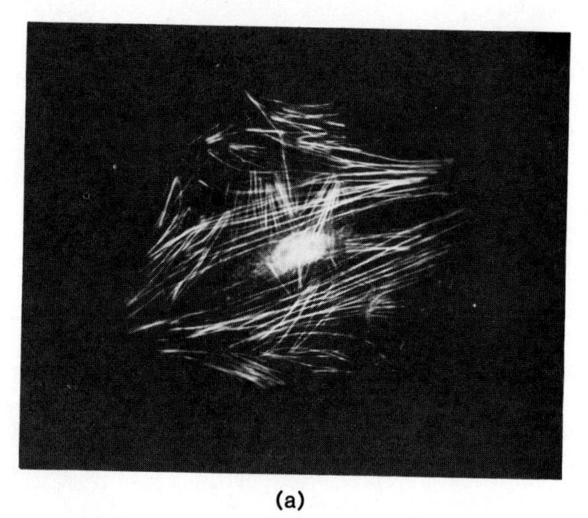

(a)

(b)

from one part of the cell to another, and for the pinching constriction that splits a cell in two during cell division.

Microtubules are thin cylinders that are several times larger than microfilaments. Microtubules are complex structures constructed of two types of slightly different spherical subunits made up of a protein called **tubulin.** These subunits unite to form the cylindrical shape of the microtubule (figure 3.22*b*). Microtubules are involved in several kinds of movement processes including movement in chromosomes during cell division. Microtubules also make up two structures, cilia and flagella, that permit some cells to move actively through their environment.

Figure 3.23 The beating action of flagella and cilia. (*a*) Flagellar movement often takes the form of waves that move either from the base of the flagellum to its tip or in the opposite direction. The motion of these waves propels the organism along. (*b*) The beat of a cilium may be divided into two phases. In the effective stroke (white), the cilium remains fairly stiff as it swings through the water. This is followed by a recovery stroke (gray) in which the cilium bends and returns to its initial position. The large arrows indicate the direction of water movement in these examples.

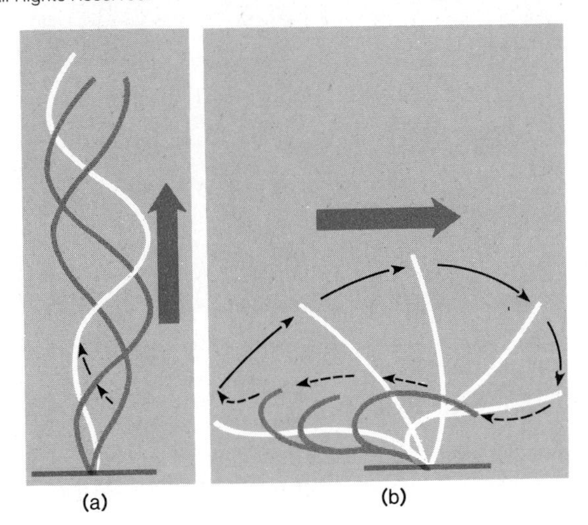

In addition to their roles in cell movement, microtubules and microfilaments are part of an interconnected network of filaments in the cytoplasmic matrix. Microtubules, microfilaments, and filaments that are intermediate in size are connected by thin filamentous bridges making up a network called the **microtrabecular lattice.** Thus, the cytoplasmic matrix, long thought to be a rather homogeneous fluid, actually contains an elaborate, highly organized meshwork of filaments.

Cilia and Flagella

Cilia (singular: **cilium**) and **flagella** (singular: **flagellum**) are long, slender, hairlike organelles associated with cell movement. They are the means by which swimming cells propel themselves. For example, sperm are propelled by flagella. In certain stationary cells, cilia or flagella move material over the cell's surface. In the epithelial cells lining human respiratory passages, for example, cilia carry mucus and dirt away from the lungs.

Cilia and flagella are similar, but they do differ from one another in two ways. First, cilia may be only 10 to 20 μm long, while flagella are much longer (about 100 to 200 μm in length). Second, the two organelles move differently (figure 3.23). A flagellum moves in an undulating fashion and acts more like a propeller. A cilium's beat, on the other hand, is more like the stroke of an oar that acts to row a cell through the surrounding fluid.

Figure 3.24 Cilia structure. (*a*) Scanning electron micrograph of ciliated cells on the gill of an amphibian embryo (magnification × 366). (*b*) Electron micrograph of the cilia of a single-celled organism cut across the cilia (cl) to show their internal structure. Note the two central microtubules surrounded by nine pairs of microtubules (magnification × 79,920).

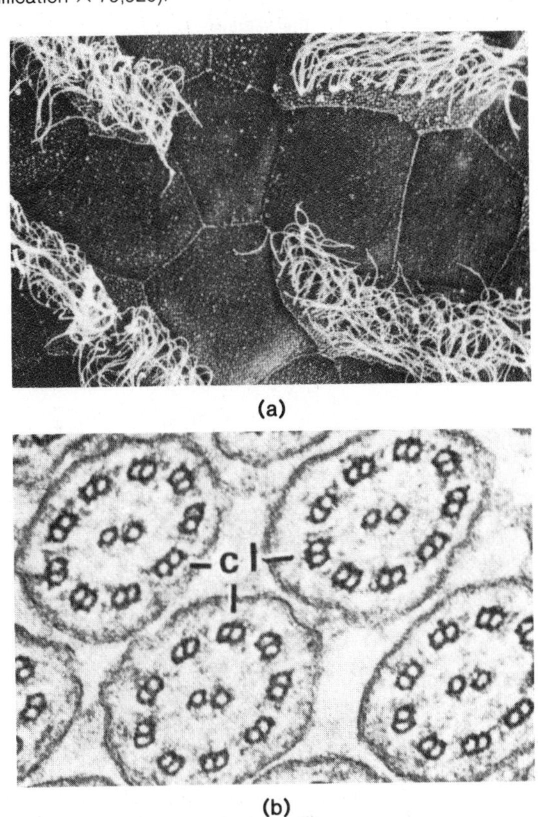

Despite these differences, the internal structure of cilia and flagella is remarkably similar. Each has an inner core that contains a very characteristic set of microtubules, which are responsible for their movement. There are nine pairs of microtubules arranged in a circle around two central tubules, all of which run through the length of the cilium or flagellum. This construction is called a 9 + 2 pattern of microtubules (figure 3.24). In terms of evolution, the 9 + 2 pattern in cilia and flagella is very interesting. Cilia and flagella of a very wide range of organisms have microtubules running through them with this same arrangement.

In the cytoplasm at the base of each cilium or flagellum lies a **basal body.** A basal body is a short cylinder also made up of microtubules. These microtubules, however, form a 9 + 0 pattern. They are arranged in a ring of nine sets of three (triplets) with none in the middle. Basal bodies play a role in assembling cilia and flagella.

Figure 3.25 Electron micrograph of a pair of centrioles. One is seen from the end and the other from the side (magnification × 142,000). The accompanying sketch shows microtubule triplets arranged in a 9 + 0 pattern in a centriole. The basal bodies of cilia and flagella have microtubules arranged in the same pattern.

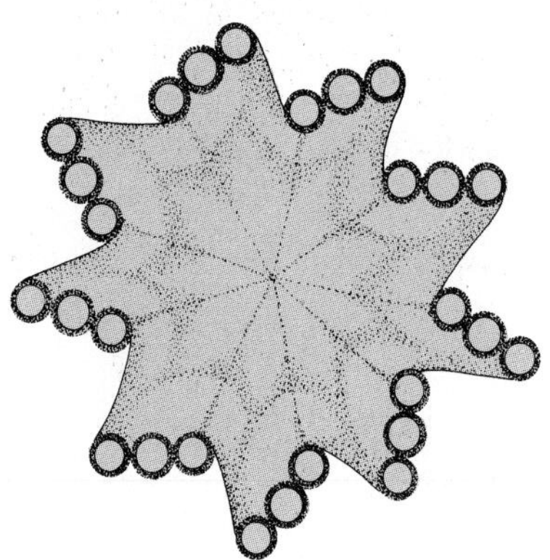

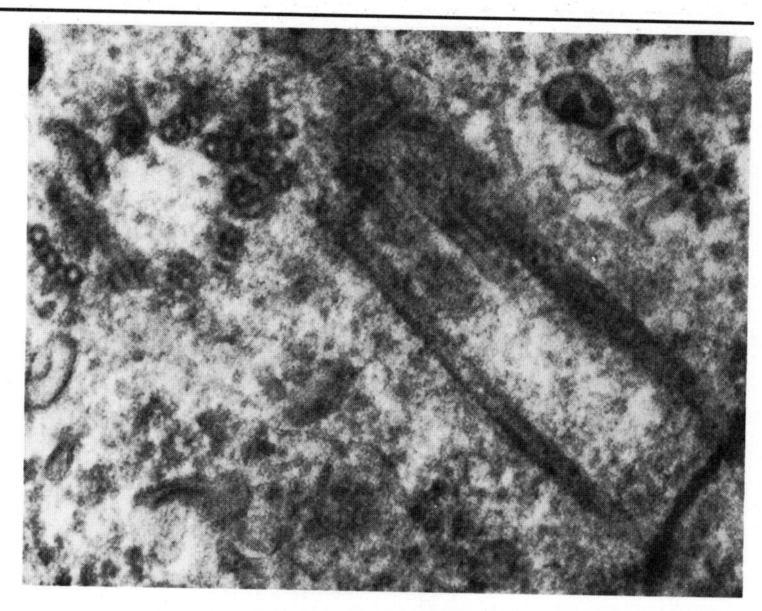

Figure 3.26 A diagrammatic sketch showing the layers of a plant cell wall.

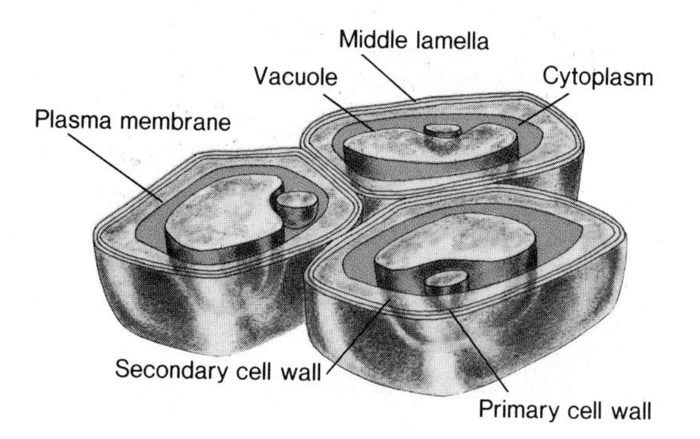

The 9 + 0 pattern of microtubule triplets is also characteristic of another organelle, the **centriole.** Centrioles have essentially the same structure as basal bodies, but they usually occur in pairs. Each centriole contains nine microtubule triplets. Centrioles are associated with cell division in most cells except those of many plants, but their actual role is not yet understood (figure 3.25).

The Plant Cell Wall

The most obvious difference between animal and plant cells is that plant cells have a **cell wall.** The cell wall is a rigid, boxlike structure that lies outside the plasma membrane and thus encloses the entire plant cell.

Cells from soft tissues, such as leaves and fruits, have cell walls composed of two layers. Each cell produces a thin **primary wall,** and the primary walls of adjacent cells are cemented together by a layer called the **middle lamella.** In addition to primary walls and middle lamellae, the cells of hard, woody tissues, such as stems and nuts, form a **secondary cell wall** (figure 3.26) between the primary wall and the plasma membrane. The secondary cell wall may have several layers; it makes the boundary around each cell much thicker and harder than it would be with a primary wall alone.

Cell walls are firm because they contain tiny fibers made up of **cellulose** molecules. Another substance called **lignin** adds to the hardness of cell walls in woody parts of plants.

The middle lamella is composed mainly of **pectin.** The consistency of a plant tissue depends heavily on the condition of the pectin in the middle lamellae between its cells. For example, as fruits ripen, the pectin of the middle lamella is converted to a more soluble form and the cells become loosely associated. You have probably tasted the effects of excessive pectin loss, for example, in overripe, mealy apples. Home canners are familiar with pectin because concentrated pectin extracts are added to fruit juices to make jelly.

Table 3.2
Comparison of Prokaryotic and Eukaryotic Cells

Property	Prokaryotes	Eukaryotes
Organization of Hereditary Material		
True nucleus	Absent	Present
DNA complexed with proteins (histones)	No	Yes
Other Cell Components		
Mitochondria	Absent	Present
Chloroplasts	Absent	Present
Plasma membrane has steroids, such as cholesterol	Usually no	Usually yes
Flagella	Submicroscopic in size; composed of one fiber	Microscopic in size; membrane bound; usually twenty microtubules in 9 + 2 pattern
Ribosomes	70S*	80S (except in mitochondria and chloroplasts)
Lysosomes	Absent	Present
Microtubules	Absent or rare	Present

*S = Svedburg unit, a measurement used to compare sizes of molecules and molecular aggregates.

Cell wall constituents from many plants are very useful economically. Wood used in construction is probably the first that comes to mind. Cotton is almost pure cellulose. Paper is a cellulose product that is manufactured from wood treated with dilute acids or bases. Lignin is extracted from wood and used to produce pigments, adhesives, vanilla flavoring, and many other items. And, as we noted earlier, pectins are important to the food industry as gelling agents used in making jams and jellies.

The Eukaryotic Cell: A Summary

Composite views of two eukaryotic cells, including many of the organelles that we have discussed, are pictured in figure 3.27. A comparison of this illustration with E. B. Wilson's sketch, done in 1925 (figure 3.3), shows how our knowledge of cells has grown since that time.

Obviously, the cell is not a simple sac of granular fluid with a nucleus at its center, as it was once visualized. It is a complex structure with a high degree of internal organization. Membranes divide the cell into a number of individual compartments. Cell processes can take place simultaneously within these compartments, but still can be independent of one another because they are separated by membranes.

Prokaryotic Cells

Near the beginning of our discussion of cell structure, we noted that there are two fundamentally different kinds of cells: eukaryotic cells and prokaryotic cells. Thus far we have focused on eukaryotic cells, cells that possess a membrane-enclosed nucleus and membranous structures in their cytoplasm. Now we turn our attention to prokaryotic cells, cells that lack a true membrane-enclosed nucleus. Prokaryotic cells are found among bacteria and cyanobacteria (blue-green algae), while all other organisms, including all plants and animals, have eukaryotic cells (figure 3.28).

Although the presence or absence of a true nucleus is the most obvious difference between these two cell types, there are other significant distinctions. As you can clearly see from table 3.2, prokaryotic cells are structurally much simpler. They lack both the diverse collection of membrane-bound organelles and the high degree of functional compartmentalization that are features of eukaryotic cells.

Despite structural differences between prokaryotic and eukaryotic cells, there are striking functional similarities at the biochemical level. For example, genetic information is stored in DNA in the same way in both types of cells. The expression of genetic information in regulation of cell function also is very similar. Many other processes also occur in virtually the same way in both prokaryotic and eukaryotic cells. Thus, beneath the significant structural and functional differences that separate prokaryotes from eukaryotes, there is an even more fundamental molecular unity that is basic to all life processes.

Figure 3.27 Diagrammatic representations of eukaryotic cells based on electron micrographs. (*a*) A generalized animal cell.

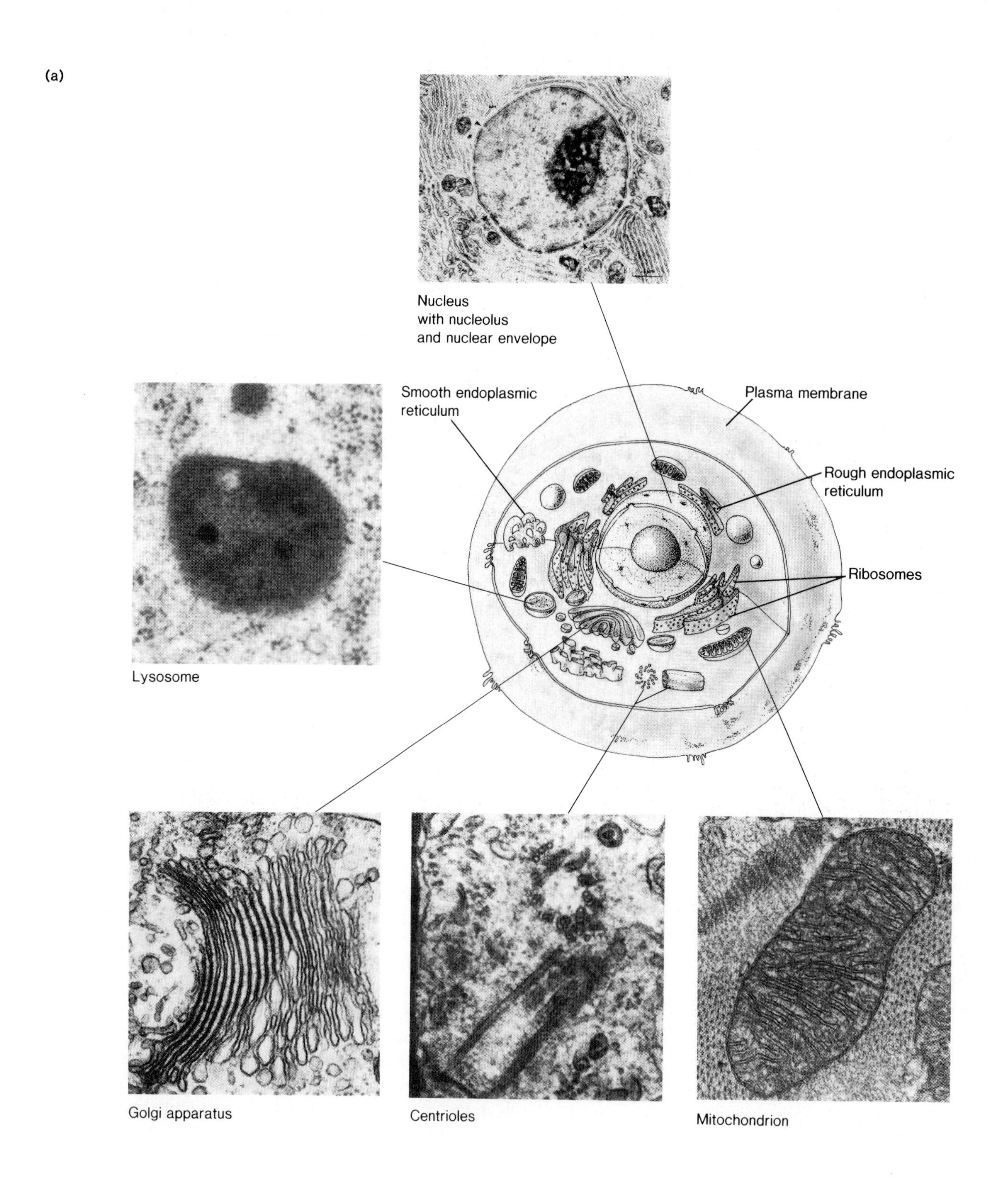

(*b*) A generalized plant cell. The vacuole actually occupies a greater percentage of the cell's volume in the average mature plant cell than it does in this sketch. Leucoplasts are noncolored plastids that store starch.

(b)

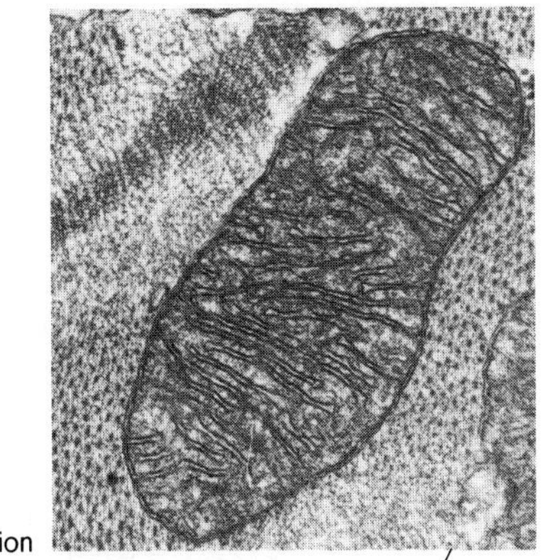

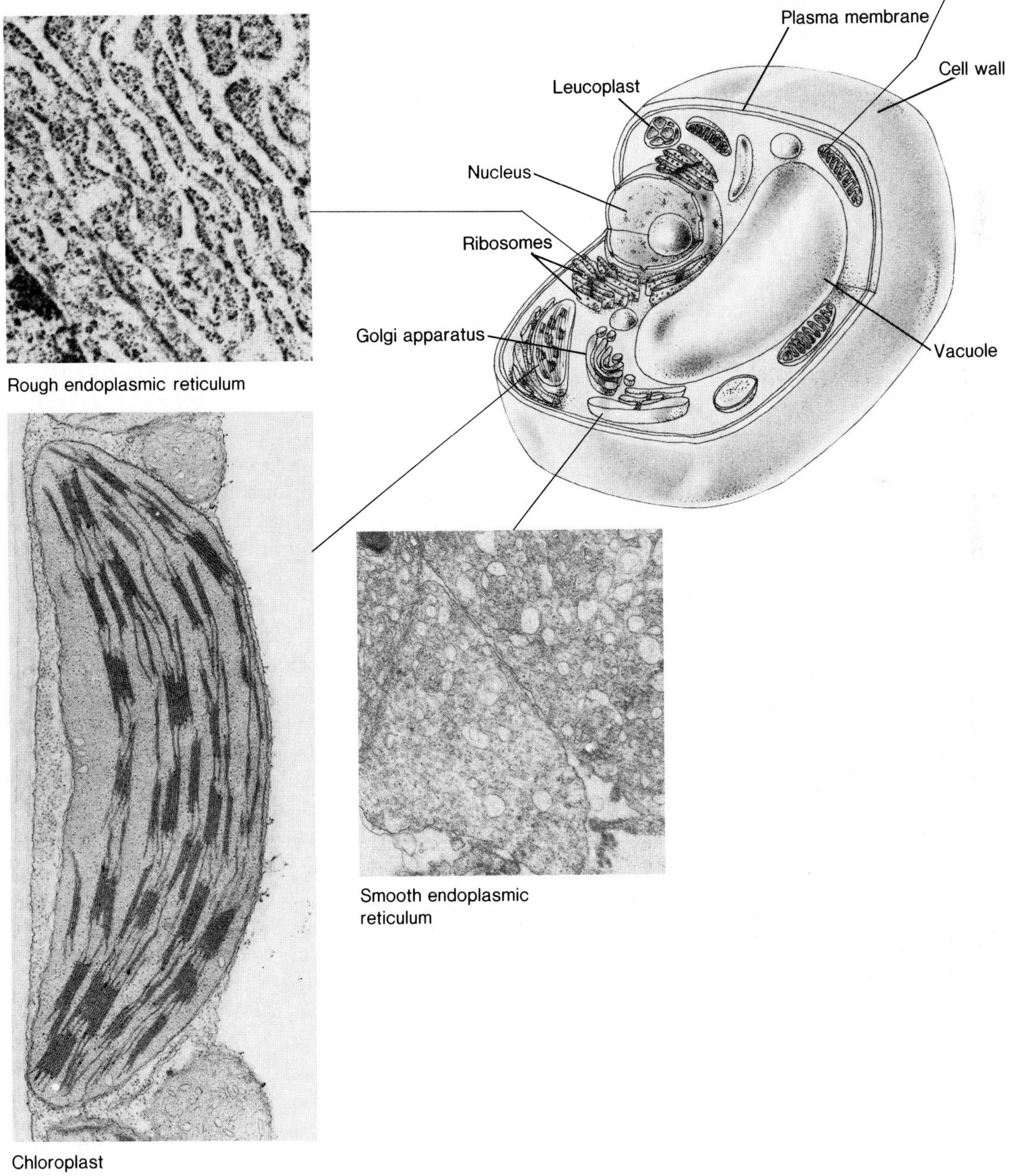

Mitochondrion

Plasma membrane

Cell wall

Leucoplast

Nucleus

Ribosomes

Golgi apparatus

Vacuole

Rough endoplasmic reticulum

Smooth endoplasmic reticulum

Chloroplast

Figure 3.28 Electron micrograph of a prokaryotic cell undergoing cell division. Parts of the cell are identified in the accompanying sketch. This micrograph is of a bacterium (magnification × 63,960). Notice that there is a nuclear body (nucleoid) rather than a membrane-enclosed nucleus like that of a eukaryotic cell, and that the cell lacks membrane-bounded organelles. Mesosomes are inward-folded portions of the plasma membrane. Mesosomes may function to increase the surface area available for absorption of nutrients. They also may be involved in cell division processes.

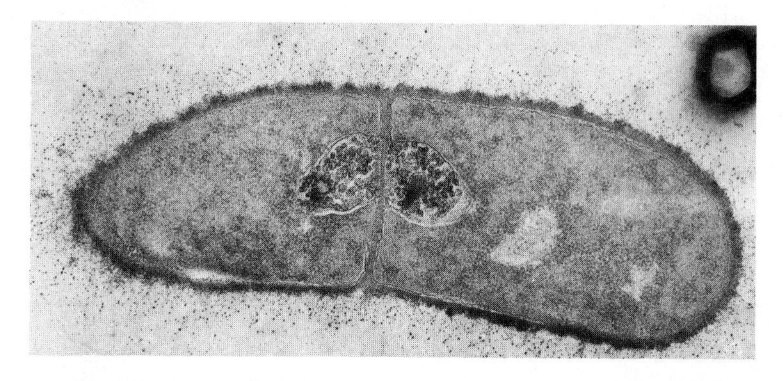

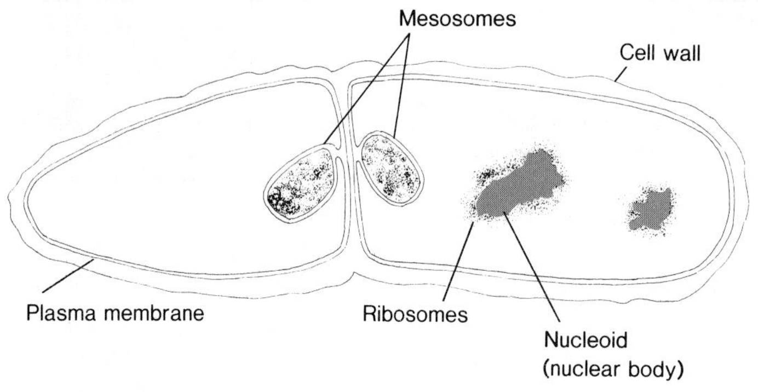

How Things Get In and Out of Cells

All living cells continually exchange materials with their environments. They must obtain some materials from and release others to the environment if they are to maintain their normal structure and functional capabilities.

Because the plasma membrane is the region of contact between the cell and its environment, one of the most important functions of the plasma membrane is the regulation of the movement of substances into and out of the cell. There are several different means by which various materials pass through plasma membranes.

Diffusion

Some substances enter or leave cells by a process called diffusion. **Diffusion** is the tendency for materials to move from an area in which they are highly concentrated to an area of lower concentration.

All molecules (and ions) are in a state of constant, random motion because of heat energy. This is called thermal agitation. The distance that molecules can move without encountering other molecules depends on the nature of the material. In a solid, movement is limited; for example, the movement of water molecules within a piece of ice is severely restrained. However, there is more freedom of movement for molecules in a liquid, which is a less orderly arrangement of molecules, and even more freedom of movement in a gas, which is still less orderly.

The process of diffusion can be demonstrated by a container filled with water and a crystal of a colored material placed in one corner of the container. The crystal will dissolve, forming a halo of color as colored particles move slowly away from the crystal and into the surrounding water. The thermally agitated particles tend to spread from regions where they are highly concentrated to regions where they are less concentrated (figure 3.29).

The colored particles will gradually move from the area of high concentration to the area of lower concentration. On the average, the particles (as they move about randomly) can

Figure 3.29 Diffusion of particles away from a crystal of colored material placed in water. A halo of color surrounds the crystal where colored particles are most concentrated. On the average, particles can move further in the direction of decreasing concentration before colliding with other particles of their own kind, so long as a concentration gradient exists. Diffusion eventually produces an equilibrium in which particles are evenly distributed throughout the water.

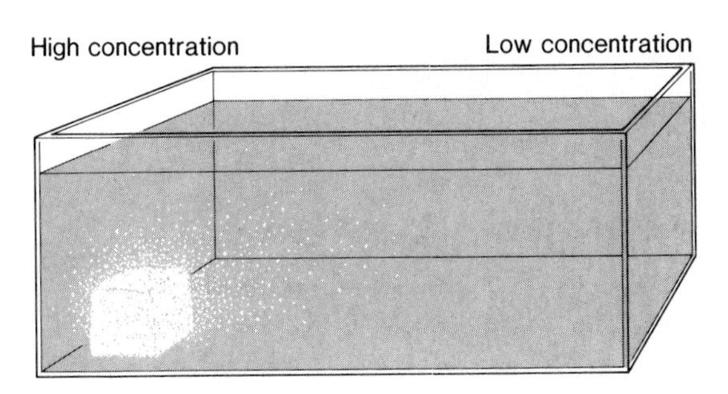

move farther in that direction before they collide with other colored particles. The eventual result will be an equilibrium in which the colored particles are evenly dispersed throughout the water.

Some substances, such as oxygen and carbon dioxide, diffuse freely through plasma membranes. Biologists call this movement **simple** (or **passive**) **diffusion** because the material enters and leaves the cell without active involvement of the cell or its plasma membrane and without energy expenditure by the cell.

Osmosis

Osmosis is a second process important in movement of material in and out of cells. Actually, osmosis is a specialized case of diffusion. Osmosis involves the passage of a solvent such as water through a selectively permeable membrane. We can demonstrate the principles of osmosis with an apparatus consisting of two compartments separated by a selectively permeable membrane—one that will allow water, but not a solute (dissolved substance), to cross it. In figure 3.30, solution 1 is pure water, while solution 2 is a mixture of sucrose (sugar) and water. The membrane will allow water to pass through, but not sucrose. In the context of osmosis, a solution is considered in terms of the concentration of solvent (water), rather than in terms of the concentration of dissolved material. Thus, the concentration of the water in solution 1 is greater than in solution 2 because of the presence of sucrose molecules in the latter. Since any material will tend to diffuse from a region of higher concentration to one of lower concentration, the water will tend to diffuse from area 1 to area 2. When water enters area 2 in figure 3.30, the volume of the solution there increases slightly and rises in the tube.

Figure 3.30 A demonstration of osmosis. Solution (1) is pure water. Solution (2) is sucrose and water. The selectively permeable membrane is permeable to water but not to sucrose. Water moves from (1) into (2). Eventually, the solution reaches a height (h) at which equilibrium is reached; that is, the weight of the column of liquid exerts exactly enough pressure to balance the tendency of water to move from (1) into (2). The pressure required to halt solvent movement across a selectively permeable membrane is called osmotic pressure.

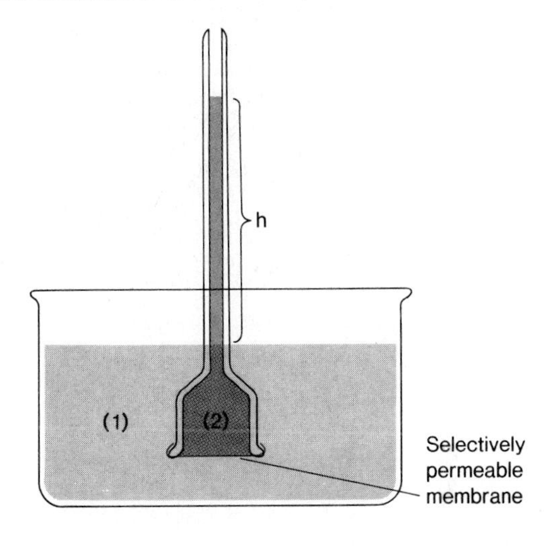

Because the plasma membrane of cells is selectively permeable, water moves in or out of cells depending upon the relative concentrations of water and dissolved material inside and outside cells. This is functionally important because cells must be in a stable water-balance relationship with their environment if they are to function normally.

The tendency for water to move in or out of cells as a result of these concentration effects is described using the terms isotonic, hypertonic, and hypotonic. An **isotonic** solution does not cause water movement in or out of a cell and, therefore, does not cause volume changes in cells (figure 3.31*a*).

A **hypertonic** solution is one in which there is a greater concentration of dissolved material in the fluid surrounding the cell than there is inside the cell. Thus, the water concentration inside the cell is higher and the cell tends to lose volume because of water loss to its environment. When a red blood cell, for example, is transferred from an isotonic solution to a hypertonic solution, water tends to leave the cell, and the cell shrinks (figure 3.31*b*). In a **hypotonic** solution, water from the surrounding solution will flow into the cell because of the relatively higher water concentration outside the cell. A red blood cell will eventually burst if it remains in a hypotonic solution for any length of time.

Figure 3.31 The effects of hypertonic and hypotonic solutions on red blood cells. (*a*) Normal cells in an isotonic solution. (*b*) In hypertonic solutions, erythrocytes lose water and their plasma membranes become wrinkled. (*c*) In hypotonic solutions, erythrocytes gain water and become swollen. Eventually, they break open and release their hemoglobin into the suspending medium. (The photo in (*c*) is a lower magnification than the photos in (*a*) and (*b*).)

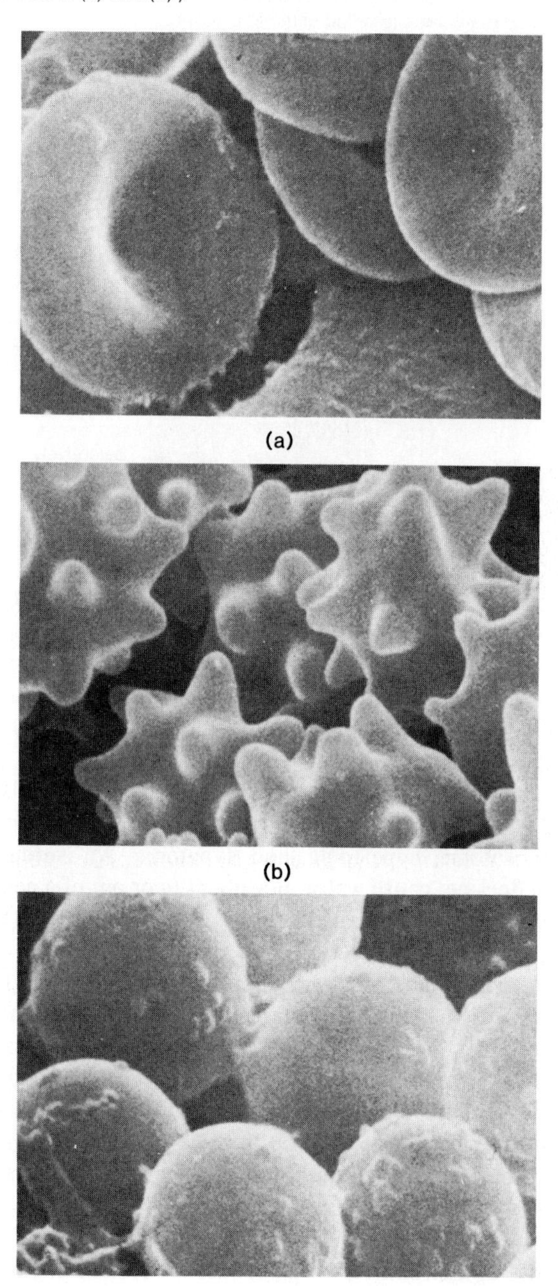

Figure 3.32 Effects of hypertonic and hypotonic solutions on plant cells. (*a*) In a hypertonic solution, the cell membrane pulls away from the cell wall because of a net water loss and shrinkage of the cell. (*b*) In a hypotonic solution, the cell is swollen, but it is prevented from bursting by the restraining force of the cell wall.

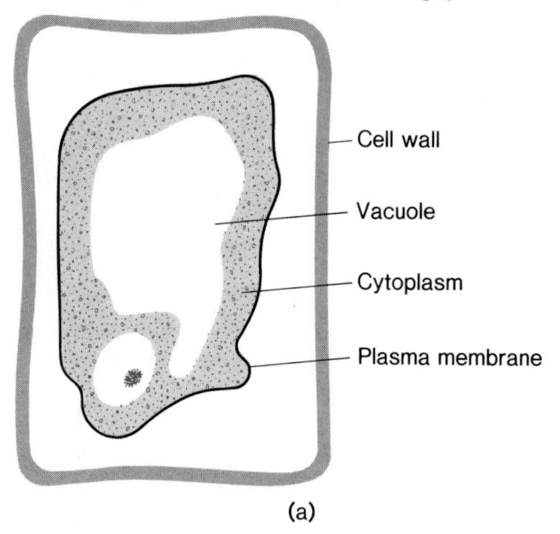

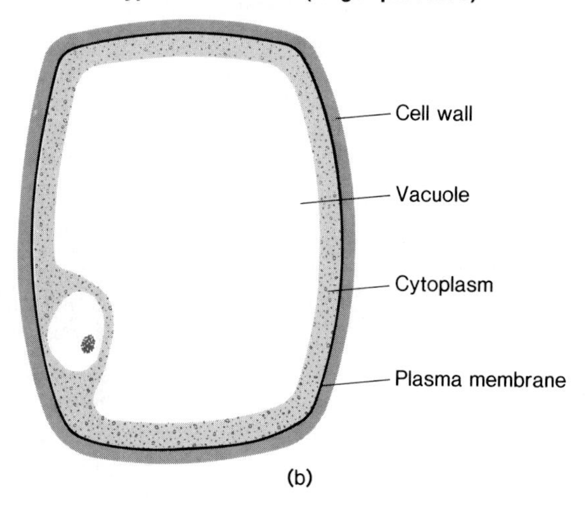

Thus far, the examples we have used to illustrate osmosis have involved animal cells. Plant cells also react to osmotic changes in their environments, but the situation is more complex in plant cells because of the presence of a cell wall around the plasma membrane. In a hypotonic solution, water enters a plant cell until the plasma membrane presses against the cell wall (figure 3.32). When the restraining force of the cell wall equals the outward push due to water entering the cell by osmosis, the net water flow ceases. The pressure exerted by plant cells against their cell walls is called **turgor pressure.** Plant cells normally are rigid from turgor pressure since the fluid in their environments is usually hypotonic.

Figure 3.33 Two different models for active transport. It is very unlikely that carrier proteins rotate as proposed in the model in (*a*). Thus, the fixed pore model (*b*) is favored by most cell biologists. Active transport differs from facilitated diffusion in that active transport requires energy (ATP) input by the cell, and active transport can and does work against concentration gradients.

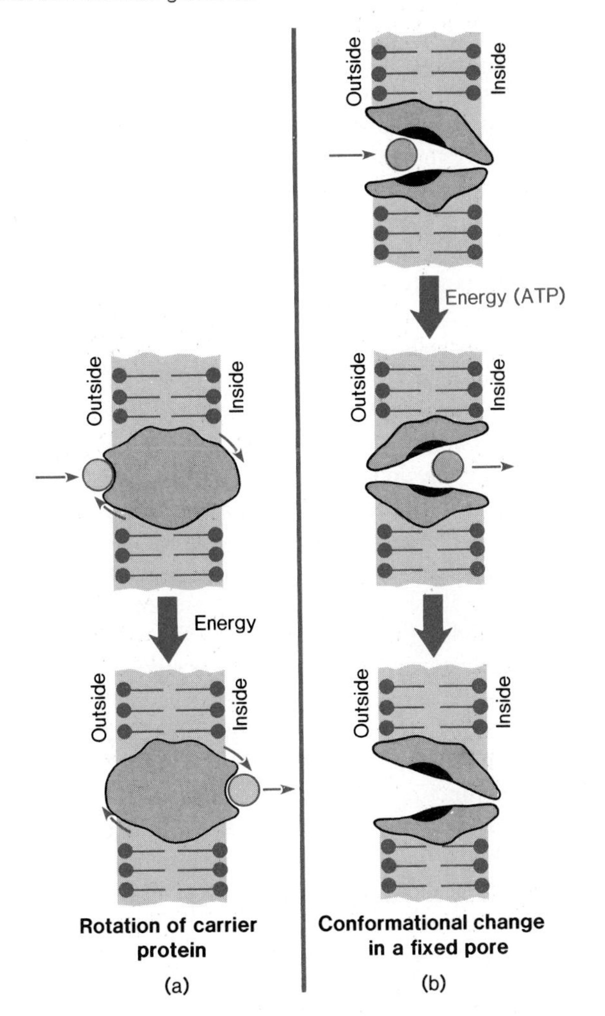

In a hypertonic environment, however, water flows out of a plant cell, causing the cell to shrink in volume and to pull away from its cell wall. This is what happens when a plant goes without water for too long. The loss of turgor pressure in many of the cells causes the plant to wilt. When the environment around a shrunken cell becomes hypotonic once again, it expands to press against its cell walls and its turgor pressure is restored.

Transport across Membranes by Carriers

Water and some other molecules diffuse through small pores in the plasma membrane, but other substances will not diffuse across the plasma membrane. These substances can traverse the membrane only by means of special protein molecules in the membrane called **carriers.** Carriers generally are very specific; that is, a given carrier normally is involved in the movement of only one type of molecule or ion.

Facilitated Diffusion

Facilitated diffusion is one of the processes involving carrier proteins that is important in movement of materials in and out of cells. In facilitated diffusion, the carrier protein provides a route into the cell for a substance that could not diffuse through the plasma membrane by itself. Facilitated diffusion only works, however, when the diffusing substance is more concentrated on one side of the membrane than the other. The diffusing substance must move from a region of higher concentration to a region of lower concentration. The carrier just provides a route through the membrane; the cell does not have to do work in the process.

Active Transport

Active transport is another process involving carrier proteins by which substances move through the plasma membranes of cells. Cells do work in active transport and can even move materials from a region of lower concentration to a region of higher concentration. Such movement requires the cell to expend energy; you could compare the process to rolling a rock uphill. Energy for this work is provided by the breakdown of molecules of **adenosine triphosphate (ATP),** which is called the "energy currency" of cells—ATP can be "spent" in order to power work in cells. We will learn more about ATP, its production, and its functions in chapter 4.

How is cellular energy actually used to move molecules across cell membranes? The substances to be transported are relatively small compared to the thickness of the membrane and must, therefore, be moved a considerable distance by the carrier protein. Several models for this transport mechanism have been suggested.

One model proposes that a carrier protein might bind the solute molecule and then rotate within the lipid bilayer so that its binding site is transferred to the other side of the membrane (figure 3.33*a*). After releasing the solute molecule, the carrier would rotate back to the original side. However, it would take a considerable amount of energy to rotate a large protein through the lipid bilayer of the membrane. There is no evidence that such carrier-molecule rotations occur.

Many biologists favor a different model called the **fixed-pore mechanism.** They suggest that a membrane may possess pores, each of which is formed by two or more protein molecules, and that these integral proteins have specific binding sites for solute molecules. After a molecule has been bound,

Figure 3.34 Exocytosis and endocytosis. These transport mechanisms involve membrane-bound vesicles and require that cells expend energy when using them to move materials across the plasma membrane.

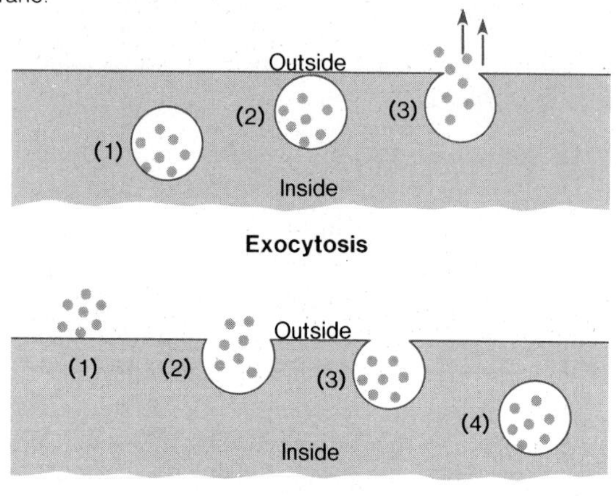

energy is used to induce a conformational (shape) change in the pore proteins (figure 3.33*b*), and this change in pore shape transports the solute molecule to the other side of the membrane, where it is released. The cycle ends when the pore proteins return to their original conformation.

Exocytosis and Endocytosis

Cells can also expend energy to use membrane-bound sacs (vesicles) in either the secretion or uptake of molecules. In **exocytosis** a vesicle approaches the inside of the plasma membrane and fuses with it to release its contents to the exterior (figure 3.34). Exocytosis is the means by which specialized secretory cells release the substances that they produce to the outside (see page 65).

In **endocytosis,** material from outside the cell is enclosed in a vesicle and taken into the cell. There are two forms of endocytosis, **phagocytosis** and **pinocytosis.** In phagocytosis solid material is ingested, while in pinocytosis the cell engulfs a small quantity of fluid.

Phagocytosis is used by a wide variety of cells. For example, in many one-celled organisms, phagocytosis serves as a basic feeding mechanism. White blood cells of vertebrates, including humans, phagocytize invading organisms such as bacteria and destroy them inside the vesicles that are formed (see figure 3.21). This is very important in resistance to infection and disease.

Pinocytosis is used by a number of cells to take in large molecules, such as proteins, that will not readily penetrate the plasma membrane.

Summary

Cells are the fundamental units of life. All organisms are either cells or aggregates of cells and cell products.

Electron microscopy makes possible high-magnification studies of cell structure not possible with the light microscope. Cell-fractionation techniques permit isolation and functional analysis of cell organelles.

Membranes are fundamental to cell structure and function. The current model describes membranes as lipid bilayers with proteins moving about in them.

The plasma membrane is a membranous enclosure around the cell. It holds cell constituents together and controls exchanges between the cell's interior and the external environment.

Eukaryotic cells contain a number of types of structurally and functionally specialized organelles. Organelles are suspended in the cytoplasmic matrix, a watery medium that contains a network of filaments and other materials.

A nuclear envelope with many pores encloses the nucleus. Chromatin is the dispersed form of the cell's genetic material. Chromatin condenses into chromosomes during cell division. The nucleolus participates in ribosome synthesis.

Ribosomes, the sites of protein synthesis in cells, are complex particles that are assembled from smaller units. Rough endoplasmic reticulum (ER) has ribosomes attached to it; smooth ER carries no ribosomes. The Golgi apparatus consists of a stack of saclike structures. It packages cell products. Lysosomes, which contain enzymes that can digest cell components, function in defense reactions, in the turnover of cell materials, and in the elimination of cell debris.

Mitochondria are complex double-membrane organelles. Important energy conversions occur in mitochondria.

Plastids are a diverse group of plant organelles. Chloroplasts are the organelles involved in photosynthesis.

Both plant and animal cells contain vacuoles, but plant cell vacuoles are much more prominent.

All cells exhibit movement of some kind. Microfilaments and microtubules are involved in maintaining cell shape and in many types of cell movement.

Cilia and flagella are both movement organelles. Their basal bodies are very similar to centrioles. Centrioles are involved in cell division.

Cell walls are multilayered enclosures around plant cells that vary in thickness and composition depending on the age and function of the cells.

The most obvious difference between eukaryotic and prokaryotic cells is that prokaryotic cells lack true nuclei and membrane-enclosed organelles. But many cell processes in the two types of cells are similar at the biochemical level.

The plasma membrane is the cell's boundary with its environment. Properties of the plasma membrane determine characteristics of a cell's exchanges with the exterior.

In diffusion, particles tend to move from an area of greater concentration to an area of lower concentration. Some materials penetrate plasma membranes by simple diffusion, but the plasma membrane is selectively permeable—some materials do not diffuse through it readily.

Osmosis is the diffusion of water through a membrane in response to solute concentration differences on the two sides of the membrane. The net osmotic movement of water through a membrane is toward the area with the greater solute particle concentration and the lower water concentration.

Cells in isotonic solutions do not show a net gain or loss of water or cell volume changes. Cells in hypertonic solutions lose water and shrink, and cells in hypotonic solutions gain water and swell.

Carrier-mediated transport involves certain specialized membrane proteins. In facilitated diffusion, the carrier increases the rate at which a material moves down a concentration gradient but does not require energy (ATP) input. Active transport requires energy (ATP) input.

Endocytosis and exocytosis are other mechanisms by which materials enter and leave cells. In phagocytosis, solid materials are incorporated into vacuoles. In pinocytosis, fluid is engulfed.

Questions for Review

1. Lens systems could be built for light microscopes that provide more magnification than commonly used light microscopes have. Explain why a light microscope with much higher magnification cannot be substituted for an electron microscope in the study of cell structure.
2. Explain the role of the Golgi apparatus in "packaging" cell products.
3. Lysosomes contain enzymes that can digest many important types of molecules. Explain why other cell constituents are safe from digestion by these enzymes and how this illustrates the significance of compartmentalization in eukaryotic cells.
4. What distinguishes smooth endoplasmic reticulum from rough endoplasmic reticulum?
5. A number of plant and animal cell organelles are similar, but there are several significant differences between animal cells and the cells of plants. List and explain at least three of these differences.
6. List three differences between eukaryotic and prokaryotic cells.
7. Animal cells cannot tolerate being in a hypotonic environment, but the normal environment of plant cells is hypotonic to them. Can you explain how plant cells can remain intact in a hypotonic environment?
8. Explain the roles of endocytosis and exocytosis in moving large molecules and even larger particles through plasma membranes of cells.
9. Why would a poison such as cyanide, which interferes with cell metabolism and ATP production, depress active transport activities in cells?

Questions for Analysis and Discussion

1. Explain why substances that interfere with the assembly of tubulin subunits into microtubules are sometimes used as part of a chemotherapy program in cancer treatment.
2. For many years, strong salt solutions (brines) have been used for food preservation. How might a very strong salt solution interfere with the life and growth of bacteria that might otherwise spoil the food? Why don't bacteria grow in honey?
3. Digestive enzymes are protein molecules. The pancreas produces digestive enzymes that go to the intestine and function there. Why would you expect a cell in the pancreas that is producing digestive enzymes to contain the following: a prominent nucleolus, a great deal of rough endoplasmic reticulum, a well-developed Golgi apparatus, and a large number of membrane-enclosed vesicles?

Suggested Readings

Books

Alberts, B.; Bray, D.; Lewis, J.; Raff, M.; Roberts, K.; and Watson, J. D. 1983. *Molecular biology of the cell.* New York: Garland.

Avers, C. J. 1981. *Cell biology.* 2d ed. New York: Van Nostrand.

Dyson, R. D. 1978. *Cell biology.* 2d ed. Boston: Allyn & Bacon.

Giese, A. C. 1979. *Cell physiology.* 5th ed. Philadelphia: W. B. Saunders.

Karp, G. 1984. *Cell biology.* 2d ed. New York: McGraw-Hill.

Sheeler, P., and Bianchi, D. E. 1983. *Cell biology.* 2d ed. New York: Wiley.

Articles

Allison, A. C. 1977. Lysosomes. *Carolina Biology Readers* no. 58. Burlington, N.C.: Carolina Biological Supply Co.

Cook, G. M. W. 1980. The Golgi apparatus. *Carolina Biology Readers* no. 77. Burlington, N.C.: Carolina Biological Supply Co.

Dustin, P. August 1980. Microtubules. *Scientific American* (offprint 1477).

Lucy, J. A. 1979. The plasma membrane. *Carolina Biology Readers* no. 81. Burlington, N.C.: Carolina Biological Supply Co.

Satir, B. October 1975. The final steps in secretion. *Scientific American* (offprint 1328).

Unwin, N., and Henderson, R. February 1984. The structure of proteins in biological membranes. *Scientific American.*

Energy, Reactions, and Life

Chapter Outline

Chapter Concepts

1. All living things must obtain energy to do work required to maintain their internal organization and carry out their life processes.
2. All energy conversions in all systems, living and nonliving, obey two basic laws. One says that energy can be neither created nor destroyed during any conversion; the other says that there is a tendency toward increasing disorder in the universe.
3. The rates at which chemical reactions proceed depend upon several factors.
4. Adenosine triphosphate (ATP) serves as a renewable cellular energy currency that provides energy for many energy-requiring processes in cells.
5. Enzymes are catalysts that increase the rates of chemical reactions in living things.
6. Many vital energy-yielding reactions in living things involve oxidation (electron loss) and reduction (electron gain).

Imagine yourself standing alongside an airport runway as a giant 747 rumbles slowly toward you (figure 4.1). Turning onto the runway, it pauses for a moment, then the rumble of its engines turns to a roar. At first it seems to lumber along, like a great silver dinosaur. But as the engines roar louder and thrust harder, the plane steadily gains speed and races along the pavement. Only seconds later, with a slight shudder, the mighty machine lifts off the ground and rises into the sky.

You have just witnessed a gigantic energy conversion. Huge quantities of jet fuel have been burned explosively to provide thrust to accelerate the plane down the runway to takeoff speed and beyond. The chemical energy of the jet fuel has been used to do work moving the massive airplane. As you read this, similar energy conversions are going on in your own body and in all living things around you. Obviously, these energy conversions are much smaller than the explosive burning of jet fuel in the 747, but they are no less important because they provide energy for all the vital processes that make life possible.

All organisms must perform work to live. Vital, work-requiring life processes, such as breathing, heartbeat, and digestion, continue even when an animal is at rest. Within cells, there are a variety of forms of chemical work being done. For example, large molecules are being made from small molecules. Cells are also constantly doing transport work, in the form of actively transporting substances into or out of various body cells. Plants, even though they do not move from place to place, must also accomplish several forms of work, including chemical work and transport work. In fact, if these various kinds of work cease in an organism, life itself will cease.

Thus, it is clear that all life depends upon the capacity to do work. We call this capacity **energy.** All living things and all of their cells require a continuing supply of energy to make this vital work possible.

The ultimate source of energy for life on earth is the sun—the radiant energy of this star bathes our planet every day. A portion of this solar energy is trapped by plants and some microorganisms as chemical energy in the form of organic chemical compounds through the process of **photosynthesis** (figure 4.2). Plants, and of course the animals that consume them, can use these compounds as both structural building blocks and as energy sources.

In order to do all of this, living things must conduct many different chemical reactions that transfer energy and convert it from one form to another. To understand those conversions, we need to understand a few things about energy and about the nature of chemical reactions in general.

Figure 4.1 Energy conversion by a 747. Huge quantities of chemical energy are used to do work as jet fuel is burned to propel an airplane.

Energy

When we consider energy in biology, we usually are interested specifically in energy changes in a particular organism, a part of an organism, or a particular process going on inside an organism. When considering energy changes, we call the focus of our attention a **system.** Therefore, for energy change purposes, we might define an entire plant or animal as a system. Or, more often, we might concentrate on an individual chemical reaction occurring in a living thing. This single reaction would then be defined as our system.

Whatever the system being studied, large or small, everything outside the system is called the **surroundings.** It is convenient to think of the immediate environment of a chemical reaction in a cell or the environment around a plant or animal as the surroundings. Technically, however, when we use the term surroundings, we are really referring to everything that exists outside the particular system being considered. All of the rest of the matter and energy in the entire universe constitutes the surroundings of any system that we define.

Two basic laws apply to all energy changes in all systems, including those that occur in living things. These laws are called the first and second laws of thermodynamics.

The First Law of Thermodynamics

The first law of thermodynamics says that energy can be neither created nor destroyed. This means that during any process, the total amount of energy in the system and its surroundings remains constant. Energy may be changed from

Figure 4.2 Radiant energy from the sun is the ultimate source of all of the energy that supports life on earth.

one form to another, as when light energy from the sun is converted to chemical energy in a green plant. But such an energy conversion does not change the *total* amount of energy present in the plant *and* its surroundings.

In any process, a system may gain or lose energy, but any change in the energy of a system is accompanied by an energy change in the surroundings. When you sit in a cold room, for example, your body loses heat energy, but your body heat loss results in an equal heat gain by your surroundings.

We will consider several chemical reactions in our discussion of energy changes in living things. These chemical reactions normally occur either with the release of energy from the reaction (system) to its surroundings or the absorption of energy from its surroundings.

When a reaction releases energy, it is called an **exergonic** reaction. A hypothetical example is:

$$\boxed{AB} \xrightarrow{\text{Energy}\uparrow} \boxed{A} + \boxed{B}$$

In this reaction, a chemical compound, AB, is broken down to its constituent parts, A and B. This reaction is exergonic because it gives off energy to its surroundings.

The opposite reaction would be:

$$\boxed{A} + \boxed{B} \xrightarrow{\text{Energy}} \boxed{AB}$$

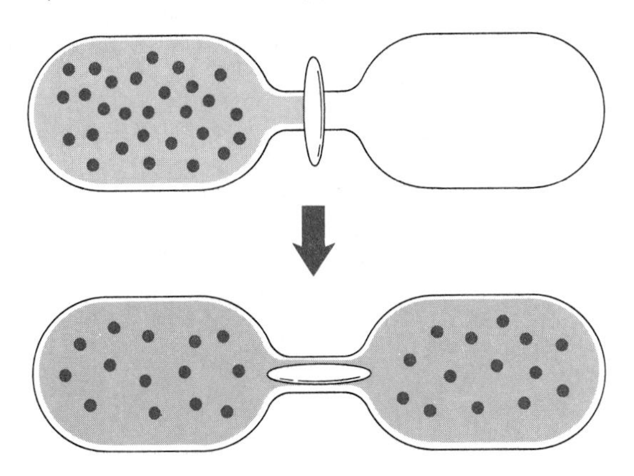

Figure 4.3 The expansion of gas into an empty cylinder when the valve is opened increases the entropy (randomness) of the system.

This reaction is **endergonic.** It absorbs energy from its surroundings. In fact, this reaction will not occur without energy being added from its surroundings.

Throughout our study of biology, we will see examples of both exergonic and endergonic reactions that occur in living things.

The Second Law of Thermodynamics

The second law states that there is a general tendency toward disorder or randomness in the universe. This disorder is known as **entropy.** All processes proceed in such a way that the amount of entropy (disorder or randomness) in the universe increases.

Let us illustrate this rather abstract idea of entropy with the example in figure 4.3. There, two cylinders are connected by a tube with a valve. One of the cylinders is filled with gas and the other is empty. If the valve is opened, gas will move into the empty cylinder until the amount of gas (pressure) in each cylinder is approximately equal. As long as one cylinder contains more gas, the net movement of gas tends to be toward the cylinder with less gas in it. Theoretically, this gas movement could be used to do work. For example, if a miniature windmill were placed in the tube between the cylinders, the movement of the gas would rotate the windmill and make possible some kind of work.

However, when the gas pressure in the two cylinders is equal, the system no longer has the potential to do work, since that potential depended upon the *difference* between the two cylinders. The system has not lost any total energy because the moving gas molecules in the two chambers still have the same total energy that they originally had, but their energy is now equally distributed between the two cylinders. The system has become less organized. Its entropy (disorder) has increased and the system has lost its potential to do work.

Figure 4.4 The relationship between an organism and its surroundings. Growing organisms are systems with increasing order (decreasing entropy). Organisms increase the entropy of their surroundings. In this sketch, S_o symbolizes the entropy of the organism and S_s symbolizes the entropy of the surroundings.

The gas movement in these cylinders is an example of the general tendency for organization and order to break down and decrease while disorder and randomness (entropy) increase. The entire universe is "running down" because of this general tendency for entropy to increase.

How does the general tendency toward an increase in entropy relate to living things? The second law of thermodynamics states only that the entropy of the universe (system plus surroundings) increases during any process. But the entropy of a specific system can actually *decrease,* and, of course, this does indeed happen. A growing organism is a good example of a system that gains increasing order and becomes more highly organized.

However, this by no means implies that living systems are exceptions to the second law. When a system gains order (decreasing its entropy), the entropy of the surroundings must increase. For instance, animals take in complex food molecules, degrade them, and release simpler waste molecules. Thus, they increase the disorder (entropy) of their surroundings, and they use the energy they obtain from nutrients to maintain or even increase their own order (figure 4.4).

Plants also must constantly obtain and use energy to minimize their entropy (maintain their organization). But in contrast with animals, they trap and use light energy to accomplish this.

You can see, therefore, that living things must obtain energy to counter the general tendency toward increasing entropy in the universe. It could be said that maintaining life is like walking in a fast-flowing river. It takes work just to stand still facing upstream in the moving current and even more work to move upstream. Life is possible in a universe in which entropy is increasing, but it takes a lot of energy to do the necessary work to live.

Figure 4.5 The structure of adenosine triphosphate (ATP). When the bonds indicated by wavy lines are broken, energy becomes available for endergonic processes.

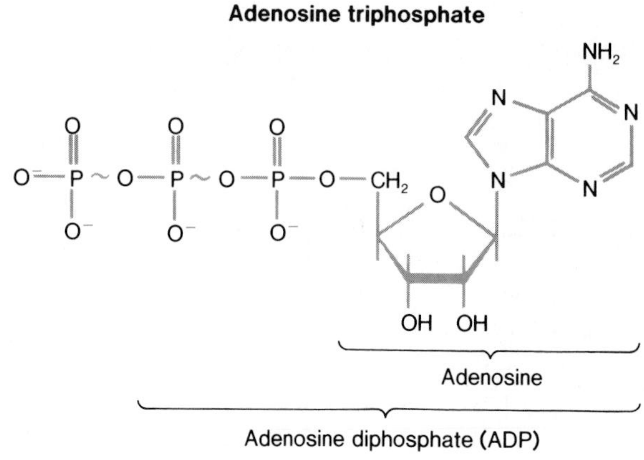

Adenosine triphosphate

Adenosine

Adenosine diphosphate (ADP)

ATP: The Cell's Energy Currency

The key to the way that living things use energy to perform work is found in **ATP.** The three letters stand for **adenosine triphosphate.** ATP provides the energy for many energy-requiring (endergonic) processes in living things, such as energy for muscle contraction and all other forms of movement. ATP also supplies energy for active transport of substances into or out of cells. It supplies energy as well for synthesis of large molecules. Because it provides energy for so many processes, it is small wonder that ATP has been called the cell's "energy currency"; that is, ATP can be "spent" to pay for a variety of important work in living things.

How then does ATP function as an energy currency? Adenosine triphosphate is an organic molecule that contains three phosphates (figure 4.5). When one of these phosphate (PO_3^{2-}) groups is removed, energy is released. When one phosphate has been removed, the molecule that remains is called **adenosine diphosphate (ADP).** We can represent this reaction as follows:

Adenosine—Ⓟ–Ⓟ–Ⓟ ⟶

Adenosine—Ⓟ–Ⓟ + Ⓟ + Energy

This exergonic (energy-yielding) reaction in which ATP is split into ADP and phosphate is the way in which ATP provides energy for many kinds of work in living things (figure 4.6).

Figure 4.6 ATP functions as a cellular energy currency. Splitting of ATP to ADP and phosphate (symbolized P_i) yields energy for many energy-requiring processes in living things.

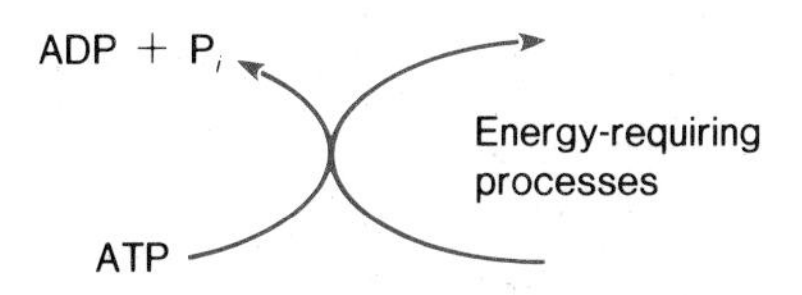

ATP is very stable in living cells and is split into ADP and phosphate only in connection with energy-requiring processes in the cell. ATP splitting (hydrolysis) is actually very closely connected (coupled) with those energy-requiring processes. ATP splitting normally does not occur in cells unless there is work to be done. Thus, ATP functions as an excellent energy currency, because it is not needlessly wasted through uncontrolled breakdown. Instead, it is used when and where it is needed to provide energy for work.

Phosphorylation

If cells are constantly using up their supply of ATP for work, how is the supply replenished? ATP can be remade from ADP and phosphate. The attachment of a phosphate to a molecule is called **phosphorylation.** Although many kinds of phosphorylations occur in cells, the phosphorylation of ADP to produce ATP is so important in living things that when biologists simply say "phosphorylation," they almost always mean the phosphorylation involved in ATP production.

Because splitting ATP into ADP and phosphate releases energy, you might expect that it would take energy to do the reverse, that is, to attach a phosphate to ADP to make ATP. And so it does:

Adenosine — (P) – (P) + (P) + Energy ⟶

Adenosine — (P) – (P) – (P)

The energy for phosphorylation of ADP to produce ATP comes from other chemical reactions that occur in living cells. Cells obtain and use various nutrients in reactions that provide energy for ATP production. Thus, ATP is a renewable energy currency (figure 4.7).

Figure 4.7 ATP is a renewable energy currency. Phosphate is added to ADP in phosphorylation to make ATP. Cells obtain and use various organic nutrients to provide energy for phosphorylation.

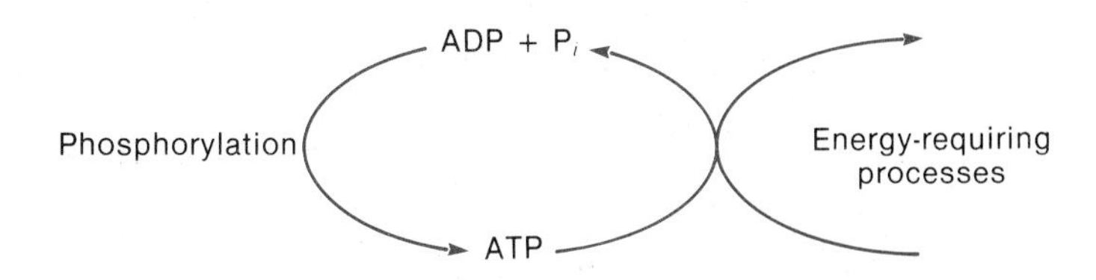

Reactions and Enzymes

If we are to understand chemical processes that occur in living things, such as the reactions that provide energy for ATP production, we need to learn a little about the ways in which chemical reactions occur.

Reaction Rates

Water is made up of hydrogen and oxygen. The reaction in which hydrogen and oxygen combine to produce water is highly exergonic; that is, it releases a great deal of energy. But if hydrogen and oxygen gases are mixed together very carefully (not an experiment to be tried in the home chemistry lab), nothing happens. However, once the reaction starts (if, for instance, a match is lit), it proceeds so rapidly that an explosion results.

How can we explain this abrupt change from no noticeable reaction to an explosively rapid reaction? Chemical reactions take place by means of collisions between **reactants,** the reacting substances. The reaction rate will never be faster than the rate at which reacting molecules collide with each other.

However, there is another factor. Reacting molecules must collide with at least a certain minimum amount of energy in order to react. This minimum energy is called the **activation energy** of the reaction. If the reactants have activation energy, the reaction proceeds quickly and products of the reaction are formed. Water, for example, is the product of the reaction between hydrogen and oxygen. Figure 4.8 illustrates activation energy. We can think of activation energy as being similar to having just enough energy to roll a boulder a short way up to the top of a hill. When we reach the top, the boulder will roll down the opposite side of the hill by itself.

Figure 4.8 Activation energy. Reacting molecules require activation energy before they can react. Once activation energy is available, the reaction proceeds. The reaction diagrammed here is exergonic because it releases energy; that is, the products contain less energy than the reactants.

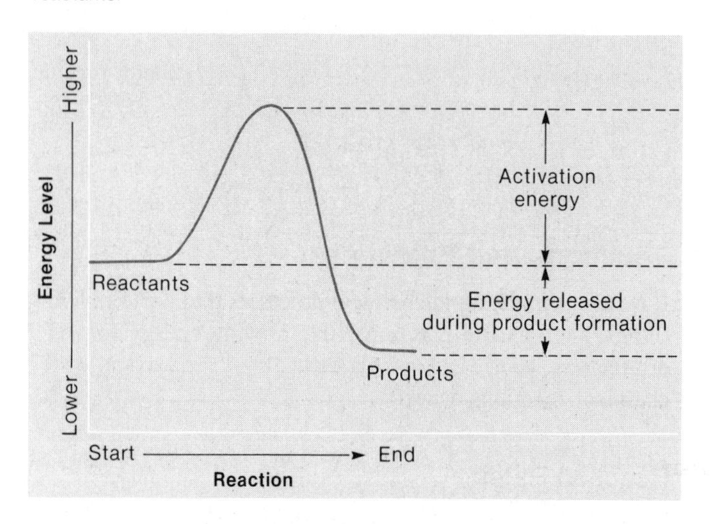

In any situation, only a certain percentage of the molecules are at the necessary energy level in order for the reaction to occur. This is because not all molecules at the same temperature have the same energy. In our example of hydrogen and oxygen gases, very few of the molecules have enough energy to react and the reaction occurs so slowly that it is virtually undetectable. But when a lighted match is added, several things happen. The heat energy speeds up the movement of hydrogen and oxygen so that more collisions occur, *and* the heat also increases the energy of the colliding reactants; that is, the match provides enough energy so that more molecules reach or exceed the activation energy, and a significant reaction begins. As we said earlier, the reaction between hydrogen and oxygen to form water is highly exergonic—it yields energy. Once the reaction is started by the match, the reacting molecules generate enough heat to raise the energy of still more molecules, and the rate of reaction increases so fast that an explosion occurs.

Catalysts

Many of the reactions that occur in living things are reactions that can occur spontaneously; that is, they are reactions that can proceed by themselves. But most of these spontaneous reactions occur so very slowly that they simply are not useful to living cells because too few of the reacting molecules have activation energy. How are these reactions speeded up so that they occur at rates that meet the needs of living

Figure 4.9 Effect of a catalyst on activation energy. A catalyst increases the rate of a reaction because it lowers the activation energy required for the reaction to occur.

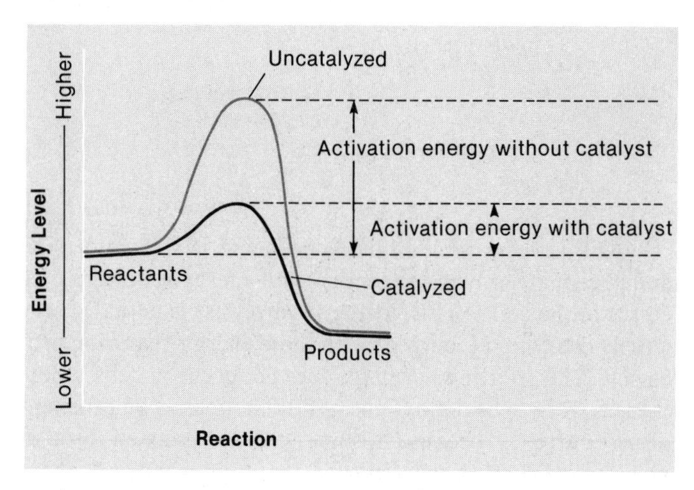

things? We know that it is possible to provide activation energy by heating up the reactants, as in the example of hydrogen and oxygen gases. But living things generally operate at moderate temperatures. However, there is another way of accomplishing the same thing under the very stable conditions that must exist inside organisms.

Reaction rates can be changed with catalysts. **Catalysts** are substances that lower activation energy by slightly changing the way in which a reaction occurs (figure 4.9). With a lowered activation energy, the reaction proceeds at a faster rate because many more molecules can react.

A catalyst enters a reaction without itself being changed by the reaction. Thus, a catalyst can function over and over again in its role of speeding up a reaction.

Very few reactions in living cells occur in the same way that they would in a chemical laboratory. In the vast majority of cases, reactions in living things involve catalysts called **enzymes** that are produced by cells.

Structure and Function of Enzymes

Enzymes, the catalysts produced by living cells, are globular protein molecules. Globular proteins have specific three-dimensional shapes (see page 45), and these shapes of enzymes are very important in their functioning as catalysts. Many different enzymes, all with different three-dimensional shapes, catalyze virtually all of the reactions that occur in living things. Each reaction has its own specific kind of enzyme catalyst.

Reacting substances in enzyme-catalyzed reactions are called **substrates.** As different enzymes catalyze different reactions, there obviously must be a specific relationship between each type of enzyme and its particular substrate molecules. What is this relationship?

Box 4.1

The Bombardier Beetle's Chemical Blaster

Many animals produce toxic substances that are used for capturing prey and for self-defense. Toxins produced by snakes, spiders, and bees are familiar examples. And when some millipedes are irritated, they release hydrogen cyanide (HCN), a deadly metabolic poison.

However, some of the most dramatic defense mechanisms are the chemical sprays that animals such as certain scorpions and some insects direct at their attackers. One of the most spectacular of these defense sprays is the hot, irritating mist produced by the bombardier beetle—a mist that is emitted in explosive blasts of fine spray at a temperature of 100°C (box figure 4.1).

The bombardier beetle in box figure 4.1, *Stenaptinus insignis,* is from Kenya. It has a pair of double-chambered glands that open at the tip of its abdomen. The inner chamber of each gland contains a mixture of hydroquinones and hydrogen peroxide (H_2O_2). The outer chamber contains enzymes that catalyze the breakdown of hydrogen peroxide to release oxygen gas (O_2) and the oxidation of hydroquinones to quinones with the release of hydrogen gas (H_2).

A bombardier beetle points its abdomen at a target and "fires" its chemical spray by squeezing the hydrogen peroxide and hydroquinones from the inner chambers to the outer chambers of its glands. This instantly sets off an explosive set of reactions as oxygen and hydrogen are released. The gases cause a dramatic pressure increase as they combine in a highly exergonic reaction ($H_2 + \frac{1}{2}O_2 \rightarrow H_2O$). The energy released from the reactions in each chamber heats the water and quinones to 100°C. Finally, the gas pressure buildup shoots the hot mixture out as a fine mist. This explosion causes an audible bang.

The bombardier beetle's spray is doubly effective in repelling attackers. Quinones are irritating to other animals, and the tremendous heat of the spray also is likely to discourage a would-be predator. The beetle can turn its abdomen in any direction to aim accurately (as it is doing in box figure 4.1, where the "attacker" is a forceps grasping one of its legs), and it can spray repeatedly in quick succession. Thus, the beetle's chemical defense system functions as a formidable repellent to would-be attackers.

Box figure 4.1 The bombardier beetle *Stenaptinus insignis.* This animal is about 20 mm long. It is shown here spraying the forceps with which an experimenter is grasping one of its legs.

Figure 4.10 A sketch of an enzyme molecule with substrate molecules fitted into its active site. Enzymes are globular proteins that are more or less spherical in shape.

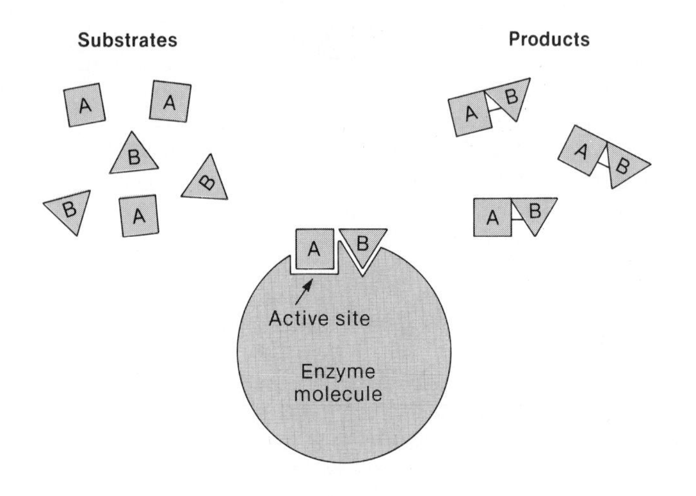

Figure 4.11 Example of an enzyme functioning as a catalyst. (*a*) Molecules that are properly oriented when they collide react with each other. (*b*) Molecules do not react if they collide in some other way. (*c*) Diagram showing reacting molecules bound in the active site of an enzyme molecule. The reacting molecules are held in exactly the right orientation to assure reaction.

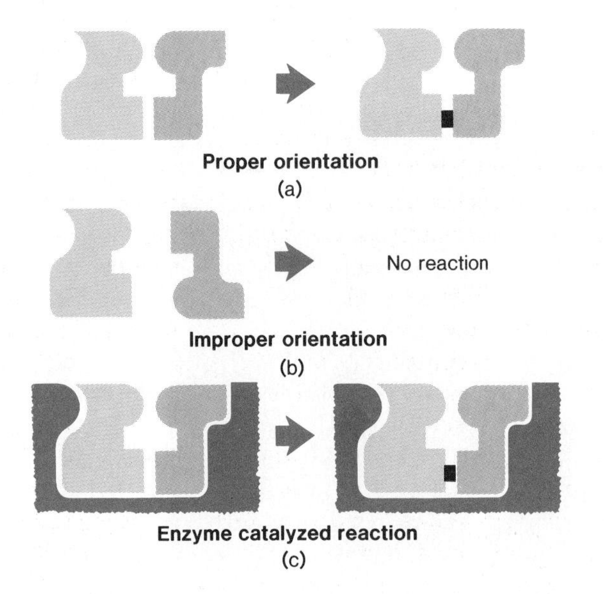

Figure 4.12 Some enzymes are flexible and change shape as they catalyze a reaction. Attachment of substrate molecules caused this enzyme to flex, bringing the substrates together for reaction. When the reaction is complete and the catalytic site is empty, the enzyme returns to its original shape and can catalyze the reaction of another set of reactants.

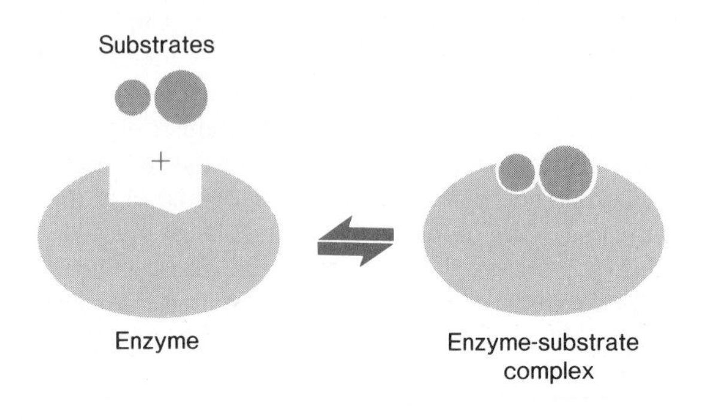

Figure 4.13 A diagram of a simple metabolic pathway with five enzyme-catalyzed reactions. The product of each reaction becomes the substrate in the next reaction. The activity of this pathway is regulated and balanced by the negative effect of product F on the activity of enzyme A. When there is more of product F, it inhibits enzyme A and slows the activity of the entire pathway. When there is less of product F, the inhibition is released and the activity of the pathway increases.

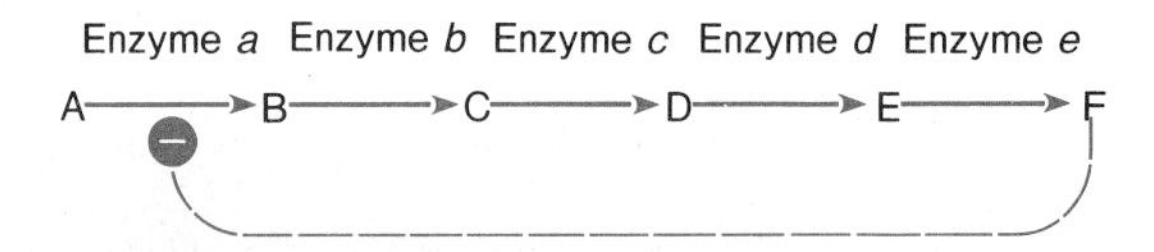

Each enzyme molecule has a specific region known as its **active site** (figure 4.10). This site is shaped so that a substrate molecule or several substrate molecules (depending on the reaction) fit into it in a very specific way and are held in place by weak chemical forces such as hydrogen bonds (page 32). The specific fit of an enzyme's active site and an appropriate substrate molecule is very much like the specific fit of a lock and its key. Other molecules may come close to fitting the active site, but they do not fit perfectly, as the substrate does.

The Mechanism of Enzyme Action

When an enzyme and its substrate molecules are bound together, they form an **enzyme-substrate complex.** In the enzyme-substrate complex, the substrates are held very close together and oriented in such a way that proper bonds can form easily (figure 4.11).

Figure 4.14 Regulation of an enzyme by attachment of an inhibitor to its regulatory site. This enzyme has two parts; one (gray) has the active site and the other (color) has the regulatory site. A normal enzyme functioning in the absence of inhibitor is shown across the top. When an inhibitor attaches to the regulatory site (1), the enzyme's shape changes (2), and the active site no longer provides a neat fit for substrate molecules.

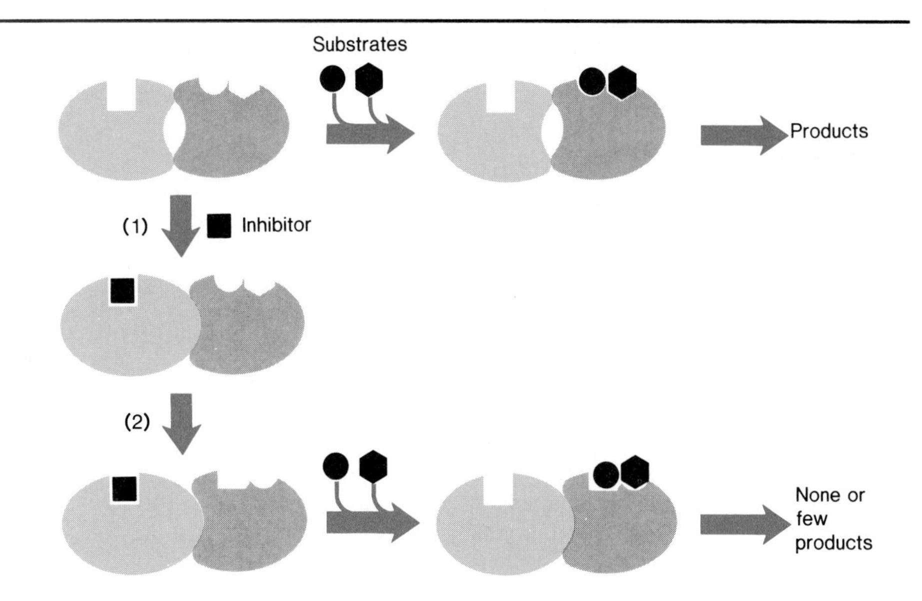

Some enzymes have flexible shapes. This flexibility means that the shape of the active site changes after the proper substrate binds to the enzyme. As the enzyme changes shape, the substrates are brought into ideal position for reaction. After the reaction is complete, the product detaches and the enzyme returns to its original shape, ready to begin the cycle again (figure 4.12).

Enzyme shape changes can also be involved in taking a molecule apart in a reaction that breaks bonds. In such reactions, flexing the enzyme strains or distorts a substrate so that the bonds break more easily.

Control of Enzyme Function

In order for cells to function smoothly and use their nutrients most efficiently, it is necessary that cellular chemical processes be regulated. Chemical reactions must proceed rapidly when products of the reactions are needed, but the reactions must be slowed down when the products are plentiful. Such regulation conserves resources in living things.

Many enzyme-catalyzed reactions are parts of **metabolic pathways,** which are series of reactions in cells. Very often in these pathways, the product of one enzyme-catalyzed reaction is a substrate for the enzyme that catalyzes the next reaction in the series (figure 4.13). In such a case, each reaction effectively removes the product of the reaction that preceded it in the sequence; and with the product constantly being removed, the reaction continues to convert substrate to product. However, it is not economical for cells to have enzyme-catalyzed reactions occurring at maximum rates until every last available molecule of substrate has reacted. There must be some way to regulate the catalytic activity of enzymes so that the output of a reaction series can be correlated with the chemical requirements of the cell.

Very often regulation of a metabolic pathway depends upon slowing down or speeding up the catalytic activity of an enzyme catalyzing just one of the series of reactions. In the metabolic pathway illustrated in figure 4.13, a final product of the reaction series, substance F, has a negative effect on the activity of enzyme *a*. When the activity of enzyme *a* is inhibited, the activity of the whole pathway slows. Thus, enzyme *a* serves as a regulatory enzyme for the pathway. The activity of enzyme *a* depends upon how much of substance F is present. When there is an excess of substance F, enzyme *a* is inhibited. When the concentration of substance F decreases, enzyme *a* and the entire metabolic pathway are released from inhibition. This kind of control of a metabolic pathway is called **feedback inhibition.**

How does feedback inhibition work? Some enzymes, especially some key regulatory enzymes, have another site to which a molecule can bind, in addition to the active site where the enzyme's substrate binds (figure 4.14). When a particular molecule binds to this second site, which is called a regulatory site, the enzyme's general shape changes. Of course, this changes the shape of the active site so that it no longer binds the substrate efficiently. When the molecule leaves the regulatory site, the enzyme returns to its original shape and resumes its normal catalytic function.

Figure 4.15 Electrons flowing through a series of electron carriers. As one carrier gives up its electrons (becomes oxidized), the next one gains electrons and becomes reduced.

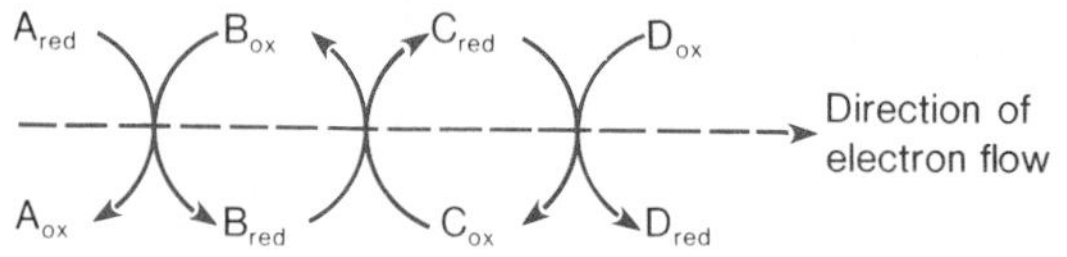

A_{red} B_{ox} C_{red} D_{ox}

Direction of electron flow

A_{ox} B_{red} C_{ox} D_{red}

Coenzymes

Many functional enzymes contain a nonprotein portion in addition to protein. These nonprotein components must be present for the enzymes to catalyze reactions. One of the most common types of nonprotein components is called a **coenzyme.** Coenzymes are small organic molecules that can attach and detach from the enzyme protein. However, the coenzyme must be present for an enzyme to play its normal role in a reaction. For instance, in various reactions, coenzymes function as carriers that transfer electrons, atoms, or molecules from one enzyme to another. Such coenzyme-mediated transfers are vital to a number of reactions in the energy exchanges of photosynthesis and in utilization of nutrients by cells.

Oxidation and Reduction

Many of the enzyme-catalyzed reactions that living things use to obtain energy for ATP production are reactions that involve oxidation and reduction. **Oxidation** is the loss of one or more electrons; **reduction** is the gain of electrons.

An iron (Fe) atom forms part of several molecules that are important in energy-yielding reactions in living things. The iron atom may be in either the reduced (Fe^{2+}) or the oxidized (Fe^{3+}) state and it is changed from one to the other by oxidation or reduction reactions. The oxidation (loss of an electron) reaction is:

$$Fe^{2+} \longrightarrow Fe^{3+} + \text{Electron}$$

After it has lost one electron with its negative charge, the iron has a greater positive charge. The opposite reaction would be a reduction in which an electron is added:

$$Fe^{3+} + \text{Electron} \longrightarrow Fe^{2+}$$

Figure 4.16 In some electron carrier series, energy is released as electrons are passed from carrier to carrier. The energy released in such exergonic electron transfers can be used for such purposes as ATP production.

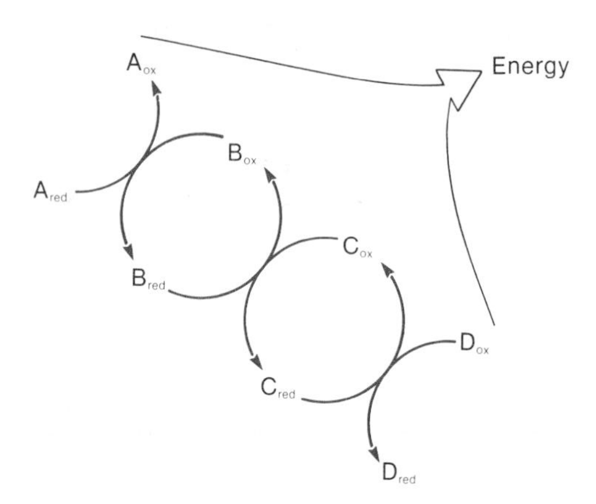

These reactions might be a little misleading, however, because electrons never really appear free from atoms. In reality, when one substance becomes oxidized, something else must become reduced. Thus, we often speak of oxidation-reduction reactions.

In living things, oxidation-reduction reactions occur in series. Electrons are passed along from one electron carrier to another. As one becomes oxidized (loses electrons), the next becomes reduced (gains electrons). Thus, electrons "flow" through a series of carriers (figure 4.15).

Electron Flow and Energy

In terms of energy, electron flow from one carrier to another in a series can be either exergonic (energy-yielding) or endergonic (energy-requiring). Another way of looking at the energy flow is to think of it as being either "downhill" (exergonic) or "uphill" (endergonic).

If the flow of electrons is "downhill" (exergonic), energy is released and can be used for other purposes. This happens when cells are using organic nutrients to obtain energy for ATP production. First, electrons are removed from the nutrients (the nutrients are oxidized). Then, the electrons pass through a "downhill" (exergonic) electron carrier series and the energy released is used for ATP production (figure 4.16).

In chapter 5, we will see that in photosynthesis, the sun's energy is used to drive electrons "uphill" through an electron carrier series. This is a key step in converting the sun's energy into chemical energy useful for all living things.

Figure 4.17 The oxidized (top) and reduced (bottom) forms of ubiquinone, an electron carrier that gains and loses hydrogen ions (H^+) along with electrons.

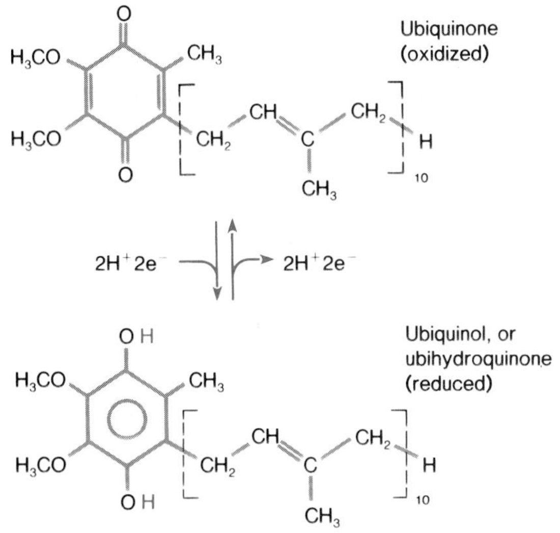

Figure 4.18 NAD, a coenzyme that accepts a pair of electrons during enzyme-catalyzed reactions. In the reduced form (bottom), one electron is attached to the nitrogen atom at position (1). The second electron attaches to a carbon atom at position (4) and is accompanied by a hydrogen ion. A supply of the B vitamin niacin is required for synthesis of this vital coenzyme.

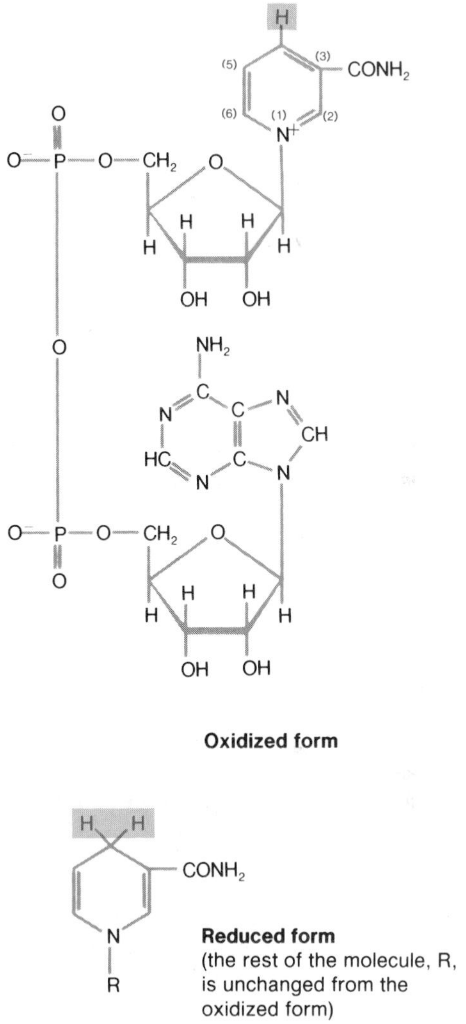

Electron Carriers

Some oxidations and reductions in living things are more complicated. In some cases, electrons are transferred in pairs, and also, electrons very often are accompanied by hydrogen ions (H^+). This is illustrated by the electron carrier ubiquinone, which is shown in figure 4.17. When ubiquinone gains or loses two electrons, it also gains or loses two hydrogen ions.

Earlier when we discussed enzymes, we mentioned coenzymes, molecules that must be present if the enzyme is to function. In many cases, the actual role of the coenzymes is to accept electrons removed from a substrate molecule in the reaction catalyzed by the enzyme. A good example of a molecule that functions as an electron acceptor for several different enzymes is the coenzyme **nicotinamide adenine dinucleotide (NAD).** See figure 4.18.

Energy and Vitamins

NAD accepts a pair of electrons in a number of important reactions in cells, and those reactions will not proceed in the absence of NAD. NAD is synthesized in the body from the B vitamin niacin, and it cannot be synthesized without a supply of niacin. Because NAD is so critically important to the function of all body cells, it is easy to understand why a niacin deficiency in the diet has such widespread harmful effects on the body. Several other vitamins are required for synthesis of other coenzymes. As a result, assuring an adequate supply of energy in an animal body requires a continuing supply of a variety of organic nutrients, including vitamins.

In chapters 5 and 6, we will explore the processes by which nutrients are supplied and used to provide energy for living things.

Summary

Energy is the capacity to do work and all living things require a continuing supply of energy to function normally. The ultimate source of energy for life is the radiant energy of the sun. Photosynthetic organisms use sunlight to provide energy for the synthesis of organic compounds. These compounds in turn provide a source of chemical energy for all forms of work in organisms.

Energy conversions are studied in systems and their surroundings. All energy changes obey two basic laws. The first law states that energy can be neither created nor destroyed. Although the energy content of a particular system may change as a result of energy exchanges with the surroundings, the total energy content of the universe remains constant. The second law states that all processes proceed in such a way that the disorder (entropy) of the universe increases. A living organism may decrease its entropy, but it does so by increasing the entropy of its surroundings.

Energy is released when adenosine triphosphate (ATP) is split into ADP and phosphate. ATP provides energy for many kinds of work in living things. The supply of ATP is replenished by phosphorylation of ADP. Energy for ATP production comes from other chemical reactions that occur in living cells.

Several factors are important in determining the rates of chemical processes in living things. The rate of collision, the orientation of the reactants, and the activation energy all affect the rates of chemical reactions. Large activation energies lead to slow reactions. Catalysts provide alternate reaction routes with lower activation energies.

Enzymes are globular protein molecules that act as catalysts in living cells. They have active sites that complement the shapes of their substrates. Nonprotein components such as coenzymes are essential to the functions of many enzymes.

Enzymes lower activation energies by binding (and thus concentrating and orienting) substrate molecules. Some enzymes change shape after binding substrates to bring the substrates into optimal orientation or to distort the substrates so that bond making or breaking is easier.

Enzyme activities are regulated so that their functions meet, but do not exceed, the needs of cells.

Oxidation-reduction reactions involve the transfer of electrons. Oxidation is the loss of electrons; reduction is the gain of electrons. Some oxidation-reduction reactions in biological systems involve the transfer of a single electron, but in many others, pairs of electrons are transferred. Often hydrogen ions accompany transferred electrons. Living cells use energy released during oxidation-reduction reactions for ATP production.

Questions for Review

1. How does the sun serve as an energy source for animals?
2. What is the difference between an exergonic reaction and an endergonic reaction?
3. Explain the role of ATP as an energy currency.
4. Define phosphorylation using the terms ADP and ATP.
5. Use the term "activation energy" in explaining how a lighted match causes an explosion in a mixture of hydrogen and oxygen.
6. Use the term "activation energy" when you explain the role of a catalyst in a chemical reaction.
7. Explain the role of the active site in formation of an enzyme-substrate complex.
8. Clearly define oxidation and reduction and explain why oxidation of one substance is always accompanied by reduction of another.

Questions for Analysis and Discussion

1. Growing organisms clearly are systems that have increasing order (decreasing entropy). Is life, therefore, an exception to the second law of thermodynamics? Explain why or why not.
2. How does feedback inhibition of enzyme action serve to conserve energy and resources in cells?
3. Can you think of some reasons why ATP provides energy for many different kinds of work in cells? Why is it better *not* to have separate "energy currencies" for different kinds of work?

Suggested Readings

Books

Baker, J. J. W., and Allen, G. E. 1981. *Matter, energy and life.* 4th ed. Reading, Mass.: Addison-Wesley.

Lehninger, A. L. 1971. *Bioenergetics.* 2d ed. New York: Benjamin.

Stryer, L. 1981. *Biochemistry.* 2d ed. San Francisco: W. H. Freeman.

Articles

Chappell, J. B. 1977. ATP. *Carolina Biology Readers* no. 50. Burlington, N.C.: Carolina Biological Supply Co.

Hollaway, M. R. 1976. The mechanism of enzyme action. *Carolina Biology Readers* no. 45. Burlington, N.C.: Carolina Biological Supply Co.

Photosynthesis

5

Chapter Outline

Chapter Concepts

1. In photosynthesis, light energy is used to produce organic compounds that provide chemical energy for all living things.
2. The leaves of plants serve as absorbers and converters of light energy.
3. Chloroplasts are the sites of photosynthetic activity in plant cells.
4. Pigments involved in the light reactions are associated with the internal membrane structures of the chloroplast.
5. Carbon dioxide incorporation occurs in the stroma matrix of the chloroplast.
6. The C_4 and CAM pathways provide alternative carbon dioxide incorporation mechanisms.

An African lioness moves silently through the tall grass and shrubs toward a herd of grazing gazelles. Several of the animals toss their heads and look about nervously, but then resume their feeding. The lioness slowly moves closer, keeping her body low to the ground and hidden from view. Then suddenly, she springs into action. The gazelles immediately take flight, but it is too late. The lioness's charge overtakes one of them and pulls it down in a cloud of dust. In a few moments, it is all over. The lioness drags the dead gazelle off to be shared with her cubs (figure 5.1), and just a short distance away, the herd resumes eating grass, almost as if nothing had happened. And all the while the hot African sun shines down on the entire scene.

Figure 5.1 An African lioness dragging a Thomson's gazelle that she has just killed.

The gazelles obtain energy for life by eating grass, but the lioness must kill and eat other animals to obtain energy for herself and for her growing cubs. The source of all this energy for the growing grass, the grazing gazelles, and the feeding lions is the sunlight that bathes the African plain.

The energy that all living things use comes directly or indirectly from the sun. Sunlight provides energy for the process of **photosynthesis,** in which green plants convert light energy to chemical energy in the form of organic chemical compounds. These substances provide energy for all of the biological world—for the green plants themselves, for organisms that eat plants, and for other organisms that feed on the plant eaters.

Photosynthesis forms organic compounds from carbon dioxide and water. Using a sugar, glucose, as an example of a product of photosynthesis, we can summarize the process as follows:

$$\underset{\text{(Carbon dioxide)}}{6CO_2} + \underset{\text{(Water)}}{6H_2O} \xrightarrow{\text{Light}} \underset{\text{(Glucose)}}{C_6H_{12}O_6} + \underset{\text{(Oxygen)}}{6O_2}$$

In photosynthesis, carbon dioxide from the air is reduced to make organic molecules; that is, electrons and hydrogen ions are added to carbon dioxide to make organic compounds such as glucose. This requires energy and the source of that energy is sunlight (figure 5.2).

Figure 5.2 Photosynthesis, respiration, and the flow of energy from the sun into living things. Organic compounds are produced in photosynthesis using CO_2, water, and light energy. They are broken down in respiration to provide energy for ATP production.

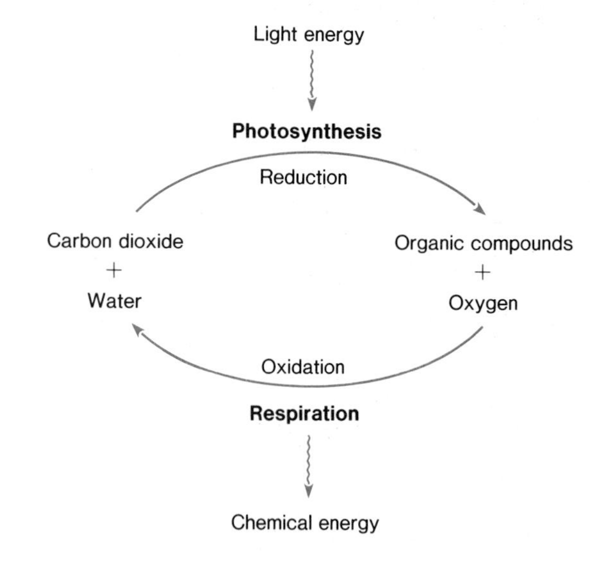

Respiration is roughly the opposite process. In respiration organic compounds are oxidized. Electrons are removed and in the process, energy becomes available to make ATP, the cellular "energy currency." Again using the sugar glucose as a representative organic compound, respiration can be summarized as follows:

$$C_6H_{12}O_6 + 6O_2 \longrightarrow 6CO_2 + 6H_2O$$
$$\searrow \text{Usable energy (ATP)}$$

In this chapter, we begin our study of the processes by which the sun's energy flows through the living world, providing energy for vital processes in all living things. We will examine photosynthesis as it occurs in the leaves of green plants.

Figure 5.3 A leaf is a broad, flat, bladelike structure with a large surface area for light absorption. Because leaves are thin, no cell is far from the supply of CO_2 in the surrounding air.

The Leaf

In most plants, photosynthesis occurs mainly in a specialized organ, the **leaf.** Usually, a leaf is a flat, thin bladelike structure that contains a large number of cells specialized to carry on photosynthesis. Leaves are well-suited to their photosynthetic function because of their shape (figure 5.3). Since they are broad and flattened, they have large surface areas for light absorption. But it is also important that leaves are thin; light must reach photosynthetic cells. Also, photosynthesis requires a supply of carbon dioxide from the air, and it is vital that the distance between the photosynthetic cells and the outside air be small enough so that CO_2 can reach the cells fairly easily.

The requirement of a supply of carbon dioxide for photosynthesis causes some problems for plants. CO_2 is present only in very small amounts in the air (about 0.03 percent), and, thus, photosynthetic cells must have good access to air. CO_2 must first dissolve in water before it enters leaf cells. Thus, each photosynthetic cell has a water film covering its surface. This means that a large amount of wet surface area is exposed to the air and water is continually lost by evaporation. This water loss is not all bad for plants, as we shall see later, but excessive water loss will cause **wilting.** Wilted leaves become limp and droopy, and they can die if the wilt continues too long.

How are plants protected against excessive water loss? The leaf surface is covered with a waxy **cuticle,** which allows very little water to pass through it.

Unfortunately, the cuticle, which protects so well against water loss, also acts as a barrier to gases. It blocks passage of the carbon dioxide required for photosynthesis. Because of this, gas exchange in the leaves of land plants takes place mainly through small pores called **stomata** (singular: **stoma**). In many plants, stomata are located mainly, or even only, on the lower surface of the leaf where they are relatively protected from the sun's heat and drying wind currents.

The diagram in figure 5.4 shows the internal structure of a leaf. The upper and lower epidermis are each protected by a waxy cuticle. Stomata are shown in the lower epidermis. Each stoma lies between two specialized cells called **guard cells.** Guard cells are smaller than ordinary epidermal cells, and they contain chloroplasts, which are not found in regular epidermal cells.

The photosynthetic cells in the leaf interior are known as **mesophyll** (meaning "middle leaf") cells. Toward the top of the leaf, mesophyll cells are lined up side by side in the **palisade mesophyll** and underneath, cells are more loosely packed in the **spongy mesophyll.** Loose packing increases the efficiency of CO_2 uptake by these photosynthesizing cells because it leaves lots of room for air in the spaces among the cells. These air spaces are connected with the outside air by the stomatal openings.

Figure 5.4 Leaf structure. (*a*) Diagram of the internal structure of a leaf showing palisade and spongy mesophyll. The vascular bundle is part of the transport system that moves materials from one part of the plant to another. Each stoma is enclosed by two guard cells. (*b*) Scanning electron micrograph showing the arrangement of mesophyll cells (magnification × 1,150).

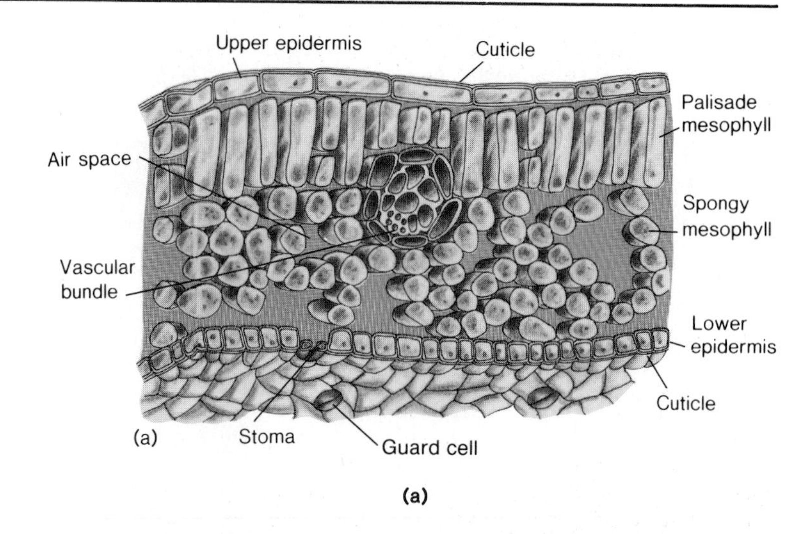

(a)

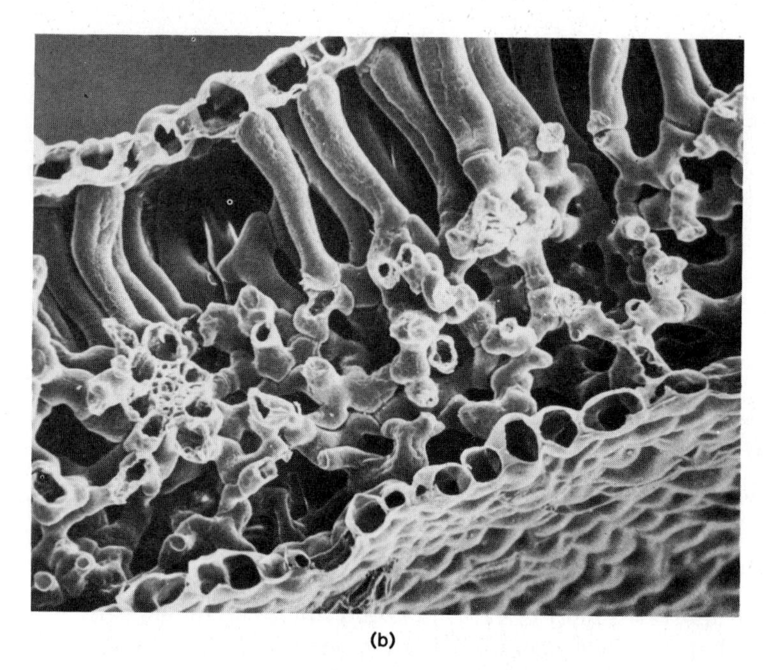

(b)

Opening and Closing Stomata

Stomata are pores that can open and close. It is important that somata open when leaf cells are carrying on photosynthesis because the cells require a continuing supply of carbon dioxide. But it is equally important that these openings to the outside air be closed at other times to prevent excessive water loss and possible wilting.

Stomata open when a leaf is carrying on photosynthesis because guard cells take up water by osmosis (see page 75) and swell against their cell walls. Although it would seem logical that swelling of guard cells would close the stomata, this is not the case. Guard cells have bands of cellulose microfibrils around them so that cell walls cannot increase in girth as the cells swell. This forces the cell walls to elongate instead.

A second structural feature of the guard cells that causes a stoma to open is that guard cells are very tightly connected to one another at their ends. When the guard cells swell and get longer, the tight fastening at their ends causes them to curve outward, thereby opening the stoma (figure 5.5).

Figure 5.5 Stomata. (*a*) Closed and open stomata on the bottom of a leaf. (*b*) Scanning electron micrograph of an open stoma showing its two guard cells (magnification × 2,380). (*c*) Guard cell movement during stomatal opening. Dashed lines indicate position of guard cells when the stoma is nearly closed. Radial arrangement of bands of cellulose microfibrils in guard cell walls is shown in color. Guard cells are fastened together tightly at their ends. K^+ uptake and osmotic entry of water cause guard cell swelling, but cells cannot increase in girth because of the cellulose microfibrils running around their walls. They do increase in length, but curve outward because of the connections at their ends. (*d*) A two-balloon model of the guard cells of a stoma. Masking tape represent the radial cellulose microfibrils. Balloons were glued together at their ends with rubber cement before inflation. Drs. Frank Salisbury and Cleon Ross who put together this two-balloon model report that it took them eight trials before this demonstration worked without breaking the balloons.

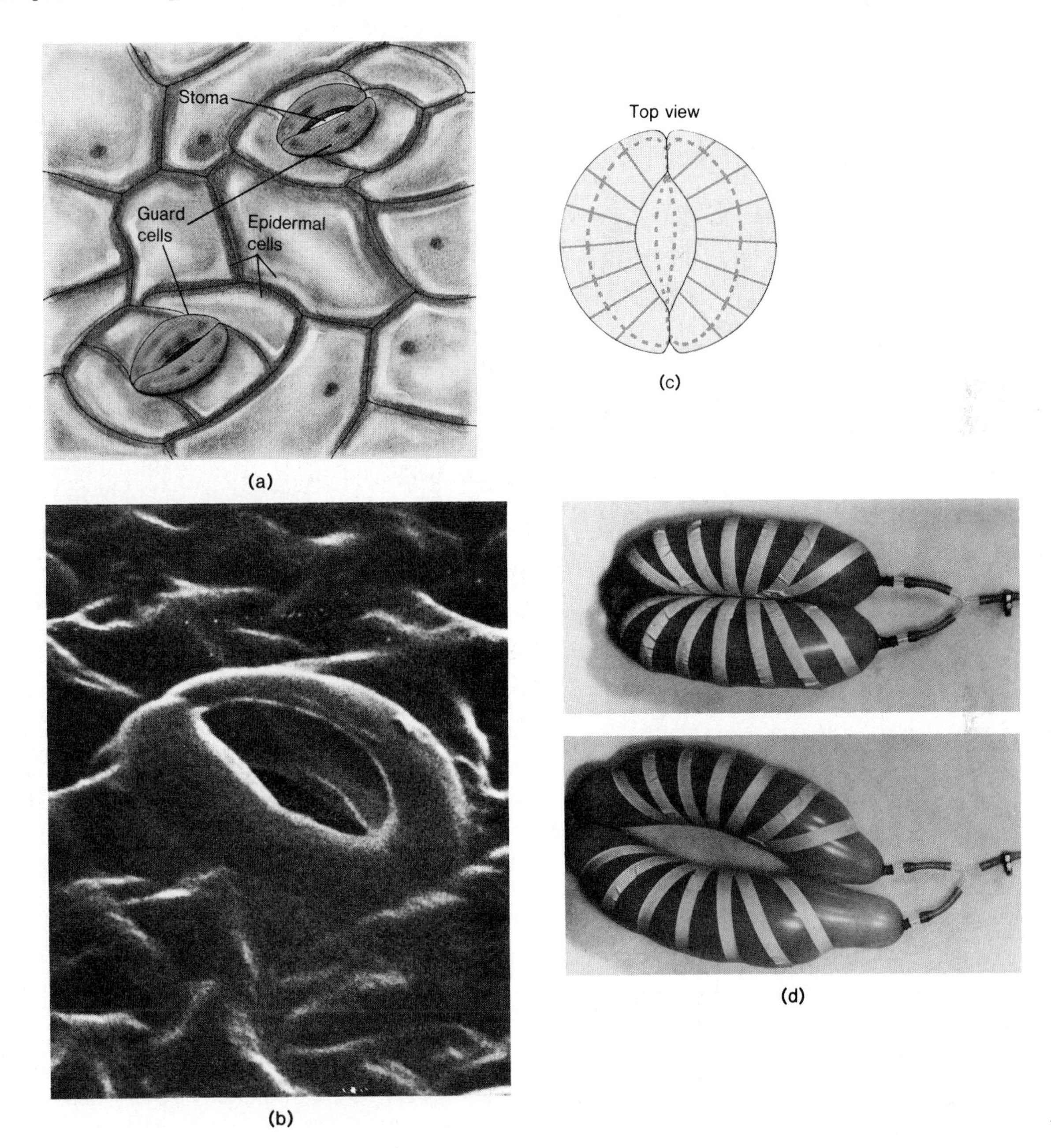

What causes the swelling of guard cells when leaf cells require carbon dioxide for photosynthesis? This is one of those intriguing questions in biology for which we have only a partial answer. Recall that guard cells have chloroplasts. This means that they carry on photosynthesis when light strikes them. Their photosynthetic activity uses up the CO_2 inside them. Somehow, decreasing CO_2 content causes guard cells to begin active uptake of potassium ions (K^+).

Guard cells with decreased CO_2 content pump potassium ions inward through their plasma membranes. When the K^+ concentration builds up in the cells, water begins to enter by osmosis. This swells the guard cells and opens the stomata.

Because it is so important for the productivity of agricultural plants and their water requirements, stomatal opening and closing is an important area of plant research.

Photosynthesis

Because of the work of a number of scientists over the years, some basic characteristics of photosynthesis were well understood by the beginning of this century. It was known, for instance, that green plants use water from the soil and carbon dioxide from the air to synthesize carbohydrates. It was also known that oxygen is a byproduct of this process and that this oxygen is released into the atmosphere. Further, it was well established that photosynthetic activity is centered in the chloroplasts of green plant cells. Beyond these basics, however, practically nothing was known about the details of the actual chemical processes in photosynthesis. In this chapter, we will consider some of the facts about photosynthesis that have been learned in this century.

The Chloroplast

The **chloroplast** is the site of photosynthesis within the cell. Chloroplasts are plastids, distinctive organelles found in the cells of green plants (see chapter 3). The structure of a chloroplast can be understood best if we think of a chloroplast as having four "layers"; some of the layers are membranes and others are spaces between membranes (figure 5.6).

The whole chloroplast is enclosed by a membrane called the **chloroplast envelope.** Raw materials for photosynthesis enter and products of photosynthesis pass out through this membrane.

The second layer is the area inside the chloroplast envelope, the **stroma matrix.** It contains a number of enzymes that catalyze important reactions in photosynthesis.

Figure 5.6 The chloroplast. (*a*) Chloroplast structure showing the four layers. (*b*) Electron micrographs of chloroplasts from a broad bean. (1) The whole chloroplast contains stored carbohydrate in the form of two starch granules (magnification × 10,500). (2) Detail to show stacking of thylakoids to form grana.

(a)

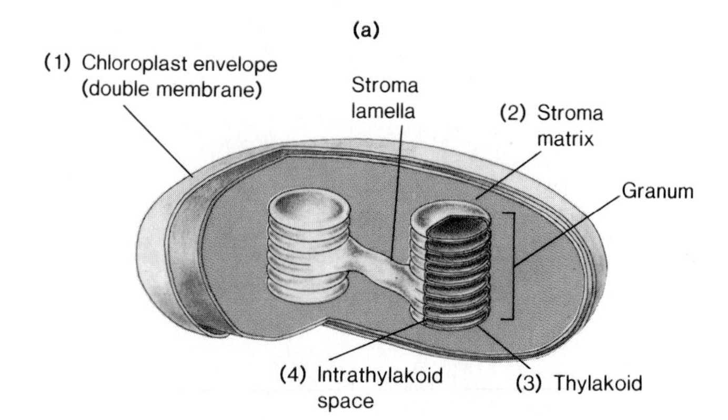

(b)

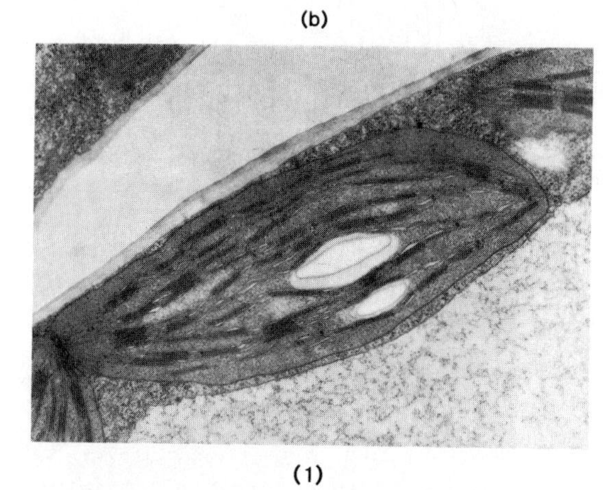

(1)

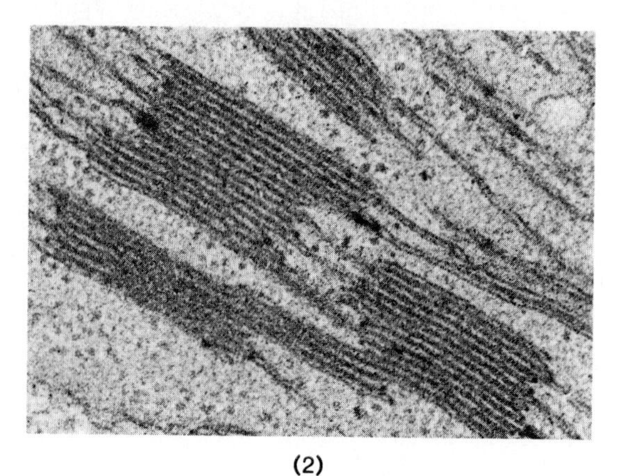

(2)

Figure 5.7 The structure of NADP that is reduced in the light reactions of photosynthesis. In the reduced form (bottom), one electron is attached to the nitrogen atom at position (1). The second electron attaches to a carbon atom at position (4) and is accompanied by a hydrogen ion. That is why reduced NADP is called NADPH.

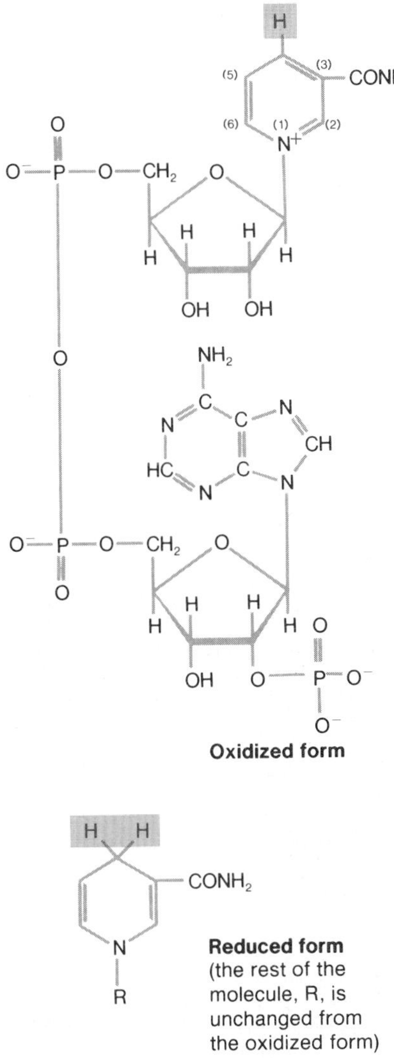

Oxidized form

Reduced form (the rest of the molecule, R, is unchanged from the oxidized form)

Inside the chloroplast, a complicated array of membranes forms the third layer. These membranes are arranged in structures that are visible by light microscopy as darker grains and are called **grana** (singular: **granum,** which means "grain" in Latin). The more detailed pictures of an electron microscope show that these grana are composed of stacks of flattened bags or discs called **thylakoids.** The various grana stacks are connected by membranous structures called **stroma lamellae.** The pigments (see page 100) that give the chloroplast and the plant their characteristic green color, as well as numerous proteins that play roles in photosynthesis, are contained in the membranes of the thylakoids and the stroma lamellae.

The space inside the thylakoids, the **intrathylakoid space,** is the fourth layer of the chloroplast.

Light Reactions and Dark Reactions

Modern analysis of the process of photosynthesis began in about 1905 when F. F. Blackman proposed that there are two basically different kinds of reactions in photosynthesis. He suggested that part of the photosynthetic process, which he called the **light reactions,** requires light, and that part of photosynthesis, called the **dark reactions,** does not require light.

Other scientists followed up Blackman's proposal. In the 1930s, C. B. van Niel and Robert Hill discovered that in the light reactions, water molecules are split to provide electrons that can be used for reducing carbon dioxide as it is incorporated into organic compounds. They found that the splitting of water releases the oxygen from the water molecules. This answered an old question about where the oxygen that is released during photosynthesis actually comes from.

Later, it was learned that the electrons removed from split water molecules in chloroplasts attach to (reduce) an electron carrier molecule called **nicotinamide adenine dinucleotide phosphate (NADP)** (figure 5.7).

Thus, the light reactions produce reduced NADP (NADPH). It was also discovered that ATP is produced when light shines on leaves.

These discoveries led Daniel Arnon to test an important hypothesis about photosynthesis. Arnon concluded that the role of the light reactions is to provide a source of electrons (NADPH) and energy (ATP) to be used to incorporate CO_2 into sugars and other carbohydrates. He reasoned that if he supplied chloroplasts with NADPH, ATP, and CO_2, they should produce organic molecules, even in the dark. He did the experiment, and the chloroplasts did indeed synthesize carbohydrates in the dark. Thus, the processes by which carbon dioxide is incorporated into organic molecules do not directly require light.

Arnon's experiment proved that Blackman had been right many years before when he said that part of photosynthesis (the light reactions) requires light, and part (the dark reactions) does not. Do not get the idea, however, that dark reactions require darkness or take place only in darkness. Dark reactions, of course, take place in leaves in the light, at the same time that light reactions are occurring. The dark reactions do not require light energy itself, but they require the products of the light reactions: reduced NADP (NADPH) and ATP.

We also know where in the chloroplast these reactions occur. The CO_2–incorporating reactions (dark reactions) take place in the stroma matrix of the chloroplast (see figure 5.6). The light reactions, which produce NADP (NADPH) and ATP, occur in the thylakoid membranes (figure 5.8).

Figure 5.8 The light and dark reactions of photosynthesis. The light reactions supply reduced NADP (NADPH) and ATP for the reduction and incorporation of CO_2 into carbohydrate molecules in the dark reactions.

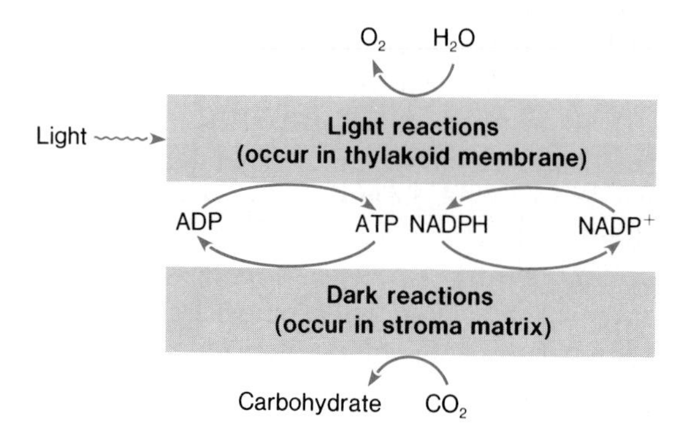

Light and Pigments

The light reactions in the thylakoid membranes of chloroplasts involve the interaction of light energy and colored chemical compounds called **pigments** that can absorb light energy and become changed as a result. Light reactions in the chloroplasts thus involve photochemical reactions, reactions between light and chemical compounds.

Our eyes also function through photochemical reactions. We can see what we call **visible light** because it is energy that can cause photochemical reactions in our eyes. Our eyes are sensitive to only part of a broad spectrum of kinds of radiant energy that there are (figure 5.9). Pigments in chloroplasts also absorb light energy in the part of the spectrum we call visible light.

Leaf Pigments

The green color of plants is due mainly to the presence of the pigment **chlorophyll.** Chlorophyll absorbs blue light and red light. The unabsorbed or reflected light appears green, thus making chlorophyll, and the chloroplasts and leaves that contain it, appear green to our eyes. There are two slightly different forms of chlorophyll, which are designated as chlorophyll *a* and chlorophyll *b* (figure 5.10). Other kinds of pigments, known as accessory pigments, also absorb light energy

Figure 5.9 Visible light makes up only a small part of the electromagnetic spectrum, a broad range of kinds of radiation with various wavelengths. Wavelengths are given in nanometers (nm). One nanometer equals 10^{-6} mm (0.000001 mm). The visible light portion of the spectrum (between 400 nm and 700 nm) is expanded in the lower part of the diagram. Color is the interpretation of what is seen when radiation of a given wavelength strikes the eyes. Thus, it is only a matter of convenience to speak of blue light or red light when describing light of a particular range of wavelengths. Bees and other insects can see parts of the spectrum that we cannot see. If we could see those parts as well, we would need to invent new names for the additional colors that we saw.

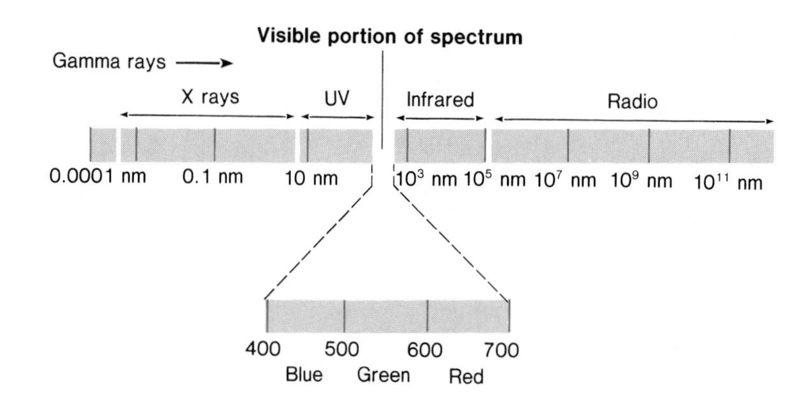

in chloroplasts. Some of these accessory pigments, the **carotenoids,** which are yellow, are familiar because they are found in other places besides green leaves. Carotenoids, for example, give carrots their orange color and make some autumn leaves yellow.

Excited Chlorophyll

When a molecule such as chlorophyll absorbs light energy, it is changed because electrons in one or more of its atoms are moved to new positions. Such an altered molecule is energy enriched and is said to be in an **excited state.** Chlorophyll (Chl) in its excited state is symbolized Chl*.

The excited state can be passed from molecule to molecule. This is what happens in chloroplast membranes where hundreds of pigment molecules are clustered together to make a **photosynthetic unit.** Each photosynthetic unit has one special chlorophyll *a* molecule called its **reaction center chlorophyll** to which the excited state eventually passes (figure 5.11). Only when this special molecule becomes excited can the light reactions of photosynthesis proceed. But absorption of light energy by any of the pigment molecules in the photosynthetic unit will lead eventually to excitement of this crucial reaction center chlorophyll molecule.

The molecules in a photosynthetic unit essentially function like an antenna system. Another good analogy is that a photosynthetic unit functions much like a large upright funnel leading to a small container set out in the rain. Raindrops land in various places in the funnel, but they all run down through the funnel into the container. Similarly, light energy absorbed anywhere in the photosynthetic unit eventually "funnels down" to the reaction center chlorophyll, which must become excited for the light reactions to proceed.

In an excited state, a molecule has different properties than the same molecule has in its normal condition. For example, excited chlorophyll (Chl*) loses electrons much more easily than Chl does. Excitement of chlorophyll molecules is the very heart of the light reactions of photosynthesis because Chl* donates electrons that are passed through several carriers to NADP. Thus, Chl* is responsible for the reduction of NADP, which we identified as one of the key results of the light reactions.

After chlorophyll has donated electrons for NADP reduction, how are they replaced? Oxidized chlorophyll (chlorophyll minus the donated electrons) picks up electrons from water molecules that have been split. Recall that this splitting also results in the release of oxygen during photosynthesis (figure 5.12).

Figure 5.10 The structure of chlorophyll. Some biologists argue that chlorophyll is the most important substance in the world because it is the key to harvesting the sun's energy for life. The basic structure of chlorophyll consists of a cluster of rings of carbon atoms (shown in gray) with a magnesium atom held in their center by four nitrogen atoms. The portion of the molecule marked R is the only part that differs between chlorophyll *a* and chlorophyll *b*.

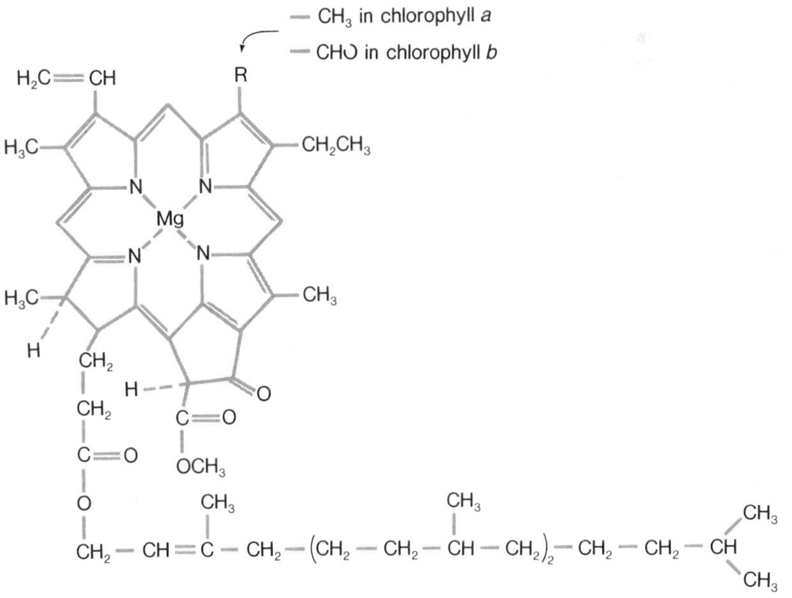

— CH_3 in chlorophyll *a*
— CHO in chlorophyll *b*

Figure 5.11 The photosynthetic unit. The pigments in the thylakoid membrane of the chloroplast are arranged in units, each comprised of several hundred molecules. Light absorbed by any one of these molecules can eventually cause the formation of the excited state of one special chlorophyll molecule, the reaction center chlorophyll. The chain of molecules shown in color is an example of the way that the excited state resulting from absorption of light energy can pass from molecule to molecule through the unit to the reaction center molecule. Some energy is lost in each transfer of the excited state, but there is still enough energy left to cause an excited state in the reaction center chlorophyll.

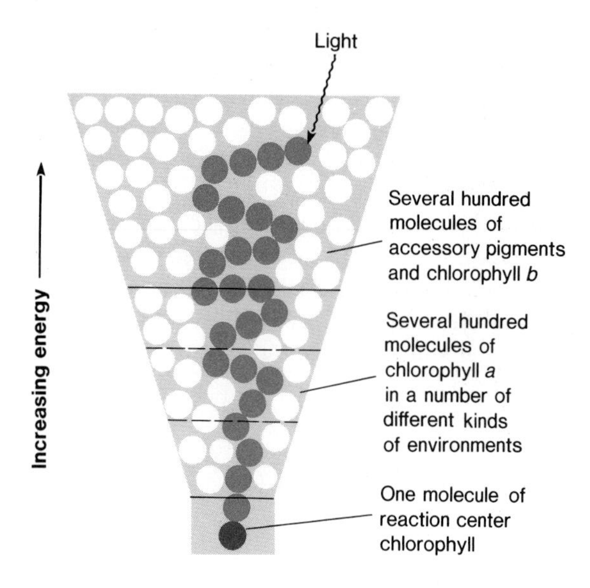

Photosystems

The overall result of the electron transfers in the light reactions is the reduction of NADP and the oxidation of water; that is, water serves as an electron donor for the light reactions in which NADP is reduced to NADPH.

This transfer of electrons from water to NADP occurs in two stages and involves two different types of reaction center chlorophyll molecules, called **Photosystem I** and **Photosystem II.**

Photosystem I is directly involved in supplying electrons to carriers that transfer them to NADP, while Photosystem II receives electrons from split water molecules and sends them through a series of carriers to Photosystem I (figure 5.13). The chlorophyll molecules that become excited in each of these photosystems are slightly different in terms of their light-absorbing properties. The reaction center chlorophyll in Photosystem I is called P700, and the reaction center chlorophyll in Photosystem II is called P680.

Figure 5.12 The splitting of water during photosynthesis supplies electrons (e^-) that combine with chlorophyll molecules that have donated electrons for NADP reduction. The upward arrow by the O_2 indicates that oxygen is given off as a gas. We will account for the fate of the hydrogen ions (H^+) later.

$$2H_2O \longrightarrow 4e^- + 4H^+ + O_2\uparrow$$

The energy boosts provided by light striking chlorophyll molecules are more than adequate to move the electrons "uphill" from water to NADP; there is energy to spare. This spare energy is released during the transfer of electrons from carrier to carrier in the series of carrier molecules associated with the photosystems. The energy released is used for ATP production. Thus, the photosystems in chloroplast thylakoid membranes also are involved in production of the other main product of the light reactions, ATP.

Our description might lead you to think that all the details of the light reactions are well understood, but this is far from true! The splitting of water molecules in Photosystem II is a critically important chemical process because it supplies electrons to the photosystems *and* because it continually replenishes the supply of atmospheric oxygen that the majority of living things require for life. However, we do not understand exactly how this crucial process occurs.

Likewise, we do not fully understand exactly how energy released during electron transfer results in ATP production, but there are some good hypotheses concerning the mechanism.

Phosphorylation

What might be the actual link between the transfer of electrons from one carrier to another in the internal membranes of chloroplasts, and the combination of ADP and phosphate to form ATP? The hypothesis that seems to fit best with observed facts is one proposed by Peter Mitchell, who was awarded a Nobel Prize in 1978 for his work on phosphorylation. This hypothesis is called the **chemiosmotic hypothesis** or simply the Mitchell hypothesis.

Mitchell proposes that as electrons are being passed from carrier to carrier through electron transport systems in thylakoid membranes, hydrogen ions (H^+) are dragged along in some of the transfers. He suggests that the carriers are arranged in the membrane so that they pick up hydrogen ions from the stroma matrix and release them into the intrathylakoid space (figure 5.14). This results in a buildup of hydrogen ions inside the thylakoids (see figure 5.6*a*).

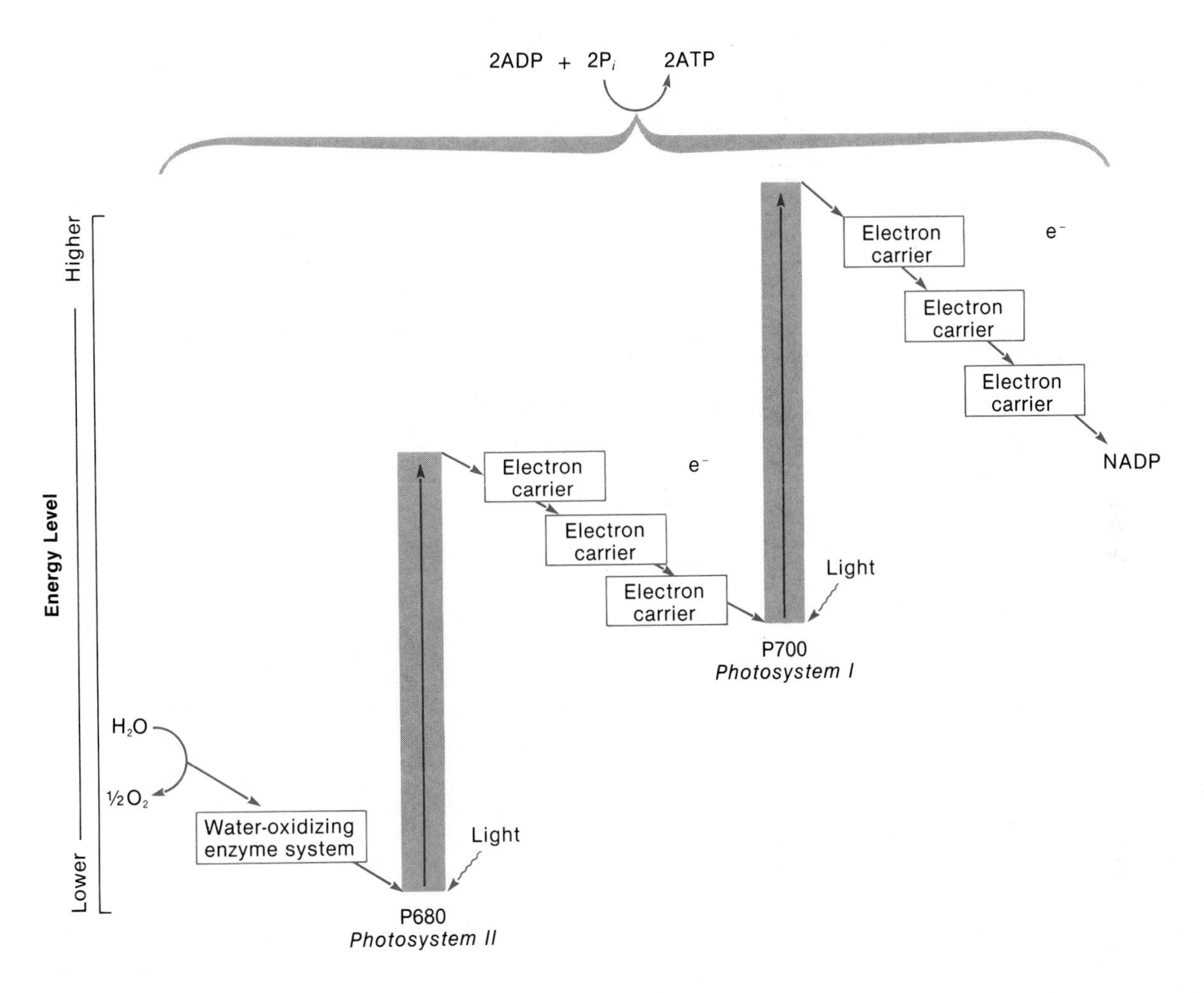

Figure 5.13 The two photosystems. Light energy excites chlorophylls P680 and P700 so that they readily donate electrons. This two-stage system uses energy from sunlight to transport electrons from water to NADP. Energy released during electron transfers is used to produce ATP.

Because H^+ ions are more concentrated in the intrathylakoid spaces, they tend to diffuse out if a way is open to them. Furthermore, there is an electrical charge difference between the inside and the outside because all those positively charged hydrogen ions have been moved to the inside. Thus, there are forces that tend to drive H^+ ions out through the thylakoid membranes. Mitchell hypothesizes that when H^+ ions do flow out through the membrane, energy is released that results in the formation of ATP by a special enzyme system also located in the membrane.

This hypothesis provides one explanation for ATP formation (phosphorylation) during photosynthesis, but not all biologists think that it works exactly this way. However it happens, though, light energy striking chloroplasts causes the movement of electrons from carrier to carrier, and the end result is the production of ATP and reduced NADP (NADPH), the two products of the light reactions that are needed in the dark reactions for carbon dioxide to be incorporated into organic molecules.

Building Organic Molecules

You will recall that Daniel Arnon proved that even in the dark, chloroplasts supplied with carbon dioxide, ATP, and NADPH would produce carbohydrates. How then are ATP and NADPH used during **CO_2 fixation,** the incorporation of carbon dioxide into organic molecules? During the 1940s,

Figure 5.14 Models of Peter Mitchell's hypothesis explaining ATP production during electron transport in chloroplast membranes. (*a*) The thylakoid membrane encloses the intrathylakoid space and separates it from the stroma matrix. (*b*) In the chloroplast, the space to the left would be the stroma matrix, the membrane would be the thylakoid membrane, and the space to the right would be intrathylakoid space. How reduction and oxidation of a carrier (Q) at opposite sides of a membrane could cause the transport of hydrogen ions across the membrane according to Mitchell's hypothesis. An electron carrier on one side of the membrane reduces Q. This requires two electrons and two protons that are taken up from the stroma matrix. The reduced carrier (QH_2) then diffuses across the membrane. On the other side, it donates its electrons to another electron carrier. Two hydrogen ions are released on that side. Finally, the oxidized Q diffuses back across the membrane to complete the cycle.

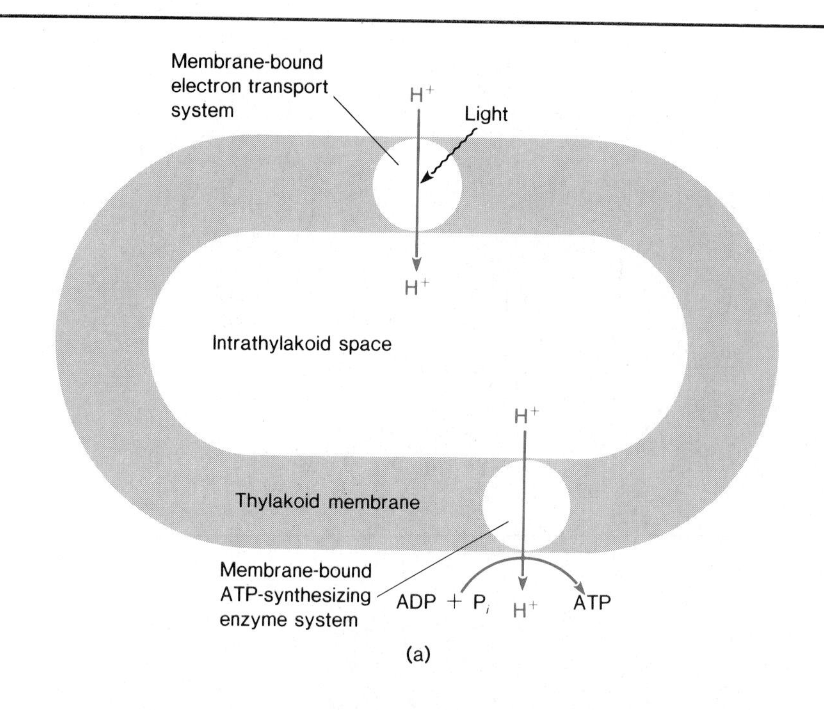

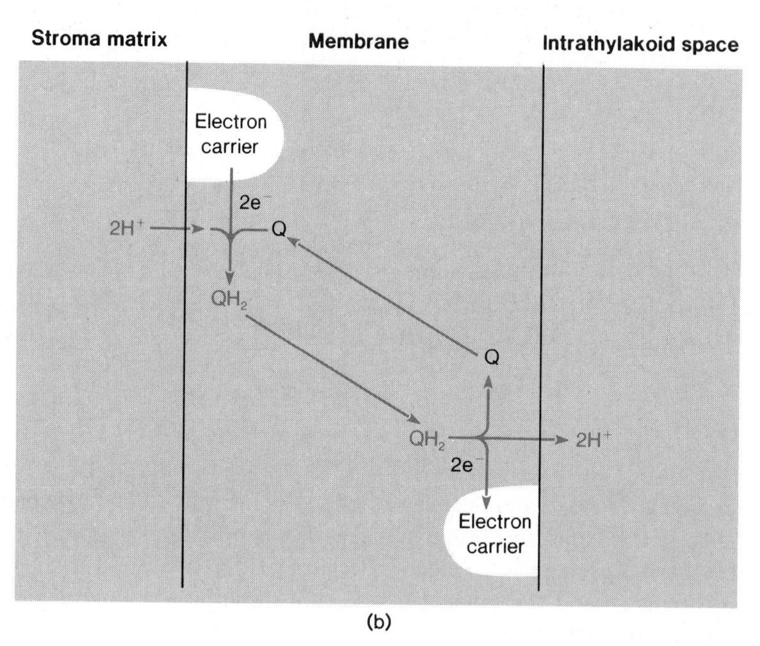

Melvin Calvin and his colleagues did experiments on CO_2 fixation using a radioactive form (isotope) of carbon, ^{14}carbon. They shined light on a suspension of cells of a unicellular alga, *Chlorella,* to which they supplied $^{14}CO_2$, carbon dioxide containing radioactive carbon. After different brief lengths of time, the algae were killed by dropping the cells into boiling alcohol, which instantly stopped all chemical reactions in the cells (figure 5.15). Carbon-containing compounds were then extracted from the cells and analyzed. Calvin and his coworkers were able to determine what compounds contained radioactive carbon after a very short time, and after longer periods of time. This told them how CO_2 becomes part of organic molecules in photosynthesis.

Figure 5.15 The apparatus used by Melvin Calvin for his classic experiments. The algae were contained in the flat flask, which was illuminated by the two lamps. $^{14}CO_2$ was added with a syringe, and the algae were killed after various periods by opening the stopcock and allowing them to fall into the beaker containing hot alcohol.

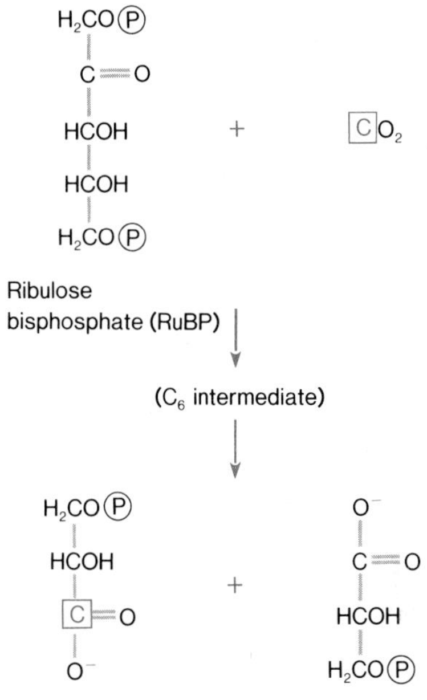

The C_3 Pathway

The first radioactive organic compounds that Calvin and his colleagues discovered, after a very short exposure to $^{14}CO_2$, contained three carbon atoms. For this reason, the set of reactions involved in CO_2 fixation is called the **C_3 pathway.** (It is also known as the **Calvin cycle.**)

Does this mean that CO_2 is bound to a two-carbon compound to make these three-carbon products? Calvin found it does not mean that. Actually, CO_2 is joined to a five-carbon (C_5) compound to make a six-carbon (C_6) compound, which exists for an extremely short time and then splits into two C_3 molecules. We can symbolize these reactions as follows:

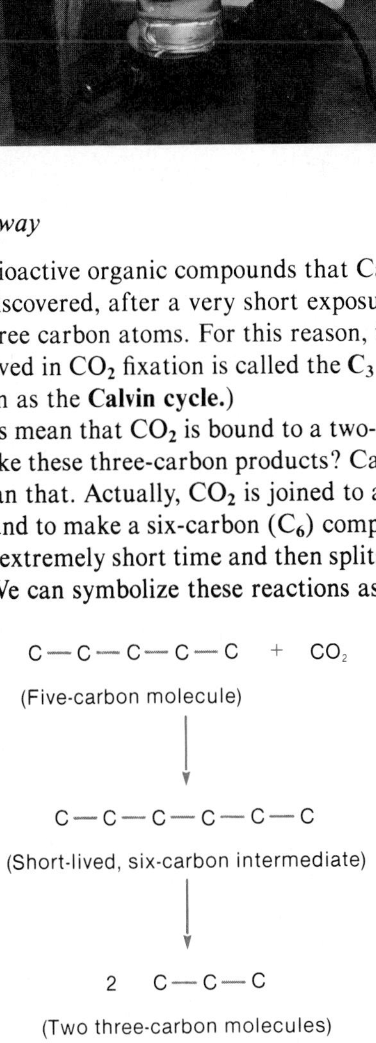

C—C—C—C—C + CO_2

(Five-carbon molecule)

↓

C—C—C—C—C—C

(Short-lived, six-carbon intermediate)

↓

2 C—C—C

(Two three-carbon molecules)

Figure 5.16 The reaction by which carbon dioxide is fixed and becomes part of an organic molecule PGA. This reaction is catalyzed by the enzyme RuBP carboxylase.

H_2COⓅ | C=O | HCOH | HCOH | H_2COⓅ + [C]O_2

Ribulose bisphosphate (RuBP)

↓

(C_6 intermediate)

↓

H_2COⓅ | HCOH | [C]=O | O^- + O^- | C=O | HCOH | H_2COⓅ

3-phosphoglycerate (PGA)

The five-carbon molecule to which CO_2 is attached is a sugar with two phosphates called **ribulose bisphosphate (RuBP).** The enzyme that catalyzes this reaction (figure 5.16) is called **ribulose bisphosphate carboxylase,** and the two three-carbon molecules produced are molecules of **3-phosphoglycerate (PGA).**

This is one of the most important of all biochemical reactions because the great majority of all carbon atoms in organic compounds in all living things are derived from CO_2 through this reaction.

We have said that ATP and reduced NADP (NADPH) are required for CO_2 fixation. How are they used? PGA is not a final product of photosynthesis that can be used to make other compounds in living cells. It must be converted to another, slightly different substance called **glyceraldehyde-3-phosphate (GAP),** and this is where ATP and NADPH come in. One ATP is used and one NADPH must give up its two electrons each time PGA is converted to GAP (figure 5.17).

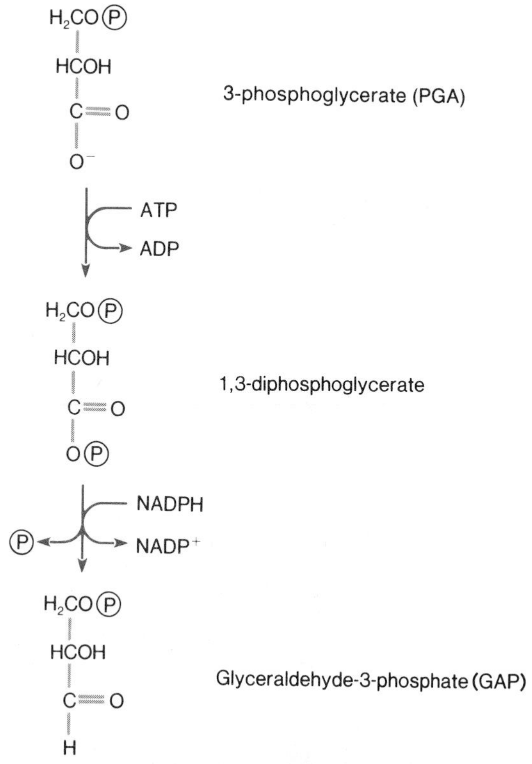

Figure 5.17 ATP and NADPH are required for reactions in which PGA is converted to GAP, the product of photosynthesis that can be used to make other important substances in cells.

H_2COⓟ
HCOH
C=O
O^-

3-phosphoglycerate (PGA)

ATP
ADP

H_2COⓟ
HCOH
C=O
Oⓟ

1,3-diphosphoglycerate

NADPH
ⓟ
$NADP^+$

H_2COⓟ
HCOH
C=O
H

Glyceraldehyde-3-phosphate (GAP)

GAP is the product of photosynthesis that can be used to make sugars, starch, fats, proteins, and all other compounds needed by the plant. But only part of the GAP produced by photosynthesis can be used for these things. Much of it must be used to maintain the process of CO_2 fixation.

Keeping the Cycle Going

If a RuBP is used each time a CO_2 is fixed, what keeps the supply of RuBP from running out? Most of the GAP produced in photosynthesis is fed into a series of reactions that produce RuBP. This is why we say that these reactions are a cycle; they come back to the same starting point. This explains the other name for the C_3 pathway: the Calvin *cycle* (figure 5.18).

Regenerating the supply of RuBP also requires ATP. This is a second place where the dark reactions use ATP produced in the light reactions.

After replenishment of the supply of RuBP, the GAP that remains is the net product of photosynthesis. This GAP is available for use for energy or for synthesis of glucose and other organic molecules. Some of the GAP is used within the leaf cells, but much of it is "exported" to other parts of the plant. Thus, products of photosynthesis in chloroplasts eventually supply the energy and raw material needs of the entire plant.

The C_4 Pathway

Over the years, photosynthesis has been studied in a great many plants, and in most, carbon dioxide is incorporated into organic compounds in the way first discovered in Calvin's classic experiments on algae. But some plants incorporate carbon dioxide differently. In those plants, the first detectable products of CO_2 incorporation are four-carbon (C_4) acids, not three-carbon acids. Such plants are called "C_4 plants." Some familiar examples of C_4 plants are corn *(Zea mays)* and sugar cane.

The C_4 plants share a number of distinctive characteristics. Most have a specialized leaf anatomy in which the vascular bundles are surrounded by distinctive **bundle sheath** cells that contain chloroplasts (figure 5.19). In cross section, this sheath looks somewhat like a wreath; the German word for wreath, *Kranz,* is applied to this type of structure. Thus, C_4 plants are said to display **Kranz anatomy.** Chloroplasts are not present in bundle sheath cells of C_3 plants.

The C_4 plants have an extra set of reactions, in addition to the Calvin cycle reactions, by which CO_2 is incorporated into organic compounds. These reactions do not replace the Calvin cycle reactions, but rather supplement them.

Ordinary mesophyll cells in C_4 plants do not carry on Calvin cycle reactions, but they do fix carbon dioxide by joining it to a three-carbon acid, phosphoenolpyruvate (PEP), to make a four-carbon (C_4) acid. This C_4 acid is converted to another form and exported to a nearby bundle sheath cell where it gives up the CO_2 for use in the Calvin cycle and things proceed in the ordinary way (figure 5.20). Calvin cycle reactions occur *only* in the bundle sheath cells of C_4 plants.

Of what value is this extra set of reactions to the C_4 plants? The function of these extra reactions seems to be providing a *concentrated* supply of CO_2 for use in the Calvin cycle in bundle sheath cells. Specifically, providing a good supply of CO_2 greatly increases the efficiency of the enzyme ribulose bisphosphate carboxylase.

Figure 5.18 The Calvin cycle. Some of the GAP produced in photosynthesis is available for use by the plant, but some of it must be used to replenish the supply of RuBP and keep the CO_2 fixation process going.

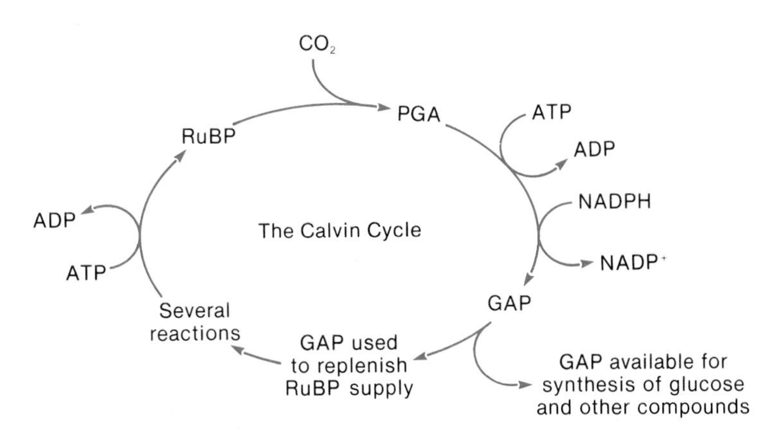

Figure 5.19 Comparison between leaf anatomy in plants using C_3 and C_4 pathways of CO_2 fixation. Scanning electron microscope photos of leaves from a C_4 plant bluestem, *Andropogon* (left) and a C_3 plant brome grass, *Bromus* (right) (magnification × 277 and 267, respectively). Accompanying sketches highlight some major structural differences. Vascular bundles are separated by only a few (usually five or less) mesophyll cells in C_4 plants' leaves. C_4 bundle sheath cells contain large numbers of chloroplasts. Mesophyll cells form a concentric ring around the bundle sheath and contribute to the "wreath" appearance. When a bundle sheath is present in a C_3 plant, its cells contain no chloroplasts, and vascular bundles usually are farther apart, separated by a number of mesophyll cells.

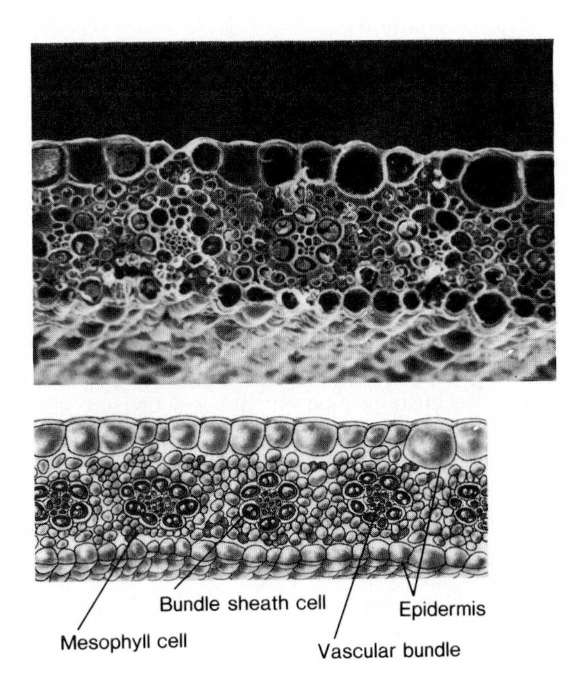

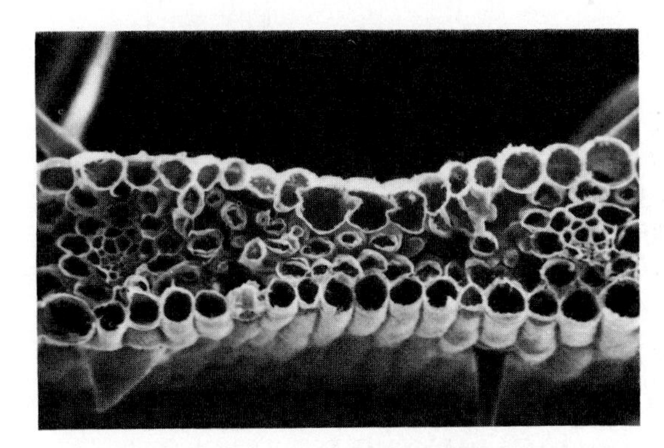

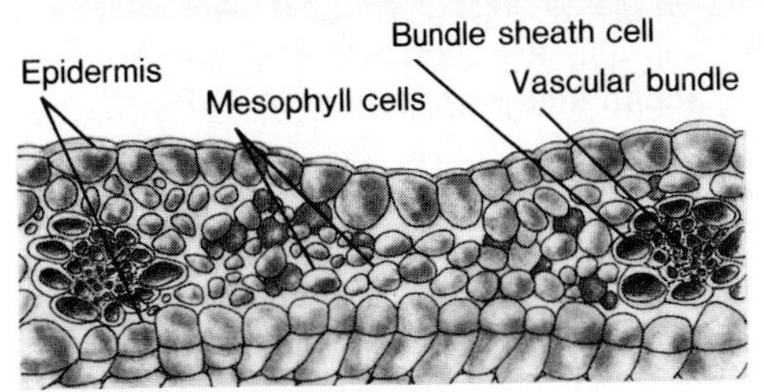

Figure 5.20 Summary diagram of C_4 photosynthesis. CO_2 is incorporated into C_4 organic acids in mesophyll cells. Organic acids are transferred to bundle sheath cells where they give up CO_2 for use in Calvin cycle reactions. This mechanism produces a high concentration of CO_2 in bundle sheath cells and increases the efficiency of the Calvin cycle reactions.

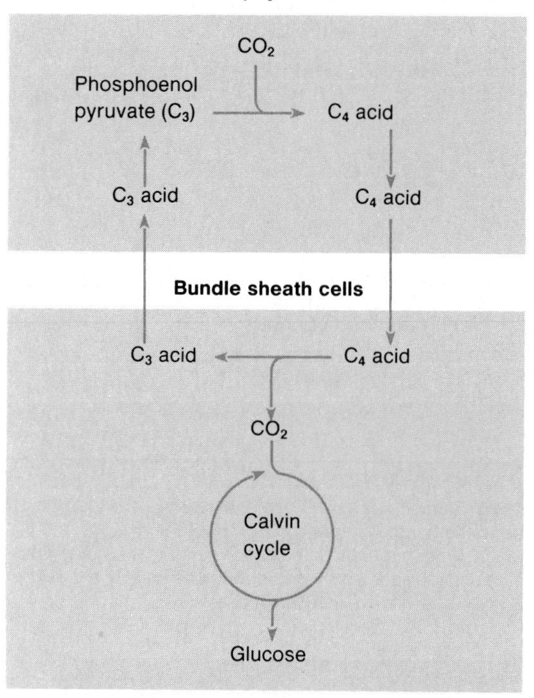

Obtaining CO_2 in this way requires energy and costs C_4 plants some ATP, but it seems to be worth it, especially to plants living in environments with high temperatures and limited water supplies. Possessing the C_4 photosynthetic mechanism helps such plants because they do not have to keep their stomata open as long as C_3 plants to obtain the necessary CO_2. This reduces their water loss by evaporation and helps to prevent water stress and wilting.

There are yet other photosynthetic mechanisms (see box 5.1) that meet the needs of plants in various environments. Many biologists are working to understand the various photosynthetic mechanisms better so that present crop plants can be made more productive and new crop plants might be developed. For example, the desert shrub *Euphorbia* contains a sap that can be distilled to produce hydrocarbons similar to those in petroleum. Melvin Calvin and others propose that *Euphorbia* might be cultivated commercially in "petroleum plantations," using land that is too dry for other commercial crops, to reduce dependence on present petroleum supplies.

To meet human society's future requirements for energy and raw materials, it may well be that more efficient harvesting of the sun's energy will become possible not through construction of bigger and better solar energy collection devices, but by learning more about, and taking advantage of, the remarkable energy conversions of photosynthesis.

Box 5.1

Night and Day Photosynthesis

The CAM System

A very interesting variation in photosynthetic mechanisms is found among plants growing in very dry environments, such as deserts. This type of photosynthesis was discovered first in plants belonging to the plant family *Crassulaceae,* and thus the photosynthetic mechanism is called **crassulacean acid metabolism** or **CAM,** for short.

CAM is similar to C_4 photosynthesis in that CO_2 is first fixed into C_4 acids, but CAM plants do not have the characteristic *Kranz* anatomy that ordinary C_4 plants have.

The biggest difference is that in CAM plants CO_2 incorporation into C_4 acids occurs *at night.* During the cool desert night, when the danger of water loss decreases, plants with CAM open their stomata and fix large amounts of CO_2 into C_4 acids. These acids are stored in high concentrations in the cell vacuoles of mesophyll cells. Succulent plants have thick, fleshy leaves because their leaves contain such large quantities of these mesophyll cells.

During the day, the C_4 acids in CAM plants are broken down, providing a source of CO_2 for fixation by the Calvin cycle enzymes. To prevent water loss, the stomata are firmly closed during the day.

Thus, the CAM mechanism permits desert plants to obtain the CO_2 that they need for photosynthesis without opening their stomata during the daytime, when high temperatures would lead to excessive and potentially dangerous water loss.

Summary

A fundamental characteristic of living organisms is their need for a continual input of energy. The energy that all living organisms use comes directly or indirectly from sunlight. Reduced organic compounds produced in photosynthesis provide chemical energy for all organisms. These reduced organic compounds are broken down in respiration, resulting in the formation of ATP.

Leaves are photosynthesizing organs. Most leaves are flat and thin with a large surface area exposed to light. Photosynthesizing mesophyll cells in the leaf interior are surrounded by air spaces that connect with the outside atmosphere through stomata. Guard cells surround the stomata and regulate their opening and closing. Stomatal control is important in regulating the CO_2 supply for photosynthesis and in preventing excessive water loss from leaves.

Chloroplasts are membrane-bound organelles containing complex internal membrane structures, the thylakoids and stroma lamellae. The stroma matrix is the fluid area surrounding these membrane structures. Pigment molecules involved in light reactions are associated with the thylakoids. Enzymes involved in dark reactions are in the stroma matrix.

Light causes specific photochemical reactions in pigment molecules that produce an "excited" state.

Pigments in chloroplasts are arranged in photosynthetic units. Excitation of a pigment molecule anywhere in the unit causes energy transfers leading to an excited state in the reaction center molecule.

Excited chlorophyll molecules readily donate electrons. Photosystem I reduces oxidized NADP. Photosystem II is directly involved in water splitting and oxygen release. Photosystem II gains electrons from water and transfers them to Photosystem I to replace those given up to oxidized NADP by the chlorophyll in Photosystem I.

Phosphorylation (ATP production) occurs as a result of energy changes during electron transport. One hypothesis suggests that hydrogen ions are moved across the thylakoid membrane during electron transport. Their return through the membrane provides energy for ATP formation.

In the Calvin (C_3) cycle, CO_2 is combined with a C_5 molecule, ribulose bisphosphate (RuBP). The first stable products of this combination are two molecules of phosphoglycerate (PGA), a three-carbon compound. ATP and NADPH are used in subsequent steps to produce glyceraldehyde-3-phosphate (GAP). Most of the GAP produced is used in regeneration reactions to resupply RuBP. The rest of the product molecules can be withdrawn from the Calvin cycle and used for energy or as raw materials for synthesis of other organic compounds.

In plants using the C_4 photosynthetic pathway, mesophyll cells incorporate CO_2 into C_4 compounds. The CO_2-concentrating mechanism in C_4 plants permits them to photosynthesize efficiently with less stomatal opening. The result is less water loss.

Questions for Review

1. Explain how swelling of a guard cell due to osmotic water uptake can result in *opening* of a stoma.
2. Define the terms granum and thylakoid and explain what we mean by intrathylakoid space.
3. How did Daniel Arnon prove that the dark reactions do not directly require light, but rather the products of light reactions?
4. Why is visible light "visible"?
5. Why is chlorophyll green?
6. What property of excited chlorophyll (CHl*) is the key to its role in the light reactions?
7. Two products of the light reactions are required for CO_2 fixation. What are those products?
8. Explain how Melvin Calvin and his colleagues did the experiments in which they discovered how carbon dioxide is incorporated into organic molecules.
9. RuBP carboxylase is one of the most important enzymes in the living world. What reaction does it catalyze?
10. Identify GAP and explain what happens to the GAP produced in photosynthesis.
11. What is *Kranz* anatomy?
12. How does C_4 photosynthesis get its name?

Questions for Analysis and Discussion

1. Can Photosystem I continue to donate electrons for NADP reduction if Photosystem II is not operating? Explain why or why not.
2. Explain why the Calvin cycle is a cycle.
3. If all photosynthetic activity were to cease, all animals would die long before food supplies ran out. What would be the cause of death?

Suggested Readings

Books

Hall, D. O., and Rao, K. K. 1977. *Photosynthesis.* 2d ed. Studies in Biology no. 37. Baltimore: University Park Press.

Lehninger, A. L. 1982. *Principles of biochemistry.* New York: Worth Publishers.

Articles

Björkman, O., and Berry, J. October 1973. High-efficiency photosynthesis. *Scientific American* (offprint 1281).

Calvin, M. 1979. Petroleum plantations for fuel and materials. *BioScience* 29:533.

Govindjee, and Govindjee, R. December 1974. Primary events of photosynthesis. *Scientific American* (offprint 1310).

Heath, O. V. S. 1981. Stomata. *Carolina Biology Readers* no. 37. Burlington, N.C.: Carolina Biological Supply Co.

Cellular Metabolism

6

Chapter Outline

Chapter Concepts

1. Reduced organic compounds, originally produced by photosynthetic activity, are broken down in a stepwise series of oxidation reactions to provide energy for ATP production.
2. Processes that do not directly require oxygen use a small portion of the potential energy of glucose molecules to produce ATP.
3. Organic molecules are broken down further by aerobic (oxygen-requiring) processes that occur in mitochondria.
4. As organic molecules are broken down, CO_2 is released, and oxidations supply electrons to mitochondrial electron transport systems.
5. Phosphorylations in electron transport systems produce a great deal of ATP.
6. Fats and proteins can be oxidized in the same metabolic pathways as carbohydrates, but they must first undergo preparatory reactions.
7. Compounds can also be withdrawn from these reactions for use in synthesis.

In a long distance race, world-class runners seem to be running at high speed as they circle the track time after time. But as they near the end of the race, they call on a reserve of strength and quicken their pace. After many minutes of running, they are still able to accelerate until they are sprinting. As they near the finish, however, some runners strain even harder and push themselves to even greater speeds (figure 6.1). Then suddenly they flash across the finish line and the race is over.

Figure 6.1 A runner expending energy to achieve a maximum effort at the end of a race. This is Sebastian Coe crossing the finish line in a world record-setting performance.

If we follow one of the runners after the race, we see that he stands still or walks slowly, almost stumbling at times, as he gulps one deep breath after another. Each breath supplies oxygen to his body and carries away carbon dioxide. His rapid breathing continues for several minutes after the race ends, but gradually he returns to a normal breathing rate.

How was energy supplied in the runner's body for the tremendous effort of the race? How is oxygen used during and after the race? Where does the carbon dioxide that the runner breathes out come from? Why does heavy breathing continue for some time after the race is over?

It may sound strange to say that the energy for running comes from the sun, but the sun is the ultimate source of energy used for movement and all other forms of work in all living things. Of course, the energy used by a runner does not come directly from the sun, but from nutrients, organic chemical compounds that the runner has eaten.

The organic compounds produced from simple, inorganic precursors in photosynthesis provide energy for all living things. The breakdown and oxidation of these reduced carbon compounds yields energy for the many energy-requiring processes that take place in the cells of living organisms (figure 6.2).

Reduced organic compounds are oxidized to yield energy. This oxidation is not like burning, an oxidation with which we are all familiar. When something is burned, it is oxidized very quickly with a rapid release of energy. In energy-yielding processes in living cells, the gradual breakdown of reduced organic compounds is the key characteristic. These processes release the chemical energy of carbohydrates and other organic molecules in "packets" of manageable size through a stepwise series of reactions. This gradual

Figure 6.2 The relationships between photosynthesis and respiration.

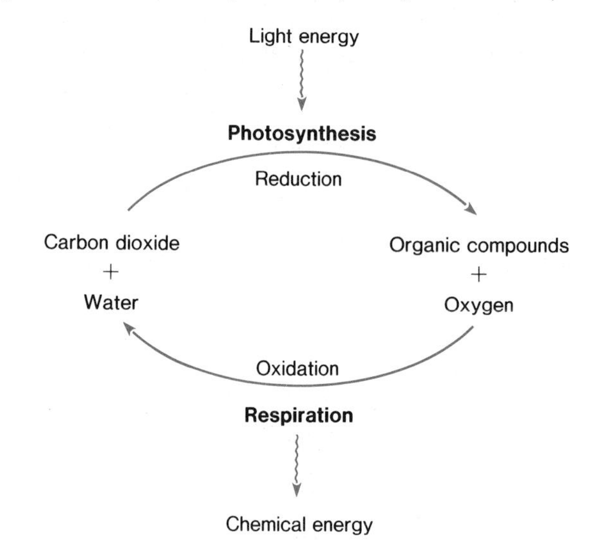

breakdown provides energy for formation of adenosine triphosphate (ATP), which can then be used to provide energy for a number of kinds of biological work.

We can summarize cellular respiration, the oxidation of organic nutrients, using glucose as an example:

$$C_6H_{12}O_6 + 6O_2 \longrightarrow 6CO_2 + 6H_2O$$

Usable energy (ATP)

The Central Role of ATP

The primary function of the energy conversions that we will consider is production of ATP, the cellular "energy currency." It is difficult to overemphasize the importance of ATP and its central role in cellular energy economy. The energy from the **exergonic** (energy-yielding) reactions of cellular respiration generally cannot provide energy directly to various **endergonic** (energy-requiring) processes in living cells (figure 6.3). Rather, ATP is the directly usable energy currency that can drive a variety of endergonic processes. Work such as synthesis, transport, and movement in living things is done using energy from ATP.

Thus, as we examine each of these stepwise breakdown processes, we will again and again ask the question, "How much ATP is produced as a result of this reaction?" As we consider these stepwise breakdown processes and ATP production, make certain that you consult all of the figures cited in the text. We believe that the figures will help you to understand and analyze these processes more easily.

Using Glucose: An Overview

It is best to begin a consideration of cellular energy conversions with the utilization of a carbohydrate, the sugar glucose. The first series of reactions in the breakdown of glucose is called the **Embden-Meyerhof (E-M) pathway** after two German scientists who studied these processes in the 1920s and 1930s. The term **glycolysis** ("glucose breakdown") is sometimes also used as a synonym for the Embden-Meyerhof pathway. The E-M pathway is a several-step process that occurs in the cell's cytoplasmic matrix. It can proceed either in the presence or absence of oxygen and, by itself, makes available only a small portion of the potential energy that can be extracted from glucose.

The formation of two three-carbon molecules of **pyruvic acid** is the final step of the E-M pathway (figure 6.4), so the process does not decrease the number of carbon atoms in organic molecules (glucose is a six-carbon molecule: $C_6 \rightarrow C_3 + C_3$). Pyruvic acid formation is an important step in utilizing carbohydrates such as glucose. Different cells further metabolize pyruvic acid in different ways.

Under **anaerobic** conditions (in the absence of oxygen), pyruvic acid is further processed through one of several types of **fermentation.** Two kinds of fermentation are identified in figure 6.4. In **alcoholic fermentation,** which occurs in many microorganisms (for example, brewer's yeast), pyruvic acid is converted to ethanol (ethyl alcohol). **Lactic fermentation,** which occurs in some microorganisms and some animal cells (for instance, in muscle cells during heavy exercise, when the oxygen supply is insufficient), produces lactic acid. Neither kind of fermentation yields additional usable energy for the cells beyond that extracted in the Embden-Meyerhof pathway.

Figure 6.3 ATP as energy currency. Oxidation of reduced organic compounds provides energy to combine ADP and phosphate to replenish the supply of ATP. Breakdown of ATP to ADP and phosphate provides energy to drive energy-requiring processes in cells.

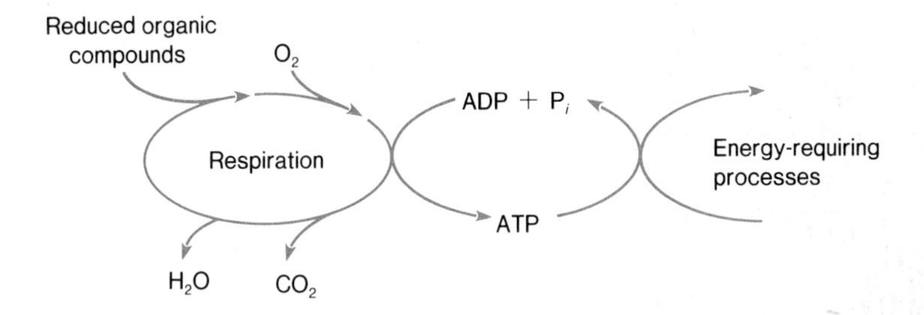

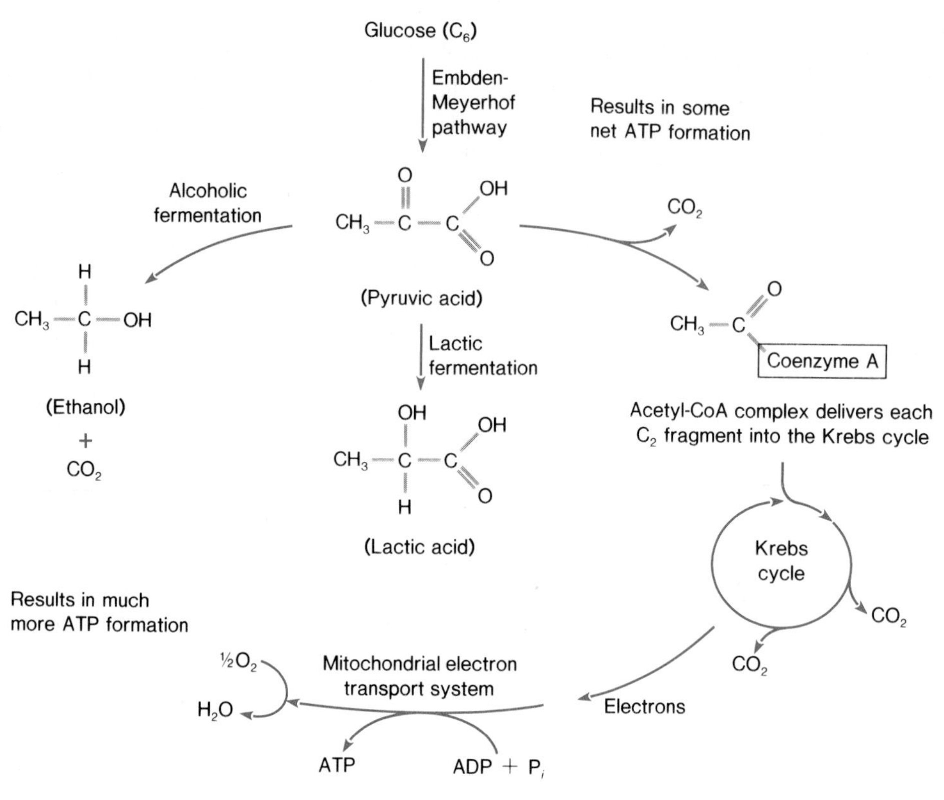

Figure 6.4 Preview of carbohydrate breakdown pathways using glucose as a starting point. For each molecule of glucose processed in the E-M pathway, two molecules of pyruvic acid are produced. Alcoholic and lactic fermentation occur in certain cells under anaerobic conditions. The Krebs cycle is part of aerobic respiration because electrons from Krebs cycle oxidation reactions must go through the electron transport system to oxygen.

Under **aerobic** conditions (in the presence of oxygen), oxygen-requiring processes collectively known as **respiration** use much more of the potentially available chemical energy to form ATP. Pyruvic acid enters another series of reactions that occur inside the mitochondria. This reaction series is the **Krebs cycle,** named for Sir Hans Krebs, who clarified the cyclical nature of the reactions involved. In the Krebs cycle, organic molecules are further degraded by the removal of one-carbon fragments as CO_2. This process of CO_2 removal is called **decarboxylation.**

Oxidations of organic molecules in the Krebs cycle provide electrons that are transferred to electron carriers in the mitochondrial membranes. There, a series of oxidation-reduction reactions occurs as electrons are passed from carrier to carrier in an **electron transport system.** Energy released during these transfers makes possible the production of considerable amounts of ATP.

These electron carriers finally donate electrons to oxygen, which combines with hydrogen ions (protons) to form water. The reduction of oxygen, which serves as the *final electron acceptor,* and the subsequent formation of **metabolic water,** is a fundamental characteristic of aerobic respiration. The formation of metabolic water is summarized in the following equation:

$$O_2 + 4e^- + 4H^+ \longrightarrow 2H_2O$$

Thus, the final products of the oxidative pathways are CO_2 and water, the CO_2 produced in decarboxylation reactions and the water produced by the reduction of oxygen. The energy made available during the oxidation of organic compounds and subsequent electron transport events is used to produce ATP.

The Embden-Meyerhof Pathway

The progressive breakdown of glucose to produce pyruvic acid in the Embden-Meyerhof (E-M) pathway occurs in a series of nine reactions. Each reacton is catalyzed by a specific enzyme. All of these details are shown in figure 6.5, but we can more easily identify the key events in the E-M pathway if we represent the molecules more simply.

To get the process started, sugar molecules must be "activated" in preparation for the reactions that follow. Since these preparatory reactions are phosphorylations (that is, phosphate groups are attached to carbohydrate molecules), they use ATP. The ATP molecules donate their terminal phosphate groups in the reactions. Thus, the E-M pathway reactions actually begin with an expenditure of ATP. But these preparatory reactions provide activation energy at the beginning to make all the subsequent reactions possible.

Figure 6.5 The reactions of the Embden-Meyerhof pathway. The name of the enzyme that catalyzes each reaction is given by each arrow. Most enzyme names end in "-ase." The positions of the carbons in the molecules are represented symbolically as the "corners" of the ring. Note that acids such as pyruvic acid actually dissociate and give up a hydrogen ion (H^+) inside cells. Thus, the pyruvate ion, which is produced by pyruvic acid dissociation, is pictured rather than the pyruvic acid molecule. (The "-ate" ending is used generally to indicate ionic forms of dissociated acids.) The oxidized form of NAD is symbolized NAD^+, while the reduced is NADH.

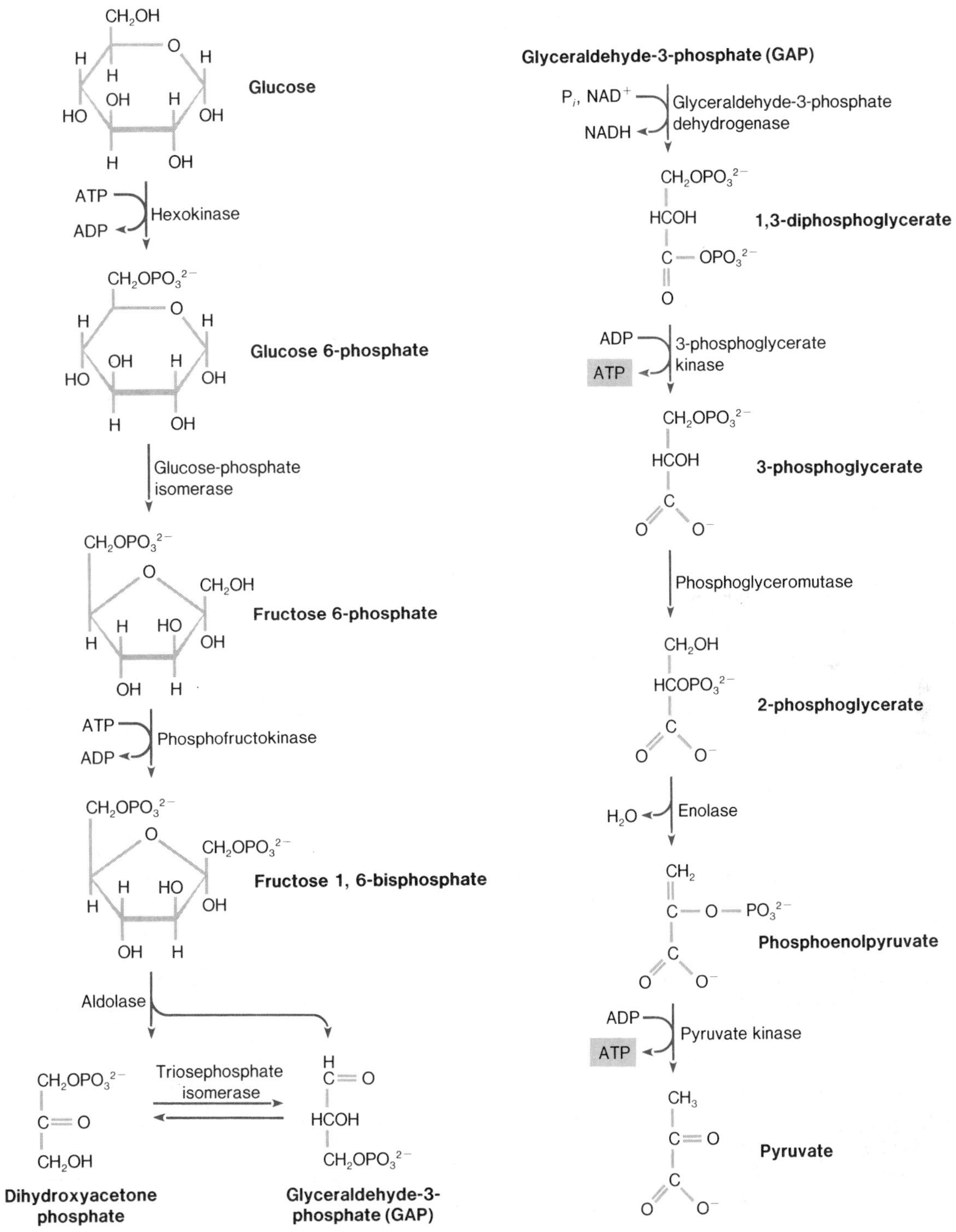

CH_2OH

H O H

H

OH H

HO OH

H OH

Glucose

ATP → ADP Hexokinase

$CH_2OPO_3^{2-}$

H O H

OH H

HO OH

H OH

Glucose 6-phosphate

Glucose-phosphate isomerase

$CH_2OPO_3^{2-}$

O CH_2OH

H HO

H OH

OH H

Fructose 6-phosphate

ATP → ADP Phosphofructokinase

$CH_2OPO_3^{2-}$

O $CH_2OPO_3^{2-}$

H HO

H OH

OH H

Fructose 1, 6-bisphosphate

Aldolase

$CH_2OPO_3^{2-}$

C=O

CH_2OH

Dihydroxyacetone phosphate

Triosephosphate isomerase

H

C=O

HCOH

$CH_2OPO_3^{2-}$

Glyceraldehyde-3-phosphate (GAP)

Glyceraldehyde-3-phosphate (GAP)

P_i, NAD^+ → NADH Glyceraldehyde-3-phosphate dehydrogenase

$CH_2OPO_3^{2-}$

HCOH

C — OPO_3^{2-}

O

1,3-diphosphoglycerate

ADP → ATP 3-phosphoglycerate kinase

$CH_2OPO_3^{2-}$

HCOH

C

O O^-

3-phosphoglycerate

Phosphoglyceromutase

CH_2OH

$HCOPO_3^{2-}$

C

O O^-

2-phosphoglycerate

Enolase → H_2O

CH_2

C — O — PO_3^{2-}

C

O O^-

Phosphoenolpyruvate

ADP → ATP Pyruvate kinase

CH_3

C=O

C

O O^-

Pyruvate

The first E-M pathway reaction results in the formation of **glucose 6-phosphate** (glucose with a phosphate attached to one of its carbon atoms).

C C C C C C + ATP ⟶

(Glucose)

C C C C C C (P) + ADP

(Glucose 6-phosphate)

Two additional preparatory reactions follow the phosphorylation of glucose. A molecular reorganization yields **fructose 6-phosphate,** and then the transfer of another phosphate group from ATP produces **fructose 1,6-bisphosphate.**

C C C C C C (P) + ATP ⟶

(Fructose 6-phosphate)

(P) C C C C C C (P) + ADP

(Fructose 1,6-bisphosphate)

To this point, then, two ATP molecules have been used up in preparatory reactions. How then does the E-M pathway actually produce an energy gain for the cell?

Subsequent reactions in the Embden-Meyerhof pathway are energy yielding, and thus the whole series of reactions does result in an ATP gain for the cell. Fructose 1,6-bisphosphate is split into two three-carbon molecules, **dihydroxyacetone phosphate (DHAP)** and **glyceraldehyde-3-phosphate (GAP).**

(P)—C—C—C—C—C—C—(P) ⟶

(Fructose 1,6-bisphosphate)

(P)—C—C—C + C—C—C—(P)

(DHAP) (GAP)

These three-carbon molecules are readily interconvertible in a reversible reaction. Since one product, GAP, reacts in the next step of the pathway and is used up, the DHAP is converted to GAP. Thus, two GAP molecules must be accounted for in determining energy relationships of subsequent reactions.

Further processing of C_3 molecules begins with an oxidation in which the transfer of electrons involves the simultaneous removal of hydrogen ions. In this particular reaction, two electrons, with an accompanying hydrogen ion, are transferred to the electron carrier **nicotinamide adenine dinucleotide (NAD).** At the same time, a phosphate group is attached to the molecule, so that products of this reaction are a C_3 molecule with two phosphate groups, **1,3-diphosphoglycerate,** and a reduced molecule of NAD (NADH).

Figure 6.6 ATP arithmetic of the Embden-Meyerhof pathway. Two ATPs must be spent to start the process. Four ATPs are produced by substrate-level phosphorylations, giving a net gain for the cell of two ATPs.

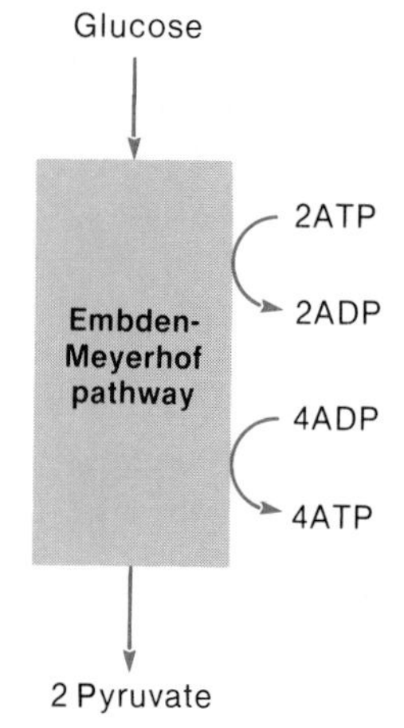

In the next reaction in the sequence, the energy change is large enough to allow a removed phosphate group to be transferred to ADP by phosphorylation, thus yielding a molecule of ATP. This is the first of two such transfers that are essential to energy retrieval in the E-M pathway. Because this phosphorylation is associated directly with a reaction involving the substrate of an enzyme in the pathway, it is called a **substrate-level phosphorylation.**

After two molecular reorganizations, there is a second phosphorylation, which occurs in the reaction that yields **pyruvate** (the ionized form of pyruvic acid in solution). Pyruvate is a key compound in cellular metabolism because it stands at the junction of several biochemical pathways.

Energy Yield of the E-M Reactions

After all these reactions, how much energy (ATP) has the cell gained? This is actually a rather tricky question because the answer depends upon whether the reactions are taking place under aerobic (oxygen present) or anaerobic (oxygen absent) conditions. Thus we can give only a partial answer now. It costs two ATPs to get the process started. Then two ATPs are produced for each of the two three-carbon molecules produced from each glucose molecule. Thus, two ATPs are spent and four are produced, leaving a net gain of two ATPs (figure 6.6).

What happens to the pyruvic acid produced by the E-M pathway reactions? This is another question that has several possible answers, depending basically on whether conditions are aerobic or anaerobic. Let us look first at fermentation, which occurs under anaerobic conditions.

Figure 6.7 Reactions in alcoholic and lactic fermentation shown side by side. The hydrogens (shown in color) show where hydrogen ions attach when electrons are donated by reduced NAD (NADH). Oxidized NAD is symbolized NAD^+.

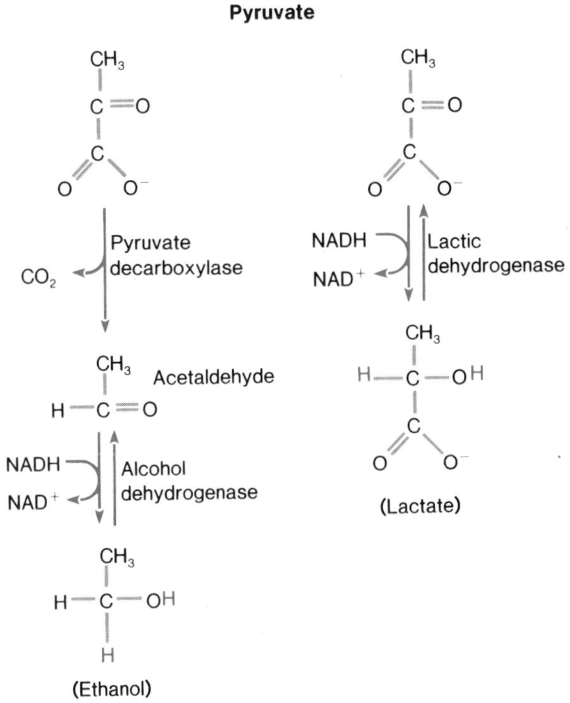

Pyruvate

CH_3–C(=O)–COO^- CH_3–C(=O)–COO^-

CO_2 ← Pyruvate decarboxylase   NADH → NAD^+ Lactic dehydrogenase

CH_3–CH=O Acetaldehyde CH_3–CH(OH)–COO^- (Lactate)

NADH → NAD^+ Alcohol dehydrogenase

CH_3–CH_2–OH (Ethanol)

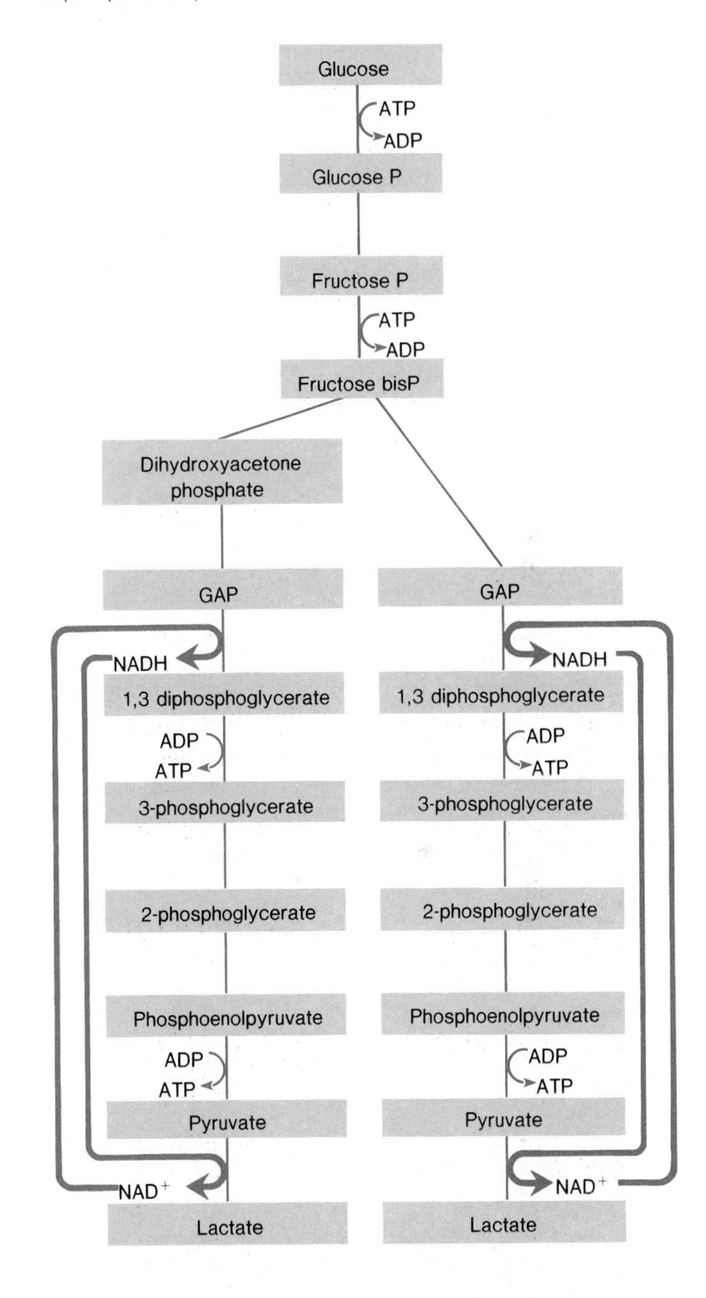

Figure 6.8 Summary of anaerobic metabolism of glucose showing the E-M pathway and lactic fermentation. Parallel reaction series are shown to account for the two C_3 chunks produced when fructose bisphosphate is split in two.

Fermentation

Under anaerobic (oxygen absent) conditions, fermentation converts pyruvic acid into one of several products, depending upon the kind of cells and the enzymes that they possess.

Some cells, such as yeast, convert pyruvic acid to ethanol (ethyl alcohol). Alcoholic fermentation is very important in the brewing and wine-making industries, as well as in baking, where carbon dioxide produced by yeast cells causes bread to "rise."

Other cells convert pyruvic acid to lactic acid. Some bacteria that spoil milk do this. Animal muscle cells, including those in human muscles, also conduct lactic fermentation during vigorous exercise.

In both alcoholic and lactic fermentation (figure 6.7), there is a reaction in which reduced NAD (NADH) donates electrons. These electrons are followed by hydrogen ions (see colored H's on ethanol and lactate molecules in figure 6.7). This is critical in keeping the Embden-Meyerhof pathway reactions going in these cells, because oxidized NAD (NAD^+) must be available or the whole process stops at the E-M reaction where it is required. The quantity of oxidized NAD available in cells is very small, so this constant recycling is necessary (figure 6.8). There must be a place for NAD to get rid of the electrons picked up in the E-M pathway reactions, and the fermentation reactions provide that place.

Aerobic Respiration

A net gain of only two ATPs for each molecule of glucose run through the E-M pathway may not seem to you to be a very large energy gain, and it really is not. Much more ATP can be produced if pyruvic acid molecules are broken down further under aerobic conditions rather than being used in fermentation. This is what happens in aerobic respiration inside mitochondria. Oxygen is used in further oxidation of organic molecules and a great deal more ATP is produced.

All of the E-M pathway reactions occur in the cytoplasmic matrix, but aerobic respiration occurs inside the mitochondria, the tiny "powerhouses" of cells.

Mitochondrial Structure

As you saw in chapter 3, a mitochondrion has a complex membrane system that consists of two parts: a smooth outer membrane and a folded inner membrane (figure 6.9). The folds in the inner membrane are called **cristae.** The **mitochondrial matrix,** the material enclosed by the inner membrane, contains a number of important enzymes.

Electron Transport, Oxygen, and ATP

In aerobic respiration inside mitochondria, pyruvate, the product of the Embden-Meyerhof pathway, is *not* converted to lactic acid or ethanol, as it would be under anaerobic conditions. Instead it is broken down further with a great increase in the ATP yield. This further processing occurs in the reactions of the Krebs cycle.

In the Krebs cycle, there are several reactions in which organic compounds are oxidized and electron-carrying coenzymes such as NAD become reduced. These electron carriers pass their electrons to oxygen, and in the process, energy becomes available for ATP production.

The reduced coenzymes do not react directly with oxygen. Instead, electrons are passed down a series of electron carriers known as the **electron transport system.** In chapter 5 we encountered a similar arrangement of electron carrier molecules in the chloroplast. In mitochondria, many of the electron carriers are membrane-bound proteins containing metal ions that can gain or lose an electron. **Cytochromes,** for example, contain an iron atom. The word *cytochrome* literally means "cell color." The name was applied because of the pronounced pink color of the reduced forms of cytochromes. The electron-transporting activity of the cytochromes depends on the alternate reduction and oxidation of the iron atoms that they contain:

$$Fe^{2+} \underset{\text{Reduction}}{\overset{\text{Oxidation}}{\rightleftharpoons}} Fe^{3+} + e^{-}$$

Another particularly important element in the electron transport system is **ubiquinone.** Ubiquinone is a small molecule of the inner membrane of the mitochondrion. Ubiquinone's oxidation and reduction involve both protons and electrons. The substance illustrates an important fact about the electron transport system: electrons are actually passed through the system not singly but in pairs (figure 6.10). All of our discussion of electron transport from now on will concern the passage of pairs of electrons from carrier to carrier in the system.

Figure 6.9 Structure of a typical mitochondrion. Note inner and outer mitochondrial membranes, cristae, and matrix.

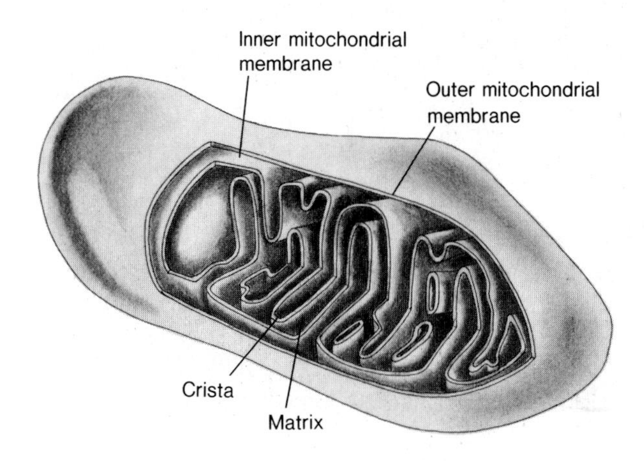

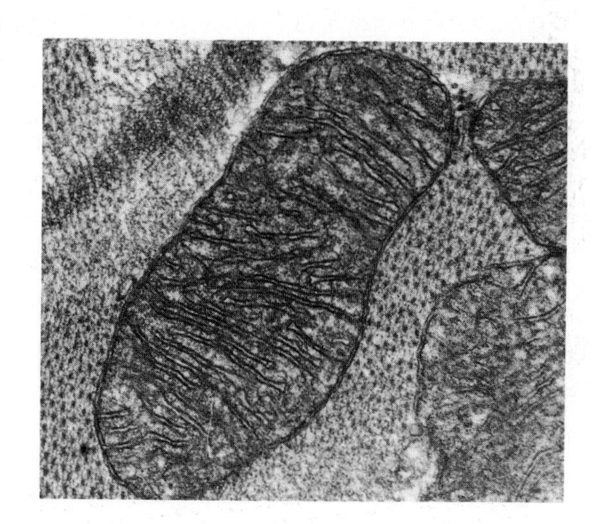

We do not know how the electron carriers of each electron transport system are actually arranged, but we do know that thousands of transport systems are located in the inner membrane of each mitochondrion. We also know quite a bit about the elements of each of these systems. An electron transport system seems to have three complexes, each containing several components, and these complexes are linked by ubiquinone and one of the cytochromes, cytochrome *c* (figure 6.11).

Other electron carriers beside NAD can donate electrons to an electron transport system. Note in figure 6.11 that another electron-carrying coenzyme, **flavin adenine dinucleotide (FAD),** donates electrons to ubiquinone.

At the end of this electron transport system, oxygen serves as the final electron acceptor. Electrons are transferred to oxygen from the third complex, the cytochrome oxidase complex, thus reducing the oxygen atoms. Hydrogen ions join the oxygen atoms along with the electrons, forming what is called metabolic water.

Figure 6.10 The oxidized and reduced forms of ubiquinone.

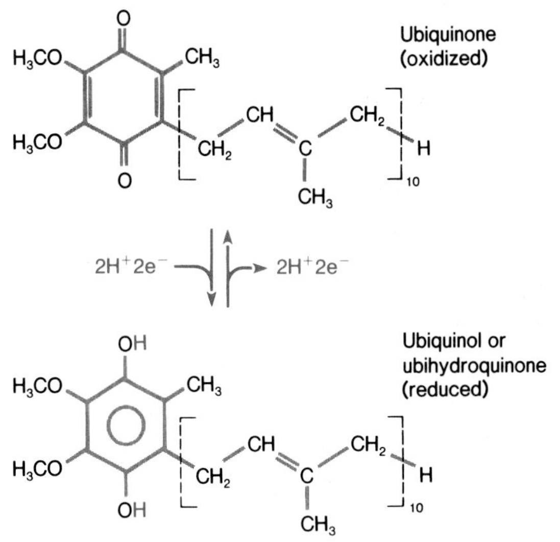

Ubiquinone (oxidized)

$2H^+2e^-$ ⇌ $2H^+2e^-$

Ubiquinol or ubihydroquinone (reduced)

Figure 6.11 The mitochondrial electron transport system through which pairs of electrons are passed to oxygen. There are three complexes joined by ubiquinone and cytochrome *c*. The third complex (cytochrome oxidase) transfers electrons to oxygen. Energy changes in the electron transport system are adequate to permit ATP formation (phosphorylation) at three levels in the system. Thus, three ATPs can be formed for each pair of electrons passed from NADH to oxygen.

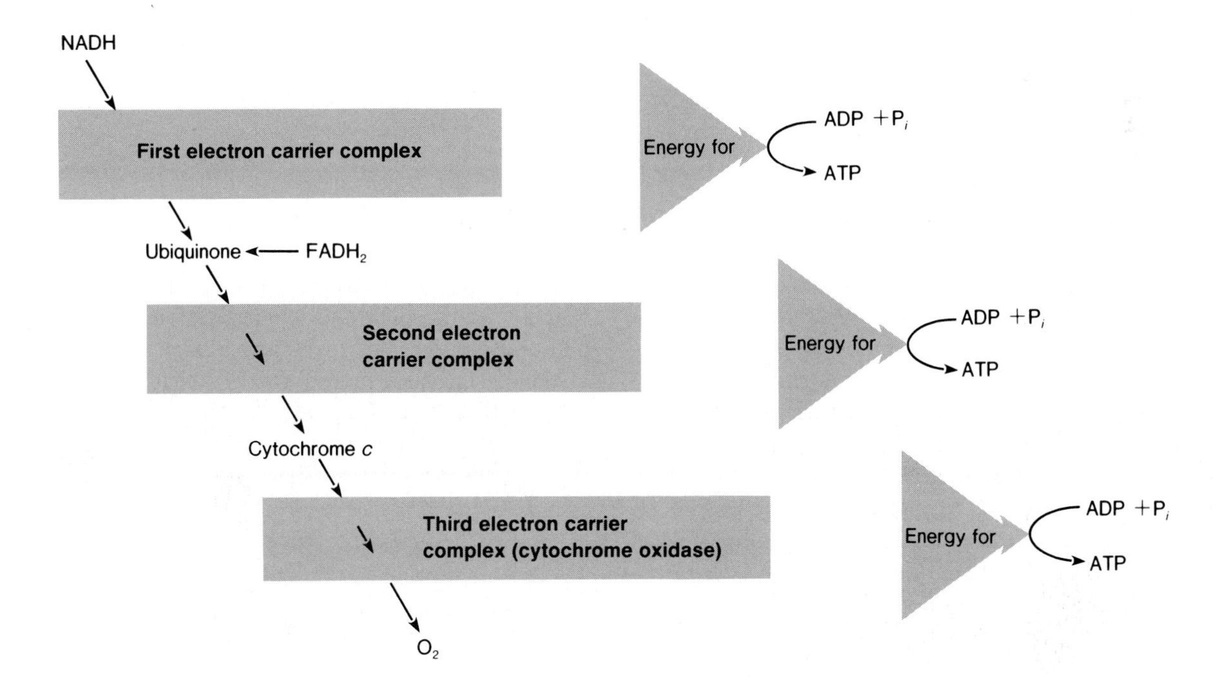

Box 6.1

Respiratory Control and Metabolic Poisons

Do cells keep on using glucose as long as there is any available to them? It only takes a moment's thought to realize that the answer must be "no." The oxidation of nutrients must be correlated with a cell's energy requirements. Therefore, cellular energy conversions must be regulated. In fact, there is a whole network of regulatory interactions that controls the rates at which the Embden-Meyerhof pathway and the Krebs cycle proceed.

One interesting way in which aerobic respiration is regulated is called **respiratory control.** This is a misleading name, but it is in common use among biologists. Respiratory control, which depends on the coupling of reactions, is a form of regulation that depends on the ATP-to-ADP ratio. Normally, electron flow through the electron transport system is tightly coupled to phosphorylation; that is, electron movement from one electron carrier to another will not occur unless ATP is being formed simultaneously. When there is a high ATP-to-ADP ratio, electron transport is slowed because there simply is not enough ADP available for the coupled reactions of electron transport and phosphorylation to proceed. On the other hand, if a great deal of ATP-requiring work is going on, a relatively large supply of ADP will be available for phosphorylation and, consequently, electron flow through the electron transport system will speed up.

However, when the electron transport system is "backed up" because ADP is not available for phosphorylation, reduced NAD and FAD cannot "unload" electrons. As a result, the entire respiratory process is slowed because all reactions requiring these coenzymes as electron acceptors are inhibited. Through this mechanism, then, the rate of ATP formation can be regulated by the rate that ATP is being used for work in the cell (box figure 6.1).

A metabolic poison, **2,4-dinitrophenol (DNP),** uncouples phosphorylation from electron transport. This uncoupling allows the metabolism to "run wild" because there is no control over the rate of electron transport. Of course, ATP production ceases when oxidative phosphorylation is uncoupled from electron transport.

Even when electron transport is coupled with phosphorylation, quite a bit of energy is released in the form of heat. When these processes are uncoupled, the remainder of the energy released in the electron transport system also is dissipated as heat. It is therefore not surprising that elevated body temperature is a major symptom of DNP poisoning.

Another metabolic poison, **cyanide,** affects the electron transport system in a different way by interfering with the cytochrome oxidase complex so that electron transfer to oxygen is inhibited. The cell's inability to reduce oxygen when poisoned by cyanide has widespread effects. Krebs cycle reactions are shut down

It is this passage of electrons through the electron carrier system that provides energy for ATP production, capturing much more of the energy potentially available in organic molecules. Although some of the energy is dissipated as heat, a good deal of it is captured in the form of ATP that is available for cell work.

For every pair of electrons passed through the electron transport system from NADH to oxygen, enough energy becomes available to cause the production of three molecules of ATP from ADP and phosphate (see figure 6.11). This form of phosphorylation, associated with the oxidation-reduction reactions in the electron transport system, is called **oxidative phosphorylation.** This energy conversion is the basic function of the electron transport system in aerobic respiration. The role of oxygen in aerobic respiration is to serve as an acceptor for electrons removed from organic molecules and passed through the electron transport system.

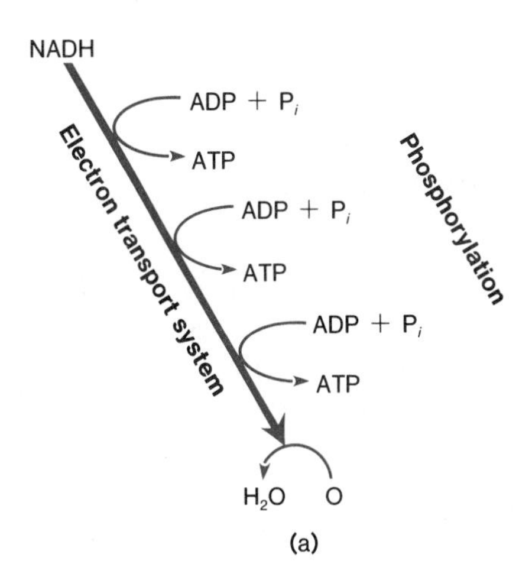

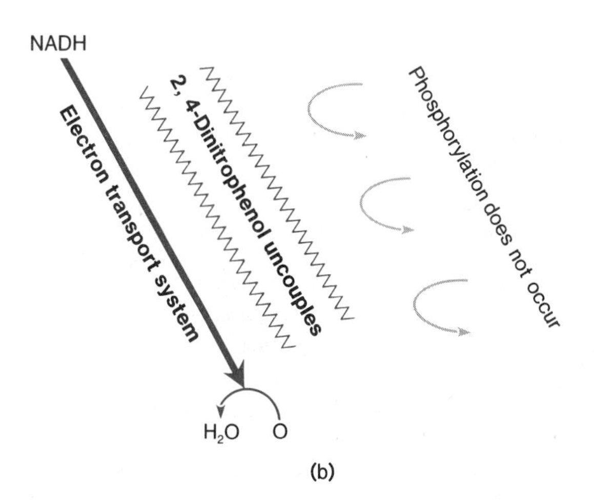

because the Krebs cycle is an aerobic process—it depends on coenzymes (NAD and FAD) that must become oxidized by donating electrons to oxygen through the electron transport system. Oxidative phosphorylation decreases sharply with cyanide poisoning, and no ATP-requiring cell work can be done because the supply of ATP is quickly diminished.

Uncoupling and electron-transport blocking agents are important research tools used to study the mechanics of electron transport systems. Many of the facts known about these important oxidation-reduction reaction series have been learned through experiments employing metabolic poisons.

Box figure 6.1 "Respiratory Control." (*a*) Electron transport and phosphorylation are coupled. ADP availability controls rate of electron flow through the electron transport system. This is a control relationship in which the cell's energy requirements adjust the rate of electron transport. (*b*) Uncoupling agents such as 2,4-dinitrophenol disrupt this control by uncoupling electron transport and phosphorylation. With its "brakes" off, the electron transport system runs at full speed, and energy is released as heat rather than being used for ATP production.

Oxidative Phosphorylation

As is the case with ATP production in photosynthesis, we are not certain exactly how ATP production in mitochondria (oxidative phosphorylation) actually results from the electron transport process. But again, as in photosynthesis, Peter Mitchell's hypothesis concerning movement of hydrogen ions across membranes during electron transport seems the best explanation at this time (see page 102).

Mitchell proposes that as electrons are passed from carrier to carrier in the electron transport system located in the inner mitochondrial membrane, hydrogen ions (H^+) picked up from the mitochondrial matrix are dragged along and are released into the space between the two mitochondrial membranes (figure 6.12). This results in a higher concentration of hydrogen ions in the space than in the matrix. There is a strong tendency for those H^+ ions to move back toward the matrix because of the concentration difference and because of the excess of positive charges in the space between membranes.

Figure 6.12 How reduction and oxidation of an electron carrier (Q) on opposite sides of a membrane could lead to transport of hydrogen ions across the membrane according to Peter Mitchell's hypothesis.

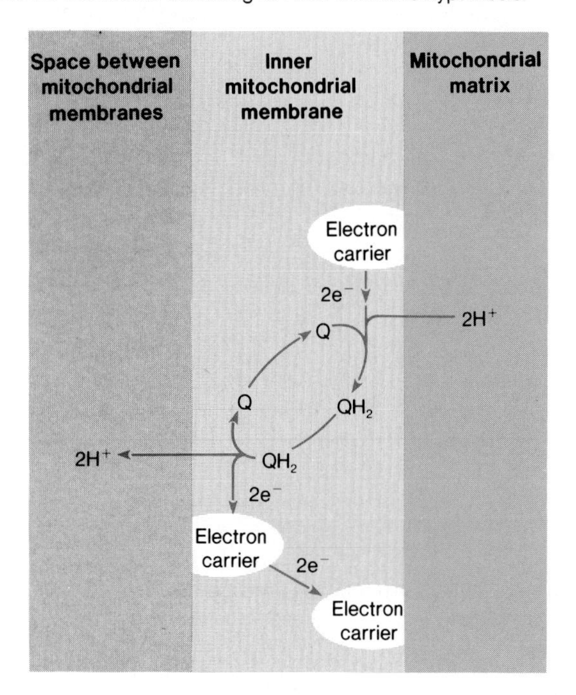

Mitchell further suggests that there are special channels through which the H^+ ions can move in response to the forces acting on them (which are both electrical and concentration differences). In some as yet undetermined way, the movement of protons back through the membrane in this way is coupled to the phosphorylation of ADP to form ATP (figure 6.13).

As important as ATP formation is to all living things, we are not yet certain if this explanation is completely correct. It does fit well with the situation that exists in both chloroplasts and mitochondria where chains of electron carrier molecules are situated in membranes that separate watery spaces. But there are still significant gaps in our knowledge about the actual production of ATP, which is, after all, the key outcome of all of the cellular energy conversions.

The Krebs Cycle

How are electrons supplied to the electron transport system? The Krebs cycle completes the oxidation of carbohydrates and supplies electrons to the electron transport system. The Krebs cycle is a set of reactions, discovered by Hans Krebs and others, catalyzed by enzymes that are located in the mitochondrial matrix.

Figure 6.13 Summary of the mechanism of oxidative phosphorylation in mitochondria according to the Mitchell hypothesis. (1) Electron flow results in H^+ moving across the membrane, leaving an excess of OH^- behind. (2) This results in differences in H^+ concentration and electrical charge on the two sides of the membrane. (3) Hydrogen ions moving back through the membrane somehow provide energy for ATP formation.

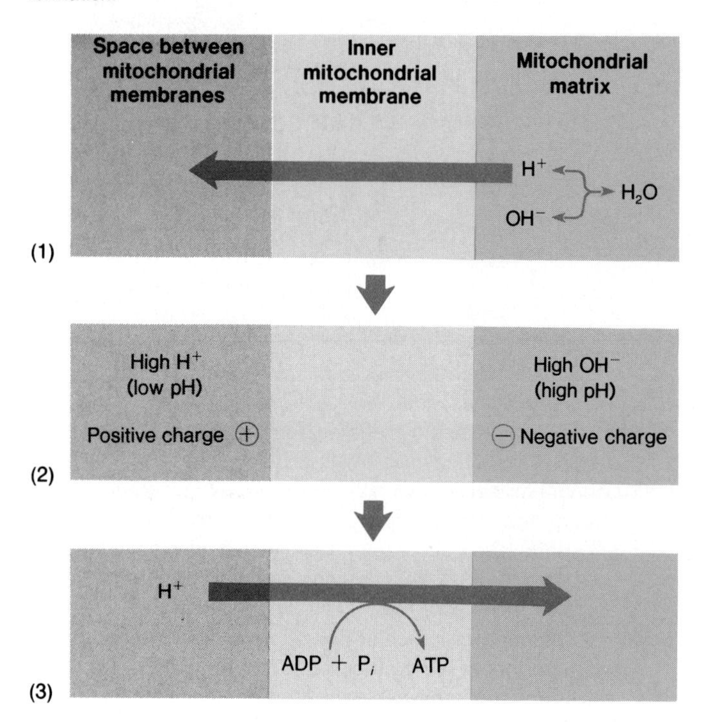

Entering the Krebs Cycle

Pyruvate, which enters the mitochondrion, does not itself enter Krebs cycle reactions. Entry into the Krebs cycle requires preparation of a two-carbon fragment called an **acetyl group.** This preparation actually takes several steps, but we have combined them in figure 6.14. During these reactions, there is a **decarboxylation;** that is, a carbon dioxide molecule is released, thereby shortening the carbon chain by one carbon. Also, a molecule of NAD is reduced. And finally, the two-carbon fragment (acetyl group) that remains after decarboxylation of pyruvic acid is attached to a carrier molecule called **Coenzyme A.** This complex, made up of the two-carbon fragment and Coenzyme A, is called **acetyl CoA** for short and is now in an "active" form, ready to be fed into the Krebs cycle.

Figure 6.14 The preparatory reactions that break down pyruvate and produce acetyl CoA, which enters the Krebs cycle. A decarboxylation results in CO_2 production. Also, NAD becomes reduced.

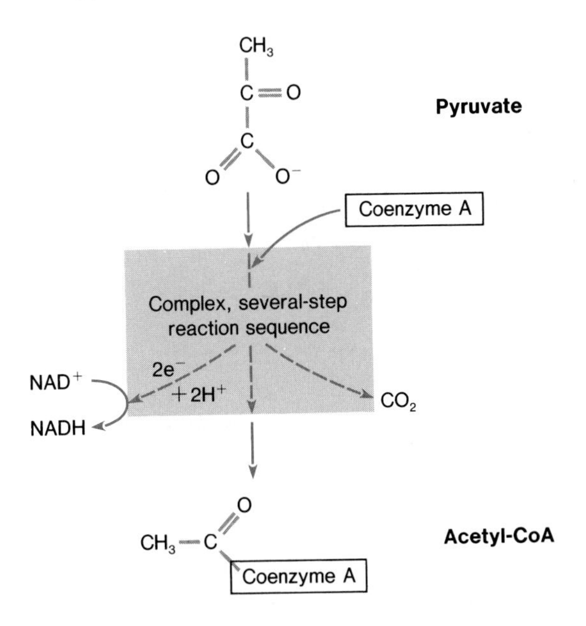

Figure 6.15 "Carbon arithmetic" of the preparatory reaction and the Krebs cycle. The names of organic acids involved in Krebs cycle reactions and the number of carbon atoms contained in each are given. Each time a decarboxylation (CO_2 removal) occurs, the number of carbons in the product is reduced, *e.g.*, from C_6 to C_5. Three CO_2 molecules are produced for each pyruvate molecule processed.

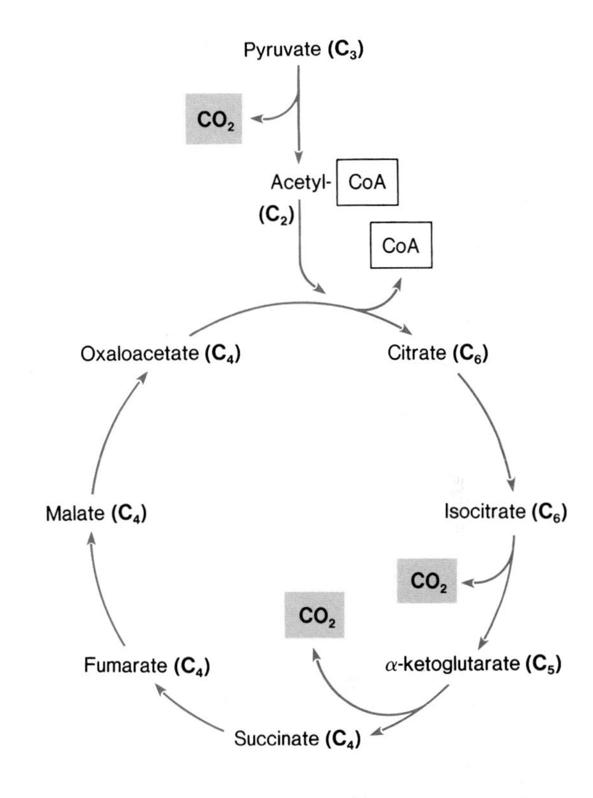

In the Krebs Cycle: Carbon Arithmetic

The Krebs cycle is a cycle because it is a series of reaction events, each catalyzed by its own enzyme, that ends up back at its original starting point. The reactions involve a series of organic acids with different numbers of carbon atoms. We will use the processing of one acetyl group in the Krebs cycle to explain the cyclical nature of the reactions. The Krebs cycle is illustrated in simplified form in figure 6.15.

An acetyl group containing two carbons (C_2) is combined with a C_4 molecule to produce a C_6 molecule. Subsequent reactions of the Krebs cycle involve two decarboxylations that produce two molecules of CO_2 and reduce the length of the carbon chains by two carbons. This brings the cycle back to the starting point, where a four-carbon molecule is available to combine with another acetyl group and to start the whole process again.

Recall that for every molecule of glucose, two molecules of pyruvate are produced. Here we see that for every molecule of pyruvate processed in a mitochondrion, one molecule of CO_2 is produced during the formation of acetyl CoA and two molecules of CO_2 are produced during the ensuing turn of the Krebs cycle. This accounts for all three carbons present in each molecule of pyruvate and for all six carbons in the original glucose molecule ($3CO_2 \times 2 = 6CO_2$). Thus, we can account for all of the carbon dioxide produced during the utilization of each molecule of glucose. But carbon dioxide production is not our main concern here. We are really more interested in how much energy is obtained in the form of ATP as a result of these processes.

In the Krebs Cycle: Energy Conversions

At four separate points in the Krebs cycle, pairs of electrons are transferred to coenzymes and eventually, to the electron transport system. In three cases, the coenzyme serving as initial electron carrier is NAD, and in the other case (the step from succinate to fumarate), the coenzyme FAD is reduced as the initial electron acceptor and passes electrons to ubiquinone (figure 6.16). Refer again to figure 6.11 where you

Figure 6.16 ATP arithmetic of the Krebs cycle. Each star is worth one ATP formed. Each reduced NAD molecule permits three phosphorylations in the electron transport system. A reduced FAD permits two phosphorylations in the electron transport system. A substrate-level phosphorylation forms GTP, which can transfer phosphate to ADP to produce ATP. Thus, one turn of the Krebs cycle produces twelve ATPs, or, starting from pyruvate and including the preparatory steps, fifteen ATPs are formed.

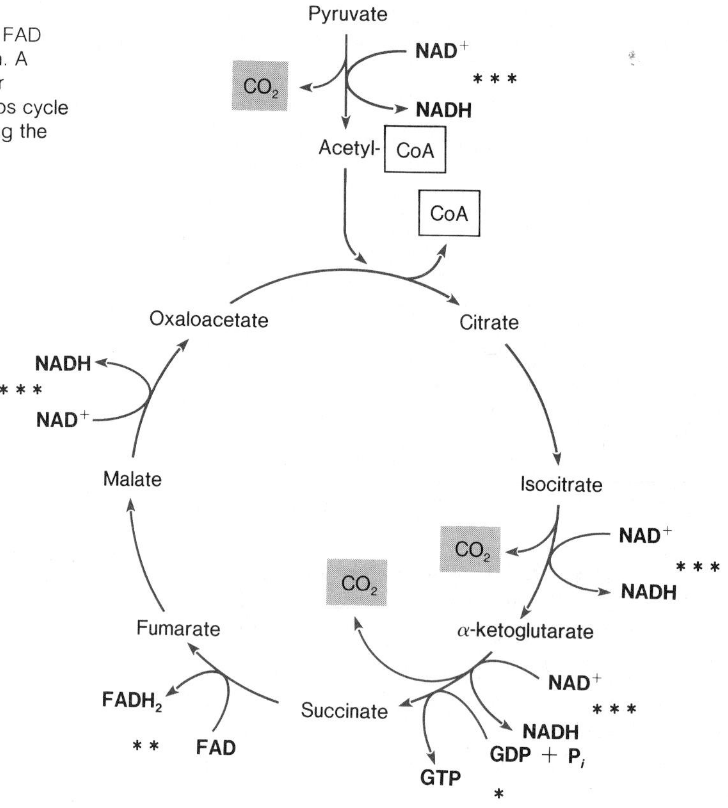

can see that for every pair of electrons passed through the respiratory chain from NADH to oxygen, three molecules of ATP are formed. However, the electron transport resulting from succinate oxidation permits just two phosphorylations because the electrons enter the electron transport system at a "lower" point. Thus, in terms of ATP formation, reduced FAD is not as energy rich as reduced NAD.

In each turn of the Krebs cycle, one substrate-level phosphorylation also occurs. Guanosine diphosphate (GDP) is phosphorylated to produce guanosine triphosphate (GTP), which can then transfer its terminal phosphate to ADP to make ATP. But this substrate-level phosphorylation accounts for only a small part of the energy conversion in the Krebs cycle because the bulk of the ATP synthesis resulting from Krebs cycle reactions is associated with the electron transport system.

Figure 6.16 summarizes ATP formation and the Krebs cycle. When the substrate-level phosphorylation, the phosphorylations occurring as a result of electron transport initiated by reduced coenzymes from the Krebs cycle reactions, and the NAD reduction in the preparatory reactions are taken into account, a total of fifteen molecules of ATP are formed for every molecule of pyruvate entering the reaction sequence inside the mitochondrion.

Overall Energy Sums

Now we can summarize the energy return in ATPs formed from the complete oxidation of a glucose molecule under aerobic conditions. We can also see how aerobic and anaerobic processes compare in terms of efficiency.

Aerobic conditions change the energy story of the Embden-Meyerhof pathway somewhat because the NADH produced during the E-M reactions can transfer electrons to

Table 6.1
ATP Formation during Aerobic Oxidation of Glucose

Embden-Meyerhof Pathway	**ATP Calculations**	**Sums**
ATP expended to phosphorylate sugar molecules	−2	
Substrate-level phosphorylations	4	
ATP formed as a result of electron transport from two NADs reduced during E-M pathway reactions (two ATP each)	4	
Net gain from E-M pathway under aerobic conditions		6
Two Pyruvate to two Acetyl CoA		
Two NAD reduced (three ATP each)	6	6
Krebs Cycle (two turns to process two Acetyl CoA)		
Six NAD reduced (three ATP each)	18	
Two FAD reduced (two ATP each)	4	
Two substrate-level phosphorylations (one ATP each)	2	
Total from Krebs cycle (two turns)		24
Total ATP Formation per Molecule of Glucose		36

the electron transport systems inside mitochondria, thereby salvaging more usable energy than is possible under anaerobic conditions, where the electrons are accepted during the production of organic end products (lactic acid or ethanol). But to do so, electrons must be transferred into the interiors of the mitochondria, where they become available to the electron transport systems that are located in the internal membranes of the mitochondria.

It turns out that reduced NAD cannot itself enter a mitochondrion directly. Thus, it cannot pass its electrons directly to an electron transport system in the inner mitochondrial membrane. Instead the electrons are passed to a shuttle carrier that takes them inside the mitochondrion and then donates them to an electron transport system. Usually the shuttle donates the electrons to ubiquinone (see figure 6.11). Thus, they do not start at the "top" of the electron transport system, and their transport results in only two ATPs being formed.

Nevertheless, being able to get some ATP production from the NAD reduced during the Embden-Meyerhof reactions makes the E-M pathway more efficient in terms of energy production when it operates under aerobic conditions. Under aerobic conditions, the E-M pathway yields a net gain of six ATP produced per glucose molecule processed (figure 6.17): the four direct (substrate-level) phosphorylations, which we accounted for when we discussed the E-M pathway, *plus* four ATPs formed because of the two NAD reductions (one NAD reduction for each three-carbon carbohydrate molecule), *minus* the two ATP molecules used initially for activation. If the ATP output from the E-M pathway is added to the ATP yield from preparing and feeding pyruvate products into the Krebs cycle, a total of thirty-six molecules of ATP is formed for each molecule of glucose oxidized completely to CO_2 and water under aerobic conditions (table 6.1).

Figure 6.17 ATP production from the Embden-Meyerhof pathway when it is associated with aerobic respiration.

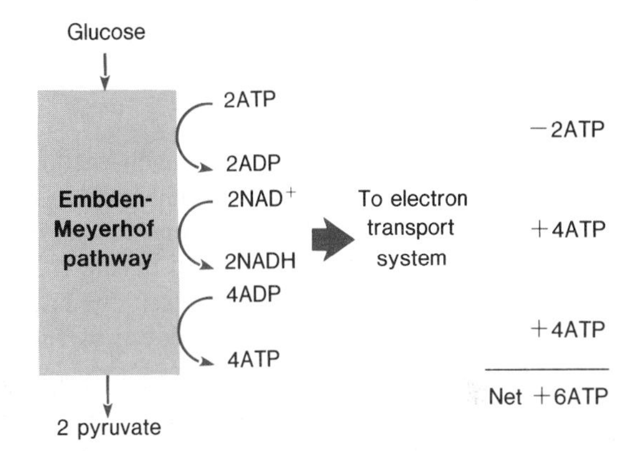

Since anaerobic processes yield a net gain of only two molecules of ATP per molecule of glucose, the aerobic process yields a net gain of eighteen times as much usable energy as did strictly anaerobic processes. Thus, aerobic respiration clearly is a much more efficient way to use glucose to obtain usable energy in the form of ATP.

As a result of stepwise breakdown involving a series of oxidations, cells derive a considerable amount of ATP from processing organic nutrient molecules. It is estimated that

Box 6.2

Oxygen Debt

Energy for the Long Run

At the beginning of this chapter, we asked several questions about runners during and after a hard race. By now you have a pretty good answer for several of those questions. You know that energy is supplied during the race by oxidations of organic nutrients in processes that result in ATP production. You also know that oxygen is used as an electron acceptor that receives the electrons removed from various organic molecules. The carbon dioxide that runners breathe out is produced in decarboxylation reactions that remove CO_2 from organic molecules.

But one key question remains unanswered. Why does heavy breathing continue for some time after the race is over? The answer is that the runners have acquired an **oxygen debt.** But what is an oxygen debt?

Efficient as the aerobic respiratory processes are in producing ATP, they cannot produce ATP fast enough to supply working muscles during heavy exercise. Muscle cells use oxygen as fast as it is supplied to them, but it simply is not enough; they must supplement aerobic respiration in order to produce enough ATP.

During heavy exercise, muscle cells not only use aerobic respiration at maximum possible rates, but they also use additional glucose to carry out anaerobic lactic acid fermentation (see page 117). This allows them to produce extra ATP. Fermentation does, however, lead to a buildup of lactic acid in muscle cells and in the blood, and this lactic acid must be processed after exercise is completed.

Some of the lactic acid is exported to the liver for processing and some is converted back into pyruvic acid when an adequate supply of oxygen again becomes available after the muscle has returned to rest. Both liver and muscle processing of the accumulated lactic acid require oxygen and thus there is indeed an oxygen debt that must be repaid after heavy muscle work ends.

about 40 percent of the chemical energy available in a carbohydrate such as the sugar glucose is converted to energy stored in ATP. Most of the remaining 60 percent of the energy available in the nutrients is released during the reactions as heat. We should not think of this heat as completely lost or wasted energy, however. Heat produced by these reactions affects the body temperature of organisms in which they occur.

A 40 percent efficiency of energy conversion actually sounds pretty good when compared with energy conversions in other places. For example, automobile engines are estimated to be only about 25 percent efficient in using chemical energy in fuel. The remaining 75 percent is lost as heat. Thus, cellular energy conversions seem to be quite efficient processes.

Oxidation of Other Nutrients

So far we have concentrated on the oxidation of carbohydrates using the sugar glucose as our primary example. Other organic compounds can be oxidized in the pathways that we have described. But in each case, there must be preparatory steps to produce molecules that can enter the pathways. We will now consider use of two other important classes of nutrients: fats and proteins.

The Oxidation of Fats

It takes several steps to prepare a common fat (**triglyceride**) for oxidation. First, an enzyme known as a **lipase** catalyzes the separation of the triglyceride molecule into one molecule of **glycerol** and three **fatty acid** molecules (figure 6.18).

In most of the triglycerides (fats) found in animals, these fatty acids have even-numbered chains of carbon atoms (that is, they are likely to be made up of chains of C_{16} or C_{18}, or some other even number, rather than C_{15} or C_{17}, or some other odd number). Keep this even-number rule in mind because you will soon see that it is important in the carbon arithmetic of fatty acid oxidation.

The glycerol derived from fat breakdown can enter the Embden-Meyerhof pathway after it is used to produce dihydroxyacetone phosphate (DHAP) (figure 6.19). Fatty acids are prepared for use through a process that cleaves their carbon chains into two-carbon fragments in a repeating series of reactions (figure 6.20). These two-carbon fragments end up as acetyl CoA, which then may enter the Krebs cycle.

This two carbon at a time utilization process explains why fatty acids in animal fats have even-numbered carbon chains.

During the process of chopping a fatty acid up into two-carbon chunks, FAD and NAD are reduced. Each acetyl CoA formed can contribute a two-carbon fragment to the Krebs cycle where still more energy is made available for ATP production.

The ability to acquire an oxygen debt is not a life and death matter for racing humans, but it certainly is for many animals. A rabbit fleeing from a fox, for example, would not survive long if its muscles could work only to the extent permitted by aerobic metabolism, which depends on the amount of oxygen delivered to the muscles.

It is possible to improve the efficiency with which the circulatory system delivers oxygen and nutrients to muscles. In fact, this is one of the main goals of physical conditioning and athletic training. But still, the ability to accumulate an oxygen debt in the form of a temporary lactic acid buildup is critically important for normal muscle functioning during heavy work or vigorous exercise.

Box figure 6.2 Runners at the end of a race breathe heavily as they repay oxygen debts acquired during the race.

Figure 6.18 Separation of a triglyceride (fat) into glycerol and three fatty acid molecules. This reaction catalyzed by the enzyme lipase is a hydrolysis because components of water molecules are added during the reaction. The "R" in each case represents an even-numbered chain of carbons.

$$3\,(R{-}CH_2{-}C(=O){-}O{-})\,C_3H_5 + 3H_2O \xrightarrow{\text{Lipase}} 3\,R{-}CH_2{-}C(=O)OH + C_3H_5(OH)_3$$

Triglyceride molecule + $3H_2O$ → (Lipase) **3 molecules of fatty acid** + **Glycerol**

Figure 6.19 The reactions by which glycerol enters oxidative pathways. DHAP enters the reactions of the Embden-Meyerhof pathway and is processed there.

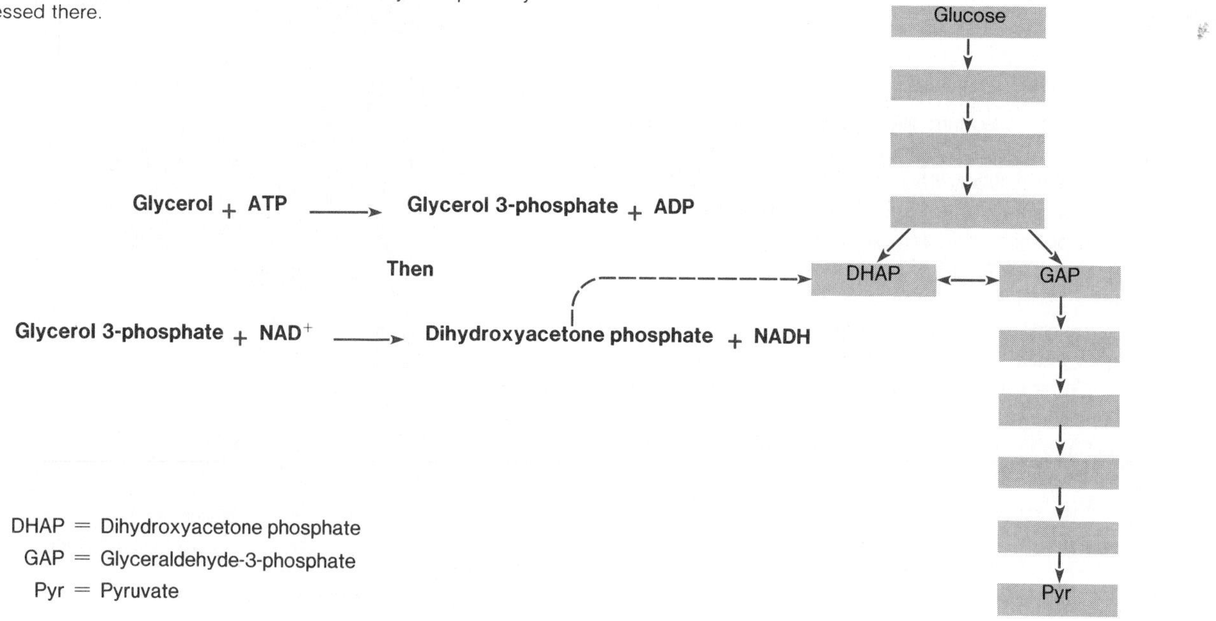

Glycerol + ATP ⟶ Glycerol 3-phosphate + ADP

Then

Glycerol 3-phosphate + NAD^+ ⟶ Dihydroxyacetone phosphate + NADH

Glucose → → → → DHAP ⟷ GAP → → → → → Pyr

DHAP = Dihydroxyacetone phosphate
GAP = Glyceraldehyde-3-phosphate
Pyr = Pyruvate

Fats, Calories, and Energy

It is clear that a great deal of energy becomes available when fats are oxidized. Because of the relatively large amount of energy that can be derived from their oxidation, fats are essential storage compounds.

Why are fats so well suited for use in energy storage? Fats are more highly reduced compounds than carbohydrates; that is, they contain more removable electrons per unit weight than carbohydrates. Thus, fats have the potential for greater energy conversion. In practical terms, this means that the oxidation of fats yields more energy per unit weight than the oxidation of carbohydrates.

In discussing energy conversion and comparing energy contents of various nutrients, the most commonly used energy unit is the **calorie.** A calorie is the amount of heat energy required to raise the temperature of one gram of water 1 °C.* Sometimes this unit is called a **gram calorie** or **small calorie** because another unit, called a **large calorie** or **kilocalorie,** also is used (1 kcal = 1,000 gram calories). Kilocalories are familiar in everyday life because these are the "Calories" that we count in diets.

How do fats and carbohydrates differ in calorie content? General average values for energy yields from oxidation are about 9 kcal per gram for fats and about 4 kcal per gram for carbohydrates. Obviously, fats are much more energy rich compounds than carbohydrates. In terms of nutrition, fats contain more than twice as many calories per gram than carbohydrates.

It is not surprising that except for some short-term storage of small amounts of carbohydrates, most animals depend more on fats than on carbohydrates for energy storage. Plants, on the other hand, use carbohydrates (especially starch) as their main storage compounds. Interestingly, some animals such as clams, which live a relatively sedentary life, do store considerable amounts of carbohydrates. Obviously, such animals do not have to carry the extra weight and can afford to store extra fuel in a bulkier form.

*Although the calorie is the traditional energy unit, you should be aware that another energy unit is coming into use. Modern international scientific usage is rapidly converting to the **joule** as the basic measure of energy because joules are much easier to convert to work measurements. One gram calorie of heat energy is equal to 4.186 joules. Because of the everyday familiarity of kilocalories, however, we will continue to use calories in our discussion.

Figure 6.20 The steps by which a fatty acid, such as palmitic acid, is cleaved into two-carbon chunks to make acetyl CoA. Each acetyl CoA can then be fed into the Krebs cycle for further oxidation.

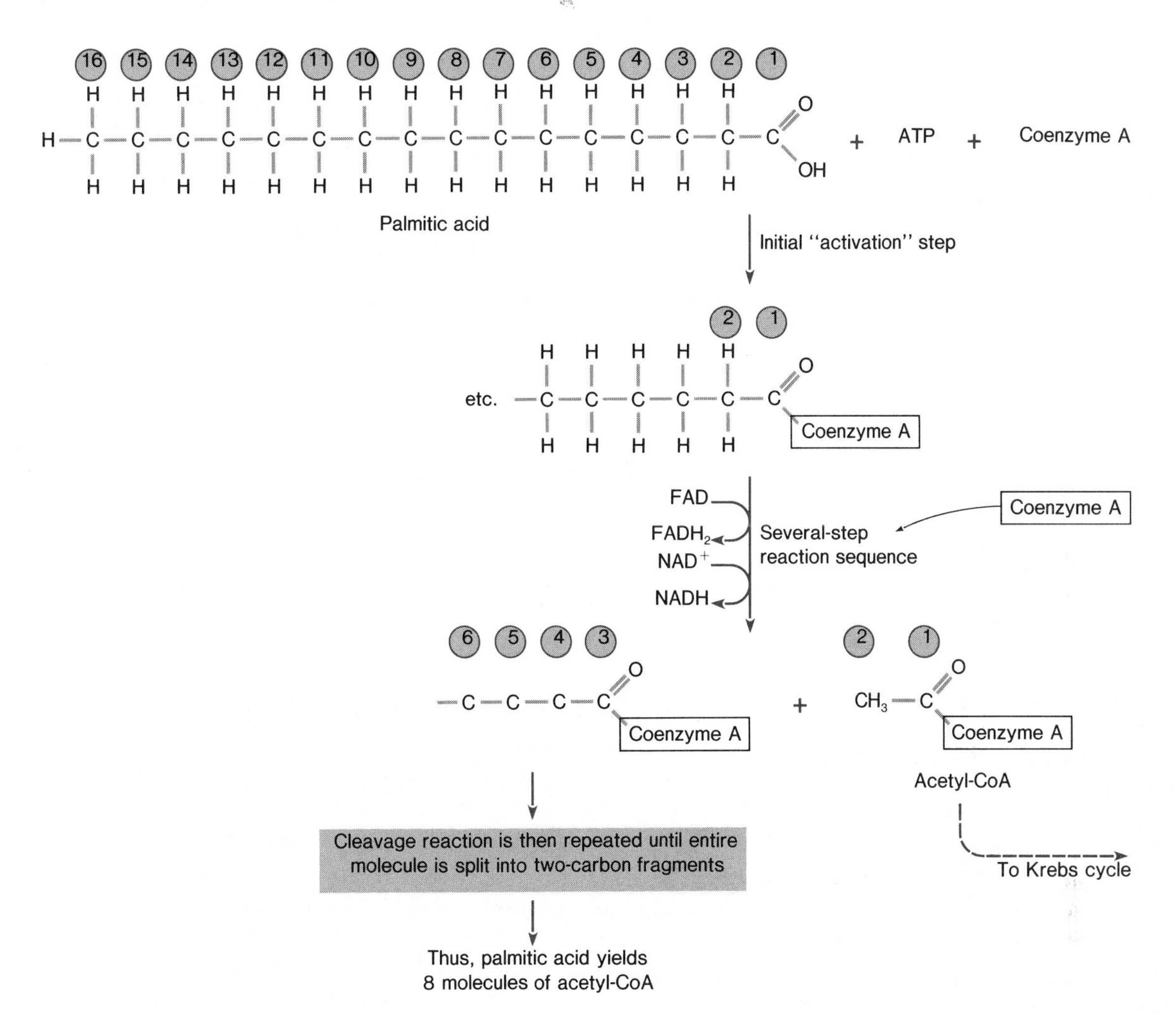

The Metabolism of Proteins

Sometimes when a fad diet is described in a magazine or newspaper, it sounds almost as if proteins contribute no calories in the diet. Some of these diets suggest that you can consume virtually unlimited quantities of proteins without gaining weight so long as you restrict your fat and carbohydrate intake. This is simply not correct.

Proteins do contribute calories in the diet because proteins can be used for energy-yielding oxidations. In the digestive tract, proteins are digested to their component amino acid molecules. Some of these amino acids are used for synthesis of new proteins in cells, but some are left over. Extra amino acids are not stored, but rather are prepared for oxidation. There is also a normal continuing protein replacement process in cells in which some protein molecules are broken down and replaced by new molecules. Some of the amino acids from this process also become available for oxidation. Finally, under starvation conditions, cells break down some of their proteins for use in vital energy-yielding processes after they run out of carbohydrates and fats.

Figure 6.21 Three amino acids with corresponding keto acids. Colored boxes highlight atoms that differ in the amino and keto acid in each case. Note that these particular keto acids can all be fed into the Krebs cycle (see figures 6.15 and 6.16). Other keto acids produced by deamination can be fed into other metabolic pathways.

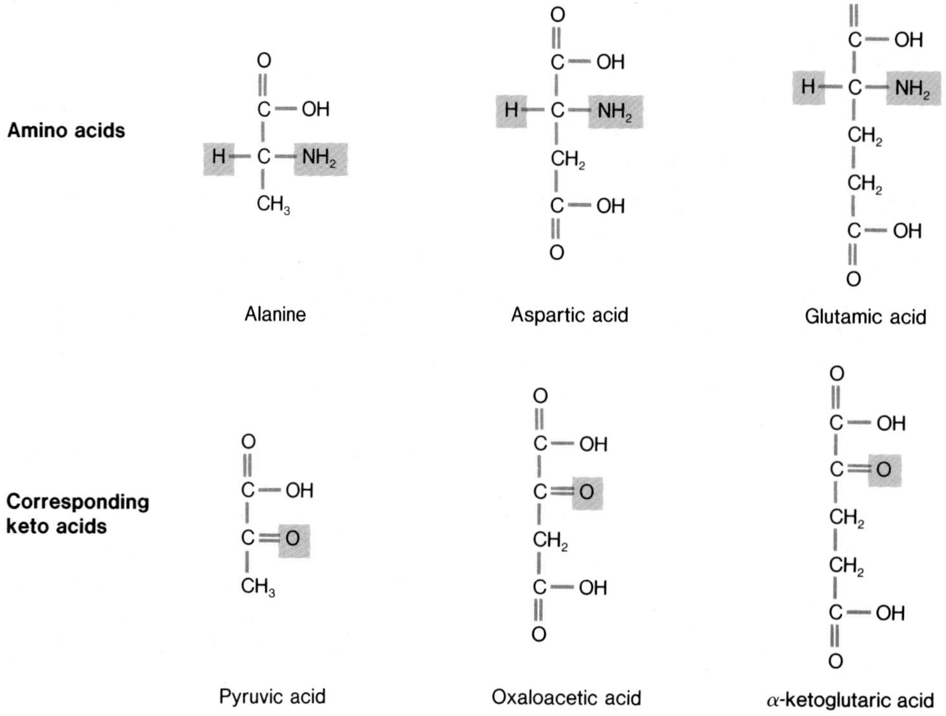

Amino acids

Alanine

Aspartic acid

Glutamic acid

Corresponding keto acids

Pyruvic acid

Oxaloacetic acid

α-ketoglutaric acid

Figure 6.22 An example of a transamination reaction in which an amino group is exchanged for an oxygen atom. A transaminase is an enzyme that catalyzes this exchange, and each different enzyme is named for the specific exchange that it catalyzes.

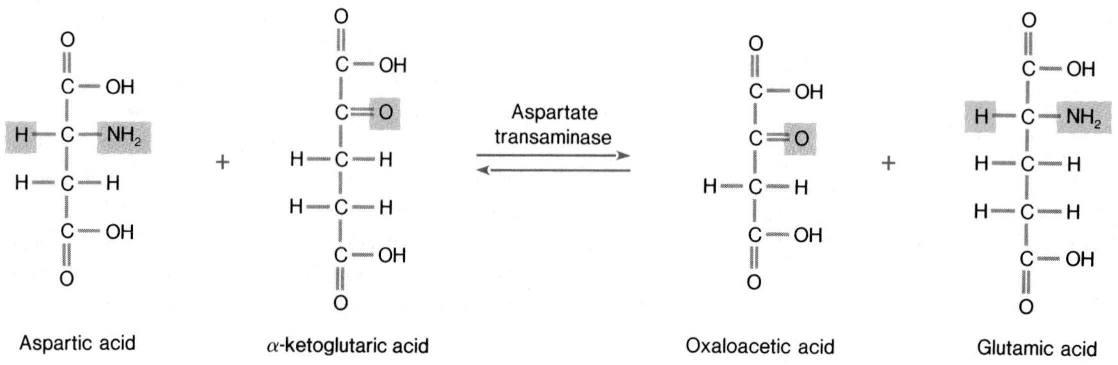

Aspartic acid + α-ketoglutaric acid ⇌ (Aspartate transaminase) Oxaloacetic acid + Glutamic acid

Figure 6.23 Deamination of glutamic acid (glutamate). In this reaction, catalyzed by the enzyme glutamate dehydrogenase, glutamate is deaminated to produce α-ketoglutarate, and ammonia (NH_3) is released.

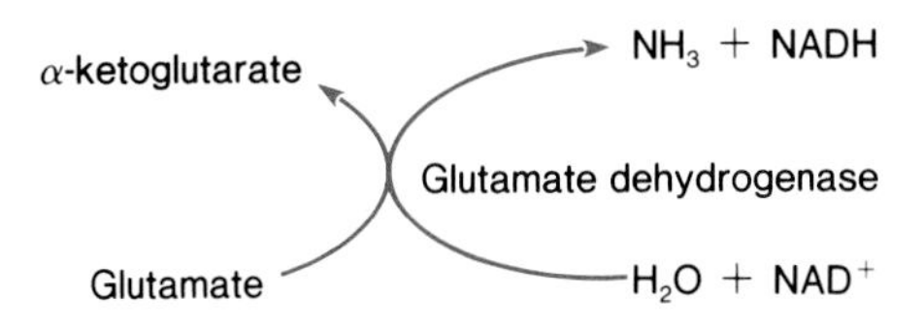

Figure 6.24 The "glutamate funnel." Other amino acids trade their amino groups to glutamate, from which the amino group is removed.

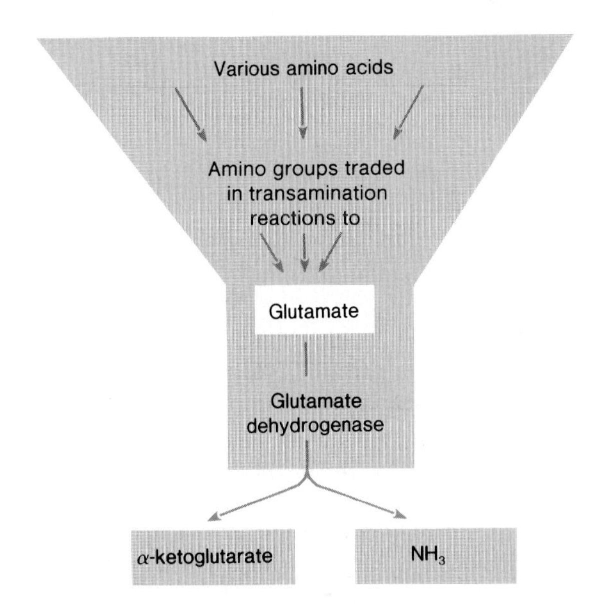

Thus, there are several ways in which amino acids become available for oxidation in metabolic pathways. How then are they utilized?

An amino acid is prepared for oxidation in the Krebs cycle or in other oxidative reactions by the removal of its amino group in a **deamination** reaction. In deamination reactions, amino groups are replaced by oxygen atoms; that is, amino acids are converted to keto acids (figure 6.21).

Most amino acids get rid of their amino groups by trading them to other molecules for oxygen atoms. This transferring of amino groups is known as **transamination.** One example of such a transamination reaction is shown in figure 6.22. These trades can be made between a number of combinations of molecules.

Figure 6.25 Urea, a nitrogenous (nitrogen-containing) waste carrier molecule. Ammonia produced by deamination is incorporated into urea, which is then eliminated from the body.

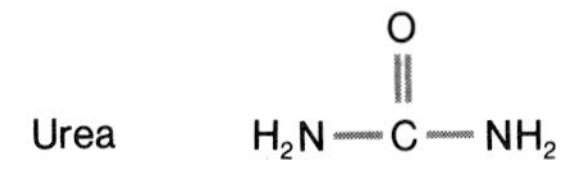

Another type of deamination is one in which the amino group is not traded off but removed directly. In animals, however, only one amino acid is deaminated in this way (figure 6.23). Glutamic acid (called glutamate in its ionized form in cells) is deaminated by a reaction in which the products are α-ketoglutaric acid and ammonia (NH_3).

This reaction is the key to using all amino acids for energy because other amino acids pass their amino groups to glutamate by transaminations. The amino groups are then removed from glutamate molecules. Because of this arrangement, amino acids are said to be "poured through the glutamate funnel" on their way to oxidative pathways (figure 6.24).

What happens to the ammonia produced during deamination of glutamate molecules? This ammonia is incorporated into waste carrier molecules such as **urea** (figure 6.25), which the body must then eliminate. In humans, glutamate deamination and urea production occur mainly inside mitochondria in liver cells.

Nutrition and the Metabolic Pool

We have seen how various substances such as fats and proteins are processed to produce compounds that can be oxidized by Embden-Meyerhof pathway and Krebs cycle reactions. This is not simply a one-way flow, however. Various products of these reactions can also be withdrawn for use in synthesis. Thus, we think of the products of the E-M pathway and Krebs cycle reactions as being part of a **metabolic pool** (figure 6.26). Material can be poured into or withdrawn from this metabolic pool.

Some examples may help to clarify this idea. At times when more than adequate supplies of organic nutrients are present in the body, fat synthesis begins. Glycerol is produced for fat synthesis from DHAP. Excess acetyl CoA makes two-carbon fragments available for incorporation into the even-numbered carbon chains of fatty acids. The components of fat molecules, glycerol and fatty acids, are thus produced from material withdrawn from the metabolic pathways.

Figure 6.26 The metabolic pool. The metabolic pathways function as an open system. Various compounds enter at different points, for oxidation, and certain products can be used for synthesis. Reactions involved in synthesis usually are not exactly the reverse of reactions preparing compounds for entry into oxidative pathways because different enzyme systems are involved.

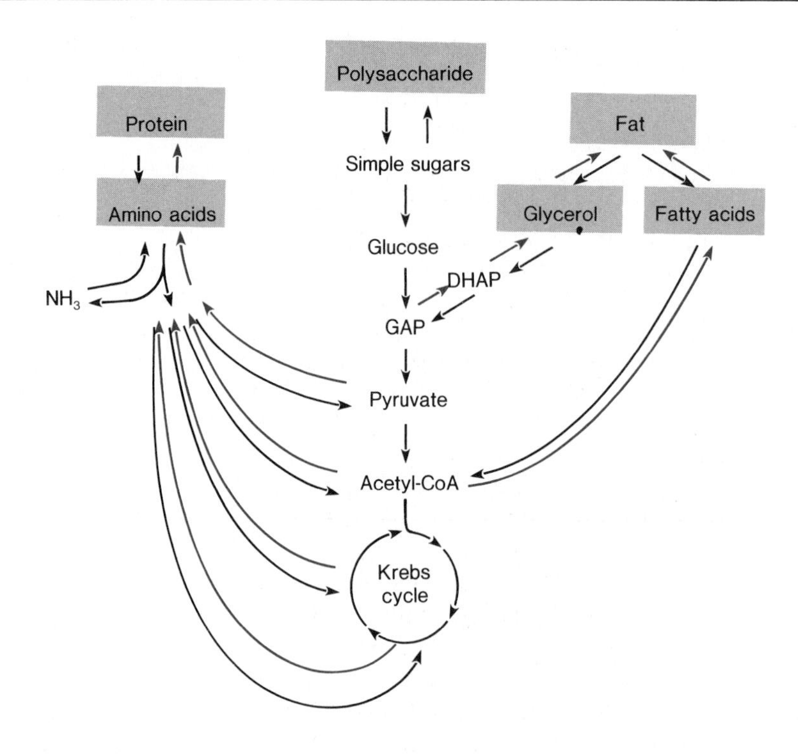

A second example of the metabolic pool concept is related to the need for amino acids to be used in protein synthesis. In animals, a large proportion of these amino acids are normally provided by the digestion of dietary proteins. But in plants, and in part in animals, intermediate products from the oxidative pathways are used for amino acid synthesis. Molecules withdrawn from the metabolic pool are used in **amination** reactions, in which the addition of an amino group produces amino acids. Amination reactions are essentially the opposite of the deamination reactions that occur when amino acids are being prepared for oxidation.

Dietary Requirements

Plant cells must be able to synthesize all of the kinds of amino acid molecules that they need for protein synthesis because they have no external source of amino acids. But in the cells of many animals, the ability to synthesize amino acids is limited; that is, these organisms lack the enzyme systems needed to synthesize certain amino acids. Those particular amino acids must be obtained from outside sources. Thus, in animals, there are certain **essential amino acids** that the animal cannot synthesize and that are therefore absolute dietary requirements; **nonessential amino acids** can be synthesized. Furthermore, because extra amino acids are not stored in the body, an animal must continually obtain essential amino acids in its diet.

Humans (and, incidentally, white rats, which are often used in nutrition research) can synthesize ten of the amino acids found in protein molecules but must obtain the other ten (the essential amino acids) through diet (see table 6.2). It is possible to eat large quantities of proteins and still suffer from deficiency symptoms if those proteins do not contain adequate quantities of the essential amino acids.

This sometimes happens to people who do not eat any animal protein (vegetarians). A number of plant proteins contain low quantities of one or more of the essential amino acids. For example, protein from corn (maize) contains very little lysine but does contain enough tryptophan, while protein from beans contains very little tryptophan but adequate lysine. This means that a person on a vegetarian diet would need to eat both corn and beans to obtain a full supply of essential amino acids. Interestingly, the two foods must be eaten at the same time in order for the diet to be adequate. It does not work well to eat corn at one meal and beans at the next because animals, including humans, simply do not store significant amounts of amino acids for future use.

Having introduced the idea of essential amino acids, we should mention that there is a similar problem associated with fat synthesis. There are certain **essential fatty acids** that cannot be synthesized and must be included in the diet. For example, there are two fatty acids, linoleic acid and linolenic acid, that must be present in human diets. Fortunately, these two fatty acids are abundant in many plant foods and in fish and poultry products, so people rarely have deficiency problems.

Table 6.2
Human Dietary Requirements Among the Twenty Amino Acids That Occur in Proteins

Nonessential	Essential
Alanine	Arginine
Asparagine	Histidine
Aspartate	Isoleucine
Cysteine	Leucine
Glutamate	Lysine
Glutamine	Methionine
Glycine	Phenylalanine
Proline	Threonine
Serine	Tryptophan
Tyrosine	Valine

The reactions of the E-M pathway and the Krebs cycle do more than provide energy for ATP production. They also constitute a metabolic pool that supplies raw materials for synthesis of many important constituents of living things. Since animals have somewhat limited synthetic abilities and cannot synthesize certain vital organic compounds, essential amino acids, essential fatty acids, and vitamins, all must be obtained from outside the metabolic pool, through a proper and complete diet.

Summary

Reduced organic compounds such as glucose, originally produced by photosynthetic activity, are oxidized in a stepwise set of reactions. These cellular oxidation-reduction reactions result in ATP production.

The first series of reactions in the oxidative breakdown of glucose is the Embden-Meyerhof pathway. The E-M pathway reactions occur in the cytoplasm. Phosphorylations at the beginning of the Embden-Meyerhof pathway require two ATP molecules per molecule of glucose. After some molecular reorganizations, a six-carbon molecule is split into two three-carbon molecules. These molecules are processed further to produce pyruvate. Pyruvate stands at an important branchpoint in carbohydrate metabolism.

Under anaerobic conditions, pyruvate may undergo either alcoholic or lactic fermentation. In the fermentation reactions, reduced NAD donates electrons to organic molecules, thus renewing the supply of oxidized NAD for Embden-Meyerhof reactions.

The net gain in ATP production produced through the E-M reactions and fermentation is only two ATP molecules per molecule of glucose oxidized. Although four ATPs are produced by substrate-level phosphorylations, two ATPs are spent in activating reactions.

Much more ATP can be produced if pyruvate is oxidized in aerobic respiration. Respiration, which occurs inside mitochondria, releases much more of the potential energy of carbohydrates. Reactions of the Krebs cycle are a key part of aerobic respiration.

Krebs cycle reactions occur in the mitochondrial matrix. Electron transport and oxidative phosphorylation are associated with the inner mitochondrial membrane. The mechanism of oxidative phosphorylation is not fully understood. However, it seems likely that electron transport produces a hydrogen ion gradient across the inner mitochondrial membrane and that the movement of hydrogen ions back through the membrane results in ATP production.

Coenzymes, reduced during Krebs cycle reactions, donate electrons to an electron transport system. During the transfers, energy changes make possible phosphorylations of ADP to produce ATP.

In addition to oxidation-reduction reactions, Krebs cycle decarboxylations produce carbon dioxide and shorten the carbon chains of carbohydrate molecules. Overall, six CO_2 molecules are produced for each glucose molecule entering the oxidative pathways.

As a result of all the phosphorylations occurring in conjunction with the electron transport system and the one substrate-level phosphorylation in the Krebs cycle, a total of fifteen ATP molecules are produced per molecule of pyruvate. Embden-Meyerhof reactions, under aerobic conditions, result in a net gain of six ATP molecules. Thus, overall, a total of thirty-six ATP molecules are gained from the aerobic oxidation of one glucose molecule.

Other organic compounds can be prepared for oxidation in the same reactions that oxidize carbohydrates. Glycerol and fatty acids from fats enter oxidative pathways as dihydroxyacetone phosphate and acetyl CoA, respectively. Fats are more highly reduced compounds than carbohydrates and thus yield more calories per gram oxidized.

Proteins are hydrolyzed into amino acids, which are then deaminated prior to entry into the oxidative pathways. Excess amino groups produced during amino acid deamination are removed from the body in nitrogenous waste carrier molecules such as urea.

Various molecules can be withdrawn from the metabolic pool for synthesis. Embden-Meyerhof and Krebs cycle intermediates are used as carbon skeletons in fat and amino acid syntheses. Not all required compounds can be synthesized, however, and therefore certain organic compounds are essential dietary requirements.

Questions for Review

1. Why is carbon dioxide a product of alcoholic fermentation and not of lactic fermentation?
2. Define the term decarboxylation.
3. What are oxidation and reduction?
4. Define the term metabolic water and explain how metabolic water is produced.
5. Explain how six molecules of CO_2 are produced for each molecule of glucose oxidized through the Embden-Meyerhof pathway and the Krebs cycle.
6. Why do the Embden-Meyerhof pathway reactions yield more ATP when they are associated with aerobic respiration than when associated with fermentation?
7. Explain what we mean by the term metabolic pool.
8. How is urea production related to protein metabolism?
9. Why is fat better than starch as a means of storing chemical energy for most animals?
10. Most fatty acids in animal fats have even numbers of carbon atoms, for example, C_{16} or C_{18}. How does this relate to the ways in which fatty acids are oxidized and synthesized?
11. Why are some amino acids essential for human diets while others are nonessential?

Questions for Analysis and Discussion

1. The human body temperature becomes elevated during extended periods of heavy exercise. You should be able to explain at least one of the causes of this temperature elevation.
2. How many molecules of metabolic water are produced as a result of coenzyme reductions and electron transport activity during the complete processing of one molecule of pyruvate in aerobic respiration?
3. Could you explain to a friend who wants to become a strict vegetarian what protein nutrition problems such a diet can produce? Could you explain how strict vegetarians might solve those problems?

Suggested Readings

Books

Baker, J. J. W., and Allen, G. E. 1981. *Matter, energy, and life.* 4th ed. Reading, Mass.: Addison-Wesley.

Lehninger, A. L. 1982. *Principles of biochemistry.* New York: Worth Publishers.

Articles

Chappell, J. B. 1979. The energetics of mitochondria. *Carolina Biology Readers* no. 19. Burlington, N.C.: Carolina Biological Supply Co.

Hinkle, P. C., and McCarty, R. E. March 1978. How cells make ATP. *Scientific American* (offprint 1383).

Nicholls, P. 1984. Cytochromes and biological oxidation. *Carolina Biology Readers* no. 66. Burlington, N.C.: Carolina Biological Supply Co.

Function and Structure of Organisms

Part 2

To maintain stable internal conditions (homeostasis), living things must selectively exchange materials with their environments. In single-celled organisms, the individual cell's plasma membrane is the boundary between the organism and the outside world. But in multicellular organisms, specialized body surfaces function as exchange boundaries between organisms and the environment. In addition, a variety of transport mechanisms move materials within organisms, further contributing to maintenance of homeostasis.

Land plants obtain water and mineral nutrients from the soil. They transport these materials through vascular systems to the leaves, where photosynthesis occurs. The products of photosynthesis are then transported to other parts of plants.

Animals obtain reduced organic compounds and other materials from their environments. They, in turn, excrete waste materials into the environment. These exchanges occur at specialized exchange boundaries that are connected with other body areas by circulatory systems.

In chapters 7 through 10 we will look at the specialized exchanges, transport mechanisms, and internal processing of materials involved in maintenance of the stable internal environments that are essential to the lives of plants and animals.

Facing page Aloe in bloom at Ngorongoro Crater, Tanzania.

Nutrition and Transport in Plants

7

Chapter Outline

Chapter Concepts

1. A vascular plant's body is specialized to obtain and transport materials.
2. Open stomata in leaves permit carbon dioxide diffusion into the leaves. At the same time there is transpiration (evaporative water loss from the leaves).
3. A stream of water continuously moves up from roots through the xylem of the vascular plant, replacing water lost by transpiration and carrying mineral ions from the roots to the rest of the plant.
4. Transporting elements of the xylem consist of cell walls left after the cytoplasmic degeneration that occurs during xylem development.
5. Evaporation of water during transpiration pulls water molecules out of plant cells. Because of the cohesion of water molecules, water is drawn up through the xylem in response to the pulling force of transpiration.
6. Phloem translocates organic molecules from one part of a plant to another.

In 1643, J. B. van Helmont, a Flemish physician, planted a tiny willow tree in a pot. Over the next five years he added only rainwater or distilled water to the pot. After five years, van Helmont found that the plant had increased in weight by over 74 kg, while the soil was only about 0.057 kg lighter than it had been at the beginning of the experiment (figure 7.1). He concluded that the plant's growth was due to the incorporation of the added water.

Until van Helmont's time, people had believed that plants grew by absorbing and incorporating soil material. His experiment showed that only a tiny fraction of the plant's weight gain actually was due to incorporation of soil material. We now know that van Helmont was mistaken when he decided that all the rest of a plant's growth was due to incorporation of the water that he added, because he did not know about the incorporation of atmospheric carbon dioxide in photosynthesis (figure 7.2).

The tiny fraction of the plant's weight gain that is due to incorporation of soil material is very important, however. Various minerals from the soil are essential for plant growth and development. Thus, plants have three fundamental requirements: carbon dioxide, water, and certain minerals from the soil. Plants are **autotrophic** ("self-feeding") organisms. They can produce all organic compounds that they require using inorganic substances obtained from the environment. But to do so, plants must have a continuing supply of carbon dioxide, water, and mineral nutrients. In this chapter, we will consider the means by which land plants obtain and transport materials to meet their fundamental needs.

Figure 7.1 Van Helmont's experiment. Van Helmont demonstrated that weight gained by a growing plant was not equaled by weight loss in the soil in which it was grown. Although he concluded mistakenly that the weight gained by the plant came entirely from the water added during its growth, he dispelled the long-established notion that plants derived all their substance from the soil.

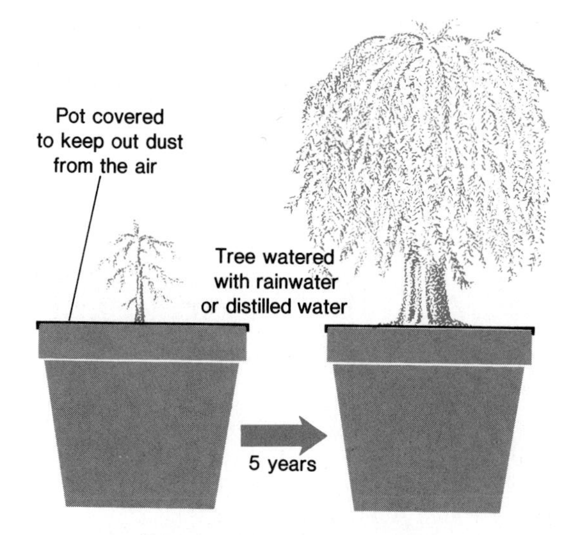

Figure 7.2 The proportions of elements making up a plant's dry weight. Photosynthesis incorporates carbon and oxygen from atmospheric CO_2, and hydrogen from water taken up from the soil. The other elements are obtained as mineral nutrients from the soil.

Functional Organization of Plants

Vascular plants have distinctive regions of specialization. **Roots** absorb water and mineral ions from the water that surrounds soil particles. **Leaves** absorb light energy and are specialized for the gas exchange required for photosynthesis. **Stems** connect roots and leaves and support the leaves in an elevated position, where the leaves are well exposed to light.

Vascular tissues run through the roots, stems, and leaves, and transport materials from one part of the plant to another. The two types of vascular tissue are xylem and phloem. Water and minerals from the soil are transported through **xylem** in only one direction, from the roots upward through the plant body. **Phloem,** on the other hand, can move dissolved material from place to place in the plant. For example, sugar produced in photosynthesizing leaf cells can be transported down through the phloem to the roots where it is stored, or it can be transported back up from the roots through the plant (figure 7.3).

Figure 7.3 Generalized sketch of a land plant. Various specialized regions of plants are connected by the vascular tissues xylem and phloem.

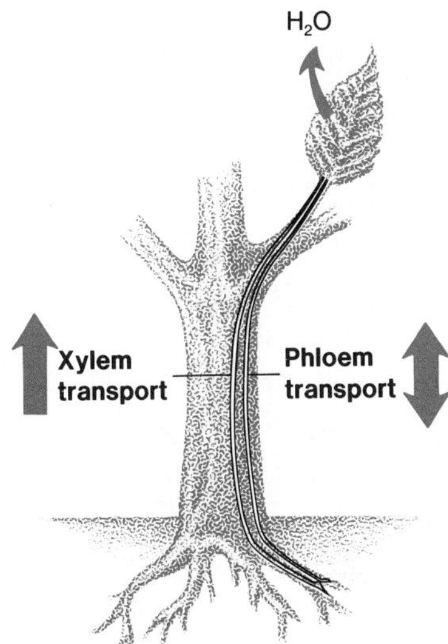

Obtaining Carbon Dioxide

Leaves are thin, delicate organs specialized for photosynthesis. They are protected against water loss by a waxy **cuticle.** Because carbon dioxide does not pass through this cuticle easily, CO_2 enters leaves through small, closable pores called **stomata** (figure 7.4). This permits CO_2 from the outside air to reach the moist surfaces of photosynthetic cells inside the leaf where it dissolves before entering the cells.

Water evaporates from the cell surfaces and diffuses out through open stomata. This evaporation of water from leaves to the outside air is called **transpiration.**

Figure 7.4 (*a*) Leaf structure. Mesophyll cells carry on photosynthesis and thus must have a supply of carbon dioxide. The cuticle contains a waxy substance that is impermeable to water and poorly permeable to CO_2. Carbon dioxide enters the leaves mainly through stomata (singular, stoma). (*b*) A stoma. This scanning electron micrograph shows the two guard cells that control the size of the stomatal opening. In transpiration, water evaporates from leaves through stomata (magnification × 2,660).

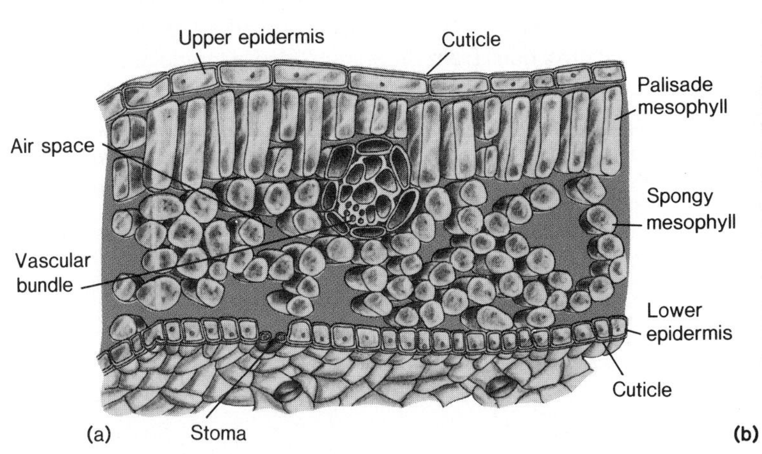

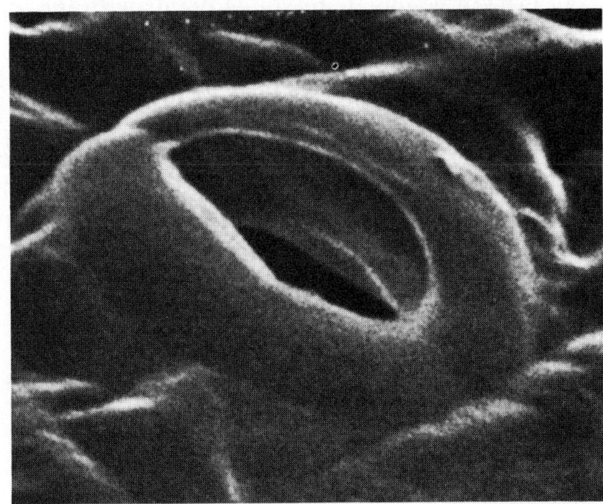

Water Loss and Replacement

Although plants continuously lose water by transpiration when the stomata are open, the water films on the surfaces of internal leaf cells are not depleted. This is because a **transpiration stream** of water continuously moves up through the xylem from the roots. The transpiration stream replaces water lost by transpiration or consumed in photosynthesis, and also supplies the water that is retained in plant tissue (figure 7.5).

Most of the water absorbed by the roots is used to replace the water being lost by transpiration. Only a very small percentage of the absorbed water is used in photosynthesis or retained in cells and tissues. For example, it is estimated that corn (*Zea mays*) plants growing in a moderately dry climate will transpire 98 percent of all the water that their roots absorb. A single *Zea mays* plant loses somewhere between 135 and 200 l of water through transpiration during a growing season. If this water loss is multiplied by the number of corn plants in a heavily planted cornfield, it is easy to understand why agricultural demands on soil water are so great.

This considerable water loss might seem like a very negative factor for plants, and it can be if excessive water loss leads to wilting. But there is another side to the story. The transpiration stream of water rising through a plant carries with it minerals absorbed from the soil. Thus, the seemingly wasteful water loss of transpiration is actually functionally important for plants. We will explore this relationship further when we consider function of the xylem.

Figure 7.5 The transpiration stream. Water moves up from the roots to replace water lost to the atmosphere by transpiration. This flow continues as long as stomata are open. A single corn (*Zea mays*) plant such as this one loses between 135 and 200 liters of water by transpiration during a growing season.

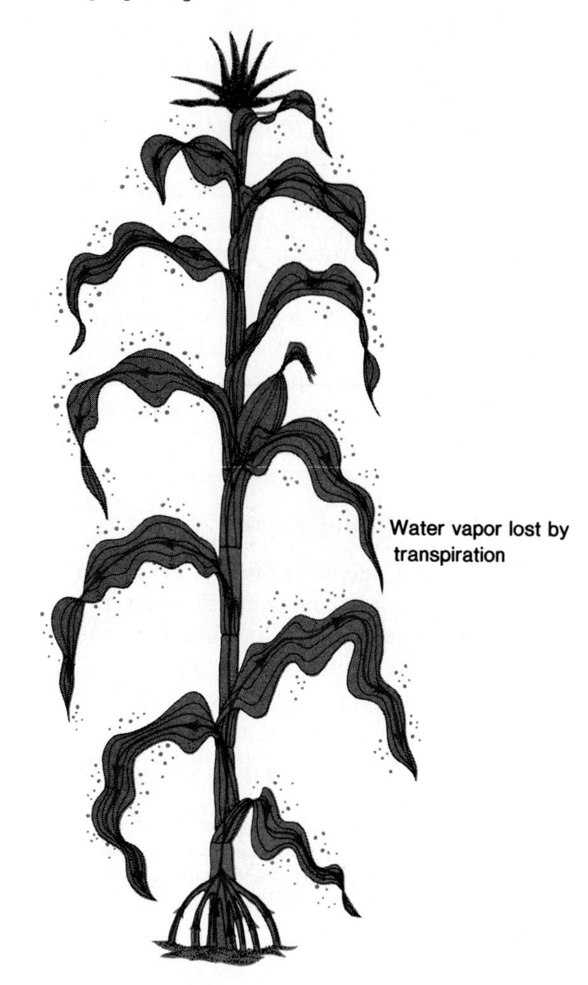

Roots

One of the most striking features of root function is that roots are not static; they grow continuously. Growing root tips constantly move through the soil and squeeze their way into new spaces among soil particles, where moisture may be more plentiful.

Root growth begins in all young plants with the growth of a single **primary root.** In some plants, the primary root grows straight down and remains the dominant root of the plant, with much smaller **secondary roots** growing out from it. This arrangement is called a **taproot system** (figure 7.6*a*). In other plants, a number of slender roots develop, and no single root dominates. These slender roots and their lateral branches make up a **fibrous root system** (figure 7.6*b*).

Root growth carries the roots far out to the sides of the plant in some cases. Corn roots spread as much as a meter in all directions from the stem. A dense tangle of intertwined root networks exists under the neatly rowed orderliness of a cornfield. The roots of other plants grow far down into the soil. Alfalfa roots reach a depth of 6 meters, and the roots of some desert plants penetrate to several times that depth.

Figure 7.6 Root systems. (*a*) The taproot system of a dock plant. (*b*) The fibrous root system of a zinnia.

(a)

(b)

Root Tips and Root Hairs

Near the tip of each growing root is an area of active cell division, the **meristematic region,** where cell division continues to produce new cells all through the life of a plant.

Some of the cells produced in the meristematic region form a protective cover over the tip of the root, called the **root cap.** Root cap cells are ground off by the wear and tear of being pushed through spaces among the rough soil particles, and they must be replaced constantly. Other cells produced by the cell division in the meristematic region are added to the length of the root itself. As these cells grow and lengthen, they push the root's tip further through the soil.

A growing root tip has **root hairs,** outgrowths of surface cells a few millimeters back from its tip. New root hairs form as the root pushes its tip further through the soil. The area where root hairs are found, the **region of maturation** or **root-hair zone,** is only a short section of the length of the root. Individual root hairs actually function for only a few days. The root-hair cycle is continuous—new root hairs are produced, they mature and function briefly, and then they wither and fall off (figure 7.7).

Figure 7.7 Root hairs on the primary root of a developing radish seedling.

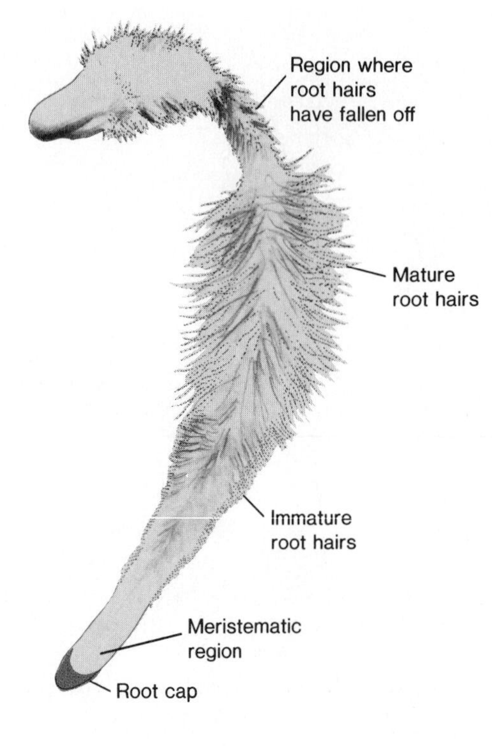

Root Structure

We can examine the internal structure of a root by looking at a cross section (a thin slice) cut across a root (figure 7.8).

The **epidermis,** the outer layer of the root, usually is a single layer of thin-walled cells. Root hairs develop as extensions of some of the epidermal cells.

Inside the epidermis is the **cortex** of the root. It contains loosely packed, multi-sided cells called **parenchyma cells.** There are open spaces among the parenchyma cells of the cortex. But where parenchyma cells touch each other, there are tunnel-like connections that join the cytoplasm of the individual cells.

The innermost layer of the cortex, the **endodermis,** forms a distinct boundary between the loosely organized parenchyma cells of the cortex and the **stele,** the central core of the root. Endodermal cells fit snugly together, and their walls are fused with one another. Since their top, bottom, and side walls contain a waxy material, water and solutes cannot pass through spaces between the endodermal cells (figure 7.9). This cell wall barrier to water and solute movement is known as the **Casparian strip.** Because of the organization of the Casparian strip, water or minerals moving from the cortex to the stele must cross a cell membrane and pass through the cytoplasm of the endodermal cells. There are no uncontrolled, easy-access routes from the cortex to the stele.

The central core of the root, the stele, lies inside the endodermis. The first layer of the stele is the **pericycle,** a layer whose cells can divide to produce branch roots (see figure 7.10).

The central portion of the stele is occupied mainly by the two vascular tissues, xylem and phloem (see figure 7.8). (The cross section shown in figure 7.8 is only a thin slice of the root. Of course, in a whole plant, the xylem and phloem are parts of continuous tubular transport systems that extend all the way up through the root and stem into the leaves.) Xylem transports water and minerals upward from the roots, while phloem moves dissolved materials from place to place in the plant.

Figure 7.8 Cross sections of a root. A photomicrograph (above) and a diagram (below).

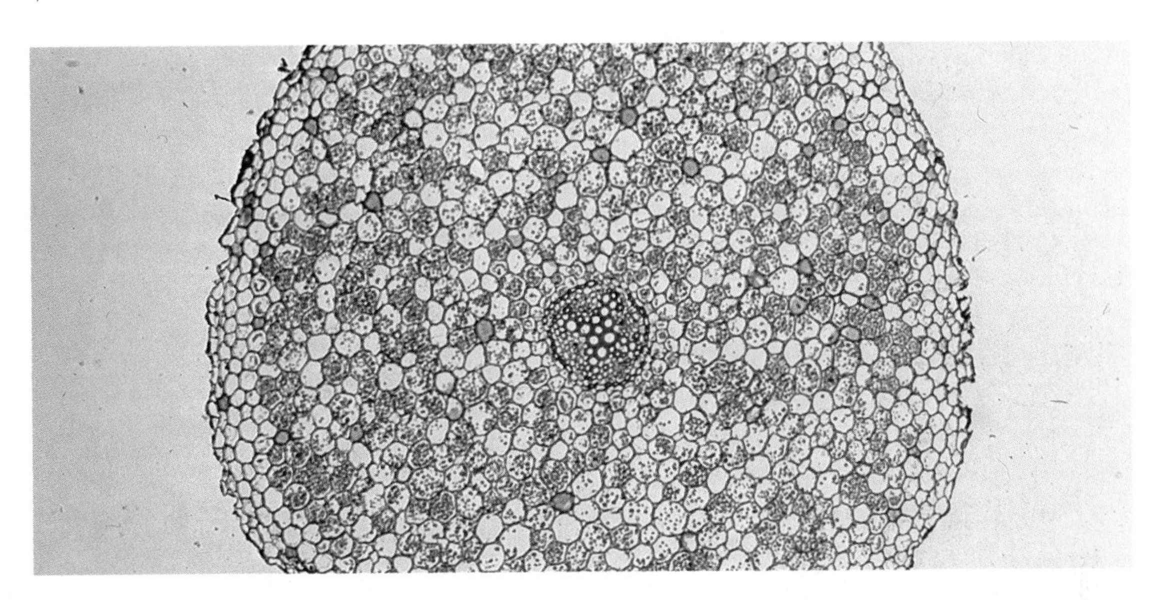

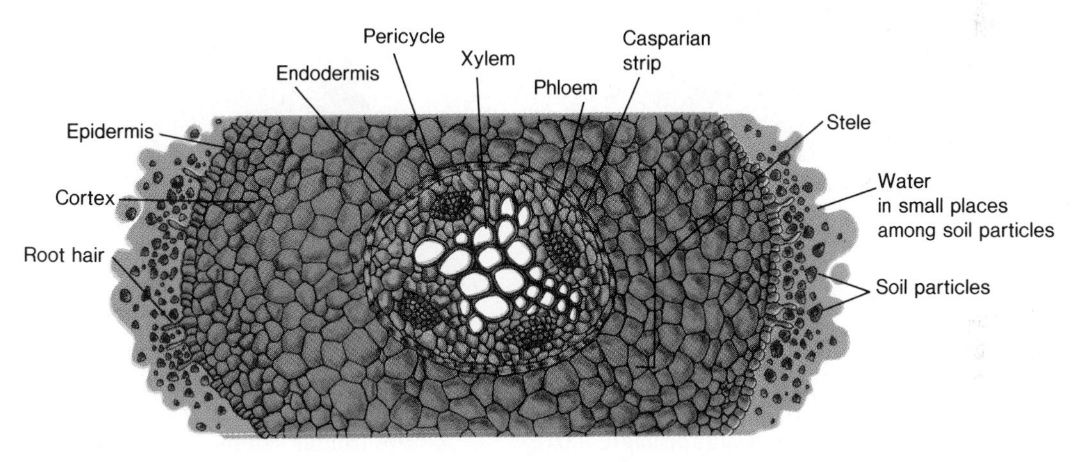

Figure 7.9 The Casparian strip. Fused endodermal cell walls impregnated with fatty material make an impermeable boundary between spaces in the cortex and in the stele. Water and solutes must pass through the cytoplasm of endodermal cells to enter the stele.

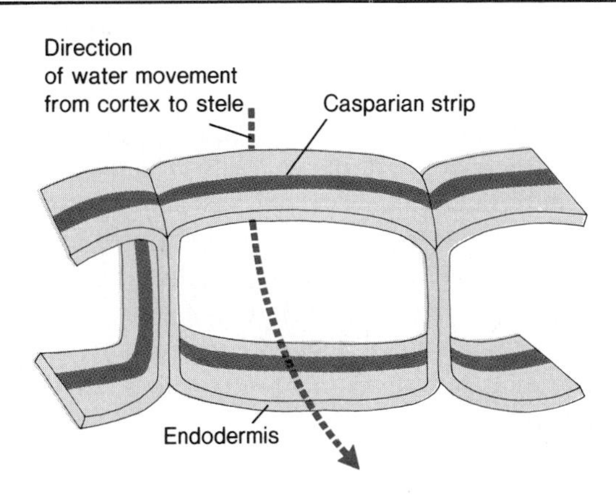

Figure 7.10 A diagram showing how a branch root develops when cell division begins in an area of the pericycle. Because branch roots originate deep inside the root, they are connected to vascular tissue of the stele from the beginning. They push physically and digest their way by enzyme action to force through other tissue to the surface and out into the soil.

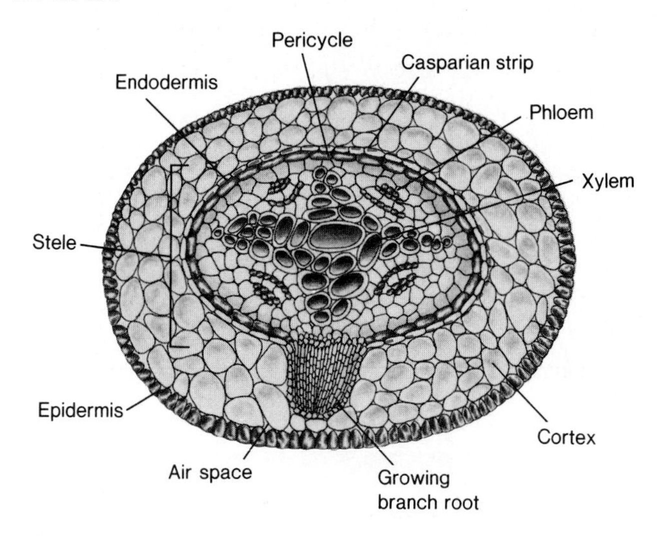

A single layer of tissue, the **cambium,** lies between the xylem and phloem. Cambium tissue is meristematic. Meristematic tissue contains cells that are able to divide throughout the life of the plant. Cell divisions in the cambium add more layers of xylem and phloem and thereby increase the thickness of roots and stems.

Figure 7.11 Water can either flow into the spaces between parenchyma cells in the cortex until it enters one of the cells (*a*). Or, water can enter epidermal cells directly from soil water spaces (*b*). Once inside a cell, water passes from cell to cell through the cortex and through the endodermis into the stele to the vessels of the xylem.

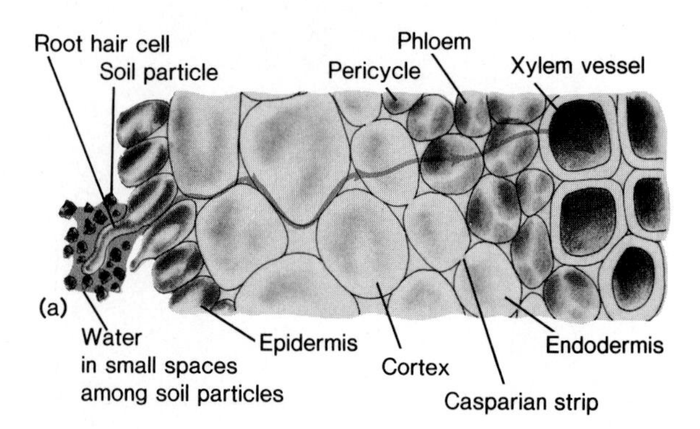

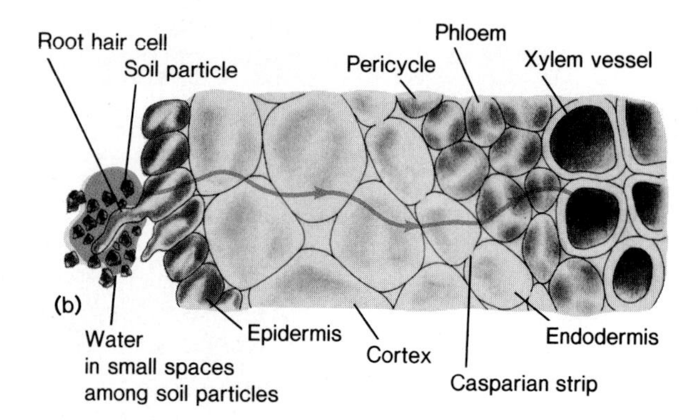

Water Movement in the Root

Water from the tiny spaces among soil particles can enter either epidermal cells or parenchyma cells of the cortex. Once water has entered a cell, it flows from cell to cell until it passes through endodermal cells into the stele and is delivered to the xylem (figure 7.11).

Water tends to flow inward from the soil because water is continually being drawn from the root and up through the plant by transpiration. Another force involved in the movement of water is the osmotic pressure difference between soil water and the cytoplasm of root cells. The cytoplasm normally contains more dissolved and suspended material than soil water, so water tends to enter the root cells osmotically.

The total mineral content of soil influences water entry into plants, and the osmotic pressure of salty soil water actually can prevent water from entering roots. High salinity can cause roots to lose water to salty soil. After their final victory over Carthage, a city state in north Africa, the ancient Romans were determined to eliminate the possibility of future trouble. So they applied this principle and salted the fields of Carthage, making them unfit for agriculture.

A modern practical problem with soil salinity is the soil salting that can occur in irrigated fields. Certain combinations of soil properties, mineral content of the irrigation water, and evaporation and runoff conditions can result in gradual buildup of salt deposits in the soil. These salt deposits threaten future crop production either because of osmotic changes that interfere with water entry into roots or toxic oversupply of some particular mineral. This is already happening in some irrigated areas in California. Thus, soil and water chemistry must be considered carefully whenever irrigation is planned.

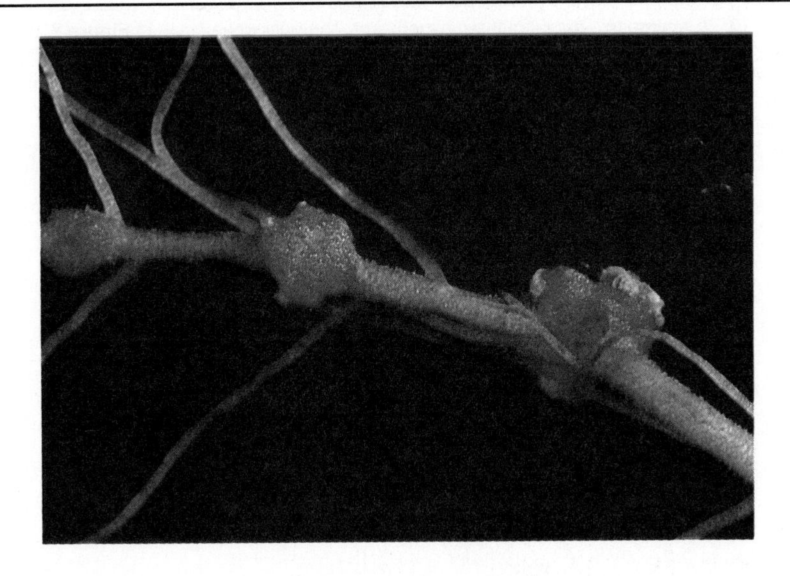

Figure 7.12 Symbiosis and mineral nutrition. Nodules on soybean roots are made up of plant tissue and nitrogen-fixing bacteria. These nodules are red because cells in the nodule produce a red pigment, leghemoglobin, during the time that active nitrogen fixation is occurring. Leghemoglobin resembles the hemoglobin found in vertebrate red blood cells.

Mineral Absorption

The mineral ions that plants absorb are found in solution in the water in spaces among soil particles. How do these minerals get into roots?

Sometimes when particular ions are very concentrated in soil water, as they can be after fertilizer application, the ions enter root cells by simple diffusion. But very often various ions are more concentrated inside root cells than in soil water and they will not enter by diffusion. How do ions enter root cells under such circumstances?

The most important means by which ions are moved from soil water or the space inside the cortex into the cytoplasmic network of root cells is by **active transport** across root cell membranes. Active transport requires energy (ATP) input. We can demonstrate this experimentally. If a respiratory inhibitor such as cyanide is applied to roots, thereby cutting off the supply of ATP, active transport of ions is slowed or stopped.

Once ions are inside root cells, they move from cell to cell through the cortex and through the endodermis into the stele. There, stele cells actively pump ions into the xylem through which they are carried to upper parts of the plant.

Symbiosis and Mineral Nutrition

Some plants get help from other organisms in obtaining mineral nutrients. Nitrogen nutrition is a good example of this.

Nitrogen is one of plants' most important nutritional requirements. Because nitrogen is a plentiful substance (about 78 percent of the air is nitrogen) you might think this requirement would not present any particular problem for plants. But plants cannot absorb molecular nitrogen (N_2) from the atmosphere. They require nitrogen in the form of nitrate (NO_3^-) or ammonium (NH_4^+) ions. Plants, therefore, are dependent on the activity of soil microorganisms that incorporate atmospheric nitrogen by the process of **nitrogen fixation.** After nitrogen fixation has occurred, usable nitrogen compounds (NO_3^- and NH_4^+) are available for both the microorganisms and the plants.

Many of these nitrogen-fixing microorganisms are free-living in the soil. In some plants, particularly legumes (such as beans and peas), plant tissue and nitrogen-fixing bacteria are found together in swellings called **nodules** that develop on the roots (figure 7.12). The nitrogen-fixing bacteria convert molecular nitrogen (N_2) into forms required by the plant. At the same time, the plant provides the microorganisms with a supply of reduced organic compounds (produced photosynthetically) that meets the energy requirements of the microorganisms. This sort of close functional interdependence between two different organisms is called **symbiosis,** which literally means "living together."

Figure 7.13 A hydroponic (liquid culture) experiment. The buckwheat plant on the left was grown in a nutrient solution without potassium. The culture on the right had a complete nutrient solution containing all required nutrients including potassium.

Figure 7.14 Elements that make up plants. Macronutrients and micronutrients together total only 4 percent of the total weight of a plant, but they are absolutely essential to the life and growth of a plant.

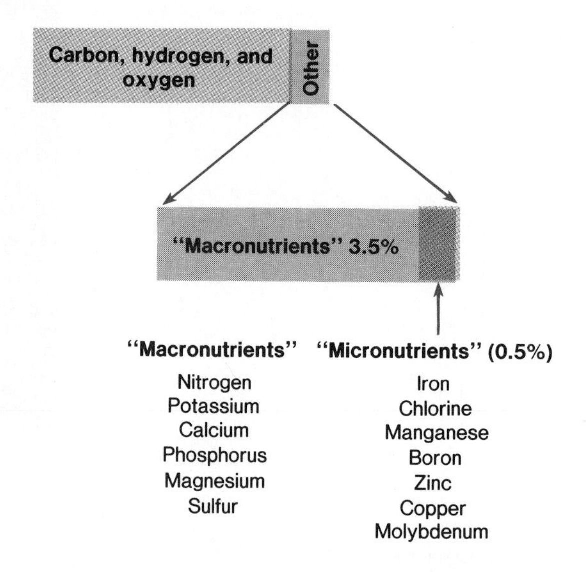

Mineral Requirements of Plants

The discovery of photosynthesis made it clear to biologists that plant bodies were not entirely constructed of material from the soil, but it also was obvious that plants cannot live and grow without certain mineral nutrients from the soil. Biologists began to ask specifically what these nutrients are and how much of each is required.

One of the most effective methods of determining the mineral nutrient requirements of plants is **hydroponics**—growing plants in water cultures containing the same ions that occur in soil. Plants' responses to addition or deletion of various elements can be used to determine the mineral requirements of the plant (figure 7.13).

During the nineteenth century, hydroponics showed that seven elements are essential nutrients for plants: nitrogen, phosphorus, potassium, calcium, magnesium, sulfur, and iron. The first six of these elements are required in the greatest quantities by most plants and are called **macronutrients.** Iron is required in far smaller quantities than any of the other six, and thus usually is now placed with a group of elements (chlorine, manganese, boron, zinc, copper, and molybdenum) called **micronutrients,** which need to be supplied to plants in much smaller amounts (figure 7.14). Several other elements, for example, sodium and cobalt, seem to be required by some plants and not by others.

Mineral Deficiencies in Plants

Mineral deficiencies cause a variety of symptoms in plants and some macronutrient deficiencies have direct bearings on agricultural production.

Because of the importance of nitrogen as a constituent of proteins, nucleic acids, chlorophyll, and several coenzymes, it is not surprising that nitrogen deficiency causes a variety of problems. Nitrogen-deficient plants are generally stunted and weak, and they have small, distorted leaves that are subject to yellowing (chlorosis). Nitrogen deficiency is a very common agricultural problem.

Figure 7.15 Symptoms of mineral nutrient deficiencies. (*a*) Calcium deficiency symptoms in tomato fruits. "Blossom end rot" forms on the side away from the stem. (*b*) Boron deficiency symptoms include early death of stem tips. Leaves become crinkled, and stems and leaf stalks crack. (*c*) Iron deficiency symptoms in tomato leaves. Chlorosis (yellowing) occurs between veins.

(a)

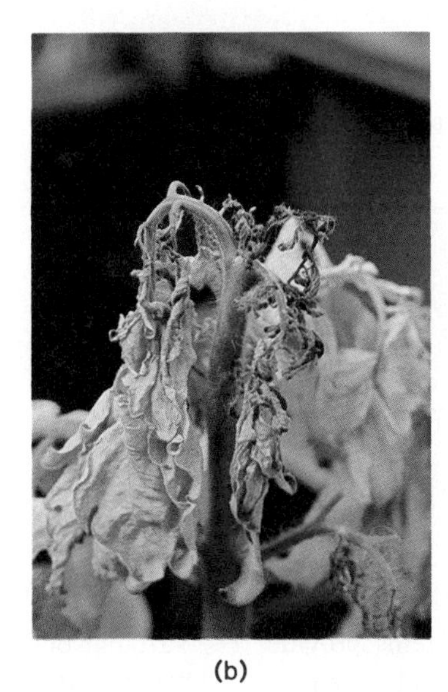

(b)

(c)

There are many macronutrient and micronutrient deficiency syndromes. They frequently have very characteristic symptoms, which are recognized readily by plant nutrition experts (figure 7.15). However, many gardeners have probably seen these symptoms without knowing what they were.

Mineral Nutrition and Modern Agriculture

Despite natural replacement processes, intensive modern agriculture depletes soil mineral elements so drastically that they must be replaced through other means to maintain productivity of the land.

Nitrogen, phosphorus, and potassium are the three macronutrients that most frequently must be supplemented by fertilization of agricultural soils. Fertilizers commonly are labeled with a formula that indicates the percentage of each of these three. A formula of 20–10–10 indicates that the fertilizer contains 20 percent nitrogen, 10 percent phosphoric acid, and 10 percent soluble potassium (as K_2O). Often, a fourth element, calcium, also is included in fertilizers, and occasionally other nutrients may be added as well.

Fertilization, however, is not the simple solution it appears to be for replenishing these elements. Heavy nitrogen fertilization is absolutely essential for the maintenance of high levels of agricultural production. But the energy input required to produce and deliver nitrogen fertilizers is becoming a critical problem. Production of nitrogen fertilizers consumes large amounts of fossil fuel, and transport from production sites to fields requires an additional heavy energy input. The constraints of soil chemistry and plant mineral nutrition coupled with increasing energy costs and decreasing supplies of fuel reserves further complicate the problem of meeting the food requirements of a growing world population.

Figure 7.16 Comparison of monocots and dicots, the two major groups of flowering plants.

Monocot

1. One cotyledon in seed
2. Vascular bundles scattered in stem
3. Leaf veins parallel
4. Flower parts in threes and multiples of three

Dicot

1. Two cotyledons in seed
2. Vascular bundles in a definite ring in stem
3. Leaf veins form a net pattern
4. Flower parts in fours or fives and multiples of four or five

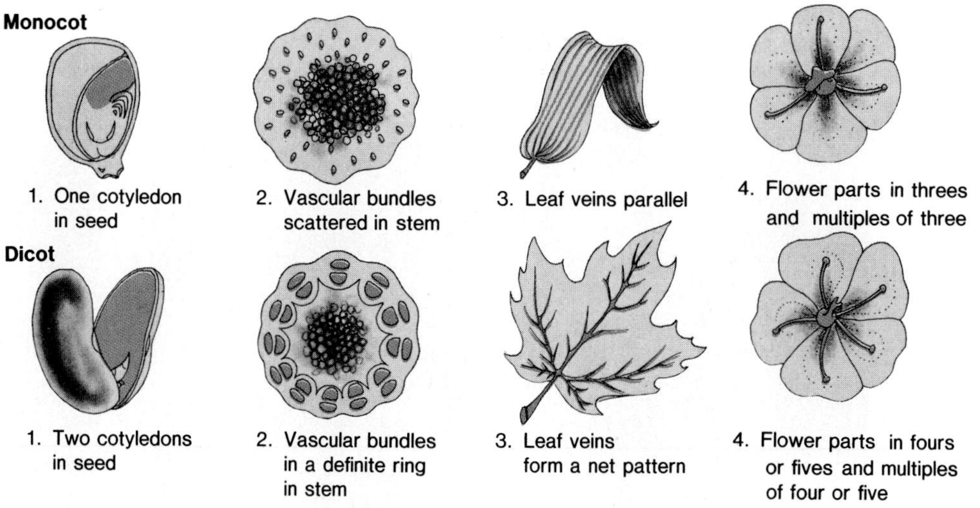

Stems

Stems have two major functions in plants. Stems provide **support** to hold branches and leaves upright, and they provide a route for **transport** of materials between roots and leaves through xylem and phloem tissues.

All stems are structurally organized to fulfill these support and transport roles, but different types of plants have different kinds of stems. There are two major groups of flowering plants, the **dicotyledonous** plants and the **monocotyledonous** plants. These two groups are often called **dicots** and **monocots** for short. Monocots have only one **cotyledon** ("seed leaf"), and dicots have two cotyledons (figure 7.16). Monocots and dicots have fundamentally different stem structures. You can see these differences in the arrangement of **vascular bundles,** the long strings of xylem and phloem tissue that run through stems.

Monocot Stems

The basic pattern of a typical monocot stem is shown by a cross section of a corn (*Zea mays*) stem in figure 7.17. Vascular bundles are scattered throughout the stem. All of the space around the vascular bundles is filled by parenchyma cells.

Stems are supported by **mechanical tissue.** Mechanical tissue is a collective name for the thick-walled cells in vascular bundles, including both xylem tissue and various fiber cells surrounding the xylem and phloem cells of the vascular bundles.

Dicot Stems

Vascular bundles in a dicot stem are arranged in a ring lying between the cortex and a large central **pith,** both of which consist of packing parenchyma cells (figure 7.18). In addition, dicot stems have cambium (meristematic tissue).

Woody Stems

Many **perennial** plants (plants that live for several to many growing seasons) have **woody** stems. Woody stems are hard stems that grow in length and diameter each growing season, year after year. They grow in length because new cells are produced each year at their tips in a growth region called the **apical meristem.**

Woody stems develop **buds** whose apical meristems become active at the beginning of each growing season and continue to add new tissue throughout the season. There are buds at the tip of the stem (terminal buds) and just above the points where leaves are attached (lateral buds) (figure 7.19). Terminal buds add to the length of the original stem, while certain of the lateral buds produce stem branches.

Each terminal bud contains the beginnings of a new section of stem. Leaves and their stalks encircle the central region of the tip, the apical meristem. As the stem lengthens, the leaves, which are closely packed within the bud, space out at regular intervals along the stem. The points of leaf attachment are known as **nodes,** and the spaces between nodes are called **internodes** (figure 7.20). All growth in length is due to growth occurring in apical regions. There is no factual basis, for example, for the popular misconception that a tree gets taller because the growth at the base of its trunk pushes it upward.

Figure 7.17 Cross section of a representative monocot stem—corn (*Zea mays*)—showing the scattered arrangement of vascular bundles. The enlargement of one corn vascular bundle shows the arrangement of tissues in a monocot vascular bundle.

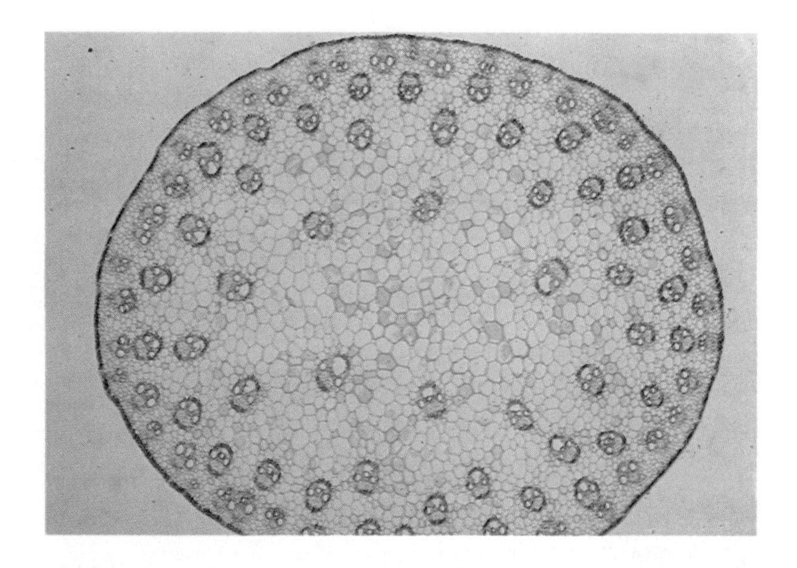

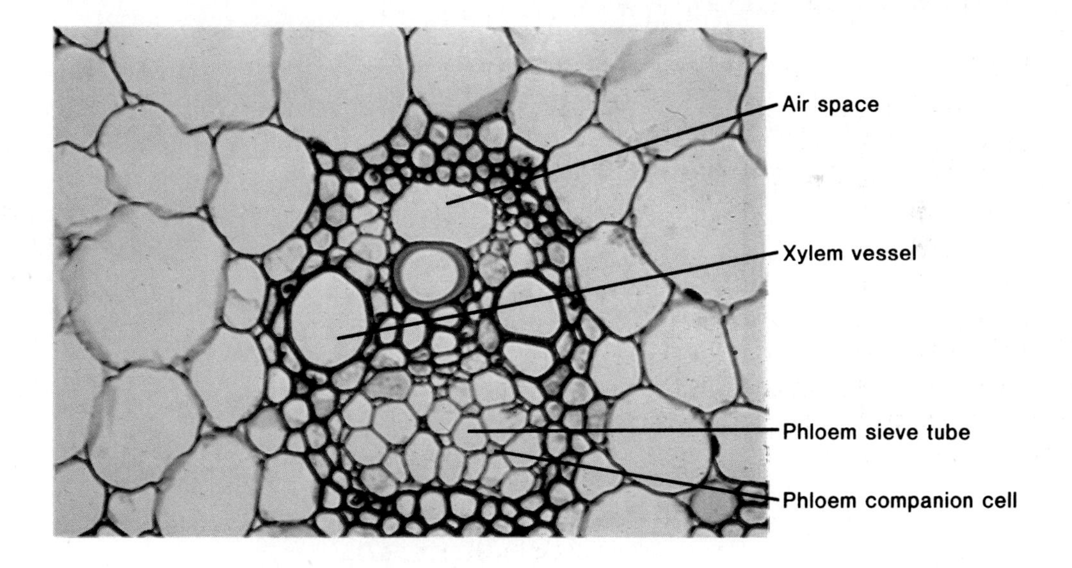

Figure 7.18 Photograph of a cross section of a representative herbaceous dicot stem—alfalfa (*Medicago*)—and a sketch of a sector of the stem.

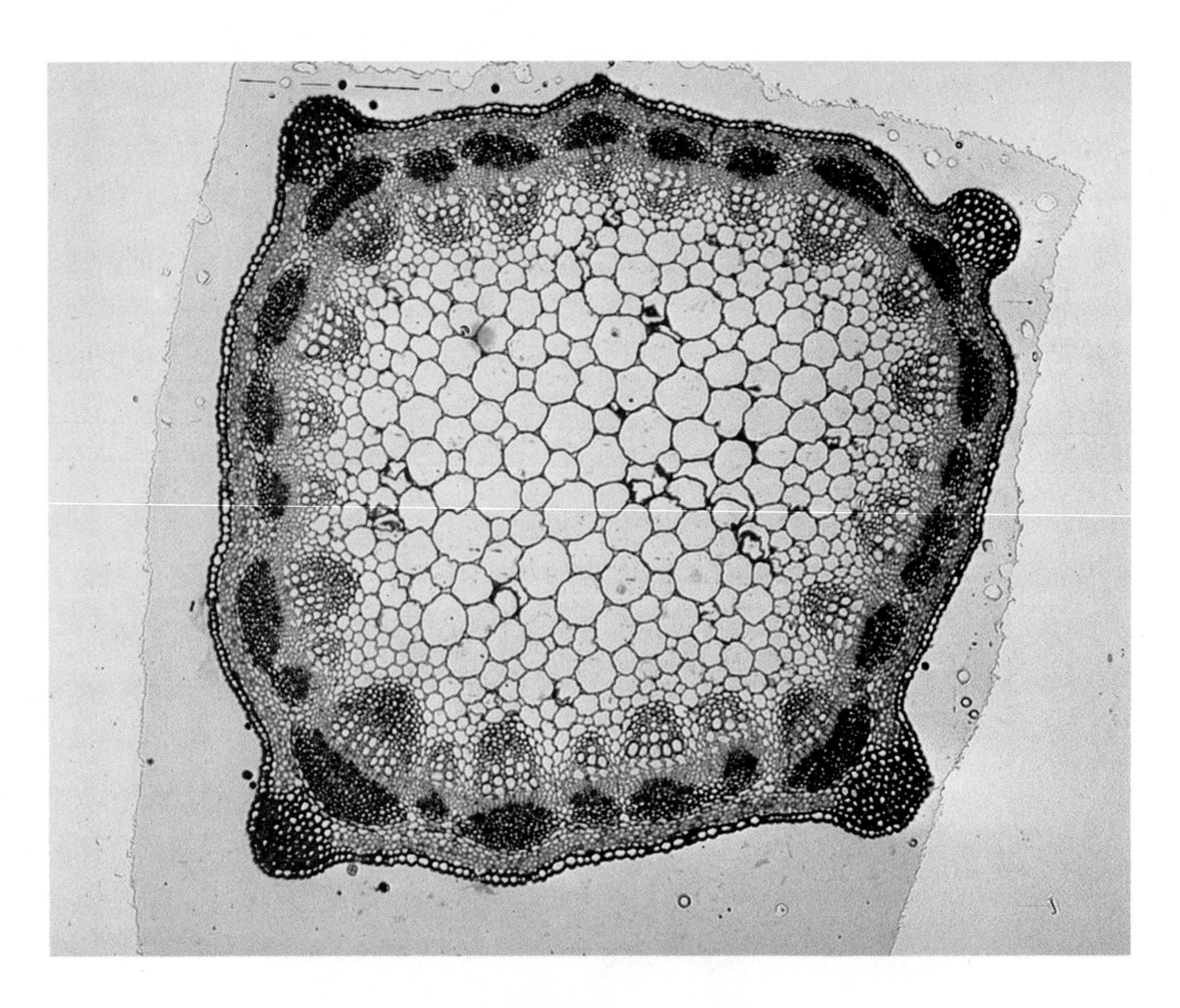

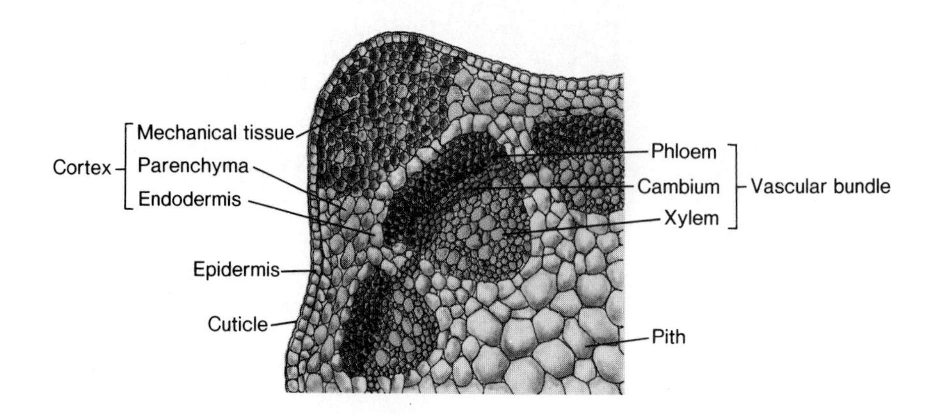

Figure 7.19 A terminal bud and two lateral buds on a shagbark hickory (*Carya ovata*) and a diagram of a longitudinal section of a terminal bud.

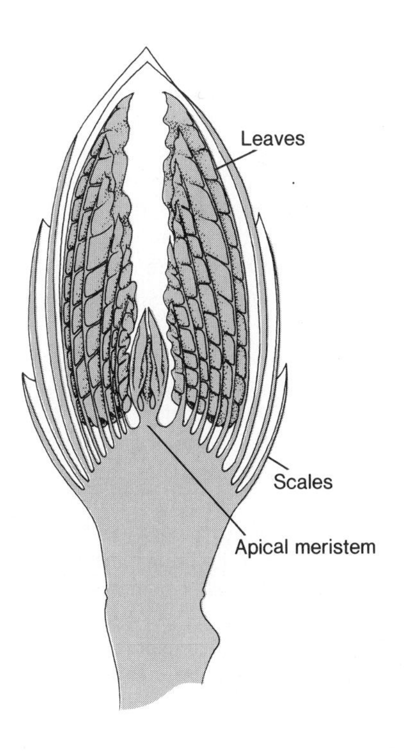

Figure 7.20 Generalized sketch of a woody stem. Terminal bud scale scars mark the beginning of each season's growth in length. Lenticels are small patches of loosely organized tissue through which oxygen and carbon dioxide diffuse.

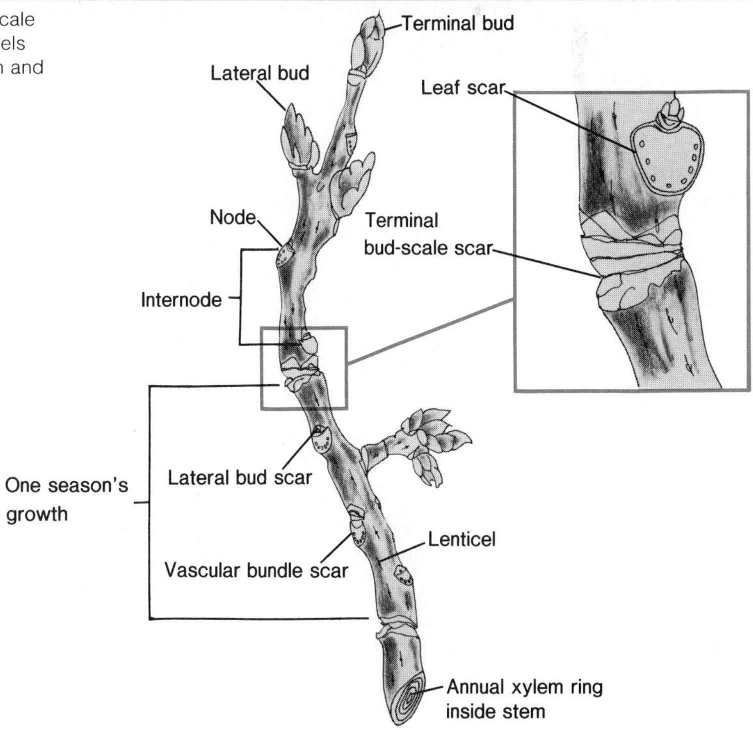

Figure 7.21 Xylem structure. Tracheids and vessels are the conducting elements of xylem. Parenchyma cells pack the space around them.

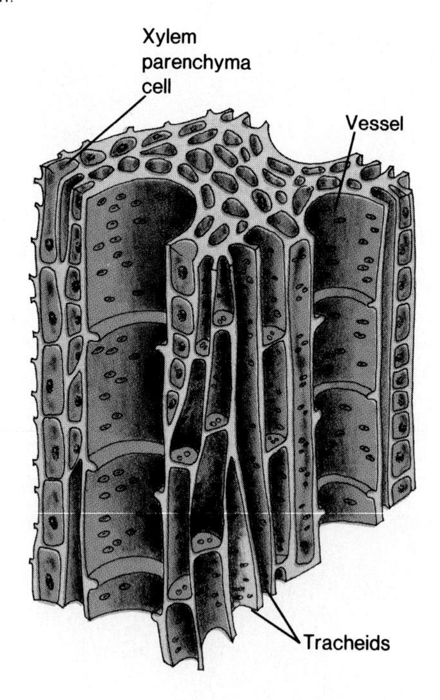

Buds form at the end of one growing season and become active in the next season. Thus, they must be adapted for overwintering. Buds are enclosed within several layers of highly modified leaves, the **bud scales,** which provide a protective coat. In the spring, buds swell and rupture the coating layer of bud scales, which eventually drop off. The area where these scales were attached to the stem is marked by the **terminal bud scale scar.** Each terminal bud scale scar identifies the beginning point of a season's growth (figure 7.20).

Cell division in a ring of cambium (meristematic tissue) produces growth in diameter in a woody stem during each growing season. Growth due to cambium activity during successive growing seasons is reflected internally as a series of **annual rings** in the xylem tissue of a woody stem (box 7.1).

Xylem

Throughout this chapter we have described the xylem as a transport tissue through which water and minerals from the soil move upward through a plant. How is xylem tissue actually constructed and how is upward transport through the xylem accomplished?

The xylem of flowering plants contains two basic types of conducting tubes, **tracheids** and **vessels** (figure 7.21). Many other plants, including evergreens and ferns, have only tracheids in their xylem tissue.

Box 7.1

How Old Was That Tree?

When a large tree is cut down, it is possible to determine the tree's age by counting "tree rings." What are tree rings? How are they produced? How do they reflect the age of the tree?

Early in its first growing season, a woody stem's basic organization is quite similar to that of an herbaceous (nonwoody) dicot stem (see figure 7.18). All of the tissues in the stem were produced by primary growth; that is, they were made up of cells produced by cell divisions in the apical meristem at the tip of the growing stem.

But in woody stems, there is a great deal of secondary growth produced by cell divisions in the cambium. A ring develops before the end of the first growing season as the areas of cambium originally present in the vascular cambium connect to produce a continuous cylinder in the stem. From then on, concentric rings of new tissue are produced each year, with secondary xylem being laid down inside the cambium rings and secondary phloem being produced outside it (box figure 7.1).

During each growing season, xylem elements produced in the spring usually are larger than those produced later in the season, so the xylem formed in one growing season can be distinguished from the adjacent growth of another growing season. These accumulating thick-walled xylem cells make up what is commonly called the **wood** of a tree or shrub. The yearly deposits of xylem constitute the annual growth rings in the wood and form an accurate record of the age of the plant.

Only the more recently formed part of the xylem, the **sapwood,** actually contains conducting xylem elements that still function in water transport. In the older, inner **heartwood,** the xylem becomes plugged with deposits of resins, gums, and other material.

Over the years, phloem tissue does not accumulate as xylem tissue does. Phloem, which consists of relatively thin-walled and delicate transport cells, is pushed outward as the diameter of the woody xylem area increases. Eventually, the outermost parts of the phloem are crushed against the bark that covers older stems. Thus, there are no obvious annual growth rings in the phloem. Only the most recently formed phloem, that portion nearest the cambium, is active transporting tissue.

Climatic conditions influence the nature and thickness of the annual rings formed each year. In this way, the tree rings record the climate of past years. Thus, the history of a tree's life is written in the woody xylem tissue.

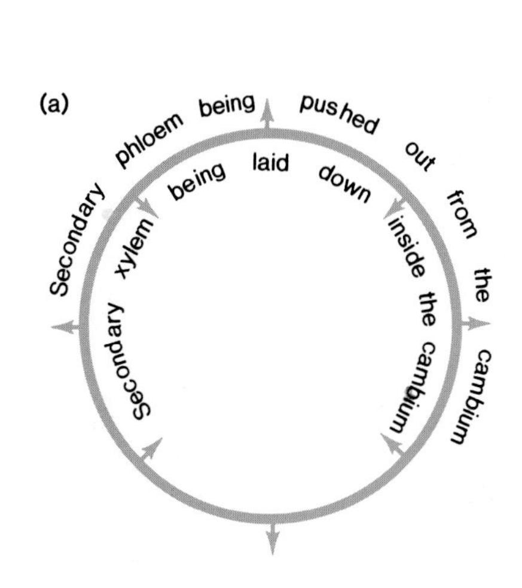

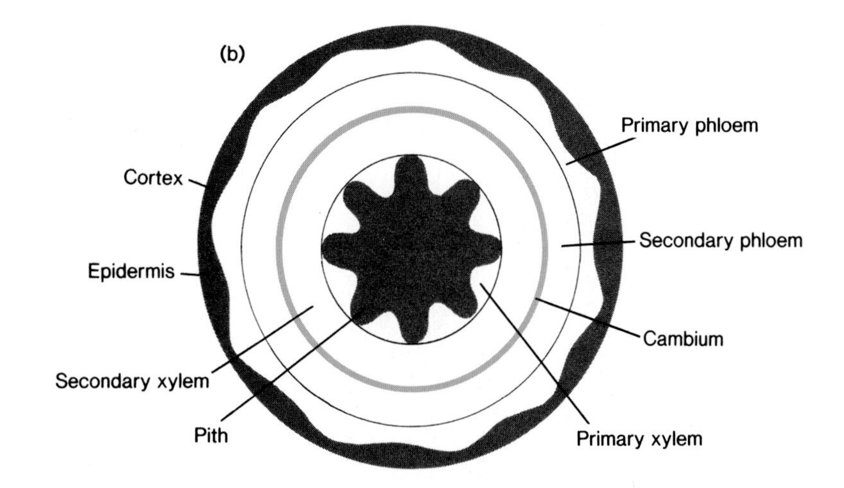

(c)

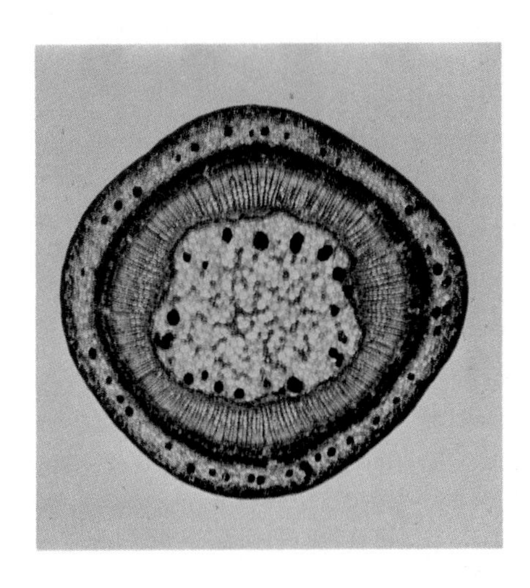

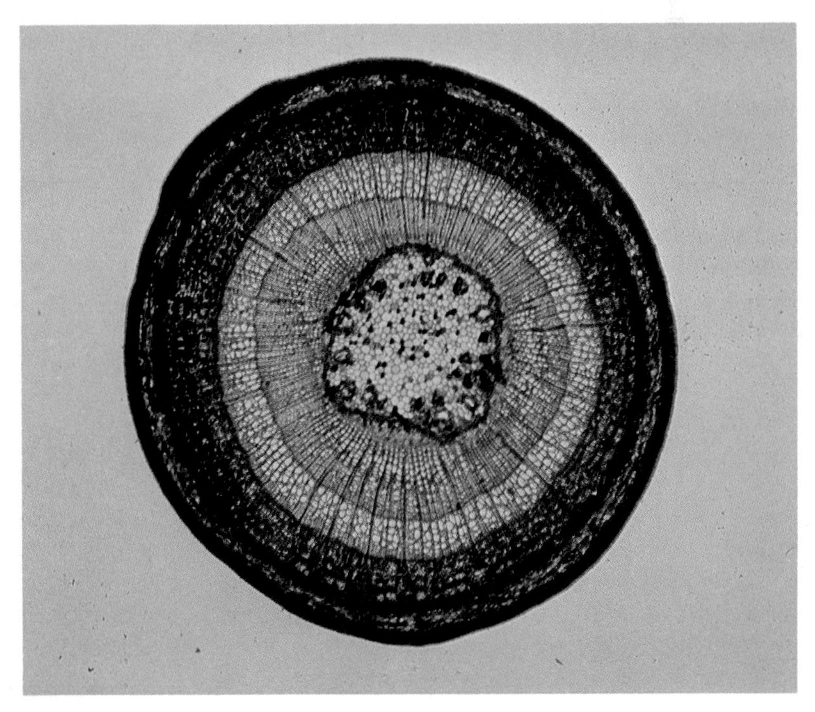

Box figure 7.1 Secondary growth. (*a*) Activity of vascular cambium (shown in color) producing secondary tissue in a woody stem. The circumference of the cambium ring increases as the stem grows. (*b*) Diagram of a woody dicot stem cross section after a season of secondary growth. (*c*) Cross sections of basswood (*Tilia*) stems after one and two growing seasons.

Figure 7.22 Comparison of tracheids and vessel elements. (*a*) Tracheids. Water must pass through the thin walls of pits to pass from one tracheid to the next. (*b*) Xylem vessel elements either have holes in their end walls (left) or lack end walls entirely (right). Thus, xylem vessels are continuous hollow tubes.

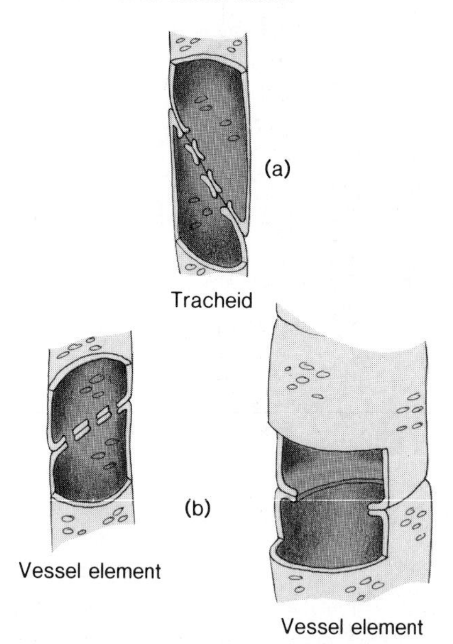

Figure 7.23 Xylem transporting elements as "cellulose pipes." Scanning electron micrograph of xylem vessels from a cucumber root. Note the sculptured appearance of the inside of these vessels (magnification × 1,050).

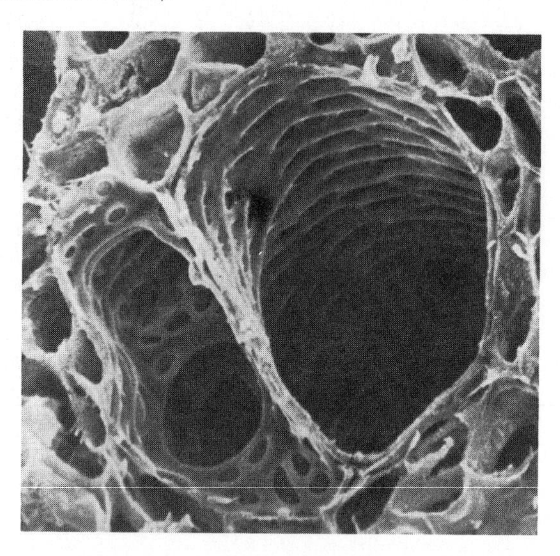

Tracheids are hollow, spindle-shaped cells that have drawn-out, flattened ends. There are a number of very thin spots (known as **pits**) in the walls of tracheids where the cells overlap one another. Water flowing upward through tracheids moves from one cell to the next by passing through these thin spots (figure 7.22*a*).

Vessels develop from end-to-end chains of cells. In some cases, the cell walls at the ends of the cells develop holes, but in other cases the end walls disappear entirely. As the end walls are thinning out and disappearing, the side walls thicken. Finally, the nucleus, cytoplasm, and membranes of the cells degenerate and disappear. This leaves the cell walls connected in hollow vessels that extend through the plant (figure 7.22*b*). Individual vessel cells are called **vessel elements.** They no longer are individual cells but are nonliving parts of a structure derived from many cells. Literally, vessels are "cellulose pipes" running through the plant. Water moving up the plant flows through these straight, open channels (figure 7.23).

Movement through the Xylem

How is **xylem sap** (water and minerals) moved upward through the hollow tracheids and vessels of the xylem? One moving force results from osmotic movement of water into roots (see page 75) from the soil. This osmotic force, called **root pressure,** is generated at the bottom of the xylem and tends to push xylem sap upward. In fact, cut ends of stems will "bleed" xylem sap for some time because of the push applied by root pressure.

Root pressure also is responsible for an interesting phenomenon that occurs when transpiration is not taking place and the soil is well watered. Drops of water are forced out of vein endings along the edges of leaves as a consequence of root pressure. This "bleeding" at leaf vein endings is called **guttation** (figure 7.24).

Transpiration Pull and the Cohesion of Water

Root pressure alone, however, simply is not strong enough to move water very far up a tall plant. If root pressure does not generate enough force to raise water to the tops of large plants, how then is it done?

Water rises through vessels and tracheids as a result of a pull applied to water at the top of the xylem and the very strong tendency of water molecules to stick to one another. This attraction among molecules of the same kind is called **cohesion.** Liquid water is not simply water molecules randomly floating around one another. There is considerable attraction between the hydrogen atoms of one water molecule and the oxygen atom of an adjacent molecule. This attraction, called hydrogen bonding (see page 32), causes water molecules to "stick together."

Figure 7.24 Young oat plants with drops of water forced out at their leaf tips by guttation.

The cohesive force among water molecules is so great that you could think of the column of water inside a xylem vessel as behaving almost like a chain of water molecules that can be pulled up through the xylem. In fact, it is easier to pull apart the molecules in fine wires made of some common metals than it is to disrupt a column of water in a small-diameter, airtight tube.

If cohesive forces among water molecules are great enough to hold a column of water together even when it is being pulled with enough force to raise it to the top of a large tree, what then supplies the actual pulling force? Biologists have come to accept an idea first suggested by an English scientist, Stephen Hales. Near the beginning of the eighteenth century, Hales suggested that transpiration provides the pulling force that draws water upward through plants.

It now seems that **transpiration pull** and the **cohesion of water** together provide a reasonable explanation of how water is drawn up through xylems of even very tall trees. Evaporation of water molecules from the surface water films of leaf cells pulls water from within those leaf cells to replenish the surface water films. This pull is transmitted in a chain reaction as water is drawn through leaf cells from the xylem in the leaf veins. Because the xylem contains continuous tubular structures that run from leaf veins down through the stem to the roots, this pull (tension) is transmitted down through the entire plant, so that whole columns of water literally are pulled up from the top.

Thus, evaporative water loss from leaves is not just a necessary evil that plants must endure in order to obtain carbon dioxide for photosynthesis. Transpiration is vital to the life of plants because it provides the pulling force that draws xylem sap up from the roots to the upper parts of the plant.

Phloem

During early spring growth, organic materials, especially sucrose, are translocated from plant storage areas to young, actively growing areas of the plant that are not yet synthesizing adequate quantities of nutrients to meet their own needs. Later in the season, organic molecules produced in mature leaves are translocated to storage tissues, which develop in modified stems or roots, or to seeds. How are these organic compounds moved from the **sources** (the production sites in the leaves) to **sinks** (the actively growing areas or storage tissues of seeds, roots, or stems)?

The phloem was identified as the translocating tissue long ago when the results of **girdling** (removal of a ring of bark) were analyzed. Girdling removes the phloem but leaves the xylem intact, and it eventually kills a tree. If a tree is girdled below the level of the majority of leaves, the bark, with the underlying phloem, swells just above the cut, and sugar accumulates in the swollen tissue. This finding led biologists to the conclusion that the phloem is the carbohydrate translocating tissue.

Figure 7.25 Phloem structure. (*a*) Arrangement of cells in phloem. Sieve tubes are the actual conducting elements in phloem. Phloem fibers are supporting cells not involved in translocation, and parenchyma cells fill spaces among the other cells. (*b*) Sketches of a sieve plate.

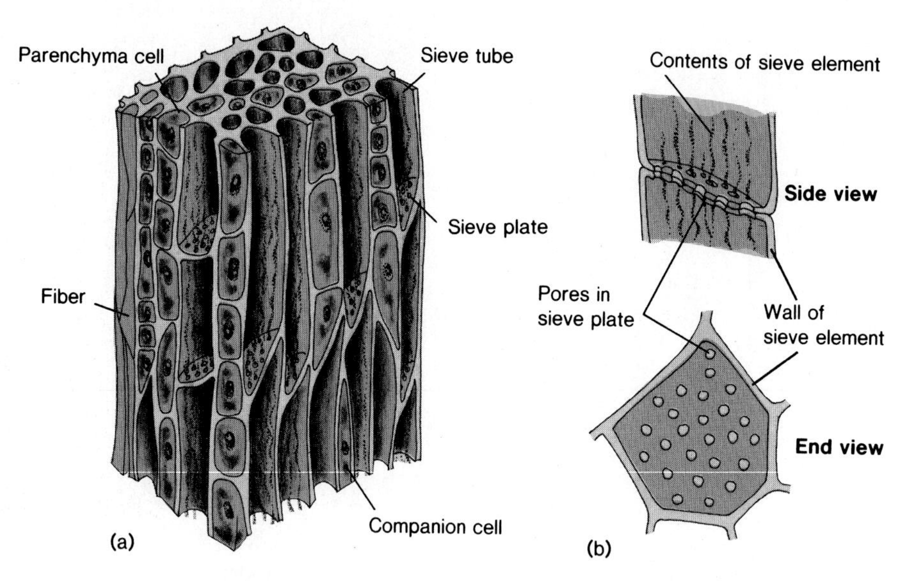

Research using radioactively labeled compounds (compounds containing a radioactive isotope) have confirmed that the phloem transports sugars, amino acids, hormones, and other substances from one part of the plant to another.

Phloem Cells

Material moved through the phloem is translocated through the **sieve tubes. Sieve tube elements,** the cells that make up sieve tubes, are highly specialized cells that are arranged in linear arrays running vertically through the length of the plant. In flowering plants, the sieve tube elements are lined up end-to-end with flat areas of contact where the end walls of cells abut one another (figure 7.25).

There are prominent holes in the end walls of sieve tube elements, and adjacent cells contact each other directly through these holes. Because of its sievelike appearance, this contact area is called the **sieve plate.** Cytoplasmic connections run through the pores of each sieve plate, connecting the cytoplasm of the sieve elements adjacent to the plate. Thus, there is cytoplasmic continuity from cell to cell in the sieve tubes.

Individual sieve tube elements are structurally different from the majority of plant cells, because a mature sieve tube element does not have either a vacuole or a nucleus. Thus, functional phloem tissue contains end-to-end rows of nonnucleated, nonvacuolated, but still living, sieve elements. This special structural arrangement is very different from that of xylem vessels and tracheids, which consist of empty cell walls.

Each sieve tube element has one or more **companion cells** adjacent to it. Companion cells have nuclei and vacuoles as well as other organelles normally found in plant cells. It seems likely that the companion cells exert some control over sieve tube functioning. And almost certainly, companion cells can transfer material into sieve tubes through pores that connect them.

Movement through the Phloem

How is material translocated from one part of a plant to another through phloem sieve tubes? It has been surprisingly difficult to answer this question because of the sensitivity of phloem cells. Most early experiments aimed at determining how phloem transport works caused a shutdown in sieve tube function, which phloem cells show in response to almost any kind of interference from the outside. But since effective techniques for studying phloem function were developed (see box 7.2), some good hypotheses have been proposed.

The most widely held hypothesis explaining phloem transport functions is the **mass flow hypothesis.** This hypothesis can best be explained using the model shown in figure 7.26*a*. This is a system bounded by differentially permeable membranes that allow water to pass freely, but do not allow solute molecules (sucrose in this case) much access. There are marked differences in the sucrose concentration at two separated points in the system.

Box 7.2

Aphids

Giving Your Body for Science

Phloem function is difficult to study because the cells involved are delicate and easily damaged. The slightest disturbance of sieve tubes leads to injury responses that shut down phloem translocation and conserve phloem sap. A slimy plug made of a substance called callose (a glucose polymer) develops in each plate pore when the phloem is injured. The callose plug apparently develops in injured phloem in nature just as it does in experiments.

Because injury responses in phloem occur quickly, researchers studying phloem function had to find a way to study phloem without injuring it. One of the most clever methods of doing this takes advantage of the feeding habits of aphids, small insects that insert long pointed feeding devices, known as stylets, into the phloem and feed on phloem sap. The aphid stylet penetrates the tissue lying over the phloem and terminates in an individual phloem sieve tube (box figure 7.2). An aphid stylet can enter a sieve tube without causing the injury response. Thus, an aphid can remain "plugged in" to a sieve tube and feed on phloem sap at leisure for hours. In experimental studies, aphids are allowed to begin feeding, then are anesthetized and cut away from their stylets. This leaves experimenters with small open channels through which phloem sap can be withdrawn without disturbing phloem function and causing an injury response.

(a)

(b)

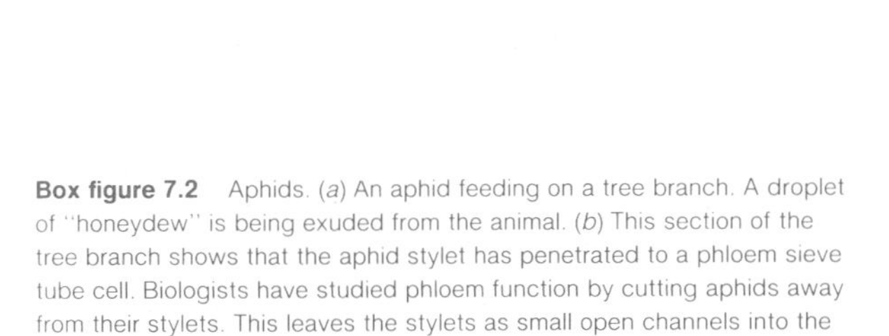

Box figure 7.2 Aphids. (*a*) An aphid feeding on a tree branch. A droplet of "honeydew" is being exuded from the animal. (*b*) This section of the tree branch shows that the aphid stylet has penetrated to a phloem sieve tube cell. Biologists have studied phloem function by cutting aphids away from their stylets. This leaves the stylets as small open channels into the phloem.

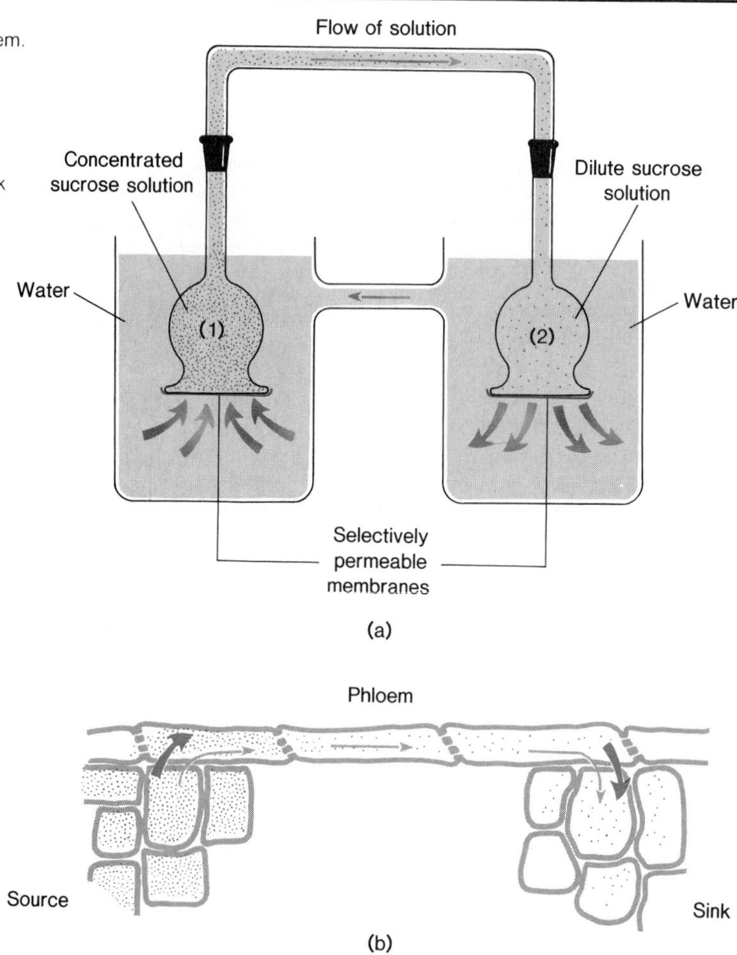

Figure 7.26 Phloem transport. (*a*) Model system in which mass flow occurs. (*b*) Proposed mass flow of water and solute through the phloem. Colored arrows represent active transport of sucrose; gray arrows represent flow of water. At a source (for example, an actively photosynthesizing leaf), sucrose is actively transported into phloem sieve tubes. At a sink (for example, in a root), sucrose is actively transported out of phloem sieve tubes. So much water enters the source end of the phloem by osmosis that water is forced out the sink end. This results in flow of water through the phloem that sweeps sucrose along with it.

When this system is submerged in distilled water, water tends to enter osmotically through both membranes, but this tendency is much stronger at membrane (1) because the concentration difference across membrane (1) is so much greater. Thus, water moves into (1) much more rapidly than into (2), and soon pressure differences in the system result in a flow of water from (1) to (2). This flow occurs because pressure actually drives water out through the membrane of (2), even though water is being forced to move against a concentration gradient. As water flows through the system it sweeps along sucrose, so that this mass flow of water tends to drive the model system toward equilibrium in a relatively short time.

These same principles can be applied to the mass flow hypothesis of phloem transport. The hypothesis states that in phloem the activities of living cells can maintain the concentration differences between different parts of the system, so that instead of approaching equilibrium as the model system in figure 7.26*a* does, the differential is maintained and mass flow continues. During vigorous photosynthetic activity, for example, the leaf can serve as a "source" of sucrose that is loaded into the phloem sieve tubes by active transport across sieve tube element membranes (figure 7.26*b*). Then at another point in the phloem, active transport processes remove sucrose from the sieve tubes and move it toward a storage "sink" in the starch-storing plastids of cells, which are in storage areas such as root cells. Thus, concentration differentials are maintained by energy-requiring active transport, and the mass flow of phloem sap continues.

The mass flow hypothesis also can account for reversal of the flow through any given chain of sieve tube elements. All that is required is a reversal of the source-sink relationship; that is, sucrose is loaded into the opposite end of the phloem and actively removed from the end that previously was the source.

Sieve tubes, therefore, are living pipes through which materials can be translocated from one part of the plant to another. The rate and direction of transport depend upon the needs of the plant at any given time.

Summary

Water and mineral ions are obtained from the soil by root systems and are moved up through vessels to reach the rest of the plant. Organic compounds and other materials are moved from one part of a plant to another through the phloem.

Carbon dioxide is obtained by diffusion through stomata. But stomatal opening results in evaporative loss of water from leaves, called transpiration. Replacement of this loss necessitates movement of large quantities of water up through the xylem from the roots. The transpiration stream also carries minerals absorbed by the roots up through the plant.

An important part of root function is the continuous growth of root tips that brings the tips into contact with new soil areas. Root structural organization is correlated with root functioning in absorption and transport.

The bulk of ion uptake from the soil is accomplished by active transport. Active transport requires ATP energy input and can work against concentration gradients.

Symbiotic relationships contribute to nutrient uptake by roots. Nitrogen deficiency is a common limiting factor in plant growth and productivity.

Dicots and monocots are two major groups of flowering plants that differ structurally. In dicot stems, vascular bundles are arranged in a circle and contain xylem and phloem separated by cambium. Monocot stems have scattered vascular bundles.

Woody stems persist from year to year. Primary growth, initiated by buds, adds to stem length each growing season. Growth, due to proliferation of cambium, increases the thickness of stems by adding new xylem and phloem tissue.

Xylem contains two basic types of conducting tubes, the tracheids and the vessels. Both consist of empty cell walls when mature.

The best current explanation of the force that causes fluid to rise through the xylem is the transpiration pull/cohesion of water hypothesis. Water evaporating from leaf cell surfaces pulls water out of leaf cells. The cohesion of water molecules causes this pulling force (tension) to be transmitted to the xylem where it pulls fluid up from the roots.

Phloem translocates material from sources to sinks through sieve tubes that consist of living but nonnucleated cells.

Characteristics of phloem transport seem to be best explained by the mass flow hypothesis.

Questions for Review

1. What are the general functions of leaves, stems, and roots?
2. Explain the significance of the Casparian strip in the movement of water and mineral ions within the root.
3. Why do plants often suffer from nitrogen deficiency when nitrogen is such a plentiful element in the environment?
4. What do the numbers 27–5–5 on a bag of lawn fertilizer mean?
5. Define meristematic tissue.
6. How could you determine the length that a twig on a tree grew during the growing season two years ago?
7. Explain the cause of root pressure.
8. Explain the relationship between the stomatal opening and the movement of xylem sap up through the xylem tracheids and vessels.
9. What do we mean by cohesion of water?
10. Define source and sink as they relate to phloem transport.

Questions for Analysis and Discussion

1. Why must new root hairs be produced continually in growing roots?
2. It has been difficult to determine whether or not certain elements are or are not required as micronutrients by certain plants. Can you suggest reasons why this might be hard to do in hydroponic experiments? Hint: Only very small quantities of some micronutrients are required.
3. What do you think would happen to xylem transport if air could leak into xylem vessels?
4. Explain the benefit for plants of the phloem injury response in which phloem transport shuts down immediately as soon as phloem tissue is damaged. What effect would applying a metabolic poison to phloem cells in the sink end of the phloem have on phloem transport? Why?

Suggested Readings

Books

Raven, P. H.; Evert, R. F.; and Curtis, H. 1981. *Biology of plants.* 3d ed. New York: Worth Publishers.

Saigo, R. H., and Saigo, B. W. 1983. *Botany: Principles and applications.* Englewood Cliffs, N.J.: Prentice-Hall.

Salisbury, F. B., and Ross, C. W. 1978. *Plant physiology.* 2d ed. Belmont, Calif.: Wadsworth.

Sutcliffe, J. F., and Baker, D. A. 1974. *Plants and mineral salts.* Studies in Biology no. 48. Baltimore: University Park Press.

Articles

Brill, W. J. March 1977. Biological nitrogen fixation. *Scientific American* (offprint 922).

Epstein, E. May 1973. Roots. *Scientific American* (offprint 1271).

Wooding, F. B. P. 1978. Phloem. *Carolina Biology Readers* no. 15. Burlington, N.C.: Carolina Biological Supply Co.

Animal Nutrition and Digestion

Chapter Outline

Chapter Concepts

1. Animals are heterotrophic organisms that must obtain organic compounds from their environment.
2. Besides organic nutrients needed for energy and synthesis, animals require vitamins, minerals, and water.
3. Animal digestive tracts are specialized to transport, process, and absorb the nutrients that must be supplied to all body cells.
4. The kind and degree of specialization in any animal's digestive tract is related to the animal's metabolic requirements and the nature of its food supply.
5. Human digestion takes place in a highly specialized, complete digestive tract. Enzymes hydrolyze large organic nutrient molecules to produce smaller absorbable molecules.
6. Despite efficient protective mechanisms, the digestive tract can be damaged by its own digestive secretions and also is subject to occasional infection.

In Michigan territory, on the 6th of June, 1822, Alexis St. Martin, a young French-Canadian fur trader, was accidentally wounded by a musket shot. He was brought to an army surgeon named William Beaumont for examination and treatment. Beaumont found a large wound through which part of St. Martin's lung and a portion of his stomach protruded. Upon further examination, Beaumont discovered that the stomach had been perforated. Stomach contents were pouring out of the wound.

After cleaning the wound, Beaumont pushed the lung and stomach back into place and proceeded with standard treatments of the time. After several months, although most of the wound was healed, the stomach still remained open to the outside. Unless pressure was maintained on the perforation, stomach contents emptied out. But St. Martin steadfastly refused to undergo surgery to close the opening.

Beaumont kept St. Martin under observation for a number of years during which St. Martin worked for Beaumont, enjoyed good general health, and fathered several more children. During this time, William Beaumont took advantage of this unique opportunity to study the functions of the human stomach. Periodically, he inserted a tube into St. Martin's stomach to collect gastric juices. He also tied tiny pieces of food to silk threads and inserted them through the hole. After varying periods of time, he withdrew the food and studied the effects of the stomach's digestive activity.

Clearly, William Beaumont was a person with an inquiring mind who seized on a great learning opportunity when it was presented to him. Through these studies, Beaumont was able to add considerably to the understanding of digestive functions.

We can separate all organisms into two categories—**autotrophic** or **heterotrophic**—on the basis of the methods by which they obtain the energy and raw materials that they require to live. Plants are the most familiar autotrophs; they use light energy, in photosynthesis, to produce reduced organic compounds. As you saw in chapter 5, plants are chemically versatile. They can produce the full range of organic compounds that they require using only CO_2, H_2O, and various mineral ions taken up from the soil as starting materials.

In this chapter, we will focus on nutrition in heterotrophic ("other-feeding") organisms, such as animals. Heterotrophs must meet their energy and raw material requirements by obtaining already-formed reduced organic compounds. Some animals (**herbivores**) accomplish this by eating only plants. Other animals (**carnivores**) eat only other animals. Still other animals (**omnivores**), such as humans, eat mixed (plant and animal) diets.

Usually, the organic nutrients that animals ingest are not in forms that can readily be absorbed into their bodies. Nutrients must be converted into absorbable forms so they can be transported across the plasma membranes of cells lining the digestive tract. **Digestion** is the sum total of physical and chemical processes that convert the food that animals eat into these absorbable forms. Thus, animals require **digestive systems** to make these conversions of food material.

Nutrients

The three major categories of nutrients are: (1) organic molecules used in energy conversions and as raw material in synthesis; (2) minerals, typically in the form of inorganic salts or ions; and (3) vitamins.

We can subdivide the first category into four groups, three of which are familiar to almost everyone. They are **carbohydrates, proteins,** and **fats.** The fourth group of organic molecules contains the **nucleic acids.** These are not required nutrients because all cells can make their own nucleic acids, but they are always present in animal diets because animals eat the cells of other organisms.

Carbohydrates

Strictly speaking, carbohydrates such as sugars and starch are not essential nutrients for most animals, since carbohydrates can be synthesized in animal cells from proteins or fats. But carbohydrates are important nutrients because they provide the bulk of the organisms' chemical energy supply. One carbohydrate, the sugar **glucose,** is a centrally important nutrient because, as you saw in chapter 6, it is used in cellular energy conversions in the cells of all organisms.

Glucose and other sugars are included in the diets of many animals. But animals also ingest carbohydrates in the form of **polysaccharides,** such as **starch, cellulose,** and **glycogen** (chapter 2). Polysaccharides are polymers of glucose; that is, they are chains of glucose molecules fastened together. Glucose and other monosaccharides are readily absorbed from the digestive tract, but polysaccharides are not. Polysaccharides must be enzymatically broken down (**hydrolyzed**) into sugar units that can be absorbed (figure 8.1).

Proteins

Proteins are important constituents of all cells. Thus, all organisms must obtain the raw materials for protein synthesis. Proteins are polymers of **amino acids,** and animals obtain amino acids by eating and digesting the proteins of other organisms. Proteins are digested to yield amino acids that are then absorbed through digestive tract linings (figure 8.2).

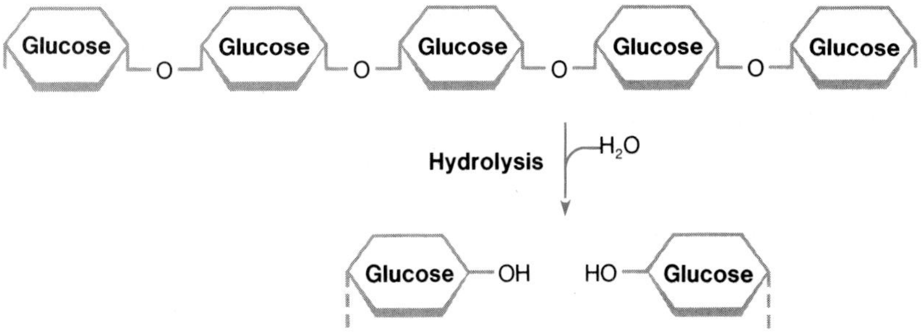

Figure 8.1 Breaking of the bond connecting glucose units in a polysaccharide such as starch. This is called hydrolysis because components of a water molecule are added during the digestive reaction.

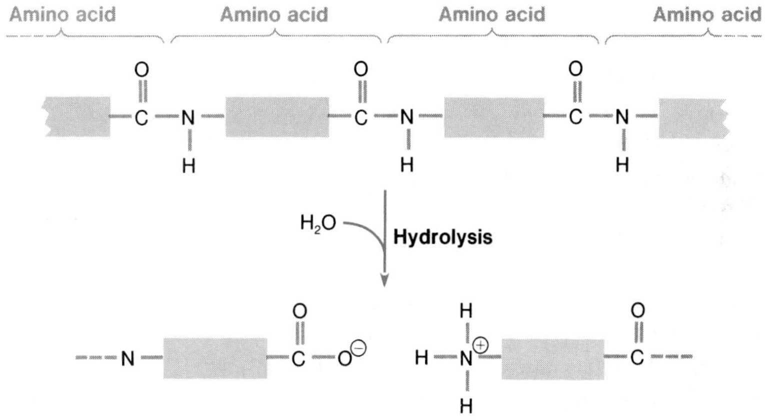

Figure 8.2 Hydrolysis of a peptide bond between two amino acids of a protein molecule in protein digestion. Products of protein digestion are absorbed either as amino acids or as dipeptides (two linked amino acids).

Most animals can use carbohydrate starting material to synthesize some of the amino acids needed for protein synthesis, but not all. The amino acids that animals cannot synthesize are known as **essential amino acids** (see chapter 6). To insure good health, essential amino acids must be present in the diet. Different animals have different sets of essential amino acids. Ten amino acids are essential in the human diet.

Many people around the world suffer from **kwashiorkor.** This is a protein deficiency disease that occurs in people who eat a mainly starchy diet, even though their diets may contain sufficient calories. Growing children with kwashiorkor become anemic and listless, and they have very low resistance to even the mildest diseases and infections. Even if they survive the childhood diseases, many of them become mentally retarded and unhealthy adults. A supply of quality dietary protein is one of the world's greatest health problems.

Table 8.1
Energy Yielded by Oxidation of Organic Nutrients

Nutrients	Average Energy Yield in Kilocalories (and Kilojoules, kJ) per Gram
Carbohydrates	4.1 (17.2 kJ)
Proteins	4.4 (18.4 kJ)
Fats	9.3 (38.9 kJ)

Fats

Fats are very energy rich because fat molecules are highly reduced compounds. They yield more energy per unit weight when they are oxidized than do carbohydrates or proteins (table 8.1).

Figure 8.3 During fat digestion, hydrolysis of fats yields glycerol and fatty acids. Most digested fats are absorbed either as fatty acids or monoglycerides, glycerol with one fatty acid still attached to it.

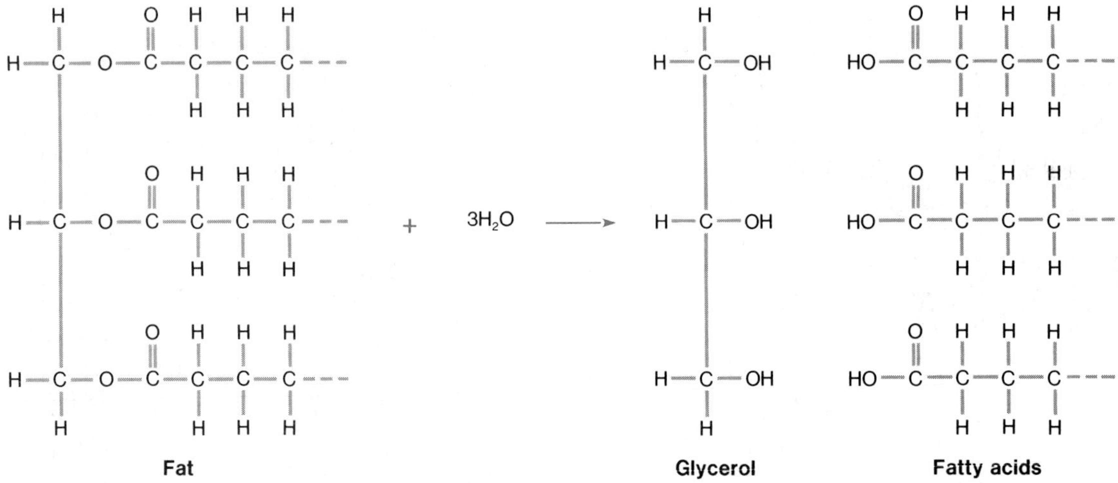

A molecule of the kind of fat commonly eaten by animals contains glycerol and three fatty acids. While some fat is absorbed intact as small droplets that pass through the digestive tract linings, much of the ingested fat is first digested to yield glycerol and fatty acids that are then absorbed (figure 8.3).

As is the case with some of the amino acids required for protein synthesis, certain animals cannot synthesize some fatty acids required for fat synthesis. These **essential fatty acids** must be included preformed in the diet. Without them, deficiency symptoms arise.

Minerals

In animals, some minerals are required in relatively large quantities. For example, sodium and potassium are vital constituents of all body fluids. Calcium and magnesium are found in large quantities in bones and teeth. Phosphorus is important in bones and teeth, but also is critical for synthesis of nucleic acids and other phosphate-containing substances such as ATP.

Other minerals are required in much smaller quantities. These minerals, known as **trace elements,** are no less important, however. For example, iron is essential for hemoglobin production and iodine is required for thyroid hormone synthesis.

Just because an element is present in the body, though, does not necessarily indicate that the element is required in the diet. Indeed, a chemical analysis of a typical human from an industrialized society probably would reveal trace amounts of mercury and lead. These elements definitely are not essential; in fact, they are poisons! However, they are not easily eliminated from our bodies and tend to accumulate in tissues such as fat and hair.

Vitamins

Vitamins are a set of relatively simple organic molecules that are required in small quantities for a variety of biological processes. Because vitamins are compounds that cannot be synthesized in the body, they must be present in small quantities in an organism's diet. The lists of organic compounds classified as vitamins are not the same for all species because of differing capabilities for synthesis. For example, rats can synthesize **ascorbic acid,** but humans cannot. Thus, ascorbic acid is a vitamin (**vitamin C**) for humans and must be included in the diet, but it is not classified as a vitamin for rats.

Vitamins play a major role in a wide variety of biological processes. For example, B vitamins are essential as coenzymes involved in cell metabolism (see page 91). Vitamin A is required for photochemical reactions in vision, and vitamin C must be present for maintenance of connective tissues and promotion of normal healing. Vitamin molecules can function over and over in these roles and, thus, are not depleted as quickly as other nutrients are. Maintenance of normal vitamin levels requires only a small, steady supply from the diet. Serious, noticeable vitamin deficiency symptoms develop only after extended periods of dietary vitamin deprivation.

Surprisingly, the need for vitamins has been recognized, at least indirectly, for centuries. The ancient Egyptians determined that certain eye problems could be cured by eating

Figure 8.4 Scurvy. Gum bleeding is one of the most characteristic symptoms of severe vitamin C deficiency.

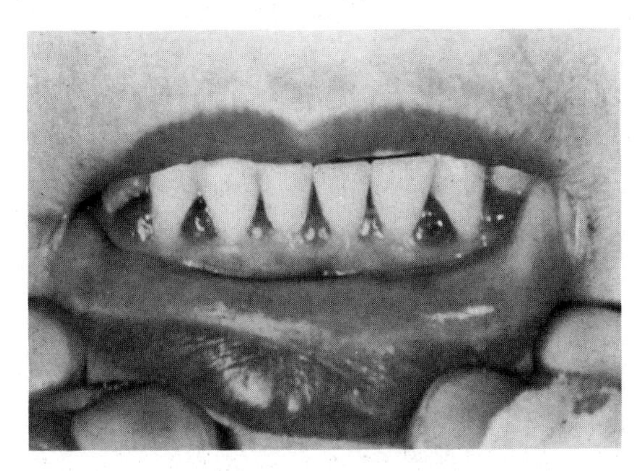

liver, which we now know contains a great deal of vitamin A. Over 200 years ago, an English physician noted that sailors who ate citrus fruits did not develop **scurvy,** a disease whose symptoms include anemia, slow healing, painful joints, gum bleeding, and eventual loss of teeth (figure 8.4). After this discovery, scurvy, formerly the scourge of sailors of the day, was prevented because the British Admiralty insisted that all ships carry limes on board to be included in the sailors' rations. This regulation accounts for the British (and especially British sailors) being called "Limeys." Vitamin C, the nutrient that had been missing from the sailors' diets, was chemically identified as ascorbic acid in 1933.

In 1886, a group of Dutch scientists was sent to investigate the crippling disease **beriberi** in the East Indies. Originally, they assumed that beriberi was caused by a microorganism. But then one of them noted that chickens developed beriberilike symptoms when they were fed the same kind of polished rice eaten by people suffering from beriberi. Polished rice is simply rice from which the hull has been removed to retard spoilage. When the chickens were fed rice with intact hulls, they quickly recovered. The same treatment proved effective for human victims of beriberi. Years later, the "antiberiberi factor" was identified as **thiamine (vitamin B_1).** Because most of the thiamine of rice is contained in the hull, and because much of that remaining in the rice kernels is lost in cooking (thiamine is water soluble), polished rice, like other milled grains, is vitamin deficient. This also explains why white flour typically must be "vitamin enriched." Virtually all the vitamins originally present in the grain are lost during the milling process, and vitamins must be added to the flour.

Over the years, the list of vitamins required for humans has grown and is today considered virtually complete, although the functions of a few vitamins (such as vitamin E) are still not well understood.

Figure 8.5 Common vitamin deficiency syndromes. (*a*) **Pellegra,** due to niacin deficiency. An obvious symptom of pellegra is dermatitis, a severe skin irritation that develops in areas exposed to sunlight. (*b*) **Rickets,** due to vitamin D deficiency. Because a vitamin D deficiency reduces the body's ability to absorb and use calcium and phosphorus, bones become soft and deformed. This child has deformed ribs (arrow), ankles, and wrists, as well as bowing of the legs, which is a characteristic symptom of rickets.

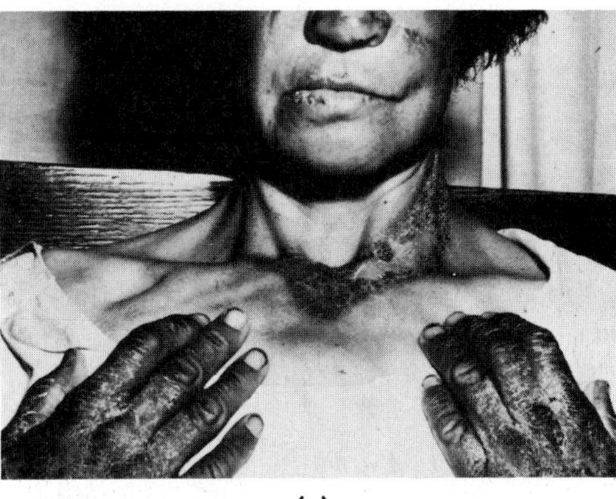

(a)

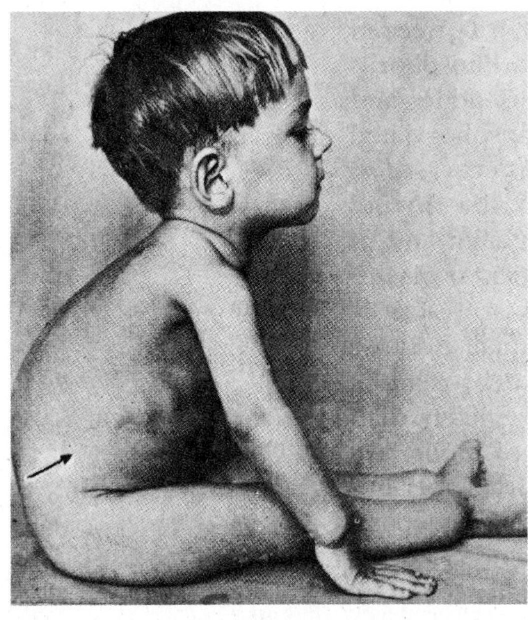

(b)

We can divide the vitamins into two major groups: those that are soluble in organic solvents (**fat-soluble vitamins**) and those that dissolve in water (**water-soluble vitamins**). Fat-soluble vitamins include A, D, E, and K. Water-soluble vitamins include the vitamins of the B complex and vitamin C. Information about vitamins required by humans is summarized in table 8.2. Two common vitamin deficiency syndromes are shown in figure 8.5.

Table 8.2
Functions, Deficiency Symptoms, and Sources of Vitamins

Vitamin	Physiological Role	Deficiency Symptoms	Some Major Food Sources
Water-Soluble			
Thiamine (B_1)	Coenzyme in carbohydrate metabolism	Beriberi, neuritis, loss of appetite, heart failure, indigestion, fatigue, edema, mental disturbance	Whole grains, organ meats, yeast, nuts, pork
Riboflavin (B_2)	Coenzyme in protein and carbohydrate metabolism, as part of FAD	Vascularization of the cornea, inflammation and fissuring of the skin, sores on the lip, swollen tongue	Milk, cheese, eggs, liver, yeast, leafy vegetables, wheat germ
Niacin (B_3)	Coenzyme in energy metabolism (part of NAD and NADP)	Pellagra, skin eruptions, neuritis, fatigue, digestive disturbances	Whole grains, yeast, liver and other meats
Pyridoxine (B_6)	Coenzyme in many phases of amino acid metabolism	Anemia, convulsions, neuritis, dermatitis, impairment of antibody synthesis	Whole grains, kidney, liver, fish, yeast
Pantothenic acid	Forms part of Coenzyme A	Neuromotor and cardiovascular disorders, impairment of antibody synthesis	Present in most foods, especially meat, whole grains, eggs
Biotin	Coenzyme in decarboxylation and deamination	Dermatitis, muscle pains (deficiency is rare)	Egg yolk, liver, yeast
Folic acid	Coenzyme in formation of nucleotides and heme	Anemia, impairment of antibody synthesis	Leafy vegetables, liver
Cobalamin (B_{12})	Coenzyme in formation of nucleic acids and proteins	Pernicious anemia	Liver and other organ meats
Ascorbic acid (C)	Vital to collagen and intercellular cement for bones, teeth, cartilage, and blood vessels; maintains resistance to infection; frees iron to make hemoglobin	Scurvy, anemia, slow wound healing	Citrus fruits, tomatoes, green leafy vegetables
Fat-Soluble			
A (retinol)	Formation of visual pigments; maintenance of normal epithelial structure	Night blindness; dry, flaky skin; and mucous membranes	Green and yellow vegetables, dairy products, egg yolk, fruits, butter, fish-liver oil
D (calciferol)	Increases absorption of calcium and phosphorus and their deposition in bones	Rickets	Fish oils, liver, egg yolk, milk and other dairy products, action of sunlight on lipids in the skin
E (tocopherol)	Antioxidant; protects red blood cells, vitamin A, and unsaturated fatty acids from oxidation	Fragility in red blood cells, male sterility (in rats)	Widely distributed, especially in meat, egg yolk, green leafy vegetables, seed oils
K	Needed in synthesis of prothrombin, which is necessary for blood clotting	Slow blood clotting and hemorrhage	Green leafy vegetables, synthesis by intestinal bacteria

Minimum daily requirements (now usually referred to as **"recommended daily allowance"** or **RDA**) for most of the vitamins have been calculated for humans. In a sense, however, these values are really only educated guesses. If you ingest the recommended daily allowance for vitamin C, you presumably will never develop scurvy. But might there be milder, less dramatic consequences of getting no more than the minimum amount? Some scientists such as Linus Pauling, winner of two Nobel Prizes, think that we actually require vitamin C in much larger doses than the recommended daily allowance. Pauling says that failure to meet this larger "requirement" does not result in any life-threatening disease, but he believes that large doses of vitamin C increase resistance to upper respiratory diseases such as the common cold.

Serious problems can develop from taking large doses of some vitamins (**megavitamin therapy**), however. Although excess amounts of the water-soluble vitamins are excreted in the urine, excess amounts of fat-soluble vitamins (especially vitamin A) accumulate in fat tissues in the body. Death by massive vitamin overdose is rare, but vitamin overdoses can cause health problems. Furthermore, there is little or no clear-cut evidence that megavitamin therapies actually produce the benefits claimed by their proponents.

Figure 8.6 The gastrovascular cavity of *Hydra* is an incomplete digestive tract because it has only one opening, a mouth, that must serve as both entry and exit. Extracellular digestion occurs in the cavity. Intracellular digestion occurs inside food vacuoles formed when phagocytic cells engulf food particles.

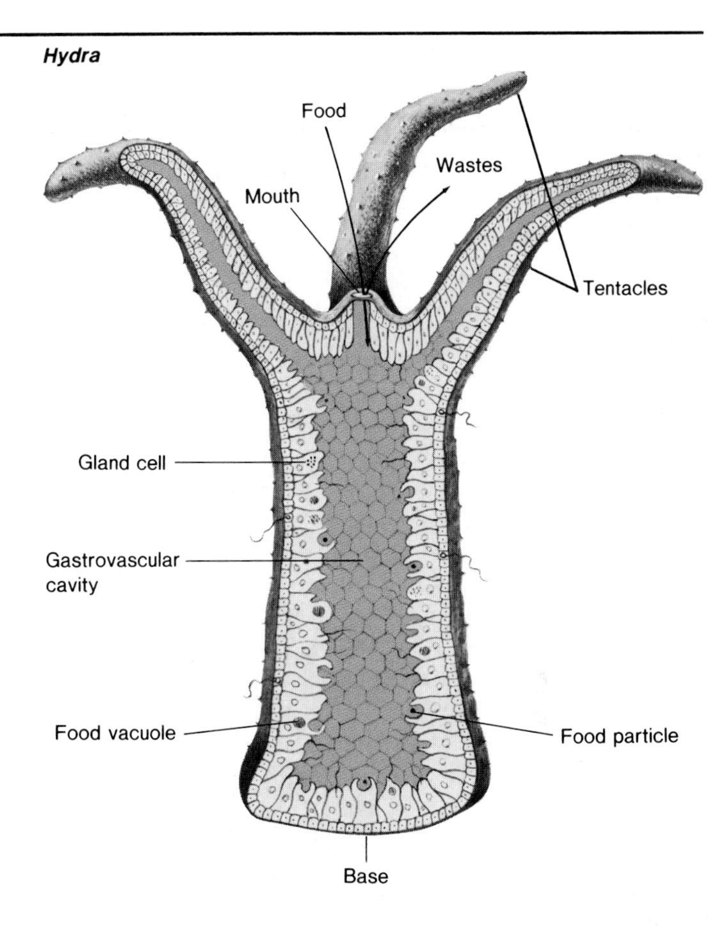

Animal Digestive Tracts

In animals, the cells that are specialized for digestion surround and line a digestive cavity (**gut**). Thus, one group of cells in one part of the body is specialized to obtain nutrients for all cells in the entire body.

In the most primitive animals, the gut is a blind (closed) sac with only one opening that must serve as both entrance and exit. Undigested residues must be ejected through the same opening through which the food originally entered. Such digestive tracts are said to be **incomplete.**

Larger, more complex multicellular animals have gut tubes that are tunnellike. Food enters at one end through a **mouth** and flows through the gut where digestion and absorption occur. Undigested residues pass out through a posterior exit, the **anus.** Such flow-through systems are called **complete digestive systems.**

Incomplete Digestive Tracts

The small freshwater animal *Hydra* has an incomplete digestive tract. Its digestive cavity (**gastrovascular cavity**) is a blind sac with only one opening, the mouth (figure 8.6). Body projections called **tentacles** capture prey organisms and push them through the mouth opening into the digestive cavity.

Some specialized cells in the cavity lining secrete digestive enzymes into the gastrovascular cavity, where digestion of a food particle or prey organism begins. Such digestion outside cells is called **extracellular digestion.** But the bulk of digestion in *Hydra* is **intracellular digestion.** Phagocytic cells in the lining of the gastrovascular cavity engulf pieces of food material and continue digestion inside food vacuoles. The phagocytic cells, in turn, relay the products of digestion to other cells of the organism.

Flatworms such as **planarians** also possess an incomplete gut (figure 8.7).

Figure 8.7 A planarian's gastrovascular cavity is more complex than that of *Hydra* because it branches extensively, but the two are functionally similar because they are both incomplete digestive tracts with only one opening. When a planarian feeds, it sticks its muscular pharynx out of its mouth and sucks in food.

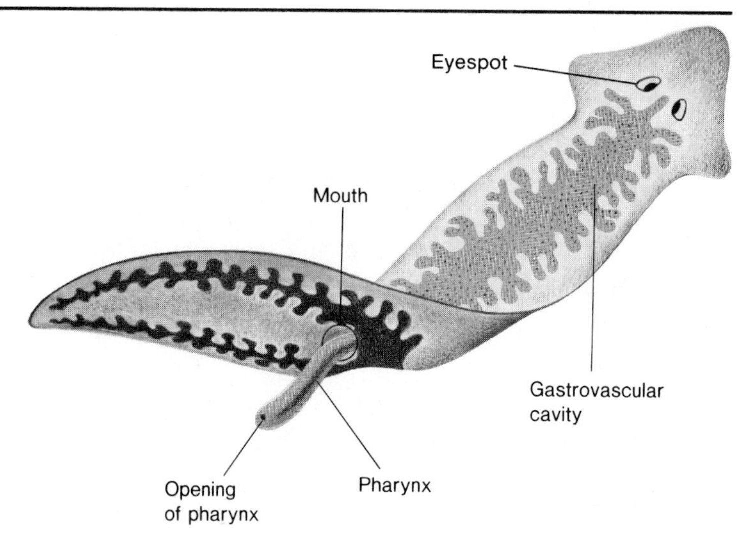

Figure 8.8 Example of a complete digestive tract, the simple digestive tract of the parasitic roundworm *Ascaris*, which lives in the intestines of various domestic animals.

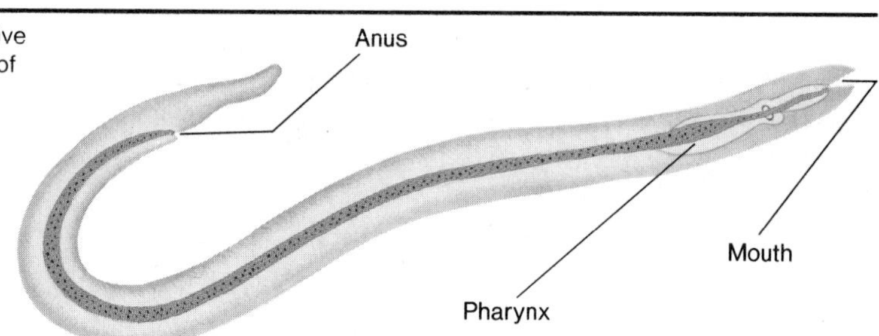

Complete Digestive Tracts

In an incomplete digestive tract, there is no effective means of segregating recently ingested food from food ingested much earlier. A complete digestive tract, however, permits one-way flow because undigested food does not have to pass back out the same way that it came in (figure 8.8).

Complete digestive tracts also have the advantage of progressive digestive processing in various specialized regions along the gut. Food can be digested efficiently in a series of distinctly different steps.

Food Storage

Many animals' complete digestive tracts have regions specialized for food storage. This permits such animals to eat relatively large meals periodically and then digest their food while doing other things.

Stomachs provide storage for large meals eaten by many animals. Another storage mechanism is the **crop.** A crop is an enlargement in the first part of the digestive tract. Crops are found in such diverse animals as earthworms, insects, and birds (figure 8.9).

Physical Breakdown of Food

Digestion is more efficient if food is broken up into small fragments, thereby exposing more surface area to the action of enzymes. **Teeth** physically break down food. Some animals' teeth are used for grinding up food, while others use their teeth for killing or for ripping chunks out of their prey (figure 8.10).

Another food grinding device is the **gizzard,** an organ found in the digestive tracts of many kinds of animals (see figure 8.9*a* and *c*). Gizzards have very muscular walls between which gut contents are rubbed back and forth vigorously, thus grinding the food. Hard particles swallowed with food aid this grinding action. Perhaps you have noticed birds eating small stones. They do this regularly to supply grinding material for their gizzards.

Many species of animals do not need any mechanism for physical breakdown of food because they are **fluid feeders** that pierce other organisms and feed on their body fluids (figure 8.11).

Box 8.1
"Predatory" Fungi

In addition to animals, one other kingdom of organisms consists entirely of heterotrophic organisms. This is the kingdom Fungi, and it includes molds, rusts, smuts, and mushrooms.

What is the source of nutrient supply for the familiar bracket fungus growing on the side of a tree (box figure 8.1*A*)? If you guessed the tree, you are correct, since the fungus lacks chlorophyll and cannot photosynthesize. The fungus secretes enzymes that digest large organic molecules of the tree. This extracellular (outside the cell) digestion produces a supply of absorbable nutrients that can be utilized by the fungus.

Fungi that live on dead organisms and derive their nutrients by participating in decay processes are called **saprophytes.** Other fungi, including many that grow on living trees, are **parasites.** Parasitic fungi grow on or inside many different living plants or animals. The invasive and digestive activities of these parasitic fungi can cause significant damage to the host organism.

Some of the most unusual of fungal nutritional adaptations are seen in a group of soil fungi that capture living animals. These "predatory" fungi produce a sort of hangman's noose (box figure 8.1*B*) that tightens around the body of a soil roundworm and traps it. Once a worm has been secured, fungal cells grow into the worm and secrete enzymes that begin to digest the worm's body. Capturing a worm in this way provides the fungus with a concentrated source of organic nutrients.

Box figure 8.1*A* A bracket fungus growing on a tree.

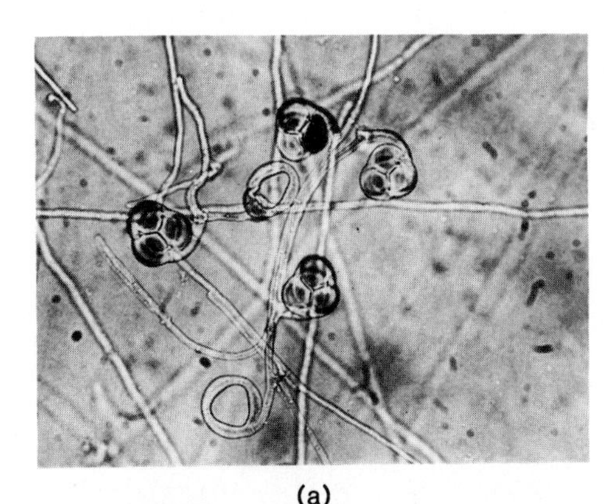

(a)

(b)

Box figure 8.1*B* (*a*) A fungus that traps roundworms (nematodes). The "nooses" consist of three fungal cells (magnification × 275). (*b*) When a roundworm enters the loop, the cells swell and the worm is held firmly in place. Fungal cells invade the body and secrete enzymes that digest it (magnification × 275).

Figure 8.9 Crops are expanded regions in the first part of the digestive tract that function as food storage organs. Animals that have crops include (*a*) earthworms, (*b*) grasshoppers, and (*c*) birds.

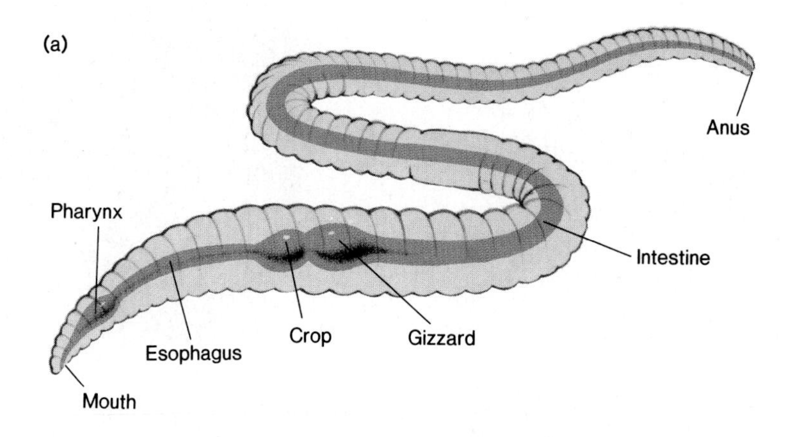

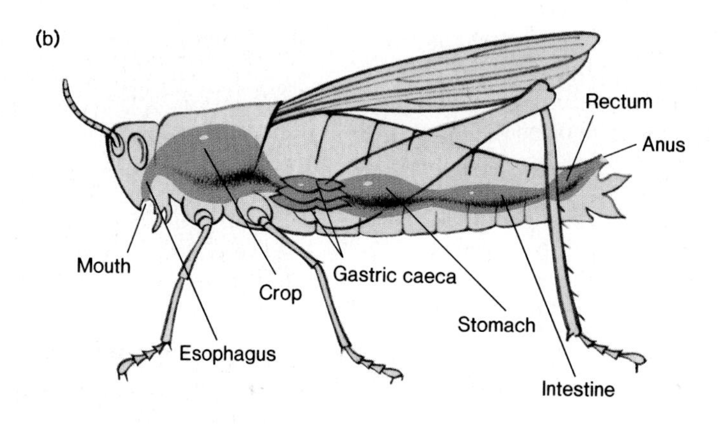

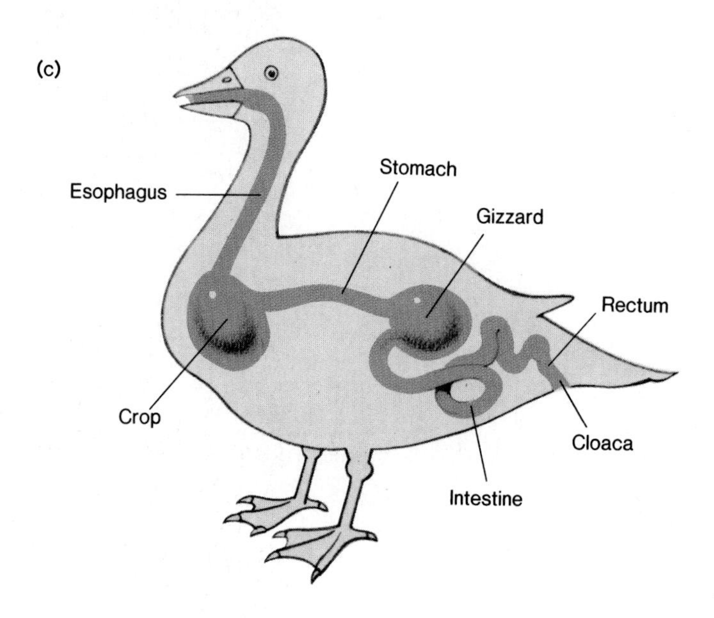

Figure 8.10 Comparison of teeth of several mammals. (*a*) Horse, a grazing herbivorous mammal that clips off plants with its incisors and grinds them with its flat molars. (*b*) Lion, a carnivorous mammal that uses its teeth for killing and for tearing off chunks of flesh that are swallowed whole. (*c*) Human, an omnivorous mammal.

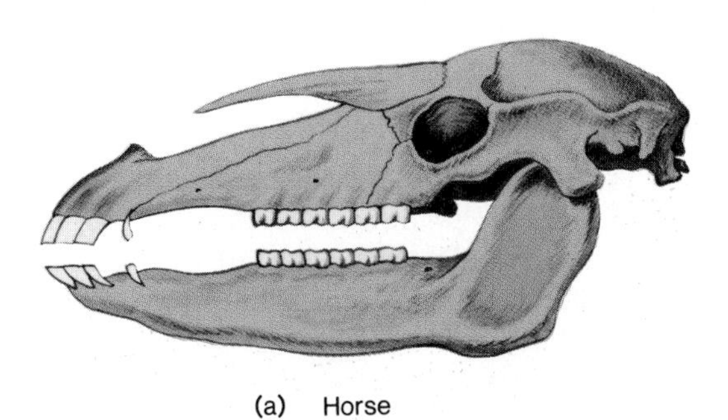

(a) Horse

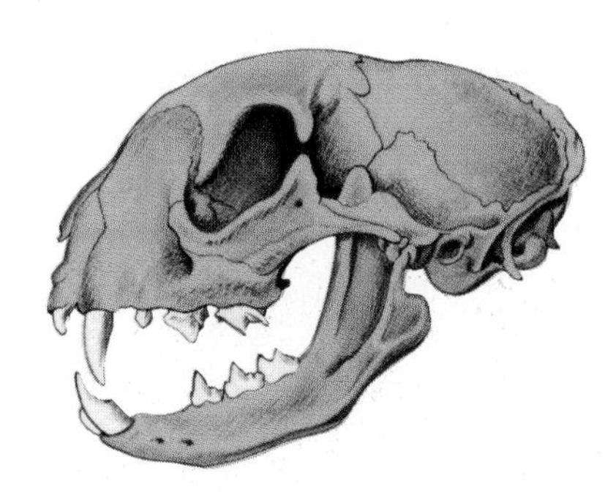

(b) Lion

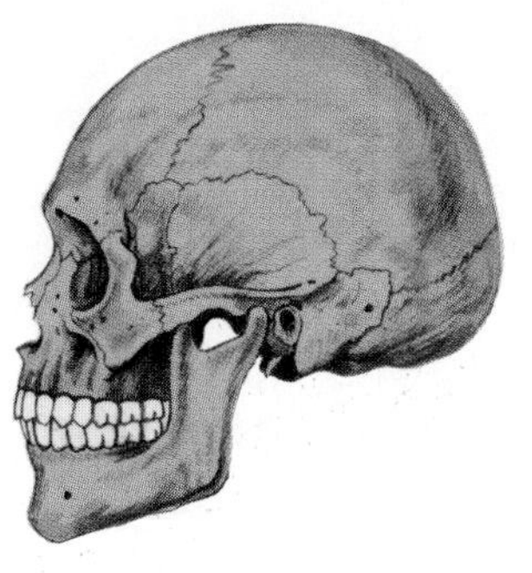

(c) Human

Figure 8.11 Fluid feeders. Leeches (*a*), vampire bats (*b*), and mosquitoes (*c*) feed on blood. Aphids (*d*) feed on plant sap.

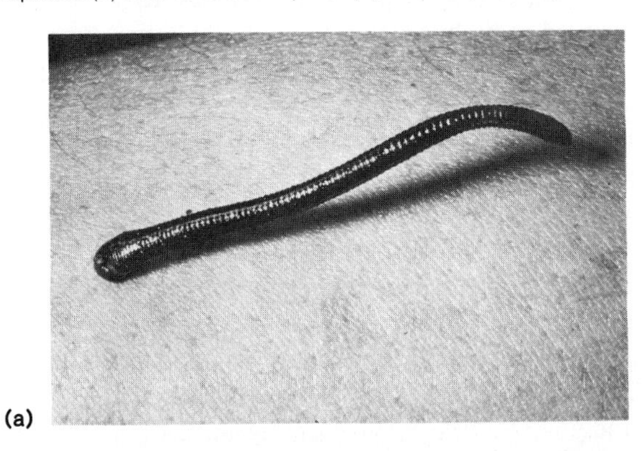

(a)

(b)

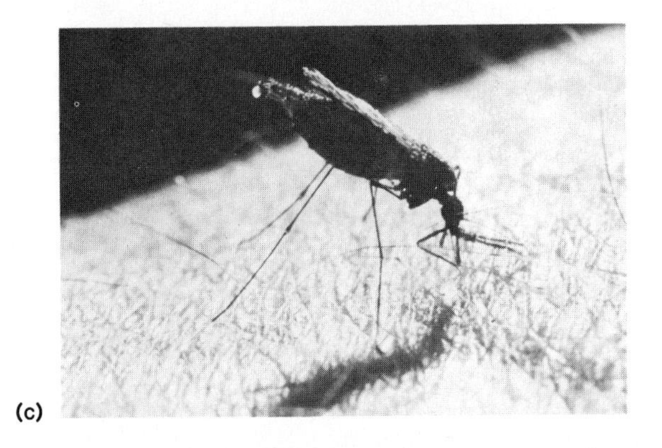

(c)

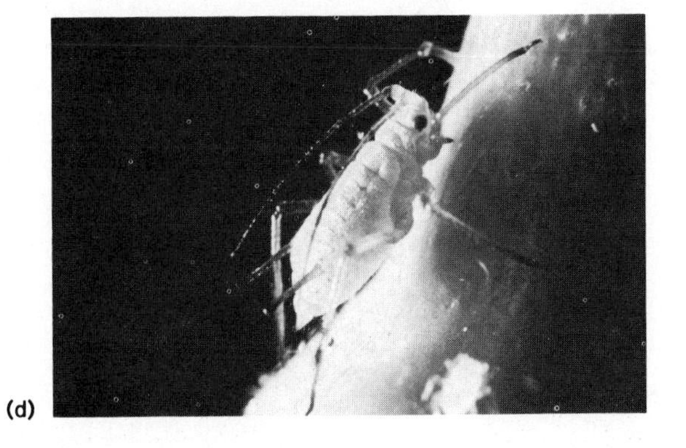

(d)

Carnivores, Herbivores, and Omnivores

The food of carnivores (flesh) is relatively easier to digest than the food of herbivores or omnivores. Consequently, carnivores tend to have relatively shorter and less complicated guts.

On the other hand, herbivores tend to have longer, more complex digestive tracts. Herbivores digest the contents of plant cells that are broken open during chewing and during further processing in the herbivores' digestive tracts. In addition, many herbivores have mechanisms that allow them to derive nutrition from the cellulose in plant cell walls. Microorganisms that can break down cellulose are housed in special saclike regions (such as the **rumens** of cows, sheep, deer, and goats) of the herbivores' digestive tract. These microorganisms produce **cellulase** (see page 42), an enzyme that hydrolyzes cellulose (figure 8.12). This symbiotic arrangement makes it possible for herbivores to derive energy from cellulose, which is abundantly available, even though their own digestive tracts do not produce cellulase.

Omnivores fall between carnivores and herbivores in a dietary sense and, not surprisingly, their gut lengths generally are intermediate between the two extremes. Omnivores do not have mechanisms for utilizing cellulose, so they can digest only the intracellular (cell contents) portion of the plant material in their diets.

Absorption Surfaces

There are several different specializations that provide increased surface area for nutrient absorption. Perhaps the simplest way to have a large surface area for absorption is to have a long absorptive section in the gut. Many different animals do indeed have long, coiled guts.

But surface areas are increased in the relatively straight, uncoiled guts of some animals by other means. For example, most earthworms have a longitudinal fold called a **typhlosole** that gives their gut nearly twice as much surface area as a simple round tubular gut of similar length (figure 8.13). Sharks have a complex folded membrane called a **spiral valve**

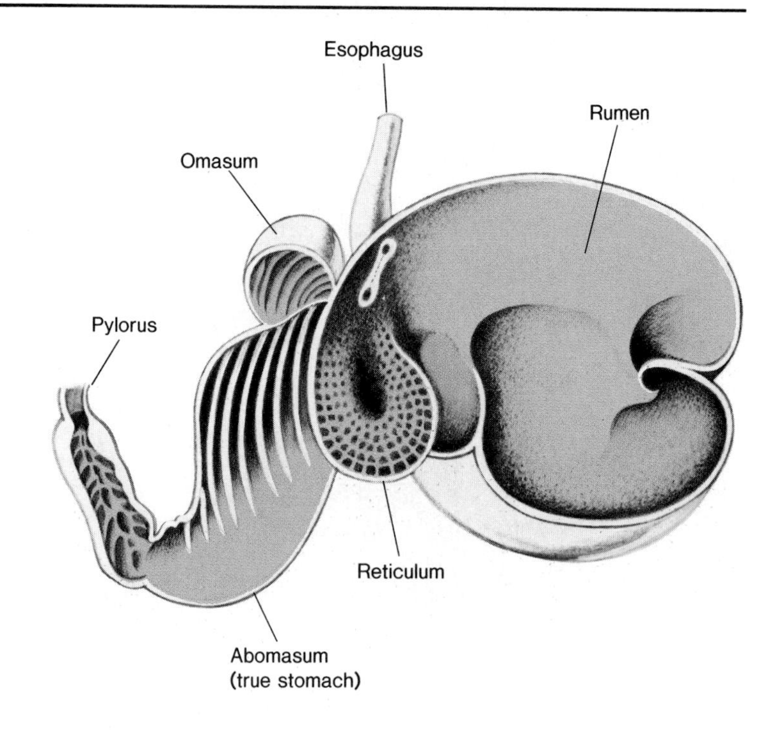

Figure 8.12 Stomach compartments of a ruminant mammal. Bacteria and protozoa, living in a dilute salt solution in the rumen and reticulum, hydrolyze cellulose in plant material. Food is regurgitated repeatedly for "cud" chewing and is swallowed again. Eventually, food is swallowed into the omasum, which churns it up before it passes into the abomasum. The abomasum is a typical vertebrate glandular stomach.

that forces food to follow a route resembling a spiral staircase as it passes through the gut. As a result, the food is exposed to a much larger surface area and more nutrients can be absorbed more easily (figure 8.14).

Further modification of the layers lining the gut has produced much larger absorptive surface areas in mammals, including humans. For example, the human small intestine is not a simple, smoothly lined tube; it has ridges and furrows that give it an almost corrugated appearance. On the surface of these ridges and furrows are small fingerlike projections called **villi.** Cells on the surfaces of villi have minute projections called **microvilli** (figure 8.15). These structural features increase the actual absorptive surface of the human small intestine to about 120 times the area of a simple, smooth tube of similar length. A smooth tubular small intestine would have to be five to six hundred meters long to have a surface area comparable to the human small intestine. Such a huge, long gut would be hard to carry around, and if the movement rate were the same as it is in a normal human gut, it would require weeks for food to traverse the small intestine.

Figure 8.13 Cross section of an earthworm showing the typhlosole, which is a longitudinal fold that greatly increases the gut's absorptive surface area.

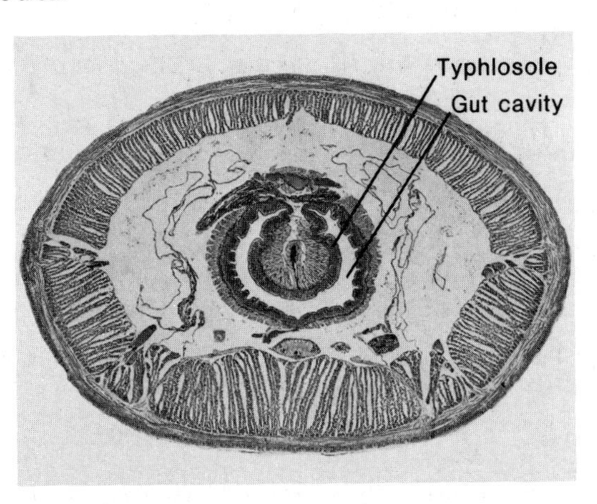

Figure 8.14 The spiral valve greatly increases the internal surface area of the shark intestine.

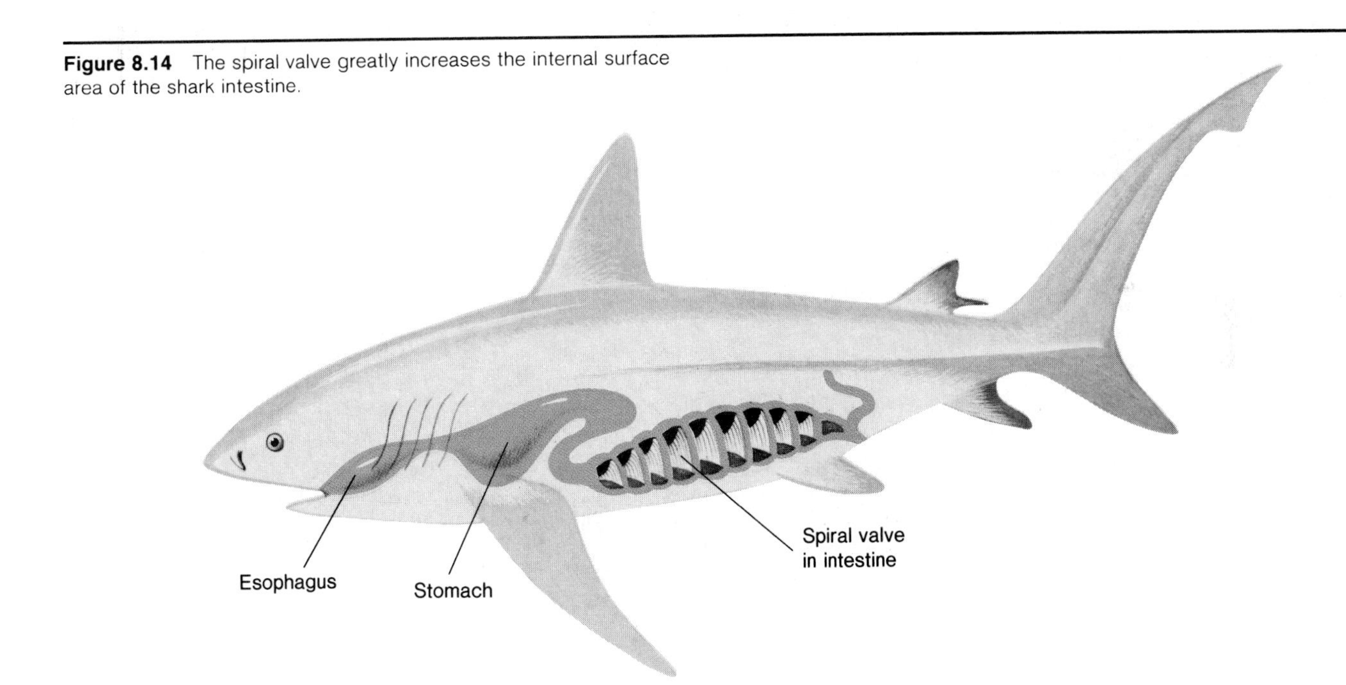

Figure 8.15 Structures that increase surface area for absorption. (*a*) Sketch of several villi in the human small intestine. (*b*) Scanning electron micrograph of microvilli (Mv), which are tiny projections on cells of villi (magnification × 10,370).

(*b*) Kessel, R. G., and Kardon, R. H. *Tissues and Organs: A Text-Atlas of Scanning Electron Microscopy.* © 1979 by W. H. Freeman and Company.

(a)

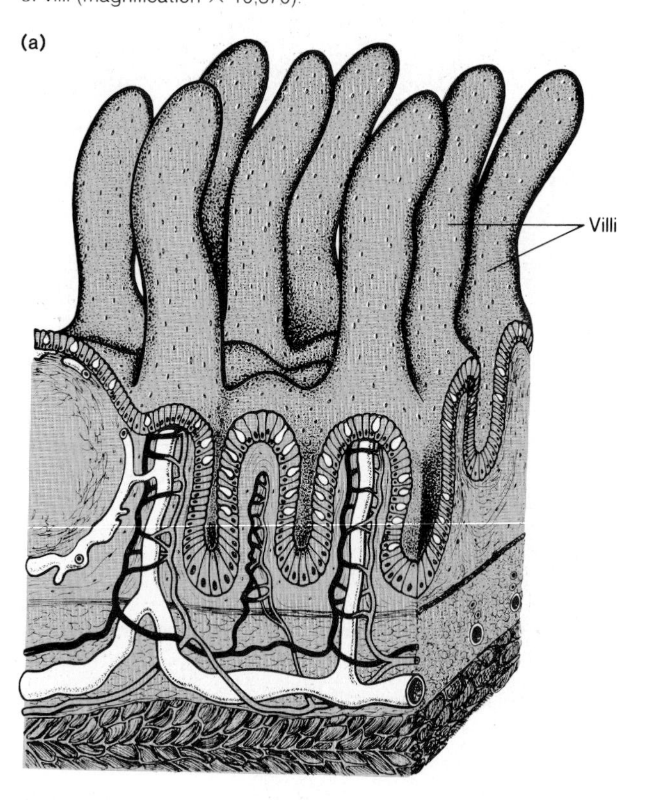

(b)

Box 8.2

Why Do Gut Lining Cells Die Young?

All along the digestive tract, cells in the layer lining the gut cavity are continuously being replaced. This replacement occurs at a surprisingly rapid rate in all parts of the digestive tract, but it is especially rapid in the small intestine. Cell divisions produce new cells that replace the millions of lost cells that shed from the tips of the intestinal villi into the gut cavity each day. Cells on the surface of villi in the small intestine, on the average, live less than forty-eight hours, a very short cell life span compared to other cell types. For example, red blood cells usually are considered relatively short-lived cells, but they live about 120 days.

Why do gut lining cells have such a short life span? The answer is not altogether clear, but there are several possibilities. Cell replacement in the gut may be a mechanism that repairs normal wear and tear caused by abrasion and the actions of digestive enzymes on intestinal lining cells, despite the protective mucous coating that covers them. It might also be, however, that only young cells are efficient enough to carry out the vital absorptive processes in the intestine. Two-day-old cells may have "aged" so much that they can no longer function adequately.

Unfortunately, the vital importance of this continuing cell replacement in the digestive tract is sometimes demonstrated in cancer patients receiving chemotherapy. These treatments suppress cell division throughout the body, both in the cancer that is the target of the therapy and in other areas where continuing cell division is necessary for normal cell replacement. One of the side effects of cancer chemotherapy can be intestinal problems associated with failure of the gut to produce enough new lining cells.

Figure 8.16 The human digestive tract. The liver is drawn smaller than normal proportional size and moved back to expose the gallbladder, ducts, and duodenum. The **common bile duct** carries bile from the gallbladder and liver to the duodenum.

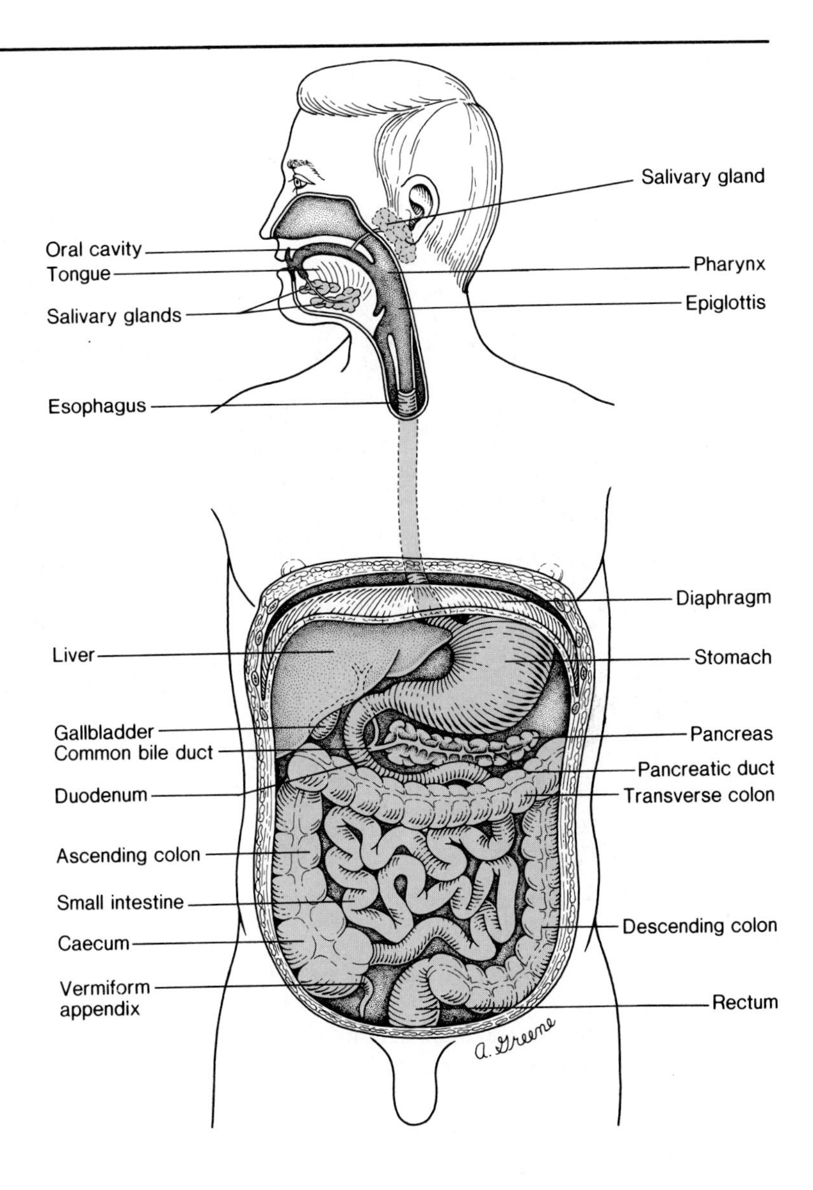

Digestion in Humans

The human digestive tract (figure 8.16) is specialized to fulfill the following basic functions of digestion: food acquisition, food storage and transport, physical breakdown, chemical digestion, absorption of nutrients, and concentration and evacuation of wastes.

The Mouth

Food enters the human digestive tract through the mouth, where it is manipulated by the muscular tongue and chewed by a set of teeth that are specialized for several different actions. The sharp **incisors** in the front of the mouth are used for biting; that is, they cut off chunks of food from larger pieces. Pointed **canine** teeth are used for tearing food, and the flattened **premolars** (**bicuspids**) and **molars** located in the posterior part of the mouth are specialized for grinding and crushing (figure 8.17*a*).

The tongue has sensory functions in addition to its food-manipulating role. It has touch, temperature, and pressure receptors similar to those in the skin, and **taste buds** are scattered over its surface. Taste buds are chemical receptors that provide sensory information about the chemical composition of food (figure 8.17*b*).

Three pairs of **salivary glands** secrete **saliva** into the mouth where it mixes with the food. Saliva is a watery liquid that moistens food so that it can move on more smoothly.

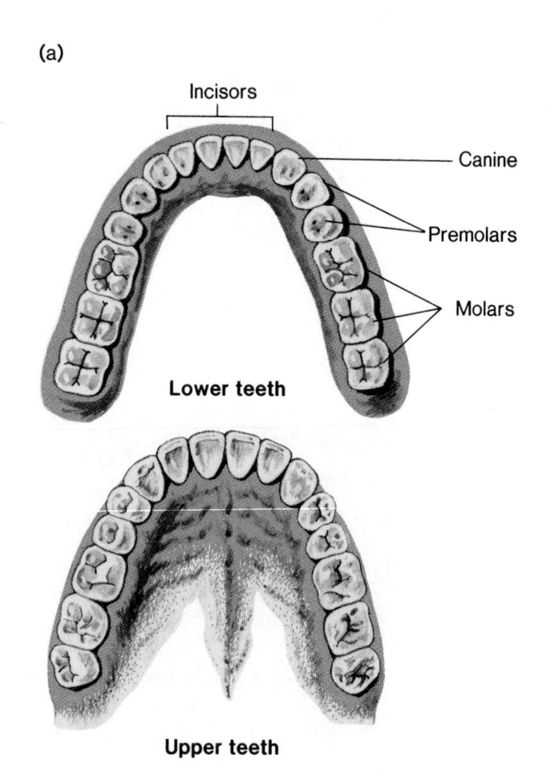

(b)

Figure 8.17 Teeth and taste buds. (*a*) Human teeth. (*b*) Papillae on the tongue. Papillae give the tongue's surface its rough texture. Taste buds are located in circular depressions below the surface of the flat-topped papillae.

Saliva also contains an enzyme, **salivary amylase,** that begins the process of starch digestion by hydrolyzing some of the bonds between the glucose units making up starches.

When food has been chewed and mixed with saliva, the tongue initiates the process of **swallowing** by pushing the food back through the **pharynx** toward the **esophagus.**

The Pharynx and Esophagus

The digestive and respiratory passages cross in the pharynx (figure 8.18). Thus, swallowing poses a potential problem because food might enter the **trachea** ("windpipe") and block the path of air to the lungs. Normally, however, swallowing involves a set of reflexes that close off the opening into the **larynx,** which lies at the top of the trachea. A flap of tissue, the **epiglottis,** covers the opening as muscles move the food mass through the pharynx into the esophagus.

Muscle contractions in the esophagus, and throughout the remainder of the digestive tract, occur in rhythmic waves. This rhythmic muscle contraction that moves material along the digestive tract is called **peristalsis** (figure 8.19).

The Stomach

The stomach can fairly easily hold up to two liters of material. You probably are well aware, however, how much your stomach can stretch to hold even more. The muscular walls of the stomach contract vigorously and mix food with **gastric secretions** (secretions of the stomach) from glandular (secreting) cells in the stomach wall. Because of the churning and mixing, stomach contents come to have a thick, soupy consistency. This thoroughly mixed suspension of food and gastric secretions is called **chyme.**

Some stomach lining cells secrete **mucus** that lubricates food as it is being mixed. Mucus also forms a protective coating over the stomach lining and prevents the stomach from being damaged by the other secretions of glandular cells in the stomach.

Other secretory cells in the stomach lining secrete **hydrochloric acid (HCl)** into the **lumen** (internal cavity) of the stomach, making stomach contents very acidic (pH of 2.0 or less). This highly acid environment kills most living cells in the food, as well as bacteria and other microorganisms taken in with the food. The acid also erodes and loosens up food material and exposes more surface area, thus facilitating further digestion.

Figure 8.18 Swallowing. Respiratory and digestive passages cross in the pharynx (*a*). During swallowing (*b*), the epiglottis covers the opening into the trachea and prevents food from entering it.

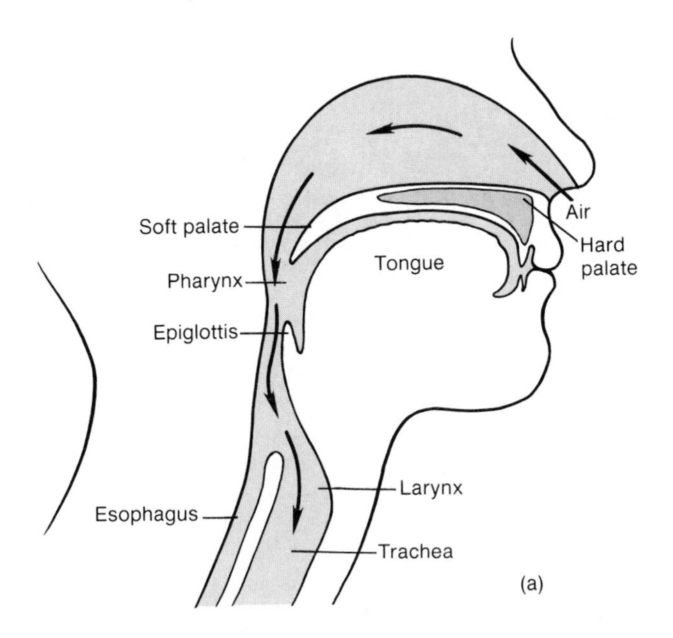

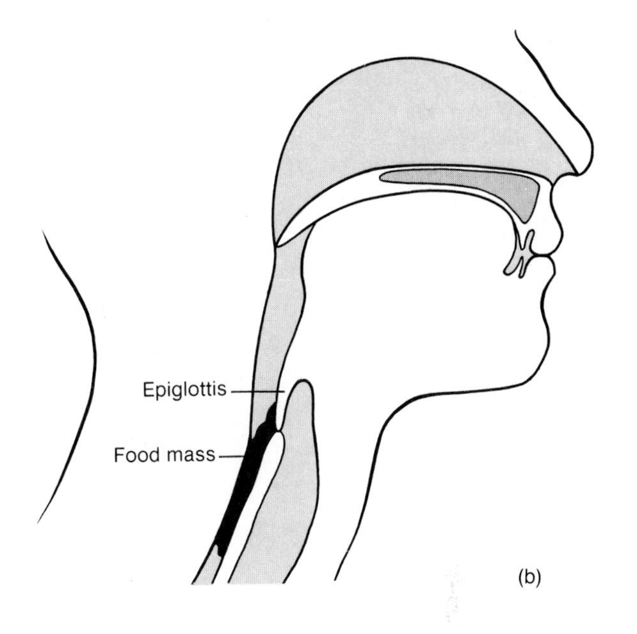

Certain secretory cells in the stomach lining secrete a protein-digesting enzyme, **pepsin,** that functions well at such low pH levels. Pepsin begins the process of protein digestion by hydrolyzing some of the peptide bonds that bind amino acids together in protein molecules. Pepsin breaks up a protein molecule into shorter chains of amino acids (peptides), but further digestion by other enzymes in the small intestine is required to yield the individual amino acids that can be absorbed through the intestinal wall.

Pepsin is an enzyme that hydrolyzes peptide bonds in many kinds of protein molecules. You might ask what prevents it from attacking the cells that produce it or from digesting the lining of the stomach itself. One protection against self-digestion is that pepsin is produced and secreted in an inactive form, **pepsinogen.** Pepsinogen does no damage inside the cells that synthesize it and is converted into the active enzyme pepsin only after it is in the stomach lumen (cavity). Normally this safeguard, together with the mucous coating, effectively protects the stomach from self-digestion by pepsin.

Chyme, the soupy mixture of food and digestive secretions, leaves the stomach through a narrow opening controlled by a ring of powerful muscles called the **pyloric sphincter.** Periodically, the sphincter relaxes as a wave of contraction moves through the stomach wall toward it. This relaxation allows a small quantity of chyme to squirt through the opening into the **duodenum,** the first part of the small intestine.

Figure 8.19 Peristalsis. Rhythmic waves of muscle contraction move material along the digestive tract. The three sketches show how a peristaltic wave of contraction moves through a single section of gut over time (left to right).

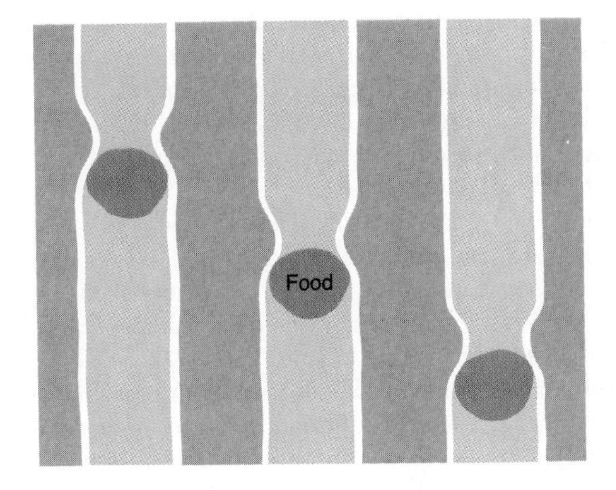

Table 8.3
Digestive Enzymes in Humans

Digestive Organs and Enzymes	Source of Enzyme	Action
Mouth		
Amylase	Salivary glands	Hydrolyzes starch to dextrins (small polysaccharides) and the disaccharide maltose
Stomach		
Pepsin (pepsinogen)	Stomach mucosa (lining)	Hydrolyzes peptide bonds and splits proteins into smaller peptides
Small Intestine		
Trypsin (trypsinogen)	Pancreas	Hydrolyzes peptide bonds in proteins and peptides
Chymotrypsin (chymotrypsinogen)	Pancreas	Hydrolyzes peptide bonds in proteins and peptides
Carboxypeptidases and aminopeptidases	Intestinal glands and pancreas	Hydrolyze peptide bonds to release individual amino acids from peptides
Lipase	Pancreas	Hydrolyzes fat to glycerides, fatty acids, and glycerol
Amylase	Pancreas	Hydrolyzes starch to the disaccharide maltose
Disaccharidases, including		
Sucrase	Intestinal glands	Hydrolyzes sucrose to glucose and fructose
Maltase	Intestinal glands	Hydrolyzes maltose to two glucose units
Lactase	Intestinal glands	Hydrolyzes lactose to glucose and galactose
Ribonuclease	Intestinal glands and pancreas	Hydrolyzes RNA to nucleotides
Deoxyribonuclease	Intestinal glands and pancreas	Hydrolyzes DNA to nucleotides

The Small Intestine

When chyme enters the duodenum, proteins and carbohydrates have been only partly digested and are not yet in absorbable forms. Fat digestion has not begun at all. Considerably more digestive activity is required before these nutrients can be absorbed through the intestinal wall.

The duodenum is the short (about 20 cm) first section of the human small intestine. Ducts carrying **bile** and **pancreatic juice,** the complex secretions of the liver and pancreas, respectively, empty into the duodenum about midway along its length. These fluids contain sodium bicarbonate, which neutralizes acid in chyme.

Bile also contains **bile salts,** which are fat-emulsifying agents; that is, they break fat globules into smaller droplets. This exposes more surface to the action of fat-digesting enzymes secreted by the pancreas.

Enzymes in pancreatic juice, together with enzymes produced by glands in the intestinal wall, digest each of the major classes of nutrients to absorbable forms (table 8.3).

Lipase hydrolyzes fat molecules to fatty acids, glycerol, and some monoglycerides (glycerol with one fatty acid still bonded to it).

Pancreatic amylase completes the process begun by salivary amylase, of splitting starch into maltose, a disaccharide made up of two glucose units. Then maltose and other disaccharides, such as sucrose, are hydrolyzed by various specific **disaccharidases** (table 8.3) to yield monosaccharides (simple sugars) that are absorbed through the intestinal lining.

Protein digestion, started by pepsin in the stomach, continues in the small intestine, although pepsin itself is inactivated by the slightly basic pH of the small intestine. Two pancreatic enzymes, **trypsin** and **chymotrypsin,** hydrolyze peptide bonds in proteins and peptides, thus producing smaller peptide fragments. These smaller peptides are digested by other enzymes secreted by intestinal glands, **carboxypeptidases** and **aminopeptidases,** that remove individual amino acids from peptides. Their action completes the process of protein hydrolysis to amino acids, the absorbable end products of protein digestion.

Trypsin and chymotrypsin, like pepsin, are general protein-digesting enzymes that hydrolyze peptide bonds in many different kinds of proteins. Thus, they present a threat of self-digestion in the pancreas and the small intestine similar to that posed by pepsin in the stomach. These enzymes, like pepsin, are secreted as enzymatically inactive molecules,

Figure 8.20 Hormonal control of digestive secretions. Colored arrows indicate stimulatory relationships. The black arrow indicates that enterogastrone inhibits stomach secretion and mobility. CCK-PZ stands for cholecystokinin-pancreozymin.

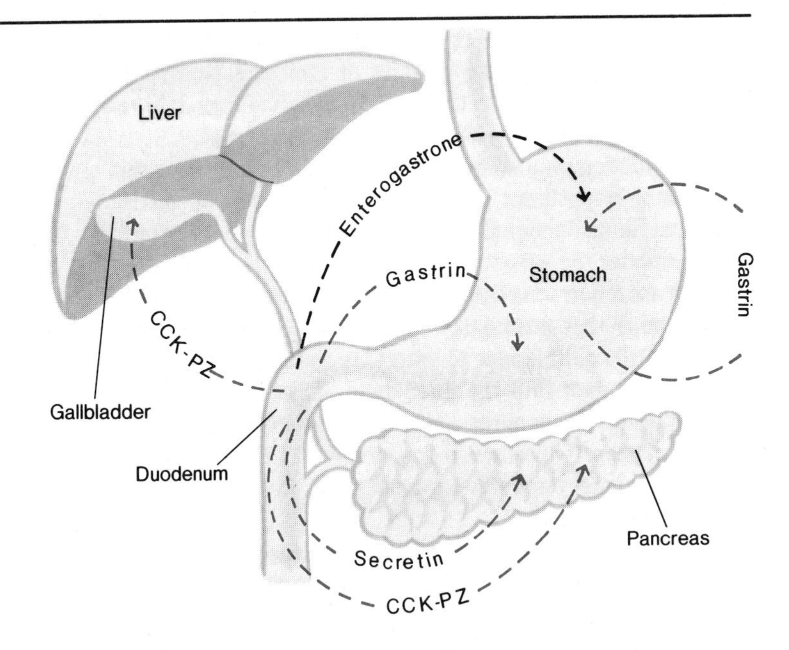

trypsinogen and **chymotrypsinogen,** that are converted to active enzymes in the lumen of the small intestine. Intestinal cells also produce mucus that lubricates intestinal contents and protects the intestinal lining from the protein-hydrolyzing action of trypsin and chymotrypsin.

Nutrient Absorption

Only a few substances such as ethanol (the alcohol in liquor) are absorbed through the stomach wall. The bulk of nutrient absorption occurs in the small intestine.

Simple sugars (monosaccharides) and amino acids are absorbed into intestinal lining cells by active transport. After entering intestinal lining cells, these sugars and amino acids are passed to blood vessels in the intestine and are carried away toward the liver.

Droplets of absorbed fat enter vessels of the lymph system (a system involved in fluid balance in your body), which carry them into the chest, where they pass into veins near the heart. Sometimes after a fatty meal, so many fat droplets enter the circulation that they give the blood an almost milky appearance.

The Large Intestine

About 10 l of water enter the digestive tract daily as a result of eating and drinking, as well as secretion by the digestive glands. About 95 percent of this water is reabsorbed into walls of the large intestine (**colon**). This water reabsorption is essential. Failure to reabsorb water can result in **diarrhea,** which can lead to serious dehydration and also ion loss, especially in children. In addition to water conservation, the large intestine functions in ion regulation. Vitamin K, vital for normal blood clotting, is produced by intestinal bacteria and also is absorbed in the colon.

Digestive wastes (**feces**) eventually leave the body through the rectum and anus. Feces are about 75 percent water and 25 percent solid matter. Almost one-third of this solid matter is made up of intestinal bacteria. The remainder is undigested plant material, dead intestinal lining cells, fats, waste products from the liver and elsewhere, inorganic material, and proteins, especially mucus.

Control of Digestive Secretions

In order to maintain homeostasis, it is important that all digestive secretions be released when food is present and not at other times.

Salivary gland secretion is controlled by the nervous system. Saliva flows in response to the taste, smell, or sometimes even the thought of food. In fact, you might notice right now that saliva sometimes flows in response just to thinking about saliva flowing.

Nervous system reflexes also stimulate secretion by glandular cells in the stomach. But a major factor in gastric secretory regulation is a hormone, **gastrin,** which is produced in the stomach itself. When proteins contact the stomach lining, gastrin-producing cells are stimulated to release gastrin into the bloodstream (figure 8.20). As soon as gastrin circulates through blood vessels and reaches the acid- and enzyme-secreting cells of the stomach lining, these cells respond by secreting large quantities of HCl and pepsin. As the stomach empties, both the neural reflexes and gastrin release subside, and less HCl and pepsin are secreted.

The duodenum also may produce some gastrin, but this is of minor importance in the overall regulation of gastric secretion. Some duodenal cells produce the hormone **secretin** that stimulates the pancreas to release pancreatic juice. Other cells release a hormone called **CCK-PZ** (formerly called **cholecystokinin**) that stimulates the release of bile. CCK-PZ stimulates the gallbladder to empty its contents through the **common bile duct** into the duodenum. This same hormone also stimulates the pancreas, especially the enzyme secretion of the pancreas. Therefore, it was renamed CCK-PZ, which stands for **cholecystokinin-pancreozymin,** in recognition of this dual role.

A hormone called **enterogastrone** is released into the circulation by duodenal cells in response to the presence of fatty food in the intestine. Enterogastrone inhibits stomach gland secretion and slows muscular movements of the stomach. This slows stomach emptying and allows more time for processing meals that contain large amounts of fat. This is important because fat is digested more slowly than other nutrients (figure 8.20).

Liver Function

The liver is an extremely important and versatile chemical-processing center. Blood leaving the intestinal tract reaches the liver by way of the **hepatic portal system,** a system of blood vessels beginning with capillaries in the intestine and ending in spaces among liver cells. This blood, carrying recently absorbed nutrients, comes into contact with liver cells, which can make a variety of exchanges with it (figure 8.21).

For example, the liver plays an important role in **blood glucose** regulation. Many cells in the body are sensitive to changes in the level of the glucose being supplied to them by the circulatory system. Brain cells, for instance, are especially dependent on a stable glucose supply, and they can be seriously damaged by unusual fluctuations in blood glucose level.

When blood sugar levels are high, the liver absorbs glucose and synthesizes glycogen for storage. When blood sugar levels fall, the liver breaks down glycogen and releases glucose into the blood. The liver continues to respond to the stimulus of lowered blood glucose even when the glycogen

Figure 8.21 Liver. (*a*) Blood leaving the digestive tract flows through the hepatic portal vein to the liver, where it enters spaces (called **sinusoids**) among the liver cells. (*b*) Scanning electron micrograph showing sinusoids (S) in the liver. Important chemical exchanges occur between blood in the sinusoids and liver, or hepatic, cells (Hc). Small channels collect bile secreted by liver cells and carry it toward the gallbladder (magnification × 592).

(*b*) Kessel, R. G., and Kardon, R. H. *Tissues and Organs: A Text-Atlas of Scanning Electron Microscopy.* © 1979 by W. H. Freeman and Company.

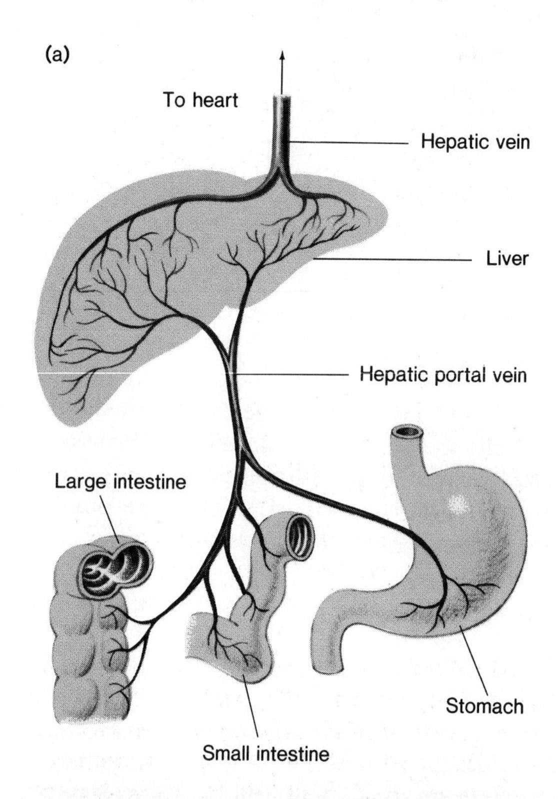

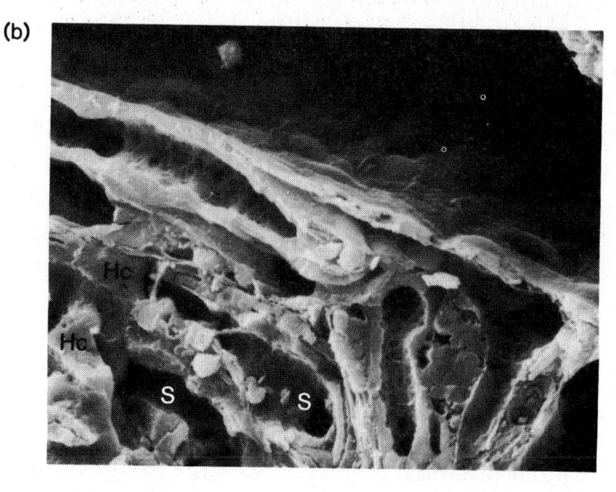

Figure 8.22 The vermiform appendix is an extension off the caecum, a blind sac at the beginning of the colon.

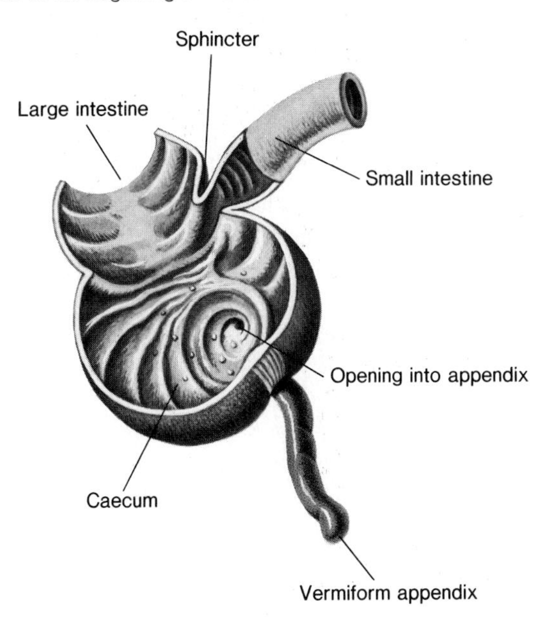

reserve has been used up. After the glycogen supply is exhausted, stored fats are used for glucose synthesis. If blood glucose levels remain very low for an extended period, as in starvation, blood glucose eventually is maintained by synthesizing glucose from deaminated amino acids (chapter 6). This depletes body proteins and seriously damages the body if continued too long.

The liver also plays an important role in processing extra amino acids. During times of adequate nutrient supply, there may be excesses of particular amino acids. Liver cells deaminate these amino acids to produce usable carbohydrates. The amino groups removed during deamination are incorporated into **urea** molecules, which are excreted from the body (see chapter 10).

In addition to these and other metabolic functions, the liver collects and breaks down many toxic substances that enter the body.

Problems with the Digestive Tract

The gut tube, for all its twists and expansions, is really just a tunnel through the body. In the strictest sense, the contents of the gut are outside the body and remain so unless they pass across cell membranes of gut-lining cells. Indeed, the contents of the gut must not come into direct contact with the blood or other interior tissues of the body because the chyme contains bacteria and other potentially hazardous components. The nature of this risk is graphically illustrated by two rather common digestive ailments—appendicitis and ulcers.

The **vermiform appendix,** or just appendix for short (figure 8.22), is a short blind sac off the beginning of the colon, just past its junction with the small intestine. The human appendix does contain lymphoid tissue and might, therefore, be involved in resistance to infection, but its function is poorly understood. The appendix all too often becomes infected and develops an inflammation called **appendicitis.** If untreated, the inflamed appendix can rupture, allowing gut contents to come in contact with the **peritoneum,** the tissue that covers all the digestive organs and lines the body cavity. The material that escapes from a ruptured appendix causes **peritonitis,** a life-threatening infection.

Ulcers are eroded areas of varying depth and position in the wall of the gut. Most are located in the duodenum and are called **duodenal ulcers.** Somewhat more rarely, **gastric ulcers** may form in the stomach itself. Ulcers are caused by the combined effect of acid and enzymes eroding the protective lining of the gut. However, what is not well understood is why in some individuals the normal mechanisms that protect the linings of the digestive tract function adequately for a lifetime, and in others they do not. It is widely believed that worry, stress, or frustration can contribute to ulcer development because gastric juice secretion is at least partly controlled by the nervous system. In fact, one of the more drastic treatments for serious ulcers is to cut the vagus nerve, the nerve that stimulates stomach secretion.

Most ulcers are really rather mild, albeit painful, abrasions in the gut lining. Healing can be rapid because the cells lining the gut tube are quickly replaced, but the causative features (often extreme stress) must be removed and the patient must switch to a diet that does not stimulate excess acid secretion. The real dangers come when ulcers begin to bleed or when they perforate (produce a hole directly through the gut wall). Peritonitis can develop following perforation unless prompt medical care is received.

Problems also can occur in the digestive glands. For example, the liver and its bile-storing structure, the gallbladder, are subject to several types of problems. One common condition is **gallstones,** which result from the precipitation (settling out) of bile salts in the gallbladder or in the bile duct. Blockage of the bile duct may cause bile to back up and ultimately leads to **jaundice,** a yellow discoloration of the skin. Jaundice develops when **bile pigments** get into the circulation. Bile pigments are breakdown products of hemoglobin from old red blood cells that are destroyed in the liver. Normally, bile pigments flow into the digestive tract as part of bile. Bile pigments that pass on through the digestive tract are at least partly responsible for the dark color of feces.

Jaundice also may be a symptom of various types of liver failure, a much more serious condition, given the overall importance of this organ. Inflammation of the liver is known as **hepatitis,** and it may result from a virus infection (viral hepatitis), poison (toxic hepatitis), or protein deficiency (deficiency hepatitis). The liver does have considerable regenerative ability, but when liver cells are destroyed, they sometimes are replaced by scar tissue. This scarring is called **cirrhosis** of the liver. All of these liver problems weaken the liver's metabolic capacities and greatly reduce its ability to deal effectively with toxins entering the body.

In addition to being susceptible to problems such as appendicitis, ulcers, gallstones, and liver disease, the digestive tract is subject to invasion by many disease organisms, from viruses to tapeworms.

Yet, despite all of these potential problems, most human digestive tracts function normally and, except for occasional minor upsets, seldom cause much concern. A normally functioning digestive tract is essential if the body is to receive an uninterrupted supply of vital nutrients.

Summary

Heterotrophic organisms must obtain organic compounds from their environment to meet their energy and raw material requirements. The organisms or parts of organisms that animals eat to provide these organic compounds are processed in digestive systems, where large organic molecules are broken down into absorbable forms.

Carbohydrates are important nutrients mainly because they provide chemical energy sources. Much ingested carbohydrate consists of polysaccharides that must be hydrolyzed to absorbable sugars.

Animals must ingest proteins to obtain amino acids for their own protein synthesis. Animals' cells can synthesize some amino acids from other compounds, but other amino acids are essential dietary constituents.

Fats are highly reduced organic compounds that are important for energy and synthesis of structural components. Certain animals cannot synthesize some fatty acids, and these fatty acids must be included preformed in the diet.

Some minerals are required in large quantities, but only trace amounts of others are needed.

Vitamins are relatively simple organic compounds that must be present in the diet to maintain health.

Incomplete digestive tracts have only one opening, a mouth. Complete digestive tracts have both a mouth and an anus. In complete digestive tracts, food can move through and be processed in assembly-line fashion in specialized regions.

In the human digestive tract, food is chewed and manipulated in the mouth. Then it is swallowed through the esophagus to the stomach. The stomach stores food and mixes it with mucus and gastric juice to produce chyme. Protein digestion begins in the stomach.

Chyme passes through the pyloric valve to the duodenum. There, bile, pancreatic juice, and intestinal gland secretions are added to the chyme and begin to act upon it. Enzymes hydrolyzing all of the organic nutrients act in the small intestine. Most nutrient absorption takes place in the small intestine, but some water and mineral ions are absorbed in the colon.

Digestive wastes leaving the colon contain water, bacteria, undigested material, fats, waste products, inorganic material, proteins, and dead intestinal lining cells.

In order to maintain homeostasis, digestive secretions are released only when food is present.

The liver is involved in chemical processing of absorbed food and in maintenance of stable concentrations of nutrients such as glucose in the blood. The liver also processes nitrogenous wastes and breaks down toxins.

Problems with the digestive tract include possible infection by a wide range of organisms, as well as erosion of the tract's protective lining by its own secretions.

Questions for Review

1. Name three kinds of polysaccharides included in animal diets that are polymers of glucose.
2. What is kwashiorkor?
3. Define the terms scurvy and beriberi.
4. Why are overdoses of vitamin C much less likely to cause serious problems than overdoses of vitamin A?
5. Explain the advantages of a complete digestive tract over an incomplete digestive tract.
6. Name five structural features that provide increased surface area for absorption in various digestive tracts.
7. Define the term amylase.
8. What normal protective mechanisms guard against self-digestion of digestive glands and the linings of the digestive tract?
9. What are the steps in digestion and absorption of starches in the human digestive tract? Of proteins? Of fats?
10. What are bile pigments and what is their relationship to jaundice?

Questions for Analysis and Discussion

1. Explain why ascorbic acid is classified as a vitamin (vitamin C) for humans, but is not essential in the diet of rats.
2. Which would you expect to be longer: the digestive tract of a two kilogram cat or the digestive tract of a two kilogram rabbit? Why?
3. Why are cows, sheep, deer, and goats able to derive nutritional benefit from the cellulose that they eat while humans are not?

Suggested Readings

Books

Eckert, R., and Randall, D. 1983. *Animal physiology.* 2d ed. San Francisco: W. H. Freeman.

Guyton, A. C. 1979. *Physiology of the human body.* 5th ed. Philadelphia: W. B. Saunders.

Schmidt-Nielsen, K. 1983. *Animal physiology.* 3d ed. New York: Cambridge University Press.

Articles

Harpstead, D. D. August 1971. High-lysine corn. *Scientific American* (offprint 1229).

Loomis, W. F. December 1970. Rickets. *Scientific American* (offprint 1207).

McMinn, R. M. H. 1977. The human gut. *Carolina Biology Readers* no. 56. Burlington, N.C.: Carolina Biological Supply Co.

Moog, F. November 1981. The lining of the small intestine. *Scientific American* (offprint 1504).

Scrimshaw, N. S., and Young, V. R. September 1976. The requirements of human nutrition. *Scientific American.*

Sherlock, S. 1978. The human liver. *Carolina Biology Readers* no. 83. Burlington, N.C.: Carolina Biological Supply Co.

Gas Exchange and Circulatory Systems

Chapter Outline

Chapter Concepts

1. Gas exchange is an urgent, continuing requirement for animal life.
2. Small animals can exchange gases directly through body surfaces, but larger animals have specialized gas exchange mechanisms.
3. Aquatic animals generally exchange gases through gills, while terrestrial animals have tracheal systems or lungs.
4. Circulatory systems move blood through animals' bodies, transporting materials between body exchange areas and internal body cells.
5. In open circulatory systems, blood leaves vessels and circulates among body organs. In closed circulatory systems, blood is continuously enclosed inside vessels.

The African lungfish is a very curious creature (figure 9.1). Lungfish have lungs and can breathe air. Being able to breathe air at the surface, they can survive in very stagnant water when they cannot obtain enough oxygen through their gills. Sometimes, a lungfish even wriggles across the ground to a new pond if its own pond is drying up.

At other times, a lungfish goes to the bottom of a drying pond, forms a cocoon around itself, and enters an inactive state in which it can survive for long periods, even for years, buried in the dried mud. When rain finally comes and the pond refills, the lungfish emerges from its cocoon and resumes its normal activity.

You might conclude, then, that lungfish are hardy animals, able to withstand very bad conditions. But when biologists went to Africa to study lungfish, they encountered peculiar problems. Many lungfish died in the nets that were used to catch them, and others that were caught alive almost invariably died, no matter how gently they were handled while they were being carried to the laboratory in buckets of water. This seemed strange because native fishermen who caught lungfish on lines and carried the fish on their backs were able to deliver lungfish alive to the market for sale. What were the biologists doing wrong?

Eventually the scientists discovered that the lungfish died in the buckets because they could not bend their bodies to reach the surface of the water. They are obligate air breathers; that is, they must breathe air to live. In the nets or buckets, lungfish were dying for lack of oxygen. In other words, African lungfish are fish that can drown!

Figure 9.1 The African lungfish, a fish that can drown.

Continuing gas exchange is a basic requirement for life in animals. This dependence is much more immediately urgent than the need for food or water. We humans can survive for several weeks without eating, or for several days without drinking, but we can live for only a very few minutes without breathing.

In animal cells, oxygen is continuously being used as an electron acceptor in metabolic oxidation-reduction reactions that produce ATP for a variety of energy-requiring processes. Another requirement for gas exchange results from the decarboxylation of carbon compounds, which yields carbon dioxide as a metabolic waste product. This CO_2 must be released to the environment as rapidly as it is produced.

In this chapter we will discuss the mechanisms by which animals accomplish this vital gas exchange. We will also see how circulatory systems function to deliver oxygen and other required substances to body cells and to remove carbon dioxide and other wastes.

Figure 9.2 Animals with small bodies in which gases must diffuse only short distances between body cells and gas exchange surfaces. (*a*) The cells of planarians and other flatworms are spread out in their flattened bodies so that no cells are far from an outer body surface. (*b*) The *Hydra* has a hollow internal cavity. Some cells exchange gases with water inside this cavity.

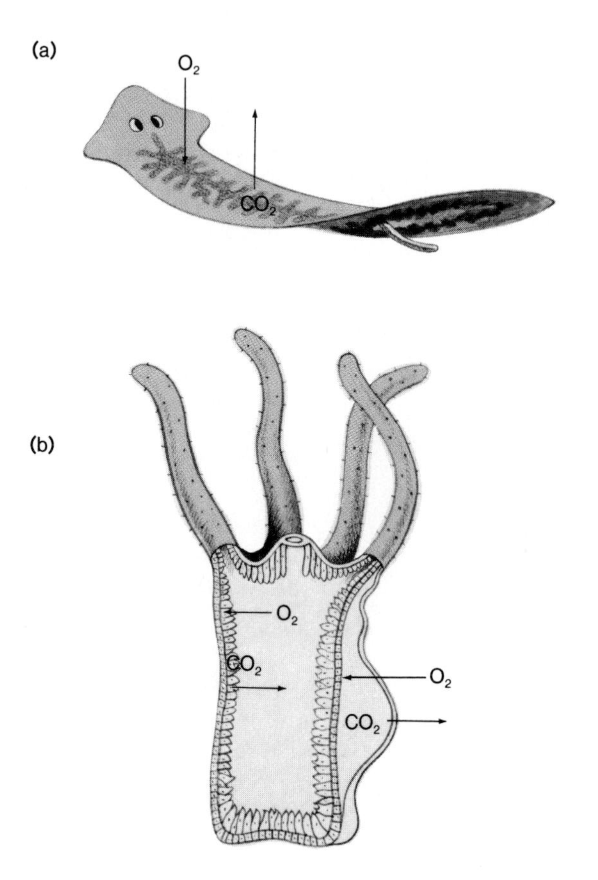

Gas Exchange

The kinds of gas exchange and transport mechanisms that animals have depend upon body sizes and shapes. Animals with small bodies in which all cells are relatively near gas exchange surfaces can depend on simple diffusion for gas movement to body cells (figure 9.2).

But diffusion alone is only adequate to meet the needs of small animals. All large animals must have arrangements that bring oxygen to internal body cells and carry carbon dioxide away. Many of these animals have **circulatory systems** with gas transport mechanisms that carry oxygen from a specialized exchange area to cells throughout the body (figure 9.3).

Most animals that depend on circulatory gas transport have specialized **respiratory (oxygen-carrying) pigments,** such as hemoglobin, in their blood. These carriers reversibly bind oxygen molecules and greatly increase the oxygen-carrying capacity of the blood.

The four basic categories of animal gas exchange specializations are integumentary (skin) exchange, gills, tracheal systems, and lungs (figure 9.4)

Integumentary Exchange

Integumentary exchange (exchange through the skin) is the principal method of gas exchange only for quite small animals and for a small number of relatively larger animals that live in moist environments. For example, earthworms have capillary networks just under their skin surfaces; they exchange gases with the air in spaces among soil particles. Since oxygen must go into solution before it can diffuse into surface cells, earthworms must maintain moist body surfaces. But by doing so, they risk serious desiccation if they enter dry soil areas or remain exposed to dry air above ground for any length of time.

On the other hand, heavy rains threaten earthworms with suffocation because soil air spaces become filled with water. To survive, earthworms must emerge from the ground and exchange gases with the air until the rainwater drains out of the upper layers of soil. This is why you see earthworms scattered on the ground and sidewalks after a heavy rain.

Figure 9.3 A circulatory system brings blood to a specialized gas exchange surface. From there the blood is carried to body tissues where cells exchange gases with it.

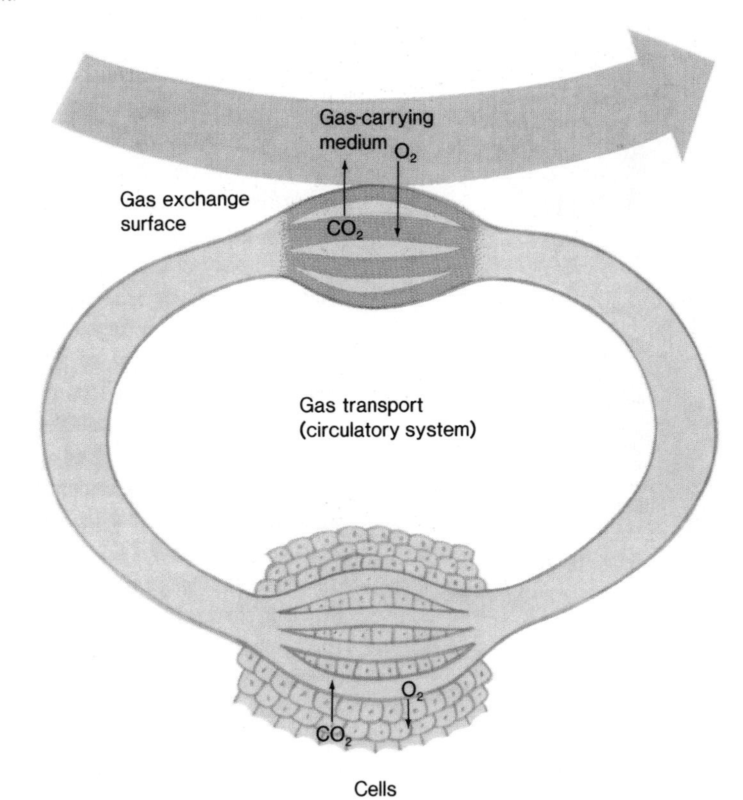

Figure 9.4 The four basic types of animal gas exchange specializations. (*a*) Integumentary (skin) exchange at the body surface. (*b*) Gills. (*c*) Lungs. (*d*) Tracheae in a tracheal tube system.

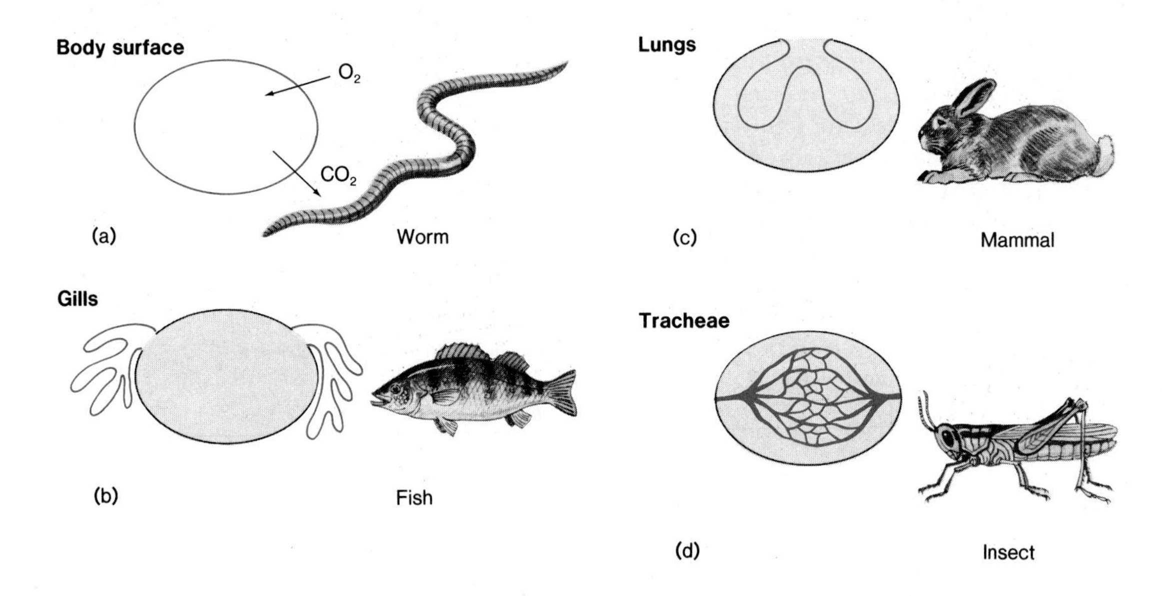

Figure 9.5 Starfish gills are small delicate projections scattered over the body surface. They are protected by spines and small pincers called pedicellariae.

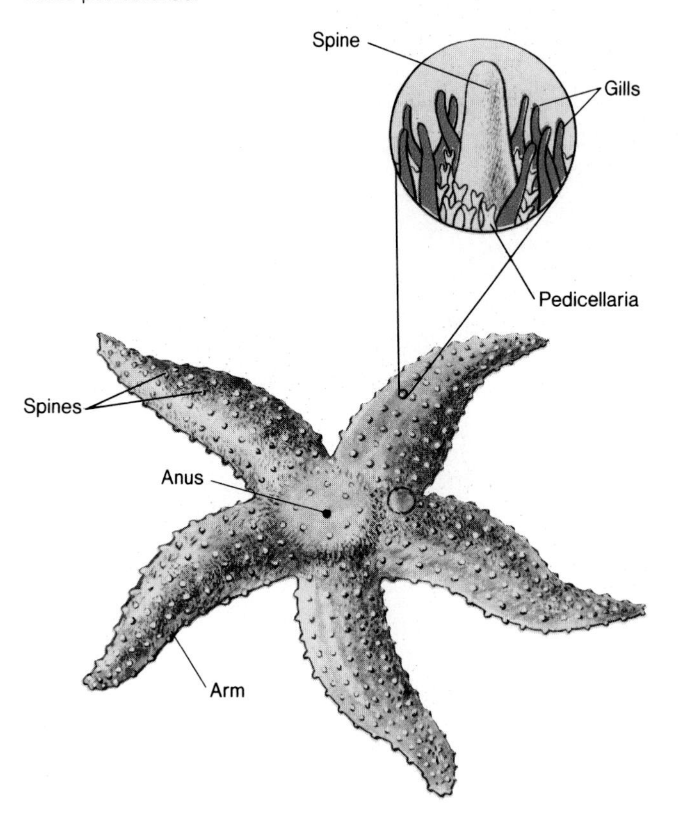

Gills

Most animals that live in water carry out gas exchange through **gills.** Gills are body surface outgrowths that are exposed to the water. The gas exchange surfaces of gills are very thin (one cell thick in many cases), so the gas diffusion distance across them is quite small.

The simplest gills are small individual projections of the skin, such as the tiny gills of starfish (figure 9.5). While starfish gills are scattered over much of the body surface, many aquatic animals have much more compact gills that are located in only one area of the body. For compact gills to have adequate surface area for gas exchange, they must be finely divided into highly branched structures (figure 9.6).

Fish gills consist of many layers of gill filaments with numerous delicate plates that protrude all along them and greatly increase the surface area for gas exchange. These plates are richly supplied with blood vessels. A hard, bony cover, the **operculum,** encloses and protects the gills. Fish force water through their gills by using muscles in the mouth and pharynx (figure 9.7).

Gas exchange in fish gills is very efficient because blood flowing through vessels and capillaries inside gills moves in the opposite direction of water flow over the gill surface. This **countercurrent** flow arrangement, described in figure 9.8, allows fish gills to absorb oxygen efficiently enough so that gills can be relatively small compared to total body size and yet meet the gas exchange requirements of the animals.

Figure 9.6 The delicate, feathery gills of crayfish are protected inside a gill chamber. The cover of the gill chamber has been removed in this diagram to show the gills. Water is moved through the chamber past the gills by a beating paddlelike structure.

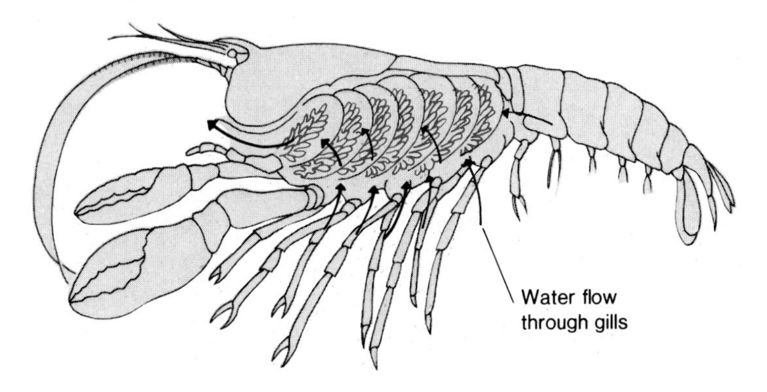

Figure 9.7 Fish gills. (*a*) The operculum (1) covers and protects several layers of delicate gills. The operculum has been removed from (2) to expose the gills. (*b*) Water is taken into the pharynx through the mouth (1). Then the oral valve closes the mouth from the inside and muscles contract, forcing water over the gills and out through the operculum (2).

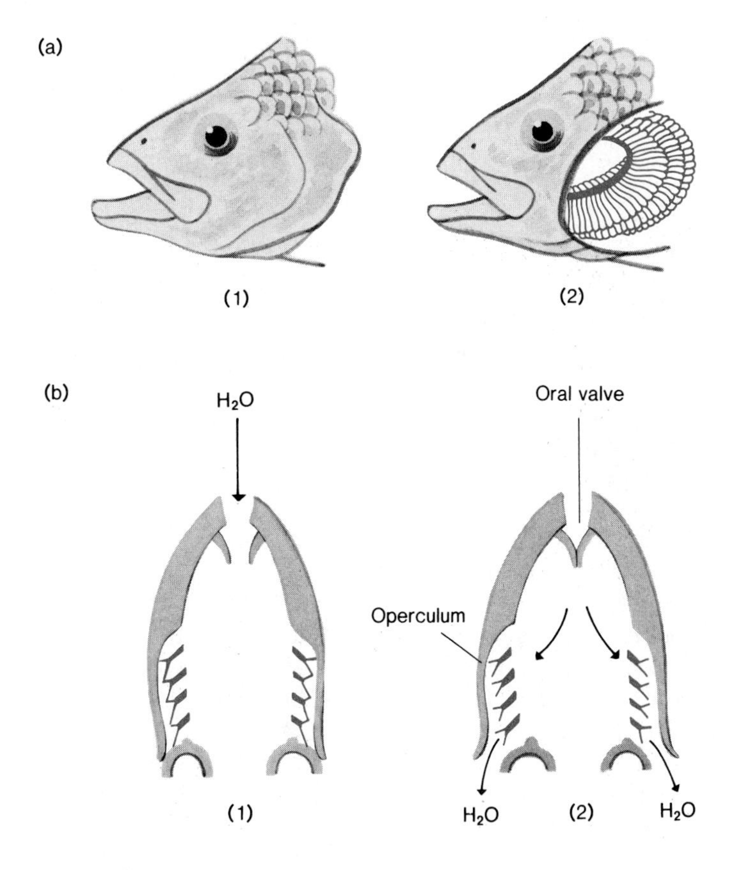

Figure 9.8 (*a*) Several gill arches are under the operculum. Each arch has two rows of gill filaments, and each filament has many thin, platelike lamellae. Gases are exchanged between blood inside the lamellae and the water that flows between them. Note that the blood inside the lamellae flows in a direction opposite to the direction of water flow. This countercurrent flow increases the efficiency of gas exchange in fish gills. (*b*) Scanning electron micrograph of the tip of a single gill filament from a trout. Note the regularly arranged lamellae. Water flows through the filaments in a direction perpendicular to the plane of this page. (See *a*.) (*c*) Diagram (1) shows how countercurrent flow permits oxygen uptake throughout the time that blood and water are in close proximity in the gill lamellae. This allows oxygen content of the blood to reach the highest possible level. Water may leave the gill having lost 90 percent of its initial oxygen content. Compare this with a hypothetical arrangement (2) in which blood and water move through gills in the same direction. No further exchange of oxygen would take place after equal concentrations were achieved.

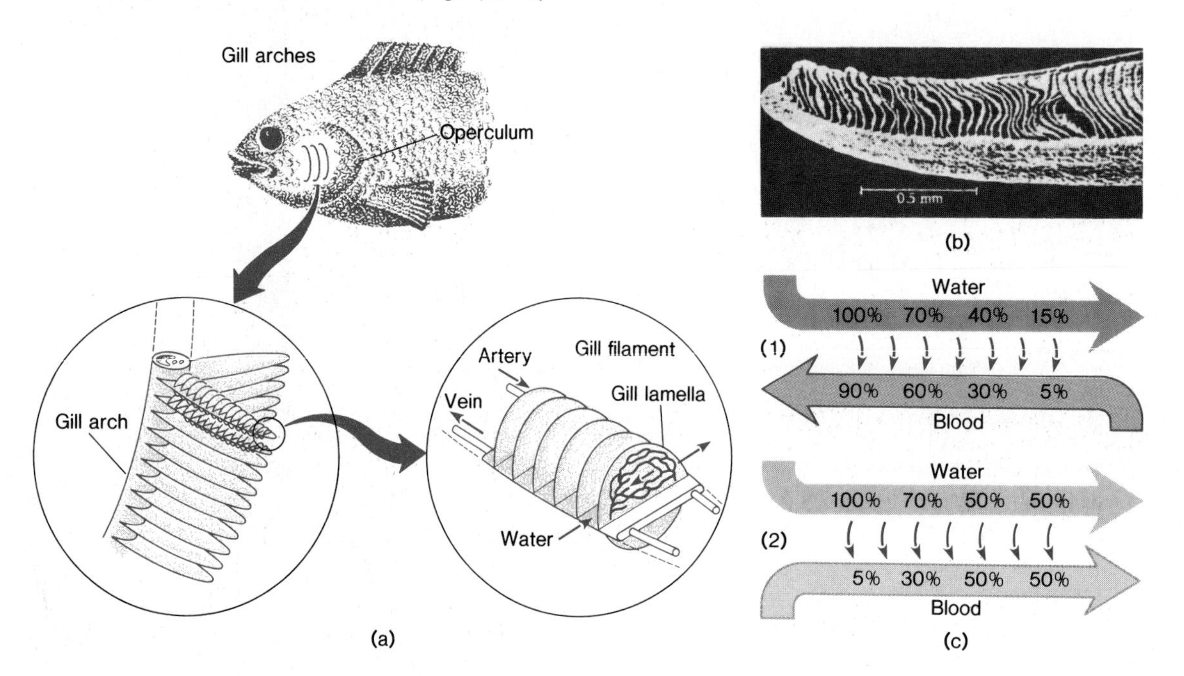

Tracheal Systems

Tracheal systems are networks of air tubes that penetrate the tissues of insects and other terrestrial arthropods (centipedes, millipedes, and some spiders). These tubes, called **tracheae** (singular, **trachea**), open to the outside air through holes in the body surface called **spiracles.** Oxygen diffuses into the tracheal system, and carbon dioxide diffuses out through the spiracles (figure 9.9). The finest end branches of the system penetrate all parts of the body, and as a result, no cell is more than one or two cells away from a tube. Thus, gas exchange occurs directly between body cells and air in the tracheal system.

Larger insects, such as bees and grasshoppers, pump air in and out of the system by muscle contractions. They have enlarged collapsible bags (air sacs) on some of their tracheal tubes, and muscular pressure on these bags aids the pumping process. Even so, the efficiency of tracheal systems is limited. Science-fiction movies aside, there probably never could be giant insects as large as animals who exchange gases by other means because tracheal systems cannot provide adequate gas exchange for very large bodies.

Lungs

Lungs are saclike internal structures that provide a moist gas exchange surface in a sheltered internal environment (see figure 9.4). Although a few kinds of invertebrate animals, such as some snails, have lungs, lungs are primarily a vertebrate adaptation.

Vertebrate Lungs

The lungs of various vertebrate animals differ depending upon the gas exchange requirements of the animals. Frog lungs are relatively simple sacs that are subdivided into compartments by partitions (figure 9.10*a*). The lungs of turtles and other reptiles are subdivided by branched and rebranched partitions (figure 9.10*b*). These lungs have significantly more gas exchange surface area than do frog-type lungs of similar size. And as you might expect, the lungs of birds and mammals are even more elaborately subdivided into small passageways and spaces.

Figure 9.9 Tracheal systems. (*a*) The tracheal system of a small insect. Tubes penetrate all parts of the body, and body cells exchange gases with the air inside the tracheal system. The tracheal tubes open to the outside air through spiracles. (*b*) Larger insects such as grasshoppers pump air through their tracheal systems. This pumping occurs when contracting muscles put pressure on collapsible bags (air sacs) that are part of the tracheal system. This sketch shows larger respiratory structures of the left side of the body. (*c*) Photomicrograph of part of an insect tracheal system. Walls of tracheae contain chitin, a polymer synthesized from glucose. Heavy deposits of chitin make structures rigid. Note the ringed appearance of the tubes. The rings are chitinous thickenings of the walls that help to prevent collapse of the tubes.

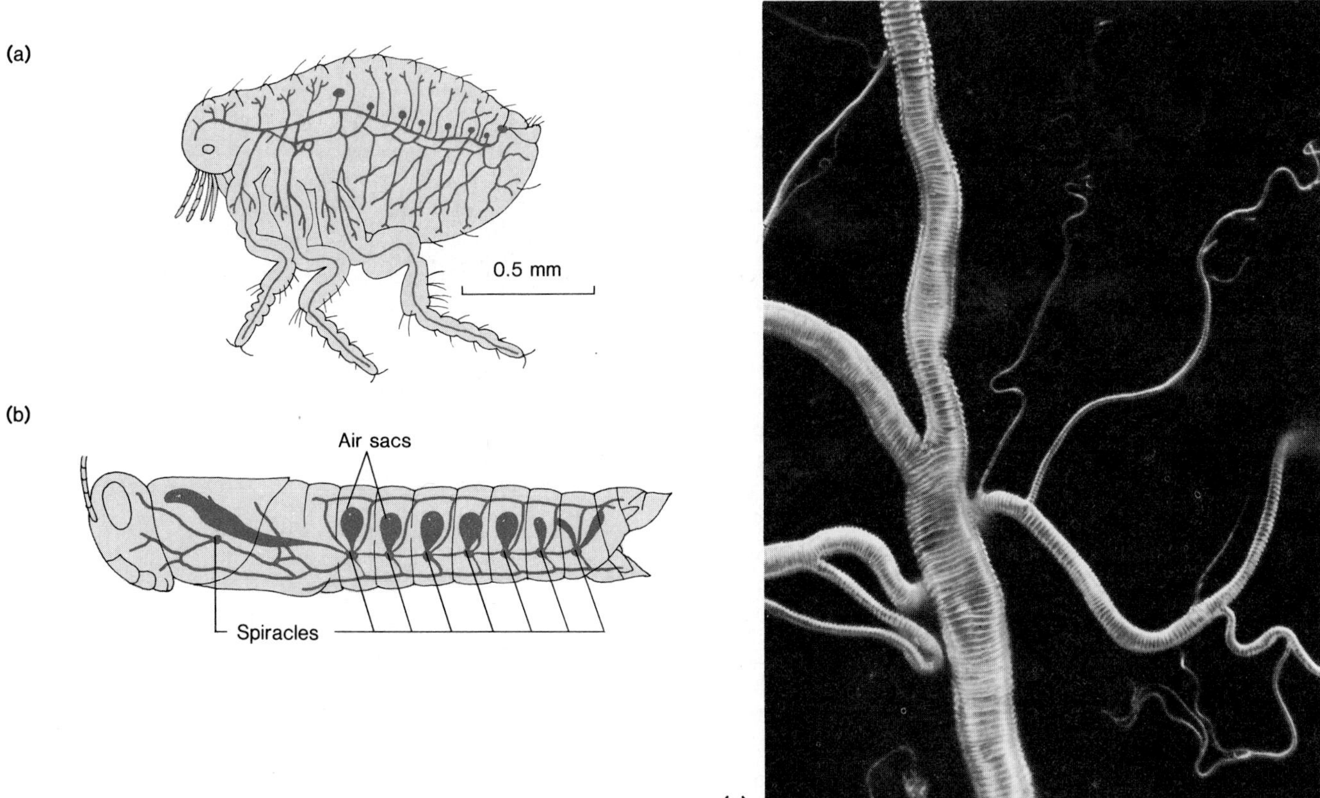

Figure 9.10 Lung types. (*a*) Frog. (*b*) Turtle.

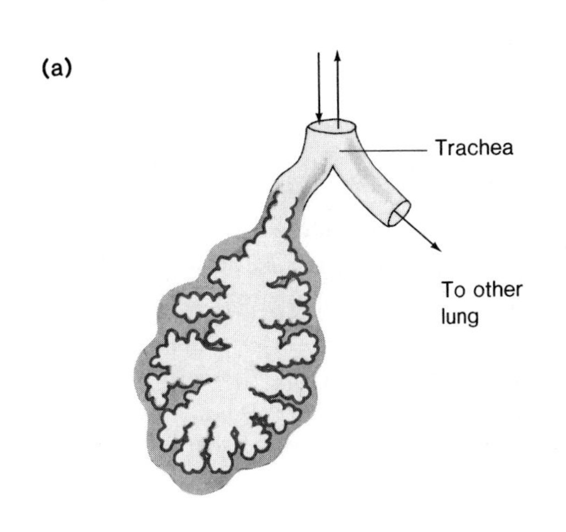

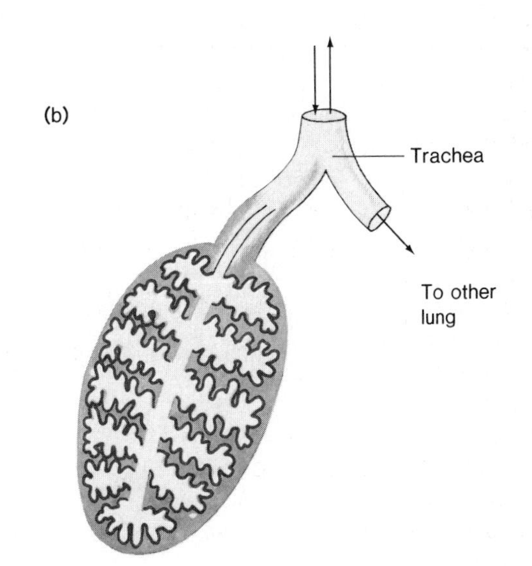

Lung Ventilation

Vertebrate animals **breathe;** that is, they **ventilate** their lungs by causing air to move in and out of them. Without regular, continuing ventilation, oxygen depletion and carbon dioxide buildup would occur quickly in the air inside the lung cavities. Even with ventilation by ebb and flow breathing, the oxygen level is lower and the carbon dioxide level higher inside vertebrate lungs than in the outside air. This is because there is never complete air exchange during breathing; the lungs are not completely emptied and refilled during each breathing cycle.

Flow-through lungs with air coming in one opening, passing over gas exchange surfaces, and leaving the body through a different opening would provide very efficient gas exchange. But such a system would result in considerable water loss because there would not be the opportunity for water recovery that reverse flow (air entering and leaving through the same passageway) makes possible. Thus, vertebrate lung structures and ventilation mechanisms represent an evolutionary compromise between two urgent requirements—the need for efficient gas exchange and the need to conserve body water.

Most vertebrate animals draw air into their lungs by a negative pressure system; that is, they **inhale** (breathe in) by suction. In mammals, the entire rib cage that surrounds the lungs can be lifted at the points where the ribs join the vertebral column. The **thoracic** (chest) cavity has a muscular floor, the **diaphragm,** that can be flattened. Together, these two movements decrease pressure around the lungs, causing inhalation as atmospheric pressure forces air into the lungs. **Exhalation** (breathing out) occurs when the muscles of the rib cage and diaphragm relax and put increased pressure on the lungs (figure 9.11).

Figure 9.11 Ventilation of human lungs, an example of breathing in a mammal. (*a*) In inhalation, muscle contractions lift the ribs up and out, and lower the diaphragm. These movements increase the size of the thoracic cavity and decrease the pressure around the lungs. Then atmospheric pressure causes more air to enter the lungs. (*b*) Exhalation follows the relaxation of the rib cage and diaphragm muscles.

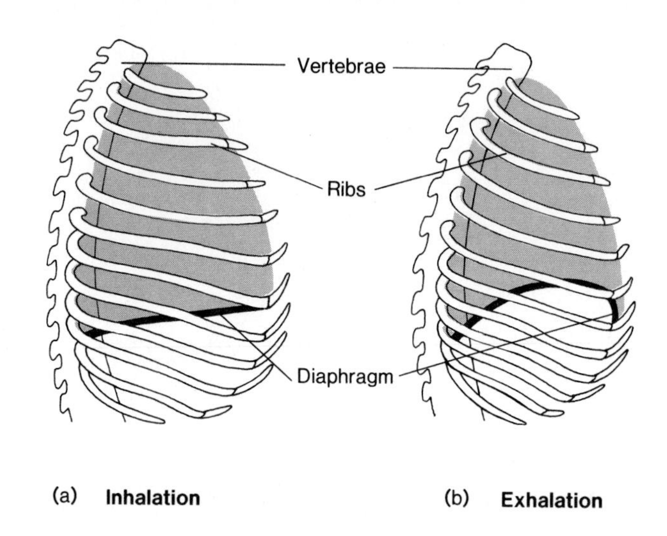

The Human Gas Exchange System

The human gas exchange system (**respiratory system**) is similar to the gas exchange systems of other mammals both in terms of basic structure (figure 9.12) and ventilation by an ebb and flow mechanism that uses negative pressure inhalation.

Most people at rest generally breathe through their nose. During inhalation, air enters through **nostrils** and passes through the **nasal cavities.** The linings of the nasal cavities near the nostrils have hairs, while the linings deeper in the cavities have cilia on their surfaces. Hair and cilia, as well as mucus on the lining surfaces, trap dust particles and other foreign material. Air is also warmed and moistened in the nasal passages. Blood vessels in the linings of the nasal cavities lose heat to the air, and water evaporates off the moist surfaces. By the time air enters the lungs, it has been cleaned and is more than 99 percent saturated with water vapor. Conversely, air leaving the lungs cools as it passes out through the nasal passages, and as it cools it loses water by condensation.

Air passes from the nasal cavities into the **pharynx,** where it crosses the path of food. Swallowing temporarily closes the way to the lungs while food is passing through the pharynx (see page 179). It may seem inefficient for food and air to pass through the same space, and certainly there is danger of choking if food accidentally enters the passage leading toward the lungs, but this arrangement does have one big advantage—it permits mouth breathing, if the nostrils or nasal cavities become blocked. It also permits greater air intake during heavy exercise when greater gas exchange is required.

Figure 9.12 The human gas exchange (respiratory) system. The structure of the smallest elements is shown in the enlarged diagram of one part of a lung. Gas exchange actually occurs in the alveoli, which are bulges in the walls of alveolar sacs.

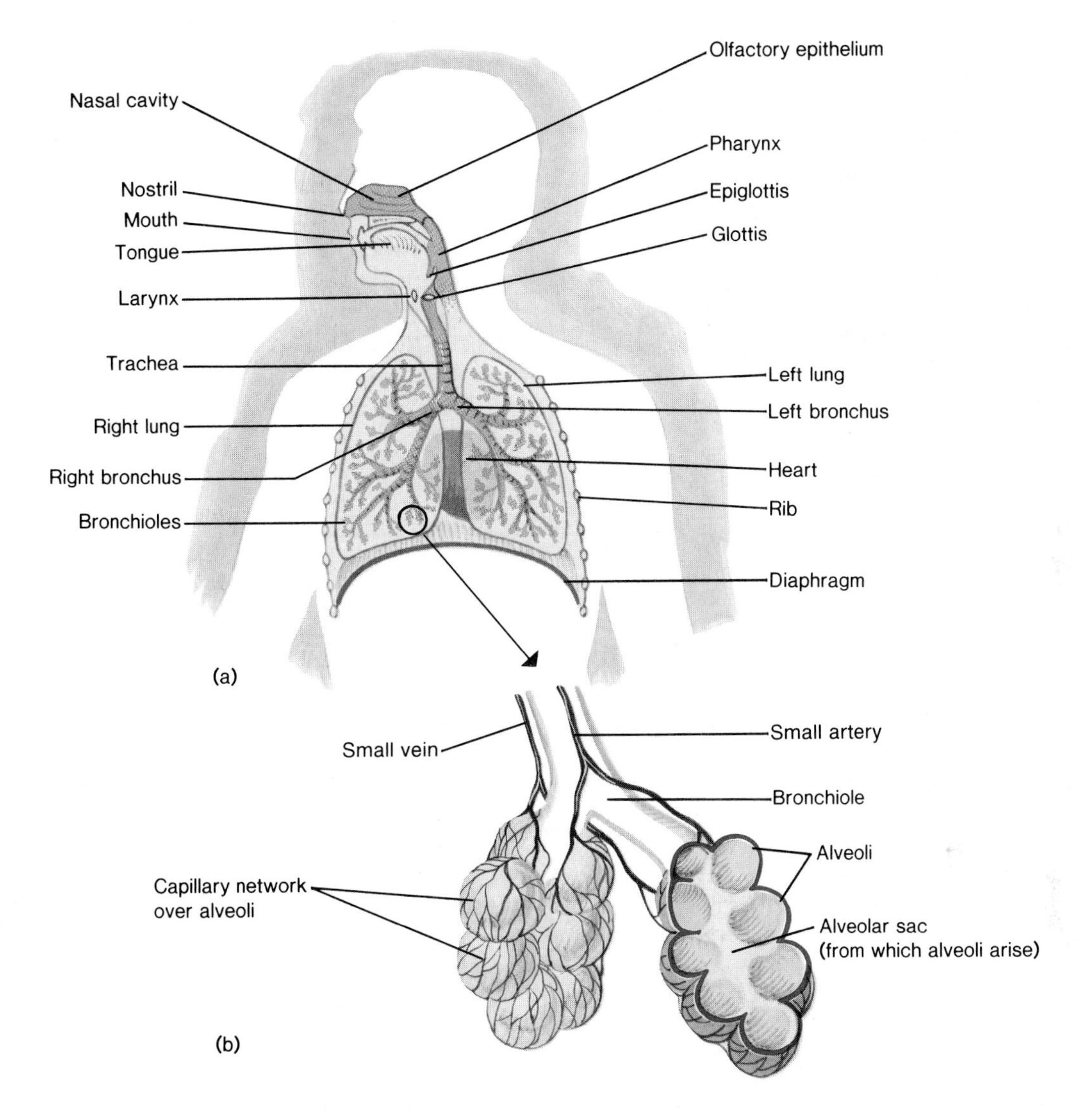

Air passes on from the pharynx through the **glottis,** an opening into the **larynx.** The larynx, or "voice box," lies at the top of the **trachea.** During swallowing, the larynx is raised and a flap, the **epiglottis,** covers the glottis, temporarily closing it. The larynx is a rigid, cartilage-supported structure. It contains a pair of elastic ridges, the **vocal cords,** that vibrate when currents of air pass between them, producing sounds.

Air passes on through the larynx into the trachea ("windpipe"). The trachea is a tube with a series of C-shaped cartilages in its wall that prevent it from collapsing. The tracheal lining is covered with cilia-bearing cells and mucus-secreting cells. The cilia beat in waves and move mucus and any trapped particles up toward the larynx (figure 9.13). This helps to protect the lungs from foreign substances. The tracheal lining cells also trap many potentially harmful microorganisms before they reach the lungs.

Figure 9.13 Scanning electron micrograph of the tracheal lining. The hairlike structures are cilia and the rounded structures are mucus-secreting goblet cells (magnification × 800).

The lower end of the trachea divides into two **bronchi** (singular: **bronchus**) that carry air to the right and left lungs. Each bronchus branches and rebranches into a network of smaller passages called **bronchioles.**

The smallest bronchioles terminate in **alveolar sacs** that have tiny bulges called **alveoli** (singular: **alveolus**) on their surfaces. Alveoli are the chambers within which gas exchange between air and blood actually occurs. Although each alveolus is very small (human alveoli are about 0.25 mm in diameter), there are hundreds of millions of alveoli in a pair of human lungs, thus making the total surface area of the alveolar linings enormous. Alveolar surface areas in human individuals range from 50 to 100 m^2 with the average probably being about 70 to 80 m^2, an area about forty times as large as the entire skin surface area of a human adult.

Box 9.1

Lung Diseases
A Threat to Life

Continued efficient gas exchange is vital to life and any health problem that interferes with lung functioning can be life threatening.

Because their warm, moist, blind sac interiors provide very favorable environments for growth of microorganisms, the lungs are quite susceptible to infection. Infections of the lungs can produce a condition known as **pneumonia.** In pneumonia, alveolar sacs fill with fluid and dead white blood cells. This reduces the amount of surface available for efficient gas exchange because gases must diffuse through this fluid to reach the alveolar surfaces. Thus, pneumonia can result in seriously reduced blood O_2 levels. Bacterial pneumonia is not so great a threat now as it once was because bacterial infections of the lungs can be treated with antibiotics. Viral pneumonia, however, still causes serious problems.

Another once-dreaded bacterial disease of the lungs is **tuberculosis.** But this degenerative disease also responds to antibiotic therapy and is a serious health problem only for people who lack adequate medical attention.

Unfortunately, reduction of the prevalence of bacterial infections of the lungs has not relieved concern about disorders of the respiratory tract. In recent years, there has been a marked increase in the incidence of

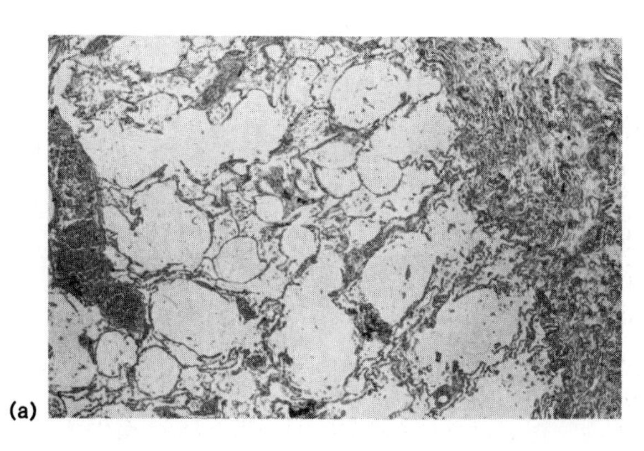

(a)

other lung disorders, such as **emphysema.** This problem seems to have increased as a result of certain features of life in modern, technological societies.

Emphysema is a progressive disease in which alveolar walls degenerate and lose their elasticity. Some alveolar walls break down and larger chambers form (box figure 9.1). Gas diffusion through the damaged membranes is reduced. Even mild exercise can cause emphysema sufferers to gasp for breath because it is very difficult for them to increase their oxygen and carbon dioxide exchange enough to meet the demands created by exercise.

Alveoli are well adapted for their role in gas exchange because more of their surface area is occupied by capillaries, the smallest of blood vessels, than by spaces between capillaries. The alveolar linings consist of a single layer of very flattened cells, as do the capillary walls. Thus, the distance that gases must diffuse between air and blood is very short, about 0.5 μm.

Human Breathing

A resting human adult (you, for example, as you read this chapter) breathes about fourteen times per minute. But during vigorous exercise, the breathing rate may rise to 100 breaths per minute. Fast, deep breathing can move nearly 100 times as much air in and out of the lungs per minute as is moved by normal breathing in a person at rest.

The breathing rate and the volume of air inhaled and exhaled are controlled by a **respiratory center** in the brain. Changes in the breathing rate and in the volume of air being breathed are made automatically by the respiratory center in response to changes in the gas content of the blood, especially increases or decreases in the amount of carbon dioxide in the blood.

Gas Transport in Blood

Most of the oxygen entering the blood combines with the hemoglobin in red blood cells to form **oxyhemoglobin.** Each molecule of hemoglobin contains four polypeptide chains, and each of these is folded around a heme unit that contains an iron atom (see chapter 2 for details). Because oxygen molecules bind to these iron atoms, each molecule of hemoglobin

(b)
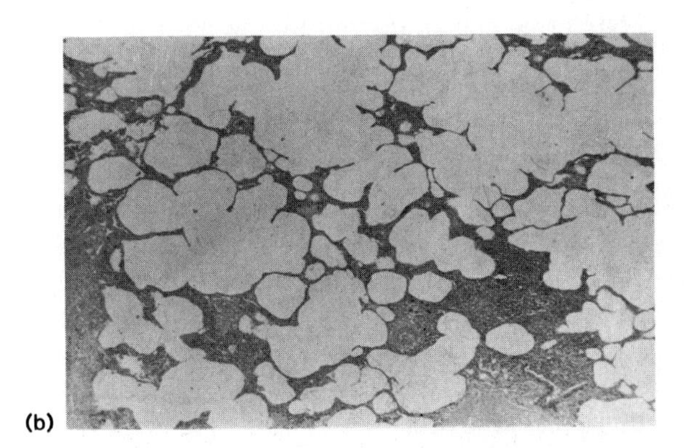

(c)
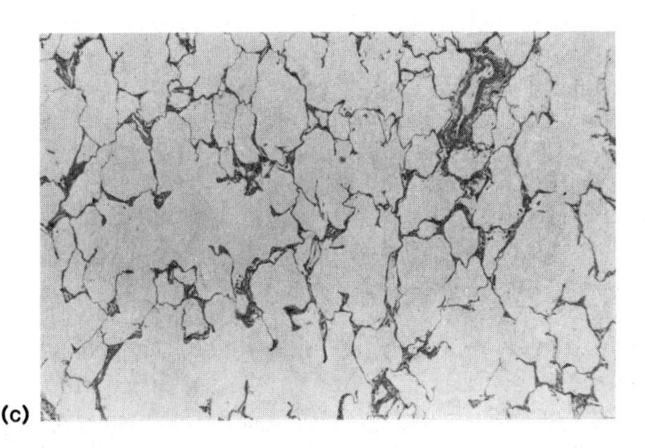

Box figure 9.1 Development of emphysema. (*a*) Normal alveoli. (*b*) Breakdown of some alveolar walls to produce enlarged chambers. Lungs lose both gas exchange surface area and elasticity as emphysema develops. (*c*) Advanced stage of emphysema in which functional alveolar tissue breaks down completely, leaving only connective tissue, thus eliminating this part of the lung as a functional gas exchange area.

Because of the loss of lung elasticity, exhalation does not occur by the normal passive mechanisms, and people with advanced stages of emphysema must make an effort to exhale. The extra work required for breathing combined with the reduced efficiency of gas exchange essentially incapacitates many emphysema victims.

Although the causes of emphysema are not clear, many scientists believe that prolonged exposure to respiratory irritants in polluted air, or tobacco smoke, or both, play a major role in the disease's development. Cigarette smoke is especially suspect because it contains irritating particulate matter and substances that inhibit the activity of the cilia that normally move debris up out of the respiratory passages.

can carry four molecules of oxygen. Since there are about 280 million hemoglobin molecules in each **erythrocyte** (red blood cell), each cell can carry more than one thousand million molecules of oxygen.

In the lungs, hemoglobin becomes loaded with oxygen which remains firmly attached as the blood passes through arteries. But when blood enters various body tissues where oxygen concentrations are lower because cells are using oxygen in respiration, oxygen quickly unloads from hemoglobin and diffuses into cells.

Blood also transports carbon dioxide from body tissues back to the lungs. A small amount of CO_2 (about 7 percent of the total) dissolves in blood plasma. About 10 to 15 percent of the CO_2 transported is bound to hemoglobin. But dissolved carbon dioxide and CO_2 bound to hemoglobin account for less than one-fourth of the CO_2 transported by the blood. Most of the CO_2 carried by the blood is transported as **bicarbonate ions** (HCO_3^-). Bicarbonate ions are produced because carbon dioxide tends to combine with water to produce **carbonic acid:**

$$H_2O + CO_2 \longrightarrow H_2CO_3$$

In blood, carbonic acid tends to be dissociated to hydrogen ions (H^+) and bicarbonate ions:

$$H_2CO_3 \longrightarrow H^+ + HCO_3^-$$

The following equation summarizes what happens when carbon dioxide enters the blood flowing through body tissues:

$$H_2O + CO_2 \longrightarrow H_2CO_3 \longrightarrow H^+ + HCO_3^-$$

This reaction lowers the pH of the blood, but body cells cannot withstand significant changes in blood pH. Therefore, serious blood pH fluctuations are prevented by buffers (see page 36). Hemoglobin and other blood proteins function as buffers because they combine in various ways with the excess hydrogen ions.

In body tissues, carbon dioxide diffuses into the blood. But in the lungs, the carbon dioxide concentration in the air inside alveoli is lower than in the blood. Therefore, CO_2 diffuses out of the blood and into the lungs where it is exhaled.

Figure 9.14 The heart of a grasshopper (color) pumps blood through an open circulatory system. Blood moves forward from the heart, through the body cavity, and back into the heart through the ostia.

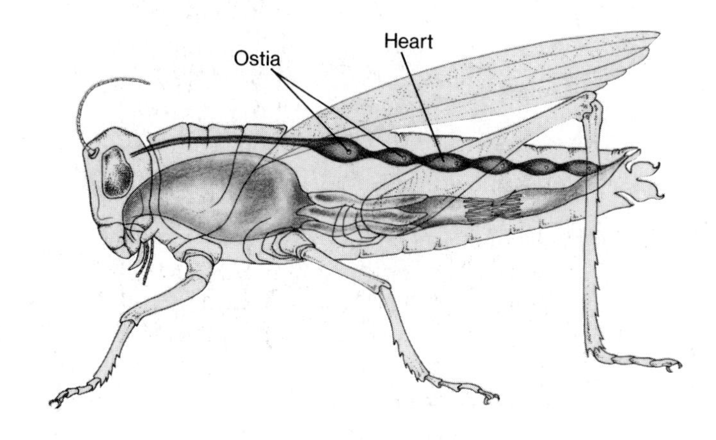

Circulatory Systems

In addition to oxygen and carbon dioxide, many other substances must be carried to and away from body cells in animals. These important transport functions are accomplished in all large animals by **circulatory systems.** A circulatory system is a transport system by which fluid is moved through the body in a regular fashion by the pumping action of a muscular **heart.**

Open and Closed Circulatory Systems

The two basic types of circulatory system arrangements are **open circulatory systems** and **closed circulatory systems.** In open systems, the heart pumps blood out into the body cavity, or at least, through parts of the body cavity, where the blood bathes organs and tissues of the body. In closed circulatory systems, blood circulates only within the confines of a system of tubular **blood vessels.**

Insect circulatory systems are open. An insect's heart pumps blood through vessels that open into the body cavity. Blood that has circulated through the body cavity enters the posterior part of the heart through openings called **ostia** (figure 9.14).

In closed circulatory systems, blood is not pumped out into the spaces around body cells as it is in open circulatory systems. Exchanges between body cells and the blood are made through the walls of the smallest blood vessels, called **capillaries.** A capillary wall is only one cell thick. Since capillaries form diffuse networks in all body tissues, no body cell is far from circulating blood.

Figure 9.15 The closed circulatory system of an earthworm (color). Hearts pump blood into a ventral blood vessel. Branches from it lead to capillary networks. Blood returns to the heart through a dorsal blood vessel.

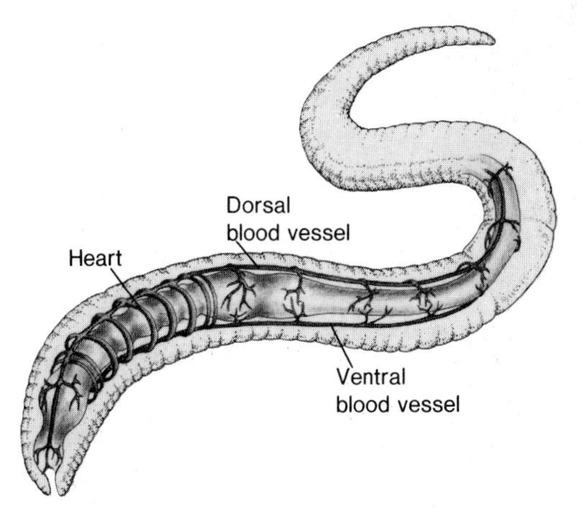

The common earthworm has a closed circulatory system. Its five pairs of pulsating hearts pump blood out to a ventral blood vessel that runs along the bottom side of the body. Branches from the ventral blood vessel carry blood into the various body tissues. After blood passes through the capillary networks in an earthworm's tissues, it collects in vessels leading to a major dorsal blood vessel that runs along the top side of the body and returns blood to the heart (figure 9.15).

Vertebrate Circulation

All vertebrate animals have closed circulatory systems. The vertebrate circulatory system has a strong, muscular heart with **valves** that prevent backward flow when the heart beats. Blood travels away from the heart through **arteries** to smaller vessels called **arterioles.** Arterioles lead to capillary beds where exchange between blood and the various body cells takes place. From capillaries, blood collects in **venules** that flow together into **veins,** the major vessels carrying blood toward the heart (figure 9.16).

Figure 9.16 Comparison of the walls of arteries, capillaries, and veins. (*a*) Diagrammatic comparison in which capillary is drawn much larger than its normal proportional size. Artery walls are more rigid and do not expand or change shape as readily as vein walls. Capillary walls consist of a simple endothelium that is only one cell thick and corresponds to the lining surface of the innermost layer of arteries and veins. (*b*) Scanning electron micrograph of an artery (A) and its companion vein (V), illustrating the difference in wall thickness and lumen size (magnification × 56). (*c*) Scanning electron micrograph showing layers in the wall of an artery. The inner layer with an endothelium (E) on its surface, the muscular layer (M), and the connective tissue layer (CT) are shown (magnification × 1,316).

Kessel, R. G., and Kardon, R. H. *Tissues and Organs: A Text-Atlas of Scanning Electron Microscopy.* © 1979 by W. H. Freeman and Company.

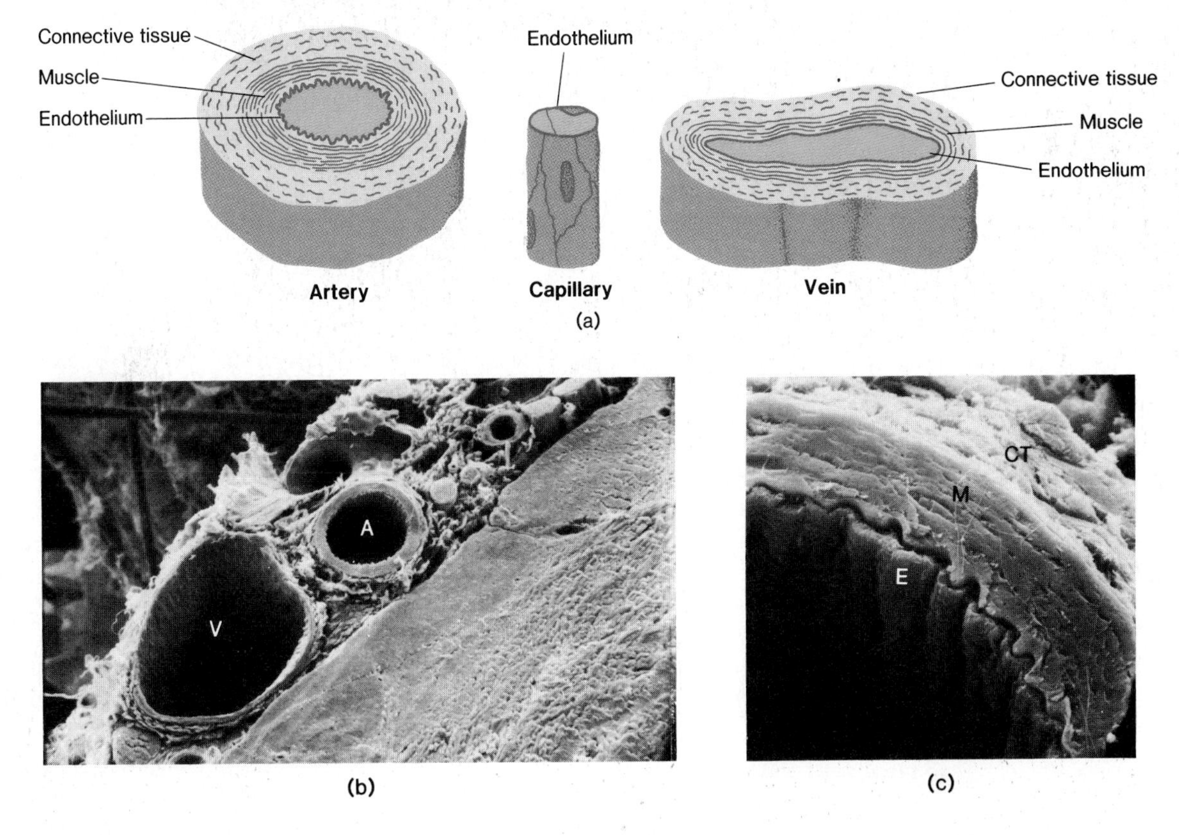

Figure 9.17 The single-circuit circulatory arrangement of fish. Blood passes from the heart, through the gill circulation, and directly on to the systemic circulation. Oxygenated blood is indicated by color. Two major parts of the heart, the atrium (A) and ventricle (V) are shown.

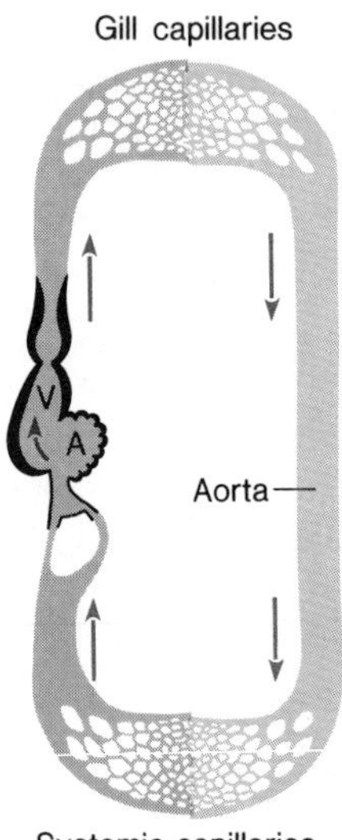

The major differences among vertebrate circulatory systems involve the relationship between circulation through the gas exchange organs (gills or lungs) and circulation to the rest of the body.

The Single-Circuit Plan of Fish

In fish, the circulatory plan involves a single "loop" through the body. Blood leaves the heart and goes forward through a large vessel, the **ventral aorta.** The ventral aorta carries blood to the gill area where smaller vessels distribute it to the gill capillary networks. In the gills, blood is **oxygenated** (it picks up oxygen), at the same time giving off CO_2. Blood leaves the gills through vessels that flow into a large **dorsal aorta.** The dorsal aorta is the main systemic (body) trunk and supplies the **systemic circulation;** that is, it carries blood to all of the capillary networks in the body tissues (figure 9.17).

After blood has traveled from the heart, through the gill circulation, and through the systemic circulation, it returns to the heart. This single-circuit circulatory arrangement has a significant drawback—it has relatively low systemic blood pressure.

Blood pressure is the force that drives blood through the circulatory system. Because blood pressure decreases as blood passes through the gill capillary network, the blood pressure in the fish circulatory system beyond the gills is reduced.

Figure 9.18 Separation of pulmonary (lung) and systemic circulations in birds and mammals. (*a*) Diagrammatic view of the circulatory arrangement in birds and mammals. (*b*) Diagram of a four-chambered heart with the chambers stretched out for clarity. Complete separation of two streams of blood is maintained because there are separate right and left atria and right and left ventricles.

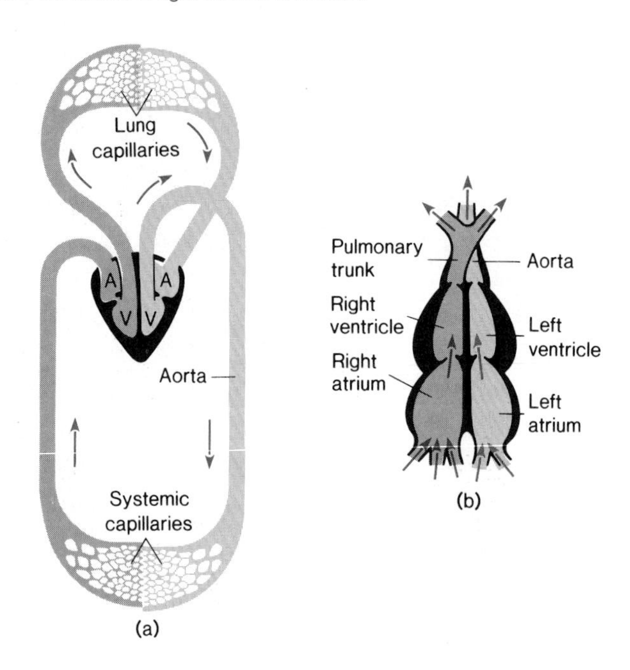

While this low pressure systemic circulation is adequate for fish, it could not deliver oxygen and nutrients and remove metabolic wastes rapidly and efficiently enough to meet the needs of other vertebrates.

The Two-Circuit Plan

Other vertebrates have a two-circuit arrangement. Blood passes through the lungs (**pulmonary circulation**), where it is oxygenated, and it returns to the heart. Then the heart pumps this oxygenated blood out through the systemic circulation. Birds and mammals have completely separate pulmonary and systemic circulatory loops (figure 9.18). Because of the separation of pulmonary and systemic circulations, the blood pressure in the systemic circulation is relatively high, making circulatory supply to body tissues more efficient than it is in fish.

The main pumping portion of a vertebrate heart is the **ventricle,** a muscular chamber that forces blood out through the blood vessels of the body. Before a vertebrate ventricle contracts, blood enters the relaxed ventricle from a thin-walled chamber called the **atrium** (plural: **atria**). The ventricle then contracts, forcing blood out of the heart into the blood vessels.

Figure 9.19 Schematic diagram of the human circulatory system. Oxygenated blood is shown in color. Deoxygenated blood is shown in gray.

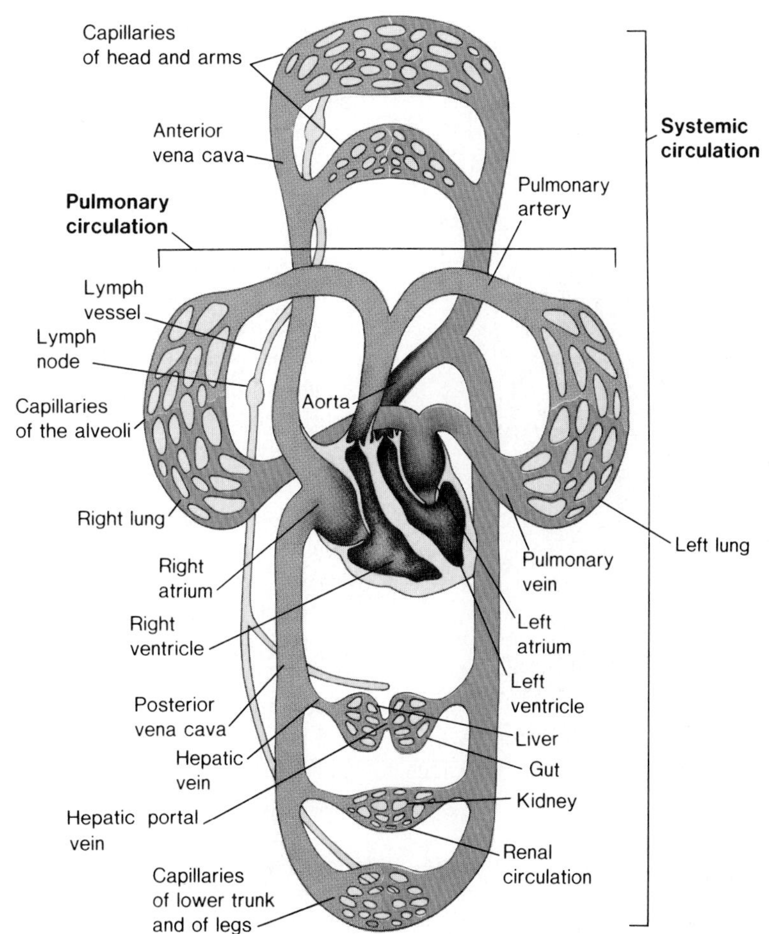

Bird and mammal hearts have right and left atria and right and left ventricles. The two atria beat simultaneously and then the two ventricles beat, pumping blood into the vessels leading to the pulmonary circuit through the lungs and to the systemic circuit. In such **four-chambered hearts,** the complete partitioning into right and left halves permits the separation of two streams of blood passing through the heart.

Human Circulation

Like other mammals, humans have a two-circuit arrangement with a four-chambered heart (figure 9.19).

Blood returns from the systemic circulation to the right atrium of the heart by way of two large veins, the **anterior (superior) vena cava** and the **posterior (inferior) vena cava.** The anterior vena cava returns blood from the head, arms, and upper trunk while the posterior vena cava returns blood from the remainder of the systemic circulation. All of this blood entering the right atrium is oxygen-poor (deoxygenated) blood. From the right atrium, blood passes into the right ventricle, which then pumps the blood out through the **pulmonary trunk.** The pulmonary trunk branches into two pulmonary arteries that carry blood to the lungs. After passing through lung capillaries around the alveoli, blood, now oxygenated, returns to the left atrium of the heart through **pulmonary veins.** This oxygenated blood passes into the left ventricle, which then pumps it out through the **aorta,** the large arterial trunk that supplies the entire systemic circulation.

The aorta sends branches to all parts of the body. The first branches off the aorta are **coronary arteries.** Because individual heart cells cannot exchange material with the blood being pumped through the chambers of the heart, coronary arteries must provide **coronary circulation** to the heart muscle

Figure 9.20 X-ray photograph of the coronary arteries taken after injection of the arteries with barium sulphate, a substance that absorbs X-rays and makes the blood vessels show up very clearly.

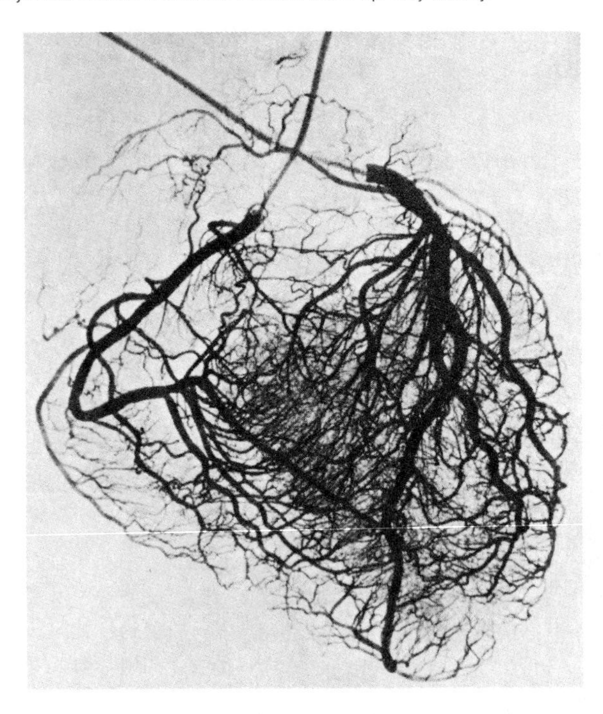

itself (figure 9.20). Normal blood flow through coronary arteries is critical for normal functioning of the heart. Blockage of any of the coronary arteries quickly results in damaged heart muscle and impaired heart function.

Anterior to the heart, the aorta bends around to the left, forming an arch, and then passes posteriorly through the body. Branches off this arch of the aorta supply anterior parts of the body. As the aorta descends through the trunk, branches run to the digestive organs, the kidneys, the body wall, the legs, and all other posterior parts of the body.

Blood returns from many of these structures directly through veins that flow into the posterior vena cava. The digestive tract, however, has a special circulatory arrangement. Blood leaving the capillary networks of the major digestive organs—the stomach and the small and large intestines—flows into veins leading to a large **hepatic portal vein.** The hepatic portal vein does not empty into the posterior vena cava. Instead, it carries blood to a network of capillarylike spaces (**sinusoids**) inside the liver. This arrangement facilitates processing by the liver of nutrients just absorbed from the digestive tract. After passing through liver sinusoids, blood flows into **hepatic veins** that carry it to the posterior vena cava. This special circulatory arrangement involving the veins of the digestive tract, the hepatic portal vein, and the liver is called the **hepatic portal system.**

The Heartbeat

The human heart is a remarkably efficient and reliable pumping device, and its regular, continual beating is essential to life. The heart can never rest or stop beating for even a short time because while some organs can survive brief pauses in circulatory supply, others cannot. The brain, for example, is critically sensitive to even short failures of circulation. If circulation to the brain stops for as little as five seconds, consciousness is lost. A four-minute break in circulation to the brain causes death of significant numbers of brain cells, and a nine- or ten-minute circulatory failure causes massive, irreversible brain damage.

During each minute of life, a normal, resting person's heart beats about seventy times and pumps out a total of about 5 l of blood, an amount approximately equal to the total blood volume in the body. All of this work is done by a relatively small organ—the average human heart weighs only 300 g, less than 0.5 percent of total body weight. This small organ sustains its heavy work load, without rest, from the time it begins to beat early in embryonic development until death.

Each heartbeat involves a cycle of contraction (**systole**) and relaxation (**diastole**) of the atria, and systole and diastole of the ventricles. Although the human heart has two completely separated streams of blood flowing through its right and left sides, the two sides contract in unison; that is, right and left atria contract at the same time, and then slightly later, right and left ventricles contract.

There is a brief pause between heartbeats when atria and ventricles both are relaxed. Blood from the atria begins to fill the ventricles during this pause. The relaxed ventricles are 70 to 75 percent filled during this period. Then atrial systole forces blood still contained in the atria into the ventricles and completes their filling.

The atria relax as the ventricles begin to contract. The beginning of ventricular systole (contraction) increases pressure in the ventricles and would force blood back into the atria except for **valves** that shut to prevent backward flow. These valves are the **tricuspid valve,** located in the canal between the right atrium and right ventricle, and the **bicuspid (mitral) valve,** located in the canal between the left atrium and left ventricle.

As ventricular systole continues, blood is forced out into the major arteries. The elastic walls of the arteries expand as blood is pushed into them. When ventricular diastole (relaxation) begins, the artery walls recoil. Blood would be forced back into the ventricles except that here again valves prevent backward flow. Sets of cuplike **semilunar valves** at the bases of the pulmonary trunk and the aorta slam shut and prevent blood from returning to the ventricles (figure 9.21).

Figure 9.21 Circulation through the human heart. (*a*) The paths of blood flow. This diagram shows the veins entering the heart, the arterial trunks leaving the heart, and the locations of valves in the heart. This is a ventral view with several parts of the wall of the heart removed.
(*b*) Valves. This is a view from above the heart with the valves exposed.

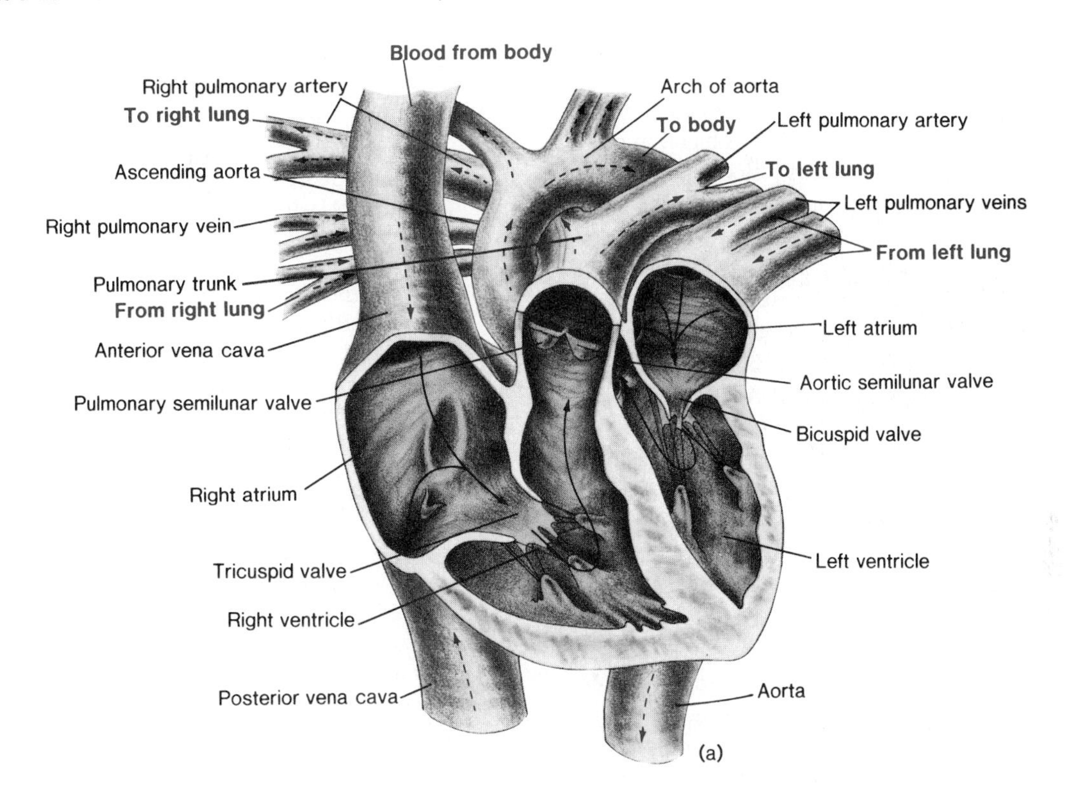

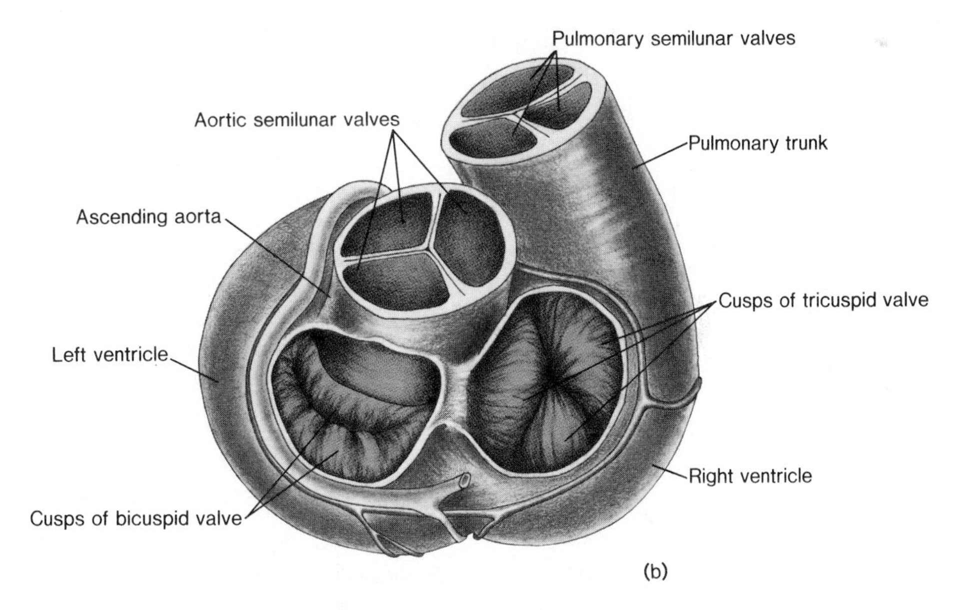

All of this takes about 0.8 second. Then there is a period of about 0.4 second between the end of ventricular systole and the start of the next atrial systole at the beginning of the next heartbeat.

Blood Pressure

Pressure generated by the heartbeat forces blood into the aorta and supplies the driving force that causes blood to move through the entire circulatory system. Blood pressure is highest during each ventricular systole, when blood is being forced into the aorta. At this time there is a resultant temporary expansion of the elastic aortic wall, which passes like a wave along the wall of the aorta and the walls of each of its larger branches. This wave of expansion is the **pulse** that you can detect in certain arteries near the body surface. Because there is one arterial pulse per ventricular contraction, the arterial pulse rate can be used to determine the heart rate. Arterial wall expansion is only temporary because arterial walls are elastic and quickly return to their original diameter. This elasticity of arterial walls is very important functionally because it helps to maintain blood pressure between heartbeats.

Thus, two different pressure levels alternate in large arteries. **Systolic pressure** results from blood being forced into the arteries during ventricular contraction (systole), and **diastolic pressure** is the blood pressure in the arteries during ventricular relaxation (diastole). Human blood pressure is measured using a device called a sphygmomanometer. A sphygmomanometer has a pressure cuff that measures the amount of pressure required to stop the flow of blood through an artery. Blood pressures normally are measured on the brachial artery, an artery of the upper arm, and are stated in millimeters of mercury (Hg). A blood pressure reading includes two numbers, for example 120/80, and these numbers represent systolic and diastolic pressure, respectively.

As blood moves through the aorta and then spreads through smaller arteries and arterioles, pressure falls. When blood reaches the capillaries, the pressure is much lower and there is no pulse, but only a smooth, steady flow of blood.

Blood pressure in the veins is very low and is not adequate to move blood back to the heart, especially from lower parts of the body. Venous return to the heart depends in part on mechanisms that apply force from outside the circulatory

Box 9.2

Circulatory System Health

The circulatory system is prone to many kinds of problems. One of the most common is **atherosclerosis,** an arterial disease that contributes to nearly one-half of all deaths in the United States. This common condition involves the formation of soft masses of fatty material in blood vessel linings. These fatty masses, called **plaques,** contain large quantities of cholesterol. They often form in arteries, and where they form, the arterial lining becomes much rougher than normal. As plaque develops, it decreases the diameter of the blood vessel and interferes with blood flow. The formation of calcium deposits in the plaque and degenerative changes in the arterial wall lead to hardening of the artery (box figure 9.2). Hardened (sclerotic) arteries lose elasticity and are especially susceptible to rupture. The rupture of hardened brain arteries, which often results in brain damage, is called a **stroke**.

Plaques also can break loose and circulate until they block a small blood vessel or their rough surfaces cause formation of a blood clot that blocks a vessel. This is especially devastating if the blocked vessel is one of the coronary arteries of the heart. The portion of the heart muscle denied a blood supply is said to be **infarcted,** and the whole process is called **myocardial** (heart muscle) **infarction.** There often are simultaneous disturbances of the impulse-conducting system so that rapid and uncoordinated heart beating occurs. These are symptoms of a heart attack. If the victim survives the initial crisis of a heart attack, there is a long recovery period during which dead muscle tissue in the infarcted area is replaced by fibrous tissue. People who recover from a myocardial infarction have a reduced ability to meet extra circulatory requirements because part of the heart muscle is permanently lost.

The circulatory system is prone to many other types of problems. Little can be done about inherited circulatory system tendencies, but positive steps can be taken to improve and maintain circulatory system health.

Dietary moderation is a good general rule. Obesity increases circulatory demands and the heart's work load. Also, people whose diets contain large quantities of fat are more likely to develop atherosclerosis than people with low-fat diets. High salt consumption can be another problem because it may be correlated with various circulatory problems, especially **hypertension** (high blood pressure).

Not smoking or stopping smoking also can preserve circulatory system health. Cigarette smoking is very highly correlated with an increased risk of heart and blood vessel disease.

Finally, sensible exercise programs can contribute substantially to maintenance of circulatory system health.

Box figure 9.2 Atherosclerosis. (*a*) An X-ray picture taken after a substance that appears opaque on X-ray photographs was injected into the bloodstream. An area where plaque formation has decreased the diameter of the femoral artery in the thigh is indicated by the arrow. (*b*) A diagram showing plaque accumulation in an artery. The circular broken line indicates the approximate size of the normal lumen. Note that the lining in the plaque area is much rougher than the normal endothelium. (*c*) Changes in arteries affected by atherosclerosis. (1) A normal artery. (2) Advanced atherosclerosis has led to wall hardening in this artery. (3) This sclerotic artery is completely blocked by a clot.

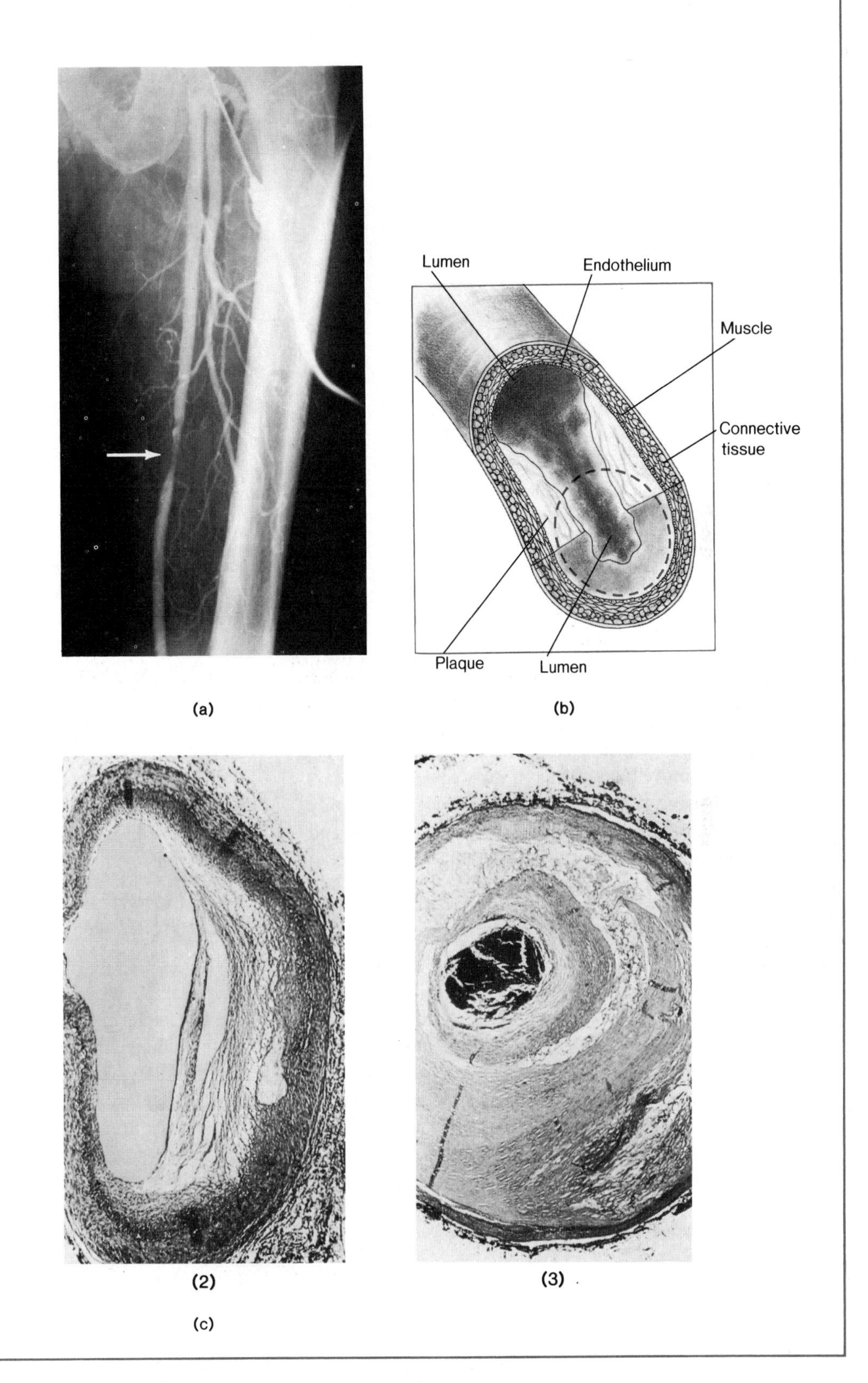

Figure 9.22 The role of valves in movement of blood through veins. (*a*) Blood pressure is low in veins. Muscle contractions near veins put pressure on the blood and force it through veins. (*b*) Blood moves only toward the heart because valves prevent backward flow when muscle pressure decreases.

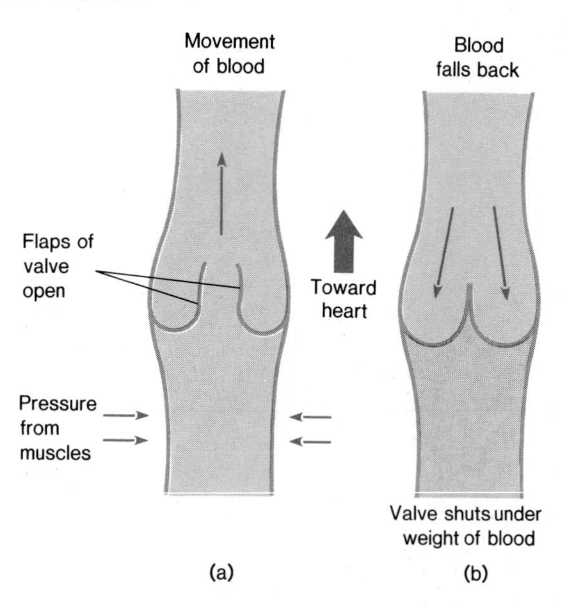

Figure 9.23 The position of a sphincter muscle that controls blood flow into a capillary bed. Nervous system control over sphincters throughout the body makes it possible to regulate the blood supply to different body regions in response to changing circulatory requirements.

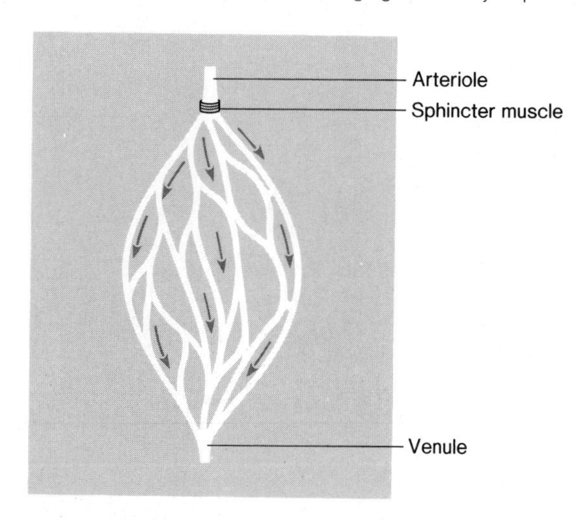

system to keep blood moving. Walls of veins are much thinner than the walls of arteries and, therefore, are quite easily collapsed. When body muscles that are near veins contract, they put pressure on the veins' collapsible walls and on the blood contained in the veins. Because veins have valves that prevent backward flow, pressure from muscle contractions effectively helps to move blood through veins toward the heart (figure 9.22). During periods of inactivity, blood moves only very sluggishly though veins in some parts of the body and can accumulate in them. For example, blood accumulates in leg veins during long periods of standing.

Capillary Circulation

Capillaries deliver materials to cells in tissues of the body and remove wastes from them. Thus, vital exchanges between blood and body cells occur in capillary beds. How is the blood supply to various tissues regulated?

Blood flows into capillary beds from arterioles. Arterioles have circular **sphincter muscles** in their walls that can contract, thus restricting flow through the capillaries. When sphincters open (dilate), they again allow blood into capillary beds (figure 9.23). These arteriole sphincters are controlled by the nervous system.

The nervous system's control over arterioles functions to meet specific circulatory requirements in specific areas of the body, and those requirements change with time. For example, in resting muscles blood flow is maintained through only a few capillary beds. But when exercise begins, many more sphincters open, and circulation begins through those arterioles and the capillary beds that they supply. At the same time, capillary circulation is decreased selectively in other parts of the body, notably in the digestive tract. Conversely, following a large meal, many arterioles open to increase the flow through capillary beds in the digestive tract, thereby increasing the blood supply for transport of absorbed products of digestion.

Capillary Function

Capillaries are distributed throughout every tissue of the body, forming very complex networks, so that no cell is far from one or several capillaries (figure 9.24).

Individual capillaries are very small, barely large enough for red blood cells to squeeze through single file. However, because they are so numerous, capillaries have a very large total surface area in body tissues. This large surface area is exposed to the fluid present in the spaces around cells, the **interstitial fluid.** Much of the exchange between blood and interstitial fluid occurs by diffusion through capillary walls, which are very thin and consist of a single layer of flattened cells.

Figure 9.24 Circulation in a small area of the retina of the human eye. This specimen has been injected with a material that makes the blood vessels more clearly visible. At the top is an arteriole with its branches, the terminal arterioles, which lead to capillaries. Capillaries form a dense network and are uniform in diameter. The capillaries flow together into the branches of a venule (bottom) (magnification × 67).

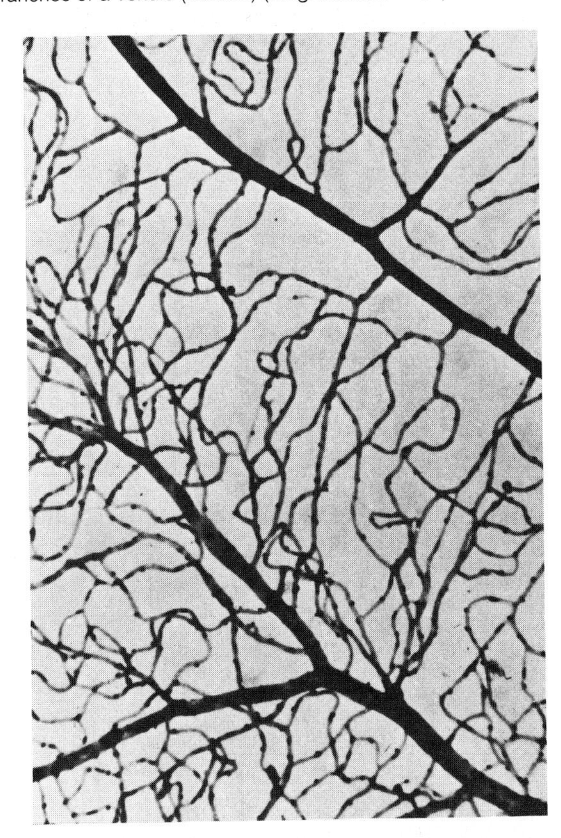

But there is another major factor in capillary exchange. Fluid actually flows back and forth through capillary walls. At the arteriole end of the capillary, water and dissolved materials are forced out through the capillary wall, while they flow into the capillary at the end nearer the venule (figure 9.25).

This regular fluid movement through the capillary wall is important because water carries nutrients and other dissolved materials from the blood to the interstitial spaces, and water returning to the capillary carries metabolic wastes. Not quite as much fluid returns to capillaries as is forced out. But as we shall see in chapter 10, this extra fluid is returned to the circulatory system by another route.

Blood and tissue fluids constantly exchange materials through capillary walls, and this close relationship between blood and tissue fluids is vital to the circulatory system's role in transporting substances to and from body cells.

Figure 9.25 Electron micrographs of a cross section of a capillary in heart muscle. (*a*) The wall of a capillary consists of a single layer of flattened endothelial cells. This picture shows parts of two endothelial cells; the nucleus and part of the cytoplasm of one cell, and a small part of the cytoplasm of another cell, in the lower right part of the capillary. The lumen is the inside of the capillary. (*b*) Enlarged view of part of this capillary wall showing the place where two endothelial cells join (arrow). Water and solutes flow through pores in these junction regions.

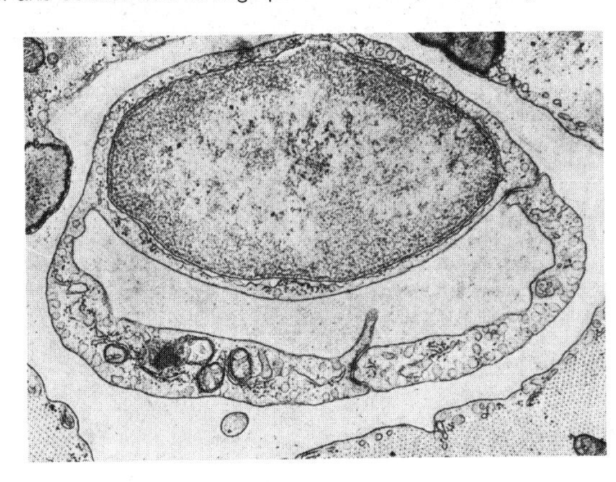

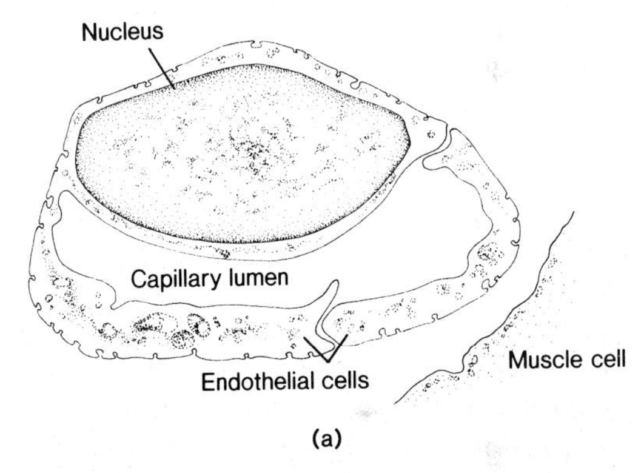

(a)

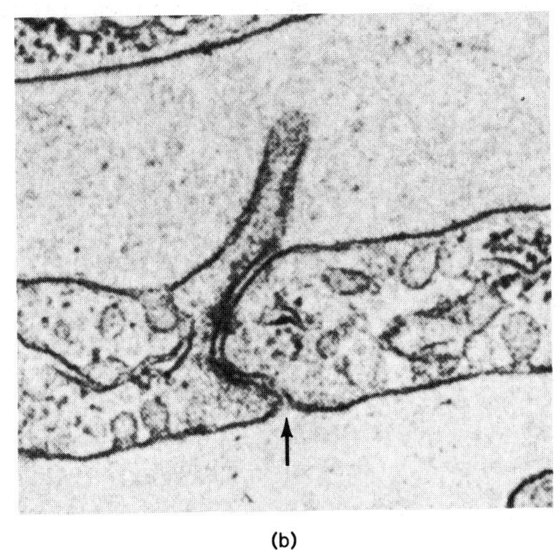

(b)

Summary

Gas exchange is an urgent requirement for animal life. All animals must continuously obtain oxygen and release carbon dioxide.

The four basic categories of animal gas exchange mechanisms are integumentary exchange, gills, tracheal systems, and lungs. Most aquatic animals exchange gases through gills. Tracheal systems are networks of air tubes that penetrate the body tissues of many small terrestrial animals such as insects. Lungs are highly vascular, thin-walled internal sacs. Lungs are found primarily in terrestrial vertebrate animals.

Because vertebrates' lungs are blind sacs connected to the outside air by narrow passages, breathing mechanisms are necessary for lung ventilation. Most vertebrates inhale by muscular movements that produce negative pressures around the lungs so that atmospheric pressure forces air into the lung.

Most of the oxygen transported in blood is bound to hemoglobin as oxyhemoglobin. Hemoglobin loads with oxygen quickly and completely in the lungs and unloads it when the blood reaches capillaries in body tissues.

Carbon dioxide is carried in the blood mainly as bicarbonate ions. The buffering action of hemoglobin and other protein molecules prevents significant pH changes in the blood.

A circulatory system is a transport system in which blood is moved through the body in a regular fashion. Complex circulatory systems have hearts that pump blood through a network of blood vessels.

In open circulatory systems, blood leaves blood vessels and moves through parts of the body cavity. In closed circulatory systems, blood does not leave the confines of blood vessels.

Circulatory systems of vertebrate animals are closed and have a heart, arteries, capillaries, and veins. In fish, blood passes through gill circulation, where it is oxygenated, and then moves on to the systemic circulation supplying the rest of the body. Vertebrate animals that have lungs for gas exchange have separate pulmonary and systemic circulatory loops. Separation of pulmonary and systemic circulatory loops is associated with a partitioning of the heart into the right and left chambers.

Humans have a four-chambered heart and separate pulmonary and systemic circuits. The human heart has a regular cycle of contraction and relaxation in which atria contract and then ventricles contract.

Blood pressure provides the driving force that propels blood through the circulatory system. Pressure falls as blood moves through capillary beds and is so low in veins that movement through veins toward the heart depends on body muscle contraction and a system of valves that prevent backward flow.

Sphincter muscles in arterioles regulate flow through capillary beds.

Individual capillaries are very small and very numerous. Much exchange between body cells and the blood is by diffusion through capillary walls, but solute-carrying fluid also regularly moves out from the capillaries into tissue spaces and returns to capillaries near their venule ends.

Questions for Review

1. Why can animals with small, flattened bodies live without specialized gas exchange structures?
2. Explain the mechanism by which fish move water over their gills for gas exchange.
3. What are respiratory pigments? Name an example of a respiratory pigment.
4. Explain what we mean by negative pressure inhalation.
5. The trachea has rigid C-shaped cartilages in its wall. Why are they important?
6. Distinguish between open and closed circulatory systems.
7. What are arterioles? Venules?
8. Explain the advantages of a circulatory system with separate pulmonary and systemic loops over a system with a single loop, such as that of fish.
9. What is coronary circulation?
10. When patients are immobilized in hospital beds, arrangements often are made to elevate their feet and legs higher than their trunks at least part of the time. What normal circulatory mechanism does this procedure replace?
11. How does an arteriole sphincter muscle control blood flow through a capillary bed?

Questions for Analysis and Discussion

1. Insect blood does not contain respiratory pigments. How do you suppose this fact relates to the insect method of gas exchange?
2. Lungs of mammals are much more elaborately subdivided into small passageways and spaces than are the lungs of frogs and reptiles. What do you think are the reasons for these differences?
3. Even with increased public awareness of the dangers of cigarette smoking, many young people still start to smoke each year. Present arguments based on what you know about respiratory and circulatory system problems that might help to convince them not to start smoking, or to quit smoking if they have already started.

Suggested Readings

Books

Schmidt-Nielsen, K. 1983. *Animal physiology.* 3d ed. New York: Cambridge University Press.

Tortora, G. J., and Anagnostakos, N. P. 1981. *Principles of anatomy and physiology.* 3d ed. New York: Harper and Row.

Articles

Hughes, G. M. 1979. The vertebrate lung. *Carolina Biology Readers* no. 59. Burlington, N.C.: Carolina Biological Supply Co.

Johansen, K. October 1968. Air-breathing fishes. *Scientific American* (offprint 1125).

Lassen, N. A.; Ingvar, D. H.; and Skinhøj, E. October 1978. Brain function and blood flow. *Scientific American* (offprint 1410).

Neil, E. 1979. The mammalian circulation. *Carolina Biology Readers* no. 82. Burlington, N.C.: Carolina Biological Supply Co.

Schmidt-Nielsen, K. May 1981. Countercurrent systems in animals. *Scientific American* (offprint 1497).

Wigglesworth, V. B. 1972. Insect respiration. *Carolina Biology Readers* no. 48. Burlington, N.C.: Carolina Biological Supply Co.

Blood and Body Fluid Homeostasis

Chapter Outline

Chapter Concepts

1. Body fluid regulation is a vital part of homeostasis.
2. Blood, which is composed of plasma, cells, and cell fragments, functions in transport and body defense reactions.
3. Excretory structures remove nitrogenous wastes and other cell wastes. They also function in maintenance of water balance and regulation of ionic composition and pH of body fluids.
4. The tubular units of kidneys are adapted to produce concentrated urine, thereby accomplishing the water conservation that is vital in terrestrial animals such as humans.
5. Some kidney diseases interfere with body fluid homeostasis. Kidney failure causes serious illness that can quickly lead to death.

Very few people have ever seen a tardigrade, but those who have are not likely to forget the comical appearance of the tiny "water bear." Tardigrades are microscopic animals that range in size between 0.1 mm and 1 mm in length. They use eight clawed legs to crawl over the water plants or wet moss on which they feed. The tardigrade's nickname, "water bear," comes both from its appearance and its ambling, bearlike gait. A tardigrade does look something like a miniature bear when you see it under the microscope (figure 10.1).

The most unusual thing about tardigrades, however, is their remarkable ability to survive very long periods of drought.

When a tardigrade's surroundings begin to dry up, instead of moving on or attempting to conserve water, it simply contracts into a barrellike shape and becomes inactive. Its body dries until it contains barely detectable quantities of water, as little as 3 percent of its predrying water content. A tardigrade can remain in this dehydrated condition for years and yet be ready to rehydrate and become active quickly when water is once again available. A museum specimen of moss that had been stored for 120 years yielded living tardigrades when it was placed in water.

This dormant state in which an animal has extremely low metabolic activity is called **cryptobiosis** (meaning "hidden life"). Cryptobiotic tardigrades are very resistant to harmful conditions, such as severe drought, that would quickly kill normal, active tardigrades, or any other animals for that matter. In their dried condition, tardigrades are not harmed by exposure to heat or solar radiation, and they are so light they can be blown about by the wind. (For more information, see box 25.1 on page 548.)

The ability to reduce metabolism and survive with very little body water remaining is a very rare phenomenon in the animal world. For the great majority of animals, including humans, of course, any significant change in quantity or composition of body fluids is a serious problem that can lead to death if not quickly corrected. Thus, stability of body fluids is a key part of homeostasis, the maintenance of a stable internal body environment. In this chapter, we will discuss body fluids of animals and consider some of the mechanisms by which body fluid homeostasis is maintained.

Figure 10.1 Tardigrades ("water bears"). (*a*) Scanning electron micrograph of a tardigrade. This is the normal appearance of the animal when it is active. Tardigrades live in very moist environments and their bodies contain about 85 percent water (magnification × 60). (*b*) Scanning electron micrograph of a tardigrade in cryptobiosis. The animal has assumed a barrellike shape and has dehydrated as its environment dried up. A cryptobiotic tardigrade's body contains very little water (magnification × 60).

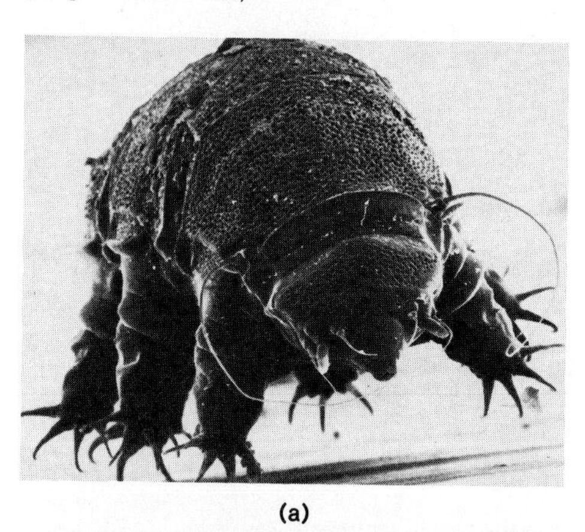

(a)

(b)

Figure 10.2 Body fluid compartments of a vertebrate animal.

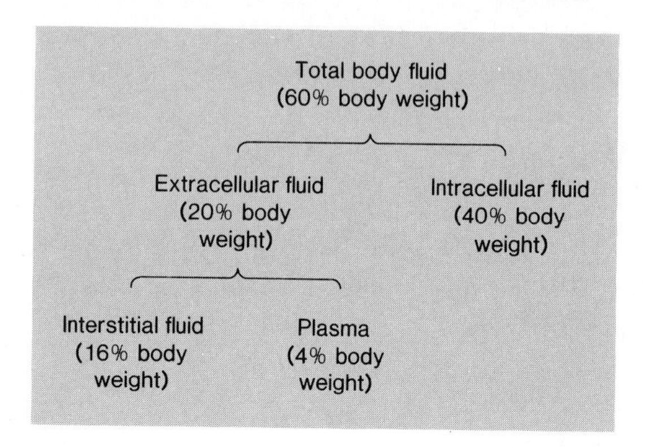

Figure 10.3 Relationships of body fluid compartments to one another and to the organs involved in body fluid regulation in a vertebrate animal. Blood plasma is the only fluid directly modified by external exchanges. Materials are exchanged with the external environment at specialized exchange boundaries (gut, lungs, kidneys, skin). Activities of the liver continuously modify plasma composition although the liver is not an external exchange organ of the body.

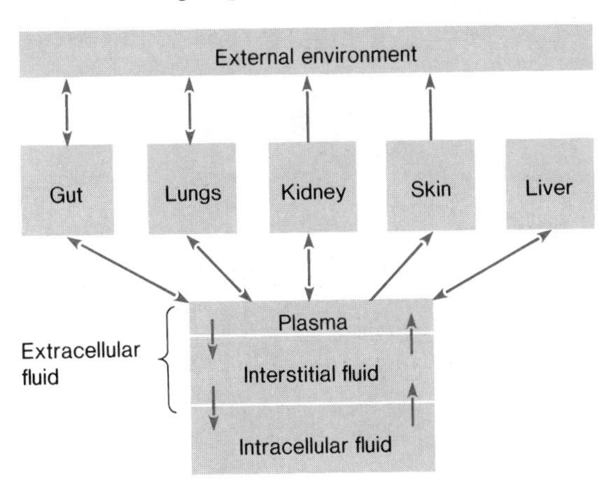

Body Fluid Compartments

Body fluids of animals are made up primarily of water containing specific quantities of ions (from dissolved salts), certain proteins, and a few other kinds of organic molecules. These body fluids are found in several fluid-containing compartments that are separated from each other by plasma membranes or by layers of cells (figure 10.2).

In vertebrate animals, for example, about two-thirds of the fluid (around 40 percent of body weight) is **intracellular fluid**—the fluid contained inside body cells. About one-third of the body fluid (around 20 percent of body weight) is **extracellular fluid**—fluid that is outside body cells. We can subdivide extracellular fluid into two categories: **interstitial fluid,** located in the spaces that surround body cells, and blood **plasma,** the fluid portion of blood.

Body fluid homeostasis is dynamic, and it involves exchanges among all the body fluid compartments. It depends on the specialized activities of several body organs and systems and on exchanges made between the body's internal environment and the external environment at specialized exchange boundaries (figure 10.3).

Fluids moving through the walls of capillaries carry materials back and forth between the blood plasma and the interstitial fluid around cells (see chapter 9). Most, *but not quite all,* of the fluid that is forced out of capillaries into spaces among cells, returns to the capillaries at the end of capillary beds. What happens to the extra fluid left in the tissues?

Figure 10.4 Photomicrograph of a valve in a lymph vessel. Lymph is forced through lymph vessels as contracting muscles exert pressure on the vessels' walls. Valves maintain one-way flow through lymph vessels by preventing backward flow (magnification × 45).

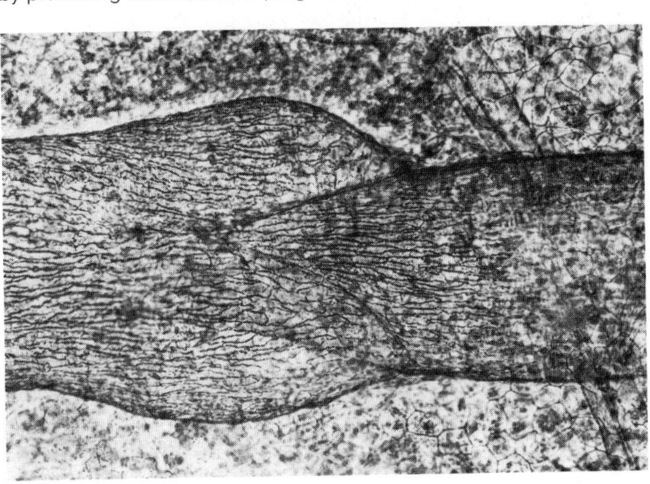

Vertebrate animals have a special system of vessels, the **lymphatic system,** that drains excess fluid from interstitial spaces in the tissues and returns it to the blood. This system maintains a balance between blood volume and interstitial fluid volume in the body.

The Lymphatic System

There are **lymph capillaries** in all parts of the body. Once interstitial fluid enters the lymphatic system, it is called **lymph.** Lymph moves through lymph capillaries and **lymph vessels** as a result of pressure applied by muscle contractions near the vessels and a system of valves that prevent backward flow (figure 10.4).

Lymph moves from capillaries into smaller vessels that unite to form larger vessels. Finally, two major **lymph ducts** empty into large veins near the heart (figure 10.5).

In addition to returning fluids from the interstitial spaces to the circulatory system, the lymphatic system has several other important functions. Lymph capillaries of the digestive tract, known as **lacteals,** are involved in transport of absorbed fats during digestion.

The lymphatic system also has a role in filtering of the lymph. **Lymph nodes** located along the lymph vessels contain a network of connective tissue fibers with **phagocytic cells** scattered among them. These phagocytic cells engulf dead

Figure 10.5 The lymphatic system. Lymph vessels flow into two main channels. The thoracic duct drains lymph vessels from all parts of the body except the upper right portion. Vessels from that area drain through the right lymphatic duct. These two large lymphatic ducts drain into the subclavian veins, thus returning fluid from the tissues to the circulatory system. Nodes filter debris and bacteria out of lymph. The inset is an enlarged sketch of part of the intestinal lining showing the location of lacteals in the intestinal villi. Products of fat digestion pass through the lacteals to lymph vessels and eventually to the circulatory system.

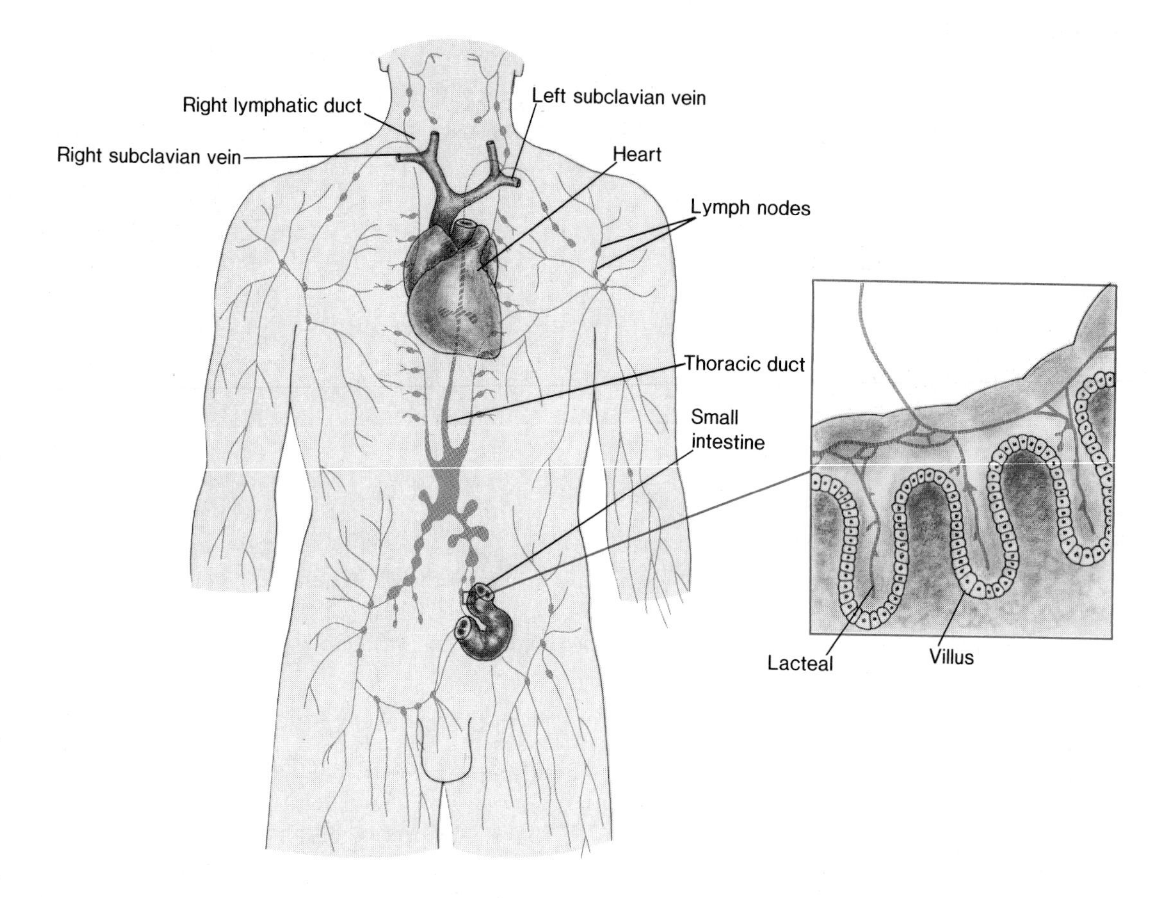

cells, cell debris, and foreign objects, such as bacteria, as the lymph filters through the lymph nodes. Because the lymph nodes are very active in defense responses, they often become swollen and tender during infections. The so-called "swollen glands" that you can often feel when you have a sore throat are active lymph nodes.

Anything that interferes with the flow of lymph through a lymph vessel or lymph node causes fluid accumulation in the part of the body drained by that vessel. Such fluid accumulation is called **edema,** and it often is visible externally as a swelling in that part of the body.

A very dramatic example of edema can be seen in the tropical disease **filariasis.** When infected mosquitoes bite humans, they transmit a parasitic worm into the bloodstream. The larvae of these filarial worms invade the lymph vessels and interfere with (or completely block) the flow of lymph. Tissues in the area of the blocked lymph vessel can swell to enormous proportions. The condition is appropriately called **elephantiasis** (see figure 29.22).

Blood

Blood is a red liquid made up of cells and cell fragments suspended in the blood plasma. Plasma makes up about 55 percent of blood, while the cells and cell fragments constitute the remaining 45 percent. These include **erythrocytes (red blood cells), leukocytes (white blood cells),** and **platelets,** which are small fragments of cells.

Figure 10.6 Erythrocytes. Human erythrocytes are flattened, biconcave discs. This side view shows the inside of an erythrocyte and reveals how thin the center is. Erythrocyte shape apparently is an adaptation for more efficient gas diffusion.

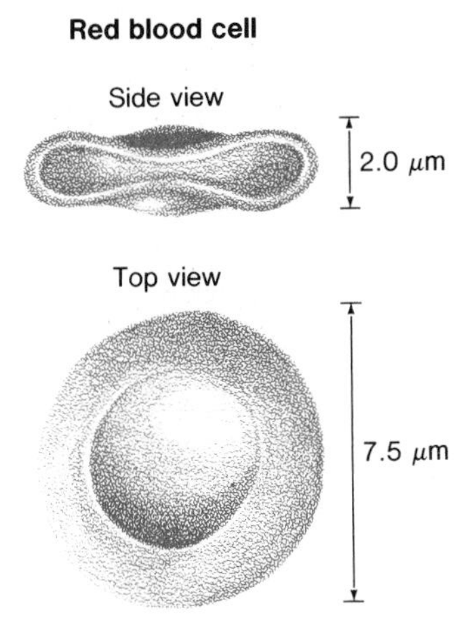

Red Blood Cells

Erythrocytes (red blood cells) are the most numerous of the blood cells. Usually there are from four to six million red blood cells per cubic millimeter of human blood. A human erythrocyte is a thin, biconcave disc with a very thin center and a thicker rim. A mature circulating erythrocyte does not have a nucleus because the nucleus degenerates during the erythrocyte's development (figure 10.6). Approximately one-third of an erythrocyte's volume is taken up by about 280 million hemoglobin molecules which bind oxygen for transport.

Erythrocytes normally are quite elastic. Even though they are bent as they pass through small blood vessels, they spring back into shape. As erythrocytes age, however, they become more fragile, and such bending can damage or even rupture them. Aged, damaged erythrocytes are withdrawn from circulation as the blood passes through the liver or spleen. How old is an "old" erythrocyte? The average life span of human erythrocytes is about 120 days. New erythrocytes must be produced as fast as old ones are destroyed, and they develop in the **bone marrow,** which is located in cavities inside bones. About two million new erythrocytes are produced in the human body per second, a rate that equals the rate of red cell destruction in a normal healthy person.

Sometimes the mechanisms that normally balance erythrocyte production and destruction fail. For example, nutritional deficiencies or diseases cause slower than normal erythrocyte production. This results in a condition known as **anemia** in which there are fewer than the normal number of erythrocytes (or, in some cases, reduced hemoglobin content in the erythrocytes). People suffering from anemia often are fatigued or chilled because their tissues lack adequate oxygen for cellular energy conversion and heat production.

Human Blood Types

The surfaces of human erythrocytes contain certain genetically determined sets of molecules, and people can be divided into groups or **blood types** depending on which sets of molecules are present. Because these sets of molecules are genetically determined, a person's blood type is inherited. Two major classifications of human blood types are based on **ABO** grouping and the **Rh** system.

ABO grouping is based on differences in the type of glycoprotein molecule (protein with carbohydrate attached—see page 62) present on erythrocyte surfaces. Type A people have A type glycoproteins on their erythrocyte surfaces; type B people have B type glycoproteins; type AB people have both of these glycoproteins; and type O people have neither of them.

The A and B glycoproteins function as **antigens;** that is, they combine specifically with **antibody** molecules. **Anti-A** antibody and **anti-B** antibody are protein molecules that occur in human plasma, and they combine specifically with the A and B antigens, respectively. We call antigens and the specific antibodies with which they combine **complementary.** When antigens combine with complementary antibodies, erythrocytes clump together. This happens because antibodies can combine with antigens on the surfaces of two different red cells, thus causing the cells to be bound together. Whole clumps of red cells can be stuck together in this way. This cell clumping is called **agglutination** (figure 10.7).

Type A people have the anti-B antibody in their plasma; type B people have anti-A antibody; type AB people have neither of these antibodies; and type O people have both antibodies (table 10.1). Knowing this, you can see why it is vital to determine someone's blood type before giving blood transfusions. Donor and recipient must be compatible, or a severe agglutination reaction can occur. For example, if type A blood is given to a type O person, the anti-A antibody in the recipient's blood combines with the A antigen on the surface of the donor cells and causes them to agglutinate. The clumps of cells formed by this reaction can block blood vessels and cause serious circulatory problems.

Whenever possible, recipients should be given blood of the same ABO type as their own, but there are possible alternatives. AB people can receive blood of any of the ABO types because their plasma has neither of the antibodies. If the correct type is not available, type O blood can be given to a person having any ABO blood type because type O erythrocytes have neither of the antigens on their surfaces.

Figure 10.7 Erythrocyte agglutination reactions. (*a*) Comparison of unreacted blood sample (left) and blood sample with agglutinated erythrocytes (right). Agglutination occurs when cells bearing particular cell surface antigens are mixed with plasma containing specific antibodies against their antigens. (*b*) A diagrammatic representation of ABO blood typing reactions. Plasma antibody types used in the tests are indicated at the top of the columns.

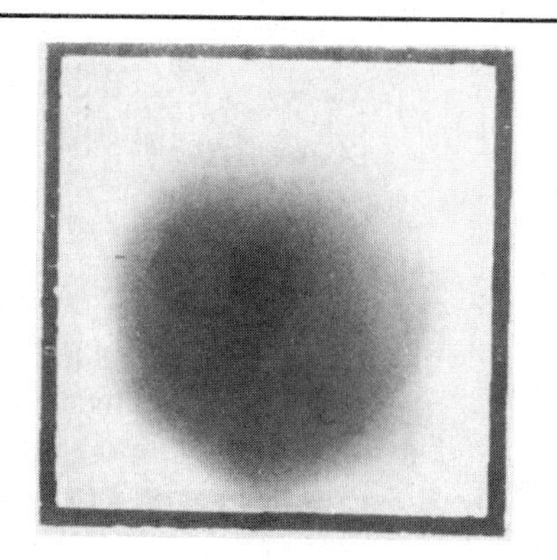

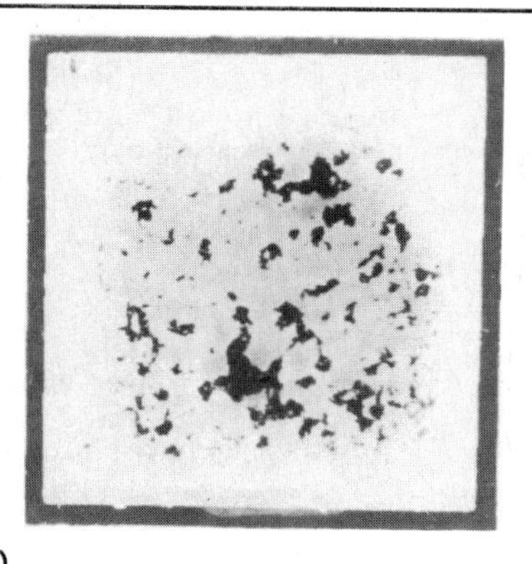

(a)

Anti A | Anti B | Type blood

O

A

B

AB

(b)

Table 10.1
Antigens and Antibodies in Human ABO Blood Groups

Blood Type	Antigen Present on Erythrocyte Membranes	Antibody in Plasma	Incidence of Type in the United States	
			Among Whites	*Among Blacks*
A	A	Anti-B	41%	27%
B	B	Anti-A	10%	20%
AB	A and B	Neither	4%	7%
O	Neither	Anti-A and Anti-B	45%	46%

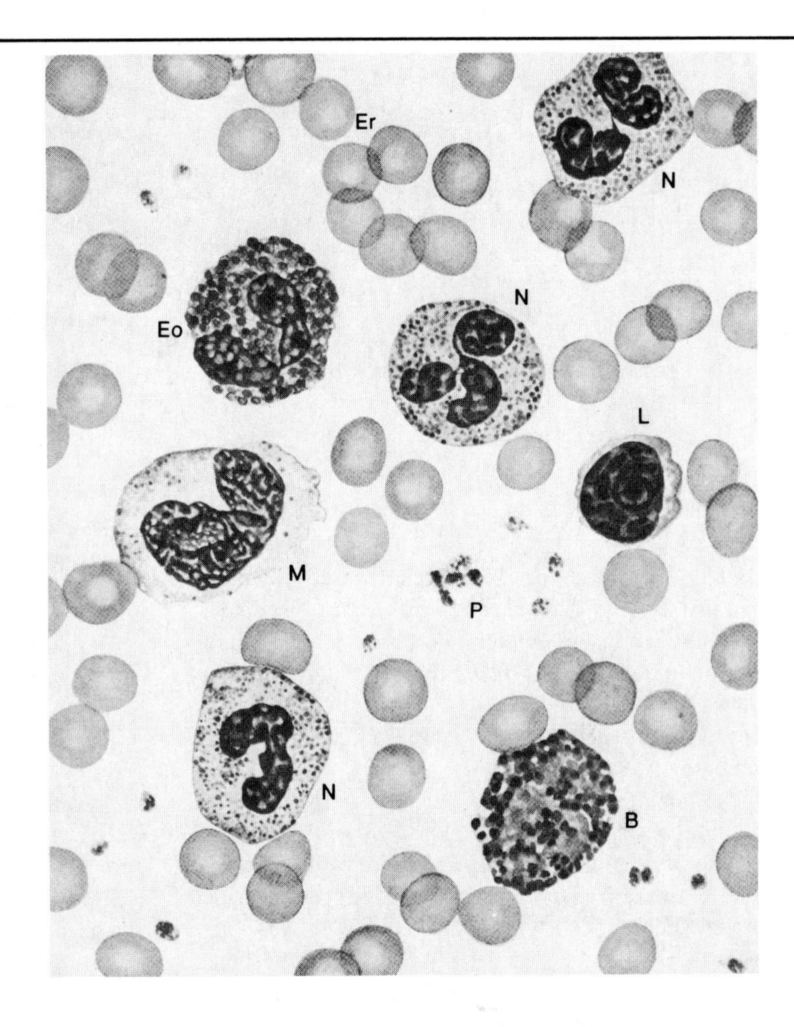

Figure 10.8 A representation of human blood cells. Shown are erythrocytes (Er), three kinds of granular leukocytes (eosinophil—Eo, neutrophils—N, and a basophil—B), a monocyte (M), a lymphocyte (L), and platelets (P). The names of the granular leukocytes are based on the different ways that the cells stain when stained blood smears are prepared. Neutrophils are the most numerous of all leukocytes. Several types of leukocytes are phagocytic cells that engulf and destroy bacteria, viruses, and scraps of damaged tissue both inside the circulatory system and in spaces around tissue cells. We will examine the roles of lymphocytes in disease resistance and immunity in chapter 21.

The second major classification of human blood types is based on the Rh type. (Rh stands for **rhesus;** this factor was first identified in the blood of the rhesus monkey.) Rh classification depends on the presence or absence of the **Rh antigen** on red blood cells. **Rh-positive (Rh^+)** people have the Rh antigen on their erythrocytes while **Rh-negative (Rh^-)** people do not. About 85 percent of all people in the United States are Rh-positive.

White Blood Cells

Leukocytes (white blood cells) contain nuclei and do not contain hemoglobin. They participate in body defense reactions. Leukocytes are capable of leaving the circulatory system by squeezing out between the cells in capillary walls. In this way, leukocytes can act both inside blood vessels and in the spaces around tissue cells.

Leukocytes are far less numerous than erythrocytes. Normally, human blood contains from 7,000 to 8,000 leukocytes per cubic millimeter. Red cells thus outnumber white cells by about 700 to 1. You can see the various types of leukocytes in figure 10.8.

Platelets

Platelets are small disc-shaped cell fragments that lack nuclei. They average from 2 to 4 μm in diameter, and there are between 200,000 and 400,000 platelets per cubic millimeter of blood. Platelets are involved in the formation of clots and in this way function in preventing fluid loss from the circulatory system. Platelets must be produced in large quantities because they have a life span of only about one week.

Figure 10.9 Major steps in blood clotting.

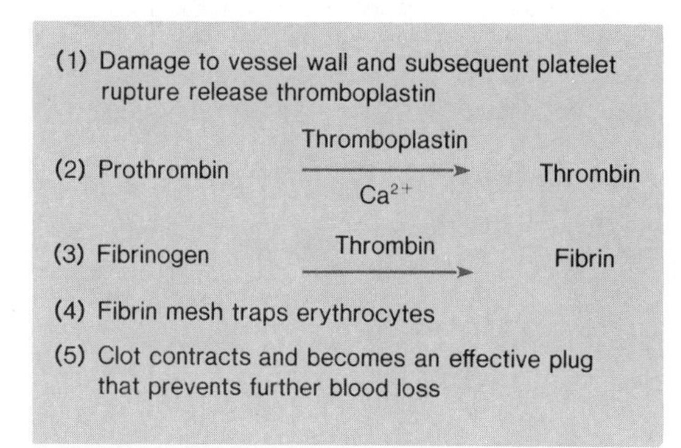

Blood Plasma

Although blood plasma is 90 to 92 percent water, it is much more than just a liquid that suspends the various blood cells. Blood plasma contains a complex mixture of ions and inorganic and organic molecules. By far the most numerous ions are sodium (Na^+) and chloride (Cl^-) ions. These and other important ions are required in different amounts for normal cell functioning.

Some of the important organic nutrients in the blood are glucose, amino acids, and various lipids. Nutrients are transported in the blood following absorption from the digestive tract. They also are transported by the blood from nutrient storage sites, such as the liver, to various body tissues.

Plasma proteins make up 7 to 9 percent of the blood plasma. The three main types of plasma proteins are the **albumins,** the **globulins,** and **fibrinogen.** Albumins are involved in regulation of fluid exchanges between capillaries and the interstitial fluid. Globulins function in transport and in immune responses to foreign antigens, such as those of invading bacteria. Fibrinogen plays a critical role in the blood-clotting process.

Blood Clotting

Any injury that breaks blood vessel walls can quickly lead to serious blood loss. This life-threatening danger of excessive blood loss is countered by the complex **clotting (coagulation)** mechanism. Clots are plugs that form temporary barriers to blood loss until vessel walls have healed.

You may have examined a small cut and noticed how bleeding slows and soon stops. While the process may seem quite simple, blood clotting involves a chain of reactions, each of which depends on completion of the reaction that precedes it in the sequence (figure 10.9).

Figure 10.10 Scanning electron micrograph showing an erythrocyte caught in the fibrin threads of a clot. Fibrin threads form a mesh that catches many blood cells. Then the clot contracts, squeezing out fluid and forming a solid plug that prevents further blood loss (magnification × 7,600).

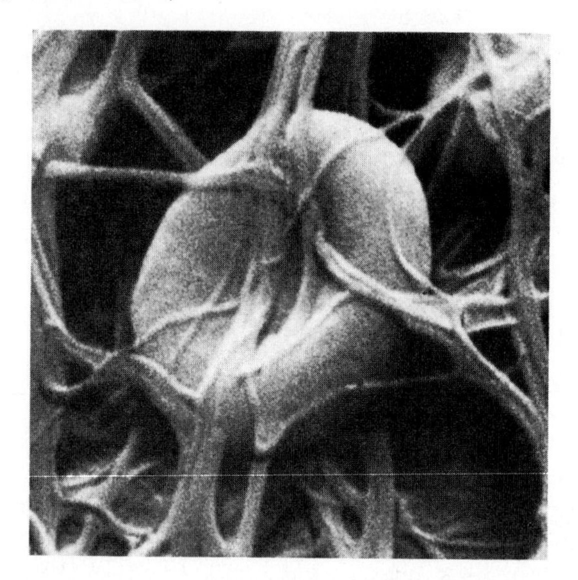

When there is damage to the wall of a blood vessel, platelets, which are very fragile, stick to the rough surfaces and break open, releasing their contents. Ruptured platelets release **thromboplastin,** which affects one of the globulin group of plasma proteins, **prothrombin,** in a very specific way. In the presence of thromboplastin and calcium ions (Ca^{2+}), prothrombin is converted to an active form called thrombin. Thrombin catalyzes the conversion of the plasma protein fibrinogen to a gellike form called **fibrin.** It is fibrin that forms the fibrous mesh of a blood clot (figure 10.10). This fibrin mesh traps blood cells and forms a plug that prevents further blood loss.

When you are injured, clots must form readily in order to prevent excessive blood loss. But on the other hand, a grave danger is posed if clots form too easily. Clots formed inside blood vessels can interfere with or even entirely shut off blood flow to an area of the body. Despite protective mechanisms, clots sometimes form inside blood vessels, most often where the inner surface of a blood vessel is roughened. A clot on the inner wall of a vessel is called a **thrombus** (figure 10.11). A thrombus can slow or block blood flow through the vessel where it forms, or it can break loose from the vessel wall and begin to flow with the blood through the circulatory system. Such a free-moving clot in the circulatory system is called an **embolus** (plural: **emboli**). Emboli generally flow along in the blood until they reach a small narrow point where they lodge in a vessel and stop flow through it. This is called an **embolism.**

Figure 10.11 Scanning electron micrograph of a thrombus, a blood clot formed inside an unbroken blood vessel. Blood cells are trapped in a mesh of fibrin threads. Erythrocytes (Er) often become misshapen (crenated) when they are in an altered environment such as the inside of this thrombus. Some crenated erythrocytes (CE) are labelled (magnification × 625).

Kessel, R. G., and Kardon, R. H. *Tissues and Organs: A Text-Atlas of Scanning Electron Microscopy.* © 1979 by W. H. Freeman and Company.

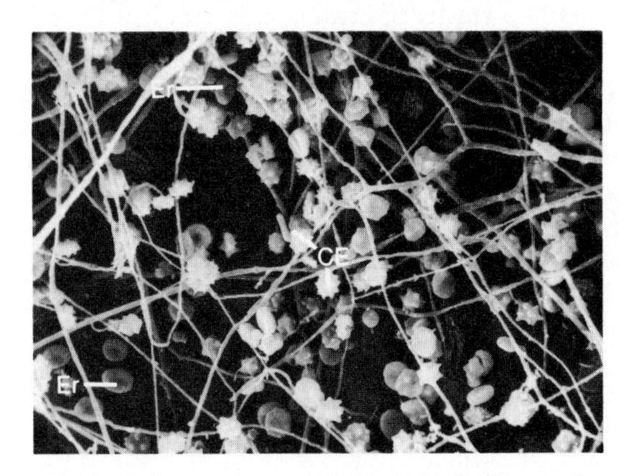

Not all problems with blood clotting involve extra, undesirable coagulation. Sometimes, clotting is so inefficient that excessive bleeding results. A vitamin K deficiency results in poor clotting because vitamin K is required for normal prothrombin synthesis by the liver, and lowered plasma prothrombin levels lead to poor clotting.

Another condition that results in excessive bleeding is **hemophilia,** a hereditary problem that occurs almost exclusively in males. Victims of hemophilia bleed excessively because they have a deficiency of one of the plasma factors required for clotting. Even the slightest injury can lead to massive blood loss (**hemorrhage**). Very often, hemophilia victims die at a young age following repeated episodes of severe bleeding. Hemophilia influenced European history during the nineteenth century because young male members of several of the royal families of Europe inherited the disorder.

Nitrogenous Wastes

In order to maintain homeostasis, an animal must rid its blood and body fluids of waste substances. **Excretion** is the process that removes wastes, excess materials, and toxic substances from the body's cells and extracellular fluids.

We have seen (chapter 9) how the metabolic waste carbon dioxide is released into the atmosphere through specialized gas exchange organs. Other major wastes produced by cellular metabolism are **nitrogenous** (nitrogen-containing) **wastes.**

Figure 10.12 Chemical formulas for the three major nitrogenous wastes of animals. Note that in solution inside animals' bodies, ammonia gains a hydrogen ion to become an ammonium ion.

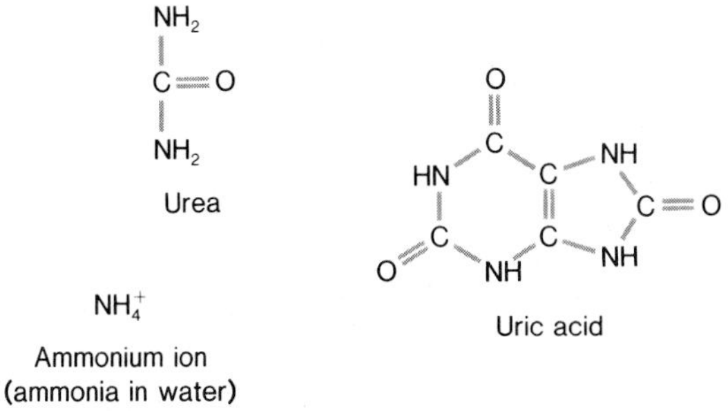

NH_2–C(=O)–NH_2

Urea

NH_4^+

Ammonium ion
(ammonia in water)

HN, C=O, NH, C, C, C=O, NH, NH (purine ring)

Uric acid

Nitrogenous waste compounds are products of cellular protein metabolism. Extra amino acids not used for synthesis of body proteins are oxidized to generate energy or converted to fats or carbohydrates, which can be stored. Before this can happen, however, nitrogen-containing **amino groups** ($-NH_2$) must be removed from these amino acids. **Ammonia** (NH_3) is the nitrogenous waste molecule produced by this reaction.

Some organisms excrete ammonia directly. Many other animals, however, incorporate ammonia into organic molecules, mainly either **urea** or **uric acid,** which are then excreted (figure 10.12). A number of terrestrial organisms, such as mammals and adult amphibians (frogs, toads, salamanders, etc.), excrete urea as their main nitrogenous waste product.

On the other hand, reptiles, birds, and insects excrete mainly uric acid. Uric acid leaves the bodies of animals that excrete it as a damp, concentrated mass of semicrystalline material. Animals that excrete uric acid recover most of the water from this semisolid waste before it leaves their bodies. This makes uric acid excretion quite advantageous for animals living in very dry environments.

Excretory Organs

Because of the importance of water in life processes, maintenance of **water balance** is vital for homeostasis in living things. An organism is in water balance when its water loss to the environment equals its water gain from the environment. Regulation of the ionic composition of body fluids is interrelated with maintenance of water balance and waste excretion.

How do animals rid themselves of wastes? In some simple multicellular animals, cell wastes simply diffuse away into the surrounding water. But most larger animals have specialized excretory structures that function in body fluid homeostasis.

Figure 10.13 The flame-cell excretory system of a planarian worm.

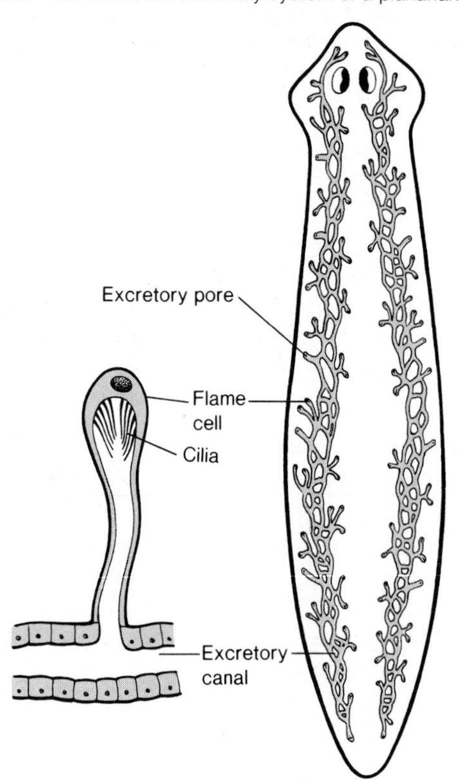

Invertebrate Excretion

Most animals' excretory systems consist of tubules or collections of tubules. Among the simplest of these are the **flame-cell systems** found in flatworms such as planarians (figure 10.13). Bulblike **flame cells** are located along tubular excretory canals. These bulbs are called flame cells because each one contains a cluster of beating cilia that looks like a flickering flame when viewed under a microscope. The beating of flame cell cilia propels fluid through the excretory canals and out of the body through excretory pores. Flame-cell systems function mainly in eliminating excess water that has entered the body osmotically.

Earthworms have more complex excretory systems than planarian worms. Earthworm excretory structures are tubules called **nephridia** (singular: **nephridium**). Each body segment has a pair of nephridia (figure 10.14).

Fluid from the body cavity enters the nephridium through an opening called the **nephrostome** and is processed as it moves through the tubule. Ions are reabsorbed and carried away by a network of capillaries that surround the tubule, and waste materials, delivered by the capillary network, are added to the fluid. The tubule leads to an enlarged bladder that empties to the outside of the body via an opening called the **nephridiopore. Urine,** the excretory fluid that leaves the body, is a dilute fluid containing metabolic wastes.

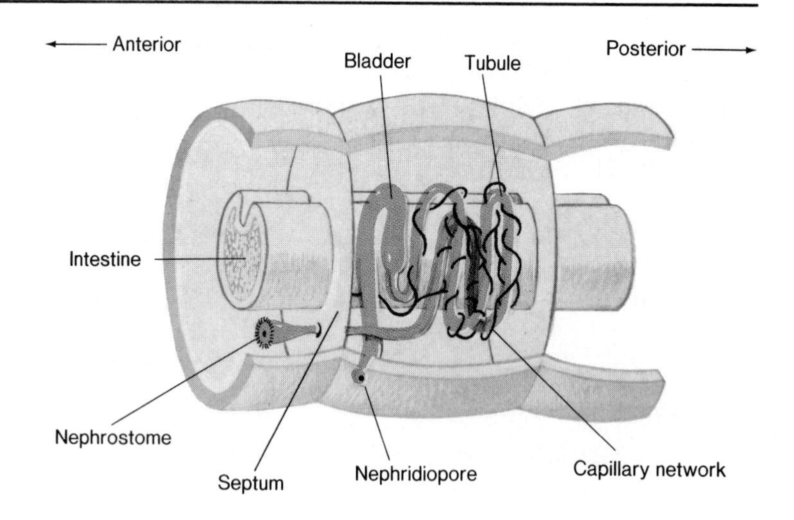

Figure 10.14 The earthworm nephridium. Fluid enters the nephridium through the nephrostome. The main tubular portion of the nephridium is coiled and surrounded by a capillary network. Urine can be stored in a bladder before being released to the outside. Note the close relationship between blood vessels and the tubules.

Insects have an excretory system that is quite different from those we have considered so far. Insects' excretory ducts do not open to the outside of the body. Instead, the system consists of a cluster of long, thin **Malpighian tubules** that are attached to the gut. The Malpighian tubules and the gut together function as the insect excretory system (figure 10.15).

Some insects living in dry environments reabsorb most of the water from their urine and excrete a dry, semisolid mass of precipitated uric acid. For example, mealworms, which are the larva of a flour beetle *(Tenebrio molitor),* live in dry flour and have no access to water or moist food throughout their entire lives. Metabolic water produced in cell respiration (see page 119) is their major source of water. Yet they manage their water balance by efficiently reabsorbing water and producing a very concentrated excretory product.

Figure 10.15 Malpighian tubules in the grasshopper. The Malpighian tubules are attached to the junction of the stomach (midgut) and the intestine (hindgut). Both Malpighian tubules and posterior portions of the digestive tract are involved in excretory functions. The Malpighian tubules are named for the Italian microscopist Marcello Malpighi (1628–1694), who discovered them.

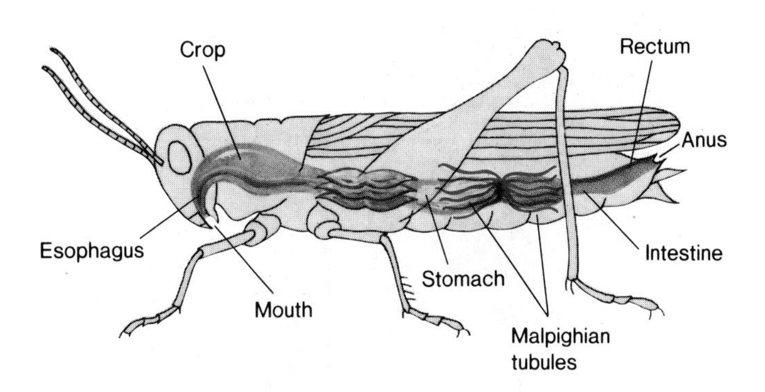

The Human Excretory System

All vertebrate animals have specialized excretory structures called **kidneys.** We will examine the human excretory system as an example of vertebrate excretion (figure 10.16).

Figure 10.16 The human excretory system. The adjective "renal" refers to the kidney. Thus, a renal artery is the artery that supplies the kidney. Ureters carry urine from the kidneys to the urinary bladder, where it is stored until emptied through the single urethra.

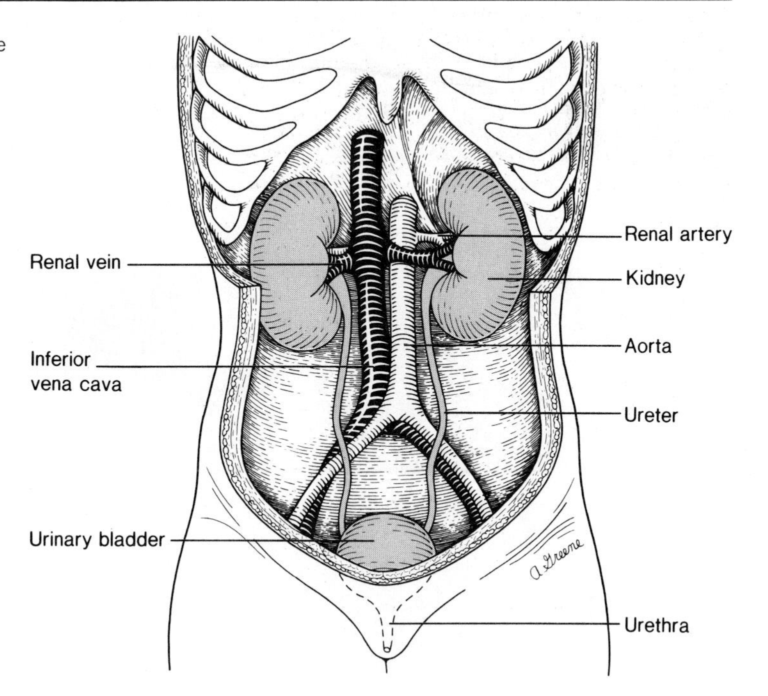

Figure 10.17 Internal structure of the human kidney. One section is enlarged to show location of tubules.

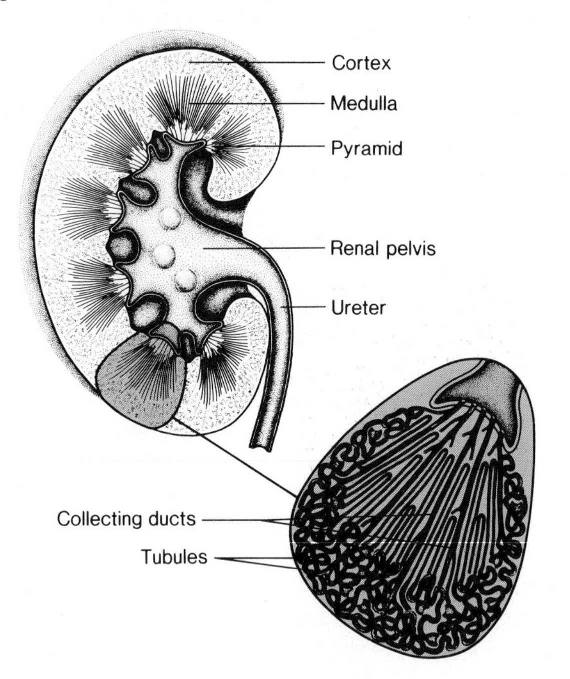

The human kidney consists of a compact mass of individual functional tubular excretory units, the **nephrons.** Each of approximately one million nephrons in an adult human kidney makes its own contribution to the total excretory function of the kidney. In human nephrons, as in those of other terrestrial vertebrates, water conservation is a critically important part of the processing of excretory fluid.

The kidney's outer region, the **cortex,** has a somewhat granular appearance (figure 10.17). The **medulla,** which lies inside the cortex, is arranged in a group of pyramid-shaped regions, each of which has a striped appearance. The innermost part of the kidney is a hollow chamber called the **pelvis.** Urine formed in the kidney collects in the pelvis on its way to the ureter.

As we mentioned earlier, the basic functional units of human kidneys are the one million or so nephrons that make up each kidney. Figure 10.18 shows more details of the anatomy of an individual nephron. Each nephron is essentially a rather complex tubule with an expanded, hollow end

Figure 10.18 The human nephron with associated circulatory structures. Each kidney contains over one million nephrons. The term proximal means nearer; distal means further. In this case the proximal convoluted tubule is proximal (closer) to Bowman's capsule and the distal convoluted tubule is further from it.

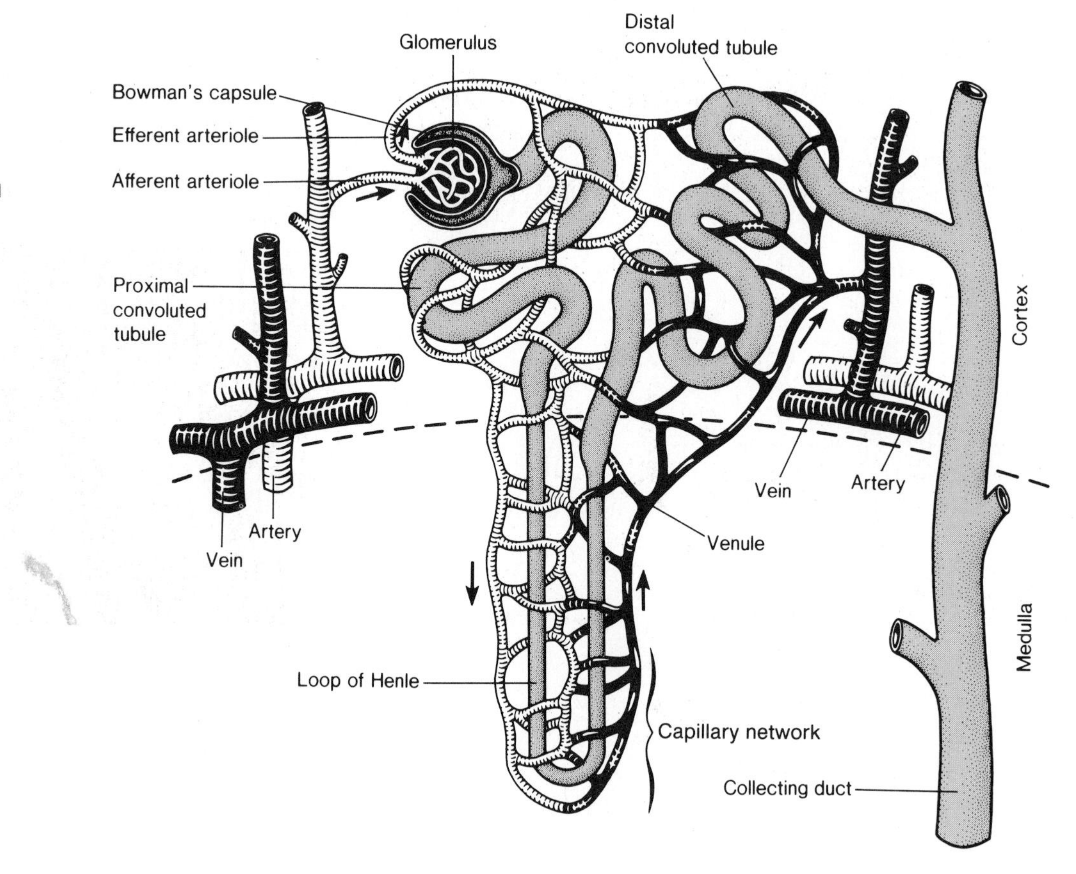

called **Bowman's capsule.** Enclosed within the Bowman's capsule is a knot of capillaries called the **glomerulus.** The remainder of the nephron consists of a tubule with three distinct regions. Adjacent to the Bowman's capsule is a coiled portion called the **proximal convoluted tubule.** The proximal convoluted tubule leads to the **loop of Henle,** a narrowed portion that plunges deep into the medulla. Another coiled portion, called the **distal convoluted tubule,** follows.

The end of each nephron drains into a **collecting duct** that recovers fluid from several nephrons in an area of the kidney and delivers urine to the pelvis (figure 10.18).

Each nephron is supplied blood by an **afferent arteriole** that branches into the capillary knot making up the glomerulus. The glomerular capillaries drain into an **efferent arteriole,** which subsequently branches into a second capillary network around the tubular parts of the nephron. These capillaries pick up materials reabsorbed from the tubules and deliver materials that are secreted into the tubules. The capillaries drain into venules, and these flow together into veins that carry blood to the single renal vein leaving the kidney.

Kidney Function

The functioning of human nephrons in body fluid regulation involves three basic processes: filtration, reabsorption, and secretion. Fluid from the blood is forced through the walls of the capillaries of the glomerulus into the space inside Bowman's capsule, a process called glomerular **filtration** (figure 10.19). This fluid (**filtrate**) is identical to the blood in composition except that it does not contain blood cells or plasma proteins.

As the fluid passes through the nephron, nutrients, essential ions, and water are reabsorbed and passed to capillaries. Most of this reabsorption takes place in the proximal convoluted tubules. This tubular **reabsorption** is vital, because the body cannot afford to lose all of the nutrients and ions that pass from the blood into the glomerular filtrate (figure 10.20).

Certain other substances, delivered to the tubule area by the capillaries, are actively secreted by cells in the tubule walls. Through this tubular **secretion,** additional wastes are added to the fluid for eventual urinary excretion.

About 125 ml of fluid are filtered in your two kidneys every minute. This means that a total of about 180 l of fluid leave your blood and enter the Bowman's capsules each day.

Figure 10.19 The glomerulus. (*a*) A diagram showing the glomerulus as a knot of capillaries inside a Bowman's capsule. Filtration forces fluid out through the walls of the capillaries into the cavity of a Bowman's capsule and on into the proximal tubule. (*b*) A scanning electron micrograph of a section of kidney cortex showing a glomerulus inside a Bowman's capsule. The holes surrounding the glomerulus are cross sections of tubules.

(a)

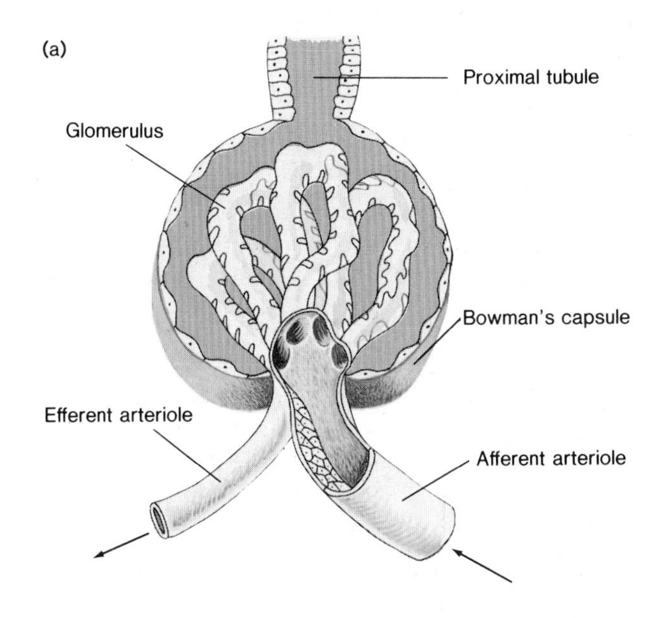

(b)

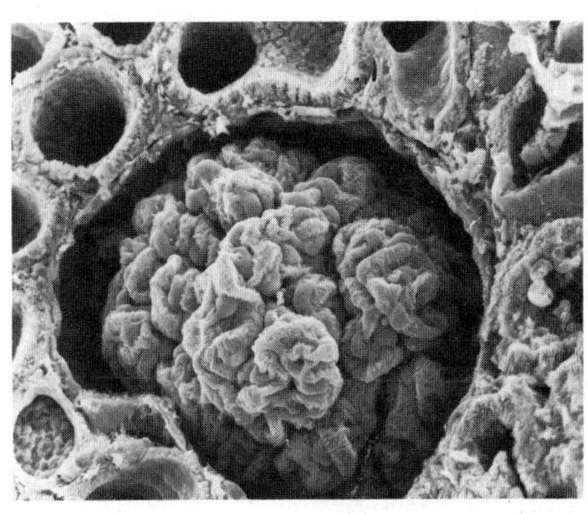

Figure 10.20 Filtration, reabsorption, and secretion in nephron function. In the glomerulus, water and solutes are filtered into the kidney tubules, while proteins and cells remain in the blood. As the glomerular filtrate passes through the proximal and distal convoluted tubules, solutes may be actively reabsorbed from the filtrate or secreted into it.

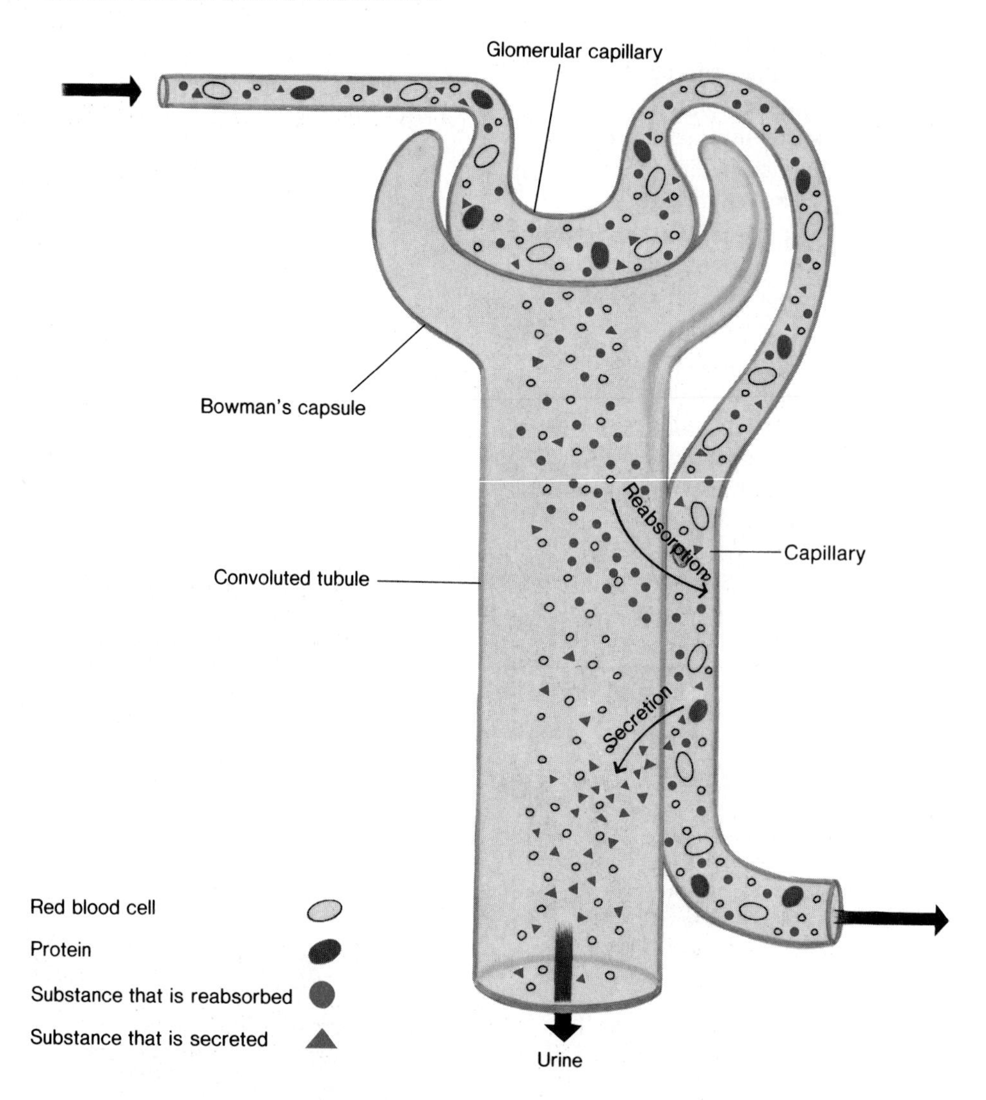

Very obviously, however, you do not produce and void 180 l of urine a day. Our kidneys recover over 99 percent of this filtrate and return it to the blood leaving the kidney.

Water Conservation

All terrestrial animals, including humans, must conserve as much water as possible, while at the same time excreting nitrogenous wastes and other materials.

How is water conservation accomplished in the human kidney? A major part of the water in the kidney tubules is reabsorbed as the fluid flows through the kidney's collecting ducts. Water is recovered because a very salty environment is maintained in the medulla of the kidney where the collecting ducts are located. Because this environment is hypertonic to the fluid in the collecting ducts, water tends to move out of the collecting ducts by osmosis. Thus, water is recovered by the body. The hypertonic environment maintained in the kidney medulla is the key to water conservation.

How is the salty environment around the collecting ducts maintained? This is the function of the loops of Henle. Fluid moves down toward the center of the kidney through the **descending limb** of the loop of Henle, and it flows in the opposite direction through the nearby **ascending limb.** There are important differences between the descending limb and the ascending limb of the loop. The descending limb does not

Figure 10.21 Water reabsorption functions of the loops of Henle and collecting ducts in the kidney. Relative salt concentrations are shown as osmolarity, a measure of osmotically effective concentrations. Sodium and chloride ions pumped out of the ascending limb of the loop of Henle diffuse into the descending limb. This cycling of ions maintains the salty environment of the fluid in the part of the medulla around the tips of the loops of Henle and around the collecting ducts.

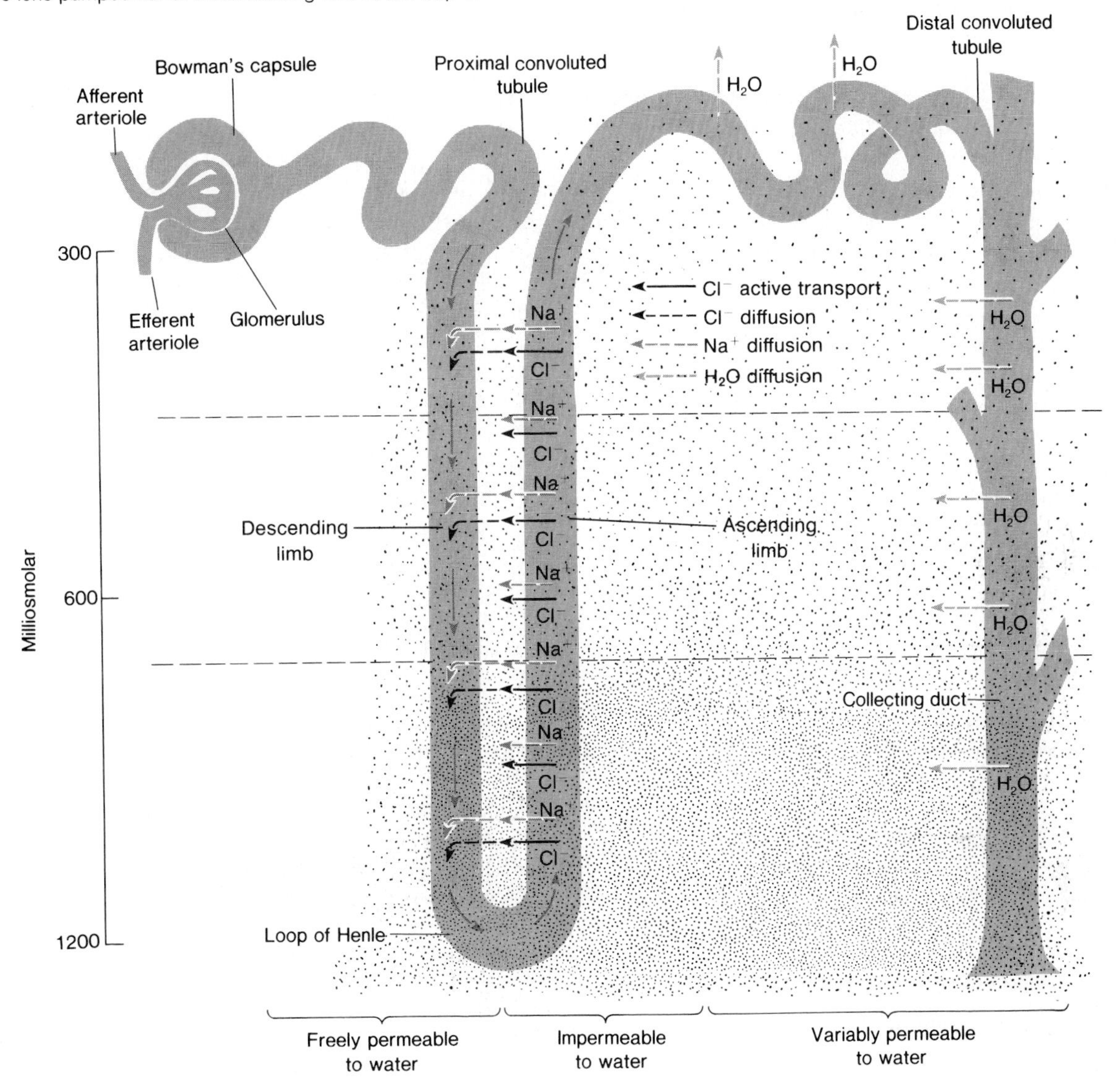

actively transport salt. It is somewhat permeable to Na^+ and Cl^-, and it is very permeable to water. In contrast, the ascending limb actively transports chloride ions outward to the interstitial space, and sodium ions follow due to electrical attraction. Also, the ascending limb is practically impermeable to water.

Thus, the ascending limb actively pumps ions out of the fluid moving up through it. Ions from the salty environment around the loop of Henle diffuse into the descending limb, adding to the Na^+ and Cl^- already in the fluid in the lumen of the tubule. This makes the Na^+ and Cl^- content of the fluid inside the tubule at the bottom of the loop of Henle very high. But most of this salt is pumped out of the ascending limb, which at the same time is impermeable to water.

The result of all this activity is that a good deal of sodium chloride is retained near the bottom of the loop of Henle where it cycles. It is pumped out of the ascending limb into the interstitial fluid, where it diffuses into the descending limb and is then carried around to the ascending limb, where it is pumped out again. Recycling sodium chloride permits the kidney to maintain a salty environment in the interstitial fluid in the area of the tips of the loops of Henle (figure 10.21).

All of this sets the stage for the final phase of water reabsorption, which takes place from the collecting ducts. As fluid in the collecting ducts passes through the salty area in the medulla, water leaves osmotically through their walls. Using this mechanism, the human kidney can produce quite concentrated urine.

Efficient water recovery and production of concentrated urine are vital for all animals, including humans, that live in environments in which the water supply is limited. But our kidneys cannot make urine as concentrated as that made by the kidneys of some other mammals that live with more severe permanent water shortages (see box 10.1).

Water Balance Regulation

In order to maintain body fluid homeostasis, what regulates the kidney so that it can change to meet either the need for water conservation or elimination of excess water?

One of the important factors regulating water recovery is a peptide hormone called **antidiuretic hormone (ADH)**. ADH opposes **diuresis,** the increased discharge of urine. Specifically, ADH does this by increasing the water permeability of the collecting duct walls. When the circulating ADH level is high, more water is recovered and the volume of urine decreases. Conversely, a decrease in ADH concentration decreases the water permeability of the collecting ducts and decreases water recovery so that urinary water elimination increases. ADH is released into the blood from the posterior lobe of the pituitary gland (chapter 11). The rate of production and release of ADH is very responsive to changes in body water balance. When the body is losing too much water, more ADH is produced and released. When excess water accumulates in the body, less ADH is produced and a larger volume of more dilute urine is excreted (figure 10.22).

This mechanism explains a disease called **diabetes insipidus,** which occurs when not enough ADH is produced. ADH deficiency causes constant elimination of large quantities of dilute urine.

A number of substances, called **diuretics,** increase urine volume either by acting directly on collecting duct walls or by inhibiting ADH production. Coffee, tea, and ethyl alcohol all act as diuretics.

Box 10.1

The Kangaroo Rat

A Mammal that Does Not Need to Drink

The kangaroo rat (box figure 10.1), which gets its name from its habit of hopping around on its hind legs, performs some remarkable feats of water balance maintenance in a very dry environment. This small rodent thrives in the desert regions of the southwestern United States, even in Death Valley where there is no drinking water available. How does this animal manage to live in its harsh, dry environment?

The kangaroo rat lives in burrows that are deep enough to be much cooler and more humid than the desert ground surface in the daytime. A nocturnal animal, the kangaroo rat stays in its burrow during the warmer hours of the day and ventures out only at night in search of the seeds that make up its diet. This avoidance of daytime heat and dryness minimizes water loss by evaporation.

Some animals living in waterless environments survive by eating succulent leaves that have a high water content. But the kangaroo rat eats mainly dry seeds. The water content in these seeds is so low that only 10 percent of the animal's daily water requirements can be supplied by the seeds. The remainder of its water gain is metabolic water (page 114) formed as a result of oxidation of food material. Oxidation of 1 g of carbohydrate yields 0.6 g of water, and oxidation of a gram of fat yields 1.1 g of water. The seeds that the kangaroo rat eats are high in fat and carbohydrate content and low in protein content. The low protein content of the seeds is important because using protein for energy requires more water for excretion of nitrogenous wastes.

Because the kangaroo rat's water supply is so meager, it must conserve water very efficiently. It reabsorbs so much water from its large intestine that its fecal material is almost completely dry. In addition, the kangaroo rat has exceptionally long loops of Henle in its kidneys. These enable the animal to produce urine that is over three times as concentrated as human urine. Thus, the kangaroo rat loses very little water through its kidneys.

Scientists demonstrated the effectiveness of a kangaroo rat's kidney by substituting soybeans, which

Figure 10.22 Regulation of body water content by antidiuretic hormone (ADH). Colored arrows indicate cause-and-effect relationships. Increase in water content causes decrease in ADH secretion, while decrease in water content causes increase in ADH secretion.

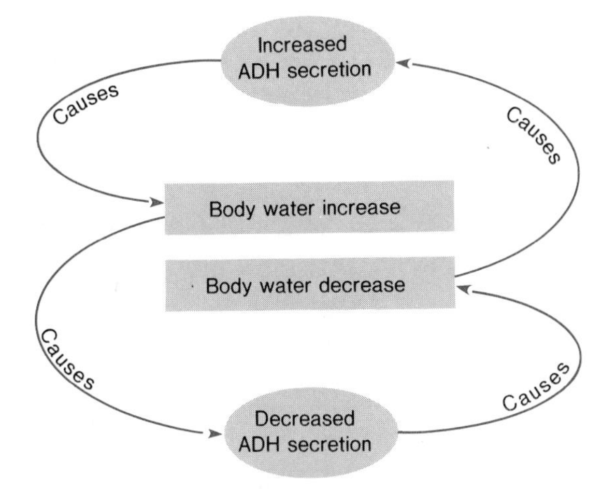

Problems with Kidney Function

The importance of the kidney for maintenance of body fluid homeostasis is vividly illustrated when there is kidney failure due to disease.

There are numerous kidney diseases which can lead to partial or complete renal (kidney) failure. The kidneys become unable to handle normal waste excretion and other homeostatic functions, and wastes build up in the blood. This condition is called **uremia,** and it will lead to death if it persists.

People suffering from either temporary or permanent renal failure can be treated with an artificial kidney machine. Blood circulates from an artery in the patient's arm to the kidney machine and returns to a vein. In this way, their blood continuously is circulated through a **dialyzing unit** in the artificial kidney. The dialyzing unit has very thin, selectively permeable membranes that separate blood flowing in one direction from a **dialysis fluid** that flows in the opposite direction (figure 10.23). Dialysis fluid used in artificial kidneys contains concentrations of various substances that are the same as those in normal blood plasma, but it lacks urea and other substances that are too concentrated in the blood

have a high protein content, for kangaroo rats' normal diet of high fat, low-protein seeds. This created a need for more water to be used in nitrogenous waste excretion and resulted in excessive water loss. If the rats were supplied with seawater to drink, however, they managed nicely. Humans cannot maintain water balance while drinking seawater because human kidneys do not produce urine that is concentrated enough to avoid extra water loss as a result of the salt intake. But the effective water reabsorption mechanism of the kangaroo rat kidney made it possible for the kangaroo rat to drink seawater as a supplemental water source in the experiments.

Most of the total water loss from a kangaroo rat's body is by evaporation, mainly from its lungs. Even this loss is kept to a minimum because the animal's long nose allows exhaled air to be cooled. This cool exhaled air carries out less precious body water than it would if it remained at body temperature.

Its minimal water needs and highly effective methods of water conservation make the kangaroo rat remarkably well adapted to life in a dry environment.

Box figure 10.1 The kangaroo rat (*Dipodomys spectabilis*).

Figure 10.23 The artificial kidney and its operation. (*a*) A schematic diagram showing the machine's construction and operation. (*b*) An enlarged view of a parallel plate dialyzer unit. Blood and dialysis fluid flow past each other in opposite directions. Body wastes diffuse from the blood across the dialysis membrane to the dialysis fluid and are carried away.

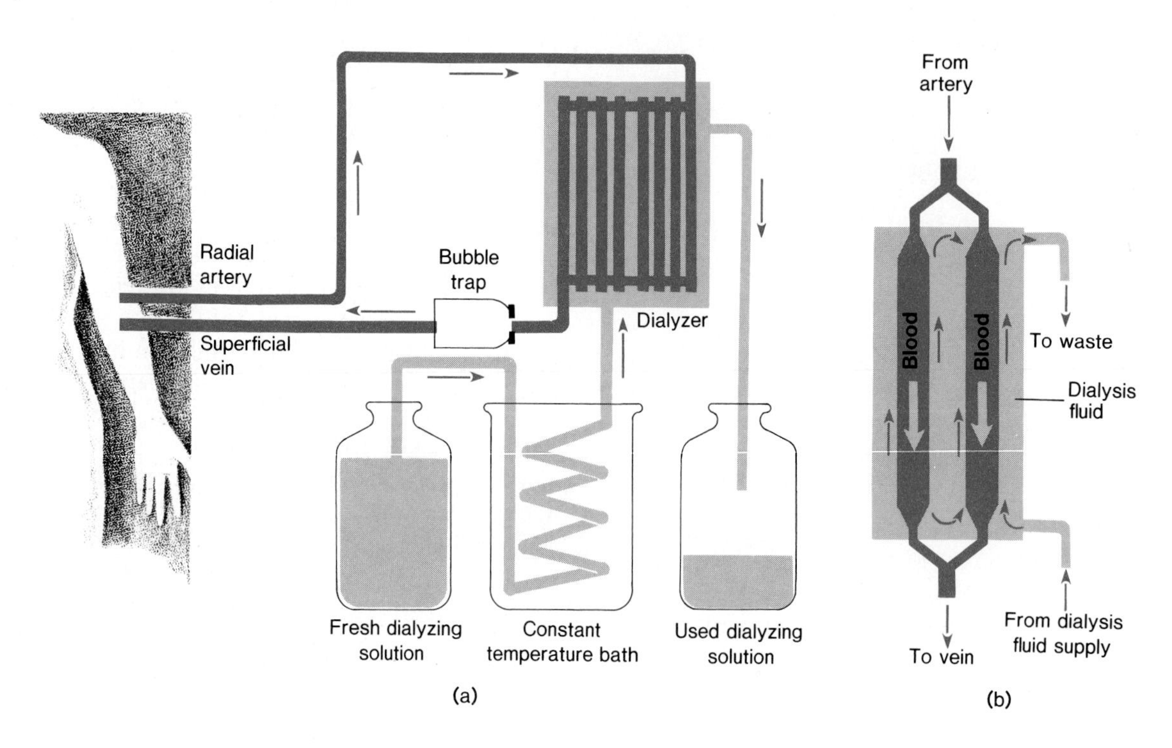

of uremic patients. The membranes separating blood and dialysis fluid are permeable to ions and small molecules such as urea, but are impermeable to blood cells and plasma proteins. Thus, as the blood and dialysis fluid flow through the dialyzing unit, the unwanted substances diffuse out of the blood through the membranes and are carried away by the dialysis fluid.

Dialysis for several hours returns a uremic person's blood to its normal homeostatic balance and removes nitrogenous waste materials as well. Kidney machines have saved the lives of many people suffering short-term kidney failure resulting from disease or toxic responses, such as mercury poisoning. The machines also prolong the lives of victims of permanent renal failure. Unfortunately, artificial kidneys and extended dialysis treatment are very costly. Recent technical advances in dialyzing unit design, however, may make treatment available to more people who suffer from these disturbances of body fluid homeostasis.

Summary

The cells of specialized multicellular organisms live in a stable internal environment. In order to maintain homeostasis, body fluids must remain stable.

About two-thirds of body fluid is intracellular (inside body cells). The remaining third is extracellular fluid (interstitial fluid and blood plasma). Body fluid homeostasis requires exchanges of materials among these compartments and regulated exchanges between the blood plasma and the external environment.

Excess interstitial fluid is drained by the lymphatic system, through which it returns to the circulation. Lymph vessels also transport fats absorbed in the intestine. The lymphatic system is especially important for its role in body defense reactions to infection.

Human blood consists of blood cells and platelets suspended in the plasma. Erythrocytes (red blood cells) are the most numerous cells and they function in gas transport.

Two major classifications of human blood types are based on ABO grouping and the Rh system.

Leukocytes (white blood cells) are much less numerous than erythrocytes. They participate mainly in body defense reactions.

Platelets are small cell fragments that are critically important for blood clotting.

Blood plasma is a watery material containing ions, organic nutrients, nitrogenous wastes, hormones, plasma proteins, and dissolved gases.

Blood clots form in response to injury of vessel walls, preventing blood loss until healing occurs. A clot forms when a series of clotting reactions leads to formation of a mesh of sticky fibrin threads, which traps blood cells.

Abnormal clot formation inside unbroken vessels can block blood vessels and cause severe tissue damage. Clotting deficiencies, such as hemophilia, allow excessive blood loss in response to even minor injuries.

Nitrogenous wastes result from the removal of amino groups from amino acids. Some organisms excrete ammonia directly, but many others excrete urea or uric acid.

Flame cells, nephridia, and Malpighian tubules are excretory structures in invertebrate animals. All of them function in waste excretion and maintenance of water and ion balance. Vertebrate animals have kidneys.

Nephron function begins with pressure filtration of fluid from the blood in the glomerulus. Some materials are reabsorbed from the fluid. Other substances are secreted into the fluid as it moves along the tubule.

Production of concentrated urine is important for maintenance of water balance. Water moves out of the collecting ducts osmotically because the loops of Henle maintain a salty environment in the region through which collecting ducts pass.

The amount of water reabsorbed from the collecting ducts depends on the ducts' permeability, which is under the control of antidiuretic hormone (ADH).

As a result of some kidney diseases, uremia develops as excess water, salts, and urea are retained in the body. Uremia can be treated with artificial kidney machines. Urea and other wastes diffuse from the blood to the dialysis fluid and are carried away, thus artificially helping to restore body fluid homeostasis.

Questions for Review

1. Define interstitial fluid.
2. Infections that affect lymph nodes sometimes produce a swollen puffiness in some parts of the body that cannot be explained by enlargements of the lymph nodes themselves. How would you explain the condition?
3. Explain why a person with any ABO blood type can be given type O blood in an emergency?
4. Briefly describe the formation of a blood clot.
5. What is the source of nitrogenous wastes in animals?
6. Name three forms in which nitrogenous wastes are excreted by various animals.
7. Name three things reabsorbed in the proximal convoluted tubules.
8. Explain the role of the loop of Henle in maintaining a hypertonic environment around the collecting ducts of the kidney.
9. What is ADH and what does it do?
10. In terms of kidney function, how would you account for "beer diuresis," a condition sometimes experienced by college students?

Questions for Analysis and Discussion

1. Anemia develops in some cancer patients being treated with radiation or chemotherapy to inhibit cell divisions of cancer cells. Why?
2. What effect would you expect elevated blood pressure to have on the volume of urine being produced? Why?
3. If a urine sample contains blood cells and plasma proteins, what kidney structures have been damaged? How did you reach your conclusion?

Suggested Readings

Books

Hill, R. W. 1976. *Comparative physiology of animals: An environmental approach.* New York: Harper and Row.

Hole, J. W., Jr. 1984. *Human anatomy and physiology.* 3d ed. Dubuque, Ia.: Wm. C. Brown Publishers.

Schmidt-Nielsen, K. 1983. *Animal physiology.* 3d ed. New York: Cambridge University Press.

Articles

Crowe, J. H., and Cooper, A. F., Jr. December 1971. Cryptobiosis. *Scientific American* (offprint 1237).

Doolittle, R. F. December 1981. Fibrinogen and fibrin. *Scientific American* (offprint 1506).

Moffat, D. B. 1978. The control of water balance by the kidney. *Carolina Biology Readers* no. 14. Burlington, N.C.: Carolina Biological Supply Co.

Schmidt-Nielsen, K., and Schmidt-Nielsen, B. July 1953. The desert rat. *Scientific American* (offprint 1050).

Zucker, M. B. June 1980. The functioning of blood platelets. *Scientific American* (offprint 1472).

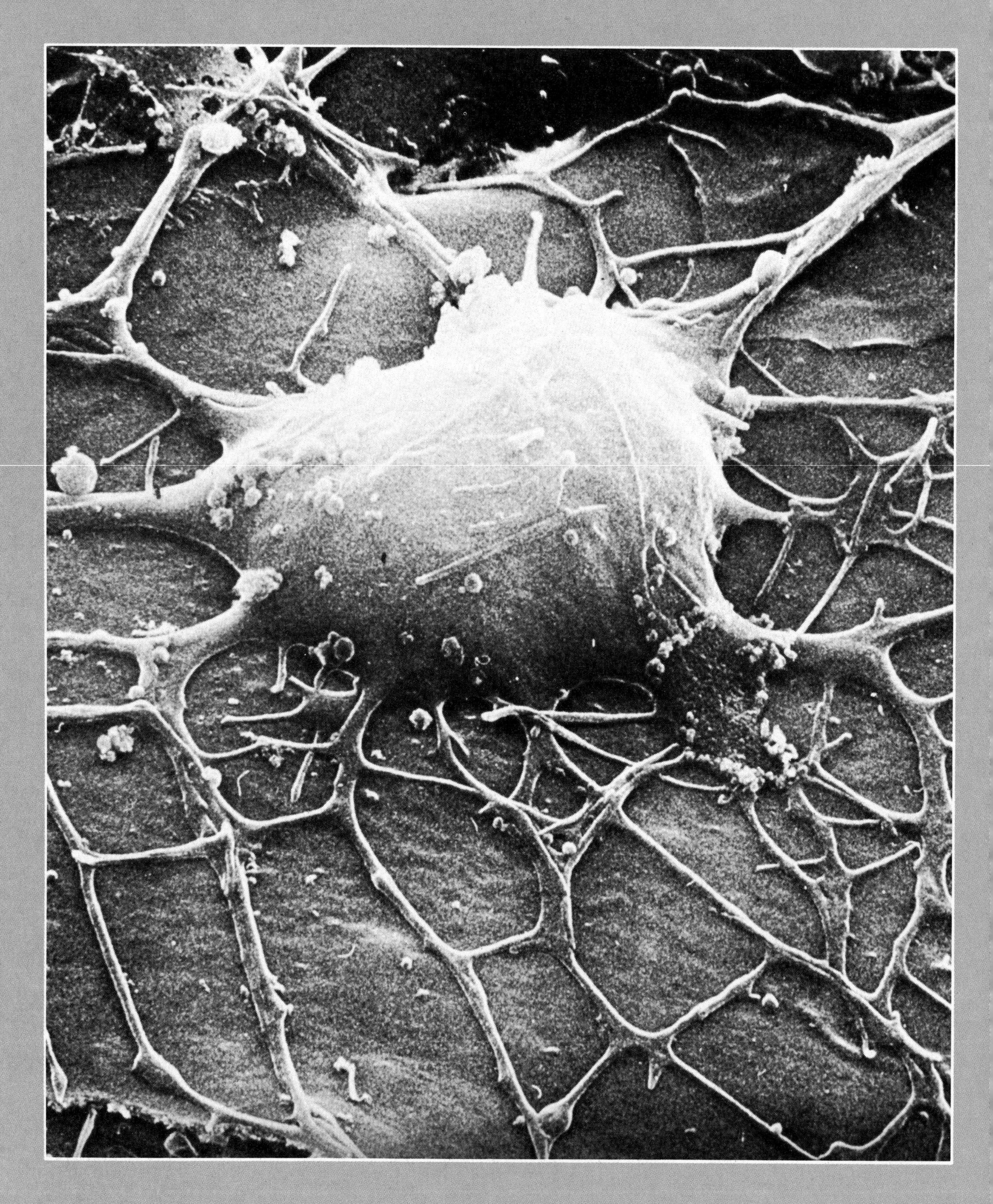

Regulation in Organisms

Part 3

Within a multicellular organism there must be effective coordination of many functional activities. This coordination requires communication among body parts that are, in many cases, widely separated from one another. Furthermore, these coordinating mechanisms must be flexible enough to permit adjustments in response to changes in the external and internal environments.

In animals, chemical control by hormones provides relatively slower but more sustained regulation. Nervous mechanisms are involved in relatively faster but shorter-lived regulatory responses.

Hormonal regulation in plants functions mainly in control of growth and developmental responses to environmental factors, the most important of which is light. These responses are part of each plant's continuing interaction with its environment. In this interaction, various hormonal and timing mechanisms function to synchronize growth and other activities with favorable environmental conditions.

In chapters 11 through 14 we will first explore hormonal and nervous control mechanisms in animals. Then we will turn our attention to the functions of various plant hormones and the timing mechanisms that coordinate plants' lives with seasonal environmental changes.

Facing page Scanning electron micrograph of cultured nerve cells from the retina of a housefly.

Animal Hormones

11

Chapter Outline

Chapter Concepts

1. Animals' bodies are chemically regulated by hormones.
2. Hormones, in addition to nerves, provide means of communication within the organism that help maintain homeostasis.
3. Substances produced by nerve cells that function as hormones form an important link between the nervous and endocrine systems.
4. Many vertebrate endocrine glands are regulated by secretions from the pituitary gland, which is in turn regulated by hormones released from the hypothalamus, a region of the brain.
5. In humans, numerous hormones control metabolism and fluid and ionic balance.
6. Cells must have receptors in order to respond to hormones.
7. There are several different mechanisms by which hormones can exercise their effects upon target cells.

P. T. Barnum was a promoter. He founded the "American Museum," an exhibition of "curios," and the "Greatest Show on Earth," a huge circus. He knew he could get people to pay for a chance to see the "largest," "smallest," or "strangest" that the world had to offer. For example, Barnum managed to buy the enormous elephant, Jumbo, from the Royal Zoological Society of London and transport it to the United States before the general public in England realized what he was doing.

But P. T. Barnum's greatest successes were his "exhibits" of people who differed from the norm. Possibly the most famous of these was the midget, Charles Stratton, whom Barnum called "General Tom Thumb." Stratton was very tiny throughout his life (figure 11.1) because he suffered a deficiency of a pituitary growth hormone during his growing years. Stratton's small stature was clear testimony to the powerful effects of a hormone deficiency and the role of hormonal regulation of growth.

Many other activities beside growth are regulated by hormones. Hormones are involved in coordination of many everyday activities that contribute to homeostasis. In multicellular organisms, there is a division of labor among the cells, tissues, and organs of the body. Each of the body regions performs specialized functions that help maintain a stable internal environment for all body cells. Obviously, all of these specialized activities must be coordinated if this stability is to be maintained.

Animals possess two major methods of internal coordination: nervous (neural) and hormonal control mechanisms. Neural control permits the rapid conduction of messages and causes quick but usually short-term responses. Hormonal control, on the other hand, is accomplished through specific chemical messengers carried in the blood. Hormonal control requires longer time periods both for the message to be transmitted and for its effects to take place.

Figure 11.1 As a result of a childhood growth hormone deficiency, "General" Tom Thumb (Charles Stratton), like other pituitary midgets, had normal body proportions despite his small size. He is shown here with P. T. Barnum, the showman who exhibited him.

Hormones, Glands, and Target Cells

Hormones are synthesized and released into the bloodstream by specialized glandular cells and are carried in the blood until they reach specific **target cells** (responding cells). Hormones cause target cells to respond by altering their activity in a specific way.

Hormones have traditionally been defined as substances secreted by organs called **endocrine glands.** The word endocrine (from the Greek *endon,* meaning within) indicates that these glands secrete their products directly into the bloodstream, while **exocrine** (from the Greek *exo,* or outside) **glands,** such as digestive glands, secrete their products through ducts. We now know, however, that defining hormones only in terms of the familiar endocrine glands (such as the pituitary, thyroid, and adrenals) barely touches the surface of hormonal regulation in the body.

We also know that the two major animal control systems (neural and hormonal) are not separate and distinct, but are very closely interconnected with one another. For example, certain specialized nerve cells called **neurosecretory cells** produce and release chemical substances (**neurosecretions**) that are carried in the blood (figure 11.2). Neurosecretions have specific effects on target cells more or less remote from the nerve cells that produce them. Thus, neurosecretory cells are nerve cells, but they do exactly what all hormone-producing cells do: they produce and release chemical messengers that are transported to target cells, on which they have a specific effect.

Figure 11.2 A comparison of chemical control by neurosecretory cells and gland cells. (*a*) Neurosecretory cells synthesize neurosecretions that pass down axons (fibrous extensions of the cells) and are released either directly into the circulation (1) or into storage cells (2) for later release into the bloodstream. (*b*) Typical gland cells secrete their product directly into the circulation.

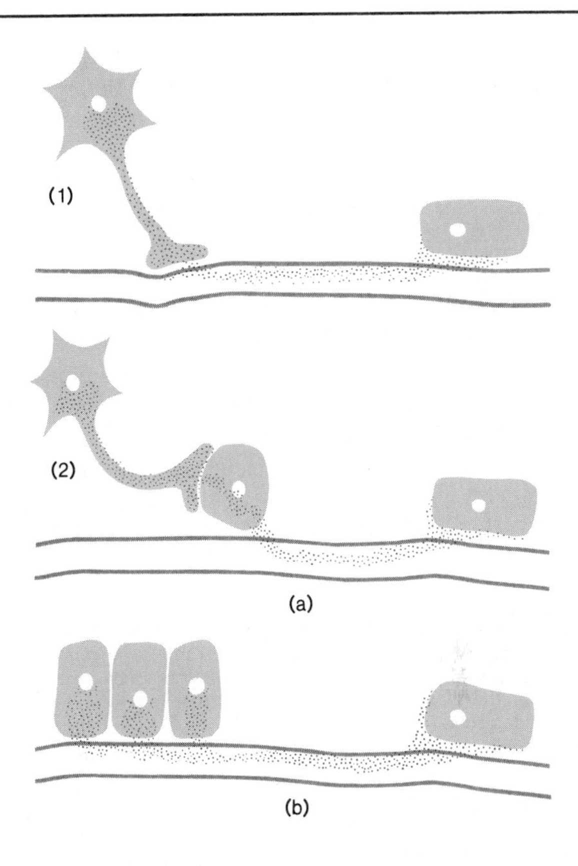

The Discovery of Hormones

The actions of some endocrine glands have been known for many years. For example, castration (removal of the testes) of young male vertebrates has long been used to prevent certain male animals from breeding or to raise animals that are more docile and that produce better meat.

However, scientific experimentation on the functioning of endocrine glands did not actually begin until 1849, when A. A. Berthold transplanted testes back into the bodies of young castrated male chickens. Berthold found that this treatment restored the male characteristics normally seen in roosters. They developed combs, gained the ability to crow, and began to exhibit the generally pugnacious behavior expected of roosters. These treated males looked and acted like roosters even when the transplanted testes were far from their normal location in the body and had none of the normal nerve connections of testes. Berthold concluded that testes must release some substance that causes the development of male characteristics. It was later learned that the testes secrete the hormone **testosterone,** which is responsible for development of many male characteristics.

Over the next few years, similar gland removal and replacement experiments helped to identify other hormone-producing structures. Correlation of disease symptoms with degenerated glands found during autopsies helped establish still more structures as endocrine glands. For example, in 1855 Thomas Addison described a set of symptoms—including poor appetite, low blood pressure, muscular weakness, and a characteristic skin discoloration—associated with the deterioration of the outer portion or cortex of the adrenal glands. The syndrome is still known today as Addison's disease.

The search is not over, however. Additional hormones are still being discovered; some are produced by the traditionally recognized endocrine glands and some by other body tissues.

Although our knowledge about hormones is steadily increasing, we still have only partial answers to many questions about hormonal regulation. How is the synthesis and release of hormones regulated? Why do hormones act on certain target cells, but not on other cells? What changes do hormones cause within target cells, and how do those changes cause the ultimate effects of the hormones? Before we examine these questions further, let us identify some additional important animal hormones.

Hormones in Invertebrates

Most of the early studies of hormonal regulation involved vertebrate animals, especially mammals. But some of the hormones that regulate functions in invertebrate animals have also been discovered. The principal hormones involved in regulation of insect growth and maturation are undoubtedly the most thoroughly studied invertebrate hormones.

Hormones and Insect Development

Insects have a rigid outer body covering called an **exoskeleton.** Because exoskeletons do not expand to allow for growth, insects must periodically shed their exoskeletons in a process known as **molting (ecdysis)**. Molting is followed by the expansion and hardening of a new, soft, larger exoskeleton that develops underneath the old exoskeleton.

In insect development, **metamorphosis** is the change in body form that converts an immature individual into an adult. Metamorphosis is either **complete (abrupt),** involving **larval, pupal,** and **adult** stages, or **incomplete (gradual),** involving a series of **nymphs.** Nymphs generally resemble adults and become somewhat more adultlike at each molt, until a fully developed adult emerges from the final molt. Newly hatched

Figure 11.3 Hormonal control of molting and metamorphosis in a moth. Ecdysone produced by the prothoracic gland and juvenile hormone released by the corpus allatum together produce a larva-to-larva molt. In the last larval stage, the juvenile hormone level decreases, and pupation occurs.

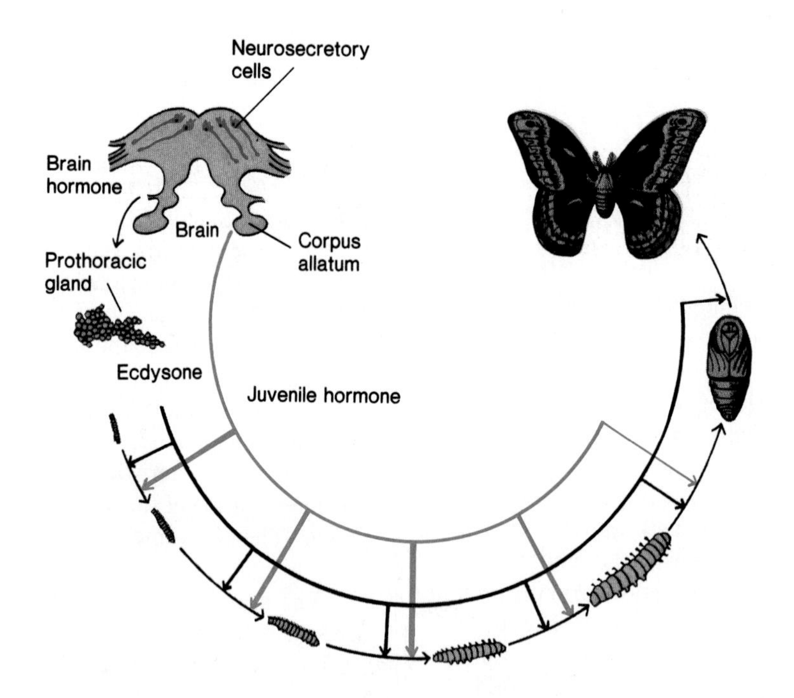

grasshopper nymphs, for example, are clearly recognizable as grasshoppers. They do not have the wormlike appearance that characterizes larvae of insects with complete metamorphosis.

Several hormones regulate insect growth, molting, and metamorphosis. Neurosecretory cells in the brain produce a **brain hormone** that stimulates the release of another hormone, **ecdysone.** Ecdysone is called the molting hormone or molt and maturation hormone because it promotes molting and tends to spur the developmental process along toward pupation and adulthood. During the larval stages, the maturing effect of ecdysone is countered by the action of yet another hormone called **juvenile hormone.** While juvenile hormone is secreted, the organism remains immature; larval molting leads simply to additional larval stages. But in the last larval stage, juvenile hormone secretion decreases dramatically, and the animal pupates and begins its transformation to the adult form (figure 11.3).

Experiments on the hormonal regulation of insect development have shown that it is possible to produce miniature adults or giant, abnormal adults (figure 11.4). The powerful effects of juvenile hormone on insect development also make this hormone an interesting subject for insecticide research. If larval insects or the food they consume could be sprayed with juvenile hormone, the larvae might develop abnormally and die, or fail to reach an appropriate stage of their life cycle in time to survive the winter. Research on juvenile hormones as insecticides seems well worth the effort because the hormones are effective at very low concentrations and are naturally occurring, biodegradable (broken down by natural decay processes) compounds that should not pose the threat of cumulative toxicity that many insecticides do.

Figure 11.4 Experiments on the role of juvenile hormone in insect development. (*a*) Normal last (fifth) larval stage, pupa, and adult of the silkworm moth *Bombyx mori.* (*b*) The source of juvenile hormone was removed from this third larval stage. Instead of entering a larva-to-larva molt, it became a pupa and developed into a miniature adult. (*c*) Juvenile hormone releasing glands from young larvae, which were still actively secreting juvenile hormone, were transplanted into a fifth larval stage. Instead of becoming a pupa, it entered a larva-to-larva molt and produced this large extra larval stage, which then pupated and developed into a giant adult.

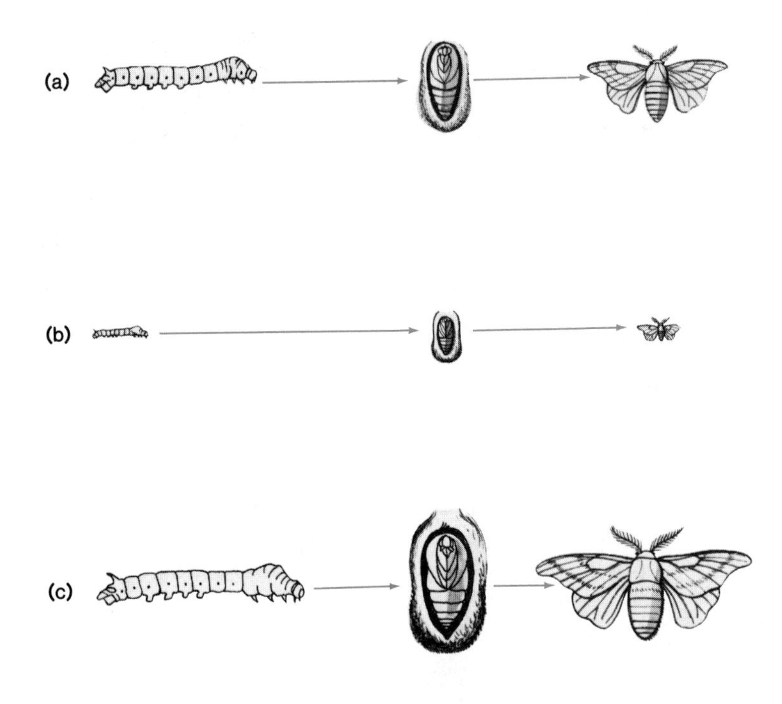

Box 11.1
"Social Hormones"

In addition to the hormones secreted within an organism that affect its own physiological processes, there are other chemical regulatory compounds that are secreted by organisms into the environment. These compounds, called **pheromones,** cause behavioral, developmental, or reproductive responses in animals of the same species. Therefore, pheromones are sometimes called "social hormones."

Well-known examples of pheromone functions include trail-marking behavior in ants, and territory marking by male dogs, deer, and antelope through pheromones secreted in urine or from scent glands. In these cases, the pheromone proclaims to other males of the species that the territory belongs to a particular individual.

An intriguing example of pheromones is the **sex attractants,** which have been most extensively studied in insects. The first sex attractant identified and synthesized was bombykol, which comes from females of the silkworm moth *(Bombyx mori).* Sex attractants have a very compelling effect on behavior. Bombykol can attract male silkworm moths several kilometers away. Sex attractant pheromones have been discovered in other moths and in many other insect species as well (box figure 11.1). For example, the queen bee releases a pheromone during her nuptial flight that entices the drones to follow her and mate.

Sex attractants have also been studied in mammals. In a number of mammalian species, females emit a special odor during **estrus,** their period of fertility and sexual receptiveness. Normally docile, well-behaved male dogs will scratch doors, break leashes, or jump fences and roam great distances to reach a female in estrus once they detect her chemical signal.

Box figure 11.1 The head of a male *Polyphemus* moth with its long, feathery antennae. The antennae bear receptors with which the moth can detect minute quantities of the sex attractant released by the female of the species.

A sex attractant has also been isolated from female rhesus monkeys. This focuses attention on primate pheromones in general and human pheromones in particular. Whether or not sex attractants and other pheromones exist in humans remains a matter of some controversy. At any rate, modern men and women seem to prefer to wash off natural body chemicals and replace them with laboratory-compounded scents. Who knows what could happen if manufacturers of these scents should succeed in isolating, identifying, synthesizing, and marketing real human pheromones?

Vertebrate Glands and Hormones

More is known about mammalian hormones than about the hormones of other vertebrates, and most of the study of mammalian hormones has focused on hormonal regulation in the human body. Figure 11.5 is a diagram of the human body showing the locations of some of the major endocrine glands: the hypothalamus, the pituitary, the thyroid, the parathyroids, the pancreas, the kidneys, the adrenals, the ovaries, and the testes. Table 11.1 lists some of the major mammalian hormones along with their sources and the class of chemical compounds to which each belongs.

As we discuss these hormones, we will explore three main topics: (1) the hormones of the pituitary gland and the relationship between the pituitary and hypothalamus, (2) the hormones regulating metabolism, and (3) the hormones regulating ionic balance in the body. Grouping hormones in this way helps emphasize the role of hormonal regulation in the maintenance of homeostasis through the coordination of specialized body functions.

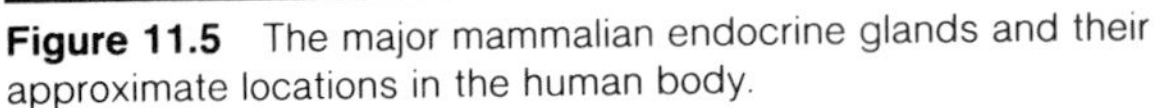

Figure 11.5 The major mammalian endocrine glands and their approximate locations in the human body.

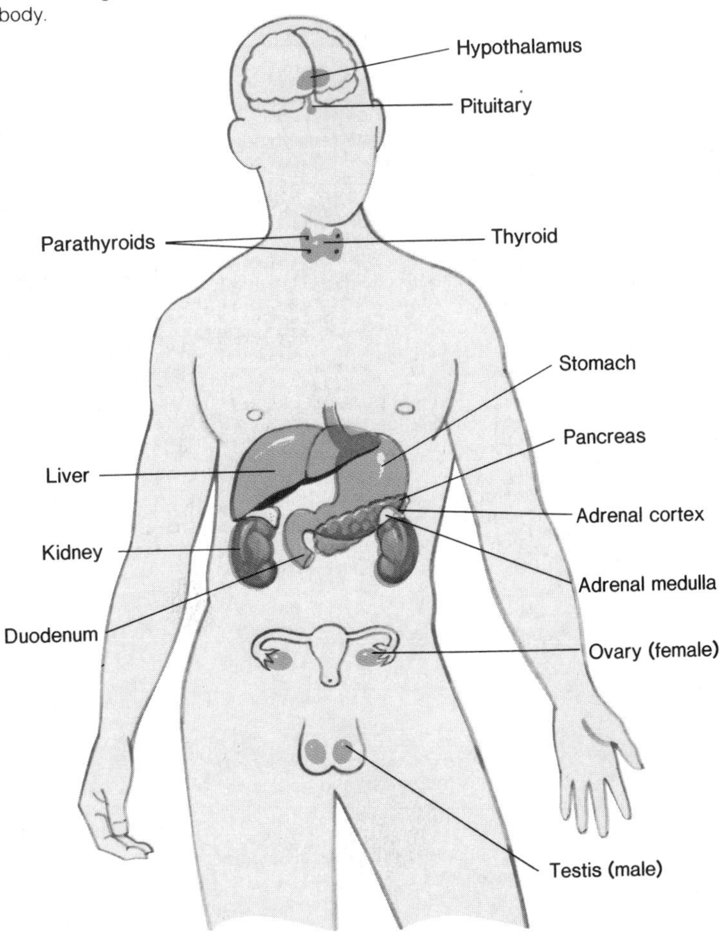

The Pituitary Gland

The **pituitary gland,** also known as the **hypophysis,** which in humans is about the size of a garden pea, is suspended on a stalk extending down from a part of the brain called the **hypothalamus** (figure 11.6). For many years, the pituitary was known as the "master gland" among the endocrine glands of the body. While this is true to the extent that pituitary hormones do regulate several other endocrine glands and their secretions, the pituitary does not function in a simple master-slave relationship with other glands.

The human pituitary is composed of two major regions, a **posterior lobe** and an **anterior lobe.** In some other vertebrates, there is an additional **intermediate lobe,** which lies between the posterior and anterior lobes. The tissue of the posterior lobe resembles nerve tissues, while the anterior lobe consists of specialized secretory cells. Although these two lobes lie close together in what appears to be a single structure, they are in reality two distinctly different glands that secrete different hormones and have different functions.

Figure 11.6 The location of the human pituitary gland below the brain. The posterior lobe has nerve connections to the hypothalamus, but the anterior lobe does not. These two lobes are distinctly separate glands that differ structurally and functionally.

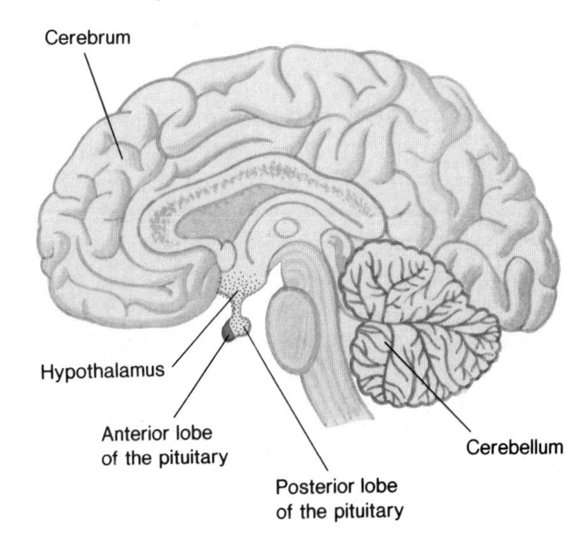

Table 11.1
Some Major Mammalian Hormones

Source	Hormone	Class of Chemical Compound
Hypothalamus	Thyrotrophin-releasing hormone (TRH)	Peptide
	Corticotrophin (ACTH)-releasing hormone (CRH)	Peptide (or several peptides)
	Luteinizing hormone-releasing hormone/ follicle stimulating hormone-releasing hormone (LH-RH/FSH-RH)	Peptide
	Somatostatin or growth hormone-release inhibiting hormone (GH-RIH)	Peptide
Hypothalamus, with storage in the posterior pituitary	Antidiuretic hormone (ADH) or vasopressin	Peptide
	Oxytocin	Peptide
Anterior pituitary	Growth hormone (GH) or somatotrophic hormone	Protein
	Thyroid-stimulating hormone (TSH) or thyrotrophic hormone	Glycoprotein
	Adrenocorticotrophic hormone (ACTH)	Polypeptide (39 amino acids)
	Follicle-stimulating hormone (FSH)	Glycoprotein
	Luteinizing hormone (LH)	Glycoprotein
	Prolactin	Protein
Thyroid	Thyroxin and triiodothyronine	Iodinated amino acids
	Calcitonin	Polypeptide (32 amino acids)
Pancreas	Insulin	Polypeptide (51 amino acids)
	Glucagon	Polypeptide (29 amino acids)
Adrenal medulla	Epinephrine (adrenalin) and norepinephrine (noradrenalin)	Modified amino acids
Adrenal cortex	Glucocorticoids (cortisol, corticosterone, cortisone, and so on)	Steroids
	Mineralocorticoids (aldosterone and so on)	Steroids
Parathyroids	Parathyroid hormone	Polypeptide (32 amino acids)
Ovaries	Estrogens and progesterone	Steroids
Testes	Testosterone	Steroid

The Posterior Lobe

Two major hormones, the **antidiuretic hormone (ADH)** and **oxytocin,** are associated with the posterior lobe of the pituitary gland (figure 11.7). As you saw in chapter 10, ADH promotes water reabsorption by kidney collecting tubules, thus maintaining water balance in the body. Oxytocin causes contraction of smooth muscles in the uterus. Because of its powerful effect on uterine muscles, doctors use oxytocin as a very reliable means of inducing labor and childbirth, when, for example, a pregnancy has continued long past term.

Oxytocin also causes the release of milk from the mammary glands of nursing female mammals. The release of oxytocin in response to suckling is a good example of a **neuroendocrine reflex.** In such a reflex, sensory information received by the nervous system leads to the secretion of a hormone by an endocrine organ.

ADH and oxytocin were originally thought to be synthesized *and* released by the posterior lobe of the pituitary. More recent studies, however, have clearly demonstrated that they are products of neurosecretory cells in the hypothalamus, and that they travel down axons (nerve cell fibers) to the posterior lobe, where they are stored until their release (figure 11.8).

Figure 11.7 Structural formulas of oxytocin and ADH, which are very similar peptides (small chains of amino acids). Note that the amino acid sequences of the two hormones differ by only two amino acids (indicated by color shading).

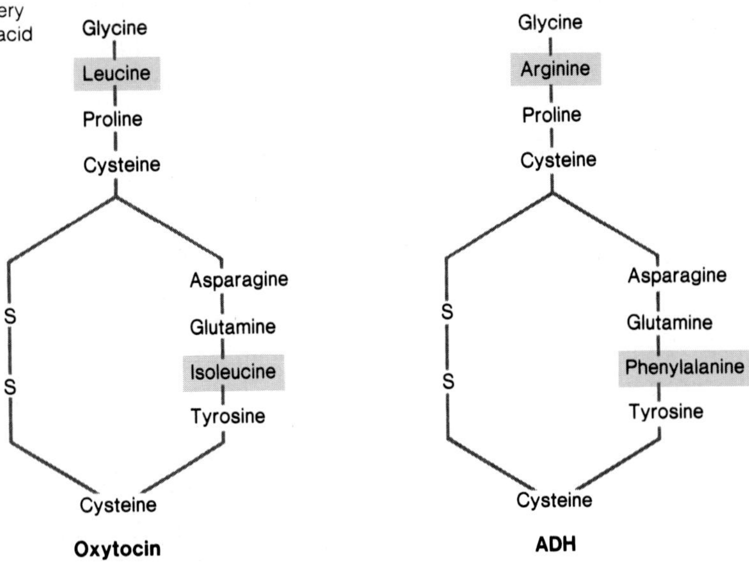

Figure 11.8 Relationship between the hypothalamus and the posterior lobe of the pituitary. Neurosecretory cells in the hypothalamus synthesize ADH and oxytocin. The hormones are transported down axons to the posterior lobe and released into the bloodstream there.

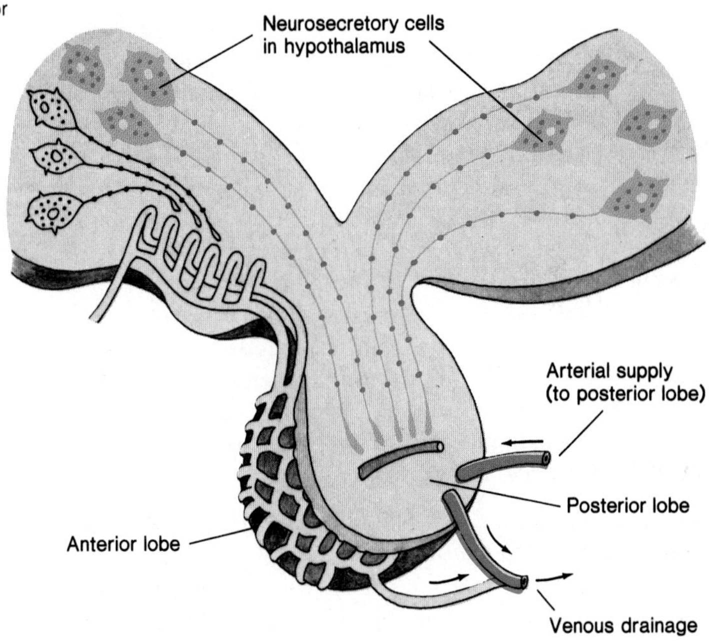

The Anterior Lobe

The hormones of the anterior lobe have a wide range of effects on many processes in developing and adult mammals. Perhaps, however, the effects of **growth hormone** are best known because they are so obvious—and even dramatic, in some cases. If the secretion of growth hormone is insufficient during childhood and adolescence, an individual becomes a pituitary midget (see figure 11.1), while excess secretion during the same period produces a pituitary giant. The oversecretion of growth hormone in an adult causes the toes, fingers, and face to continue or resume growth. This condition, called **acromegaly,** is characterized by distorted facial features, including an enlarged jaw and thick, heavy eyebrow ridges.

Growth hormone is also known as **somatotrophic hormone,** or **STH.** The suffix *trophic* comes from a Greek word meaning to feed or cause to grow, and is usually used to describe hormones that stimulate their target organs to secrete other hormones. The anterior lobe also produces **thyrotrophic hormone,** more commonly called **thyroid stimulating hormone (TSH),** and **adrenocorticotrophic hormone (ACTH).** TSH stimulates the thyroid gland to produce its hormones, while ACTH stimulates the adrenal cortex to produce a group of steroid hormones.

Other trophic hormones from the anterior pituitary include the gonadotrophic hormones, **follicle stimulating hormone (FSH)** and **luteinizing hormone (LH),** which stimulate sex hormone secretion and egg or sperm production by ovaries or testes. We will examine the gonadotrophic hormones in chapter 20 when we discuss human reproduction.

The last anterior lobe hormone that we will mention is **prolactin.** Interestingly, prolactin is produced in the anterior lobes of both female and male mammals, but its function is clearly understood only in females. Prolactin acts on the mammary glands, causing them to produce milk. (Do not confuse this hormone with oxytocin, which is necessary for milk release from the glands.) More than thirty functions have been proposed for prolactin in male vertebrates, but for the most part, these are yet to be proven.

The Hypothalamus and the Pituitary

It has long been known that stress can cause the pituitary to secrete more ACTH, thereby causing stimulation of the adrenal glands. But certainly the pituitary itself does not perceive and respond to stress. Similarly, in some species, the mere sight of a potential mate stimulates the release of pituitary gonadotrophins, which in turn stimulates the production of sex hormones by the gonads. Again, how does this happen?

These responses are additional examples of neuroendocrine reflexes. There is a close relationship between the hypothalamus of the brain and the posterior lobe of the pituitary; the two structures are connected by neurosecretory fibers. But years of careful anatomical studies failed to show similar nervous connections between the hypothalamus and the anterior pituitary. However, in 1936, researchers discovered blood vessels extending from the hypothalamus through the pituitary stalk to the anterior portion of the pituitary (figure 11.9). This special network of capillaries and vessels has since been named the **hypothalamic-pituitary portal system.**

In 1945, G. W. Harris proposed that neurosecretory cells in the hypothalamus release substances into the blood vessels of the portal system and that these substances stimulate the anterior pituitary to secrete its hormones. In 1955, scientists reported that extracts of the hypothalamus could, indeed, stimulate the anterior pituitary to secrete ACTH. Since then, in addition to this **corticotrophin releasing hormone (CRH),** a number of different hormones from the hypothalamus that are carried through the portal system and act upon the anterior pituitary have been discovered. Roger Guillemin and Andrew Schally shared the Nobel Prize in 1977 for the isolation and characterization of hypothalamic factors with Rosalyn Yalow, who developed analytical techniques that have dramatically changed hormone research and diagnosis.

Thyrotrophin releasing hormone (TRH) was the first of the releasing factors from the hypothalamus to be isolated and purified. TRH is a small peptide composed of only three amino acids. It causes the anterior pituitary to secrete thyroid stimulating hormone (TSH), which, in turn, acts upon the thyroid to stimulate the release of thyroxin. TRH was isolated in 1969 by Schally's research team using pig hypothalami and by Guillemin's team using sheep brains. As has been the case throughout the history of research on hormones, this was no easy task. Guillemin collected 500 tons of sheep brains to remove 7 tons of hypothalami, and his team put in four years of intense work to produce a single milligram of TRH!

Since that time other releasing hormones have been discovered (see table 11.1), but not all the hormones from the hypothalamus are releasing hormones. Some inhibit rather than stimulate secretion of pituitary hormones.

Thus, the anterior lobe of the pituitary, which regulates secretion of several other glands, is itself regulated by hormones produced by neurosecretory cells in the hypothal-

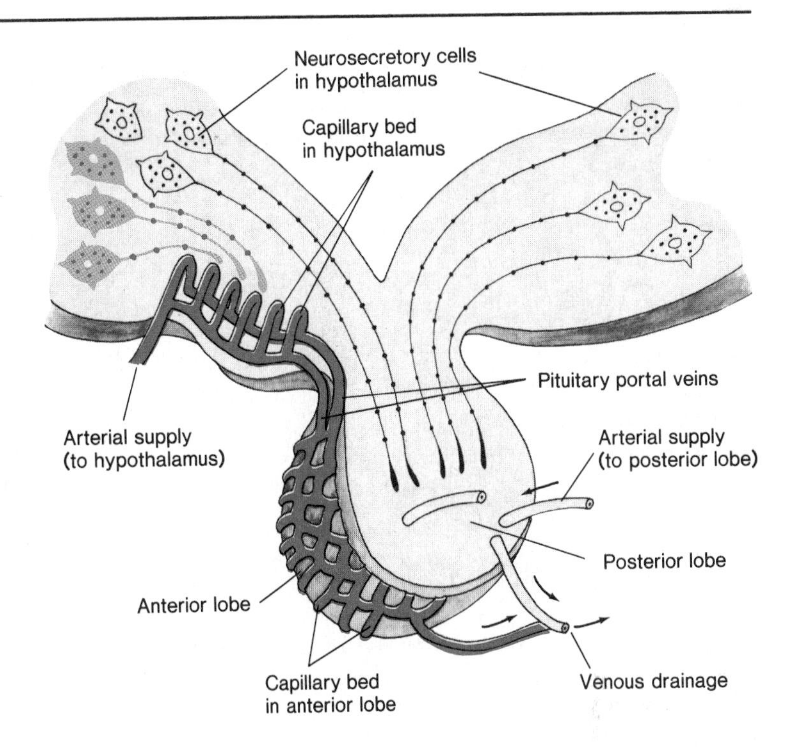

Figure 11.9 The hypothalamic-pituitary portal system. Neurosecretory cells in the hypothalamus produce releasing and inhibiting factors and secrete them into a capillary bed in the hypothalamus. Vessels carry them from the hypothalamus to capillaries in the anterior lobe of the pituitary where they control secretion of anterior lobe hormones.

amus. The hormonal and neural regulatory systems are intimately connected by this functional relationship between a part of the brain, the hypothalamus, and the pituitary gland (figure 11.10).

Hormones Controlling Metabolism

Metabolism can be defined as the sum total of all the biochemical processes that take place within an organism, and the **metabolic rate** is the rate at which these reactions proceed. In order for homeostasis to be maintained, all of these processes in various cells of the body must be regulated and balanced with each other. Several hormones are involved in control of metabolism.

Thyroid Hormones

Thyroid hormones, which are produced by the thyroid gland located at the base of the neck (see figure 11.5), are the principal hormones controlling metabolic rate. There are two major thyroid hormones, **thyroxin (T_4)** and **triiodothyronine (T_3)**, which differ only in that thyroxin contains four iodine atoms while triiodothyronine contains only three (figure 11.11).

Figure 11.10 The hypothalamus and pituitary and the control of secretion of endocrine glands.

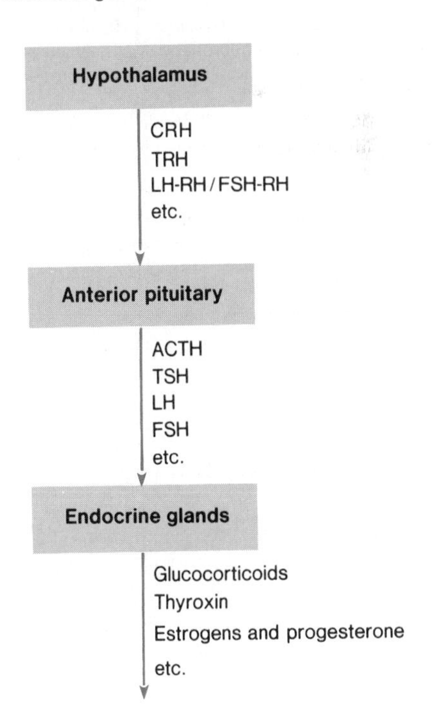

Figure 11.11 The thyroid. (*a*) The structures of the two major thyroid hormones, thyroxin (T_4) and triiodothyronine (T_3). They differ by only one iodine atom. (*b*) Thyroid tissue. The thyroid consists of follicles, hollow balls of follicle cells, which secrete thyroid hormones. The hormones are stored until release in the colloid that fills the inside of follicles.

Thyroxin (T_4)

Triiodothyronine (T_3)

(a)

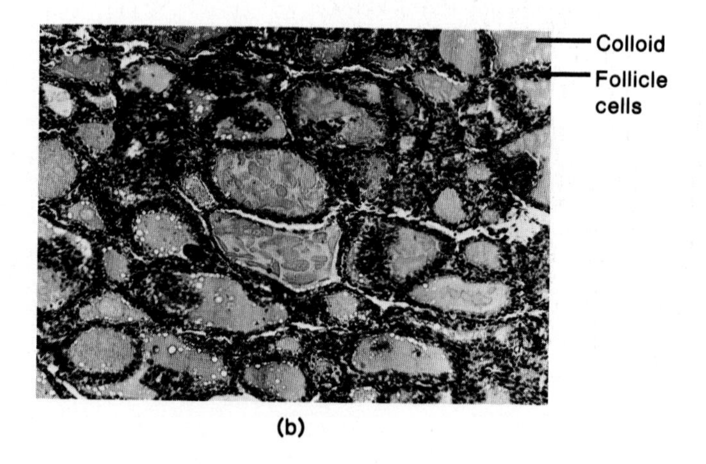

(b)

Figure 11.12 Negative feedback control mechanism that regulates secretion of the thyroid hormones. Increased thyroxin (T_4) and triiodothyronine (T_3) cause decreased TRH release from the hypothalamus and inhibit TSH release from the anterior lobe. When T_4 and T_3 in the blood decrease, inhibition of the hypothalamus and anterior lobe is relieved. More TSH is released, and the thyroid is stimulated to release more T_4 and T_3. Thus, the negative feedback system normally balances the secretions of all elements and keeps the concentration of T_4 and T_3 in the blood within the proper range.

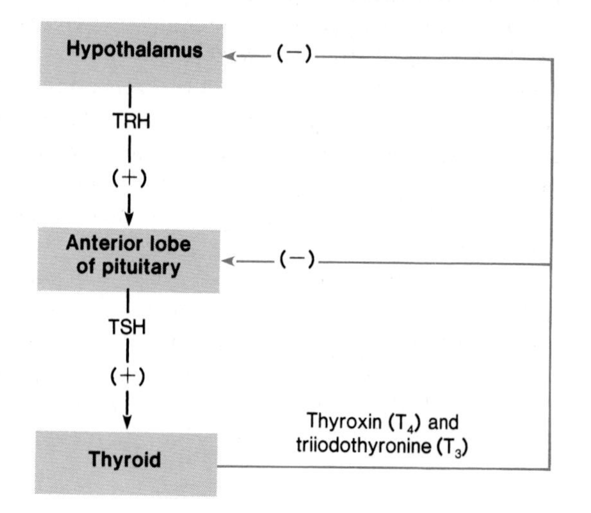

The thyroid hormones powerfully affect all body cells. T_4 and T_3 increase the metabolism and oxygen consumption of cells throughout the body. Thyroid hormones are also instrumental in promoting normal growth and development largely through the stimulation of normal protein synthesis. In humans, thyroid deficiency during infancy and childhood causes a dwarfed condition known as **cretinism,** in which the victim never matures sexually and suffers severe mental retardation. Fortunately, cretinism can be prevented by administering hormones to children with thyroid deficiencies.

As we saw earlier, TSH from the pituitary's anterior lobe controls the production and release of thyroid hormones. But the rate of TSH production also must be regulated so that only normal, adequate stimulation of the thyroid is achieved. TSH secretion is affected by the level of thyroid hormones in the blood. When thyroid hormone concentrations in the blood increase, TSH secretion temporarily slows down. When thyroid hormone concentrations in the blood decrease, TSH secretion temporarily increases. This relationship in which a substance (in this case, thyroid hormone) negatively affects the secretion of another substance that regulates its own production (TSH) is called a **negative feedback control mechanism** (figure 11.12). Many such feedback mechanisms control the secretion of various hormones and are very important for maintenance of homeostasis.

Figure 11.13 A woman suffering from goiter, an enlargement of the thyroid gland that often accompanies hypothyroidism. Continued TSH stimulation causes massive overgrowth of thyroid tissue.

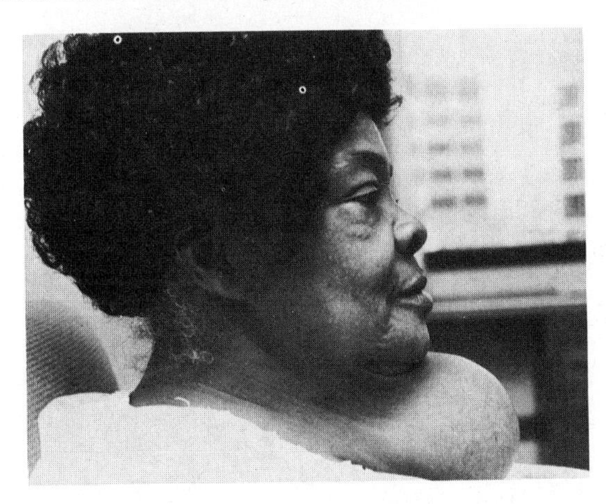

Figure 11.14 An islet of Langerhans in a section of pancreas tissue. The islets are separate areas of endocrine tissue surrounded by the tissue that secretes the digestive enzymes of the pancreas.

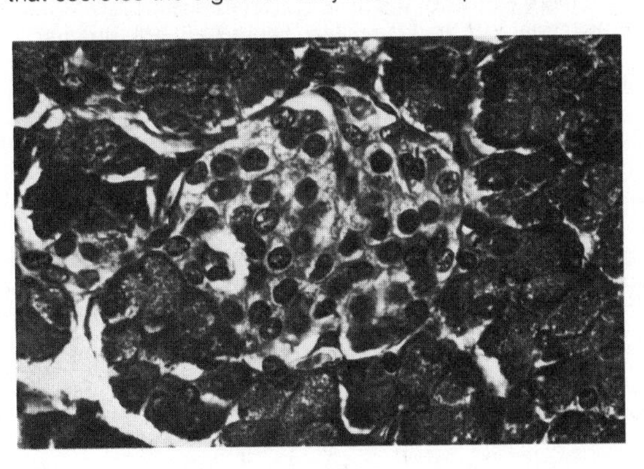

The feedback control relationship between thyroid and pituitary secretion is mediated by the hypothalamus. Increased levels of T_4 and T_3 in the blood inhibit TRH production by the hypothalamus, while decreased T_4 and T_3 levels reduce inhibitions to TRH production and release, thereby causing increased TSH secretion.

The delicately balanced feedback control of thyroid secretion prevents excess thyroid secretion (hyperthyroidism), which causes weight loss, elevated body temperature, profuse perspiration, and nervous irritability. On the other hand, thyroid deficiency (hypothyroidism) leads to a lowered metabolic rate, obesity, physical sluggishness, mental dullness, and abnormal skin and hair conditions.

Clear evidence that the normal feedback mechanism is not functioning properly is seen when a condition called **goiter** develops. Goiter is a tremendous expansion of the thyroid gland that is externally visible as a swelling at the base of the neck (figure 11.13). Low thyroid secretion rates, caused by hormone synthesis problems or by a deficiency in dietary iodine, result in increased TSH secretion through the normal feedback relationship. The excess growth of the thyroid tissue, producing a goiter, occurs in response to continuing TSH stimulation.

Hypothyroidism and goiter are rare in humans in seacoast areas because dietary iodine is usually adequate when a considerable amount of seafood is included in the diet. But chronic thyroid deficiency problems can develop inland, where dietary iodine levels may be lower. Fortunately, this problem can be prevented by using iodized salt, table salt to which very small quantities of iodine have been added.

Pancreatic Hormones

Maintenance of a stable supply of energy to body cells is one of the most important aspects of homeostasis. This energy supply is assured through maintenance of stable blood sugar (glucose) levels. The task of storing glucose (as glycogen, a polymer of glucose) and making it available to other body tissues falls primarily to the liver (see page 182). In large part, glucose storage and blood sugar levels are regulated by two major hormones, **insulin** and **glucagon,** produced in the pancreas.

The cells producing insulin and glucagon are located in the **islets of Langerhans,** patches of tissue scattered among the portions of the pancreas that secrete digestive enzymes (figure 11.14). The cells of the islets are glucose sensitive—they respond quickly to changes in blood glucose concentration. When glucose is abundant in the blood, insulin secretion is stimulated. Insulin lowers blood sugar levels mainly by promoting the uptake of glucose into body cells and the use of glucose in cells. It also stimulates glucose storage in the liver. When there is a decrease in blood glucose, glucagon secretion is stimulated. Glucagon promotes glucose release from the liver and thus raises blood sugar levels. Thus, glucagon and insulin are **antagonistic** hormones; they oppose and balance one another. A normal, adequate supply of glucose for body cells and tissues depends to a large extent on this antagonistic relationship between insulin and glucagon.

Figure 11.15 The adrenal gland. (*a*) Cross section of an adrenal gland of a mouse showing the inner medulla and outer cortex. Human adrenal glands are more flattened than this mouse adrenal because they lie against the surface of the kidney (see figure 11.5). (*b*) Structural formulas of the two hormones of the adrenal medulla, epinephrine (adrenalin), and norepinephrine (noradrenalin). (*c*) Cortisol (hydrocortisone), the principal glucocorticoid produced by the adrenal cortex. Most of the carbons and hydrogens in the molecule are not shown because the adrenal cortical steroids share a common skeleton and differ only in side chains. But these side chain differences can result in very different functional properties (for example, compare the structure of cortisol with aldosterone, a mineralocorticoid shown in figure 11.16).

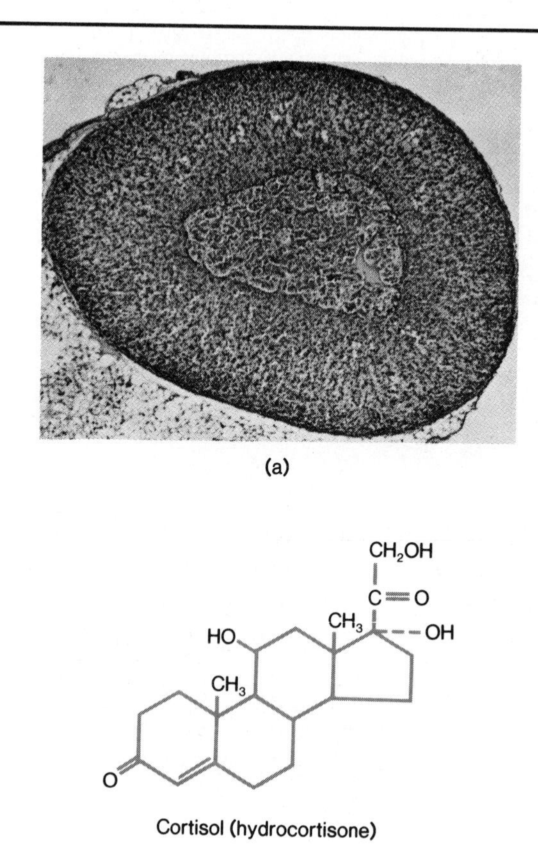

HO — (ring) — CH — CH_2 — N(H)(CH_3)
OH OH

Epinephrine

HO — (ring) — CH — CH_2 — NH_2
OH OH

Norepinephrine

(b)

A deficiency in insulin secretion results in the condition called **diabetes mellitus.** In diabetes mellitus, glucose storage decreases and there is inefficient glucose absorption and utilization in the body cells. As a result, blood glucose levels rise so high that glucose is excreted in the urine. At the same time, more water is lost and the diabetic becomes dehydrated. Despite blood glucose excesses, body cells cannot meet their energy needs, and they therefore metabolize proteins and fats. These metabolic alterations lead to a general weakness and susceptibility to disease.

Insulin can now be produced by genetic engineering techniques (see chapter 18), and as a result, research on diabetes and success in its control are progressing rapidly. But the battle is still far from won. There are still problems because prolonged insulin therapy itself may cause problems for diabetes mellitus patients.

Adrenal Hormones

Some of the adrenal gland hormones also affect blood glucose levels and metabolism in general. The adrenal glands, one of which is located above each kidney, are really two glands in one, the inner **medulla** and the outer **cortex** (figure 11.15).

Table 11.2
Effects of Epinephrine Secretion

Effects
Increased blood sugar level
Increased heart rate
Increased force of cardiac contractions
Increased respiratory rate
Increased blood flow to brain
Increased blood flow to liver and heart
Increased blood flow to skeletal muscles
Increased blood pressure
Dilation of pupil of the eye
Dilation of bronchioles of the respiratory tract
Decreased blood flow to digestive tract
Decreased digestive function

The adrenal medulla secretes the hormones **epinephrine** and **norepinephrine** (also called adrenalin and noradrenalin). The effects of the two hormones are similar, though not identical. In general, epinephrine is secreted in response to stressful or emergency situations and prepares the organism to cope with such circumstances. The secretion of epinephrine causes an increase in blood sugar levels by stimulating the liver to release glucose. Epinephrine also has many other effects not directly related to blood sugar (table 11.2).

The cortex of the adrenal gland secretes numerous hormones—perhaps as many as fifty different compounds—and all of them are steroids, molecules whose structure includes several rings of carbon atoms. Some of these steroids, the **glucocorticoids,** also are involved in the regulation of blood sugar. The principal glucocorticoid is **cortisol (hydrocortisone).** Glucocorticoids also play a role in the body's adaptation to long-term stress situations.

Hormones Controlling Fluid and Ionic Balance

Animal blood and tissue fluids are essentially dilute salt solutions containing mostly sodium and chloride ions with smaller amounts of calcium, potassium, phosphate, magnesium, and other ions. Maintenance of homeostasis requires that these ions be kept under strict control. Several hormones are involved in regulating ionic balance in the body cells and in tissue fluids. The **mineralocorticoids,** a second class of steroid hormones from the adrenal cortex, are involved in ion regulation in the body (figure 11.16).

Figure 11.16 Aldosterone, the principal member of the class of adrenal cortical hormones called mineralocorticoids, which are involved in ionic regulation in the body. Aldosterone promotes sodium reabsorption by the kidney.

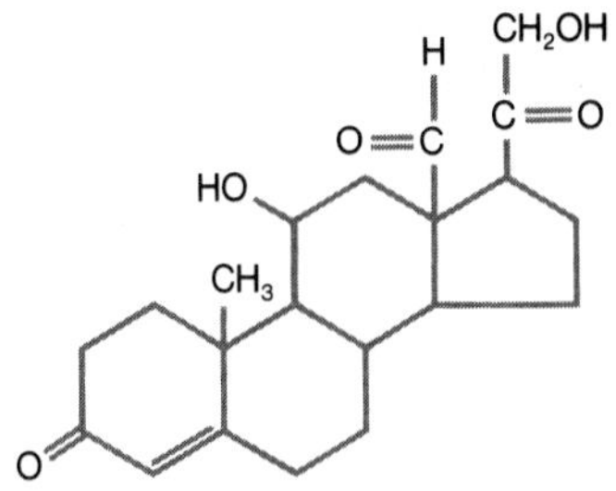

Control of Calcium Balance

Calcium is especially important to the functions of nerve and muscle tissue. Death due to calcium imbalance can occur very quickly through massive nervous disruption or disturbance of heart muscle contraction. Consequently, calcium levels in blood are closely regulated within a very narrow range. The **parathyroid glands** play a major role in this regulation. They produce **parathyroid hormone,** which raises blood calcium levels and lowers blood phosphate levels. Parathyroid hormone acts on bone cells, causing them to remove calcium ions from bones and transfer them into the bloodstream.

For many years parathyroid hormone was thought to be the only hormone controlling calcium balance in the body. But in the early 1960s, a second calcium-regulating hormone, **calcitonin,** was discovered. Calcitonin is produced by a special group of cells in the thyroid, the C cells. It is an antagonist to parathyroid hormone; that is, it has the opposite physiological effect. Calcitonin lowers the blood calcium level by inhibiting bone cells from releasing calcium into the blood. Calcitonin secretion is stimulated by a rise in blood calcium, and it acts quickly to lower the blood calcium to a proper level.

Yet a third hormone is now known to affect calcium levels in body fluids. This recently discovered hormone, **1,25-dihydroxyvitamin D_3,** is a derivative of vitamin D_3, a common form of vitamin D. Vitamin D_3 is either ingested in the diet or synthesized in the skin under the influence of ultraviolet radiation in sunlight. Vitamin D_3 is itself a **prohormone;** that

Figure 11.17 Steps in the conversion of vitamin D_3 into a molecule that is active as a hormone. The carbon and hydrogen skeleton of the molecule does not change during these steps so it is not shown in detail. Vitamin D_3, which is synthesized in the skin under the influence of ultraviolet radiation or ingested in the diet, is converted to 25-hydroxyvitamin D_3 in the liver and finally to 1,25-dihydroxyvitamin D_3 in the kidney, thus producing the active form of the hormone.

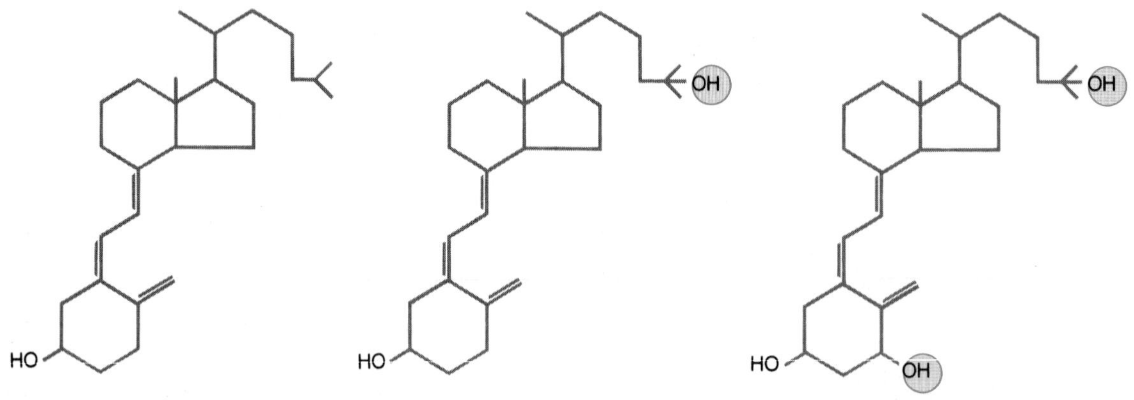

Vitamin D_3 (calciferol)

25-hydroxyvitamin D_3 (25-hydroxycalciferol$_3$)

1,25-dihydroxyvitamin D_3 (1,25-dihydroxycalciferol$_3$)

is, it is a molecule that can be converted into an active hormone form. The first step in converting it into a hormone occurs when liver cells change the vitamin to 25-hydroxyvitamin D_3 (figure 11.17). This substance is then transported to the kidney, where enzymes in kidney cells change it to the active hormone, 1,25-dihydroxyvitamin D_3 (symbolized 1,25-$(OH)_2D_3$). The major effect of this hormone is to stimulate the cells of the intestine to absorb more calcium, thus raising blood calcium levels.

Hormone Synthesis

Some of the most interesting recent discoveries in research on hormones have concerned the synthesis of polypeptide hormones, the hormones that consist of chains of amino acids. When biologists began to study synthesis of these hormones using the sophisticated techniques of modern molecular biology, they got some surprises. It soon became apparent that, in several cases, the polypeptide first synthesized in gland cells is *not* identical to the polypeptide hormone that the cells secrete. What are the differences between these originally synthesized polypeptides and the actually secreted hormone molecules, and how is one converted into the other?

Prohormones

A number of polypeptides have been isolated and described as prohormones. These prohormones are larger molecules than the hormones, and they do not act on target cells in the same way. They must be converted into the smaller, active hormone form. For example, the single-chain polypeptide **proinsulin** is converted to insulin in pancreas cells (figure 11.18).

More recently, even larger molecules, which are precursors of the prohormones, have been discovered. These preprohormones make the story of polypeptide synthesis even more complicated.

Figure 11.18 Conversion of proinsulin to insulin. The polypeptide chain of the proinsulin molecule is longer (eighty-four amino acids) than the double polypeptide chain of the insulin molecule (fifty-one amino acids). The active hormone is produced by removal of a peptide fragment that contains thirty-three amino acids.

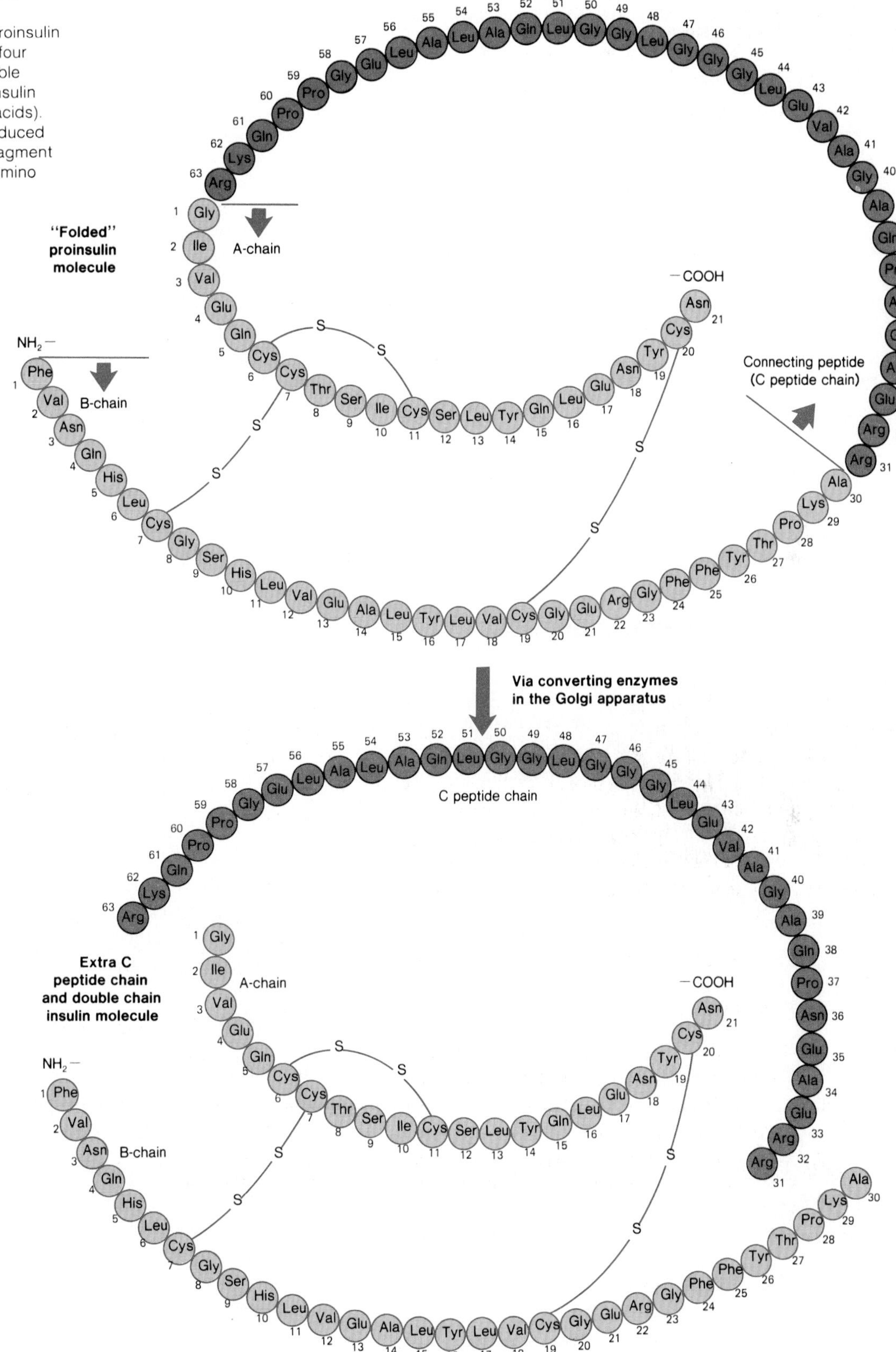

Box 11.2
Why Can Morphine Kill Pain?

The powerful effects of opiates (opium, morphine, and related drugs) have been known for many years. Once the concept of receptor molecules was developed, scientists hypothesized that the powerful pain-killing effects of the opiates must involve specific opiate receptors on cells in the nervous system. In fact, such receptors were found. But this raised a new question: why would cells carry receptors for substances not normally found in the body? Perhaps, it was suggested, the opiates combine with receptors that normally bind naturally occurring regulatory molecules in the nervous system. Further research isolated small peptides called **enkephalins** (Greek for "in the head") that do bind to opiate receptors. Larger peptide molecules called **endorphins** ("internal morphine") that include the enkephalin peptide sequences were also found to bind with the opiate receptors.

The discovery of enkephalins, endorphins, and even larger polypeptides that include these smaller molecules within them has opened a new field of study in chemical regulation. It may now be possible to determine the basis for physical drug addiction. These peptides might even serve as nonaddictive pain killers. Nervous system disorders might also be treated with enkephalins, endorphins, or other peptide regulatory substances. Science has barely scratched the surface in understanding these and other peptides that act as regulators throughout the body.

Hormone Receptors and Target Cells

How does a hormone interact with its target cells, and what determines which cells are target cells for a particular hormone and which cells are not? A target cell has the appropriate receptors for a hormone and responds functionally to that hormone. A receptor is a molecule or molecular complex that is part of the target cell and that can specifically recognize a particular hormone and bind with it. Receptors for polypeptide and adrenal medulla hormones appear to be located on the cell surface, that is, on the plasma membrane. Receptors for steroid hormones, on the other hand, appear to be located in the interior of the cell, in the cytoplasm.

Nontarget cells presumably do not have the appropriate receptors for a particular hormone and, hence, do not respond to it. Different types of cells are assumed to have different hormone receptors, and cells that respond to several hormones must have receptors for each one.

Effects of Hormones Inside Cells

Hormones exercise their effects on target cells in several major ways. (1) Some hormones alter the transport of substances across the cell membrane. The prime example of this type of action is the hormone insulin, which increases the transport of glucose into cells by stimulating glucose transport systems in the plasma membrane. (2) A second major mode of hormonal action is the alteration of enzyme activity to produce **cyclic AMP** (see below), which can in turn alter numerous other enzyme systems inside the cell. (3) Finally, a hormone can exert control by acting directly on the genetic material to stimulate the synthesis of messenger RNA, which leads to production of a particular protein or set of proteins.

Cyclic AMP, The Second Messenger

A nucleotide, **cyclic adenosine monophosphate (cAMP)** acts as a **second messenger** in responses of target cells to certain hormones. Earl W. Sutherland received the Nobel Prize in 1971 in recognition of his discovery of the importance of cyclic AMP in hormone activity. At the time, Sutherland was studying the effects of epinephrine on liver cells. Epinephrine stimulates the release of glucose into the blood by causing an increase in the rate at which glycogen is broken down in the liver cells.

Figure 11.19 The enzyme adenylate cyclase catalyzes the production of cyclic AMP from ATP. Note that the phosphate group in the colored area is part of a ring structure with oxygens and carbons.

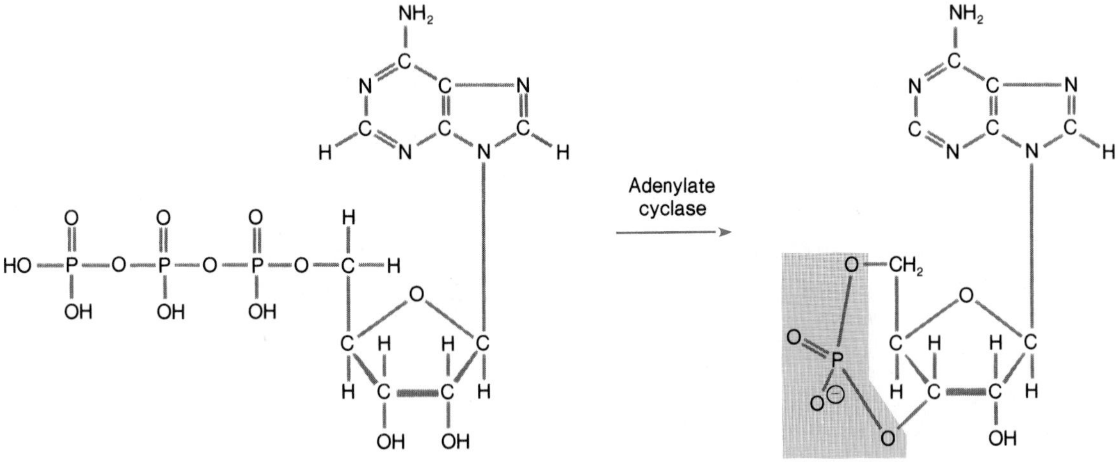

Adenylate cyclase

Adenosine triphosphate

Cyclic adenosine monophosphate

Although epinephrine binds to receptors on the surfaces of target cells, it does not enter the cytoplasm. How, then, does this hormone exert its influence on glucose release? Sutherland and his colleagues determined that epinephrine stimulation of liver cells causes an increase in the production of cyclic AMP in the cytoplasm. They demonstrated that cyclic AMP functions as a "second messenger" in stimulating glucose release. The hormone epinephrine is the first messenger because it brings a message to the plasma membrane of a target cell. There it stimulates production of the second messenger, cyclic AMP, which actually causes the response inside the target cell—in this case, activation of glucose release.

Cyclic AMP is formed from ATP in a reaction catalyzed by the enzyme **adenylate cyclase** (figure 11.19). Adenylate cyclase is associated with receptors in the plasma membrane. As the hormone binds to the receptors, the activity of adenylate cyclase increases. Adenylate cyclase then produces cyclic AMP from ATP. Hormone receptors, adenylate cyclase, and cyclic AMP are linked with the intracellular effects of a number of hormones.

There is one final, critical stage in the cyclic AMP mechanism. No regulatory system can function efficiently if the regulator substance accumulates in excessive amounts. The regulator substance must be disposed of fairly quickly after it has had time to act. In the case of cyclic AMP, an enzyme called **phosphodiesterase** catalyzes the conversion of cyclic AMP to an inactive form of AMP (figure 11.20).

The Actions of Steroid Hormones on Genes

Although the hormones of the adrenal medulla (epinephrine and norepinephrine) and the polypeptide hormones act through cyclic AMP, the steroid hormones interact with cells through a different mechanism.

Steroid hormones are relatively small molecules that appear to enter the cell with ease. Once in the cell, a steroid hormone acts on the genetic material in a series of reactions.

First, the hormone binds to a receptor in the cytoplasm. The hormone-receptor complex is then able to act on the cell's nucleus. After it enters the nucleus, the hormone-receptor complex binds to specific parts of the chromosome, where it activates certain genes, which lead to the cellular response characteristic of that hormone. All of this is summarized in figure 11.21.

Figure 11.20 A model of the role of cyclic AMP as a second messenger in the effects of epinephrine on a liver cell. Cyclic AMP stimulates activation of enzymes that release glucose into the bloodstream. Phosphodiesterase prevents accumulation of cyclic AMP by converting it to an inactive form of AMP.

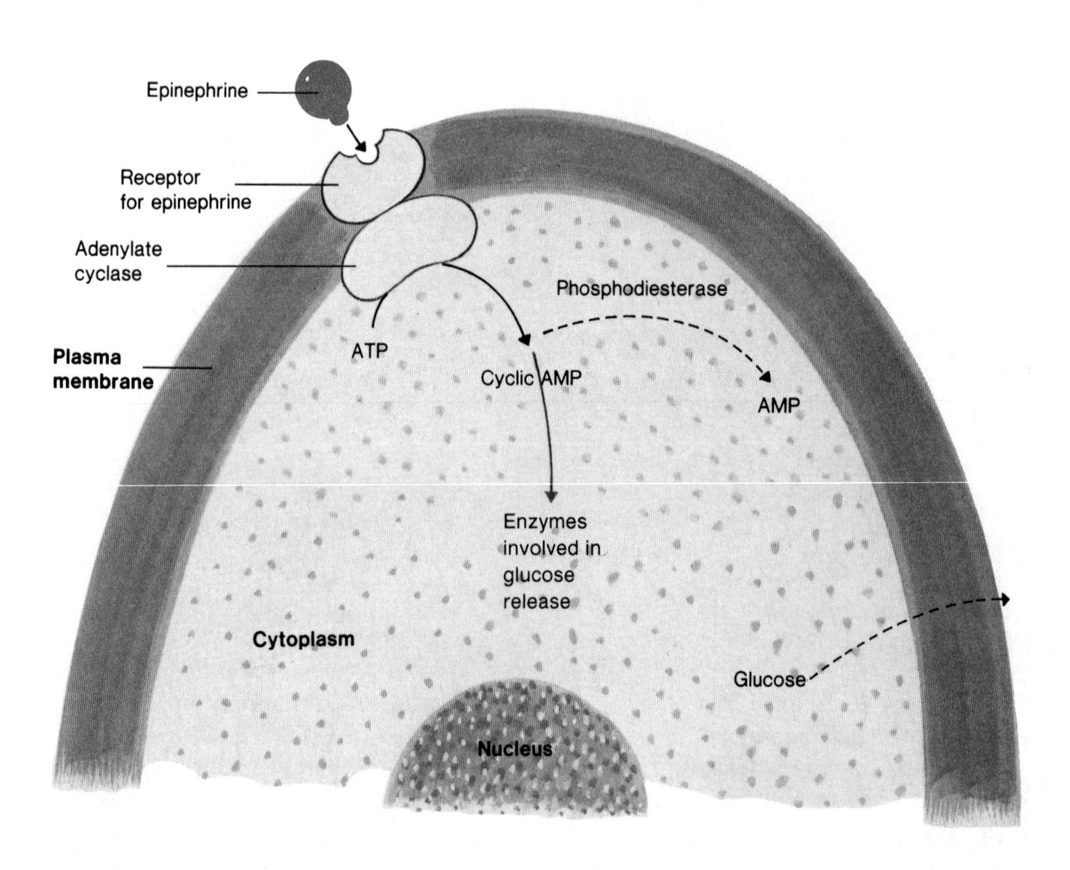

Prostaglandins

The **prostaglandins** are a family of twenty-carbon fatty acids formed in the plasma membranes of a wide variety of tissues. They are produced in small amounts, are extremely potent, and exist for only a short time (figure 11.22).

The effects of the prostaglandins are widespread and varied. Some of them even have opposing effects. For example, one lowers blood pressure, but another raises it. Some stimulate the contraction of the smooth muscle of the uterus. Careful administration of prostaglandins can be helpful in childbirth. Prostaglandins also are possible abortion-inducing agents, not only because they cause the uterus to contract but also because some may cause a reduction in the secretion of progesterone, a hormone necessary for the maintenance of pregnancy. It is possible that some of the prostaglandins may prove useful for birth control in the future.

Prostaglandin research has already provided a possible explanation of the way in which aspirin acts. Aspirin has been used as a drug for several hundred years without complete understanding of what it does. But we now know that aspirin blocks the synthesis of prostaglandins. This discovery seems to be a step toward explaining some of the effects of this most commonly used of all drugs.

Figure 11.21 A model of the mechanism of action of a steroid hormone on genes in a target cell. The hormone passes through the plasma membrane and binds with a receptor in the cytoplasm. The hormone-receptor complex is modified in some way that activates it, and then it enters the nucleus, where it facilitates transcription of specific genes to produce messenger RNA (chapter 18). This leads to protein synthesis, an important step in the cell's response to the hormone.

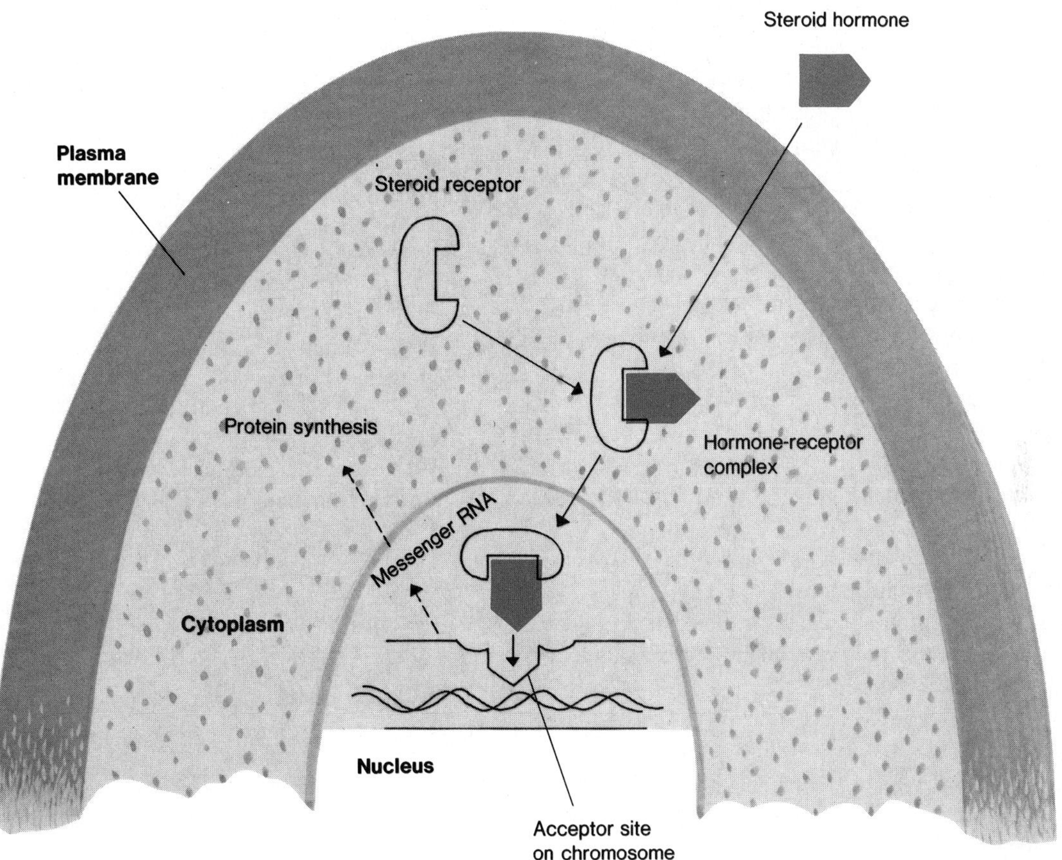

Figure 11.22 One of the prostaglandins, PGE_2. The broken line denotes a bond that extends below the plane of the ring portion of the molecule. Only a skeleton of the molecule is shown, without details of most carbon and hydrogen atoms.

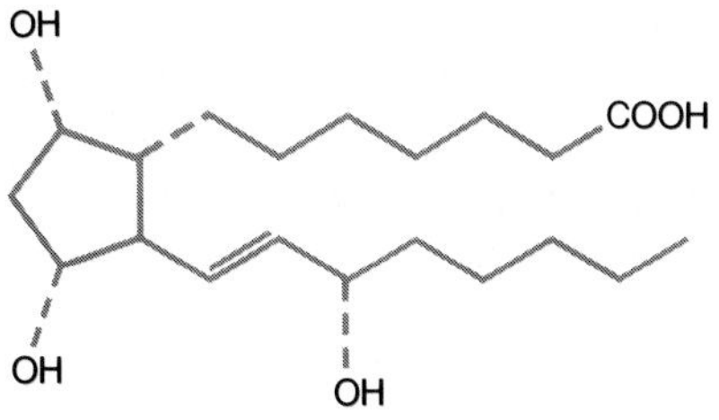

OH

COOH

OH OH

Summary

Multicellular organisms must have coordinating mechanisms for their cells, tissues, and organ systems. The endocrine and nervous systems are the two major methods of cellular communication within an animal. Both are necessary for the maintenance of homeostasis in the organism.

Endocrine glands are ductless glands that secrete substances called hormones. Hormones are carried in the bloodstream to target cells, which respond to them.

Insect development is controlled by brain hormone from the neurosecretory cells in the brain, by ecdysone, and by juvenile hormone.

In humans and other vertebrates the major endocrine glands are the pituitary, the thyroid, the parathyroids, the pancreas, the adrenals, the ovaries, the testes, the hypothalamus of the brain, and the kidneys. Many other tissues also secrete chemical regulatory substances.

Oxytocin and ADH, hormones of the posterior pituitary, are produced by neurosecretory cells in the hypothalamus and released into the blood from the posterior lobe. Anterior-lobe hormones stimulate the growth and development of the entire organism, secretion by other endocrine glands, and milk production by the mammary glands.

Hormone secretion by the anterior lobe is regulated by releasing hormones and release-inhibiting hormones delivered from the hypothalamus via the hypothalamic-pituitary portal system.

Metabolism is under the control of several hormones, notably the thyroid hormones. Carbohydrate metabolism is affected by insulin and glucagon from the pancreas, epinephrine and norepinephrine from the adrenal medulla, and glucocorticoids from the adrenal cortex.

Three hormones are involved in the regulation of blood calcium levels: parathyroid hormone, thyrocalcitonin, and 1,25-dihydroxyvitamin D_3.

Research on hormone synthesis has led to the discovery of prohormones and preprohormones.

A receptor for a hormone binds with it and initiates the chain of events that lead to the cell's ultimate physiological response to the hormone.

Some hormones appear to affect their target cells by altering intracellular levels of a substance known as cyclic AMP. This chemical then acts within the cell to produce the final effect of the hormone. Thus, cyclic AMP is sometimes known as the second messenger.

Steroid hormones exert their effects upon target cells by entering the cell, combining with a receptor, entering the nucleus of the cell, and stimulating protein synthesis.

Prostaglandins are powerful hormonelike substances found in a wide variety of tissues.

Questions for Review

1. Define the terms exocrine gland and endocrine gland.
2. What are neurosecretory cells?
3. Explain how juvenile hormone might be used as an insecticide.
4. Describe the anatomical and functional connections between the hypothalamus of the brain and the posterior lobe of the pituitary.
5. Explain how the secretions of the anterior lobe of the pituitary are controlled by the hypothalamus.
6. What is a neuroendocrine reflex?
7. Liver cells that are involved in glucose storage and release have receptors for several different hormones. Why? What are some of the hormones?
8. Describe how proper calcium levels are maintained in human blood.
9. In what way does cyclic AMP act as a second messenger in hormone action?

Questions for Analysis and Discussion

1. American Indians called the famous mountain man Jim Bridger "Big Throat." Can you speculate on what Jim Bridger's problem might have been and on a possible environmental cause for that problem in a person who spent his life wandering the central part of the North American continent?
2. Weight loss is a common symptom of diabetes mellitus. Why do body cells of diabetics oxidize vital proteins and fats when there is actually excess glucose available around them?
3. Why would it have been impossible for researchers studying hormones to have determined the role of the kidneys in producing 1,25-dihydroxyvitamin D_3 using the methods that Berthold used to discover the hormonal activity of the testes?
4. Caffeine inhibits phosphodiesterase activity. Can you explain why and how caffeine might therefore affect many kinds of cells?

Suggested Readings

Books

Gorbman, A.; Dickhoff, W. W.; Vigna, F. R.; Clark, N. B.; and Ralph, C. L. 1983. *Comparative endocrinology.* New York: Wiley.

Turner, C. D., and Bagnara, J. R. 1976. *General endocrinology.* 6th ed. Philadelphia: W. B. Saunders.

Articles

Baile, C. A.; Della-Fera, M.A.; and Krestel-Rickert, D. 1985. Brain peptides controlling behavior and metabolism. *Bio Science* 35:101.

Blake, C. A. 1984. The pituitary gland. *Carolina Biology Readers* no. 118. Burlington, N.C.: Carolina Biology Supply Co.

Guillemin, R., and Burgus, R. November 1972. The hormones of the hypothalamus. *Scientific American* (offprint 1260).

O'Malley, B. W., and Schroeder, W. T. February 1976. The receptors of steroid hormones. *Scientific American* (offprint 1334).

Snyder, S. H. March 1977. Opiate receptors and internal opiates. *Scientific American* (offprint 1354).

Wigglesworth, V. B. 1983. Insect hormones. *Carolina Biology Readers* no. 70. Burlington, N.C.: Carolina Biological Supply Co.

Nerve Cells and Nervous Systems

Chapter Outline

Chapter Concepts

1. The nervous system is specialized to provide coordination and communication in the animal body. This system senses environmental changes and organizes appropriate responses to them.
2. Neurons are the functional units of all nervous systems, which range from simple, diffuse networks of cells to highly specialized systems including large, complex brains.
3. Nerve impulses are electrical charges that sweep rapidly along neuron fibers.
4. Neurons communicate with one another by way of specialized connections called synapses.
5. Neurons are specifically arranged in functional pathways, the simplest of which are reflex arcs.
6. The central nervous system (CNS) is composed of the brain and the spinal cord.
7. The central nervous system has a number of different functionally specialized areas.

It is very interesting to watch a *Hydra* eat. Under a microscope, a *Hydra* looks like a small, flexible, white object fastened to the bottom on one end, swaying gently in the water with a ring of tentacles extending from its free end. It hardly looks like a predator that captures and eats other animals.

But if a small animal such as a water flea should blunder against one of those tentacles, the result is dramatic. One moment the water flea is all action with its tiny swimming appendages beating as fast as the eye can follow. Suddenly, all movements stop abruptly—the water flea is almost instantly paralyzed by poison from stinging cells on the *Hydra's* tentacles.

Then the *Hydra* begins a very impressive performance. Its tentacles begin to work together to move the water flea into position over its mouth. Moments later, the water flea disappears into the mouth and becomes nothing more than a bulge in the *Hydra's* body (figure 12.1). The *Hydra* then resumes its vigil, waiting for the next unlucky creature to fall prey to its poison tentacles.

Even if you flood the *Hydra* with large numbers of prey animals at the same time, it manages quite nicely. Many of the animals are paralyzed in short order. Some drop off the tentacles and in time, the area around the *Hydra* looks like a miniature battlefield with paralyzed animals scattered all around. But the *Hydra* continues to feed and its tentacles work in coordination to move certain of the captured prey into its mouth while others are ignored for the time being.

Figure 12.1 A *Hydra* containing a recent meal.

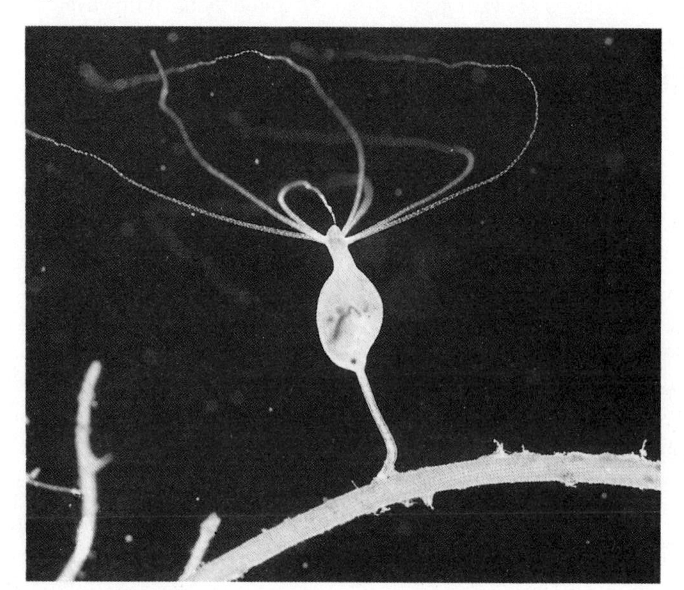

This complex feeding behavior is possible because *Hydra,* like the vast majority of animals, has a **nervous system** which, though it is very simple as nervous systems go, coordinates all the *Hydra's* actions.

A complex division of labor among body parts is necessary to maintain homeostasis in multicellular organisms. The diverse activities of the various body parts of animals are coordinated by hormonal and neural regulation, both of which are vital to maintaining homeostasis. In chapter 11, we saw that hormonal regulation involves transport of chemical messengers, called hormones. It takes minutes, hours, or even longer for a hormonal message to produce an effect. Neural regulation, however, involves much more rapid communication. Neural messages can be passed from one part of the body to another in only thousandths of a second.

An animal's nervous system gathers and processes information concerning both the external and internal environments of the body. It also organizes appropriate responses and actions by various parts of the body. This work requires rapid transmission of messages from one place to another by the individual nerve cells (**neurons**) that make up a nervous system.

Figure 12.2 shows the structure of one of the many types of neurons found in a nervous system. This neuron consists of three basic parts. The **dendrites** are slender, branched extensions that form a major receptor area of the neuron; that is, they receive messages from other cells. The **nerve cell body** contains the nucleus and the other organelles normally found in cells. Finally, the **axon** is a long process extending from the cell body that carries messages (nerve impulses) on to the next cell.

Figure 12.2 Structure of a motor neuron in a vertebrate animal. Note the branched dendrites and the long single axon that branches only near its tip. The myelin sheath is an insulating structure that is present around many neuron processes.

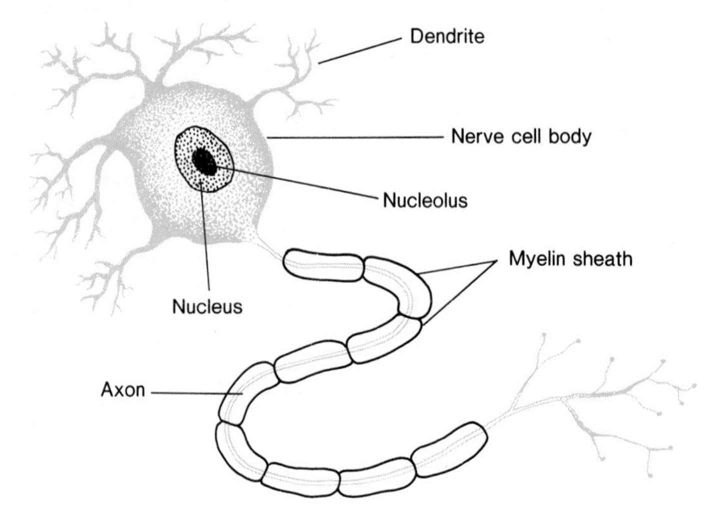

Nervous Systems

How are nerve cells organized into nervous systems in animals? There are a great many answers to that question. Our *Hydra* has one of the simplest kinds of nervous systems, the **nerve net.** A nerve net consists of neurons interwoven into a diffuse network that extends through all parts of the body (figure 12.3). Messages can be carried in all directions through this net beginning at any point. Such a nervous system is adequate for a *Hydra,* but since it has no central processing unit (brain), its functions are limited.

As you might expect, more complex animal bodies have more complex nervous systems. Such complex systems have clusters of nerve cells that make up a **central nervous system (CNS** for short). In humans, the central nervous system includes the brain and spinal cord. The central nervous system is connected to the rest of the body by bundles of neuron fibers. These bundles, called **nerves,** make up the **peripheral nervous system.**

In more complex nervous systems, we can divide neurons into three broad categories based on their functions. Some are **sensory neurons** that either receive stimuli directly or respond to specific changes that occur in other specialized receptor cells. Others are **motor neurons** that direct the organism's response to stimuli. And still others are **interneurons,** acting as intermediates between sensory and motor neurons. A large proportion of the neurons in the central nervous system are interneurons, and the more complex the nervous system, the greater the number of interneurons present.

Figure 12.3 A very simple nervous system, the nerve net of *Hydra.*

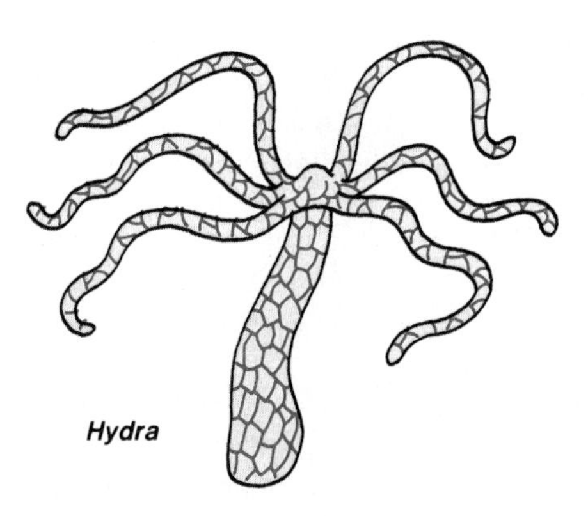

Nervous systems of annelids (segmented worms) such as earthworms illustrate some of the features of more complex nervous systems (figure 12.4). A concentration of nerve cells in the head constitutes the **brain,** and a large, solid ventral **nerve cord** runs through the body. Along this nerve cord are clusters of nerve cell bodies called **ganglia** (singular: **ganglion**), arranged with one ganglion in each body segment. The brain, nerve cord, and ganglia make up the central nervous system, while as we said earlier, bundles of neuron fibers (nerves) make up the peripheral nervous system. These nerves, which contain both sensory and motor neuron fibers, connect all parts of the body with the central nervous system.

Figure 12.4 The nervous system of an earthworm, showing the central nervous system and peripheral nerves. Animal brains characteristically are located near sense organs in the head.

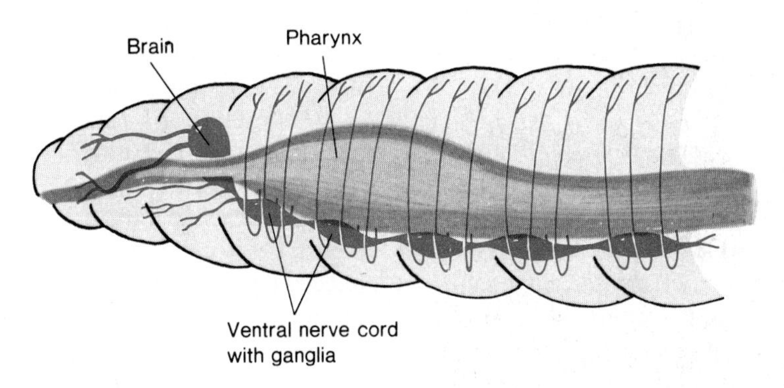

Figure 12.5 An example of a vertebrate nervous system. The human nervous system showing the central nervous system (brain and spinal cord) and the nerves of the peripheral nervous system.

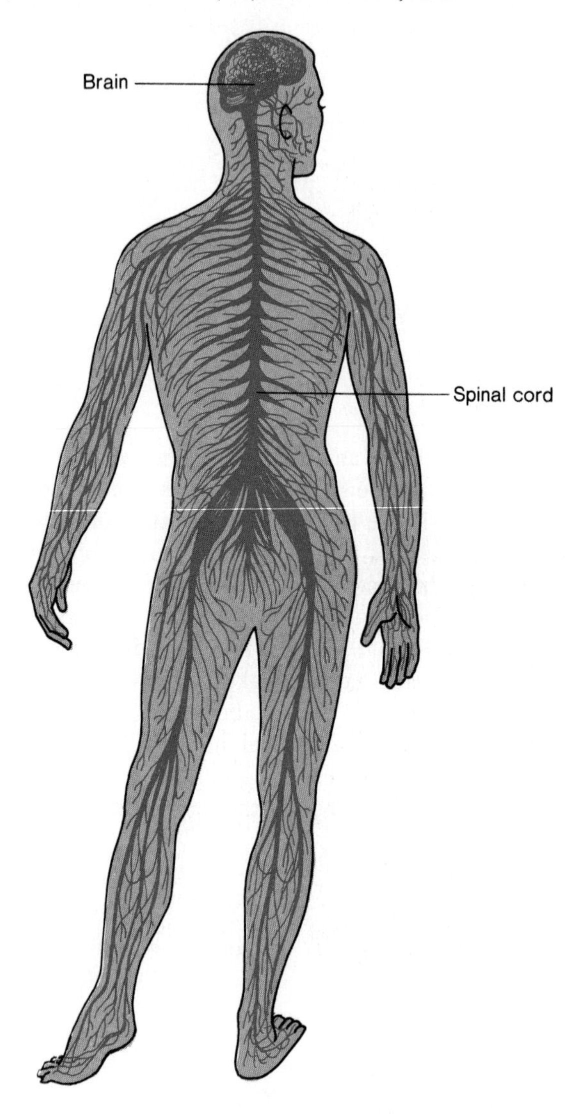

While the nervous system of a complex invertebrate (an animal without a backbone) such as an insect may contain a total of about one million neurons, the nervous system of a vertebrate (an animal with a backbone) contains many thousands of times that number. In vertebrates, huge numbers of interneurons are organized into brains that function as elaborate information-processing centers (figure 12.5). Such centers can generate much more complex responses and behavior patterns than those of invertebrate animals.

Nerve Cell Functions

Nervous systems are made up of neurons, each playing its specific role in the total functioning of the system. Neurons can participate in neural functions because they possess several fundamental properties. Every neuron can undergo specific electrical changes in response to appropriate stimulation, and, by doing so, it can communicate information to other cells. Each neuron, therefore, is itself a small message receiving and transmitting device.

Irritability and Neuron Functions

How do neurons respond to stimuli to receive information, and how do they carry messages? All living cells are **irritable;** that is, they can respond to stimuli. But irritability is especially well developed in neurons, which are extremely sensitive and responsive. Appropriate stimulation of a neuron causes an electrical change, which can travel along the entire length of the cell, even in neurons that are several meters long.

How exactly do neurons carry messages? They do not simply function as small wires like those in telephone or telegraph cables. The message-carrying ability of neurons is quite complex. The generally small size of neurons made research on the nerve impulse difficult until biologists found that some animals have neurons with very large axons. Giant axons from squid (figure 12.6*a*), for example, are large enough so that a tiny electrode can fit inside them (figure 12.6*b*). Using these giant axons, biologists were able to investigate neural functions in much more detail.

Electrical measurements show that there is a difference in electrical charge (an **electrical potential**) between the inside and the outside of a neuron's plasma membrane. Electrical potentials are usually measured in **volts,** but the electrical charge differences across neuron membranes are so small that they are measured in **millivolts** (thousandths of a volt). Nevertheless, these small differences in electrical charge are the key to neuron functions.

The interior of a neuron is negative relative to its exterior. Another way of describing the situation is to say that the neuron's plasma membrane has a negative side (inside) and a positive side (outside) (figure 12.7*a*). This charge difference is always present in a normal, living neuron and is called the **resting membrane potential,** or simply **resting potential.**

When a neuron responds to a stimulus, the electrical potential across its membrane changes. The charge difference reverses, so that for a short time the outside is negative relative to the inside (figure 12.7*b*). This membrane potential

Figure 12.6 Giant axons. (*a*) The squid *Loligo*. Giant axons from the squid have been used to study neuron functions. (*b*) Electrodes in position to measure electrical charge differences across the plasma membrane of an axon.

(a)

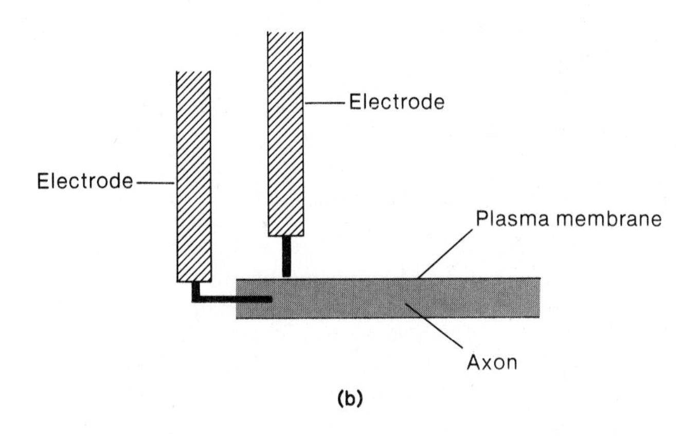

(b)

Figure 12.7 Neuron membrane potentials. (*a*) The resting potential. The outside of a neuron's plasma membrane is normally positive relative to the inside. (*b*) The action potential involves a temporary reversal of the charge relationship in part of the membrane. (*c*) A nerve impulse is a chain reaction series of action potentials that sweeps rapidly along a neuron's membrane (see arrows in 1–4).

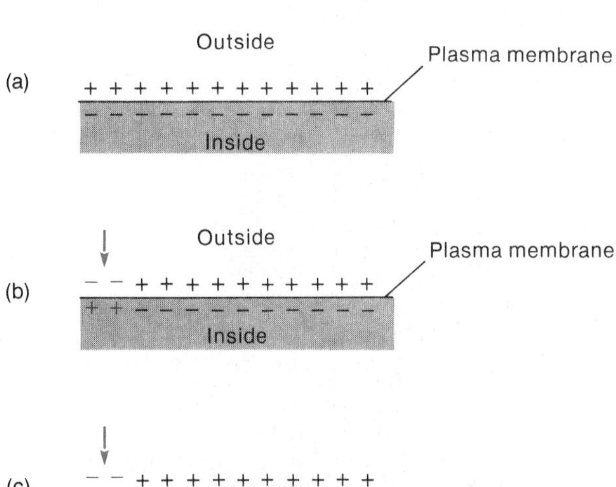

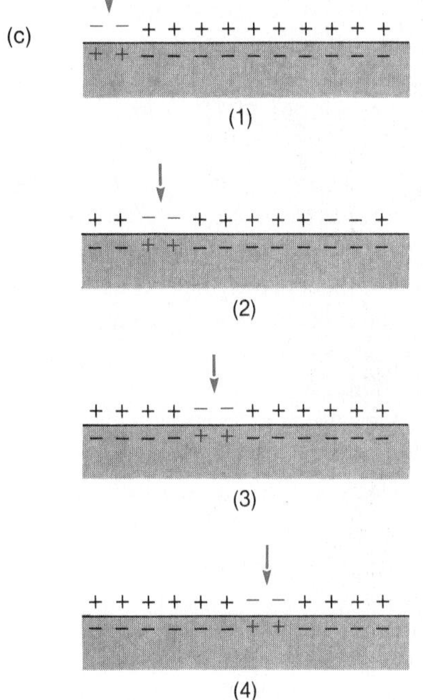

change is called an **action potential.** An action potential is a short-lived, temporary event, but it is the key to a neuron's ability to react to stimuli and conduct information over distances quickly. A **nerve impulse** is a chain reaction series of action potentials running the length of a neuron's membrane. Nerve impulses sweep along a neuron's membranes very quickly and permit rapid, long-distance transmission of electrical messages (figure 12.7*c*).

The Resting Potential

The irritability of a neuron depends upon maintenance of the resting potential across its plasma membrane. Why is there a resting potential, and how is it maintained?

Several properties of the plasma membrane establish and maintain a neuron's resting potential. Normally, there is a small excess of positively charged ions outside the plasma

Table 12.1
Concentrations of Ions Outside and Inside the Plasma Membrane of a Cat Nerve Cell

	Extracellular Fluid	Intracellular Fluid
Na^+	150 mM*	15 mM
K^+	5 mM	150 mM
Cl^-	125 mM	10 mM
	Fewer negatively charged proteins	More negatively charged proteins

*millimolar

Modified from A. J. Vander et al., *Human Physiology*, 3d ed. (New York: McGraw-Hill, 1980).

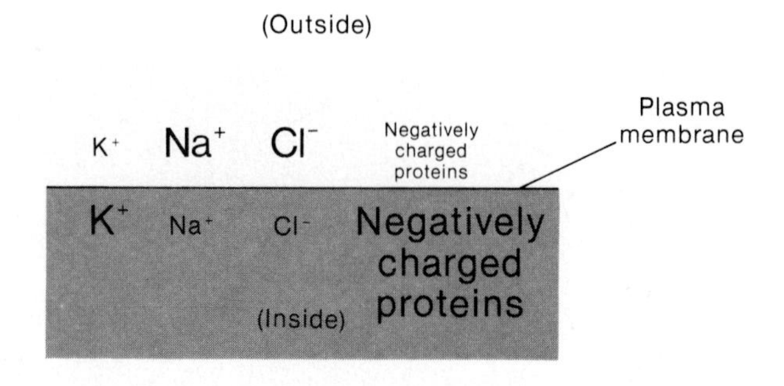

Figure 12.8 Diagram representing concentration differences between the inside and the outside of the neuron's plasma membrane.

membrane and a small excess of negatively charged ions inside the cell. This is because there are unequal distributions of certain ions on the inside and outside of the plasma membrane (table 12.1). Note that the external fluid that bathes the neuron contains more sodium ions (Na^+) and chloride ions (Cl^-) than the neuron's interior, while the neuron's cytoplasm contains more potassium ions (K^+) and more negatively charged protein molecules than the cell's external environment (figure 12.8). *The net effect of these ion concentration differences is an electrical charge difference across the neuron's plasma membrane.*

Plasma membranes, especially neuron plasma membranes, do not permit various ions to move with equal freedom into or out of cells. The neuron's plasma membrane is not very permeable to Na^+ ions but is 50 to 100 times more permeable to K^+ ions. Therefore, K^+ ions, which are more concentrated inside the cell, tend to leak out through the plasma membrane fairly easily. Only a relatively few Na^+ ions diffuse from the outside (where they are in higher concentration) to the inside of the cell (where they are in lower concentration).

But if there is a little leakage of Na^+ ions and considerable leakage of K^+ ions, how are the concentration differences maintained? There is an ATP-driven "pump" mechanism that actively transports Na^+ ions out of the cell and K^+ ions to the inside. The action of this **sodium-potassium pump** helps to maintain the differences in Na^+ and K^+ concentrations inside and outside the plasma membrane.

These factors—a difference in the membrane permeability for Na^+ and K^+ (resulting in greater outward leakage of K^+ in response to its concentration gradient); the presence of more large, negatively charged molecules inside the neuron; and the sodium-potassium pump—account for the resting membrane potential.

The Action Potential

As we mentioned earlier, a neuron's response to an appropriate stimulus is an action potential. An action potential is a specific and characteristic set of changes in a membrane's electrical charge (see figure 12.6*b*) and permeability characteristics, which occur very quickly (these changes can actually be completed within one millisecond). Let us look at these changes more closely.

During an action potential, the electrical charge difference of the membrane (inside relative to outside) changes from its normal −60 to −70 millivolts to +30 or even +55 millivolts and then returns to its original condition (figure 12.9).

Each of these electrical changes results from ion movements that take place as the plasma membrane undergoes a series of alterations in its permeability characteristics. First, the membrane temporarily becomes much more permeable to Na^+ ions, and a number of them rush in. Because they bring additional positive charges to the interior, the inside of the cell temporarily becomes positive relative to the outside. Then, a fraction of a millisecond later, the sodium permeability of the membrane returns to its regular low level, but the membrane's permeability to potassium ions suddenly increases. Potassium ions flow out of the cell. The outward flow of K^+ returns the membrane to its resting condition.

How rapidly can action potentials actually follow one another? The cell membrane is unresponsive to further stimulation for a certain length of time after an action potential has occurred. This time is called the **refractory period.** Although this limits the number of action potentials a cell can have, the fact that action potentials and refractory periods are completed in milliseconds means that many nerve cells can produce up to hundreds of action potentials per second.

Figure 12.9 The action potential. Diagram shows timing of electrical changes and relative permeabilities of membrane to Na^+ and K^+.

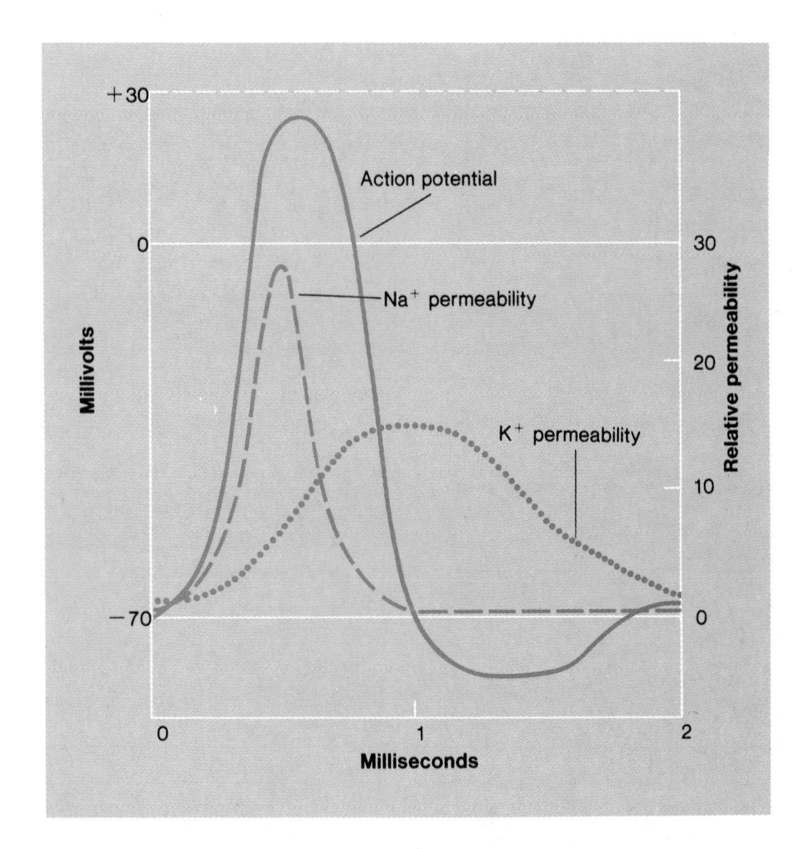

Local anesthetics, such as the novocaine or xylocaine your dentist administers before drilling near the very sensitive nerve endings in your teeth, seem to act by preventing action potentials in the nerve cells of the "deadened" area. Such anesthetics prevent the increase in membrane permeability to sodium that is necessary for an action potential to occur.

Nerve Impulse Conduction

The action potential is a series of events occurring in a small area of a membrane. But action potentials form the basis of the nerve impulse, the passage of an electrical signal from point to point along the fibers of neurons. How then are action potentials involved in nerve impulse conduction, which permits neurons to transmit information long distances in small fractions of a second?

In impulse conduction, each action potential triggers a new action potential right next to it in the plasma membrane. As each action potential generates a new one, the electrical change sweeps along the membrane. A nerve impulse is thus a chain reaction series of action potentials along a neuron's membrane.

Figure 12.10 pictures impulses being transmitted along an axon. At any one point where an action potential occurs, the inside of the axon membrane becomes positive and the outside negative because of the inward flow of Na^+ ions. This triggers a change in permeability at the next point along the axon, starting an action potential there. In this way, the impulse travels along the axon from the nerve cell body to the end of the axon.

Figure 12.10 Impulse transmission. A nerve impulse is a self-propagating series of action potentials. (*a* and *b*) The progress of a single impulse along an axon. (*c*) A second impulse (left) following the first.

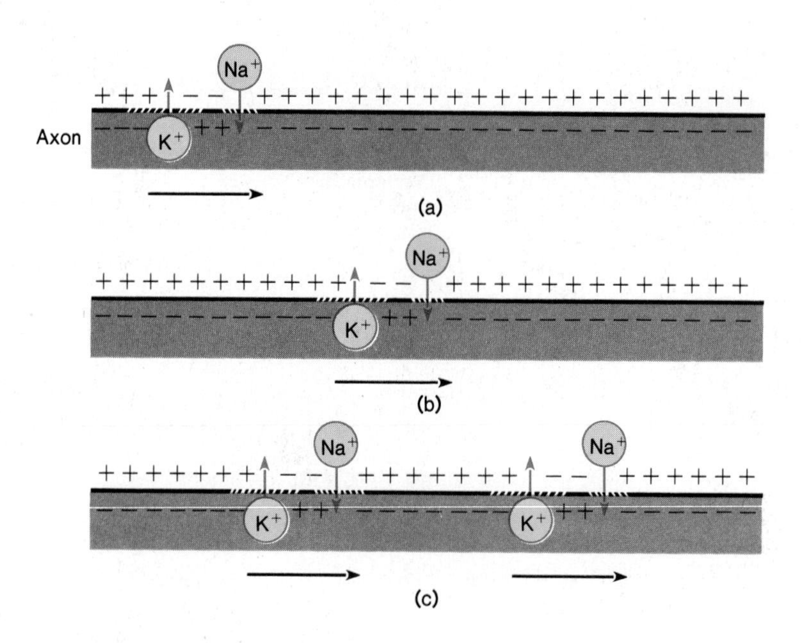

Synaptic Transmission

Conduction of nerve impulses in individual neurons is a key part of nervous system functioning, but it is only part of the story. Neurons have to be able to communicate with each other to make a system. Communication between neurons occurs at special junctions called **synapses.**

Some synapses, especially in central nervous systems, involve direct electrical effects of one nerve cell on another nerve cell where they come into contact. But most synapses operate by chemical transmission. A chemical messenger, called a **neurotransmitter,** is released by one nerve cell and causes an action potential in another nerve cell.

The cell sending the signal at a synaptic junction is the **presynaptic neuron,** and the cell receiving the signal is the **postsynaptic neuron.** The **synaptic cleft,** a very small space about 20 nm across, separates presynaptic and postsynaptic neurons. A small molecule can diffuse across the cleft in about a tenth of a millisecond.

The presynaptic portion of a synapse is usually the end of an axon, called the **axon terminal** (synaptic knob). The postsynaptic neuron has a special membrane area where it receives messages from the presynaptic neuron (figure 12.11).

Figure 12.11 is quite simplified compared to situations that actually exist in nervous systems. A postsynaptic neuron may have thousands of synaptic junctions on its dendrites and cell body. Some neurons in the human spinal cord have as many as fifteen thousand synapses. In the brain, a single neuron may receive even more synapses—perhaps as many as one hundred thousand synapses per nerve cell. Since the brain contains as many as 10^{10} neurons, its complex circuitry contains an astronomical number of synapses.

Figure 12.11 Synaptic connections between neurons. Note the knob at the end of the axon called the terminal and the thickening of the membrane just under the synapse, the postsynaptic membrane.

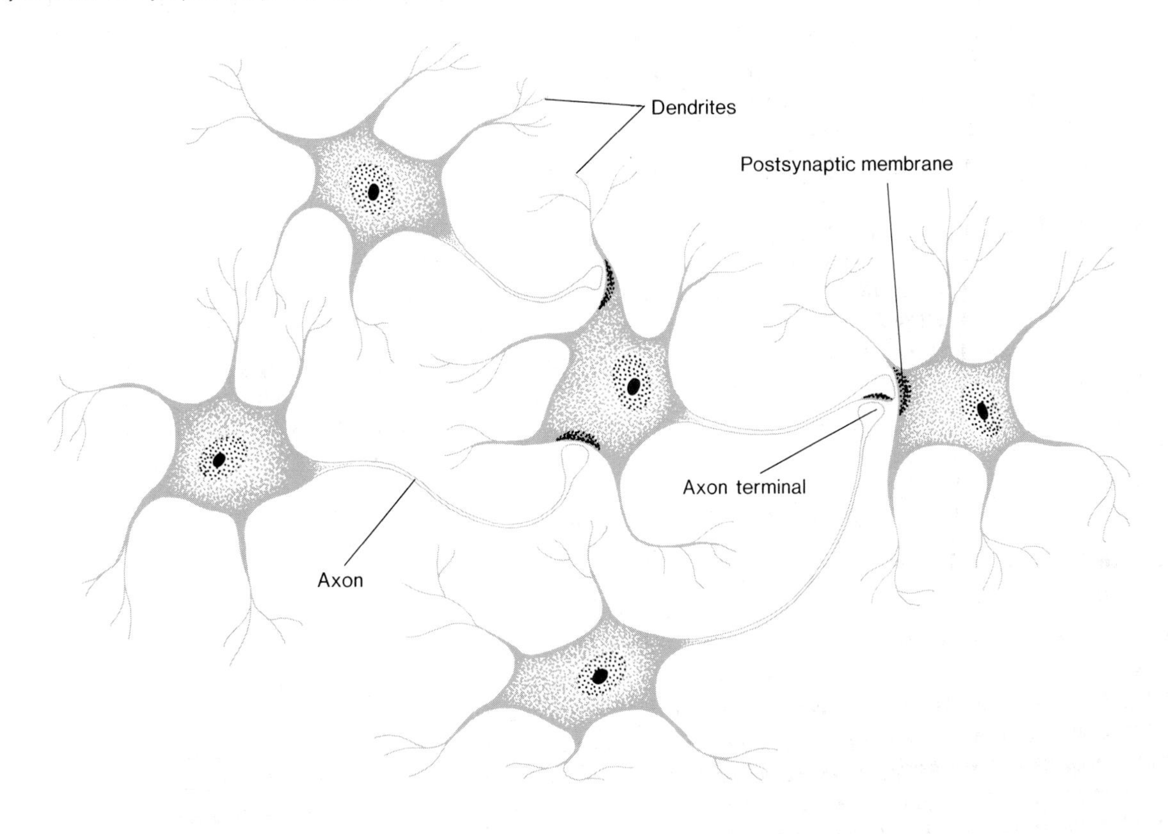

The Function of the Synapse

How do nerve cells communicate with one another in a synapse? Details of a synapse, as magnified greatly by the electron microscope, are diagramed in figure 12.12. The axon terminal (synaptic knob) contains small membrane sacs called **synaptic vesicles.** Each vesicle is about 20 nm in diameter and contains neurotransmitter molecules. The arrival of a nerve impulse transmitted down the presynaptic neuron causes the release of neurotransmitter into the synaptic cleft. Hundreds of synaptic vesicles fuse with the cell surface and empty their contents into the synaptic cleft in a very short time (figure 12.13).

Once neurotransmitter molecules have diffused across a synaptic cleft, they attach to special receptor molecules in the postsynaptic membrane. When neurotransmitter molecules attach to these receptor molecules, specific changes occur in the postsynaptic membrane. The interaction of neurotransmitters and receptor molecules causes ion channels in the membrane to open. Sodium ions flow in and an action potential is underway—the message has passed from one neuron to another.

We have just described an excitatory relationship in which synaptic transmission promotes action potential occurrence in a postsynaptic neuron. We should mention that other synaptic transmissions are inhibitory; that is, they act to inhibit action potential occurrence in postsynaptic neurons. Both kinds of synaptic functions are vital for normal nervous system operation.

Figure 12.12 Synapse structure. (*a*) A nerve cell body with several axon terminals (synaptic knobs) on it. (*b*) Diagram of the structure of the synapse as revealed by the electron microscope.

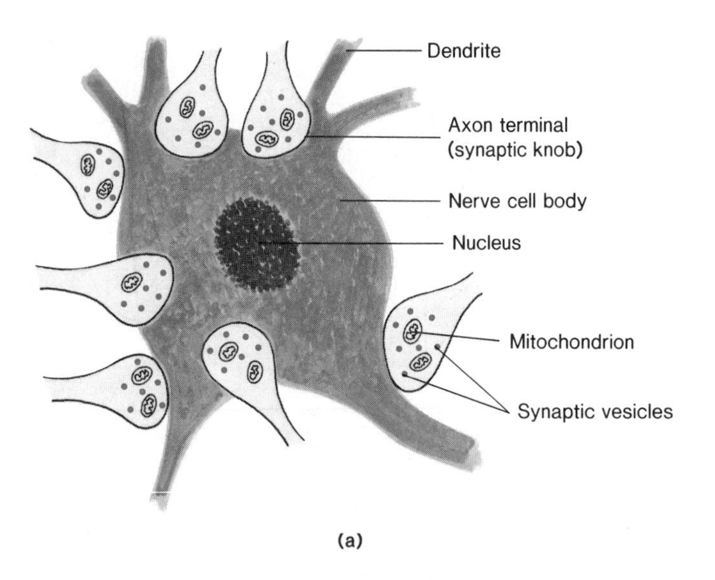

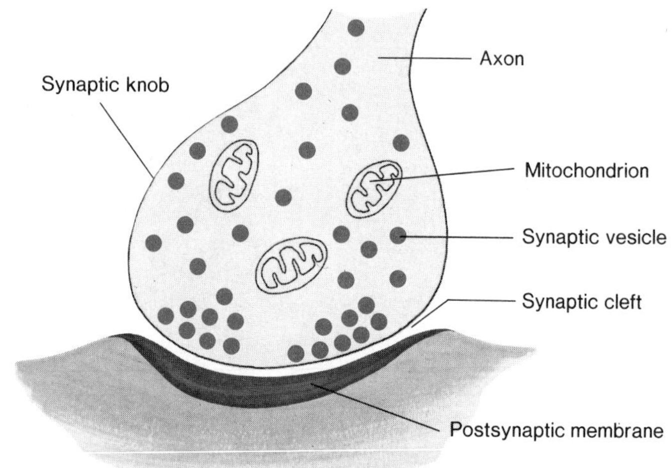

Figure 12.13 Synaptic transmission. (*a*) Electron micrograph showing synaptic vesicles apparently opening into the synaptic cleft (magnification × 104,000). (*b*) Proposed interpretation of (*a*). A synaptic vesicle (1) fuses with the membrane of the synaptic knob (2) and releases its contents into the synaptic cleft (3). Neurotransmitter molecules fuse with receptor molecules in the postsynaptic membrane. This causes opening of ion channels, sodium ion entry, and the beginning of an action potential.

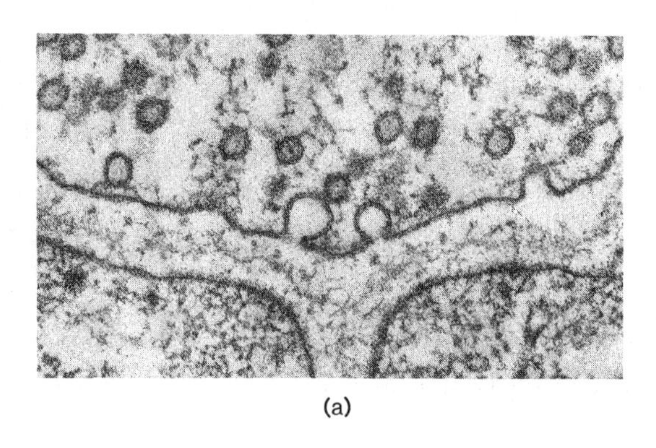

(a)

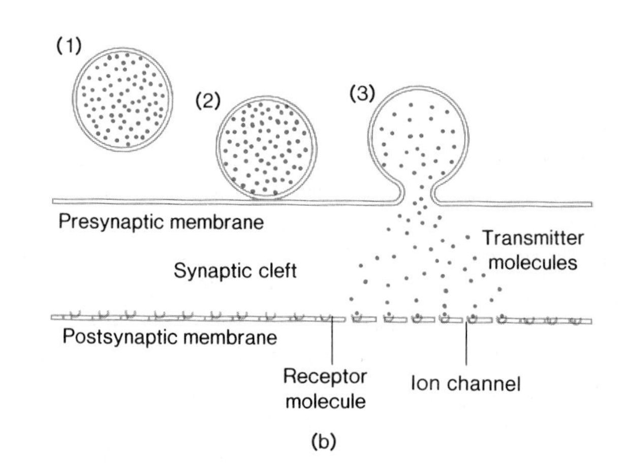

(b)

Figure 12.14 Some neurotransmitters. Glutamic acid and glycine are two of the amino acids that function as neurotransmitters. Many other substances not shown here are known or suspected to be neurotransmitters.

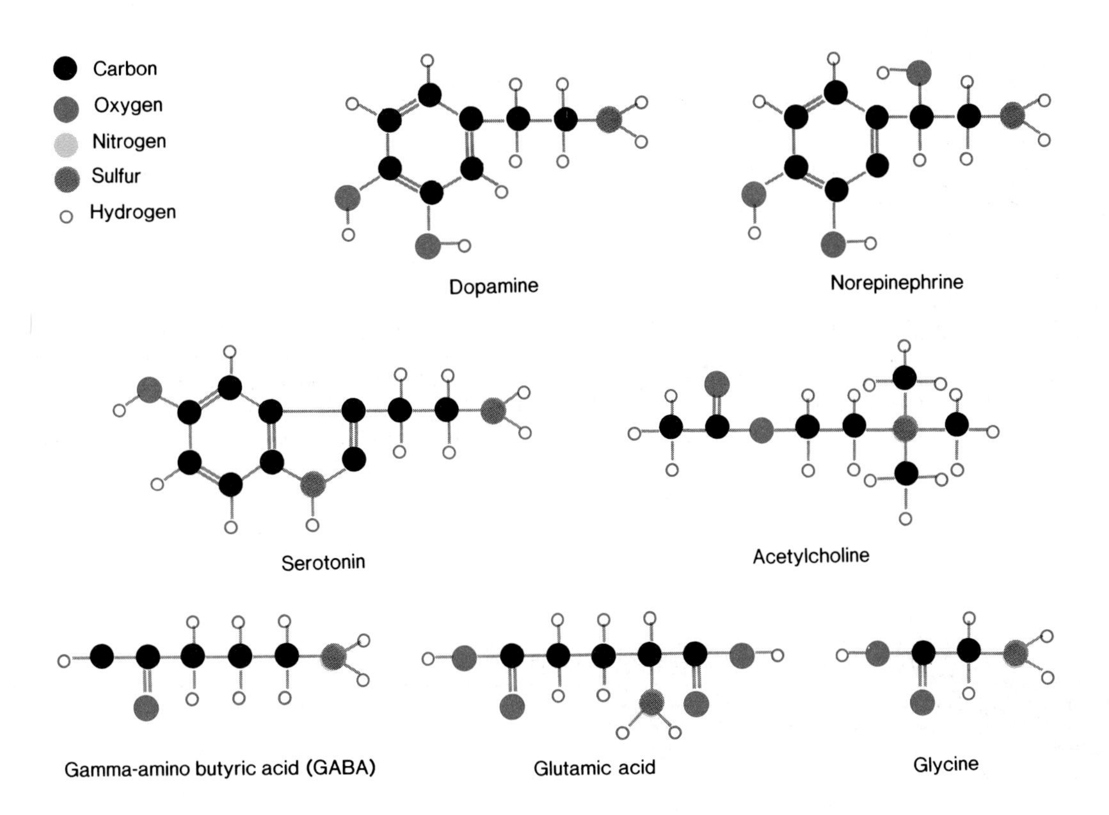

Neurotransmitters

An individual neuron uses only one kind of neurotransmitter to send signals from its axon terminal, but a number of chemicals are known to act as neurotransmitters. Thus, various neurons release different neurotransmitters. Some of the most thoroughly studied neurotransmitters include (1) **acetylcholine,** the first neurotransmitter to be discovered; (2) **dopamine** and **norepinephrine;** and (3) **gamma-amino butyric acid (GABA),** which appears to be the most abundant neurotransmitter in the brain. Several amino acids also function as neurotransmitters (figure 12.14).

Some relatively recently discovered neurotransmitters are peptides such as the enkephalins and endorphins we discussed in chapter 11 (page 250). These molecules may act as both hormones and neurotransmitters, illustrating the overlapping functions of hormones and the nervous system.

The Destruction of Neurotransmitters

Obviously, neurotransmitter molecules cannot remain permanently bound to receptor molecules, otherwise the postsynaptic neuron would not be capable of receiving subsequent messages. The neurotransmitter acetylcholine, for example, is broken down by an enzyme, **acetylcholinesterase,** to simpler chemicals that are unable to activate receptors. Other neurotransmitters are broken down by different enzymes or removed from synaptic clefts by other means so that neurotransmitters do not accumulate in synaptic clefts under normal circumstances (see box 12.1).

Box 12.1
Sacred Cactus and the Synapse

When the Spaniards invaded Mexico in the sixteenth century, they discovered rich and diverse cultures that included many practices foreign to the European mind. One of these was a religion that focused on the cactus known as peyote *(Lophophora williamsii)*. Peyote is a small, spineless cactus that has a substantial carrotlike root, but only a small exposed top (box figure 12.1).

Indians cut the peyote tops and dried them to make "buttons," which were then eaten during religious rituals. After eating the buttons, users experienced hallucinations in which they saw visions in brilliant colors and heard voices, said to be the voices of ancestors giving them advice and guidance.

The Spaniards tried to stamp out peyotism, the religion of the "devil's root," as they called it. But the religion persisted despite Spanish opposition and is, in fact, practiced to this day in some parts of Mexico. It also has spread to the United States, where it is practiced by almost a quarter of a million members of an organized religion called the Native American Church. This religion has members among more than 60 Indian tribes across much of the United States. Its followers have won the legal right to use peyote in their religious rituals. This marks one of the few cases in which the use of a hallucinogenic (hallucination-producing) substance is permitted outside of psychiatric treatment.

Peyote contains thirty or more different nitrogen-containing chemicals known as alkaloids, which cause intoxication when eaten. The major member of this group of compounds is a substance known as mescaline. Mescaline affects the action of the neurotransmitter serotonin on brain cells involved in vision and emotions, and thereby causes hallucinations.

Research on synapses and the neurotransmitters that carry messages between nerve cells has provided insights into how mescaline and other psychoactive drugs affect the nervous system. Some drugs interfere with receptor molecules, while some cause changes in synthesis of a transmitter. Others affect packaging or release of transmitters, while still others offset the breakdown of neurotransmitters or their removal from the synaptic cleft.

Tranquilizers such as valium apparently interact with GABA receptors. Stimulants such as cocaine and amphetamines block the removal of certain transmitters, leaving them in the synaptic cleft. Hallucinogens such as LSD interact with the receptors for the neurotransmitter serotonin. The list could go on, but the point is that understanding the synapse enables biologists to understand better the actions of many of these drugs. Some of the drugs that act on synapses may provide new tools for treating neurological disorders and perhaps even various types of mental illnesses.

Already, Parkinson's disease, a serious disease of the central nervous system, is treated with L-DOPA, a precursor of the transmitter dopamine. Myasthenia gravis, a disorder characterized by skeletal muscle weakness and fatigue, involves a decrease in acetylcholine receptors and is treated by inhibiting acetylcholinesterase. Schizophrenia is treated by drugs that block dopamine receptors.

We hope that with increased understanding of the functions of neurotransmitters and receptors, and of the substances that affect them, we can do even more in the fight against disorders of the central nervous system.

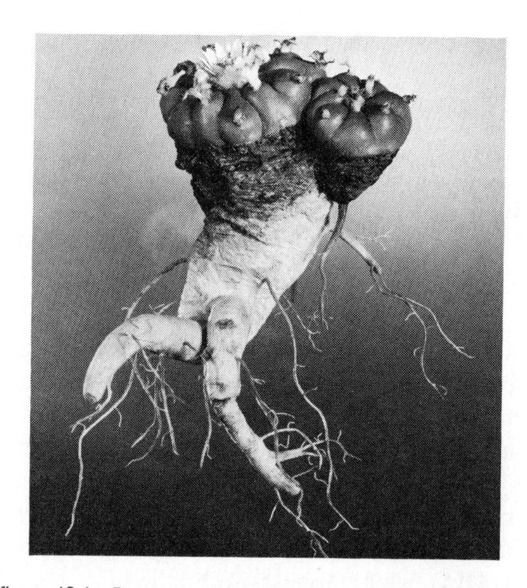

Box figure 12.1 Peyote (*Lophophora williamsii*). The small, spineless, above ground portion of the peyote cactus is attached to a substantial carrotlike root. The above ground parts are harvested and dried to produce "buttons."

Figure 12.15 The pain withdrawal reflex involves three neurons and two synapses. Note the interneuron within the spinal cord. The flexor muscle bends the forearm, withdrawing the hand from the painful stimulus.

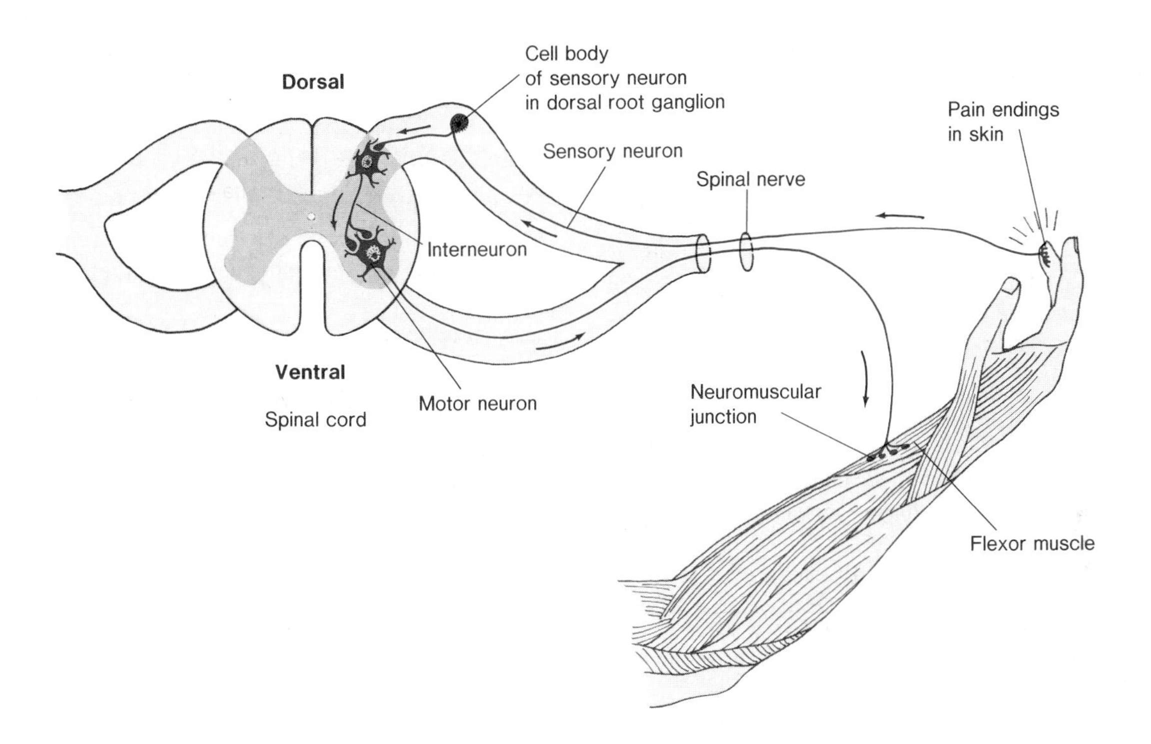

Reflexes

The functioning of a nervous system depends upon messages being passed from neuron to neuron through functional neural pathways. The simplest neural pathways are those involved in **reflexes,** which are automatic responses to specific stimuli. The messages involved in a reflex travel over a pathway called a **reflex arc.** A reflex arc links a receptor, which senses a stimulus, and an **effector,** which carries out an appropriate response to the stimulus. Some of the simplest reflexes (but most important for survival) involve only the spinal cord and peripheral nerves, not the brain.

Almost everyone is familiar with the pain withdrawal reflex. If you were to touch a hot stove, you would automatically flex your forearm, withdrawing your hand from the hot surface. This happens in a fraction of a second and does not require a conscious decision by your brain. Pain receptors in your skin are stimulated, and the message is relayed by sensory neurons to the spinal cord. There interneurons transmit impulses to the appropriate motor neurons, and then impulses travel over these motor neurons to muscles. The proper muscles contract, your arm flexes, and your hand is withdrawn from the stove (figure 12.15).

There are many other examples of reflexes involving special sensory receptors, sensory neurons, interneurons, motor neurons, and skeletal muscles acting as effectors. But let us turn our attention to another type of reflex, one that involves the internal organs of the body.

The Autonomic Nervous System

The skeletal muscles involved in the reflexes we just described are under the control of the **somatic** motor division of the nervous system. Somatic motor activities can generally be controlled voluntarily (for example, you can flex your arm however you wish, just by deciding to do so). The motor activities of the **autonomic nervous system,** however, are not normally under voluntary control. The autonomic nervous system controls reflexes involving internal organs. These are called **visceral reflexes** (figure 12.16), and the effectors are glands or the involuntary muscles of organs.

The sensory neurons that function in visceral reflexes enter the spinal cord on the dorsal side, and the nerve cell bodies of the sensory neurons are located in **dorsal root ganglia.** The difference between somatic motor innervation and autonomic innervation is in the motor neurons of each system.

Somatic motor control depends on the axons of *single* motor neurons that reach from the central nervous system to the muscles being stimulated. But autonomic innervation involves *two motor neurons arranged in sequence* to transmit stimulation from the central nervous system to the effector. These two cells are called the **preganglionic motor neuron** and **postganglionic motor neuron,** respectively. The axon of the preganglionic motor neuron leaves the CNS, but it synapses with the postganglionic motor neuron in an **autonomic ganglion.** The axon of this second motor neuron reaches to the effector.

The Parasympathetic and Sympathetic Systems

We can divide a mammal's autonomic nervous system into two parts on the basis of structure and function: the **parasympathetic** system and the **sympathetic** system (figure 12.17).

Many organs are innervated by both sympathetic and parasympathetic neurons (figure 12.18). But, in most cases, sympathetic and parasympathetic neurons act antagonistically—one set of neurons *stimulates* the organ, and the other *inhibits* it (table 12.2). For example, sympathetic nerves

Figure 12.16 A visceral reflex. Sensory information from the intestine is carried to the spinal cord via a sensory neuron. In the spinal cord it synapses with the first motor neuron (the preganglionic motor neuron). This, in turn, synapses in an autonomic ganglion with the postganglionic motor neuron, which goes to the muscle in the wall of the intestine.

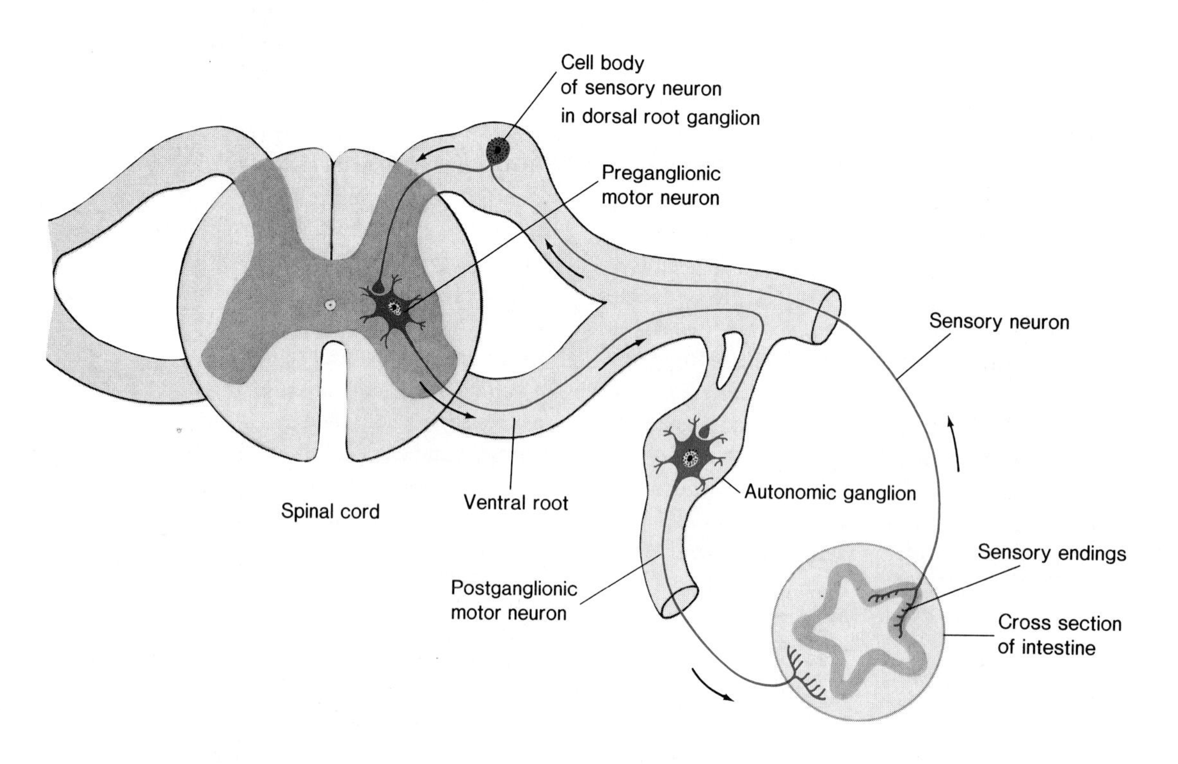

Figure 12.17 Motor elements of the nervous system.

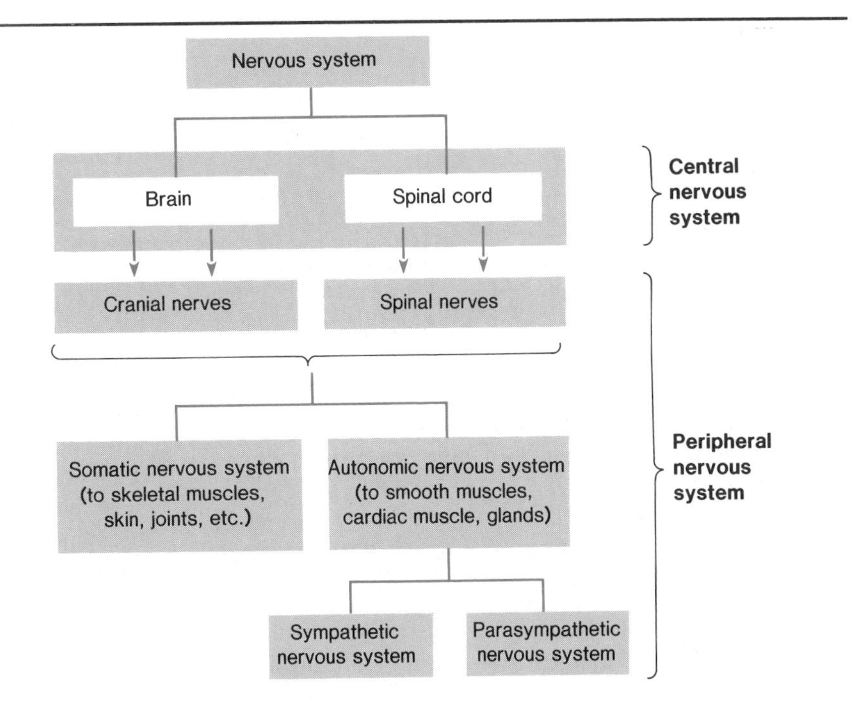

Figure 12.18 The sympathetic and parasympathetic divisions of the autonomic nervous system and the organs they innervate.

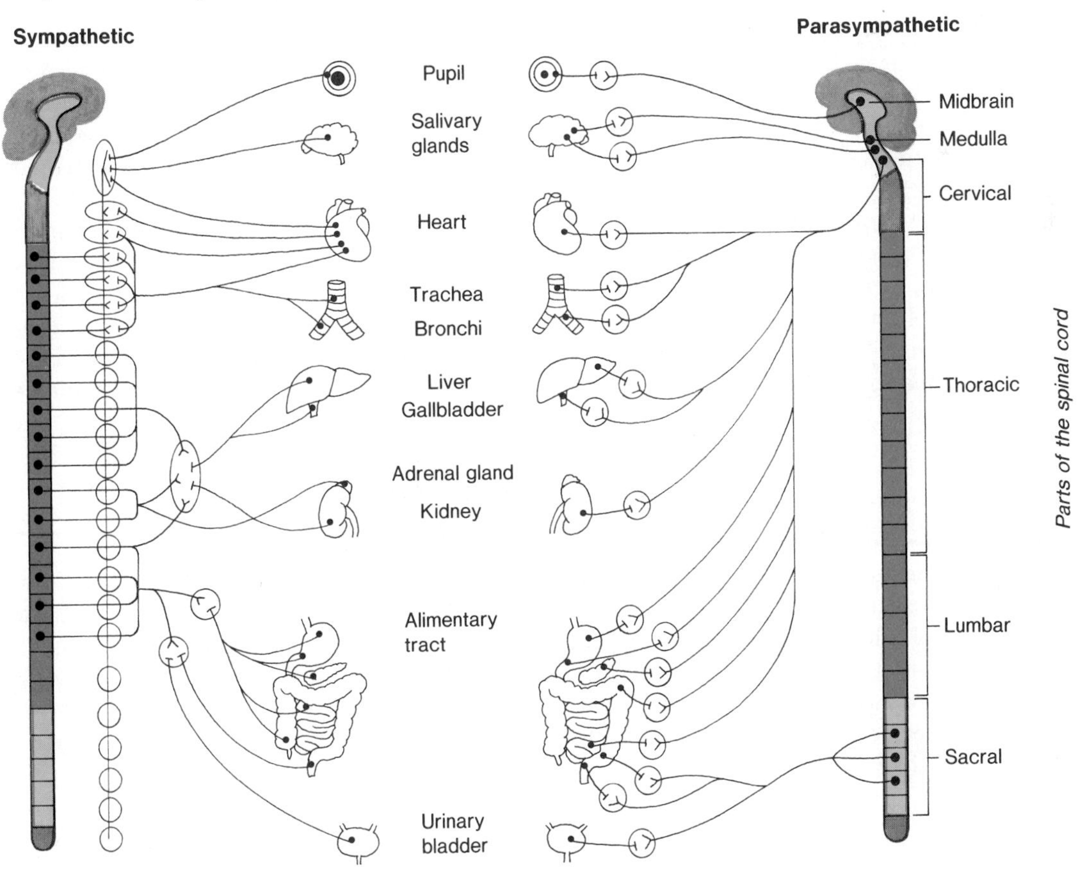

Table 12.2
Functions of the Autonomic Nervous System*

Sympathetic Division	Parasympathetic Division
Dilates pupils of the eyes	Constricts pupils
Increases heart rate	Decreases heart rate
Increases strength of heart contractions	Decreases strength of heart contractions
Dilates bronchioles of respiratory system	Constricts bronchioles
Decreases peristalsis of digestive tract	Increases peristalsis
Probably inhibits the secretion of digestive enzymes	Stimulates the secretion of digestive enzymes
Relaxes urinary bladder	Contracts urinary bladder
Constricts blood vessels to genital organs	Dilates blood vessels to genital organs (causes erection)

*This list is not exhaustive. There are several other effects, particularly on the blood supply to various organs.

stimulate the heart to beat faster and stronger. Parasympathetic nerves cause the heart to contract more slowly and with less strength. On the other hand, the contractions (peristalsis) of the muscles in the digestive tract are speeded up by the parasympathetic system and slowed or even stopped by the sympathetic system.

The autonomic nervous system is essential to the maintenance of homeostasis, the dynamic, steady state of the organism's internal environment. In general, the parasympathetic system stimulates everyday activities and functions (digestion, urination, and so on). Generally, the sympathetic system inhibits these activities. On the other hand, the sympathetic system promotes the ability of the body to meet emergencies, often called the "fight-or-flight" response. Stimulation by the sympathetic system increases the heart rate (pumping more blood), dilates the pupil of the eye (letting in more light), dilates the bronchioles of the respiratory tract (increasing gas-exchange efficiency), and inhibits digestion, which is not immediately essential during an emergency.

The Brain and Neural Reflexes

Although many somatic reflexes involve only the spinal cord, you do become conscious of what has happened to your body after the reflex has occurred. This is because other interneurons have transmitted the information to the areas of your brain that are responsible for awareness. On the other hand, while many visceral reflexes actually involve the brain directly, we are not consciously aware of them.

But the brain's functions go far beyond simply becoming aware of sensory information coming in from receptors and performing reflex responses. The brain is the seat of many complex functions of the nervous system, including consciousness, thought, learning, memory, and emotions.

The Central Nervous System

Of all scientific endeavors, one of the most fascinating is the study of the structure and function of the central nervous system. This awesome cellular circuitry, with billions of interconnected neurons, regulates the body's internal environment and allows meaningful interactions with the outside world.

The central nervous system is made up of the brain and the spinal cord. These two structures are actually continuous tissue, not separate organs. Together they look like a thick-walled tube that expands at one end. The brain is enclosed and protected by the bony cranium (skull). The spinal cord, which extends from the base of the skull, is protected by the vertebral column (backbone) (figure 12.19).

The Spinal Cord

Functionally, the spinal cord and brain play very different roles. The brain carries out the most complex functions of the central nervous system. The spinal cord directs certain reflex responses and acts as a liaison between many elements of the peripheral nervous system and the brain. It brings sensory information to the brain and carries motor commands from it.

Spinal nerves containing sensory and motor fibers are attached to the spinal cord. In humans there are thirty-one pairs of spinal nerves. Near the spinal cord, each spinal nerve splits into two branches—a **dorsal root** and a **ventral root,** carrying sensory and motor information, respectively. There also are twelve pairs of **cranial nerves** connected directly to the brain.

The spinal nerve connections and the internal structure of the spinal cord are shown in figure 12.20. The interior of the spinal cord is divided into **gray matter** and **white matter.** The center of the cord is gray because it is composed mostly of nerve cell bodies. The surrounding area is white because it is made up mainly of nerve cell fibers, many of which are enclosed in fat-containing sheaths. (Fat has a white or light yellow color.)

Figure 12.19 The human brain encased in the bony cranium and the spinal cord encased in the vertebral column. Note the central canal in the spinal cord and ventricles, which are cavities inside the brain. Both the spinal cord's central canal and the brain's ventricles are filled with cerebrospinal fluid.

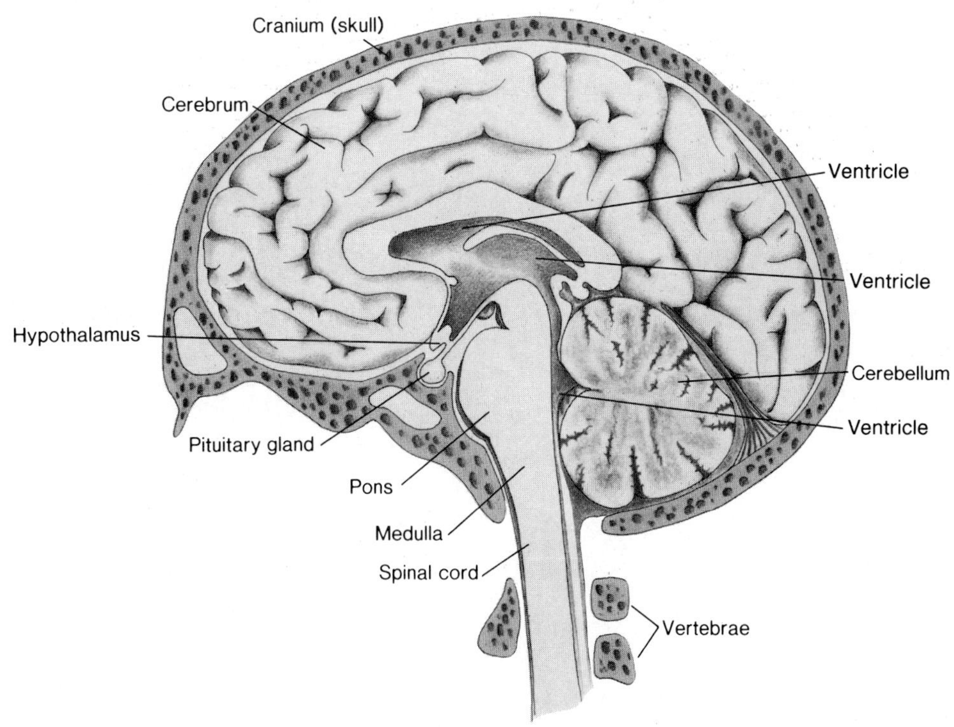

Figure 12.20 Interior of the spinal cord showing white and gray matter. Note also the spinal nerve, dorsal root, and ventral root. Arrows indicate directions of nerve impulse conduction—toward the spinal cord via sensory fibers and away from the spinal cord via motor axons.

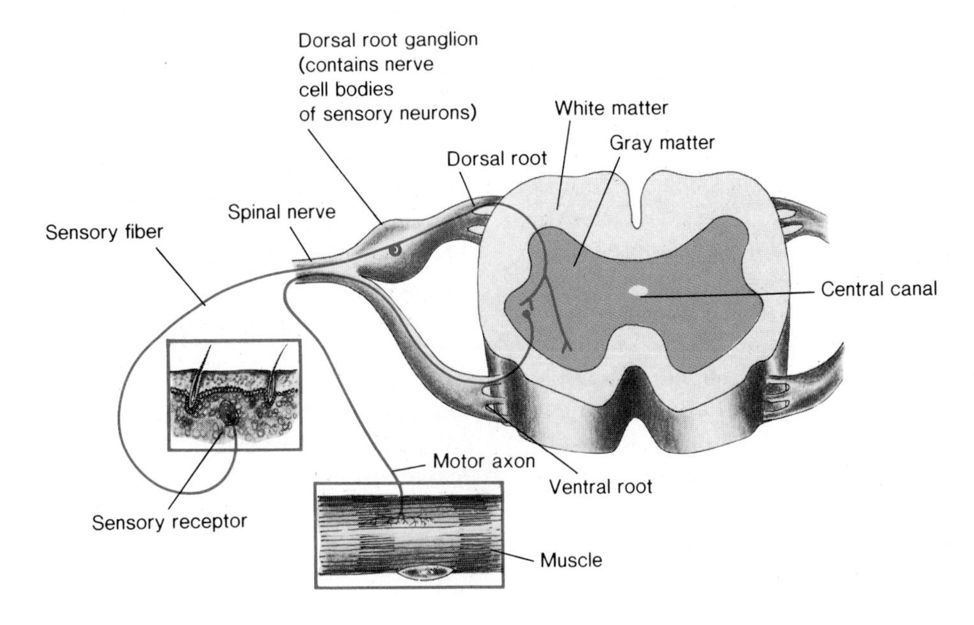

Figure 12.21 Organization of the brain. (*a*) The major brain regions—forebrain, midbrain, and hindbrain—as they appear early in the development of a vertebrate embryo. (*b*) Structures present in the major brain regions of the fully developed adult human brain. The hindbrain is composed of medulla, pons, and cerebellum. The tectum and tegmentum comprise the midbrain. The forebrain is composed of hypothalamus, thalamus, and cerebrum.

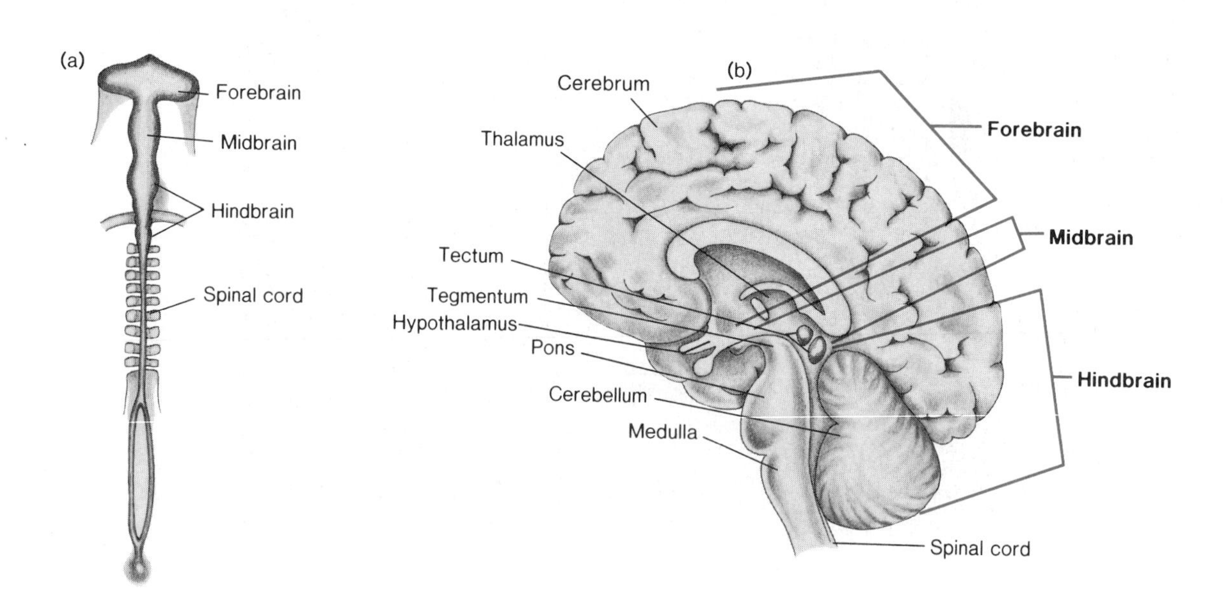

In the very center of the spinal cord is a hollow canal, running throughout the central nervous system. In the brain, the canal expands into cavities called **ventricles** (see figure 12.19). The central canal of the spinal cord and the ventricles are filled with **cerebrospinal fluid.** This fluid carries nutrients and chemical signals and also protects the central nervous system from damage by providing an inner cushion.

The Brain

Even without the aid of a microscope, it is obvious that the brain is a highly complex structure. It has characteristic grooves and ridges on its surface and is divided into several clearly distinguishable parts. Let us examine each of these parts briefly and consider their contributions to the overall functioning of the brain.

The major brain parts are separated into regions or groups on the basis of their development. Early in the life of a vertebrate embryo, three swellings arise in the developing brain region. These embryonic divisions form the basis for naming the general brain regions: **hindbrain, midbrain,** and **forebrain** (figure 12.21). The hindbrain is composed of the medulla, pons, and cerebellum; the midbrain is composed of the tectum and tegmentum; and the forebrain includes the hypothalamus, thalamus, and cerebral cortex.

The Hindbrain

The first structure within the brain, beginning just above the spinal cord, is the **medulla.** In addition to relaying signals to and from the forebrain, neurons in the medulla play an important role in control of the heart rate, digestion, and breathing—basic functions that are essential to life. In fact, life can often continue following damage to other brain regions, but damage to the medulla usually causes death very quickly.

The **pons** is the next structure upward. It functions as a relay center connecting various parts of the nervous system. Many fibers pass through it into the cerebellum, the third major part of the hindbrain.

Figure 12.22 A photograph of a human brain showing the many ridges and grooves on the surface of the cerebrum. The cerebellum is visible in the lower right portion of the photo.

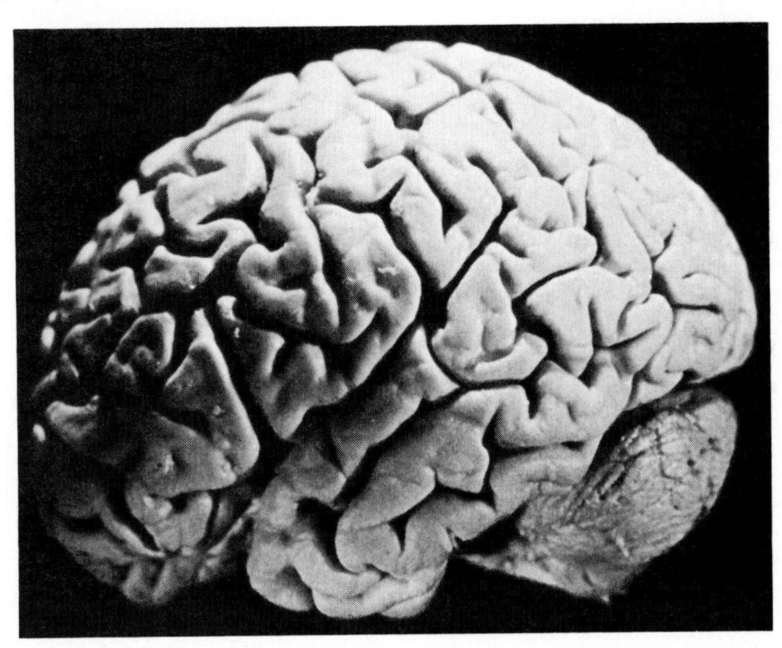

The **cerebellum** is especially important in the control of movement. It is a bulblike structure extending upwards from the pons and contains two halves or **hemispheres.** The cortex, or outer portion, of the cerebellum is gray matter, which means that nerve cell bodies are located close to its outer surface. The cerebellum is one of only three regions of the brain that have a gray cortex. In all other parts of the brain, nerve cell fibers lie near the outside and nerve cell bodies are confined to the interior, as in the spinal cord.

The cerebellum is responsible for fine tuning muscle activity and is especially important in controlling rapid movements. The cerebellum functions like a computer to keep track of movements initiated by the forebrain. It compares actual movement with desired movements and provides correctional signals as needed. If the cerebellum is damaged, jerky body movements result because the corrections necessary for coordination are no longer made.

The Midbrain

The midbrain is a small part of the central nervous system that contains the **tegmentum,** a relay center connecting the hindbrain region with the forebrain region, and the **tectum.** The tectum, which has a gray cortex, receives sensory information from the eyes and ears. All auditory (hearing) information enters the tectum before being sent to the cerebrum where sensory information about sound is interpreted.

The Forebrain

The forebrain, the largest part of the brain in humans and other mammals, is responsible for many important functions. The **hypothalamus,** the **thalamus,** and the **cerebrum** are the major parts of the forebrain.

Although the hypothalamus is small, weighing only about 4 grams in humans, it fulfills several major functions. The hypothalamus regulates the autonomic nervous system. It controls the release of hormones from the pituitary gland, as well as regulating hunger, thirst, osmotic balance, body temperature, metabolic rate, and circulation. The thalamus is another of the major relay centers that connect various brain regions with one another.

By far the largest and most striking part of a mammal's brain, however, is the cerebrum. In the human brain, it accounts for over half of the total mass of the brain. Its outer area, the **cerebral cortex,** is gray matter made up of billions of nerve cell bodies. It is divided into halves, the **cerebral hemispheres,** whose surfaces are covered with many ridges and grooves (figure 12.22). Within a species, the locations of the ridges and grooves are always essentially the same. Thus, the surface of the cortex can be mapped.

Figure 12.23 A map of a human cerebral hemisphere showing the approximate locations of a number of functional areas, including both sensory and motor functions.

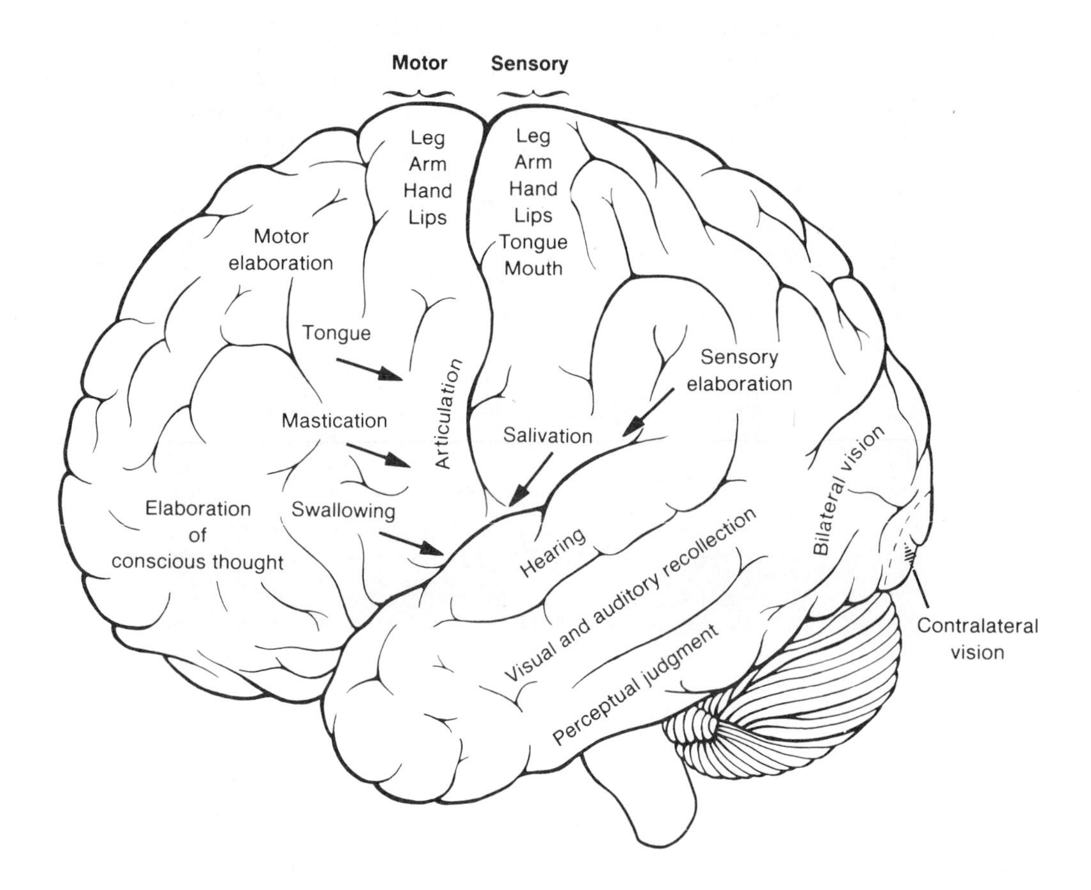

The functions of various parts of the cerebral hemispheres have been partially determined by studying neurological problems caused by damage to specific areas. In humans, it has been possible to observe people who have suffered strokes or gunshot wounds that damage small, localized areas. Furthermore, scientists have identified specific areas of the cerebral hemispheres that interpret information from various sense organs. There are regions, for example, that process visual information, auditory information, and other kinds of sensory input (figure 12.23).

These functions have been mapped only because they are fairly easy to identify. However, much of the vast complexity of the cerebral cortex is yet to be unraveled. We still have a great deal to learn about the locations of such functions as thought, learning, and memory. Considering the complexity of the cerebral cortex, with its astronomically large numbers of nerve cells and interconnections among the cells, it is clear that further building of our understanding of the cerebral cortex will require much painstaking research. But such research seems well worthwhile because, at least in the case of humans, the neural functions that interest us most, especially those functions that distinguish us from other animals, reside in the cerebral cortex.

Summary

The nervous system provides a means of rapid communication in the animal body. It gathers information from both the external and internal environments, stores the information, and directs appropriate responses of the organism.

Neurons are the functional units of any nervous system. Neurons receive inputs, respond with electrical changes, and generate outputs to other cells. Many neurons have long fibrous processes along which nerve impulses are conducted.

Sensory neurons carry information toward the central nervous system, while motor neurons carry information away from the CNS toward effector cells. Interneurons are intermediate components involved in information processing.

Unequal concentrations of charged particles (ions) inside and outside a neuron's membrane result in a charge difference (electrical potential) across the membrane. The resting membrane potential exists because of the selective permeability of the membrane and the action of the sodium-potassium pump.

During an action potential, the membrane becomes much more permeable to sodium, which rushes inward along its concentration gradient. This results in a temporary reversal of the membrane's polarity. Then an outward flow of potassium restores the normal resting potential.

A nerve impulse involves a chain reaction series of action potentials sweeping along the neuron's plasma membrane.

Synapses transmit information from neuron to neuron. In chemical synapses, neurotransmitters carry a message across the synaptic cleft.

Neurotransmitter molecules are released from synaptic vesicles. They then diffuse across the synaptic cleft and attach to receptor molecules in the postsynaptic membrane, where they either promote or inhibit action potential occurrence, depending upon the type of synapse.

The simplest neural pathways are those involved in reflexes, which are automatic responses to stimuli. Visceral reflexes occur below the level of conscious control and operate by autonomic innervation of internal effectors.

Autonomic innervation requires two motor neurons arranged in sequence. The parasympathetic and sympathetic nervous systems usually act antagonistically; one stimulates a given process or activity while the other inhibits it.

The central nervous system is made up of the brain and the spinal cord. The nerves of the peripheral nervous system connect to these two areas. Nerves contain sensory and motor neuron fibers.

The spinal cord has gray matter, composed mostly of nerve cell bodies, in its center, and white matter, made up of nerve cell fibers, surrounds the gray matter. The ventricles in the brain and the hollow central canal in the spinal cord contain cerebrospinal fluid.

The forebrain, the midbrain, and the hindbrain are the three major regions in the embryonic brain. Each region produces a specific set of adult brain structures.

The cerebral cortex with its two cerebral hemispheres is a forebrain derivative. It contains large quantities of gray cortex material arranged in ridges and grooves. Specific regions of the cerebral hemispheres are identified with specific functions.

Questions for Review

1. Define the terms central nervous system and peripheral nervous system.
2. What are interneurons?
3. Explain the electrical and chemical changes that occur during an action potential.
4. Why is a nerve impulse described as a chain reaction series of action potentials?
5. What is a refractory period?
6. Explain the difference between gray matter and white matter.
7. People who have suffered severe damage to a large portion of the cerebral cortex are mentally impaired, but they often remain alive. Significant damage to the brain stem, however, especially to the medulla, almost always causes death in a very short time. What aspects of brain function explain this difference?
8. List three areas of the brain that have a gray cortex (nerve cell bodies at the surface).
9. Name three derivatives of the embryonic forebrain and make a general statement about the function of each.

Questions for Analysis and Discussion

1. There are drugs that inhibit the function of the enzyme acetylcholinesterase. Can you explain why these drugs cause widespread problems in functioning of the nervous system?
2. Can you explain how you withdraw your hand from a hot surface well before you become consciously aware of the pain?
3. Make some suggestions regarding the evolutionary significance of ridges and grooves in the cerebral cortex of mammal's brains, especially human brains.

Suggested Readings

Books

Kuffler, S. W.; Nicholls, J. G.; and Martin, A. R. 1984. *From Neuron to Brain.* Sunderland, Mass.: Sinaur Associates, Inc.

Shepherd, G. H. 1983. *Neurobiology.* New York: Oxford University Press.

Articles

Adrian, R. H. 1980. The nerve impulse. *Carolina Biology Readers* no. 67. Burlington, N.C.: Carolina Biological Supply Co.

Gray, E. G. 1977. The synapse. *Carolina Biology Readers* no. 35. Burlington, N.C.: Carolina Biological Supply Co.

Iversen, L. I. September 1979. The chemistry of the brain. *Scientific American* (offprint 1441).

Nauta, W. J. H., and Feirtag, M. September 1979. The organization of the brain. *Scientific American* (offprint 1439).

Reception and Response

Chapter Outline

Chapter Concepts

1. Sensory receptors respond to environmental changes and their responses cause nerve impulses to be transmitted to the central nervous system.
2. Each type of sensory receptor responds best to a specific kind of stimulus.
3. Sensory information is interpreted by the central nervous system and appropriate responses are initiated through effectors.
4. Skeletal muscles are effectors that produce limb movements and other whole body movements.
5. Body movement requires interaction of muscular and skeletal elements.
6. Muscle contraction is produced by the energy-requiring interaction of sliding filaments containing specialized contractile proteins.
7. During vigorous muscle activity, skeletal muscles use fermentation to supplement aerobic respiration in the production of ATP, which is necessary for muscle contraction.

You may recall a physical examination during which the doctor gave you a sharp tap just below the kneecap with a small mallet. Immediately your leg swung out, extending in response to the tap. This response is very fast—it occurs in about 50 msec. The kneejerk response is a **reflex** similar to those mentioned in chapter 12. This reflex involves only two neurons, a sensory neuron and a motor neuron, and the synapse between them (figure 13.1).

The kneejerk reflex is one of many **stretch reflexes.** Receptors in muscles detect stretching and transmit impulses to the spinal cord. There, neurons respond by sending impulses to muscles causing them to contract. This pattern of sensing by receptors followed by muscles responding to the sensory information goes on all the time without conscious awareness, helping to maintain our posture and body position.

A variety of other kinds of receptors also gather information about conditions in the body and in the environment and feed it into the central nervous system. The nervous system processes the information and directs proper responses by effectors (responding structures). In this chapter we will examine some receptors and the ways in which they gather and transmit information. Then we will turn our attention to an important group of effectors, the muscles.

Receptors and Senses

In animals, specialized **receptor cells** sense conditions in the environment. These cells respond to specific kinds of environmental stimuli. Some cells are very sensitive to changes in temperature, while others respond to mechanical energy (movement), light energy, or chemicals.

Each type of receptor cell is sensitive to one type of stimulus and responds best to that particular stimulus. However, receptor cells can respond to stimuli other than the one to which they are most sensitive. For example, the flashes of light that you see when your eyes are rubbed hard (or punched!) are caused when receptor cells that normally respond to light energy respond instead to the mechanical energy of pressure on the eye. Light receptors are very sensitive to light, but it takes a great deal of pressure to cause this light flash response.

The same is true of all receptor cells. They are specialized to be very sensitive and responsive to one kind of stimulus and are much less responsive to others. Receptors usually are classified by the type of stimulus to which they respond. **Photoreceptors,** for example, respond to light. Some **thermal receptors** respond to heat, others to cold. **Mechanoreceptors,** such as touch receptors in the skin or the receptors in the ear

Figure 13.1 A stretch reflex. Note the sensory neuron coming into the dorsal side of the spinal cord. It synapses inside the spinal cord with the nerve cell body of the motor neuron. The axon of the motor neuron leaves the spinal cord on the ventral side and goes to the skeletal muscle. Neuromuscular junctions are the contacts between the axon of a motor neuron and the muscle cells that it controls.

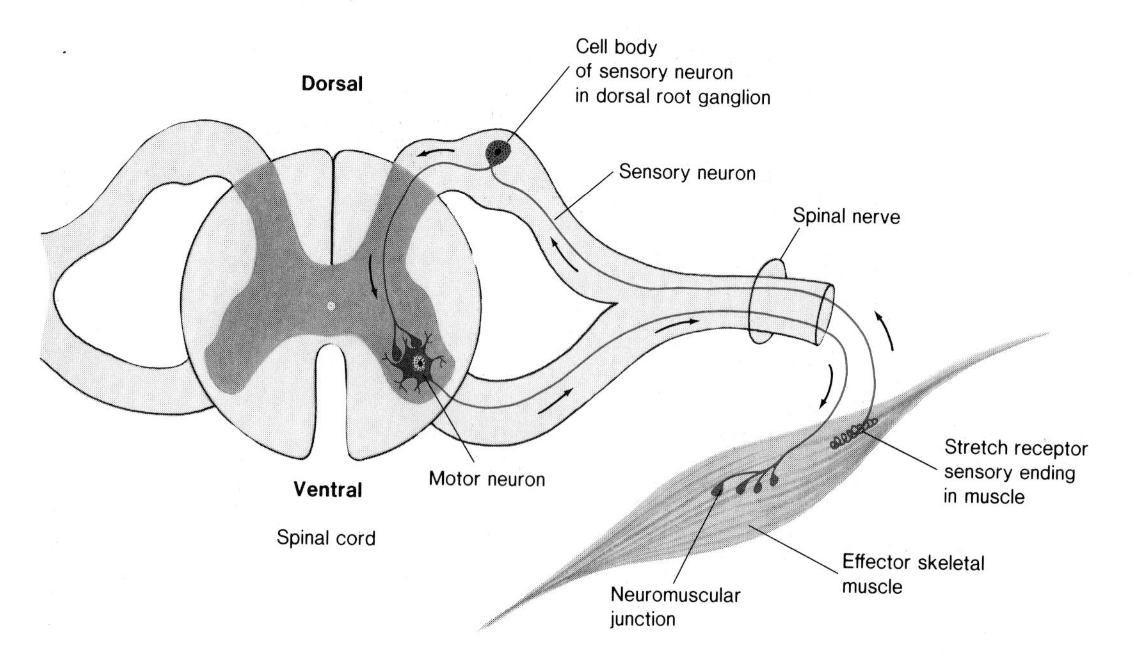

that pick up vibrations in the air, respond to mechanical (movement) stimuli. **Chemoreceptors** in the tongue and the nasal passages are fundamental to the senses of taste and smell.

Most people are familiar with the standard list of five basic senses: touch, smell, taste, hearing, and sight. But as you can see in table 13.1, human beings have many more than five kinds of receptors. Both internal and external environments are monitored, some at a conscious level and others below the level of consciousness.

How Receptors Function

There are two basic categories of receptors, based on the specific responding cell that detects the stimulus. When a sensory neuron itself acts as a receptor cell, it is called a **primary receptor.** One type of primary receptor in humans is a pain receptor. Pain receptors are simply the ends of fibers of certain sensory neurons.

But there are other cases in which there is another specialized type of receptor cell *in addition to* a sensory neuron. These receptor cells function in cooperation with sensory neurons and are called **secondary receptors.** We will see that secondary receptors are important in the sensory functions of the eye, the ear, and in taste buds.

Several general properties apply to essentially all sensory functions. The response of sensory receptors can vary with the **intensity** of the stimulus. When a receptor is stimulated more strongly, nerve impulses are transmitted to the brain more frequently. The frequency of impulses is a code that the brain uses to determine the strength and importance of the stimulus.

A second general property concerns interpretation of sensory information. It is the brain that interprets the stimulus originally detected by a receptor. In the strictest sense, you do not see with your eyes, you "see" when specific areas of the brain receive information about responses to light energy by receptor cells in the eyes. The brain interprets information coming from those receptors in only one way, as light. Remember that a punch in the eye can be "seen" as a flash of light! Thus, receptors *detect,* but the brain *interprets.*

Another interesting property of sensory reception is **sensory adaptation.** In sensory adaptation, the response of receptor cells declines, even though the stimulus is sustained at the same level of intensity. Some receptors adapt relatively quickly, while others adapt quite slowly. For example, touch receptors of the skin adapt quickly. This is fortunate if you do not wish to be continually reminded of the shirt on your back or the rings on your fingers. Pain receptors, however, adapt quite slowly. Slow adaptation to pain is an important safety device. If pain were easy to ignore, its function as a warning system would be much less effective.

Table 13.1
Types of Sensory Receptors

Sense	Receptor	Location
Vision	Rods and cones	Retina of the eye
Hearing	Hair cells	Organ of Corti in the inner ear
Smell	Olfactory neurons	Lining of the nose
Taste	Taste receptor cells	Taste buds in the mouth
Rotational movement	Hair cells	Semicircular canals of the inner ear
Touch	Nerve endings	Skin
Heat	Nerve endings	Skin
Cold	Nerve endings	Skin
Pain	Naked nerve endings	Throughout body
Joint position and movement	Nerve endings	Joint
Muscle length	Muscle spindle	Muscles
Muscle tension	Golgi tendon organ	Tendons
Arterial blood pressure	Stretch receptors	Carotid sinus and aortic arch
Venous blood pressure	Stretch receptors	Venae cavae and right atrium
Inflation of lung	Stretch receptors	Lung tissue
Internal body temperature	Nerve endings	Hypothalamus of the brain
Osmotic pressure of blood	Osmoreceptor cells	Hypothalamus
Blood sugar level	Glucostatic cells	Hypothalamus
Blood oxygen levels	Chemoreceptor cells	Carotid and aortic arteries

Figure 13.2 Sensory receptors in the skin for touch, pressure, warmth, cold, and pain. All are ends of sensory neuron fibers. Only part of each fiber is shown, but these fibers actually are very long as they extend all the way to cell bodies in dorsal root ganglia near the spinal cord.

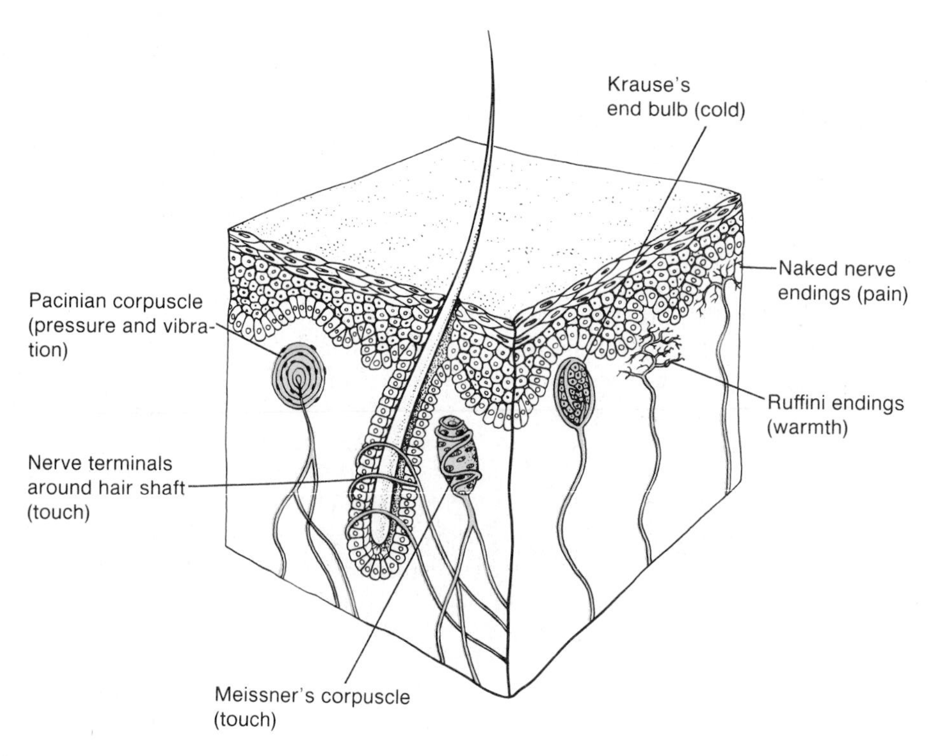

Sensory Receptors in the Skin

The skin has receptors for a number of senses, including touch-pressure, pain, warmth, and cold. Sensory receptors of the skin actually are the ends of nerve cell processes. These sensory neuron ends may be encapsulated by specialized structures, or they may be naked nerve endings (figure 13.2).

Smell

The sense of smell, or **olfaction,** employs chemoreceptor cells, cells that are responsive to specific chemical substances. These cells are located in part of the lining of the nasal cavity inside the nose. Receptor cells are mixed in among other cells, some of which secrete mucus, a liquid important in the sensory functioning of these receptors. The receptor cells have little extensions, resembling cilia, that project into the layer of mucus on the surface of the membrane (figure 13.3).

Various chemicals from the air dissolve in the mucus and stimulate the receptor cells. These cells synapse with other nerve cells that form part of the olfactory nerve connecting to the brain. Thus, stimulation of the receptor cells by chemicals results in nerve impulses being conducted to olfactory regions in the brain where they are interpreted as odors.

The sense of smell is much more important in many other mammals than it is in humans, and correspondingly, the olfactory regions comprise a much larger fraction of the total brain mass of these mammals. Still, the human sense of smell is quite acute and can be made even more so with practice and training. For example, trained perfumers can discriminate among hundreds of fragrances, some of which are very similar to one another.

Figure 13.3 The olfactory receptors located in the lining of the upper nasal passages. Special structures known as olfactory cilia extend into the mucus that covers the lining. The cilia are the sites of response to chemical stimuli. The olfactory bulb is part of the brain.

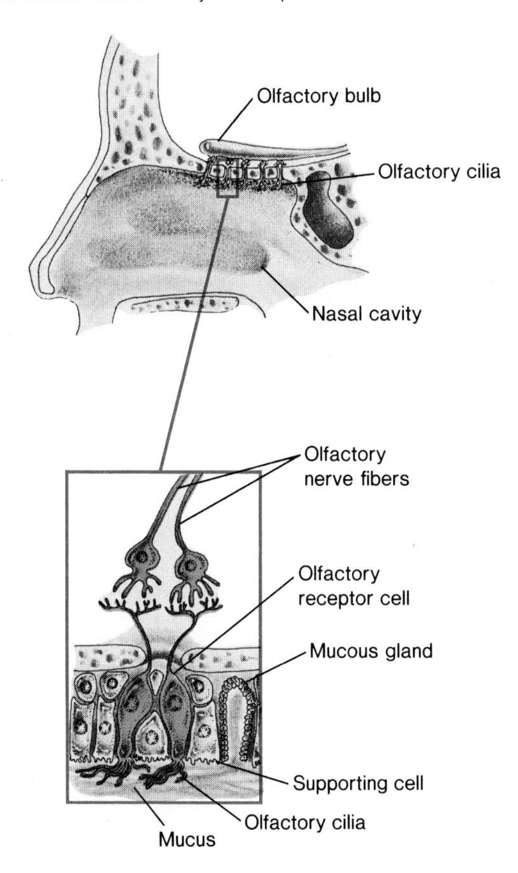

Taste

Taste "buds" in the tongue contain chemoreceptors specialized for registering sensations of sweet, salt, bitter, and sour. However, the sense of taste is neither so sensitive nor so diverse as the sense of smell. Indeed, a great deal of the "taste" of food actually comes from olfactory sensations received during eating. The sense of taste is dulled when smell is impaired, as you probably have experienced while suffering from a common cold.

The structure of a taste bud is shown in figure 13.4. Receptors for taste are secondary sense cells, not neurons. Chemicals in food stimulate receptor cells. Responses of these receptor cells cause neurons surrounding the receptor cells to send impulses to the brain, where the impulses are interpreted as "tastes."

Sense Organs of the Ear

The ear is considerably more complex than the other sensory structures that we have discussed so far. It houses two quite different senses, **hearing** and **balance.** The organs for these senses are found in the inner part of the ear and are associated with several fluid-filled chambers. One of these chambers is the **cochlea,** which contains the **organ of Corti,** in which specialized cells initiate the hearing response. Three other chambers, the **semicircular canals,** respond to movements of the head and play a role in the sense of balance.

Hearing

The human ear has three general areas: the outer, middle, and inner ears (figure 13.5). The outer ear, which includes the **pinna** (plural: **pinnae**) and the **auditory canal,** plays little direct role in the actual sensory functioning of the human ear, but some other animals, such as dogs, can direct their pinnae for better hearing. You may have noticed resting dogs or cats lying almost motionless, moving only their ears (pinnae) as they track the various sounds of activities going on around them.

The **eardrum (tympanic membrane)** separates the outer ear and the middle ear. Sound waves cause vibrations in this membrane. Within the middle ear, the vibrations of the eardrum are passed on to three miniature bones (**ossicles**). These three bones, commonly known as the **hammer (malleus), anvil (incus),** and **stirrup (stapes)** because of their shapes, cause the vibrations of the eardrum to be reproduced on another membrane located at the **oval window.** When the eardrum moves inward, the stirrup, connected to the oval window, also moves inward. The system is arranged so that vibrations of the large eardrum are fully transmitted to the much smaller oval window.

The middle ear is protected against rupture of the tympanic membrane, which could result from pressure differences between the middle ear and outer ear. The **eustachian tube,** which connects the middle ear cavity with the pharynx (throat), permits pressure equalization on the two sides of the membrane. Because of this passageway, for example, the discomfort from pressure changes during airplane takeoffs and landings can be relieved by yawning and swallowing.

The oval window separates the middle ear from the inner ear. Vibrations in the oval window, initiated by pressure from the stirrup, induce movement of the fluid in the cochlea. The

Figure 13.4 Approximate locations of the four primary tastes on the human tongue. The diagram of a taste bud on the right shows the receptor cells lying beneath a pore through which chemical stimuli reach the cells. Sensory neurons that synapse with the taste receptor cells carry signals to the brain.

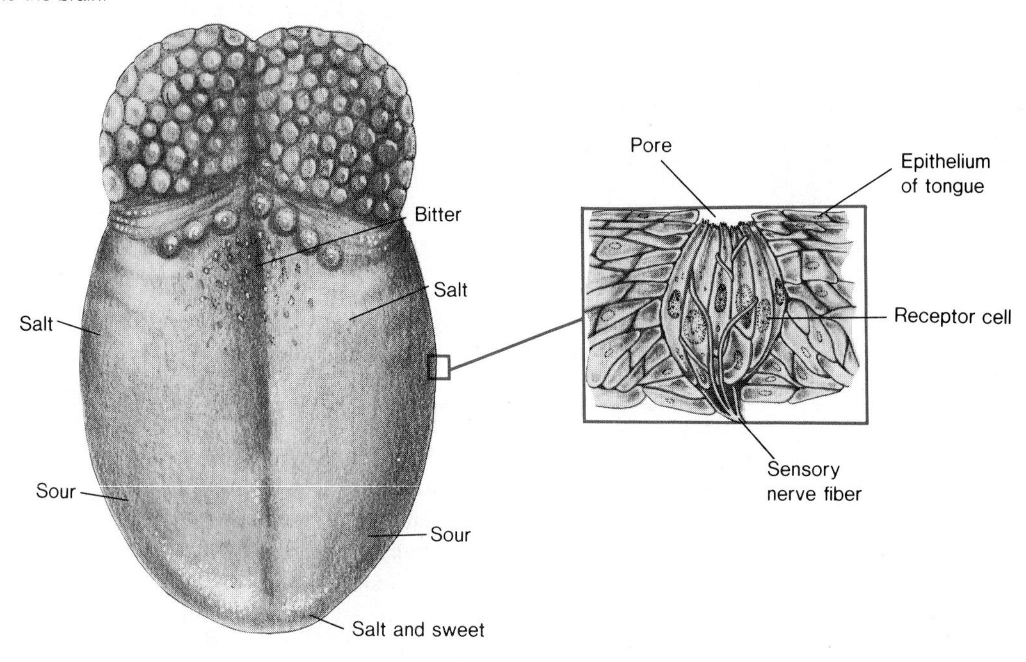

Figure 13.5 Structure of the human ear. Note the outer, middle, and inner ear. Part of the inside of the snail-shaped cochlea is drawn in detail at the right. It shows the three fluid-filled compartments and the location of the organ of Corti.

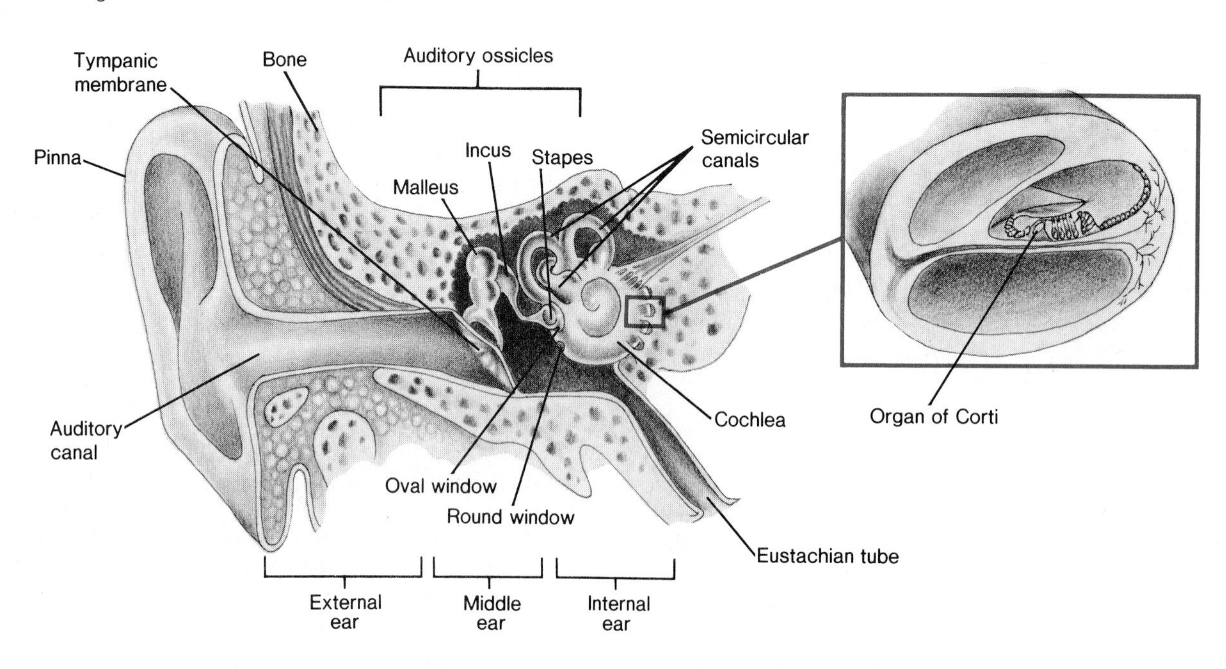

Figure 13.6 Structure of the cochlea with the organ of Corti. Note the hair cells, the sensory neurons with which they synapse, and the overlying tectorial membrane.

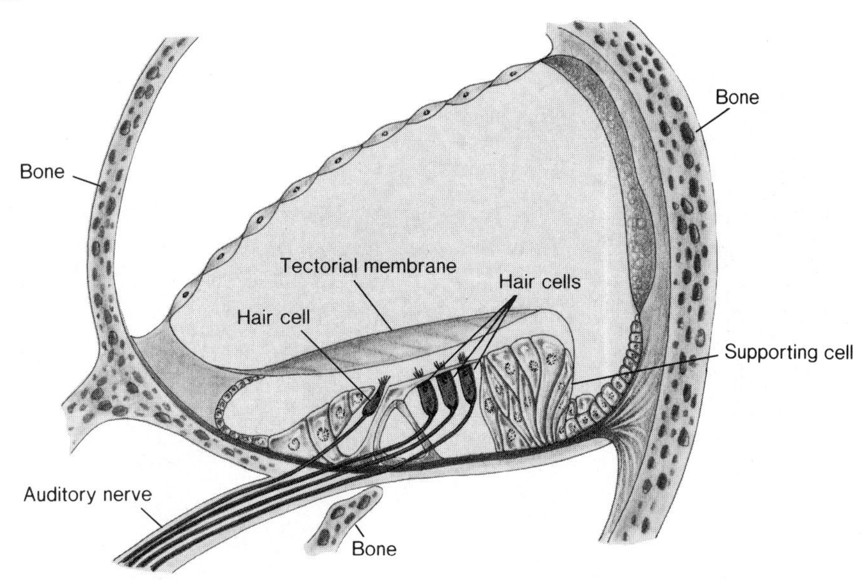

three-dimensional structure of the cochlea with its three internal compartments looks something like a snail shell (figure 13.5). The movement of the fluid within the cochlea is possible because of the expandable membrane at the **round window.** Stretched across the round window, this membrane bulges into the middle ear just as the membrane of the oval window bulges into the inner ear.

The sensory cells for hearing, called **hair cells,** are located inside the organ of Corti where they are covered by a membrane called the **tectorial membrane** (figure 13.6). When the oval window vibrates, the fluid in the cochlea vibrates. This vibration causes the tectorial membrane to vibrate against hair cells. When hair cells respond to vibration, sensory neurons that touch them are stimulated to send impulses to the brain via the auditory nerve. Finally, these impulses are interpreted by the brain as sounds.

Balance

The semicircular canals are fluid-filled chambers in the inner ear that have similar properties to the cochlea, but their function is related to movement and balance rather than hearing (see figure 13.5). The three semicircular canals are oriented so that movements in any direction are detected. Hair cells within these canals are stimulated by movement of the fluid in the canals. Pressure occurs because, as the head turns, the hair cells move with the head but the fluid lags slightly behind. The brain uses information it receives from the three canals to determine the direction and rate of movement. This information is used, together with information about position sensed by other receptors in the inner ear, in maintaining balance. Very rapid or prolonged movements of the head, such as a continual rocking motion in a boat, may cause uncomfortable side effects, typically dizziness or nausea. Seasickness is a reaction of the body to continual activation of receptors in the semicircular canals.

Vision

The reception of visual information begins with cellular responses to light, which take place in photoreceptor (light receptor) cells in the **retina** of the eye. This light sensitive part is in the back of the eye (figure 13.7). The rest of the eye functions essentially as an optical instrument that focuses light on photoreceptors in the retina.

We can observe three features of the eye directly: the **sclera** (the white part of the eye), the **iris** (the colored part of the eye), and the **pupil** (an opening that appears as the dark center of the iris). Light passes into the eye through the pupil, and reflex-controlled smooth muscles in the iris increase or decrease the diameter of the pupil in response to the brightness of the light. The **cornea** is a piece of transparent tissue curving over the front of the eye. Between the cornea and the iris is a compartment filled with fluid called the **aqueous humor.** A larger fluid compartment, filled with **vitreous humor,** is between the lens and the retina, and gives the eye its shape (figure 13.7).

Figure 13.7 A diagram of the human eye. Note that the image reaching the retina is inverted.

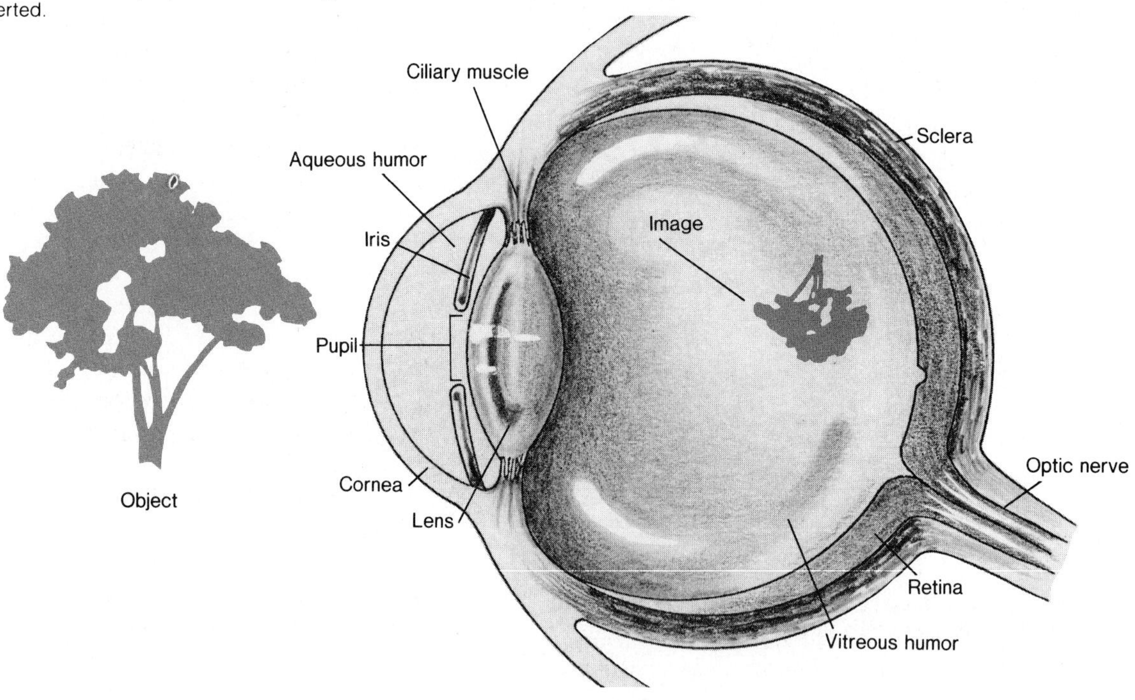

Figure 13.8 Structure of the retina showing the three layers of cells. Light must pass through two layers of nerve cells to get to the rods and cones at the back of the retina.

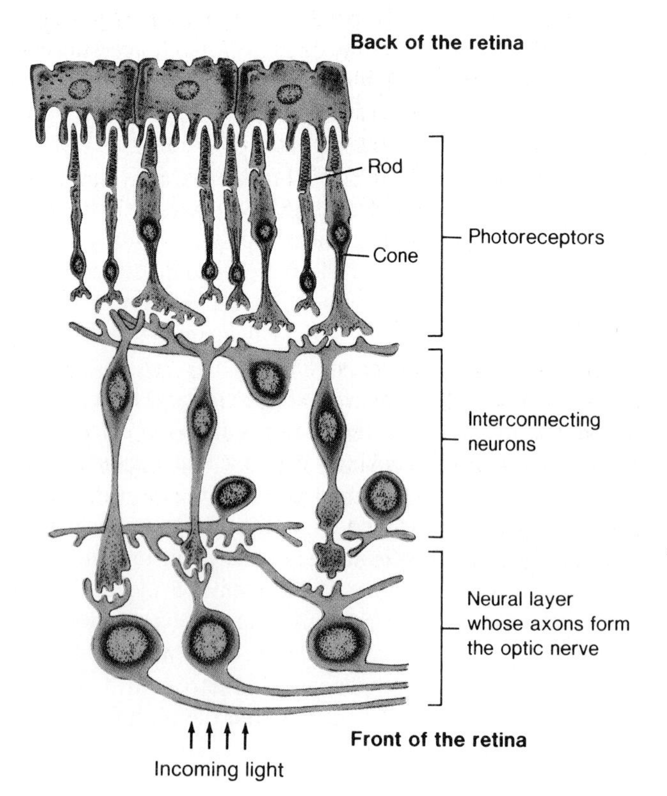

Light is focused by the cornea and the **lens.** The lens changes shape when acted on by **ciliary muscles.** These muscles alter the curvature of the lens in order to focus light coming from different distances on the retina.

Light passing through the cornea and lens is refracted (bent). As a consequence, the actual image that arrives at the retina is upside down and right-left reversed. However, the brain's processing of sensory information inverts the images, thus compensating for the refraction and yielding a correct perception of the physical environment.

Events in the retina represent the real beginning of visual sensation. The retina is a thin sheet of tissue that contains three important layers of cells (figure 13.8). The layer at the back of the retina holds the **photoreceptor cells.** These special photoreceptor cells are called **rods** and **cones** because of their characteristic shapes.

Rods and cones have different functions. Rods are important for *visual sensitivity* and *night vision* and cones are important for *sharpness* of vision. Nocturnal (night-active) animals have a much higher ratio of rods to cones than do day-active animals. Rods function well in dim light and are extremely light sensitive. Another important difference between rods and cones is that rods do not detect color differences while cones do. Thus, color vision depends on the functioning of cones. Humans and other primates probably are the only mammals with color vision. In fact, most other mammals probably see only shades of gray. Food dyes added to give pet foods a bright, red "meaty" color mean more to pet owners than cats and dogs, who cannot see them anyway! But insects, fish, some reptiles, and most birds also see in color.

Light reception by rods and cones is based on **photochemical reactions,** reactions in which light induces changes in certain chemical compounds. In rods, the molecule involved in photochemical reactions is **rhodopsin,** and part of the rhodopsin molecule is derived from vitamin A. These photochemical reactions are the key to photoreceptors' responses to light. The reaction between light and molecules such as rhodopsin leads to electrical changes in receptor cells, which in turn cause nerve impulses to be transmitted along fibers in the **optic nerve** which connects the eye to the brain.

Responses and Effectors

Stimulated sensory receptors cause nerve impulses to be sent to the central nervous system. These impulses are interpreted to provide the central nervous system with information about conditions in the external environment around the body and in the body's internal environment. The central nervous system processes this incoming information and directs appropriate responses to changes in either of these environments. Nerve impulses go out from the central nervous system along motor neurons to effectors, which carry out the responses directed by the nervous system.

This pattern of sensing, processing information, and carrying out appropriate responses through effectors is repeated thousands of times each day. Many of the responses are made by internal effectors and include secretion by some glands, contraction of muscles in the walls of blood vessels and digestive organs, and adjustment of the rate of heartbeat. These motor functions are automatic. The activities of these internal effectors are controlled by the autonomic nervous system (see page 268) and they are not subject to conscious control. For example, it ordinarily is not possible for a person to will a change in heart rate or the rate at which peristaltic waves of muscular contraction move along the digestive tract.

Many other responses, however, involve body movements produced by the actions of **skeletal muscles,** the muscles attached to the supporting skeleton of the body. Skeletal muscles are responsible for major movements of body parts and of the body as a whole. The activities of this important group of effectors are under conscious, voluntary control. Animals can make decisions about body movements that require skeletal muscle action and consciously act on these decisions.

In the next few pages we will consider muscles, especially vertebrate skeletal muscles, and their roles as effectors carrying out responses directed by the central nervous system.

Muscles, Skeletons, and Body Movements

Muscles are made up of contractile cells—cells that can shorten forcefully—and muscles exert force as a result of simultaneous shortening of many of these cells. To bring about body movements, however, the force of muscle contractions must be specifically directed against other body parts. This is the role of skeletons in animal movement. Muscles cause movement by pulling against skeletons.

The animals with the most highly developed arrangements for body movement are the arthropods (insects and their allies) and the vertebrates. Both groups have skeletons made of rigid supporting elements to which muscles are attached. Arthropods have an **exoskeleton,** an external skeleton that doubles as a hard, protective skin. Vertebrates have an **endoskeleton,** an internal skeleton that provides support from the inside.

Shortening is the only way that muscles can exert force; they cannot exert force by active stretching. Muscles lengthen by relaxing. This has important consequences for the design of the skeletal-muscular units that produce body movements. Movements usually are produced around **joints,** movable junctions between parts of the skeleton. Because muscles can exert force only by shortening, each joint is supplied with a double set of muscles (or groups of muscles) that function as an **antagonistic pair.** One muscle (or group of muscles) passively stretches while the other actively contracts. For example, in limbs, one muscle of an antagonistic pair brings the limb toward the body, bending the joint. This muscle is called a **flexor.** The other member of the antagonistic pair, called an **extensor,** straightens the joint. Figure 13.9 illustrates the arrangement of antagonistic muscle pairs and compares the actions of muscles in the limb of a vertebrate, where they are attached to elements of an endoskeleton, with the muscles in the limb of an arthropod, where they are attached to an exoskeleton.

Vertebrate Skeletons

The vertebrate skeleton provides rigid or semirigid support for the vertebrate body as well as the attachment points for the skeletal muscles that move the body. Vertebrate skeletons are made up of two kinds of tissue: cartilage and bone. **Cartilage** is a somewhat rubbery tissue that provides fairly firm support yet allows for a degree of flexibility, while **bone** provides much more rigid support.

Figure 13.9 Antagonistic muscle pairs. Muscles can exert force only by shortening, and movable joints are supplied with double sets of muscles that work in opposite directions. (*a*) The human arm showing an example of antagonistic muscles attached to a vertebrate endoskeleton. The biceps is a flexor; it contracts and flexes the arm; that is, it bends at the elbow (colored arrow). The triceps passively stretches while the biceps contracts. When the triceps contracts, the lower arm is extended (gray arrow). The biceps passively stretches while the triceps contracts. (*b*) Muscles in an insect's leg as an example of antagonistic muscles attached to the inside of an arthropod exoskeleton. Note the relative positions of flexor and extensor when muscles are inside the skeleton as opposed to the vertebrate arrangement where muscles are attached to the outside of the skeleton.

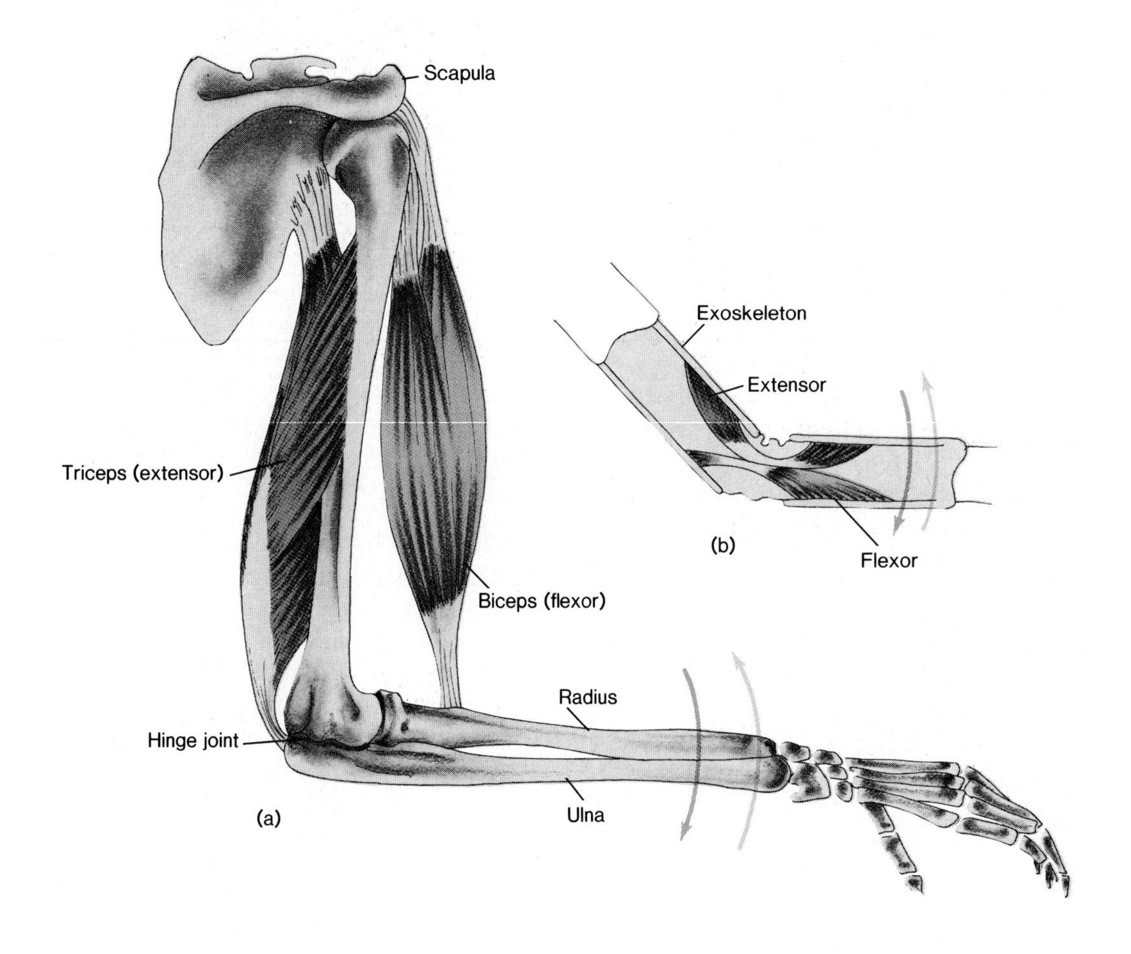

Cartilage makes up most of the skeleton during development of vertebrate embryos. Some vertebrates—notably sharks, skates, and rays—have a cartilaginous skeleton throughout life. But in most other vertebrates much of the cartilage of the embryonic skeleton gradually is replaced by bone during subsequent development. In adult humans, for example, cartilage remains only in such flexible structures as the ears and the nose tip, and it also provides a form of padding for the areas where bones lie adjacent to one another in movable joints.

Bone is a more rigid supporting tissue that occurs in two forms—spongy bone and compact bone (figure 13.10). **Spongy bone** is composed of a latticelike network of bars. The irregular spaces among the bony bars, as well as the hollow cavities inside many bones, contain **bone marrow.** Bone marrow is the site where replacement blood cells develop in adult vertebrates. **Compact bone** is more solid than spongy bone and provides stronger support in such sites as the shafts of the longer bones of the body.

Figure 13.10 Bone structure. A drawing and a scanning electron microscope picture showing compact and spongy bone. Spongy bone contains irregularly shaped bony bars with marrow spaces among them.

(b) Kessel, R. G., and Kardon, R. H. *Tissues and Organs: A Text-Atlas of Scanning Electron Microscopy.* © 1979 by W. H. Freeman and Company.

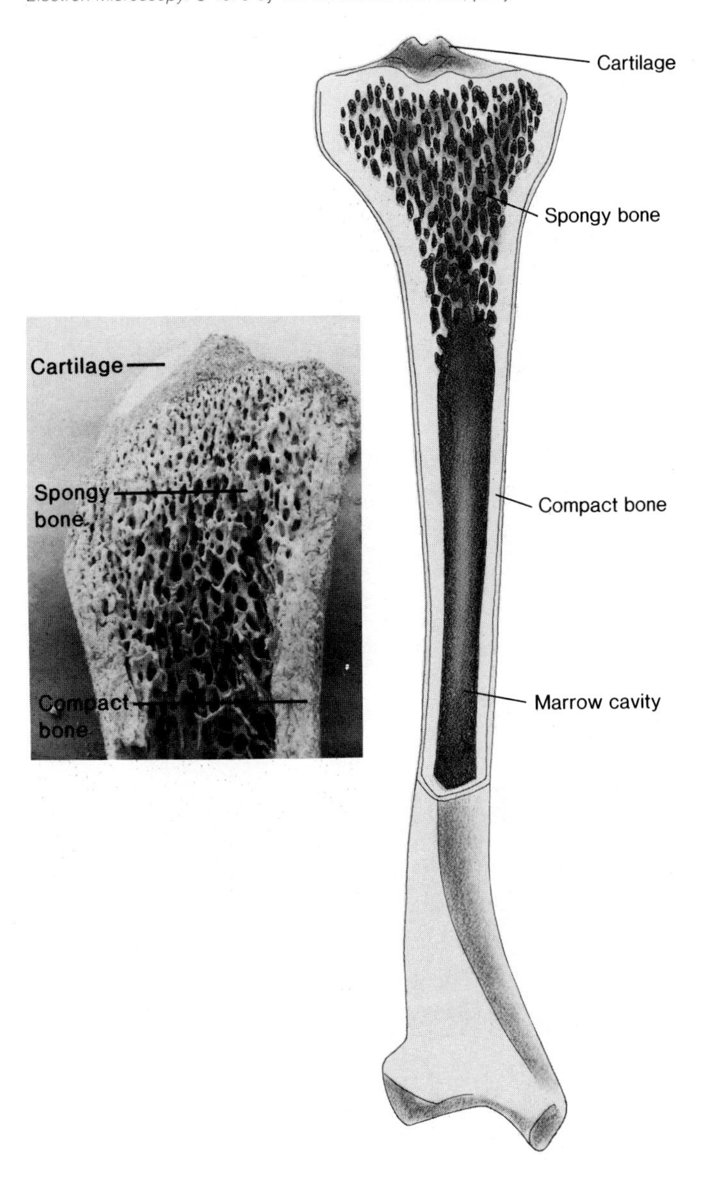

Figure 13.11 The human skeleton.

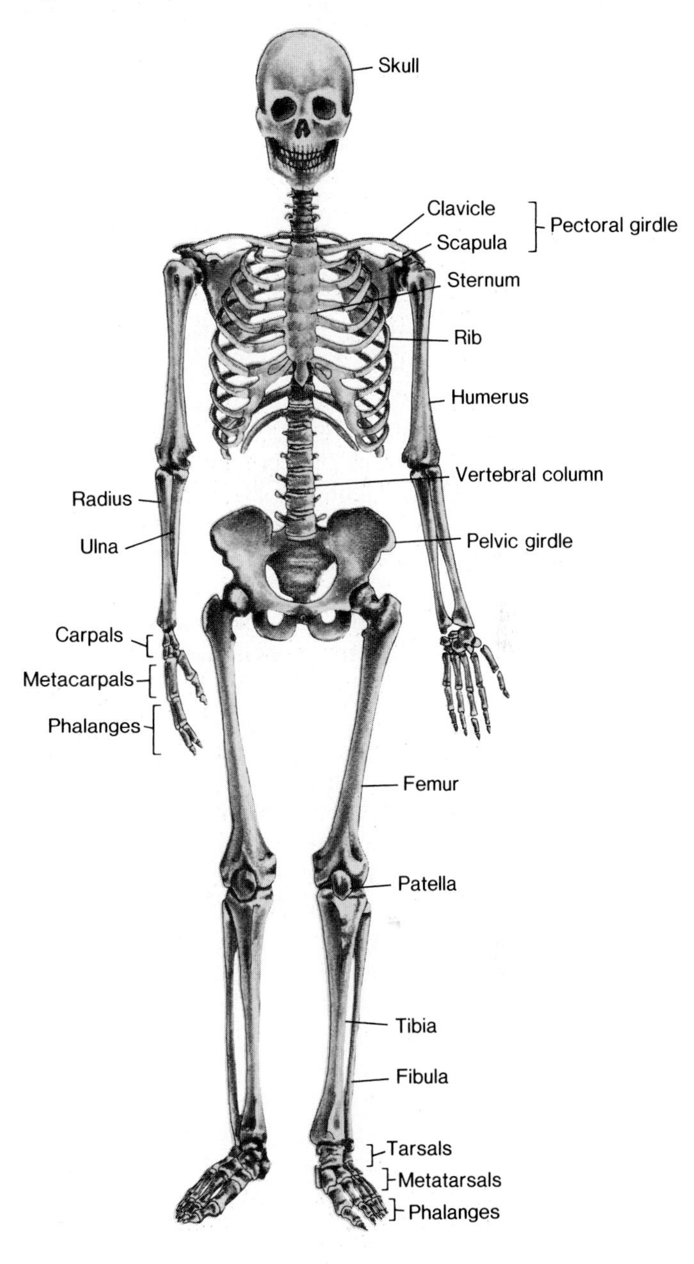

The Human Skeleton

The human skeleton, like all vertebrate animals' skeletons, can be divided into two basic parts (figure 13.11). The **axial skeleton** supports the main body axis—the head, neck, and trunk. It includes the **skull,** which surrounds and protects the brain; the **vertebrae,** a series of bones making up the **vertebral column** or backbone; and the **ribs,** curved bones that are attached to vertebrae and enclose and protect vital organs, such as the heart and lungs, located in the upper body trunk.

The **appendicular skeleton** includes the bones of the paired appendages (the arms and legs), as well as the bones of the pectoral and pelvic girdles that support the arms and legs, respectively, and connect them to the axial skeleton.

Some bones, such as those of the skull, are connected to one another by immovable joints so that they are permanently fixed in position relative to one another. But many other bones meet in movable joints that are held together by **ligaments.** Ligaments are tough, but flexible, connecting straps between bones. Movable joints are flexible enough to permit body movements in response to contractions of skeletal muscles.

Vertebrate Muscles

Vertebrate animals have three types of muscle tissue: **skeletal muscle, smooth muscle,** and **cardiac muscle.** Each of these three types of muscle tissue has different structural and functional characteristics.

Skeletal muscle tissue is the contractile tissue of the skeletal muscles involved in limb movements and other whole body movements. It consists of tubular, multinucleate (more than one nucleus) cells, called **muscle fibers,** that are held together in bundles by connective tissue (figure 13.12*a*). Skeletal muscle tissue is also called **striated muscle** because of the alternating light and dark bands (**striations**) that cross its fibers. Nuclei in a skeletal muscle fiber lie near the edge of the tubular cell, just inside the plasma membrane.

Smooth muscle tissue is found in many sites, including the walls of the digestive tract, the urinary bladder, the uterus, and encircling blood vessels, where smooth muscle contraction regulates blood flow to specific body regions. Smooth muscle tissue is different from skeletal muscle in that it contains individual contractile cells with *single* nuclei. Smooth muscle cells lack the obvious striations seen in skeletal muscle (figure 13.12*b*). Cardiac muscle is the specialized contractile tissue of the heart (figure 13.12*c*).

Many skeletal muscle fibers are innervated by motor neurons that are under conscious (voluntary) control. Smooth muscle tissue and cardiac muscle tissue are involved in movements in internal body structures that are not under voluntary control. Their contractions are controlled by nerves of the autonomic division of the nervous system.

Figure 13.12 Vertebrate muscle fibers. (*a*) Skeletal (striated) muscle fibers. Note striations and the peripheral location of nuclei in the multinucleate fibers. (*b*) Smooth (visceral) muscle fibers. Smooth muscle fibers have a single nucleus and lack striations. (*c*) Cardiac muscle fibers. Note the branching of the fibers, the central position of the nuclei, and the presence of intercalated discs, which are the complex junctions between adjacent individual cells.

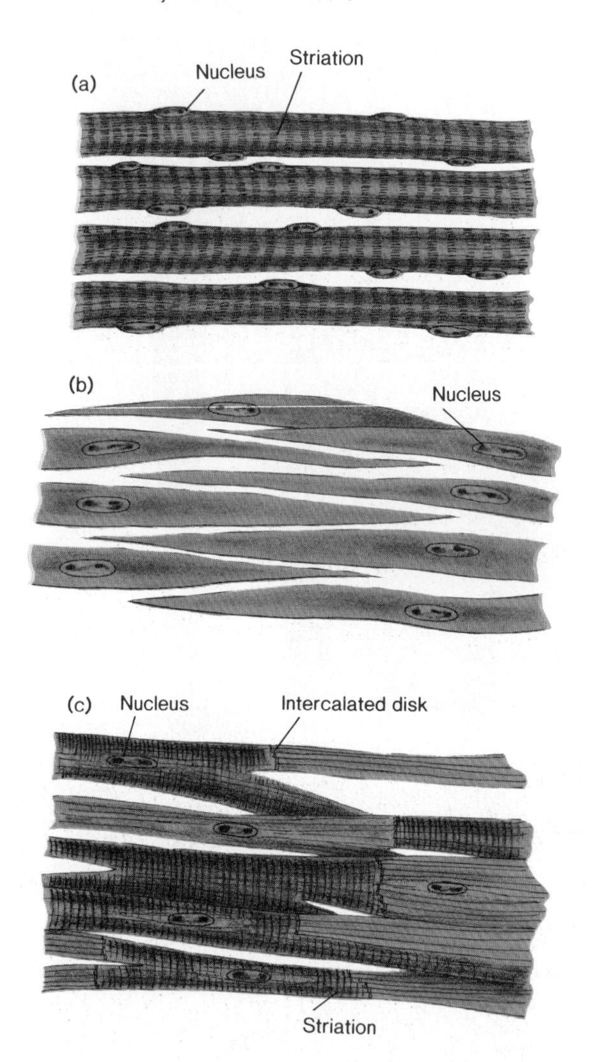

Skeletal Muscle

Skeletal muscle is the most abundant muscle type in the body, constituting about 80 percent of the body's soft tissue.

A skeletal muscle is made up of many bundles of skeletal muscle fibers held together by connective tissue. All of these bundles are surrounded by a tough outer connective tissue sheath with a smooth surface that reduces friction between the muscle and adjacent structures during muscle contraction and relaxation.

Most skeletal muscles taper at their ends and are connected to bones by **tendons.** Figure 13.13*a* shows the gastrocnemius muscle in a frog's leg. The gastrocnemius muscle is attached by the Achilles tendon to a bone in the frog's foot. When the gastrocnemius muscle contracts, it produces a powerful straightening of the foot that is an important part of the frog's jump. The nerve innervating the frog gastrocnemius is the sciatic nerve, which passes through the muscles of the thigh (figure 13.13*b*).

Box 13.1

Muscle that Stimulates Itself

Cardiac muscle has a number of unique characteristics associated with its role in the pumping action of the heart. One of the most interesting features of cardiac muscle is its inherent rhythmicity. This continuous rhythmic beating of the heart is controlled by an internal **pacemaker.**

Each individual heartbeat begins with spontaneous action potentials in cells in a specialized region, called the **sino-atrial (S-A) node.** Excitation spreads from the sino-atrial node, over a network of fibers, to all parts of the heart and stimulates a heartbeat. Next there is a short pause, during which membrane potentials are restored to their resting level in cells of the S-A node and the fibers of the conducting network. Then action potentials develop again in the S-A node and the stimulus for the next heartbeat goes out to all of the heart muscle.

This internal stimulus generation means that individual heartbeats do not depend at all upon neural stimulation from the central nervous system, though the nervous system does adjust the beating rate and the strength of individual beats. Thus, cardiac muscle generates the stimulus for each of its own contractions. This is distinctly different from skeletal muscle, which contracts only when it receives external stimulation from the nervous system.

The rhythm of internally-generated beats alternating with periods of relaxation begins in the embryo long before birth and continues throughout life. As a matter of fact, the embryonic heartbeat begins even before the heart has any nerves connecting it to the central nervous system. Cardiac muscle truly is muscle that stimulates itself.

Figure 13.13 Muscle, tendons, and nerves. (*a*) Dissection of a frog's leg that shows the gastrocnemius muscle and the Achilles tendon that attaches it to a bone in the foot. (*b*) Dissection of a frog's thigh that exposes the sciatic nerve, which contains axons that provide motor innervation for the gastrocnemius muscle as well as those supplying other leg and foot muscles.

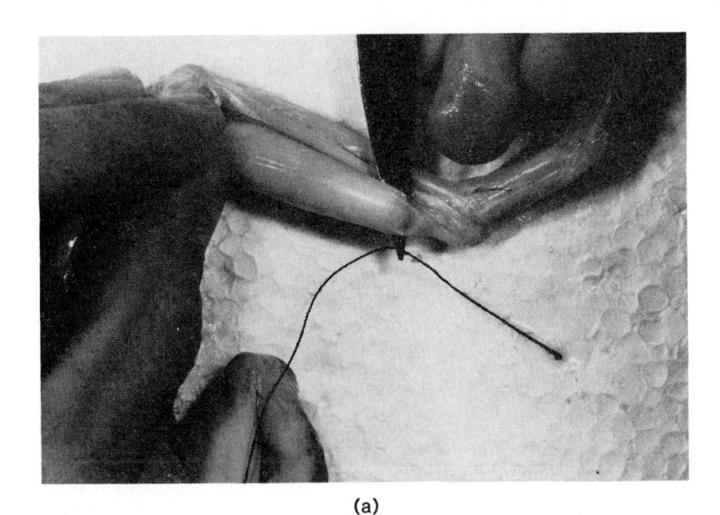

(a)

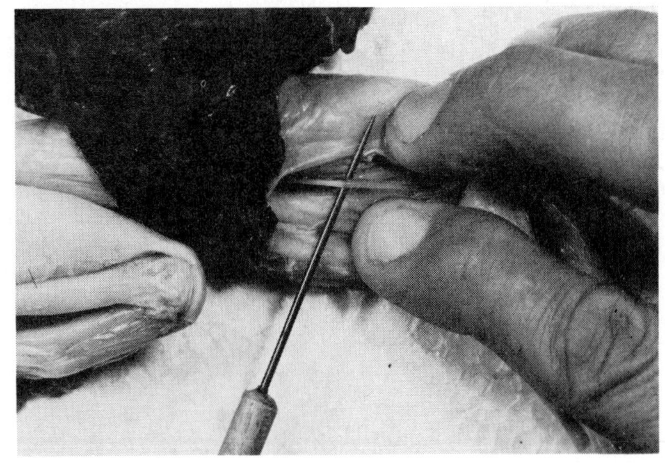

(b)

Motor nerve cells' axons running through nerves such as the sciatic nerve contact muscle cells at special contact points called **neuromuscular junctions.** The arrival of a nerve impulse at the end of one of these axons causes the muscle cell (or cells) that it contacts to contract.

The strength of the muscle contraction depends on the number of motor neuron axons that conduct impulses to the muscle fibers that they contact. Also, motor axons carry series of impulses that cause repeated contractions of muscle cells. The frequency of repetition of these impulses is a major factor in determining the strength of a whole muscle's contraction. Thus, the strength of muscle contractions can be adjusted to the work load at hand. For example, you would use the same set of muscles to lift a piece of paper from the edge of a table or to lift the edge of the table itself. The nervous system adjusts the strength of the muscle contractions involved in the two jobs by varying the number of motor units activated and the frequency of impulse conduction by motor axons.

Muscle Contraction

How do muscle cells contract so that muscles can exert a pulling force? This question has long intrigued biologists and has inspired a great deal of research on muscles and muscle cells.

One of the key discoveries concerning muscle contraction was made in the 1940s by Albert Szent-Györgyi. He extracted and purified two proteins, **actin** and **myosin,** the major constituents of muscle cells. He then mixed actin and myosin in solution so that they produced a complex called actomyosin, which formed tiny threads or fibers.

Szent-Györgyi found that fibers of actomyosin, in a solution containing appropriate ions, actually shortened when he added ATP to the solution. Neither pure actin fibers nor pure myosin fibers would shorten by themselves. Thus, Szent-Györgyi demonstrated that actin and myosin indeed are contractile proteins that can shorten physically, but that they do so only as actomyosin, a complex of the two proteins. He further showed that ATP can provide energy directly to this molecular contraction process.

Sliding Filaments

Szent-Györgyi's demonstration that threads of actomyosin could shorten if supplied with ATP focused attention on the two contractile proteins actin and myosin and on the actual physical arrangement in muscle fibers that permits them to interact during contraction. Much of the present understanding of these relationships has resulted from the research of two English scientists: A. F. Huxley, who studied muscle contraction mainly with the light microscope, and H. E. Huxley, who studied the fine structure of muscles with the electron microscope. They and their coworkers discovered how actin and myosin are arranged inside intact muscle cells and how they actually work together to cause the shortening of muscle cells, which is the basis of muscle contraction and movement in animals.

Each skeletal muscle fiber (cell) is striated (see figure 13.12); it has a pattern of alternating light and dark bands. The striation of whole fibers arises from the alternating light and dark bands of the many smaller, tubular **myofibrils** contained in each muscle fiber (figure 13.14). Please refer frequently to the diagrams in figure 13.14 and the electron micrograph in figure 13.15 as we discuss the internal structure of muscle fibers (cells) and the myofibrils inside them. The myofibrils have dark bands, called **A bands,** that alternate with lighter bands, called **I bands.** Each A band has a region in its center that is lighter than the rest of the band. This lighter, central region is called the **H zone.** Finally, the middle of each I band has a distinctive, thin, dark line called the **Z line** (figure 13.15). The banding patterns of myofibrils reflect their functional organization, and the portion of a myofibril running from one Z line to the next is a single functional contracting unit called a **sarcomere.**

Relatively thicker filaments, which are made up of myosin, run through the A band. Relatively thinner filaments, which contain actin (and several other proteins), run through the I band and also overlap with the thick myosin filaments in part of the A band. The portions of the A band in which the thick and thin filaments overlap are darker than the H zone. The H zone is the central portion of the A band that contains only the thick myosin filaments (see figure 13.14). The Z line is a structure to which the smaller, actin-containing filaments are anchored.

The two Huxleys proposed a model for the molecular events in muscle contraction that is now called the **sliding filament theory.** This theory says that in response to a stimulus these filaments slide by one another and increase the amount by which they overlap. This sliding draws the two Z lines of each sarcomere closer together. The H zone in each

Figure 13.14 The structure of skeletal muscle. Each muscle (1) contains many muscle fibers (2 and 3). Muscle fibers contain myofibrils (3 and 4), each of which consists of many sarcomeres (5). The banding of sarcomeres arises from the arrangement of thick and thin filaments inside them (6). The cross section sketches (7) show the very precise arrangement of the thin and thick filaments in small areas of the I and A bands. Each sarcomere actually contains hundreds of thick and thin filaments.

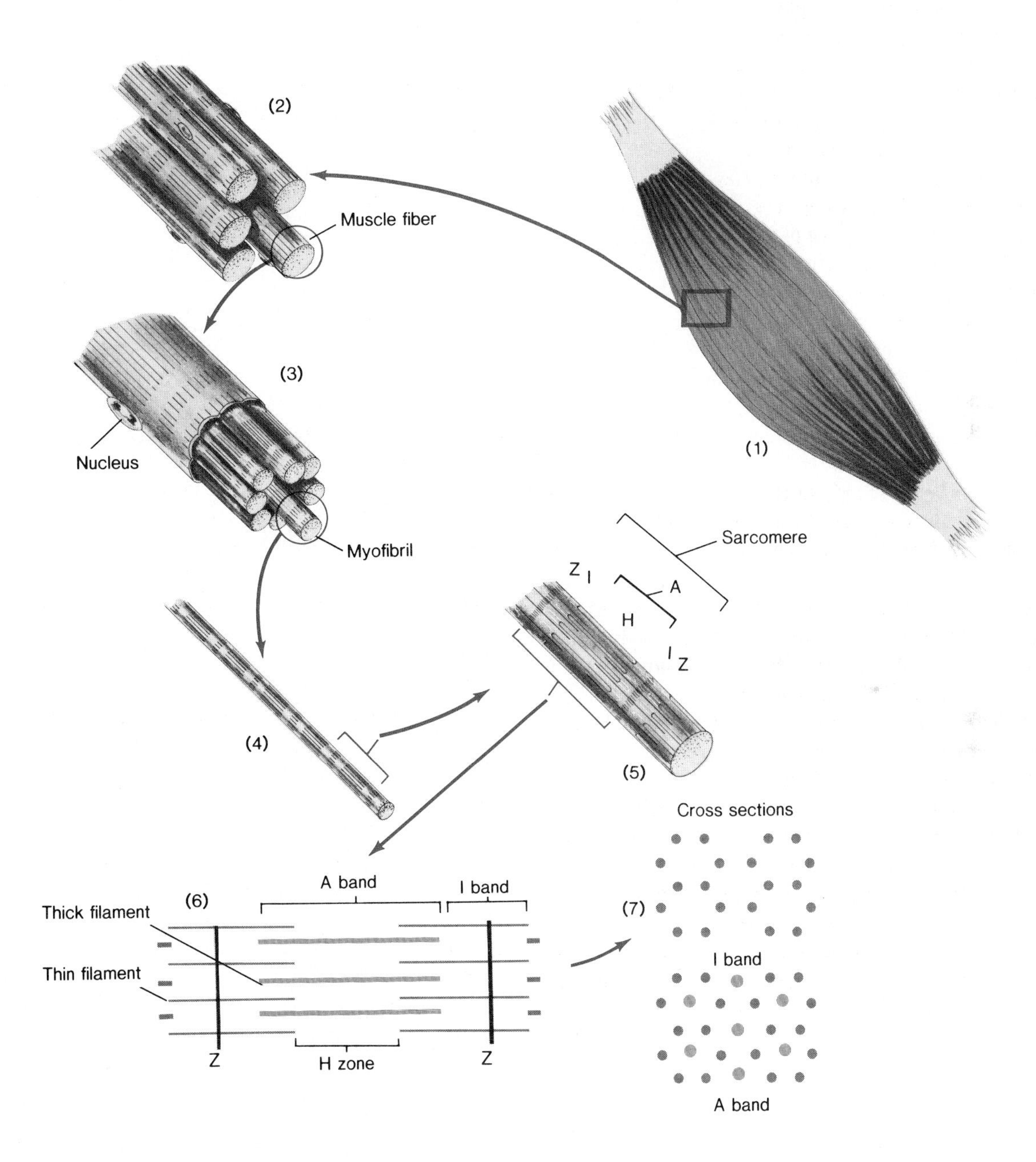

Figure 13.15 An electron micrograph of several myofibrils in a skeletal muscle fiber of a rabbit showing the appearance of the various regions represented diagrammatically in figure 13.14. A band (A), I band (I), H zone (H), and Z line (Z) are identified (magnification × 760).

sarcomere essentially disappears, and the I bands all along the sarcomere become narrower while the widths of the A bands remain constant (figure 13.16). A shortening of all of the sarcomeres within each myofibril shortens the entire myofibril, and simultaneous shortening of all of the myofibrils in a muscle fiber shortens the whole fiber. Of course, it is the shortening of many muscle fibers that causes the whole muscle to contract and exert a pulling force.

How then do these thick and thin filaments slide and bring about muscle contraction? Filament sliding involves temporary connections that form between myosin in the thick filaments and actin in the thin filaments. The connections are flexible, temporary cross-bridges that are established when the globular "heads" of myosin molecules attach to binding sites on actin molecules in the thin filaments (figure 13.17). Once a cross-bridge forms, it bends, thus exerting a pulling force on the thin filament that slides it by the thick filament.

The original binding between actin and myosin requires the presence of an ATP molecule. During cross-bridge binding, the ATP molecule breaks down to ADP and phosphate (P_i) and provides energy for the bending. Then, the linkage between actin and myosin is released. The myosin "head" then returns to its prebending position and is ready to form a new cross-bridge with an actin molecule further along the thin filament.

This cross-bridging cycle is repeated over and over by each of a huge number of myosin molecules during each contraction. At any given instant there are numerous links between myosin molecules and actin molecules, and filament sliding is the composite effect produced by bending of many of these cross-bridges.

Figure 13.16 Sketches showing how filaments in an individual sarcomere slide during muscle contraction.

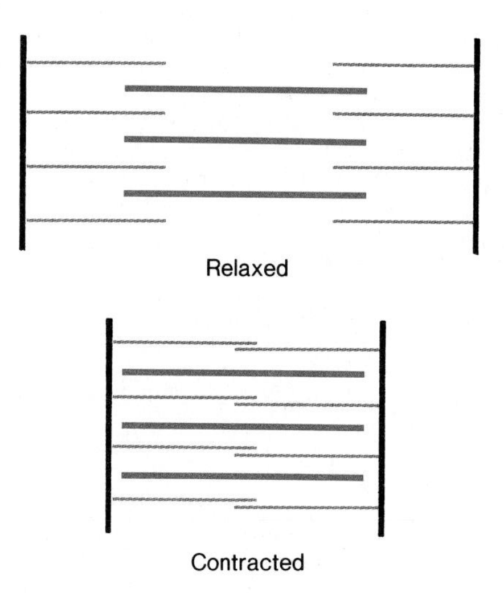

After the contraction of a muscle fiber is complete, it relaxes. During relaxation, the filaments slide back to their original positions relative to one another, and all of the changes in myofibril and muscle fiber length are reversed. Then the muscle fiber is again ready to respond to stimulation and contract again.

Control of Muscle Contraction

Now that we have described the mechanism by which muscle cells contract, we can ask what makes them contract in response to signals from nerve cells. The ends of motor axons form neuromuscular junctions with muscle fibers that are very much like the synapses between nerve cells (see page 262). The arrival of a nerve impulse in the axon initiates an action potential (electrical change) in the muscle cell's plasma membrane. Then a chain reaction series of action potentials passes along the plasma membrane just as happens during impulse conduction in a nerve cell.

But how do these electrical events in plasma membranes cause the filament sliding inside muscle fibers that we have just described? Part of the answer lies in the structure of the plasma membrane of muscle fibers. It is not simply a smooth envelope, but juts inward into the cytoplasm in the form of hollow depressions called **T tubules** (for transverse tubules). The T tubules come into close contact with membranous sacs inside the muscle fiber (figure 13.18*a*). These sacs serve as storage sites for calcium ions (Ca^{2+}) in the cell.

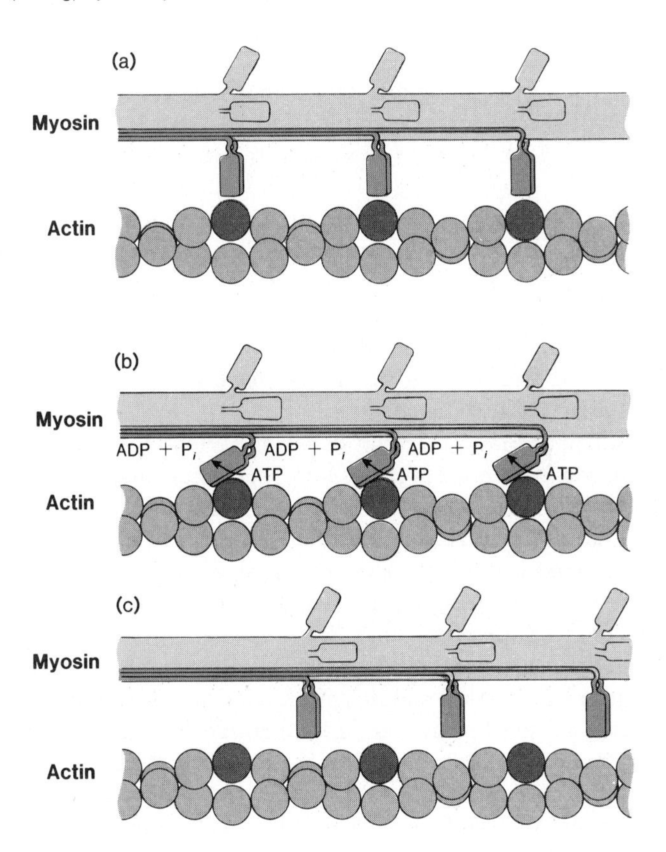

Figure 13.17 The proposed mechanism of filament sliding during contraction. (*a*) Thick filaments contain myosin molecules, which have globular double "heads" that form temporary connections with actin molecules. Thin filaments consist of two chains of spherical actin molecules that are intertwined in a double helix. (*b*) Myosin "heads" bind with actin molecules and then flex, thus exerting a sliding force on the thin filament. ATP is required for binding and flexing of the myosin "heads." (*c*) After flexing, the myosin heads release from actin, return to their original shape, and then bind with other actin molecules further along the filament. Note the position of the actin molecules (darker) originally contacted by the myosin heads. The actin filament is moving (sliding) by the myosin filament.

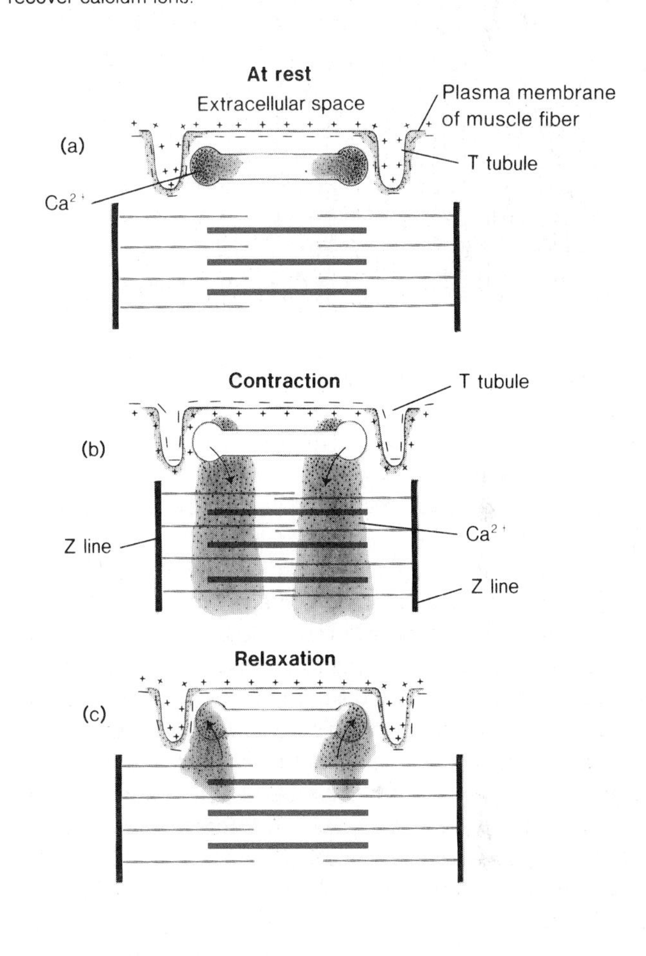

Figure 13.18 Simplified diagram of the control of actin and myosin filament sliding by calcium release and recovery. (*a*) T tubules, which are continuous with the plasma membrane, actually branch extensively and penetrate all parts of the muscle fiber. (*b*) When action potentials pass down the T tubules, calcium is released from storage sacs, and filament sliding occurs. (*c*) Relaxation occurs when the storage sacs recover calcium ions.

When a motor axon stimulates a nerve cell and electrical changes pass over its plasma membrane, they also occur in the T tubules. Somehow electrical changes in the T tubules cause calcium ions (Ca^{2+}) to be released from the storage sacs (figure 13.18*b*). It is this flood of calcium ions that actually causes filament sliding and contraction of the cell. Then, calcium ions are taken back into the storage sacs during relaxation, and the whole cell is ready to respond again when the next nerve impulse comes along the motor axon (figure 13.18*c*).

Energy for Contraction

We have seen that ATP is used up during filament sliding in muscle contraction. When a muscle is contracting vigorously, all of the ATP in the muscle fibers is used up quickly and the cells begin to oxidize glucose to replenish the supply of ATP. But if heavy muscle exercise continues, problems develop.

Oxygen-requiring cell respiration is the most efficient means of producing ATP, and muscle fibers employ it, but its usefulness to a working muscle is limited by the amount of oxygen that can be delivered to the muscle. Physical conditioning improves the efficiency of oxygen delivery to muscles, but even muscles in an individual in excellent physical condition suffer an oxygen delivery deficit as heavy muscle exercise continues. Thus, ATP produced by oxygen-requiring respiration falls short of the muscle's needs because of a shortage of available oxygen.

It would not do, of course, for an animal to be unable to continue heavy muscular work for extended periods. How then is the additional required ATP produced? Muscle cells have the ability to carry on lactic fermentation, which permits them to produce additional ATP anaerobically, that is, without additional oxygen. (See page 117 if you have forgotten what we mean by fermentation.) As a result of this fermentation, lactic acid builds up in muscle fibers and in the fluid around them during heavy and extended muscle work, but an adequate supply of ATP to power continuing muscle contractions is maintained.

An accumulation of lactic acid in contracting muscles must be dealt with when vigorous contraction stops. The results of this temporary employment of anaerobic metabolism amount to an **oxygen debt,** a chemical deficit that must be made good when the muscle returns to rest (see page 126). Some of the lactic acid built up in working muscles is exported to the liver, where it is processed. The remainder of the lactic acid is processed in the muscle fibers themselves when extra oxygen becomes available after vigorous exercise stops.

The anaerobic metabolic capabilities in skeletal muscle fibers have great adaptive value. The ability to continue vigorous muscle contractions and to accumulate a temporary oxygen debt can be a matter of life and death for most organisms. Eventually, however, all oxygen debts must be paid back so that muscles are ready to function normally again when they are called upon to act as effectors in response to nervous stimulation.

Summary

Sensory receptor cells respond to specific stimuli. Receptor cells are neurons or are associated with neurons that carry impulses to the central nervous system.

Sensory receptors of the skin and muscles are the ends of neuron processes. Various receptors are specialized to respond to touch, pressure, pain, and temperature changes.

Olfactory receptors are embedded in nasal passages. They detect chemicals present in fluid on the surface of mucous membranes.

Receptors for taste are secondary sense cells in the taste buds. Responses of these cells cause impulse transmission in nerve cells that synapse with them.

The ear houses the senses of hearing and balance. Sound strikes the tympanic membrane as a series of vibrations. Vibrations are transmitted via the middle ear ossicles to the cochlea, where fluid movements trigger responses in hair cells of the organ of Corti. Hair cells communicate with neurons that conduct impulses to the brain.

Body movements are sensed by hair cells in the semicircular canals that respond to fluid pressure changes inside the canals.

The eye is an optical device that focuses light on the retina, where photoreceptor cells are located. Rods are receptor cells that are very light sensitive and can function in dim light. Cones require bright light and are associated with sharp image formation. Cones also detect color differences.

The central nervous system sends nerve impulses over motor axons to various effectors.

Skeletal muscles are effectors that exert force by contracting and pulling against skeletal elements. Because muscles exert force only by shortening, skeletal muscles usually are arranged in antagonistic pairs.

The two major skeletal tissues in vertebrates are cartilage and bone. Cartilage has some flexibility, but bone provides very rigid support.

The human skeleton has two main parts. The axial skeleton supports the head, neck, and trunk of the body. The appendicular skeleton includes the bones in the limbs and the bones that support the limbs.

Muscle contraction depends on filaments of contractile proteins inside the myofibrils of muscle fibers. Thick myosin filaments overlap thin, actin-containing filaments. During contraction, myosin molecules in the thick filaments form temporary, bendable cross-bridges with actin molecules in the thin filaments. Energy provided by ATP permits bending of these cross-bridges. The bending pulls the thin filaments along so that they overlap further with the thick filaments. Filament sliding shortens the myofibrils and whole muscle fibers. During relaxation, the filaments return to their original positions.

Nerve impulses coming over motor axons set off electrical changes in the plasma membranes of muscle fibers and in the T tubules that end near membranous sacs that contain calcium ions. This triggers calcium release and causes filament sliding.

Continuing contraction soon exceeds the ability of aerobic respiration to supply ATP because oxygen delivery to the muscle is a limiting factor. The muscle then uses anaerobic lactic fermentation to supplement ATP production. This causes lactic acid buildup, and the muscle accumulates an oxygen debt that must be repaid when muscle work stops.

Questions for Review

1. What is the difference between primary receptors and secondary receptors?
2. How is intensity of stimulation communicated?
3. What is a mechanoreceptor?
4. Explain the physical basis for a "seasickness" response.
5. Why is vitamin A necessary for good vision?
6. What characteristic of muscle function necessitates the arrangement of skeletal muscles in antagonistic pairs?
7. Compare and contrast the three kinds of vertebrate muscle tissue.
8. Distinguish between the terms muscle fiber and myofibril.
9. Explain why H zones seem to disappear in contracting myofibrils.
10. What is oxygen debt?

Questions for Analysis and Discussion

1. Explain why a sharp blow to the side of the head can make your ears "ring."
2. Why might you assume that Ca^{2+} ions had to be included in the medium that Szent-Györgyi used when he successfully induced shortening of actomyosin threads by adding ATP?
3. What is the functional significance of the fact that T tubules extend deep into muscle fibers and branch extensively?

Suggested Readings

Books

Bullock, T H.; Orkand, R.; and Grinnell, A. 1977. *Introduction to nervous systems*. San Francisco: W. H. Freeman.

Wilkie, D. R. 1976. *Muscle*. 2d ed. Studies in Biology no. 11. Baltimore: University Park Press.

Articles

Buller, A. J., and Buller, N. P. 1980. The contractile behaviour of mammalian skeletal muscle. *Carolina Biology Readers* no. 36. Burlington, N.C.: Carolina Biological Supply Co.

Friedmann, I. 1979. The human ear. *Carolina Biology Readers* no. 73. Burlington, N.C.: Carolina Biological Supply Co.

Horridge, G. A. July 1977. The compound eye of insects. *Scientific American* (offprint 1364).

Hudspeth, A. J. January 1983. The hair cells of the inner ear. *Scientific American.*

Lazarides, E., and Revel, J. P. May 1979. The molecular basis of cell movement. *Scientific American* (offprint 1427).

Parker, D. E. November 1980. The vestibular apparatus. *Scientific American* (offprint 1484).

Weale, R. A. 1978. The vertebrate eye. *Carolina Biology Readers* no. 71. Burlington, N.C.: Carolina Biological Supply Co.

Regulation in Plants

Chapter Outline

Chapter Concepts

1. A plant continually interacts with its environment.
2. Plant hormones mediate these interactions and cause specific developmental responses in plants.
3. Light is the dominant environmental factor to which plants respond. Light determines growth patterns of plants and coordinates the timing of important events in their lives.
4. Changes in the twenty-four hour cycle of alternating light and dark are particularly important in timing flowering and preparation for dormancy.
5. Internal clocks are necessary for time measurement in plant responses to day length changes.
6. Internal clocks also time daily rhythms in many functions, but the nature of the basic cellular clock remains obscure.

Imagine a walk through a mountain meadow in spring or early summer. You see flowers everywhere. In fact, the ground is virtually a carpet of flowers (figure 14.1). And yet, if you return to the same meadow only a week or two later, everything is green and growing, but there is hardly a flower in sight. The flowering period is over and the plants have moved on to other phases of their seasonal activity.

Every phase of a terrestrial vascular plant's growth, function, and reproduction involves interactions with a changing environment. Because plants remain in a single place throughout their lives, they cannot avoid adverse changes in their environment by moving to more favorable locations. They are directly exposed to all of the seasonal changes in light, temperature, and available moisture, as well as to the possibility of being eaten by animals. They must adjust to seasonal changes in conditions around them, and occasionally, to the loss of parts of their bodies.

Plants develop in an organized way. Flowering and seed production are completed during a specific part of the growing season, and then plants prepare for winter dormancy. How is all of this coordinated? Throughout the year, changes in light conditions are the most reliable indicators of changing seasons. Seasonal day length and light intensity changes are the same every year, while temperature and moisture conditions can vary considerably from year to year. Thus, light is the dominant external factor affecting seasonal plant responses. The orderly and properly timed responses of plants to light (and to other environmental factors, as well) involve the actions of plant hormones.

Plant Hormones

A plant hormone is like an animal hormone in that it is produced in one part of the organism and transported to other parts of the organism, where it causes a response, even though it is present only in a very low concentration. But there are also some important differences between hormonal regulation in plants and animals. The most significant difference is that almost all hormone effects in plants modify growth patterns. Plant hormones are regulators of the lifelong growth of plants and are not as closely associated with short-term, reversible physiological adjustments, as many animal hormones are.

Figure 14.1 Events in the lives of plants are precisely regulated in a continuing interaction with the environment. This picture of a mountain meadow illustrates the simultaneous flowering of many individual plants that occurs during only a brief period out of the entire growing season.

Auxins

Auxins were the first of the plant hormones to be identified. Their discovery resulted from their role in various tropisms (turning responses to environmental stimuli), especially **phototropism** ("light turning"). Many plants respond to directional light by a specific growth curvature. House plants that "turn toward the light" are good examples of plants showing phototropic responses. Auxins are now known to be involved in a great variety of plant cell responses.

Phototropism

Scientific analysis of phototropism began with the experiments of Charles Darwin and his son Francis around 1880. The Darwins used grass seedlings to investigate the bending of plants in response to unidirectional light. They found that seedlings failed to bend toward the light after their tips were cut off or covered with black caps. A seedling would bend toward the light, however, if its tip was exposed while the rest of the seedling was buried in fine black sand. The Darwins concluded that curvature of grass seedlings in response to light depended on some "influence" transmitted from the seedling tip to the rest of the seedling (figure 14.2*b*).

Figure 14.2 The coleoptile and the discovery of auxins. (*a*) A photograph of two oat (*Avena*) seedlings, one with its husk removed (left), showing the coleoptile and other seedling parts. The first leaves are rolled up inside the coleoptile. (*b*) A diagrammatic summary of some of the important experiments that demonstrated the existence of auxins. Both the Went and Paàl experiments were conducted in the dark.

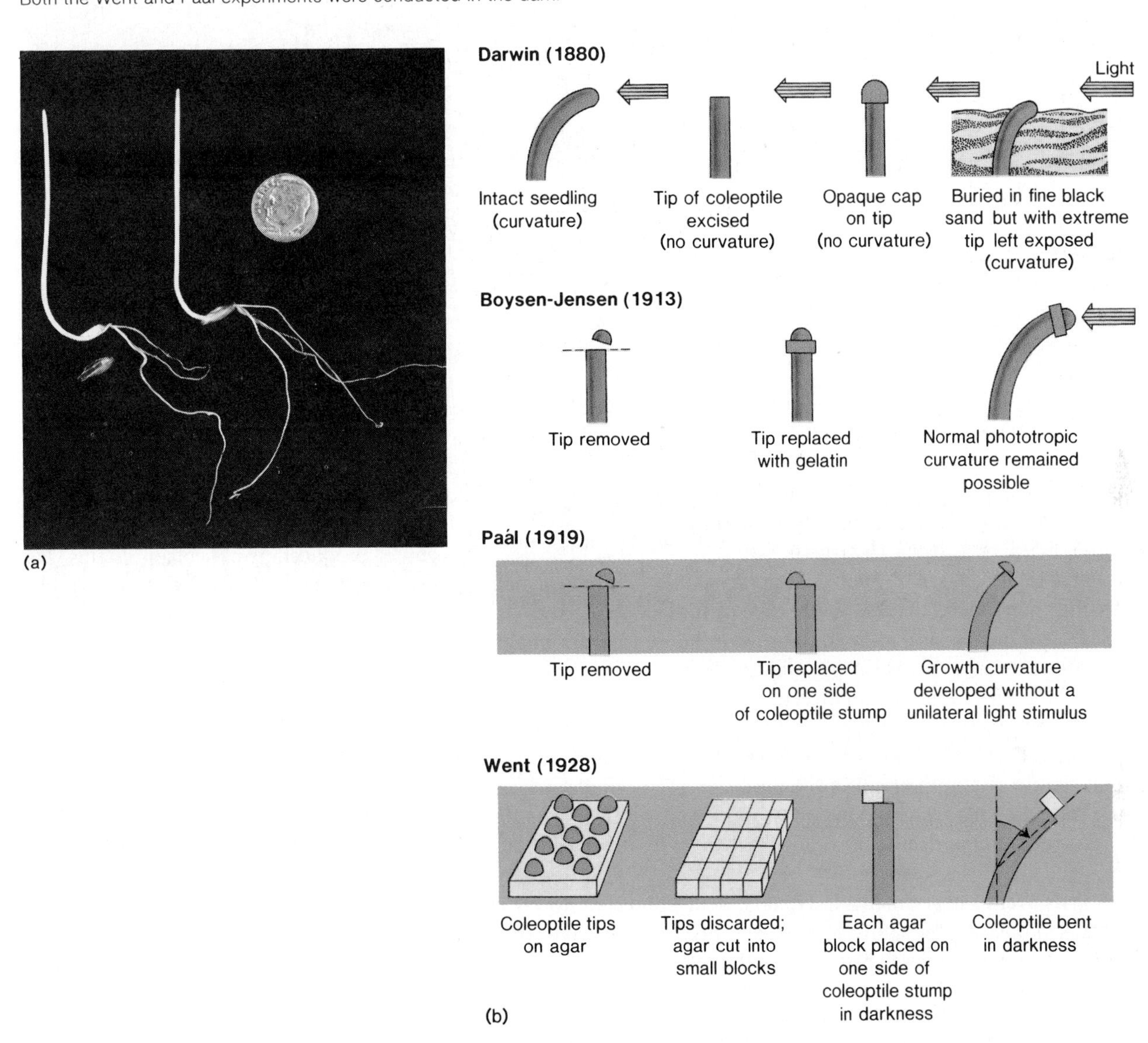

Work on phototropism then focused on this role of the seedling tip. Usually, oat (*Avena*) seedlings were used in the experiments. An oat seedling has a sheath, called the **coleoptile,** that encloses the first leaves, and the tip of this coleoptile is the key to the phototropic response of the young seedling (figure 14.2*a*).

P. Boysen-Jensen, working in Denmark about thirty years after the Darwins, experimented with cutting off the tips of the oat coleoptiles. When he did this to a coleoptile, it stopped growing. Next he placed a piece of gelatin on the cut surface and set the cut tip on top of the gelatin. A short time later

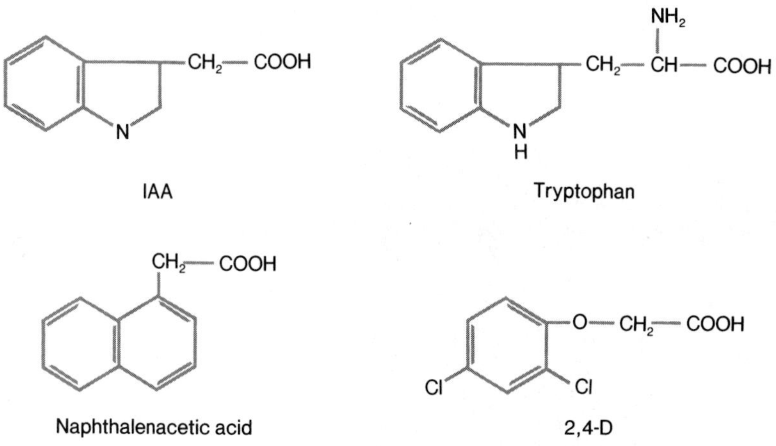

Figure 14.3 Indoleacetic acid (IAA) is the principal naturally occurring auxin. The formula of the amino acid tryptophan is given to show its close similarity with IAA. Naphthaleneacetic acid and 2,4-dichlorophenoxyacetic acid (2,4-D) are synthetic auxins.

IAA

Tryptophan

Naphthalenacetic acid

2,4-D

the coleoptile resumed growing, and in this condition it would also show the normal phototropic response to light from one side. Boysen-Jensen's experiments demonstrated that the influence of the coleoptile tip did not depend on the tip being in its normal place on the plant because the influence of the tip could pass through a gelatin block (figure 14.2*b*).

A few years later in 1918, Arpad Paàl in Hungary cut off coleoptile tips and then placed the tips to one side of the cut surfaces. When these plants were allowed to grow in the dark, Paàl observed a curvature that was very similar to the normal response of intact seedlings to unidirectional light (figure 14.2*b*). Thus, asymmetrical placement of the tip imitated the effect of light shining on the plant from one side. Paàl reasoned that the tip of a normal coleoptile produces a growth-promoting substance that travels downward, and that light must make it travel asymmetrically, thus causing greater growth on the shaded side of the seedling.

Then in 1926 Frits Went's experiments in Holland demonstrated that Paàl's proposal was right: a substance produced in the coleoptile tip causes growth responses in the rest of the coleoptile. Went cut off coleoptile tips, placed them on a gelatinlike material called agar, and left them for varying periods of time. He then cut up the agar gel and put pieces of it on "decapitated" coleoptiles that were kept in the dark. A piece of this agar placed squarely on top of a cut coleoptile caused the seedling to grow straight upward, but if Went placed the agar to one side of the top of the coleoptile, the seedling curved just as it did in Paàl's experiment with the tip itself. This seedling curvature response in the dark, and in the complete absence of the coleoptile tip, proved conclusively that the effect of the coleoptile depends not on the actual physical presence of the tip, but on a chemical substance that can accumulate in the agar (figure 14.2*b*).

The growth hormone produced by oat coleoptile tips was given the name auxin, from the Greek word *auxein* meaning "to grow." Biologists next set out to determine the chemical nature of this growth hormone.

Nature and Action of Auxins

Early researchers thought that they were working with a single substance in their experiments on growing coleoptiles. Tests of substances extracted from a variety of organisms showed that auxin could be isolated not only from plants, but also from a number of other seemingly unlikely sources, including various molds and yeasts, and even animals. In fact, the first auxin to be identified chemically was not isolated from plants, but from human urine! This auxin is **indoleacetic acid (IAA),** a molecule whose molecular structure is very similar to that of the amino acid tryptophan (figure 14.3).

Evidence now indicates that IAA is the principal naturally occurring auxin in plants. Other compounds isolated from plants also act as auxins, as do a number of laboratory-synthesized compounds. What all of these compounds have in common is that they cause plant cells to elongate. Elongation is the major effect of an auxin.

Figure 14.4 Greater cell elongation on the shaded side, a response to greater auxin concentration, causes curvature in phototropism.

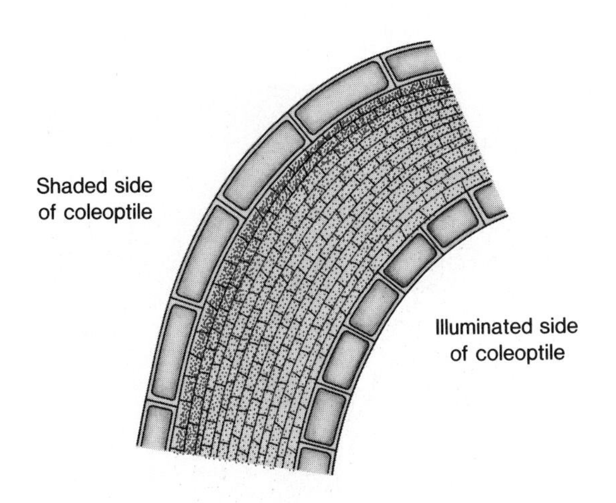

Some of the synthetic auxins are economically important. For example, **naphthaleneacetic acid** is widely used in horticulture for plant propagation because it promotes root development on stem cuttings (figure 14.3). It also is used in orchards to prevent early fruit drop. A commonly used **herbicide** (plant killer), **2,4-dichlorophenoxyacetic acid (2,4-D),** acts as a synthetic auxin (figure 14.3). It causes abnormal growth and metabolic responses in dicots (broadleaf plants) at concentrations that do not harm monocots, such as lawn grasses. This selective killing of "broadleaf weeds" makes 2,4-D and similar synthetic auxins practical for weed control in lawns and especially in growing cereal grain fields. Although these synthetic auxins have been used for many years as weed killers, it still is not known why a given concentration of 2,4-D kills broadleaf plants and leaves grasses unharmed.

Cell Elongation and Plant Responses

As you have seen, elongation on the side of a plant away from a unidirectional light source causes the curvature response in phototropism. Many years ago, it was proposed that cells elongate more on the shaded side because the auxin concentration is greater there (figure 14.4). How does the auxin concentration differential that causes the greater expansion on the shaded side come about? In growing coleoptile tips, auxin is *transported* to the shaded side. This lateral transport of auxin establishes a difference in auxin concentration and in turn causes differential cell elongation, which results in the curving growth response. Unfortunately, very little is known about the actual mechanism by which this lateral transport takes place.

Figure 14.5 Geotropism. Stem and roots grow out of a seed (*a*) and respond to gravity (*b*) with stem growing up and roots growing down. Stem curvature results from differential auxin concentration. Root curvature involves several factors.

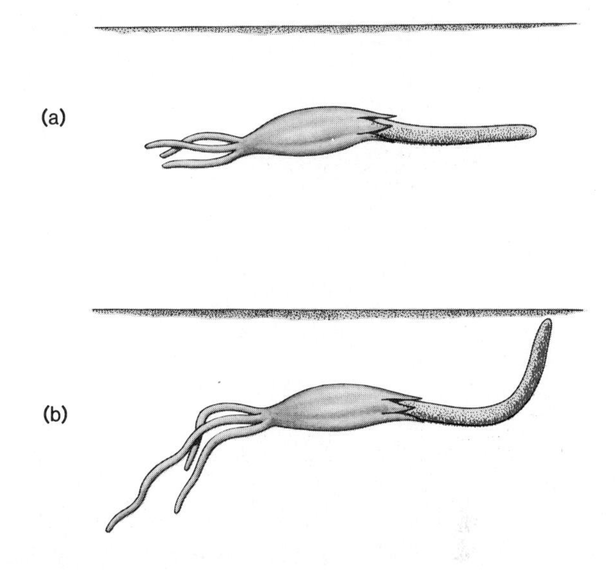

Light is not the only factor affecting the growth direction of seedlings. Growing seedlings also bend in response to gravity. This response, called **geotropism,** is very important in early seedling growth because it gets seedlings properly and uniformly oriented in the soil. In geotropism, stems grow upward and roots grow downward. Thus, stems are **negatively geotropic** because they grow away from the earth's gravity center, and roots are **positively geotropic** because they grow toward it (figure 14.5). Auxin is involved in the cell elongations that produce geotropic responses.

Apical Dominance and Other Auxin Effects

The general growth pattern of any dicot (broadleaf) plant depends on the relative growth at the tip (apex) of the main stem and the growth of lateral buds, which produce branches. Plants with a dominant main stem and very little lateral growth have a straight, narrow shape, while plants with extensive lateral growth are bushy. You can see this difference in growth pattern quite dramatically in trees (figure 14.6).

Figure 14.6 General growth patterns of plants depend on relative growth of main stem and lateral branches. Compare the growth of trunk and branches and the shapes of a poplar (*a*) and a box elder (*b*).

(a)

(b)

Auxin produced in the apical meristem (the region of cell division at the stem tip) is mainly responsible for **apical dominance** (dominance of the main stem over lateral branches). The auxin concentration that promotes cell elongation in the main stem simultaneously inhibits development of lateral buds. Clearly, the nature of the response to a given concentration of auxin depends on the type and condition of the responding tissue.

Removal of the stem tip (shoot apex) removes the source of the auxin that had maintained apical dominance and releases the lateral buds from inhibition. Lateral buds then grow actively and produce branches. This procedure is put to practical use in horticulture and everyday gardening when stem tips are "pinched off" to make plants such as *Coleus* develop a fuller, bushier growth pattern by increasing the number of lateral branches. Experimentally, apical dominance can be maintained following shoot apex removal if auxin is applied to the cut surface of the stem (figure 14.7).

Figure 14.7 Experiments showing growth of lateral (axillary) buds when the stem apex is removed and role of auxin in apical dominance. IAA applied to cut apex inhibits growth of lateral buds.

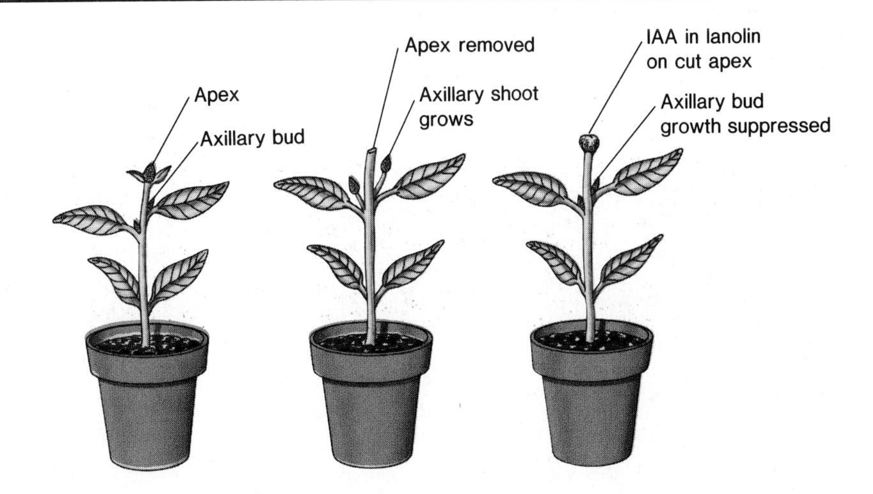

Figure 14.8 Auxin released from seeds influences fruit growth. (*a*) Normal strawberry fruit. (*b*) Fruit of the same age from which all seeds had been removed. (*c*) Magnified view of a fruit with all seeds except one removed. There is a normal area of fruit development around this one seed. (*d*) A strawberry fruit that developed after all seeds were removed and replaced by a lanolin paste containing a synthetic auxin.

Auxin also causes a variety of other responses, both alone and in conjunction with other hormones. For example, in many plants, fruits develop in response to auxin produced by embryos in developing seeds. Early removal of seeds causes abnormal fruit development (figure 14.8).

Ethylene

Ethylene is the only one of the currently known plant hormones that is a gas at normal environmental temperatures and pressures (figure 14.9). It has been known for many years that plants produce ethylene, and as early as 1935, there were suggestions that it might function as a hormone. However, the very small quantities of ethylene involved in plant responses were very difficult to detect and measure until some years after that.

The best-known response to ethylene is fruit "ripening." Of course, "ripeness" is more of a subjective human judgment than a specific physiological state in fruits. Basically, we say fruits are ripe when they are ready to be eaten. Ethylene promotes hydrolysis (breakdown) of starch, with a resulting increase in sugar concentration. Ethylene also stimulates production of **cellulase,** an enzyme that hydrolyzes cell wall cellulose and thus generally softens the fruit tissue during ripening.

Ethylene production is "contagious"; that is, when a fruit releases ethylene, it stimulates other fruits to begin to produce it as well. Thus, the old saying, "one bad apple spoils the barrel," is very accurate. To counteract this contagious ripening, apples now are stored in sealed compartments with a high concentration of atmospheric carbon dioxide, because CO_2 blocks the action of ethylene. On the other hand, ethylene can be used to stimulate ripening in fruits that are picked green and shipped long distances.

Figure 14.9 The structure of ethylene.

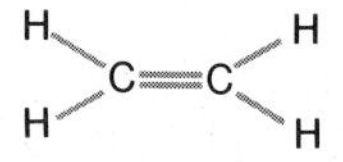

Ethylene

Cytokinins

In the early 1940s, biologists attempted to grow plant tissues in culture vessels using media containing auxin and all known plant nutrients. But something seemed to be missing. Plant cells in these cultures would enlarge, often to spectacularly large sizes, but cell divisions were rare. Thus, a search began for substances that would promote cell division in cultured plant cells. Eventually, by trial and error, it was discovered that coconut milk (a liquid nutrient storage tissue from the inside of a coconut) would greatly stimulate cell division in cultured cells when added to a culture medium.

Other preparations, such as crude yeast extracts, also could provide the necessary stimulus for cell division. But coconut milk and yeast extracts are complex mixtures of many substances. It took years of research to isolate and identify the specific chemical compounds that can cause cell division in cultured plant cells. When they were finally discovered, they were named **cytokinins** (from *cytokinesis,* meaning cell division). Chemical analysis revealed that cytokinins are structurally similar to the purine adenine, which is a component of DNA and RNA (figure 14.10).

In cultured plant tissue, there is a complex interaction between auxin effects and cytokinin effects on cells (figure 14.11). When auxin and cytokinin concentrations are balanced, the tissue grows as an undifferentiated mass called a **callus.** When the auxin to cytokinin ratio is increased, roots develop; when the auxin to cytokinin ratio is decreased, shoots and leaves develop. Similarly, in intact plants, regulation of various growth processes involves changing ratios of auxin and cytokinin concentrations.

Figure 14.10 Cytokinins. All cytokinins are structurally similar to the purine, adenine. Kinetin is an artificial cytokinin that is often used in research. Zeatin is a natural cytokinin isolated from corn (*Zea mays*) seeds. Zeatin riboside occurs, along with zeatin, in coconut milk. Note that zeatin riboside has the same basic structure as zeatin but has a sugar (ribose) attached.

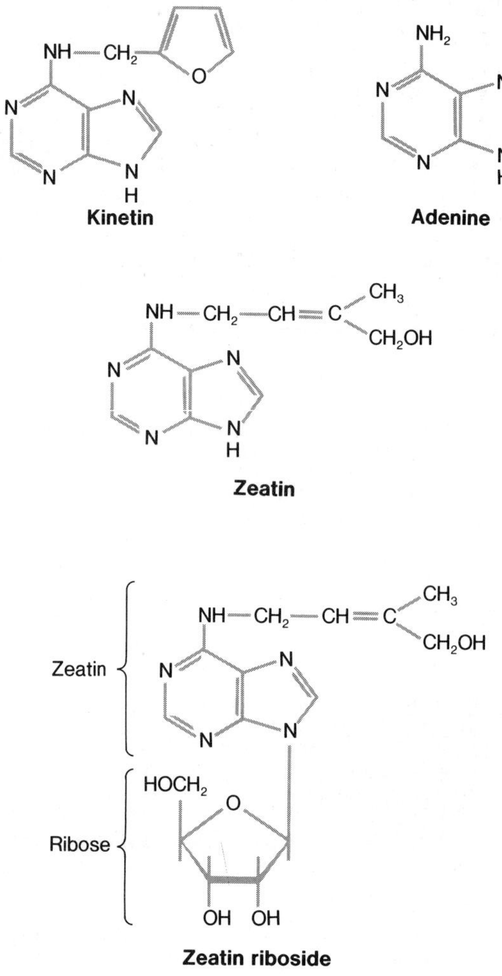

Kinetin

Adenine

Zeatin

Zeatin riboside

Gibberellins

Gibberellins were first found as a result of research on a rice plant disease called "foolish seedling disease." Affected rice seedlings become unusually tall, but they are spindly and weak, and usually break and fall over before they produce ripe rice grains. Japanese scientists learned that the disease is caused by a fungus, *Gibberella fujikuroi.* In 1926, E. Kurosawa showed that a substance that caused excessive growth in rice plants could be extracted from the fungus. Because of World War II, however, biologists in other countries did not become aware of this work until the 1950s. Then biologists began to study the effects of this extracted substance, **gibberellic acid (GA),** on vascular plants and to search for

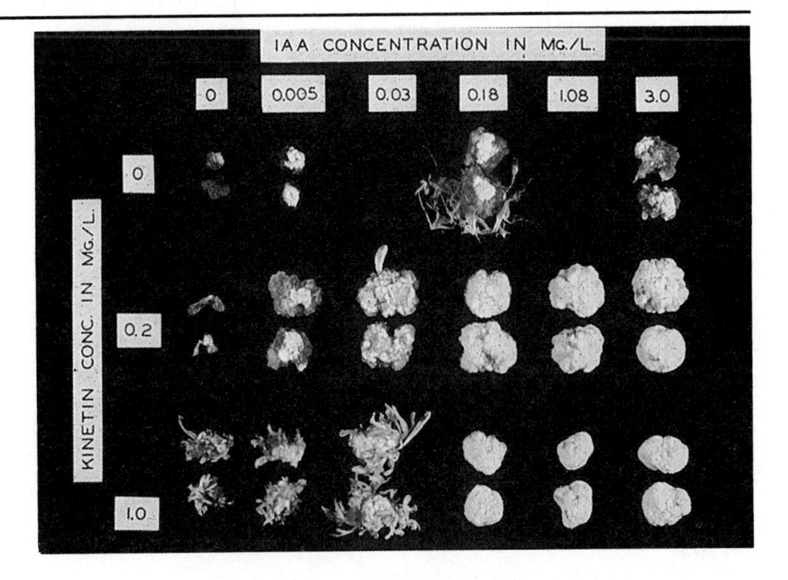

Figure 14.11 Interactions between auxin (IAA) and cytokinin (kinetin) in cultures of tobacco tissue. When IAA and kinetin concentrations are balanced, cultures grow as undifferentiated calluses (middle rows). Higher IAA to kinetin ratios result in root growth (upper right). Lower IAA to kinetin ratios result in shoot growth (lower left).

similar naturally occurring compounds. Eventually, chemically similar compounds that caused such growth responses were found in vascular plants, as well as in algae and in other fungi. Collectively, these compounds are known as **gibberellins** (figure 14.12).

One of the most striking effects of gibberellic acid (GA) is the stem elongation that it causes in dwarf varieties of certain plants. When GA is applied to such dwarf plants, they grow to normal size (figure 14.13).

Another interesting effect of gibberellic acid is the response it causes in plants growing as rosettes. A rosette is a compact growth form where leaf attachments (nodes) are very close together on the stem. Cabbage, for example, is a plant that grows in a rosette form (the familiar cabbage "head") during its first growing season. During its second season, however, cabbage grows tall, flowers, and produces seeds. In order to grow tall during the second growing season, a cabbage plant normally must experience chilling during the winter between its first and second growing seasons. Such chilling is the normal stimulus for stem elongation. GA applied to a first-year plant, however, can replace chilling and cause a rapid elongation of the stem known as bolting (figure 14.14).

Gibberellins also regulate important changes in germinating seeds. Inside seeds, embryos are in a metabolically slowed state called **dormancy.** At germination, the embryo resumes its growth, using nutrients stored in the seed, particularly starch. Starch must be broken down by enzymes to make glucose available to the growing embryo. When germination begins in barley seeds, for example, the embryo

Figure 14.12 Structure of gibberellic acid.

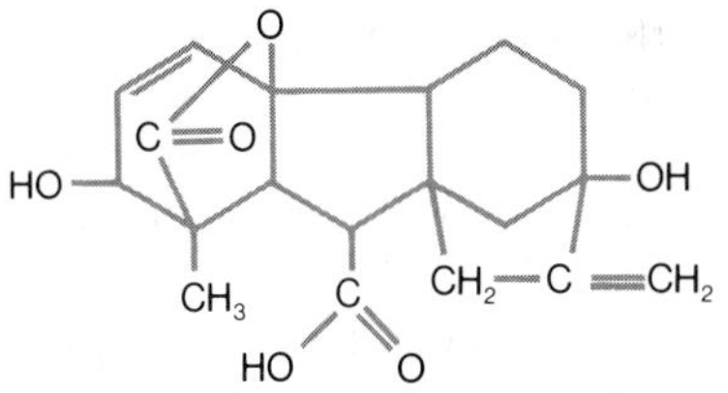

O
C=O
HO
OH
CH_3
C
CH_2—C =CH_2
HO
O

Gibberellic acid

Figure 14.13 Effect of a single application of gibberellic acid (GA). The normal plants at the left received no treatment. The normal plants second from the left received GA, but GA has little effect on the length of normal plants. The dwarf plants third from the left received no treatment. The plants on the right are dwarf plants that received a GA treatment. They have grown to the same length as normal plants.

Figure 14.14 Bolting caused by gibberellic acid. The cabbage plants on the left are untreated. Treatment with gibberellic acid has caused bolting and flowering in the plants on the right.

Figure 14.15 Structure of abscisic acid.

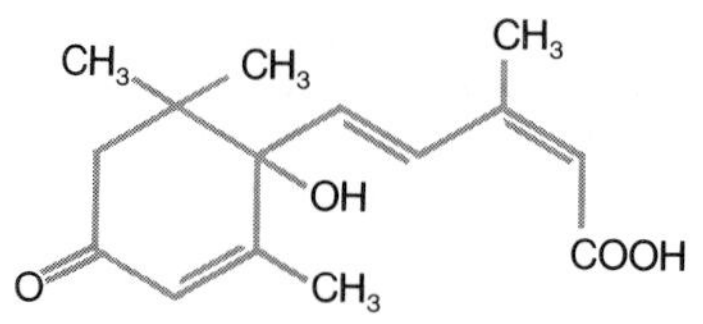

CH_3 CH_3 CH_3 OH COOH O CH_3

Abscisic acid

produces GA, which stimulates enzyme production and starch digestion. This reaction is particularly important in the brewing industry because barley seeds are germinated in the malting process, an early step in beer production. This makes sugar available for alcohol production by yeast during brewing. GA is used to speed up the malting process.

Abscisic Acid

The plant hormones that we have considered so far generally are stimulators of various processes in plants. Another group of plant hormones play very different roles in plants because they act basically as inhibitors. The most widely studied of these inhibitors is **abscisic acid** or **ABA** (figure 14.15).

Abscisic acid, which is produced by aging leaves in autumn, causes bud scales to form around delicate buds that must be protected during their long winter dormancy. In this way, ABA plays an important role in plants' adjustments to the changing seasons. ABA also causes fruits of a number of plants to come loose and fall off. This process is called abscission, and although ABA does much more than simply cause fruit to fall, it is where we get the name abscissic acid.

Photomorphogenesis

Plants show a number of kinds of responses to light in their environment. Some of them involve turning toward light (phototropism) while others depend on the length of the light period during each day (photoperiodism).

Seedlings grown in the dark provide a good example of a third category of responses, known as **photomorphogenesis.** These seedlings have a peculiar, spindly appearance and they have extremely long internodes (the spaces between leaf attachments) (figure 14.16). No new leaves develop beyond those present in the seed. The plants also have a very pale color because they have colorless **etioplasts** instead of green chloroplasts. These symptoms, which develop during growth in the dark, are caused by abnormally high auxin levels and excessive ethylene accumulation. Such dark-grown plants are said to be **etiolated.**

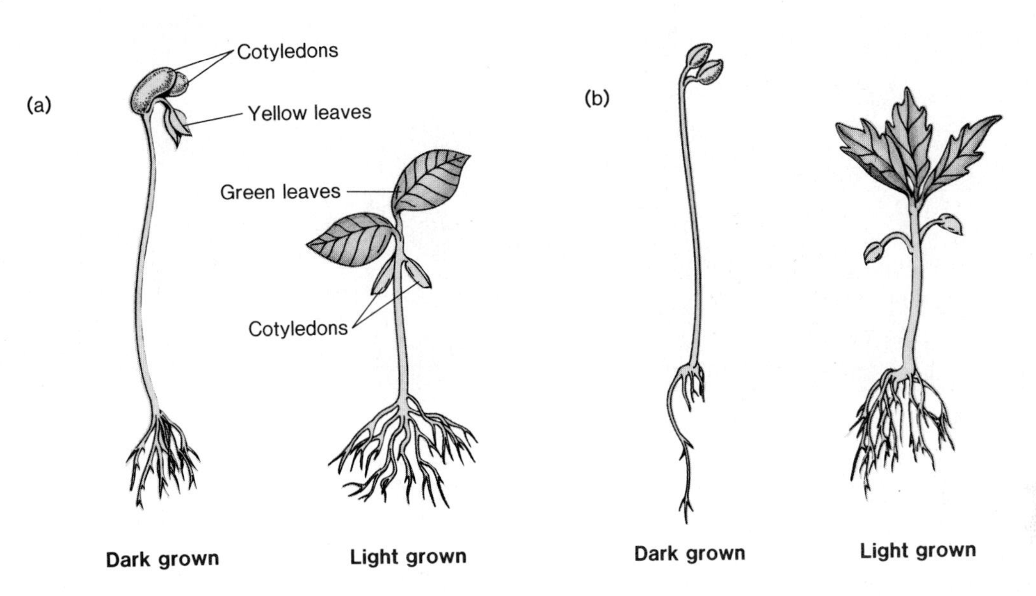

Figure 14.16 Etiolation in dark-grown seedlings of (*a*) bean and (*b*) mustard. Note the elongate stems in both and the small, colorless leaves of the bean. Leaves are lacking from the etiolated mustard seedling, while the etiolated bean has only the leaves that were already present inside the seed. Cotyledons are thickened leaves that store nutrients in seeds.

It is interesting, however, that very dim, continuous light or even a single, brief exposure to brighter light each day prevents etiolation of a growing plant. Experiments have determined the quantity and quality (wavelength) of light required to switch etiolated seedlings to normal growth. A dark-grown pea seedling, for example, switches to normal growth if exposed for only five minutes to red light (660 nm wavelength). But, interestingly, if such a red light exposure is followed by a five-minute exposure to far-red light (730 nm wavelength), etiolated growth continues. This rather intriguing result leads to a fundamental question: How does light actually exert control over plant growth processes?

Phytochrome

In many plant responses to light, a specific kind of pigment (light-absorbing) molecule called **phytochrome** ("plant color" or "plant pigment") plays a key role. Phytochrome exists in two forms that are interconvertible in **photochemical** (light-chemical) reactions. Phytochrome molecules are reversibly converted from one form to the other when they absorb light energy. Red light (660 nm wavelength) converts the red-light-absorbing form of phytochrome, called P_r, to the far-red-light-absorbing form, called P_{fr} (figure 14.17). And P_{fr} is converted back to P_r by far-red light (730 nm wavelength).

Figure 14.17 Photochemical reactions of phytochrome. Red and far-red light interact with the phytochrome molecule to convert it back and forth from one form to the other.

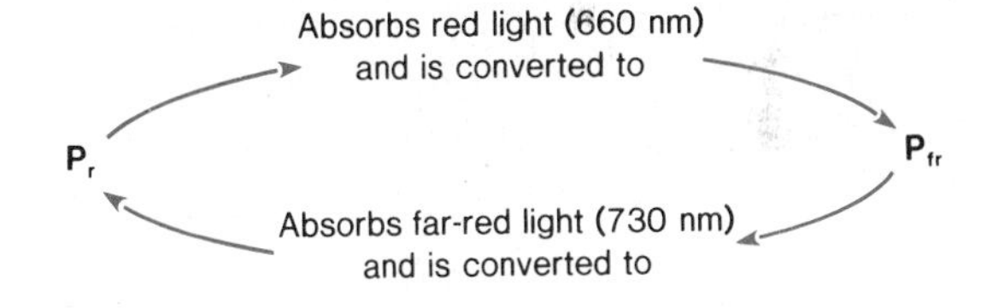

This red/far-red reversibility of the forms of phytochrome was very clearly demonstrated in research in the 1930s on germination of lettuce seeds of the Grand Rapids variety. First, scientists discovered that wavelengths of light most effective in promoting lettuce seed germination were in the red region of the light spectrum, and that far-red light depressed the germination rate, even below the low rate found in seeds kept in total darkness. Later it was shown that if

Figure 14.18 Red/far-red reversibility of light-stimulated germination of lettuce seeds. D: Seeds kept in dark. R: Received three minutes red light. R/F: Received three minutes red light followed by three minutes far-red light. See also table 14.1.

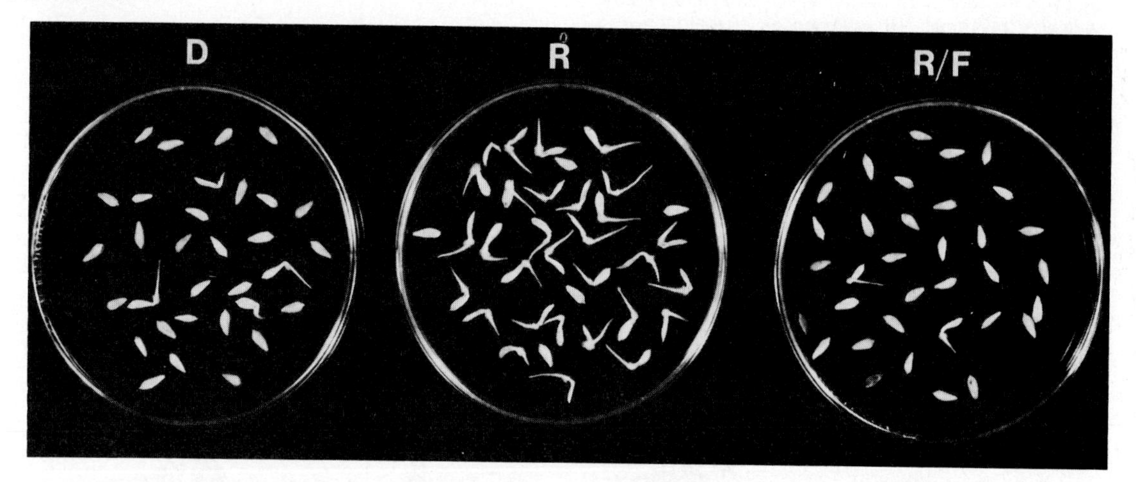

seeds stimulated by red light were then exposed to far-red light before germination had actually begun, germination was inhibited. Scientists concluded, then, that far-red light reverses the effect of red light on Grand Rapids lettuce seeds (figure 14.18).

In fact, the germination process can be "switched on or off" over and over again using first one kind of light and then the other to change the form of phytochrome in the seeds (table 14.1). You can see in table 14.1 that the percentage of seeds that germinate depends almost entirely on which form of light they were exposed to *last*.

Phytochrome seems to play a role in many kinds of plant responses to light in the natural environment as well, but there is still much to be learned about its functions.

Table 14.1
Data on Lettuce Seed Germination Percentages Following Exposure to Red and Far-Red Light

Irradiation	Percentage Germination
Red	70
Red/Far-red	6
Red/Far-red/Red	74
Red/Far-red/Red/Far-red	6
Red/Far-red/Red/Far-red/Red	76
Red/Far-red/Red/Far-red/Red/Far-red	7
Red/Far-red/Red/Far-red/Red/Far-red/Red	81
Red/Far-red/Red/Far-red/Red/Far-red/Red/Far-red	7

Source: H. A. Borthwick, et al., "A Reversible Photoreaction Controlling Seed Germination," *Proceedings of the National Academy of Sciences 38*(1952): 662–66.

Photoperiodism

Experienced gardeners can plan their flower gardens so that the advance of spring is marked by a very orderly series of floral displays. Perhaps you have noticed a neighbor's garden in which new blossoms appear on a regular basis—flowering border plants are followed by blooming vines, which, in turn, are succeeded by flowering shrubs and trees. Considering the variability of springtime temperature and moisture conditions, maybe you have wondered how gardens can be programmed to provide a regular series of flowering events. Such programming is possible because many plants flower in response to photoperiodic stimuli, that is, changes in the twenty-four hour cycle of alternating light and dark. Because different species respond to different day lengths, gardeners can select flowers so that they bloom in sequence as the light-dark cycle changes.

Flower gardeners are simply taking advantage of a system, which has evolved in many plant species, that assures that all members of a species in a given area flower at the same time. You can understand the adaptive value of this precise photoperiodic timing of flowering. It ensures simultaneous flowering and thus increases the opportunities for reproductive success.

In addition to flowering, other aspects of plant life also are regulated photoperiodically. For example, the photoperiodic stimuli of shorter days and longer nights in the fall stimulate plants to prepare for dormancy, so that they are ready to withstand the rigors of winter (figure 14.19). The

Figure 14.19 A leafless, dormant tree in midwinter. Dormancy preparations are made in response to photoperiod changes that provide reliable information about the time of year.

advantage of photoperiodic timing is that photoperiodic changes occur in the same regular fashion year after year. It is of great adaptive value for many species of plants that such critical parts of their lives as flowering and entry into dormancy are controlled by photoperiodic changes rather than by other, less regular, seasonal environmental changes, such as temperature and moisture changes.

Types of Photoperiodic Responses

Research on photoperiodism began in the 1920s when W. W. Garner and H. A. Allard made a series of soybean plantings over a period of several weeks. In late summer, they observed the flowering time of the plants in the various groups. Despite age differences due to the different planting times, all of the soybeans flowered surprisingly close to the same time. All of the soybeans flowered in late summer as the days shortened. Further study showed that flowering occurred when days shortened below a critical length.

Soybeans and other plants that flower when day length decreases below a critical point are called **short-day plants.** On the other hand, **long-day plants** show an opposite response: they flower when day length increases beyond a particular point. **Day-neutral plants** (for example, the tomato) do not seem to be influenced by photoperiod. They flower in response to other environmental factors or simply when they become mature (figure 14.20).

Figure 14.20 The general categories of photoperiodic responses to the daily light-dark cycle in flowering plants. (*a*) Short-day plants. (*b*) Long-day plants. (*c*) Day-neutral plants.

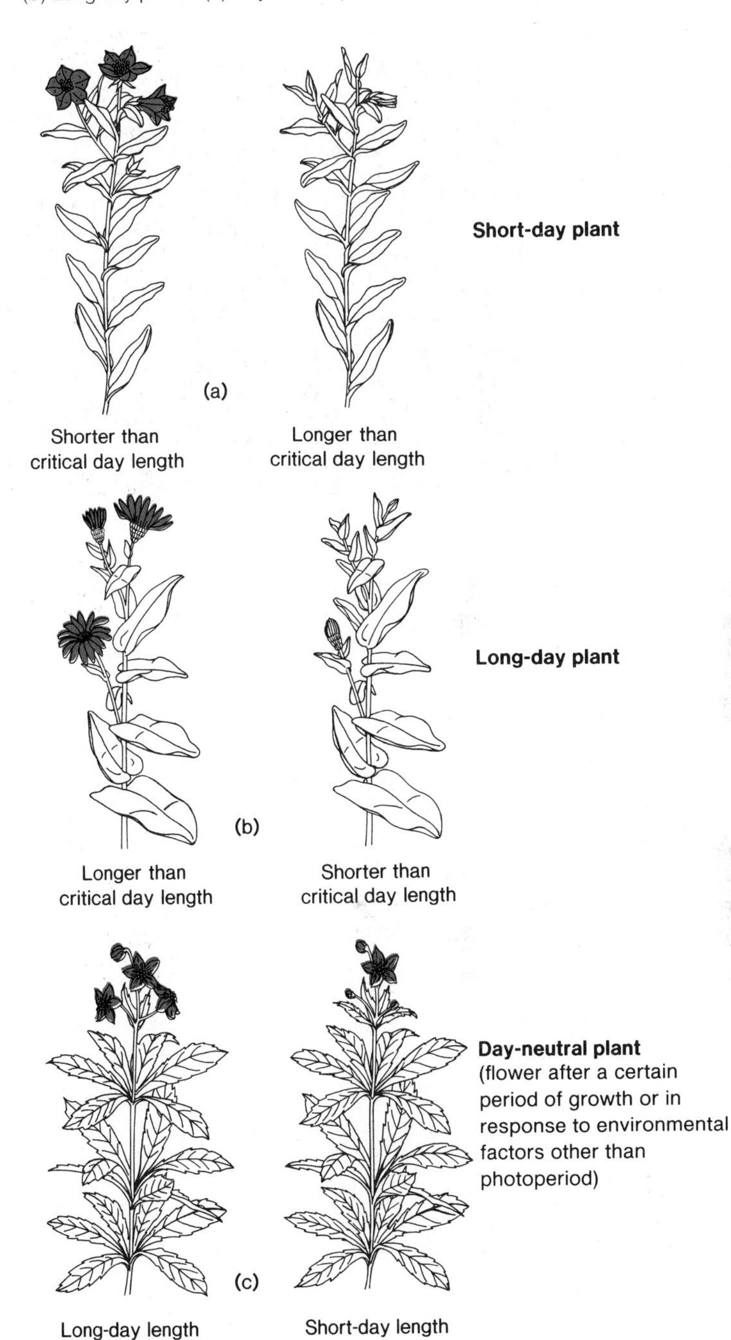

Generally, short-day plants (for example, cocklebur, ragweed, and asters) grow and mature during spring and early summer and become ready to flower in response to the shortening day length of late summer and fall. Long-day plants (for example, radish, clover, and wheat) flower in response to increasing day length in the spring and early summer.

The real distinction between short-day and long-day plants, however, is the way that each type of plant responds to its own **critical photoperiod,** a specific ratio of day and night length. Short-day plants flower when the light period becomes shorter than their critical photoperiod, however long that critical photoperiod might be. Long-day plants flower when the light period becomes longer than their critical photoperiod, however long that critical photoperiod might be. For example, *Xanthium* (cocklebur) is a short-day plant because it flowers when day length decreases below its critical photoperiod in late summer. The critical daily light-dark cycle for *Xanthium* turns out to be fifteen hours of light and nine hours of darkness. Although this is not a particularly short day by everyday human standards, it is the critical photoperiod for this particular short-day plant.

Short Days/Long Nights

In the 1930s, K. C. Hamner and James Bonner experimented further with the relative roles of light and dark periods by testing the effects of a variety of light-dark combinations on the cocklebur *Xanthium.* As we said earlier, *Xanthium* normally flowers when it has fifteen hours of light and nine hours of darkness. But Hamner and Bonner found that flowering also occurred when they subjected *Xanthium* to artificial, experimentally lengthened daily cycles with light periods longer than fifteen hours, *as long as the dark period was nine hours or more.* In fact, a critical dark period of longer than nine hours seemed to be the primary determining factor in *Xanthium's* short-day response. Thus, short-day plants such as *Xanthium* might more properly be described as "long-night plants."

Another experiment underscored the importance of the dark period. If the dark period was interrupted by a short period of intense light, even a period as brief as one minute, short-day plants failed to flower (figure 14.21). This is further evidence that short-day plants respond to an uninterrupted dark period of a critical length.

Figure 14.21 The importance of the dark period in the flowering response of short-day plants. The dark periods in cycles 2 and 3 are the same length, but the brief light interruption of cycle 3 prevents the flowering response. This principle has practical applications for greenhouse operations. It is cheaper to use a brief light period to interrupt the night than to keep lights on for several more hours each day if short-day plants must be prevented from flowering.

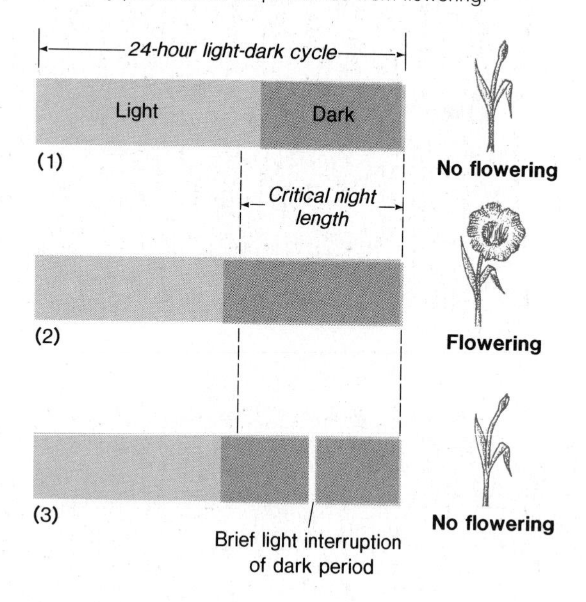

Experiments with long-day plants have shown that night length also is critical in their photoperiodic responses. In the case of long-day plants, nights that are *shorter* than a particular critical length are necessary to cause flowering. Nights longer than the critical length inhibit the flowering of long-day plants, even if they are given very long days in artificial, experimentally lengthened daily cycles. If the longer than critical-length night is interrupted by a brief light period, however, long-day plants will flower. Clearly, the length of the *uninterrupted* night period is also a key factor in the photoperiodic responses of long-day plants.

Time Measurement in Photoperiodism

Phytochrome plays some part in the time measurement necessary for photoperiodic responses, but its exact role is not fully understood. Plants clearly can measure the lengths of light and dark periods very accurately, but the phytochrome system alone does not adequately explain this time-measuring ability. Time seems to be measured by an "internal clock" mechanism that is separate from the phytochrome system. A short-day plant measures night length using this clock mechanism in a way that is as yet undetermined.

Let us now turn our attention to timekeeping in living things.

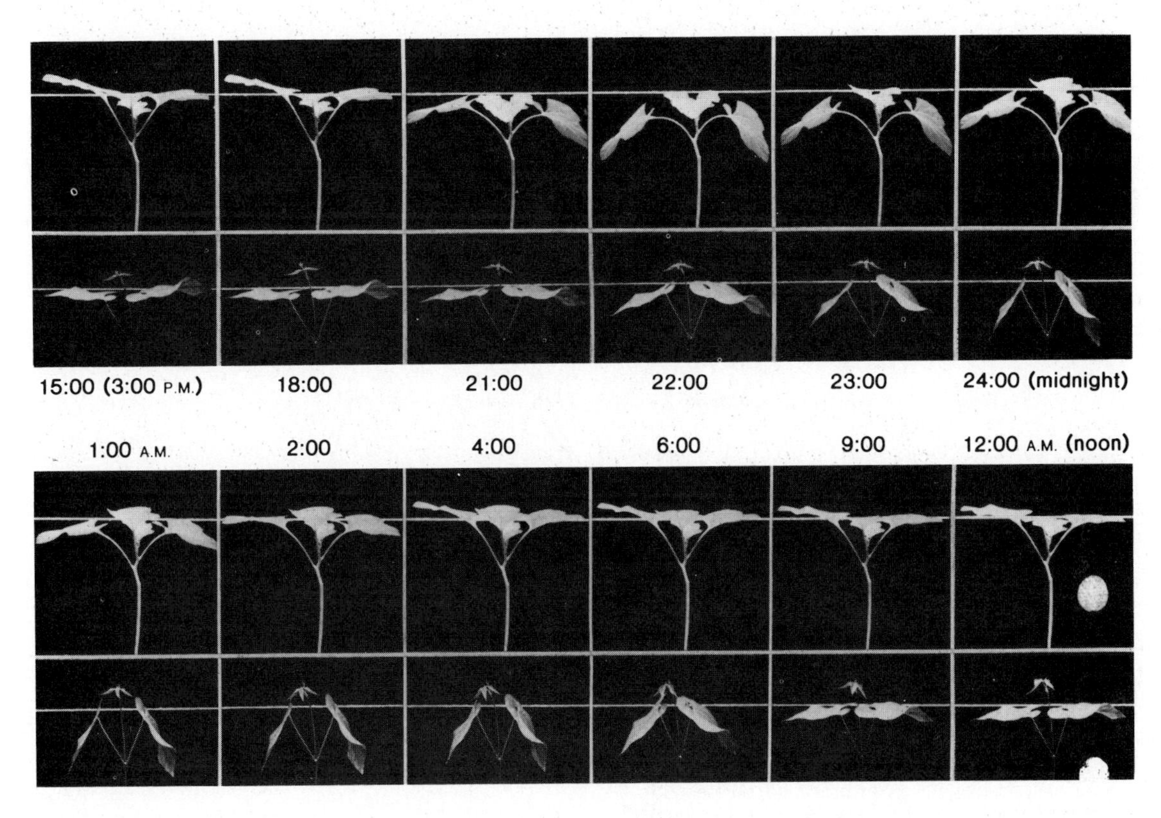

Figure 14.22 A photographic record of rhythmic leaf movements in cocklebur (top row) and bean (second row). The plants were photographed at hourly intervals from noon to noon. Note that bean leaves drop more sharply and later in the evening than the cocklebur leaves.

Biological Rhythms

The nature of time measurement in living organisms is one of the most intriguing and puzzling questions in modern biology. Obvious daily cyclic fluctuations (**rhythms**) in many physiological processes are expressions of this biological timekeeping.

Daily Rhythms

Plants in nature show obvious daily cycles. For example, beans and many other plants spread their leaves horizontally in the daytime, exposing a greater leaf surface area for light absorption. At night their leaves fold up or down in what has been called a sleep movement (figure 14.22). In addition to such externally obvious daily cycles or rhythms, plants have many subtle, rhythmic, internal physiological fluctuations. For example, some enzyme activity levels, certain ion concentrations in internal fluids, and sensitivities to many drugs and other chemicals change throughout the day in a rhythmic fashion. All plants and all animals, including humans, have such daily rhythms. Of course, it is not surprising that life processes vary in rhythmic manner with the time of the twenty-four-hour **solar day** because all organisms have evolved in a decidedly rhythmic world where light and darkness cycle with great regularity. Other physical factors, such as temperature and barometric pressure, also fluctuate, but not with nearly the same regularity as the daily light cycle.

Figure 14.23 Measurement of bean plant sleep movements under constant laboratory conditions. Leaf is attached to a lever with a pen that traces on paper fastened to a rotating drum. Results of several days of recording are shown.

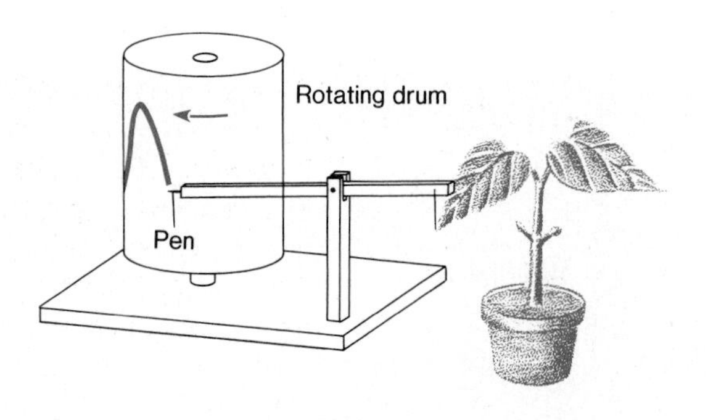

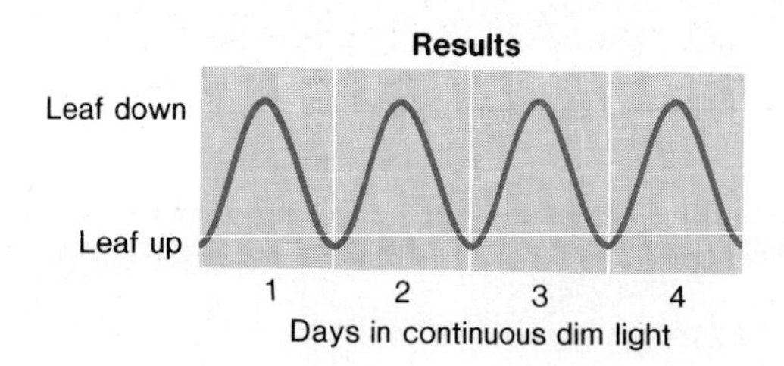

Figure 14.24 Biological rhythms and temperature. (*a*) Period of rhythm is the time from a point in one cycle to the same point in the next cycle. Amplitude is the difference between the high point and the low point within one cycle. (*b*) Temperature relationships. Increased temperature would be "expected" to change *both* period length and amplitude of rhythms. However, amplitude changes are as predicted, but period length changes little if at all in response to raising the temperature.

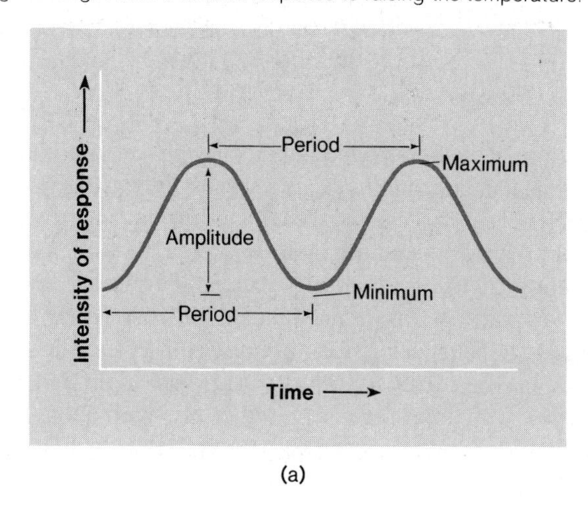

(a)

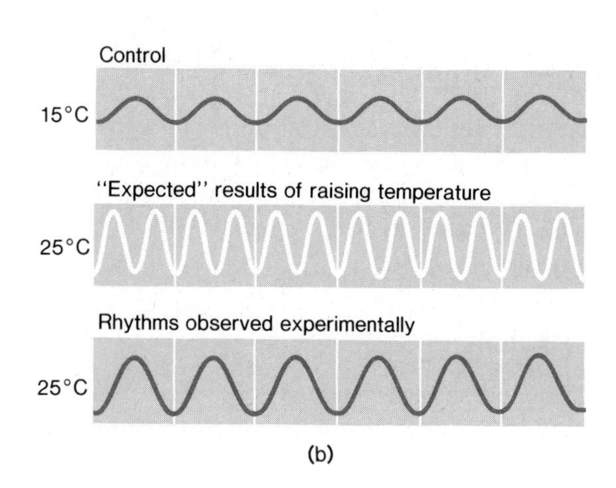

(b)

Figure 14.25 Comparison between an exact twenty-four-hour daily rhythm and a circadian rhythm. (*a*) An exact twenty-four-hour solar day rhythm, the characteristic pattern of all daily rhythms under natural entrainment and a few rhythms even under laboratory constant conditions lacking normal entraining mechanisms. (*b*) A circadian rhythm with a period length less than twenty-four hours. This slightly exaggerated model shows how peaks of rhythmic activity occur earlier during each actual solar day. Such circadian rhythms are observed under laboratory constant conditions when organisms are deprived of normal entrainment, such as the daily light-dark cycle.

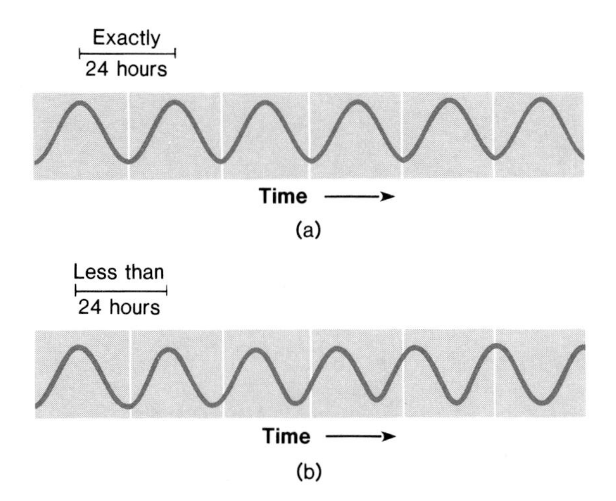

What *is* surprising is that the solar-day rhythms of many living organisms continue even when the organisms are deprived of obvious external information about the time of day. The cycles continue when organisms are maintained in laboratories in constant light, constant temperature, and even constant barometric pressure. For example, bean plant sleep movements continue for days when a bean plant is left undisturbed under constant conditions (figure 14.23). Biologists have concluded that this persistence under constant conditions must mean that organisms possess internal clocks that time rhythmic processes even in the absence of obvious external cues about the time of day.

Another surprising result in this research is that in many cases, temperature change has very little effect on the rates at which rhythmic processes go through their daily cycles. This result was unexpected, because rates of many biological processes are strongly affected by temperature increases. Scientists demonstrated that the **period length,** which really is a measure of the rate at which a biological rhythm "runs," is quite temperature independent in most cases (figure 14.24).

Another feature of most rhythms in organisms kept under constant conditions in the laboratory, especially in constant light or constant darkness, is that period lengths usually are slightly longer or slightly shorter than exactly twenty-four hours. We call these **circadian** (literally "about a day") **rhythms** (figure 14.25). Sometimes, biologists use this name much more generally and speak of all clock-timed rhythmic phenomena as circadian rhythms or circadian clocks. But, technically, the term circadian describes only rhythms that have period lengths slightly different from twenty-four hours when measured under constant conditions. Circadian rhythms are seen only under laboratory conditions, where organisms are deprived of information about normal light/dark cycles. In nature, the daily light/dark cycle keeps period lengths of daily rhythms exactly twenty-four hours long. This external control by environmental factors that keeps rhythms precise is called **entrainment,** and the environmental factors that entrain rhythms are called **zeitgebers** (from the German *Zeitgeber,* meaning "time-giver"). Thus, rhythms become circadian in the laboratory only when deprived of normal entrainment by environmental zeitgebers.

Rhythms and Clocks

Most or maybe all of the biological rhythms, such as plant sleep movements, that scientists have described and studied are overt, external expressions of timekeeping by an underlying **biological clock.** You should think of rhythms as representing the hands of the clock—they are resettable just as the hands of an ordinary clock can be reset to any time setting. In all plants and animals there seems to be an underlying clock, and the resettable rhythms (the hands) are linked to it. Rhythms are reset by environmental zeitgebers.

Interpretation of the results of biological rhythms experiments is very difficult. Experimental results might relate to a property of only the clock's hands (observable, overt rhythms), the clock's basic timing mechanism, the linkage between the two, or some combination of all three. One certainty, though, is that there are some challenging unanswered questions about both the underlying biological clock and its relationship to the overt rhythms that play such a large part in the lives of all plants and animals. Possibly the most fundamental of these questions involves the nature of the clock (or clocks) and its location (or locations) in the organism (see box 14.1).

Box 14.1

Biological Clocks

Internal or External Timing?

The search for biological clocks, the underlying timekeepers in living organisms, has generated considerable controversy. There are two opposing, but not entirely different, viewpoints on the nature of the biological clock. One hypothesis proposes that the clock is an entirely *internal,* biochemical oscillator mechanism operating at the cellular level. This proposed cellular clock works like the pendulum that measures time in a pendulum clock. Of course, pendulum clock hands (rhythms) can be reset to any time on the face of the clock. The hypothesis also proposes that the oscillator mechanism may be regular, cyclical changes in some cellular process, such as a complex enzyme reaction series, differences in membrane permeability, or possibly even repeated transcription (see page 379) of a segment of a DNA molecule.

The other hypothesis also suggests that the cellular clock is internal, but that its basic timing information comes from the *external* environment. This proposed cellular clock is like the motor of an electric clock. Again, the hands (rhythms) of an electric clock are resettable, but basic time measurement is by an electric motor driven by alternating current supplied from the outside. Scientists who support this hypothesis think that regular daily fluctuations of physical environmental forces, such as magnetic or electrostatic field intensities, cosmic radiation, or some combination of these or other physical forces, provide the basic timing information. They propose that these forces may override all efforts of investigators to maintain constant laboratory conditions.

Either of these hypotheses can be used to explain almost every result of experiments on biological rhythms and almost every inferred property of biological clocks. It is difficult to conceive of experiments that would clearly distinguish between the two hypotheses regarding the nature of basic clock timing. All such studies are handicapped because all available experimental results provide us with information about the behavior of only the clock "hands," that is, the observable, overt rhythmic processes. The search for the underlying biological clock (or clocks) of living things is one of the more difficult and challenging areas of biological research.

Box figure 14.1 Two models of the internal cellular clock. The "hands" can be reset (entrained by environmental zeitgebers) in either case, but what is the basic clock mechanism? (*a*) Is it an internal biochemical oscillator that measures time like a pendulum? (*b*) Or is it like an electric motor clock, where time measurement depends on alternating current from the outside?

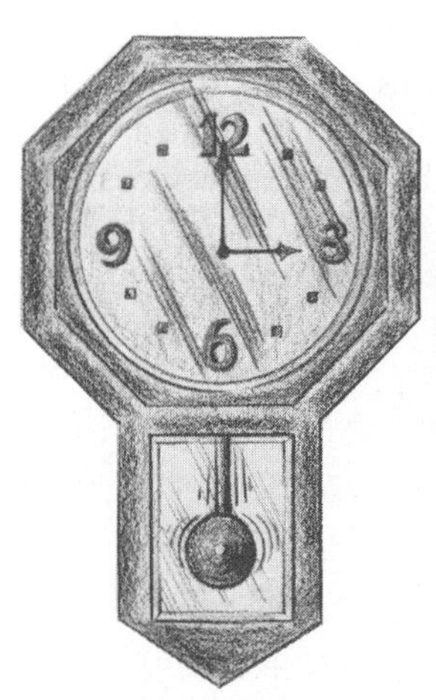

(a) Pendulum clock model—timing comes from biochemical oscillator

(b) Electric clock model—external timing cues reach clock

Figure 14.26 The abscission zone. (*a*) Abscission zones (color) may be located at the base of a petiole, at the junction of a petiole and a leaf, or even at several different points. (*b*) Internal structure of an abscission zone at the junction of a petiole and the stem of a silver fir.

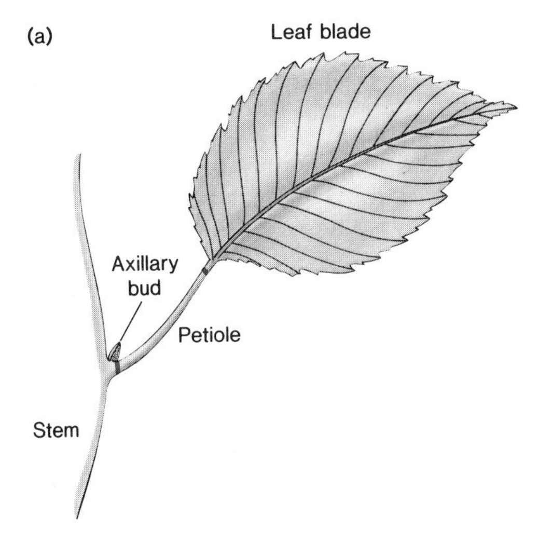

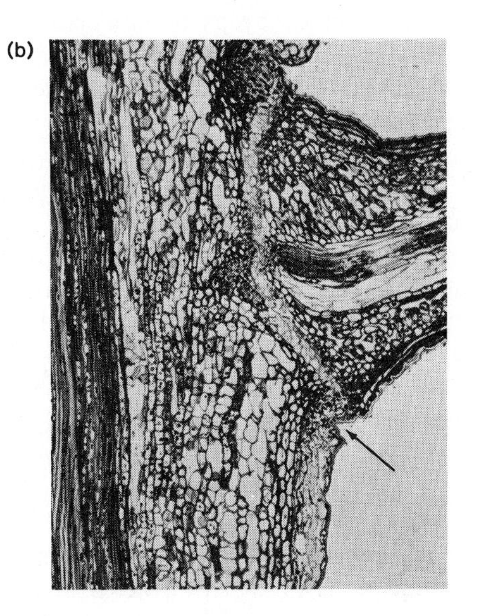

Season-Ending Processes

Perennial plants (plants that are active during several to many growing seasons) cannot survive freezing winter temperatures in their active, growing, summertime condition. What preparations do perennial plants need to make for winter dormancy?

In late summer and early fall, nutrients are transported to underground storage sites such as roots, and tough scales or other protective devices form around the buds that will begin the next season's growth.

To most people, however, changing leaf color and the eventual loss of leaves are the most obvious and familiar signs of approaching winter in plants. Leaves contain relatively soft tissue that would be very difficult to protect from winter damage, and preparation for winter dormancy includes **senescence** (aging) and **abscission** (detachment) of leaves.

During leaf senescence, many components of leaves are broken down and the essential materials, such as mineral ions, are exported from the leaves to other parts of the plant. In this way, plants conserve valuable resources. Eventually, as senescence nears completion, the leaves are prepared for abscission.

Abscission

While senescence changes most of the cells of a leaf, abscission is due to changes in a very narrow band of cells located, depending on the type of plant, either at the base of the leaf petiole (leaf stalk) or at the point where the blade of the leaf joins the petiole. This band of cells forms an **abscission zone,** the breaking point where the leaf eventually detaches from the stem (figure 14.26). During abscission, enzymes break down pectins, which are important components of the material between cells, and the cellulose of cell walls is also weakened. These enzyme actions cause the layers of the abscission zone to separate, and the leaf falls. Corky material then covers the abscission zone and forms a **leaf scar.**

Control of abscission, like so many other processes in plants, involves complex interactions and changing balances of several plant hormones. Thus, the whole life of a vascular plant, from seed germination to aging and senescence, involves continuing interactions with the environment, and plant hormones mediate the plant's responses at every step along the way.

Summary

A terrestrial vascular plant remains in one place and interacts with its changing environment throughout its life. These interactions are mediated by plant hormones. Plant hormones act mainly as growth regulators.

Auxin was discovered as a result of experiments on phototropic responses of grass seedlings.

Phototrophic responses result from lateral transport of auxin to the shaded side of seedlings, where greater cell elongation occurs. Auxin causes this cell elongation. Auxin also is responsible for apical dominance because auxin produced in the shoot apex inhibits the growth of lateral buds. The principal naturally occurring auxin is indoleacetic acid, but other substances, including some laboratory-synthesized substances, act as auxins.

Ethylene, a gas, is an important fruit-ripening agent. It promotes hydrolysis of starch to sugar and stimulates cellulase activity, which results in fruit softening.

Cytokinins are cell-division promoters that act along with auxins in general regulation of growth responses.

Gibberellins, which were first extracted from a fungus, stimulate stem elongation in dwarf plants and probably play a role in stem elongation in normal plants as well. Gibberellic acid, a gibberellin, causes bolting in rosette plants. Gibberellic acid is important in seed germination because it stimulates synthesis of enzymes, which hydrolyze stored starch, thus making sugar available to the growing seedling.

Abscisic acid, an inhibitor, is involved in preparation for and maintenance of dormancy. It also promotes abscission of various fruits and seeds.

Many light responses are promoted by red light but opposed by far-red light. This generalization led to the discovery of phytochrome, a pigment that exists in two forms that are interconvertible in photochemical reactions.

Photoperiodism is particularly important in timing of reproductive activity (flowering) and in onset of dormancy. Short-day plants respond when nights are longer than a particular critical length, while long-day plants respond when nights are shorter than a particular critical length.

Photoperiodic responses involve time measurement that depends on an internal clock mechanism. The clock mechanism also times rhythmic daily changes in various physiological functions. But the nature of the underlying cellular clock mechanism remains obscure.

Plants prepare in several ways for winter dormancy. Nutrients are stored underground, and protective scales form around delicate buds. But the most obvious sign of preparation for winter is leaf senescence and abscission. These processes also are under hormonal control, as are all of the environmental responses in the lives of vascular plants.

Questions for Review

1. Define phototropism and geotropism.
2. Explain the role of auxin in apical dominance.
3. Why does "one bad apple spoil the barrel"?
4. Describe an etiolated plant and explain how etiolation occurs?
5. What is P_{fr}?
6. Explain the adaptive value of using photoperiodic changes rather than seasonal temperature changes to time events in plants' lives.
7. Distinguish between short-day and long-day photoperiodic responses.
8. What are circadian rhythms?
9. How do biological rhythms relate to biological clocks?

Questions for Analysis and Discussion

1. Gibberellins are not required for the growth of the fungus that causes "foolish seedling disease" in rice. What adaptive value do you suppose gibberellin production has for the fungus?
2. Why do you think some short-day plants near the northern edge of their range are able to produce mature seeds and fruits during some growing seasons, but not during others?
3. Explain jet lag in terms of biological rhythms and suggest how the clock "resetting" that cures jet lag may occur.

Suggested Readings

Books

Palmer, J. D. 1976. *An introduction to biological rhythms.* New York: Academic Press.

Saigo, R. H., and Saigo, B. W. 1983. *Botany: Principles and applications.* Englewood Cliffs, N.J.: Prentice-Hall.

Stern, K. W. 1982. *Introductory plant biology.* 2d ed. Dubuque, Ia.: Wm. C. Brown Company Publishers.

Thimann, K. V. 1977. *Hormone action in the whole life of plants.* Amherst, Mass.: University of Massachusetts Press.

Wareing, P. F., and Phillips, I. D. J. 1981. *Growth and differentiation in plants.* 3rd ed. New York: Pergamon.

Articles

Cleland, C. F. 1978. The flowering enigma. *BioScience* 28:265.

Hendricks, S. B. 1980. Phytochrome and plant growth. *Carolina Biology Readers* no. 109. Burlington, N.C.: Carolina Biological Supply Co.

Hillman, W. S. 1979. Photoperiodism in plants and animals. *Carolina Biology Readers* no. 107. Burlington, N.C.: Carolina Biological Supply Co.

Satter, R. L., and Galston, A. W. 1981. Mechanisms of control of leaf movements. *Annual Review of Plant Physiology* 32:83.

The Continuity of Life

Part 4

The ability to reproduce is a fundamental property of living things. All multicellular organisms age and eventually die. But for the species, reproduction provides a potential means of increasing numbers and results in continuing replacement of aging individuals with young, vigorous individuals. In addition, sexual reproduction produces individuals with new genetic combinations.

Reproductive processes involve specific activities of individual cells. In unicellular organisms, one individual divides mitotically to produce two individuals. Multicellular organisms produce specialized individual reproductive cells such as eggs and sperm. Via these specialized cells, characteristics are transmitted from one generation to the next by specific genetic mechanisms. Each individual offspring has a new set of genes that is somewhat different from each of the parents.

Expression of the genetic complement of a new individual occurs through developmental processes. Development continues throughout life in the form of growth, maturation, and aging. Developmental processes are also involved in continuing replacement of body cells, in healing of injuries, and in specific responses to infection and disease. But developmental processes gone awry as abnormal growth can be life-threatening.

In chapters 15 through 21 we will examine the cell division mechanisms involved in reproduction and development and the means by which genetic information is transmitted and expressed. Then we will look at some details of reproductive patterns in various organisms, and the nature of developmental processes that occur throughout life.

Facing page Scanning electron micrograph of the cells in a frog embryo.

15

Cell Division

Chapter Outline

Chapter Concepts

1. The cell cycle involves growth, duplication of genetic information, preparation for division, and division.
2. Mitosis is the process whereby the duplicated genetic information in the nucleus of a cell is precisely divided to produce two genetically identical nuclei.
3. The cytoplasm of a dividing cell is separated by cytokinesis, a process that differs in plant and animal cells.
4. A sexually reproducing organism must reduce its chromosome number by half at some point in the life cycle. This prevents a doubling of the normal chromosome number each time fertilization occurs.
5. Meiosis is a type of cell division by which this reduction in chromosome number occurs.

You are not the same person that you were yesterday. Literally millions of your body cells have been replaced in the past twenty-four hours. Flattened, dead cells are always wearing away from the surface of your skin, and those cells are constantly being replaced by new cells produced by cell divisions below the surface. As we saw in chapter 10, massive numbers of red blood cells are replaced all the time as millions of aging cells are taken out of circulation and equal numbers of new cells enter your bloodstream to take their place. In your intestine, lining cells are sloughed off the surface and carried out of the body as part of the feces. These cells are replaced by cell divisions in the gut lining.

In each of these cases, continuing cell division is vital for replacement of lost or worn-out cells if homeostasis is to be maintained in your body.

But, of course, the importance of cell division goes far beyond just cell replacement in adult organisms. All living things are composed of cells and materials produced by cells. All of those cells are produced by the division of previously existing cells. Thus, cell division is also the key to reproduction and growth of all living things.

The genetic information that determines the characteristics of cells and controls cells' many functions is contained in chromosomes in cell nuclei. During cell division, duplicate copies of this genetic material are distributed to the new cells with great precision, thus ensuring that the genetic information is accurately transmitted from cell to cell. The process by which the genetic material is distributed is called **mitosis.** Mitosis results in two new nuclei, each genetically identical to the original nucleus, because each receives a set of chromosomes identical to that of the original nucleus. The cytoplasm is divided, more or less equally, by a separate process known as **cytokinesis.**

In some types of organisms, specialized cells, produced by mitotic cell division, may develop directly into new individuals that are genetically identical to the parent. This type of reproduction is called asexual reproduction (see chapter 19).

The other principal method by which organisms reproduce is sexual reproduction. Sexual reproduction requires a genetic contribution from two different cells and involves a special type of cell division called **meiosis.** After we have discussed mitosis, we will consider meiosis and how it plays its key role in sexual reproduction.

The Cell Cycle

For a long time the emphasis in studies of cell division was almost entirely on the movement of chromosomes during mitosis. Detailed, accurate descriptions of that part of cell division were made in the late 1800s. These mitotic processes were studied because they could be observed with the light microscope. But events in the remainder of cells' life spans, the periods between visible mitotic events, were largely unknown. In fact, the period when mitotic activity is not occurring was viewed as a resting stage, an "interphase."

Figure 15.1 The cell cycle. Mitosis (M) and the synthesis of the genetic material (S) are separated by two gap phases (G_1 and G_2).

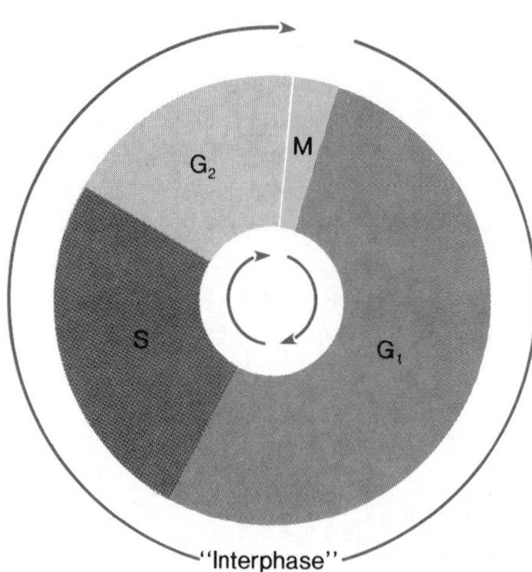

It was not until much later, with the development of increasingly sophisticated biochemical techniques beginning in the 1950s, that biologists were able to discover and describe many of the processes occurring in cells between mitotic divisions. They learned that the period between divisions is actually a very important and active time for the cell. The entire cycle—from the formation of a new cell until the cell divides—is called the **cell cycle,** and mitosis can be fully understood only in the context of this cell cycle.

After it was confirmed that DNA is the primary genetic material (see chapter 18), it was clear that the genetic information contained in DNA must be copied (replicated) before cell division. Biologists set out to learn exactly when DNA replication takes place. They learned that the DNA of a cell is replicated during a specific period of the so-called resting stage, frequently hours before mitosis occurs. The discovery of the timing of DNA replication led to a description of a cell cycle divided into four distinct, functional phases (figure 15.1).

The phase in which the DNA is synthesized is designated the S (for synthesis) phase. The period of time after a cell is produced by mitosis and before it enters the S phase is called the gap 1 (G_1) phase. (The G_1 phase is a "gap" only in the sense that it does not include major cell reproductive events involving DNA.) During this G_1 phase, growth and the synthesis of many compounds other than DNA occur. After the genetic material is duplicated in the S phase, another gap occurs (G_2). Not much is known about the G_2 phase except that it apparently involves the synthesis of various elements necessary for mitosis. The fourth phase is the nuclear division process itself, mitosis (M phase).

Box 15.1

Controlled and Uncontrolled Cell Division

Actively dividing populations of cells are necessary for normal continuing cell replacement processes in the body. The actual number of dividing cells varies now and then to meet changing needs, but it remains relatively stable over the long term.

On the average, one out of every two daughter cells produced in such replacement processes becomes a functioning, nondividing body cell. One-half of the cells remain capable of dividing again. Thus, the process is regulated so that the size of the population of cells capable of cell division remains stable.

Occasionally, however, body cells become transformed so that they divide in an uncontrolled manner. These transformed cells divide and divide again without the restraint that normally functions in cell replacement processes. Thus, both the cell population size and the number of dividing cells increase. This is characteristic of the behavior of cells in the growths that we commonly call tumors.

A popular definition of abnormal growth in a tumor might well include the phrase "rapid, uncontrolled cell division." While some abnormal growths do contain rapidly dividing cells, this is certainly not true of all. In fact, some normal cell replacement processes, such as replacement of aging red blood cells, involve cells that divide more frequently than cells in some tumors. The important difference is in the fate of the cells produced by the cell divisions. In tumors, both the cell population size and the number of dividing cells increase. The normal controls over the cell cycle and cell division fail to operate, and growth usually continues until it creates problems when the tumor interferes with some normal function. We will examine the transformation of normal cells into tumor cells in chapter 21.

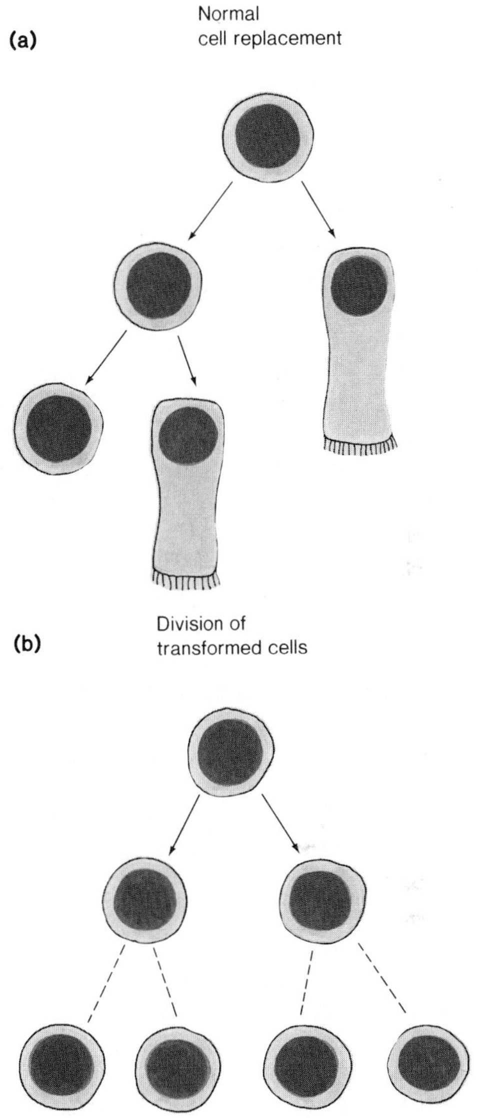

(a) Normal cell replacement

(b) Division of transformed cells

Box figure 15.1 Difference between normal cell replacement (*a*) and cell division of transformed cells in a tumor (*b*). In tumor growth, the number of dividing cells increases continually.

The transition from the G_1 phase to the S phase is a critical point in the cell cycle. However, the stimulus that causes a cell to enter the S phase is unknown. If the cell does not enter the S phase, it does not divide, and the majority of cells in adult organisms remain in the G_1 phase, permanently blocked from entering the S phase. However, as we noted earlier, some cells in mature adult organisms continue to divide periodically to replace cells that are lost or cease to function. Such cells that are going to divide do make the critical transition from G_1 to the S phase. Once a cell enters the S phase, it is usually committed to proceed through mitosis.

A better understanding of exactly what occurs at the critical point between G_1 and the S phase might offer clues for treatment and control of abnormal cell division. Cancer cells, for example, are cells that grow and divide without restraint or control. They probably are originally derived from cells that are normally blocked from entering the S phase; that is, cells that normally do not divide. But some factor, external or internal, transforms these cells so that they continue to enter the S phase in each cell cycle and thus, are committed to uncontrolled cell division (see box 15.1). Much

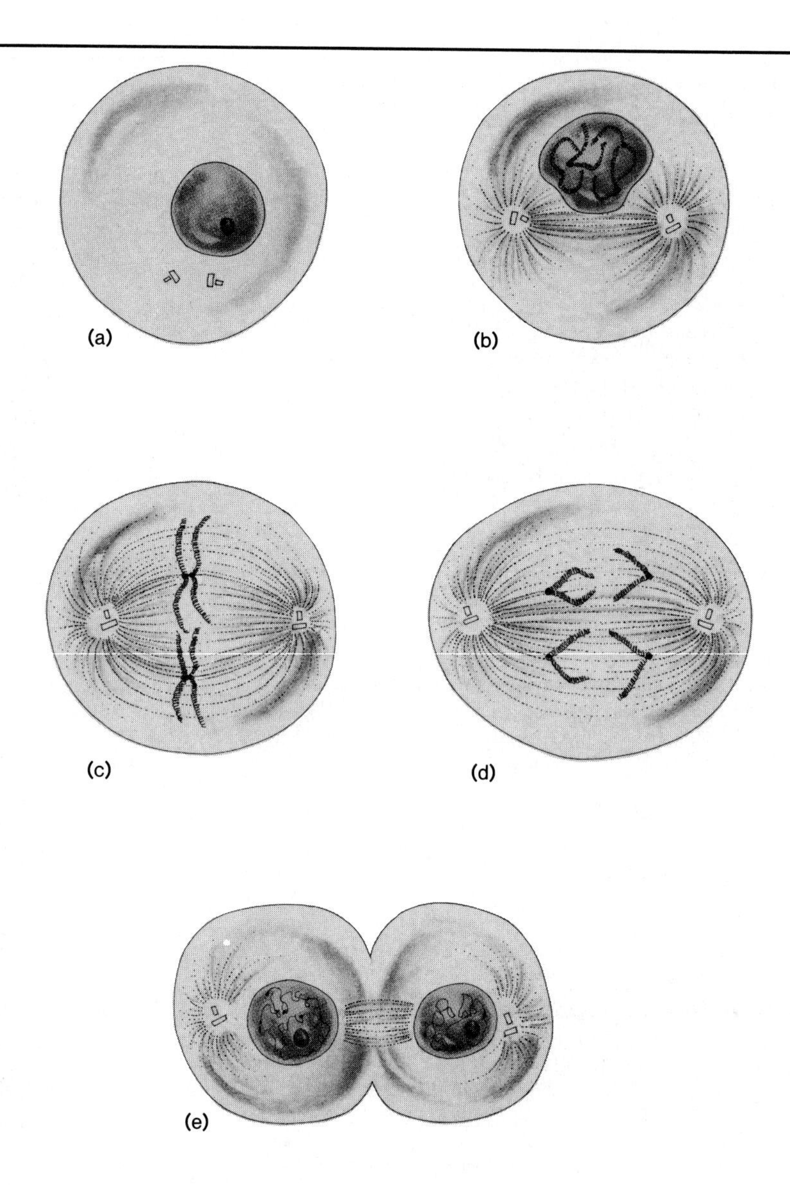

Figure 15.2 Mitosis. (*a*) An animal cell just before entering mitosis. (*b*) Prophase. (*c*) Metaphase. (*d*) Anaphase. (*e*) Telophase and the beginning of division of the cytoplasm.

of the basic research on cancer is now aimed at determining why cancer cells are not under the same restraints that prevent normal cells from dividing in an uncontrolled fashion.

Mitosis

Mitosis is the division of the nucleus in eukaryotic cells (cells having a membrane-enclosed nucleus) that results in equal distribution of duplicated genetic information to the two new cells (called **daughter cells**). This division is very precise, and the end result of mitosis is two nuclei with genetic information identical to that of the parent cell's nucleus.

Stages of Mitosis

Mitosis is a continuous process that we can arbitrarily divide into four stages for convenience of description. As you read about the stages of mitosis, look carefully at the figures. To help you to keep the big picture in mind, refer repeatedly to figure 15.2, which shows all four stages in a diagrammatic form.

The first stage of mitosis is called **prophase.** A typical cell entering prophase from G_2 has an intact nuclear membrane and may have visible nucleoli, but chromosomes are not visible with a light microscope (figure 15.3). During the early stages of prophase, chromosomes begin to condense.

Figure 15.3 The nucleus of a peony cell before entering mitosis. Doubled chromosomes are not visible, and the nuclear membrane is intact (magnification × 1,046).

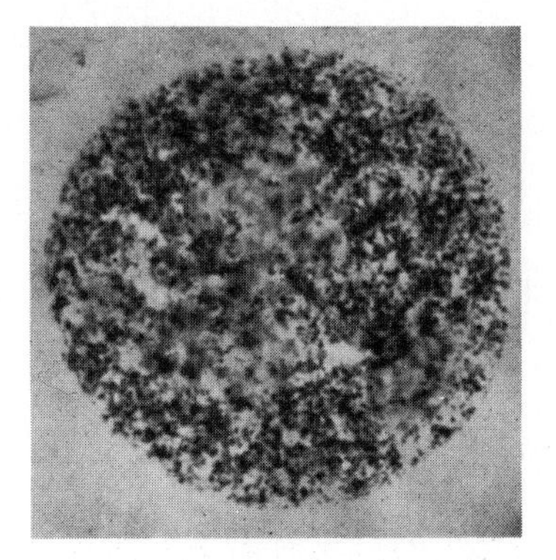

Figure 15.4 Mid-prophase in the peony. The chromosomes are now visible (magnification × 995).

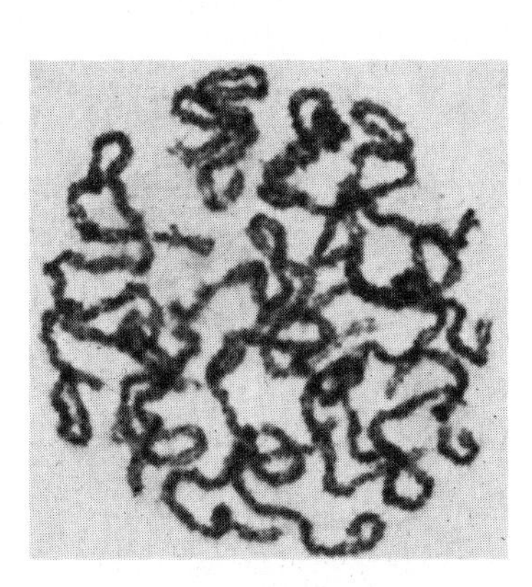

This apparent condensation is actually a coiling process that results in a shortening and thickening of each chromosome. This coiling results in chromosomes that are easily visible with a light microscope after they have been stained.

The chromosome condensation that occurs at mitosis is temporary because the chromosomes uncoil at the end of each cell division. This is necessary because chromosomes must be uncoiled for normal genetic functioning between divisions.

In the cells of most organisms, the nuclear membrane begins to break down during prophase. Later, this allows for the chromosomes to be apportioned to each of the two new nuclei that will be formed. Any nucleoli present in the dividing cell also disappear during the early part of prophase.

By midprophase (figures 15.2*b* and 15.4), you can clearly see stained chromosomes under the light microscope as double-stranded structures. There are two strands at this time because the chromosomes were duplicated before the beginning of mitosis, during the S phase of the cell cycle. Each strand of the doubled chromosome is called a **chromatid** (figure 15.5), and the two chromatids of each chromosome are joined in the regions of their centromeres. The **centromere** of each chromatid is a small platelike structure located on a constricted area of the chromatid.

Also during prophase, the **spindle apparatus** begins to form. The spindle apparatus consists of a set of microtubules (see chapter 3 for a discussion of microtubules), which are associated with chromosome movement later in mitosis. Another obvious occurrence at this time, at least in animal cells, is the movement of the **centrioles,** which were duplicated earlier during the S phase of the cell cycle. During prophase,

Figure 15.5 A replicated human chromosome. The two chromatids are connected in the region of their centromeres (magnification × 33,824).

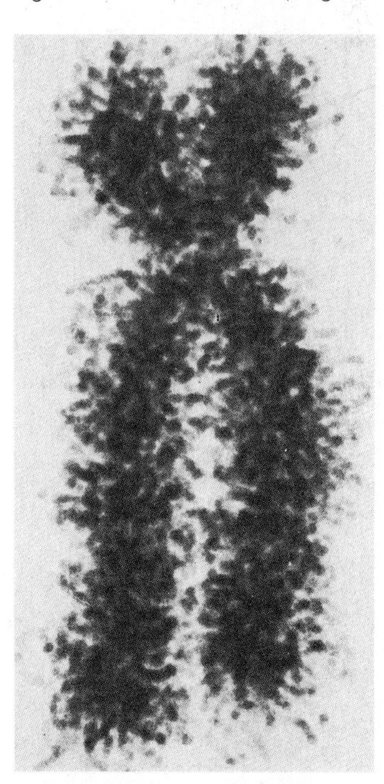

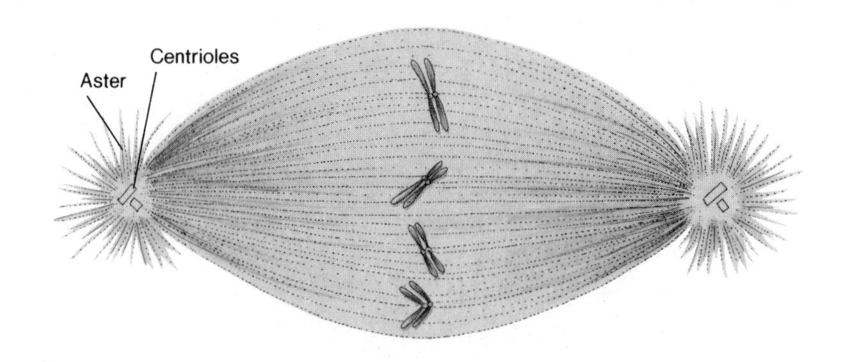

Figure 15.6 The spindle apparatus in a dividing animal cell. Some of the microtubules are attached to the chromosome centromeres. Others overlap to extend from one end of the spindle apparatus to the other. The centrioles are surrounded by short microtubules, which make up the aster.

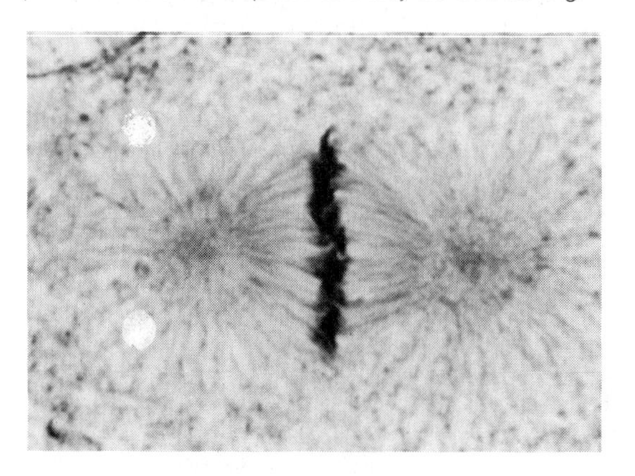

Figure 15.7 Metaphase in the whitefish. Duplicated chromosomes are lined up in the middle of the spindle as if they are on a flat ring.

Figure 15.8 Anaphase in the onion. The centromeres have divided, and the chromatids have separated, becoming independent chromosomes.

one pair of centrioles moves to one side of the nucleus and the other moves to the opposite side. The centriole pairs are accompanied during this movement by a halo of microtubules called the **aster,** part of which remains around each of them in their new locations on opposite sides of the nucleus. The spindle apparatus develops between these pairs of centrioles (figure 15.6).

Although centrioles are very prominent in dividing animal cells, cells of many other organisms do not have centrioles. For example, certain amoebae, some algae, some gymnosperms (such as pines), and all angiosperms (flowering plants) lack centrioles altogether. Thus, it appears that the centriole movement that occurs during animal cell division is not an integral part of the mitotic process. Instead, it may be a mechanism that assures that each cell produced receives centrioles so that they will be available for other functions in the daughter cells. In fact, removal of centrioles does not prevent cells that normally have centrioles from forming a spindle apparatus and dividing.

During late prophase, the chromosomes migrate toward the center of the cell where they line up and remain during the next relatively short phase of mitosis, **metaphase.** During metaphase the chromosomes are lined up across the middle of the cell as if on a flat ring that lies at right angles to the direction of the spindle microtubules (figures 15.2*c* and 15.7).

The beginning of the next phase of mitosis, **anaphase** (figures 15.2*d* and 15.8), is marked by the separation of the two chromatids of each chromosome, so that each chromatid becomes an independent chromosome. During anaphase, one chromosome from each of these newly separated pairs of chromosomes moves toward each end of the cell.

Although the mechanism of chromosome movement during anaphase is not fully understood, it is generally agreed that some of the microtubules of the spindle pull chromosomes toward the ends of the cell. This apparent pulling results in the characteristic shapes of anaphase chromosomes.

Some of them assume a V shape or a J shape as they move, with the apex of the V or J being the centromere, which is at the point of spindle microtubule attachment.

When the chromosomes arrive at each end of the cell, **telophase,** the final phase of mitosis, begins (figures 15.2*e* and 15.9). During this phase the chromosomes begin to uncoil and lengthen, and the two nuclei organize. This nuclear reorganization involves formation of nuclear membranes and the reappearance of the nucleoli.

Cytokinesis

So far our discussion of cell division has been limited to the activities of the nucleus. The rest of the material in the cell divides during a process called cytokinesis. Animal cells divide by a constriction process (figure 15.10). As cytokinesis begins, the cell membrane constricts and forms a furrow around the middle of the cell due to the activity of microfilaments. The furrow progressively deepens until the cell is actually pinched into two new cells. This constriction cuts through the cell exactly where the middle of the spindle apparatus was located.

Cytokinesis in plants is a very different process because of the rigid cell wall that surrounds each cell. In plant cells, division of the cytoplasm occurs from the inside to the outside instead of outside to inside as in the cleavage of animal cells. A **cell plate** begins to develop in the center of the spindle

Figure 15.9 Telophase. (*a*) Early telophase in the onion. The chromosomes have reached the poles of the spindle. Note the line in what was the middle of the spindle. This is the cell plate that is the first step in formation of a cell wall between the two newly formed cells. (*b*) Late telophase in the onion. Nuclear membranes have formed and nucleoli have reappeared. A cell wall separates the two cells.

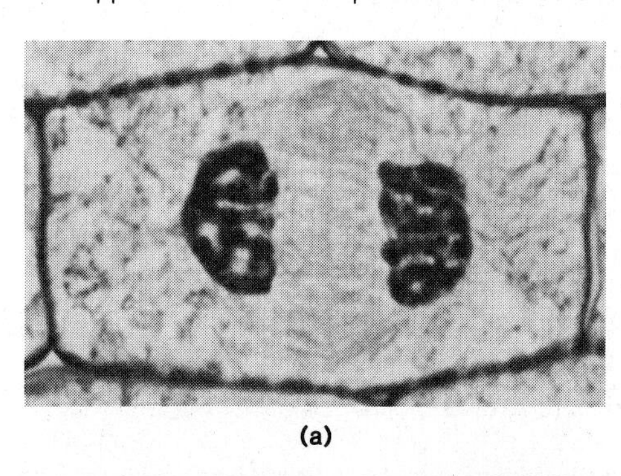

(a)

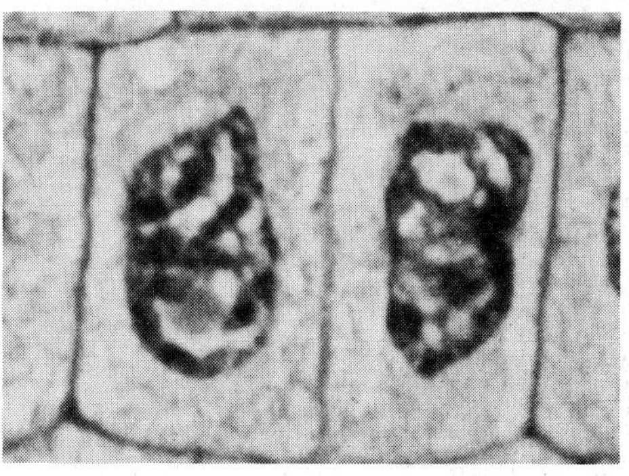

(b)

Figure 15.10 Cytokinesis in animal cells (sea urchin eggs).

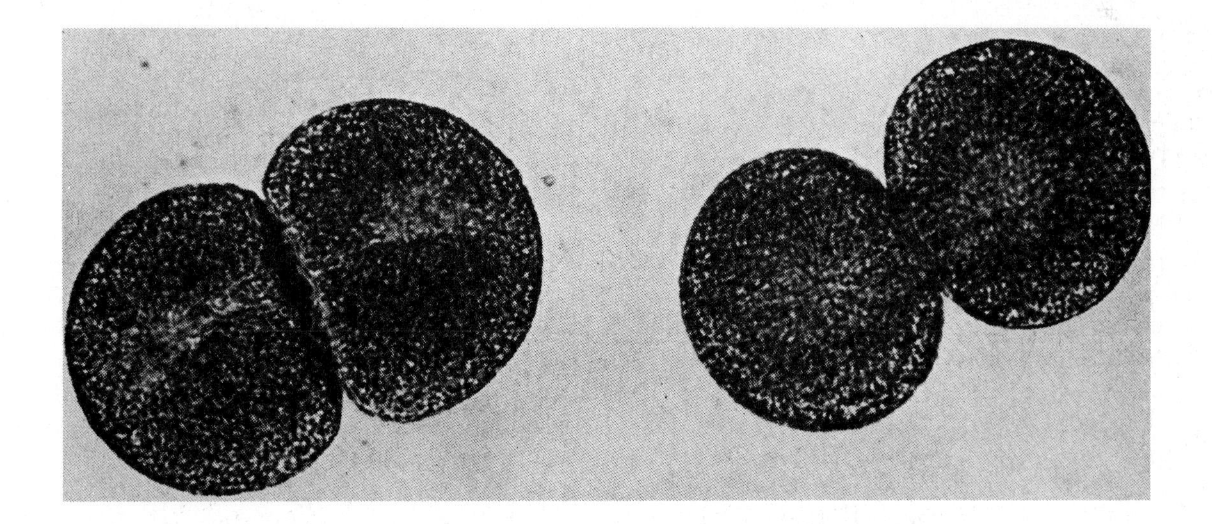

Figure 15.11 Cytokinesis in plant cells. Electron micrograph of cell plate formation in the root tip of a soybean. The cell plate forms by fusion of membranous spheres derived from the Golgi apparatus of the original cell. The process begins in the center of the spindle and spreads outward. Note the presence of some spindle microtubules and the re-forming nuclear envelopes of the daughter cells (magnification × 6,650).

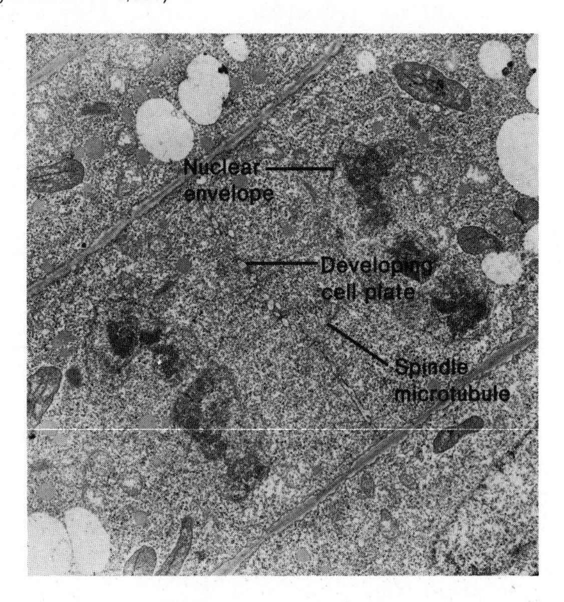

(see figure 15.9*b*). This plate forms by fusion of membranous spheres derived from the Golgi apparatus (figure 15.11). More and more spheres are added to the growing cell plate, and it increases in size until the cell is divided down the middle by two plasma membranes with a space between them. These membranes become continuous with the old plasma membrane, thus dividing the cell into two daughter cells. Then, new cell walls develop between them, and at the end of the process, two plant cells, each surrounded by its own cell wall, exist where there had been one.

Meiosis

In sexual reproduction, two specialized reproductive cells, called **gametes,** combine to form a single cell called a **zygote,** the first cell of the new organism. Fusion of two gametes to form a zygote is called **fertilization.** In fertilization, each of the two gametes contributes half of the genetic information for the new organism.

Before fertilization, organisms that reproduce sexually must produce gametes with half the normal chromosome number of ordinary body cells. If this did not occur, the fusion of the two gametes would result in a zygote with twice the parents' chromosome number, and the number of chromosomes would continue to double with each generation. Reduction in chromosome number is accomplished by a type of cell division called meiosis. The doubling of the chromosome number by fertilization and the halving of the chromosome number by meiosis ensure that the number of chromosomes is stable from generation to generation.

Most cells are **diploid,** which means that they have two of each kind of chromosome. In other words, diploid sets of chromosomes contain several to many chromosome pairs. Members of these pairs of chromosomes are **homologous** chromosomes. Homologous chromosomes are similar in appearance, and the two chromosomes in a pair contain genetic information controlling the same characteristics.

Meiosis reduces the number of chromosomes to one half of the diploid number. Cells that contain this reduced number of chromosomes are said to be **haploid** or to have a haploid number of chromosomes. For example, in an organism having a diploid chromosome number of eight, the haploid number of chromosomes contained in the gametes is four. Human body cells have forty-six chromosomes, while human eggs and sperm, which are haploid, have twenty-three chromosomes.

Haploid cells have only one of each "kind" of chromosome (only one member of each of the homologous pairs of chromosomes found in diploid cells). The haploid number of chromosomes normally is indicated by the designation **N,** and the diploid number is designated **2N.**

The Role of Meiosis in Life Cycles

Meiosis occurs in all sexually reproducing organisms, but the timing of this process in the life cycle varies in different types of organisms. In most animals, reduction in the number of chromosomes from the diploid number to the haploid number occurs during gamete production.

In plants, meiosis occurs during the production of spores. These haploid (N) spores then develop into multicellular structures made up entirely of haploid cells. Gametes develop directly from some of these haploid cells. Thus, meiosis is not directly involved in gamete production in plants, but occurs elsewhere in the plant life cycle. We will explore the role of meiosis in plant and animal reproduction more fully in chapter 19.

Figure 15.12 Stages of meiosis. Meiosis involves two successive cell divisions and produces cells with half the number of chromosomes contained in the original cell. The organism used in this example has a diploid number of four, and meiosis produces cells with two chromosomes, the haploid number for the organism.

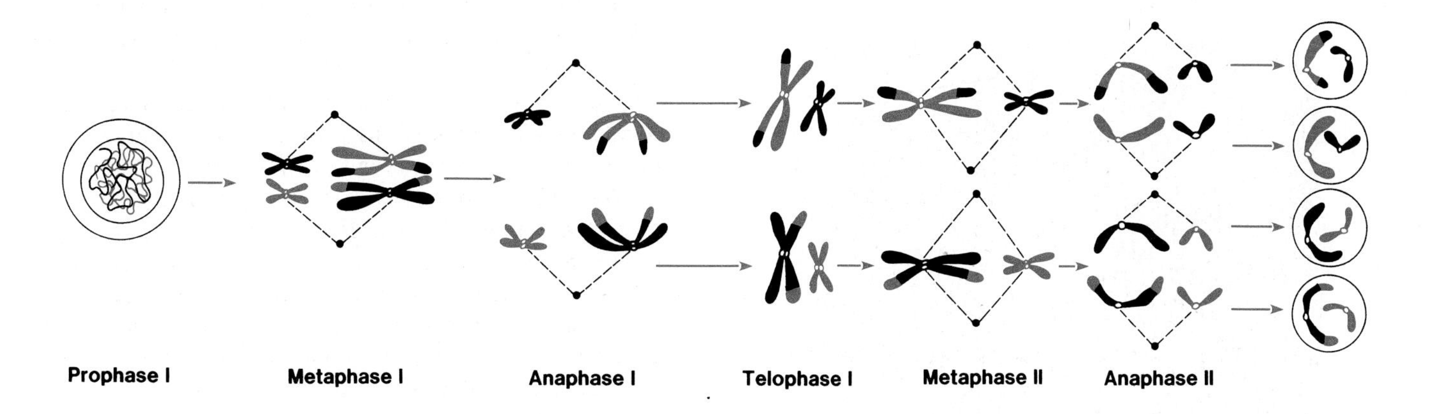

Stages of Meiosis

Meiosis consists of two successive divisions (figure 15.12). The first division reduces the *actual number of chromosomes* to half. But at the end of this first division, each of the chromosomes is still doubled because chromosomes were replicated in the S phase preceding meiosis, and chromatids do not separate in the first division of meiosis; that is, at the end of the first division, each chromosome still consists of two chromatids attached in the areas of their centromeres. The second division follows immediately without any DNA replication. It resembles mitosis in that the centromeres separate and the chromatids part to become independent chromosomes. In order to keep an overview of meiosis as you examine individual phases, we recommend that you refer repeatedly to figure 15.12.

As with mitosis, meiosis is divided into arbitrary stages for descriptive purposes. **Prophase I** is the first stage of the first division. During early prophase I (figure 15.13), the nuclear membrane and the nucleoli remain intact. The chromosomes appear as very long threads with beadlike thickened areas. Although these chromosomes have been replicated, they appear to be single strands; chromatids are not distinguishable at this stage.

Figure 15.13 Early prophase I in corn (*Zea mays*). The large dark round body is a nucleolus.

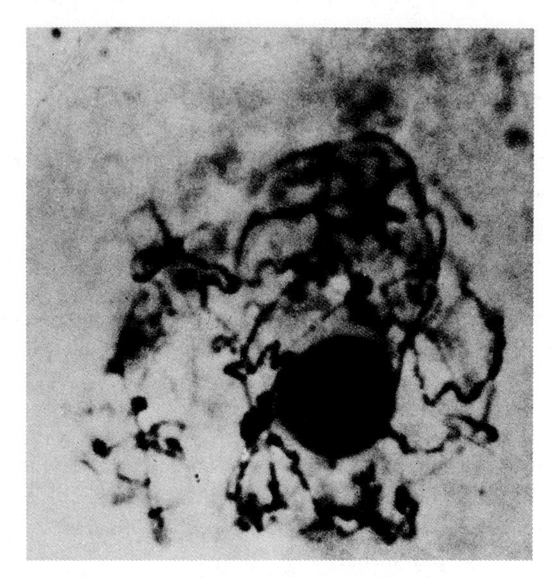

Figure 15.14 Synapsis. This photograph of barley chromosomes shows the pairing of the fourteen chromosomes into seven pairs of homologous chromosomes.

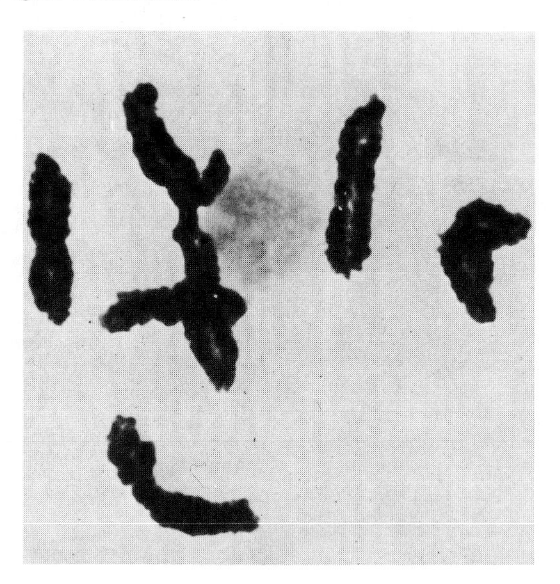

Figure 15.15 A diagram showing chromosome segment exchange during crossing over. (*a*) Paired homologous chromosomes before crossing over. (*b*) Two of the chromatids exchanging segments. (*c*) The chromosomes after crossing over has been completed.

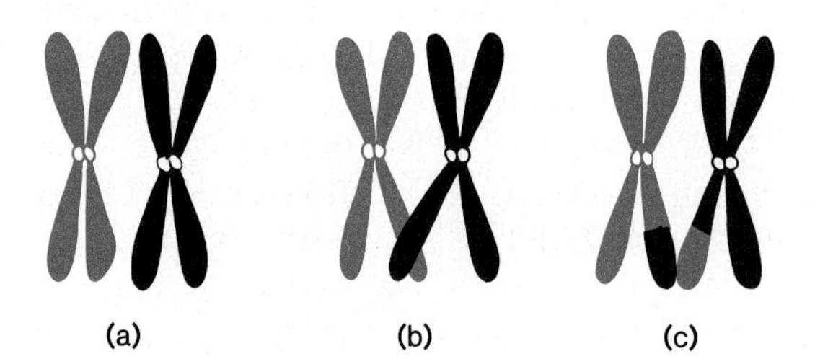

Figure 15.16 Paired chromosomes held together at points where crossing over has occurred. These chromosomes are in a newt cell. A newt is a tailed, aquatic amphibian (magnification × 343).

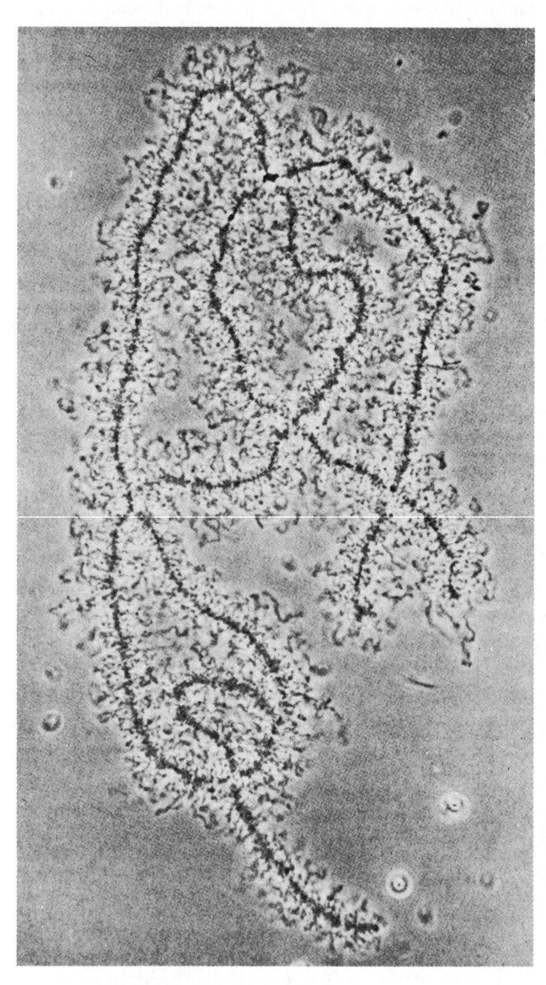

Next, homologous chromosomes begin to come together and pair with each other. *This pairing process, called* **synapsis** (figure 15.14), *has no counterpart in mitosis.* Each of these synaptic pairs, therefore, consists of two chromosomes (four chromatids) in close proximity to one another.

The chromosomes gradually become more condensed, and an important phenomenon called **crossing over** occurs. During crossing over, chromatids of the homologous chromosomes exchange small segments (figure 15.15). This effectively redistributes genetic information among the paired homologous chromosomes and produces new combinations in the sets of genes borne by the various chromatids. After crossing over, there is a slight separation of the homologous chromosomes. These chromosomes, however, remain attached at the places where crossing over previously has occurred (figure 15.16). At the end of prophase I, the nuclear membrane breaks down, the nucleoli disappear, and the chromosome pairs move toward the center of the cell.

Metaphase I is similar to the metaphase of mitosis in that the chromosomes are lined up in the center of the cell (figure 15.17).

Figure 15.17 Metaphase I in corn (*Zea mays*).

Figure 15.18 Anaphase I in corn (*Zea mays*).

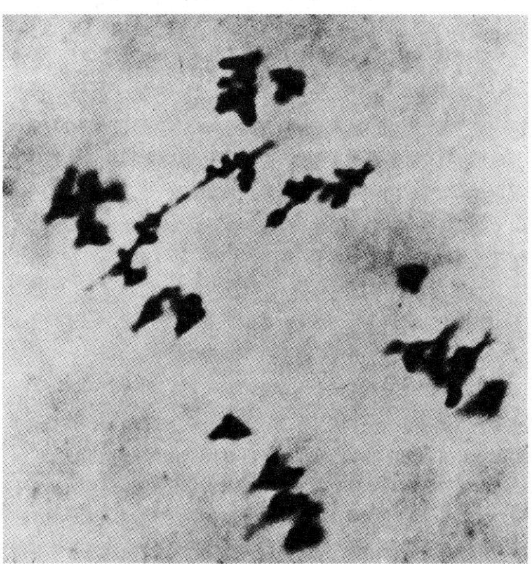

Anaphase I begins when homologous chromosomes separate and begin to move toward each pole (figure 15.18). Chromatids do not part at this stage because there is no centromere separation in the first meiotic division. Because the chromosomes of each homologous pair separate and move in opposite directions, each daughter cell receives one chromosome from each homologous pair. However, each of these chromosomes still consists of two chromatids.

At the end of the first meiotic division, the number of chromosomes has been reduced to half of the diploid number, but each chromosome still consists of two chromatids. One more division is needed to separate the chromatids. We remind you once more that there is *no* DNA replication (S phase) between the two meiotic divisions.

In the second division of meiosis, the phases are referred to as **prophase II, metaphase II, anaphase II,** and **telophase II** (see figure 15.12 again). During anaphase II, centromere separation finally occurs, and the chromatids part to become independent chromosomes. At the end of telophase II, the final products of the two divisions of meiosis are four new cells. Each of these four cells is haploid; that is, it has half the diploid number of chromosomes and half the DNA content of diploid cells. Because of the crossing over that has occurred and because of the assorting of homologous chromosomes into different associations, the meiotic products are genetically different, not only from each other, but also from the original parental cell. (See box 15.2 for more information on assortment of chromosomes in meiosis.)

Meiotic Errors

Errors in mitosis ordinarily do not have serious consequences for organisms unless they occur early in development so that genetic or chromosomal problems are perpetuated in many body cells. But errors in meiosis are another matter because they result in genetic problems in gametes. These problems are perpetuated in all the cells of an individual produced as a result of fertilization involving a genetically abnormal gamete. A common error in meiosis is **nondisjunction,** the failure of chromosomes to become distributed normally during anaphase. Nondisjunction produces cells that have an extra chromosome and cells that are short a chromosome.

Nondisjunction can occur, for example, as a result of both chromosomes of a homologous pair moving to the same pole at anaphase I (figure 15.19). This means that, after meiosis is complete, some cells will have N + 1 chromosomes and some will have N − 1 chromosomes. If these cells ultimately participate in fertilization by combining with normal haploid gametes, the resulting organisms have either one extra chromosome (2N + 1) or are missing one chromosome (2N − 1). A chromosome number that varies from the

Box 15.2

How Are There So Many Different Types of Individuals?

Part of the answer to the question of individual diversity among living things is that there are so many types of gametes. Every individual produces many different types of gametes because of the nature of the process of meiosis. You may be interested to know that you, as a human being, can produce gametes that can have any one of over 8 million chromosome combinations. Let us see how this is possible.

All body cells are descended by mitotic cell divisions from a single cell, the zygote. Each cell has, just as the zygote had, a diploid (2N) number of chromosomes made up of a characteristic (N) number of homologous pairs. In each of these homologous pairs, one chromosome is a copy of a chromosome received from the individual's mother; for purposes of convenience, it can be identified as a "maternal chromosome." The other member of the homologous pair can be identified as a "paternal chromosome," as it is a replicate of a chromosome received from the individual's father. Of course, these "maternal" and "paternal" chromosomes differ from one another genetically.

Because the haploid cells produced by meiosis have only a single representative of each homologous pair of chromosomes, how are these different chromosomes distributed to daughter cells during the meiotic process? This is a key genetic question because the cells produced by meiosis become gametes directly (in animals) or develop into multicellular haploid structures that produce gametes (in plants).

Box figure 15.2 Key parts of meiosis in an organism with a hypothetical diploid (2N) number of 4 and a haploid (N) number of 2. "Maternal" chromosomes are shown in color, and "paternal" chromosomes are shown in black. These diagrams show that there are four possible outcomes of the meiotic process in terms of the chromosome combinations produced. For each additional pair of chromosomes possessed by an organism, the number of possible chromosome combinations in the gametes is doubled.

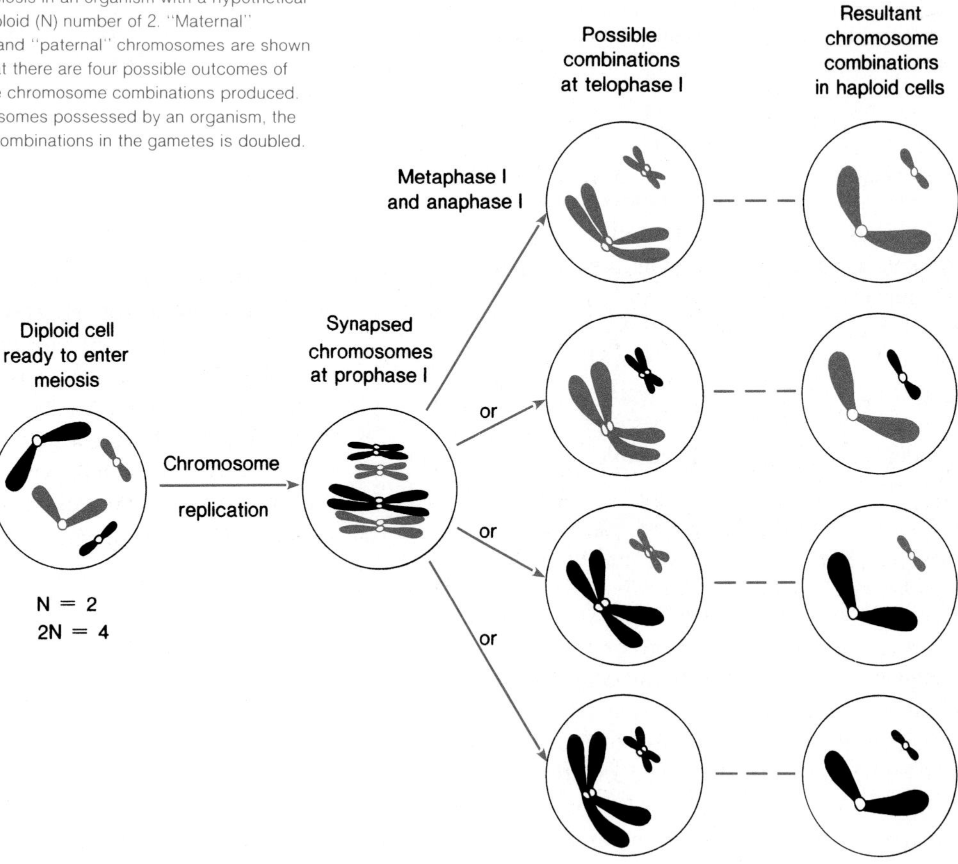

The distribution of the members of any homologous pair of chromosomes to daughter cells during the first meiotic division is totally independent of the distributions of members of all other homologous pairs of chromosomes. *Each type of chromosome assorts independently of all others.*

What does this mean in terms of the meiotic process in cells? Box figure 15.2 diagrammatically shows some key parts of meiosis in an organism with a hypothetical diploid (2N) number of 4 and a haploid (N) number of 2. "Maternal" chromosomes are shown in color and "paternal" chromosomes are shown in black. Chromosomes are shown following chromosome replication and the synapsis of chromosomes that occurs in prophase I.

The key to understanding independent assortment of chromosomes is understanding the possible outcomes of anaphase I when members of homologous chromosome pairs separate from one another. As box figure 15.2 shows, the fate of each homologous pair during this separation is totally independent of the fate of each other homologous pair. It is a random process that can produce, in the example, four possible outcomes. As a result of this independent assortment, four chromosome combinations are possible for the haploid nuclei produced when chromatid separation in the second meiotic division is complete.

What about organisms with larger chromosome numbers? Meiosis in an organism with six chromosomes ($2N = 6$, $N = 3$) can produce twice as many possible chromosome combinations as the organism with a diploid number of 4 in our example. The number of possible meiotic outcomes doubles with each additional pair of homologous chromosomes. Another way of saying this is that the number of possible outcomes is equal to 2^N (2 to the Nth power), where N is equal to the number of homologous chromosome pairs (same as the haploid number, N). As we saw in the example, when $N = 2$, there are four (2^2) possible combinations. When $N = 3$, there are eight (2^3), when $N = 4$, there are sixteen (2^4), etc. In humans, where $N = 23$ (humans have forty-six chromosomes), the number of possible chromosome combinations produced by meiosis is a staggering 2^{23}, or 8,338,608!

Independent assortment, together with crossing over, results in a great variety of genetic combinations in the haploid cells produced by meiotic cell division.

Figure 15.19 An example of nondisjunction in meiosis. Only the homologous pair of chromosomes involved in nondisjunction are shown in this illustration. All other chromosomes are presumed to behave normally. Some gametes have N + 1 chromosomes and some N − 1 as a result of nondisjunction.

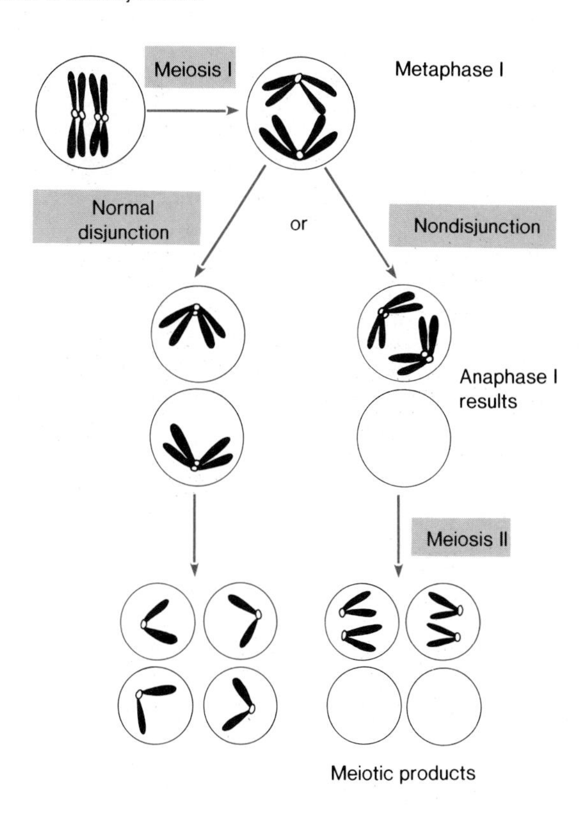

normal chromosome number of the species by a small number of chromosomes is called **aneuploidy.** The most common of human birth defects, Down's syndrome, is the result of aneuploidy. Affected individuals have an extra chromosome in one of the sets of homologous chromosome pairs (chromosome no. 21), and they suffer a number of physical problems as well as mental deficiencies that range from relatively mild impairment to severe retardation. Clearly, meiotic errors can have broad and significant consequences.

Figure 15.20 A diagrammatic summary comparing mitosis and meiosis. Meiosis is simplified here by omission of crossing over, which is a normal part of the process.

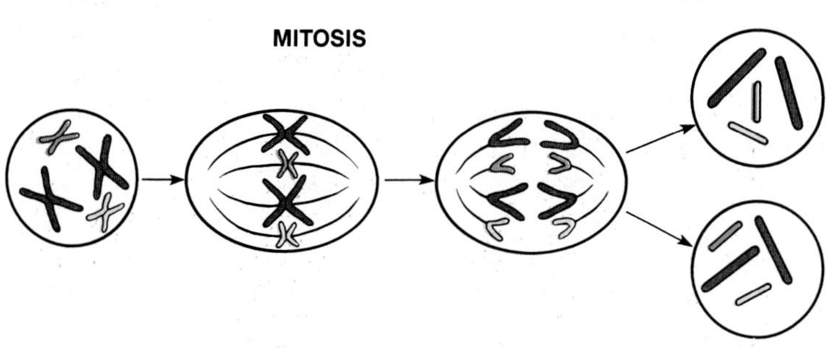

Prophase
Four replicated chromosomes, each consisting of two chromatids joined in the centromere region.

Metaphase
Replicated chromosomes line up in the middle of the spindle.

Anaphase
Centromeres part and chromatids separate to become independent chromosomes, which move to opposite poles of the spindle.

Telophase
The end of mitosis. Two daughter cells, each with exactly the same chromosomal complement as the original cell.

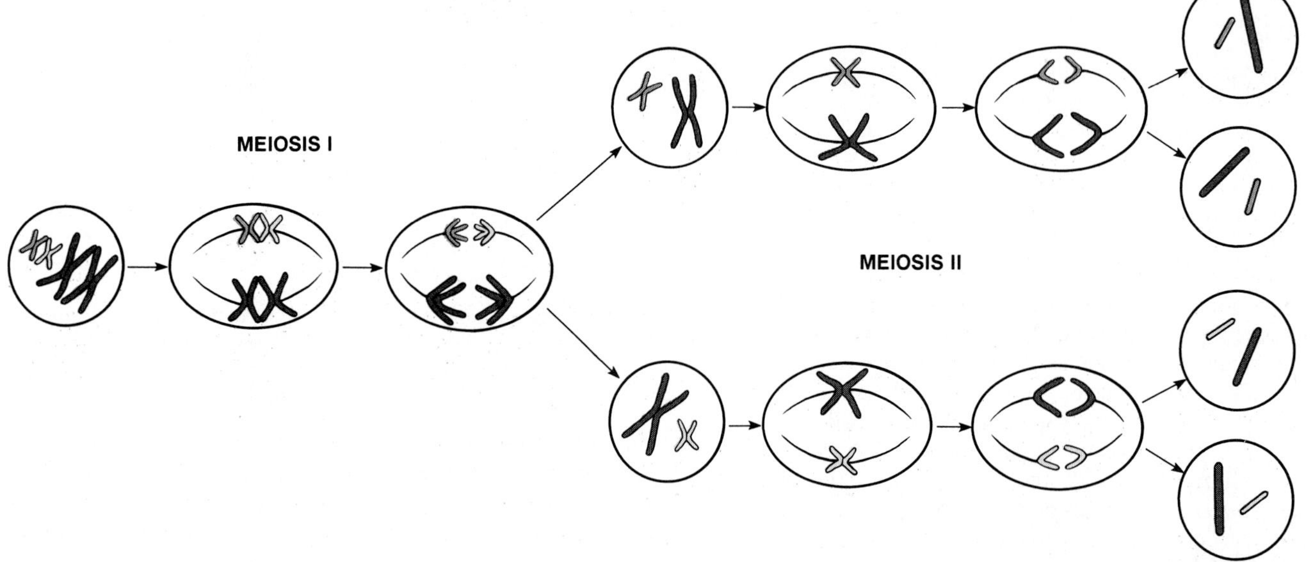

Prophase I
Four replicated chromosomes; homologous pairs are synapsed.

Metaphase I
Homologous pairs line up in the middle of the spindle.

Anaphase I
Centromeres do not separate and chromatids remain together. Homologous chromosomes separate.

Telophase I
End of Meiosis I. One member of each homologous chromosome pair is present in each daughter cell. Chromatids remain attached at the centromere.

Metaphase II
In each of the two cells, chromosomes line up in the middle of the spindle.

Anaphase II
The centromeres finally part, allowing chromatids to separate as independent chromosomes, which move toward the poles of the spindle.

Telophase II
The end of meiosis: four cells, each with one-half the number of chromosomes in the original cell.

A Comparison of Mitosis and Meiosis

Mitosis and meiosis are similar in that they are both forms of cell division, incorporating complex mechanisms that ensure that the genetic information is reliably passed to each new cell (figure 15.20). These two methods of division play different roles in the life cycle of the organism.

The outcome of mitosis is simply production of new cells from previously existing cells. This is the method by which multicellular organisms grow and also replace worn-out cells. Meiosis is necessary for the reduction of chromosome numbers in the life cycle of organisms that reproduce sexually.

Mitosis may occur in either haploid or diploid cells, and it does not result in any change in chromosome number. Meiosis always begins with a diploid cell and produces haploid cells.

The process of mitosis takes place in one division with four stages. Meiosis requires two divisions: one to separate the homologous chromosomes and one to separate the replicated chromatids.

Finally, the products of the two processes differ. Mitosis produces two new cells, each of which has a set of chromosomes exactly like that of the parental cell. Meiosis produces four cells, each of which has half as many chromosomes as the parental cell.

Summary

Cells arise from preexisting cells by a process of cell division. The life span of a cell from the time that it has been formed by cell division until it too divides can be described in terms of four distinct stages that make up the cell cycle. After division, a cell goes through a process of growth and maturation. This is the gap 1 (G_1) phase of the cell cycle. If the cell is going to divide, it moves into the synthesis (S) phase. During this phase the genetic material is replicated in preparation for division. After the S phase, the cell enters a gap 2 (G_2) phase. During this phase the cell synthesizes the materials necessary for division. The fourth phase, mitosis (M), is nuclear division. It is accompanied by cytokinesis, the division of the cytoplasm.

Mitosis is the division of the nucleus of a eukaryotic cell so that two new nuclei are formed. For descriptive purposes, mitosis is divided into four stages. During the first stage, prophase, the chromosomes shorten and thicken, and the nuclear envelope and nucleoli disappear. The doubled chromosomes line up in the center of the cell during the second stage of mitosis, metaphase. Anaphase, the third stage, occurs when chromatids of each doubled chromosome move to opposite ends of the cell. Each chromatid becomes an independent chromosome at the time of separation. The fourth stage, during which the two new nuclear membranes are formed and the nucleoli reappear, is called telophase.

The cytoplasm of the cell is divided by cytokinesis. In most animal cells cytokinesis occurs by a constriction of the cell from the outside. The constriction continues until the cell is pinched in two. Plant cells have a rigid cell wall, and direct constriction to produce two cells is not possible. The cytoplasm of plant cells is divided by the formation of a cell plate across the center of the cell. This cell plate enlarges and materials are added until a new plasma membrane and cell wall separate the two newly formed cells.

Meiosis produces cells that have half the diploid number of chromosomes. This reduction in chromosome number is accomplished in two successive divisions. The first division separates the pairs of homologous chromosomes. The two cells that are produced have the haploid number of chromosomes, but each of these chromosomes consists of two chromatids. The second division of meiosis separates these chromatids and produces cells that have a haploid (N) number of chromosomes.

Questions for Review

1. Does a cell duplicate most of its component parts before or during mitosis? Explain.
2. What is a chromatid, and when does a chromatid become a chromosome?
3. Define cytokinesis.
4. What aspects of cell division are common to both plant and animal cells? What aspects are different?
5. Explain the significance of meiosis in the life cycles of organisms that reproduce sexually.
6. What are homologous chromosomes?
7. Define synapsis and crossing over.
8. Why is nondisjunction in meiosis much more likely to cause serious problems than nondisjunction in mitosis?
9. List four differences or distinctions between mitosis and meiosis.

Questions for Analysis and Discussion

1. When early microscopists studied cell division, they concentrated on the visible events of mitosis and named the period in which there was no visible activity between mitotic events the "interphase." They thought of the interphase as a "resting stage." Is the period between cell divisions (interphase) a resting stage? Justify your response.
2. Explain why we say that at the end of the first meiotic division the chromosome number of the daughter nuclei is haploid, but their DNA content is the same as that in diploid cells. How does this statement relate to the fact that there are two successive divisions in meiosis?

Suggested Readings

Books

DeRobertis, E. D. P., and DeRobertis, E. M. F., Jr. 1980. *Cell and molecular biology.* 7th ed. Philadelphia: Saunders College.

Karp, G. 1984. *Cell biology.* 2d ed. New York: McGraw-Hill.

Kemp, R. 1970. *Cell division and heredity.* Studies in Biology no. 21. Baltimore: University Park Press.

Wolfe, S. 1981. *Biology of the cell.* 2d ed. Belmont, Calif.: Wadsworth.

Articles

John, B., and Lewis, K. R. 1980. Somatic cell division. *Carolina Biology Readers* no. 26. Burlington, N.C.: Carolina Biological Supply Co.

———. 1983. The meiotic mechanism. *Carolina Biology Readers* no. 65. Burlington, N.C.: Carolina Biological Supply Co.

Prescott, D. M. 1978. The reproduction of eukaryotic cells. *Carolina Biology Readers* no. 96. Burlington, N.C.: Carolina Biological Supply Co.

Rensberger, B.; Bush, H.; and Blonston, G. September 1984. Cancer: The new synthesis. *Science 84.*

Sloboda, R. D. 1980. The role of microtubules in cell structure and cell division. *American Scientist* 68: 290.

Patterns of Inheritance

16

Chapter Outline

Chapter Concepts

1. The science of genetics began with Gregor Mendel's fundamental research in the mid-1800s.
2. Mendel's first law, the law of segregation, states that genes for each trait come in pairs and that these genes segregate so that the gametes have only one gene for each trait.
3. Mendel's second law, the law of independent assortment, states that different genes segregate independently of each other.
4. Mendel's laws can be used to predict the proportions of offspring types resulting from specific crosses.
5. Interactions between different genes and between genes and the environment may affect inheritance patterns.
6. Some traits are determined by many genes, with each gene having a slight effect.

Ancient records indicate that when Alexander the Great invaded India in 329 B.C., the Saluki (figure 16.1*a*) was clearly recognized as a distinctive breed of dog. Much earlier records in the form of pyramid pictures in Egypt dating to 2100 B.C. clearly show dogs with the characteristics of Salukis. In fact, still older pictures from Sumerian civilization, which flourished between 4000 and 6000 B.C., also show dogs that seem to be Salukis. Thus, the Saluki may be the oldest of all the recognized breeds of dogs.

Throughout history, dogs with certain desired characteristics have been bred together to enhance expression of those features. This selective breeding has resulted in establishment of many recognized breeds, all of which are descended from only a few relatively similar types of wild dogs. This selective breeding process continues even today with some new breeds having been established very recently. For example, only in 1956 did the American Kennel Club recognize Belgian Tervurens (figure 16.1*b*) as a breed separate from other closely related Belgian shepherds.

The story of domesticated dogs is only one of many stories of the selective breeding of domesticated animals and plants. Throughout most of history, selective breeding was done without understanding of the underlying mechanisms by which characteristics of living things are passed from generation to generation.

In sexually reproducing organisms, both parents contribute certain traits to their offspring. **Genetics** is the branch of biology that deals with this inheritance of characteristics. Although people had known about the concept of **heredity** (the transmission of characteristics from parents to offspring) for centuries, very little was known about the nature of the genetic process until 1866, when two of the basic laws of heredity were reported. Before that time, the prevailing idea of heredity was a notion of blending. The hereditary material was thought of as something resembling a fluid, perhaps something in the blood of animals or the sap of plants. The traits of offspring were thought to be a combination of the parents' traits due to the blending of these fluids.

Furthermore, the difficulty in determining the precise manner in which genetic information is transmitted from generation to generation was not due to a lack of study by scientists, but to the fact that most scientists were attempting to study simultaneously the transmission of a large number of characteristics. This made it much more difficult to understand and explain what was happening.

Modern genetic analysis really began with a series of experiments reported by an Austrian monk named Gregor Mendel in the 1860s. Mendel's discoveries were the result of remarkably thoughtful analysis of simple but careful experiments. Modern biologists recognize the significance of Mendel's conclusions and refer to the study of certain types of inheritance patterns as **Mendelian genetics.**

Figure 16.1 Some examples of selective breeding in dogs. (*a*) The Saluki, probably the oldest of recognized breeds of dogs. (*b*) The Belgian Tervuren, one of the more recently recognized breeds of dogs.

(a)

(b)

Mendel's work went unappreciated until 1900 when it was rediscovered by three different geneticists—Hugo DeVries in Holland, Carl Correns in Germany, and Erich von Tschermak in Austria—each working independently. Since that time, the science of genetics has expanded rapidly, and the last decades have seen especially significant advances in the understanding of the molecular aspects of genetics. Mendel's basic conclusions, however, remain a fundamental part of genetic theory.

Figure 16.2 Gregor Mendel.

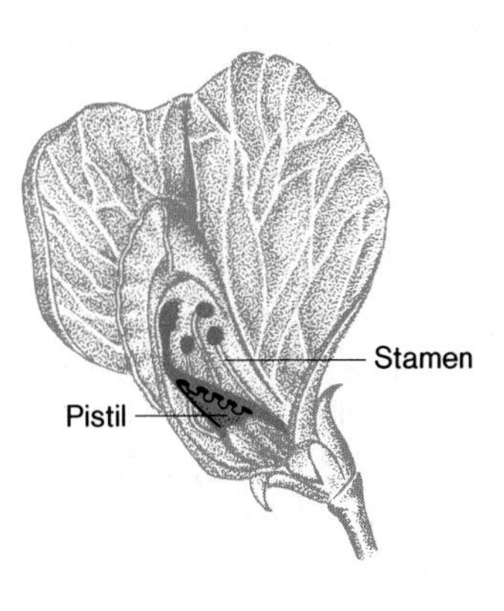

Figure 16.3 A cross section of a garden pea, *Pisum sativum*, flower. The petals that normally surround the reproductive organs have been cut away. These surrounding petals ensure self-fertilization under natural conditions. Pollen grains that later produce sperm nuclei are produced by the stamen. Haploid tissue that produces the egg develops in the pistil.

The Work of Mendel

Gregor Mendel (figure 16.2) had two years of university training before entering the monastery. While at the university, he became familiar with the genetic theories of the time. He also was very aware of the interest of many scientists in plant hybridization (crosses between two differing types).

Mendel began his own experiments around 1856 with the garden pea (*Pisum sativum*). This plant was well suited to Mendel's experiments for two reasons. First, the structure of the pea flower is such that it normally self-fertilizes; that is, the female gametes unite with male gametes from the same flower (figure 16.3). If a cross with a different plant is desired, the pollen-bearing structures (stamens) can be removed. Pollen from another plant, designated as the male parent, can then be placed on the appropriate part of the pea flower. Eventually, the male gamete contained in the pollen unites with the female gamete within the ovule. A second reason why the garden pea was well suited to Mendel's experiments is that genetic experiments require the testing of large numbers of individuals. Sufficient numbers of peas could be grown in the limited garden space available to Mendel at the monastery.

Table 16.1
The Seven Characteristics Used by Gregor Mendel in His Studies of the Garden Pea

Seeds	Round vs. wrinkled
Seeds	Yellow vs. green
Flowers	Purple vs. white
Flowers	Axial vs. terminal
Pods	Inflated vs. pinched
Pods	Green vs. yellow
Stem	Tall vs. short

Mendel was not the first to study hybridization in garden peas, but his approach was different. While others chose to examine the simultaneous inheritance of a large number of characteristics, Mendel chose, at the beginning of his experiments, to examine one characteristic at a time. He also chose characteristics with alternate forms that were easily recognized.

Mendel made a number of crosses involving each of seven different characteristics (table 16.1). In one experiment he crossed peas with tall stems and peas with short stems (figure 16.4). The plants he used for the first parents were plants

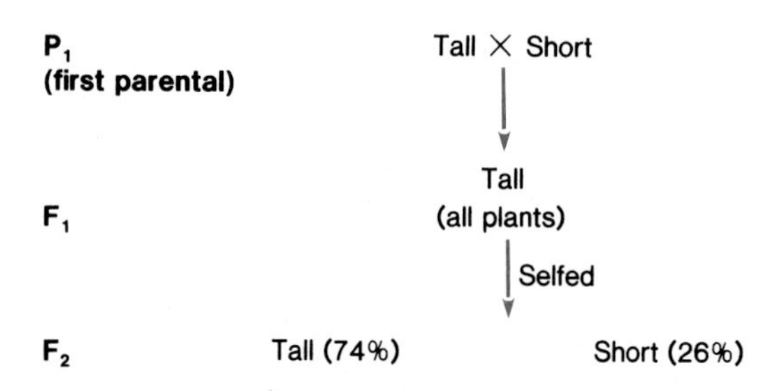

Figure 16.4 Results of Mendel's experiments on the inheritance of stem length in pea plants.

derived from a number of generations of selfing (self-fertilization). Mendel considered these plants to be "true breeding" in the sense that offspring of selfed tall plants always were tall and offspring of selfed short plants always were short. In his experiment, Mendel crossed a true breeding tall plant with a true breeding short plant. He collected all the seeds (peas) resulting from this cross and planted them. The plants that developed from these seeds were all tall. We call the first generation of offspring in a genetic experiment such as this the **first filial** or the F_1 generation. Next, Mendel allowed all of these tall plants to self, forming the **second filial (F_2)** generation, and they produced a total of 787 tall plants and 277 short plants. This works out to a ratio of about 3 tall : 1 short. Mendel did the same type of crosses with each of the six other characteristics in table 16.1 and obtained approximately the same ratios each time.

Segregation

In interpreting these results, Mendel stated what is now called Mendel's **law of segregation.** Mendel concluded that characteristics such as tall or short stems are determined by discrete factors. These factors we now call **genes.** Mendel stated that *each organism contains two factors for each characteristic and that the factors segregate during the formation of gametes so that each gamete contains only one of each pair of factors.* Mendel was able to draw these conclusions from his experimental results even though he was unaware of the fact that his factors were located on chromosomes. He was also unaware of meiosis, which, as you recall from chapter 15, is responsible for the segregation of chromosomes (and therefore genes) before the formation of gametes.

Figure 16.5 shows how Mendel's theory can be used to interpret the results of his experiment on stem height in peas. Each pea plant has two copies of the gene controlling the height of the stem. In this example, each copy may be either of two possible forms. Alternate forms of a gene are called

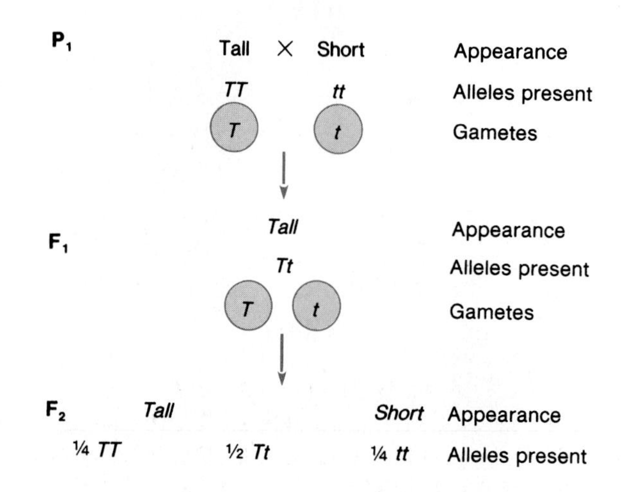

Figure 16.5 Explanation of the genetic basis for the results of Mendel's experiment on stem length (figure 16.4). Genetic makeup of all plants and gametes produced are shown.

alleles. Thus, in this example there is an allele for tallness and an allele for shortness. The original parents used in Mendel's cross, referred to as the P_1 generation, were "true breeding." Therefore, the tall plants had two copies of the allele for tallness and the short plants had two copies of the allele for shortness. When an organism has two of the same kind of alleles, it is **homozygous** for that gene. Because the first parents were homozygous, all gametes produced by the short plant contain the allele for shortness, and all gametes produced by the tall plant contain the allele for tallness. In the resulting F_1 generation, all the individuals have one allele for tallness and one for shortness. When an organism has two different alleles of a gene, it is **heterozygous** for that gene. Although the plants of the F_1 generation have one of each type of allele, they are all tall. When expression of only one allele in a heterozygous individual is observed, this allele is a **dominant** allele. The allele whose expression is masked in a heterozygote is a **recessive** allele. In modern genetic notation, we indicate the dominant allele with an uppercase letter and the recessive allele with a lowercase letter.

Genotype and Phenotype

From these results you can now see that two organisms with different allele combinations for a trait may have the same outward appearance (*TT* and *Tt* peas are both tall). For this reason it is important for us to distinguish between the genetic constitution and the outward appearance of an organism. The genetic constitution of an organism is its **genotype.** The genotype is determined at fertilization and does not change during the lifetime of the organism. The **phenotype** of an organism is a set of observable characteristics of

Figure 16.6 This diagram indicates why a 3:1 phenotypic ratio is expected from a cross between two heterozygotes. It shows the genotypes that can result when the possible types of male and female gametes combine.

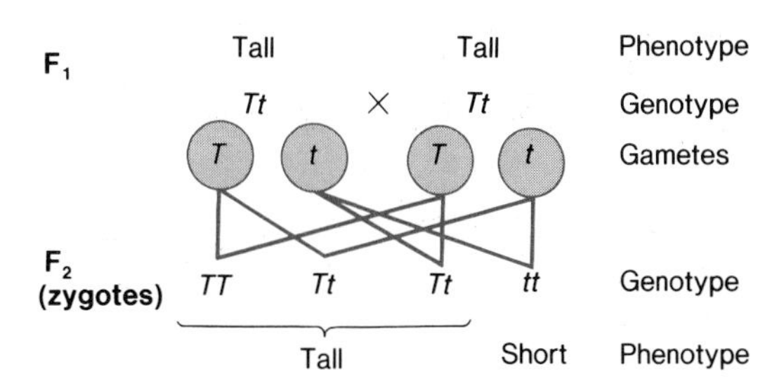

Figure 16.7 These results convinced Mendel that the genes for seed shape and color assort independently of each other.

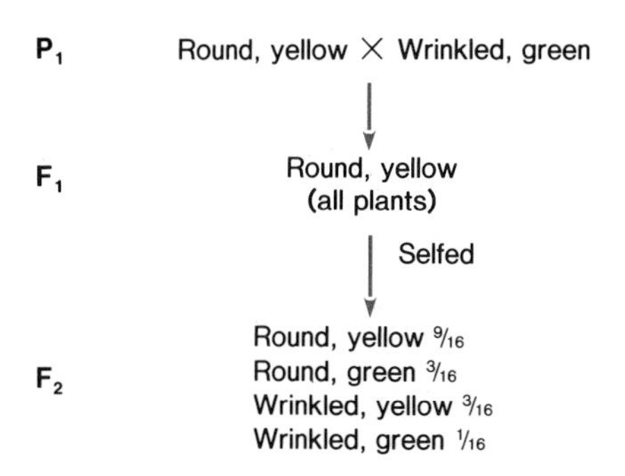

that organism. In the pea example, both *TT* and *Tt* are genotypes that give the tall phenotype. As you continue, it is important for you to keep the distinction between genotype and phenotype in mind.

When the heterozygous F_1 pea plants are selfed, the results are the same as those obtained when two heterozygous individuals (heterozygotes) are crossed. Approximately half of the gametes (both male and female) have the allele for tall stems and half have the allele for short stems. When these combine to form the F_2 zygotes (figure 16.6), the results are one-fourth homozygous tall plants, one-fourth homozygous short plants, and one-half heterozygotes. Because the heterozygotes display the tall phenotype, 75 percent of the F_2 generation are tall and 25 percent are short, thus producing the 3:1 phenotypic ratio.

Independent Assortment

Mendel next conducted experiments in which he traced the inheritance patterns of two different genes at the same time. He crossed plants producing round, yellow seeds with plants producing wrinkled, green seeds. The round and yellow alleles are dominant; the wrinkled and green alleles are recessive. Mendel had shown previously that both the seed shape and color traits obeyed the law of segregation.

When the two genes were traced simultaneously, the F_1 generation showed, as expected, that all individuals had the dominant characteristics—round and yellow seeds. Mendel then hypothesized that if the F_1 individuals were selfed, there were two possible outcomes for the F_2 generation.

If the two traits, seed shape and seed color, were inherited together, there would be just two kinds of seeds in the F_2 generation. The seeds would be either round and yellow, or wrinkled and green because these two sets of alleles would be inherited together. The two kinds would be in a 3:1 ratio with 75 percent of the seeds being round and yellow and the remaining 25 percent being wrinkled and green.

If the two traits were inherited independently of each other, however, the F_2 generation would show four possible combinations of seed color and shape: round and yellow (both dominant characteristics), round and green (dominant, recessive), wrinkled and yellow (recessive, dominant), and wrinkled and green (both recessive).

The F_2 generation conformed to the second of these hypothetical outcomes (figure 16.7); that is, Mendel found all four different genetic combinations among the F_2 generation plants. The combinations occurred in a ratio of 9 round, yellow : 3 round, green : 3 wrinkled, yellow : 1 wrinkled, green.

We will see later (page 343) why these particular ratios were obtained. For now, however, the important thing to remember is that the results of these experiments led Mendel to formulate a second genetic principle, called the **law of independent assortment.** This principle states that the members of one pair of genes segregate independently of other pairs.

These two laws, the law of segregation and the law of independent assortment, form the basis for modern genetic analysis. But when Mendel published his conclusions in 1866, no one realized their significance. One probable reason for this was that most investigators continued to study inheritance of large numbers of characteristics simultaneously. This led to such complex results that the underlying principles were virtually impossible to detect. These investigators may have thought that Mendel's experiments, based on the study of one or only a few traits, were a gross oversimplification of natural processes.

In his later years at the monastery, Mendel devoted most of his time to administrative battles with the local government, and he died in 1884 without the satisfaction of having the significance of his work understood and appreciated. Finally though, in 1900, De Vries, Correns, and von Tschermak discovered that Mendel's earlier conclusions stated very clearly what they themselves were discovering from their own experiments.

Figure 16.8 A Punnett square showing the results of a cross between two heterozygous pea plants.

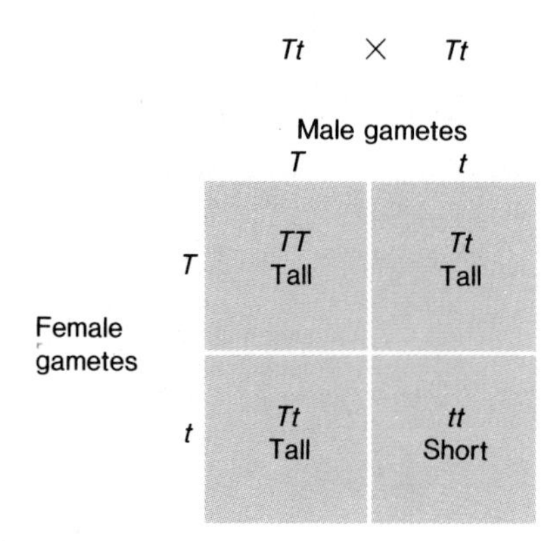

Figure 16.9 A test cross. The possible outcomes of the cross are diagrammed with Punnett squares.

Tall × Short

T? *tt*

If the tall plant is a heterozygote:

	t	
T	*Tt*	50% tall
t	*tt*	50% short

If the tall plant is a homozygote:

	t	
T	*Tt*	All offspring are tall

Mendelian Genetics

Now let us see how the principles of Mendelian genetics—the law of segregation and the law of independent assortment—can also be applied to present-day genetic analysis of hereditary patterns.

Monohybrid Crosses

A **monohybrid cross** is a cross between parents that differ in a single characteristic. The individuals involved usually differ in other characteristics, but the analysis is concerned with only one. Mendel's experiment with the tall and short pea plants is an example of a monohybrid cross. Using the law of segregation we can calculate the expected proportions of genotypes and phenotypes if we know the genotypes of the parents.

One way to determine this is to use a **Punnett square.** A prominent poultry geneticist, R. C. Punnett, introduced this simple method in the early 1900s. Figure 16.8 shows how the Mendelian cross between two types of peas can be diagrammed in a Punnett square. All the possible kinds of male gametes are placed across the top of the square, with all the types of female gametes along the left side. The ways in which these gametes may combine are determined by filling in the cells of the square. The results show that the expected proportions of offspring are 3 tall : 1 short for the phenotypes, and 1 *TT* : 2 *Tt* : 1 *tt* for the genotypes.

You should bear in mind that these are only approximate expected proportions, however, not absolute numbers. For example, if the cross shown in figure 16.8 produced 200 offspring, approximately 150 of them would be tall and approximately 50 would be short. In terms of the genotypes, approximately 50 of the plants would be *TT*, about 100 would be *Tt*, and the remaining 50 or so would be *tt*.

In order to make such predictions about the genetic results of crosses between different individuals, however, we must be able to determine genetic characteristics of the various types of gametes the individuals can produce.

We must know the genotypes, not just the phenotypes, of an individual in order to determine the types of gametes that the individual will produce. This is because different genotypes may have the same phenotypes. Remember, for example, that heterozygous (*Tt*) pea plants have the same appearance (phenotype) as homozygous tall (*TT*) plants. Mendel solved this problem by using plants that had been selfed for many generations. He knew these "true breeding" plants were homozygous. But even if the genetic history of an individual is not known, the individual's genotype can still be determined by a genetic means that involves making a specific kind of cross, called a **test cross,** which takes advantage of certain predictions made by the law of segregation.

An example of a test cross is diagrammed in figure 16.9, again using tall and short pea plants. In a test cross, the tall individual being tested for homozygosity or heterozygosity is crossed with an individual that is known to be homozygous for the recessive characteristic; that is, an individual with an unknown genotype is crossed with an individual with a known genotype. The homozygous recessive individual can produce only one type of gamete (*t*) as regards this particular gene. We do not know the genotype of the tall plant in figure 16.9, but if the plant is a heterozygote, half the offspring will be tall, and half will be short. If the tall parent is a homozygote, all of the offspring will have at least one *T* allele and will exhibit the dominant (tall) characteristic. But if any short offspring are produced, the parent with the unknown genotype must be a heterozygote.

Figure 16.10 A Punnett square for a dihybrid cross and the resulting phenotypic ratios. Note that each gamete has an allele for each gene being studied. Identical genotypes are connected by dashed or solid lines for convenience. In the summary below the Punnett square, the (−) symbol is used to indicate that, due to dominance, it does not matter which second allele is present.

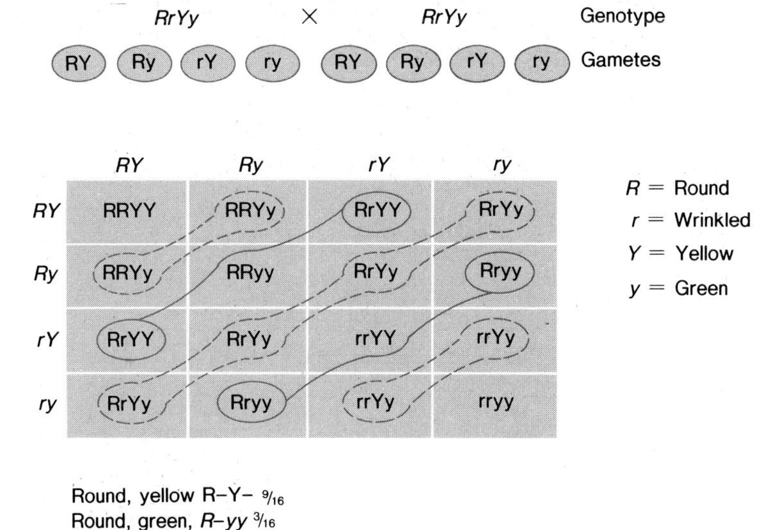

Dihybrid and Trihybrid Crosses

The same principles used for the analysis of monohybrid crosses can be used when analyzing the inheritance of the combinations of two or more different characteristics. Mendel's law of independent assortment states that the members of one pair of genes segregate independently of other pairs.* This makes it possible to predict the genotypes of the offspring of a cross involving more than one gene.

A **dihybrid cross** is a cross between individuals differing in two pairs of genes. Mendel's experiment with seed shape and color in peas is an example of a dihybrid cross. Figure 16.10 shows the Punnett square method of determining the genotypic and phenotypic ratios in the offspring. The method is basically the same as with the monohybrid cross. The major difference is that each gamete type now is shown with one allele of each of two different genes.

Another means of working out the results of a dihybrid cross is to use a **dendrogram,** a branching, treelike (dendro = "tree") diagram. The dendrogram technique can be used, for example, to predict phenotypic ratios in a dihybrid cross (figure 16.11). The probability of producing offspring with round seeds is 3/4 (1/4 homozygous round + 2/4 heterozygous). The probability of the green phenotype is 1/4 because only the homozygous recessive individuals have this phenotype. Therefore, the probability of producing an individual with round, green seeds is 3/16 (3/4 × 1/4). You can work out the other probabilities in the same way to show that a dihybrid cross between two heterozygous parents yields a ratio of 9 dominant, dominant:3 dominant, recessive:3 recessive, dominant:1 recessive, recessive. The dendrogram technique also can be used to predict genotypic ratios.

Figure 16.11 The dendrogram method of predicting phenotypic ratios from a dihybrid cross.

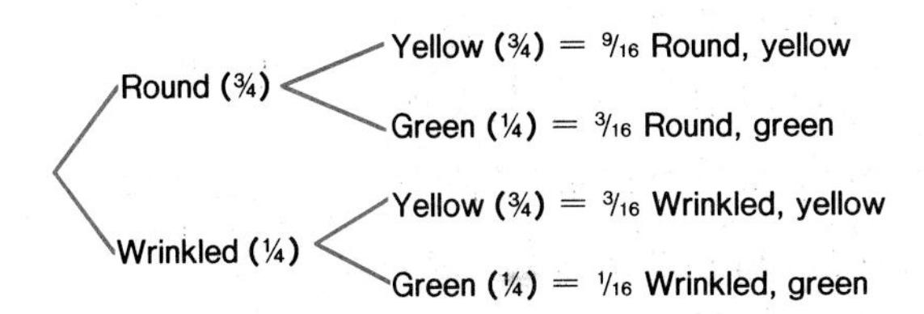

A **trihybrid cross,** which involves three characteristics, follows the same principles as a dihybrid cross. However, using a Punnett square for a trihybrid cross is a large job because each trihybrid parent can produce eight different types of gametes (figure 16.12). The resulting Punnett square has sixty-four cells!

*There actually are numerous exceptions to the law of independent assortment. We will discuss the reasons for these exceptions and their importance in chapter 17.

Figure 16.12 An organism that is heterozygous for three genes (trihybrid) may form gametes with any one of eight different genetic combinations.

Genotype	*Aa Bb Cc*
Gamete types	*ABC*
	ABc
	Abc
	AbC
	aBC
	abC
	aBc
	abc

Figure 16.13 How the three blood type alleles combine to produce the four possible blood types.

Blood type (phenotype)	**Possible genotypes**
A	I^AI^A, I^Ai^O
B	I^BI^B, I^Bi^O
AB	I^AI^B
O	i^Oi^O

Figure 16.14 How an offspring of A and B parents could have any of the four posible blood types if both parents happen to be heterozygous.

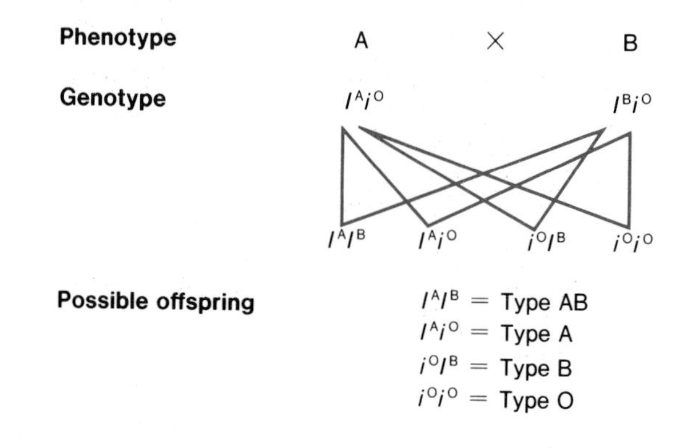

Table 16.2
Number of Different Genotypes Possible for Different Numbers of Alleles

Alleles	Possible Number of Genotypes
1	1
2	3
3	6
4	10
5	15
N	$\frac{N(N + 1)}{2}$

Multiple Alleles

All of the examples we have discussed so far have concerned genes that have two alleles. But you should not get the idea that this is the only kind of genetic pattern that exists. Our examples were simply the easiest ones to use in explaining basic genetic principles. There certainly are other patterns. In some populations, there may be only one form of a gene for a particular trait. In other cases, a gene can have three or more alleles. However, any single diploid individual can have a maximum of only two alleles per gene. (Remember the paired nature of genes and of the chromosomes that bear them.)

One of the best known examples of a three allele gene is the one that controls ABO blood types in humans. There are four possible ABO blood types (phenotypes): A, B, AB, and O. These phenotypes are produced from combinations of three different alleles: I^A, I^B, and i^O. The i^O allele is recessive, and the I^A and I^B alleles are both dominant. Figure 16.13 shows the different combinations of alleles that form the different phenotypes. Types O and AB are each produced from a single genotype, but types A and B may each be either homozygous or heterozygous.

It is possible, within limits, to predict the blood types of children from those of their parents. This relationship is sometimes used, for example, in cases of disputed paternity. Paternity cannot be proven from blood types, but it is possible to eliminate certain individuals as possible fathers. For example, two parents with blood type O will have only type O children. Children whose mother is type O and whose father is type A could be neither B nor AB. On the other hand, a type A parent and a type B parent could possibly have children with any of the four blood types, if both of them are heterozygous; that is, if blood type A parent has the I^A i^O genotype and blood type B parent has the I^B i^O genotype (figure 16.14).

An important point to note from this example is that the addition of a third allele increases the number of possible genotypes from three to six. In general, if there are N alleles, there are $N(N + 1)/2$ genotypes.

Figure 16.15 Crosses involving the frizzle feather abnormality demonstrate the phenomenon of incomplete dominance. (*a*) General pattern of inheritance. (*b*) Data from actual experimental crosses.

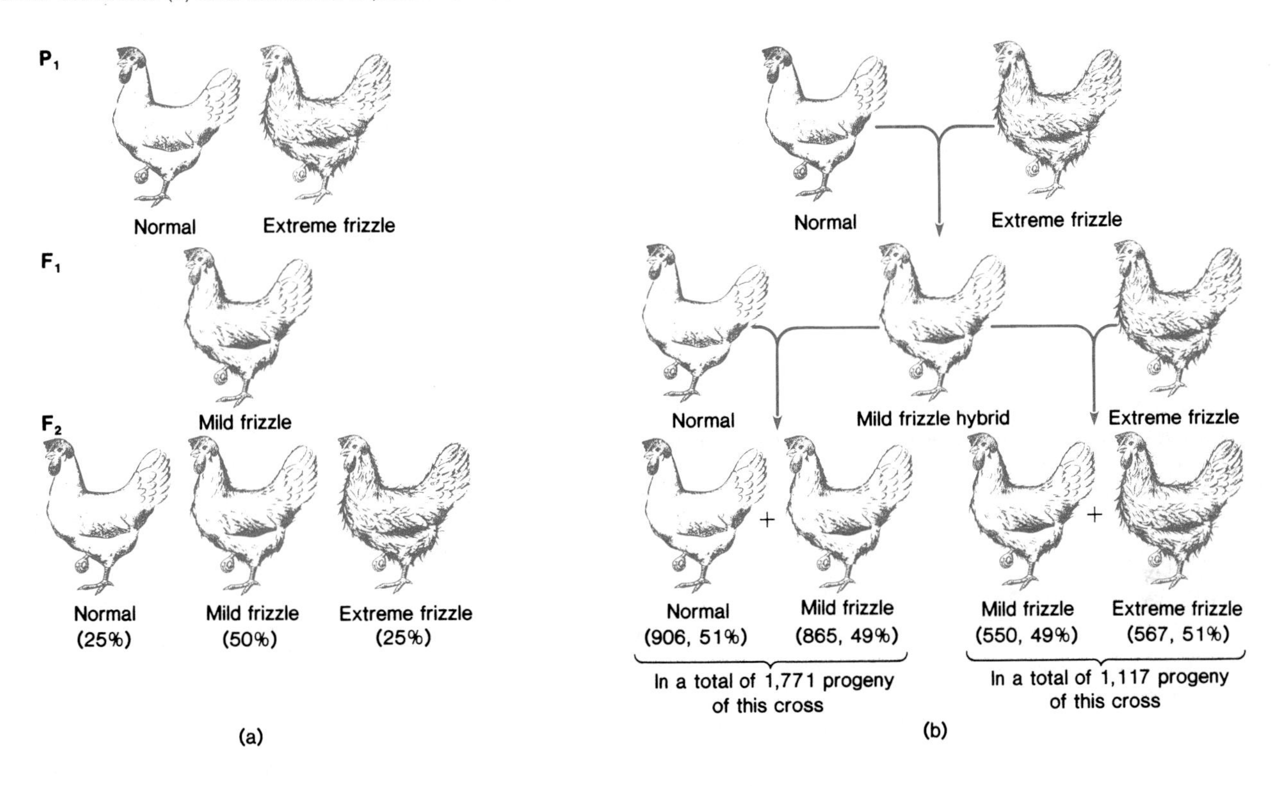

Other Inheritance Patterns

The pattern of human blood type genotypes and phenotypes, which we used in the previous section as an example of multiple alleles, differs in another important way from the inherited characteristics that Gregor Mendel studied. All of the characteristics of garden peas that Mendel reported studying display simple dominant-recessive relationships. Both individuals with the homozygous dominant genotype and individuals with the heterozygous genotype express the dominant phenotype, while individuals with the homozygous recessive genotype express the recessive phenotype. In other words, only two phenotypes are expressed. Human blood group inheritance, however, shows four phenotypes. Many other inherited traits also show several possible alternative phenotypes.

Intermediate Inheritance

Alleles are not always either dominant or recessive to each other. In chickens, for example, there is a mutation affecting feather structure descriptively called "frizzle." This condition produces bristly feathers that wear off easily. Poultry geneticists have noted that some chickens have a more extreme form of the frizzle condition than others. If a chicken with extreme frizzle is mated with a normal feathered chicken (figure 16.15), all of the F_1 offspring have a mild frizzle condition. If two of these mild frizzle chickens are mated to produce an F_2 generation, 25 percent of these offspring have extreme frizzle, 50 percent have mild frizzle, and the remaining 25 percent have normal feathers. With the segregation of parental types in the F_2 generation, it becomes obvious what has happened. The intermediate form occurs in the F_1 because neither of the two alleles exerts dominance, and an intermediate phenotype is produced. This phenomenon is called **incomplete dominance.**

Incomplete dominance also occurs in snapdragons. If red-flowered snapdragons are crossed with white-flowered snapdragons, all of the offspring have pink flowers. If these are selfed, the F_2 ratio is 25 percent red, 50 percent pink, and 25 percent white. Thus, the only way to maintain a garden with all pink snapdragons is to replant F_1 seeds each year.

Another type of intermediate inheritance occurs when the traits of both parents are distinctly expressed in the F_1 as opposed to forming an intermediate phenotype. The genetic control of blood types in humans, which we described earlier, is an example of this. If one parent contributes an I^A allele and the other contributes an I^B allele, the child will have blood type AB. The A and B alleles are said to be **codominant.**

Modifier Genes

In cases where dominance is complete, incomplete, or there is codominance, we see the effects of interactions between alleles of the same gene. Sometimes the situation is a little more complex, however, because the expression of certain genes is affected by entirely different genes, genes that are not their alleles. One example of this occurs in the genetic determination of coat pattern in cattle. The basic spotted pattern in Holstein cattle is determined by a single gene, but the relative amounts of black and white are determined by a different series of genes. Genes such as these are called **modifier genes,** because they modify the expression of other genes that are not their alleles.

Actually, there are many genetic combinations in which the phenotypic expression of a given gene depends upon the genetic makeup of the individual (genetic environment) in which the gene functions. These interactions among genes complicate genetic research because genes can actually be present without being expressed phenotypically. It should be clear to you by now that Gregor Mendel was very lucky indeed (or very perceptive) in choosing to study a set of genes that are inherited and expressed simply and directly.

Box 16.1

When Is a Colored Chicken White?

Genes can actually be present without being expressed phenotypically because other genes cover up their expression entirely. This is called **epistasis** ("covering up"). There is, for example, a color gene in fowl designated *C*. A *cc* genotype produces a white phenotype, a *Cc* or *CC* genotype is expected to have color.

But there is another gene *I*, which covers up the expression of the *C* gene. A genotype of *Ii* or *II* produces a white fowl, whether or not a *C* gene is present. Thus, the fowl must have a homozygous recessive genotype *ii* if it is to have color.

The white leghorn breed of chickens has the genotype *IICC*. The white wyandotte breed has the genotype *iicc*. Thus, the explanation for white feather color is different in the two breeds. The white leghorn is white because of the presence of the *I* gene. White leghorns actually carry the dominant *C* gene for color, but it is covered by the *I* gene. White wyandottes are white because they are homozygous recessive (*cc*) for the color gene. A white wyandotte could be colored if it had one or two *C* genes because wyandottes are homozygous recessive for the masking gene *ii*.

What happens when members of these two white breeds are crossed with each other? The results of such a cross are shown in box figure 16.1. All members of the F_1 generation are white because of the presence of the *I* gene, which masks the expression of genes for color. But in the F_2 generation, some colored chickens indeed are produced.

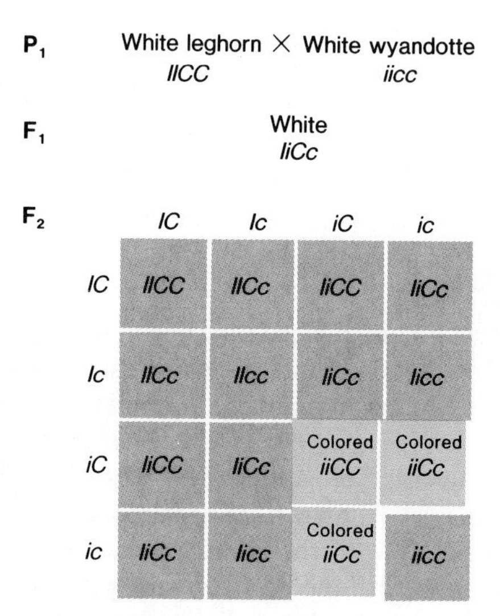

Box figure 16.1 Results of crosses between two breeds of white chickens, white leghorns and white wyandottes. The *I* gene masks expression of the *C* gene in the F_1 generation, but colored chickens are produced in the F_2 generation.

Pleiotropy

We have now seen some cases in which several genes affect the phenotypic expression of a single character. There is an entirely different kind of situation in which a single gene affects several different phenotypic characteristics. This is called **pleiotropy.**

There are many known examples of pleiotropy. In humans, for example, there is a gene which, in the homozygous recessive condition, results in phenylketonuria (PKU). Untreated individuals with this problem have unusual amounts of the amino acid phenylalanine in their blood, and also lower mental ability, somewhat larger heads, and lighter hair color. All of these characteristics are determined by a single gene.

Other examples of pleiotropy include the following: blue-eyed white cats are generally deaf, and humans with Eddowe's syndrome have both brittle bones and a blue color to the normally white sclerotic coat of the eye. An extreme example is a gene appropriately called polymorph (polymorph means "many form" or "many shape") in *Drosophila,* the fruit fly. This single gene affects eye color, body proportions, wing size, wing vein arrangement, body hairs, size and arrangement of bristles, shape of testes and ovaries, viability, rate of growth, and fertility.

The fact that pleiotropy is very common reflects the way in which genes act through effects on biochemical processes. An alteration in a single biochemical pathway, resulting from the expression of a single gene, may affect many developing characteristics.

Quantitative Genetics

The alternate traits that we have discussed up to this point, such as tall vs. short plants, green vs. yellow seeds, A vs. B blood types, etc., are clearly different from each other. This kind of variation is called **discontinuous variation,** because the variants fall into discrete, nonoverlapping sets. With regard to other traits, however, organisms may exhibit **continuous variation** so that different individuals may vary from each other only slightly. Differences among these individual phenotypes are *quantitative*—differences in degree or amount requiring precise measurement—rather than *qualitative*—differences requiring only simple observation for sorting of phenotypes into discrete groups. The study of the inheritance patterns of these traits is called **quantitative genetics.**

Many sets of alleles may be involved in quantitative inheritance of certain characteristics, and the environment may play an especially large part in determining the phenotype. Because large numbers of genes may be involved, quantitative inheritance is sometimes called **multiple factor** or **polygenic** ("many genes") **inheritance.** Phenotypes of quantitatively inherited characteristics frequently are distributed in a bell-shaped (normal) curve. Human height, for example, is a quantitatively inherited trait determined by many genes, and the distribution of human heights follows a normal distribution (figure 16.16).

Figure 16.16 Height distribution in 175 men recruited for the Army around 1900.

Number of individuals	1	0	0	1	5	7	7	22	25	26	27	17	11	17	4	4	1
Height in inches	58	59	60	61	62	63	64	65	66	67	68	69	70	71	72	73	74

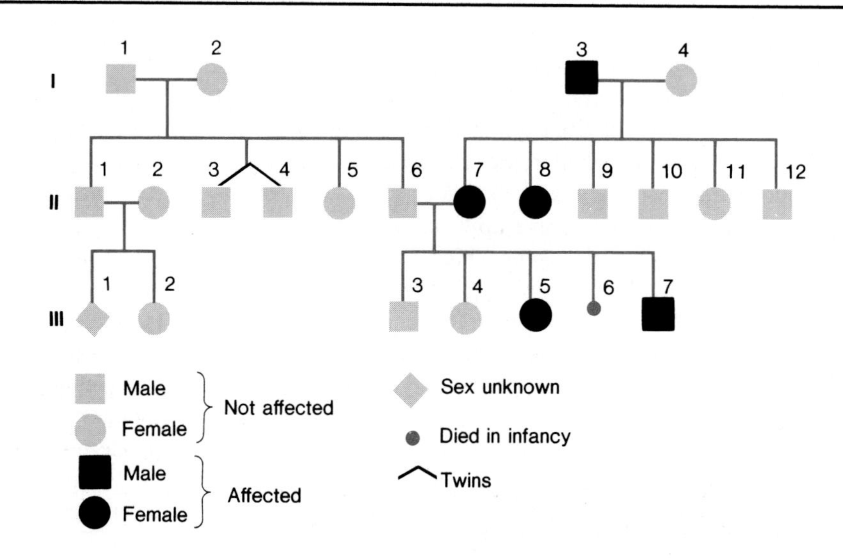

Figure 16.17 A typical human pedigree with the standard notation. For purposes of reference, Roman numerals to the left identify generations, and each individual within a generation has his or her own number. For example, II-7 is the seventh person in the second row.

Many of the economically important characteristics of agricultural plants and animals are inherited in a quantitative fashion. Some examples are milk production in dairy cattle and yield in corn. For this reason, those involved in breeding agricultural plants and animals must have a thorough understanding of quantitative genetics.

Many of the geneticists of Mendel's time were unknowingly studying polygenic inheritance. Because quantitative inheritance patterns resemble a blending pattern, these scientists were unable to recognize the laws of segregation and independent assortment or to appreciate the significance of Mendel's work.

Human Genetics and Genetic Counseling

Over the years, considerable progress has been made in the study of inheritance of many human characteristics. A list of Mendelian traits in humans compiled in 1978 includes 736 known dominants, 753 possible dominants, 521 recessives, 596 possible recessives, and 107 traits on the X chromosome (these are called X-linked genes and are discussed in the next chapter). Most of these identified traits are concerned with various diseases because, as you might imagine, there is much more medical incentive to study inherited diseases than the inheritance of various normal human characteristics.

Couples frequently are concerned with questions such as, "Will one of these genetic disorders, which we may display or may be present in a relative, or in previous children, also be likely to occur in future children?" Answers to such questions should be sought from a qualified genetic counselor. A genetic counselor (frequently a physician) is trained in genetic principles and is familiar with the inheritance patterns of human diseases.

The patterns of inheritance of a trait or characteristic in a particular family are commonly analyzed by drawing a **pedigree.** Figure 16.17 shows a sample pedigree and definitions of some of the symbols used. If the information is complete enough, a genetic counselor may be able to reconstruct the genetic history of a particular trait. After studying the pedigree, the counselor can inform the parents of the risk of having an offspring with the particular genetic defect. Once the risk has been determined, most genetic counselors encourage the prospective parents to decide for themselves whether or not to have children.

Modern Genetic Counseling

Many pedigrees are very complex and risks are often difficult to estimate. Sometimes, the risk must be estimated from past statistics rather than from the knowledge of the nature of inheritance of a particular trait. But some factors make the job of a genetic counselor easier. With some traits, heterozygotes can be detected by medical tests. For example, individuals heterozygous for Tay-Sachs disease or for sickle cell anemia can be detected by appropriate tests. When such testing is available, risk to offspring of certain marriages can be much more precisely determined.

Table 16.3
Mortality in Offspring of First Cousin Marriages Compared with That in Offspring of Unrelated Parents

Trait	Unrelated	First Cousins
Stillbirths and neonatal deaths	0.044	0.111
Infant and juvenile deaths	0.089	0.156
Juvenile deaths	0.160	0.229
Postnatal deaths	0.024	0.081
Miscarriages	0.129	0.145

Adapted from *An Introduction to Human Genetics*, Third Edition by H. Eldon Sutton. Copyright © 1980 by Saunders College/Holt, Rinehart and Winston. Reprinted by permission of CBS College Publishing.

Marriage between Relatives

The risk of genetic defects in offspring is important not only when there is a history of genetic disease, but also when close relatives are considering marriage. The offspring of two individuals who are genetically similar show a decrease in the number of genes that are heterozygous and an increase in the number of genes that are homozygous. This increase in homozygosity may have serious consequences because most individuals contain at least a few recessive alleles that, if homozygous, would be harmful, perhaps lethal, to the individual. Because the expression of these alleles normally is masked in a heterozygous condition, their presence causes no harm to heterozygous carriers.

The increase in homozygosity in the offspring is proportional to how closely related the parents are. Mating between parents and children or between sisters and brothers is called incest. There is a taboo against incest in most societies, and most states and countries have laws forbidding this type of marriage. A more common type of marriage is between first cousins. The frequency of first cousin marriages varies from society to society. In the United States, almost half the states have laws forbidding such a marriage, while other states permit it. However, in Japan, first cousin marriages are encouraged and occur in up to 10 percent of the marriages in some areas.

Table 16.3 shows some of the effects of the increased level of homozygosity due to first cousin marriages. The presence of harmful recessives is referred to as a type of **genetic load.** Unfortunately, the genetic loads of close relatives are more likely to include the same harmful recessive genes than are those of unrelated or more distantly related individuals. Clearly, there is a sound genetic basis for the ancient taboos against incest.

Heredity and Environment in Genetic Expression

One of the most basic, and also most often forgotten, principles of genetics is that the genotype and the environment interact to produce the phenotype. It is frequently quite difficult to determine how much of a phenotype is determined by heredity ("nature") and how much by environment ("nurture"). An example of the complex interaction of heredity and environment is seen in the determination of human height. Many genes affect height. Some genes involved are genes controlling growth hormone, digestive enzymes, rate of calcium deposition in the bones, and many others. In addition, environmental factors, such as the presence or absence of adequate nutrition during critical growing periods, may strongly influence an individual's height. A combination of all these factors ultimately determines how tall a person is.

In humans, twins occasionally provide an opportunity to study the relative roles of heredity and environment. There are two kinds of twins: **monozygotic** (identical) and **dizygotic** (fraternal). Dizygotic twins are produced from two different zygotes and are therefore no more similar genetically than brothers and sisters of different ages. Monozygotic twins are produced from the splitting of a single zygote and, as a result, are genetically identical. Usually, twins are raised in the same environment so that there is similarity in both the genetic and environmental components. But what happens when monozygotic twins are reared apart from each other? In these situations, the individuals have identical genetic components, but different environmental components. A comparison of the phenotypes of such twins provides an estimate of the relative genetic and environmental contributions to certain traits. Cases of identical twins raised separately, in different environments, are so rare, however, that progress is very slow in this research. The relative contributions of heredity and environment to such complex traits as intelligence are difficult to determine. Needless to say, interpretation of the results of research on such subjects often leads to controversy.

Summary

The modern science of genetics was founded by Gregor Mendel in the mid–1800s. From a series of experiments with garden peas, Mendel derived two basic laws of genetics that still are valid today.

Mendel's first law, the law of segregation, states that organisms contain two discrete factors (genes) for each characteristic and that these genes segregate so that a gamete has only one of the pair of alleles (forms of the gene) for each trait. Mendel's second law, the law of independent assortment, states that genes affecting different traits segregate independently of each other. These two laws can be used as a basis to predict the proportions of different types of offspring from a specified cross.

The types of genes present in an organism constitute that organism's genotype. Interaction of the genotype with the environment results in the outward appearance or phenotype of an individual.

Sometimes, one form of a gene, or allele, may mask the expression of another allele in the same organism. This effect, called dominance, causes organisms with different genotypes to have the same phenotype. In other cases, both alleles of a pair are expressed in the phenotype, resulting in various types of intermediate inheritance. Pleiotropy is where single genes have multiple effects.

A number of genes, each having a small effect, can combine to form a particular characteristic. The study of this kind of inheritance is called quantitative genetics.

The application of genetic principles to human populations has resulted in genetic counseling services. Couples may, in many cases, obtain estimates of the risk of having offspring with certain genetic diseases.

Both genotype and environment contribute to an organism's phenotype. The relative contributions of each, however, are difficult to resolve.

Questions for Review

1. What are alleles?
2. Briefly explain Mendel's law of segregation and his law of independent assortment.
3. Define the terms homozygous and heterozygous.
4. Explain what we mean by the terms genotype and phenotype.
5. How many different alleles can be present for a single gene in a diploid individual? Is there any limit to the number of alleles that might be present in a population of individuals?
6. How many types of gametes may be formed in an organism with the genotype A^1 A^2 B^1 B^2 C^1 C^1? *A, B,* and *C* are genes located on separate chromosomes.
7. In cattle, polled is dominant to horned. A polled bull is crossed with a horned cow. Of the four offspring, one is horned and the other three are polled. Is the bull homozygous or heterozygous for this gene?
8. In humans, assume that right-handedness is dominant to left-handedness. Explain how two right-handed people might have a left-handed child. If a right-handed person who had one left-handed parent married a left-handed person, what predictions could you make about their offspring?
9. Some genes are lethal when present in the homozygous condition. In chickens, when a gene known as "creeper" is present in the homozygous condition in a developing embryo, the spinal cord and vertebral column develop abnormally and the embryo dies inside the eggshell. When the creeper gene is present in the heterozygous condition, the chicken hatches, but it has skeletal abnormalities and walks with a peculiar creeping, stumbling gait. This set of characteristics gives the gene its name. Interpret the results of the following crosses:

   Normal × Normal—96 normal
   Normal × Creeper—51 normal, 48 creeper
   Creeper × Creeper—54 creeper, 26 normal

10. In crosses between two crested ducks, only about three-quarters of the eggs hatch. The embryos in the remaining quarter of the eggs develop nearly to hatching and then die. Of the ducks that do hatch, about two-thirds are crested and one-third have no crest. What results would you expect from a cross between a crested and a noncrested duck?
11. A cross is made between two parents with genotypes *AaBB* and *aabb*. If there are thirty-two offspring, how many of them would be expected to exhibit both dominant characteristics?

12. In rabbits, black color is due to a dominant gene *B* and brown color to its recessive allele *b*. Short hair is due to the dominant gene *S* and long hair to its recessive allele *s*. In a cross between a homozygous black, long-haired rabbit and a brown, short-haired one, what would be the nature of the F_1 generation? Of the F_2 generation? If one of the F_1 rabbits was mated with a brown, long-haired rabbit, what kinds of offspring in what ratio would you expect?
13. In radish plants, the shape of the radish produced may be long, round, or oval. Crosses among plants that produced oval radishes yielded 121 plants that produced long radishes, 243 that produced oval radishes, and 119 that produced round radishes. What type of inheritance appears to be involved? What results would you expect from a long with long cross? A round with round cross?
14. A man of blood type A and a woman of blood type B produce a child of type O. What are the genotypes of the man, the woman, and the child? If the couple were to have other children, what possible blood types could be produced? Indicate genotypes as well as phenotypes.
15. Define the term pleiotropy.
16. Explain why the study of quantitative genetics is important both in human and agricultural genetics.

Questions for Analysis and Discussion

1. Can you think of an experiment that would determine whether or not the differences between two populations of plants are due to genetic or strictly environmental influences? Remember that most plants can be propagated asexually and transplanted.
2. We saw in the discussion of cell division in chapter 15 that it is not the genes themselves that segregate during meiosis but the chromosomes on which the genes are located. If two genes were located next to each other on the same chromosome, do you think that they still would segregate independently? (We will discuss this question further in chapter 17.)
3. Do you think the genetic risk to offspring of first cousin marriages is significant enough that such marriages should be legally discouraged? Justify your answer.

Suggested Readings

Books

Ayala, F. J., and Kiger, J. A., Jr. 1980. *Modern genetics.* Menlo Park, Calif.: Benjamin Cummings.

Crow, J. F. 1976. *Genetics notes.* 7th ed. Minneapolis: Burgess.

Gardner, E. J., and Snustad, D. P. 1984. *Principles of genetics.* 7th ed. New York: John Wiley.

Singer, S. 1978. *Human genetics.* San Francisco: W. H. Freeman and Company.

Suzuki, D. T.; Griffiths, A. J. F.; and Lewontin, R. C. 1981. *An introduction to genetic analysis.* San Francisco: W. H. Freeman and Company.

Articles

Cavalli-Sforza, L. L. 1983. The genetics of human races. *Carolina Biology Readers* no. 21. Burlington, N.C.: Carolina Biological Supply Co.

Crow, J. F. February 1979. Genes that violate Mendel's rules. *Scientific American.* (offprint 1418).

Fincham, J. R. S. 1983. Genetic recombination in fungi. 2d ed. *Carolina Biology Readers* no. 2. Burlington, N.C.: Carolina Biological Supply Co.

Holden, C. November 1980. Twins reunited. *Science 80.*

Chromosomes and Genes

Chapter Outline

Chapter Concepts

1. Genes are located in linear arrays along chromosomes.
2. Species differ from each other in the number of chromosomes that they have.
3. Chromosome number changes may involve increases or decreases in entire sets of chromosomes or in individual chromosomes.
4. In some organisms, the different sexes differ from each other with respect to a single pair of chromosomes.
5. Genes on sex chromosomes have an inheritance pattern that is different from that of other genes.
6. Chromosomes may vary not only in number but in structure.
7. Genes close together on the same chromosome do not segregate independently during meiosis.
8. It is possible to map the relative locations of genes on chromosomes.
9. Some genes are located outside the nucleus in mitochondria and in chloroplasts. The inheritance patterns of these genes do not follow Mendel's laws.

Jeff is a fine young man. He has a ready smile for all of his many friends wherever he encounters them. He is almost always positive and cheerful. Unlike many teenagers, he feels few problems in his relationships with his parents and his brother and sister. He is warm-hearted and kind in all of his interactions with family and friends.

Although he does not play basketball, Jeff is a great basketball fan. He eagerly cheers for the high school and college teams in his hometown. Like other teenagers, he is delighted when his favorites win and very disappointed when they lose. He remembers stars of earlier years and enjoys comparing them with current varsity members.

Jeff is a special person because of his delightful personality, and because of his deep interest and involvement in many aspects of teenage life. But Jeff is also a special person because he has problems that he shares with a relatively small percentage of teenagers. Jeff is able to read only simple material written for children much younger than he is. He can write only his name, his address, and a few other words. He may be able to learn a little more, but learning is an agonizingly slow process for him and for his teachers. You see, Jeff has Down's syndrome. Jeff and other people who have Down's syndrome share certain facial features (figure 17.1) and other physical characteristics, and most suffer some degree of mental retardation. In addition, many people with Down's syndrome have other physical problems including certain heart problems.

Jeff has an extra chromosome (chromosome no. 21) in the nucleus of every cell of his body. It is the presence of this extra chromosome that results in the set of problems we recognize as Down's syndrome.

How can an extra chromosome have such profound effects on growth and development? The answer to that question is not yet clear, but we do know that each chromosome bears a set of genes, and thus is involved in determining many different characteristics of the individual.

Since the time of Gregor Mendel, much progress has been made in working out the relationship between chromosomes and genes. Mendel thought that his "factors" (we now call them genes) were independent, discrete units within the cell. Later, scientists learned that genes are organized in linear arrays as part of chromosomes and that each gene occupies a specific site on the chromosome. In this chapter, we will examine chromosomes and the relationships between genes and chromosomes.

Figure 17.1 A young man who, like Jeff (not his real name), has Down's syndrome, a condition that results from the presence of an extra chromosome in all body cells.

Chromosomes

At the time that Gregor Mendel published his results in the 1860s, it was known that all living things are made up of cells. It was also known that plant and animal cell nuclei contain rodlike structures called **chromosomes.** But Mendel did not know enough about the behavior of chromosomes in dividing cells to realize the significance of chromosomes for his theories of inheritance.

When Mendel's work was rediscovered in 1900 (see page 341), a great deal more was known about the behavior of chromosomes in dividing cells. In 1902, two investigators, Walter S. Sutton of the United States and Theodor Boveri of Germany, suggested independently that the genetic material is contained in the chromosomes. This idea has developed into the **chromosome theory of heredity.**

Sutton and Boveri had several reasons for believing that Mendel's independent factors, or genes, were located on chromosomes. They knew that within the nucleus, chromosomes exist in pairs. This coincides with Mendel's theory that organisms have two copies of each gene. According to the chromosome theory of heredity, one copy of each gene is located on each one of a pair of chromosomes. Thus, an individual with a heterozygous genotype (for example, *Aa*) has one allele (*A*) on one chromosome of a given pair and another allele (*a*) on the other. It had also been noted that *gametes* have only *one* of each pair of chromosomes. The separation

Figure 17.2 Nucleosomes. (*a*) This photograph shows uncoiled chromatin material from chicken cells. The "beads" in the photograph are the nucleosomes and the "string" is the DNA between nucleosomes. (*b*) A drawing of a portion of a chromosome, showing the nucleosomes (spheres of aggregated histone molecules) around which the DNA (lines) is wrapped. The boxed area shows how the DNA is coiled around the nucleosomes and how it connects nucleosomes. The upper left portion of the diagram (outside the box) shows the coiled form in which the entire nucleosome chain sometimes is found. This coiled form may represent the condition of the chromatin in chromosomes when they are most condensed, as during metaphase of mitosis.

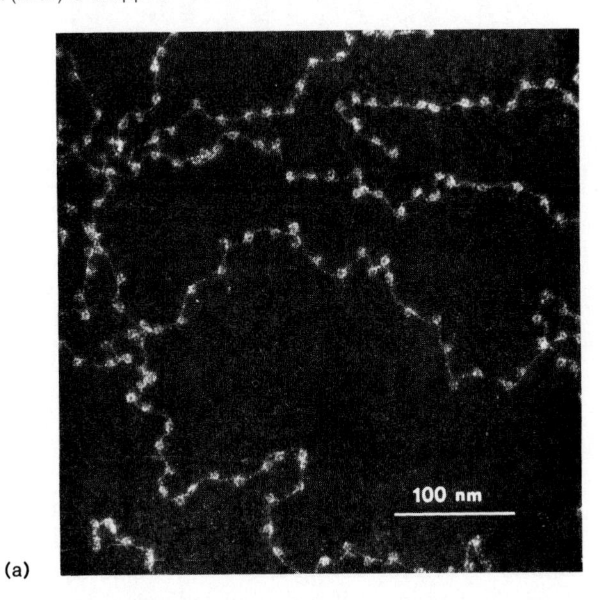

(a)

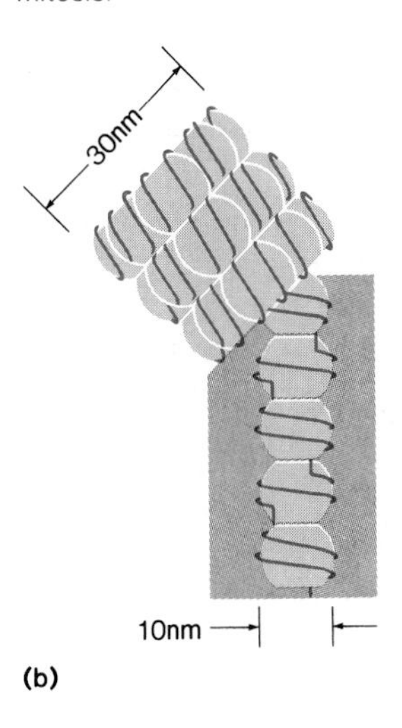

(b)

of pairs of chromosomes during meiosis (see chapter 15) explains how the alleles of a gene segregate and then recombine in the formation of zygotes at the time of fertilization. However, before we consider the arrangement of genes on chromosomes, we need to learn a little more about the structure and function of chromosomes.

Chromosome Structure

The chromosomes in most eukaryotic cells (cells with nuclei) are rod shaped and stain quite strongly with certain dyes when they are condensed during cell division. Biologists call the stainable fibrous material that makes up chromosomes **chromatin.** Chromatin, however, is not a single substance, but a complex, highly organized set of substances. A chromatin fiber consists of a long molecule of DNA associated with basic protein molecules called histones, along with quantities of other proteins.

The DNA and proteins in the chromatin are organized in a very specific way. When chromatin is isolated, spread out, and examined with the electron microscope, it appears to consist of a series of "beads" that are about 10 nm in diameter and arranged in a chain (figure 17.2*a*). Each of these "beads" is a basic structural unit of a chromosome called a **nucleosome.** Each nucleosome is an aggregate of eight histone molecules with a specific length of DNA wrapped in a helical coil around the protein aggregate (figure 17.2*b*).

Nucleosome "beads" in chromatin appear to be held together by DNA, which, in addition to coiling around each nucleosome, stretches from one nucleosome to the next. Finally, the whole nucleosome chain itself sometimes is found in a coil, measuring about 30 nm in diameter.

Number of Chromosomes

While the basic structural units in chromosomes, the nucleosomes, seem to be remarkably similar in all the eukaryotic cells studied so far, there are major differences among the chromosome complements of various organisms. One of the most obvious differences is the number of chromosomes present.

Differences among Organisms

The number of chromosomes found in different organisms varies a great deal. The number ranges from 4 chromosomes found in one plant, *Haplopappus gracilis,* and in several insects and other invertebrate animals, to nearly 500 chromosomes in some ferns (table 17.1). The number of chromosomes present is usually, but not always, consistent within a species and is not correlated with either the size of the cell or the evolutionary "advancement" of a species.

Table 17.1
Chromosome Numbers of Some Common Plants and Animals

Species	Common Name	Chromosome Number
Triticum aestivum	Wheat	42
Zea mays	Corn	20
Lactuca sativa	Lettuce	18
Lycopersicon esculentum	Tomato	12
Gossypium barbadense	New World cotton	52
Phaseolus vulgaris	Kidney beans	22
Glycine max	Soybeans	40
Pyrus communis	Apple	34
Ipomea batatus	Sweet potato	90
Ophioglossum vulgatum	Fern	480
Apis mellifera	Honeybee	16
Bombyx mori	Silkworm	56
Cyprinus carpio	Carp	100
Rana pipiens	Green frog	26
Alligator mississippiensis	Alligator	32
Anas platyrhynchos	Mallard duck	78
Bos taurus	Cattle	60
Equus caballus	Horse	64
Felis felis	Cat	38
Canis familiaris	Dog	78

Polyploidy

Up to this point, we have discussed mainly organisms that are diploid; that is, organisms that have two of each kind of chromosome. However, there are other arrangements. When an organism has more than two sets of chromosomes, the condition is called **polyploidy.** Polyploid organisms are named according to the number of sets of chromosomes that they have: triploids (3N) have three of each kind of chromosome, tetraploids (4N) have four sets of chromosomes, pentaploids (5N) have five sets, etc. Polyploidy is quite common in plants but relatively unusual in animals. Many ornamental plants are bred as polyploids because they tend to have larger leaves and flowers than diploid plants of the same species.

Aneuploidy

While polyploidy involves the number of sets of chromosomes, **aneuploidy** is a general name for conditions where the chromosome number varies from the normal number for the species by a small number of chromosomes. Usually, aneuploidy involves an excess or a deficiency of an individual

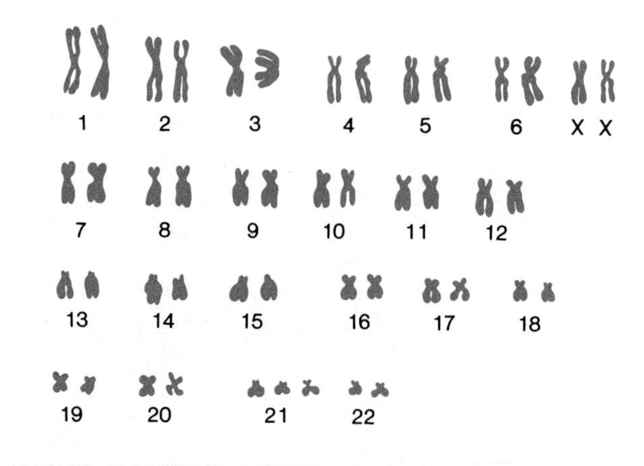

Figure 17.3 Chromosomes of an individual with Down's syndrome. Note the extra copy of chromosome no. 21.

chromosome in an otherwise normally diploid cell. For instance, if a diploid individual has one extra copy of a chromosome, the condition is called **trisomy** and is designated as 2N + 1. Recall that 2N symbolizes the normal diploid number and thus in humans (with a normal chromosome number of 46), for example, 2N + 1 would equal 46 + 1, or 47 chromosomes. If a diploid individual is missing one chromosome, the condition is called **monosomy** and is indicated by 2N − 1.

As noted in chapter 15, aneuploidy usually arises because of nondisjunction of homologous chromosomes during cell division. If it occurs during meiosis, it can have drastic consequences in the development of a zygote produced as a result of fertilization involving a gamete with a chromosome excess or deficiency.

A well-known example of aneuploidy is the human condition **Down's syndrome,** which we mentioned earlier. This condition, sometimes referred to as "mongolism," results from the presence of an extra copy of the chromosome designated no. 21 (figures 17.3 and 17.1). The extra copy of the chromosome results from nondisjunction during sperm or egg formation; either the duplicated pair of chromosomes fails to separate during the first division of meiosis or the chromatids of the duplicated chromosome fail to separate during the second division. In humans, the chances of producing an aneuploid child grow larger as the parents, especially the mother, grow older. This is an important consideration when a woman over age 35 considers having a child.

Down's syndrome and other chromosomal disorders now can be detected in early pregnancy by the use of a technique called **amniocentesis.** A needle is inserted through the mother's abdomen into the uterus, and a sample of the amniotic fluid surrounding the developing fetus is withdrawn. Cells

Figure 17.4 The procedure for preparing karyotypes of cultured human white blood cells. Because chromosomes are best observed in metaphase, colchicine, a substance that stops mitosis at metaphase, is added to the culture after some cell growth has occurred. This provides the person doing the work with many metaphase cells.

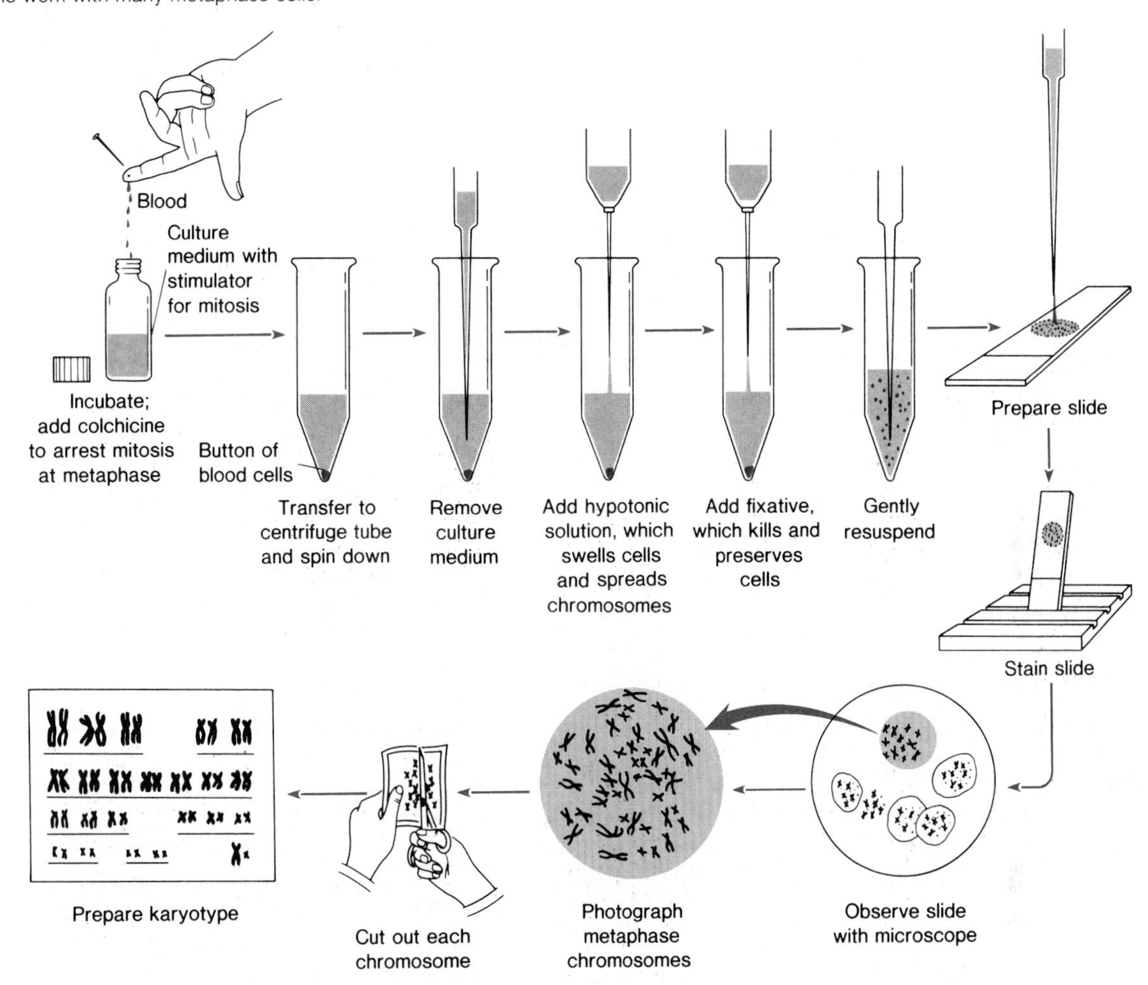

from the amniotic fluid can then be cultured and examined for the presence of the extra no. 21 chromosome, or other problems. Detection of a chromosome problem does not solve the problem, however. Rather, it presents the parents with some terribly difficult decisions to be made.

Human Chromosomes

Although we are now familiar with pictures of sets of human chromosomes such as that shown in figure 17.3, virtually all knowledge of human chromosomes has been obtained since the 1950s. Even the exact number of chromosomes in humans was not determined with precision until 1956. This seems extraordinary in light of the fact that chromosomes in other organisms had been observed and described for over a hundred years. But the chromosomes of humans are quite difficult to prepare in a manner in which consistent chromosome counts can be made, and routine techniques for studying human chromosomes have been available for only a relatively short time.

Preparations of human chromosomes are usually made from certain white blood cells (lymphocytes). Figure 17.4 diagrams the procedure. When dividing cells are observed, metaphase chromosomes are photographed. The individual chromosomes are cut out of the photograph and lined up with one another in matching homologous pairs. This ordered arrangement of all the chromosomes in the nucleus is called a **karyotype.**

Figure 17.5 Photograph of the chromosomes of a human male in a cell at metaphase of mitosis.

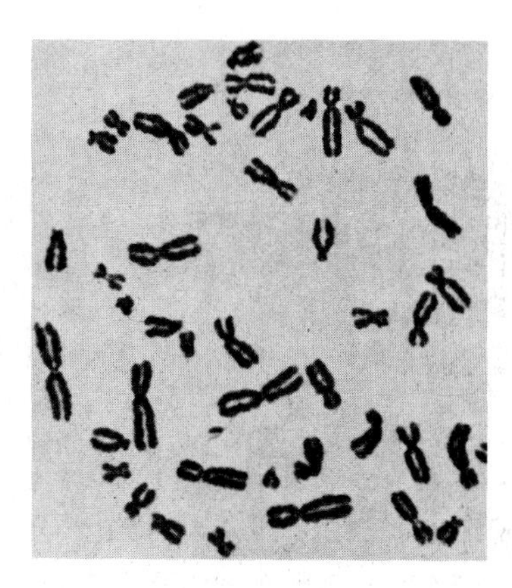

Figure 17.6 A human male karyotype. A human female karyotype would differ only in that it would have two X chromosomes rather than an X and a Y chromosome.

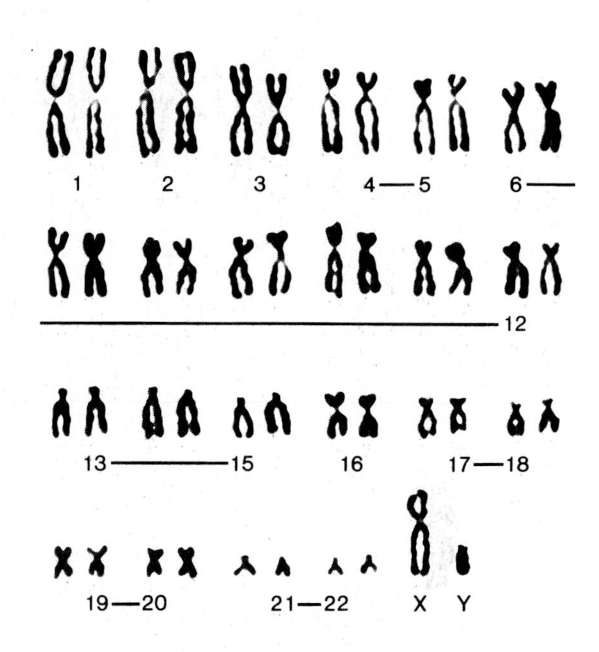

Figure 17.5 is a photograph of the metaphase chromosomes of a human male. Figure 17.6 is a karyotype made from the chromosomes such as those shown in figure 17.5. Such karyotypes show the characteristic diploid chromosome number of human body cells, which is forty-six (twenty-three pairs). The last pair of chromosomes in the karyotype, labeled X and Y, are not identical. These are the **sex chromosomes.** The human male has the differing X and Y chromosomes in this pair, while the human female has two X chromosomes.

The study of human chromosomes was greatly advanced in the 1970s with the introduction of new methods for chromosome staining. These new methods produce distinct patterns of dark and light stained bands. Figure 17.7 shows one such type of banding, referred to as G banding. The bands are called **G bands** because of the Giemsa stain used to prepare the chromosomes for examination. The visualization of this type of banding pattern is extremely important because it permits reliable identification of individual chromosomes. Many of the chromosomes are otherwise very similar and can be sorted only into groups with common general appearance when routine karyotyping procedures are used without Giemsa staining (as they are in figure 17.6). Such staining also allows for easier identification of structural abnormalities in the chromosomes and helps to detect changes in the number of chromosomes. The mapping of the locations of particular genes (page 368) on specific chromosomes also is aided by the study of G bands.

Sex Chromosomes

In the karyotype in figure 17.6, you can see that the human male has an X chromosome and a Y chromosome, and twenty-two other pairs of chromosomes. The karyotype in Figure 17.7 shows that the genetic makeup of the human female differs from the human male only in that the female has a pair of X chromosomes and no Y chromosome. As we said earlier, the X and Y chromosomes are called sex chromosomes. The other chromosomes, the ones that are common to both males and females, are called **autosomes.**

Sex Determination

The pair of sex chromosomes segregates at meiosis so that each gamete receives just one member of the pair. In humans, each egg has an X chromosome, but a sperm cell can have either an X chromosome or a Y chromosome (figure 17.8). Therefore, the sex of the new individual is determined at the time of fertilization, and it depends on the type of sex chromosome present in the sperm. If the sperm cell has a Y chromosome, the zygote (XY) will become a male. If the sperm cell has an X chromosome, the zygote (XX) will become a female.

Figure 17.7 A karyotype of a normal human female, showing the different G-band patterns that permit reliable sorting of individual chromosomes. Note that the sex chromosome pair in females consists of two identical X chromosomes.

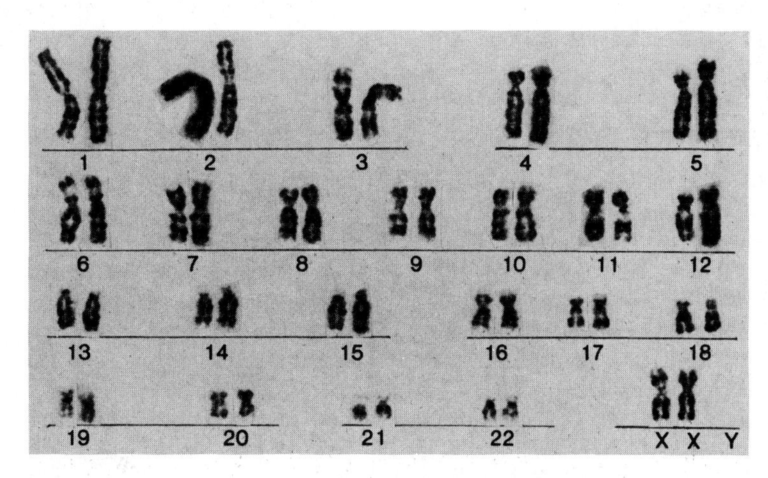

Figure 17.8 Sex chromosomes and sex determination in humans. The number of male zygotes produced should equal the number of female zygotes.

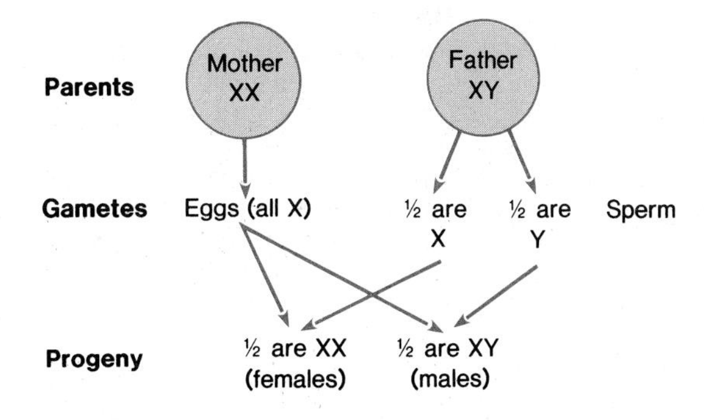

Many animals have the same sex determination system as humans. An organism with two X chromosomes is a female, and an organism with an X and a Y is a male. But different sex determination patterns also exist. In birds, moths, and butterflies, the situation is opposite of the human pattern because the male has two X chromosomes, while the female has only one X chromosome. Y chromosomes may or may not be present, depending on the species. In ants, wasps, and bees, there is yet another pattern. Some eggs develop parthenogenetically (without fertilization) into males, while the fertilized eggs develop into females. In these animals, all cells in males are haploid (N), and all cells in females are diploid (2N). The development of different types of females, such as queens and workers in bees, is due to differences in nutrition, not to differences in chromosome complement.

Box 17.1
Sex Ratios

The **sex ratio,** the ratio of males to females among offspring, is approximately one to one in many organisms with separate sexes; that is, the number of females born is approximately equal to the number of males. This is because of the segregation of the sex chromosomes in the organism with the XY genotype. The two kinds of gametes, X and Y, should be produced in equal proportions, and therefore, the sex ratio of the offspring should be roughly 1:1 (see figure 17.8).

The sex ratio in humans, however, shows that the number of males does not equal the number of females. Among Caucasians in the United States, there are approximately 106 boys born for every 100 girls. This ratio, the ratio at the time of birth, is the secondary sex ratio. The sex ratio at the time of fertilization is the primary sex ratio. If there is a different survival rate between fertilization and birth for the two sexes, the primary and the secondary sex ratios may be different. Is the higher number of males at birth due to a higher death rate among developing females? This apparently is not the case. Studies of unborn fetuses actually show an even higher ratio of males to females. In fact, it is estimated that the human primary sex ratio is about 114 males to 100 females.

It has been suggested that the sperm cells bearing the lighter Y chromosome are able to swim faster than those bearing the X chromosome, and that for this reason they are able to reach the egg sooner. Another possibility is that the environment of the female reproductive tract favors the survival of more Y-bearing sperm than X-bearing sperm. The actual reason for the slightly distorted sex ratio in humans, however, remains unknown.

Even though there are more human males than females at birth, this ratio ultimately reverses itself because more males than females die at every age. Females begin to exceed males as early as age eighteen in some countries or as late as age fifty-five in others. The reasons for this differential survival also are unknown.

The sex ratio conceivably could be altered if parents were given the option of choosing the sex of an offspring. The technical methods of determining the sex of a fetus already exist, through amniocentesis followed by karyotyping. Selective abortion at this point could alter the sex ratio, although this approach would raise grave ethical problems. However, animal breeders now are able to separate X and Y sperm outside the body of the animal; either type can then be introduced into a female by artificial insemination. In years to come, many ethical and social problems could be raised by the introduction of parental choice regarding sex of human offspring.

Figure 17.9 T. H. Morgan's experiments with *Drosophila* eye color.

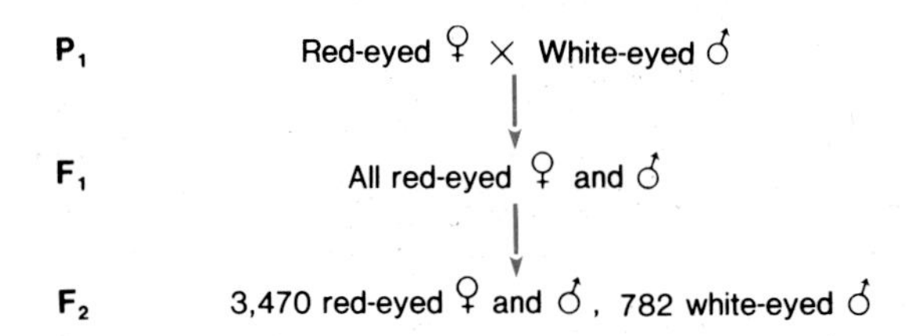

Sex Linkage

Genes occupy specific sites on chromosomes. The specific part of a chromosome that is the site of a particular gene is known as its **locus** (plural: **loci**). During this century, **chromosome mapping** studies (the determination of relative positions—loci—of individual genes on specific chromosomes) have yielded extensive maps of gene loci on chromosomes of many organisms, including humans. All of this work on determining the positions of gene loci on chromosomes began with a discovery that involved sex chromosomes.

Thomas Hunt Morgan and his colleagues at Columbia University were the first biologists to associate a specific gene with a specific chromosome when they published the results of the following experiment in 1910. In one of their many genetic experiments on the fruit fly (*Drosophila melanogaster*), they crossed female flies that had red eyes with males that had white eyes (figure 17.9). In the F_1 generation, all

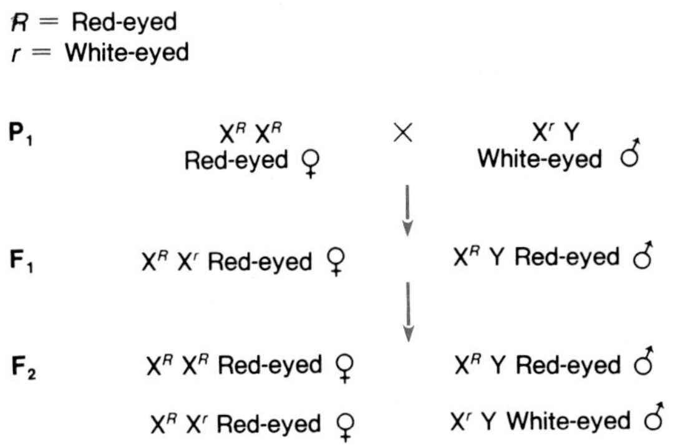

Figure 17.10 How Morgan explained his results. The gene for eye color is found on the X chromosome and is therefore sex linked (or X linked). The Y chromosome does not carry a gene for eye color.

R = Red-eyed
r = White-eyed

P_1	$X^R X^R$ Red-eyed ♀	×	$X^r Y$ White-eyed ♂
F_1	$X^R X^r$ Red-eyed ♀		$X^R Y$ Red-eyed ♂
F_2	$X^R X^R$ Red-eyed ♀		$X^R Y$ Red-eyed ♂
	$X^R X^r$ Red-eyed ♀		$X^r Y$ White-eyed ♂

Figure 17.11 Color vision in humans as an example of sex-linked inheritance.

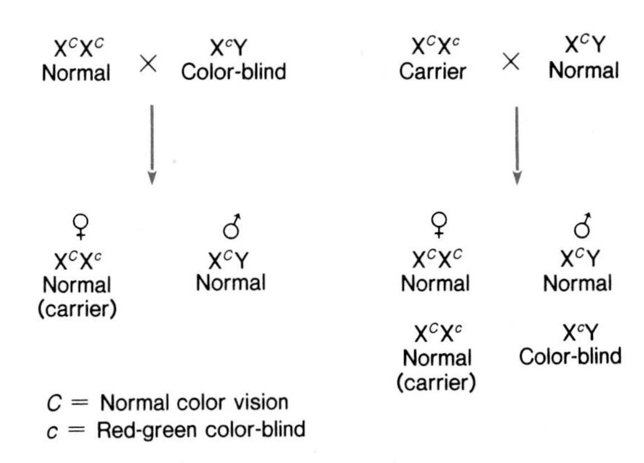

the flies had red eyes. This was not a surprising result. It would be expected if red eyes were dominant to white. When crosses were made between members of the F_1 generation, a ratio of 75 percent flies with red eyes to 25 percent with white eyes was expected among the offspring. The F_2 generation did have 3,470 flies with red eyes and 782 with white eyes, roughly a 4:1 ratio, not too far from the expected ratio of eye colors. But the experimenters noticed a peculiarity in their results. *All* the flies with white eyes were males, while there were both males and females among the red-eyed flies.

Because of this seemingly unusual result, Morgan and his colleagues did further experiments and carefully took sex differences into account in interpreting their results. They knew that *Drosophila* males have an X and a Y chromosome, while the females have two X chromosomes. Eventually they concluded that the gene for eye color is located on the X chromosome and that the Y chromosome bears no gene for eye color. Thus, female fruit flies have two genes for eye color while males have only one. Figure 17.10 illustrates this explanation of Morgan's results.

Any gene, like the genes for red and white eye color in *Drosophila,* that occurs on a sex chromosome, is called a **sex-linked gene.** The vast majority of sex-linked genes occur on the X chromosome, as opposed to the Y chromosome. When analyzing any cross involving a sex-linked gene, it is important to keep track of the X and Y chromosomes and to treat the sexes separately. In humans (and in *Drosophila* and other animals with the same sex chromosome pattern), all male offspring receive the Y chromosome from their father and thus cannot receive any gene located on the father's X chromosome. All female offspring receive the father's X chromosome and one of the two maternal X chromosomes.

In humans, an example of a sex-linked gene is the one responsible for common red-green color blindness. The gene for this condition is carried on the X chromosome. The allele for color blindness is recessive to the allele for normal vision. If a male receives an X chromosome from his mother that has the allele for color blindness, he will be color blind because the Y chromosome does not carry an allele for normal vision. A female, because she has two X chromosomes, may be heterozygous for the alleles. She would have normal vision but would be a "carrier" of the allele for color blindness. Figure 17.11 shows a cross between a color-blind male and a female that is homozygous for normal vision. None of the male offspring can possibly inherit the allele for color blindness. All of the female offspring are carriers because they all get an X chromosome with the allele for color blindness from their father and one of the two X chromosomes with the normal allele from their mother. If one of the carrier daughters marries a normal male, approximately half of the male offspring will have red-green color blindness. None of the female offspring will be color blind, although they may be carriers. This is because half the males receive the allele for color blindness from their mother, and all of the females get a normal allele from their father. Again, the important points in working any problem concerning sex linkage are to diagram the crosses in terms of X and Y chromosomes and to look at the phenotypic ratios in both sexes in the offspring.

Figure 17.12 A pedigree showing the sex-linked inheritance of hemophilia in European royal families. The members of the present British royal family are unaffected.

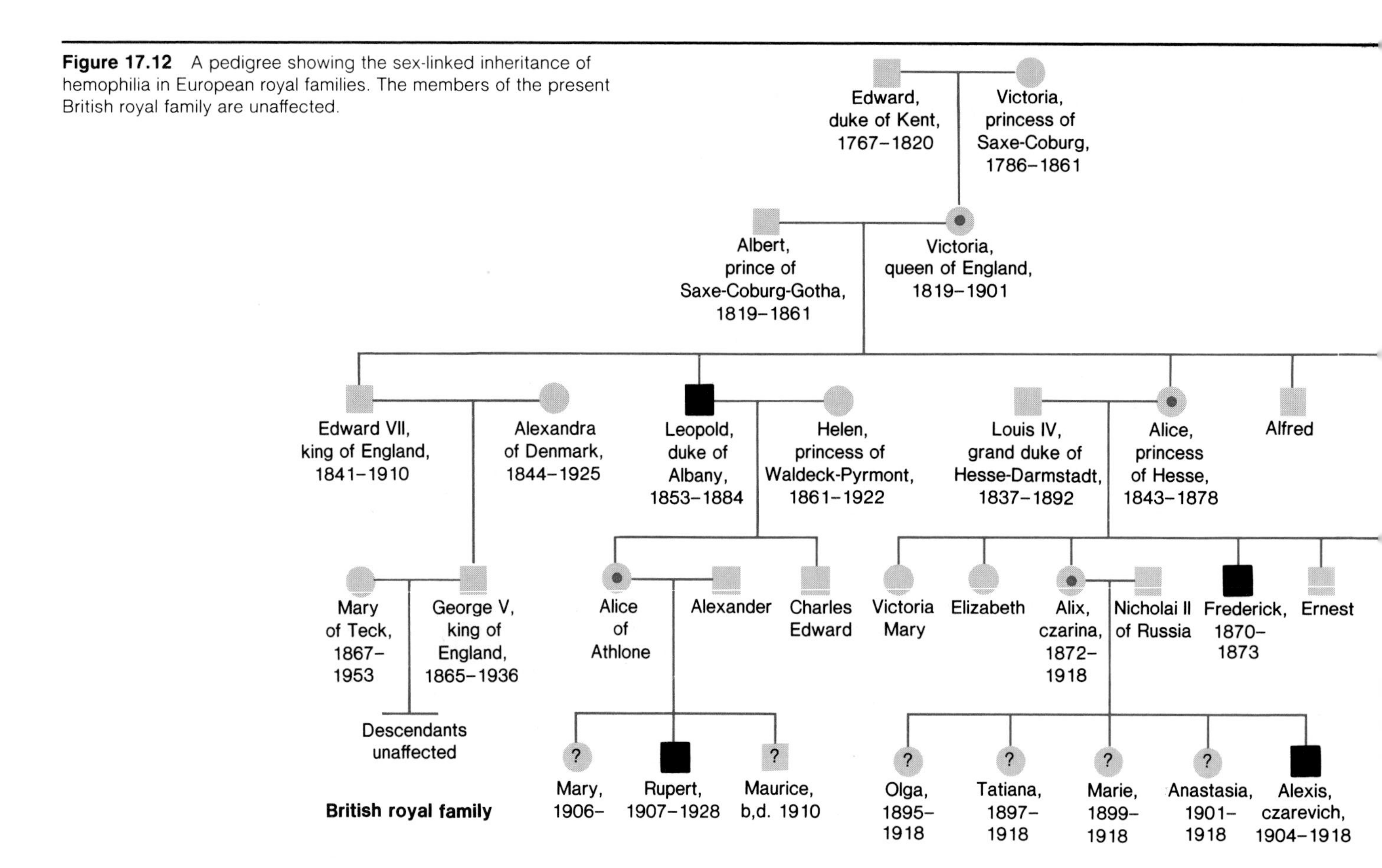

Figure 17.13 These three brothers have hairy ear rims. The gene for this condition is located on the Y chromosome.

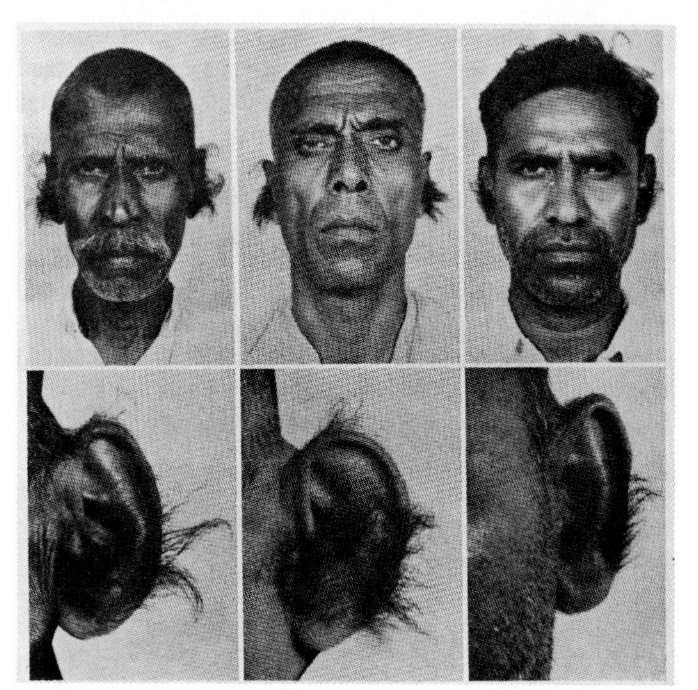

Another example of a sex-linked trait in humans is a recessive allele that results in hemophilia, "bleeder's disease." This disease played a role in history because it has been found in royal families in Europe that are related to Queen Victoria of Great Britain. Hemophilia resulted in or contributed to the early death of male members of several royal families (figure 17.12).

Genes borne on the Y chromosome are very rare. In humans, for example, only a very few genes are known to occur on the Y chromosome. One of them is a gene with an allele for hairy ear rims (figure 17.13). Genetically, a gene on the Y chromosome is often easy to detect because if the father has a particular allele, all of the sons and none of the daughters have it.

Sex Influence

Some genes not located on the X or Y chromosomes are expressed differently in the two sexes, due probably to the hormonal differences between the sexes. This is called **sex influence.** An example of a sex-influenced characteristic in humans is pattern baldness, which is far more common among men than women. The allele for pattern baldness is dominant

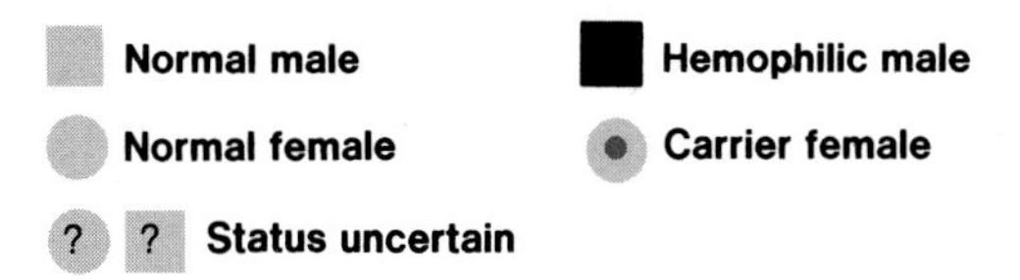

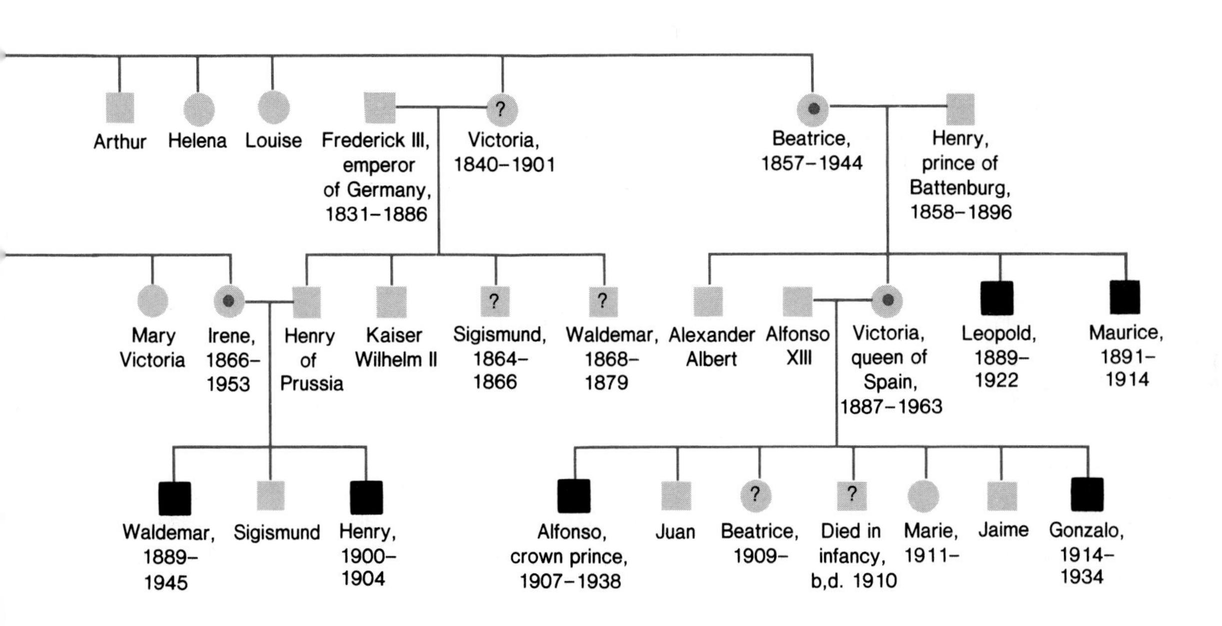

Figure 17.14 The inheritance of pattern baldness shows that it is a sex-influenced characteristic.

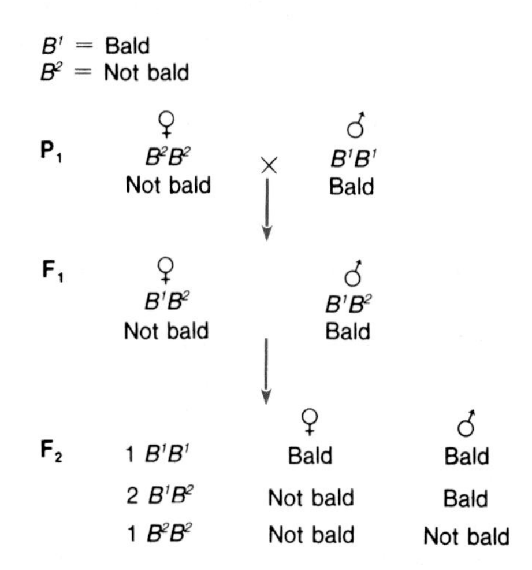

Figure 17.15 Another sex-influenced trait is the relative lengths of the index finger and fourth finger of the hand. Both individuals in the photograph are heterozygous for the trait. The hand on the right is that of a female, and the hand on the left is that of a male.

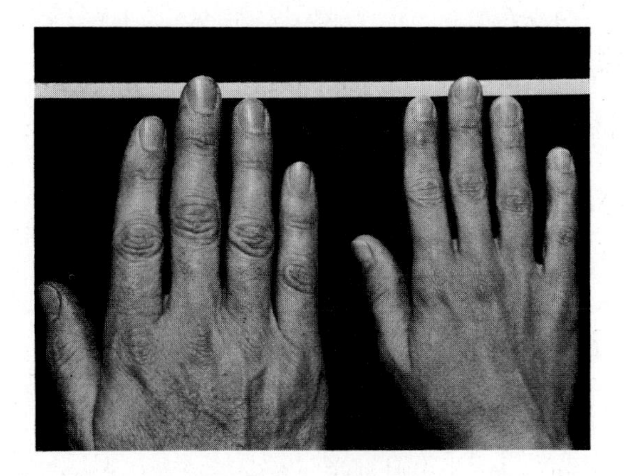

in males but recessive in females (figure 17.14). Another example of sex-influenced inheritance in humans involves the comparative lengths of the index finger and the fourth finger of the hand. The index finger being equal to or longer than the fourth finger is dominant in females and recessive in males (figure 17.15).

Chromosomal Abnormalities

Changes in chromosomal structure can have important genetic consequences, and such changes often occur as a result of chromosome breakage. Chromosomes may be broken by radiation, various chemicals, or even viruses. Sometimes fragments of broken chromosomes are simply lost. Loss of a segment of a chromosome is called a **deletion.** Often, however, chromosomal fragments rejoin in a different arrangement.

Changes in the genetic material of a cell that can be passed on to daughter cells are called **mutations.** Deletions and chromosomal reorganizations are called **chromosomal mutations** or **macromutations** to distinguish them from the gene mutations, changes in individual genes, which we will discuss in chapter 18.

Figure 17.16 Chromosomal mutations. Arrows show points of chromosome breakage, and displaced chromosome fragments are colored. In deletions, the broken chromosome fragment does not reattach and is lost. In duplication, a broken segment from one chromosome attaches to its homologous chromosome. An inversion involves breakage and reattachment to the same chromosome in a reversed position. A translocation is a transfer of a chromosome fragment to a nonhomologous chromosome. Radiation damage to cells increases the frequency of these problems because radiation causes greatly increased chromosome breakage.

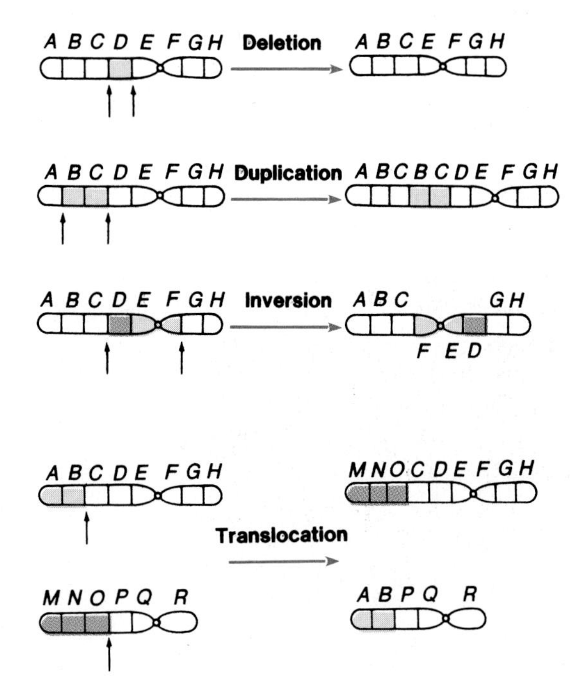

Figure 17.16 illustrates deletions and various other types of chromosomal mutations. A **duplication** results in the presence of the same chromosome segment more than once in the same chromosome. An **inversion** is when a segment of a chromosome is turned around 180°. **Translocations** result from the actual interchange of blocks of genes between two nonhomologous chromosomes.

Chromosomal mutations cause problems in meiosis. When homologous chromosomes come together in synapsis

Box 17.2

Active Y Chromosomes and Inactive X Chromosomes

Although the Y chromosome in mammals carries few genes, it has a powerful effect on sex determination. One intriguing gene thought to be borne on Y chromosomes is identified as $H\text{-}Y^{+}$. Even if the $H\text{-}Y^{+}$ gene is not actually borne on the Y chromosome, possibly it is a gene on another chromosome whose expression is regulated by a gene located on the Y chromosome. In any event, this gene directs synthesis of a specific kind of protein molecule, **H-Y antigen,** which is present in the membranes of virtually all cells in male mammals but not in those of females. The H-Y antigen may play a role in determining that the gonads of a developing embryo become testes rather than ovaries. Once that determination is made, hormone differences cause other body parts to develop in a male direction. In addition to a possible role in that first commitment to maleness in the developing embryo, the presence of the H-Y antigen may make male cells different from female cells for a lifetime, in ways that are as yet undetermined.

Another question about sex chromosomes that had interested biologists for some time is how males manage with only one "dose" of the genes present on the X chromosomes (sex-linked genes), while females apparently have a double "dose" of such X-linked genes. That question turned out to have a rather unexpected answer. Apparently, cells in females also have only one functional X chromosome. How does this come about?

Years ago, M. L. Barr observed a consistent difference between nondividing cells from female and male mammals, including humans. Females have a small, darkly staining mass of condensed chromatin present in their nuclei (box figure 17.2). Males have no comparable spot of chromatin in their nuclei. This darkly staining spot in female nuclei is now called the **Barr body** or **sex chromatin body.**

during prophase I, they cannot line up properly with one another if one of them is abnormal as a result of a chromosomal mutation.

Chromosomal mutations also cause difficulties in normal genetic functioning of chromosomes, resulting in serious problems for the individual carrying them. For example, deletions usually are lethal if both homologous chromosomes are deficient. While a deletion in one of the chromosomes of a homologous pair may not be lethal, it can have drastic effects on the development and function of the individual. An example in humans is the **cri-du-chat syndrome** (cat's cry syndrome). This syndrome results from a deletion in the short arm of chromosome no. 5. Infants who are heterozygous for this deletion have a high-pitched mewing cry and usually have severe growth abnormalities and mental retardation.

Finally, chromosomal mutations are very frequently observed in cancer cells. It appears that chromosome breakage and reconnecting of chromosome segments in new (abnormal) arrangements may be closely connected to the conversion of normal body cells into abnormally growing tumor cells.

In 1961, Mary Lyon, a British geneticist, proposed that the Barr body represents a condensed, inactive X chromosome whose genes are not expressed in cells of females, and that only genes on the other X chromosome are expressed. She further proposed that X chromosome inactivation occurs sometime during development and that once inactivated in a cell, a given X chromosome remains inactive in that cell *and* in all of its descendants.

This **Lyon hypothesis,** as it is called, has been tested and found to be valid. Early in development, X chromosomes are inactivated in cells of female embryos, but the inactivation is not a consistent inactivation of a single one of the two X chromosomes; that is, one X chromosome is inactivated in some cells, while the other is inactivated in others. As Mary Lyon proposed, once inactivation of a particular X chromosome occurs in an embryonic cell, that X chromosome forms a Barr body in each and every cell descended from the original cell. Because many sex-linked genes are heterozygous in any given individual, this X chromosome inactivation has interesting genetic consequences for body cells. Some cells will have one allele of a gene being expressed while other cells will have another allele being expressed. The female body, therefore, is a mosaic with "patches" of genetically different cells. It has been demonstrated, for example, that blood cells from heterozygous females belong to two separate subpopulations that produce either one or the other of two different forms of certain enzymes, depending on which allele of sex-linked genes is being expressed. Cells from skin and other tissues have also been shown to be divided into two subpopulations with regard to expression of sex-linked genes. Thus, it now seems quite certain that female mammals, including human females, are genetic mosaics with reference to expression of sex-linked genes.

Having a mixture of cells with different X chromosomes inactivated allows heterozygous females to escape the effects of certain harmful, or even lethal, genes. Although harmful alleles may be the only ones expressed in some body cells, normal alleles are expressed in other body cells and these "cancel out" the effects of the harmful alleles.

The connection between inactivated X chromosomes and Barr bodies also has implications for genetic testing. Human cells have one less Barr body than the number of X chromosomes they contain. Thus, normal males have no Barr bodies. Normal females have one. Metafemales may have two or more Barr bodies because they have three or more X chromosomes. This correlation between the numbers of X chromosomes and Barr bodies is useful in genetic analysis because it provides a simple and direct means of detecting possible sex chromosome problems. Any suspected problems can then be pursued by preparation of a full karyotype.

Sex chromatin (Barr body) tests are sometimes used in other contexts as well. For example, they are sometimes applied to entrants in Olympic and other sporting events that are open only to one sex.

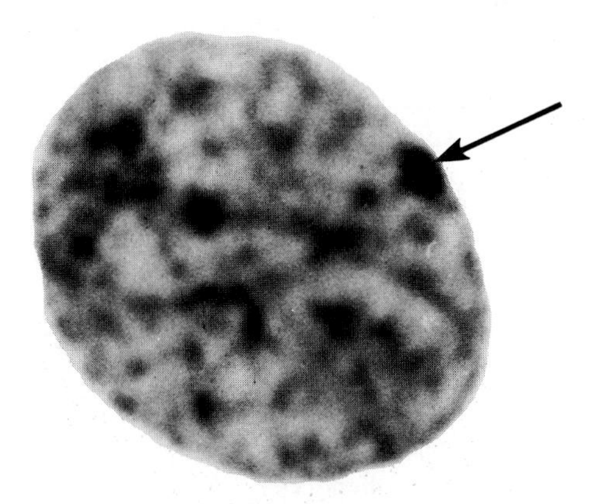

Box figure 17.2 The Barr body (sex chromatin). The inactivated X chromosome appears as a dense, darkly staining mass known as a Barr body in the nucleus of a cell from a female mammal. This nucleus is in a cell taken from the lining of a normal woman's mouth.

Chromosome Number Problems in Humans

We have already discussed Down's syndrome, which is an example of aneuploidy in humans. There are several other conditions that result from aneuploidy in the autosomes, but we will focus our attention on abnormalities that result from aneuploidy in the sex chromosomes.

An individual with **Turner's syndrome** has only one X and no Y or second X chromosome. People with this condition are sterile females with no ovaries and little development of the secondary sex characteristics. Usually, mental deficiency is not associated with this syndrome.

Metafemales have three or more X chromosomes. These females have only limited fertility and usually are mentally retarded.

An individual with **Klinefelter's syndrome** has a Y chromosome with two or more X chromosomes. Individuals with this chromosomal condition are sterile males with some tendency toward femaleness.

Studies of these human chromosomal abnormalities teach us some important things about the mechanism of sex determination in humans. Sex determination in *Drosophila* had been studied thoroughly during the early years of genetics before human chromosome studies began. Results of those studies indicated that sex is essentially determined in *Drosophila* by the number of X chromosomes present, whether or not a Y chromosome is present. Presence of only one X chromosome produces a male. Thus, either an XY or an XO (a lone X chromosome with no other sex chromosome) genotype produces a male. A *Drosophila* individual with an XXY genotype becomes an essentially normal female. These results indicate that the Y chromosome is a basically neutral factor in *Drosophila* sex determination.

When chromosomal studies began on human sex determination, it was assumed that the chromosomal mechanism of human sex determination might be similar to the *Drosophila* model because both have the X-Y sex chromosome system. *But this clearly is not the case.* Turner's syndrome subjects have an XO genotype and an at least partly female phenotype. Klinefelter's syndrome subjects have an XXY genotype and a basically male phenotype. Clearly the Y chromosome is not a neutral factor in human sex determination, as it seems to be in *Drosophila.* Rather, it is a male-producing factor. Among the few genes that the Y chromosome bears are genes whose expression contributes to development of the male phenotype.

Figure 17.17 Example of the kind of deviation from Mendelian predictions that might be seen when genes *A* and *B* are close together on the same chromosome. Genotypes are recognizable in this experiment since any offspring showing dominant phenotype must be heterozygous.

AaBb × *aabb*

↓

Genotypes	Expected	Observed
AaBb	25%	42%
Aabb	25%	8%
aaBb	25%	8%
aabb	25%	42%

Figure 17.18 The chromosomes involved in the cross in figure 17.17, showing the chromosomal rearrangement that results from crossing over. Linkage between genes *A* and *B* explains the observed ratio in the offspring.

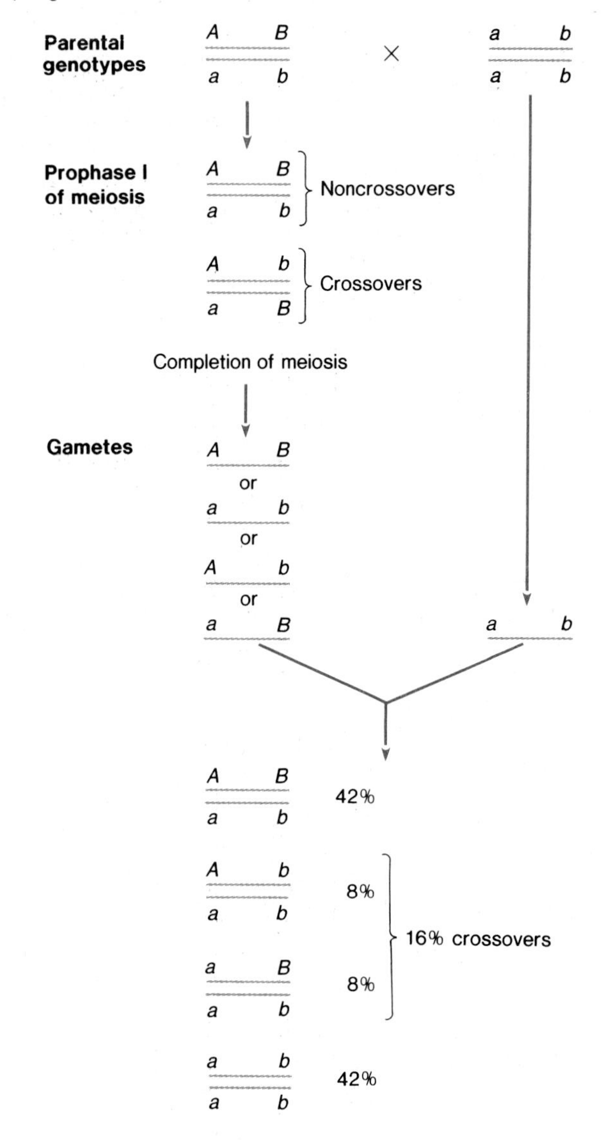

Box 17.3
How Did Mendel Do It?

Gregor Mendel reported experimental crosses involving seven characteristics of pea plants, and the genes determining each of these characteristics assorted independently of the genes for each of the other characteristics that he tested. Independent assortment occurs optimally only if the genes in pairs being studied are on separate chromosomes, or are far apart on the same chromosome. This makes Mendel's reported results rather amazing and raises an interesting question about Mendel and his work. Was Mendel almost incredibly lucky in choosing seven characteristics to be studied genetically, or did he study other characteristics, which did not assort independently, and then choose to ignore them when he reported his results? It is very unlikely that we will ever know for certain.

Linkage and Chromosome Mapping

During meiosis, the chromosomes assort at random into the gametes. The gametes then combine to form the zygotes of the next generation. Mendel's law of independent assortment states that each gene segregates independently of other genes. Mendel assumed that genes behaved as independent particles in cells. Genes, however, are not independent particles, but are attached parts of linear arrays on chromosomes. It is actually the chromosomes that segregate independently during meiosis. Thus, if two gene loci are close together on the same chromosome, they tend to assort together during meiosis, and they do not obey Mendel's law of independent assortment. This tendency of certain genes to assort together is called **linkage.**

Linkage

An example of linkage is given in figure 17.17. If a test cross is made between an individual heterozygous at each of two genes and another individual that is homozygous recessive for both of those genes, each of the four possible genotypes is expected in the offspring in equal ratios. But in the example shown in figure 17.17, there is a strong deviation from this expected ratio.

We can explain this deviation from Mendelian expectations by assuming that the two gene loci involved are on the same chromosome and thus cannot assort independently of one another; that is, the genes are linked. But how can the particular ratio of genotypes found among the offspring be accounted for? The answer is that even when gene loci are linked because they are located on the same chromosome, linkage is not demonstrated 100 percent of the time because of the chromosome fragment exchange that occurs as a result of crossing over (see page 330) during meiosis.

Thus, both linkage and crossing over are involved. The diagrammatic representation of the chromosomes and alleles in figure 17.18 shows what happens. The homozygous recessive parent has the *a* and *b* allele on each chromosome of the pair. All gametes produced by that parent would contain a chromosome with *a* and *b* whether or not crossing over occurred. Given the ratios of genotypes in the offspring (figure 17.17), most of the gametes from the other parent had either *A* and *B* on the same chromosome, or *a* and *b* on the same chromosome. These would represent gametes resulting from a meiotic sequence during which no crossing over occurred between the two loci.

Cross-over events produce gametes in which *A* and *b* are on the same chromosome and *a* and *B* are on the homologous chromosome. The results show that a combined total of 16 percent of the offspring resulted from gametes formed after crossing over between the two genes in the heterozygous parent. The other 84 percent were formed by gametes resulting from meiosis during which no crossing over occurred between the two genes. When two or more gene loci are close enough on the same chromosome so that there is little crossing over between them, as the loci in this example are, Mendel's law of independent assortment does not hold.

Figure 17.19 The relative positions of genes *A, B,* and *C* as determined by the analysis of the crossover percentages. The numbers given are the map units.

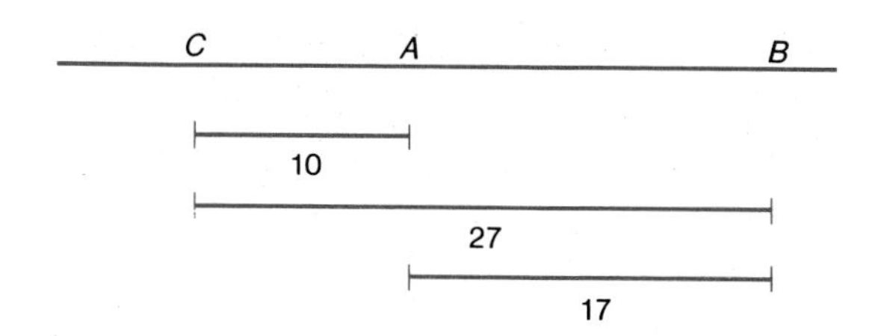

Figure 17.20 Another gene, *D,* is added to the chromosome map begun in figure 17.19.

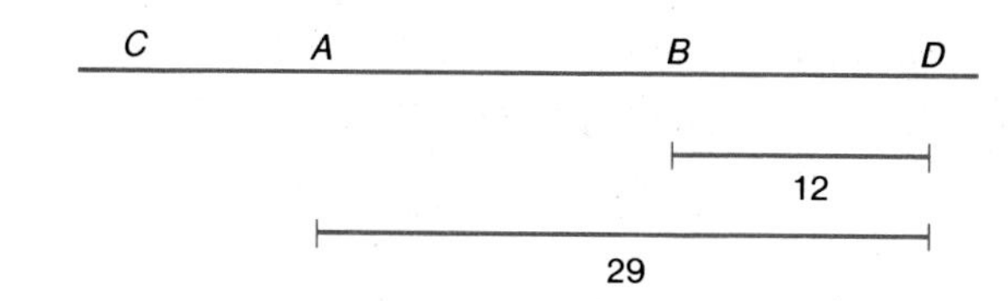

Chromosome Mapping

The study of linkage and crossing over has been used to prepare **chromosome maps,** which are schematic representations of the relative positions of various gene loci on chromosomes.

In synapsis during meiosis, chromatids are breaking and rejoining all up and down their length. Sometimes broken fragments rejoin the same chromatid; sometimes they cross over to join a chromatid of the homologous chromosome in exchange for a broken piece of that other chromatid. The further apart two loci are, the greater the chance that one of these breaks will occur between them. This also means that the amount of crossing over between loci is proportional to the distance between them; that is, crossing over occurs more frequently between loci that are farther apart along the chromosome than between those that are closer together.

Working under these assumptions, it is possible to determine the arrangement of the different genes on a chromosome by examining their crossing over frequencies. For example, if there is a 10 percent crossing over between genes *C* and *A,* 27 percent between *C* and *B,* and 17 percent between *B* and *A,* there is only one possible arrangement of genes along the chromosome (figure 17.19). If another gene, *D,* is found to have 12 percent crossing over with *B* and 29 percent with *A,* its location can be added to the map (figure 17.20). By this method, a map of the entire chromosome can be made.

The distances between the genes along the chromosome are designated in terms of **map units.** One map unit is equal to 1 percent crossing over. Therefore, in the example, *C* is ten map units from *A, A* is seventeen map units from *B, C* is thirty-nine map units from *D,* and so forth.

Mapping techniques permit biologists to construct chromosome maps that show the relative positions of specific genes on chromosomes. For example, the chromosome map in figure 17.21 shows the relative locations and distances of many of the known genes on the chromosomes of the fruit fly, *Drosophila melanogaster.*

Extranuclear Inheritance

In this chapter we have concentrated on the relationships between chromosomes and the genes that they bear. But it is important to point out that this nuclear genetic material, organized in linear arrays of gene loci on chromosomes, does not represent the only place where genetic information is carried in eukaryotic cells.

Genetic information located outside the nucleus can affect the phenotype of a cell or an organism. For example, mitochondria contain DNA and functioning genes, as do chloroplasts in plant cells. These genes direct synthesis of certain polypeptides found in those organelles that are important in their structure and function. Mapping studies indicate that these extranuclear genes usually are part of a single circular molecule of DNA in each organelle.

The inheritance patterns of these genes differ from normal Mendelian inheritance. During the formation of gametes, so little cytoplasm is included in the male gametes, such as sperm cells, that most extranuclear genes, the genes contained within the mitochondria and chloroplasts, are transmitted through the female parent. Thus, in terms of mitochondrial genes, you probably are much more your mother's child than your father's.

Figure 17.21 A chromosome map showing three of the four linkage groups of *Drosophila melanogaster.* The map of the tiny fourth chromosome is not included. The linkage groups are connected with the appropriate chromosomes (inset) by dashed lines. By convention, one end of each chromosome is designated as the zero end. Crossover map units are shown to the left of each map, and names and genetic symbols of various mutant alleles are shown to the right. When an entire chromosome is mapped, map units are determined from the gene closest to one end of the chromosome to the gene adjacent to it. Then the map units between the second gene and a third gene are determined, and so forth. In constructing a chromosome map, larger values are given as sums of shorter intervals, but there actually cannot be more than 50 percent crossing over between any given pair of genes on a single chromosome.

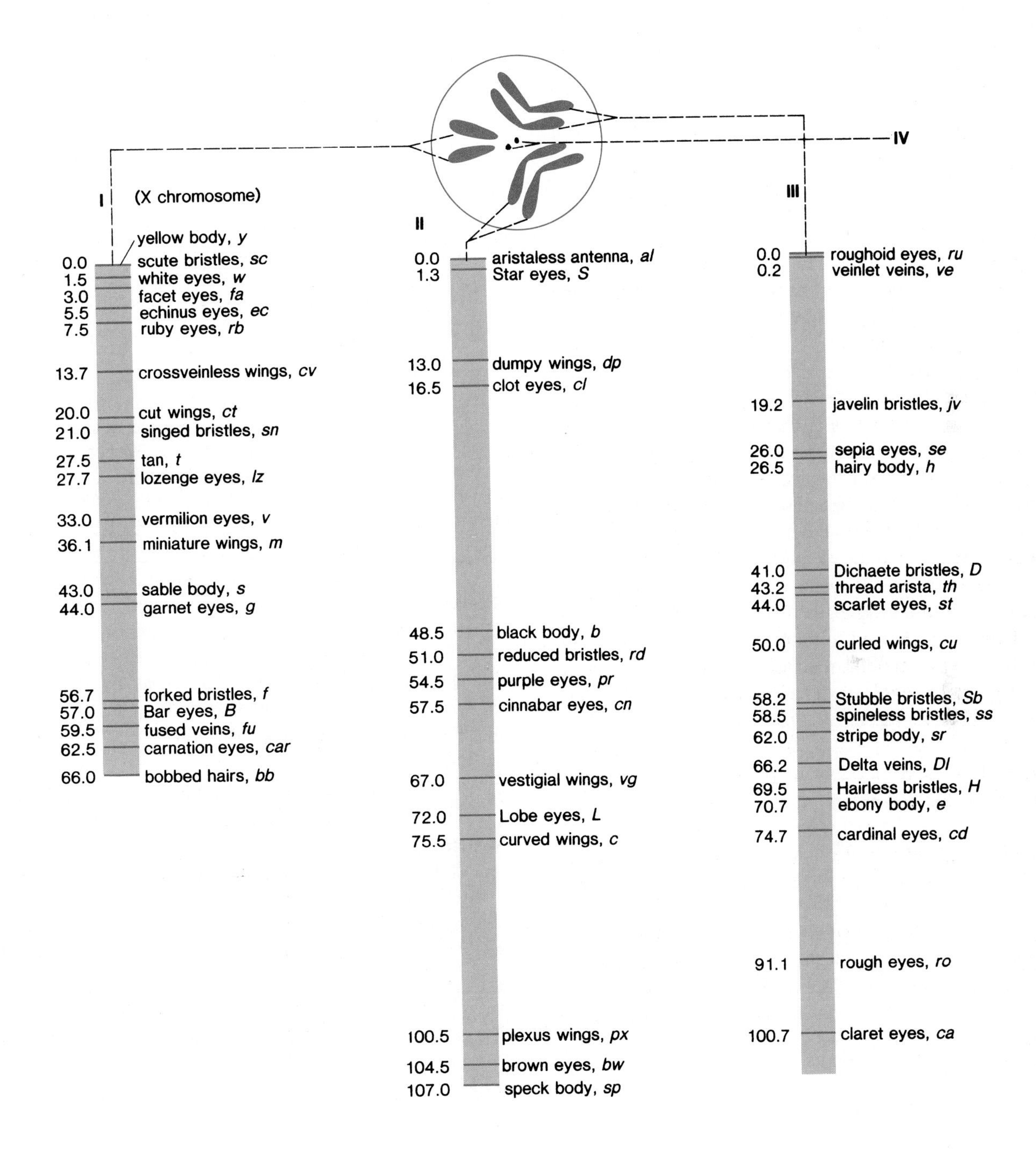

Summary

Genes are located in linear arrays on chromosomes.

Species differ widely in the number of chromosomes that they contain. The number of chromosomes in an organism may be increased or decreased. In polyploidy, the increase or decrease is in terms of entire sets of chromosomes. In aneuploidy, the increase or decrease is in terms of individual chromosomes.

Many organisms have sex chromosomes. In humans, an individual with two X chromosomes is a female, and an individual with an X chromosome and a Y chromosome is a male. Genes on these chromosomes are called sex-linked genes, and their patterns of inheritance are different than the patterns of inheritance of genes on autosomes. Sex-influenced inheritance is the differential expression of autosomal genes due to hormonal differences between the sexes, with one allele dominant in one sex and the other allele dominant in the other.

Changes in chromosomal structure include deletions, which occur when a segment of a chromosome is lost; duplications, which occur when a chromosome is repeated; inversions, which involve a reversal of part of a chromosome; and translocations, which involve an exchange of material between nonhomologous chromosomes.

Genes that are close together on the same chromosome deviate from Mendel's law of independent assortment. The amount of crossing over between any two genes on the same chromosome is proportional to the distance between them. This principle can be used to map genes on chromosomes.

Genes also are present in circular pieces of DNA in mitochondria and chloroplasts. These genes are usually inherited from the female parent.

Questions for Review

1. What are nucleosomes?
2. Account for all the possible genotypes and phenotypes in offspring of a *Drosophila* cross between a heterozygous red-eyed female and a white-eyed male.
3. Two normal-visioned parents produce a red-green color-blind son. What are the genotypes of the parents?
4. What is the difference between sex-influenced inheritance and sex linkage?
5. What is a karyotype?
6. Why do chromosomal mutations cause difficulties during prophase I of meiosis?
7. How many Barr bodies would you expect to observe in the cells of individuals with each of these genotypes: XO, XYY, XXX, XXXY?
8. Explain why a *Drosophila* that has an XXY genotype is a normal female while a human with an XXY genotype has sexual development problems (Klinefelter's syndrome).
9. If there is 3 percent crossing over between genes 1 and 2, 12 percent crossing over between genes 2 and 3, and 15 percent crossing over between genes 1 and 3, what are the relative positions of these three genes along the chromosome?
10. Why are the extranuclear genes inherited almost exclusively from the female parent?

Questions for Analysis and Discussion

1. Why are fewer instances of polyploidy found in organisms with sex chromosomes?
2. How do you suppose that the visualization of bands in chromosomes assists in the genetic mapping of the chromosomes?
3. In cats, the genotype *BB* is yellow, *Bb* is calico, and *bb* is black. The gene is on the X chromosome. If a calico female is crossed with a black male, what types and frequencies of offspring would be expected? Could there ever be a calico male? Explain.

Suggested Readings

Books

Elseth, G. D., and Baumgardner, K. D. 1984. *Genetics.* Reading, Mass.: Addison-Wesley.

Mange, A. P., and Mange, E. J. 1980. *Genetics: Human aspects.* Philadelphia: Saunders College.

Mittwoch, U. 1967. *Sex chromosomes.* New York: Academic Press.

Sutton, H. E. 1980. *An introduction to human genetics.* 3d ed. New York: Holt, Rinehart & Winston.

Articles

Fuchs, F. June 1980. Genetic amniocentesis. *Scientific American* (offprint 1471).

Jinks, J. L. 1978. Cytoplasmic inheritance. *Carolina Biology Readers* no. 72. Burlington, N.C.: Carolina Biological Supply Co.

Kornberg, R. D., and Klug, A. February 1981. The nucleosome. *Scientific American* (offprint 1490).

McKusick, V. A. April 1971. The mapping of human chromosomes. *Scientific American* (offprint 1220).

Mittwoch, U. July 1963. Sex differences in cells. *Scientific American* (offprint 161).

Upton, A. C. February 1982. The biological effects of low-level ionizing radiation. *Scientific American* (offprint 1509).

Molecular Aspects of Genetics

18

Chapter Outline

Chapter Concepts

1. Genetic information is coded in the DNA molecule.
2. In DNA replication, the two strands of the molecule are split apart, and these strands are used as patterns to form new complementary strands.
3. Genetic information coded by the nucleotide sequence in DNA is transcribed into a complementary segment of RNA and then translated in the cytoplasm through the use of ribosomes into specific amino acid sequences in polypeptides.
4. Structural genes code information for synthesis of structural proteins or enzymes. Other genes, called regulatory genes, regulate the activity of structural genes.
5. Recombinant DNA techniques permit insertion of genes from one organism into the DNA of another organism.
6. Point mutations are changes in the nucleotide sequence of DNA molecules.

In July of 1980, seventeen healthy volunteers in London and Liverpool were given doses of human insulin that had been produced by bacteria. The trial was a success because the injected insulin caused lowered blood sugar levels in the volunteers. More clinical trials followed in Britain and in the United States. Late in 1982, the United States Food and Drug Administration (FDA) approved this new form of insulin for medical use.

This insulin, produced by bacterial cells, holds great promise for treatment of human diabetes. It will be cheaper to produce than the insulin extracted from animal tissues that are obtained from slaughterhouses. Extraction costs for animal insulin are high because about 3000 kg of pig or cow pancreas must be processed to yield one kilogram of insulin. Furthermore, the human insulin produced by bacteria should be less likely to provoke allergic responses in patients than animal insulin preparations.

You may be wondering how bacterial cells can be made to produce a human protein such as the hormone insulin. This is possible because of the development of genetic engineering techniques (see page 387). Human genes actually can be introduced into bacterial cells. These cells are then induced to produce the desired substance, in this case, insulin (figure 18.1).

We have barely scratched the surface of the potential offered by genetic engineering. Soon bacteria may be producing commercial quantities of human growth hormone and a variety of substances that are badly needed for medicine and other purposes. Someday, scientists very likely will use variations of these techniques to alter certain genetic characteristics of agriculturally important plants and animals. We will come back to genetic engineering later in this chapter.

In the 1950s, several fundamental discoveries were made concerning the chemical nature of genes and the way that genes function in cells. Let us first consider some of those critical discoveries. Then we will examine some of the basic principles of the science of molecular genetics.

DNA as the Genetic Material

The first task of molecular geneticists was to determine which molecular substance actually makes up the genetic material. After Sutton and Boveri suggested the chromosomal theory of inheritance in 1902 (see page 354) and T. H. Morgan and his colleagues proved that genes are linear arrays in chromosomes, biologists knew that the genetic material is contained within the chromosomes. Chemical analysis of chromosomes showed that they are made up of DNA and protein. It then became a question of which of these substances was the genetic material.

Figure 18.1 Crystals of purified human insulin made by bacteria into which the human insulin gene was introduced by genetic engineering (recombinant DNA) techniques.

At the time that this question was being argued most vigorously, the chemical structure of DNA was poorly understood. Many biologists thought that DNA's chemical structure was too simple to encode genetic information. Proteins, on the other hand, were known to be quite complex. Biologists suspected that the sequence of amino acids in certain proteins in chromosomes might form a genetic code and that DNA simply might be a structural material that provided a supporting framework for the protein molecules that carried the genetic information.

However, several important studies in molecular genetics indicated that DNA, and not protein, is the carrier of genetic information.

Transformation

The first of these studies was published by Fred Griffith in 1928. Griffith's experiments involved the process of **transformation.** Transformation is a change in a bacterial strain that results when the bacteria receive genetic information from another strain of bacteria.

Griffith's experiments were done with strains of the bacterium *Streptococcus pneumoniae.* Some strains of this organism cause a fatal pneumonia in mammals. These pathogenic (disease-causing) bacteria produce a polysaccharide capsule that protects the bacteria from the defense mechanisms of the infected animal. Another strain of *S. pneumoniae* cannot synthesize capsules. The two strains also differ in that they produce different types of colonies when grown in culture. Colonies made up of cells with capsules have a smooth appearance. These cells are called strain-S cells. Cells without a capsule produce colonies with a rough

Figure 18.2 Griffith's transformation experiments. (*a*) Mice injected with strain-S bacteria died of pneumonia. (*b*) Mice injected with strain-R bacteria survived. (*c*) Mice injected with heat-killed strain-S bacteria survived. (*d*) Mice injected with living strain-R bacteria and heat-killed strain-S bacteria contracted pneumonia and died. Furthermore, it was possible to extract strain-S bacteria from the blood of these sick mice.

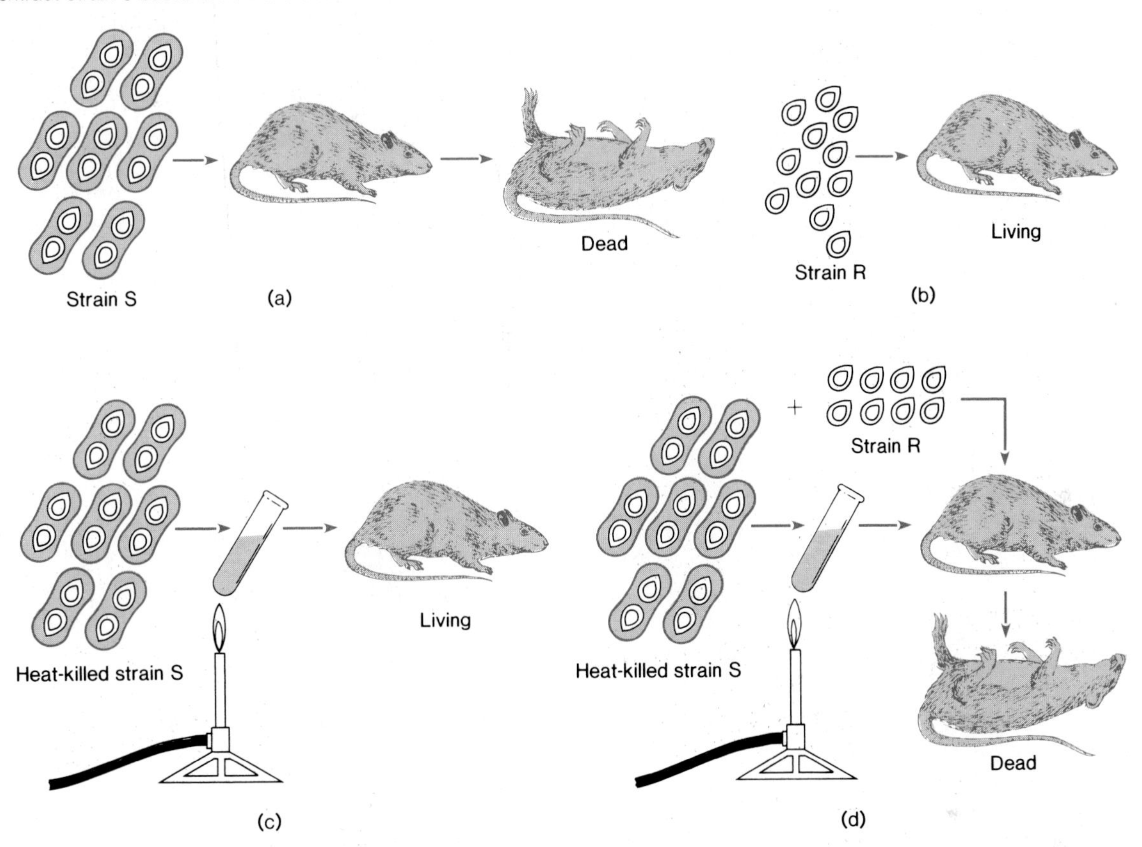

appearance and are called strain-R cells. The strain-R cells do not cause pneumonia in animals that they infect; the strain-S cells do.

Griffith performed several experiments designed to reveal more about the differences between these two strains and the role of the capsule in infection. When he injected living strain-S cells into a mouse, the mouse died. Thus, strain-S cells were pathogenic (disease-causing). However, living strain-R cells were relatively harmless. Griffith then injected heat-killed strain-S cells into a mouse. This time the mouse did not die. In short, the disease was caused only when living strain-S cells were present in the mouse. In a later experiment, Griffith injected live strain-R bacteria (normally relatively harmless) together with heat-killed strain-S cells (figure 18.2). Surprisingly, the mice injected with this combination died of bacterial pneumonia! Somehow, the relatively harmless strain-R cells had been transformed so that they had acquired the disease-causing ability of strain-S cells.

Other studies showed that transformation does not depend on the living host animal. When heat-killed strain-S cells are added to a culture of living strain-R cells, some of the strain-R cells are transformed into capsule-producing cells that form smooth colonies. Even a cell-free extract of strain-S would transform strain-R cells into strain-S cells. Some chemical factor from the strain-S cells transforms living strain-R cells into strain-S cells. Furthermore, these transformed bacteria retain their capsule, colony, and pathogenic characteristics for generation after generation. Thus, transformation results in a stable genetic change.

After years of chemical investigation, O. T. Avery and his colleagues Colin MacLeod and Maclyn McCarty demonstrated conclusively that it is the DNA extracted from the heat-killed strain-S bacteria that is the transforming factor.

Figure 18.3 The Hershey-Chase experiments. Results of two experiments are combined in summary form in this diagram. Hershey and Chase labeled the protein coats of one batch of bacteriophage with ^{35}S. (Sulfur is found in protein but not in nucleic acids.) They allowed the phages to attach to bacteria for a time, and then they were agitated in a blender. Radioactivity measurements indicated that little radioactive material entered bacterial cells; the ^{35}S labeled protein remained outside. But the infection proceeded, and new virus particles were formed and released. Hershey and Chase labeled the DNA in another batch of phages with ^{32}P. (Phosphorus is an important constituent of nucleic acids but not of proteins.) Again, after a period of attachment, phages were separated from bacterial cells in a blender. This time the bulk of the radioactivity was found inside the host cells, indicating that it is DNA that enters host cells and infects them. These important experiments helped to prove that the primary genetic material is DNA, not protein.

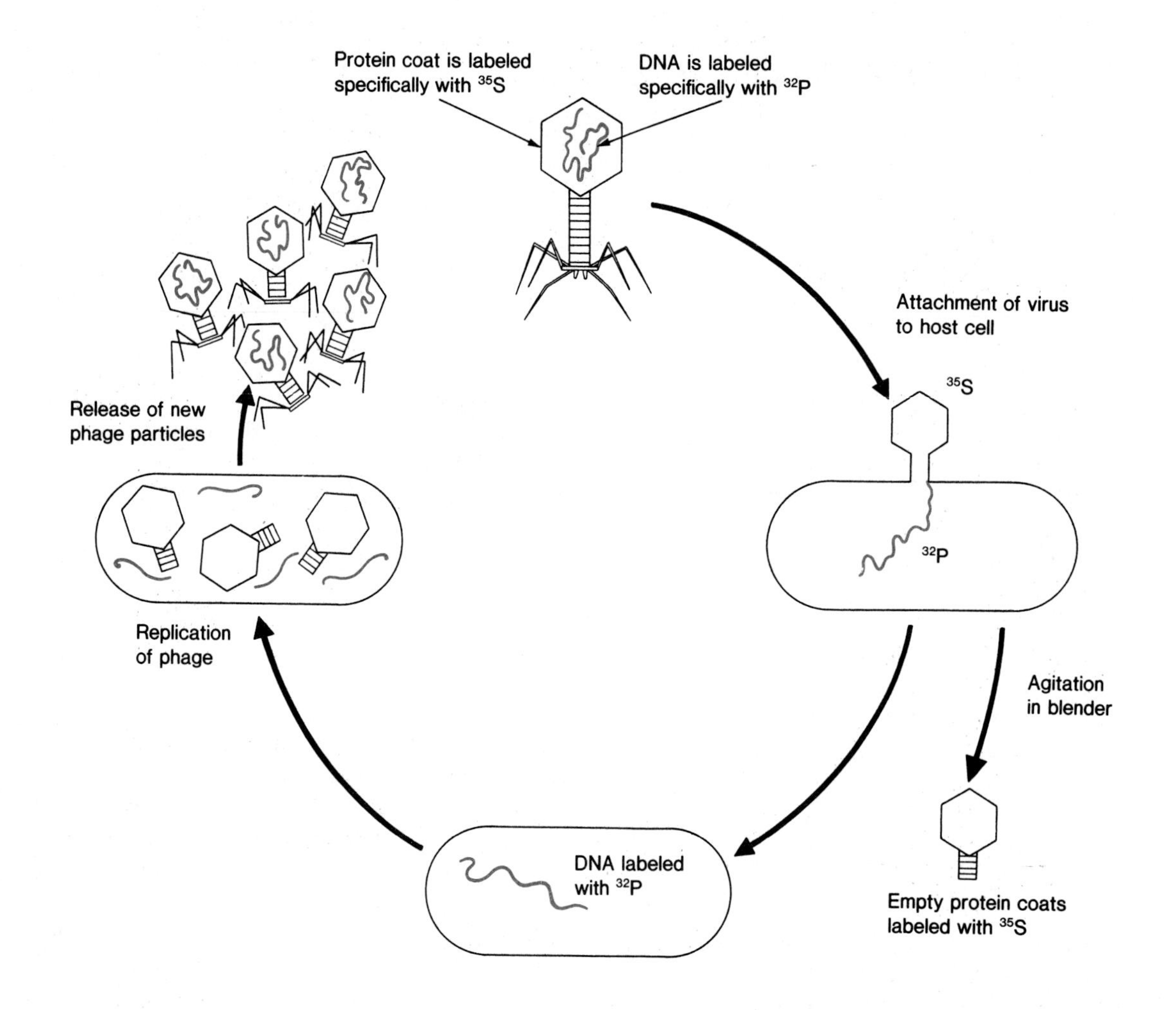

DNA and Virus Infection

More evidence that DNA is the genetic material came from a series of experiments with bacteriophages published by A. D. Hershey and Martha Chase in 1952. A **bacteriophage** (also called **phage** for short) is a virus that infects a bacterial cell and causes the cell's genetic machinery to produce more viruses. The T2 bacteriophage that Hershey and Chase used consists of a protein coat and a core of DNA. Hershey and Chase labeled the protein coat of one batch of bacteriophage with radioactive sulfur (^{35}S) and labeled the DNA of another batch of bacteriophage with radioactive phosphorus (^{32}P). The incorporation of these radioactive elements into the two parts of the virus enabled Hershey and Chase to trace which material actually enters the bacterium. They found that it is DNA that enters and infects the cell. The protein coat remains on the outside of the cell. Once the infection has been started, the empty protein coat can be removed from the cell without affecting the progress of the infection (figure 18.3). It is clear, then, that DNA is the substance inserted by the T2 bacteriophage to take over the genetic mechanisms of the cell. Thus, it must carry the genetic information of the virus.

The Double Helix Model of DNA

The third major line of evidence that confirmed DNA as the genetic material came when the structure of the DNA molecule was finally revealed by Watson, Crick, Franklin, Chargaff, and Wilkins. (You might want to reread the story of their efforts and the description of the structure of the DNA molecule in chapter 2.) When biologists realized that DNA consists of two complementary strands of nucleotides, arranged in the now familiar double helix, it became obvious that each of these two strands could serve as a template (pattern) for the synthesis of a precise copy of the partner strand. This would provide a mechanism for the duplication of genetic information during the S phase of the cell cycle. Also, it was pointed out that DNA molecules have a great deal of complexity and variety. The base pairing from one strand to another is restricted; that is, adenine (A) always pairs with thymine (T), and guanine (G) always pairs with cytosine (C). But there is no structural restriction on the order of nucleotides in the molecule, and it became clear that their sequence could code genetic information. This work, published in a brief article in the journal *Nature* in 1953, marks the point at which it was finally confirmed to nearly everyone's satisfaction that the genetic code is carried by DNA molecules.

There always seem to be a few exceptions to every rule, however. In some genetic systems, RNA is the primary genetic material, not DNA. Certain viruses, for example, contain cores of RNA rather than DNA. But for the great majority of living things, DNA is the primary genetic material, and our examination of molecular genetics will focus on those organisms.

Figure 18.4 The two strands of a DNA molecule separate and each serves as a template for the formation of a new complementary strand. The strands of the original molecule are gray; the new strands are colored. Adenine (A) always pairs with thymine (T), and guanine (G) always pairs with cytosine (C). (Details of DNA structure are discussed in chapter 2.)

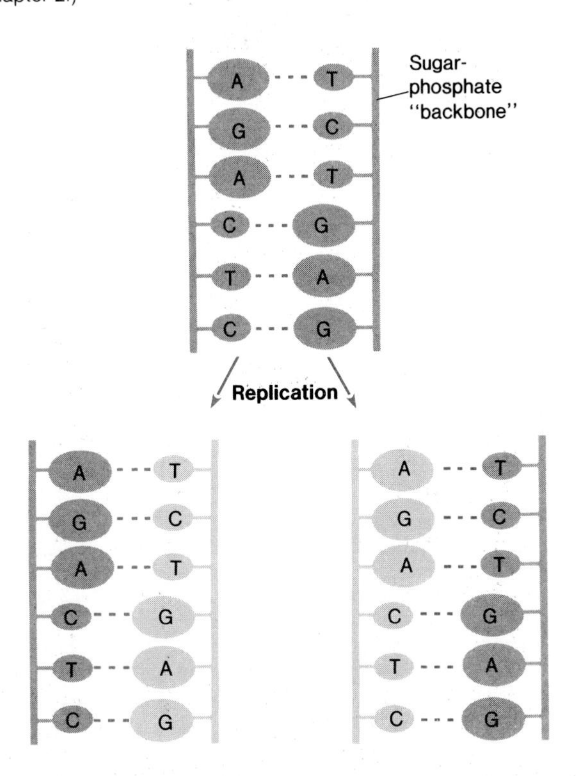

DNA Replication

As we said above, discovery of DNA structure led investigators to predict a mechanism of DNA replication (copying) based on complementary base pairing. The proposed mechanism was that a double-stranded molecule of DNA would uncoil, the two strands would separate, and each single strand would serve as a template (pattern) for the formation of a new complementary second strand (figure 18.4). For example, where there is an A in the existing strand, a T is incorporated into the forming strand; where there is a C in the existing strand, a G is incorporated into the forming strand, and so forth. This type of replication, in which half of each new DNA molecule is carried over (conserved) from the previous generation and the other half is newly synthesized, is called **semiconservative replication.**

Soon biologists began to make progress in understanding the process of DNA replication and in identifying enzymes that are involved in its various steps. One of the most important of these discoveries was made by Arthur Kornberg and his colleagues who demonstrated that an enzyme called **DNA polymerase** catalyzes the formation of new strands of DNA. They showed that in the presence of this enzyme a single strand of DNA can, by itself, act as a template for formation of a new strand of nucleotides. Kornberg and his coworkers prepared a mixture of the four nucleotides found in DNA and DNA polymerase. Then they added a single strand of DNA. The result was new, double-stranded DNA. The length of the new DNA was determined by the length of the single-stranded DNA originally present in the mixture. When the nucleotide content of the new strand was analyzed, the new strand was found to be complementary to the original strand.

Chromosome Replication

At first it was thought that all DNA molecules must be replicated beginning with separation of the two strands at one end of the molecule, leading to assembly of a new strand complementary to each of the original strands. Scientists assumed that this process must progress from one end of the molecule to the other. But then it was discovered that the

Figure 18.5 Replication of DNA in a eukaryotic chromosome. Nicks (N) in a single strand serve as swivel points as the double helix unwinds during replication. Replication begins simultaneously at a number of initiation sites (I) along the linear chromsome and proceeds in both directions from each initiation site. When the short, replicated segments have joined one another, the entire DNA molecule is replicated, and there are two molecules where there had been one.

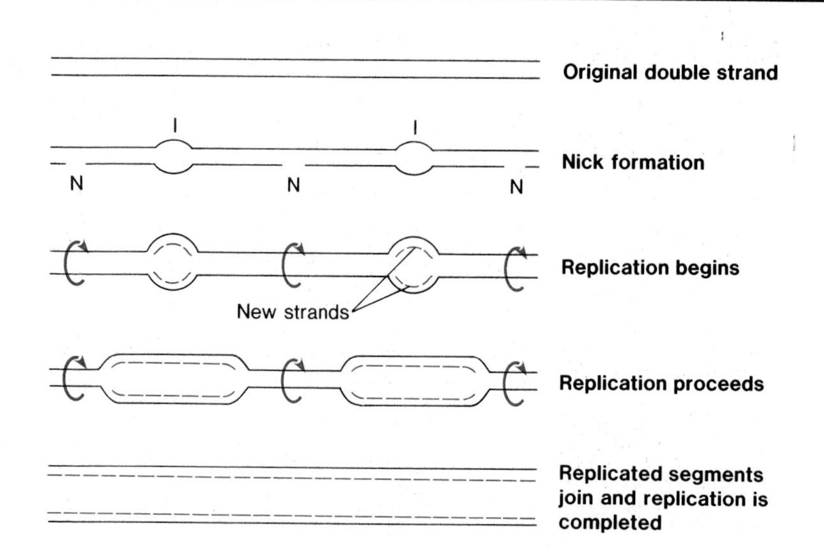

Figure 18.6 Alkaptonuria and related metabolic problems. The amino acids phenylalanine and tyrosine are obtained from dietary protein. Normally, they are used in a variety of reactions, each of which is catalyzed by one or more specific enzymes. But genetic problems affect some of these reactions. Colored bars on arrows with numbers indicate reactions that may fail to occur in various individuals because a genetic problem results in deficiency in enzyme function. The enzyme (homogentisic acid oxidase, site 2 in diagram) deficiency studied by Garrod is just one of several possible hereditary problems that can be associated with metabolism of these compounds. (1) Enzyme deficiency here leads to phenylketonuria (PKU). (2) Enzyme deficiency here leads to alkaptonuria. (3) Enzyme deficiency here leads to albinism because normal pigmentation cannot develop. (4) Enzyme deficiency here leads to thyroid deficiency.

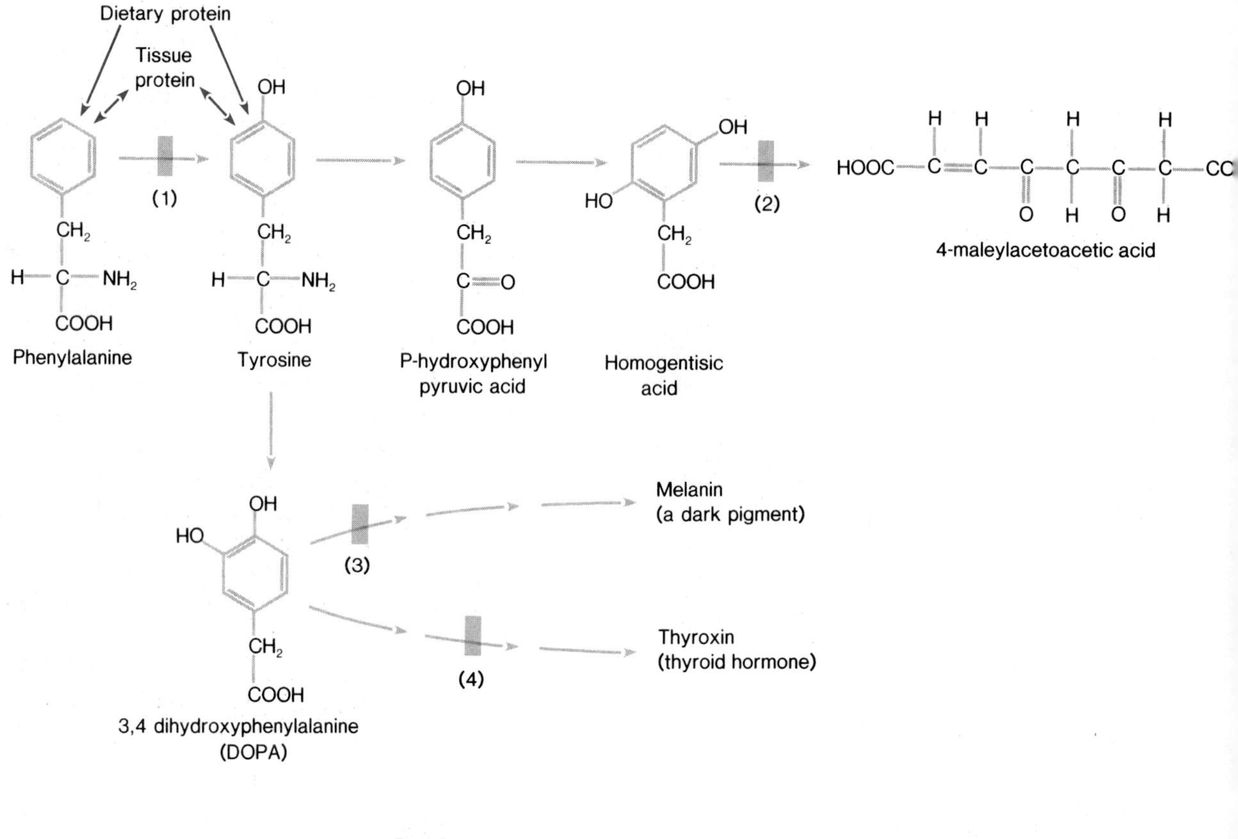

Dietary protein

Tissue protein

(1)

(2)

Phenylalanine

Tyrosine

P-hydroxyphenyl pyruvic acid

Homogentisic acid

4-maleylacetoacetic acid

(3)

(4)

Melanin (a dark pigment)

Thyroxin (thyroid hormone)

3,4 dihydroxyphenylalanine (DOPA)

replication of the very long DNA molecules in the chromosomes of eukaryotic cells happened much more quickly than it could if it was a single process moving from one end of the molecule to the other. How does this relatively rapid replication occur?

It turns out that DNA replication in eukaryotic cells is initiated at hundreds of sites, almost simultaneously, along each chromosome. A short segment from a eukaryotic cell's chromosome showing only two of many such sites is shown in figure 18.5. First, one strand of the double helix is broken ("nicked") by enzymes at various places along the chromosome. These breaks or nicks serve as swivel points as the double helix unwinds and its two strands separate. As new complementary strands form along each of the previously existing strands, two double helixes form where there had been one. Unwinding and synthesis occur simultaneously, followed by rewinding of the helix, up to the nicked region. When the nicks are rejoined, replication is complete.

Chromosome replication is an extremely complex process involving many steps and enzymes. In addition to DNA polymerase (the enzyme that catalyzes the actual joining of nucleotides into strands), there are enzymes that make nicks in the sugar-phosphate backbones, enzymes that "heal" the nicks, and proteins that assist in the unwinding and rewinding processes. The description we have given is greatly simplified, and if you want additional details of the process you should consult some of the readings we suggest at the end of this chapter.

Further reactions
that yield products
that can be oxidized
by reactions in
fatty acid metabolism
and the Krebs Cycle

Genes and Enzymes

Until now we have considered some aspects of replication of the genetic material, DNA. Now let us turn our attention to another important question in modern genetic research. How do genes determine the characteristics of living things? More precisely, how is the genetic information, coded as a sequence of bases in DNA molecules, used in the cells of living things?

Many types of genes function by providing coded information that directs synthesis of enzymes or other proteins. Enzymes are the catalysts involved in virtually all biochemical reactions in living cells. Thus, many genes function by determining the characteristics of these vital catalysts, thereby determining the nature of key functions in cells and the organisms that they make up.

The discovery of this relationship between genes and enzymes has an interesting history. The relationship was first suggested not by a biologist working on experimental organisms in a research laboratory, but by the English physician Archibald Garrod who studied human metabolic diseases in the early 1900s.

In 1902 Garrod reported his findings on the disease **alkaptonuria.** Individuals with alkaptonuria usually have arthritis and characteristically produce urine that turns black when it is exposed to air. Garrod observed that people with alkaptonuria excreted, in their urine, all the homogentisic acid produced in their bodies. Homogentisic acid is a substance that is produced in cells by reactions involving the amino acids tyrosine and phenylalanine. Normal healthy individuals can break down homogentisic acid; people with alkaptonuria cannot. Garrod hypothesized that people with alkaptonuria lack an enzyme necessary to catalyze the conversion of homogentisic acid to another compound. Figure 18.6 shows the reaction that is catalyzed by the enzyme missing in alkaptonuria. Because the disease is inherited in a simple Mendelian fashion and appeared to be controlled by a single recessive gene, Garrod proposed that there was a relationship between a gene and an enzyme. Garrod went on to propose similar mechanisms for three other human conditions. He referred to these biochemical hereditary diseases as "inborn errors of metabolism" and published a book with that title.

Figure 18.7 Beadle and Tatum's method for detecting nutritional mutants in *Neurospora crassa.* A single spore is placed in a complete medium that supplies a variety of nutrients, including vitamins, amino acids, and others. Then a sample of the growing culture is transferred to minimal medium (left), in which *Neurospora* normally can grow, because it can synthesize all the compounds that it requires beginning with material present there. If the mold fails to grow in the minimal medium, it is a nutritional mutant; that is, it cannot make one of the compounds that normal mold can make. The nature of its deficiency can then be determined by transferring other samples from the culture growing in complete medium to various cultures containing minimal medium, each of which is supplemented with a different single nutrient. In this example, the mutant strain is able to grow on minimal medium supplemented with pantothenic acid (a B vitamin). This indicates that this strain lacks an enzyme involved in pantothenic acid synthesis.

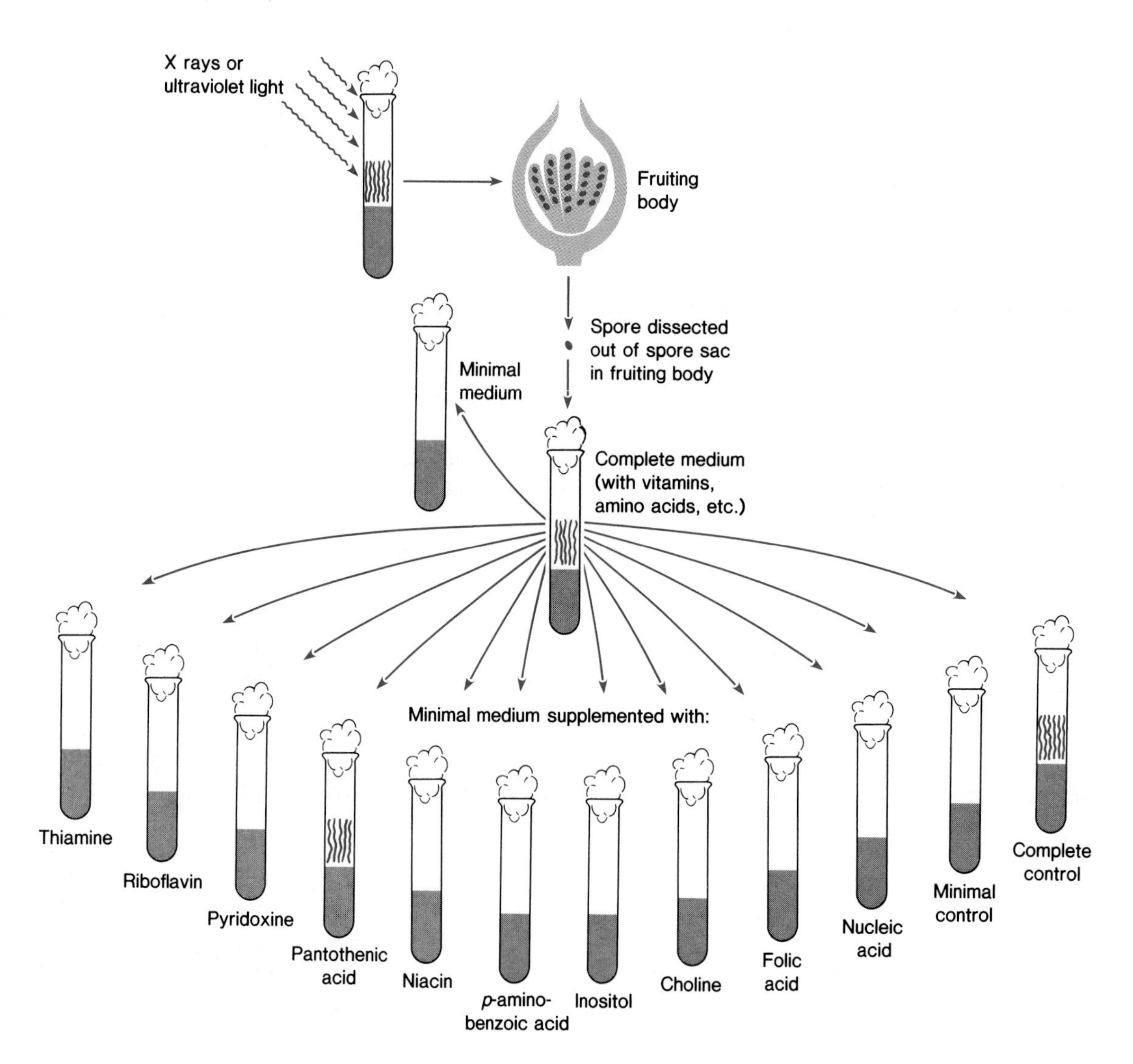

It was many years, however, before Garrod's hypothesis was verified experimentally. In 1951, G. W. Beadle and E. L. Tatum conducted a series of experiments using *Neurospora crassa,* the red bread mold. They were able to formulate some precise relationships between genes and enzymes. In their experiments, Beadle and Tatum used X rays to induce mutations in the mold. They isolated a series of mutant molds, each of which was unable to grow without the addition of a specific nutrient to the growth medium (figure 18.7). These were nutrients that normal molds are able to synthesize for themselves. In each case, however, the mutant lacked an enzyme that catalyzed a step in the synthesis of the nutrient. Different mutant strains required different nutrient supplements because they lacked different functional enzymes.

Figure 18.8 The first seven amino acids of the β polypeptide chain of human hemoglobin. The substitution of a single amino acid—valine substituted for glutamic acid—at the sixth position results in sickle-cell anemia. This substitution of only one of the 146 amino acids in the β polypeptide chain drastically alters properties of hemoglobin and the red blood cells that contain the altered hemoglobin.

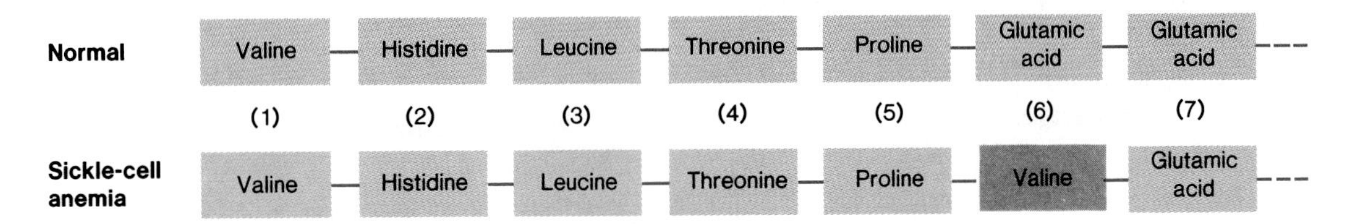

Beadle and Tatum found a one-to-one correspondence between a genetic mutation and the lack of a specific enzyme required in a biochemical pathway. They proposed, therefore, that each gene specifies the synthesis of one enzyme. This is called the one gene/one enzyme hypothesis. We now know that the relationship is more complex than that, but the work of Beadle and Tatum clearly proved the relationship between genes and enzymes that Garrod had proposed so long before.

In 1957, V. M. Ingram reported that the hemoglobin in people with sickle-cell anemia differs from normal hemoglobin by a single amino acid in the β (beta) polypeptide chain (figure 18.8). (Hemoglobin is made up of two β and two α (alpha) polypeptide chains; see page 45.) This indicated that genes actually specify the amino acid sequence of proteins. They also code specific information for assembly of all proteins (polypeptide chains), including both enzymes and other proteins such as hemoglobin. By renaming the one gene/one enzyme hypothesis the one gene/one polypeptide hypothesis, Ingram stressed that genes code for proteins that are single polypeptides and also for the individual polypeptides that aggregate in complex proteins such as hemoglobin. The genetic information for such complex proteins, therefore, is coded in not one gene, but several. The genes that code information for synthesis of polypeptide chains are called **structural genes.**

Genetic Expression

One of the keys to understanding the role of genes is found in their expression in the assembly of specific kinds of polypeptides. We need to learn, therefore, how genetic information coded as a sequence of nucleotides in a DNA molecule is translated into a sequence of amino acids in a polypeptide. This process can be divided into two phases: **transcription** and **translation** (figure 18.9).

Figure 18.9 Transcription and translation. The message in a sequence of DNA nucleotides is copied as a sequence of nucleotides in a messenger RNA (mRNA) molecule that leaves the nucleus and enters the cytoplasm. Messenger RNA associated with ribosomes directs assembly of polypeptides made up of specific amino acid sequences.

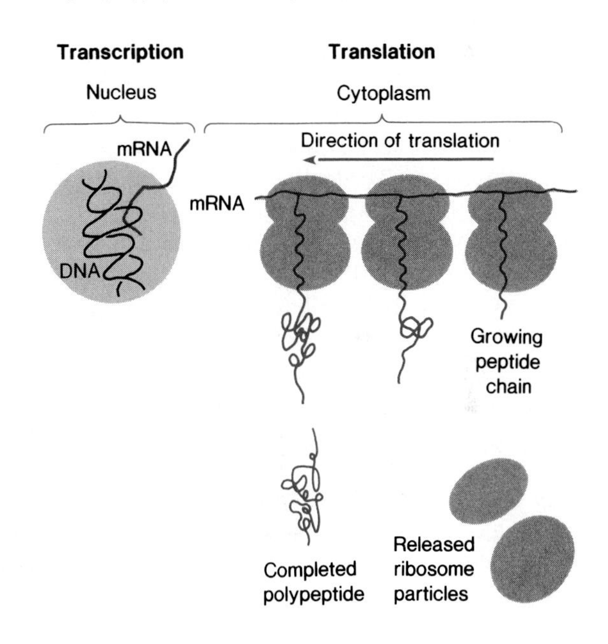

Transcription is the process of actually "reading and rewriting" the genetic code of the DNA molecule. In transcription, the sequence of nucleotides in a segment of DNA is transcribed into a complementary strand of RNA. In the process, the sequence of nucleotides in the DNA segment specifies the sequence of nucleotides in the RNA strand. The RNA strand produced is called **messenger RNA (mRNA).** It is a messenger in the sense that it carries a message about genetic information contained in nuclear DNA out into the cytoplasm where the actual assembly of amino acids into polypeptides takes place.

Figure 18.10 Transcription. (*a*) A representation of mRNA production. Transcription involves assembly of a strand of mRNA that is complementary to part of one of the strands of DNA. The DNA strands must temporarily separate in order for transcription to occur. The enzyme RNA polymerase catalyzes the linking of nucleotides into a chain. Note the complementarity of the bases and that uracil (U) is present in RNA rather than thymine (T). (*b*) Diagrammatic representation of the sense strand and replication strand in a segment of DNA. Only the sense strand is transcribed; mRNA produced during transcription is complementary to the sense strand. The replication strand is, of course, necessary for DNA replication.

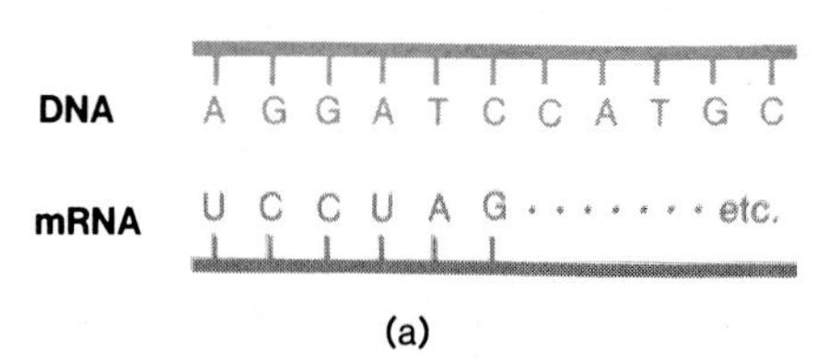

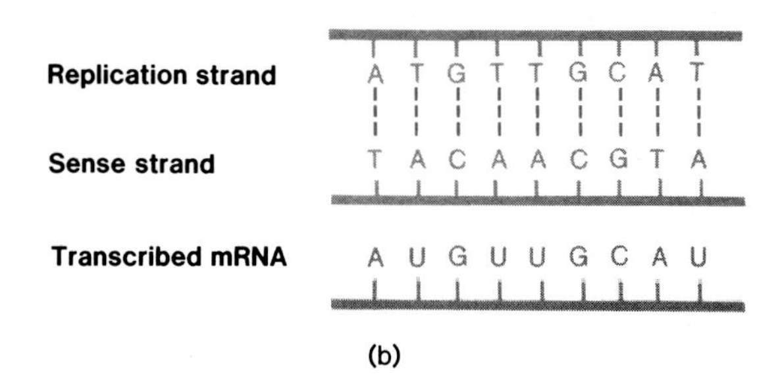

The use of the information, now encoded in mRNA, to produce a polypeptide is called translation. The mRNA strand moves out into the cytoplasm, where it becomes associated with groups of ribosomes. Each ribosome moves along the mRNA strand and reads the genetic code. When it reads a three-base sequence along the mRNA strand (the code is a "three-letter" code, as we shall see shortly), a specific **transfer RNA (tRNA),** bearing a specific amino acid, attaches to the ribosome. As the ribosome moves further along the mRNA strand, the tRNA leaves the ribosome, leaving behind its amino acid. The ribosome reads the next part of the coded information along the mRNA strand, binds another tRNA with its particular amino acid, and the process is repeated. Thus, sequentially coded information in mRNA specifies which tRNA molecules are bound and, therefore, what amino acids are added to the growing chain in what sequence. As each ribosome moves down the mRNA strand, it produces a chain of amino acids, a polypeptide. When the ribosome reaches the end of the mRNA strand or reaches a coded signal to stop, it breaks away and the completed polypeptide is released.

Mechanisms of Protein Synthesis

Transcription and translation are complex processes and not fully understood, but the molecular biology of genetic expression is one of the most active areas of biological research. In this section, we will explore the transcription and translation processes in greater detail.

Table 18.1
Complementary Base Pairing between DNA and Messenger RNA.

DNA	RNA
Adenine	Uracil*
Cytosine	Guanine
Guanine	Cytosine
Thymine	Adenine

*Note that RNA contains uracil rather than thymine. Wherever an adenine appears in a DNA strand being transcribed, uracil is incorporated into the mRNA molecule produced.

Transcription

The production of mRNA during transcription is somewhat similar to the replication of DNA. The formation of a mRNA strand requires the presence of an enzyme, **RNA polymerase,** that catalyzes the joining of nucleotides into a polynucleotide strand (table 18.1). One of the DNA strands acts as a template for the assembly of a complementary mRNA strand (figure 18.10). The transcript is made from only one of the two strands of DNA; this strand is called the **sense strand.** The other DNA strand is called the **replication strand.** Although the DNA replication mechanism depends on the double-stranded nature of the molecule, only one strand's coded information is transcribed.

Figure 18.11 Transfer RNA (tRNA). Base sequences in tRNA are arranged so that complementary base pairing occurs within the molecule. Hydrogen bonds between base pairs stabilize the tRNA molecule in its characteristic cloverleaf shape. This "cloverleaf" actually twists into a more complex three-dimensional shape. The point of attachment of an amino acid is symbolized aa. Each type of tRNA molecule specifically attaches one and only one kind of amino acid and has a specific base triplet (anticodon) that is complementary to a specific codon in mRNA. The attachment of the appropriate amino acid to the correct tRNA molecule is catalyzed, in each case, by a specific enzyme.

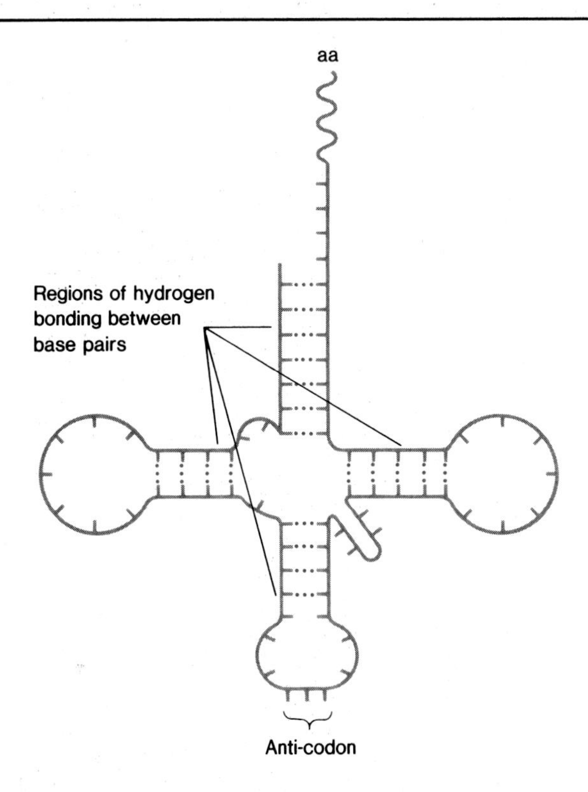

Transfer RNA molecules have a shape that resembles a cloverleaf (figure 18.11). Each tRNA molecule attaches to one specific amino acid, and only to that kind of amino acid. This specific binding of particular amino acids to particular tRNA molecules is absolutely essential for the assembly of normal protein molecules during the process of translation.

Translation

Messenger RNA moves out from the nucleus into the cytoplasm and becomes associated with the ribosomes. Ribosomes become attached to the mRNA strand and move along the strand, "reading" its message. Usually, several ribosomes attach to the same mRNA strand so that a number of identical polypeptides are being assembled at the same time.

Each ribosome is made up of two subunits, one large and one small. In bacteria, these are designated 50S and 30S. S is the Svedberg unit, which is a measure of mass. The larger the particle, the higher the number. The ribosomal subunits in eukaryotic cells are a little bit larger, consisting of 60S and 40S units. Both ribosomal subunits are made up of protein and nucleic acid. The nucleic acid is a third kind of RNA, called **ribosomal RNA (rRNA).** The subunits are free in the cytoplasm when they are not directly involved in protein synthesis. When they attach to a mRNA molecule during protein synthesis, however, the two subunits join together to form a functional ribosome.

Some details of the translation process are illustrated in figure 18.12. Sequences of three nucleotides in mRNA are called **codons.** Each codon specifies an amino acid. The three nucleotides on the tRNA that will base-pair with each codon are called the **anticodon.** The specific pairing between each mRNA and the correct anticodon of a tRNA molecule brings the correct amino acid into exactly the right place in a polypeptide. Obviously, each tRNA molecule must carry the appropriate amino acid if this is to occur correctly.

Peptide bonds form between amino acids as they are brought into place by tRNA molecules (see chapter 2 for more details concerning peptide bonds). Once its amino acid has been joined to the growing polypeptide chain, a tRNA molecule falls away from the codon on the mRNA. Movement of the ribosome along the mRNA molecule exposes one codon after another. Appropriate tRNA molecules bind, and the polypeptide chain grows in length.

Figure 18.12 The translation process. Transfer RNA molecules bring appropriate amino acid molecules into place as the ribosome moves along the mRNA molecule from codon to codon. Each amino acid is attached to the growing polypeptide chain. As it gives up its amino acid, a tRNA first moves aside and then detaches from the mRNA. This makes room for attachment of the next tRNA, which is specified by the next codon in the mRNA molecule. That tRNA brings its attached amino acid into place for its addition to the growing polypeptide chain. The enlargement shows how anticodons and codons are complementary to each other. Note that tRNA molecules are drawn here in a form that symbolizes the three-dimensional shape that the "cloverleaf" shown in figure 18.11 actually assumes in cells.

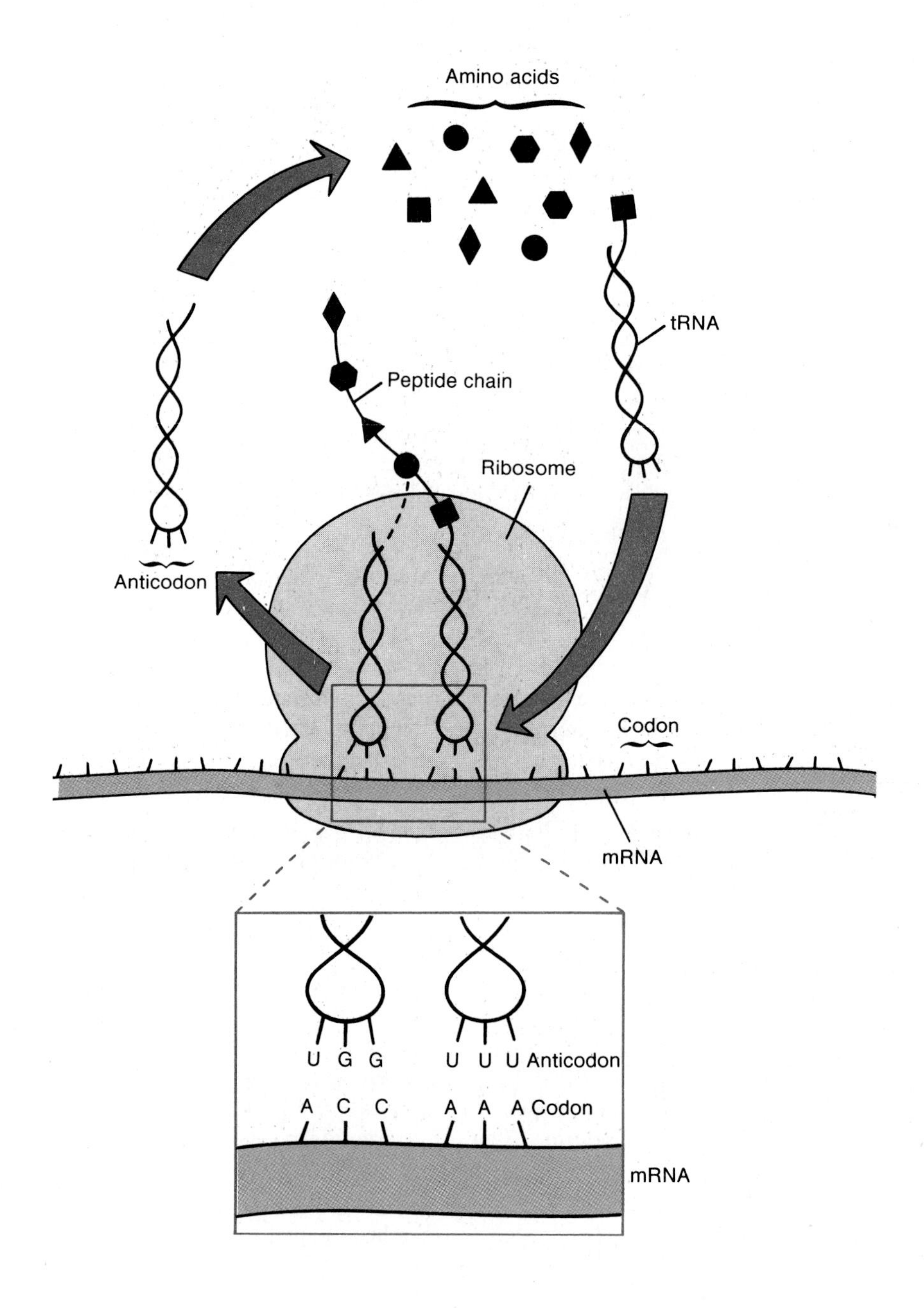

Figure 18.13 Ribosomes and mRNA. Several ribosomes are attached to an mRNA molecule at a time and each of them is involved in some stage of assembly of a polypeptide as it moves along the mRNA. When the polypeptide is complete, the ribosome detaches from the mRNA and separates into its two subunits. The polypeptide is released at the same time.

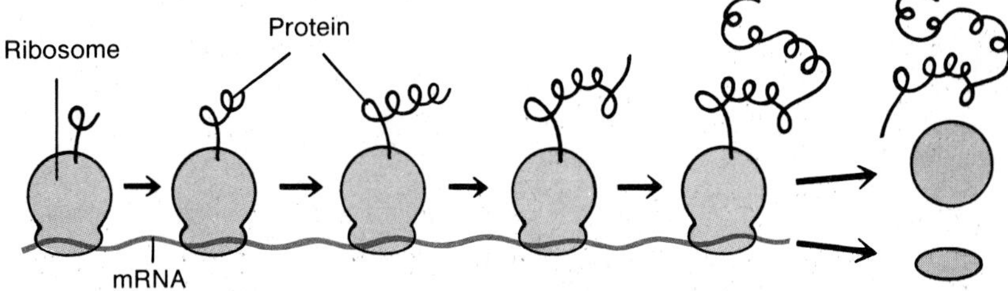

This process continues until the end of the message, at which time a newly completed polypeptide chain is released. At the same time that the polypeptide is released, the two subunits of the ribosome break apart (figure 18.13).

This, then, completes a sequence of molecular processes involved in genetic expression in cells. The genetic message coded as a sequence of bases in DNA has been expressed, through the processes of transcription and translation, as a specific type of protein molecule consisting of a particular sequence of amino acids.

The Genetic Code

As we saw in the previous section, codons of mRNA and anticodons of tRNA are sequences of three bases. The DNA molecule itself also has three-base sets that code information. But how was it actually proven that this code depends on base sequence, and how was the code "cracked"?

The first step involved some simple mathematical reasoning. Because twenty amino acids are incorporated into naturally occurring proteins, the codon has to be at least three bases long. Why is this so? If the codon were a single base, only four amino acids could be coded because there are only four different bases in the nucleic acid molecules. In other words, you cannot write nearly enough one-letter words with this four-letter alphabet to code for the twenty different amino acids found in proteins. What if the codon were two bases long? Only sixteen amino acids could be specified. This is because you can write only sixteen different two-letter words using a four-letter alphabet.

But if the codon were three bases long, sixty-four amino acids could be coded; that is, sixty-four different three-letter words can be constructed using a four-letter alphabet. Therefore, three is the smallest number of bases that provides enough combinations to code for twenty different amino acids. It has been demonstrated experimentally that the three-base (triplet) code, the simplest code possible, is indeed the form of genetic coding found in living things.

However, because only twenty different amino acids are used in the assembly of naturally occurring proteins and there are sixty-four different possible three-base combinations in nucleic acid molecules, we must ask a critical question: Could more than one three-base codon code for the same amino acid? Either there is more than one codon for some amino acids, or some codons do not code for an amino acid at all. Research shows that all but three of the sixty-four possible codons do specify amino acids, and that most amino acids are coded for by two or more different codons.

How was it determined which codons specify which particular amino acids? One of the keys to this determination was an enzyme found by Severo Ochoa that could be used to link nucleotides together into RNA molecules in the absence of DNA. This enzyme permitted biologists to produce artificial RNA. For example, in the presence of this enzyme, uracil triphosphate (UTP) is converted into long chains of RNA that contain only uracil as a base. The composition of the RNA produced in this way is described as poly U.

Biologists soon discovered how such artificial RNA could be combined with material extracted from cells to produce amazing results. This specially prepared RNA could actually function as artificial messenger RNA. In 1961, Marshall Nirenberg discovered that when artificial mRNA that

Figure 18.14 The genetic code. This figure shows all sixty-four combinations of nitrogenous bases in mRNA and the amino acid for which each combination codes. Most of the combinations code for amino acids (note that there is more than one combination for most amino acids). There are three terminator codons (labelled "stop") that signal the end of a polypeptide. When a terminator codon is present, translation stops and the polypeptide is released from the ribosome that is reading the message of an mRNA molecule.

First position	Second position: U		C		A		G		Third position
U	UUU	Phe	UCU	Ser	UAU	Tyr	UGU	Cys	U
	UUC		UCC		UAC		UGC		C
	UUA	Leu	UCA		UAA	STOP	UGA	STOP	A
	UUG		UCG		UAG		UGG	Trp	G
C	CUU	Leu	CCU	Pro	CAU	His	CGU	Arg	U
	CUC		CCC		CAC		CGC		C
	CUA		CCA		CAA	Gln	CGA		A
	CUG		CCG		CAG		CGG		G
A	AUU	Ile	ACU	Thr	AAU	Asn	AGU	Ser	U
	AUC		ACC		AAC		AGC		C
	AUA		ACA		AAA	Lys	AGA	Arg	A
	AUG	Met	ACG		AAG		AGG		G
G	GUU	Val	GCU	Ala	GAU	Asp	GGU	Gly	U
	GUC		GCC		GAC		GGC		C
	GUA		GCA		GAA	Glu	GGA		A
	GUG		GCG		GAG		GGG		G

contained only uracil (poly U) is added to a cell extract, a particular type of polypeptide is made. The single unusual polypeptide produced in the reaction mixture consists entirely of the amino acid phenylalanine. Poly U directs production of polyphenylalanine, a polypeptide containing only one phenylalanine after another. Thus, the codon specifying phenylalanine must be UUU. Using this discovery, three large laboratories headed by Nirenberg (National Institutes of Health), Ochoa (New York University), and H. G. Khorana (University of Wisconsin) used similar methods to determine the remaining codons. This tremendous effort resulted in the construction of a simple but history-making RNA/amino acid dictionary (figure 18.14). The genetic code was "cracked"!

After years of study on a multitude of different organisms, one fact has become quite clear: the genetic code appears to be universal; that is, in virtually all living things, from bacteria and even viruses to human beings, the same codons specify the same amino acids. This discovery powerfully supports the idea that at the level of molecular genetics, there is a very fundamental unity of life that indicates a common evolutionary origin.

Regulatory Genes

So far, we have discussed genes that provide coded information for synthesis of proteins (polypeptides). As we mentioned before, these genes are called structural genes because they determine protein structure. There is strong evidence indicating that another class of genes, called **regulatory genes,**

Box 18.1

Split Genes and Jumping Genes

"What is true for *E.coli* is true for the elephant." These words became one of the popular mottos of molecular biology during the twenty years that followed the determination of the structure of DNA in 1953. Using prokaryotic organisms, especially the human intestinal tract bacterium, *Escherichia coli,* molecular biologists broke the genetic code and delved into the mechanisms by which genetic information, encoded in DNA molecules, is expressed through the mechanisms of transcription and translation.

Optimistic pronouncements about *E.coli* and elephants became commonplace when it was determined that the genetic code is universal. That is, the same three-nucleotide codons specify the same amino acids in virtually all cells, from the prokaryotic cells of *E.coli* and other bacteria to the eukaryotic cells of complex plants and animals. It seemed quite logical to assume that similar parallels would be found in the organization of genes and in the mechanisms of genetic expression.

But, in the late 1970s, came the startling discovery that genes in eukaryotic cells are organized differently from genes in prokaryotic cells. In prokaryotic cells, the amino acid sequence in a polypeptide is a direct reflection of the information in the structural gene that codes for it. The nucleotide sequence of the structural gene is transcribed in a mRNA molecule and then translated directly into an amino acid sequence in a polypeptide.

This is not always the case in eukaryotic cells. Many genes in eukaryotic cells are split. Stretches of DNA that specify parts of the amino acid sequence in the protein product are separated by intervening nucleotide sequences that are not translated. The information-containing sequences are called **exons,** and the intervening sequences are called **introns.** An entire section of the DNA molecule, including both exons and introns, is transcribed, but only the exons are represented in the mRNA that is finally exported to the cytoplasm for translation. There is a splicing process by which the introns are cut out and the exons are connected in a single uninterrupted sequence.

Split genes seem to be widespread, possibly even commonplace in the genetic material of eukaryotic cells, while they have not been observed in prokaryotic cells. Thus, in terms of gene organization, what is true for *E. coli* is not necessarily true for the elephant after all.

Split genes were not the only surprise that geneticists got from detailed analysis of the structure and organization of the genetic material. In the late 1940s and early 1950s, Barbara McClintock did genetic crosses and developmental studies on corn (maize) plants. The results led her to conclude that during development of various cells, genes move around and become rearranged in new positions relative to one another.

When McClintock first reported these results, biologists either did not understand exactly what she meant or did not think that such genetic instability could be possible. At that time, virtually everyone was certain that genes were arranged in fixed positions on chromosomes and that the only time genes became rearranged was during crossing over in meiosis.

But then in the 1970s, it became clear to biologists able to use techniques of molecular genetics, which were not available to Barbara McClintock, that rearrangement of **movable gene segments** (also called **transposons**) is indeed widespread. In fact, such genetic rearrangement seems to be a normal part of development in many organisms, just as Barbara McClintock had first suggested in the early 1950s.

Sometimes recognition comes too late to those who make discoveries that are so advanced that they are not fully understood or accepted by contemporary scientists. Possibly, Gregor Mendel is the most famous example of late appreciation. Fortunately, Barbara McClintock has received her well-deserved recognition including the 1983 Nobel Prize for Medicine and Physiology, given for her pioneering discovery of "jumping genes."

exists. Regulatory genes regulate the activity of structural genes. For example, it is known that specific mRNAs and their resulting proteins are synthesized at specific times in development of any individual organism. This fact implies that certain structural genes are somehow "turned on" and then "turned off" at specific times, apparently via the action of regulatory genes.

In bacteria, genetic elements that turn genes on and off have been described in some detail (see box 18.2), but mechanisms involved in control of gene expression in eukaryotic cells are not understood as well. There is, however, a great deal of research currently being done on the subject and we can expect much progress in the next few years.

Box 18.2

Turning Genes On and Off

In bacteria, genetic elements that turn genes on and off have been described in great detail. These regulatory elements, together with the structural genes that they regulate, are called **operons.** The most thoroughly studied operon is the **lactose** or **lac operon.** Based on their work with the lac operon, François Jacob and Jacques Monod in France proposed a general scheme for control of genetic expression that is now known as the Jacob-Monod model.

As you can see in box figure 18.2, there are five genetic units in the lac operon: two regulatory units (P and O) and three structural genes (Z, Y, and A). The P site, or promotor, is a segment of DNA to which RNA polymerase binds. Under appropriate conditions, RNA polymerase catalyzes transcription of mRNA from the structural genes Z, Y, and A. These genes code for enzymes that are involved with the breakdown of the

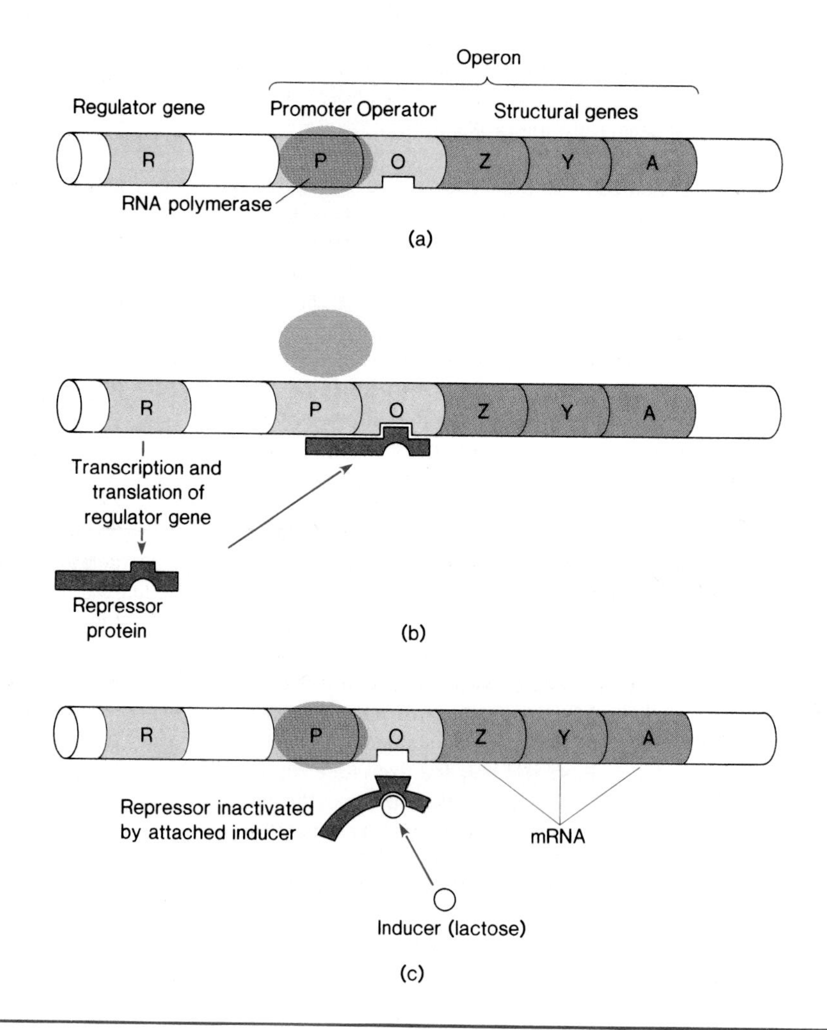

Box figure 18.2 The Jacob-Monod model of structural gene regulation as illustrated by function of the lac operon of the bacterium *Escherichia coli.* (*a*) Elements of the operon with RNA polymerase in place on the promoter site of operon, as it must be for transcription to occur. (*b*) The lac operon in its repressed (inactive) condition with the repressor molecule binding to the operator gene and preventing positioning of RNA polymerase on the promoter site. (*c*) The lac operon in its derepressed (induced or active) condition. Lactose binds to its site on the repressor molecule, thus altering the repressor so that it no longer blocks position of RNA polymerase on the promoter site, and transcription of the structural genes proceeds.

sugar lactose. Another genetic unit, R, a regulator gene, is not a physical part of the operon but is involved in functional control of the operon. The regulator gene codes for production of a protein called the **repressor,** which has two binding sites, one for lactose and one for the operator gene O.

The repressor is an **allosteric** protein molecule, a molecule in which the shape of one binding site is affected by the condition of the other binding site. In this case, the lac operon repressor can bind with the operator (O) gene when it does not have its lactose binding site occupied. When its lactose binding site is occupied, the repressor no longer binds with the operator.

In the lac operon, when the repressor is binding the O gene, the polymerase cannot bind and function at the P site, and the structural genes Z, Y, and A are not transcribed; they produce no messenger RNA coding for enzyme synthesis. This is the situation when there are low levels of lactose in the medium. If lactose levels rise, lactose binding alters the conformation of the repressor so that it uncouples from the O gene, the polymerase binds at the P site, and the structural genes are transcribed (mRNA is produced). When the enzymes synthesized by translation of the messenger RNA have broken down all the lactose, the repressor returns to the O gene, shutting down enzyme synthesis. In this system the presence of lactose can, in effect, turn on genes (induce transcription), whereas the absence of the molecule turns off the genes automatically. In the Jacob-Monod model, any substance that acts as lactose does is called an **inducer molecule.**

In addition to inducible genes, such as those of the lac operon, which normally are "off" and can be turned "on," there also are repressible genes, which normally are "on" and can be turned "off." The value of gene regulation to the cell is that the enzymes are produced only when needed; thus, the cell's resources are conserved.

In eukaryotic organisms, however, the details have not been worked out as well. Biologists who have studied control of gene expression in eukaryotic cells conclude that the control mechanisms involved are somewhat more complicated than those found in prokaryotic cells and described by the Jacob-Monod model.

Recombinant DNA

Now that we have explored some fundamentals of the molecular aspects of genetics, we can return to the topic with which we began this chapter—the idea of genetic engineering. How has knowledge of molecular events involved in genetic replication and expression allowed us to develop techniques for modification of the genetic makeup of living things?

The method that has been developed for inserting a small segment of DNA from one species into the genetic material of another unrelated species is called the **recombinant DNA** (gene-splicing) technique. Let us see how the method works.

Recombinant DNA technology is based on two sets of discoveries in molecular genetics. The first of these was that there is extrachromosomal (outside the chromosomes) DNA in bacterial cells. One form of extrachromosomal DNA in bacteria is the **plasmid,** a relatively small circular DNA molecule that can replicate inside bacterial cells and that also can be transferred to other bacterial cells. Resistance to certain antibiotics, for instance, is passed from cell to cell by plasmid transfer.

The second major finding that made recombinant DNA technology possible was the discovery of enzymes known as **restriction endonucleases.** These enzymes have the capacity of recognizing and binding to a specific pattern of base sequences in a DNA molecule, and splitting the molecule in a very characteristic fashion where that base sequence occurs (figure 18.15). When a restriction endonuclease cuts (nicks) the two strands of a DNA molecule in this way, it leaves fragments that have overlapping free ends. DNA from most species of organisms has sites that can be split by the same restriction endonucleases that nick bacterial plasmids.

Figure 18.15 The action of a restriction endonuclease. A restriction endonuclease recognizes and attaches to a specific sequence of nucleotides in DNA. It then nicks (cuts) the single strands at specific points, indicated here with colored arrows (*a*). This leaves DNA fragments that have free single-stranded ends (*b*).

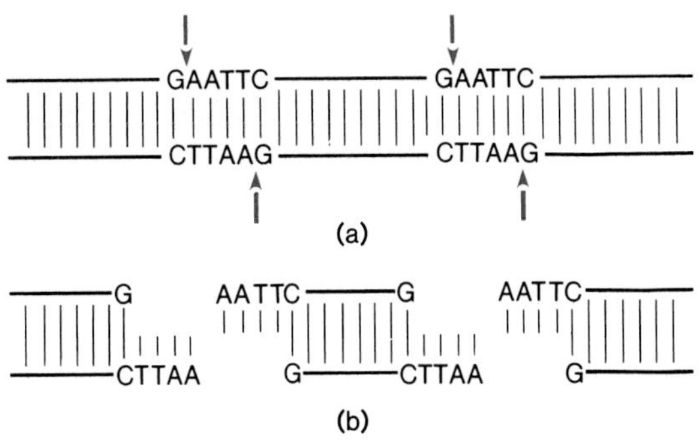

Figure 18.16 Diagrams outlining the recombinant DNA (gene-splicing) technique. Once the plasmid containing the desired gene has been introduced into bacterial cells (5), millions of cells can be grown in culture (6). Each of these cells contains one to many copies of the introduced gene.

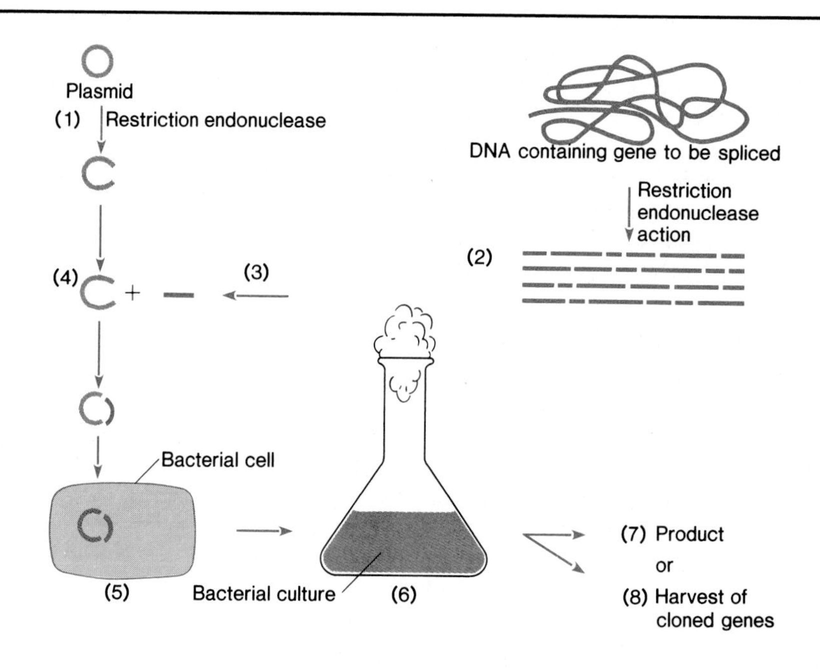

Figure 18.17 Electron micrographs taken at stages during recombinant DNA experiments. (*a*) Bacterial plasmid ring has been nicked by restriction endonuclease. Fragment of foreign DNA to be inserted is at lower right. (*b*) Plasmid after fragment of foreign DNA has been incorporated. (*c*) Plasmid adjacent to bacterial cell into which it can be introduced.

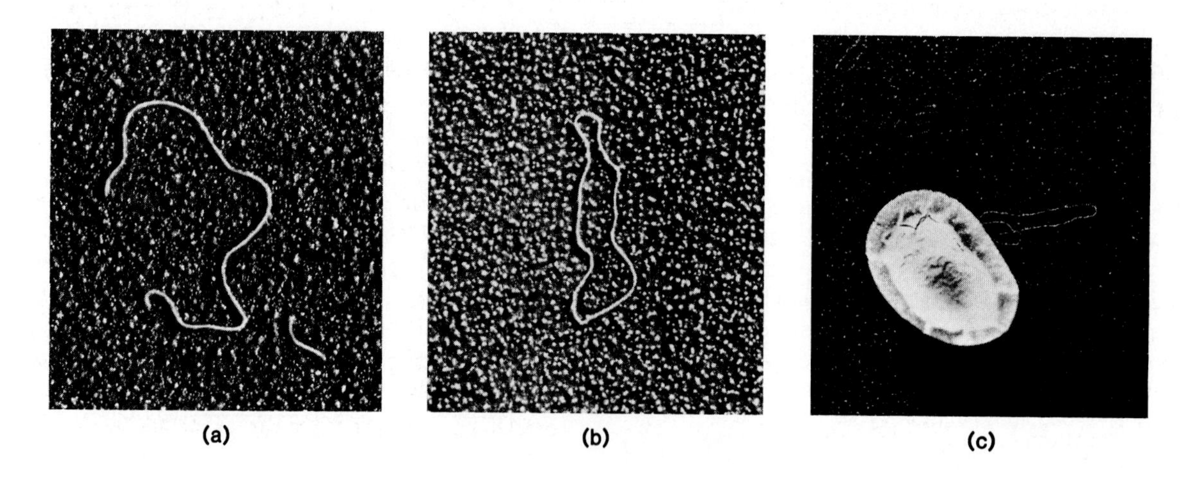

How, then, does the recombinant DNA technique work? The actual methods used were devised by Stanley Cohen, Herbert Boyer, and others. Specific plasmids containing a single restriction endonuclease site are harvested from their host cells and cut open with restriction endonuclease (figure 18.16). The foreign DNA to be inserted is extracted and purified. Then it is cut up into many fragments using the same restriction endonuclease (part 2 in figure 18.16). A specific fragment, known to contain the desired gene or genes, is selected according to weight. The overlapping ends cut by a restriction endonuclease are complementary; molecular geneticists say they are "sticky." Thus, a specific recognition and association can take place between the cut ends of the foreign DNA fragment and the cut ends of the plasmid. The sugar-phosphate backbones of the double helix are joined using an enzyme known as a **ligase** (part 4 in figure 18.16), thereby producing the normal closed circle of the plasmid into which the foreign gene is now incorporated. The recombinant plasmid composed of the plasmid DNA and the foreign DNA fragment (that is, recombinant DNA) is introduced into a bacterial cell (part 5 in figure 18.16). Cells containing the recombinant plasmid are allowed to reproduce until there are large numbers of cells (part 6 in figure 18.16), thus producing many more copies of the recombinant DNA. The many exact copies of the plasmid make up a **clone** of recombinant DNA. Certain strains of plasmids also replicate many times inside each cell, further amplifying the number of **cloned genes.** Figure 18.17 shows electron micrographs of several steps in the recombinant DNA technique.

Either the products of the expression of the cloned genes or the cloned genes themselves then can be isolated. For gene products to be isolated, the gene must be transcribed in the bacterial cell, where mRNA can be produced. Then the message contained in the mRNA must be translated into polypeptides with specific amino acid sequences. If this can be accomplished, it is possible to design bacteria that can "manufacture" a specific polypeptide product.

A cloned gene can be isolated by harvesting the plasmids from the cells, purifying them, and using the same restriction endonuclease to cut up the plasmid. The fragment, now in numerous copies, can then be purified.

One of the most important reasons for cloning genes is economic. For example, the protein hormone insulin, used for treatment of diabetes, has been isolated, in an expensive operation, from the pancreas of animals. But as we noted at the beginning of this chapter, it now has become possible to produce insulin using recombinant DNA techniques. Copies of the human insulin gene have been obtained. The gene is inserted into a plasmid, and the plasmid is introduced into bacteria that can be grown in large quantities. Transcription and translation occur in bacteria containing the plasmid. Thus, bacteria can now be made to synthesize human proinsulin, the larger polypeptide precursor of insulin (see page 248). This proinsulin can be converted into active insulin and purified.

It now seems very likely that a number of other polypeptides that are medically important will soon be in commercial production using recombinant DNA techniques. Some of these polypeptides, such as human growth hormone, can now be obtained only from human tissues and thus have been so scarce that supplies have fallen far short of medical needs. We hope that polypeptide production using recombinant DNA techniques soon will make needed treatments available to many people previously not able to obtain them.

Gene cloning has also found valuable applications in genetic research. In detailed study of genes, such as determination of base sequences, large quantities of purified genes are required. Already, base sequences of a number of cloned genes have been determined, and it looks as if the technique will be even more useful in the future.

Recombinant DNA and gene-cloning experiments may, however, have potential dangers. The possibility of scientists creating a "monster" strain that might escape from the lab and wreak havoc to life caused considerable public discussion of such work in the mid–1970s. Since *Escherichia coli* (an inhabitant of the human intestine) is the organism most commonly used in recombinant DNA studies, there has been concern that dangerous genes or groups of genes might somehow be transferred to humans. For example, what dangers would be in store if *E.coli* containing cloned genes of a cancer-causing virus were to escape the lab and infect humans? Would it cause a new cancer epidemic? As a result of such fears, federal guidelines were devised to regulate research in this area. However, experiments on potential dangers have so far yielded negative results, and research using recombinant DNA techniques is proceeding with caution.

Looking further into the future, there is the distinct possibility that gene-splicing techniques may be modified for use in treatment of human hereditary diseases in what amounts to "gene therapy." It is also possible that these techniques may allow desirable traits to be put together in useful new combinations in agriculturally important plants and animals.

Mutations

We introduced the concept of mutation in chapter 17 with a discussion of stable heritable changes in chromosome structure that are transmitted to all of the cells descended from any cell in which such a change occurs. Such changes are **chromosomal mutations (macromutations)**.

Much smaller changes in the genetic material are called **point mutations.** Macromutations affect large pieces of the genetic material, in most cases involve breaks in the chromosome, and frequently can be detected using the light microscope. But point mutations generally involve a single base change in a DNA molecule and, as a result, are not microscopically detectable.

Thus, the definition of mutation can be broadened to include any relatively stable change in the genetic material of a cell that is transmitted to all cells descended from it. Although mutation is defined to include both general types of genetic change, biologists often tend to use the word mutation specifically to mean point mutation because they are the more common genetic changes.

There are several kinds of **mutagens** (agents that cause mutation). A number of chemicals are **mutagenic** (mutation causing), as are ultraviolet light and various forms of ionizing radiation (X rays, gamma rays, etc.).

Although point mutations—changes in single bases in the long sequences of bases in DNA molecules—may seem small in chemical terms, they are very important biologically. The effect of a single-base change can be dramatic, and it may spell the difference between life and death for an organism whose cells all contain an inherited point mutation. Let us see why this is so.

Types of Point Mutations

There are several classes of point mutations. The first is the **missense** mutation. A missense mutation is a single-base substitution that changes a codon specifying one amino acid to a codon specifying another amino acid. The sickle-cell hemoglobin mutation is a classic example of a missense mutation. The change from the glutamic acid codon to a valine codon in messenger RNA results from a single-base substitution: GAG → GUG (see figure 18.14).

When a gene containing a missense mutation is expressed, there is an amino acid substitution in the polypeptide formed (valine substituted for glutamic acid in the sickle-cell hemoglobin mutation, for example). How can substitution of a single amino acid among many in a polypeptide have such drastic consequences?

Figure 18.18 A segment of mRNA showing the effect of frame-shift mutation resulting from insertion of an extra base into the DNA strand from which the mRNA was transcribed. During translation, the mRNA transcribed from a gene that is frame-shifted will direct incorporation of incorrect amino acids because all codons beyond the addition point (G) are altered.

Before	AUG	CCA	UAC	UGG	
	Methionine	Proline	Tyrosine	Tryptophan	
After	AUG	CCA	(G)UA	CUG	G—
	Methionine	Proline	Valine	Leucine	

If the amino acid substituted is at a critical region, the molecule could be nonfunctional. In the case of the sickle-cell hemoglobin, the substitution does occur at a critical position, and properties of the entire hemoglobin molecule are abnormal as a result. The structure of the molecule is unstable in solutions of low oxygen concentration. Structural changes in the molecules are related to a structural change in red blood cells (they assume the sickle shape). These sickle-shaped cells interfere with circulation through small blood vessels and cause a host of problems in body tissues. This whole chain of events is initiated by a missense mutation, a single base substitution in a DNA molecule.

There are other missense mutations in the hemoglobin gene and in other genes that do not adversely affect the function of the molecules. Occasionally, a missense mutation actually improves the function of the molecule produced as a result of expression of the altered gene. Such mutations are very important in terms of evolution because they provide a source of new favorable characteristics.

The second class of point mutations is the **frameshift** mutation. As you know, messenger RNA transcribed from DNA is a sequence of codons in which each set of three bases constitutes a separate reading unit or frame. If an extra base is inserted into the base sequence in DNA, all of the codons after the insertion are changed or shifted to codons calling for different amino acids. This drastic change is transcribed into mRNA and translated into an altered polypeptide. Figure 18.18 shows a segment of an mRNA molecule that illustrates the effect of a frameshift. By inserting a guanine between the second and third codon, the third and fourth amino acids and all subsequent amino acids in the polypeptide produced during translation will be changed. If such a mutation occurs anywhere but at the very end of the gene (the last codons transcribed and translated), a completely nonfunctional protein results. The deletion of a base has a similar frameshift effect.

Often, mutations have a different kind of effect. There are codons (terminators) to indicate where reading of mRNA stops (see figure 18.14). No amino acid-carrying tRNA attaches at a point in an mRNA molecule where there is a terminator codon. Because this interrupts assembly of the polypeptide chain, a terminator codon functions as a period in the genetic code. If a frameshift or a base substitution produces a terminator codon (for example, the mRNA transcribed might contain UAA instead of AAA), then during translation, a peptide bond does not form between the amino acids called for by the codons straddling the terminator codon introduced by mutation. As a result, the translation terminates prematurely, producing an abnormally short polypeptide. If such a mutation occurs near the end of the message, less harm is done, since almost all of the protein has been formed. However, if the mutation occurs earlier in the message, translation does not yield a functional polypeptide.

Causes of Mutations

Some mutations arise accidentally as a result of chemical errors in replication of the DNA, but many are caused by outside agents.

Chemical Mutagens

Some mutations are caused by chemical mutagens. An example of a chemical mutagen is the very reactive chemical nitrous acid (HNO_2). Nitrous acid converts adenine into a compound that resembles guanine. As a result of HNO_2 treatment, adenines are converted into guaninelike compounds that pair with cytosine during DNA replication. This change is permanent because DNA molecules produced in subsequent replications will have C-G base pairs instead of A-T pairs in the positions where the original changes occurred. A variety of other chemical compounds also act as mutagens.

Physical Mutagens

Several types of electromagnetic radiation also cause mutation. Although wavelengths of the electromagnetic spectrum longer than those of visible light, such as infrared and microwaves, also are thought to be mutagenic, the best understood types of mutagenic radiation are ultraviolet light, X rays, and gamma rays.

Ultraviolet light is most mutagenic at the wavelength that is absorbed most by DNA—260 nm. An atom in a DNA molecule becomes excited by ultraviolet light. This results in a rearrangement of electrons. In this excited state, the atom is chemically reactive, allowing it to bind covalently with another atom. One frequent effect of ultraviolet light is the formation of a thymine dimer, where a thymine molecule bonds to an adjoining thymine in the same strand of the DNA molecule (figure 18.19). Thymine dimers do not pair properly during DNA replication. As a result, formation of a thymine dimer can cause a permanent change in DNA that will be passed on through subsequent replications.

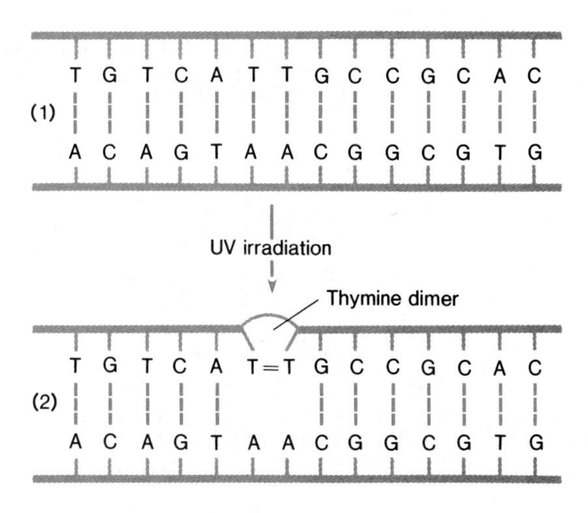

Figure 18.19 Formation of a thymine dimer (covalent bonding between adjacent thymine molecules within one strand of a DNA molecule) caused by ultraviolet light. Once a thymine dimer has formed, the thymines no longer function normally during DNA replication.

If thymine dimers do occur, most cells can repair the damage by breaking the dimer with a specific enzyme. Humans that have the genetic disease *xeroderma pigmentosum* are very sensitive to ultraviolet light, have a high degree of freckling, and often suffer from skin cancer. This is because cells in such individuals are unable to repair thymine dimers, apparently because they lack the functional repair enzymes. Thus, the ultraviolet light in ordinary sunlight can cause severe damage to their skin and lead to development of numerous skin tumors (figure 18.20).

The other important type of mutagenic radiation is ionizing radiation, such as X rays, gamma rays, and high energy particles, such as protons and electrons. This type of radiation is called ionizing radiation because when it hits a molecule directly, electrons are dislodged completely. Such a hit can result in a broken covalent bond, and an ion pair can form at the break point. The ions at a break point are chemically reactive. If it is a base in DNA that is ionized, a point mutation may result. If an ion pair is produced in the sugar-phosphate backbone of the double helix, a break in the helix may occur, resulting in a broken chromosome.

Figure 18.20 *Xeroderma pigmentosum,* an inherited disease in which cells are less efficient at repairing damage to DNA caused by exposure to ultraviolet light. The basis for the deficiency seems to be less efficient functioning of repair enzymes. This boy suffers from the extensive skin tumors that characteristically develop in the disease.

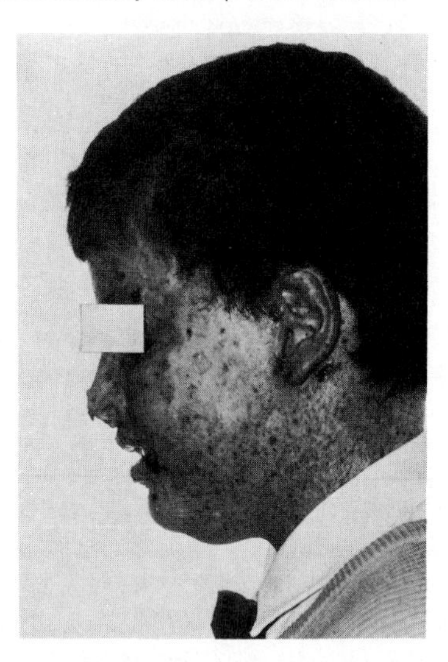

All living things have been subjected to naturally occurring physical mutagens in the environment throughout the history of life on earth. But modern use of radioactive materials in industrial, medical, and energy conversion processes has significantly increased human exposure to ionizing radiation. It is not yet known how serious the long-term effects of this increased exposure will be. But increased exposure to ionizing radiation together with increased exposure to many potentially mutagenic chemicals does increase the frequency of alterations in the heritable set of chemical instructions for life known as the genetic code.

Summary

At the molecular level, genetic information is contained within DNA molecules or, in the case of some viruses, an RNA molecule. DNA is double stranded with complementary base pairing, and it goes through a semiconservative form of replication.

Structural genes code for the formation of polypeptides. The genetic code is transmitted to the cytoplasm of the cell by mRNA. Ribosomes read the code, and tRNA molecules bring the appropriate amino acid to be attached along the developing polypeptide chain. The genetic code itself is determined by the sequence of nitrogenous bases in the DNA. It is a nonoverlapping, triplet code.

Regulatory genes regulate the activity of structural genes.

The use of enzymes called restriction endonucleases has led to the development of a technology for transferring genes from one organism to another. Pieces of DNA can be removed from one organism and spliced into the DNA of another. This is called the recombinant DNA technique.

A number of chemical and physical factors can produce changes, called point mutations, in the base sequence of DNA. A point mutation can cause a change in the code because of insertion of a single different amino acid in the polypeptide, a change that results in production of a completely abnormal polypeptide chain, or a chain that is terminated far short of completion.

Most cells have repair mechanisms that correct some of the DNA alterations caused by certain mutagenic agents.

Exposure to mutagenic agents has increased with development of modern industrialized societies.

Questions for Review

1. Explain how Griffith's transformation experiments and Hershey and Chase's bacteriophage experiments provided evidence that DNA is the primary genetic material.
2. What is DNA polymerase?
3. Distinguish between the terms transcription and translation.
4. Why is it essential that there be a different kind of tRNA molecule for each kind of amino acid?
5. Define the term anticodon.
6. What are structural genes? Regulatory genes?
7. Take the following hypothetical stretch of DNA and diagram the base sequence in the mRNA transcribed from it (use table 18.1) and the anticodon sequences of the tRNAs involved in translation. Then refer to the mRNA genetic code in figure 18.14 to determine what amino acids this DNA would ultimately code for.

   G C G G T G C A C T T T

8. The ends of DNA molecules cleaved by restriction endonucleases are sometimes described as being "sticky." How do you interpret that description?
9. How are plasmids used to insert genetic material into bacterial cells?

Questions for Analysis and Discussion

1. Why did the structure of DNA have to be known before it was finally accepted as the hereditary material?
2. Why must gene activity be regulated? (Remember that all cells within an organism contain the same genetic information.)
3. Why is it possible that certain point mutations do not actually result in the insertion of a different amino acid in the polypeptide produced when the gene is expressed?

Suggested Readings

Books

Goodenough, U. 1984. *Genetics*. 3d ed. New York: Holt, Rinehart & Winston.

Lewin, B. 1983. *Genes*. New York: Wiley.

Suzuki, D. T.; Griffiths, A. J. F.; and Lewontin, R. C. 1981. *An introduction to genetic analysis*. 2d ed. San Francisco: W. H. Freeman.

Watson, J. D. 1968. *The double helix*. Norton Critical Reader, Stent, G. S. (ed.), 1980. New York: W. W. Norton.

Articles

Chambon, P. May 1981. Split genes. *Scientific American* (offprint 1496).

Chilton, M. D. June 1983. A vector for introducing new genes into plants. *Scientific American*.

Denhart, D. T. 1983. Replication of DNA. *Carolina Biology Readers* no. 120. Burlington, N.C.: Carolina Biological Supply Co.

Fox, E. F. October 1981. McClintock's maize. *Science 81*.

Gilbert, W., and Villa-Komaroff, L. April 1980. Useful proteins from recombinant bacteria. *Scientific American* (offprint 1466).

Heylin, M., et al. 1984. Genetic engineering report. *Chemical and Engineering News* 62(33):1.

Hopwood, D. A. September 1981. The genetic programming of industrial microorganisms. *Scientific American*.

Lake, J. A. August 1981. The ribosome. *Scientific American* (offprint 1501).

Novick, R. P. December 1980. Plasmids. *Scientific American* (offprint 1486).

Travers, A. A. 1978. Transcription of DNA. *Carolina Biology Readers* no. 75. Burlington, N.C.: Carolina Biological Supply Co.

Reproduction and Development

Chapter Outline

Chapter Concepts

1. All living things arise from other living things by reproductive processes.
2. Asexual reproduction permits production of new individuals from only one parent. It does not, however, produce new genetic combinations among offspring, as does sexual reproduction.
3. Gametes are highly specialized, haploid reproductive cells that fuse in fertilization to produce diploid zygotes.
4. Development of zygotes involves repeated cell divisions (cleavage), specific spatial organization of multicellular bodies (gastrulation), and structural and functional cell differentiations.
5. Meiosis occurs during gamete formation in animals, but in plants, meiosis occurs during spore formation.
6. Cell differentiation (specialization) results from different uses of a common store of genetic information by various cells in different parts of the body.

Which came first, the chicken or the egg? It should not surprise you to hear that biologists have, on occasion, turned this old riddle into a biological argument. But biologists do not really come much closer to a final solution to this question than children arguing in a schoolyard after a playmate poses the riddle.

Everyone, however, agrees on one fundamental fact. You cannot have chickens without eggs and vice versa. All living things arise from other living things by reproductive processes. In fact, the most distinctive general characteristic of all living things is the capacity for **reproduction.** This ability to produce offspring is vital to the continuation of species. In each species, new, vigorous, young organisms replace the aging and dying organisms, and the cycle of life continues.

We are most familiar with reproduction in humans and common domesticated animals, which have a rather simple kind of life cycle. Mature adults produce specialized reproductive cells, called **gametes.** These gametes—relatively large **eggs** in females and small **sperm** in males—join to produce **zygotes,** each of which is a single cell that begins the life of a new individual organism. The process of gamete fusion to produce a diploid zygote is called **fertilization.**

But there are many other types of reproductive patterns. Many of these diverse reproductive strategies involve one form or another of **asexual reproduction,** the production of new individual organisms without the fusion of gametes.

Asexual Reproduction

The most common and direct form of asexual reproduction is mitotic cell division (see page 324). This is how unicellular organisms multiply. One individual simply divides to produce two new individuals. Genetically, of course, each of these new individuals is identical to the original individual (cell).

Some multicellular organisms reproduce by fragmenting or subdividing their bodies. New individuals grow from multicellular pieces of the original individual (figure 19.1). Once again, in genetic terms, cells of the new individuals produced by these fragmentation processes are genetically identical to the cells of the original parent individual.

Vegetative reproduction in plants is another kind of asexual reproduction. A new individual starts when some plant part, which is not a specialized reproductive structure, simply begins to grow into a new plant. The rooting response of strawberry plant runners in a spreading strawberry bed is

Figure 19.1 Asexual reproduction by fragmentation or direct outgrowth from the body. (*a*) Buds on the surface of a sponge. Each bud can eventually become an independent new individual. (*b*) Gemmae cups on the liverwort *Marchantia*. Multicellular bodies (gemmae) produced in these cups can develop directly into new plants when they are washed out on the ground.

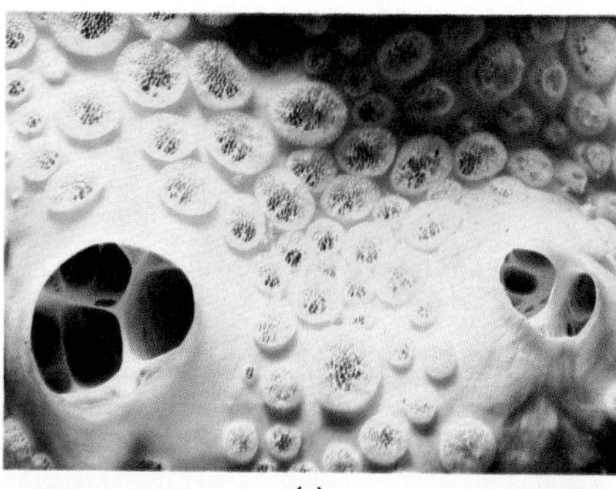

(a)

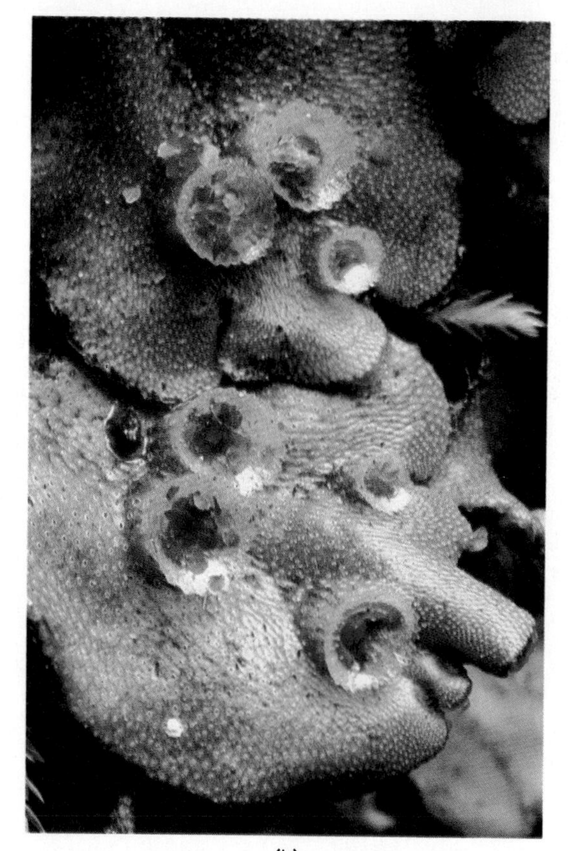

(b)

Figure 19.2 Rooting by strawberry plant runners produces a new plant wherever adventitious roots are put down.

a familiar example of the proliferation of individuals by vegetative reproduction (figure 19.2). Vegetative reproduction is economically important because an individual plant with especially desirable characteristics can be easily propagated. Small pieces (cuttings), especially pieces of stem, are treated so that they produce roots and develop into complete new individuals. Often, a group of plants can be grown from cuttings of one parent plant. Such groups of organisms are called **clones.** Members of a clone are derived directly from a single parent organism and are genetically identical with it and with one another.

Asexual reproduction also has some economic applications in animals. For example, sponge producers chop sponges into small pieces and place them where each piece can grow into a new sponge mass. At one time, oyster fishermen chopped starfish to pieces to keep the starfish from preying on oysters. But they stopped doing so when they found that chopping a starfish to pieces does not kill it at all. Instead, each of the fragments grows into a complete starfish, thereby increasing the total number of starfish!

Sexual Reproduction in Animals

Despite the many forms of asexual reproduction, **sexual reproduction,** which requires the presence of two parents, is the dominant reproductive pattern in multicellular organisms.

Reproductive Cycles in Animals

Most animals reproduce during only a relatively short time each year. During that period, the majority of individuals in a species come into breeding condition simultaneously. This increases the chances that fertilization will occur and that offspring will encounter favorable conditions for their development and growth (figure 19.3).

Many animals produce eggs that contain relatively small quantities of yolk or other stored nutrient material. Such eggs usually develop quickly into free-living individuals that can begin feeding for themselves. In many cases, this free-living form is a **larva,** an individual that is structurally different from the adults of the species, but capable of independent existence. For example, a tadpole is the larval stage of a frog.

Figure 19.3 Annual reproductive cycles are timed so that members of animal populations come into breeding condition simultaneously at times that are favorable for reproductive success. (*a*) Amplexus, simultaneous gamete shedding by female and male frogs. Frogs breed when conditions are favorable for larval feeding and growth. (*b*) Female white-tailed deer with her young fawn. Offspring of many wild mammals are born in favorable seasons after an extended period of internal development (gestation).

(a)

(b)

Figure 19.4 A typical animal life history.

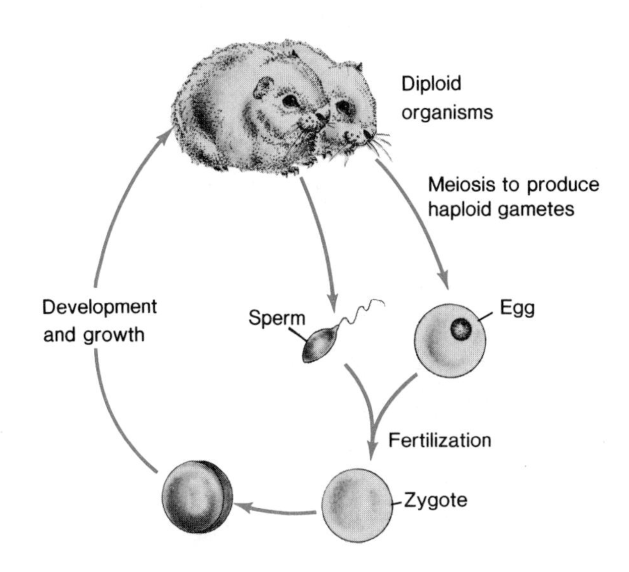

Following a period of independent life, a marked structural modification known as **metamorphosis** converts the larva into an adult. Reproductive cycles in such larva-producing animals usually are timed so that breeding occurs when food will be immediately available for the larvae.

In mammals, developing offspring are retained inside the female's body for a relatively long period of time. Breeding in wild mammals is timed so that the end of the period of internal development (**gestation**) comes when conditions are favorable for the appearance of the newborn animal (figure 19.3*b*).

Gamete Formation

Animal body cells normally have a diploid (2N) chromosome number. Each species of animal has a characteristic diploid chromosome number that remains the same from generation to generation. Because fertilization involves a fusion of two cells and brings together two sets of chromosomes, there must be a mechanism that reduces the chromosome number or it would double every generation. As we saw in chapter 15, meiosis reduces the chromosome number. In animals, meiosis occurs during gamete formation. Thus, gametes are haploid (N), and gamete fusion during fertilization establishes the diploid chromosome number in the zygote (figure 19.4).

Most animals have specialized reproductive organs, called **gonads,** within which meiosis takes place as part of **gametogenesis,** the structural and functional development of gametes. Sperm develop in the **testes** (singular: **testis**) of males. Sperm are relatively small cells that are specialized for motility (movement). Eggs develop in the **ovaries** of females. Eggs are relatively large, nonmotile cells that, in many species, contain stored materials that are used during the early development of the zygote.

Spermatogenesis

A population of specialized cells called **spermatogonia** (singular: **spermatogonium**) are the ancestors of sperm cells. In sperm formation (**spermatogenesis**), spermatogonia grow to become **primary spermatocytes.** These larger cells enter meiosis. The cells produced by the first of the two meiotic divisions are known as **secondary spermatocytes.** Each of these divides again in the second meiotic division. At the completion of meiosis, then, there are four haploid cells, called **spermatids,** for each spermatogonium that began the process (figure 19.5*a*). Spermatids develop into functional sperm cells.

Conversion of a spermatid into a sperm cell involves several important changes in the cell (figure 19.5*b*). The nuclear genetic material (chromatin) of the spermatid becomes condensed in the rounded **head** of the developing sperm. Much of the cytoplasm of the spermatid is lost as the sperm assumes a streamlined shape, an adaptation for swimming.

Mitochondria become tightly packed in the **middle piece** of the sperm cell. These mitochondria provide energy for the beating **tail** that propels the sperm when it swims. Energy for swimming is limited, however, because the metabolic reserves of sperm cells are small. The active, functional life of animal sperm cells is quite short.

Each sperm cell has two centrioles (figure 19.5*c*). One is located near the chromatin in the sperm head, and the other lies at the base of the tail. This second centriole is the basal body for the sperm tail. In this way, tails of sperm cells resemble the flagella and cilia of other motile cells.

Finally, a caplike covering, the **acrosome,** develops over the front tip of the sperm's head. The acrosome functions in the fertilization process.

Oogenesis

Meiosis in egg formation (**oogenesis**) is somewhat different from meiosis in spermatogenesis. The cytoplasmic divisions during oogenesis are very unequal. At each division of an **oocyte** (developing egg), virtually all of the cytoplasm goes to

Figure 19.5 Spermatogenesis. (*a*) A spermatogonium grows to become a primary spermatocyte. Meiotic divisions produce four spermatids, which become sperm cells. (*b*) Stages in the structural modification of a spermatid to produce a functional sperm cell. (*c*) An animal sperm with details discovered with the electron microscope. The sperm tail has the same general organization found in flagella of all kinds (see page 69).

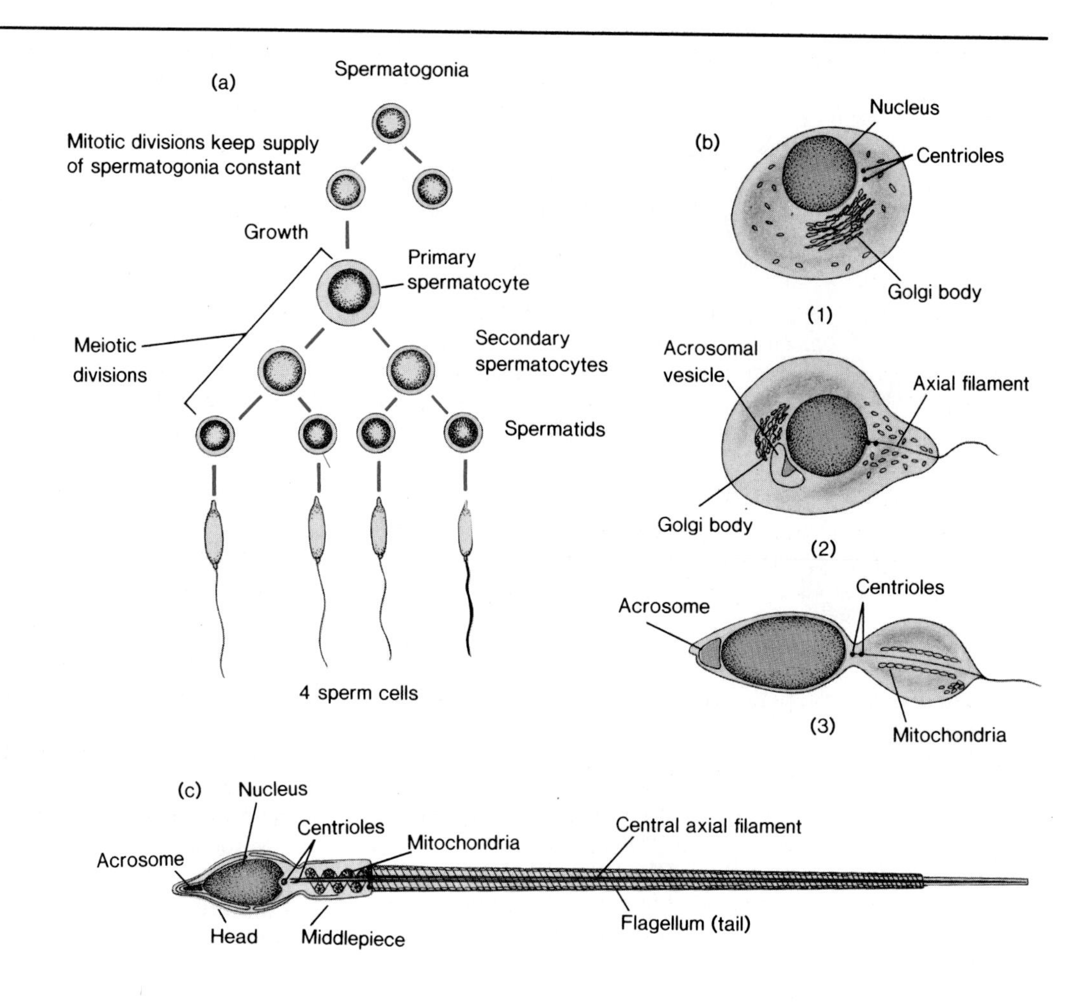

Figure 19.6 Meiosis in oogenesis with unequal cytoplasmic division that produces one egg cell and several polar bodies.

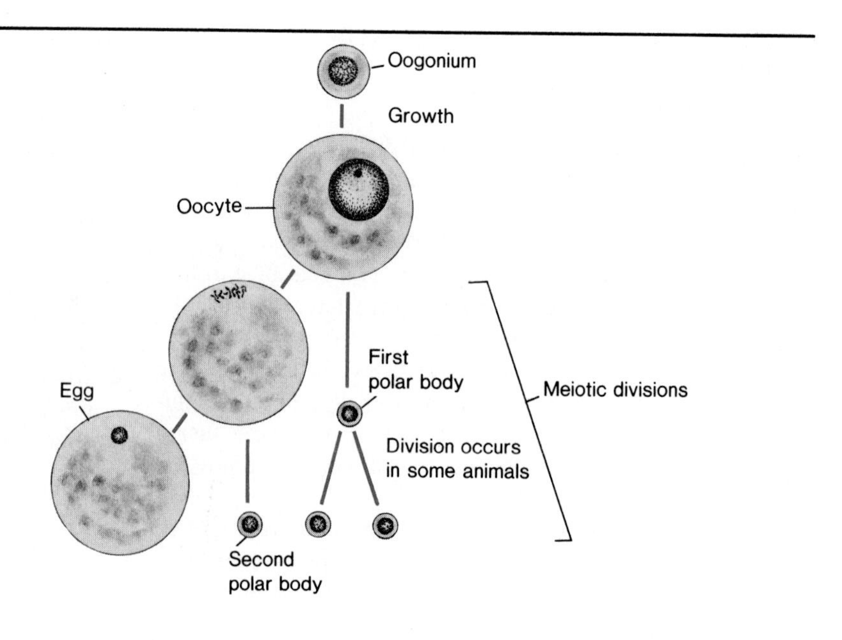

one of the daughter cells, while the other daughter cell receives a nucleus and only a minimal amount of cytoplasm. These tiny daughter cells are called **polar bodies** and are nonfunctional by-products of meiosis (figure 19.6). As a result of these unequal divisions, only one functional egg cell is produced from each **oogonium** (egg-producing cell). Cytoplasmic materials used during early development of the zygote are thus concentrated in the future egg cell.

Spermatogenesis in the majority of animals is direct and may require only a matter of days or weeks. However, the meiotic divisions in oogenesis may include pauses over long portions of a female's lifetime. For example, meiotic divisions of prospective egg cells begin in the ovaries of various young female vertebrate animals even before they are hatched (or born). Meiosis begins, but then is arrested in the prophase of the first meiotic division until the animal reaches maturity. Meiosis is resumed only at the time that eggs are being prepared for release from the ovary (**ovulation**). In human females the meiotic pause in some cells can last forty years or more.

Sexual Reproduction in Echinoderms

To illustrate some additional principles of sexual reproduction in animals, let us focus our discussion on the reproduction and development of one group of animals, the echinoderms. We will emphasize sea urchins and sand dollars.

In many animals (for example, humans) there is definite **sexual dimorphism;** that is, males and females are recognizably different in appearance and, in many species, in size as well. But male and female echinoderms look alike. The only effective way to determine sex in sea urchins is to examine the gametes that each animal sheds (figure 19.7).

Sea urchins tend to live in clusters. They each release large numbers of gametes into the seawater around them, but still, it is a big ocean! Thus, as with so many other animals, most species of sea urchins have definite reproductive cycles. Individuals of a species in any given area generally become reproductively ready ("ripe") and shed gametes at the same time of year, thus increasing the chances of fertilization.

Fertilization

We often define fertilization only in genetic terms because we emphasize the fusion of the haploid sets of chromosomes of the gametes to produce the diploid zygote. But fertilization is much more than just the coming together of chromosomes. At the time of fertilization, the egg is a complex developmental system, poised and waiting for fertilization to stimulate it to develop further. Sperm entry **activates** the egg and sets off a host of physical and biochemical changes in the egg cell.

Figure 19.7 Adult sea urchins. This is the purple urchin, *Stronglyocentrotus purpuratus*. Male and female sea urchins look alike.

Figure 19.8 Scanning electron micrographs of a sea urchin egg and sea urchin sperm during fertilization. (*a*) Sperm attached to membranes on the surface of an egg (magnification × 1,920). (*b,c,d*) Stages of sperm penetration through membranes and into the egg. The fertilization membrane was removed before micrographs (*c*) and (*d*) were made (magnification × 10,000, 4,355, and 6,750, respectively).

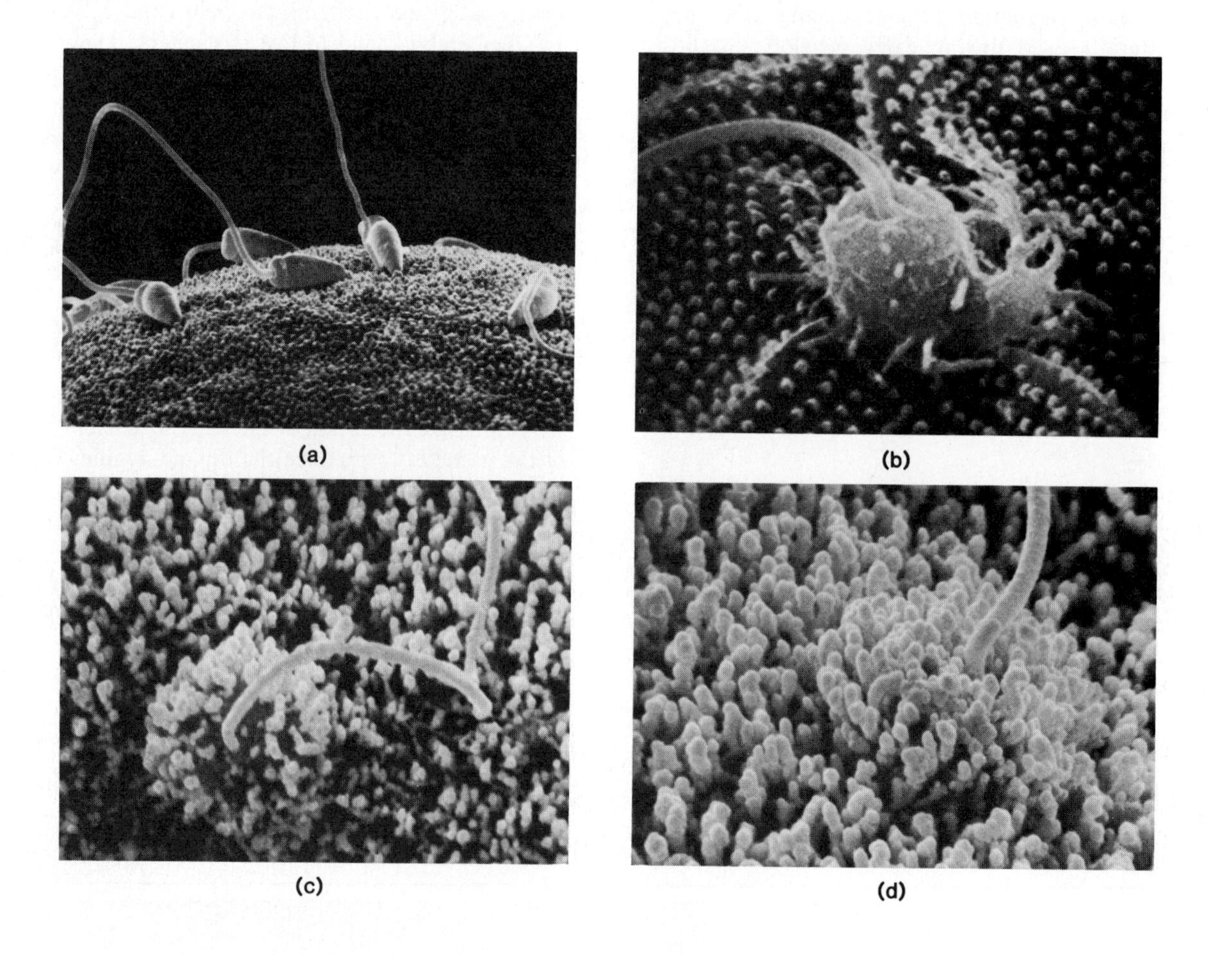

Complex interactions begin when a sea urchin sperm succeeds in penetrating the coats around the egg and contacts the egg's plasma membrane. Contact sets off an elaborate membrane fusion and reorganization process that permits the sperm to enter the egg (figure 19.8).

But this is true of only the first sperm to contact the egg. After the initial contact, the egg surface quickly changes and is no longer responsive to additional sperm cells. This apparently prevents **polyspermy,** the entry of more than one sperm into a single egg, which can cause abnormal or arrested development. A tough protective envelope, the **fertilization membrane** (figure 19.9), forms within a minute or so following the initial sperm contact. Fertilization membrane development begins where the first sperm makes contact, and spreads peripherally until it covers the egg's entire surface. The fully developed fertilization membrane is a virtually impenetrable barrier to sperm cells and a permanent block to polyspermy.

Following sperm entry into the egg, the **pronuclei** (haploid nuclei of egg and sperm) migrate toward the center of the egg cell. There they join on the mitotic spindle, which has been assembled there in preparation for the first of the repeated cell divisions of the zygote.

Figure 19.9 Development of the fertilization membrane. (*a*) A sand dollar egg before fertilization. The colored spots around the egg are pigment granules in the jelly that surrounds the egg when it is shed by the female. (*b*) An egg just after sperm entry. Note the fertilization membrane which looks like a halo around the egg.

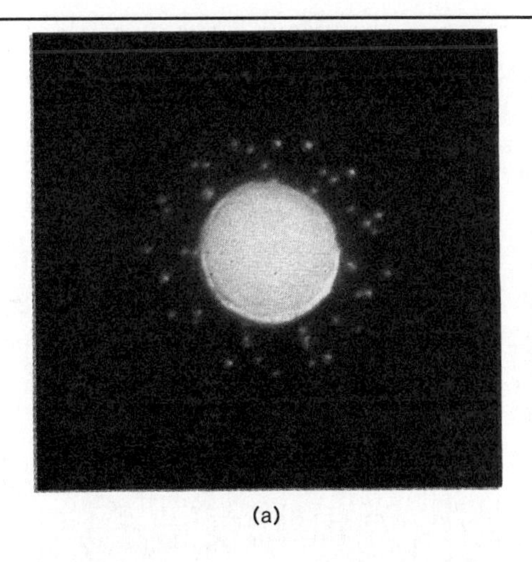

(a)

(b)

Figure 19.10 Cleavage of sand dollar embryos. (*a*) First cleavage. (*b*) Second cleavage. (*c*) Early blastula stage. Cells are becoming arranged as a hollow sphere surrounding the fluid-filled blastocoel.

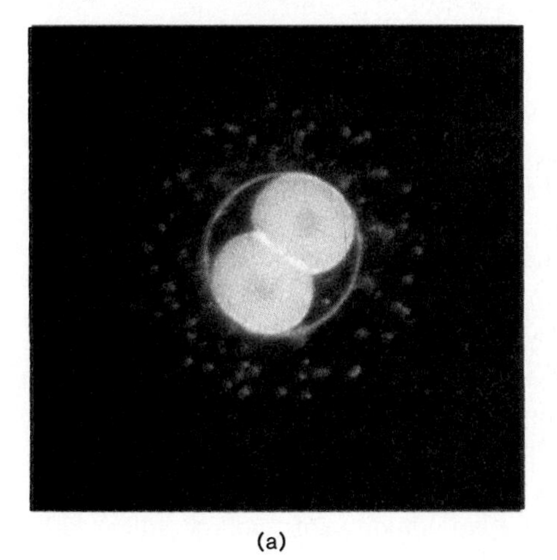

(a)

(b)

(c)

Postfertilization Development

The echinoderm zygote (like the zygotes of all other organisms) carries a genetic blueprint specifying the characteristics of the adult organism that will develop. It also carries a developmental program, a set of genetic instructions for the step-by-step construction of the new individual. Each step must be completed successfully if the new individual is to develop normally.

The postfertilization developmental program begins with a series of mitotic cell divisions known as **cleavage.** Cleavage produces a cluster of smaller cells. Next, the general structure of the body is organized by a series of cell movements and the segregation of groups of cells. These processes are known collectively as **gastrulation.** Finally, cells become differentiated and organized into groups that will carry out specific functions. Let us now examine each of these phases of development in a little more detail, using photographs of developing sand dollar embryos.

Cleavage

Cleavage converts the unicellular zygote into a multicellular aggregate. During cleavage, each mitotic cell division produces two cells that divide again, and their daughter cells divide again, and so forth, all without growing between divisions. This cycle of repeated cell divisions continues until a spherical mass of cells known as the **morula** stage of development is formed (figure 19.10*b*).

Figure 19.11 Gastrulation in sand dollar embryos. (*a*) The beginning of gastrulation. Note the indented area on one side of the spherical embryo and the primary mesenchyme cells lying loose in the internal cavity of the embryo. (*b*) Mid-gastrula stage of sand dollar. (*c*) Pluteus larva. The pluteus is the swimming, feeding larval stage of echinoderm development. The body is supported by skeletal spicules and has a complete digestive tract with mouth and anus.

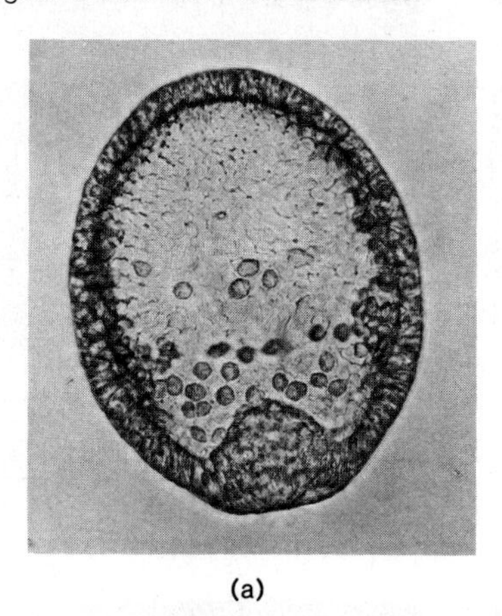

(a)

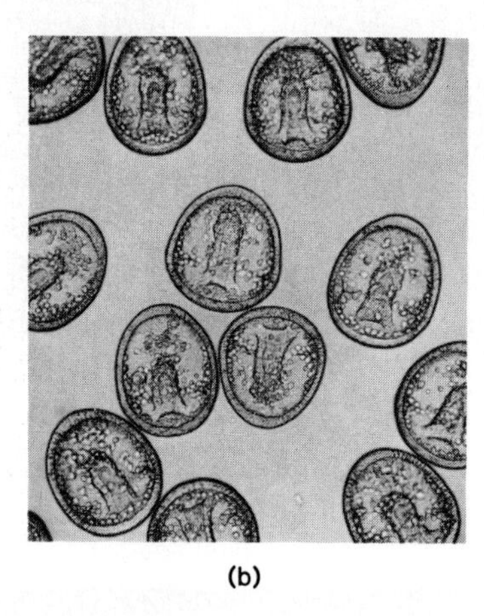

(b)

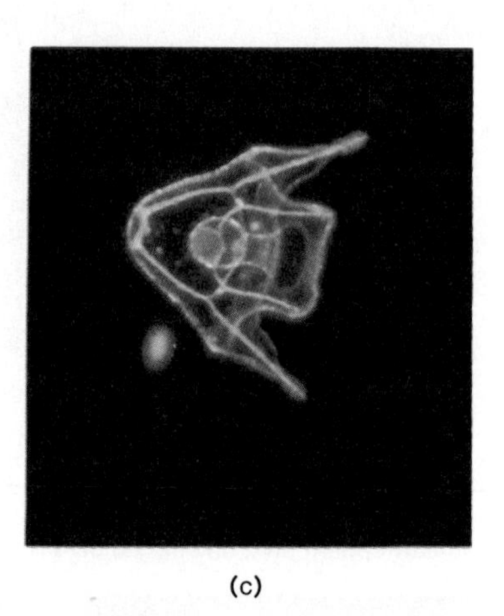

(c)

Figure 19.12 A diagram of a fully developed pluteus larva. The mouth has formed, and three basic body layers have differentiated. Skeletal spicules are produced by mesoderm (primary mesenchyme cells).

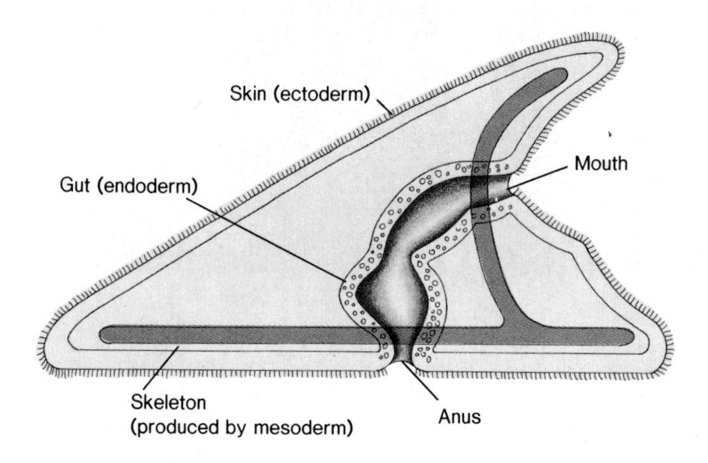

As cleavage divisions continue, the cells organize themselves into a hollow ball that surrounds a fluid-filled cavity known as the **blastocoel.** This hollow-sphere embryo, which develops at the end of cleavage, is called a **blastula** embryo (figure 19.10*c*).

Gastrulation

Gastrulation converts the simple, single-layered sphere of the blastula into a several-layered body with structurally and functionally specialized areas. These changes in the embryo's organization, which depend on cell movements and cell shape changes, are part of **morphogenesis** (meaning "development of form"). Morphogenesis is the progressive development of pattern and form of the developing embryo. Gastrulation in echinoderms begins when one side of the blastula wall sinks in to form a pit. You can visualize this process by imagining the indentation formed if you push your finger into one side of a balloon or a soft, hollow rubber ball (figure 19.11).

As this sinking process is beginning, some of the cells in the depressed area pull loose and come to lie free inside the blastocoel cavity. Soon, these cells move out to strategic locations in the embryo, where they produce the skeletal spicules that support the body of the developing larva. The cells that set off individually on this specialized developmental course are known as **primary mesenchyme cells.**

The blastula wall continues to sink in until the pit resembles the finger of a glove. This lengthening hollow structure is the developing **gut,** the future digestive tract. The gut soon contacts another point in the wall of the still hollow embryo where a breakthrough opens the gut to the outside. This new opening becomes the **mouth,** and the opening into the original pit on the side where the sinking process began becomes the **anus.** Thus, as a result of gastrulation, the hollow sphere of the blastula stage embryo has been converted into a body that has several layers and includes a complete gut, with a mouth and an anus.

Figure 19.13 Amphibian eggs. Eggs of the wood frog, showing the darkly pigmented animal pole and the light-colored, yolky vegetal pole.

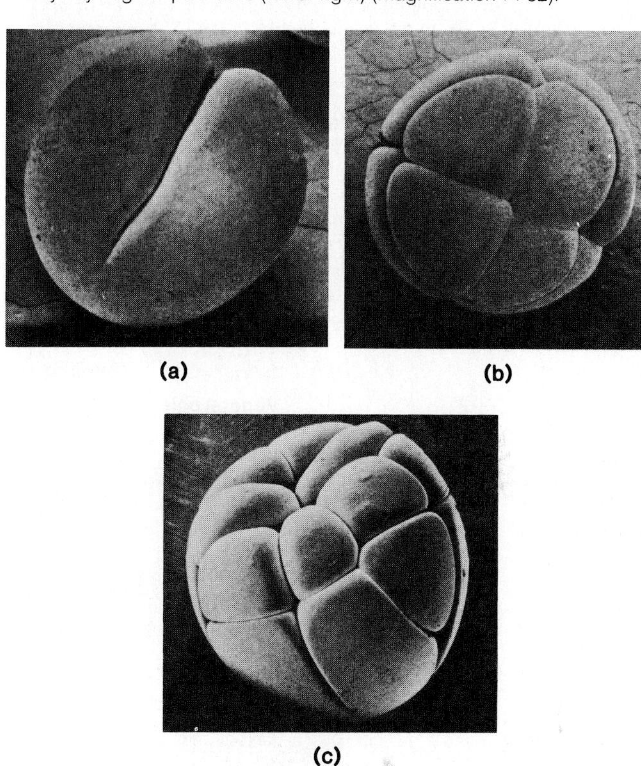

Figure 19.14 Scanning electron micrographs of frog (*Rana pipiens*) embryo cleavage stages. (*a*) The first cleavage division. The cleavage furrow cuts through the animal pole (upper right) quickly and then slows in the yolky vegetal pole (magnification × 29). (*b*) The eight-cell stage at the end of the third cleavage viewed from above. Four smaller cells lie above four larger, yolky cells (magnification × 32). (*c*) The sixteen-cell stage showing the size difference between animal pole cells (upper left) and yolky vegetal pole cells (lower right) (magnification × 32).

Body Organization

The fundamental body layers, or **germ layers,** that characterize the basic body plan of multicellular animals are fairly simply illustrated in the echinoderm embryo (figure 19.12).

Cells remaining on the surface after gastrulation make up a body surface layer known as **ectoderm.** Ectodermal cells form the "skin" of the echinoderm embryo.

Gut cells constitute the innermost body layer, the **endoderm.** Endoderm is the layer that produces the digestive system lining. The larval echinoderm gut quickly becomes functional, and the larva begins to feed on small organisms in the surrounding water.

Tissue between ectoderm and endoderm forms the middle body layer, the **mesoderm.** Mesodermal cells produce the larval skeleton. Mesodermal tissue also lines the body cavity or **coelom,** the space between internal organs and the outer body wall. Eventually, in the adult echinoderm, mesoderm produces a number of structures that make up a considerable part of the body mass.

Vertebrate Development

The same three basic body layers (germ layers) that develop in echinoderms—ectoderm, mesoderm, and endoderm—also develop in vertebrate (animals with backbones) embryos. Let us look at some of the developmental processes that are involved.

Amphibian Development

Amphibian eggs contain considerable yolk. Much of it is concentrated in one hemisphere of the egg in a yolky area known as the **vegetal pole.** The opposite side of the egg, called the **animal pole,** is darkly pigmented and has much less yolk (figure 19.13).

When an amphibian zygote divides, the cleavage furrow that cuts the cytoplasm in half moves quickly through the animal pole but is slowed by the bulky yolk in the vegetal pole. A second set of cleavage furrows divides each of the first two cells to produce four cells. At the third cleavage the division plane changes (figure 19.14). This third division yields two distinctly different sets of cells—four smaller cells lying above four larger, yolky cells. Cleavage continues, and soon the embryonic cells organize into a blastula with a blastocoel cavity (figure 19.15*a*).

Figure 19.15 Amphibian gastrulation. (*a*) A section cut through a frog blastula-stage embryo showing the location of the blastocoel, which is displaced toward the animal pole side. (*b*) A section of a frog embryo at the beginning of gastrulation. Small animal pole cells move downward to surround and enclose the yolky vegetal pole cells. Some animal pole cells leave the surface and migrate to the interior through the blastopore (colored arrow). When they reach the interior, they migrate away from the blastopore and organize the embryonic mesoderm. (*c*) A scanning electron micrograph showing the surface of a frog embryo at the time that cells begin to move into the blastopore. The first part of the blastopore to form is called the dorsal lip (DL) (magnification × 35). (*d*) Scanning electron micrograph of a later stage of frog gastrulation. As gastrulation proceeds, the blastopore becomes circular with cells migrating inward all around it. Small animal pole cells have completely enclosed the yolky vegetal pole cells except for the yolk plug (YP), made up of vegetal pole cells that lie in the center of the blastopore (magnification × 31).

(*b* and *c*) Kessel, R. G., and Shih, C. Y. *Scanning Electron Microscopy in Biology.*

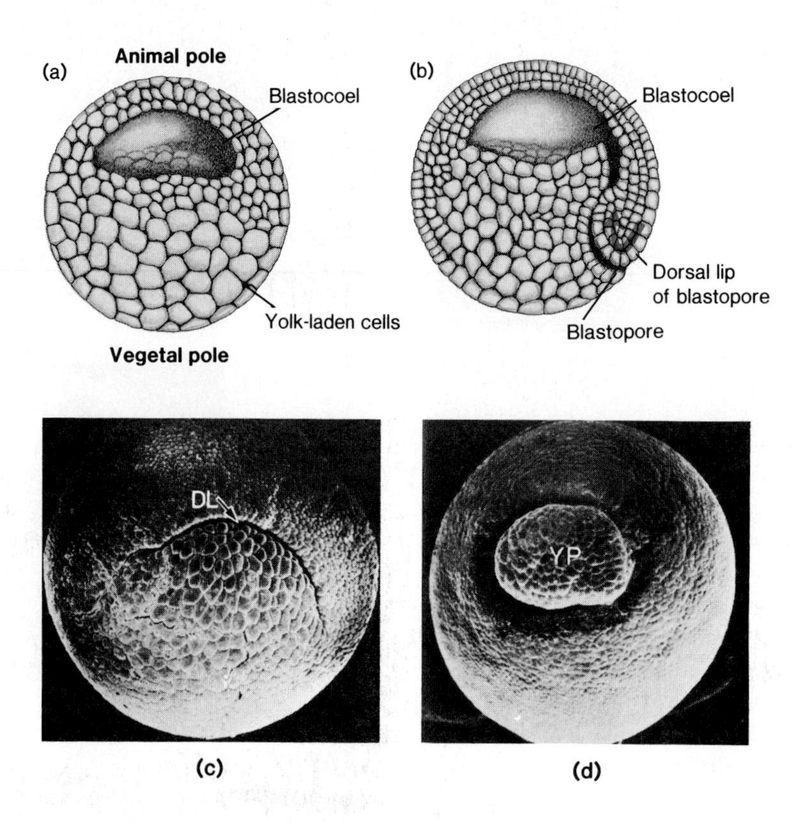

Gastrulation in the amphibian embryo involves extensive movement by all of the cells in the blastula (figure 19.15*b*). Smaller animal pole cells move downward to envelop the larger, yolky cells of the vegetal pole. Some of the animal pole cells from the surface of the embryo turn inward and move to the interior through a structure called the **blastopore.** These cells form the mesoderm of the embryo. Cells that remain at the surface produce the ectoderm. Yolky vegetal pole cells rearrange themselves to produce the gut. They become the endoderm.

Bird Development

The avian (bird) egg proper is what we call "yolk" in everyday terminology. The albumen ("egg white"), the shell, and the shell membranes are accessory structures that surround and protect the embryo.

Patterns of cleavage and gastrulation in birds are very different from those of amphibians. The large mass of yolk is not cleaved during development of the avian zygote. Cleavage is restricted to a small quantity of developmentally active cytoplasm on one side of the yolk. The cleavage divisions there produce a flat disc of cells called the **blastoderm** (figure 19.16).

The cells in the blastoderm then sort into two distinct layers—an upper **epiblast** and a lower **hypoblast.** During gastrulation, certain cells of the epiblast migrate toward and down into a shallow depression called the **primitive streak.** After moving downward into the space between the epiblast and hypoblast, many of these migrating cells move outward to produce the mesoderm of the avian embryo, located between the two previously existing layers.

Figure 19.16 Bird cleavage and gastrulation. (*a*) The avian (bird) egg. The "yolk" is the egg proper, and development is restricted to an area of developmentally active cytoplasm at one side of the egg cell. Albumen, shell membranes, and shell are accessory structures that enclose and protect the developing embryo. (*b*) Cleavage divisions (1–4) in birds produce an embryonic disc (the blastoderm) at one side of the yolky egg. (*c*) Cells of the blastoderm (1–3) sort into two separate layers, an upper epiblast and a lower hypoblast. (*d*) Some of the cells of the epiblast migrate toward, down into, and away from the primitive streak to become the mesoderm of the avian embryo.

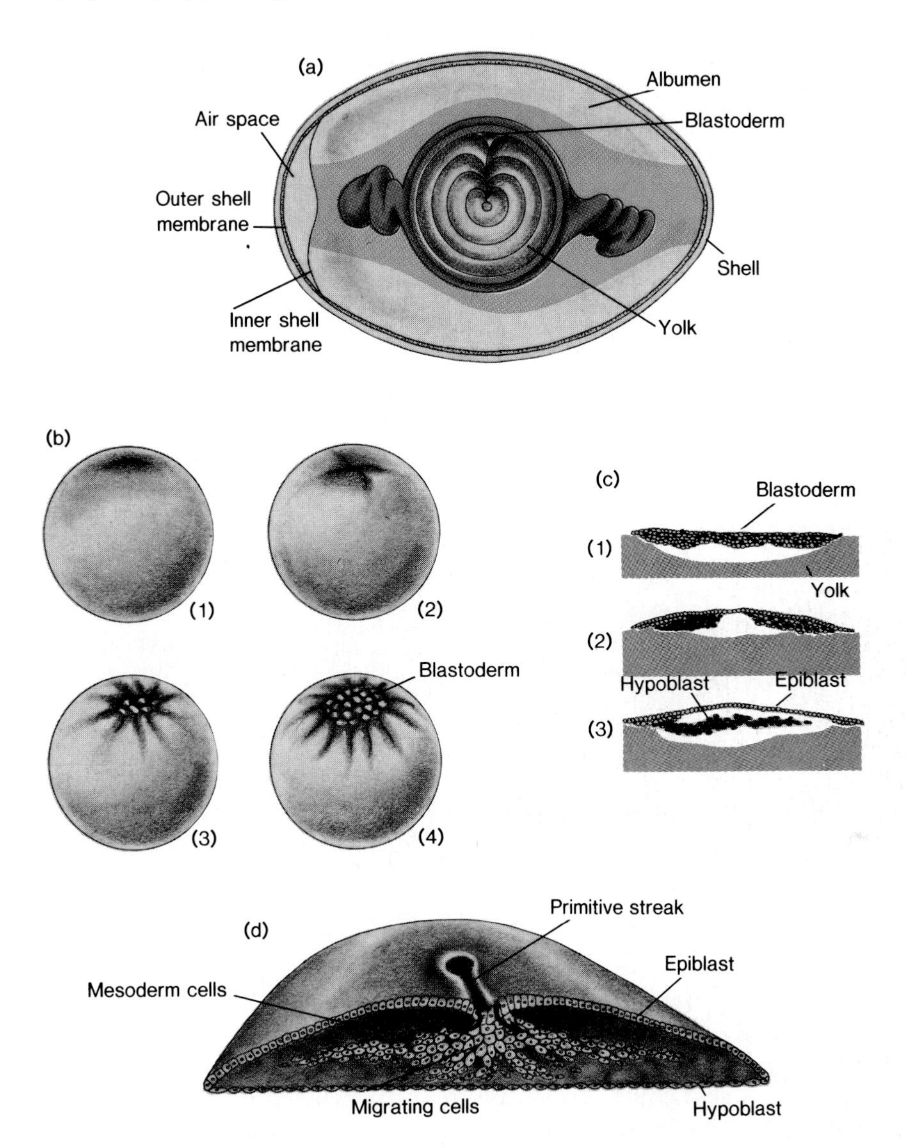

Epiblast cells that remain in the top layer constitute the ectoderm, while the endoderm of the avian embryo is made up of the original hypoblast plus some cells that come down from the primitive streak to join it.

Eventually the flat, three-layered avian embryo rolls up to produce a tubular body. This tubular body form is characteristic of all vertebrate embryos.

During all of these changes, the large yolk mass of the avian egg is not incorporated into the embryo. Rather, a structure known as the yolk sac, which is attached to the embryo, encloses the yolk and absorbs nutrients from it. These absorbed substances are transported through blood vessels of the yolk sac into the embryo, where they are used to support the embryo's growth and development.

Development of Mammals

It is not surprising that reptiles, who also have large yolky eggs, develop much as birds do. It is surprising, however, that early embryonic development in mammals also closely resembles the pattern seen in birds, despite the fact that the great majority of modern mammals produce eggs that contain very little yolk. This developmental similarity with reptiles and birds strongly suggests that mammals are descendants of animals that had much larger eggs containing much more yolk than do the eggs of present-day mammals. We will learn more about development in mammals when we examine human development in chapter 20.

Nervous System Development

Development of the nervous system is a good example of development of an organ system in a vertebrate embryo. The processes involved in establishing the nervous system are called **neurulation.** Neurulation is very similar in all vertebrate embryos.

In neurulation, the ectoderm thickens along what will be the dorsal midline of the body to form a **neural plate.** Cells in the plate change shape, causing the neural plate to roll up first into a pair of neural folds and then into a **neural tube** (figure 19.17). The entire central nervous system (brain and spinal cord) of a vertebrate embryo develops from the neural tube.

Figure 19.17 Nervous system development. (*a*) Scanning electron micrographs of frog embryos. In (1), the outlines of the neural plate can be seen as the edges of the plate begin to roll up as neural folds (magnification × 23). In (2), neural folds approach one another as neural tube formation is nearly completed (magnification × 30). (*b*) Scanning electron micrograph of part of the developing nervous system of a chick embryo showing neural folds (magnification × 120). (*c*) A series of cross sections showing the process of neural tube formation in the chick embryo.

(*a*) Kessel, R. G., and Shih, C. Y. *Scanning Electron Microscopy in Biology.* © Springer-Verlag, 1976.

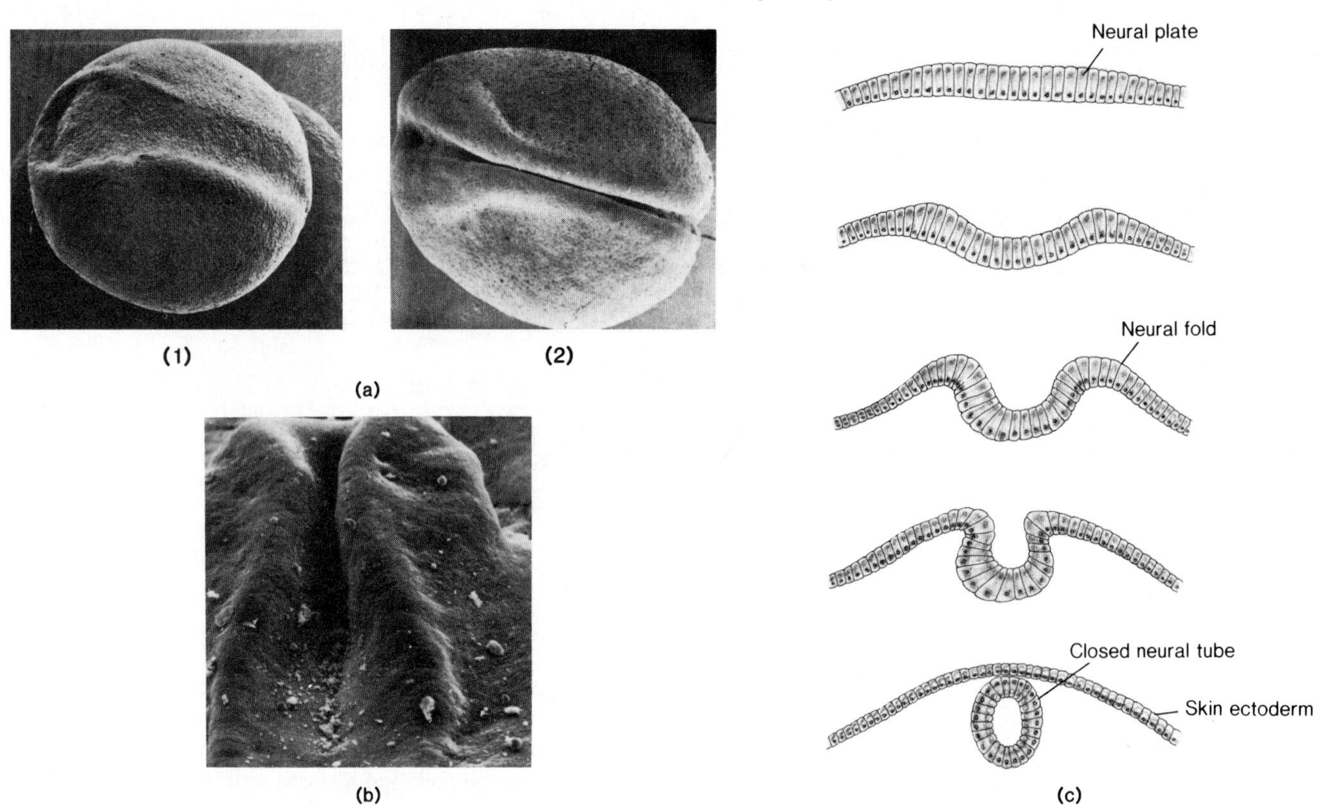

Parthenogenesis

Some animals have reproductive processes that are very different from those that we have seen thus far. For example, **parthenogenesis** is development of eggs without fertilization.

Male honeybees develop by parthenogenesis. During the single mating on her nuptial flight, the queen bee receives a lifetime supply of sperm cells. She stores these sperm and controls fertilization of the eggs that she lays. When she permits fertilization, her eggs develop into females which become workers, and the few young queens. When she withholds sperm, thus preventing fertilization, the eggs develop into males (drones) (figure 19.18). These male bees retain the haploid chromosome number of the egg cell in all of their body cells, including the testis cells that eventually produce gametes. Therefore, meiosis is not required for sperm production. Haploid testis cells can be converted directly into sperm cells.

Figure 19.18 Parthenogenesis plays an important part in the life cycle of bees. Queens and workers (also females) develop from ordinary zygotes, but males develop parthenogenetically. Sperm are produced directly from haploid cells in testes.

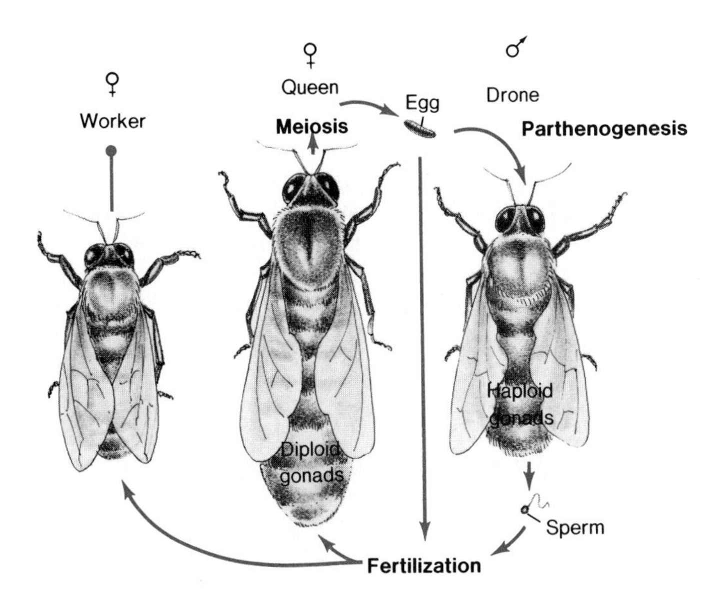

Sexual Reproduction and Plant Life Cycles

Plant life cycles are fundamentally different from the majority of animal life cycles. In plants, meiosis usually is not directly involved in gamete production. Meiosis in plants produces spores, and these haploid spores develop into haploid bodies (plants consisting entirely of haploid cells). Gametes are produced by direct differentiation of already haploid cells within specialized areas of these bodies. Then gametes fuse to produce diploid zygotes, which grow into diploid bodies (plants whose body cells are all diploid). The life cycle is completed when meiosis, occurring in specialized areas of these diploid bodies, produces haploid spores (figure 19.19).

Development of haploid spores into multicellular haploid structures means that plants have two body forms that alternate with one another. This is called **alternation of generations.** The diploid, spore-producing plant body is known as the **sporophyte generation,** and the haploid, gamete-producing plant body is the **gametophyte generation.**

Figure 19.19 Plant life cycles, illustrated here by a fern life cycle. Most plants form a multicellular haploid structure. Meiosis is not involved directly in gamete formation.

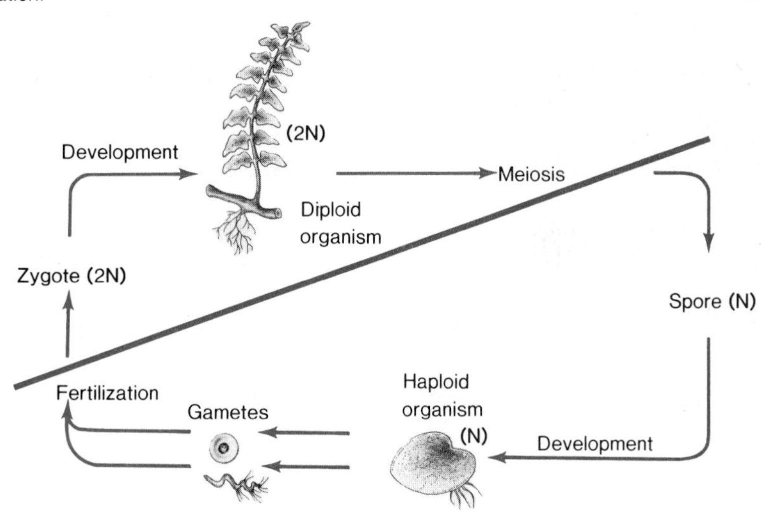

Figure 19.20 Structure of a flower.

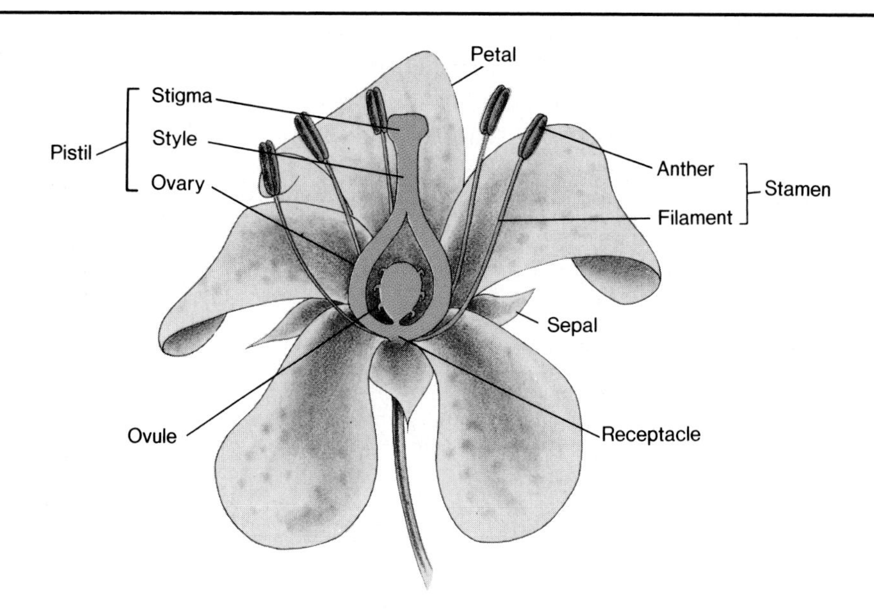

While this general life history description can apply to all multicellular plants, there are many variations on the basic plan. In chapter 28 we will see how variations in reproductive strategies and details of life histories are related to habitats and general biology of the various kinds of plants. In this section, we will focus only on the highly specialized flowering plants known as **angiosperms.**

Flowers

The majority of angiosperms are terrestrial (land-dwelling) plants, and their reproductive patterns are tied to their habitat. The delicate, unicellular spores and gametes of angiosperms are enclosed and protected inside **flowers.**

A flower is a specialized shoot with a cluster of highly modified leaves. It forms a protective envelope around the areas where cellular reproductive events take place. The modified leaves are arranged in concentric rings, or **whorls,** which are attached to a modified stem tip, the **receptacle.** The modified leaves in the lowest, outermost whorl are called **sepals,** and the next whorl inward consists of **petals** (figure 19.20). Sepals frequently are green and quite similar to ordinary leaves, while petals often are large and colorful. Sepals and petals, while not directly involved in reproductive processes, attract insect or bird pollinators in those plant species that depend on this process for reproduction.

The whorl inside the petals consists of **stamens,** and the innermost whorl is the **pistil.** Stamens are highly modified and not very leaflike in appearance. The pistil of most flowers consists of several very highly modified leaves that are fused into a single unit.

Many, but certainly not all, flowers contain both stamens and pistils. In some plants, separate flowers with stamens but no pistils (**staminate flowers**) and flowers with pistils but no stamens (**pistillate flowers**) are borne on the same plant body. In other plants, staminate and pistillate flowers are borne on entirely separate plants.

Spores and Gametophytes

Flowering plants produce two different kinds of spores. **Megaspores** develop into female gametophytes within the pistil. **Microspores** become **pollen grains.** Pollen grains actually are very small male gametophytes that are released and carried by the wind or by animal pollinators.

The enlarged base of the pistil, the **ovary,** contains from one to many **ovules** depending on the plant species. Each ovule is the site of production of a functional megaspore (figure 19.21). Actual megaspore production begins with a **megaspore mother cell** that divides meiotically to produce four haploid megaspores. Only one megaspore will eventually produce a female gametophyte structure, while the other three will degenerate without developing further.

Female gametophyte (**megagametophyte**) development also takes place inside the ovule (figure 19.21). The functional megaspore cell expands, and three successive mitotic divisions produce eight nuclei in this one expanded cell. Three nuclei are clustered at each end of the cell, and the other two nuclei, the **polar nuclei,** lie in the middle of the cell. Cell walls develop around the nuclei at the ends to form clusters of three small cells. The polar nuclei are left in the middle in a larger,

Figure 19.21 Development of female and male gametophytes, pollination, and double fertilization in an angiosperm.

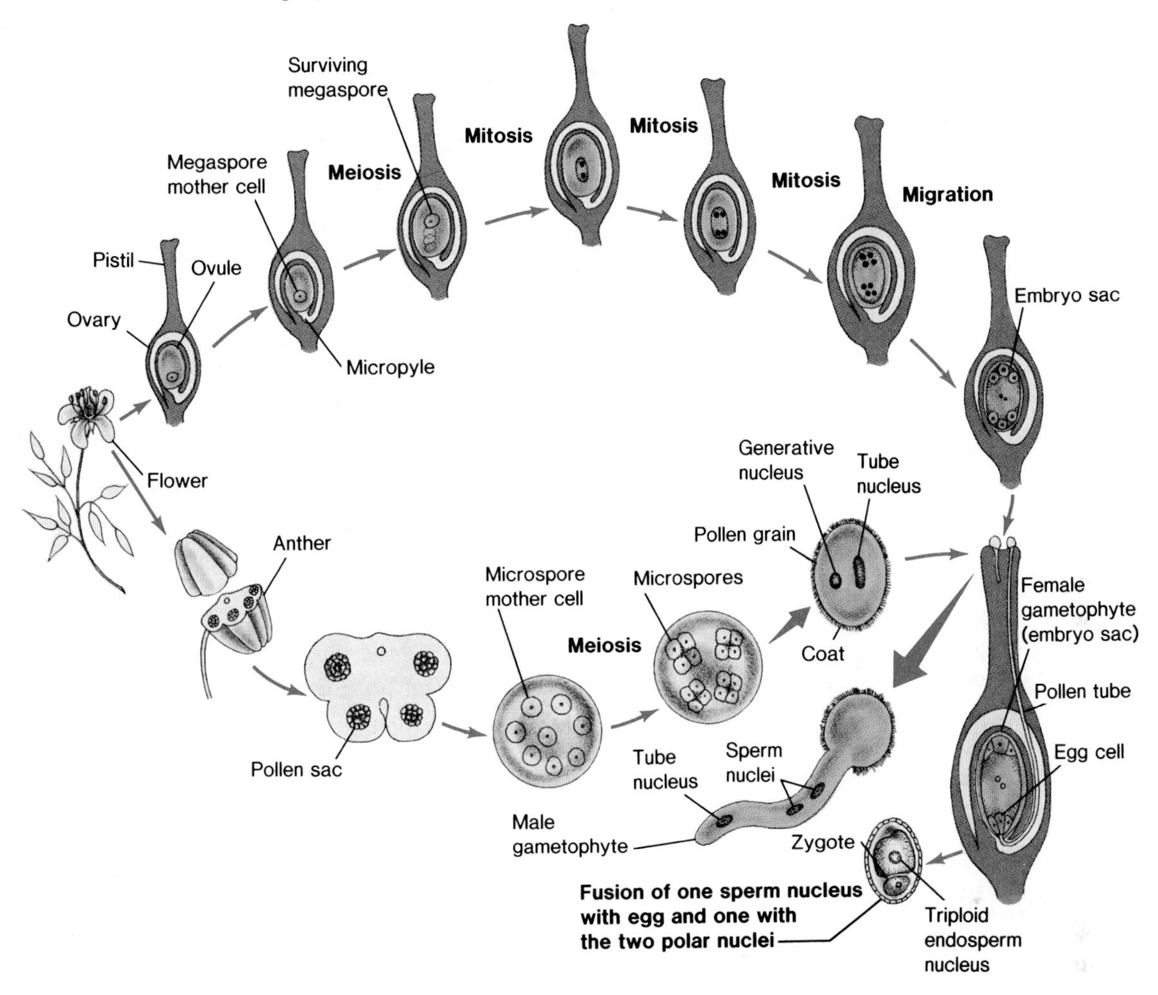

seventh cell. This seven-celled structure is called the **embryo sac,** and it is the fully developed female gametophyte. Thus, the haploid generation develops completely while still enclosed inside the ovule. Clearly, the haploid generation is a very reduced and inconspicuous part of the flowering plant's life history.

One of the cells at one end of the embryo sac becomes the functional egg cell. The other two cells at the egg-cell end and the three cells at the opposite end all degenerate without playing any obvious role in reproduction. The large cell in the middle of the embryo sac with its two polar nuclei is involved in the formation of an important accessory nutrient storage structure in the seed called the **endosperm.**

Microspores are produced in pollen sacs in the **anther,** which sits atop the **filament** at the tip of the stamen (figure 19.21). Specialized **microspore mother cells** enter meiosis, and four microspores are produced per microspore mother cell. External ornamentation develops on the cell wall as the microspore becomes a functional pollen grain. Internally, the haploid nucleus of the developing pollen grain divides mitotically to produce two nuclei, the **tube nucleus** and the **generative nucleus.** Pollen grains are released in this condition, and they develop further only if **pollination** occurs; that is, if they land on the appropriate part of the pistil. There they germinate and continue male gametophyte development.

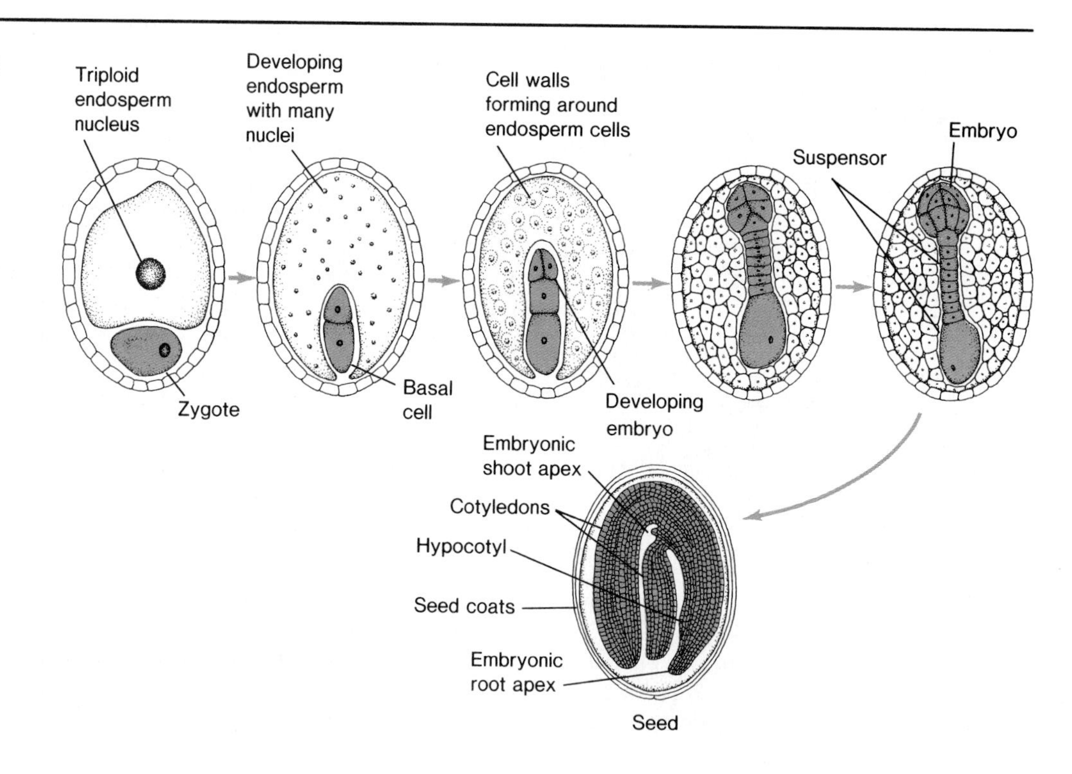

Figure 19.22 Stages in the development of a dicot embryo and seed.

Pollination and Fertilization

Pollen grains land on the **stigma,** which is the sticky upper tip of the pistil. Contact with the stigma induces pollen grains to germinate (figure 19.21). A long outgrowth from the pollen grain, the **pollen tube,** grows down through the **style** that connects the stigma with the ovary. During pollen tube growth, the two nuclei of the pollen grain enter the tube. The tube nucleus remains near the tip of the growing tube. The generative nucleus divides mitotically to produce two haploid **sperm nuclei.** The sperm nuclei are located just back from the tube tip, which enters the ovule through an opening called the **micropyle** and approaches the embryo sac.

Flowering plant fertilization actually involves two separate nuclear fusions. One of the two sperm brought into the embryo sac by pollen tube growth fuses with the egg cell to produce the diploid zygote, while the other sperm fuses with the two polar nuclei. This latter fusion brings together three haploid nuclei (the sperm nucleus and the two polar nuclei) and produces a triploid (3N) nucleus known as the primary endosperm nucleus. This primary endosperm nucleus eventually will divide repeatedly to establish a multicellular endosperm, which stores nutrients in the developing seed.

Early Zygote Development

Zygote development begins with a transverse cell division that produces two cells, a terminal cell and a larger basal cell near the micropyle. Additional divisions in the basal cell produce a linear chain of cells, the **suspensor,** which is an embryo attachment structure (figure 19.22).

Terminal cell divisions produce a flat plate of cells, the **embryo.** You can see in figure 19.22 that the embryo body consists of several major areas. A **hypocotyl** develops into the lower part of the stem and the underground parts of the plant. At the opposite end of the embryo body, seed leaves (**cotyledons**) develop. Some angiosperms (the dicots) have two cotyledons and some (the monocots) have only one. As the description "seed leaves" implies, the cotyledons are temporary structures that will be lost later. In some angiosperms, the cotyledons are relatively small. But in others the cotyledons may be very large and serve as major nutrient storage sites in the seed. You have probably seen the large, fleshy cotyledons of bean seeds when beans split open while they are being soaked or cooked. Near the base of the cotyledons is the **shoot apex.** While the shoot apex may seem a fairly inconspicuous part of the embryo, it is actually the future growing source of practically all of the permanent aboveground parts of the plant.

Figure 19.23 Seventeenth-century biologists proposed that preformed bodies existed inside sperm cells. They made sketches such as this and called the miniature body a homunculus ("little man"). Arguments even arose over the possibility that bodies of all future generations were still smaller and were inside the homunculus.

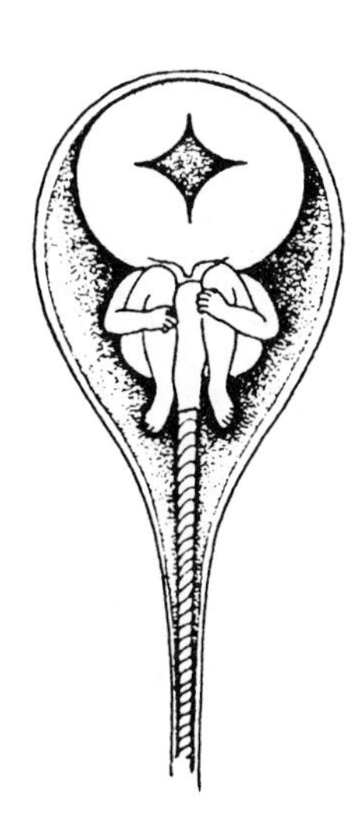

The growth centers of the stem and root are established in the embryo. **Meristematic regions** (centers of continuing cell division) are located in the shoot apex and also in the **root apex** at the tip of the hypocotyl. When the seed germinates, active proliferation in these regions lengthens the shoot and root of the growing seedling.

Seeds

The embryo, packed in with the endosperm, is surrounded by ovule tissue. Parts of the ovule harden to form tough, protective **seed coats.** Most seeds then dry out until their water content falls to very low levels (5 to 20 percent), and they maintain only a minimal maintenance level of metabolism. In many species, the remainder of the ovary develops into a **fruit** around the seed.

Embryos within seeds resume development only after seed **germination.** Germination occurs when a dry seed, in a favorable site, takes up water and swells. Then, metabolic activities of the embryo increase and rapid growth resumes. The seedling grows out of the seed and develops into a mature, diploid plant body (sporophyte).

Some kinds of seeds are **dormant;** that is, they are not physiologically responsive even when subjected to conditions that would normally cause germination. Such dormant seeds require exposure to some specific environmental stimulus (often a period of chilling) before they can germinate.

Development and Differentiation

Few natural phenomena are so impressive as the normal development of a complex, multicellular organism from a single cell, the zygote. Development has attracted interest throughout the history of biology. During the seventeenth and eighteenth centuries, biologists thought that eggs contained miniature bodies. After fertilization, these transparent, preformed bodies supposedly unfolded and began to develop. However, this **preformation** hypothesis was complicated by the discovery of sperm cells. Some biologists then began to argue that the preformed body was in the sperm rather than in the egg (figure 19.23). They proposed that the preformed body contained in the sperm could develop only when the sperm entered the hospitable environment of the egg.

Then, in 1759, Caspar Friedrich Wolff published detailed observations of chick embryo development. In his studies, Wolff had found no evidence of a preformed body. Instead, he had seen "granules" that organized into layers and then folded into a body. Wolff concluded that the body was assembled from simpler, less organized material. This concept, that there is progressive organization of a body from less organized material, was called the **theory of epigenesis.**

The Problem of Differentiation

Modern studies of development center on a very basic, but very complex question. If all of the cells in the body of a multicellular organism receive a complete and accurate copy of the zygote's nuclear genetic information, how can cells with identical genetic makeup become **differentiated** (specialized) in all the different ways needed to produce the specialized parts of the body?

Totipotency in Plant Cells

One proposed explanation for cell specialization during development was that cells differentiate because their developmental ability becomes progressively restricted, so that each can only become one kind of cell; that is, their developmental potential might be progressively narrowed until they can develop in only one way and no other. There is, however, some good evidence from studies on plants that this is not the case.

You already know that many plants can be propagated from cuttings of stems or leaves. Cuttings will eventually take root and produce whole growing plants. In plant tissue culture studies, whole plants can be grown from small clusters

Figure 19.24 Steward's experiments on totipotency of carrot cells. Embryoids develop from small clusters of cells or single cells broken off cultured phloem tissue. Some embryoids can go on to produce whole normal plants.

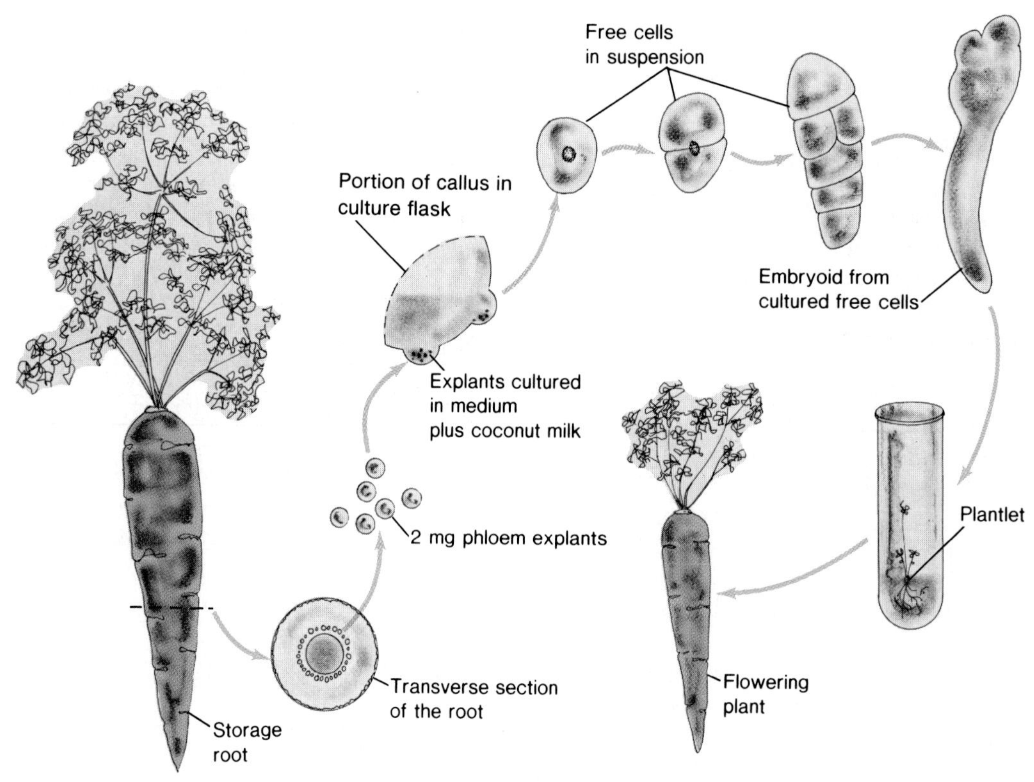

of cells broken off **calluses** (masses of undifferentiated tissue that can be started from almost any part of a vascular plant). Thus, under appropriate conditions, whole plant bodies can grow from cells originally descended from only one part of a plant.

If small clusters of cells can grow into a whole plant, could a single cell also do so? F. C. Steward and his colleagues broke up callus cultures of carrot root phloem tissue into small pieces, some of which contained a few cells and some of which probably were individual cells. When these pieces were placed in a medium containing coconut milk, some began to divide and produced cell clusters that resembled early embryos. These clusters (**embryoids**) developed into little plants that were transferred to solid cultures (figure 19.24). Later, other researchers showed that single tobacco plant cells, isolated from calluses, could divide to produce embryoids that eventually grew into completely normal tobacco plants.

Since then, embryoids have been obtained in cultures of cells from mature tissues of many species of plants. This ability of single plant cells to develop into whole plant bodies is known as **totipotency.**

As a result of these studies on totipotency, it is clear that nuclei of some differentiated plant cells retain the ability to express all of the genetic information needed to direct development of a complete body. In other words, they can develop into an entire plant just as the original zygote could. Their developmental potential has not been restricted. Totipotency of this sort has not been demonstrated for whole cells of animals, but some remarkable experiments on nuclear transplantation in animals have demonstrated that at least some nuclei of differentiated animal cells retain their full developmental potential (box 19.1).

Box 19.1
Nuclear Transplantation and Animal Clones

It has not been possible to demonstrate totipotency in animal cells as it has in plants; that is, individual cells from adult animals will not respond to treatments intended to make them divide and develop into an organism that is genetically identical to the individual from which the cell came. But there is another way to achieve a similar result, at least in some animals. This is the technique of **nuclear transplantation.**

Nuclei of animal cells can express a wide range of developmental potentials, but testing their potential requires that they be placed in appropriate new cytoplasmic environments. This was originally demonstrated by Robert Briggs and T. J. King who transplanted nuclei taken from various embryonic cells of the frog *Rana pipiens* into eggs whose nuclei had been removed (**enucleated eggs**). These combinations tested the capacity of nuclei from cells at advanced stages of development to interact with mature egg cytoplasm to produce the full range of normal developmental processes (box figure 19.1*A*). Briggs and King found that nuclei from cells of blastula-stage embryos could interact with enucleated eggs to direct normal development. Thus, even in blastula embryos, which have 8,000 to 16,000 cells, it appears that nuclei have not lost any of their developmental capacity; they still are able to direct the entire range of developmental processes—from egg to tadpole and even on through metamorphosis to adulthood. Briggs and King found in further experiments, however, that if nuclei from cells from still later stages of development were transplanted into enucleated eggs, very few normal tadpoles developed. At the time, Briggs and King concluded that as development proceeds beyond a certain point, nuclei of embryonic cells become restricted in their developmental potential.

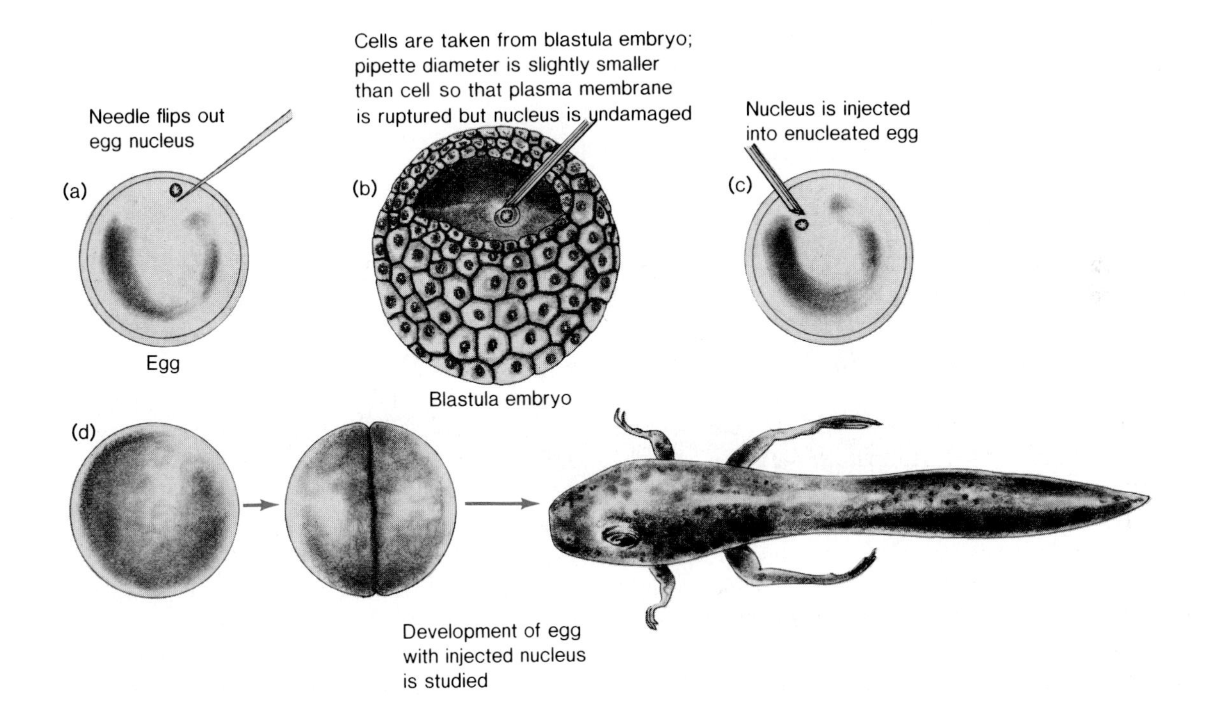

Box figure 19.1*A* Nuclear transplantation technique used by Briggs and King. Combinations of blastula nuclei with enucleated eggs developed into normal embryos in 60 to 80 percent of the cases, but they had less success with the nuclei taken from older embryos.

Box 19.1
Nuclear Transplantation and Animal Clones (continued)

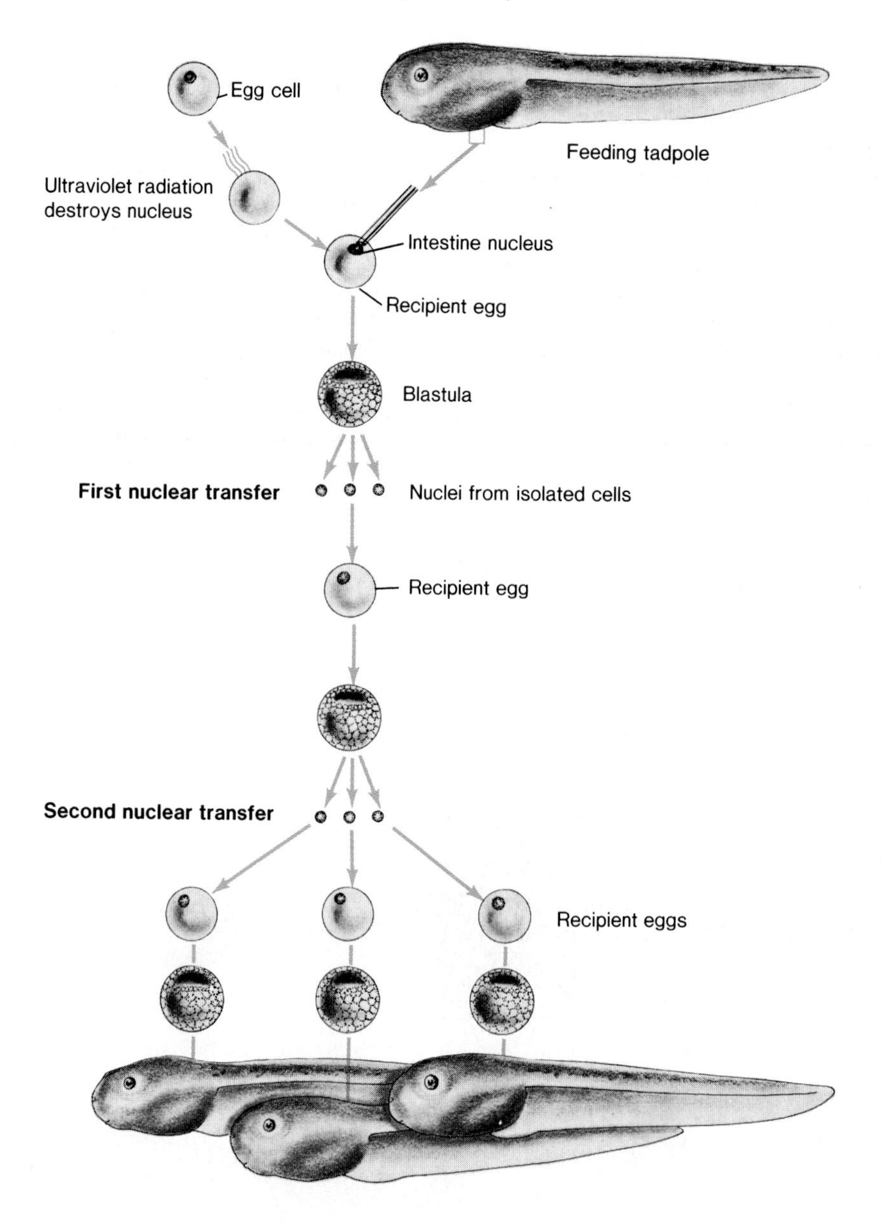

Box figure 19.1*B* Gurdon's serial nuclear transplants using *Xenopus* embryos. After a second transfer, nuclei mitotically descended from an original gut cell nucleus can interact with an enucleated egg's cytoplasm to produce normal development. Tadpoles produced are a nuclear clone because they bear identical nuclear genetic information to one another and to the individual from which the first transplanted nucleus was taken.

However, J. B. Gurdon obtained very different results when he did two successive nuclear transplantations using nuclei of the clawed frog *Xenopus*. Gurdon transplanted nuclei from advanced embryos into enucleated eggs. After these combinations reached the blastula stage, he transplanted nuclei from cells of these blastulae into other enucleated eggs (box figure 19.1*B*). This several-step process overcame a problem of basic timing incompatibility. Nuclei from slower-dividing cells of older embryos are not immediately compatible with enucleated eggs, which are geared up for the relatively rapid divisions of early cleavage. Using this serial transplantation technique, Gurdon found that even nuclei originally transplanted from differentiated gut cells of feeding tadpoles produced normal development. Thus, when tested under appropriate experimental conditions, nuclei from even differentiated cells of advanced embryos do not show signs of restricted developmental capacity. Eventually, Gurdon and his colleagues cultured adult frog skin cells and transplanted their nuclei into enucleated eggs. Even some of these combinations produced normal embryos (box figure 19.1*C*).

This work has attracted widespread interest because it seems to open the way for cloning of adult animals. Even mammals, including humans, possibly could be cloned. Experiments with mammals pose greater technical difficulties because mammalian eggs are much smaller than amphibian eggs and their nuclei are more sensitive to handling. But there appear to be theoretical barriers to nuclear transplantation and even cloning in mammals if the technical problems can be solved.

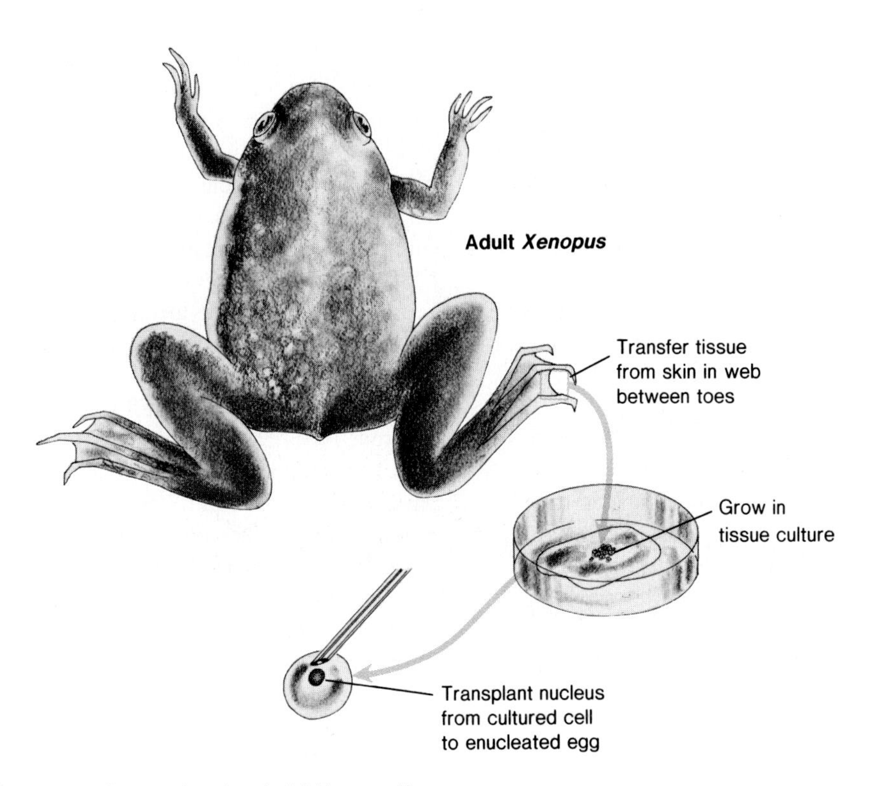

Box figure 19.1*C* Technique for transplantation of adult *Xenopus* skin cell nucleus to an enucleated egg. Cells taken from adult skin are placed in a culture dish where they divide to produce a growing population of dividing cells. Then a nucleus from one of the cells in the culture is transplanted into an enucleated egg.

Figure 19.25 Giant polytene chromosomes from a *Drosophila* larval salivary gland cell stained to show banding pattern.

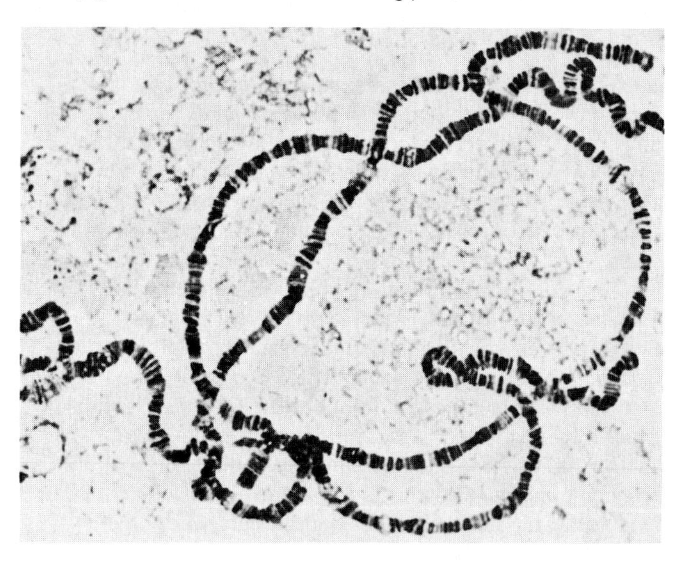

Genes and Differentiation

If differentiation is not explained by progressive restriction of developmental capacity, how then can we explain it? How do cells actually become structurally and functionally specialized for the division of labor that occurs in complex, multicellular organisms? Cell differentiation (specialization) must result from different uses of a common store of genetic information by various cells in different parts of the body.

During the process of differentiation, cells must actively express specific parts of their total store of genetic information (genome), while other parts are kept inactive (repressed); that is, some portions of the genome are used to produce one type of specialized cell (for example, a brain cell), while different portions of the genome are used to produce another type of specialized cell (for example, a kidney cell). But this does not necessarily mean that the unused parts of the genome are destroyed or lost. They simply are not expressed.

Cell differentiation, then, results from *differential gene expression.* The genetic program for development directs a series of precisely timed and positioned sets of these differential gene activations.

Chromosome Puffing

We know that differential gene expression must occur in the development of all multicellular organisms because we can see its outcome—the development of many different kinds of differentiated cells. It is possible, however, to observe evidence of differential gene expression directly in special chromosomes in the cells of certain tissues of some insects. These cells grow very large, and as they grow, their genetic material replicates repeatedly. Up to ten sequential replications produce nuclei that are 1024N (recall that normal diploid body cells are 2N). This repeated replication without mitosis produces giant **polytene** (multistrand) **chromosomes** because the replicated DNA and associated chromosomal protein of each chromosome remain together in one structural unit. Each polytene chromosome, when it is stained for microscopic examination, has characteristic patterns of light and dark bands (figure 19.25). Genetic analysis has correlated various bands with mapped locations of genes on the chromosome. Thus, specific bands are considered to be sites of genes (gene loci).

Genetic expression in polytene chromosomes involves a specific change in chromosome organization. The many strands of a particular region of a chromosome loosen up and loop out. This local expansion produces a **puff** on the chromosome. Function of the puffs can be demonstrated by supplying radioactively labeled uridine (a component of RNA, but not DNA) to cells with polytene chromosomes. Radioactivity accumulates selectively on the chromosome puffs (figure 19.26). This result implicates the puffs as sites of genetic transcription (messenger RNA synthesis).

Because their gene loci and sites of genetic transcription can be seen directly, these giant polytene chromosomes are remarkably useful model systems for studying gene activation during development. Puff patterns should be different in different types of specialized cells where polytene chromosomes are found. Some puffs indeed are found in certain cell types but not in others. Thus, the giant polytene chromosomes clearly illustrate the principle of differential gene expression during development.

Although the handy tool of giant polytene chromosomes is not available in other organisms, differentiation and its control are being studied intensively in a variety of organisms. Most biologists think that when we gain a clear understanding of cell differentiation and its control, we will not only understand normal development better, but we will also have the means to analyze and possibly prevent problems that arise when developmental processes go wrong.

Figure 19.26 Puffing in insect polytene chromosomes. (*a*) A short section of the same chromosome taken from two different cells of the fly *Trichocladius*. Puffs are regions in which the chromosome strands loosen up and loop out. The bands are numbered for reference to show that different parts of the chromosome are puffed in different cells. This indicates that different parts of the genetic material are being transcribed in each cell. This is clear evidence of differential gene expression during development. (*b*) RNA synthesis in polytene chromosomes from the salivary gland of the larva of the midge *Chironomus tentans*. Radioactively labeled uridine, indicated by the black spots, accumulates selectively on a puffed part of the chromosome indicating that puffs are sites of RNA synthesis.

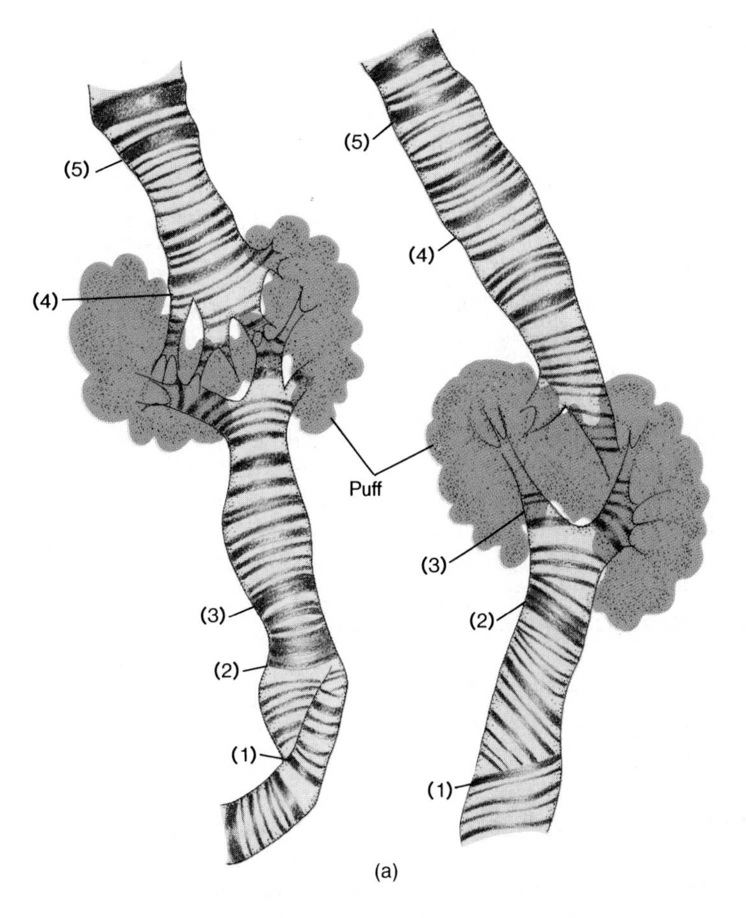

(a)

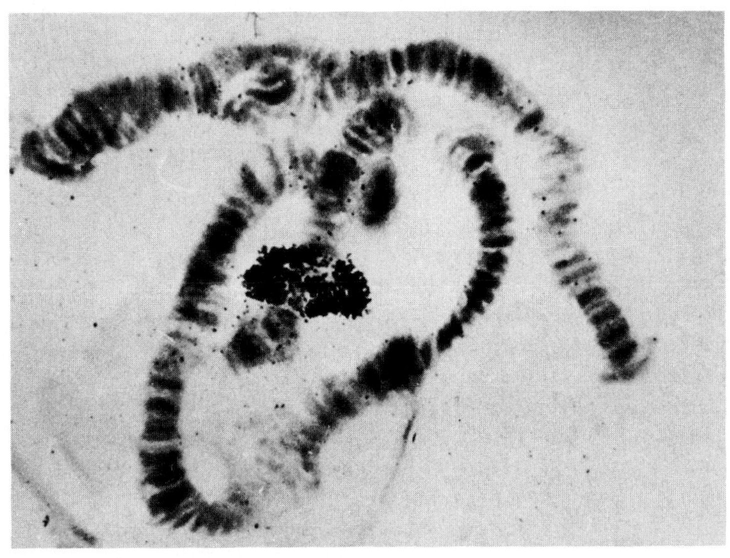

(b)

Summary

All organisms have the capacity to reproduce. Many organisms can reproduce asexually, but the majority of multicellular organisms reproduce sexually with specialized gametes fusing in fertilization to produce zygotes.

Most organisms reproduce seasonally. Their periods of reproductive activity are timed to occur when reproductive success is most likely.

Gametes are highly specialized cells that are well adapted for their reproductive function. During animal gametogenesis, meiosis reduces the chromosome number from diploid to haploid.

Animal fertilization is a complex set of interactions between egg and sperm. In most animal species, the egg becomes unresponsive to contact with additional sperm once it has been contacted by one sperm.

Zygotes, single relatively large cells, are converted into multicellular embryos by cleavage, a series of mitotic divisions without intervening growth. These cells become organized according to a basic body plan during gastrulation. Finally, cells become structurally and functionally differentiated to perform diverse, specialized functions in the body.

Some animals, such as honeybees, reproduce by parthenogenesis as well as by sexual reproduction with gamete fusion.

Plant life cycles include two separate generations—a haploid gametophyte generation and a diploid sporophyte generation. Gametes are produced by differentiation of already haploid cells of the gametophytes.

Flowers are clusters of highly modified leaves that protect reproductive structures. Meiosis occurs in the development of spores. Megaspores develop into embryo sacs (female gametophytes), and microspores develop into pollen grains. Pollen grains complete development into male gametophytes if pollination occurs.

Mitotic divisions of a plant zygote produce an embryo that is enclosed, along with the storage tissue of the endosperm, inside seed coats. Following seed maturation, embryo development pauses until germination of the seed, when growth resumes and proceeds toward development of a mature sporophyte.

All cells in a multicellular organism are descended from a single cell, the zygote, by mitosis. They all carry the same set of nuclear genetic information, but they become structurally and functionally specialized in many different ways. This differentiation depends on differential gene expression.

Questions for Review

1. Define the term clone.
2. Compare and contrast the cell divisions of meiosis in spermatogenesis with those of oogenesis.
3. What is polyspermy? How is polyspermy normally prevented?
4. How is the pattern of division without intervening cell growth related to the basic function of cleavage in animals?
5. Name the three germ layers that develop in an animal embryo.
6. Define parthenogenesis.
7. How is spermatogenesis in male bees different from the "average" situation in animals?
8. What do we mean by alternation of generations in plant life cycles?
9. What are the two separate nuclear fusions that occur in fertilization in flowering plants?
10. Contrast animal and plant life cycles generally with reference to relationships of meiosis and gamete formation.
11. Differentiating cells are characterized by differential gene activation. Explain this statement.
12. What are puffs on polytene chromosomes?

Questions for Analysis and Discussion

1. Suggest some advantages of sexual reproduction and of asexual reproduction in terms of reproductive success of individual organisms and in evolutionary terms.
2. Explain how the demonstration of totipotency in plant cells provides evidence against the hypothesis that differentiation involves progressive restriction of the developmental potential of nuclei.
3. Some biologists say that our present views of development lie somewhere between the historic ideas of preformation and epigenesis. What do you think? (Clue: Include coded genetic information in DNA in your consideration.)

Suggested Readings

Books

Browder, L. W. 1984. *Developmental biology,* 2d ed. Philadelphia: Saunders College.

Johnson, L. G., and Volpe, E. P. 1973. *Patterns and experiments in developmental biology.* Dubuque, Iowa: Wm. C. Brown Company Publishers.

Karp, G., and Berrill, N. J. 1981. *Development,* 2d ed. New York: McGraw-Hill.

Saunders, J. W., Jr. 1982. *Developmental biology* New York: Macmillan.

Stern, K. R. 1982. *Introductory plant biology,* 2d ed. Dubuque, Iowa: Wm. C. Brown Company Publishers.

Articles

Blair, J. G. January/February 1982. Test-tube gardens. *Science 82.*

Epel, D. 1980. Fertilisation. *Endeavour* 4:26.

Gurdon, J. B. 1978. Gene expression during cell differentiation. *Carolina Biology Readers* no. 25. Burlington, N.C.: Carolina Biological Supply Co.

Northcote, D. H. 1980. Differentiation in higher plants. *Carolina Biology Readers* no. 44. Burlington, N.C.: Carolina Biological Supply Co.

Shepard, J. F. May 1982. The regeneration of potato plants from leaf-cell protoplasts. *Scientific American.*

Human Reproduction

20

Chapter Outline

Chapter Concepts

1. Human reproductive systems are specialized so that delicate reproductive cells and small, fragile developing individuals are enclosed and protected within sheltered internal body environments.
2. The pituitary, the ovaries, and the uterus are all involved in the complex human female monthly reproductive cycle.
3. Intricate hormonal interactions control reproductive processes and pregnancy responses.
4. Embryonic development establishes basic body organization, and growth and maturation proceed during fetal development.
5. The placental relationship sustains the developing infant throughout its development in the uterus.
6. After birth, a human infant is physiologically independent but still weak and helpless.
7. Fertility regulation and the increasing incidence of sexually transmitted diseases are worldwide social problems associated with human sexuality and reproduction.

Using ultrasound equipment, it is possible to visualize a human fetus inside its mother's uterus so clearly that an eyeblink can be detected (figure 20.1). This ultrasound imaging, as it is called, allows quite close inspection of a fetus without harming or disturbing it. Ultrasound imaging and other techniques such as amniocentesis (page 440) now allow us to learn a great deal about the developing human individual before its birth. Despite extensive past research and these recent technical achievements, there is still much to be learned about the development of a human infant. Most of us find this topic fascinating because each of us has come into being through these very developmental processes.

Furthermore, the study of reproductive processes goes far beyond scientific and medical questions into important personal and social issues. Human sexuality is much more than just a reproductive capacity; it is a powerful psychological and social force in human life. In this chapter, we will examine both functional aspects of human reproduction and some related, socially important issues.

The basic design of the human reproductive system is a result of adaptation to life in a terrestrial environment. It is specialized so that delicate reproductive cells and small, fragile, developing individuals are enclosed and protected within moist internal body environments. For internal fertilization and development, sperm must be delivered to the inside of the female body. This internal delivery is achieved when the male **penis** deposits fluid containing sperm cells in the moist internal environment of the female **vagina.** Should fertilization occur, the developing zygote is maintained inside the female reproductive tract, and development proceeds inside the **uterus,** an organ highly specialized for maintenance of the developing embryo. The embryo becomes embedded in the uterine wall in a process known as **implantation,** and a **placenta** then develops in the implantation site. The placenta functions in exchange of materials between the circulatory systems of mother and child during embryonic development and the long **fetal period** of growth and maturation until the time of birth.

The Male Reproductive System

The male gonads are paired testes that are suspended in a saclike structure, the **scrotum.** Testes have a dual function; in addition to producing sperm, they also synthesize and release male sex hormones.

Figure 20.1 An ultrasound image of a human fetus at eleven weeks.

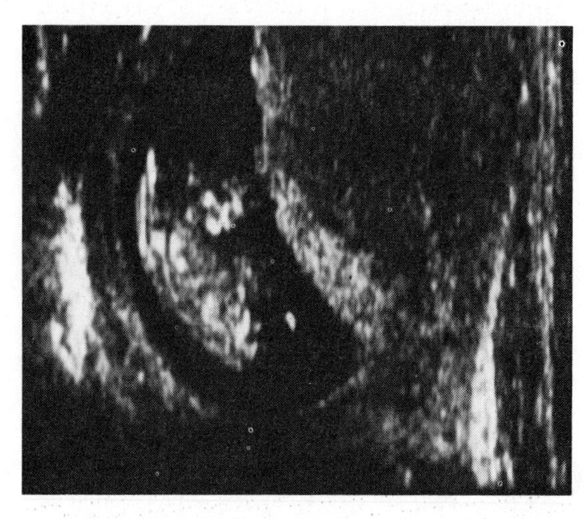

Sperm production occurs in the **seminiferous tubules** of the testes. Each testis contains about one thousand of these small, highly coiled tubules. The total length of the seminiferous tubules in an average human testis is estimated to be as much as 250 m. Spermatogenesis begins in the outer part of the seminiferous tubule, and the developing cells move toward the center of the tubule. Mature sperm cells become detached and lie in the central cavity (**lumen**) of the tubule (figure 20.2).

Male sex hormones (chiefly testosterone) are produced by the **interstitial cells** scattered in the spaces among the seminiferous tubules (figure 20.3). Testosterone is responsible for development of **secondary sex characteristics,** such as general male body form and muscle development, male hair distribution, and voice deepening in maturing males. If the testes are surgically removed (an operation called castration) before the time of sexual maturation (**puberty**), male secondary sex characteristics do not develop.

Two hormones from the anterior lobe of the pituitary gland regulate testis functions. **Follicle-stimulating hormone (FSH)** controls spermatogenesis, while **luteinizing hormone (LH)** stimulates testosterone production by interstitial cells. These hormones, along with several other pituitary hormones, are classified as trophic hormones (see chapter 11) because they stimulate secretion of other glands. Because they are trophic hormones associated with the gonads, they are called **gonadotrophic hormones** or **gonadotrophins.**

As with other anterior pituitary hormones, gonadotrophin secretion is controlled by releasing factors produced in the hypothalamus. Maturation of this control system is involved in the activation of reproductive function at the time of puberty. In adult human males, this hypothalamus-pituitary link is part of a feedback relationship that controls and balances the level of testosterone in the blood (figure 20.4).

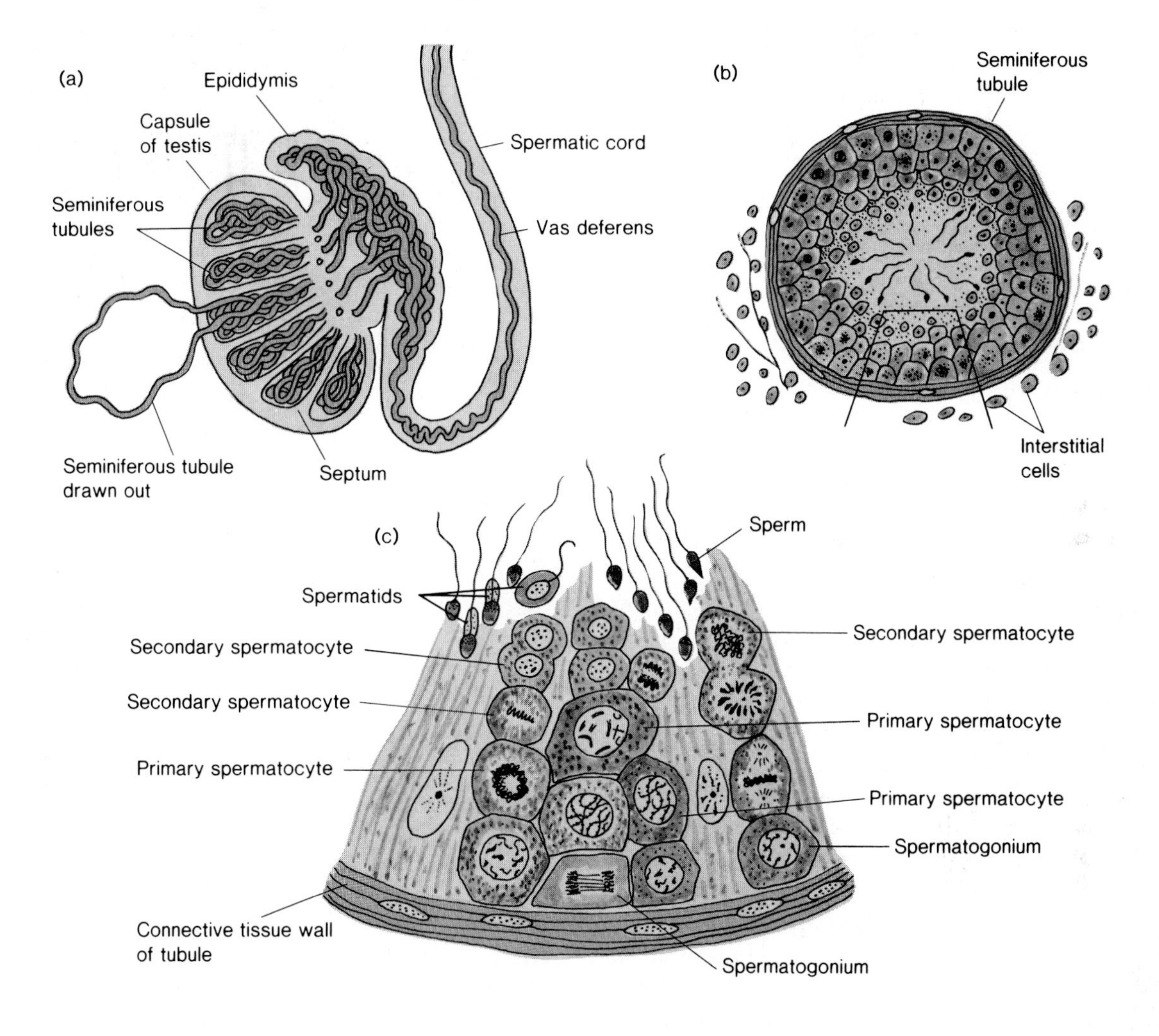

Figure 20.2 Human testis organization and sperm production. (*a*) Schematic diagram of testis organization. Packed-in, coiled seminiferous tubules produce sperm, which are transported to the epididymis and through the vas deferens. (*b*) A diagrammatic cross section showing how sperm production proceeds as cells move from the periphery toward the lumen in the center of a tubule. Interstitial cells produce male sex hormones. (*c*) Stages of spermatogenesis in one small area (outlined in *b*). Mature sperm become detached in the lumen and are moved away toward the epididymis.

Figure 20.3 The structure of the steroid hormone testosterone. Carbons and hydrogens of the basic steroid molecular skeleton are not shown. Testosterone is the main male sex hormone, and it causes development of male secondary sex characteristics.

OH
CH_3
CH_3
O

Testosterone

Figure 20.4 The hypothalamus-pituitary-testis control relationship (see chapter 11). Testosterone acts on various body tissues and balances the amount of releasing hormones being sent to the pituitary. This, in turn, affects gonadotrophin secretion by the pituitary. FSH controls spermatogenesis, and LH regulates the amount of testosterone produced.

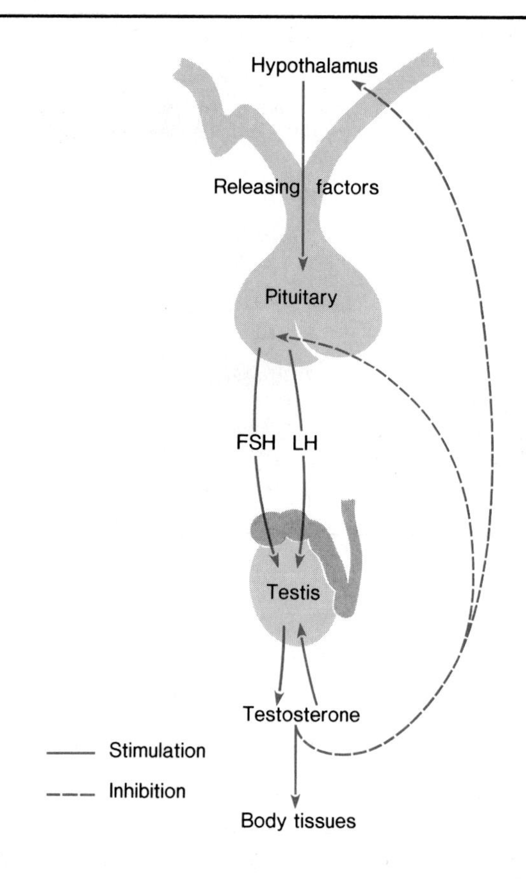

Figure 20.5 The human male reproductive system. (*a*) Frontal view showing location of reproductive organs. (*b*) Relationships of male reproductive structures and the pathway of sperm transport (indicated by arrows).

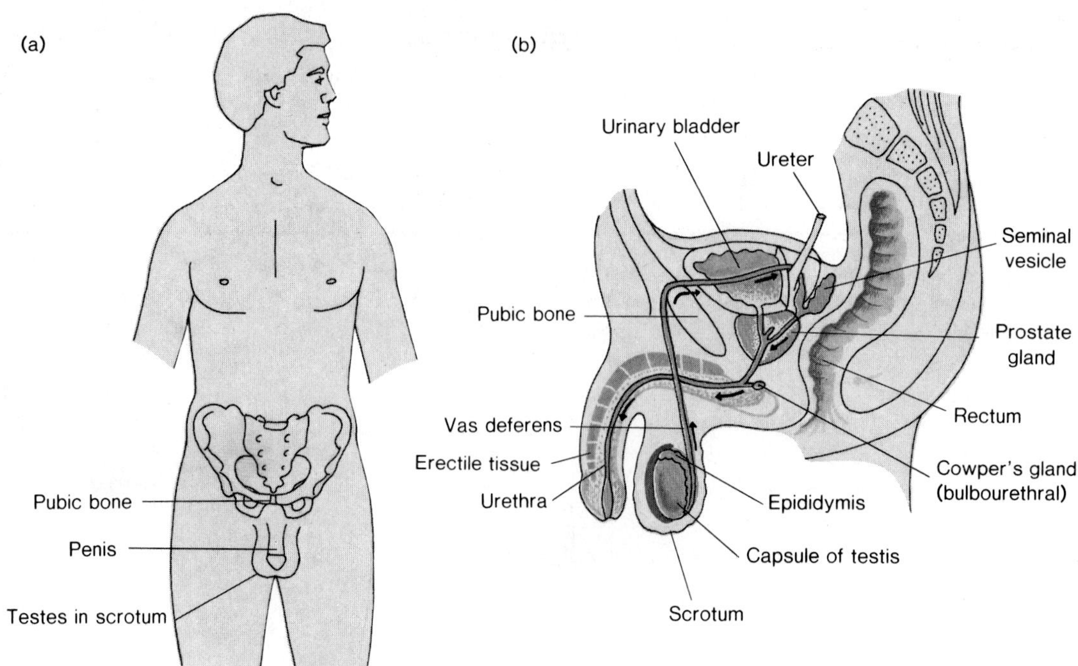

Figure 20.10 A highly diagrammatic section of a human ovary showing follicles in all stages of development. Of course, a single follicle goes through all of these stages in one place in the ovary.

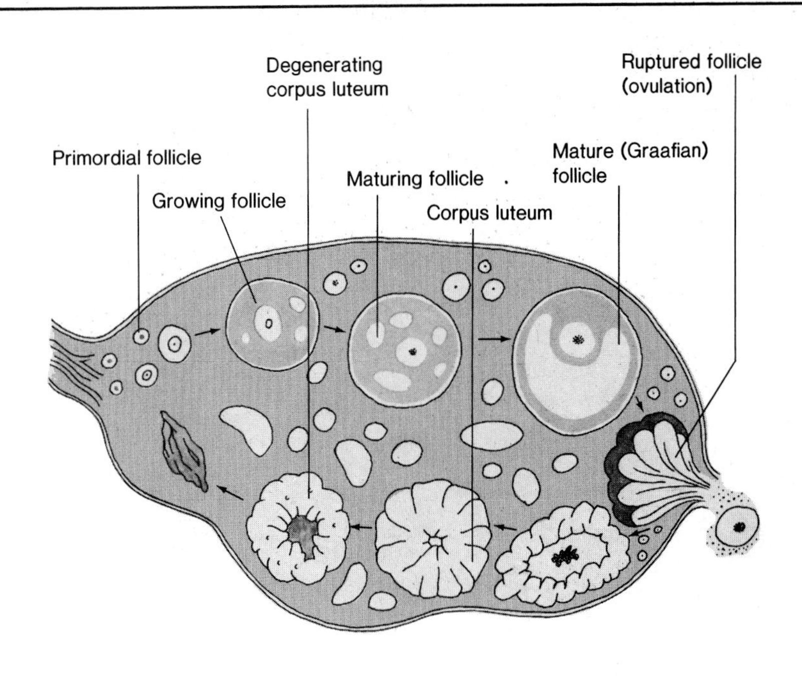

The Menstrual Cycle

Changes in the uterine lining, the **endometrium,** are coordinated with egg maturation and ovulation so that the uterus is prepared to accept an implanting zygote. The timing is such that the uterus is fully ready for implantation just as the zygote is ready to implant, if fertilization and early development have proceeded on schedule. Cyclical changes in hormone levels time these processes during the human female **menstrual cycle.** The name menstrual cycle is derived from **menstruation,** the periodic shedding of blood and tissue from the endometrium.

The uterine cycle runs concurrently with an ovarian cycle, and the two cycles are intimately related to one another. The ovarian cycle involves maturation of the future egg and development of the **follicle,** a specialized area of ovarian tissue around the maturing egg (figure 20.10). Following ovulation, the follicle changes and develops into another type of tissue, the **corpus luteum.** Changes during the uterine cycle are responses to hormones secreted by the ovary during the various phases of its cycle. Ovarian hormone production, in turn, is regulated by a feedback relationship involving the hypothalamus and pituitary (see chapter 11).

The day on which menstruation begins commonly is called "day one" of the menstrual cycle. During the early days of the cycle, **follicle-stimulating hormone (FSH)** from the pituitary causes a follicle containing a prospective egg cell to grow and proceed with meiosis in preparation for ovulation. Growing ovarian follicles secrete increasing amounts of estrogens that stimulate growth of the lining of the uterus. Layers of the endometrium that were shed during the last menstruation are rapidly replaced. This phase of endometrial repair lasts for eight to ten days after the end of menstruation.

At about the midpoint of the twenty-eight or twenty-nine day menstrual cycle, the pituitary gland produces and releases an increased amount of **luteinizing hormone (LH).** This increased LH sets in motion changes in the follicle and, subsequently, ovulation. After ovulation, continued LH stimulation causes the follicle to develop into the corpus luteum and to continue hormone secretion. The maturing corpus luteum secretes some estrogens, but it also secretes increasing quantities of another steroid hormone, progesterone. Progesterone acts on the uterus by stimulating vascularization (blood vessel development), glandular development, and glycogen accumulation in the endometrium. All of these progesterone-induced changes are necessary to make the uterus ready for the potential arrival of an implanting zygote (figure 20.11).

Figure 20.11 A composite diagram showing the time relationships of pituitary hormone secretion, the ovarian cycle, ovarian hormone secretion, and the condition of the uterine lining during the human menstrual cycle. Note that different units and different scales are used for the various hormones. Do not be as concerned with the units and actual amounts of hormones as with the relative hormone levels at different points in the cycle.

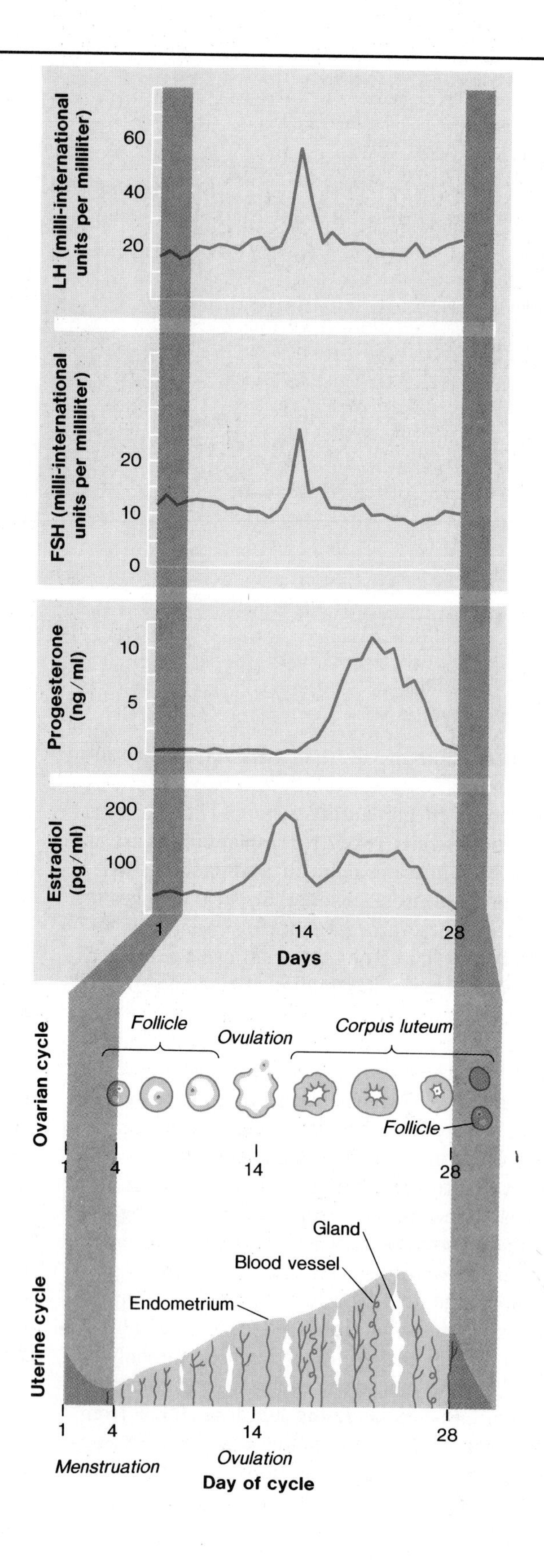

Figure 20.12 Human ovulation. The surface of the ovary swells and breaks open. Fluid from the follicle pours out, and the oocyte emerges surrounded by the zona pellucida and several layers of follicle cells.

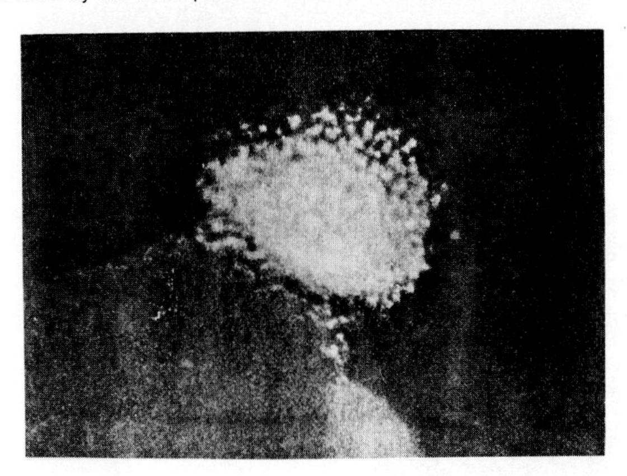

In a normal menstrual cycle, when a pregnancy has not begun, the hypothalamus and then the pituitary respond to the high levels of steroid hormones put into circulation by the active corpus luteum. This leads to a decrease in LH production, and falling LH levels cause the corpus luteum to begin to degenerate by about the twenty-second or twenty-third day. As the corpus luteum is degenerating, its hormone output decreases.

Maintenance of the fully developed uterine endometrium depends on the relatively high progesterone levels present while the corpus luteum is most active. Thus, when circulating progesterone levels decrease as a result of decreased secretion by the corpus luteum, the uterine linings start to degenerate and slough off, and menstruation begins. Before long, however, the hypothalamus and pituitary respond to hormone changes in the blood; FSH production rises again, follicle growth and maturation are stimulated, and the whole complex cycle starts over.

Menstruation is a clear signal that pregnancy has not begun during that particular menstrual cycle. The beginning of a pregnancy, however, greatly modifies hormonal and uterine events. In pregnancy, the corpus luteum is maintained in response to a hormonal influence from the developing embryo, and it continues production of adequate levels of progesterone to keep the uterine lining intact. We will see later how this works.

The Oocyte and Ovulation

Meiotic divisions of prospective egg cells begin in the ovaries of human female fetuses. Thus, a human female has initiated the process of egg production even before she is born. Some 400,000 prospective egg cells enter the first stage in the two fetal ovaries of each human female, but only 400 or so cells will proceed to ovulation during the reproductive period of an average adult woman.

During each menstrual cycle, one (rarely two or more) prospective egg resumes meiosis and proceeds as far as metaphase of the second meiotic division. The cell, now properly called a secondary oocyte, pauses again and remains at that meiotic stage through the time of ovulation. Meiosis is completed later if the cell is contacted by a sperm (chapter 19). Without sperm cell contact, meiosis proceeds no further, and the oocyte degenerates.

Ovulation occurs in response to the midcycle increase in LH, and it involves a rupture of the ovarian surface over the mature (**Graafian**) follicle (figures 20.12 and also 20.10). Fluid accumulated inside the mature follicle escapes and carries the oocyte with it. The oocyte, as it leaves the ovary, is enclosed by two covering membranes and several layers of cells that surround them. Immediately over the surface of the oocyte is the **vitelline membrane,** which is secreted by the oocyte itself. Just outside the vitelline membrane is another covering layer, the **zona pellucida,** which is produced by cells in the follicle. Around the zona pellucida are several layers of follicular cells, which accompany and enclose the oocyte and its membranes.

Fertilization and Early Development

Sperm must be present in the upper one-third of the oviduct if fertilization is to occur, since the oocyte must encounter sperm early in its three-day passage through the oviduct. Certain "aging" changes seem to affect the oocyte beginning about a day after ovulation, and the oocyte soon loses its ability to participate normally in fertilization reactions.

An enzyme, **hyaluronidase,** released from the acrosome portion of sperm cells, loosens up the follicular cells around the oocyte by hydrolyzing **hyaluronic acid,** a substance that cements the cells together. This allows sperm to gain access to the egg surface. The need for adequate quantities of hyaluronidase may explain why a large number of sperm must be present in the oviduct for fertilization to occur, even though only one sperm actually interacts with the oocyte in the fertilization process.

Contact with a sperm cell activates the oocyte, and it resumes the second meiotic division, which has been arrested since before ovulation. The second polar body is extruded, leaving the functional egg with a haploid nucleus (**pronucleus**). Then egg and sperm pronuclei migrate into the center of the cell where they meet. Chromosomes of both pronuclei replicate and the first of the mitotic cell divisions of cleavage begins (figure 20.13). This first cleavage division is completed within about a day after fertilization, and subsequent

Figure 20.13 Human fertilization and early development of the human embryo. (*a*) A human oocyte with its enclosing layers. Note sperm cells around it and in the follicle cells that surround the oocyte. (*b*) A human embryo at the two-cell stage. (*c*) The six-cell stage. Cell divisions are not synchronous so there is not a regular progression of two, four, eight, sixteen cells, and so on. (*d*) A human embryo at the morula stage of development.

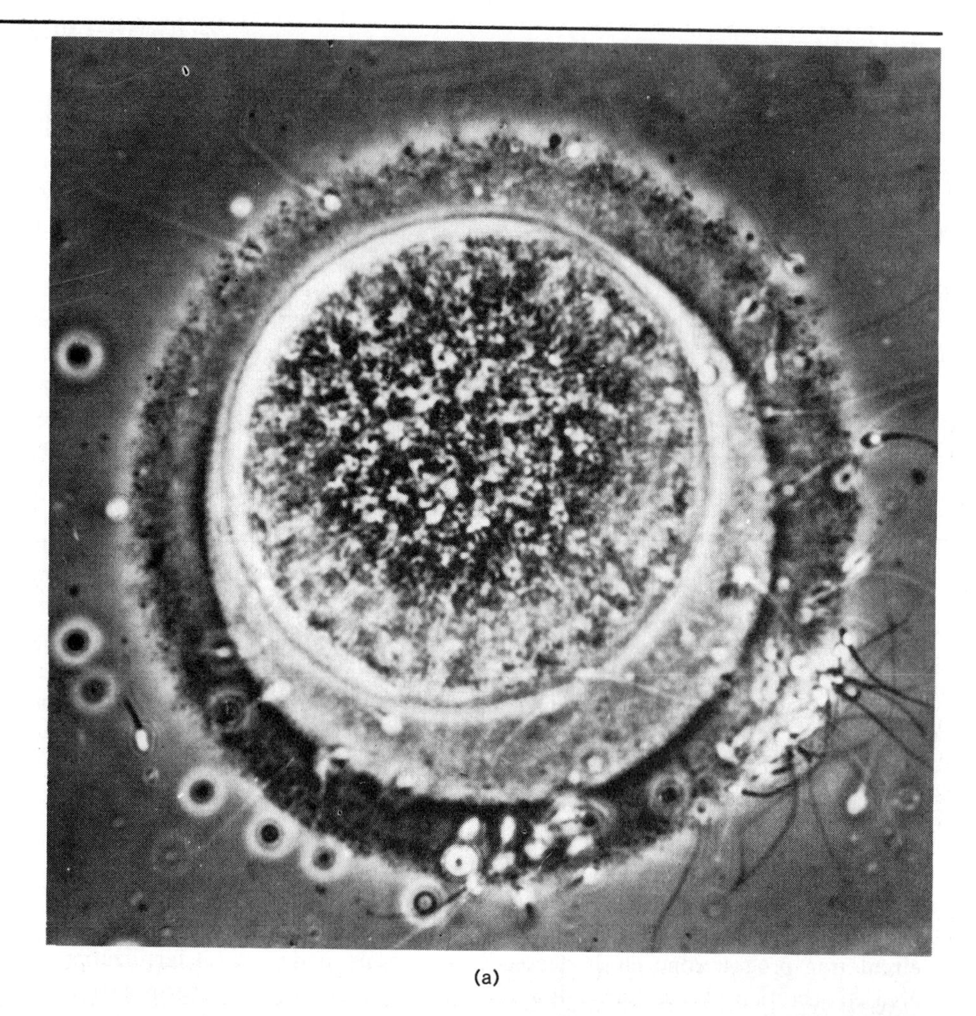

(a)

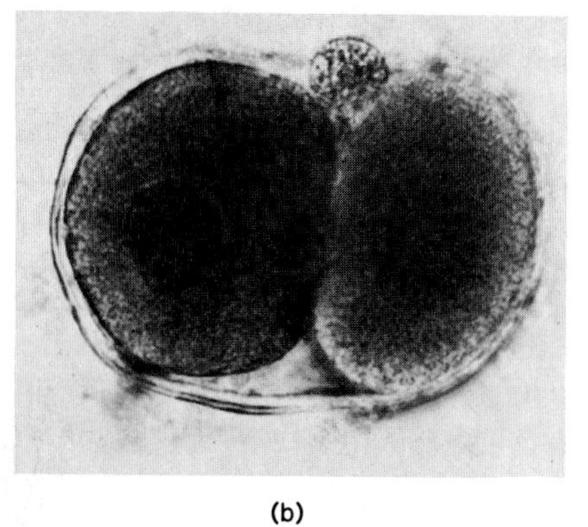

(b)

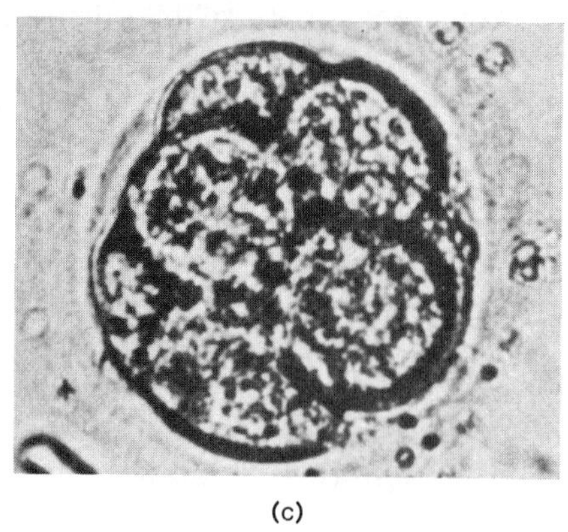

(c)

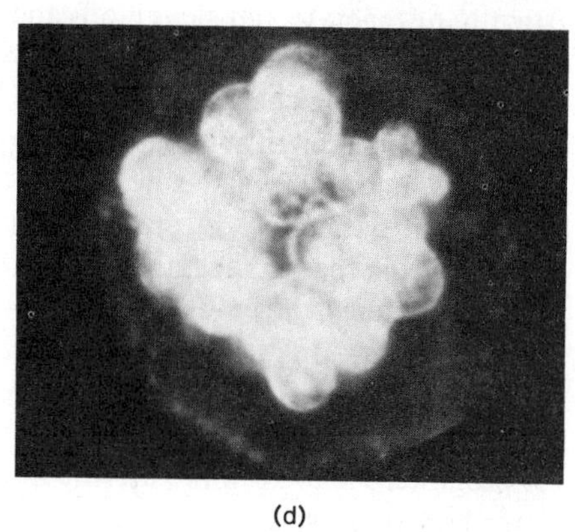

(d)

Figure 20.14 The human female reproductive tract with the uterus and one oviduct opened up to show progress of the developing embryo at various stages. The embryo normally becomes implanted in the lining of the uterus, as shown here. Implantation in other areas causes pregnancy complications.

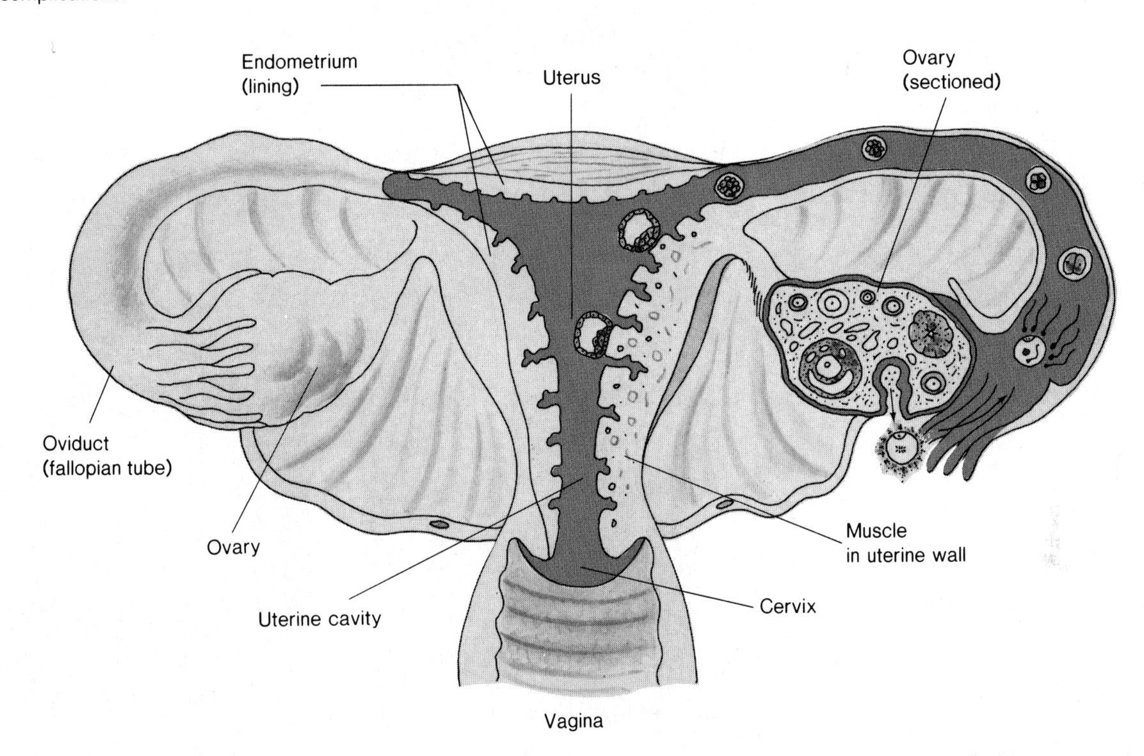

divisions occur at intervals of from eight to ten hours as the early embryo moves down through the oviduct toward the uterus (figure 20.14). Passage of the embryo through the oviduct takes from three to three and one-half days. During this time, the follicular cells around the developing embryo are lost completely, but the cells of the embryo still are enclosed and held together by the zona pellucida.

Implantation

While the embryo lies free in the uterine cavity, further cell divisions produce a rather loose aggregate of much smaller cells, the **morula.** During the next two or three days, these cells segregate themselves into two distinctly different groups as they organize the characteristic structure of the **blastocyst** stage of development (figure 20.15). The blastocyst consists of a hollow sphere of small, flattened cells, the **trophoblast,** surrounding a fluid-filled cavity. A mass of larger, rounded cells, the **inner cell mass,** is situated to one side of the cavity.

Cells of the trophoblast layer play no part in the formation of the embryo itself but produce an extra-embryonic membrane, the **chorion,** and contribute to placenta development. The body of the developing embryo is produced entirely within the inner cell mass.

The blastocyst stage is a critical point in human development because the blastocyst must implant in the wall of the uterus. At least 25 percent and possibly as many as 40 percent or more of all developing blastocysts fail to implant in the wall of the uterus. In the absence of implantation, there is no sign of pregnancy, and the embryo simply dies and is lost. Some of these implantation failures undoubtedly involve developmental abnormalities that produce blastocysts incapable of implanting.

Normally, the inner cell mass side of an implanting blastocyst contacts the endometrium first. The initial contact involves a specialized tissue on the surface of the trophoblast. This specialized surface tissue is syncytial; that is, it is a multinucleate tissue not divided into individual cells. Properties of this syncytial tissue are very important because, in a sense, implantation of the human blastocyst into the uterine wall is the acceptance of a "graft" of foreign tissue. A better understanding of this special tolerance of the endometrium for an implanting blastocyst could provide information that would be helpful in improving transplantation and organ grafting techniques.

Figure 20.15 The blastocyst and implantation. (*a*) A human blastocyst. Note the inner cell mass and the individual trophoblast (developing chorion) cells. (*b*) A section of a monkey blastocyst attached to the endometrium, which shows how the blastocyst approaches and enters the endometrium inner cell mass side first. (*c*) Section of a human implantation site showing the implanted blastocyst and the glands and blood vessels of the fully developed endometrium. (*d*) A surface view of an implantation site twelve days after fertilization. Uterine tissue grows over the surface and heals the point of entry.

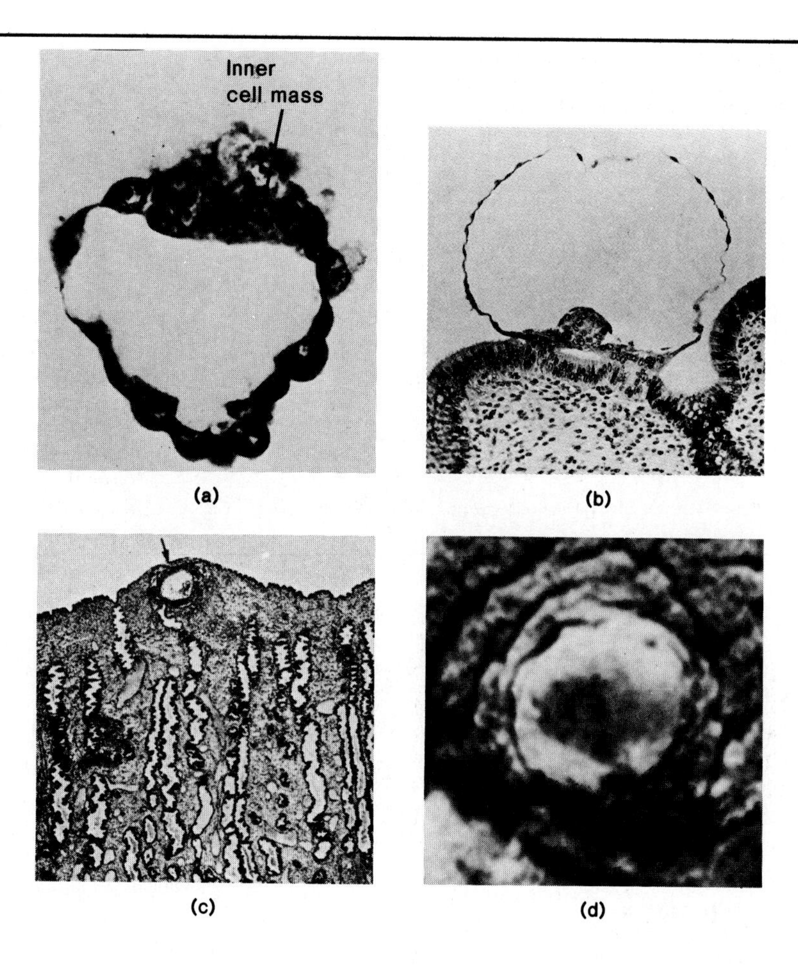

Enzymes secreted by trophoblast cells erode uterine tissue as the blastocyst sinks into the endometrium. After the blastocyst has entered the endometrium, uterine tissue grows over the surface and heals the implantation site.

Hormones and Pregnancy

During a normal menstrual cycle, a chain of events involving lowered LH production, regression of the corpus luteum, and the consequent decrease in circulating progesterone levels leads to menstruation. If implantation occurs and pregnancy is to proceed, this normal sequence must be interrupted.

An implanted blastocyst actually produces a hormone that prevents menstruation. The developing chorion (which is produced by the trophoblast and surrounds the developing embryo) begins very early to produce a hormone called **chorionic gonadotrophin** (abbreviated **HCG** for human chorionic gonadotrophin). Functionally, HCG replaces LH and keeps the corpus luteum active. The resulting continued progesterone secretion by the corpus luteum keeps the uterine lining intact.

Chorionic gonadotrophin is produced in such large quantities that pregnant women excrete considerable amounts of it in their urine. Thus, chemical pregnancy tests can be based on detection of urinary chorionic gonadotrophin (figure 20.16).

Occasionally, corpus luteum activity may decrease due to illness of the mother or a failure of the chorion to continue adequate HCG production. The resultant drop in progesterone level initiates menstruationlike breakdown of the endometrium and loss (**miscarriage**) of the implanted embryo. Miscarriage during the first month or so of pregnancy may result in a blood flow somewhat heavier and longer than normal menstruation, but the embryo is still so small that the entire brief and abruptly terminated pregnancy simply may be mistaken for a somewhat delayed menstrual period.

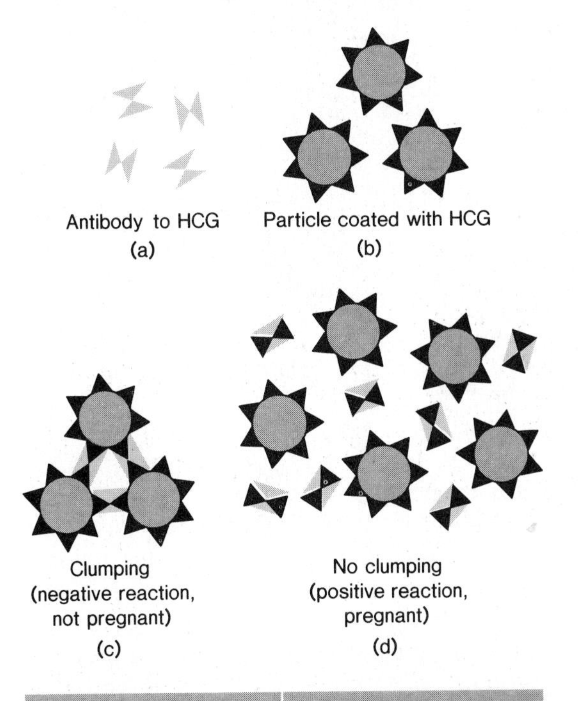

Figure 20.16 The basis for the commonly used chemical pregnancy tests. (*a*) Antibodies to human chorionic gonadotrophin (HCG) are obtained by immunizing animals against HCG. This is one reactive agent in the test. (*b*) Inert particles coated with HCG are the second reactive agent for the test. (*c*) If these two are mixed together, a reaction causes the particles to clump. (*d*) Urine from a woman being tested is first mixed with the antibody to HCG. If the woman is pregnant and HCG is present in her urine, it will combine with the antibodies. Thus, mixing this already bound antibody with the particles coated with HCG in the second step cannot cause clumping. Such chemical tests have replaced older tests that depended on reproductive tract responses of rabbits or frogs to injected urine or blood samples.

Nonpregnant woman	Pregnant woman
1. Urine + a 2. Urine + a + b	1. Urine + a 2. Urine + a + b
Agglutination (clumping)	No agglutination
Negative reaction	*Positive reaction*

In a normal pregnancy, chorionic gonadotrophin maintains vital steroid hormone production by the corpus luteum until the ninth or tenth week of pregnancy. By then, the placenta itself is well established as a hormone-producing organ. It produces both progesterone and estrogens. These placental steroids replace ovarian hormones in the maintenance of the uterine linings during the remainder of pregnancy.

The Major Stages of Prenatal Development

Development during the first two weeks following fertilization includes cleavage, blastocyst development, implantation, and very early postimplantation development. From the third through the eighth weeks, basic body organization is established. **Rudiments** (the first visible evidences of development) of all major organ systems are produced, and by the end of this period, the body is recognizably human. From the ninth week until birth is the **fetal period.** The **fetus** grows larger, and cell and tissue differentiation take place in the rudimentary organs originally laid down during the embryonic period. Various organs mature enough to permit termination of functional dependence on the placenta by the time of birth. The total period of time involved in all of these stages—from fertilization until birth—is called the **gestation period.**

The Embryo During Early Pregnancy

During the first days following implantation, cells of the inner cell mass reorganize. This cell activity establishes a flat, round, two-layered disc known as the **blastoderm,** which develops into the embryo (figure 20.17). Actually, the blastoderm is the flattened area of contact between two hollow spheres of cells formed from the originally solid ball of inner cell mass cells. One of these spheres of cells, the **yolk sac,** is so named because of its obvious similarity to yolk-digesting structures in other vertebrates, such as chick embryos. But the human yolk sac does not function in yolk digestion because the small amount of yolk originally present in the human egg is used up well before the yolk sac forms. One side of the yolk sac is one of the layers of the blastoderm. That layer, the **hypoblast** ("lower layer"), is pressed against a second layer, the **epiblast** ("upper layer"), which lies at the bottom of another spherical space, the **amniotic cavity.**

All parts of the embryo are produced from this flat, two-layered disc after it is converted into the tubular body of the embryo. But first, a third layer is added to the original two. Mechanics of mesoderm formation in the human embryo are virtually identical to those described for bird and reptile embryos (page 406). Some cells of the epiblast migrate downward through a primitive streak to establish the mesoderm of the human embryo. The three primary body layers (germ layers) of the human embryo, from top (amnion side) to bottom (yolk-sac side), are ectoderm, mesoderm, and endoderm.

Figure 20.17 A section of a human embryo showing the yolk sac and the amniotic cavity. The blastoderm is the two-layered disc where they are in contact with one another. These derivatives of the inner cell mass are attached to the inside of the chorion by the small body stalk. The body stalk becomes a major part of the umbilical cord.

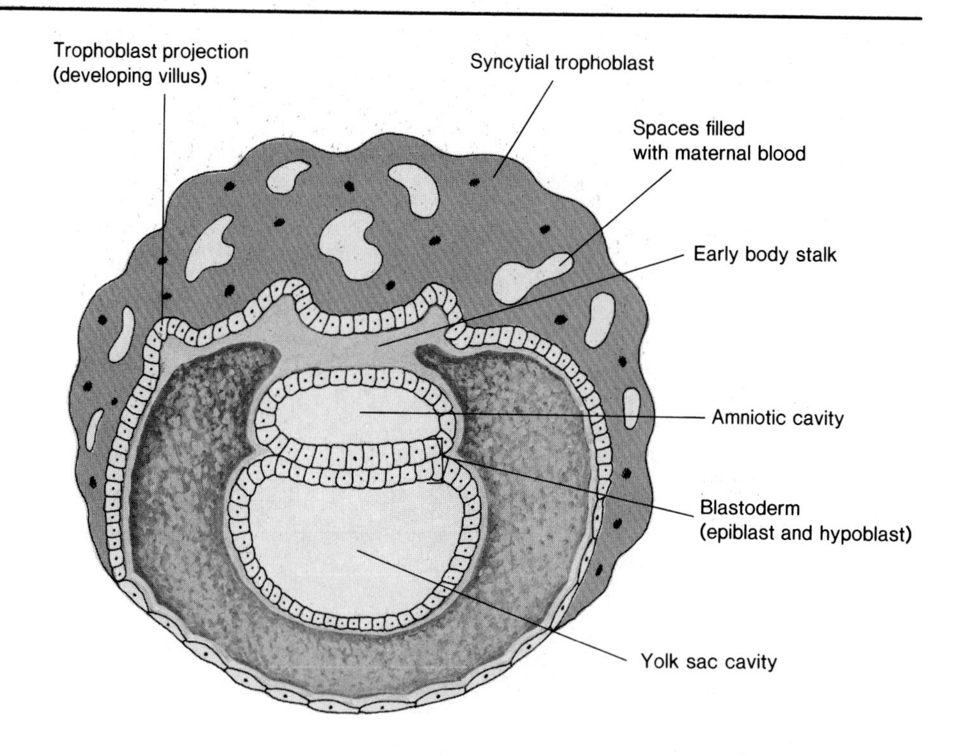

Table 20.1
Germ Layer Sources of the Major Functional Parts of Various Body Tissues, Organs, and Systems

Ectoderm
Nervous tissue
Epidermis of skin
Mesoderm
Dermis of skin
Skeleton
Muscle
Circulatory system
Excretory system
Reproductive system
Connective tissue
Endoderm
Digestive system linings
Digestive glands
Lung and respiratory tract linings

Later, the flat disc rolls up into a tubular three-layered body. This body folding brings the germ layers into their permanent relationships with one another: ectoderm on the outside, endoderm on the inside, and mesoderm between the other two. As development proceeds, primary functional parts of major organ systems develop from each of the germ layers (table 20.1).

Placenta Development and Function

While the embryo's body is being organized within the inner cell mass, a placental relationship is developing between the embryo and its mother.

A series of fingerlike outgrowths, the **villi** (singular: **villus**), develop over the chorion's surface (figure 20.18*a*). Villi on one side of the chorion are destined to grow, branch elaborately, and participate in development of the placenta, while those scattered over the remainder of the chorion regress and later disappear. Capillary beds develop inside the branching villi in the placenta area, and the villi receive a blood supply from vessels growing out from the embryo's body. These are called umbilical vessels because they pass through the **umbilical cord,** which develops from the **body stalk,** a narrow connection between the embryo and the developing placenta (figure 20.18*b* and *c*). Two **umbilical arteries** carry blood to the villi, and one **umbilical vein** carries blood back to the body from the placenta (figure 20.18*d*).

Figure 20.18 Stages in the development of extraembryonic membranes and the placenta in humans. (*a*) Syncytial portion of trophoblast (figure 20.17) and uterine tissue are not shown. Note villi projecting on the surface of the chorion. The chorion has developed from trophoblast. (*b*) The body stalk is producing an umbilical cord. Note the regression of the chorionic villi in the region farthest from the developing umbilical cord. (*c*) Final relationships of embryo and extraembryonic structures. The umbilical cord with umbilical blood vessels is well developed and serves as the route for exchange between the fetus and its mother. (*d*) Circulatory relationships within the placenta. Maternal blood pours into open spaces and bathes the villi.

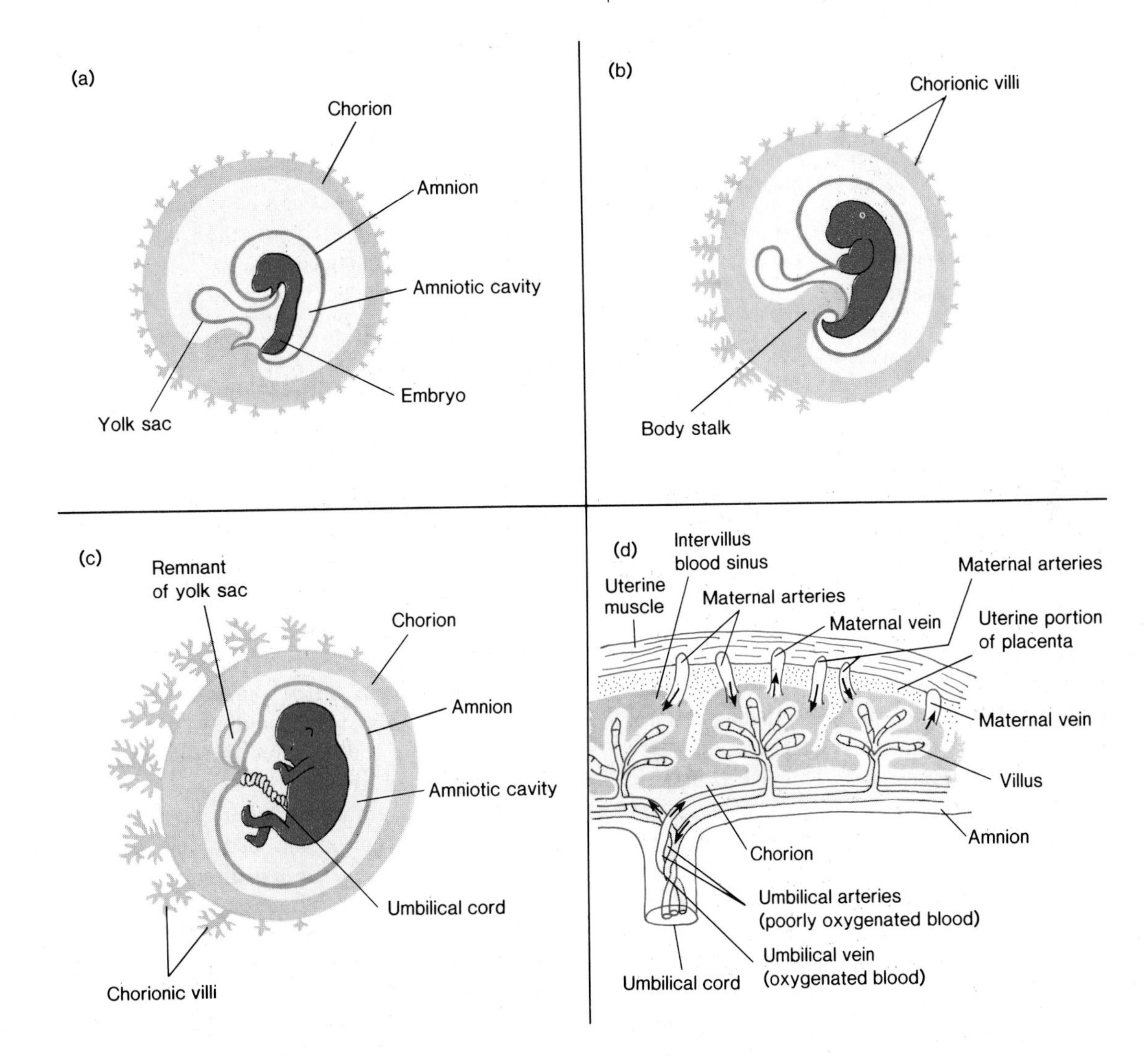

Chorionic tissue grows against and fuses with uterine tissue so that the placenta is literally a single organ constructed from tissues of the mother and the embryo. A very special circulatory arrangement develops in the spaces among the chorionic villi. Maternal arteries open up directly into these **intervillus** spaces (figure 20.18*d*). Maternal blood flows through these spaces, completely bathing the villi, before draining into veins that carry blood away from the placenta. This open circulation of maternal blood around the villi makes exchange by diffusion between the two circulations quite efficient.

Several layers of tissue normally separate the maternal bloodstream and the embryonic bloodstream. These layers are known collectively as the **placental barrier.** Properties of the placental barrier are of great interest to embryologists because the barrier is selective in allowing substances to cross from one side to the other. Some drugs and other chemicals cross; some do not. Certain everyday substances, such as ethanol (alcohol), that cross the placental barrier are mildly toxic to adults but *very* toxic to developing embryos and fetuses. Physicians now recognize a **fetal alcohol syndrome**

Figure 20.19 The placental barrier. Arrows that end at the barrier indicate that the substance normally does not cross the placental barrier. When the barrier is intact, there is no direct mixing between the two bloodstreams. Small breaks in the barrier apparently occur in a significant percentage of pregnancies without causing major problems.

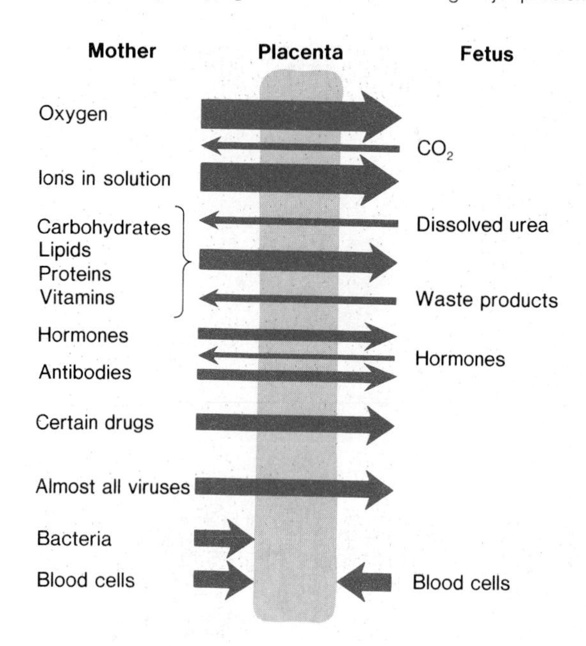

characterized by reduced growth, possible mental impairment, and other symptoms. The syndrome often is seen in infants born to mothers who drink more than a minimal amount of ethanol during pregnancy. Bacteria cannot cross the placental barrier, but viruses can. Thus, viral diseases, such as measles, can be transmitted from mother to child across the placental barrier. Nutrients and oxygen cross from the maternal side of the barrier to the fetal side, while carbon dioxide and nitrogenous wastes pass in the other direction. Maternal hormones and placental hormones also can pass from one side of the barrier to the other. All of these placental functions are essential to the welfare of the developing infant, and they are maintained until birth (figure 20.19).

The Amnion and Its Functions

The **amnion** (amniotic sac) originally encloses the amniotic cavity, the space above the embryo's body. Later, this amniotic cavity spreads and enlarges to enclose and line the whole space within which the developing fetus remains throughout development.

Amniotic fluid secreted by amnion cells fills the cavity and lubricates its surfaces. This lubrication prevents delicate tissues from sticking to one another and cushions the developing fetus by absorbing shocks when the mother moves quickly or is bumped. In later stages of development, the fetus drinks and inhales amniotic fluid. Thus, amniotic fluid bathes the linings of mouth and nasal passages as well as the skin.

Cells sloughed off body surfaces float in the amniotic fluid. Thus, samples of amniotic fluid can provide valuable information about both the progress of development and the genetic and biochemical characteristics of cells from the fetus. Amniotic fluid samples can be withdrawn for analysis using a needle inserted directly into the amniotic cavity. This technique is called **amniocentesis.**

The First Month

As we saw earlier, cleavage and blastocyst formation occur during the first week.

Implantation continues during the second week, and uterine lining cells heal over the surface of the implantation site. By the very end of the second week, the primitive streak develops, and mesoderm development begins.

Mesoderm formation continues during the third week, and the nervous system begins to develop as a pair of thickened ridges of ectoderm that rise up, meet, and fuse with each other into a hollow tube (figure 20.20*a*). Paired chunks of mesoderm tissue, called somites, appear alongside the developing nervous system. Somites later give rise to vertebrae, muscles, part of the ribs, and the dermis of the skin. Heart formation begins at this time.

Despite the embryo's rather extensive development, the mother's menstruation has been delayed by only about a week. Even though the embryo already has a developing heart and central nervous system, the mother probably is not even certain that she is pregnant!

During the fourth week, the embryo develops the beginnings of a digestive system. Strong, coordinated heart beating initiates circulation through the embryonic body and through the umbilical vessels supplying the placenta. The first circulating blood cells are produced in special areas of the yolk sac known as **blood islands.** Thus, the first blood cells that develop come from a source outside the body of the embryo itself. Only later on, during the second month, does blood cell formation begin inside the body, when the liver becomes the second temporary source of blood cells. Finally, the bone marrow matures and takes over as the third, and permanent, source of blood cells.

At the end of the first month, the embryo is still only about 5 or 6 mm in length (figure 20.20*c*).

Figure 20.20 Embryonic development. (*a*) Embryo near the end of the third week of development. Ectodermal folds (neural folds) are closing to produce brain and spinal cord. Somites are forming. (*b*) A twenty-four-day human embryo. Note that body folds have produced a tubular body from the previously flat disc. (*c*) A thirty-four-day human embryo.

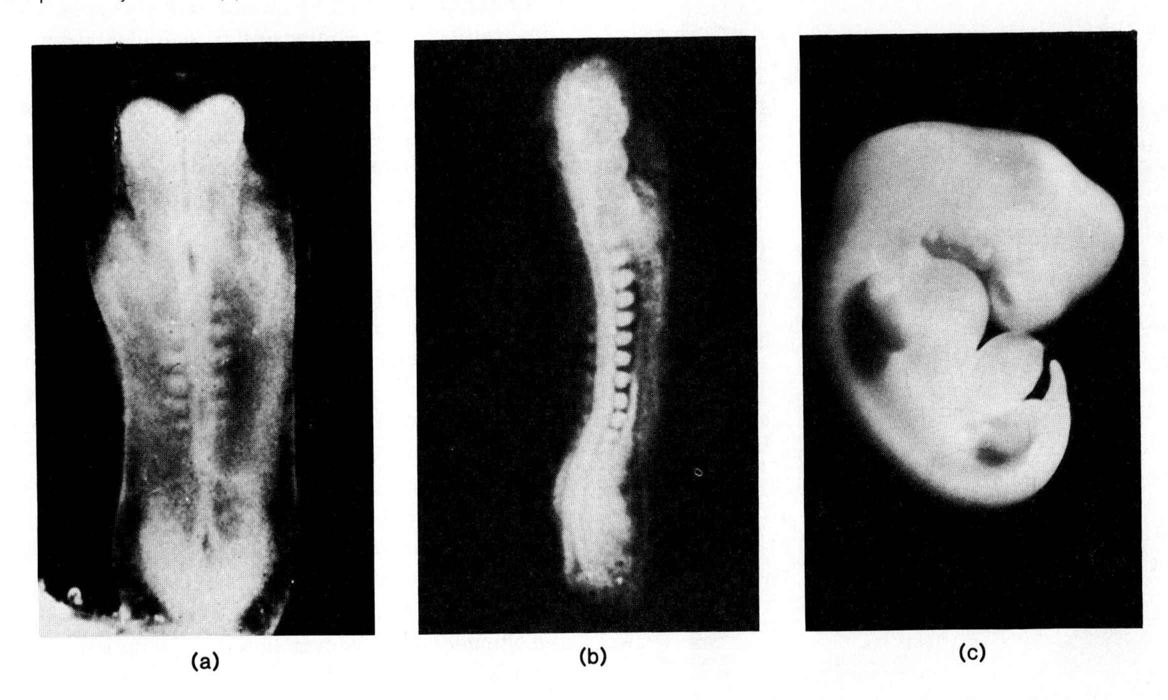

The Second Month

During the second month, rudiments of all the rest of the major internal body organs are produced in the embryo, which assumes an increasingly human appearance. Paddlelike **limb buds** grow out from the sides of the body and gradually transform in shape into recognizable arms and legs. In the head, eye and ear development proceeds, and the face differentiates. The embryo now looks like a human being and can easily be distinguished from other vertebrate embryos, which formerly it resembled very closely (figure 20.21).

Excretory and reproductive system elements develop extensively during the second month, but at first male embryos cannot be distinguished from female embryos because the developing reproductive system at this stage contains rudiments of both female and male structures. This "indifferent condition" of reproductive system development persists until about the seventh week of development, when signs of sex determination begin to appear.

The establishment of human appearance externally and all of the major organ rudiments internally marks the end of the embryonic period of development and the beginning of the fetal period. This is an important milestone in the development of the individual, and it has broader implications for the pregnancy, as well. The fetus generally is much less sensitive to harmful external influences than the embryo. Embryonic development includes a whole series of **sensitivity periods** during which delicately balanced developmental processes involved in organ formation are subject to disruption by chemical agents or disease. Harmful agents still can have damaging effects on the fetus, but they are much less likely to cause gross developmental disturbances.

The Third Month

During the third month of pregnancy, the reproductive system development proceeds to the point that the sex of the fetus is externally apparent. Overall fetal growth is considerable because the weight of the fetus increases from less than 2 gm to about 12 gm. While the body and limbs may move slightly, the fetus is so small that the movements are wholly undetectable by the mother.

Figure 20.21 A fifty-six-day human fetus. From the end of the eighth week on, it is proper to call the developing individual a fetus because rudiments of all major organs and systems are present. The fetus is clearly human in general body form.

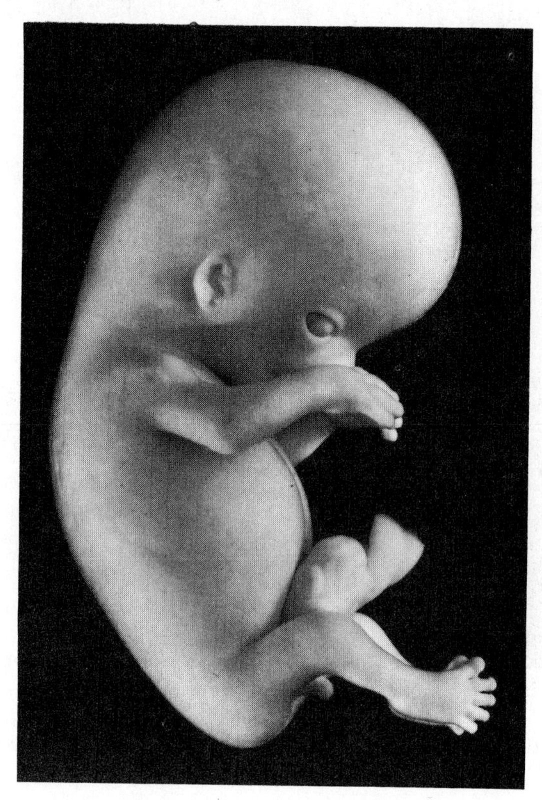

Four to Six Months (The Second Trimester)

During the second third of pregnancy (the second **trimester**), rapid fetal growth continues. For the first time, the mother can feel fetal movements, usually late in the fourth month or during the fifth month. Hair develops on the head, and a very fine hair, the **lanugo,** forms a downy covering over the body. Waxy, almost cheeselike secretions cover the skin surface and apparently protect it during its long immersion in the watery amniotic fluid. The fetal brain develops impressively during the second trimester. Sense organs become functional, and the fetus shows reflexes that clearly indicate that it is becoming responsive to changes in its environment. The fetal heartbeat can be detected fairly readily with a stethoscope because of its distinctive sound and relatively rapid rate (120 to 150 beats per minute).

Especially during the sixth month, the fetus gains weight rapidly, and by the end of the second trimester, the average fetal weight is about 630 gm. Still, this is less than 20 percent of normal average birth weight. Babies born prematurely at the end of the second trimester sometimes survive, but they invariably require intensive care under strictly controlled conditions. Infants born this early very frequently suffer from respiratory distress due to lung immaturity, and they must be carefully protected from environmental stresses, such as temperature changes.

The Final Trimester

During the last trimester, fetal growth continues with an average weight gain of about 25 gm per day. The mother's uterus continues to expand to contain the growing fetus and its membranes. At the end of gestation, the uterus weighs more than twenty times as much as it did before pregnancy began.

Nutritional research indicates that the third trimester is a critical period in brain development and that normal nervous system development is very dependent on the mother's nutritional status. Protein deficiencies in the mother's diet, for example, can have adverse effects on nervous system development that may even limit the infant's future mental capacities.

Within the ovaries of female fetuses, primary oocytes enter prophase of their first meiotic division and then pause until after puberty. Testes of male fetuses begin to move during the seventh month, and they descend out of the abdominal cavity to their final location in the scrotum. This descent is necessary for normal reproductive functioning after puberty. Testes that remain in the abdominal area (a condition known as **cryptorchidism;** literally, "hidden testes") are too warm to produce normal, functional sperm cells. Normal spermatogenesis occurs only at the slightly lower temperatures of the scrotum.

Other systems progress toward their full-term condition. Especially because of rapid lung maturation, survival chances of prematurely born infants increase dramatically during the final trimester.

Maternal antibodies cross the placental barrier and enter the fetal circulation so that the newborn infant (**neonate**) carries immunities to bacteria, viruses, and other foreign materials. This passive immunity, "borrowed" from its mother, serves the infant only temporarily. Its own immune system begins to produce antibodies shortly after birth.

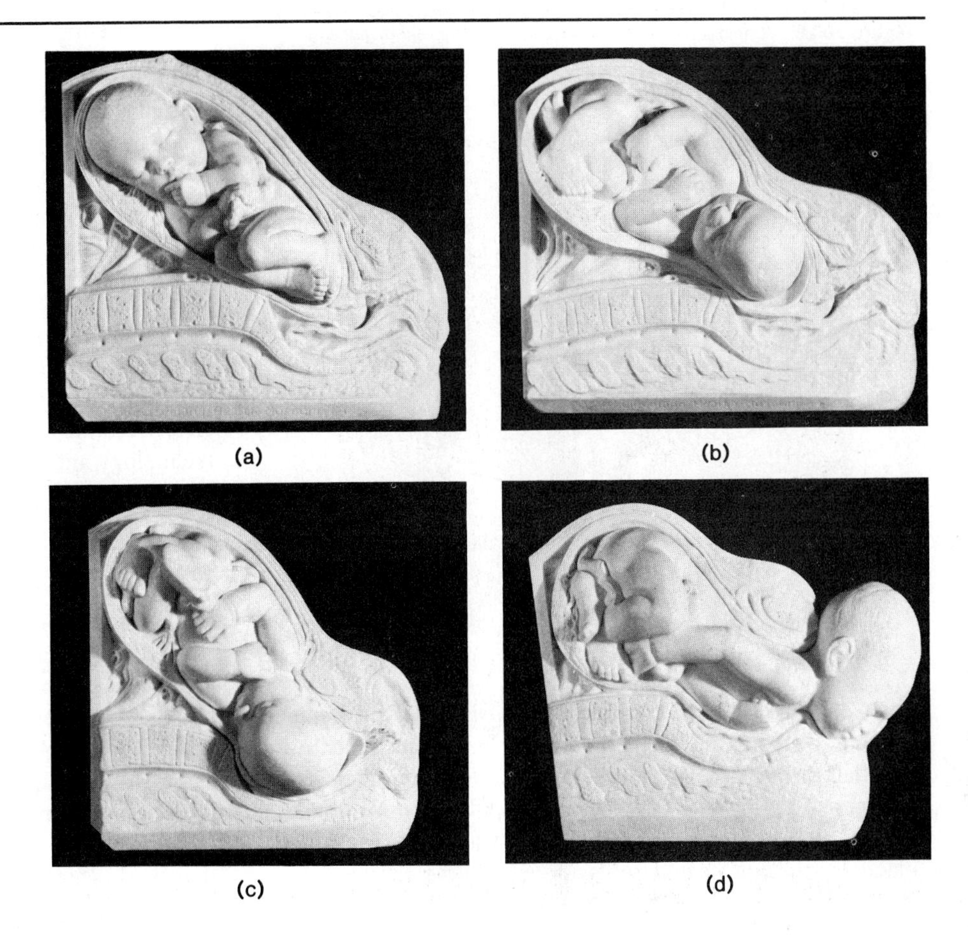

Figure 20.22 Models of the birth process. (*a*) Cutaway side view of a fetus in the uterus near the end of pregnancy. (*b*) Position of the fetus at beginning of labor. (*c*) Early in the second stage of labor as the infant moves into the birth canal. (*d*) Later in the second stage, the head has emerged.

Birth

Human birth (**parturition**) occurs after an extended period of rhythmic uterine muscular contractions known as **labor.** Birth normally takes place about 280 days after the beginning of the last regular menstrual period or about 266 days after fertilization.

Progesterone production decreases during the last two months of pregnancy (as you recall, progesterone's role has been to keep the uterine lining intact). Estrogen production increases gradually throughout the same period and then shows a sharp increase just before birth (estrogens increase the irritability and contractility of the uterine muscles). But the most direct hormonal stimulus probably comes from **oxytocin,** a hormone from the pituitary's posterior lobe, that directly stimulates contractions of uterine muscle. The cause of the increase in oxytocin level and whether the increase is absolutely essential for normal birth are not known. But oxytocin's effect on uterine muscles is reliable enough that the hormone is used routinely to induce labor artificially when the birth process must be hurried along.

Labor is a progressive three-stage process that leads to the delivery of the fetus and the afterbirth (the placenta and the fetal membranes).

Uterine contractions during early labor usually are somewhat irregular and may be mistaken for the gas pains or other intestinal discomforts of late pregnancy. But when the contractions become regular and occur at intervals of twenty minutes or less, they signal the onset of labor, although they may still stop and resume some time later.

During the first stage of labor, a mucous plug that has blocked the cervical opening of the uterus during pregnancy comes loose and is shed as the cervix begins to change shape. The cervix shortens and flattens so that it does not protrude into the vagina so far. The tiny uterine canal expands to a diameter of about 10 cm. By the end of this first stage of labor, uterine contractions have become much more forceful and frequent. Usually, the fetal membranes rupture early in labor, allowing the amniotic fluid to flow out through the vagina.

During the second stage of labor, expansion of the uterine canal is completed, and the infant is delivered through the opened uterine canal and the vagina (figure 20.22).

Figure 20.23 A newborn human infant (neonate) just after delivery.

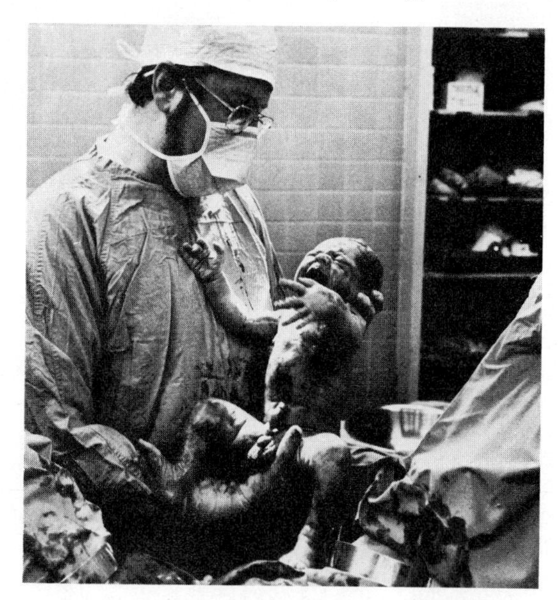

Continued uterine contractions, during the third stage of labor, dislodge the placenta, and normally the placenta is expelled through the birth canal about twenty minutes after delivery of the infant. The placenta is pulled loose from the uterine wall as a unit. Thus, in the separation of the human placenta, there is considerably more bleeding than there is at delivery of the placenta in many other mammals, where there are looser bonds between fetal and maternal tissue in the placenta.

Newborn infants (neonates) must make rapid respiratory and circulatory adjustments during the transition from complete dependence on the placenta to independent life outside the uterus (figure 20.23). Circulation to the placenta normally shuts down after delivery. This deprives the neonate of placental gas exchange and leads to a falling oxygen concentration and a rising CO_2 concentration in its blood. Brain respiratory centers respond to this increasing level of CO_2 by sending impulses to chest and abdominal muscles, thereby stimulating the contractions needed to initiate breathing if the infant has not already gasped or begun to cry in response to pressure after delivery.

But still, a newborn human infant is helpless and immature in comparison to other neonatal mammals. Baby pigs or calves, for example, are standing up, walking around, and attempting to feed within minutes following birth. Human neonates are weak and are only poorly able to regulate their body temperatures. They are completely dependent on parental care.

Regulating Human Fertility

While population growth rates are falling toward zero in the United States and several other industrialized countries, this is not the case in many underdeveloped and developing countries. Population growth either causes or aggravates practically all international social and political problems today. Effective means of dealing with population growth problems depend on development and distribution of safe and biologically effective birth control techniques.

In addition to being an international social problem, regulation of fertility is a pressing individual concern. Family size can be an urgent personal and financial problem, and effective family planning depends on adequate birth control technology and information.

Birth Control

The prevention of pregnancy is called **contraception** (literally "against conception," though not all forms of birth control actually prevent fertilization). Contraceptive techniques are evaluated on the basis of several criteria. Are they effective in reliably preventing pregnancy? Are they reversible so that when couples do wish to have children, they can stop using the technique and successfully initiate a pregnancy? Are they safe and relatively free from physiological side effects? Are they acceptable on personal and social grounds; that is, do they interfere with sexual enjoyment for one or both partners, or do they place an unfair burden of responsibility for contraception on one member of the pair? The birth control techniques currently in use all fall short of these criteria for one reason or another. Thus, the available techniques must be judged on their relative merits and weighed against the chances of unwanted pregnancy.

Birth control techniques fall into the following categories: (1) abstinence from intercourse, especially during the portion of the menstrual cycle when conception might occur; (2) suppression of egg or sperm production or release; (3) prevention of contact between egg and sperm by use of physical barriers; (4) prevention of blastocyst implantation; and (5) abortion (termination of pregnancy by removal of the embryo after it has implanted in the uterine wall). Techniques from all of these categories are used to control human fertility, but their desirability and effectiveness vary considerably (see table 20.2).

Box 20.1

Infertility, Sperm Storage, and "Test-Tube Babies"

On a personal level, infertility can represent as urgent a problem as excessive population growth can on an international scale. A significant percentage of couples wishing to have children are unable to do so. Some of this infertility is irreversible, but research into human reproductive physiology also has yielded some solutions.

One cause of male infertility is production of inadequate numbers of sperm. The average human male ejaculation produces 3 or 4 ml of semen and, with 120 million sperm per ml, a total sperm count of nearly 500 million. Because only one sperm can fuse with an egg in fertilization, 500 million seems to be an excessive number. But, in fact, when a man's sperm count is as low as 40 to 50 million per ml, he is likely to be infertile. Sperm storage technologies have provided help in such cases because semen from several ejaculations over a period of time can be stored, and the sperm can be concentrated and combined. This preparation can then be introduced into the vagina (by **artificial insemination**) at an appropriate point in the menstrual cycle, and pregnancy sometimes results. Of course, this technique does not help in cases where no sperm are produced, where sperm counts are extremely low, or where structurally abnormal sperm are produced.

A common cause of female sterility is blockage of the oviducts. This blockage prevents eggs from moving down and eliminates any possibility of fertilization. A technique for treating this condition was pioneered in England by Patrick Steptoe and Robert Edwards. In this technique, women are given pituitary hormone injections (the same treatment given other women who fail to ovulate because of pituitary gonadotrophin deficiencies), which cause maturation of one or several follicles. An oocyte is then removed through a small abdominal incision and placed in culture medium. Sperm are added, and fertilization occurs. After a blastocyst develops in culture, it is introduced into the uterine cavity at what would be the normal time of implantation if fertilization had occurred in the oviduct and the zygote had developed normally on its way down the oviduct to the uterus. If implantation occurs, development can proceed normally in some cases.

Such medical treatment raises moral problems because some people feel that it is not right to overcome "natural" infertility. Also, during the early experiments that laid the groundwork for these treatments, human zygotes were cultured and discarded. Some people question whether this can be morally justified. And yet, the techniques that resulted from these experiments permit the couples involved to become parents, which they could not have done without treatment. The borderlines are rather fuzzy when the issue involves what is "natural" and what is "not natural" in human medicine.

These techniques for *in vitro* (literally, "in glass," but generally meaning cultured outside the body) fertilization and early embryo development have also opened other new possibilities. For example, it is now possible to freeze embryos obtained in this way and store them. After a time, an embryo can be thawed and introduced into its mother's uterus.

Soon a couple who wish to do so could store their entire family and space pregnancies as they wish. This might be a great advantage if one or both are exposed in their work to radiation or chemicals that might over a period of time result in increased risk of mutations that could cause embryos to develop abnormally. Embryo storage would also allow a young couple to choose safe, but potentially irreversible birth control treatments such as vasectomy or tubal ligation without waiting years to "complete their family." But what would be done with any stored embryos in the event of a divorce?

It is also possible that embryo transfer could become a new form of adoption. Consider a couple who cannot conceive their own children. The woman, however, is fully capable of bearing a child. Might they not elect to "adopt" an embryo and thus have the pregnancy experience? But how would embryo donors be chosen? Would such procedures deprive many already born babies of potential adoptive parents? Most important of all, who decides the rules to be applied to these new patterns of human reproduction?

Table 20.2
Summary Information about Various Birth Control Techniques. Effectiveness is presented as the average number of women who become pregnant in a population of 100 sexually active women using the technique for one year. Undesirable side effects occur in only a small percentage of individuals in the case of most techniques.

Method	Effectiveness (Pregnancies per 100 Women in One Year)	Required Medical Services	Possible Undesirable Side Effects
Tubal ligation	0.04	Surgical procedure	Infection during surgery
Oral pill (21-day administration)	0.07	Prescription, regular physical exams	Early: water retention, breast tenderness, nausea Late: blood clots, hypertension
Vasectomy	0.15	Surgical procedure	Infection, possibly an autoimmune reaction
Intrauterine device (IUD)	1.5–3.4	Insertion	Menstrual discomfort, increased menstrual flow, uterine infection
Diaphragm*	12.0	Sizing, instruction	None known
Condom	14.0	None	Decreased sensation and sexual pleasure
Withdrawal	18.0	None	Decreased sexual pleasure
Spermicides alone	20.0	None	Usually none, occasional irritation
Abstinence during fertile period (rhythm method)†	24.0	None, physician counseling recommended	Psychological and physical frustration
Morning-after pill‡		Prescription	Breast swelling, nausea, water retention, cancer

*Effectiveness improved when used in combination with spermicidal jelly
†Effectiveness improved when monitored with cyclical temperature changes
‡Not frequently used as a form of birth control anymore because DES, the drug in the morning-after pill, is considered to be quite dangerous

Abstinence

Usually, total abstinence is not psychologically acceptable, so abstinence is used in the context of the so-called "rhythm" method of birth control. Couples refrain from intercourse during what is judged to be the fertile period, that is, for several days around the time of ovulation in the middle of the menstrual cycle. But the method is not totally reliable because variations in the length of menstrual cycles and days of ovulation are very common.

Suppression of Gamete Production

The most popular birth control technique in industrialized countries is the use of hormonal contraceptive pills that prevent ovulation. Characteristically, these pills contain small quantities of estrogens and larger quantities of a progesteronelike compound. These hormones suppress pituitary release of FSH and LH by interfering with the normal feedback relationship between the ovary and the hypothalamus and pituitary, thus preventing ovulation. Hormonal contraceptive pills also affect the remainder of the reproductive tract. They change the lining and motility of the oviducts. They also change the way in which the uterine endometrium grows and differentiates so that it probably would not be capable of accepting a normal implantation even if a blastocyst should reach it.

The "pill," as it is commonly called, is extremely effective in preventing pregnancy. It is taken from the fifth to the twenty-fifth days after the onset of the last menstruation. When a woman stops taking the pill on the twenty-fifth day, the endometrium begins to slough off in what approximates normal menstruation. But taking the pill on a daily basis, with no break, prevents menstruation completely. Some women may find this convenient or desirable, but they should be warned that physiologists have little idea of the long-term consequences of this abuse, which obliterates the natural cycle of menstruation.

Some unpleasant side effects of birth control pills experienced by certain women are nausea, dizziness, headaches, and vomiting. These and other symptoms similar to those of early pregnancy can make the pills undesirable for some women. But far more serious are problems of increased tendency toward blood clot (**thrombus**) formation within the

Figure 20.24 Surgical techniques that prevent normal movement of gametes. (*a*) Tubal ligation. (*b*) Vasectomy.

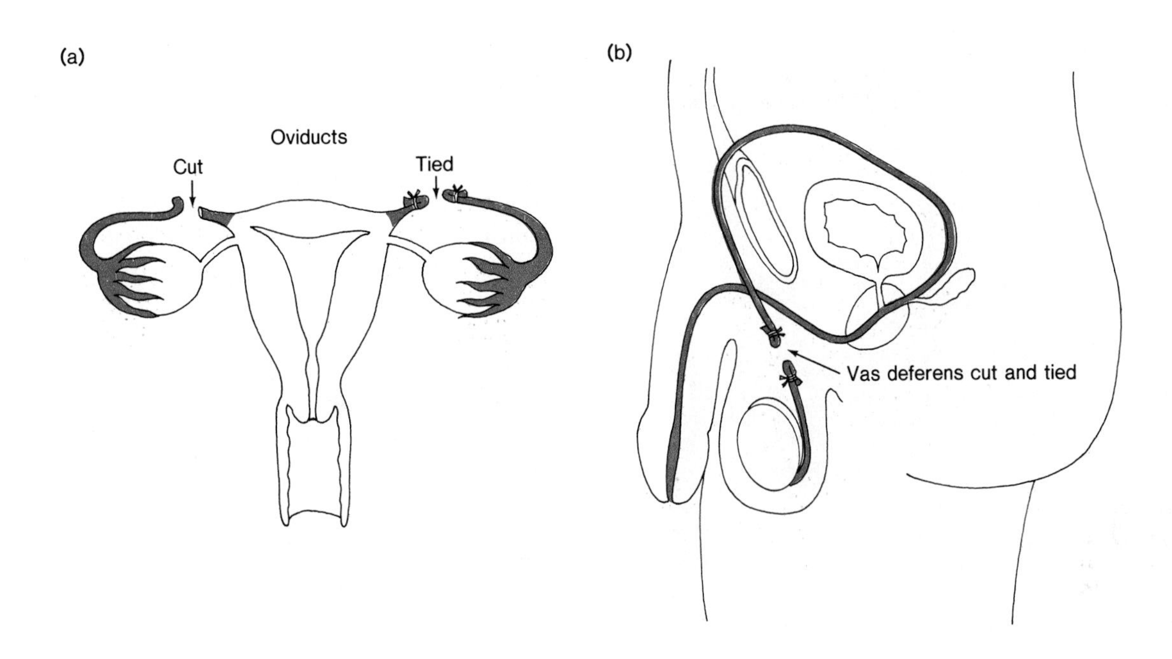

circulatory system of some women who take hormonal contraceptive pills. Occasionally, these clots come loose and circulate, and such a circulating clot (**embolus**) can lodge in a blood vessel somewhere else in the body and block a vital circulatory route. These problems must be weighed against the effectiveness and convenience of the pill, as well as the fact that the immediate risks of taking the pill are smaller than the risks that would accompany all of the pregnancies that would occur in the same population in the absence of contraception.

One unknown factor is the lifelong effect of this type of extended hormone therapy, and further study is needed in this area. Research currently is being done on hormonal contraceptives that contain much lower doses of hormones and might therefore cause fewer side effects. Another possibility for the future is development of implantable capsules that would release small quantities of hormones over a long period of time and relieve the necessity of taking pills daily.

Research on suppression of sperm production is far behind research on suppression of ovulation. Some compounds do cause reversible suppression of spermatogenesis, but the range of their possible side effects has not been completely explored. Already it is known that men taking these compounds must avoid any form of alcohol consumption because alcohol causes an unpleasant or even dangerous reaction.

Prevention of Egg-Sperm Contact

Some very old techniques and some newer surgical techniques fall into this category. Physical barriers in the vagina that prevent sperm from moving into the uterus and oviduct have been used for years. The most widely used current version of these devices is the **diaphragm.** Diaphragms are latex domes that are inserted into the upper part of the vagina to cover the cervix. They usually are used with a **spermicidal** ("sperm-killing") **jelly** that is smeared around the edge of the diaphragm. This combination is quite effective.

Another physical barrier device is the **condom,** a rubber or plastic sheath that is pulled over the erect penis just before it is inserted into the vagina. The condom traps semen and quite effectively blocks sperm transfer into the vagina.

Surgical techniques can prevent egg-sperm contact by blocking gamete transport in the reproductive system. **Vasectomy,** the cutting and ligature (tying off) of the vas deferens is the male version of this procedure, and **tubal ligation,** the cutting and ligature of the oviduct, is the female version (figure 20.24). Although microsurgical procedures can reverse some vasectomies and tubal ligations, both of these techniques should be considered irreversible.

Figure 20.25 Two types of intrauterine devices: the Copper 7 (left) and the Lippes loop (right). Apparently, there is no special shape requirement. An effective IUD must remain in place and induce the necessary reaction of the uterine endometrium.

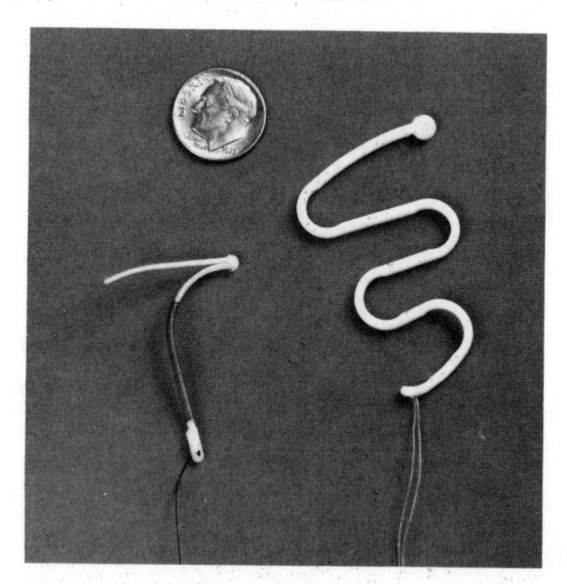

Both tubal ligations and vasectomies are done routinely, but each involves an element of risk. Tubal ligation does not interfere with normal ovulation or other ovarian functions, but the procedure requires opening the abdominal cavity, which always entails some risk, no matter how small the incision. Risks of infection in vasectomy are much lower, and the procedure is simpler, but some long-term effects are still unknown. Thus, tubal ligations and vasectomies entail some established risks and some possible risks, which must be weighed against their virtually 100 percent effectiveness in preventing pregnancy.

Prevention of Blastocyst Implantation

Intrauterine devices (IUDs) are placed inside the uterus, where they prevent implantation of blastocysts, apparently by causing slight inflammatory responses in the endometrium. Current intrauterine devices include pieces of plastic in various shapes and small copper devices, which seem more effective (figure 20.25). Intrauterine devices are inserted through the cervical opening and left in place in the uterus. They have the advantages of not requiring some action before or during sexual activity and not involving continuing hormone therapy. But they cause excessive menstrual bleeding and cramps in some women and produce uterine damage in a small percentage of cases. Sometimes, they are spontaneously expelled from the uterus, and pregnancy occurs before the loss is detected. Properly inserted IUDs that remain in place are very successful in preventing pregnancy and are cheap enough to provide hope for effective pregnancy control even in developing countries.

Blastocyst implantation also can be prevented by a hormonal treatment. The so-called "morning-after" pill is a large dose of an estrogenic substance such as diethylstilbestrol (DES). DES is a synthetic compound that produces abnormal growth of the endometrium, making normal implantation impossible. This treatment can be used in an emergency to prevent pregnancy resulting after an incident of unplanned or forced intercourse without contraceptive protection. But it must be given for several days after intercourse, and its side effects, such as nausea, vomiting, and heavy vaginal bleeding, are unpleasant.

If DES is given mistakenly to a pregnant woman during a specific sensitivity period, it has an important effect on a developing female baby. Daughters of women treated with DES during pregnancy show a markedly increased incidence of certain types of cancer of the reproductive tract. This tendency does not express itself until these women reach their twenties or thirties, and only in the last few years has it been statistically connected with DES. The ideal morning-after pill of the future, if it is developed, will not be DES, with its carcinogenic (cancer-causing) potential.

Abortion

Artificially induced abortions provide means of terminating pregnancy after implantation, but abortion is a controversial method of regulating fertility in most societies because of moral and legal concerns.

During the first three months of pregnancy, abortion can be induced by expanding (dilating) the cervical opening and scraping the uterine lining, including the implantation site or by an aspiration technique that removes the embryo by sucking it out through a tube. These techniques lead to menstruationlike bleeding.

New techniques for induction of abortion are being developed using some of the **prostaglandins,** a group of seemingly ubiquitous chemical messengers (chapter 11). Prostaglandins induce uterine contractions that lead to menstruationlike bleeding. In the future, a monthly vaginal prostaglandin insert might replace other forms of birth control. The insert would be used to induce menstruation each month, and the users would have no idea whether or not pregnancy had begun. Clearly, this sort of technique will also raise moral questions.

Sexually Transmitted Diseases

Another social problem related to human reproductive function is that of **sexually transmitted (venereal) diseases.** There are a dozen or more known sexually transmitted diseases, and probably more to be discovered.

Gonorrhea is a bacterial disease (figure 20.26*a*). In men, gonorrhea is usually readily detected because pus is discharged from the penis, and burning sensations during urination develop within a few days after infection. But in women, the infection concentrates in the cervical canal and produces very mild symptoms, if any at all. Thus, women may unknowingly develop extended infections in the reproductive tract. The most serious result of such extended infections in women is infection and inflammation of the oviducts, which can lead to partial or complete blockage of the ducts and eventually cause sterility. Gonorrhea routinely has been treated with large doses of penicillin, but penicillin-resistant strains have become more common, especially since the Viet Nam war, and treatment is more complex than it once was.

Syphilis, caused by a microorganism called a spirochete, has a rather long incubation period. On the average, the first symptom, a hard painful ulcer called a **chancre,** develops at the site of infection after about three weeks. Even if the disease is left untreated, the chancre disappears after a short time, and no more sign of disease is seen until two to four months later. At this time the disease enters its secondary stage as a generalized skin rash, and infections of various organs sometimes develop. Then the disease goes into a latent (inactive) period. This latent period can last throughout life, but in some people the disease can enter a tertiary phase that produces severe nervous system or circulatory system damage and even death. Syphilis also usually can be treated with penicillin, but as in the case of gonorrhea, increasingly common antibiotic resistance is a growing problem.

Several virus diseases are sexually transmitted (figure 20.26*b*). The best known of these is caused by the **Type 2 herpes simplex virus,** which is very similar to the virus that causes cold sores and fever blisters on the lips (Type 1 herpes simplex). Recently, however, genital infections with Type 1 herpes simplex also have become quite common. The incubation period is variable, but between two and twenty days after infection, blisters develop on the affected area. After the blisters rupture, the painful surface ulcers left by the blisters heal, and the disease becomes dormant. But as with cold sores, the infection remains and these blisters can reoccur repeatedly at variable intervals. Thus, herpes infections are particularly distressing because there presently is no cure, and because the painful symptoms return periodically.

Figure 20.26 Agents that cause two common venereal diseases. (*a*) Human white blood cell containing *Neisseria gonorrhoeae*. This cell was found in a smear of urethral discharge. Note the small dark bacterial cells. (*b*) Herpes simplex virus particles inside a cell.

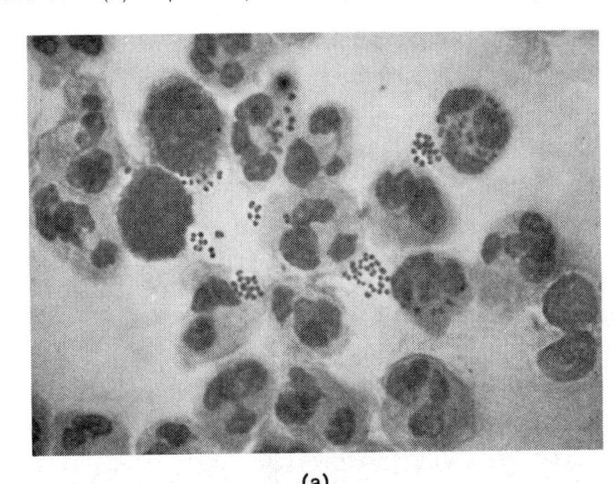

(a)

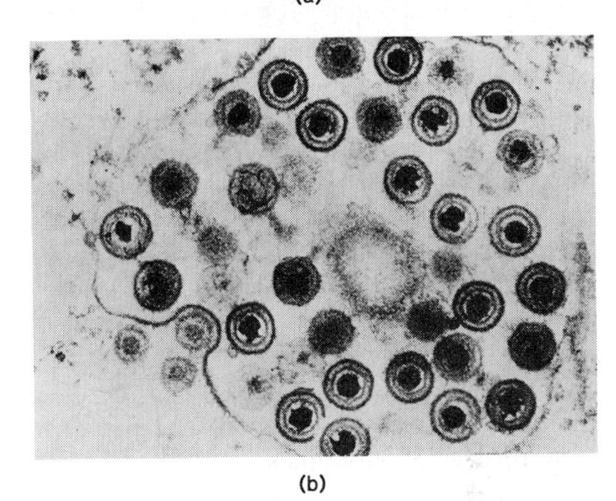

(b)

A variety of other diseases caused by viruses, bacteria, yeast, protozoa, and even small lice also are sexually transmitted.

Because it is becoming more and more common for organisms that cause sexually transmitted diseases to be resistant to antibiotic treatment, and because herpes infections currently are incurable, interest in immunization programs is increasing. You can imagine the perplexing social problems that would arise, however, by the development and experimental testing of vaccines for sexually transmitted diseases. And the publicity campaigns that would accompany these general vaccination programs undoubtedly would inspire considerable public debate.

Summary

The basic design of the human reproductive system is an adaptation to life in a terrestrial environment. Fertilization takes place inside the female reproductive tract, and there is a long gestation period of sheltered development.

Testes produce sperm and testosterone, the male sex hormone responsible for development of secondary sex characteristics. Sperm are transported to the epididymis, where they remain until ejaculation carries them through the vas deferens and urethra to the outside at the tip of the penis. During this passage, the secretions of several glands are added to the sperm to produce semen.

Ovaries produce oocytes and two kinds of steroid hormones—estrogens and progesterone. At ovulation, an oocyte is released and carried into the oviduct.

Copulation in humans involves transfer of seminal fluid into the vagina. Male orgasm is essential for reproductive success, but female orgasm is not.

Timing of reproductive events in the female is accomplished through the menstrual cycle. Pituitary hormones regulate maturation and ovulation of the oocyte, as well as hormone production by the ovary. Ovarian hormones, in turn, control cyclical events in the uterus.

Fertilization occurs in the upper part of the oviduct. Development begins in the oviduct. The blastocyst stage, which is reached in the uterus, is capable of implanting in the uterine endometrium.

During most of gestation, the placenta is the site where materials are exchanged selectively between mother and infant across a placental barrier. The placenta also functions as a hormone-producing organ.

The developing infant is enclosed and protected within the amnion and bathed with amniotic fluid throughout development.

Development of the embryo proceeds during and following implantation. Rudiments of all body organs are produced during the embryonic period, which ends at eight weeks. All body parts differentiate and grow during the long fetal period, which occupies the remainder of gestation.

Birth occurs about 266 days after fertilization. Uterine contractions expel the infant through an expanded cervix and vagina. Continued contractions result in expulsion of the afterbirth.

Regulation of human fertility is an important social and scientific concern. A variety of birth control techniques are in use, and each has advantages and disadvantages. The hormonal birth control pill is the most widely used technique in industrialized societies.

The incidence of sexually transmitted diseases is increasing. Antibiotic resistance to bacterial venereal diseases is becoming a common problem, and venereal diseases caused by viruses are being transmitted to many people.

Questions for Review

1. What are secondary sex characteristics?
2. List three glands that contribute to the seminal fluid that suspends sperm.
3. How does luteinizing hormone (LH) affect the ovary?
4. Distinguish clearly between the terms "oocyte" and "egg" in the context of human meiosis and fertilization.
5. Fertilization involves one egg and one sperm, and yet human males with sperm counts below 50 million per ml are likely to be infertile. Suggest a possible explanation.
6. Define the term inner cell mass.
7. Discuss implantation as a critical point in the continuing development of an embryo and as an example of "tolerance" in terms of the mother's normal defense mechanisms.
8. How is the production of progesterone by the corpus luteum maintained early in pregnancy?
9. Explain the distinction between the terms embryo and fetus as applied to human development.
10. Distinguish among the first, second, and third stages of labor.
11. List some advantages and disadvantages of intrauterine devices in birth control.
12. What is tubal ligation?

Questions for Analysis and Discussion

1. Can you suggest a hormonal basis for the physical and psychological letdown that some women experience just before menstruation? (Hint: Progesterone has powerful general metabolic effects.)
2. Many older legal systems make a distinction between developing humans during the first two months of pregnancy and those that have developed for more than two months. What developmental factors can be correlated with these traditional viewpoints?

Suggested Readings

Books

Jones, R. E. 1984. *Human reproduction and sexual behavior.* Englewood Cliffs, N.J.: Prentice-Hall.

Moore, K. L. 1982. *The developing human: Clinically oriented embryology,* 3rd ed. Philadelphia: Saunders.

Nilsson, L. 1973. *Behold man: A photographic journey of discovery inside the body.* Boston: Little, Brown.

Articles

Beaconsfield, P.; Birdwood, G.; and Beaconsfield, R. August 1980. The placenta. *Scientific American.* (offprint 1478).

Chedd, G. January/February 1981. Who shall be born? *Science 81.*

Edwards, R. G. 1981. Test-tube babies. *Carolina Biology Readers* no. 89. Burlington, N.C.: Carolina Biological Supply Co.

Epstein, C. J., and Golbus, M. S. 1977. Prenatal diagnosis of genetic diseases. *American Scientist* 65:703.

Hart, G. 1984. Sexually transmitted diseases. *Carolina Biology Readers* no. 95. Burlington, N.C.: Carolina Biological Supply Co.

Rhodes, P. 1976. Birth control. *Carolina Biology Readers* no. 4. Burlington, N.C.: Carolina Biological Supply Co.

Wallis, C. 1984. The new origins of life. *Time.* September 10.

Lifelong Developmental Change

Chapter Outline

Chapter Concepts

1. Developmental change continues throughout life.
2. Some homeostatically important developmental changes are involved in disease resistance.
3. Some mechanisms involved in resistance to infection and disease are general and nonspecific. Others are very specific and depend on precise recognition of foreign antigens.
4. Developmental processes that replace worn-out cells and repair damaged tissues continue throughout life.
5. Abnormal growth by transformed cells involves unregulated cell division and eventually interferes with maintenance of homeostasis.
6. Aging changes weaken organisms' abilities to resist infection and to make physiological adjustments required for maintenance of homeostasis.

How long does development last? You might first think that, in animals, development lasts until birth (or hatching, as the case may be). But after a moment's thought you would realize that your own development clearly continued through the teenage years, and even at this point in your life, you are not a finished product who will change no further. In fact, development continues throughout adulthood and even in old age. Some early developmental changes are sweeping and dramatic, while later ones, such as aging, are much more gradual. Aging is not a process found only in very old people. The onset of some biochemical changes, which become more obvious parts of the aging process later, can first be detected in humans before age twenty! Thus, development, defined in terms of continuing change, clearly is a lifelong process.

Some developmental changes are important for maintenance of homeostasis in organisms. For example, development of resistance to infection by various **pathogenic** (disease-causing) organisms is essential to well-being and continues throughout life.

Homeostasis also depends on continuing replacement of worn-out and lost cells, as well as repair of injuries, but all normal repair and replacement processes are regulated so that the number of cells produced is adequate for replacement and no more.

When the mechanisms controlling normal cell division fail, however, abnormal growth of cell populations threatens the well-being and even the lives of organisms. With advancing age, control over cell division breaks down more frequently and the incidence of abnormal growth increases.

A cycle of life, from vigorous and efficient activity to eventual aging and death, is all part of lifelong developmental change. In this chapter, we will consider several aspects of this change, beginning with defense against infection and disease.

General Defenses Against Foreign Organisms

Some general defenses against disease-causing organisms are simply barriers that keep potentially dangerous organisms away from parts of the body where the organisms might establish themselves and do damage. Other general defenses involve cell activities that are effective against a variety of foreign cells.

Surface Barriers

Land-dwelling plants usually have waterproof, tough, relatively impenetrable surfaces that prevent entry of some of the pathogenic bacteria and fungi that can grow and thrive only when they reach plants' internal tissue environments.

Animal body surfaces also serve as barriers to entry of microorganisms (figure 21.1). This is important because outside body surfaces are often densely populated with microorganisms, some of which are potentially pathogenic. This **normal flora** of the human skin, for example, usually causes no special problems, even though there may be as many as 1.5 to 2 million bacteria per square centimeter of surface, in addition to an assortment of yeasts and other fungi.

Antibiotics

An **antibiotic** is a chemical substance that is produced by a microorganism that can kill or inhibit growth of other microorganisms. We tend to focus on those antibiotics used in treating human diseases or diseases of domestic animals, but what is the significance of antibiotic production for the producing organisms themselves? The adaptive value of antibiotic production by microorganisms is that the antibiotics inhibit other organisms that compete for resources in their immediate environment.

Literally thousands of antibiotics have been discovered. Only a few of them are valuable as medicine, however, because to be safe and effective they must be selectively toxic; that is, they must be much more toxic to disease-causing microorganisms than they are to animal cells.

How are new antibiotics found? Potential antibiotic producers are isolated and grown in pure culture. Then discs of the culture medium containing the organisms are transferred to culture dishes seeded with bacteria. If the organism produces an antibiotic effective against the test bacterium, an inhibition zone develops around the disc (figure 21.2). Some antibiotics are effective against only a limited number of other organisms, but **broad-spectrum antibiotics** are effective against a wide range of organisms.

Chemical Inhibitors in Plants

Some plants' cells normally synthesize organic compounds, such as creosote, that interfere with the metabolism of invading microorganisms. Other inhibiting compounds, called **phytoalexins,** are produced in some plants only when an invading bacterium or fungus is present. Once phytoalexin production has begun in some plant tissues, growth of invading organisms is very strongly inhibited (figure 21.3).

Figure 21.1 Surface barriers as passive defense against infection in animals. Skin is relatively impermeable. It has fatty acids on its surface and therefore has a rather low pH. Digestive system linings are barriers, stomach acid kills some bacteria, and the normal bacterial flora of the intestine inhibits growth of other organisms. Cilia along air passages move materials away from lungs, which are more susceptible to infection. Mucus secretion helps to keep bacteria away from surfaces, which they might otherwise penetrate.

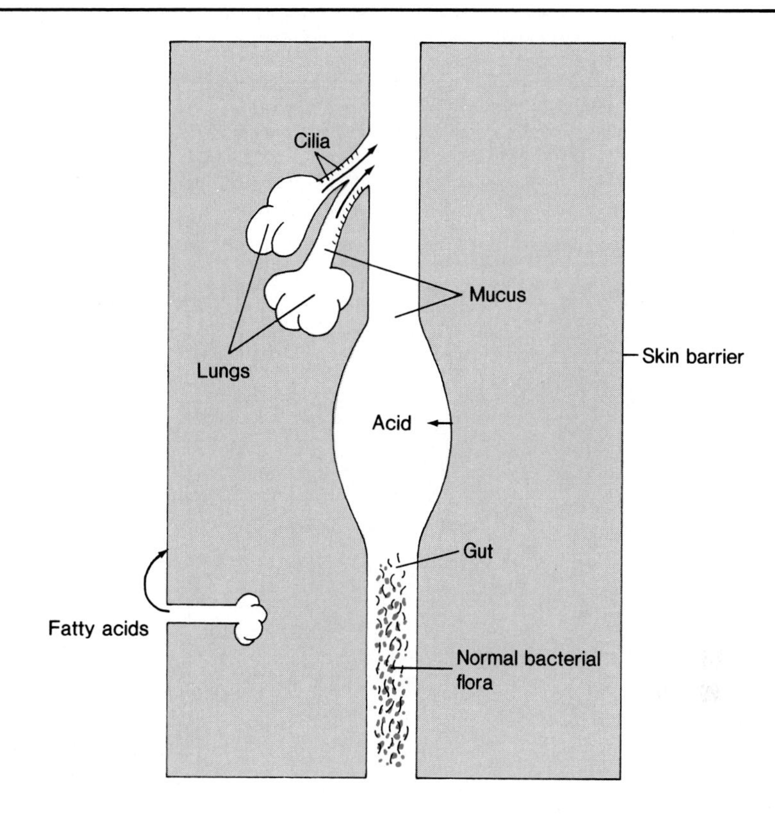

Figure 21.2 Laboratory tests for antibiotic production. (*a*) A disc of culture medium containing an organism being tested for antibiotic production is transferred to a plate seeded with bacteria. If the antibiotic is released, bacteria fail to grow in a zone around the disc. (*b*) A simultaneous test of nine different possible antibiotic producers against a particular type of bacteria. Note the inhibition zones around four of the discs.

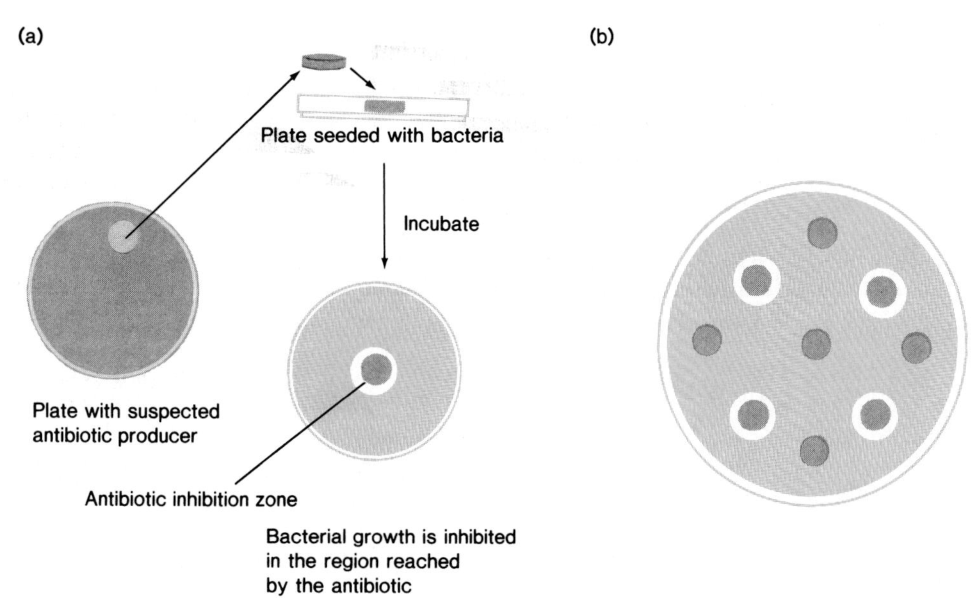

Interferon

Interferon is an antiviral substance produced by animal cells when the cells are infected by certain types of viruses. Interferon was discovered as a result of studies on the **interference phenomenon,** the observation that infection by one virus makes cells resistant to infection by additional viruses.

When viruses infect cells, they take over the cells' synthetic apparatus and cause the cells to make more virus particles. When these new virus particles are assembled in a host cell, the cell bursts and releases them. These many new particles can then infect additional cells, where the multiplication cycle is repeated. If this process continues unchecked, huge numbers of host cells soon are destroyed, with disastrous consequences for the host organism.

Interferon breaks this chain of cell infection and reinfection. While infected cells are synthesizing components of new virus particles, they also synthesize interferon, which they release into their environment. Interferon subsequently protects other cells against virus infection and thus breaks the chain of infection (figure 21.4).

Interestingly, interferon is not specific for types of viruses, but for the host cells. In response to flu viruses, interferon produced by chickens, for example, inhibits multiplication of other types of viruses in chicken cells. But it has little or no effect on multiplication of viruses—even flu viruses—in the cells of other vertebrate animals.

Figure 21.3 Diagram of an experiment on phytoalexins and plant resistance to disease that was done by Müller and Börger in 1941. Areas of cut surfaces (in gray) of potato tubers were first inoculated with an avirulent (weakly infective) form of a blight fungus that grew only a little. Later, the entire surfaces were inoculated with a virulent form that grew everywhere (colored) except in the previously inoculated areas. The "protection" came from phytoalexin produced in response to the first inoculation.

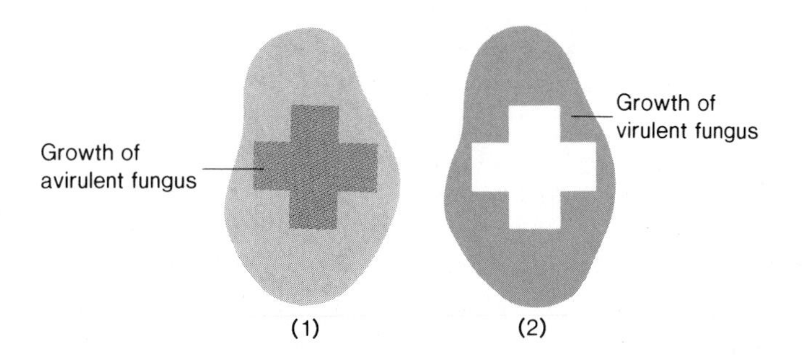

Figure 21.4 Alick Isaacs and Jean Lindemann discovered interferon in 1957. They showed that a material left in culture medium after viruses and virus-infected cells were removed from it would protect a new batch of healthy cells from virus infection. Since then, it has been shown that cells infected by virus particles (*a*) produce and release interferon (*b*) as they also produce new virus particles. These first cells rupture (lyse) and release new virus particles (*c*). Interferon then prevents virus multiplication in other cells (*d–f*).

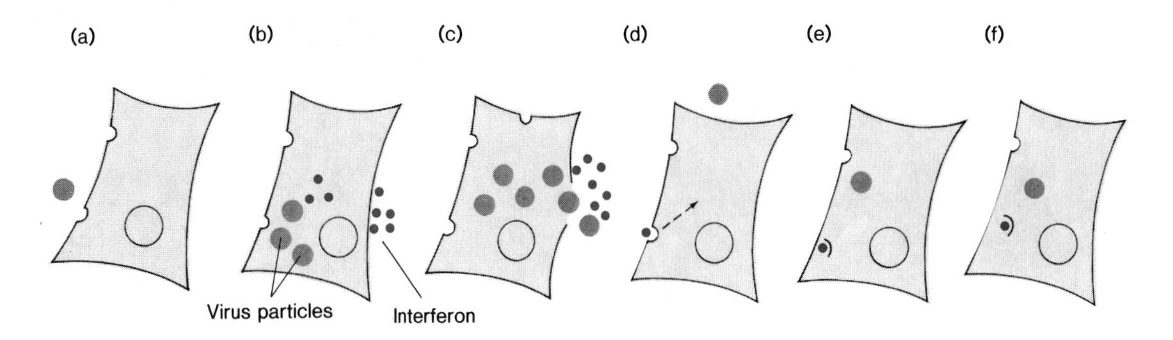

The host cell specificity of interferon relates to its mode of action. Interferon released from infected cells binds with specific cell surface receptor molecules on other cells. These interferon-receptor complexes then enter the cells and cause responses that prevent viral replication and further infection. Host cells must have receptors that bind specifically with interferon molecules if interferon is to be effective.

Interferon might seem to be the ultimate weapon against virus infection, and it could well become that in the future. However, interferon is difficult to purify, and until recently, it had to be produced by cells of the species in which it was to be used; that is, human cells had to be used to produce human interferon. Now genetically engineered bacteria can be made to synthesize human interferon in large quantities because the human interferon gene has been isolated and cloned. Large-scale production of interferon will accelerate research and clinical trials on interferon's antiviral actions.

Interferon might hold promise of being more than a new and powerful treatment for virus diseases. It also is a cell division inhibitor in normal cells, and especially, in tumor cells. Some very optimistic predictions have been made about future roles for interferon in cancer treatment. Interferon research will be one of the most active and interesting areas of biomedical investigation for years to come.

Phagocytosis

Phagocytosis is the process by which amoeboid cells in animal bodies engulf and destroy microorganisms, other cells, and various foreign particles. Phagocytosis has been recognized as a general defense mechanism for about 100 years, ever since Elie Metchnikoff found that cells accumulated around splinters in human skin. One afternoon, as he was watching motile cells within transparent starfish larvae, it occurred to him that amoeboid cells might protect other organisms against harmful invaders. He set out to test this hypothesis by picking rose thorns from his garden and inserting them into the starfish larvae. The next morning he found that the intruding thorn tips were surrounded by clusters of the amoeboid cells. We now know that phagocytosis is one of the weapons in the defense arsenal of virtually all animals.

In humans, for example, there are several important types of phagocytic cells. **Polymorphonuclear leukocytes,** sometimes also called **granulocytes** because they contain distinctive granules, are found mainly in the bloodstream (figure 21.5*a*). They generally are involved in relatively quick, focused, short-term responses to infection. **Monocytes** are cells

Figure 21.5 Phagocytic cells. (*a*) Two types of phagocytic cells. This granulocyte (polymorphonuclear leukocyte) is actually a neutrophil (chapter 10). Red blood cells average 7 μm in diameter so they can be used for a size comparison. (*b*) Phagocytosis by a monocyte. Bacterium (color) is digested by material from a lysosome. Remains are expelled from cell (5).

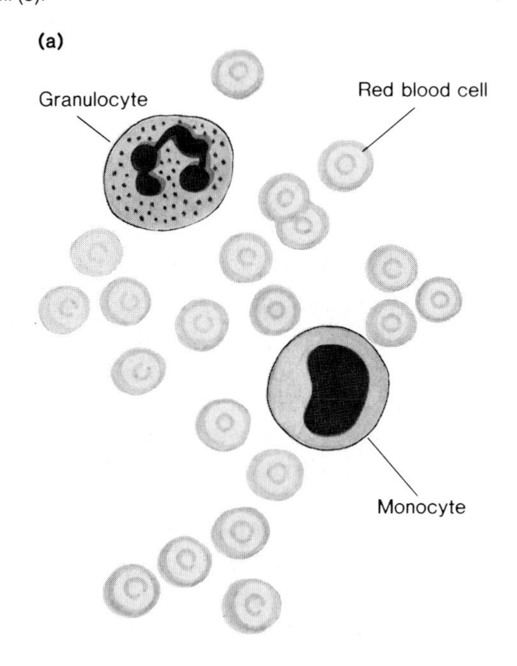

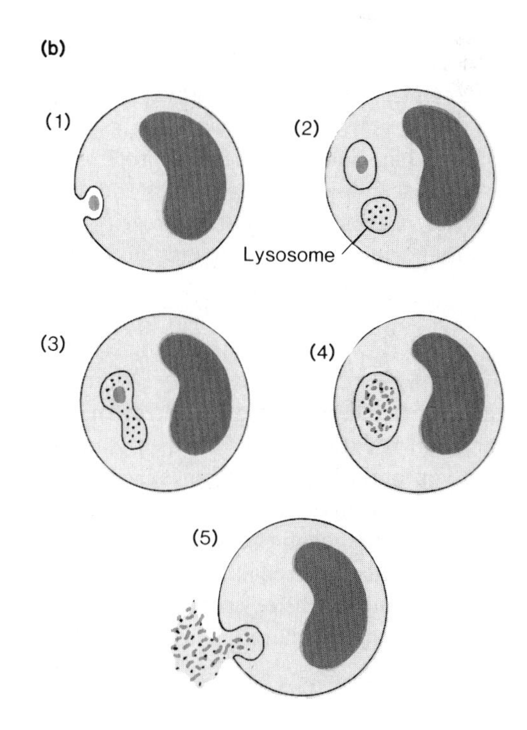

Figure 21.6 Multiple holes in a bacterial (*Escherichia coli*) cell wall caused by complement proteins (magnification × 176,000).

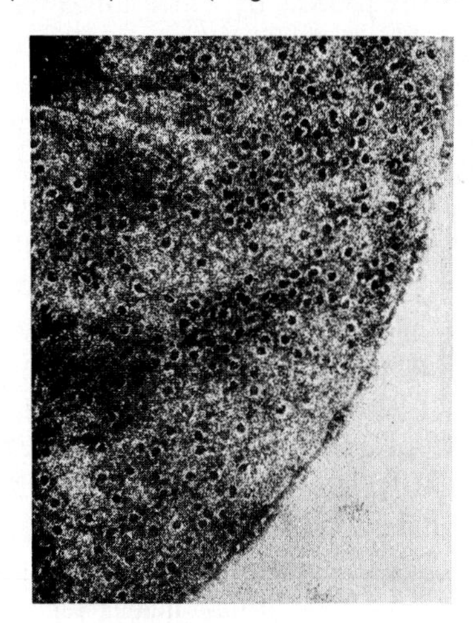

that regularly move freely between blood vessels and tissue spaces. In the tissue spaces they sometimes enlarge and remain as **macrophages.** Macrophages generally are responsible for long-term phagocytic activities, such as cleaning up dead microorganisms and host cells at the end of a successful battle against infection (figure 21.5*b*).

After foreign cells have been taken in by a phagocytic cell, they are digested by enzymes contained in lysosomes, and finally they are expelled from the cell.

Complement

Certain blood serum enzymes known collectively as **complement** or the **complement system** are always present in the bloodstream. When activated, they work against a variety of invading foreign cells. Some complement proteins make foreign cells more susceptible to phagocytosis and actually attract phagocytes. Others destroy the integrity of the cell membrane so that cytoplasm leaks out and the cell dies (figure 21.6).

Inflammation

Inflammation is a tissue and blood vessel response to infection or injury that is externally visible as a swollen, warm, and often, as you probably know from experience, painful area. Inflammation involves movement of fluid, plasma proteins, and especially phagocytic cells out of the blood vessels into tissue spaces. Chemical regulators facilitate this movement by causing contractions of cells in the walls of blood vessels, making passage of materials easier (figure 21.7*a*).

Figure 21.7 Inflammation. (*a*) During inflammation, cells in vessel walls shrink, allowing easier movement of phagocytic cells (such as the polymorphonuclear, PMN, leukocytes) into tissue, as well as increased movement of fluid and plasma proteins. (*b*) Histamine, a powerful inflammation promoter.

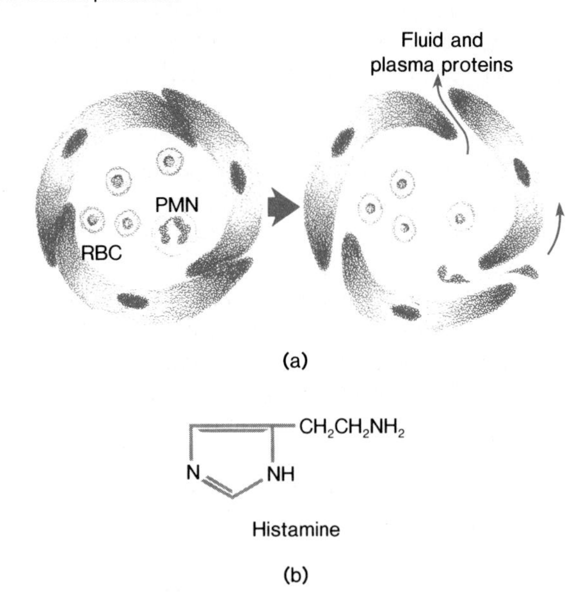

$CH_2CH_2NH_2$

N NH

Histamine

(b)

One of the best known of these chemical regulators is **histamine,** a substance released from **mast cells** (figure 21.7*b*). Complex inflammation responses are involved in several different types of normal reactions to foreign material. Some **allergy reactions,** however, can involve excessive and widespread expression of inflammatory-type responses. Sometimes the response becomes so extreme that a drop in arterial blood pressure, bronchial spasms, and even death can result. Drugs known as antihistamines are used to combat excessive inflammatory responses.

Specific Recognition and Defense Mechanisms

It has been known for centuries that once people have recovered from certain diseases, they are not likely to get them again—they have become resistant (**immune**) to subsequent infection. Edward Jenner, in the late eighteenth century, proved it was not necessary to have a disease to acquire immunity to the disease. Jenner observed that farm workers who had been exposed to cows with cowpox did not contract smallpox, which at the time was a common and extremely serious disease. Jenner purposely infected people with cowpox pus by scratching it into their skin. This treatment caused a very mild disease with only one pock in humans, and such **vaccinated** (from the Latin *vacca,* meaning cow) people did not catch smallpox. Later, Louis Pasteur and a long line of other workers developed vaccinations for many other infectious diseases. Through proper vaccinations, some diseases have been almost eliminated in developed areas of the world. Interestingly, in the late 1970s, smallpox, the first disease to be prevented by vaccination, became the first disease to be proclaimed completely eradicated by the World Health Organization.

Specific immunities to diseases are only part of a much broader system of specific chemical recognition that normally makes distinctions between parts of the body ("self") and foreign cells and their cell products ("nonself"). For example, certain cells of vertebrate animals recognize foreign proteins and carbohydrates, especially those on the surfaces of microbial cells, and respond with defense reactions aimed at removing or chemically neutralizing those cells. Substances that provoke such specific responses are called **antigens.**

Specific immune responses fall into two categories. One category involves direct attacks by certain white blood cells (**lymphocytes**) on foreign antigens. This is **cell-mediated immunity (CMI).**

The second response category involves other lymphocytes, which specialize to become **plasma cells.** Plasma cells produce **antibodies** (protein molecules that combine specifically with antigens). This **humoral antibody synthesis** is important in body defense in general and absolutely vital for acquired immunity to specific disease-causing microbes.

B Cells, T Cells, and Cellular Immunology

In the late 1950s, Bruce Glick was studying the role of the **bursa of Fabricius,** a saclike structure in birds (figure 21.8). He was bursectomizing (removing the bursas from) young chickens to test the importance of the bursa for normal growth and sexual maturation. When a colleague asked Glick for some chickens for a laboratory demonstration of antibody production, Glick supplied him with some of the bursectomized chickens. They were surprised to find that, in response to antigen injection, bursectomized chickens produced antibodies only poorly or not at all.

This result, along with Noel Warner's work on thymectomized chickens (chickens whose thymuses were surgically removed), led Robert Good and Max Cooper to do a series of experiments in which they surgically removed either the thymus or bursa from young chickens and studied the chickens' immune responses. All of this research eventually proved that the thymus and the bursa each are responsible for production of a distinct group of small lymphocytes, now called **T cells** and **B cells,** which are involved in cell-mediated immunity and humoral antibody synthesis, respectively.

Figure 21.8 Location of the thymus (in the neck) and the bursa of Fabricius of a young chicken.

It soon became clear that specific immune responses of mammals, including humans, show a similar two-partedness. In birds, lymphocytes become T cells because they pass through the thymus, and B cells develop under the influence of the bursa of Fabricius (figure 21.9). In mammals, T cell development seems the same. But since mammals do not have a bursa of Fabricius, immunologists must speak of a hypothetical "bursa-equivalent" when discussing B cell differentiation in mammals.

Cell-Mediated Immunity

As we said earlier, T cells produce responses collectively known as cell-mediated immunity (CMI). CMI is also known by an older name—delayed hypersensitivity—because the responses take from one to several days to develop as opposed to immediate hypersensitivity, which is caused by circulating antibodies and develops in minutes. CMI is involved in many allergic reactions to bacteria, viruses, and fungi, as well as in contact dermatitis (skin irritation) resulting from sensitization to certain chemicals. It also causes the hardening and reddening reaction of skin in the Mantoux test for tuberculin sensitivity and is responsible for rejection of transplanted tissue.

Figure 21.9 Processing of stem cells (unspecialized cells) to produce cells capable of functioning as T cells or B cells.

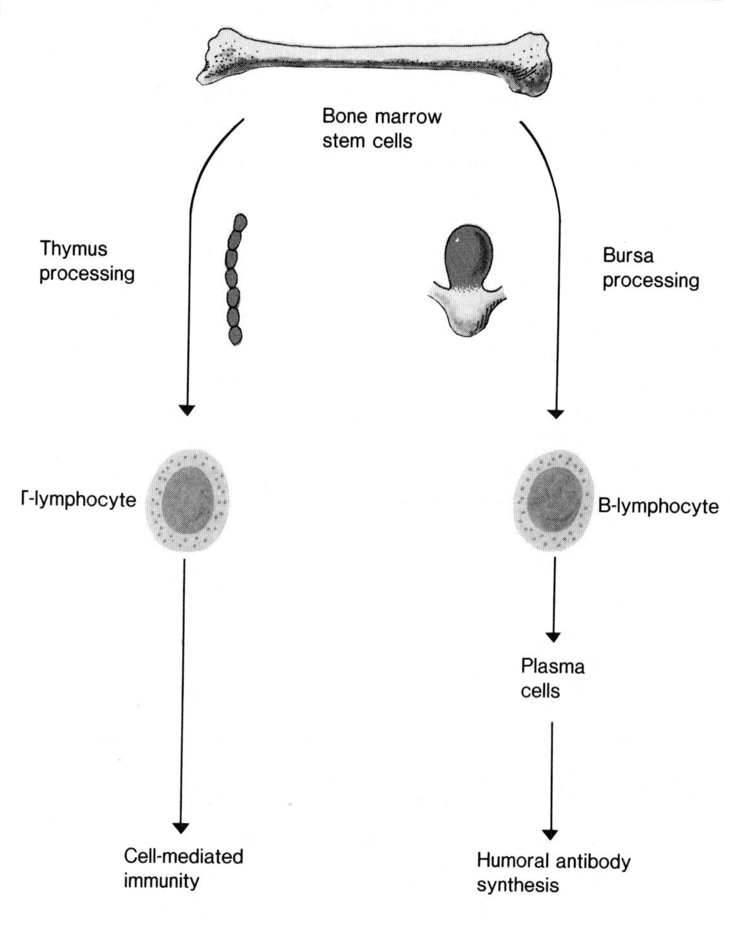

Figure 21.10 T cells divide after antigen stimulation and specialize for roles in cell-mediated immunity.

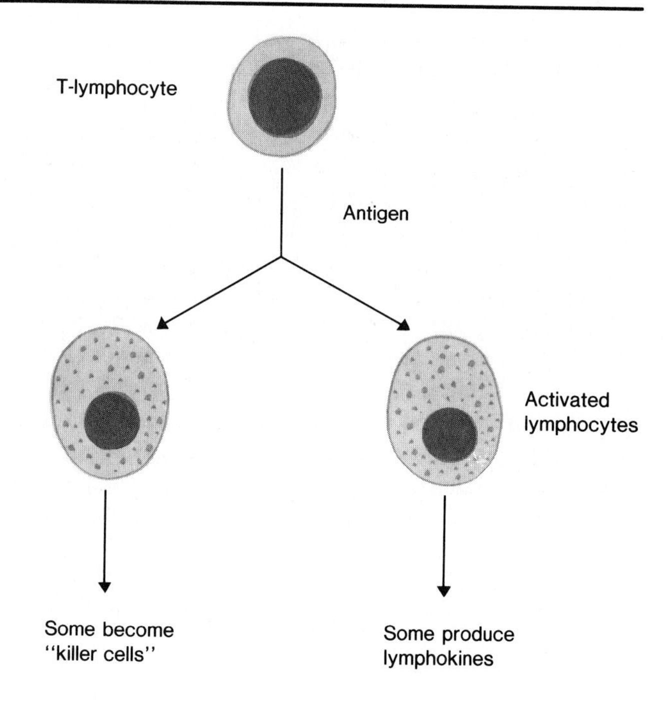

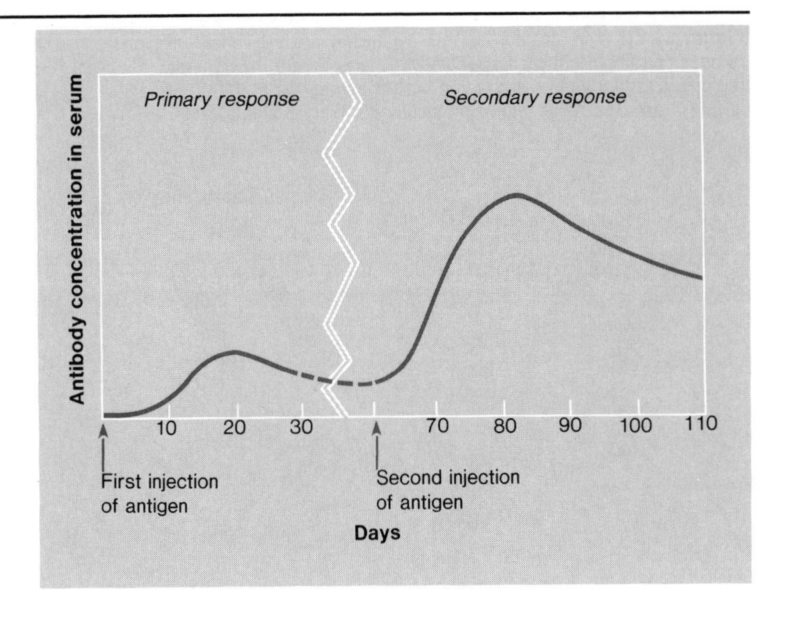

Figure 21.11 The results of two exposures to an antigen showing the difference between primary and secondary responses.

A cell-mediated immune response begins when some of the T cells become activated by a specific foreign antigen and begin to divide mitotically. Some of the T cells produced develop cytotoxic powers and become "killer cells" that attack and kill foreign cells directly, while others release substances called lymphokines (figure 21.10). Some lymphokines attract phagocytic cells, while others activate phagocytic cells and make them more active ("angry") in their attack on bacteria.

Humoral Immunity

B cells are responsible for humoral antibody production. In response to a specific foreign antigen, certain B cells develop into plasma cells that produce and release an antibody that binds specifically with the antigen. Other B cells become **memory cells,** which somehow retain a special responsiveness to the specific antigen so that when the system is challenged again by the same antigen, antibody production is much quicker and more efficient than it was the first time (figure 21.11). Thus, "memory" accounts for the difference in humoral antibody production between **primary response** (following initial exposure to an antigen) and **secondary response** (following subsequent exposures to the same antigen). This difference is the basis for acquired immunity to infectious organisms.

Immunoglobulins

Antibodies are **immunoglobulins,** a class of protein molecules present in blood, tissue fluids, and certain secretions, such as mucus, tears, and saliva. They are identified by letters: IgG (for immunoglobulin G), IgA, IgD, and IgE. About 80 to 85 percent of the circulating antibodies belong to the IgG class. These circulating immunoglobulins are found in what is called the gamma globulin fraction of blood serum.

Structure of IgG Molecules

R. R. Porter and Gerald Edelman shared a Nobel Prize in 1972 for their work on the structure of antibodies. They determined that the IgG antibody consists of four polypeptide chains. Two of these chains are called **light chains,** while the other two larger chains are called **heavy chains.** You can visualize the molecule as two heavy chains forming a Y shape, with a light chain located along each side of the arms of the Y (figure 21.12). Disulfide bonds hold the chains in place relative to one another, but the antibody is flexible at two hinge areas of the heavy chains so that the arms of the Y can open as wide as 180°.

Figure 21.12 The structure of IgG. V_H and V_L are the variable regions of heavy and light chains respectively. C_H and C_L are the constant regions. Each IgG has two antigen-binding sites, one at the tip of each arm of the Y. The hinge allows molecular flexibility during antigen binding.

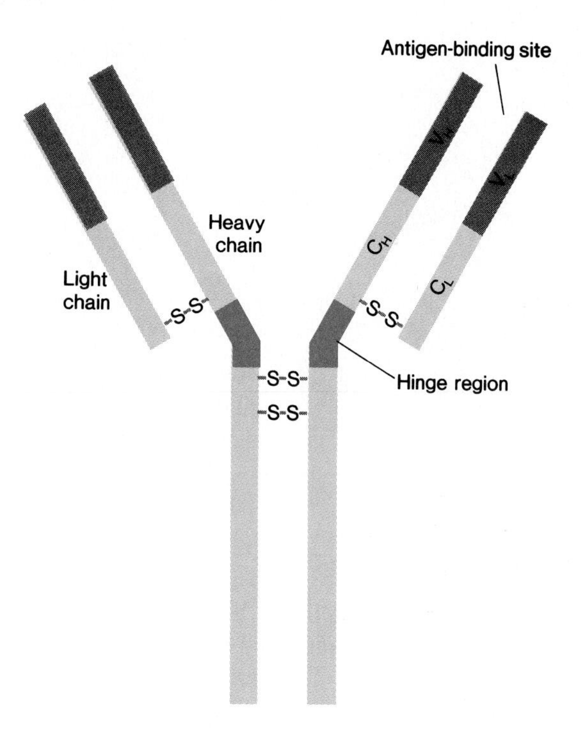

The sequences of amino acids in the heavy and light chains have been studied in detail, and it turns out that part of every heavy chain in every IgG molecule in a given animal always contains the same amino acids arranged in the same sequence, as does part of every light chain in every IgG molecule. These regions are called **constant regions.** The ends of heavy and light chains toward the tips of the arms of the Y, however, contain different amino acids in different IgG molecules and are known as **variable regions.** It is not surprising that the actual antigen-binding sites of antibody (IgG) molecules also turn out to be at the tips of the arms of the Y, the place where the IgG molecules differ from each other.

It is estimated that an individual organism is able to make antibodies for at least one million different antigens. How can this enormous range of antibody diversity be accounted for? How can so many different kinds of polypeptide chains be produced? There is not enough DNA to have separately coded instructions for synthesis of all of those individually different IgG polypeptide chains in addition to all the other information that must be coded in DNA.

Figure 21.13 Diagram of the "gene shifting" (transposition) believed to occur during differentiation of antibody-producing cells. Example shows only genes coding for the light chain, but genes for the heavy chain seem to behave in the same way. Any one of several genes (left) coding for different versions of the variable portion of the light chain may combine with the gene for the constant portion of the chain (right). Once this "gene shifting" has occurred, the genome stabilizes, and all descendants of each of the resulting cells produce identical light chains.

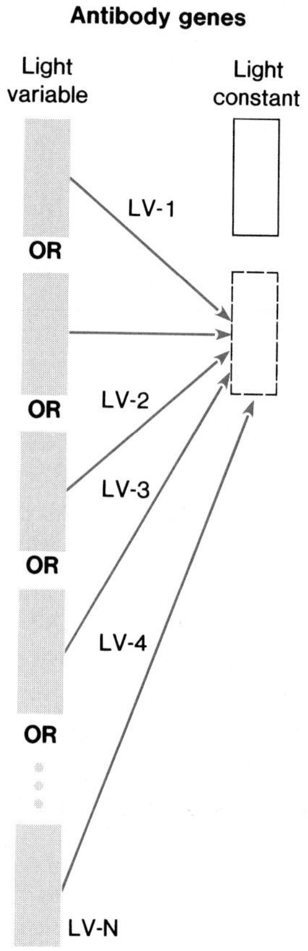

Antibody genes

Light variable

Light constant

LV-1

OR

OR LV-2

LV-3

OR

LV-4

OR

LV-N

It is interesting that in the mouse, at least, C genes (which contain genetic information for assembly of the constant region) and V genes (which contain genetic information for assembly of the variable region) are separate in embryonic cells. Later, in maturing cells, they appear to be joined. Therefore, there is a gene shifting during development (figure 21.13). This results in various C-V combinations in different developing cells, which then are maintained in those cells and their descendants. We now recognize that gene shifting (transposition) during development may be a fairly common phenomenon (see page 385). But gene shifting alone does not seem to be an adequate explanation for the observed polypeptide diversity.

Figure 21.14 Stimulated B cells divide to produce plasma cells and memory cells. Clonal selection involves selective stimulation of a certain subpopulation of B cells, the cells that produce an antibody molecule that binds specifically with that particular antigen.

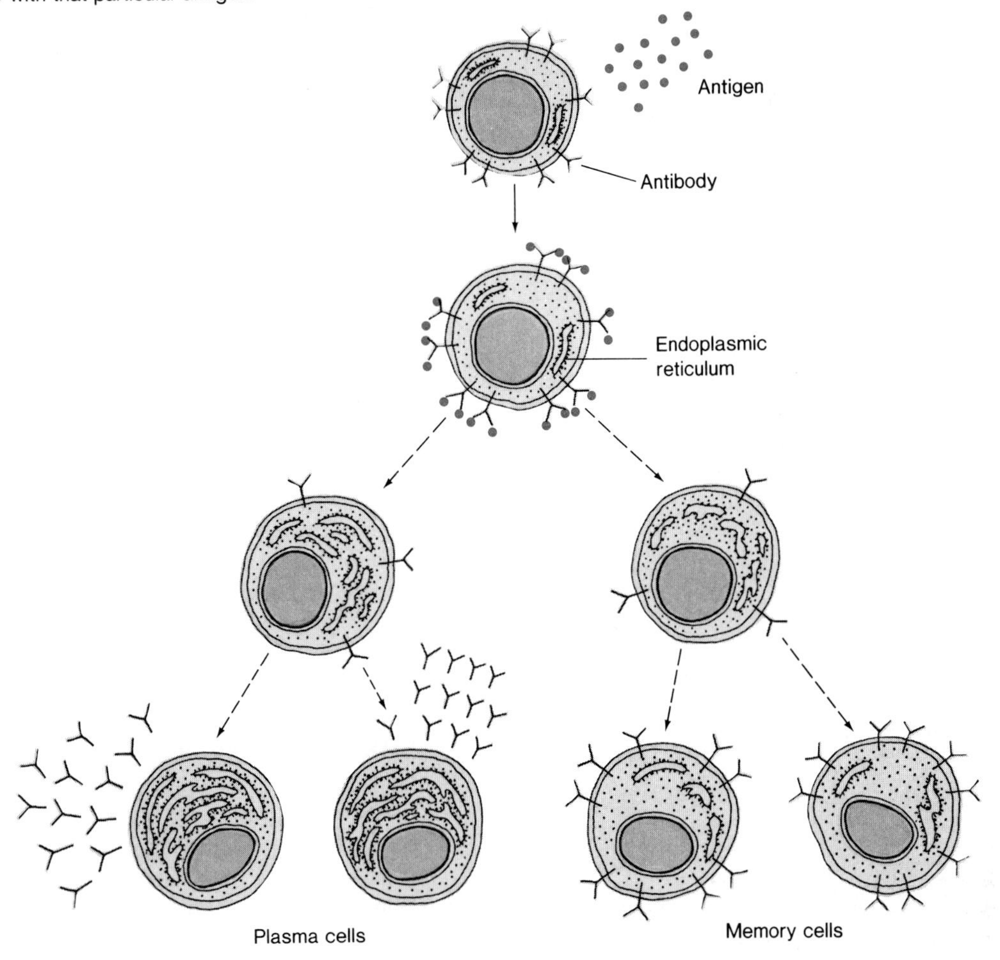

Another possibility is that there is a period in the development of B cells during which replication of the V genes is very error prone. Thus, a number of copying mistakes produce many different gene versions during cell multiplication. This hypothesis presumes a temporary "relaxation" of the normally very accurate copying mechanisms involved in DNA replication. Following this "relaxation" period, replication is presumed to become accurate once again. Subsequently, many subpopulations of cells with different V genes are maintained among the B lymphocytes.

Clonal Selection and Antibody Production

A widely accepted model, called the **clonal selection model,** states that there are many homogeneous subpopulations of B cells, each of which contain cells that produce one antibody with unique, specific, antigen-binding characteristics. Each group of B cells may be descended from a single cell (or a small number of identical cells) that became stabilized as a producer of IgG molecules with a particular set of variable region amino acid sequences. Thus, these are clones of B cells, and all the members of each clone produce the same antibody. It is an important part of this hypothesis that these antibody-producing clones arise by genetic means and are produced before the organism ever has been exposed to the antigens that its antibodies can bind.

Antibody production results when an antigen selectively stimulates the group of B cells bearing appropriate antibody molecules (figure 21.14). These B cells divide to produce a larger population and differentiate into either plasma cells or memory cells. The plasma cells produce and release immunoglobulin molecules of their own unique type, which are, of course, complementary to the stimulating antigen molecules. Memory cells, however, do not release immunoglobulin molecules; rather, they are prepared for subsequent exposures to the same antigen and production of a secondary antibody response when next that antigen enters the body.

Transplantation and Tolerance

Cells, tissue, or an organ grafted from one adult animal (the **donor**) to another (the **host**) usually grow successfully for only a short time. Then a **rejection response** sets in, and grafts between individuals who are not genetically identical are destroyed and sloughed off. The rejection is the work of T cell lymphocytes and the macrophages they stimulate in cell-mediated immunity. There is, however, a **tolerance** to "self" that permits grafts from one part of an individual's body to another part of its body. This tolerance also permits grafts between genetically identical individuals, such as identical twins.

Graft acceptance or rejection depends mainly on cell surface antigens known as **histocompatibility antigens.** Tissue matching in humans depends especially on a set of histocompatibility antigens known as the **HLA** (for human leukocyte antigen) system, which is the major histocompatibility complex in humans. When the best possible match of HLA and other antigens has been made between prospective donor and prospective graft recipient, it is usually still necessary to give **immunosuppressive drugs** to prevent rejection, but this intentional weakening of immune responses places the transplant recipient at a much greater risk from infectious disease.

Autoimmune Diseases

Usually, the body has appropriate mechanisms to prevent formation of antibodies capable of reacting with "self" components. But when tolerance of some body component fails, **autoimmune disease** can result, and components of the immune system then attack the body in various ways. Autoimmune diseases may be organ-specific diseases, such as glomerulonephritis, a degenerative disease of the kidney. Or they may be nonorgan-specific, such as rheumatoid arthritis, which can attack connective tissues in a number of body areas at the same time.

AIDS

In the late 1970s and early 1980s a distressing new disease appeared on the American scene. This disease is known as **acquired immunodeficiency syndrome (AIDS)** because it is somehow transmitted from person to person (acquired) and its victims become progressively less able to resist any form of infection (immunodeficient).

Many AIDS victims become ill with an unusual pneumonia caused by a protozoan, *Pneumocystis carinii,* but they are susceptible as well to many viral, bacterial, and fungal infections. Also, a rare form of cancer called Kaposi's sarcoma is common among AIDS victims. Even if successfully treated for one disease, AIDS victims soon fall ill with other diseases because they eventually have virtually no disease resistance.

About 75 percent of AIDS victims are homosexual men, and the disease almost certainly is transmitted by sexual contact. Many other victims are intravenous drug abusers, who have a history of sharing injection needles, thus passing the disease. A few AIDS victims are Haitian immigrants who are neither homosexuals nor drug abusers, and their cases are the most puzzling of all. Finally, a very few people apparently acquire AIDS when they are being treated for hemophilia (blood-clotting deficiency) with clot-promoting proteins extracted from donor blood. It seems likely that AIDS victims must have donated some of that blood.

AIDS is the subject of a large-scale research effort, and it appears to be closely linked with a virus called HTLV–III, which is similar to a group of viruses associated with some rare cancers (page 470). However, much remains to be learned about the cause, prevention, and treatment of this frightening disease.

Cell Replacement and Body Repair

Cell division must continue for normal cell replacement in many parts of the body throughout life. For example, it is estimated that about two million worn-out red blood cells are removed from circulation per second. This means that, on the average, about two million cell divisions must be completed in the bone marrow every second to replace those cells. Cells lining parts of the vertebrate intestine live only a few days and must be replaced. Skin cells are constantly dying and sloughing off the epidermal surface. They must be replaced by cell divisions taking place deeper in the epidermis.

All of these cell replacement processes are regulated to provide constant cell replacement and no more. On the average, each cell division produces one cell that differentiates to become a functional replacement for a lost cell, while the other remains a dividing cell (figure 21.15). Both the number of functional cells and the number of dividing cells remain constant.

Figure 21.15 Outcome of a normal replacement cell division. One cell differentiates to become a functional cell; the other retains division potential. Thus, the number of dividing cells does not change.

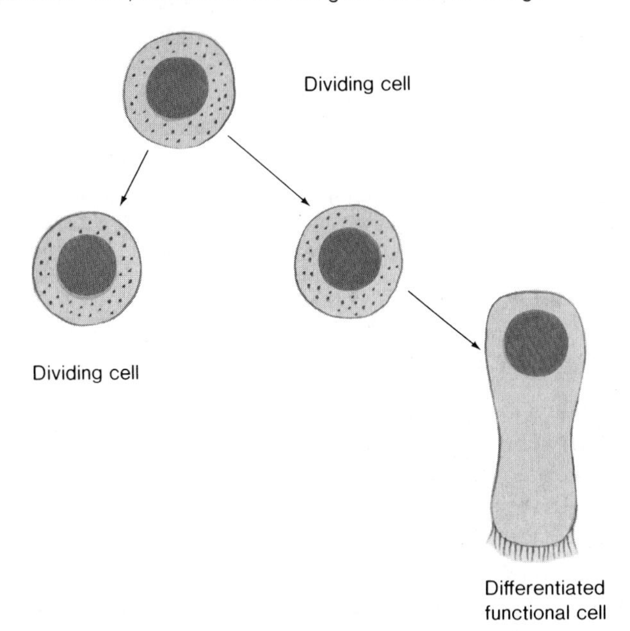

Figure 21.16 Regeneration of a planarian cut in half to produce two individuals.

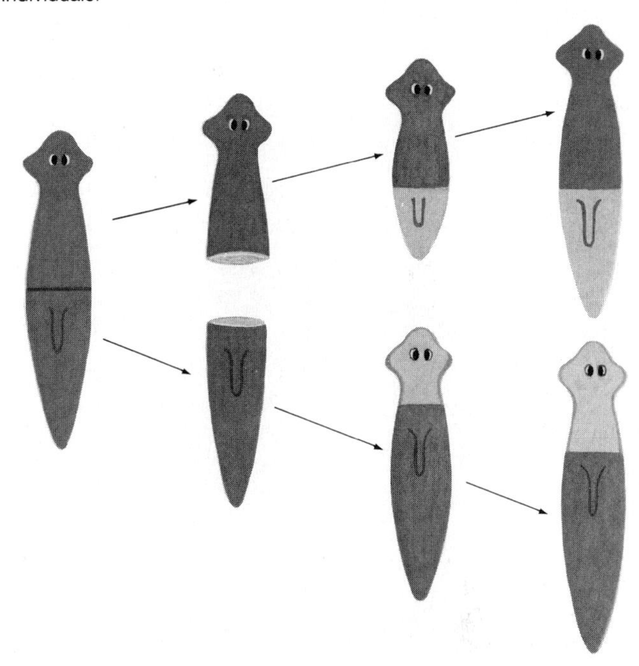

Other types of cells, however, probably never multiply during adult life. In mammals, nerve cells, voluntary muscle cells, and heart muscle cells do not divide. Damage to such tissues results in the formation of only **scar tissue.** Scar tissue is a connective tissue containing a network of tough, fibrous, protein strands. It holds the tissue together, but it does not replace lost functional cells. Thus, even relatively minor damage to these tissues causes significant reduction in the number of functional cells in the organ.

Still other types of cells that normally do not divide can divide under certain conditions. For example, bone cells (osteocytes) normally do not divide. But, if a bone is broken, osteocytes in the vicinity of the break begin to divide within a few hours and participate in the repair process. Once damage is repaired, the bone cells return to their nondividing condition.

Thus, there are clear differences in the responses of various tissues to injury. Wound healing ends with scar formation in some tissues but proceeds to functional cell replacement in other tissues. Finally, there is a fascinating extension of repair processes called **regeneration,** which results in complete replacement of lost body parts.

Regeneration

While all organisms must have wound-healing capacities, the ability to regenerate lost body parts varies greatly among groups of animals.

Planarian worms—small, free-living flatworms—have extensive regenerative ability. When a planarian worm is cut in half, regeneration begins at each cut surface. Cell division continues until an adequate population of replacement cells is produced. When the process is complete, two normal worms have been produced from the one original worm (figure 21.16).

Among the vertebrate animals, the greatest regenerative capacity is found in urodele (tailed) amphibians, such as newts and salamanders, where entire limbs can be regenerated when they are lost. Near the cut surface, cells revert to an essentially embryonic state by losing their structural and functional specializations in a process called **dedifferentiation.** Then they begin to divide. Regrowth continues as long as needed to replace the lost parts (figure 21.17).

The control mechanisms that turn off cell divisions when regeneration is complete are not understood. It also is not yet clear why the regenerative ability of other amphibians, such as frogs and toads, is different. Frog tadpoles can regenerate lost limbs, but adult frogs cannot; they simply form scar tissue on the cut surface.

Figure 21.17 Amphibian limb regeneration. Stages in regeneration of limbs of a urodele amphibian (a newt) after amputation through lower part (left) and upper part (right) of forelimb. The limb is shown before it is cut on day 0 in each case. Subsequent numbers indicate days after amputation.

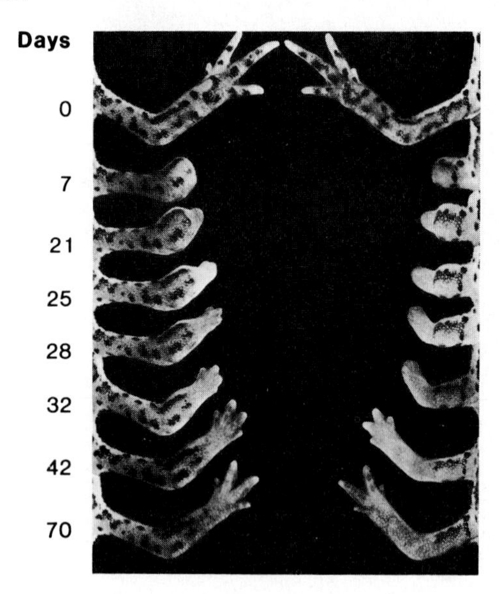

Regeneration in adult humans and other mammals is limited essentially to the liver and the skin. However, children (usually children younger than age eleven) can actually regenerate fingertips, including normal fingernails (figure 21.18). But this regeneration occurs only if the bone is not trimmed and the cut surface of the finger is not closed surgically by drawing the skin over the wound.

Possibly, further research will help to clarify the factors that limit mammalian regeneration, but the long-cherished dream of regrowing new human body parts to replace lost or badly damaged ones still seems very remote.

Cancer and Other Neoplastic Growth

Regulated populations of dividing cells are necessary for normal continuing cell replacement processes in the body. Occasionally, however, body cells become **transformed** so that they divide in an uncontrolled manner. These unregulated cell divisions disrupt homeostasis and threaten life itself. Transformed cells give rise to populations of cells committed to repeated cell divisions (figure 21.19). Thus, both the cell population size and the number of dividing cells increases. These characteristics distinguish the growth of **neoplasias** (literally, "new growths") from regulated cell divisions in normal replacement processes. Such neoplastic growths often are called **tumors,** especially when they produce a definite mass or nodule of tissue.

Figure 21.18 Regeneration of the tip of a child's middle finger. (*a*) The third finger was accidentally amputated just above the nail when the child was 22 months old. (*b*) Eleven weeks after the accident, the entire fingertip, including bone and nail, were regenerated.

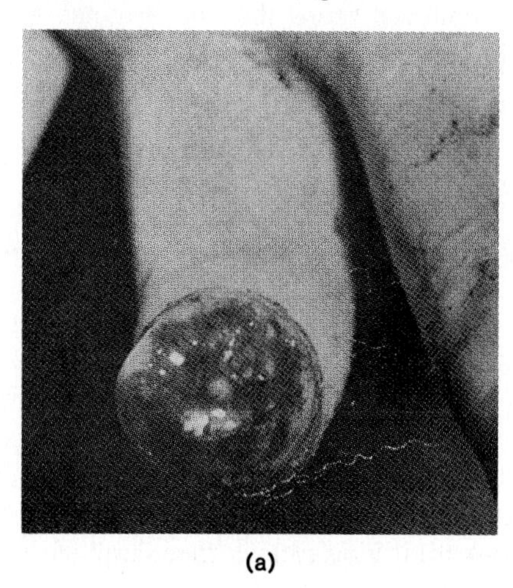

(a)

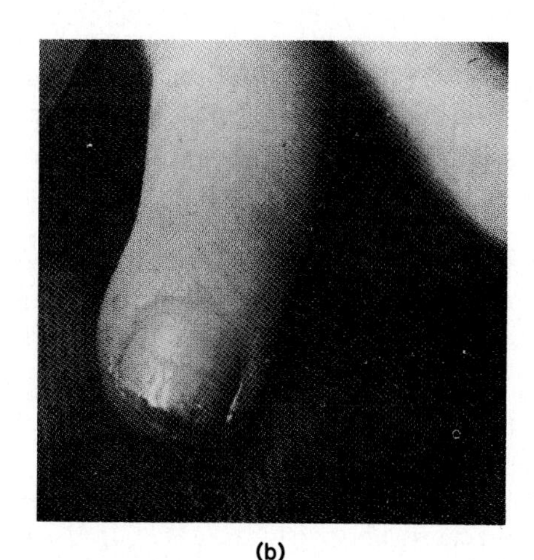

(b)

Tumors may be **benign;** that is, their cells remain within their site of origin, especially when they are enclosed in a capsule of connective tissue. But the term benign definitely does not mean harmless. Benign tumors can grow until they severely interfere with normal function simply by their bulk, by diversion of blood supply, or by other means.

Other neoplastic growths are **malignant.** They are much more likely to cause death because their cells **metastasize;** that is, they spread to new locations in other parts of the body where they establish colonies that grow into additional tumors (figure 21.20). Medically, the primary (original) tumor often can be treated quite effectively by surgery, local radiation, chemotherapy, or some combination of the three. But the colonies established by metastasizing cells are harder to detect and treat successfully. Malignant neoplastic growths are called **cancers.**

Virtually all body tissues and organs can be sites for neoplastic growths, and tumors are classified according to the tissue of origin. **Carcinomas** arise in epithelia (the lining and covering tissues), such as gut, skin, and glandular tissues. **Sarcomas** are sometimes called solid-tissue tumors because they arise in mesodermal tissues, such as bone, muscle, and connective tissue. Finally, the neoplastic growths of blood-forming tissues make up a special class called **lymphomas** or **leukemias.** Some other specialized names are applied to neoplasias with special characteristics, such as the term **melanoma** for heavily pigmented tumors.

Neoplastic Cells

You may have heard tumors described as having "rapid and uncontrolled cell division." Cell division rates in some neoplasias may be rapid, but others have cell division rates slower than those in some normal cell replacement processes. The most obvious and devastating property of neoplastic cells is uncontrolled commitment to cell division leading to a continually increasing population of dividing cells.

Microscopically, neoplastic tissues usually look different from tissues around them because they do not contain structurally differentiated cells (figure 21.21). Tumor cells have an unspecialized appearance that makes many tumor cells resemble embryonic cells. Differentiated cells have relatively more cytoplasm than undifferentiated (embryonic) cells. Chemically, tumor cells also resemble embryonic cells in several ways.

Many, though not all, types of tumor cells have abnormal chromosomes. Some have broken or missing chromosomes while others have extra copies of many chromosomes. However, the relationship between specific chromosomal abnormality and transformation is not yet clear.

Figure 21.19 Difference between normal cell replacement (*a*) and cell division of transformed cells in neoplasia (*b*). In neoplasia, the number of dividing cells increases continually.

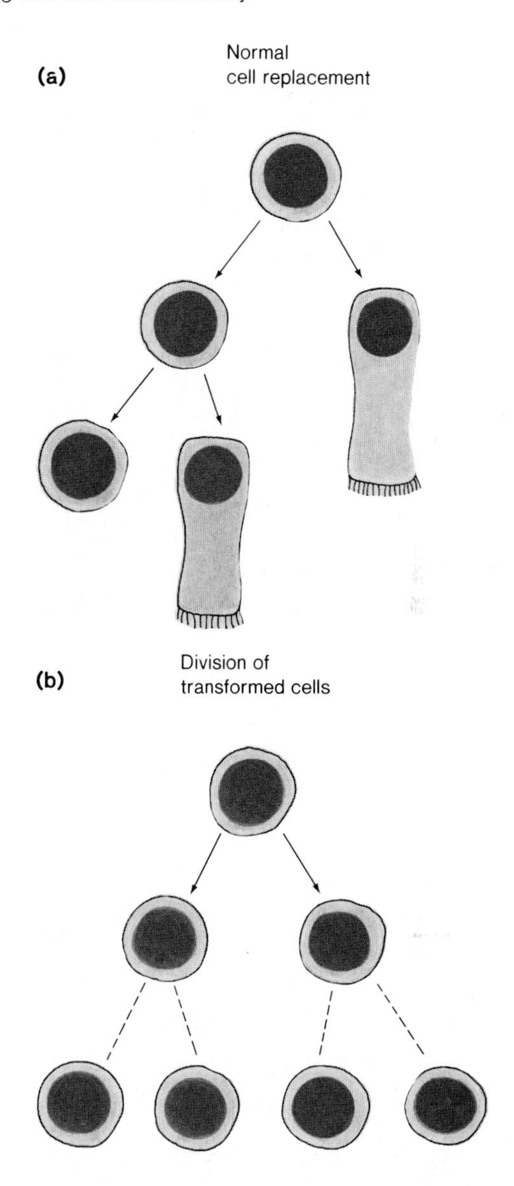

Neoplastic Transformation

Transformed cells continue to divide repeatedly. This indicates a failure of the normal regulation of cell division seen in both developing and adult organisms. How does this change come about?

One important hypothesis concerning transformation says that this permanent loss of control over cell division is a "somatic mutation"; that is, an inheritable genetic change

Figure 21.20 Metastasis. Cells from a primary tumor spread through blood vessels and lymphatic vessels and establish secondary tumors in other body organs.

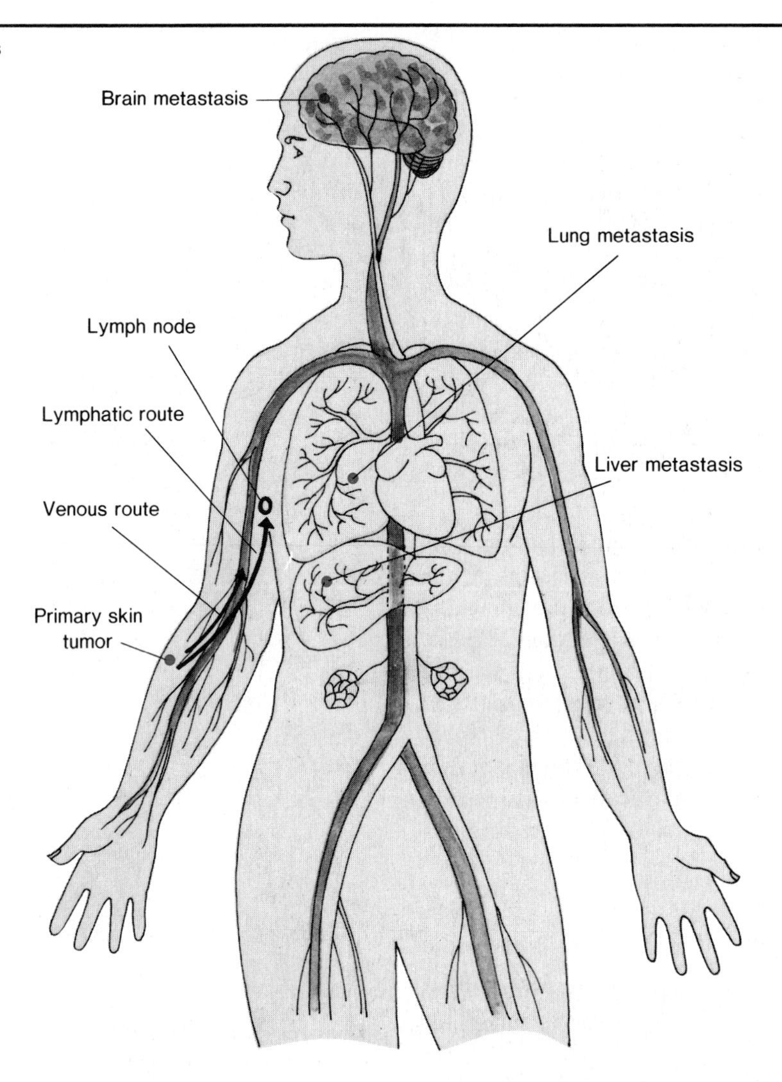

Figure 21.21 Neoplastic cells and neoplastic tissue. (*a*) Photomicrograph of a Pap smear, a sample of cells that were scraped off a human cervix. The more darkly stained cells clumped in the center are neoplastic cells. Note that they have larger nuclei and much less cytoplasm than the normal cells around them (magnification $\times$ 450). (*b*) Photomicrograph of a section of tissue from a human colon (large intestine). The left part of the picture contains normal colon tissue with prominent glands separated from other tissue beneath them by a definite boundary layer. The disorganized abnormal tissue in the right half of the picture is characteristic of colon cancer. Note the absence of functional glandular tissue and the lack of separation between tissue regions (magnification $\times$ 30).

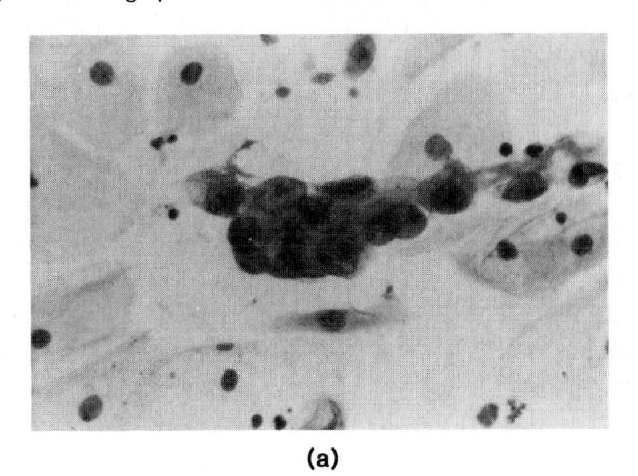

(a)

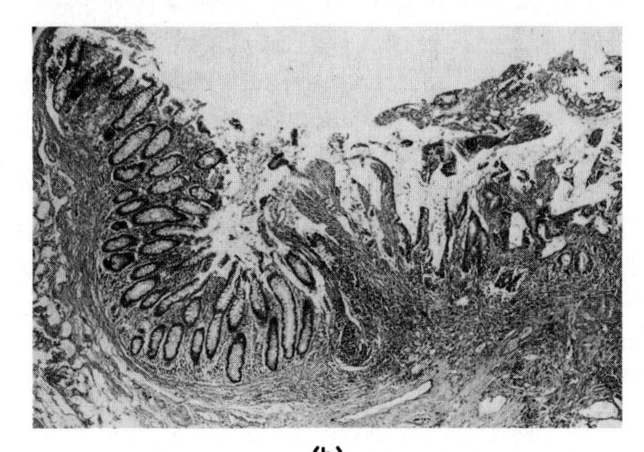

(b)

Figure 21.22 Tumor growing on the tail of a tadpole after it received a small graft of tissue from a virus-induced adenocarcinoma of an adult frog kidney. The oncogenic virus that induces this tumor is a large DNA virus.

that is transmitted from cell to cell during divisions. This hypothesis is supported by several general observations: (1) the widespread occurrence of chromosome damage in transformed cells, (2) the **carcinogenic** (cancer-inducing) potential of radiation and mutagenic (mutation-causing) chemicals, and (3) the apparent permanent loss of specialized functional characteristics.

Immune Surveillance

The immune surveillance hypothesis proposes that transformed cells are being produced continually, and that tumors develop only when the immune system does not recognize the abnormal properties of transformed cells as nonself and fails to destroy them. But there is considerable controversy about this hypothesis. A great many cancer patients, upon thorough examination, do have immune deficiencies or immune system problems of one sort or another. But it is not clear which came first—the cancer or the immune deficiency.

Etiology of Neoplastic Growths

Now that we have discussed some properties of transformed cells, we will explore a more complex problem—**etiology** (study of causes) of transformation to the neoplastic state. The facts are not yet clear here, and interpretations are conflicting at almost every point.

Viruses and Tumors

In 1910, Peyton Rous showed that viruses extracted from chicken sarcomas could induce tumor formation when reinjected into tumor-free animals. This **Rous sarcoma virus** is one of the most intensively studied of **oncogenic** (tumor-causing) **viruses.** Rous sarcoma virus is a small virus containing a single strand of RNA surrounded by a protein coat, and it is one of several small RNA viruses associated with animal tumors. DNA-containing viruses are associated with other animals' tumors, such as adenocarcinoma in frogs (figure 21.22).

Some benign human tumors, such as skin warts, are caused by viruses. Other tumors contain viruses and viruslike particles, but these viruses do not cause transformation of cultured human cells. Thus, ultimate proof that they are oncogenic viruses would depend on ethically unacceptable experiments in which they were injected into human subjects. Some of the excitement about the discovery of what appear to be virus particles in human tumor cells faded anyway when electron microscopists reported sighting similar viruslike particles in a variety of normal cells, especially embryonic cells.

The human tumor most closely associated with virus induction is **Burkitt's lymphoma,** which affects children in certain portions of Africa. The associated virus is called **Epstein-Barr virus.** It is very similar to the common herpes viruses, one of which causes cold sores. Epstein-Barr virus is found

Figure 21.23 Distribution of Burkitt's lymphoma. This rare children's cancer occurs in southern Africa but is seldom found in cooler areas where temperatures fall below about 18°C (64°F) (shaded areas). Some interaction between the Epstein-Barr virus and chronic malaria may cause the disease.

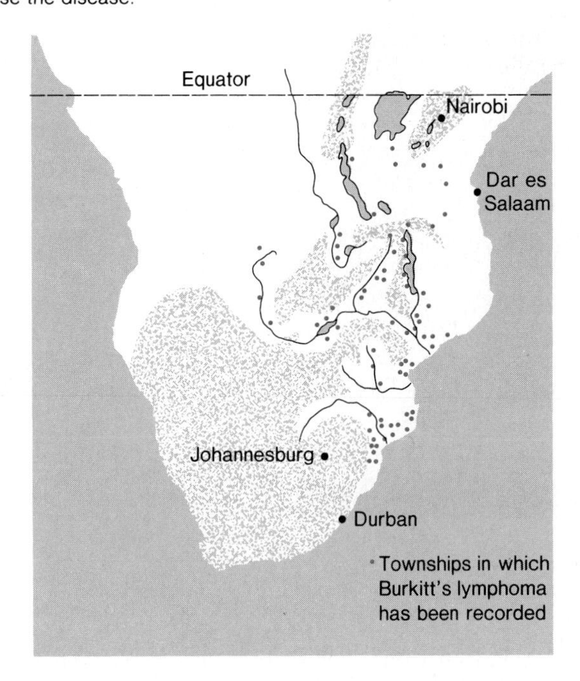

in many humans around the world, but Burkitt's lymphoma occurs mainly in areas of southern Africa with certain rainfall and temperature patterns (figure 21.23). Its absence in nonmalaria areas indicates some complex interaction between chronic malaria and the viral infection, if the virus does indeed cause Burkitt's lymphoma.

There is a similar suggestive relationship between viruses known as **human T-cell leukemia viruses (HTLV)** and a fairly rare form of cancer. HTLV infection is virtually always detectable in T cells from patients with adult T-cell leukemia-lymphoma. Furthermore, when normal lymphocytes are grown in culture with cancer cells that are releasing HTLV particles, the normal cells become infected and transformed.

Even if one or several viruses are proven to be linked with human tumors, it still must be determined how virus infection leads to transformation in some individuals and not in others. The role of environmental factors in transformation is critically important and must be studied carefully.

Natural Carcinogens

Chronic exposure to certain parts of the spectrum of sunlight and other natural background radiation does induce mutations, some of which may not be repaired by normal repair mechanisms. Some chemical constituents of the normal diet or their breakdown products may also be mutagenic. According to the somatic mutation hypothesis, some of these mutation events might cause cell transformations. Increasing incidence of neoplasia in older individuals would then be correlated with accumulation of this natural damage or with reduction of the ability to repair it.

Carcinogens in an Industrial Society

Persistent irritation contributes to transformation, and some irritants, such as asbestos dust, are widespread in industrial societies. Other airborne particles, such as coal dust and ashes, cause irritation and contain carcinogenic chemicals as well. Other carcinogenic compounds are some common organic solvents, additives to fuels, and the combustion products of a number of normally harmless organic compounds (for example, charred foods).

Screening of literally hundreds of chemical compounds is difficult because the standard animal tests are expensive and time consuming, but there is a fairly close correlation between causing mutations and being carcinogenic. The degree of mutagenic potential is easier to measure, and it gives an estimate, at least, of carcinogenic potential. Bacterial tests of mutagenesis, such as the very important **Ames test,** are very useful in identifying suspected carcinogens, which can then be tested further.

Cancer Therapy and the Cancer Cure

Considerable progress has been made in the detection and treatment of tumors over the past quarter century. But there have been no dramatic breakthroughs to compare, for example, with the revolution that antibiotics caused in the treatment of bacterial infections.

For now, the best hope for a reduction in the incidence of tumors seems to involve controlling environmental insults and reducing dietary intake of known carcinogens. The best example of potential for control using currently available knowledge is in the case of lung cancer, which is a high incidence human neoplasia (figure 21.24).

Lung cancer is so closely correlated with cigarette smoking that only the most resolutely closed-minded individuals could doubt that cigarette smoking increases the incidence of lung cancer. Yet cigarette smoking, especially among young people, has not decreased much since discovery of the relationship.

Box 21.1
Cancer Genes

A very exciting addition to the list of differences between normal cells and tumor cells is the discovery that there are specific genes that are especially active in tumor cells. It seems that activation of these **oncogenes** (cancer-causing genes) is part of the transformation process.

In animal tumors caused by viruses, an oncogene is carried into a normal cell by the virus. Once there, the oncogene is activated and plays a role in transforming the cell into a tumor cell. When oncogenes were examined closely, however, it was determined that they are not actually virus genes at all. Rather they are animal genes that are picked up by viruses as they develop in tumor cells and carried to the normal cells that they infect.

This finding is related in an intriguing way to discoveries made concerning human tumors. There also are oncogenes that are active in human tumor cells. Remember, however, that there is still no conclusive proof connecting viruses and human tumors. Human oncogenes are part of the human genome and are present in all cells, though we as yet do not understand their function in normal, nontumor cells.

How do oncogenes become active in human tumor cells? Several answers have been suggested. In some cases the oncogenes active in tumor cells are not identical with the same genes in normal cells, but the difference is small, maybe as little as one nucleotide out of the 5,000 or so nucleotides that make up the gene. Thus, the conversion from a normal gene, called a **proto-oncogene** to an oncogene may be a single point mutation (see page 390). It is true that carcinogenic substances usually also cause mutations.

In other cases, a proto-oncogene may become an active oncogene when the part of the chromosome where it is located breaks and becomes attached to a new chromosome location. This process may combine the proto-oncogene with a gene in another part of the genome that converts it into an active oncogene. Again, remember that broken and reattached chromosomes are found in many kinds of tumor cells.

More research will be needed to clarify the role of oncogenes in transformation. It is known, for example, that transformation is a two-step process. In one step a cell that is being transformed acquires a sort of "immortality"; that is, it is freed from a normal limitation on the number of generations of mitotic cell divisions that can occur in the descendents of any ordinary cell. The second step commits the cell to begin the repeated cell divisions that characterize neoplastic growth. It is not yet clear how the action of one or more oncogenes might be related to this two-step process. Further research on the role of oncogenes in transformation will expand our understanding of abnormal growth and may yield information that can help in cancer prevention.

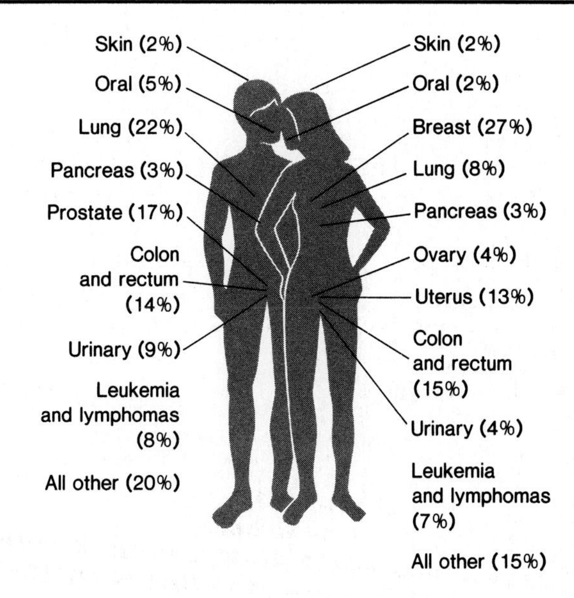

Figure 21.24 Incidence of cancer in men and women (excluding nonmelanoma skin cancer and some carcinomas). The incidence of lung cancer in women increased dramatically during the 1970s. If that increase continues, it may soon erase the sex difference in lung cancer incidence.

One suggested explanation for this failure to heed warnings about carcinogens is that somehow many people have embraced the optimistic idea that a general "cure" for cancer is just around the corner and that it will be available by the time they need it. Cancer researchers, on the other hand, tend to view neoplastic growth as a complex of problems in cell and developmental biology, and they are less optimistic about finding a generally effective cure anytime in the very near future.

Aging

Aging is a continuation of lifelong developmental processes. The outward signs of aging may be obvious, but our understanding of the underlying bases for aging changes is poor. **Gerontology,** the study and analysis of aging, is in its infancy as a science.

Aging in Populations

Each species has its own characteristic, genetically determined range of life spans. Illness and accidents cause death of some young individuals, but progressive aging changes eventually result in a marked increase in the death rate of a population made up of aging individuals (figure 21.25).

Aging changes leading eventually to death are normal developmental processes. While the average human life expectancy has been increased, for example, by eliminating many early deaths due to infectious diseases, the maximum expected life span itself has not been altered.

Expected life spans vary among individuals in a species, but each species has its approximate maximum expected life span. The maximum recorded life spans for some species of mammals are given in table 21.1.

Table 21.1
Maximum Age Attained for Selected Mammals

Species	Common Name	Maximum Life Span (in Years)
Mus musculus	House mouse	3.5
Canis familiaris	Dog	20
Equus caballus	Domestic horse	46
Elephas maximus	Indian elephant	70
Pan troglodytes	Chimpanzee	44.5
Pongo pygmaeus	Orangutan	50+
Homo sapiens	Human	118+

Figure 21.25 Survival curves and distribution of ages at death. (*a*) Male fruitflies, *Drosophila melanogaster*. (*b*) Human males.

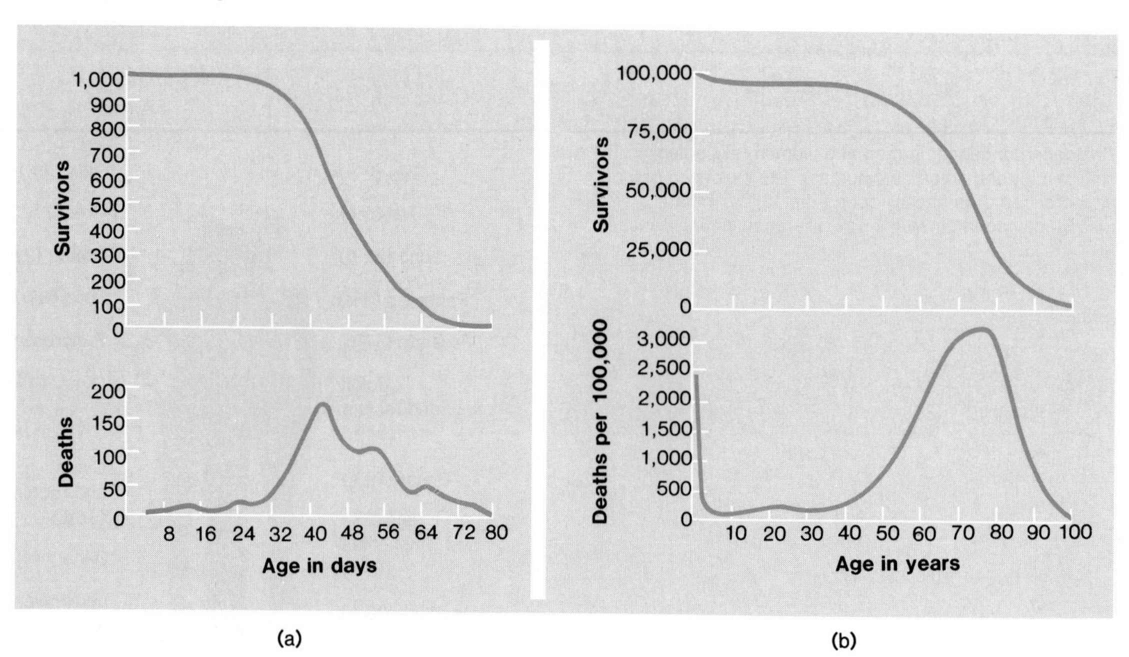

Functional Aging Changes

Many functional and structural features of organisms show signs of loss or decreased function with advancing age, and these changes, singly and in various interactive combinations, weaken an organism's homeostatic responses (table 21.2).

Cellular Aging

Tissues without significant cell turnover and replacement lose cells with aging. The nervous system, for example, suffers a progressive loss in cell numbers throughout life (figure 21.26).

Several possible causes of cell death during aging have been suggested. Certain metabolic wastes accumulate in aging cells. Somatic mutation, the accumulation of DNA damage, causes disturbances in cell functions.

Even within those tissues where dying cells are replaced early in life, cell replacement seems to become less efficient with age and eventually there may not be enough cells to maintain certain vital functions at essential levels.

The Immunological Hypothesis of Aging

The immunological hypothesis of aging is intriguing because it suggests how cellular aging might cause a wide spectrum of age-related problems. The hypothesis says that production of T cells and B cells decreases progressively with age.

This would explain decreasing resistance to infection with advancing age, and possibly, the increasing incidence of neoplastic growth in older individuals. Finally, loss of balanced control between the parts of the specific immune response may increase risk of autoimmune diseases, such as rheumatoid arthritis. All of those problems are seen in aging individuals, but the immune hypothesis is by no means established as the fundamental cellular cause of aging.

Abnormal Aging

For most people aging is a slow process that only gradually reduces certain functional capabilities and eventually weakens resistance to various health problems. Some people, however, suffer dramatic aging changes.

Table 21.2
Functional Losses Accompanying Aging

Type of Decline or Loss	Remaining Functions or Tissues* (Percentages)
Weight of brain	56
Number of spinal nerve fibers	63
Velocity of nerve impulse	90
Flow of blood to brain	80
Adjustment to normal blood pH after displacement	17
Output of heart at rest	70
Number of glomeruli in kidney	56
Glomerular filtration rate	69
Number of taste buds	36
Maximum oxygen uptake during exercise	40
Maximum breathing capacity	43
Hand grip	55
Basal metabolic rate	84
Water content of body	82
Maximum work rate for short burst	40

From Volpe, E. Peter, *Biology and Human Concerns,* 3rd ed. © 1975, 1979, 1983 Wm. C. Brown Publishers, Dubuque, Iowa. All Rights Reserved. Reprinted by permission.

*Figures are the approximate percentages of functions or tissues remaining in the average 75-year-old male, taking the value found in the average 30-year-old male as 100 percent (based on studies by Nathan W. Shock).

Figure 21.26 Cell counts for two areas of the human cerebral cortex. There is no cell replacement in nervous tissue so cell deaths result in permanent reduction of the total number of cells.

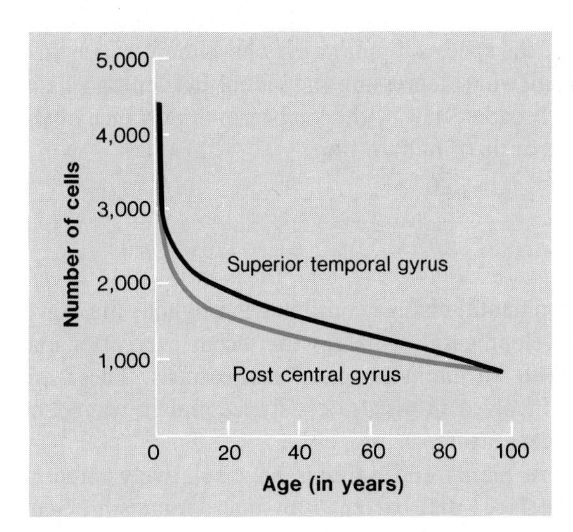

A common form of accelerated aging is a condition known as **Alzheimer's disease,** which affects as much as five to ten percent of all people over age 65. Alzheimer's disease is a degenerative brain disorder that results in severe memory loss and eventually in loss of other brain functions so that some victims become almost helpless before they die.

The brains of Alzheimer's disease victims differ from normal brains in several ways, including the death of many cells in certain brain centers. But further research will be needed to discover the causes and possible means for preventing (or delaying) these brain changes.

Future Prospects

Until fundamental aging changes are identified and understood, scientists can only work to increase the number of people who live out more of their expected life span in good health. But, if the primary cellular aging processes are identified and the life span is expanded, a new dilemma of sorts will have to be faced. Aging and death have always made room in populations for new, young individuals. Aging and death are part of humans' natural biological heritage and, from an evolutionary point of view, are a positive force for the welfare and survival of the species.

Everyone individually rebels at the notion of aging and death, and longs, along with people down through the ages, for indefinite extension of youth and vigor. But for the overall good of the species, humans may have to find ways to accept this somewhat depressing individual burden and balance it with a broader view of the long-term importance of the continuing cycle of human life.

Summary

Developmental change continues throughout life. Several of the developmental processes that occur even after maturity contribute to maintenance of homeostasis. These are processes involved in disease resistance and in wound healing and regeneration.

Both plants and animals have relatively impermeable body surfaces that bar entry by microorganisms. Some microorganisms produce antibiotics that suppress growth of other nearby microorganisms. Plants produce phytoalexins, which also are inhibitors of microbial growth. Animal cells produce and release an antiviral substance called interferon when they are infected by viruses. Interferon prevents virus replication in other cells and thus breaks the cycle of infection, replication, and virus release by cell lysis.

Phagocytosis is the process by which specialized cells engulf and destroy invading microorganisms, other cells, and various foreign particles.

The complement system is a set of nonspecific antibacterial factors in the blood.

Inflammation reactions provide easier access for phagocytic cells and complement factors to infected areas.

Specific immune responses fall into two categories. Cell-mediated immunity is based on direct attacks by T cells on foreign cells. Humoral immunity is based on certain B cells that, in response to foreign antigens, develop into plasma cells that produce and release specific antibodies (immunoglobulins).

Each immunoglobulin G molecule has two heavy chains and two light chains. Antigen-binding sites are in the variable regions near the ends of the arms of the Y-shaped molecule. Entrance of an appropriate antigen stimulates cells in one of the clones of B lymphocytes to multiply and differentiate either as plasma cells that produce and release antibodies, or as memory cells.

Grafts between individuals who are not genetically identical result in a rejection response.

Autoimmune diseases develop when the immune system mistakenly begins to recognize some body cells as foreign and attacks them.

Cell replacement continues throughout life. Regeneration is complete replacement of lost body parts.

Transformed cells are different from cells involved in all normal developmental processes because their divisions are not regulated.

The immune surveillance hypothesis proposes that transformed cells should be recognized as nonself and destroyed, but that failures of the immune system allow them to live and divide.

Some animal tumors clearly are virus induced. But no such definite connections have been made for human neoplasias.

Aging changes continue throughout adult life, and they progressively weaken homeostatic responses.

Questions for Review

1. What is the benefit of antibiotic production for microorganisms?
2. What are broad-spectrum antibiotics?
3. Explain why interferon produced by cultured cells of other animals is not effective in treatment of human virus diseases.
4. What role does inflammation play in resistance to infection?
5. Distinguish between cell-mediated immunity and humoral immunity.
6. Explain the importance of memory cells for acquired immunity to disease.
7. Why are recipients of organ grafts often susceptible to infectious diseases?
8. Define the term dedifferentiation and explain the role of dedifferentiation in regeneration.
9. List several characteristics that distinguish transformed cells from normal cells.
10. How does a test for mutagenesis help in identification of potential carcinogens?

Questions for Analysis and Discussion

1. If testosterone is injected into chicken eggs at an appropriate stage of incubation, the embryos that develop either lack a bursa of Fabricius or have a very poorly developed one. What effect would you expect this treatment to have on immune system functioning of the chick after hatching?
2. How does the increased incidence of neoplastic growths in older individuals provide indirect support for the immune surveillance hypothesis of neoplastic transformation and growth?
3. Explain why immunosuppressive drugs are of limited value in treating rheumatoid arthritis and other autoimmune diseases.

Suggested Readings

Books

Cairns, J. 1978. *Cancer: Science and society*. San Francisco: W. H. Freeman.

Hammond, S. M., and Lambert, P. A. 1978. *Antibiotics and antimicrobial action*. Studies in Biology no. 90. Baltimore: University Park Press.

Lamb, M. J. 1977. *Biology of ageing*. New York: John Wiley.

Roitt, I. M. 1980. *Essential immunology*, 4th ed. Oxford: Blackwell Scientific Publications.

Articles

Devoret, R. August 1979. Bacterial tests for potential carcinogens. *Scientific American* (offprint 1433).

Henle, W.; Henle, G.; and Lennette, E. T. July 1979. The Epstein-Barr virus. *Scientific American* (offprint 1431).

Rensberger, B.; Bush, H.; and Blonston, G. September 1984. Cancer: The new synthesis. *Science 84*.

Rose, N. R. February 1981. Autoimmune diseases. *Scientific American* (offprint 1491).

Shodell, M. October 1984. The clouded mind. *Science 84*.

Weinberg, R. A. November 1983. A molecular basis of cancer. *Scientific American*.

West, S. March 1983. One step behind a killer. *Science 83*.

Willoughby, D. A. 1978. Inflammation. *Endeavour* 2:57.

Part 5

Evolution, Organisms, and the Environment

All living things are both products of and participants in a continuing evolutionary process.

Current concepts of evolution still rest on principles of natural selection proposed in the nineteenth century by Darwin and Wallace. Modern evolutionary theories have also incorporated the principles of population genetics. Modern evolutionary research takes advantage of techniques that help biologists determine the age of fossil organisms and make detailed molecular evolutionary studies of living organisms. Newly discovered fossils are providing more information on our human origins.

Chapters 22 and 23 deal with the origins of evolutionary theory, some modern concepts of evolution, and human evolution.

In previous parts we have discussed how organisms continually interact with their physical and biological environments. Now, in chapters 24 through 26, we will go one step further and look at some behavioral interactions among organisms and some of the other interactions between organisms and their environment. There are many complex relationships among organisms, and we can fully understand living things only in the context of populations, communities, and ecosystems.

Facing page Crabs on an Indian Ocean beach at Malindi, Kenya.

Development of Evolutionary Theory

22

Chapter Outline

Chapter Concepts

1. The theory of uniformitarianism provided a conceptual framework for the explanation of earth's history and indicated that the earth was much older than had been originally estimated.
2. Charles Darwin's observations during his voyage around the world on *H.M.S. Beagle* led him to conclude that the environment plays a key role in the modification of existing species and the formation of new species.
3. Artificial selection can change the characteristics of a population through the selection of breeding stock.
4. The theory of natural selection suggests a mechanism for evolutionary change in natural populations.
5. The fossil record is the physical evidence of the history of life on earth. When used with relative and absolute dating techniques, it provides considerable evidence about the evolutionary process.
6. Modern studies of evolution involve many scientific disciplines, ranging from biogeography to comparative anatomy.
7. Results of modern molecular evolution studies correlate well with conclusions drawn from other kinds of studies concerning lines of evolutionary descent.

How old is the earth? If you could have asked that question early in the nineteenth century, most of the answers probably would have been in the range of five to ten thousand years. The most common belief at that time was that the earth and all living things on it had been created at one time and that they remain essentially as they were created. By the end of that century, however, many scientists were convinced that there is good evidence to conclude that the earth is many millions of years old.

It had been suggested, even as early as ancient Greece, that modern living things might have arisen through **evolution,** a process of change over a long period of time. By the early 1800s, several scientists supported the idea that an evolutionary process had occurred. But there was the problem of time. It did not seem that the earth was old enough for such changes in living things to have occurred. Then certain discoveries in geology made it clear that the earth was very much older than previously thought. These discoveries were to have a tremendous impact on biology.

With the new views on the age of the earth, it became possible to think seriously about the idea of evolutionary changes in living things over long periods of time. These ideas were not widely accepted, however, until Charles Darwin provided the foundation upon which the modern **theory of organic evolution** has been built. In 1859 Darwin published *On the Origin of Species,* a thoughtful explanation of his views on the evolution of life on earth. In the more than 100 years that have elapsed since the publication of *On the Origin of Species,* the theory of organic evolution has gained almost unanimous acceptance among scientists and has been greatly strengthened by evidence obtained from a wide variety of scientific disciplines.

Charles Darwin and the *Beagle* Voyage

Charles Darwin (figure 22.1) was born in England into a wealthy family. All his life Darwin loved nature. As a young man, he walked the fields and woodlands of the English countryside, studying plants and animals, and observing rock formations. This interest continued into his college years. Although his father wanted him to become either a physician or a clergyman, Darwin preferred to make friends with professors of biology and geology and continue his study of nature.

In 1831, Darwin was offered the opportunity to serve as a naturalist on H.M.S. *Beagle,* a ship commissioned to sail around the world on a survey mission. In December of that year, Darwin embarked on a voyage that was to change the history of biological thought (figure 22.2).

Lyell, Hutton, and Uniformitarianism

Darwin was not a good sailor and spent much of his time in his bunk suffering from seasickness. During the early part of the voyage, he occupied his time by reading the first volume of Charles Lyell's *Principles of Geology,* in which Lyell further developed James Hutton's uniformitarian theory of geology. Hutton contended that the earth is not a static, unchanging sphere but that the present features of the earth's surface are the result of a continuous cycle of erosion and uplift. Hutton saw evidence that the earth's continents had been continually worn away by the agents of erosion, but that "subterranean forces" operated to uplift material from below sea level to form new land surfaces. Hutton's concept of uniformitarianism stated that present forces had acted at the same rate throughout the earth's history to produce similar geological events. If the earth itself is dynamic and its present features are the result of a long process of gradual change, Darwin reasoned, could it not be possible that the biological world was also dynamic rather than static?

Fossils of South America

Darwin found his first evidence of change in the biological world in South America. Along the east coast of Argentina, in the mud and silt of river deposits, Darwin discovered a large number of fossil bones literally sticking out of the loose sediment. Among these bones he identified the remains of a giant ground sloth, a huge hippopotamuslike animal, and a species of horse. All three of the fossil forms represented **extinct species,** species not represented by any living individuals. Darwin noted close resemblance of these fossil forms to living species of sloths, hippopotamuses, and horses. Was it possible, he wondered, that these fossil forms might have been ancestral to species found on earth today? If so, the implication was that new species appear on earth as a result of descent from earlier species. Darwin later wrote that his first ideas of evolution began with his observations of these South American fossils.

The Galápagos Islands

From Argentina, the *Beagle* slowly progressed southward, rounded Cape Horn, and worked its way northward along the coast of South America to a tiny group of islands called the Galápagos. The Galápagos were formed by volcanic eruptions, and they are isolated by some 950 km of ocean from the South American mainland. Ancestors of the plants and animals that inhabit the Galápagos had come from the South American mainland. Possibly they were blown to the islands by strong offshore winds or rafted on floating vegetation.

Figure 22.1 Charles Darwin (1809–1882).

Figure 22.2 The route followed by H. M. S. *Beagle* on its surveying trip around the world (1831–1836).

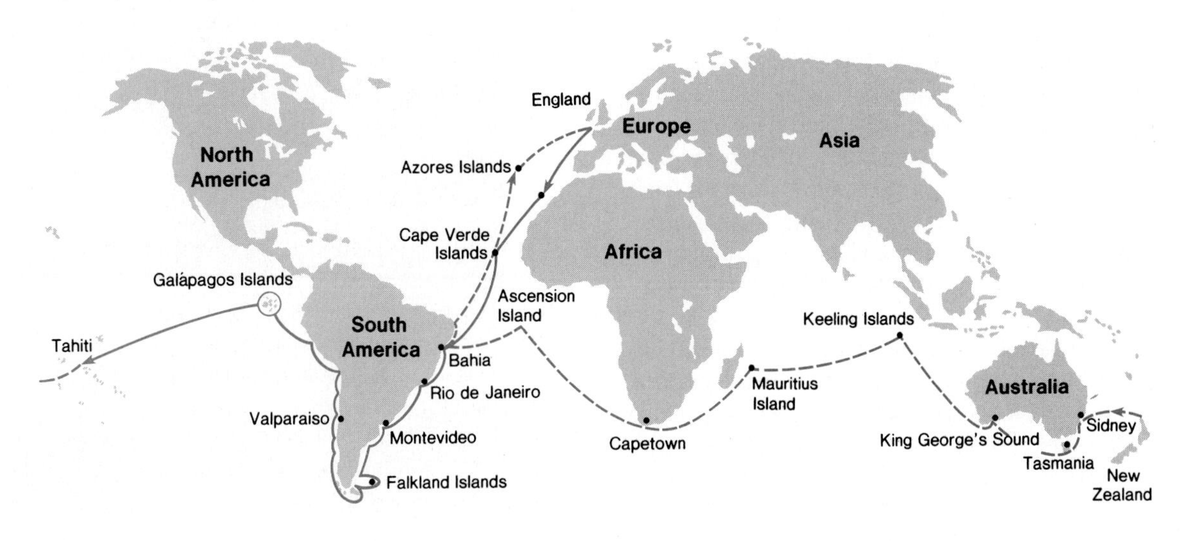

Adaptive Radiation of the Tortoises and Finches

When Darwin arrived in the Galápagos in 1835, he found a variety of plants and animals, many of which are found nowhere else in the world (figure 22.3). Among the most interesting of the Galápagos creatures were the giant tortoises that gave the islands their name (*galápago* in Spanish means tortoise). Darwin's interest in these animals was aroused when the governor of the islands chanced to remark that he could tell from which island a tortoise had come by observing the shape of its shell. Upon closer inspection, Darwin noticed that not only did the shape of the shell vary from one island to another but that other features of the tortoise varied as well. For example, the necks of the tortoises living in dry areas were longer than the necks of the tortoises living in moist areas (figure 22.4). Why, Darwin wondered, should so many distinct forms of tortoises appear in such a limited geographical area?

Darwin later hypothesized that very few tortoises could have arrived at the Galápagos from the mainland, but once there they found little competition for food and no natural predators. Under such conditions, the tortoises became established on the islands and gradually increased in numbers, spreading out wherever conditions were favorable. Because the physical conditions for survival varied from island to island, the tortoise population of each island developed, over time, individual characteristics that distinguished it from the tortoise populations inhabiting neighboring islands. Long-necked tortoises inhabited dry areas where food was scarce, since the longer neck was helpful in reaching high-growing foliage. In moist regions with relatively abundant foliage, short-necked tortoises fared well. Thus, Darwin concluded that the major factor in interisland variation among the tortoises was the environmental variation that existed from one island to the next.

Darwin also observed a wide range of variation among a group of small birds that lived on the Galápagos Islands. These relatively inconspicuous birds have gained fame in the annals of biology as "Darwin's finches," and their evolution has been studied intensively since the time of Darwin. Like the tortoises, finches were introduced to the Galápagos from the mainland of South America. Also like the tortoises, finches found little competition for food and no natural predators. Under these conditions, the finch population quickly became established.

The original finches were seed eaters, and for a time there was plenty of food and they thrived. But the very factors that led to the initial success of the finches also led to a major problem. Finches are prolific breeders, and with an initial abundance of food and no predators to control population size, the population increased to the point that competition for food and nesting sites became severe. During this period of population increase, the finches spread to all of the various islands.

Figure 22.3 Many species of plants and animals are found only in the Galápagos Islands. (*a*) Drab-colored marine iguanas climbing over rocks on Narborough (Fernandina) Island. Marine iguanas are found only in the Galápagos. (*b*) The colorful iguana found only on Hood (Española) Island. Note the large claws on its feet, which are well adapted to a life spent clinging to rocks along the shoreline. Galápagos iguanas eat seaweed exposed at low tide. (*c*) The rare flightless cormorant found only on two of the Galápagos Islands.

Figure 22.4 Tortoises of the Galápagos Islands. (*a*) Sketches of Galápagos tortoises from three different islands illustrating neck differences. Longer-necked species live in relatively dry areas and feed on the high-growing vegetation of cacti. Shorter-necked species live in moister regions with more abundant ground vegetation. (*b*) Map of the Galápagos Islands with their English names first and their current Spanish names in parentheses. (*c*) Photographs of tortoise shells illustrating the different shapes that the islands' governor pointed out to Darwin. The round shell on the left comes from James (Santiago) Island and the other comes from Hood (Española) Island.

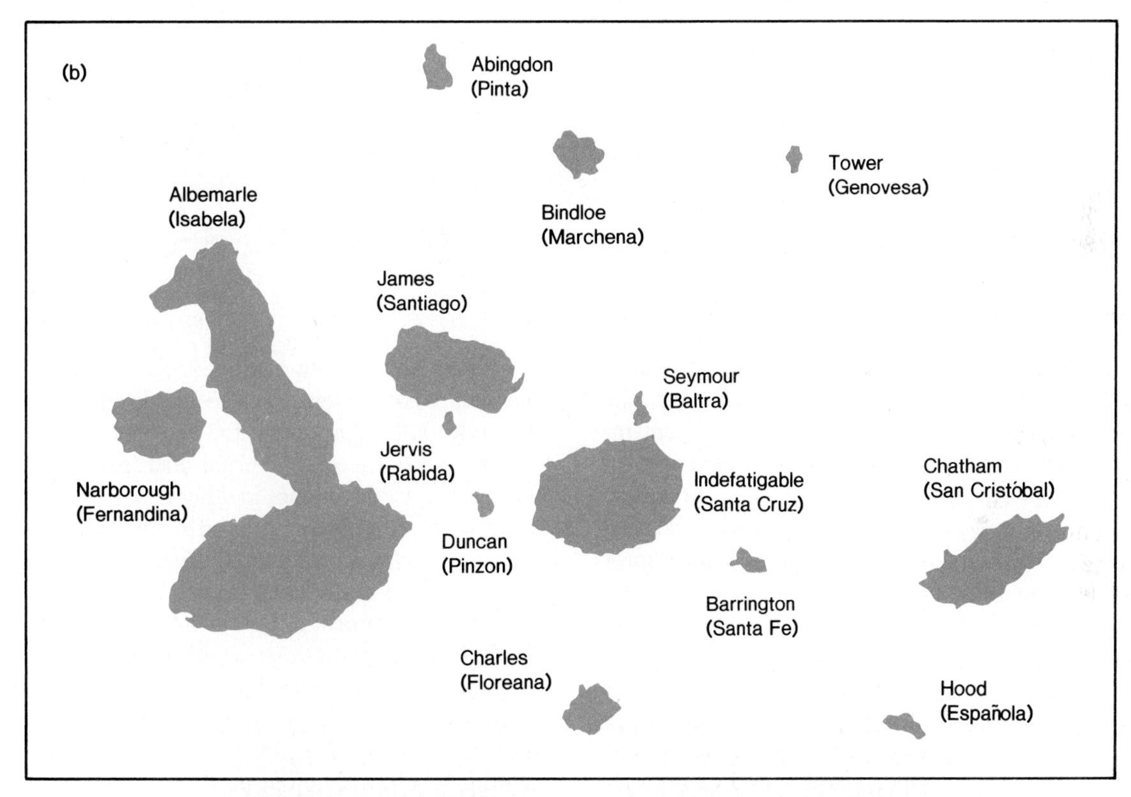

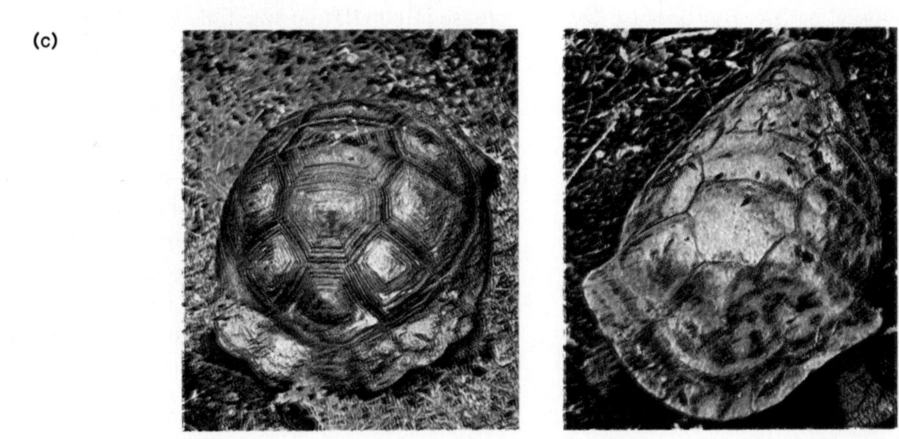

Figure 22.5 Some species of Galápagos finches. All of the species are descended from a common ancestor, but they differ markedly in appearance, habitat, and food sources. Ground finches (lower right) eat cactus flesh; warbler finches (upper left) eat insects; others eat seeds or have mixed diets. The most remarkable of the Galápagos finches is the woodpecker finch (lower left). It pecks tree bark as a woodpecker does, but it lacks the characteristically long woodpecker tongue. Instead, it uses cactus spines as tools to draw insects out of the holes that it makes.

On the different islands, the finches underwent evolutionary modification in several directions. Although the finches were primarily seed eaters, they would also eat insects. Those finches that could not obtain sufficient seed to stay alive were compelled to turn to this alternate food source. By doing so, they decreased their competition with the seed eaters and increased their chances for survival and reproductive success. Over time, finches evolved that were insect eaters rather than seed eaters. An important ingredient in the success of this transition was the lack of competition from other bird species for the insect food supply. In most ecological situations, there are bird species that feed specifically on seeds, on insects, on fruits, and so on. Under such conditions it would be very difficult for members of a seed-eating species to make the transition to insect eating. But because the Galápagos finches had little competition from other bird species over long periods of time, such evolutionary transitions were possible. Some of the finches became specialized to other habits, such as cactus eating.

After periods of isolation on separate islands, various groups of finches came to be very different from one another. And these differences involved more than variations in food habit and appearance. These groups of birds, which were descended from common ancestors, also had changed genetically during the period of geographic isolation. They could no longer interbreed. They had become reproductively isolated from one another.

There are thirteen distinct species of finches occupying the Galápagos Islands. Each species is adapted to a different life-style. These finches fill the ecological roles of warblers, parrots, woodpeckers, and other bird species that would normally occupy these niches in other geographical areas. There are distinct differences among the thirteen Galápagos finch species in such traits as plumage, nesting sites, body size, and size and shape of the beak (figure 22.5).

The development of the tortoise and finch populations of the Galápagos Islands are examples of a process modern biologists call **adaptive radiation.** Adaptive radiation occurs when members of a single species enter environments in which there is little initial competition for resources. Under such conditions, after an initial period of adjustment, a rapid increase in population size follows. This leads to intense competition for resources. This competition may result in the fragmentation of the population. After a period of isolation a number of new species develop. Each of these new species occupies a new and distinct niche, as in the example of Darwin's finches.

Profoundly influenced by his observations in the Galápagos Islands and elsewhere on his journey, Darwin concluded that species are not unchangeable and that new species are formed through the gradual modification of existing traits in response to the pressure of both the environment and biological competition.

Darwin had come to accept the idea of organic evolution—the concept that all species of plants and animals have arisen through descent with modification from previously existing species. However, not every species continues to interact successfully with its environment, especially when there are changes in the environment. Thus, some species have become extinct (died out). Change is the rule of nature, and the history of the earth has been filled with many geological and climatic changes. In fact, in the history of life many more species have become extinct than now inhabit the earth.

The Theory of Organic Evolution

After his return to England in 1836, Darwin spent the next twenty years doing biological research, reading, and developing the idea that living species are the product of a series of gradual changes that have occurred through a long period of geological time. The concept of organic evolution was not original with Charles Darwin, by any means, but he stated it clearly and directly, and systematically gathered a large quantity of evidence in support of the idea.

Darwin soon realized that he needed an explanation for the mechanism by which new species were formed. The ideas of Jean Baptiste Lamarck, though later proven wrong, provided Darwin with some food for thought.

Lamarck and Inheritance of Acquired Characteristics

Lamarck accepted variation as part of the natural world and believed that the environment played a role in the origin of new species. He thought that new traits were acquired by an organism in response to a *need* imposed by the environment.

Lamarck stated that traits could be acquired through the use or disuse of an organ; that is, a body part not used by an organism atrophies and is lost to future generations, while a body part used extensively is amplified. For example, the Lamarckian explanation for the long neck of the giraffe is that elongation occurred because the ancestors of the giraffe constantly reached into the trees to feed on high-growing vegetation. This repeated stretching of the neck over time resulted in the modern giraffe with an exceptionally long neck. A similar argument would be that a son born into a family of weight lifters should have large muscles because his father, grandfather, and great grandfather had developed bulging biceps lifting weights.

Lamarck's concept of inheritance of acquired characteristics never gained wide acceptance in scientific circles. And with the eventual rediscovery of Mendel's principles of inheritance (chapter 16), it became apparent that acquired characteristics, which are not genetically based, cannot be inherited.

Artificial Selection

Because he lacked information about genetics, Darwin was not able to provide an adequate explanation for biological variation. He did, however, develop a clear explanation of a mechanism by which evolution of new species can occur.

An activity that had been practiced by the human race for thousands of years provided a hint in Darwin's search for an explanation of evolution. For centuries humans have selectively bred fruits and vegetables, cattle, chickens, and dogs in an effort to obtain greater yields of corn, a better milk cow, more productive hens, or a dog more adept at sheepherding or rabbit hunting. These breeders observed the variation appearing in their crops and herds, and selected for breeding those individuals that best represented the properties they desired in future generations (figure 22.6). The success of this **artificial selection** was undeniable, and Darwin saw in artificial selection a model for change in the natural world. What puzzled him was how selection worked in nature. Human beings directed artificial selection, but what force directed selection in nature?

The Theory of Natural Selection

An essay on human population written by Thomas Malthus gave Darwin a clue to the solution to this problem. Malthus suggested that human populations have a general tendency to increase in size, but that such factors as war, plague, and famine serve to keep the human population of the earth more or less in check. If it were not for these factors of population control, the human population of the earth would increase explosively (a prediction that agricultural, medical, and technological advances of the last century have proven to be accurate).

Darwin reasoned that *every* species produced more young than could survive to reproductive maturity. Of the many new individuals produced by all species in nature each year, most are eaten by predators, succumb to disease, or are unable to obtain an adequate food supply or habitat in which to live. There is a constant struggle for survival. Only some members of any species live to reproduce, and only the members of any generation that succeed in reproducing contribute to the evolutionary future of the species. Darwin recognized that the same factors he had encountered in the Galápagos Islands operated throughout the biological world: nature in the form of environmental variation and biological competition operates as the selective factor in natural populations.

Figure 22.6 Variation among breeds of the domestic pigeon, showing the results of artificial selection. Darwin saw artificial selection, as practiced in the breeding of pigeons and agricultural plants and animals, as a model for evolutionary change in nature.

From these hypotheses Darwin derived his major theory, the **theory of natural selection.** We can summarize the theory of natural selection as follows: All populations of organisms possess the potential to increase at a very rapid rate, but this rate of increase is very seldom observed among natural populations. Instead, the size of populations remains almost the same from one year to the next. Thus, Darwin reasoned, there is an intense, constant struggle for food and other resources among the many young born every generation. And those individuals with the most adaptive traits, traits that make them better suited for life in the environment, are the most likely to survive and reproduce. Thus, generation by generation, adaptive traits become more frequent in the population, while nonadaptive traits tend to be eliminated.

Over long periods of time, natural selection produces species of organisms that are finely adapted to the set of environmental conditions in which they must compete. Each species is not only adapted to its physical environment but to its biological interaction with other species.

Alfred Russel Wallace

Early in 1858 Charles Darwin received a copy of an essay written by the young English biologist Alfred Russel Wallace titled "On the Tendency of Varieties to Depart Indefinitely from the Original Type." Wallace had taken an exploratory journey up the Amazon River Valley, had extensive field experience in the East Indies, and had formulated his ideas as a direct result of reading Malthus's essay on population. Wallace wanted Darwin to review the essay

Figure 22.7 Major biogeographical regions of the world. Different biogeographical regions generally have distinctively different floras and faunas. Some natural barriers that separate the regions are indicated in gray. (1) The Sahara and Arabian deserts. (2) Very high mountain ranges, including the Himalayas and Nan Ling mountains. (3) Deep water marine channels among islands of the Malay Archipelago. (A. R. Wallace recognized and wrote about this barrier, which has been called Wallace's Line). (4) The transition between highlands in southern Mexico and the lowland tropics of Central America.

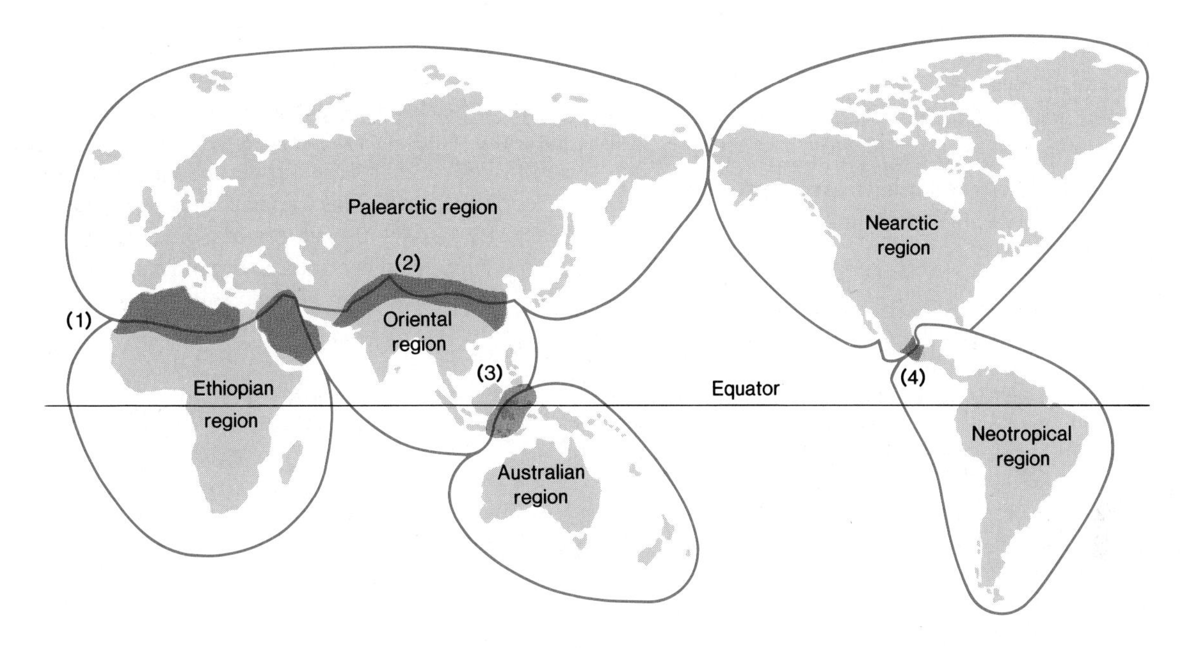

and to decide whether it merited further critical reading by Charles Lyell. Darwin was startled to find in Wallace's short essay the basic elements of his own theory of natural selection. Wallace's ideas were so similar to his own that Darwin felt that he should put off publication of his own work and let Wallace have credit for the idea of natural selection.

However, both Lyell and the botanist Sir Joseph Hooker knew of Darwin's long efforts in the development of his theory. They also knew that he had written his own ideas on the origin of species and natural selection in an abstract sent to Lyell in 1844 and in a letter to the famed American botanist Asa Gray in 1857. They persuaded Darwin to have these statements of his theory, along with Wallace's paper, presented to a meeting of the Linnean Society in the summer of 1858. Then Darwin hurried on with preparation of a large book on his ideas of organic evolution and natural selection, which was called *On the Origin of Species* and was published in November of 1859.

Evolutionary Studies after Darwin

New scientific theories often stimulate a great deal of controversy. This has a beneficial effect on scientific progress because scientists with opposing views work intensively seeking evidence to support their ideas. The theory of organic evolution and Darwin's theory of natural selection initiated such a period of intense activity, and the search for data that would either support or refute these theories began. The weight of scientific evidence over the past 100 plus years strongly supports Darwin's basic precepts, and such areas of biological specialization as biogeography, paleontology, comparative anatomy, genetics, and molecular biology have provided especially strong support.

Biogeography

Biogeography is the study of the geographic distribution of plants and animals. It attempts to explain why species are distributed as they are. As Darwin toured the world on H.M.S. *Beagle,* he discovered that plant and animal species are not distributed as broadly as are their potential habitats. Studies in biogeography since the time of Darwin have confirmed this over and over again. For example, plants of the cactus family (*Cactaceae*) are found in the deserts of southwestern North America and the high deserts of the Andes, but nowhere else. Comparable desert habitats in Africa are occupied by members of a different group, the spurge family (*Euphorbiaceae*). In fact, each major geographic region of the world has its own characteristic **flora** and **fauna** (figure 22.7). In addition, the fossil record of each region reflects an evolutionary sequence of biological events that is distinct from that of all other regions.

These observations reinforce the concept that the environment is the major force that has molded modern species. It also shows that each major biogeographical region has its own distinctive evolutionary history.

Paleontology

Information about the history of life on earth is contained in the fossil record, a collection of the remains of extinct forms. Incomplete as it is, the fossil record does permit scientists to consider events and processes that have taken place through that long history. The study of fossils and the fossil record is called **paleontology.**

Fossil Formation

Of all the organisms that have lived, only a relatively few have become fossilized, and many of those that were fossilized have later been destroyed by various geological processes.

The great majority of dead organisms are devoured by scavengers and decomposed by bacteria and fungi. Exposed bones are reduced to dust by the combined action of water, sun, and wind. The organisms preserved in the fossil record are those few that were buried by loose sediments very soon after their death. This happens most often underwater and least often in dry upland regions. For this reason, the fossil record contains more aquatic organisms, and organisms that lived near water, than land-dwelling organisms.

Box 22.1

Continental Drift

Modern geologists accept Hutton's proposals that the continents are continually worn away by the processes of erosion and that the accumulated debris of the erosion cycle is compacted and cemented into sedimentary rocks. But the earth is much more dynamic than even Hutton thought.

Early in this century Alfred Wegener in Germany proposed that at one time in the past all the continents of the earth were united in one large continental mass that he called Pangaea. Wegener thought that Pangaea began breaking apart millions of years ago, undergoing a series of changes that have resulted in the configurations and positions of the continents found today. Although Wegener presented a variety of data in support of this theory, it was not widely accepted by geologists until the second half of this century. It was given strong support by H. H. Hess of Princeton University in the early 1960s, and since that time, a great deal of evidence has been brought forward in support of the concept, which is now called the theory of **continental drift.**

According to this theory, the continents are relatively thin plates of lower density rock that float atop a denser rock that forms a "skin" around the earth. As the light continental plates move over the denser layer beneath, they are subjected to stresses that cause cracking and crumbling of the plates—stresses that geologists believe are responsible for faulting, earthquakes, volcanic activity, and mountain building. The relatively new subdiscipline of geology that is concerned with these matters is called **plate tectonics.**

The theory of continental drift has been substantiated by a variety of geological observations, and from their studies, geologists have been able to piece together the following story of the earth's history: About 225 million years ago all of the continental plates were united in a supercontinent called Pangaea. Then Pangaea began breaking up. It first split into two huge continental masses—Laurasia and Gondwana. Gradually, Laurasia and Gondwana broke up into the continental plates familiar today. About 65 million years ago the last of the previously existing plates separated, and since that time, the plates have been drifting slowly to the positions in which they are now found (box figure 22.1).

The fossil record reflects these geological events. During the time when the continents were united as Pangaea, many species enjoyed worldwide distribution—a phenomenon hard to explain if the continents were separated by ocean barriers as they are today. Even during the existence of Laurasia and Gondwana, species could be very widely distributed. But as the continents separated and drifted apart, the oceans became physical barriers to plant and animal distribution, and the flora and fauna of each continent became isolated. The plant and animal species of the individual plates have developed in geographic isolation, and over the many millenia, each continent has developed its own flora and fauna. This sequence of events is also reflected in the unique fossil record of each biogeographical region.

Generally, in order to become fossilized, an organism must possess hard body parts. Organisms without hard body parts are rare in the fossil record.

Actual fossilization can occur in one of several ways. Water may slowly dissolve the water-soluble calcium of shell or bone and leave another mineral deposited in its place. In this process, called mineralization, a durable fossil replica of the original material is left in the sedimentary rocks.

Sometimes the original shell or bone is completely dissolved away and removed from the sediment layers, leaving behind a perfect mold of the organism. At times, a cast is formed when minerals are later deposited in the hollow chamber of the mold. Both molds and casts provide good fossil replicas of body surfaces.

Interpretation of the Fossil Record

Paleontologists have done a remarkable job of reconstructing the history of life. The first (oldest) life forms to appear in the fossil record were primitive single-celled organisms resembling modern bacteria. In time, more complex cellular forms appeared, followed by multicellular plants and animals. The increase in the complexity of life forms is seen as evidence of an evolutionary sequence—the development of new life strategies continually being put to the test of natural selection. Throughout the long evolutionary process, the great majority of species failed to meet the challenge of adaptation and became extinct (figure 22.8).

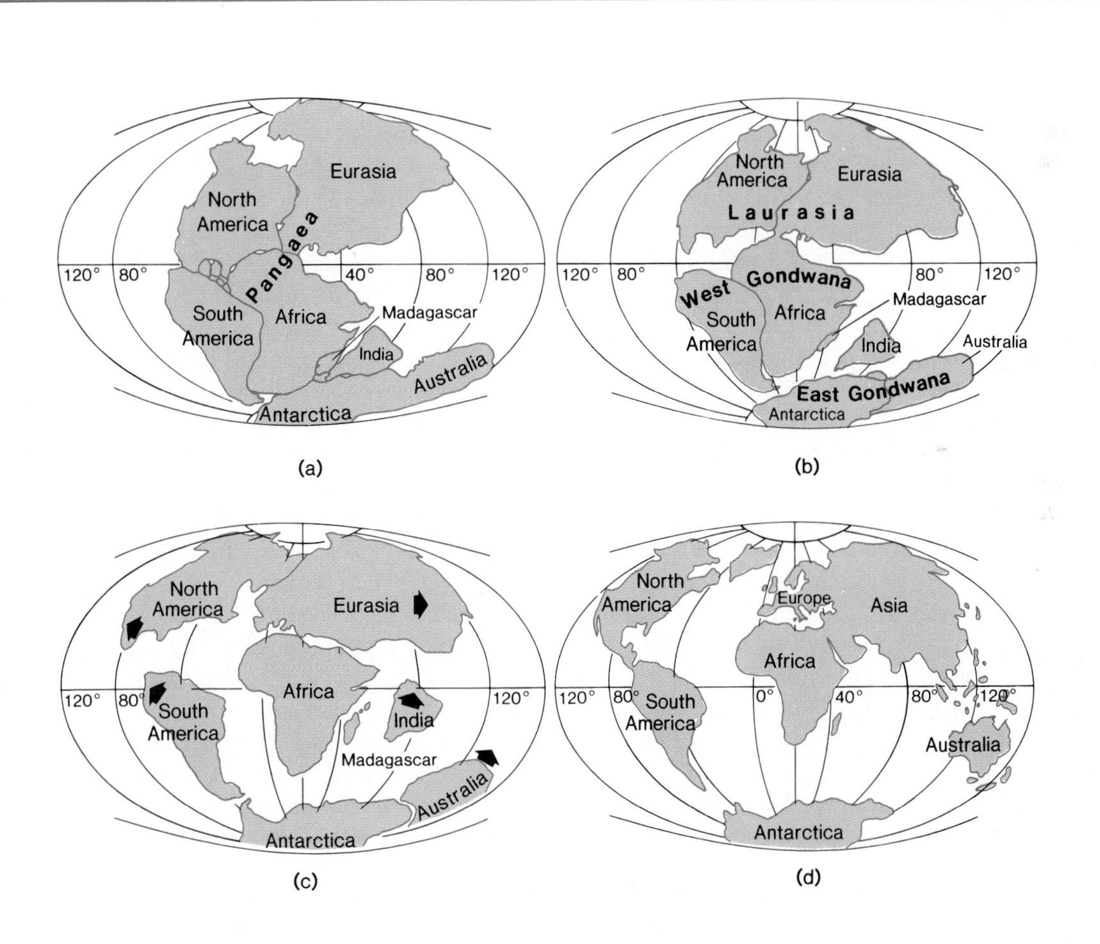

Box figure 22.1 Continental drift and the origin of the continents. (*a*) About 225 million years ago, early in the Mesozoic era, all of the earth's land existed as a single massive continent, which geologists have named Pangaea. (*b*) About 135 million years ago, Pangaea had broken up into a northern supercontinent called Laurasia and a southern supercontinent called Gondwana, which itself split into east and west parts. (*c*) About 65 million years ago, Laurasia remained a single continent, but Gondwana's breakup was complete, and its parts were drifting apart toward their present locations (arrows indicate direction of motion of the various plates). (*d*) The present arrangement of the continents.

Figure 22.8 A trilobite fossil from central Utah. Trilobites have been extinct for millions of years and are only known from the fossil record, but trilobites were once very numerous in terms of numbers of species and numbers of individuals. While the fossil record often is described as being fragmentary or incomplete, a sufficient number of fossils have been found to permit identification of nearly 4,000 species of trilobites and allow study of juvenile forms of some of those species.

The fossil record also shows clear evidence of gradual change within single lines of descent. A good example of such change is the evolution of horses. The fossil record of horses covers a period of some 60 million years (figure 22.9). The earliest fossil horse was the dog-sized *Hyracotherium* (also known as *Eohippus*), a foot-high animal with four toes on its front feet and three on its hind feet. Based on the structure of its teeth, *Hyracotherium* is thought to have fed on forest underbrush. From the variety of horses appearing in the fossil record, we can identify certain trends in horse evolution. First, the molar teeth gradually enlarged and became flatter and higher crowned, thus making horses better suited for feeding on grasses. Second, the leg bones became fused, providing increased strength in running, and legs also lengthened. Third, the digits of both the forelimb and hindlimb were gradually reduced in number. In the modern horse one digit remains, and the hoof is formed from the nail of this digit. In short, the horse became specialized for the life of a grassland grazer. These changes were gradual, and they occurred in a sequence of events, which are depicted in the fossil history of the horses.

The fossil record clearly indicates that the major groups of organisms appeared on earth in a sequential pattern. The reasonable assumption is that the older groups gave rise to the younger (more recent) groups, but there are some "missing links"—forms intermediate between major groups of organisms that are missing from the fossil record.

However, some intermediate forms have been found. A well-known example of such a transitional form is the fossil species *Archeopteryx,* a pigeon-sized animal that possessed true wings with feathers, the beak of a bird, and a skeleton that partially resembled that of the modern bird. But *Archeopteryx* had reptilian claws, and its skeleton, while bird-like, also had reptilian features (figure 22.10). Specific forms, such as *Archeopteryx,* that have been found may not be the precise species that gave rise to the modern forms, but they illustrate evolutionary change.

We should mention that there is some debate as to the tempo of evolutionary change. Many biologists think that the gradual transition in form represented by the fossil record of the horse is typical of evolutionary modification. If this view is correct, then eventually a number of intermediate forms should be found in the fossil record. However, other biologists think that there were long periods with little change. According to this view, new species form very rapidly, during periods of environmental stress. Such relatively rapid production of new forms would leave few transitional, intermediate forms in the fossil record (see the discussion of **punctuated equilibrium,** chapter 23) and would make it highly unlikely that fossil "missing links" between certain groups would ever be found.

Figure 22.9 Evolution of the horse. The fossil history of horses dates back to early Cenozoic times, about 60 million years ago. (The Cenozoic era is divided into periods, which are subdivided into epochs—see table 22.1 for more information on the geologic time scale.) The modern horse has evolved from the dog-sized *Hyracotherium* (also called *Eohippus*) via a number of intermediate forms. Note the gradual reduction in the number of digits of the limbs. A number of branches of horse evolution have ended in extinction.

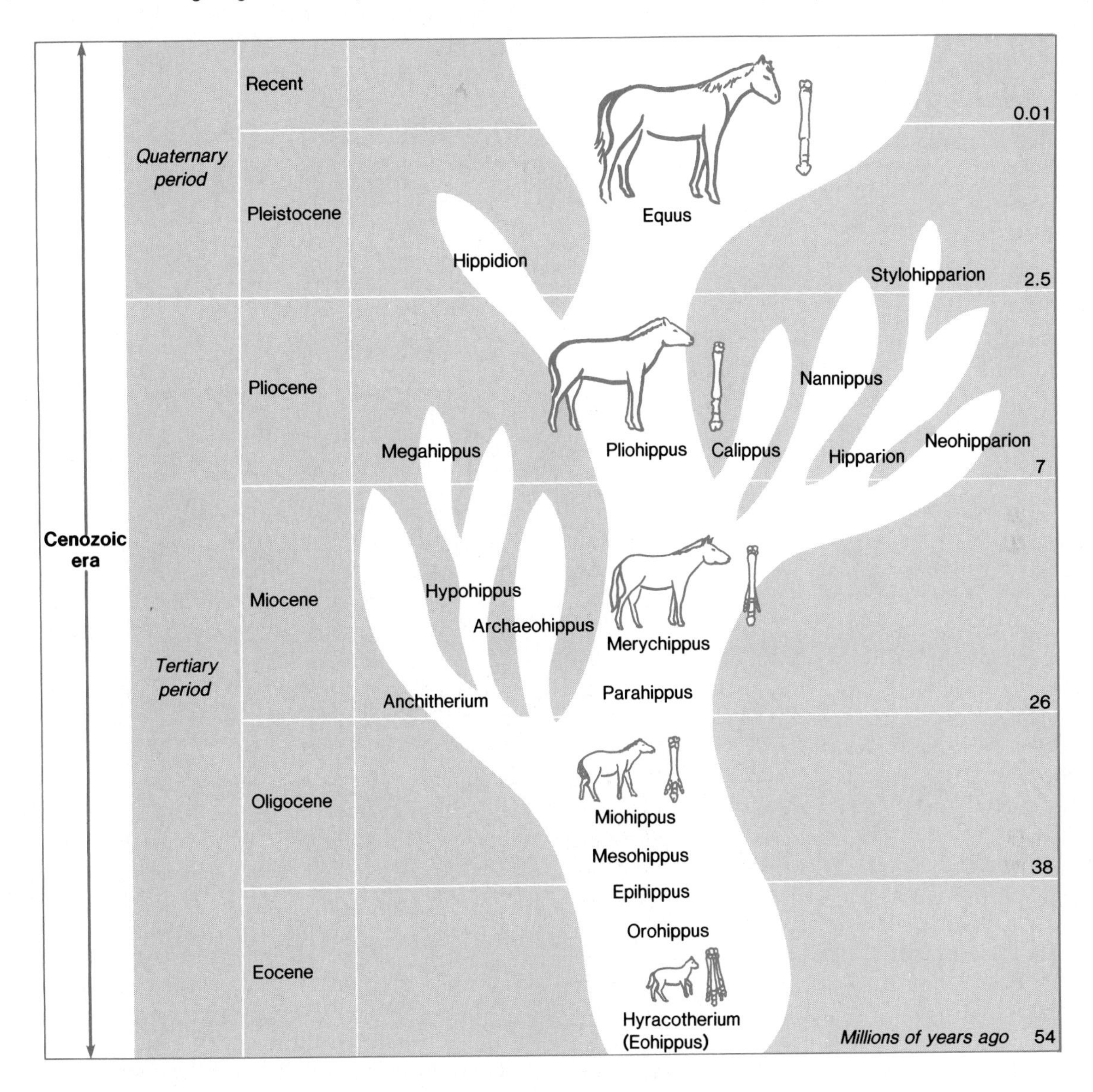

Relative Dating Techniques

To use the fossil record effectively to represent sequential changes that have occurred in the history of life on earth, we must be able to assign actual dates to fossil organisms. Paleontologists use two types of dating to determine the age of fossil-bearing rock layers—relative dating and absolute dating.

Relative dating techniques depend on the principle of **superposition.** This principle states that in any undisturbed sequence of sedimentary rocks, the bottommost layer is the oldest, and the topmost layer is the youngest. This interpretation follows from the manner in which sedimentary layers are formed. When a river reaches the ocean, it deposits its

Figure 22.10 (*a*) Comparison of a modern bird and a representation of *Archeopteryx* based on fossils. Like modern birds, *Archeopteryx* had feathers on its body and wings, but it had jaws with teeth, clawed digits at the margins of its wings, and various skeletal features that were intermediate between those of reptiles and modern birds. (*b*) A fossil of *Archeopteryx*.

(a)

(b)

sediment load on the ocean floor, and over a number of years, hundreds of meters of sediments can accumulate. Eventually, these sediments may be converted into sedimentary rock, and the oldest sediments (and hence the oldest rock) are at the bottom of the rock sequence. Therefore, in any undisturbed sequence of sedimentary rock, fossils found in the bottommost layer are older than fossils found in the layers above.

A second method of determining the relative age of fossils was developed in the 1700s by William Smith, a British geologist, surveyor, and engineer. Smith's interest in geology was a practical one, since he was concerned with the physical properties of the rock layers he encountered in his road construction work. Smith wanted to know whether sandstone that he encountered in one place was of the same origin, and thus possessed the same physical properties, as sandstone found in another area. With this goal in mind, Smith searched for a method to correlate strata from one geographic region with another.

Smith noted that in some sequences of sedimentary rock, the kinds of fossils in the bottom layers differed from those in higher layers. Because he had encountered similar fossil groupings in sedimentary layers occurring in widely separated parts of Britain, Smith proposed that all rocks containing similar fossils were of the same geological age. If so, then the fossil assemblage of a particular rock layer would provide the means of correlating rock layers occurring in widely separated geographical regions.

Smith's idea proved to be a fruitful one and became known as the principle of **fossil correlation,** a valuable tool used by geologists to this day.

These two principles—superposition and fossil correlation—were the only dating techniques available to geologists before the twentieth century. They were used to painstakingly piece together the fossil record of life in the form of a set of correlations between contents of fossil-bearing rock layers (strata) and their relative ages. No one area on earth contains a continuous sequence of rock strata from the beginning of the earth to the present. Neither does any one location contain a continuous sequence of fossils. The task confronting geologists and paleontologists was to place scattered rock layers in the proper time sequence.

By carefully examining sedimentary layers from all parts of the world and correlating the fossils they contained, a continuous sequence of rock strata and fossil life was attained. This record became known as the geological time scale (table 22.1). But there still was no way of assigning an absolute age to any single rock layer or fossil. With the discovery of radioactivity and the characteristics of radioactive decay, however, a means was devised to assign specific dates to certain rock layers.

Absolute Dating

Radioactive atoms have unstable atomic nuclei. When these nuclei break down (decay), they emit characteristic particles or rays. The end result of this radioactive decay is that another kind of atom is formed. For example, **potassium-argon dating** is a method used for determining the age of geological deposits. Radioactive potassium (^{40}K) decays to form ^{40}argon (^{40}Ar). If a rock sample contained ^{40}potassium and no ^{40}argon, all of the ^{40}K atoms would eventually be converted to ^{40}Ar, but completion of this process would take a very long time. In fact, it takes 1,300 million years for one-half of the ^{40}potassium to change to ^{40}argon. This period of time, called the **half-life,** is the time interval commonly used in radiometric dating, dating based on radioactive decay rates. Every radioactive substance has its own unique half-life.

How is this information about radioactive decay and half-lives used in radiometric dating? Let us use potassium-argon dating to see how it works.

If a rock sample contains ^{40}K and ^{40}Ar, an exact measurement of the proportion of ^{40}K to ^{40}Ar yields the absolute age of that rock sample. For example, if the rock sample originally contained 1,000 grams of potassium, after 1,300 million years it would contain 500 g of ^{40}K and 500 g of ^{40}Ar. In another 1,300 million years, half of the remaining potassium would change to ^{40}argon. Thus, if a rock sample being studied contains 250 g of ^{40}K and 750 g of ^{40}Ar, it would have passed through two half-lives and would be 2,600 million years old.

The best samples of radioactive materials for dating are found in rocks of volcanic origin (igneous rocks). This causes some problems as fossils are very rarely found in igneous rocks because the heat of the molten material from which these rocks are formed destroys organic matter rather than preserving it. Thus, the dating of fossils often depends on finding a volcanic rock layer in close association with a sedimentary rock formation. When this occurs, an absolute date can be assigned to a particular fossil species found in one location, and the fossil correlation can be used to assign an absolute age to all strata containing the same species.

Carbon dating is a means of dating fossils that still contain some carbon material from the living organism. The radioactive isotope ^{14}carbon (^{14}C) has a half-life of 5,760 years, and it breaks down to form a stable isotope of nitrogen (^{14}N). ^{14}Carbon is formed in the earth's atmosphere through the action of cosmic radiation from the sun on nonradioactive nitrogen, which is converted to ^{14}C. The ^{14}C in the atmosphere, in the form of $^{14}CO_2$, is incorporated into organic

Table 22.1
The Geological Time Scale. Major Divisions of Geological Time with Some of the Major Evolutionary Events of Each Geological Period.

Era	Period	Millions of Years Ago	Major Biological Events	
			Plants	*Animals*
Cenozoic	Quaternary		Rise of herbaceous plants	Age of humans
		2.5		
	Tertiary		Dominance of the angiosperms	First hominids Rise of modern forms Mammals and insects dominate the land
		65		
Mesozoic	Cretaceous		Spread of angiosperms Decline of gymnosperms	Extinction of the dinosaurs
		130		
	Jurassic		First flowering plants (angiosperms)	Age of dinosaurs First mammals and first birds
		180		
	Triassic		Land plants dominated by gymnosperms	First appearance of the dinosaurs
		230		
Paleozoic	Permian		Land covered by forests of primitive vascular plants	Expansion of the reptiles Decline of the amphibians Extinction of the trilobites
		280		
	Carboniferous		Land covered by forests of coal-forming plants	Age of amphibians First appearance of reptiles
		350		
	Devonian		Expansion of primitive vascular plants over land	Fishes dominate the seas First insects First amphibians move onto land
		400		
	Silurian		First appearance of primitive vascular plants on land	Expansion of the fishes
		435		
	Ordovician		Marine algae	Invertebrates dominate the seas First fishes (jawless)
		500		
	Cambrian		Primitive marine algae	Age of invertebrates Trilobites abundant
		600		
Precambrian			*Aquatic Life Only* Origin of the invertebrates Origin of complex (eukaryotic) cells Origin of photosynthetic organisms Origin of primitive (prokaryotic) cells Origin of life	
		4,600		

molecules in living tissue through the process of photosynthesis (see chapter 5), and when plants are eaten by animals, the ^{14}C is incorporated into the animals' tissues. When an organism dies, carbon exchange with the environment ceases. Any subsequent change in the ratio of ^{14}C to stable carbon isotopes in its tissues is due to radioactive decay of ^{14}C to nitrogen. The particles (beta particles) that are emitted by ^{14}carbon in a sample can be counted, the total carbon content of the sample determined, and the age of the sample calculated. Age is a function of the ratio of ^{14}C to total carbon content; the smaller the ratio of ^{14}C to total carbon content, the older the sample. One problem associated with the use of the ^{14}C technique is that the relatively short half-life of ^{14}C makes this technique valid only for fossil materials dating back to about 50,000 years ago.

Figure 22.11 Homology in the vertebrate forelimb. All of these forelimbs are built on a fundamentally similar framework of the same set of homologous bones. This indicates that they share a common ancestry even though the limbs now look very different from one another and are adapted to a variety of functions. Homologous bones are indicated as follows: humerus (upper arm)—solid color; radius (forearm)—black; ulna (forearm)—gray; carpals (wrist)—white; metacarpals (palm) and phalanges (digits)—colored dots. The digits are numbered, beginning with the first digit (thumb).

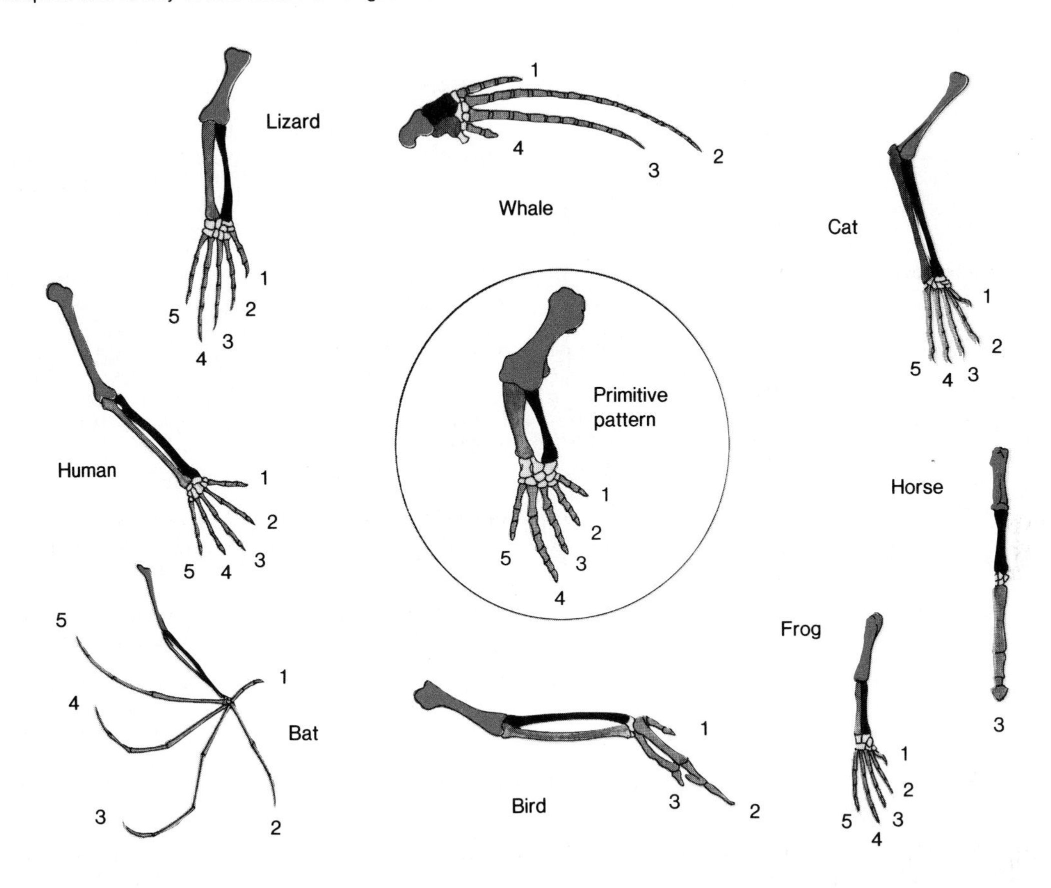

Using radiometric dating, geologists have estimated that the age of the earth is 4,500 to 5,000 million years, and reasonably accurate dates have been assigned to many of the major events of earth history and to the majority of fossil species appearing in the fossil record. The oldest known fossils, primitive bacterialike organisms, date to approximately 3,000 to 3,500 million years ago.

Comparative Anatomy

Another valuable approach to interpretation in paleontology is **comparative anatomy.** Comparative anatomists try to find similarities and differences among the fundamental structures of living organisms. Their studies have revealed certain basic plans upon which groups of organisms are structured. For example, all vertebrate animals share certain basic structural similarities, but there are important variations that distinguish the major groups of vertebrate animals from one another.

Since all species of organisms have arisen from previously existing species, it follows that those species that share a recent common ancestor are more similar genetically. Over time, the number of genetic differences increases, and the species become genetically and structurally more different.

Fundamental similarities in structure that have arisen through descent from a common genetic ancestor are called **homologies.** A good example of homology is the vertebrate forelimb (figure 22.11). All vertebrates are thought to have evolved from a common ancestor, and despite the fact that bird, whale, and human forelimbs serve different functions, they all possess the same fundamental bone structure.

Figure 22.12 Comparison of a bird wing (*a*), which has feathers attached to skin covering a skeletal framework, and the stiff, membranous insect wing (*b*) as an example of analogy. Structures that have evolved from very different evolutionary precursors and have different basic structures serve the same functon. This independent evolution of wings as an adaptation for flight is an example of convergent evolution.

(a)

(b)

Analogy is another important concept derived from comparative anatomy. Analogous structures are similar in function and appearance but differ in their fundamental structural plan. The wings of birds and insects are analogous structures. In both cases, the organs are used for flying, and each has the broad, flattened shape, but the two kinds of wings are different. The insect wing is a stiffened membrane supported by hard chitinous veins, while the bird wing has an internal bony skeleton covered by skin and feathers.

Analogous structures have arisen separately in the evolution of life forms. The ability to fly evolved completely independently in birds and in insects. Superficial similarities in the structures of their wings are due to the fact that a broad, flat surface is a prerequisite for flight. Thus, the ability to fly has appeared several times in the history of life. Analogous structures illustrate **convergent evolution**—evolution toward a common adaptation to a similar life-style (figure 22.12).

Vestigial Structures

Vestigial structures are those that appear in a simple, poorly developed form in some organisms but in a fully developed, functional form in other, closely related animals; or they are poorly developed structures thought to have been fully functional in ancestors of the modern form. Some examples of vestigial structures are: the rudimentary, functionless eyes of cave-dwelling animals; the wisdom teeth of humans; and the three to five caudal vertebrae present in humans (remnants of the tail structure found in most other mammals).

One explanation for the decline of certain structures is that natural selection favors those individuals in which the unneeded structure is reduced because they expend less energy on its development and maintenance. For example, there is selection against full expression of the genes for eye development in cave-dwelling animals. Very slowly, over long periods of time, natural selection may completely remove certain traits from populations.

Molecular Evolution

In addition to homologies found in such anatomical structures as the vertebrate forelimb, biochemists have found many homologies at the molecular level. Virtually all living things possess the same genetic material (DNA), use the same genetic code, and conduct energy conversions involving the same molecule (ATP). Many other molecules, including a number of enzymes, are found widely distributed among living forms.

The fact that practically all living things possess DNA as genetic material implies that this trait developed very early in the evolution of life and has been passed down to all organisms that have followed.

Protein homologies have been of special interest to the scientists studying evolution. Modern laboratory techniques permit biochemists to determine the exact amino acid sequences of protein molecules. Because proteins are the products of gene expression, we can infer that two organisms that possess similar proteins must also possess similar genes. Possession of similar genes is indicative of common ancestry.

We can use amino acid sequences of certain proteins to determine the closeness of evolutionary relationships among different species. If the amino acid sequences in proteins of two species are very similar, we can assume that these two species have only recently evolved from a common genetic ancestor. If, on the other hand, the proteins have very different amino acid sequences, the two species probably have been separated from a common genetic ancestor for a long period of time.

We can make these assumptions because gene mutations that change DNA base coding sequences are expressed as variations in the structure of proteins (changes in amino acid sequences). Because gene mutations occur at relatively constant rates, the accumulation of protein dissimilarities through gene mutation is used to estimate when ancestors of present-day species diverged from one another. Let us use data from studies on cytochrome *c*, an important protein in mitochondria, to illustrate this principle. The amino acid sequences in cytochrome *c* from humans and chimpanzees are identical, indicating relatively recent divergence of ancestors. Humans and chimpanzees differ from rhesus monkeys by only one amino acid out of the 104 that constitute cytochrome *c*, but they differ from dogs by 8 amino acids, from rattlesnakes by 12, and from dogfish sharks by 24.

Using known mutation rates of various genes, scientists can estimate the time since divergence. In general, the pictures of evolutionary descent developed using this "molecular clock" agree well with those derived from the fossil record and radiometric dating techniques.

Now there are techniques for direct examination of the makeup of individual genes, and this research has added an exciting new dimension to molecular evolution studies. In the next few years we can expect additional insights to be added to the understanding of evolutionary processes that has been built since Charles Darwin's pioneering work on evolution and natural selection.

Summary

The theory of organic evolution states that all species of plants and animals are the modified descendants of previously existing forms. Charles Darwin formalized the theory of organic evolution in 1859 in his book *On the Origin of Species*. Darwin's work was based on a long period of observation, experimentation, and contemplation, beginning with his five-year tour of duty as a naturalist aboard the H.M.S. *Beagle*.

Darwin also envisioned the process by which organic evolution took place—a process he called natural selection. Each species produces more young than can survive, and among the range of biological variants appearing in any population, some individuals are better adapted than are others. These better-adapted individuals make up the greater percentage of the breeding population of any generation, and thus it is likely that succeeding generations will contain a higher percentage of individuals with the adaptive traits.

The bulk of the scientific evidence gathered over the years has strongly supported the theory of organic evolution and Darwin's theory of natural selection.

Biogeographic evidence indicates that different environments have exerted different selective pressures during the evolution of the living things that occupy them.

Paleontology is the study of fossils and the fossil record, which can be used to reconstruct a general history of life on earth and to trace development within certain lines of descent.

Relative dating of fossils can be done using the principle of superposition and the fossil correlation of strata. Absolute dating is based on known rates of radioactive decay of various unstable isotopes that are found in fossil organisms or in layers near them.

Comparative anatomy and the study of vestigial structures provide data on evolutionary relationships.

Studies of molecular evolution are based on comparisons of molecular structures, mainly amino acid sequences in proteins, found in present-day organisms. Amino acid sequences are most similar in species that have relatively recently evolved from a common ancestor and much less similar in species that have been separated from a common genetic ancestor for a long period.

Questions for Review

1. What is the theory of uniformitarianism? How did it influence Darwin's thinking about the history of life on earth?
2. What is adaptive radiation? Under what conditions does it occur?
3. Describe Lamarck's theory of inheritance of acquired characteristics.
4. How did Malthus's ideas about the potential growth of human populations influence Darwin's thinking about natural populations?
5. What are the major points of Darwin's theory of natural selection?
6. What is the essential difference between artificial selection and natural selection?
7. How is the principle of superposition used in determining the age of fossils?
8. What is the principle of fossil correlation?
9. Explain what we mean by the term half-life.
10. What is the difference between homology and analogy?

Questions for Analysis and Discussion

1. Charles Darwin is often credited with developing the basic idea that living things have evolved, but he definitely was not the first to propose such an idea. Why do you suppose that his ideas attracted so much more attention than earlier ideas about evolution?
2. How is it that Darwin and Wallace conceived identical theories of natural selection at the same time without ever talking to each other about it?
3. Why is it that even if we could find all of the fossils buried in the earth (which we obviously cannot) it is extremely unlikely that we would even come close to having a complete fossil record of evolutionary history?

Suggested Readings

Books

Darwin, C. 1859. *On the origin of species*. Reprint. 1975. London: Cambridge University Press.

Miller, J., and VanLoon, B. 1982. *Darwin for beginners*. New York: Pantheon Books.

Stebbins, G. L. 1977. *Processes of organic evolution,* 3d ed. Englewood Cliffs, N.J.: Prentice-Hall.

Stone, I. 1980. *The origin.* New York: New American Library.

Articles

Ayala, F. J. 1983. Origin of species. *Carolina Biology Readers* no. 69. Burlington, N.C.: Carolina Biological Supply Co.

Grant, P. R. 1981. Speciation and the adaptive radiation of Darwin's finches. *American Scientist* 69:653.

Mayr, E. September 1978. Evolution. *Scientific American* (offprint 1400).

Nelson, G., and Platnick, N. 1984. Biogeography. *Carolina Biology Readers* no. 119. Burlington, N.C.: Carolina Biological Supply Co.

Valentine, J. W. September 1978. The evolution of multicellular plants and animals. *Scientific American* (offprint 1403).

The Evolutionary Process and Human Origins

Chapter Outline

Chapter Concepts

1. The modern theory of organic evolution is a synthesis of the theory of natural selection with modern concepts of genetics.
2. Evolution involves changes in gene frequencies in populations. Natural selection is the major force that causes gene frequency changes.
3. New species are produced when populations subjected to different selective pressures become reproductively isolated from one another.
4. The primates first evolved as tree-dwelling animals. Humans are descended from primates that became secondarily adapted to life on the ground.
5. Conclusions about the course of human evolution often are controversial because of disagreements concerning interpretation of the fossil record.
6. Cultural evolution is a key factor in the recent history of the human species.

It is really too bad that Charles Darwin had no way of knowing what you know about genetics. If he had known as much as you know, he could have developed his ideas about natural selection much further.

Darwin recognized that natural populations are variable. However, Darwin did not know about genes or alleles, the different versions of a single gene. Nor did he know anything about how different sets of characteristics (phenotypes) develop as a result of expression of different sets of genes (genotypes).

Because we understand more about genetics, we are able to relate genetics and natural selection in a modern view of evolution. Through natural selection certain genes are favored for transmission to future generations. In fact, natural selection can best be explained in terms of its effect on the genetic composition of a population. Going one step further, we can say that a good definition of evolution is any change in the frequency with which various gene alleles occur in a population.

Population Genetics

A **population** is an interbreeding group of individuals that occupy a specific geographical area. Evolutionary studies concentrate not on the genotypes of individual organisms, but on the **gene pool,** the sum total of all genes available for reproduction in a population. Therefore, the science of **population genetics** studies the events taking place in gene pools.

Genetic Equilibrium in Populations

After rediscovery of Gregor Mendel's principles of inheritance in 1900, biologists began to ask questions about the relationship of those principles to the evolutionary process. Might recessive alleles of genes tend to be overwhelmed by dominant alleles and disappear from the gene pool of a population? Might all relatively rare alleles represent genes that are disappearing and being replaced in an evolving population?

G. H. Hardy and W. Weinberg examined these questions mathematically and in 1908 proposed an answer that is now called the **Hardy-Weinberg law.** Basically, their answer was that in a large model population, under defined conditions, gene frequencies would tend to remain constant; that is, the relative frequencies of the alleles of a single gene, considered by itself, would remain stable (in equilibrium) from generation to generation (see box 23.1). And their reasoning applied not just to alleles of a single gene, but could be extended to all alleles of all genes in a population.

Box 23.1

Why Don't Recessive Genes Disappear?

How did Hardy and Weinberg convince biologists that the genetic makeup of a population would remain stable from generation to generation in a population that was not evolving? They did some elementary genetic reasoning and then constructed a simple, but very convincing mathematical model.

A hypothetical example can illustrate the Hardy-Weinberg law. To make things simple, let us consider a gene that has only two alleles, a dominant allele *A* and a recessive allele *a,* present in an entire population of sexually reproducing organisms. In this situation, three genotypes—*AA, Aa,* and *aa*—would be possible in the population. Two of those genotypes, *AA* and *Aa,* would produce the same phenotype.

Keep in mind that the Hardy-Weinberg law examines population gene pools, not individual genotypes, and it is concerned with frequencies of occurrence of alleles in the total gene pool.

We will arbitrarily set the frequencies of our hypothetical alleles *A* and *a* at 0.6 and 0.4, respectively. Gene frequencies are always given as decimals such as 0.6 and 0.4 rather than as the corresponding percentages (60 percent and 40 percent). In every case, the total of allele frequencies equals 1.0 (100 percent). In our example, $0.6 + 0.4 = 1.0$, where allele *A* makes up 60 percent of the total of the two alleles in the population, while allele *a* constitutes 40 percent of the total.

If allele *A* has a frequency of 0.6 in the entire population, it follows that 60 percent of all sperm cells and all egg cells produced in the population carry allele *A*. Similarly, allele *a,* which has a frequency of 0.4, is represented in 40 percent of all gametes produced in the population. These gamete allele frequencies can then be used in a Punnett square to determine the outcome of mating in this hypothetical population (box figure 23.1*A*).

Because allele *A* is dominant, 84 percent of the offspring (all those with genotypes *AA* or *Aa*) produced by mating in this population display the phenotype produced by expression of the dominant allele, while only 16 percent display the phenotype produced by expression of the recessive allele.

Will the more frequently occurring allele, allele *A* (which is also dominant in this case), tend to occur even more frequently in subsequent generations, at the expense of allele *a*? Do less frequently occurring alleles tend to become lost from gene pools over long periods of time? The answer to these questions provided by the Hardy-Weinberg law is that if no evolutionary change is occurring in the population, there would be no change in

Box figure 23.1*A* Allele frequencies and the Hardy-Weinberg law. (*a*) Punnett square illustrating the effects of totally random mating in a parental population in whose gene pool allele *A* has a frequency of 0.6 and allele *a* has a frequency of 0.4. (*b*) Summary of genotypes and phenotypes in the offspring generation. A simple dominant-recessive relationship between these alleles is assumed. (*c*) Analysis of allele frequencies in the gene pool of the offspring generation that demonstrates the equilibrium described by the Hardy-Weinberg law in which relative allele frequencies remain the same from generation to generation.

Eggs \ Sperm	0.6*A*	0.4*a*
0.6*A*	0.36*AA*	0.24*Aa*
0.4*a*	0.24*Aa*	0.16*aa*

(a)

Offspring genotypes	**Offspring phenotypes**
AA = 0.36	*A*– = 0.84
Aa = 0.48 (0.24 + 0.24)	
aa = 0.16	*aa* = 0.16

(b)

Contributions to frequency of alleles in gene pool in offspring generation

	Allele *A*	Allele *a*
AA individuals	0.36	0
Aa individuals	0.24	0.24
aa individuals	0	0.16
Totals	0.60	0.40

(c)

relative frequencies of alleles in populations. Even very rare alleles can continue to occur at the same constant low frequencies generation after generation and are by no means doomed to disappear from the gene pool simply because they are rare.

For example, if we analyze the frequencies of the two alleles, *A* and *a*, in the gene pool of the generation produced by the hypothetical mating in box figure 23.1*A*, we see that the relative allele frequencies are the same in the offspring generation as in the parental generation. We could go through this exercise over and over again, producing one hypothetical generation after another, and the relative frequencies of the two alleles *A* and *a* would never change.

The Hardy-Weinberg law describes this type of equilibrium mathematically. In a general mathematical statement, the frequency of one allele is represented by the letter p and the frequency of the other by the letter q, and $p + q = 1$. We can insert these symbolic gene

Box figure 23.1*B* Two demonstrations of the general model of allele frequency equilibrium described by the Hardy-Weinberg law. One demonstration uses the familiar Punnett square, and the other uses simple algebra.

Gametes	p	q
p	pp (p^2)	pq
q	pq	qq (q^2)

$$(p + q)^2 = p^2 + 2pq + q^2 = 1$$

frequencies into a Punnett square as gamete frequencies, or we can obtain the same results by using some simple algebra (box figure 23.1*B*). You probably recall that the expression $(p + q)^2$ is equal to $p^2 + 2pq + q^2$. Because the sum of genotypes in any gene pool is 1, the following equation can serve as a model of the equilibrium described by the Hardy-Weinberg law:

$$p^2 + 2pq + q^2 = 1$$

We can illustrate this further by substituting the frequencies of the two alleles (0.6 and 0.4) from the hypothetical gene pool for p and q respectively:

$$p^2 + 2pq + q^2 = 1$$
$$(0.6)(0.6) + 2(0.6)(0.4) + (0.4)(0.4) = 1$$
$$0.36 + 0.48 + 0.16 = 1$$

Thus, the terms of this equation indicate frequencies of genotypes that are identical with those obtained by the Punnett square method in box figure 23.1*A*:

$$p^2 = \text{Frequency of } AA = 0.36$$
$$2pq = \text{Frequency of } Aa = 0.48$$
$$q^2 = \text{Frequency of } aa = 0.16$$

Calculations become more complex in situations where more than two alleles are present, but the Hardy-Weinberg law applies equally well in those situations.

This model of equilibrium in gene frequencies can be applied to all alleles of all genes in the gene pool of a population.

The take-home message of the Hardy-Weinberg law is that relative gene frequencies in the gene pool of any population would remain in equilibrium if only ordinary genetic processes were acting on it. As gene frequencies do change in evolving populations, however, we can use the Hardy-Weinberg law as a jumping off point to begin consideration of the forces that bring about this evolutionary change.

Gene Frequencies and Evolution

The gene pools of natural populations, however, are not found in the stable condition calculated by Hardy and Weinberg. As we said earlier, the genetic composition (gene pool) of a natural population changes from generation to generation, indicating that these natural populations are evolving. Evolution involves changes in gene frequencies.

What are some forces that bring about these evolutionary changes in gene frequency? We can identify four main forces:

1. If a population is quite small, changes in gene frequency can occur by chance alone. This is called **genetic drift.**
2. **Migration** of individuals in or out of the population produces change in the gene pool that is not due to reproductive processes.
3. **Mutation** processes cause changes in gene frequencies.
4. Most important of all, **natural selection** may be operating; certain individuals are more likely to contribute their genes to the gene pool of the next generation.

Let us examine each of these change-inducing factors.

Genetic Drift

Genetic drift is a change in gene frequency that is not closely tied to natural selection and the adaptiveness of the traits involved, and it is a phenomenon that occurs in relatively small populations. Sometimes certain alleles are completely lost from small populations by genetic drift, while other alleles are "fixed" in the gene pool and become much more common.

We can illustrate genetic drift by describing a special case of genetic drift known as the **founder effect.** In nature, a few members of a parent population may migrate to a new area and establish a small, interbreeding population. It is highly unlikely that the gene frequencies of these few individuals are the same as those of the entire population from which they were drawn. It therefore follows that the gene pool of these individuals' descendants will reflect the gene frequencies of the founder organisms rather than those of the entire population.

An interesting example of the founder effect in human populations involves the Dunkers, a religious sect of 200 people who immigrated to Pennsylvania from western Germany in the 1700s. Since their arrival in the United States, the Dunkers have maintained strict marriage customs that prohibit marriage outside the sect. But they do permit the use of modern medical care and use modern technology. Thus, they probably are subject to the same selection pressures as members of the general population.

Table 23.1
Blood Types in Populations (Percentages)

	A	B	AB	O
United States	40	11	4	45
Dunkers	59	3	2	36
Western Germany	45	10	5	41

Under these conditions—that is, a small initial population, genetic isolation from other populations, and no obvious difference in selection pressure—significant differences in gene frequencies between the Dunkers and the United States population as a whole can be ascribed to genetic drift through the founder effect.

In genetic studies, the Dunker population was compared with the general population of the United States and western Germany. The data show significant differences in the frequencies of the genes conferring the ABO blood types (table 23.1). Blood type A occurs more frequently among the Dunkers than in the other two populations while the O type is somewhat more rare among the Dunkers. The B and AB types occur very infrequently.

Certain other physical traits were also examined (figure 23.1). Frequencies of middigital hair patterns, distal hyperextensibility of the thumb, and attached earlobes were significantly lower among the Dunkers than in the surrounding population. On the other hand, Dunkers have essentially the same incidence of left-handedness as that found in the surrounding population.

We can attribute the special set of gene frequencies found among the Dunkers to the founder effect and genetic drift.

Migration

Unless there are complete barriers to movement of individuals from one population to another, **gene flow** occurs as a result of migration and interbreeding. Gene flow is a factor that adds to the variability of natural populations, and it upsets the type of equilibria described by the Hardy-Weinberg law.

Mutation

As you know, variation in natural populations is a fundamental part of natural selection. The main source of variation is sexual reproduction. In each generation, sexual reproduction segregates sets of parental genes during meiosis and recombines the genes in new combinations in offspring.

Figure 23.1 Some inheritable physical traits included in the study of gene frequencies in the Dunker population in Pennsylvania. Traits studied included nature of earlobes, hyperextensibility of the thumb, middigital hair pattern, and left- versus right-handedness. Frequencies of some of these traits and ABO blood types among Dunkers differed significantly from those in the surrounding population. Apparently, these gene frequency differences result from the founder effect and genetic drift.

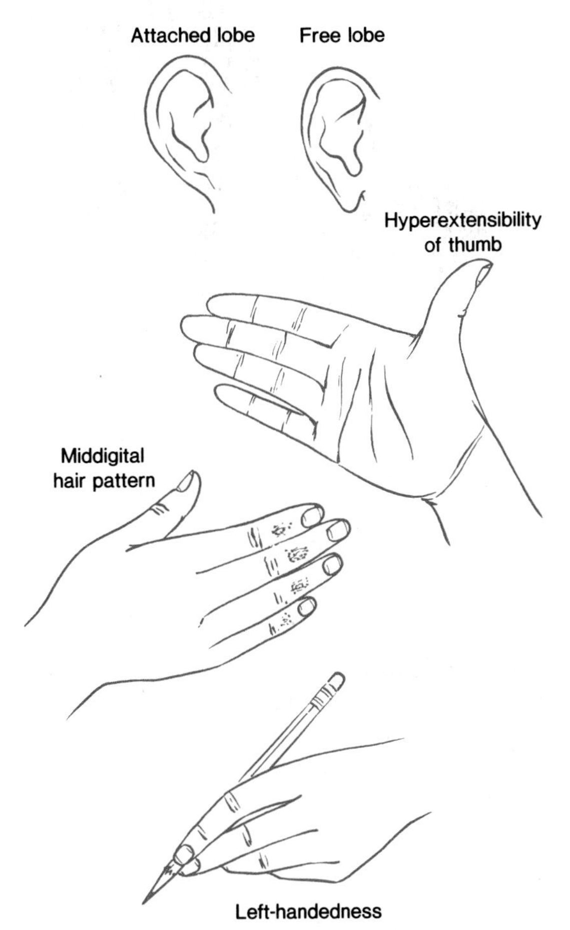

However, all of this recombination involves already existing genes. The process of mutation can contribute *new* genes to gene pools. Most often, mutations are specific changes in individual genes, which result from base substitutions in nucleic acids (chapter 18). The frequency with which these changes occur varies because each gene locus has its own characteristic mutation frequency.

Mutation rates of different alleles for the same character rarely are in equilibrium; that is, the rate of forward mutation, which is mutation from the more common allele to the less common allele, seldom is the same as the rate of back mutation, which is mutation in the reverse direction. The difference between the two is a mutation pressure that tends to produce a very slow change in allele frequencies. But evolution, slow as it is, actually moves more quickly than it would if it depended solely on change caused by mutation pressure.

Figure 23.2 Electrophoresis is a technique employing an electric current flowing through a gel for separation of similar molecules from one another. Electrophoresis of hemoglobins extracted from red blood cells demonstrates differences between hemoglobin from a normal individual and a person suffering from sickle-cell anemia. It also demonstrates that both types of hemoglobin are produced in heterozygous individuals.

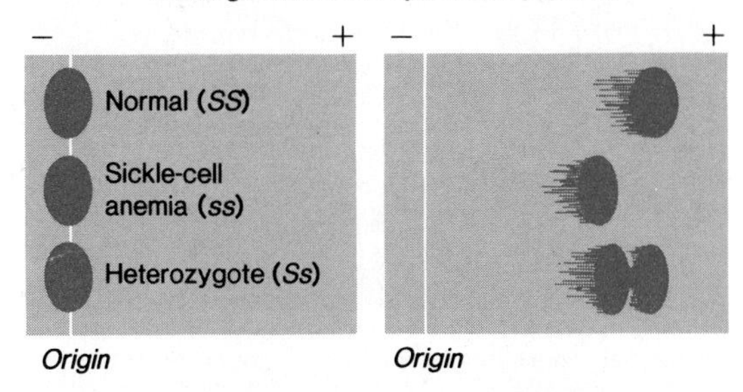

The real importance of mutation is that it is the only mechanism by which new genetic material enters the gene pool, as *some* mutations produce new alleles. Since the gene pool of any population is the product of a long period of natural selection during which genes producing more adaptive traits have increased in frequency and those producing less adaptive traits have decreased in frequency, it is not surprising that completely random mutational events usually produce alleles that are less adaptive than existing ones. But most mutational events produce alleles that are recessive to the original allele. Such recessive mutant alleles are not expressed as phenotypes until they occur in the homozygous condition. Thus, new recessive mutant alleles are not immediately exposed to the effects of natural selection.

Maintenance of recessive alleles in a gene pool is important because this reservoir of genetic variability may prove advantageous in the future. Should environmental conditions change, the adaptive value of a specific allele may also change.

Sickle-Cell Anemia

The human condition known as sickle-cell anemia is a good example of the difference in survival value of a specific allele under varying environmental conditions.

Sickle-cell anemia is an inherited disease that results from the formation of abnormal hemoglobin molecules (see page 379) in the red blood cells of afflicted individuals. Expression of the *S* allele produces normal hemoglobin, while expression of the *s* allele produces abnormal hemoglobin. Individuals with the *Ss* genotype have approximately 50 percent normal and 50 percent abnormal hemoglobin in their red blood cells (figure 23.2).

Figure 23.3 Electron micrographs of normal red blood cells (top) and sickled red blood cells (bottom) (magnification × 4,350 and 3,375, respectively).

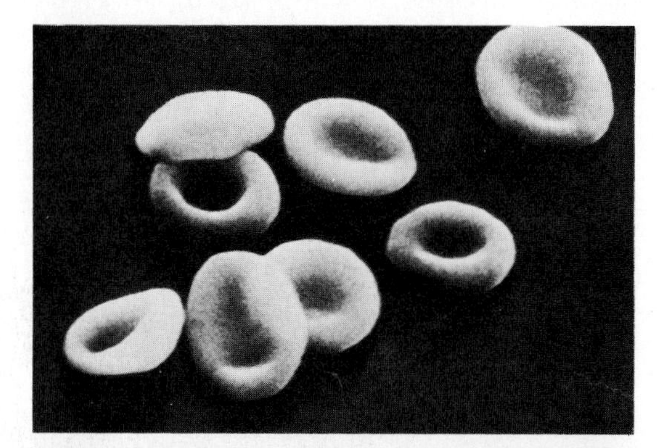

Figure 23.4 Geographic distribution of the sickle-cell condition, shown as percentages of the population afflicted with sickle-cell anemia. The frequency of the *s* gene is highest in those parts of the world in which malaria is common.

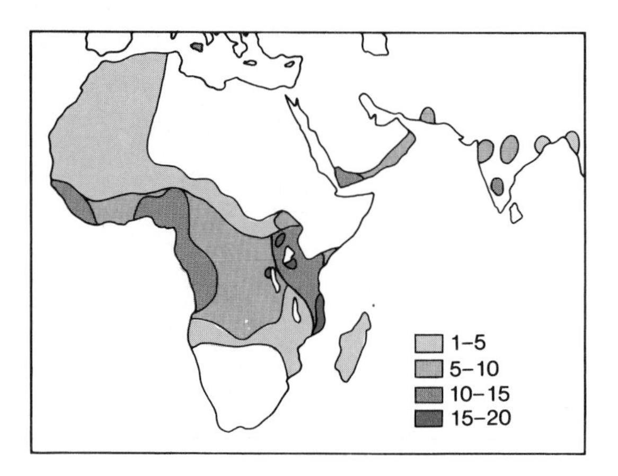

Individuals with the *ss* genotype suffer from sickle-cell anemia. When they are physically active, their red blood cells collapse, and the cells assume a characteristic shape resembling a sickle (figure 23.3). These sickled cells can block small blood vessels and cut off circulation to a tissue in a particular region of the body. Individuals with the *ss* genotype have difficulty with even mild exercise and are generally weak. Many *ss* individuals die before maturity, and very few of them live to age thirty. Heterozygous (*Ss*) individuals sometimes experience some problems when they exercise vigorously, but they are generally healthier than *ss* individuals.

You might expect that the *s* gene would have a very low frequency in the gene pool. Many *ss* individuals die young and heterozygous (*Ss*) individuals also are seriously enough affected at times to be at a selective disadvantage to individuals who are homozygous for the normal hemoglobin allele (*SS*). However, when geneticists studied populations of black Africans, in which the disease is most common, they found that the recessive allele had a surprisingly high frequency (0.2 to as high as 0.4 in a few areas).

Further research showed that another important factor affects the frequencies of these alleles. In regions where malaria occurs, the heterozygous (*Ss*) genotype had an adaptive advantage over both the *SS* and the *ss* genotypes (figure 23.4). Because the malarial parasite, which enters red blood cells, does not inhabit cells that contain the abnormal form of hemoglobin produced by expression of the *s* gene, individuals with the *SS* genotype suffer a higher death rate from malaria than do the individuals of the *Ss* genotype. And since a relatively high proportion of *Ss* individuals reproduce under these conditions, the frequency of the recessive gene remains greater than would be expected in the absence of the malaria factor. As you might expect, among American black populations who are not exposed to malaria, the frequency of the *s* allele is lower than in black populations inhabiting malarial regions.

The sickle-cell anemia example is an illustration of an evolutionary phenomenon called **balanced polymorphism.** This is a situation in which several very different phenotypic expressions are maintained in a population without one increasing in frequency at the expense of the others. In the case of the sickle-cell phenomenon in Africa, heterozygotes have a strong selective advantage over either of the two types of homozygotes. Even though both homozygotes are selected against, neither gene is eliminated because reproduction of the favored heterozygous individuals contributes both genes in equal quantities to subsequent generations. Such heterozygote superiority strongly favors balanced polymorphism.

Figure 23.5 Dark and light forms of the peppered moth *Biston betularia*. Dark and light forms on (a) a lichen-coated tree in an unpolluted region, and (*b*) on a soot-darkened tree near Birmingham, England.

(a)

(b)

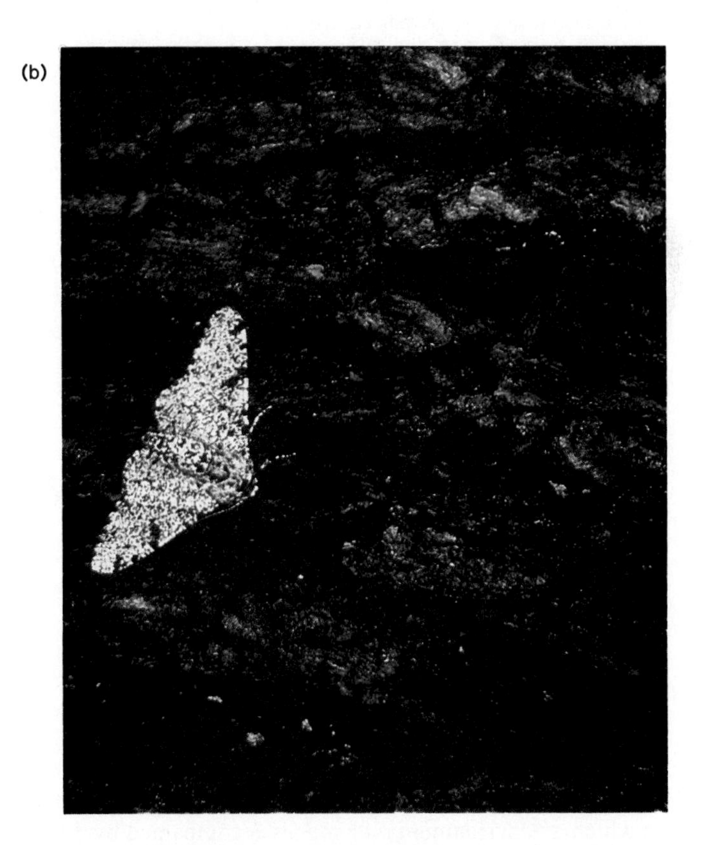

Natural Selection

Natural selection is the most important factor that changes gene frequencies in gene pools of natural populations. Every other factor that causes evolutionary change must be viewed in the context of natural selection. This is because, as Charles Darwin rightly surmised, natural selection is the major driving force that causes evolutionary change in living things.

Directional Selection

When the environment changes or when organisms migrate into new environments, natural selection operates to select those traits that are adaptive under the new set of environmental conditions. Because these changes represent a progressive adaptation to a changing environment, this type of selection is called **directional selection.**

Directional selection is clearly illustrated by **industrial melanism,** a progressive change in the average color of moths that has occurred in industrial areas, particularly in Britain and continental Europe. The best-documented example of industrial melanism occurred in the population of peppered moths (*Biston betularia*) in England between the mid 1800s and the early 1900s.

In the mid 1800s, the peppered moth population was almost entirely made up of a light-colored body form. A dark-bodied (melanic) form appeared only rarely, apparently because predatory birds had a harder time seeing the lighter moths against light-colored vegetation.

In 1850, the moth population consisted of approximately 99 percent light-colored individuals and 1 percent dark individuals. During the latter half of the nineteenth century, great quantities of soft coal were burned in industry. This produced a large quantity of air pollution and soot from smokestacks, which settled and darkened the vegetation.

As the vegetation became progressively darker, the adaptive advantage of the light-colored phase of the peppered moth was lost, and natural selection now favored the darker, melanic form. Against the soot-darkened tree trunks, the light-colored moths now stood out clearly, while the melanic form enjoyed the advantage of protective coloration (figure 23.5). Because of this change in selective advantage, by 1900 the peppered moth population of industrialized areas

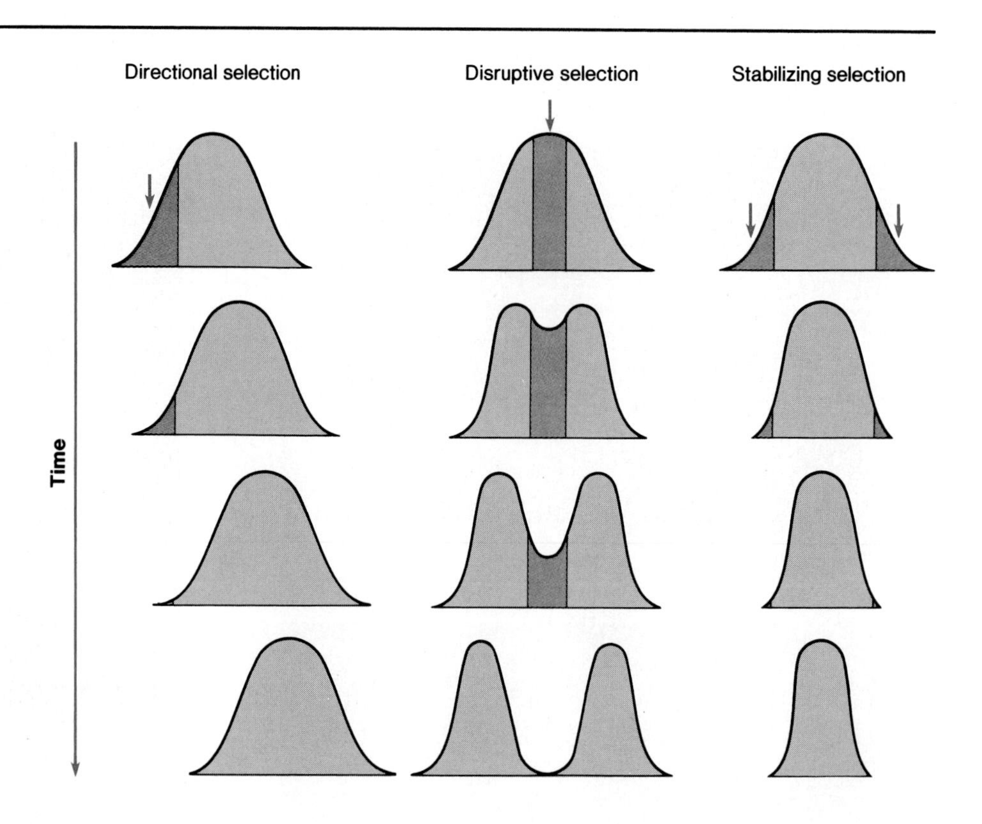

Figure 23.6 Three different types of selection. Each set of curves symbolizes the status of a particular set of characteristics in a population. Horizontal axes indicate the ranges of variability in the populations and vertical axes the number of individuals distributed in different parts of the ranges. Directional selection acts against individuals possessing one of the phenotypic extremes. It tends to result in elimination of one phenotype and proportional increase in others. Disruptive selection favors two different extreme phenotypes and acts against intermediate ones. It produces two divergent subpopulations with very different gene pools. Stabilizing selection operates when conditions remain stable for long periods, and it tends to make the gene pool more homogeneous.

of England was 90 percent melanic and 10 percent light-colored. Industrial melanism clearly shows directional selection in which an environmental change is accompanied by differential selection of phenotypes, resulting in changes in gene frequencies in the population.

Other Types of Selection

Directional selection generally results in change in one direction in response to changed environmental conditions. But natural selection can have other effects on gene pools in other situations (figure 23.6).

In **disruptive selection,** both categories of phenotypic extremes are favored over the average phenotypes in a population. This happens when a population previously exposed to a homogeneous environment becomes exposed to very different conditions in different parts of its area. Disruptive selection tends to divide a population into two contrasting subpopulations.

Another type of selection is **stabilizing selection,** which functions when the conditions under which a population lives remain constant over a period of time. Stabilizing selection works against phenotypic variation and helps to conserve the adaptive fit of the population to its environment by selecting against phenotypes produced by expression of new genetic combinations. Thus, selection is not always an agent of change.

Adaptations

An **adaptation** is a characteristic of an organism that increases the organism's fitness for life in its environment. Fitness is a measure of the likelihood that an organism will live and succeed in reproducing. Fitness, therefore, is a measure of the odds that an organism will make a genetic contribution to the next generation.

Each living thing possesses a battery of adaptations that contribute to its fitness. We can illustrate the complexity and diversity of adaptation by examining some special types of adaptations.

Figure 23.7 Cryptic coloration. For cryptic coloration to provide effective concealment, cryptically colored organisms must be behaviorally adapted to remain absolutely motionless when danger threatens. (*a*) A leaf katydid from Brazil. (*b*) A stonefish.

(a)

(b)

Cryptic Coloration and Mimicry

One of the most urgent and continuing threats to the survival and eventual reproductive success of most organisms is the danger of attack by predators. Some adaptations associated with predator avoidance are truly remarkable. Some organisms have **cryptic coloration** ("hidden coloration"), which makes them virtually undetectable when they are in position against their normal background (figure 23.7).

In **mimicry,** some organisms, instead of being hidden from the eyes of potential predators, present a showy but misleading appearance that resembles some other organism that predators avoid (figure 23.8). The similarity between the mimic and the model is often so close that predators avoid both model and mimic to nearly the same extent.

Coevolution

Coevolution is a situation in which the mutual evolutionary interaction between species is so intense that each exerts a strong selective influence on the other. Perhaps the best-known products of coevolution are the relationships between certain flowering plants and the animals upon which they depend for pollen transport from flower to flower. Adaptations of the flowers have evolved along with adaptations of the pollinators.

Flowers pollinated by bees usually are showy and bright, and they are always open during daytime hours, when worker bees do their foraging. Bee-pollinated flowers tend to be blue or yellow, but usually are not red, since bees are blind to red colors. They often have sweet, aromatic odors because bees depend on a well-developed sense of smell to locate nectar-containing flowers. Bee-pollinated flowers also have petal arrangements that provide bees a place to land before they push into the flower in search of nectar (figure 23.9).

Hummingbirds, on the other hand, see red well but blue only poorly. Because hummingbirds have a poor sense of smell, aroma is not an important factor in attracting them to flowers. Hummingbirds, like bees, forage in the daytime, but they require no landing perch. Characteristically, hummingbird-pollinated flowers usually are red or yellow and often are odorless.

Figure 23.8 Mimicry. (*a*) A lacewing (below) that mimics a wasp (above). The lacewing presents no threat to predators, but is avoided along with the stinging wasp. (*b*) A robber fly (left), a mimic of the bumblebee (right).

(a)

(b)

Figure 23.9 Pollen that sticks to a bee as it pushes into a flower in search of nectar is carried to other flowers that the bee visits subsequently.

Species and Speciation

Any change in gene frequencies is considered to be evolutionary change. The changes we have considered so far are changes that occur within the gene pools of populations. On occasion, however, sufficient genetic change has accumulated so that a new species has been formed.

Species Defined

A **species** is a population of organisms that may display a range of phenotypic (and genotypic) variation, but that still represents a biological entity distinct from all other populations. One definition of a species says simply that all members of a species are more like one another than they are like members of any other species.

But the concept of **reproductive isolation** gives a more satisfactory definition of a species. It says that a species is an interbreeding population whose members do not crossbreed with the members of any other species.

Speciation

A very important mechanism involved in speciation, the formation of new species, is **geographic isolation,** which occurs when natural barriers physically separate a species into two or more **allopatric** populations (populations between which gene flow is prevented).

Figure 23.10 hypothetically illustrates the way in which geographic isolation produces allopatric populations. Changes in these allopatric populations may lead to permanent reproductive isolation of two new species.

Figure 23.10 Geographic isolation of allopatric populations. *Stage 1:* Uniform environmental conditions. *Stage 2:* Three separate environments have been produced. These are: cool, semiarid (dry), and warmer. There are three subpopulations that are adapted to local environmental conditions (A, A_1, and A_2). *Stage 3:* The expansion of the desert wipes out A_1 and produces a geographic barrier to gene flow between A and A_2. In time, changes in these two populations may produce reproductive isolation, meaning that new species have been formed.

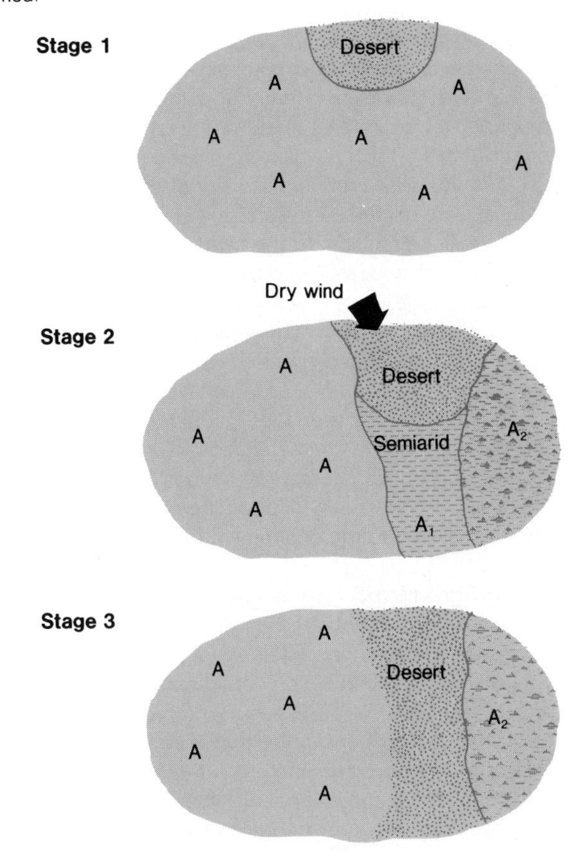

The Tempo of Evolution

Biologists have long thought that species are formed very slowly through processes that require many thousands or even millions of years. This is the concept of **phyletic gradualism.** Recently, however, a number of evolutionists, notably S. J. Gould and Niles Eldredge, have advanced the hypothesis of **punctuated equilibrium,** which suggests that evolution is "concentrated in very rapid events of speciation." They argue that throughout the greater part of its existence, a species displays very little change. This comparative equilibrium may be quite suddenly interrupted by environmental events that result in production of new species within a relatively short period of time by rather large steps. They propose further that the fossil record supports their view of the history of life because it shows long periods during which species remained essentially the same. They also point out the lack of transitional forms between taxonomic groups and suggest that transitional forms are not found because they never existed—that species formation occurs so rapidly that the long sequence of intermediate types predicted by the gradualistic concept simply were not a part of species formation.

Unfortunately, some observers outside the field of biology misinterpret arguments such as that over phyletic gradualism and punctuated equilibrium. They mistakenly conclude that biologists "are beginning to have doubts about evolution." These biologists, however, are not expressing doubts about the overwhelming evidence indicating the occurrence of an evolutionary process. Rather, they are arguing about the way that the evolutionary process proceeds. This debate actually stimulates biologists to reexamine existing data and to seek new information regarding the process of evolution.

Box 23.2
Evolutionary Controversies

The controversy between the phyletic gradualism and punctuated equilibrium viewpoints (page 509) is by no means the first such controversy in evolutionary theory. Even Darwin and Wallace, who together formulated the theory of natural selection, had disagreements about the actual role of natural selection in the evolutionary process.

The recent discussion about punctuated equilibrium must be viewed in the context of the "modern synthesis" of evolutionary theory, which developed in the first half of this century following the discovery of some of the basic mechanisms of inheritance and the growth of the science of population genetics. At the heart of the modern synthesis is the idea of long periods of accumulating gradual change, usually associated with geographic isolation, and the eventual emergence of new and reproductively isolated species. The concept of punctuated equilibrium challenges this central idea of the modern synthesis. It proposes instead that evolution of any particular line is characterized by long periods of relative stability punctuated by periods of abrupt change, change that occurs in periods of time that are relatively short compared to the enormity of the geologic time scale.

Disagreements such as the arguments about phyletic gradualism and punctuated equilibrium are not unusual happenings in science. Unfortunately, however, people who do not think that life on earth has an evolutionary history cite such controversies as indications of some fundamental weakness in the theory of organic evolution. Such controversies are inherent in the nature of science and characterize most, if not all, fields of scientific endeavor. This vigorous debate is by no means a weakness in the theory of organic evolution. Rather it is a strength, an expected characteristic of an active field of scientific endeavor. Scientific progress is made through careful examination that leads to acceptance or rejection of competing conceptual schemes. The theory of organic evolution remains one of the central unifying themes of modern biology, and it provides one of the basic conceptual frameworks for the interpretation of biological phenomena.

Views of the history of life on earth, possibly more than any other set of scientific concepts, engender emotional responses and controversy of another sort. The great majority of scientists think that available evidence indicates a very long history of life on earth that is measured in thousands of millions of years and characterized by evolutionary descent. But some individuals believe that the history of life on earth is much shorter, possibly as short as 10,000 years, and that it is characterized by a series of divine creation events. Their view of life is essentially compatible with literal interpretation of the biblical creation stories of the Judeo-Christian religious tradition.

Other scientists, however, do not find the idea of a long evolutionary process incompatible with their religious faith and experience. They do not feel that their faith is compromised because they interpret the creation stories in terms of modern scientific understanding. They recognize that the biblical creation stories in the first two chapters of Genesis were written in a form compatible with the experience of people living several thousand years ago. Possibly their view of faith and life can best be summarized with the words that Charles Darwin used following his summary of the theory of natural selection at the very end of *The Origin of Species:* "There is a grandeur in this view of life, with its several powers, having been originally breathed by the Creator into a few forms or into one; and that, whilst this planet has gone cycling on according to the fixed law of gravity, from so simple a beginning endless forms most beautiful and wonderful have been, and are being evolved."

Human Evolution

In the introduction to his book *The Descent of Man,* Charles Darwin wrote: "During many years I collected notes on the origin or descent of man, without any intention of publishing on the subject, but rather with the determination not to publish, as I thought that I should thus add to the prejudices against my views." Darwin was right, as many of his contemporaries were hostile to the notion that humans had descended from other earlier organisms and had not always been as they are now.

To a large extent, that original controversy about the evolution of human beings has passed; at least it is not a significant issue among modern biologists. But much still remains to be learned about our evolutionary history.

Human Origins

You are a mammal. Human beings belong to a group of vertebrate animals that are included in a taxonomic category called the **class Mammalia** (table 23.2).

Figure 23.11 A tree shrew. Modern tree shrews have many characteristics that are intermediate between those of modern insectivores (shrews and moles) and primates. Tree shrews may resemble the ancient ancestors of modern primates.

Since the mass extinction of the dinosaurs near the end of the Mesozoic era, 65 million years ago, the mammals have achieved dominance among terrestrial vertebrates.

Humans are members of a group of mammals classified in the **order Primates** (table 23.3). The primates first evolved as **arboreal** (tree-dwelling) animals, although not all modern primates live in the trees.

Primates have five functional digits on their hands and feet. In many primates, thumbs are **opposable;** that is, they close to meet fingertips and function efficiently in grasping.

Primate shoulder joints permit much more extensive forelimb rotation than the shoulders of other mammals allow. Other adaptations to life among the tree branches include relatively short snouts and eyes placed at the front of the head. This results in excellent binocular stereoscopic (three-dimensional) vision that permits primates to make very accurate judgments about distance and position, judgments that are essential for animals that swing and leap from branch to branch.

Most primates have a pair of mammary glands and produce only one offspring per pregnancy, whereas a great many other mammals produce litters, batches of several offspring born at the same time. Primates have relatively long infancies and develop strong, long-lasting mother-infant bonds.

Fossil remains of the earliest primates are scarce, but it generally is thought that they may have been quite similar to the living tree shrews of Southeast Asia (figure 23.11). By 50 million years ago, however, primates that grasped limbs with their digits, as modern primates do, had evolved. Their eyes faced forward rather than to the sides, as the eyes of their shrewlike ancestors had.

Table 23.2
Examples of Taxonomic Categories

Categories	Human	Domestic Dog
Kingdom	Animalia	Animalia
Phylum	Chordata	Chordata
Subphylum	Vertebrata	Vertebrata
Class	Mammalia	Mammalia
Order	Primates	Carnivora
Family	Hominidae	Canidae
Genus	*Homo*	*Canis*
Species	*sapiens*	*familiaris*

Table 23.3
Classification of the Primates

- Order Primates
  - Suborder Prosimii (lemurs and tarsiers)
  - Suborder Anthropoidea
    - Superfamily Ceboidea
      - Family Cebidae (New World monkeys)
    - Superfamily Cercopithecoidea
      - Family Cercopithecoidae (Old World monkeys and baboons)
    - Superfamily Hominoidea
      - Family Pongidae (apes)
      - Family Hominidae (humans)

Figure 23.12 Some living prosimians. (*a*) A ring-tailed lemur. (*b*) A tarsier.

(a)

(b)

Figure 23.13 Hypothetical "family tree" of the living primates.

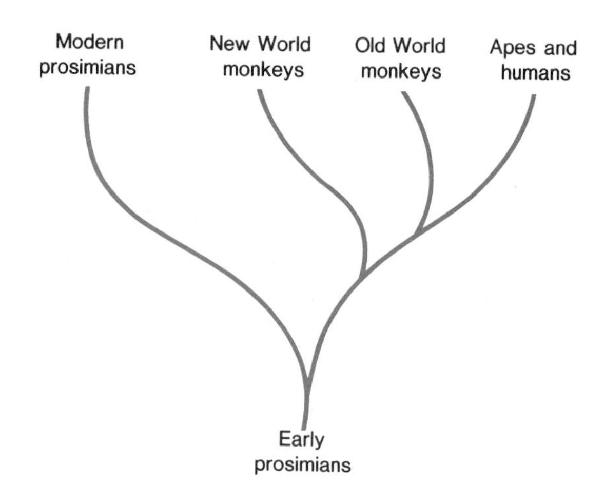

The Primate order is divided into two suborders, the Prosimii and the Anthropoidea, which includes monkeys, apes, and humans (see table 23.3).

Prosimians

Fossils of prosimians are found in many parts of the world, but modern prosimians, which include lemurs and tarsiers, are very restricted in their ranges. Today, in fact, lemurs (figure 23.12*a*) are found only on the island of Madagascar, and tarsiers (figure 23.12*b*) occur only in the Philippines and the East Indies.

Monkeys

By the Oligocene epoch, which began about 38 million years ago (table 23.4), the earliest members of the suborder Anthropoidea had diverged from the prosimians. Later two separate lines developed among the primitive anthropoids (figure 23.13). One group of anthropoids included ancestors of the **New World monkeys,** and the other group included ancestors of **Old World monkeys** (figure 23.14), apes, and humans.

Dryopithecines

During the Miocene epoch, which began about 25 million years ago, apes became abundant and widely distributed in Africa, Europe, and Asia. Members of the ape genus *Dryopithecus* may well have been the ancestors of modern chimpanzees, gorillas, orangutans, and humans.

Figure 23.14 Monkeys. (*a*) The spider monkey, an example of a New World monkey. Note that it uses its prehensile (grasping) tail to aid its movements among the branches. Old World monkeys do not have prehensile tails. (*b*) A baboon, a ground-dwelling Old World monkey. (*c*) Macaques. Some macaques are very important subjects in medical research.

(a) (b) (c)

Table 23.4
Periods and Epochs of the Cenozoic Era

Era	Period	Epoch	Millions of Years before Present*
	Quaternary	Recent	0.01
		Pleistocene	2.5
Cenozoic	Tertiary	Pliocene	7
		Miocene	25
		Oligocene	38
		Eocene	54
		Paleocene	65
Mesozoic			230

*These are approximate dates of the beginnings of these intervals.

Figure 23.15 A chimpanzee "knuckle-walking."

Figure 23.16 An African savanna, part of the Serengeti plain in Tanzania. During the late Miocene epoch, the climate became cooler and drier, lush forests dwindled, and savannas spread.

Figure 23.17 Adaptations for erect posture. (*a*) Comparison of a human skeleton with the skeleton of a "knuckle-walking" ape (a gorilla). Note differences in proportions of hindlimbs and forelimbs, differences in shape of the pelvis (color), and differences in the angle and position of attachment of the vertebral column to the skull. (*b*) Comparison of an ape foot and a human foot. The opposable toe of the ape foot is better suited for tree branch gripping, but the human arrangement is better suited to walking on the ground. (*c*) Comparison of an ape's vertebral column and a human vertebral column. Note the greater curvature in the human vertebral column, which is better suited to erect posture.

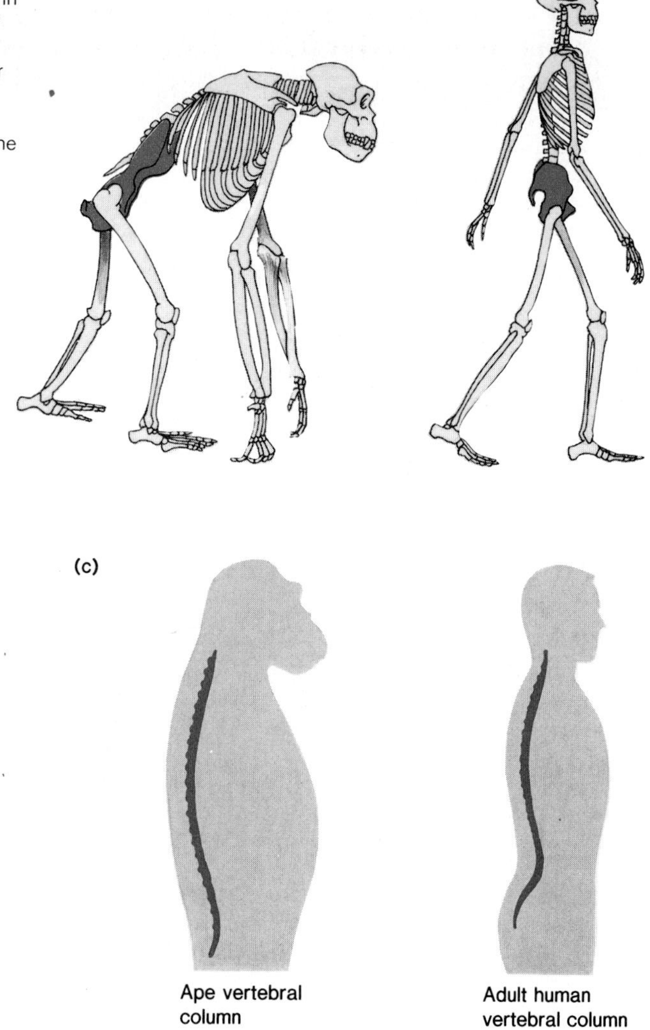

(b)

Ape foot **Human foot**

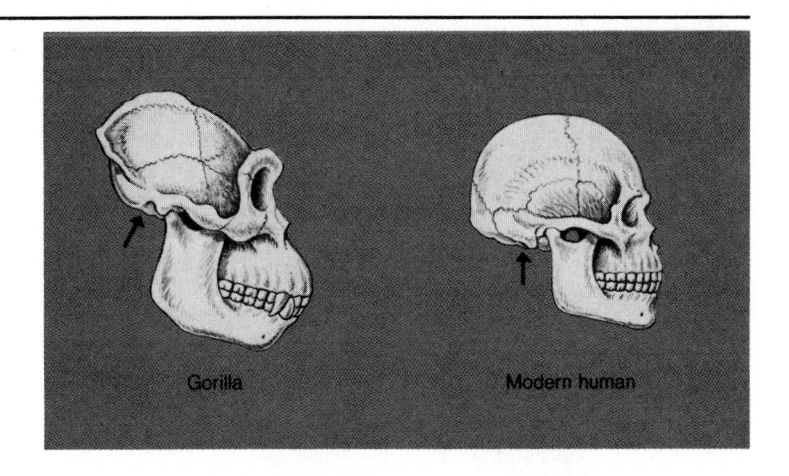

Figure 23.18 Comparison of an ape (gorilla) skull and a modern human skull. Ape skulls have prominent supraorbital ridges, which human skulls lack. The ape face projects forward, forming a muzzle, while the human face is flat. Apes have larger, heavier teeth and have especially pronounced canine teeth. The gorilla's brain capacity is only about 450 cc, while that of this modern human skull is about 1,500 cc. Note also the difference in the position of the foramen magnum (arrows).

Dryopithecine apes were forest dwellers who probably spent most of their time in the trees, but their foot skeletons indicate that they also may have spent some time on the ground, "knuckle-walking" as modern apes do (figure 23.15). The dryopithecine skull had a low, rounded cranium and **supraorbital ridges** (bony ridges that protrude above the eyes). The dryopithecine face and jaws projected forward.

Later in the Miocene epoch, the climate became progressively cooler and drier. The lush forests dwindled, and vast **savannas** (grasslands with occasional clumps of trees) spread (figure 23.16). Some late Miocene relatives of the dryopithecines, however, were not restricted to the shrinking forests. Among these late Miocene primates, which came out of the trees, stood partially upright, and moved onto the savannas, were the ancestors of humans.

The Origin of the Hominids

What evolutionary steps were involved in the transition of early hominids (members of the human family) from forest dwellers to grassland dwellers?

Erect posture and **bipedal locomotion** (walking on two legs) were two important ingredients for success on the savanna. Erect posture provided the ability to see predators in time to flee to safety, freed the hands from the necessity to support the body, thus permitting their use for carrying food or using tools, and enhanced ability to spot prey. Bipedal locomotion, an adjunct of erect posture, permitted efficient movement while both hands were free to perform other functions. The transition to erect posture required a number of individual adaptive changes in the skeleton (figure 23.17).

Was the transition to life on the savannas a gradual process that occurred as the savannas spread and the forests receded? Or were there primates that already possessed these characteristics, which would prove to be adaptive for life on the savannas, before the climatic changes set in? As with several other aspects of human evolution, there are as yet no clearcut answers to these questions. There is, however, a fossil representative of the organisms that lived on the edge of the savannas. But, as is so often the case, study of this fossil creature, *Ramapithecus,* raises almost as many questions as it answers.

Ramapithecus

Fossils of the genus *Ramapithecus* are found in the late Miocene strata (14 to 10 million years ago) in India and Africa. Only a few specimens of teeth, jaws, and parts of the face have been discovered. Even so, a number of researchers consider *Ramapithecus* either an immediate ancestor of the hominids, or possibly even a member of the hominid family.

Because there are borderline cases such as *Ramapithecus,* we need to know what characteristics are used to distinguish pongids (members of the ape family) from hominids (members of the human family) (figure 23.18).

Figure 23.19 Dental arcades of (*a*) a chimpanzee, (*b*) an australopithecine, and (*c*) a modern human. The two sides of the chimpanzee's jaw are roughly parallel, giving its dental arcade a rectangular shape. Its canine teeth are much larger than adjacent teeth, as they are in the jaws of all pongids (apes). The human and the australopithecine jaws both curve gently to give a parabolic dental arcade. Note, however, that the large grinding molars (three rear teeth on each side) of australopithecines are much sturdier and broader than human molars.

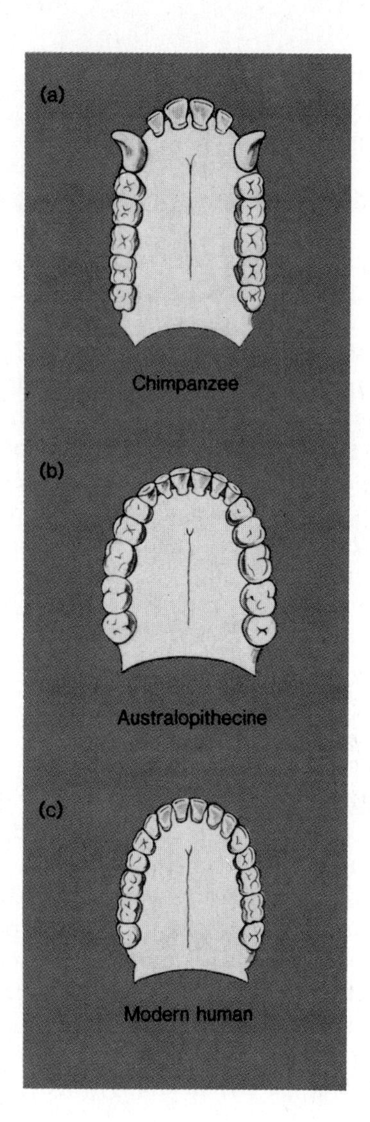

Table 23.5
A Classification of the Family Hominidae

Family Hominidae
Genus *Australopithecus*
A. afarensis
A. africanus
A. robustus
Genus *Homo*
H. habilis
H. erectus
H. sapiens neanderthalensis
H. sapiens sapiens

In apes, the **foramen magnum,** the opening through which the spinal cord attaches to the brain, is well to the rear of the skull. In humans, the foramen magnum is almost directly in the bottom center of the skull, an arrangement well suited for erect posture. Apes have pronounced supraorbital ridges. In apes, the plane of the face projects forward, forming a muzzle, while in humans the plane is flat. Apes' canine teeth are much larger than the adjacent teeth, while the canines of humans are approximately the same size as the adjacent teeth. The two sides of the ape jaw are roughly parallel, giving a rectangular shape to the **dental arcade.** In humans, the tooth pattern curves gently and continuously, giving a broad U shape to the arcade (figure 23.19).

A more recently discovered fossil form, *Sivepithecus,* may shed more light on the evolutionary significance of *Ramapithecus* and the early evolution of the hominids. But there is still a very intriguing gap in the fossil record of hominid evolution between *Ramapithecus* and the first fossils that are definitely hominid. These fossils come from strata that are nearly 10 million years more recent than the date of *Ramapithecus.*

The Australopithecines

A small skull discovered in 1924 in a limestone quarry in South Africa was sent to the anatomist Raymond Dart in Johannesburg for study. Dart named the skull *Australopithecus africanus* (southern ape of Africa) and proclaimed it to be a form intermediate between ape and human, an upright-walking member of the family Hominidae (table 23.5).

Many physical characteristics of *A. africanus* are transitional between apes and humans. The dental arcade was generally U-shaped, and the canines were reduced (figure 23.19). The face was flatter than that of the apes, and the cranium was higher and rounder. The leg and pelvis characteristics indicated that their locomotion was bipedal.

Figure 23.20 Skulls of *Australopithecus africanus* and *Australopithecus robustus*.

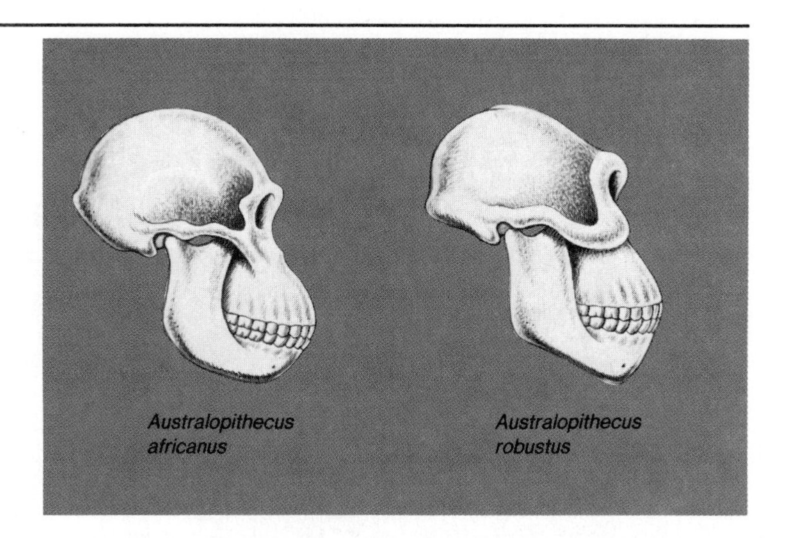

Another species of *Australopithecus, A. robustus,* has also been found in South Africa (figure 23.20). This hominid was larger than *A. africanus,* and its large jaw and strong teeth suggest that its diet was of rougher vegetation. Both species lived during the period from about 4 million years ago to about 1 million years ago, although *A. africanus* is a somewhat older species and may actually have given rise to *A. robustus.*

In 1974, near Hadar in Ethiopia (figure 23.21), Donald Johanson and his colleagues discovered a group of surprisingly complete specimens that they have named *Australopithecus afarensis.* Skeletal evidence indicates that this hominid walked fully erect some 3 million years ago. Some biologists have suggested that *A. afarensis* (if it is indeed a separate species) was ancestral to both *A. africanus* and *A. robustus* and that *A. africanus* gave rise, in turn, to *Homo,* the genus of modern humans. Johanson, however, thinks that *A. afarensis* may have given rise to two separate lines, one of which led to the other australopithecines and the other to the genus *Homo.*

Figure 23.21 Location of the Rift valley of Africa. The Rift valley, which varies in width, lies within the lighter area of this map. It is part of a long line of depressions running down from Turkey, through the Jordan River and the Dead Sea area of Israel and the Red Sea, and on through eastern Africa. The Rift apparently has been produced by movement of continental plates (see page 488). Several very important sites of hominid fossil discovery are located in or near the African Rift Valley.

Source: © M. E. Challinor/*Science 81*.

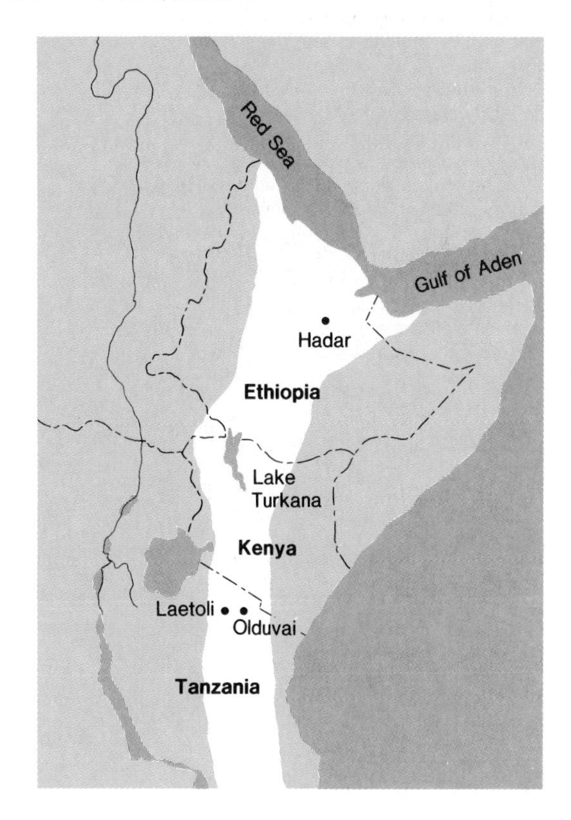

The Leakey Hypothesis

There are, however, other fossils and still other interpretations. One of them comes from the "first family" of hominid paleontology, the Leakeys (figure 23.22), who have hunted hominid fossils at Olduvai Gorge and other sites in the Rift Valley of Africa for many years.

In addition to their many other important discoveries, Louis and Mary Leakey found an especially interesting hominid skull in 1964. The Leakeys considered this 2-million-year-old skull to be much more advanced than any of its australopithecine contemporaries, and they named it *Homo habilis* ("handy man" or "man who is able to do or make").

Figure 23.22 The search for the early hominids. (*a*) Mary and Louis Leakey at work in Olduvai Gorge. Louis Leakey died in 1975, but Mary Leakey has continued the work on fossil hominids, as has their son Richard. (*b*) Richard Leakey and Donald Johanson examine fossils at the Kenya National Museum in Nairobi, Kenya.

(a)

(b)

Figure 23.23 Examples of "family trees" of the hominids. (*a*) A version of hominid relationships with *Australopithecus africanus*, or a similar form, giving rise to both the australopithecine and *Homo* lines. (*b*) A version that includes Johanson's views regarding *A. afarensis* as an ancestor of all of the hominids. There are several points of conflict. Many researchers, including the Leakeys, would include *Homo habilis* in the main line of evolution of modern humans. And the Leakey hypothesis (not pictured here), for which there is considerable evidence, proposes that the divergence between australopithecines and the genus *Homo* occurred much earlier and that the two groups have descended as separate evolutionary lines.

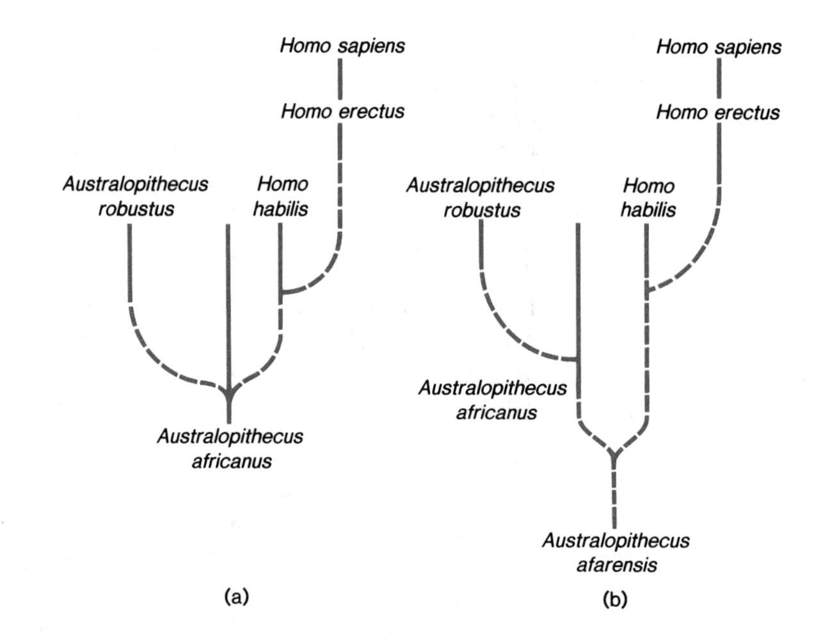

Figure 23.24 Examples of the earliest human tools. These simple tools probably were used as choppers in food preparation.

Figure 23.25 Skull of *Homo erectus. Homo erectus* was a widely distributed species, and many important human fossil finds (including "Java man" and "Peking man") are now assigned to this species.

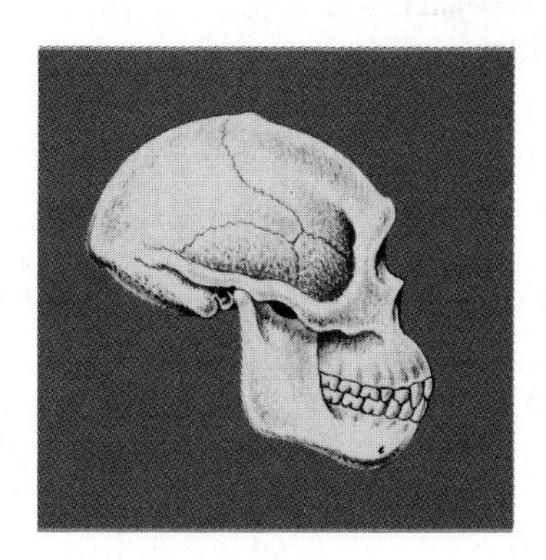

The Leakeys have maintained that the genus *Homo* is much older than previously thought, essentially as old as the australopithecines and, therefore, not descended from them. But there is much disagreement over this conclusion. Even though the skull of *Homo habilis* does have a comparatively large cranial capacity (table 23.6), some researchers prefer to classify it with the australopithecines.

Arguments about relationships of the hominid fossils are far from settled and there are a number of suggested "family trees" of the hominids (figure 23.23). Strong support for the Leakey hypothesis of the early origin of the genus *Homo* came when Richard Leakey (son of Mary and Louis Leakey) discovered an exceptionally interesting fragmented hominid skull in the Lake Turkana region of Kenya. This "skull 1470," as it is called, has been dated at 2.6 million years ago. Although there are some disagreements about the dating of the skull, its large cranial capacity and early date support the Leakey hypothesis of the early origin of the genus *Homo.*

Another very significant set of finds from eastern Africa, especially from Olduvai Gorge, has been a number of small stones dating to some 2.5 million years ago. These stones appear to have been intentionally chipped so that they could be used for cutting, pounding, and scraping; that is, they were made and used as tools (figure 23.24).

Who made these tools more than 2 million years ago? One answer is that the australopithecines might have made tools. But another answer is that while australopithecines may have found and used various objects as tools, the design and fabrication of tools is and has been a strictly human enterprise that was limited at that time to the work of *Homo habilis* or possibly other members of this genus.

Table 23.6
Cranial Capacity of Hominids

Name	Cranial Capacity (in cc)
Australopithecus afarensis	500
A. africanus	300–600
A. robustus	300–600
Homo habilis	650
H. erectus	750–1,300
H. sapiens	1,400–1,700

The Genus Homo

The earliest fossils that are accepted by all paleontologists as being members of the genus *Homo* are from a series of finds in Africa, Asia, and Europe. Collectively, these fossils are called *Homo erectus.*

Homo erectus

The oldest of the *Homo erectus* fossils dates to about 1.5 million years ago and the youngest to about 300,000 years ago. Members of this species were bipedal and fully erect in posture. Their cranial capacity was large, but *Homo erectus* differed from modern humans in that it retained some primitive skull features: the face was more projecting than that of modern humans, and the cranial bones were thicker. The teeth were large, and the lower jaw sloped back so that there was no distinct chin (figure 23.25).

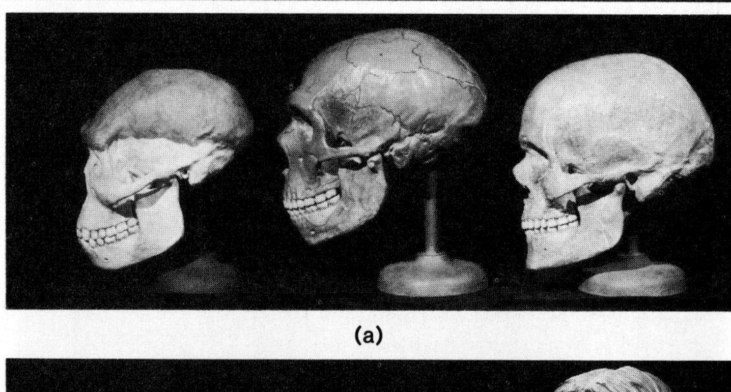

(a)

(b)

Figure 23.26 Early humans. (*a*) Restored skulls of *Homo erectus*, a Neanderthal man, and a Cro-Magnon man. There are significant differences in shape and size of craniums, supraorbital ridges, and sizes and shapes of chins and jaws. (*b*) Artistic reconstruction of possible facial features of the three types.

The oldest fossil remains of *H. erectus* have been found in Africa and younger ones in other parts of the world, a pattern that suggests that the species originated in Africa and migrated outwards to other suitable habitats. Tools of *Homo erectus* are much more elaborate than the older, more primitive tools found at Olduvai. These tools included heavy, wedge-shaped choppers, small hand axes, and various small tools.

Homo sapiens

Modern humans, *Homo sapiens,* display a number of advances in skull structure over *H. erectus*. These changes include an increase in cranial capacity, a decrease in the size of the teeth and jaws, flattening of the plane of the face, and a rounding of the cranium. *Homo sapiens* probably emerged between 250,000 and 200,000 years ago.

About 100,000 years ago a distinctive group of humans called **Neanderthals** emerged (figure 23.26). The name Neanderthal comes from the Neander Valley in Germany, where the first fossil of this type was unearthed. Actually, the Neanderthals were widespread geographically, with fossils being found in much of Europe and parts of Asia and Africa.

Fossils of fully modern humans appear in the fossil record near the end of the Neanderthals' time. These earliest modern humans, the **Cro-Magnon** people, are classified as *Homo sapiens sapiens* (as are present-day humans) to distinguish them from the Neanderthals *(Homo sapiens neanderthalensis)*. After a brief period of coexistence with the Cro-Magnon people, the Neanderthals disappeared, and from about 40,000 years ago onward, no more Neanderthal fossils occur in the fossil record.

The rapid (in terms of geological time) disappearance of the Neanderthals is somewhat of a mystery. Were they eliminated as a result of general competition with the Cro-Magnons? Or, did the two groups interbreed to such an extent that distinctive Neanderthal characteristics were no longer recognizable, and the Neanderthals were simply incorporated into the general *Homo sapiens* population?

Cultural Evolution

Also about 40,000 years ago, a new tool industry spread through the human population. The major characteristic of this industry was the production of the blade, a tool with roughly parallel sides (figure 23.27). Increasingly complex tools are physical evidences of the growth and spread of human culture. People taught other people how to use and make tools. This and other information was transmitted by

Figure 23.27 Blades produced by Cro-Magnon people. Some Cro-Magnon blades were as much as 30 cm long and only ½ cm thick. The blades represent the finest of stone tool manufacture. (Compare them with the crude choppers in figure 23.24.) The next major step in human toolmaking occurred with the advent of metal working.

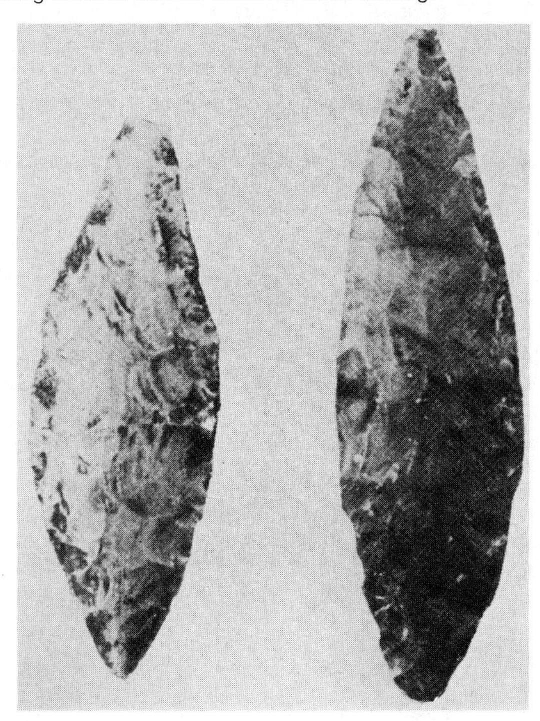

personal communication. There was now a human **culture,** a body of information transmitted from generation to generation by means that did not depend on genetic mechanisms. Human evolution had entered a new and important phase, a phase involving cultural evolution, that has continued at an accelerating pace to the present time.

Culture is passed from one generation to the next through teaching and example, and it is dynamic since it is continually modified as it is transmitted. Cultures do not evolve in the biological sense, but they do display an overall progressive change with occasional large-scale advances that take place in relatively short periods of time.

Humans were basically hunters and gatherers who sought their food in nature until about 12,000 to 15,000 years ago when agriculture became widespread. There was relatively little change in the agriculture-centered culture until about 200 years ago when the Industrial Revolution set in motion sweeping changes in human life that continue even now.

Present human culture, however, has the power for disruption and destruction that far exceeds that of any other stage of human development. The burgeoning world population is placing an ever greater strain on natural resources, and the waste products of industry are poisoning our environment. Industrial societies are producing weapons of destruction that threaten our species and others with annihilation. The very future survival of human beings, and the many other organisms that human activities affect, may depend on entry into a new cultural stage, one in which the emphasis is placed on the ability of humans to live in harmony with their environment and with each other. It seems that if this does not occur, the quality of human life must inevitably decline in the future, and cultural evolution, which represents the pinnacle of our humanness, could lead to our destruction.

Summary

Modern biologists explain the process of organic evolution in terms of changes in gene frequencies occurring in gene pools. Gene frequency changes may be caused by genetic drift, migration, and mutation, but the major force is natural selection.

Each organism possesses a set of adaptations to its environment. Some adaptations are particularly striking because of their involvement in interactions between organisms of different species. In coevolution there is an intense mutual evolutionary interaction between species in which each exerts a strong selective influence on the other.

A species is a population of organisms that is reproductively isolated from members of other populations. Geographic isolation is a major factor in species formation.

A current debate concerns the tempo of species formation. Does speciation occur as a result of long-term, gradual, steady accumulation of adaptive traits? Or are there relatively short periods of rapid formation of new species separated by longer periods of comparative equilibrium?

Humans are primates. Primates evolved as arboreal animals, and primate characteristics generally reflect adaptations for arboreal life. The oldest primates were the prosimians. Modern prosimians include lemurs and tarsiers. Monkeys, apes, and humans belong to the suborder Anthropoidea. Monkeys have diverged into New World and Old World monkeys.

The dryopithecines were ancestors of apes and humans. Climatic changes that led to forest shrinkage and produced vast savannas provided habitats for ground-dwelling hominid ancestors such as *Ramapithecus*.

The australopithecines, which lived from about 4 million years ago to about 1 million years ago, were hominids, and some of them were on, or near to, the line of human ancestry. *Homo habilis* was a contemporary of the late australopithecines.

Homo erectus emerged about 1.5 million years ago and spread rapidly over large areas of Africa, Europe, and Asia.

Modern humans, *Homo sapiens,* emerged between 250,000 and 200,000 years ago. From 100,000 years ago until about 40,000 years ago, Neanderthal humans lived in much of Europe and Asia.

Recent human evolution has been marked by rapid cultural evolution.

Questions for Review

1. What is a gene pool?
2. What factors can cause changes in gene frequencies within a gene pool?
3. Define the founder effect.
4. What is the difference between stabilizing and directional selection?
5. Define the term coevolution.
6. Explain the importance of the concept of reproductive isolation in the definition of a species.
7. List some skeletal adaptations associated with erect posture.
8. What characteristics would you expect a gorilla jaw to have?
9. Explain the fundamental disagreement between the Leakey hypothesis and the views of some other paleontologists regarding the early history of the genus *Homo.*
10. What are some differences between the skulls of *Australopithecus* and modern humans? between *H. erectus* and modern humans?

Questions for Analysis and Discussion

1. How do proponents of the punctuated equilibrium hypothesis explain the scarcity of transitional forms in the fossil record? How does their explanation differ from that offered by proponents of phyletic gradualism?
2. Why is it useful for development of an understanding of human evolution to study chimpanzees, gorillas, and baboons in the wild?
3. Explain why accurate dating is essential to understanding of evolutionary relationships of early hominids using the example of "skull 1470."

Suggested Readings

Books

Dobzhansky, R.; Ayala, F. J.; Stebbins, G. L.; and Valentine, J. W. 1977. *Evolution.* San Francisco: W. H. Freeman.

Johanson, D., and Edey, M. 1981. *Lucy: The beginnings of humankind.* New York: Warner Books.

Leakey, R. E. 1981. *The making of mankind.* New York: E. P. Dutton.

Simpson, G. G. 1967. *The meaning of evolution,* rev. ed. New Haven, Conn.: Yale University Press.

Articles

Ayala, F. J. 1983. Genetic variation and evolution. *Carolina Biology Readers* no. 126. Burlington, N.C.: Carolina Biological Supply Co.

Day, M. H. 1984. The fossil history of man. *Carolina Biology Readers* no. 32. Burlington, N.C.: Carolina Biological Supply Co.

Hammond, A. L. November 1983. Tales of an elusive ancestor. *Science 83.*

Lewontin, R. C. September 1978. Adaptation. *Scientific American* (offprint 1408).

Rensberger, B. April 1982. Evolution since Darwin. *Science 82.*

Behavior

24

Chapter Outline

Chapter Concepts

1. Behavior allows animals to respond rapidly to changes in their environment.
2. Organisms are genetically programmed through natural selection to behave in an adaptive way.
3. In some species, there are rigid, relatively unvarying responses to a given situation, while in others behavior is highly modifiable through experience (learning).
4. Hormones control behavior by altering the internal motivational states of organisms.
5. Sociobiology has provided a new way of looking at the genetics and evolution of social behavior, but application of sociobiological principles to human behavior has caused some controversy.

Walking through a gull colony can be an exciting if somewhat baffling experience. Gulls fly overhead in all directions, calling, swooping down, sometimes buffeting your head with their feet. Gulls and their young are in constant danger from predators. Thus, it is not surprising that gulls attack humans just as they do any other potential marauders that wander near their nests.

But if you hide in a blind, the gulls soon calm down and return to their normal activities. Some individuals stand alert by their nests or raise their heads in the raucous long-call so characteristic of their species. At some nests, pairs are courting, jerking their heads up and down in a chokinglike motion, while at other nests the birds are fighting with their neighbors. Other gulls are settled on their nests, incubating their eggs, while their mates stand vigilantly nearby. Like houses in a suburban development, there is a regular spacing between neighboring nests. Each gull staunchly defends the area (**territory**) around its nest, allowing only its mate into the territory. The female lays several mottled, greenish-brown eggs that are well camouflaged against the nest background. The two gulls take turns incubating the eggs for three weeks, and then they collaborate in the demanding task of feeding the young.

All of these actions are part of the gulls' behavior. **Behavior** is what an organism does (usually by movement) in response to various situations.

Analysis of Behavior

Shortly after the time of hatching, parent black-headed gulls (figure 24.1) systematically pick up empty eggshells from around the nest and remove them. If one of these discarded shells falls by the nest of a neighbor, the neighbor, in turn, also removes it.

This apparently trivial piece of behavior intrigued Nobel prize-winning ethologist Niko Tinbergen. Ethologists analyze behavior patterns in nature and are particularly interested in the adaptive significance of behavior. Why should the black-headed gull and other ground-nesting gulls take such pains to remove the shells?

There seemed to be several possibilities. For one, the sharp edges of the broken shell might be harmful to the chick or uncomfortable for the brooding parent, yet many other kinds of birds, including cliff-nesting gulls called kittiwakes (figure 24.2), do not systematically remove eggshells from their nests. Another possibility was that the behavior might reduce the risk of predators attacking the chicks. Only the outside of the gull's eggshell is camouflaged; the inside is white. The empty shell with its jagged edge and white interior might flash like a beacon to any passing predators, especially to crows flying overhead.

Figure 24.1 A black-headed gull.

Figure 24.2 Kittiwakes are gulls that nest on steep cliffs. Thus, they suffer less predation than gulls that nest on flat ground. Kittiwakes do not systematically remove eggshells from their nests.

Tinbergen and his colleagues did a series of experiments to test these ideas about predation and eggshells. They put dummy plaster eggshells in assorted colors, sizes, and shapes into the nests of wild gulls. The scientists then hid in a blind nearby and observed gulls' responses to the shells. When a gull returned and settled down on its nest, it pecked at the dummy eggshell for a few moments and then picked it up and discarded it from the nest. The more the dummy looked like the conspicuous inside of the gull's egg, the more likely the gull was to pick it up and carry it from the nest.

But how conspicuous are shells to real would-be predators? Tinbergen and his colleagues put broken eggshells at varying distances from mock nests with gull's eggs in them. Exactly as their hypothesis had predicted, the nearer the eggshells were to the nest, the more likely it was that the eggs in the nest would be eaten (figure 24.3). It seems, then, that a parent gull saves its chicks from being eaten by removing the empty eggshells from around the nest.

But one puzzling feature of the behavior remained. Despite the great risk from predation, the parent gulls did not remove the eggshells immediately after their chicks hatched. If the nests were so predictably destroyed when eggshells were present, why did the gulls wait several hours after the chick had hatched before removing the broken shell? Tinbergen made a fascinating, if somewhat gruesome, observation. When parent gulls flew up and left a chick that had not yet dried off after hatching, the chick was very likely to be gobbled up by a neighboring gull. But after a chick had time to dry and gain a little strength, it could run and hide in the grass if the parents flew up from the nest, and thus could escape cannibalistic neighbors. Tinbergen reasoned that there were really two factors affecting eggshell removal. The parents need to remove the eggshells quickly, but as long as the chicks are still wet, it is better that the parents not leave the nest, even for the few minutes required to carry off a broken eggshell.

Figure 24.3 In an experiment with mock black-headed gull nests, predation was more likely if eggshell fragments were located close to the nest.

Eggs in nest ← Distance → Broken eggshell	Percentage of eggs destroyed by predators
← 15 cm →	42
← 100 cm →	32
← 200 cm →	21

While it appears that parent gulls are thoughtfully considering the welfare of their offspring, the parent gulls are simply responding mechanically and automatically to a **stimulus** in the environment. It is as if gulls are programmed by their genes to behave in this manner when they encounter eggshells or other conspicuous objects in or around their nests. In evolutionary terms, the complex combination of genes that causes gulls to remove objects from their nests is more likely to survive (and thus be passed on to future generations) than other genetic combinations that produce other behavior patterns. If a gull did not remove eggshells, chances are that all of its eggs would be eaten, and its genes would not be passed on to another generation of gulls. There is **natural selection** for eggshell removal behavior, and **selection pressure** resulting from predation maintains the genes for the behavior in the population.

Simultaneously, however, there is also selection against the gulls leaving wet chicks unattended, even for a few moments. This is an example of the delicate balance between complex factors resulting in the evolution of the finest details in the timing and form of behavioral responses.

In discussing how a trait such as eggshell removal behavior might have evolved, we must assume that the trait was **inherited;** that is, the offspring of parents with the behavioral trait would be more likely to show the behavior than offspring of parents without the trait. Somehow, the behavior must be programmed by the genes, and those genes must be passed in an essentially unaltered form to the next generation. Genes and the behavior they program are preserved because the individuals in which they occur are more likely to survive and pass their genes on to the next generation.

Genes and Behavior

Behavior patterns that clearly are inherited have been analyzed in a number of organisms, ranging from unicellular organisms such as the protozoan *Paramecium* to complex animals.

Swimming in *Paramecium*

The single-celled *Paramecium* moves through the water in a spiral fashion, propelled by its beating cilia. When it hits something, it backs up and moves away. We now understand much of the detailed mechanism of this **avoidance response.** Running into an object causes a depolarization (a change in electrical charge) of the plasma membrane at the point of

Figure 24.4 The normal avoidance response of *Paramecium.* When the organism swims into an object (1), it backs up by momentarily reversing the direction of its beating cilia (2). After the *Paramecium* is positioned at a different angle (3–5), the cilia resume their normal movement and the organism once again moves forward (6).

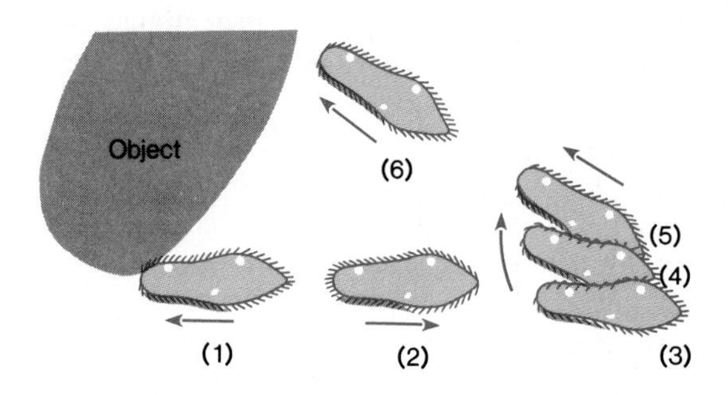

contact, which spreads quickly over the rest of the cell's plasma membrane. This process is very similar to the way that impulses pass along the nerve cells of animals (see page 259). The depolarization of the *Paramecium's* plasma membrane causes a reversal in the direction of movement of the cilia, and the *Paramecium* backs up (figure 24.4). Within a few seconds, the membrane becomes repolarized (returns to its normal electrical condition), and the *Paramecium* resumes its forward path.

A number of intriguing *Paramecium* mutants that differ in their swimming behavior have been found in the laboratory. One of these is a mutant named "paranoic." When a "paranoic" *Paramecium* runs into an object, it backs up as usual but continues swimming backwards much longer than a normal *Paramecium* would. Because of a problem in the cell's plasma membrane, the normal process of repolarization fails. This "mistake" in membrane structure is caused by a change in a single gene, which, in turn, has a direct effect on the behavior of the organism.

Learned Components of Behavior

Behavior is not limited to predictable automatic responses that are genetically programmed. Behavior can be altered by **learning,** which is the modification of behavioral responses by experience. When biologists analyze behavior, it is often difficult to distinguish clearly between innate (predictable, automatic, genetically determined) behavior and learned behavior. But some behavioral patterns do have recognizable learned components. Let us look at two examples.

Gull Chick Pecking

Parent gulls have the arduous task of trying to keep up with the feeding demands of their chicks. The parent carries the food back to the nest in its crop (a sac in the first part of the bird's digestive tract). When an adult lands at the nest, the chick pecks at the parent's bill (figure 24.5). This behavior causes the parent to regurgitate the food it is carrying. All chicks, even newly hatched ones, peck at the parent's bill. This is an example of an innate, or **instinctive behavioral pattern.** The young gulls respond automatically to a **visual communication signal.**

Outside of the nest newly hatched gull chicks will peck at a wooden model of a gull's head (figure 24.6*a*). Different rates of pecking at gull heads of different shapes and colors are reliable measures of how like the parent the chick finds various models.

The development and improvement in pecking behavior has been thoroughly studied in the American laughing gull. Newly hatched chicks are not very accurate in their pecking, but their aim improves markedly after only a couple of days in the nest. They also develop the ability to identify details of their parents' bills specifically. Eventually they will peck only at heads that look like their own parents' heads (figure 24.6*b*).

Gull chick pecking behavior was once regarded as a classic example of an innate and **fixed action pattern.** Young chicks need no prior experience to show the behavior, clumsy though they may be. They respond automatically by pecking at a specific stimulus. However, the realization that chicks got better at pecking as they grew older cast doubt on whether this behavior could be regarded as strictly innate. Not only did chicks become more coordinated, but even the form of the behavior and especially the stimulus that elicited the pecking changed as the chicks grew older.

Improvement in pecking accuracy apparently results from the fact that the chicks become more mature (for example, they become more steady on their feet), and some improvement comes from the experience of pecking. By a kind of learning known as **trial and error,** the chicks apparently *learn* how far to stand from the parent's bill in order neither to fall short nor to overshoot.

It is very difficult to make a clear distinction between the innate (instinctive or genetically determined) and learned components of behavior. All behavior at each stage of development is the result of an interaction between the effects of the animal's genes and the effects of its environment.

Figure 24.5 A gull chick pecks at the parent's bill for food.

Figure 24.6 Pecking behavior in gull chicks. (*a*) A gull chick pecking at a model of an adult gull's head and bill. (*b*) Responses of **newly hatched** (white bars) and **older** (black bars) laughing gull chicks to various models. Note that newly hatched chicks do not discriminate among the very different models (white bars of similar length), while older chicks prefer models that resemble the parent birds (black bars). The model that most closely resembles the parent laughing gull is second from the top.

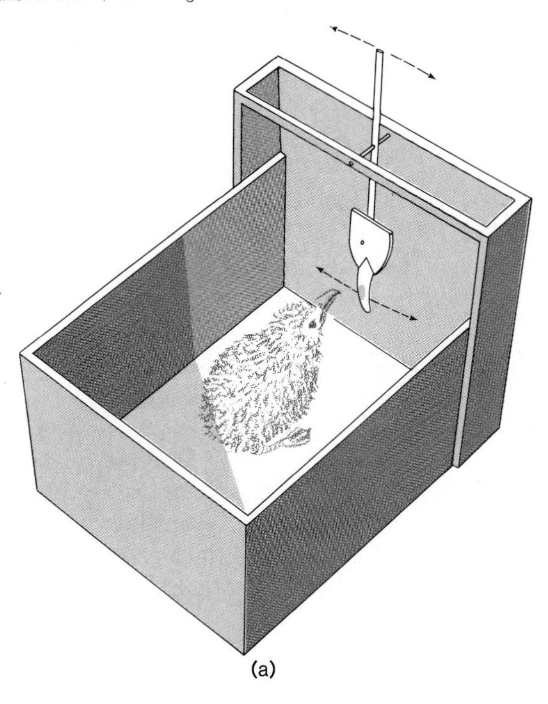

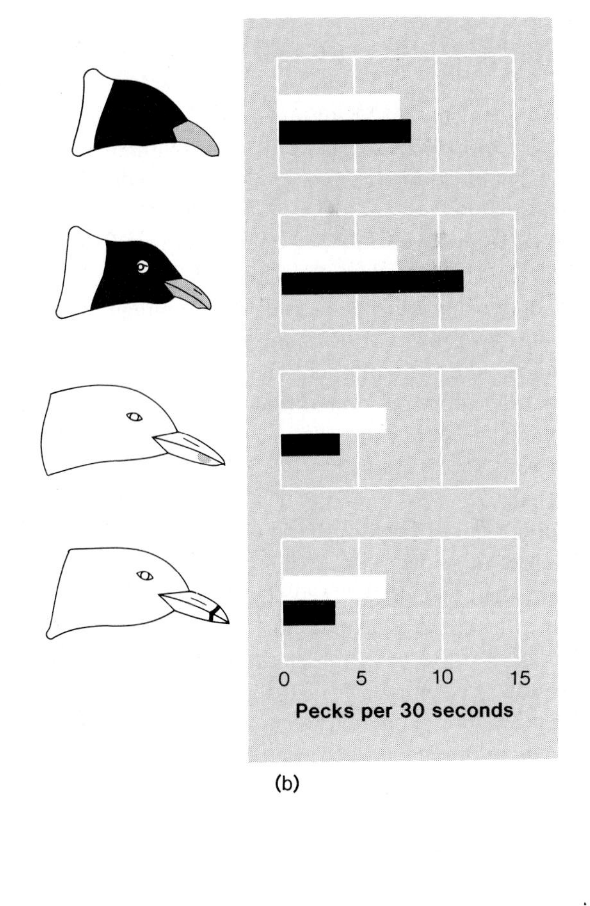

Figure 24.7 Imprinted geese following Konrad Lorenz.

Learning to Follow

You might expect that an animal's ability to recognize members of its own species and to distinguish them from other species should be innate, that is, purely instinctive. But is it, in fact, innate?

Once young birds such as ducklings have dried off and rested from the arduous task of hatching, they begin to take notice of moving objects in their environment and follow whatever moves away from them. It helps if the moving object quacks like their mother. The famous Austrian ethologist Konrad Lorenz discovered that the process of recognizing and following members of their own species, or **imprinting,** occurs very rapidly during a definite **critical period** after hatching.

Lorenz raised ducklings and goslings and allowed them to follow him as he walked along quacking (figure 24.7). When he then introduced the young birds to adult members of their own species, the ducklings and goslings ran from the adults and hurried back to Lorenz. Later in life, these imprinted ducks often paid little attention to members of their own species, preferring the companionship of humans. As adults, they sometimes even tried to mate with humans!

Many birds and mammals do not innately "know" how to recognize a member of their own species. Instead, many animals *learn* early in life to make this distinction. Birds, therefore, define a member of their own species as that which looks like what they learned to follow just after hatching. Normally, birds encounter their mother during the critical period. However, the imprinting mechanism works just as effectively in experiments where the moving object is not the mother bird, but a biologist interested in animal behavior.

Hormones and Behavior

Most temperate zone birds sing only in the spring. The expression of singing behavior, like the expression of many other behaviors, depends on **internal motivational state,** the set of conditions that exist in an animal's body at any given time. In birds, motivational states are closely related to changes that occur at different times in the quantities of circulating hormones in the birds' bodies. During active reproductive periods when sex hormone levels are high, behavioral responses are much different than at other times of the year.

Behavioral responses can be altered by permanent changes in motivational state, as by castration (testis removal) in male animals. Permanent lowering of the circulating male sex hormone level produces significant behavioral changes. Castrated male dogs, for example, do not show normally high levels of aggressive behavior toward other male dogs. And they do not respond frantically to female dogs that are sexually active (in heat). Many people consider these to be desirable behavioral changes in pet dogs because castrated male dogs remain vigilant as watchdogs, yet make more even-tempered family pets.

The Adaptive Significance of Behavior

Behavior has adaptive significance; that is, a behavior evolves through time because it confers some benefit on the organism exhibiting it. In this section, we will examine the adaptive significance of several behavior patterns.

Communication in Bees

Honeybees communicate information about new food sources by completing a movement known as the waggle dance. In the waggle dance a worker who has located a new food source moves around in a figure-eight pattern. When she is on the straight run between the loops of the eight, she moves her abdomen from side to side, hence the name waggle dance.

The direction of the straight run in the worker's dance tells other workers the exact direction of the food source, while the tempo of the dance communicates information about the food's distance away from the hive.

The significance of the waggle dance was first discovered by the Austrian ethologist and Nobel prize winner Karl von Frisch. He observed that when a worker returns to the hive, she often does the waggle dance, sometimes on a horizontal surface outside the hive or sometimes on the comb inside the hive. Fellow workers follow the dancer's movements for a few minutes and then fly out to the same feeding area that she has been using. In this way, foraging bees can quickly recruit other bees to harvest particularly rich sources of food.

If the dance is completed outside the hive, the direction of the straight run in the worker's dance tells other workers the exact direction of the food source. The other workers need only fly in the same direction as the straight run, using the sun as a compass (figure 24.8*a*).

This pattern of bee flight, which is specifically oriented with reference to a light source (in this case, the sun), is an example of a **phototaxis.** A **taxis** is a body movement in which an animal assumes a specific orientation to a stimulus source.

If the dance is done inside the hive, the workers cannot see the sun and so cannot use it directly as a compass. In a clever series of experiments, von Frisch demonstrated that foraging bees use a symbolic representation for the direction of the sun. If the forager dances the straight run part of the dance directly up the vertical comb surface, the food source is on a straight line from the hive toward the sun (figure 24.8*b*). If the dance is straight down the comb, then workers know they can find the food by flying away from the sun. If the forager dances 20° to the left of straight up, then the workers know to fly 20° to the left of the direction of the sun, and so forth. The direction to the food source is symbolized by the dancer's movement with respect to gravity (called a **geotaxis**).

Even though the sun moves across the sky during the day, the bees can compensate for this in the darkened interior of the hive because they possess an accurate **biological clock** (see box 24.1). A dancing bee inside a dark hive gradually changes the direction of its dance with the passage of time. In fact, the straight run of the dance rotates on the surface of the comb like the hands of a clock (only counterclockwise).

This complex system of communication is adaptive because it permits more workers to reach food sources more quickly, thus enhancing the food-gathering efficiency of the hive.

Figure 24.8 The waggle dance of the honeybee. When the bee moves through the straight run of the dance, she vibrates her abdomen back and forth (waggles). At the conclusion of the straight run, she circles back to the same starting position. Other workers gather information about the location of a food source from the straight run. If the bee performs the dance outside the hive (*a*), the straight run points directly toward the food source. If she performs the dance inside the hive (*b*), she orients herself using gravity as a reference, and the point directly overhead represents the direction of the sun. For example, in (*b*), because the food source is X degrees to the right of the direction of the sun, the straight run is X degrees to the right of straight up.

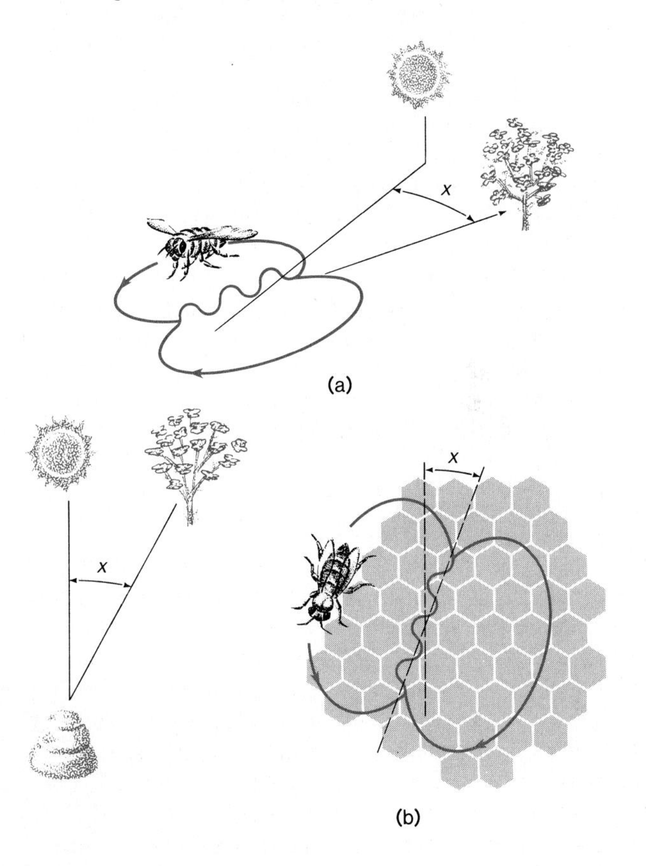

Box 24.1
Solar-Day and Lunar-Day Rhythms

Modification of the bee dance through the course of the day is only one of many behavioral patterns that change rhythmically with time. In fact, the total activity level of most animals is rhythmic; that is, animals are active at certain times of the day and inactive at others. For example, bats, rats, and moths are **nocturnal** (active at night), while dogs, butterflies, frogs, and lizards tend to be active in the daytime.

Solar-day (twenty-four-hour) rhythms are very widespread among living things (see chapter 14). Many such solar-day rhythms continue even when organisms are placed under constant conditions in the laboratory, where they are deprived of obvious information about the time of day.

However, solar-day rhythms are not the only rhythms expressed in animal behavior. Many animals that live in the intertidal zones along ocean shores display rhythmic changes in activity that correlate with the ebb and flow of tides.

Fiddler crabs of the genus *Uca,* for example, become active and feed during low tide periods (box figure 24.1), but plug their burrows from the inside and remain hidden during high tide periods. Since tides occur mainly in response to variations in the moon's gravitational pull on the earth during the 24.8-hour lunar day, tidal ebbs and flows occur about fifty minutes later each day than they did on the preceding day. Because fiddler crabs emerge from their burrows and forage on exposed mud flats during each low tide, their activity periods must begin fifty minutes later each day.

Frank A. Brown, Jr., and his colleagues discovered that these tide-related activity periods also persist under constant conditions in the laboratory. Fiddler crabs maintained in the laboratory are deprived of direct information about the ebb and flow of tides. But they are inactive during the times of high tide of their home beach, and they become active and move around when the tide is low on their home beach. Such tidal rhythms of alternating activity and inactivity also are called **lunar-day** (moon-day) **rhythms** because of the lunar periodism of the tides. Some tidal rhythms persist for many days in the laboratory, just as many solar-day rhythms do. In fact, in the absence of a tide table for fiddler crabs' home beach, it is possible to determine accurately the times of the tides on the beach (which may be quite distant from the laboratory) simply by observing the crabs' activity cycles!

Daily rhythms, and the underlying cellular clocks that time the rhythms, are important adaptations because rhythmic changes in physiology and behavior help to ensure continuing successful interactions with a fundamentally rhythmic environment.

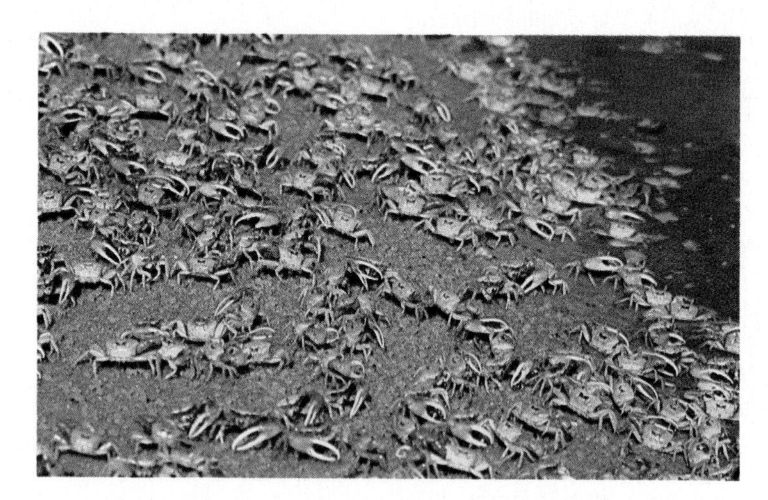

Box figure 24.1 Fiddler crabs (genus *Uca*) show a definite tidal rhythm. Here fiddler crabs are moving over a muddy beach at low tide. During high-tide periods, fiddler crabs remain inactive in their burrows.

Box 24.2

Animal Compasses

One of the most remarkable aspects of animal behavior is that animals have what might be called a sense of place. Foraging animals move out, sometimes over great distances, from their homes and are able to find their way back. Some of these animals might be using landmarks to find their way along familiar paths. But many animals demonstrate much more impressive abilities to find their way from place to place, often traversing hundreds or even thousands of kilometers in the process of migration.

For example, young green sea turtles are found on the east coast of South America, where they feed and grow. Every two or three years adult females set off swimming eastward across the Atlantic against the prevailing current. Their sea journey covers more than 2,200 km and ends on the beaches of tiny Ascension Island where they lay their eggs. They leave the eggs buried in the sand and head out to sea on their return voyage. When hatching turtles dig their way out of the sand, they scramble across the beach to the water. They apparently are carried westward by the current to the coast of South America, where the females live until they set out across the Atlantic on their own journey to Ascension Island.

Green sea turtles, however, do not hold the record for long-distance migration. Many birds migrate across large parts of a continent or even from continent to continent. Probably the champion long-distance migrant is the arctic tern. In the autumn, the arctic terns that nest in the North American arctic regions migrate across the Atlantic and travel southward along the west coasts of Europe and Africa until they reach the tip of South Africa. Then they cross the South Atlantic to Antarctica, where they live along the shore during the Antarctic summer. In this journey they cover a distance of about 18,000 km.

Some species of birds migrate in flocks in which young birds and mature adults travel together, and thus young birds travel with individuals who have made the trip before. In other species, however, older birds migrate first, leaving young birds to make their first migratory journey without the company of experienced adults.

How do birds find their way on these long journeys? The use of familiar landmarks might be a factor near the beginning and end of a journey, but long-distance navigational capabilities must exist as well. Some birds use the sun and/or the stars for orientation. Such celestial navigation requires an accurate time sense because animals using the sun as a compass, for example, must continually change their orientation relative to the sun to stay on course in a single direction. Experiments have shown that birds do indeed have and use accurate time information in their celestial navigation.

However, the migratory flights of many birds do not stop on overcast days, when celestial navigation is not possible. Birds continue their journeys, sometimes actually flying through clouds as they go. Early in the study of migration, some biologists suggested that migrating animals might use a magnetic sense for navigation, but the idea was dismissed because it was assumed that it was not possible for animals to sense the low energy levels of the earth's magnetic field.

Finally, during the 1970s, the idea that animals can sense the earth's magnetic field and can use it in orientation and navigation became widely accepted. Some of the most convincing evidence came from studies of homing pigeons. Homing pigeons were carried some distance from their home loft and released to find their way home. Observers then recorded their vanishing bearing—the direction in which they disappeared as they flew away from the release point. The birds tended to orient themselves so that they disappeared in the direction of their home loft whether the day was clear or cloudy.

W. T. Keeton, Charles Walcott, and others have demonstrated, however, that a pigeon's homing ability on a cloudy day depends on its being able to sense the earth's magnetic field. A pigeon's orientation is confused on cloudy days if it has a small bar magnet attached to its back or if it carries a small battery-powered coil that induces a uniform magnetic field in its head. It appears that use of a magnetic compass for orientation functions as a backup to the sun compass, which cannot be used on cloudy days. Magnets attached to the pigeons do not disturb the pigeons' orientation on clear days.

How do organisms actually perceive magnetic fields? This question has been only partially answered. Some organisms possess grains of an iron compound called magnetite (Fe_3O_4). For example, bacteria that show specific orientation in magnetic fields contain chains of magnetite particles. Magnetite particles ("magnetosomes") also have been reported in a number of animals, including tuna, dolphins, whales, pigeons, and green turtles. In animals, movement of magnetite particles in response to magnetic fields might be detected by adjacent sensory receptors.

We now know that organisms can respond to relatively weak magnetic fields. Our next task is to determine how organisms respond and to investigate the adaptive significance of their responses.

Figure 24.9 A swarm of honeybees on a fire hydrant in San Francisco. Swarming bees assemble in masses (usually on tree branches), while individual workers scout for a suitable new hive site.

This efficient communication system is also used when bees **swarm.** Swarming is a process by which a queen and a large number of worker bees leave an established hive and start a new hive. The bees fly out and settle on a tree branch or another surface in a mass. Then individual workers fly out and scout the area for appropriate new hive sites (figure 24.9). When a scout bee locates a potential hive site, it returns to the swarm and does a waggle dance to communicate the distance and direction to the site.

The waggle dance is a form of symbolic communication because each dance is really a miniaturized, symbolic version of the journey to a specific point. It is difficult to reconstruct the evolution of a complex behavior such as the waggle dance, but it is easy to understand that natural selection has favored the evolution of behavior that increases the efficiency of such vital processes as food gathering and location of appropriate living sites.

The Seeing Ear

How do animals that are active in almost total darkness find their way? If they are not only active, but also are trying to catch elusive and fast-moving prey, they require a very accurate system for finding their way around. One solution has been the evolution of **sonar systems,** where animals make loud noises and listen to the returning echoes. Using such a system requires that an animal be able to overcome many technical problems, but this is exactly what bats, dolphins, and a few other organisms have done in a variety of intriguing ways.

Donald Griffin clearly demonstrated that bats **echolocate.** He allowed them to fly in a room that was specially equipped with fine wires running from the ceiling to the floor. He soon learned that even blindfolded bats could find their way around these wires in total darkness.

However, plugging a bat's ears or covering its mouth seriously affected its performance. Such a bat was very reluctant to fly at all. It would hang on its perch and groom its ears and mouth vigorously, trying to remove the plugs. If the bat was forced to fly, it flew "blindly," running into wires and crashing into walls.

Although these facts strongly suggested that the animal was using sound to find its way, researchers were still puzzled because bats rarely make audible sounds. However, improved technology and the invention of what are now called "bat detectors" solved the mystery. Bats produce a more or less constant stream of clicks, but these clicks are all outside the range of human hearing; that is, they are in the **ultrasound** range. Humans can hear sounds to a maximum **frequency** (**pitch**) of about 20 KHz (20,000 cycles per second). This means that humans can hear only the lowest sounds that bats occasionally make because bats' cries usually fall between 30 and 50 KHz. Bats locate objects and capture moving prey with the echoes of these ultrasonic cries (figure 24.10).

Some bats cry through their mouths, but others emit their cries through their noses. Such bats have bizarre leaflike structures around their noses that focus or beam the sound forward much as a megaphone does (figure 24.11).

The principle of the sonar system is that a sound bounces off an object and returns as an echo to the bat's ears (figure 24.12). Bats must cry very loudly if they are to receive loud echoes from objects at any distance. Bat cries have been measured at intensities of 60 to 100 dynes per cm^2. If these sounds fell within the range of human hearing, bats would sound like jet airplanes screaming through the night. These sounds are so loud, in fact, that researchers wondered why bats do not deafen themselves.

Bat ears have to be enormously sensitive to hear the faint echoes returning from their calls. How can they possibly hear these echoes when they are making such loud noises? Some

Figure 24.10 A greater horseshoe bat catching a moth with the aid of its left wing. Bats use an echo location system to catch prey.

Figure 24.11 Outlines of the heads of five bats. (*a*) African yellow-winged bat. (*b*) Horseshoe bat. (*c*) Funnel-eared bat. (*d*) Mouse-eared bat. (*e*) Free-tailed bat. Two of these (*a* and *b*) emit sounds through their nostrils, and they have nose-leaves that act as a megaphone to focus and direct their ultrasonic cries. The others emit sounds through their mouths.

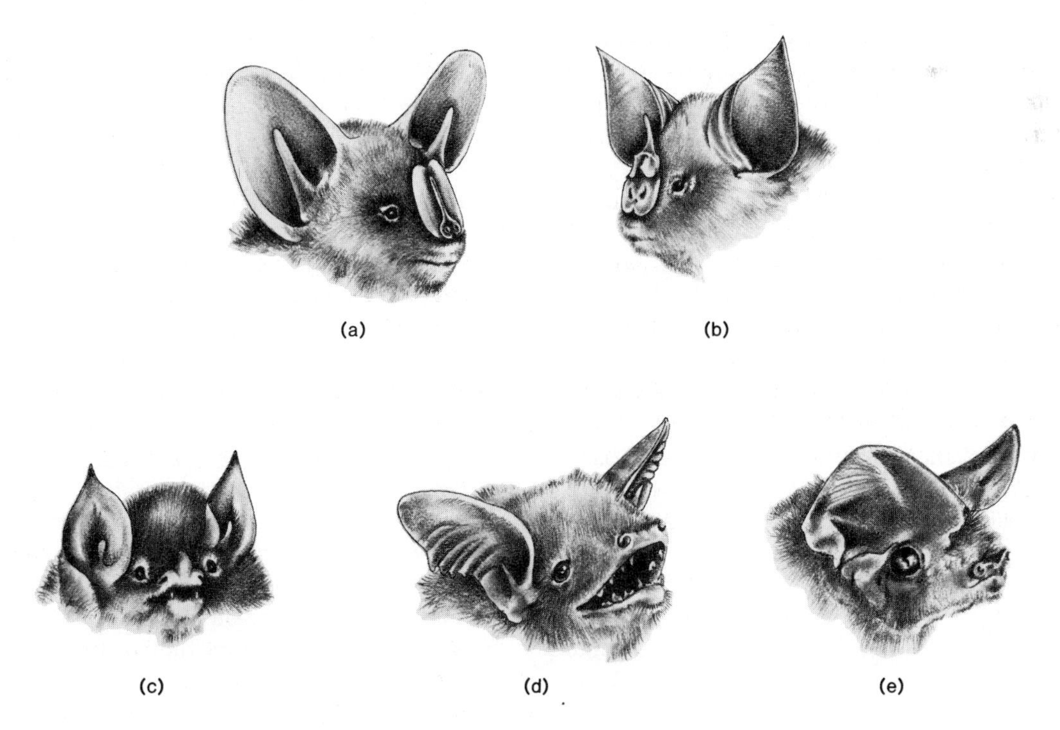

Figure 24.12 Principle of the bat sonar system. A bat emits a series of sound pulses and detects echoes that return when the sounds bounce off objects.

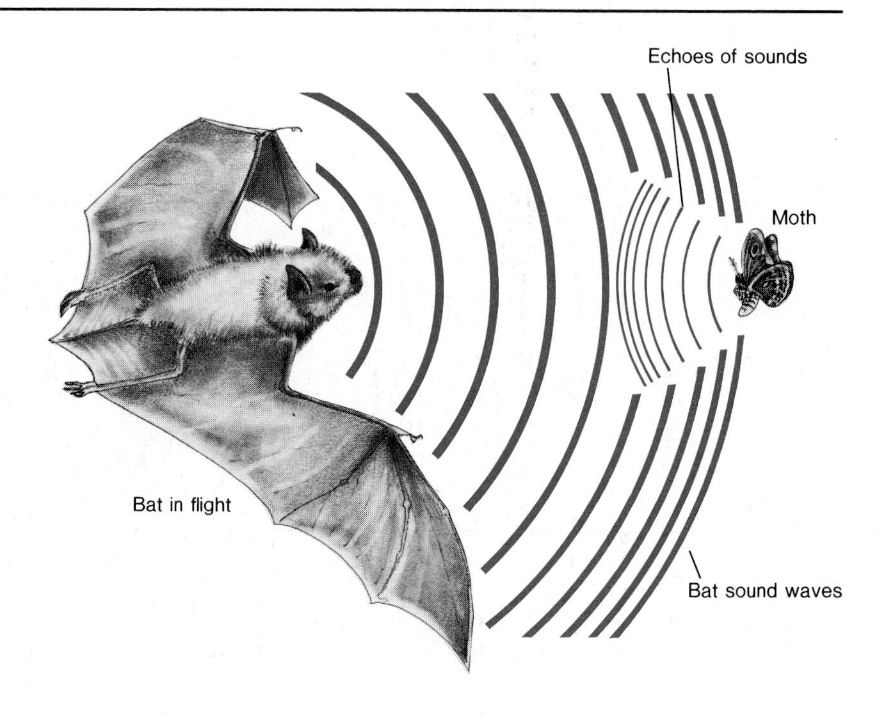

bats can temporarily shut off their hearing mechanism. They have a muscle with which they disconnect the ear ossicles while a sound pulse is going out and then reconnect them in time to hear the echo.

Bats' favorite prey have evolved ways of avoiding capture. Some insects, including many of the common night-flying moths, have ears that are extremely sensitive to ultrasound. When moths hear batlike sounds, they respond quickly. If you jingle a set of keys (this makes some sounds in the ultrasonic range) near a moth that is clinging to a wall, the moth will let go and drop to the ground. It was just such a chance observation that first puzzled Kenneth Roeder.

For several summers, Roeder spent every warm evening in his floodlit backyard playing sounds he could not hear to moths that happened to be passing by. Their behavior was striking. When a moth was 20 to 30 m from the loudspeaker, it turned and flew rapidly in the opposite direction. But when it was very close to the sound, it dived or looped.

By watching interactions between moths and local bats, Roeder and his associates quickly accounted for the different kinds of moth behavior. A moth that is 20 m away from a bat is still "invisible" to the bat; that is, it is out of sonar range. If it turns quickly, it can get completely out of the bat's range before it is even detected. However, when a moth is only 10 m or so away (within sonar range), it is in acute danger because bats can fly faster than moths. Only by executing evasive action does the moth escape.

Behavior of predators and their prey usually includes measures and countermeasures, such as those seen in bats and various insects (figure 24.13). This is a good example of coevolution (page 507) because interactions between predator and prey are so intense that they have influenced each other's evolution. In most cases, a balance has been reached through the course of evolution.

The Courtship of the Stickleback

We have examined several examples of adaptive behavior used for gathering food. Behavioral mechanisms also are involved in many other phases of animals' lives, such as reproduction.

The three-spined stickleback is a small, freshwater fish found in ponds or slowly moving streams throughout most of Europe and along both coasts of North America. The male stickleback defends a territory to which the female comes to lay her eggs. He is the sole guardian of the eggs. A male builds a nest by carrying mouthfuls of sand away from his chosen nest site until a small pit is formed in the sandy bottom. Next he collects bits of vegetation in his mouth and spits them into the pit (figure 24.14). When he has a little pile, the male glues the bits together with a secretion from his kidney. Finally, he forcefully wriggles and burrows his way through the pile of glued vegetation, making a tunnel.

Figure 24.13 Interactions between a bat and green lacewings recorded in photographs taken by Lee Miller and his colleagues. The camera shutter is kept open as bat and lacewing fly in the dark, and a series of brief flashes freeze their movements. (*a*) A bat captures an insect. The numbers indicate the position of each animal as each flash goes off. This bat swings its tail up, catches the lacewing against it, and flies off with the insect (arrow) in its mouth. (*b*) A lacewing evades the bat by folding its wings and diving before the bat reaches it. (*c*) The bat enters from the upper left, but the lacewing takes evasive action and the bat misses. This time, however, the bat swings around for another try.

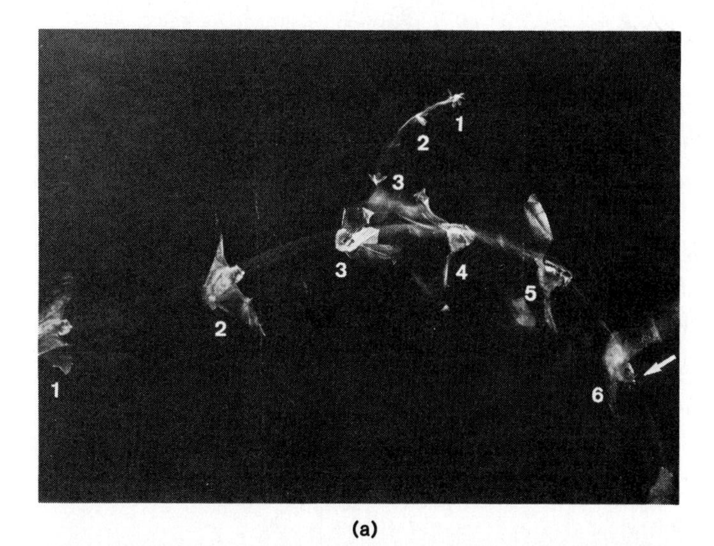

(a)

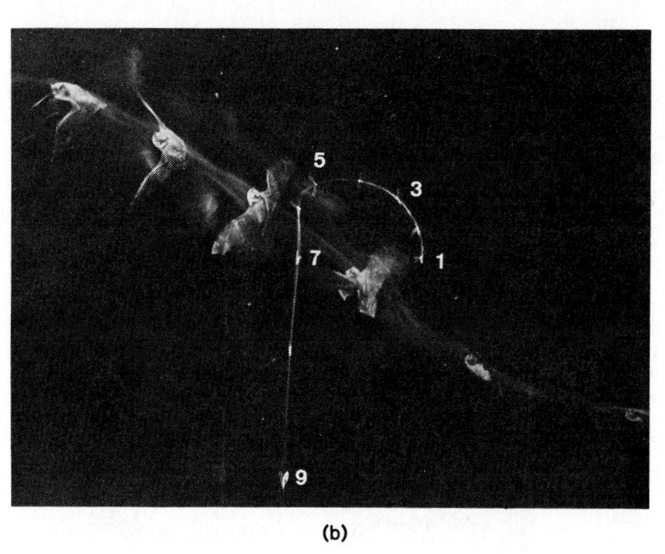

(b)

(c)

The male vigorously defends his nest site. A male stickleback recognizes another male by the bright red color of his belly (this is another example of a visual communication signal). When a male sees a spot of red in his territory, he charges out to expel it. It is easy to fool these aggressive little fish with a piece of cardboard of almost any shape, as long as it has red along the bottom side (figure 24.15*a*).

If an intruder does not flee after the initial rush, the two males may circle, or they may stand on their heads in the water, jerking up and down rapidly in gestures of **threat** (figure 24.15*b*), rather like a dog baring its teeth in a snarl or a man shaking his fist in an opponent's face.

Figure 24.14 A male stickleback selects a nest site, excavates it by removing sand and gravel and then constructs a nest over it. The completed nest contains a tunnel in which the female lays her eggs.

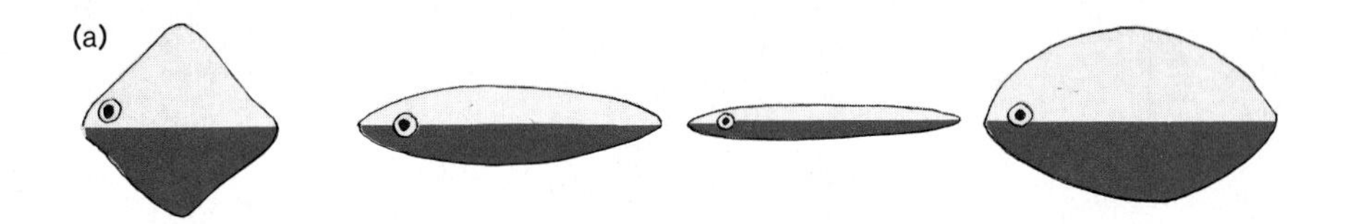

Figure 24.15 (*a*) Male sticklebacks attack models of various shapes such as these, so long as the "belly" is red. (*b*) Two male three-spined sticklebacks fighting.

When two males' territories are side by side, fighting may take on a rather bizarre appearance. Each male is **dominant** in his own territory, but **subordinate** when in the territory of another. Therefore, a stickleback that is fearlessly pursuing a rival will suddenly turn tail and flee when he discovers that he has crossed the territory boundary.

When a female enters the male stickleback's territory, his behavior becomes very different. Males recognize a female ready to lay eggs by her swollen belly. (Again, it is easy to fool him for he will respond to any object of about the right size and shape.) He launches into an elaborate course of zigzag swimming and jumps in front of her. If she follows him, he leads her to his nest and sticks his snout into the entrance. The female may then enter and wriggle into the nest tunnel. If the female does enter the nest, the male nuzzles her tail, and as he does so, she spawns her eggs and then swims out the other side of the tunnel. The male then immediately enters and sheds sperm on the newly laid eggs (figure 24.16).

The female stickleback may break off this **chain of behavior** at any point and swim from the territory. This only causes a renewed course of zigzag swimming by the male. A neighboring male also may court the female and attempt to lure her off to his own nest. He may even sneak in behind a spawning female and shed sperm on eggs in another male's nest!

Figure 24.16 Sexual behavior of the three-spined stickleback. When a female enters a male's territory, he courts her with zigzag swimming. If she responds, he leads her to his nest, where he adopts a special posture with his snout in the entrance. Then if the female enters the nest, the male nuzzles her tail, and as he does so, she spawns her eggs (inset). After she spawns, the female swims out of the nest and the male enters and sheds sperm on the eggs.

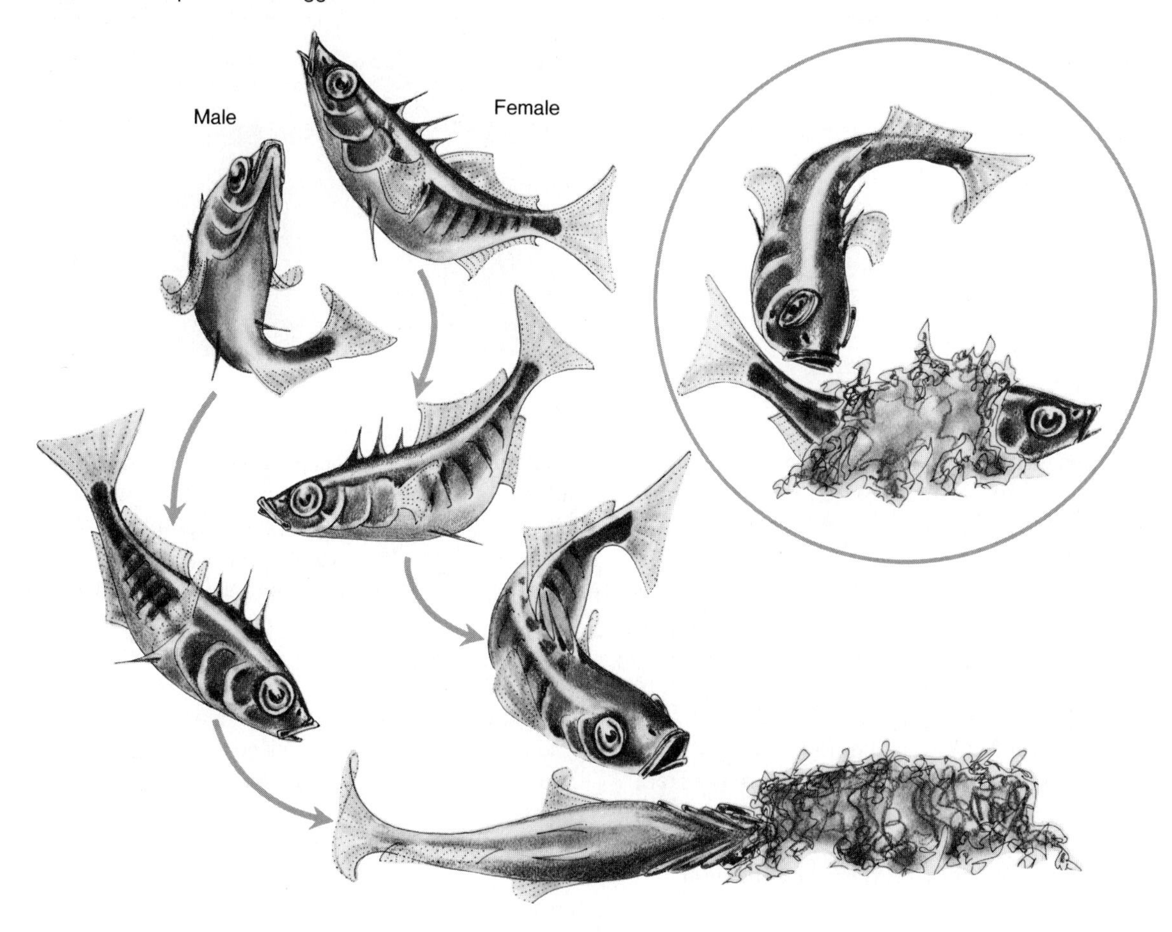

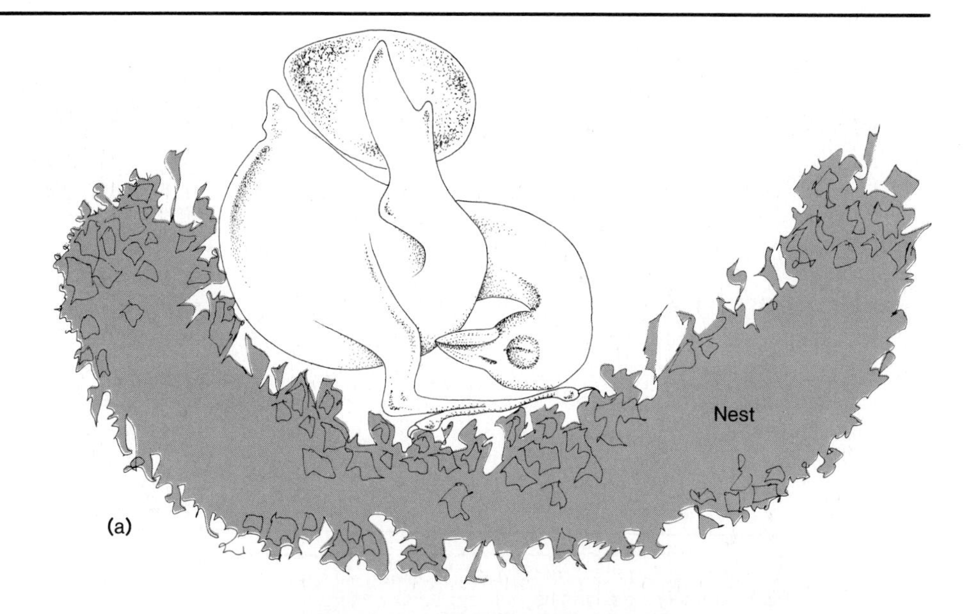

Figure 24.17 Brood parasitism by cuckoos. (*a*) A newly hatched cuckoo pushes its host's eggs out of the nest. (*b*) A hedge sparrow feeds a young cuckoo that is much larger than its foster parent.

After fertilization, the male's behavior changes dramatically. He drives the female from his territory. Then he arranges the egg mass, repairs any damage to the nest, and also lengthens the nest to make more room. Within an hour he again courts any females who come into his territory.

As the embryos develop, however, they require more of his time, and he is less likely to court. He swims near the nest and beats his fins, forcing a current of water through the nest. This fanning behavior is essential because it makes adequate gas exchange possible for the embryos.

When the young fish begin to hatch, the male continues to hover over them for several days. If any stray away from the nest, he sucks them up into his mouth and spits them back into the nest. Finally, when they are large enough, they swim off, or the male simply deserts them and the nest.

The reproductive behavior of this little fish illustrates a number of important concepts of ethology. First, the male fish is not responding generally to the female or a rival male. Only a specific part of the overall stimulus (for example, the red belly of another male) elicits a response. A specific stimulus such as another male's red belly is called a **sign stimulus,** and it is said to **release** the male's **aggressive behavior.**

Second, there are progressive changes in the male's **motivation** or likelihood to behave in a certain way. For example, just prior to mating he vigorously courts the female, and just afterwards he chases her from the territory as though she were a trespasser. It is the presence of freshly laid eggs in his nest that causes him to drive away all other fish, including the female, and to tend the eggs closely. Understanding the bases of motivational changes such as these is an important part of understanding and predicting animal behavior.

The Adaptiveness of Courtship Behavior

In many animals, copulation is preceded by elaborate patterns of courtship behavior. Members of some species are normally aggressive, and courtship helps to resolve the conflict between the drive to be hostile and the drive to mate.

In many species, courtship behavior may be necessary to resolve the conflict involved in animals touching each other, as they must do during mating. This is not a trivial problem because many adult animals respond very negatively to being touched by any other animal. This general response is understandable if you consider that, in most contexts, being touched can well mean being captured and killed. The negative response is so strong that it includes even members of the animal's own species. Courtship behavior resolves this conflict and animals are able to come into close contact for mating and nesting or other parental care activities.

Foster Parents

Many kinds of animals expend a great deal of energy on the care and feeding of their young offspring. Behavior that assures care of helpless offspring is adaptive because it is essential for reproductive success. However, some animals avoid the burdens of parental care. They trick other animals into caring for their young.

Some birds, known as **brood parasites,** lay their eggs in the nests of other species with the result that birds of the host species take care of the parasite's eggs and nestlings, along with their own.

Female cuckoos spend most of their time patiently watching the nest-building activities of prospective host species. When the female cuckoo is ready to lay an egg, she quietly flies to the host's nest, moves one of the host's eggs, slips onto the nest, and lays one of her own in its place. In a matter of seconds she is gone again, taking a stolen host egg with her.

The incubation period of the cuckoo egg is slightly shorter than that of the host's eggs. Thus, the nestling cuckoo generally hatches first. A day after hatching the apparently helpless cuckoo becomes totally intolerant of anything else in the vicinity and begins the strenuous task of evicting everything from the nest. Supporting an egg or nestling on its back and holding it with its wings, the young cuckoo pushes it along the rim of the nest and out over the edge (figure 24.17*a*). After this, the foster parents devote all of their attention to the growing cuckoo. The baby cuckoo grows quickly, and in some cases, even becomes larger than its foster parents (figure 24.17*b*).

There is an evolutionary "race" between the cuckoo and its host. Selection favors hosts that avoid taking care of cuckoo young. Hosts often desert nests that have been disturbed by cuckoos or remove eggs that do not look like their own. Hosts mob cuckoos and try to chase them away whenever they see them. But there is also pressure for cuckoos to become more successful at getting hosts to accept their eggs. This is a balanced relationship, and neither host nor parasite really wins the "struggle."

Figure 24.18 A Mexican jay feeding young in a nest.

Sociobiology: A New Look at the Evolution of Social Behavior

During the 1970s, a new emphasis developed in behavioral research. **Sociobiology** focuses on certain genetic and evolutionary explanations for social behaviors. Advocates of sociobiology maintain that many complex social behaviors, including human social behaviors, could be better understood if they were reexamined in the context of sociobiology. Because many biologists disagree with some of its basic assumptions, a great deal of controversy surrounds sociobiology.

Here is an example of a sociobiological analysis and interpretation of behavior patterns.

Helpers at the Nest

Mexican jays live in the Arizona mountains in permanent flocks of five to fifteen birds. Only one or two pairs of birds in the flock build nests at any one time. All the eggs in a given nest are laid by one female and incubated by a single pair of jays. But the unusual thing about these jays is that after the nestlings have hatched, all the birds in the flock feed the young, even the parents from the other active nest! It is a sort of bird commune (figure 24.18).

Such cooperative behavior in caring for offspring seems unusual, although it is by no means unique. Such apparently unselfish interest in the welfare of others is called altruism, and behavioral biologists call this **altruistic behavior.** But how did altruistic behavior evolve? How could natural selection favor taking care of another individual's offspring with the same care that would be given to one's own?

It appears to be very difficult for young Mexican jays to establish themselves in a new flock. Therefore, they usually remain in the same flock throughout their unusually long lifetime. Young jays do not begin breeding until they are three years old, so a flock usually contains several of these young birds. Because most jays remain within the same flock, most of the individuals in a flock are likely to be close relatives. The flock actually is an extended family unit. Thus, if an animal helps another member of the flock, it is almost certainly helping a close relative. Many examples of altruistic behavior actually turn out, like this one, to be cases of animals helping relatives. But the question still remains, how could such behavior evolve?

In the early 1960s, W. D. Hamilton developed a hypothesis that has generated much interest and controversy among biologists who study behavior. Hamilton pointed out that a parent caring for its young is just a special case of the more general phenomenon of animals helping genetic relatives. Close genetic relatives, in addition to parents and offspring, have genes in common. Thus, an animal showing altruistic behavior toward a close relative is promoting survival and transmission of at least some genes that are the same as its own. Hamilton proposed that natural selection favors such altruistic behavior toward relatives.

Does Hamilton's hypothesis provide any insight into the evolution of helpers at the nest in Mexican jays? Since members of a flock are almost always close relatives, helping to feed another individual's nestlings means that a jay is helping a close relative. Since individuals often cannot breed for several years, the best way to promote one's own genes is by helping relatives in the flock that are able to breed and transmit those same genes.

This is not to suggest that animals "know" in any conscious sense who their relatives are. However, animals who behave in this way have more surviving relatives that share and transmit the same genes, including genes for altruistic behavior, than those who do not. Even altruistic behavior can be self-serving in an evolutionary sense if the altruism is directed toward close relatives with whom the altruists have genes in common.

Sociobiology and Human Behavior

Sociobiologists insist that ideas about genetic programming of much of behavior apply to human behavior as well as to the behavior of other animals. It is on this point that sociobiology has generated a great deal of controversy. People who disagree with some of its basic assumptions argue that sociobiology puts too much emphasis on genetic aspects of behavior, and when applied to human beings, promotes a kind of **biological determinism.** They feel that sociobiology leads to the conclusion that much of human behavior, including some of its most negative aspects, is at least in part genetically determined and thus is inevitable and unalterable. Opponents of sociobiology feel that this determinism can lead people to an acceptance of the social status quo that can condone racism, sexism, and class determinism. They argue that culture and the social environment are of primary importance in shaping human behavior.

Sociobiologists answer this criticism by saying that it is not dehumanizing to investigate the roles of genes in establishing human behavior. They think that the prospect of discovering that human behavioral patterns are more a part of human biological nature than originally thought should not make people uneasy. They argue that it does not detract from one's humanity to discover that humans are different from one another in terms of behavior and that some of these differences have genetic roots deep in humans' evolutionary past.

The questions raised by the sociobiologists are intriguing, and we can certainly expect to hear more about these arguments in the future.

Summary

Adaptive behavior increases the probability that animals, or their offspring, live to reproduce successfully so that their genes are perpetuated in future generations.

The innate (genetically determined) nature of certain behavior is clear. Behavioral mutants show altered behavior patterns that are inherited by offspring of the mutant organisms.

Learned behavior results from an organism's experience with its environment, and the influence of learning on behavior patterns begins very early in life.

Many animals learn to recognize members of their own species and distinguish them from other organisms because they become imprinted on their parents at an early critical age.

Behavioral expression depends on internal motivational state. Physiological changes, such as changing hormone levels during reproductive cycles, result in changing motivation. Other motivational changes depend on completion of specific sequences of behavioral expressions.

Adaptive behavior patterns play roles in diverse aspects of animals' lives, including such vital activities as obtaining food and finding proper living places. Bee communication through the waggle dance increases the efficiency of food gathering and facilitates location of new hive sites during swarming. Bats have highly specialized sensory and behavioral mechanisms that permit them to fly safely at night and to locate and capture prey.

Courtship behavior is adaptive because it brings animals together with prospective mates of their own species and resolves potential behavioral conflicts between males and females. In many cases, courtship behavior also is part of a chain of behavior leading to parental behavior that provides care for helpless offspring. Brood parasites such as cuckoos shift a large part of the burden of their reproduction to members of other species, whom they trick into incubating their eggs and feeding their young.

Sociobiology focuses on genetic and evolutionary explanations for social behavior. Among other things, sociobiology attempts to explain how altruism directed toward close relatives promotes transmission of at least some genes that are the same as those of the altruist.

Use of the concepts of sociobiology as explanations for human social behavior has produced some controversy.

Questions for Review

1. Use the example of ground-nesting gulls removing eggshells from their nests to discuss the evolution of behavioral patterns.
2. What does an ethologist do?
3. How does imprinting affect the behavior of ducks, geese, and some other animals throughout their lifetimes?
4. What do behavioral biologists mean by motivational state?
5. Define the term phototaxis.
6. How does the waggle dance of bees communicate information about direction and distance?
7. Explain why humans cannot hear the cries of hunting bats.
8. What is a sign stimulus? Give an example.
9. What is a brood parasite?
10. Explain the sociobiological reasoning leading to the conclusion that altruistic behavior can be adaptive.

Questions for Analysis and Discussion

1. Explain why it is very difficult to conclude that a behavioral pattern is completely innate (instinctive) and does not have any learned components.
2. Although brood parasitism seems to be highly adaptive for some animals, there are only a few species that use it as part of their reproductive behavior. Why do you suppose that this is true?
3. Many people feel that great care must be taken in applying the principles of sociobiological analysis to human behavior. Explain their concern about biological determinism in terms of making future social and political decisions.

Suggested Readings

Books

Alcock, J. 1979. *Animal behavior,* 2d ed. Sunderland, Mass.: Sinauer Associates.

Fossey, D. 1983. *Gorillas in the mist.* Boston: Houghton Mifflin Co.

Gould, J. L. 1981. *Ethology.* New York: Norton.

Lorenz, K. 1961. *King Solomon's ring.* New York: Crowell.

Wilson, E. O. 1980. *Sociobiology: The abridged edition.* Cambridge, Mass.: Belknap Press.

Articles

Fenton, M. B., and Fullard, J. H. 1981. Moth hearing and the feeding strategies of bats. *American Scientist* 69:266.

Goodenough, J. 1984. Animal communication. *Carolina Biology Readers* no. 143. Burlington, N.C.: Carolina Biological Supply Co.

Hailman, J. P. December 1969. How an instinct is learned. *Scientific American* (offprint 1165).

Keeton, W. T. December 1974. The mystery of pigeon homing. *Scientific American* (offprint 1311).

Maugh, T. H., II. 1982. Magnetic navigation, an attractive possibility. *Science* 215:1492.

Palmer, J. D. 1984. Biological rhythms and living clocks. *Carolina Biology Readers* no. 92. Burlington, N.C.: Carolina Biological Supply Co.

Scheller, R. H., and Axel, R. March 1984. How genes control an innate behavior. *Scientific American.*

Smith, J. M. September 1978. The evolution of behavior. *Scientific American* (offprint 1405).

Ecology
Organisms and Environment

Chapter Outline

Chapter Concepts

1. Ecology deals with the relationships that exist between an organism and its environment.
2. Two important physical factors in an organism's environment are temperature and water. Organisms respond to environmental temperature in several different ways. Water is vital for all living organisms.
3. Populations have certain properties and characteristics, and respond in a predictable way to various factors.
4. When resources in an environment are nonlimiting, populations grow exponentially, but as resources run out, population growth slows and eventually stops.
5. Based on reproductive strategies, populations are divided into two groups, consisting of opportunistic and equilibrium species.
6. Two species that have the same ecological requirements cannot coexist indefinitely.
7. Predator and prey systems are the outcome of coevolution.
8. Parasitism and mutualism are two symbiotic relationships that exist between pairs of species.

"No man is an island, entire of itself; every man is a piece of the continent, a part of the main. . . ." These lines, written by John Donne in 1624, express an important truth about the way human lives touch one another.

In a biological sense, no living thing stands alone. An individual organism, like an oak tree or a rabbit or a human being, never exists in complete isolation. It is influenced by a large number of external factors. In the last chapter we were particularly interested in the behavioral responses of individual organisms to certain external influences. In this chapter we will examine a number of the external factors themselves.

The factors that influence an individual organism include physical factors (such as light, temperature, humidity, gravity, and nutrient availability), individual organisms of the same species, and individuals of other species. All of these factors, taken together, constitute an organism's environment. The science of **ecology** deals with the relationships that exist between an organism and its environment.

As we begin our consideration of ecology, we need to introduce two basic ecological concepts. An organism's **niche** is the particular way in which that organism interacts with its environment. How does the organism obtain nutrients? What other organisms might obtain nutrients by consuming the organism? How is the organism affected by changes in physical factors in its environment? These interactions are specific to a given kind of organism and have their origins in its evolutionary history. They enable that organism to survive and reproduce in its environment. It is important not to confuse the concept of niche with the concept of **habitat.** An organism's habitat basically decribes where an organism is to be found, the kinds of plants and animals with which it lives, and the sorts of physical factors with which it has to deal. Thus, to say that a mallard duck lives in a marsh describes its habitat and to say that it eats submerged vegetation describes one aspect of its niche.

Physical Factors

We have already discussed a number of physical factors that are important components of an organism's environment. For example, we examined the importance of light in photosynthesis (page 98), photoperiodism (page 308), and sensory perception (page 278). In this section we will examine two additional physical factors: temperature and water.

Temperature

An important physical factor in an organism's environment is temperature. Organisms respond to changes in environmental temperature in different ways.

As we saw in chapter 6, only about 40 percent of the energy available in organic nutrients is used for ATP production in the cells of living things. Most of the remaining 60 percent is emitted as heat, which, in all plants and most animals, is quickly lost to the environment. The body temperatures of these organisms are influenced by the environmental temperature. When the environmental temperature rises, their body temperatures tend to rise, and when the environmental temperature falls, their body temperatures tend to fall. Organisms with environmentally influenced, variable body temperatures are called **poikilothermic,** which means "changeable temperature" (the Greek word *poikilos* means changeable).

Some animals, however, specifically birds and mammals, conserve body heat and regulate their metabolisms so that they maintain stable, relatively constant body core temperatures. These animals are called **homeothermic** ("same temperature").

Poikilotherms

The sum of all the cellular processes that occur in an organism constitute the **metabolism** of that organism. Because the rates at which chemical reactions proceed in living organisms are temperature dependent, environmental temperature changes and subsequent body temperature changes have a significant effect on the metabolism of poikilothermic organisms.

Figure 25.1 shows how changing temperature affects the metabolic rate of a hypothetical poikilothermic organism. As you can see, there is a temperature range across which the organism is able to exist without serious difficulties, even though its metabolic rate changes as the temperature changes. This is the **range of thermal tolerance.**

Beyond the lower and upper extremes of this acceptable temperature range, some key processes are adversely affected and slow down below a critical rate, depressing the activity and the metabolism of the whole organism. **Cold torpor** and **heat coma** are names given to the inactive states near the lower and upper extremes of the range of thermal tolerance. **Cold death** or **heat death** will occur when temperatures are extreme enough to cause extensive, irreversible damage.

Figure 25.1 The relationship between temperature and the metabolism of a poikilothermic organism.

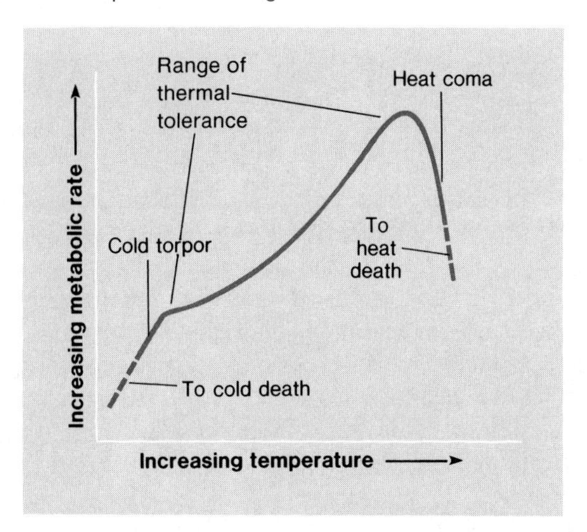

However, not all poikilotherms respond to temperature change in the same way. Different temperature responses have evolved in various organisms. Consider, for example, the difference in the effects of various water temperatures on an arctic fish and a tropical fish. In water at 4°C, an arctic fish thrives and carries out its life activities normally. A tropical fish placed in water at 4°C would be paralyzed or worse by the cold. On the other hand, a tropical fish lives normally in 20°C water. An arctic fish placed in 20°C water would suffer heat death.

Figure 25.2 Behavioral thermoregulation. (*a*) A lizard basks in the sun during cool morning hours. (*b*) During the heat of the day, the same lizard enters a cool underground burrow when its body temperature rises above a certain critical point.

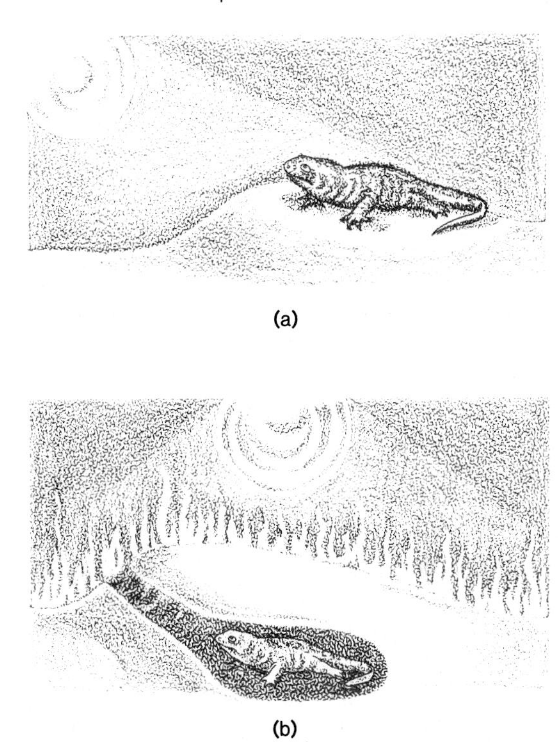

Behavioral Thermoregulation

Are poikilothermic organisms completely at the mercy of their thermal environment or are some forms of adjustment available to these organisms that face hourly temperature changes on a daily basis? Some poikilothermic animals behave in ways that adjust their body temperatures. This is called **behavioral thermoregulation.**

On cool mornings, snakes and lizards that are made very sluggish by low temperatures bask in the sunshine until their body temperatures rise to the point where they can carry out their normal activities. Later in the day, they seek shelter in cool burrows if their body temperatures rise above a certain critical point (figure 25.2).

Another form of behavioral thermoregulation is the warm-up period that some insects go through on a cool morning. Butterflies spread their wings and vibrate them slowly until the heat generated by the metabolism of their muscle cells raises their temperature. When its body temperature is high enough to permit the intense muscular activity needed for flight, a butterfly can take off. Such forms of behavioral thermoregulation help poikilothermic animals cope with changing environmental temperatures.

Homeotherms

As we mentioned earlier, a different strategy for dealing with environmental temperature fluctuations is to maintain a constant internal body temperature. Animals that do this are called homeotherms. Temperatures in the areas near the body surface may vary, but the temperature deep inside a homeotherm's body, the **core temperature,** remains relatively constant. Thus, major internal organs function in an environment where the temperature changes very little.

Homeotherms (that is, birds and mammals) have relatively higher metabolic rates than poikilothermic organisms, and many of them also have insulating layers of fat, feathers, or fur to help conserve body heat. Furthermore, they have nervous and hormonal mechanisms that regulate their metabolic rates in response to body temperature changes. The temperature-maintenance strategies of homeothermic animals use "a furnace rather than a refrigerator"—the body temperature is set and kept near the high end of the organism's normal environmental temperature range. The reason

Figure 25.3 Human responses to temperature change. Within the range between 27°C and 31°C, small adjustments in heat loss from superficial blood vessels are adequate to regulate temperature. Outside this range other physiological mechanisms come into play.

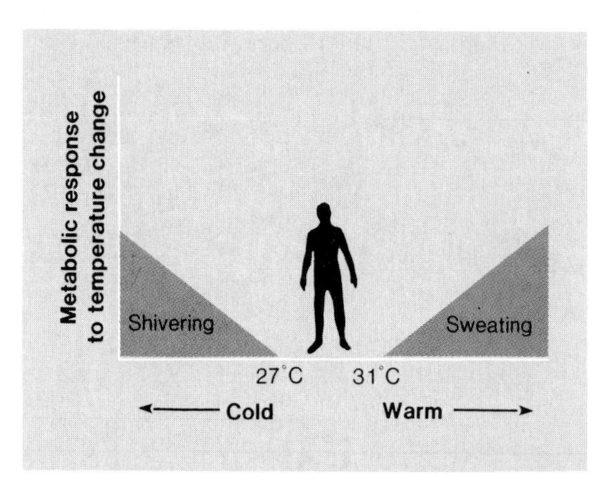

for this seems to be that regulation by heat conservation and generation is very efficient, while animals' cooling capacities are usually much more limited. Normal body temperatures of mammals usually range from 36° to 39°C, while birds' body temperatures generally are somewhat higher, in the range from 40° to 43°C.

Homeothermic animals can make some adjustments to changing environmental temperature by regulating the amount of blood that reaches body surface areas. When environmental temperature begins to fall below body temperature, blood vessels near the body's surface constrict, thus decreasing the amount of heat energy lost from the blood. When the environmental temperature begins to rise, the vessels are dilated so that there is greater blood flow near the body surface and greater heat loss to the environment.

When environmental temperature falls to a point such that adjustments in blood flow are no longer effective in maintaining body temperature, the nervous system responds by initiating **shivering** (figure 25.3). Shivering is a series of involuntary muscle contractions. These contractions produce heat that helps to compensate for heat loss. For example, when you become very chilled, shivering can be quite vigorous, even to the point that your teeth "chatter" as jaw muscles become involved. On the other hand, when the environmental temperature rises above a certain value, your body initiates **sweating,** which dissipates a great deal of body heat by evaporative cooling. Both shivering and sweating require energy expenditures and raise the metabolic rate.

Figure 25.4 Roosting vampire bats. During daytime hours, bats allow their body temperatures to fall near the temperature of their environment.

Hibernation and Other Avoidance Strategies

Considering the great metabolic cost of maintaining a high, constant body temperature, it is not surprising that some homeotherms avoid the struggle to maintain body temperature at least part of the time. Bats, for example, let their regulatory processes "slip" on a daily basis. They fly about actively and feed at night, then return in the day to roost in cool places such as caves, letting their daytime body temperatures fall to within a few degrees of the environmental temperature (figure 25.4). This normal, daily period of cold torpor seems to be a mechanism for energy conservation.

Some mammals let their body temperature and metabolism fall for much longer periods of time. These long periods of **hibernation** usually occur in winter. During hibernation, the organism's temperature may drop to just a couple degrees above the environmental temperature. Hibernating mammals can live for long periods on the metabolic reserves they accumulated before entering hibernation.

During hibernation, an animal's heart rate may fall to only a few beats per minute, and its oxygen consumption is only a fraction of the normal rate (sometimes as little as 1 percent). However, a hibernating animal has not entirely stopped regulating its temperature. Sharp decreases in environmental temperatures that could threaten to freeze body tissues cause quick metabolic increases. Sometimes, a large temperature drop will even arouse hibernating animals.

Figure 25.5 A female black bear peering out of its den in northern Minnesota in mid-April. The cub with her was born in January. While a bear's body temperature may drop a few degrees during its winter sleep, it is not hibernating.

Some mammals, such as bears, accumulate huge nutrient reserves in heavy fat deposits and simply sleep the winter away. While a bear's body temperature may drop several degrees and it may be difficult to awaken, a bear is not in an entirely altered metabolic state, as hibernating animals are. Bears and other winter sleepers do not become torpid; woodchucks, ground squirrels, and other true hibernators do (figure 25.5).

Water

Because of the importance of water in life processes, maintenance of **water balance** is vital for all living organisms. We say an organism is in water balance when its body fluids contain appropriate quantities of water, and when the organism's water loss to its environment equals its water gain from the environment.

In addition to specific functional roles, ions in body fluids contribute to osmotic properties of the organism. Thus, regulation of the ionic composition of body fluids is interrelated with maintenance of water balance. What then are some of the osmotic and ionic problems faced by animals living in various environments, and what are some solutions to these problems?

Marine Fish

Marine bony fish (teleosts) have body fluids that are about one-third the ionic concentration of seawater. Because these fish live in a hypertonic environment, they continuously lose

Figure 25.6 Water and salt balances in teleost fish. This figure illustrates the water and ion balance problems faced by marine and freshwater teleosts and also summarizes the ways in which teleosts meet these problems. The hollow arrows represent water and ion gains or losses resulting from the nature of the fish's environment. The solid arrows depict the active functions of the fish that counteract environmental pressures. Marine fish excrete some ions via their kidneys, but the primary solution to the problem of excess body salts, caused by drinking seawater, is the active excretion of Na^+ and Cl^- ions through their gills.

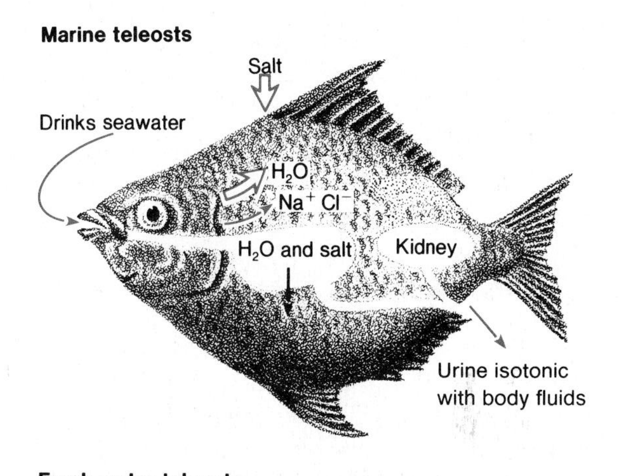

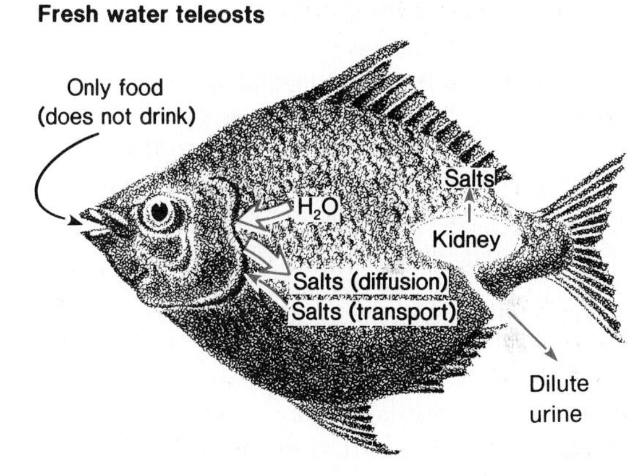

water to their environment by osmosis, especially through their gills. They replace lost water by drinking large quantities of seawater. The average marine teleost swallows an amount of water estimated to be equal to about 1 percent of its body weight every hour. This is approximately equivalent to an average-size human male drinking about 700 ml of water every hour around the clock.

Because they drink so much seawater, a great deal of salt enters marine teleosts' bodies, and they face the problem of continuous, unavoidable salt gain. To solve this problem, the gills of marine teleosts actively transport sodium (Na^+) and chloride (Cl^-) ions outward into the surrounding seawater. This active ion "pumping" is the basis of marine teleosts' ability to drink seawater to replace the water lost by osmosis (figure 25.6).

Freshwater Fish

Animals living in fresh water face a different set of problems because they live in hypotonic environments. They tend to gain water osmotically, and they lose body salts because ions diffuse out through their body surfaces. Thus, freshwater bony fish face a set of problems opposite to those of marine fish.

Freshwater fish are adapted to meet the problem of excess water gain. They never drink water and they eliminate excess water through production of large quantities of very dilute urine. They can discharge a quantity of urine equal to one-third of their body weight each day.

However, even though their urine is quite dilute, freshwater fish still lose a significant amount of salt in their urine. This urinary salt loss, coupled with salt loss by diffusion through body surfaces, represents a serious ion balance problem. These fish obtain some salt in their food, but most of the necessary salt gain is accomplished by the gills. The gills of freshwater fish actively transport ions inward. These ions, which are in low concentration in the surrounding water, are taken up by energy-requiring active transport. Thus, active movement of ions in the gills of freshwater fish is in the opposite direction from that occurring in marine bony fish (figure 25.6).

Box 25.1

Cryptobiosis

The Hidden Life of the Water Bear

A fascinating array of adaptations for dealing with environmental water deficiencies has evolved in the animal kingdom. While some animals migrate to new habitats where water is available, others handle the problem in a different way.

For example, frogs and toads that live in desert areas burrow down into the bottom mud as their ponds dry up. There they enter a resting state called **estivation.** Animals in this "summer sleep" have very low metabolic rates and can remain inactive for long periods. When rain falls and refills their ponds, they quickly become active and resume their normal lives. Sometimes this includes a flurry of reproductive activity followed by very rapid development of a new generation so that the young individuals are mature enough to enter estivation when the ponds dry up again.

But possibly the most remarkable adaptations to periods of water deficiency are those of the microscopic tardigrades (box figure 25.1*A*). These animals, which are between 0.1 and 1 mm long, are found in almost every environment—marine, fresh water, and terrestrial. They have four pairs of stumpy legs and move with a lumbering, bearlike motion. Hence, they have been called "water bears."

When a tardigrade's surroundings begin to dry up, instead of moving on or attempting to conserve water, it simply contracts into a barrellike shape and becomes inactive. Its body dries until it contains barely detectable quantities of water, as little as 3 percent of its predrying water content. A tardigrade can remain in this dehydrated condition for years and yet be ready to rehydrate and become active quickly when water is once again available. A museum specimen of moss that had been stored for 120 years yielded living tardigrades when it was placed in water. Clearly, tardigrades do not age at the normal rate while they are dehydrated (box figure 25.1*B*).

This dormant state in which the animal has extremely low metabolic activity is called **cryptobiosis** (meaning "hidden life"). Not only do cryptobiotic tardigrades somehow escape normal aging, they also are very resistant to harmful conditions that quickly kill normal, active tardigrades, or any other animals for that matter. For example, cryptobiotic tardigrades can survive for several minutes at 151°C or for long periods at temperatures near −273°C.

The ability to become cryptobiotic has great adaptive value because it allows tardigrades to survive drought conditions that would destroy many other organisms. Furthermore, tardigrades are not harmed by heat or solar radiation to which they are exposed while dry. In the dried condition they are very light and can be blown about by the wind. Thus, cryptobiosis also serves as a dispersal mechanism that permits tardigrades to enter new habitats.

Terrestrial Animals

Terrestrial animals (those that live on dry land) face the constant threat of dehydration. Relatively little water is available in terrestrial environments to make up for terrestrial animals' body water losses from nitrogenous waste excretion and evaporation. Thus, water conservation is a key requirement for fluid homeostasis in terrestrial animals.

Terrestrial animals have adapted in several ways to conserve body water. Their body surfaces usually are fairly impermeable to water. This reduces evaporation, although there is always at least some water loss from the thin moist surface areas of lungs. Many terrestrial animals conserve body water by producing very concentrated urine (see page 224). Behavioral patterns also reduce water loss. Many terrestrial animals avoid situations in which water loss is greatest. For example, certain desert animals are active during cool nighttime hours, but they rest in cool underground burrows during the hottest daytime hours.

Despite these adaptations, dehydration remains a serious threat to terrestrial animals, including human beings. Water availability is a major determining factor in the distribution of various animals in terrestrial environments.

There still are many features of this adaptation to adverse conditions that are not well understood. How is it that tardigrades, some small worms, and a few other small animals can shut down their metabolism and survive with practically no body water? What protects their cells and tissues from complete disruption during drying and rehydration? Why are other animals not able to dry out and recover in the same way? Further research will be required to answer these and other questions about cryptobiosis.

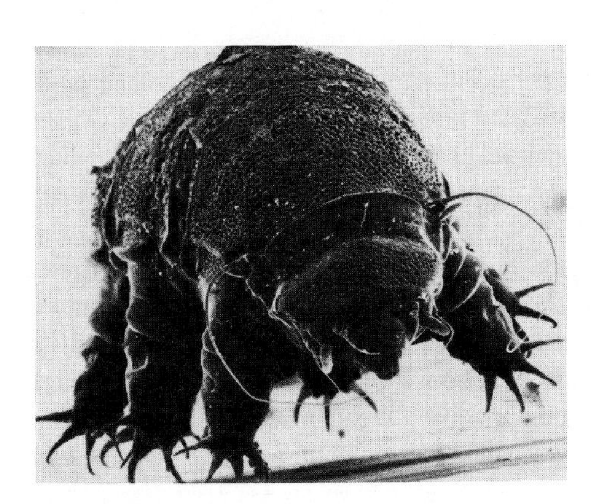

Box figure 25.1*A* Scanning electron micrograph of a tardigrade, *Echiniscus arctomys*. This is the normal appearance of the animal when it is active. Tardigrades live in very moist environments and their bodies contain about 85 percent water (magnification × 60).

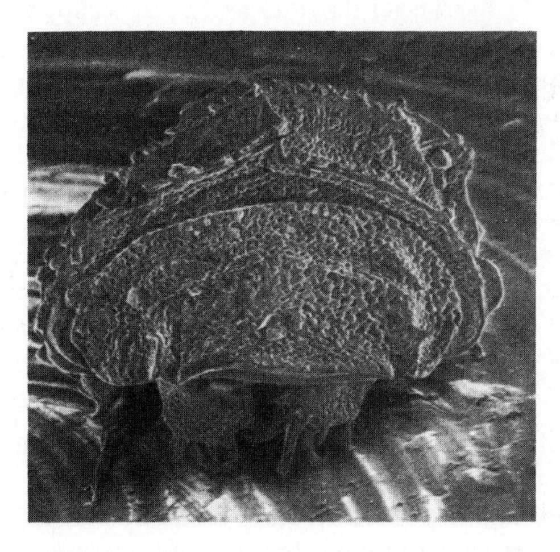

Box figure 25.1*B* Scanning electron micrograph of a tardigrade in cryptobiosis. The animal has assumed a barrellike shape and has dehydrated as its environment dried up. A cryptobiotic tardigrade's body contains only about 3 percent water (magnification × 60).

Figure 25.7 The exponential growth curve plotted (*a*) arithmetically and (*b*) logarithmically.

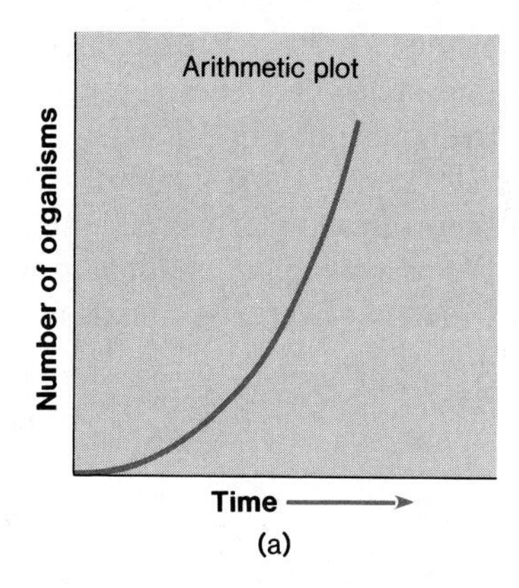

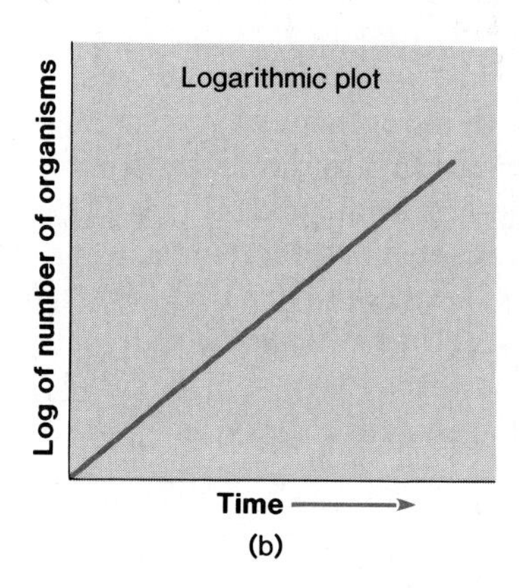

Other Organisms in the Environment

In their environment, living things interact both with members of their own species and with organisms of other species.

Interactions with Individuals of the Same Species

Individual organisms interact in a variety of ways with other individuals belonging to the same species. We can study these interactions in populations. A **population** is a group of individuals of the same species that interbreed and occupy a given area at a given time. Populations have certain properties and characteristics and respond in a predictable way to various factors. A population consists of many individuals reproducing and dying over a period of time; thus, a population has a birth rate and a death rate. Because a population usually consists of individuals of different ages, it also has an age structure.

Population Patterns in Time

Populations are variable in time. For example, the populations of birds in a woodland in the northern United States in summer are quite different from those found in the same woodland during the winter. Robins, wood thrushes, and vireos are present in the summer but leave for warmer places in the winter. During the winter populations of pine siskins and evening grosbeaks inhabit the woods.

Since animals are mobile, the seasonality of some animal species, such as many bird species, is due to local and long-range migration. However, other species, like insects, are active in summer, but spend the winter in a state of dormancy.

Plant populations also change seasonally in a woodland. In early spring, flowers such as large-flowered trillium, yellow violet, and spring beauty grow and flower. But in the fall, they are nowhere to be seen. White wood aster, gray-stemmed goldenrod, and white snakeroot are flowering instead. Among plants, such changes usually occur because different species under the influence of changing day length become dormant at different times. Dormant plants spend the nongrowing season as seeds or bulbs, or as rosettes or buds buried beneath the leaves.

Population Growth

Over a period of time, populations increase, decrease, or become stationary (remain the same size). One way to learn about animal population size changes is to introduce a small number of reproducing individuals of some species into a vacant habitat with an abundance of food and other resources.

As long as the total environment continues to be optimal (ideal for population growth), the population will grow **geometrically (exponentially).** The number of individuals added to the population would become larger for each successive increment in time. We can see this from a graph, using an arithmetic scale, where population size is plotted against time (figure 25.7). Our graph shows that the population grows slowly at first, and then the curve becomes steeper because increasingly large numbers of individuals are added to the population.

In this situation, reproductive rates are as high as they can be and death rates are as low as they can be. The optimal environment permits the population to grow at the maximum rate possible for organisms of the species being studied. Populations growing in such a manner are growing at their **biotic potential.**

Figure 25.8 The logistic or S-shaped growth curve. The population initially grows exponentially and then begins to slow down as resources become limiting. Population size remains stable when the carrying capacity of the environment is reached.

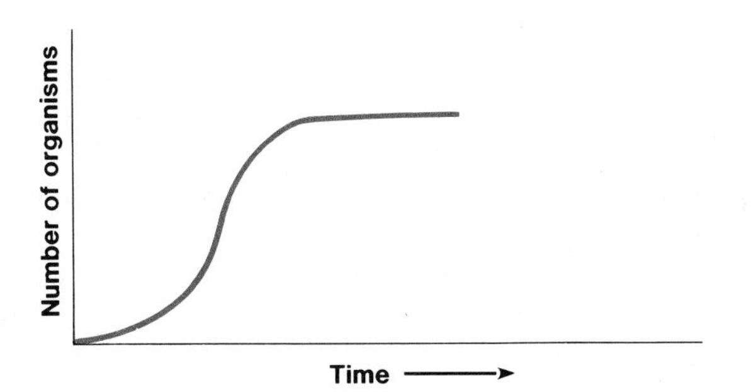

Figure 25.9 Population fluctuations in a bobwhite quail population. Open bars represent spring population sizes, and colored bars represent fall population sizes. Note that quail populations increase from spring to fall and decrease from fall to spring. But spring populations are relatively constant from year to year, reflecting the wintertime carrying capacity of the habitat.

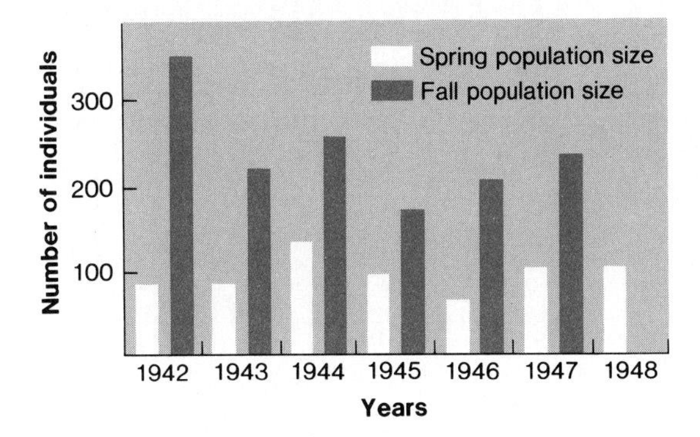

There have been many instances of exponential growth in real-life populations. These have occurred, primarily, when individuals of a species have been introduced into areas or habitats (or laboratory flasks) where the factors that normally limit population growth are not affecting the population. Two examples of very rapid population growth are the starling and the house mouse, whose populations increased dramatically following their introduction into the United States. Their populations grew exponentially at a rate near, but not quite equal to, their biotic potential, because even when conditions are very good, environments are seldom totally nonlimiting.

As a population that initially grows in exponential fashion increases, resources have to be shared by an increasing number of individuals. Eventually the environment becomes limiting. The population then increases at a rate less than that predicted by its biotic potential. The environment exerts its depressing effect on animal populations by influencing such things as the number of times that animals reproduce per year, the number of young produced each time, the age at which reproduction first occurs, and the length of time during which individuals are fertile. As a group, the factors that serve to depress population growth are known as **environmental resistance.**

When we plot population size against time for a population in a limiting environment, the resulting curve is S-shaped and is called a **logistic curve** (figure 25.8). It is a modification of the exponential growth curve and includes a factor to account for eventual limitation by the environment.

Note that populations whose growth is described by an S-shaped curve eventually reach a time when the population becomes stationary (remains the same). Population size at this point represents the maximum number of organisms a particular environment can support on a continuing basis and is known as the **carrying capacity.**

Carrying capacity varies from year to year. During favorable periods, when resources such as space or food may be more than adequate, a population may increase its size. In other periods, the environment may be less favorable, and the population decreases in size. As a result, populations tend to fluctuate from year to year (figure 25.9). The average of the series of ups and downs that occur from year to year represents the carrying capacity.

In some cases, when a population exceeds a resource such as food, environmental resistance is applied suddenly and the population undergoes a sharp decline. (Such population growth curves are sometimes described as J-shaped.) An example is the reindeer herd introduced on St. Matthews Island in Alaska. The population experienced a rapid expansion, and then declined just as quickly as the food resources became depleted (figure 25.10).

Age Structure

So far we have seen that we can make predictions about changes in population size by analyzing population growth curves. We can also make predictions on the basis of the **age structure** of a population. Age structure refers to the relative proportions of individuals that belong to each of several age categories.

Figure 25.10 The St. Matthews Island reindeer herd grew exponentially for several decades and then underwent a sharp decline as a result of overgrazing the available range.

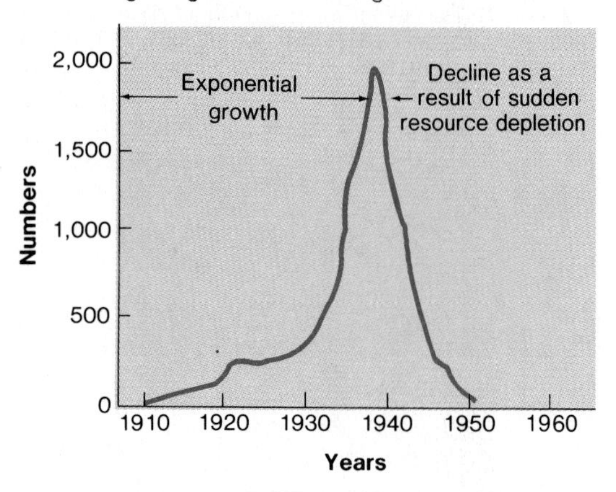

We can visualize the age structure of a population if we plot the proportions of individuals in each age class as a bar graph. This produces an age pyramid (figure 25.11). The shape of this pyramid can be used to make predictions about future changes in population size.

We can divide all age classes in a population into three main groups: prereproductive, reproductive, and postreproductive. The proportions in each are a function of the birth rate and the death rate. If the birth rate is high and mortality is low, most of the individuals in the population will fall in the prereproductive and early reproductive classes (see figure 25.11). This results in a pyramid with a broad base and a narrow top—indicating very small old age or postreproductive age classes. Because large numbers of prereproductive individuals will later be entering the reproductive age classes, such pyramids suggest that the population will grow rapidly. If the pyramid shows that the numbers of individuals in each of the age classes are about the same, except for the oldest age classes, the population probably will not experience significant growth. Finally, if the number of individuals in the prereproductive age classes is small and most of the individuals are in the midreproductive and postreproductive age classes, the population will probably decline.

Population Regulation

Monitoring of population size in some animals, such as the meadow vole, cottontail rabbit, or robin, shows that in some years the species are quite abundant and in other years relatively scarce. Over a long period of time, however, the average size of the population changes little (see figure 25.9). In general, population size is a result of the interaction between biotic potential and environmental resistance, but what are the actual factors that regulate population size?

This question has intrigued ecologists for years. Many population ecologists have argued that factors within the population related to density are of primary importance in regulating population numbers. They argue that if the population goes much over the carrying capacity, certain **density-dependent** mechanisms operate to decrease the reproduction rate or increase mortality, thereby slowing the population's growth. If the population falls below the carrying capacity, density-dependent factors function to decrease mortality and increase the reproduction rate. Other population ecologists, however, have argued that populations are controlled or influenced by mechanisms outside of the population that do not necessarily relate to population density. These **density-independent** mechanisms exert their influence in a manner that is independent of the size of the population. Today most population ecologists agree that the fluctuations of populations of most organisms result from the interaction of both types of mechanisms.

Examples of density-dependent regulatory forces include intraspecific (within-a-species) competition, stress, and dispersal. Weather and human activity are examples of density-independent factors.

Strategies for Growth

Still another way for us to understand population growth in a particular species is to examine growth characteristics in terms of the kind of environment the species inhabits. Environments range from being unoccupied, unstable, and unpredictable to occupied, predictable, and stable. Species that typically occupy the first kind of environment are said to be **opportunistic species,** while those in the second are known as **equilibrium species.**

Ragweeds in gardens and along roadsides and tent caterpillars in cherry and apple trees are opportunistic species. Opportunistic species usually are relatively short-lived, produce a large number of offspring usually in a single reproductive effort, and expend minimal energy on each individual produced. Among opportunistic animal species, for example, the eggs or young receive little or no parental care. Opportunistic species often possess efficient dispersal mechanisms that enable them to reach and exploit new food sources or newly available habitats that are relatively free of competitors. The strategy of opportunistic species is to produce large numbers of offspring, a few of which quickly disperse to colonize "open" habitats.

Equilibrium species, such as oak trees and whales, are relatively large, long-lived, and slow to mature. They produce a small number of well-developed offspring and expend considerable energy on each individual produced. Among

Figure 25.11 Types of human age pyramids. Human age pyramids are divided into dependency periods, rather than reproductive periods. Dependency periods, an economic rather than biological division, have more meaning for human populations. The divisions are 0–20, young dependents; 21–64, working periods; 65+, retired dependency periods. (*a*) The age pyramid for India (1970) is one of a rapidly expanding population. Note the broad base of young that will enter the reproductive age class (15–45). (*b*) Age pyramid for the United States (1970). The pyramid shows a constricted shape. The youngest period or class is no longer numerically the largest. This type of age structure reflects declining fertility. (*c*) Age pyramid for Sweden (1970). The population is nearly equally distributed over all age classes. This pyramid is typical of a population that is either aging and possibly declining or that has achieved zero population growth. (*d*) Age pyramid for a coal mining region in Appalachia (1960). In this pyramid the middle is pinched, reflecting a heavy emigration of younger age classes as they enter the economically active period. This emigration resulted in a high ratio of dependent young and old to the economically active. It also reflects a declining population.

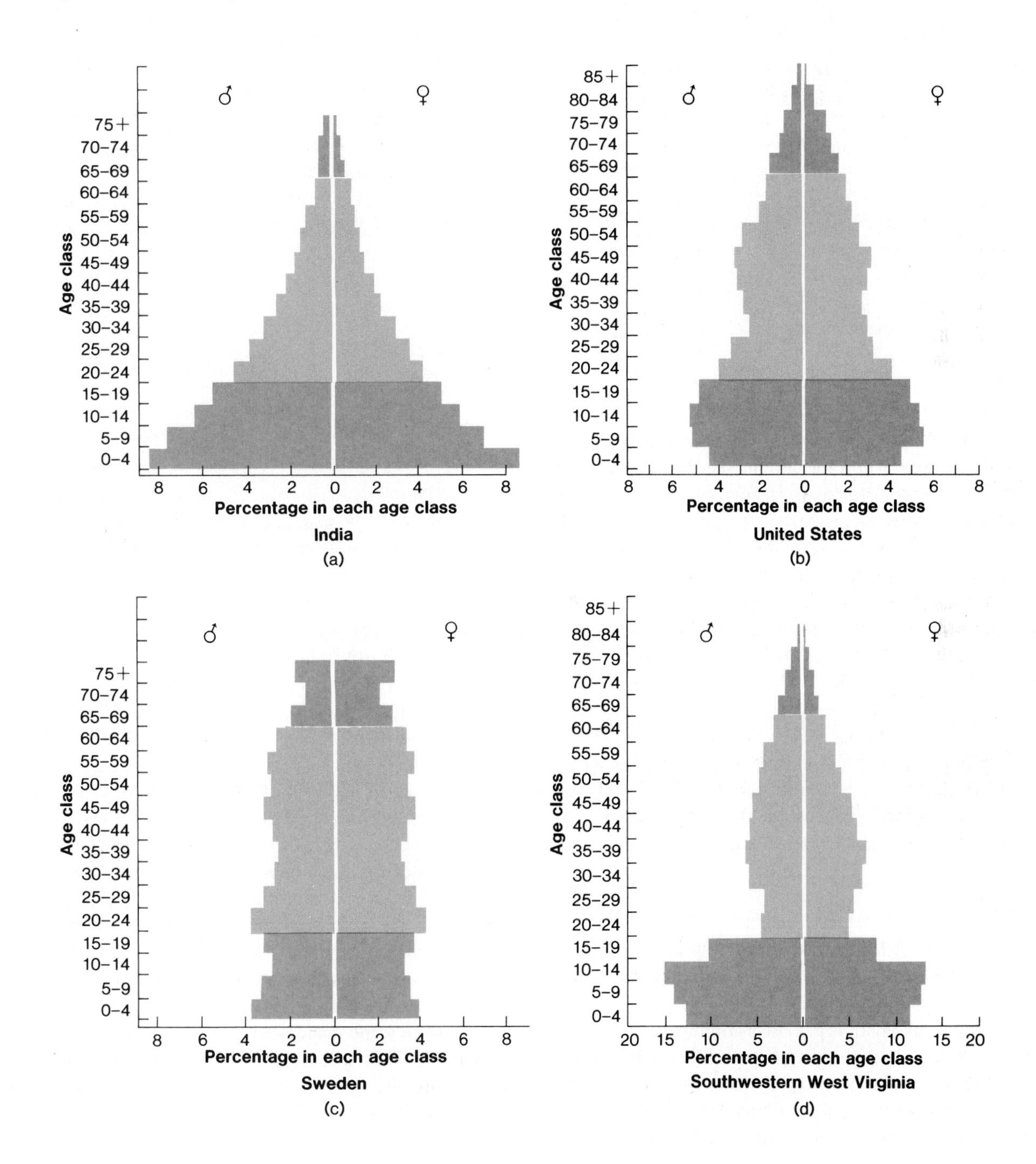

Figure 25.12 Competition between two species of *Paramecium*. When grown alone in pure culture, *Paramecium caudatum* and *Paramecium aurelia* exhibit S-shaped growth. When the two species are grown together in mixed culture, *P. aurelia* is the better competitor and its population increases. *P. caudatum* cannot grow in the presence of a growing population of *P. aurelia*, and it dies out.

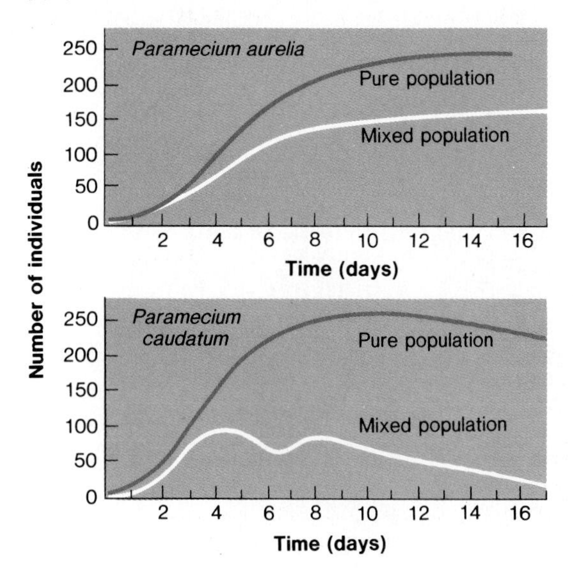

animals of equilibrium species, the young receive parental care. In plants of equilibrium species, the seeds contain a large amount of stored energy to ensure a vigorous start for seedlings. Equilibrium species are strong competitors and once established can dominate or exclude opportunistic species. They are habitat specialists rather than colonizers, and their population growth curves are logistic (S-shaped). Such populations are mostly stable and are at or near the carrying capacity of the environment.

Interactions with Individuals of Other Species

As an individual organism exists in its own particular environment, it interacts with individuals of other species as well as with individuals of its own species. These interactions include competition, predation, parasitism, and mutualism.

Competition

Competition among individuals of different species is known as **interspecific competition.** (Recall that competition within a species is intraspecific competition.)

The Russian ecologist G. F. Gause conducted a series of laboratory experiments with several species of protozoans in the genus *Paramecium.* These experiments helped to clarify some of the relationships that exist between competing species. The two species *P. caudatum* and *P. aurelia* have similar food requirements and therefore are strong competitors.

Figure 25.13 Resource partitioning by three species of annual plants in an abandoned field one year after cultivation ended. Each exploits a different part of the soil resource. Bristly foxtail has a shallow fibrous root system that exploits a variable supply of moisture. Indian mallow has a sparsely branched taproot extending to intermediate depths, where moisture is adequate during the early part of the growing season. Smartweed has a taproot that is moderately branched in the upper soil layer and develops mostly below the rooting zone of the other species, where the plant has a continuous supply of moisture.

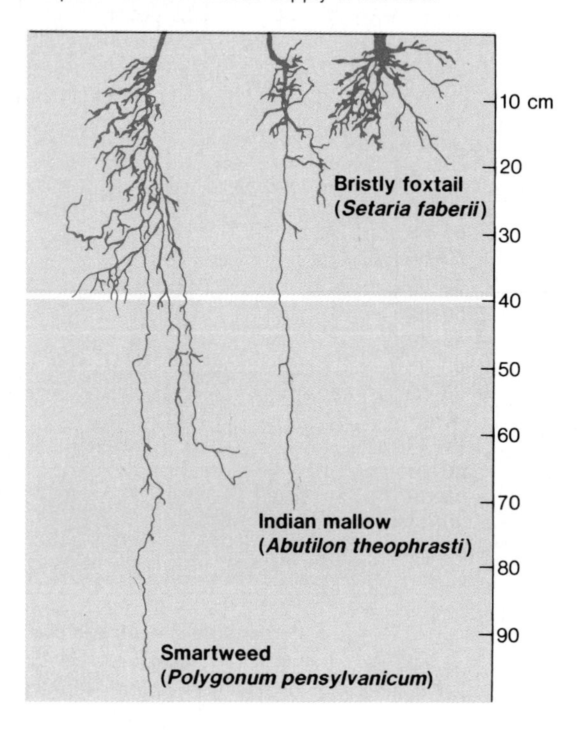

When Gause grew them separately in culture, both *P. caudatum* and *P. aurelia* increased and eventually reached the carrying capacity of the culture medium. When the two species were grown together, *P. aurelia* eventually replaced *P. caudatum* (figure 25.12).

In another experiment Gause raised *P. bursaria* (another *Paramecium* species) with *P. aurelia.* Although *P. aurelia* also outcompetes *P. bursaria, P. bursaria* was able to survive because it occupied the lower part of the culture medium, where *P. aurelia* did not occur. In this case, coexistence was possible because *P. bursaria* occupied a kind of spatial refuge.

The outcomes of Gause's experiments led to the formation of **Gause's principle,** which states that two species with exactly the same ecological requirements cannot coexist indefinitely. This idea is also known as the **competitive exclusion principle.**

You can understand, then, that coexisting species necessarily have ecological requirements that are different in some way. That is not to say that coexisting species never

Figure 25.14 Resource partitioning among five species of coexisting warblers. The time each species spent in various portions of spruce trees was determined. Each species spent more than half its time in the stippled gray zones.

compete for the same resource. They often do, and competition becomes especially sharp when that resource is in short supply. Competition for a single resource sometimes becomes especially intense when a species that is not native is introduced into a habitat. For example, the starling, which was introduced to North America, competes strongly with bluebirds and flickers for nesting sites. Such competition occurs much less often among coexisting native species. Over time, native species have evolved means of partitioning or dividing up resources. For example, they might feed in different areas, utilize different food types, or occupy different parts of the habitat (figure 25.13).

Robert MacArthur demonstrated how five species of North American warblers partition resources in their northern spruce forest habitat (figure 25.14). He divided spruce trees into zones and recorded the length of time each species spent in the different zones. In this way he was able to determine where each species did most of its feeding. He discovered that different species used different parts of the trees, although some overlap did occur. This partitioning of resources permitted all five species to feed in the same forest without seriously competing with each other.

Figure 25.15 A predator is any organism that feeds on another organism. Examples of some predators are (*a*) a barn owl feeding on a lizard and (*b*) a mountain lion burying its kill.

(a)

(b)

Figure 25.16 Orange and black stripes break up the body outline of this tiger, making it more difficult to locate in the grass. This is an example of disruptive coloration.

Predation

Predation is another way in which individuals of one species interact with individuals of another species. All organisms must extract energy from their environments. Because for many organisms, available energy occurs in the form of other organisms, energy acquisition involves a series of interactions between prey organisms and their predators. We can define predation as one living organism feeding on another, such as a cougar feeding on a deer or a robin feeding on an earthworm (figure 25.15).

While predators have evolved strategies to secure a maximum amount of food with a minimal expenditure of energy, prey organisms have evolved strategies to escape predation. Natural selection operates constantly to refine these strategies. Such refinements are coupled; that is, as predators' capture efficiency increases, there is a corresponding increase in the ability of prey to escape capture. If this were not the case, prey would be hunted to extinction, or predators would be eliminated by starvation. Recall that when two species interact so closely that evolutionary changes in one influence the direction of evolutionary changes in the other, we call the process coevolution (see page 507).

The hunting abilities of predators are constantly tested. Individual predators that are slow, lack strength, or fail to react quickly to prey movements are eliminated from the population (usually through starvation). Through natural selection predators have acquired such "tools" as claws, talons, sharp beaks, and shearing teeth that enable them to capture, hold, kill, and process their prey. Some predators have improved their hunting success by acquiring coloration that permits them to blend into their surroundings. Other predator species show **disruptive coloration** (figure 25.16). Lines and marks tend to break up the outline of the animal and make it less visible to prey.

Color patterns in some prey species also are important in enabling them to escape detection by predators. Some prey species closely resemble background objects. For example, some katydids look much like green leaves (figure 25.17), the walking stick resembles a twig, and some moths may look like bark on a tree.

Some prey species avoid predation because there is something about them that is noxious to predators. An example is the monarch butterfly, which lays its eggs on the leaves of the milkweed, a plant rich in cardiac glycosides. Cardiac glycosides are powerful poisons used in minute

Figure 25.17 The broad-winged katydid, an insect of the treetops, is an example of cryptic coloration and shape. Both its green color and its body shape and posture give the katydid the appearance of a leaf.

quantities to treat heart disease in humans. They also activate nerve centers that cause vomiting in many animals. The green caterpillar of the monarch consumes the leaves of milkweed and is able to incorporate the glycosides into its own tissue, where the glycosides remain through the monarch's pupal stage and into the adult stage. Thus, the monarch gains a chemical defense that causes it to be vomited and rejected by a predator. Of course, some individual monarchs are going to be sacrificed during the time predators learn to avoid the prey, but the monarch butterfly population as a whole escapes any significant predation.

Unpalatable prey are often highly colored; bright coloration warns potential predators that such organisms possess noxious qualities and should be avoided. Once a predator has experienced such prey, it associates the bitter taste or unpleasant experience with the color pattern and avoids the animal. This is advantageous both to the predator and to the prey. The prey is attacked less frequently, and predators do not waste time catching unpalatable prey.

Another strategy for escaping predation is to closely resemble a species that is protected by some defense mechanism. Such mimics, known as **Batesian mimics,** gain protection by looking and behaving like their models. An example of a mimic is the viceroy butterfly, a palatable species that feeds on willows and other nonpoisonous plants. The viceroy closely resembles the monarch butterfly, which as we said earlier, is unpalatable to predators (figure 25.18).

In another type of mimicry, **Müllerian mimicry,** two or more protected species mimic each other. Such mimicry gives survival advantage to each species involved because the predator soon learns to associate a particular pattern with distastefulness or danger without trying all of the species involved. Such mimicry is found among the wasps with yellow and black bands, many of which have potent stings.

Figure 25.18 Batesian mimicry. The unpalatable monarch butterfly (above) and its mimic, the viceroy butterfly (below).

Figure 25.19 Thorns of an African acacia tree. Many plants have adaptations such as thorns or spines that discourage grazing herbivores.

Just as prey animals have developed defenses against their predators, plants have evolved defenses against herbivores (animals that eat plants). Many plants possess certain chemical substances, such as tannin, that affect the herbivores that eat them. When tannin is combined with plant protein, the protein cannot be digested by the herbivore. Tannin also imparts a bitter taste to leaves and twigs and in this way discourages herbivores from consuming them. Other plants possess toxic chemicals, such as nicotine, that disrupt the normal function of the nervous system. In addition, a number of plants develop spines and thorns that discourage grazing animals (figure 25.19). Thistles in pastures are avoided by cattle, and spiny cactuses are avoided by desert herbivores. Tough seed coats of the seeds of many plants discourage animals from eating them. Other plants produce such an abundance of seeds over a wide area that seed eaters are unable to utilize the whole crop and some seeds will be left to grow.

Parasitism

Parasitism is another kind of interaction that exists between individuals of different species. Parasitism is one of several kinds of symbiotic relationships. (In general, relationships in which one organism lives in intimate association with another are known as **symbiotic** relationships.) Parasites, like predators, gain energy for life from other organisms. They are different from predators, however, in that they live in close association with a **host** for extended periods. Parasites are smaller than their hosts, they live in or on their hosts, and they utilize only a portion of the host's energy content. Parasites usually do not kill their hosts. To do so would mean that the parasite itself could no longer survive.

Figure 25.20 An example of an ectoparasite. A blood-sucking leech on a human arm.

Some parasites, known as **ectoparasites,** live on the body surface of their hosts (figure 25.20). Ectoparasites, such as lice, fleas, and ticks, usually possess some kind of specialized organ by which they remain attached to their hosts. On the other hand, **endoparasites** live within the bodies of their hosts. Some parasites utilize one host throughout their life cycle, while others require more than one host.

Mutualism

Another important symbiotic relationship between two organisms is **mutualism.** Mutualism differs from parasitism in that in this case, both organisms benefit from the association. Some mutualistic relationships are so close that neither organism can survive without the other. This is **obligatory** mutualism.

An example of obligatory mutualism is the relationship between bullthorn acacias and a certain species of ants in Central America. The large hornlike thorns of this acacia have a tough, woody covering and a pithy interior. A queen ant makes room for her brood by boring a hole into the base of a thorn and cleaning out the pith. As the colony grows, an increasing number of thorns on the plant become filled with ants. Eventually, the ant colony grows quite large, numbering in the tens of thousands of individuals. In addition to finding shelter in the acacia, the ants also obtain food from nectar found at the bases and tips of the leaves. The ants bite and sting any animals that attempt to feed on the acacia. In this way, the ants and the acacia tree are of mutual benefit to each other. The relationship is so close that apparently neither ant nor acacia can survive without the other.

Summary

The science of ecology deals with the relationships that exist between an organism and its environment. In addition to physical factors we have examined previously, two additional physical factors that affect an organism in its environment are temperature and water.

Organisms respond to environmental temperature changes in many ways. Animals whose body temperatures vary and are influenced by environmental temperature change are poikilothermic; those whose body core temperatures remain constant are homeothermic.

Water balance is vital for all living organisms. Animals living in different environments have different strategies for maintaining water balance.

A population is a group of interbreeding organisms occupying a given area at a given time. Under ideal conditions populations grow geometrically. However, as environmental resources become limiting, growth eventually slows down. This produces an S-shaped growth curve. In stable environments population size fluctuates around the carrying capacity of the environment. In unstable environments, a population may grow to exceed the carrying capacity and then sharply decline as some critical resource becomes depleted.

The size of a population can be influenced by forces from within and also from outside. A major influence from within is competition.

Species that are well adapted to occupy stable environments are known as equilibrium species, while those adapted to produce large numbers of offspring in unstable, unpredictable environments are known as opportunistic species.

Several types of interactions occur between populations of different species. One of these is interspecific competition. Coexistence is impossible for two species that have exactly the same ecological requirements.

Another interaction between species is predation. Predators and their prey have coevolved. As predators evolved ways of improving hunting efficiency, prey species evolved ways of avoiding predation.

Parasitism exists when one organism derives its nourishment from a host organism, usually without killing the host. Successful parasitism is another example of coevolution.

Mutualism is an association in which both organisms benefit. Some mutualistic organisms are obligatory—neither organism can survive without the other.

Questions for Review

1. Distinguish between the ecological concepts of niche and habitat.
2. What is the fundamental difference between poikilotherms and homeotherms? Give an example of behavioral thermoregulation in poikilotherms.
3. Why do marine bony fish drink so much seawater? What problem does this create for the fish?
4. What is a population and what are some of its characteristics?
5. When can a population grow exponentially?
6. What is carrying capacity? If resources change in the environment, how does this affect the carrying capacity of a population?
7. Distinguish between density-dependent and density-independent mechanisms in influencing population size.
8. Distinguish between opportunistic and equilibrium species.
9. What is interspecific competition, and how does it relate to the competitive exclusion principle?
10. What is parasitism? Mutualism?

Questions for Analysis and Discussion

1. Homeothermic animals must expend a great deal of energy to maintain a constant body core temperature. What advantages do they gain as a result of meeting this large energy cost?
2. Explain this statement: Actual population size represents a compromise between biotic potential and environmental resistance.
3. Explain how you would show someone who had never studied biology the importance of age structures in predicting future population sizes.

Suggested Readings

Books

Kormondy, E. J. 1976. *Concepts of ecology,* 2d ed. Englewood Cliffs, N.J.: Prentice-Hall.

McNaughton, S. J., and Wolf, L. L. 1979. *General ecology,* 2d ed. New York: Holt, Rinehart & Winston.

Odum, E. P. 1983. *Basic ecology.* Philadelphia: Saunders.

Smith, R. L. 1980. *Ecology and field biology,* 3d ed. New York: Harper and Row.

Articles

Bekoff, M., and Wells, M. C. April 1980. The social ecology of coyotes. *Scientific American* (offprint 726).

Bell, R. H. V. July 1971. A grazing ecosystem of the Serengeti. *Scientific American* (offprint 1228).

Bergerud, A. T. December 1983. Prey switching in a simple ecosystem. *Scientific American.*

de Beer, G. 1978. Adaptation. *Carolina Biology Readers* no. 22. Burlington, N.C.: Carolina Biological Supply Co.

Myers, J. H., and Krebs, C. J. June 1974. Population cycles in rodents. *Scientific American* (offprint 1296).

Ecology
Ecological Systems and Environmental Problems

26

Chapter Outline

Chapter Concepts

1. A community consists of interacting populations of plants and animals living in the same area at the same time. It is characterized by structure, dominance, and species diversity.
2. Succession is a change in the structure and species composition of a community through time.
3. Ecosystems have both structure and function. Structure involves physical nature and species composition. Function involves energy flow and nutrient cycling.
4. The sun is the source of energy that enters an ecosystem. Energy flows one way through an ecosystem. Energy and nutrients are passed along through the food chain from one trophic or feeding level to another.
5. Aquatic ecosystems are categorized on the basis of how much dissolved salt they contain. Terrestrial ecosystems are classified on the basis of the type of climax communities they support.
6. Biomes are characterized by sets of physical factors, including temperature, precipitation, and growing season length, and by the types of organisms that are adapted to live in them.
7. Worldwide population growth, pollution, and depletion of the earth's natural resources have created serious ecological problems that affect all life on earth.

The hot African sun blazes down in sharp contrast to the calm, cool stretch of blue-green ocean that lies ahead. The three Kenyan boys piloting the boat out into deeper water hand you snorkel, mask, and fins. Looking down into the clear water, you see the bottom changing—green vegetation and sand abruptly merge with coral. Suddenly the boat stops. Adjusting your mask, you slip over the edge of the boat and dive.

The first experience exploring a coral reef is an unforgettable one (figure 26.1). Schools of small, brilliantly colored fish are everywhere, swimming and feeding in all directions. The variety seems endless—parrot fish, butterfly fish, file fish—all sparkling jewels darting here and there between the stony fronds of coral in the clear sunlit waters. On the bottom small crabs scurry along, easily outdistancing the slower-moving starfish and snails. And further out to sea, where the reef begins to drop off into deeper water, a ray, body flattened and streamlined, cuts effortlessly through the warm blue depths.

Hanging weightlessly in these waters off the coast of Kenya, you are surrounded in all directions by a vast array of different populations of organisms—hundreds of species of plants and animals. Together these organisms form an ecological community, and they live their lives as part of the ocean (marine) ecosystem. In this chapter, we will consider the kinds of interactions that occur among species in communities and how these interactions shape the structure of ecological systems.

Figure 26.1 A coral reef with some of its brightly colored inhabitants.

Communities

An ecological **community** consists of all the populations of different species that inhabit a given area at a given time. Various communities are organized in different ways. These different patterns of organization make up the structure of communities, and as we shall see, community structure changes with time through the process of ecological succession. Let us have a look at community structure and succession.

Community Structure

A careful study of two communities reveals that there are many differences in community structure. A grassy field, for example, is obviously quite different from a forest. Even a casual observer can point out that the field is dominated by grasses while the forest is dominated by trees. The animal life of the grassland includes relatively fewer species than are found in the forest, and these species spend most of their time close to the ground surface. For example, many bird species are ground nesters and feed on the ground or in vegetation close to the ground. Similarly, grassland mammals live on or below the ground. In the forest, where there are relatively more species, some birds and mammals live in and on the ground, while others live in the shrubby plants below the trees, in the lower parts of the trees, or in the treetops.

An important difference, then, has to do with the way the two communities are stratified (organized in layers). The grassland essentially has two strata: a layer of litter on the ground and a layer of vegetation growing above the ground. A well-developed forest may have as many as five or six strata, including a layer of litter at the soil surface, a layer of herbaceous plants just above the ground surface, a layer of woody shrubs, a layer of small trees, and a canopy layer made up of the tops of the trees. Stratification in any habitat is important because as the number of strata or "floors" increases, so does the number of habitats available to organisms.

Dominance and Diversity

One of the concepts that is useful in making distinctions between one community and another is that of **dominant species.** A dominant species exerts more control over the character of the community than do the other species. Dominant species are usually characterized by greater numbers or greater **biomass** (living weight) than other species in the community. For example, in the northern evergreen forest,

Figure 26.2 Succession on a sand dune, an example of primary succession. The beach grass shown here is a pioneer species. Shrubs and trees replace the grass on older dunes after the dunes become stabilized and some organic matter accumulates.

spruce and fir dominate the community. They severely reduce the amount of light that reaches the forest floor, produce a litter that is not easily decomposable, and create acidic conditions that affect the character of the soil. Only those species of plants and animals able to live under the conditions created by spruce and fir can inhabit those forests.

Another important community concept is that of **species diversity.** Species diversity as calculated by ecologists considers both the number of species present (**richness**) and the relative abundance of each species (**evenness**). The greater the number of species and the more evenly the individuals are distributed among the species, the greater the species diversity.

Species dominance and species diversity are related. Species diversity is low in communities in which a single species expresses dominance. In communities in which no one species is truly dominant, the total number of individuals is equally distributed among all species, and consequently species diversity is high.

Succession

Natural communities are constantly changing. Old fields become forests. Ponds fill in to become meadows. In mature forests, trees die and are replaced by others. As the vegetational components of a community change, so do its animal components. Communities undergo a predictable series of changes in structure and function that eventually result in the emergence of a relatively stable community known as a **climax community.** We call the process by which a climax community is achieved **succession.**

Figure 26.3 Secondary succession on an old field site. The shrubs and invading trees indicate that this field is moving toward a forest community climax.

Ecologists distinguish between primary succession and secondary succession. **Primary succession** in terrestrial (land) communities takes place on areas that are devoid of soil and have not previously supported a community. Primary succession occurs on sand dunes (figure 26.2), lava flows, and rock outcrops. **Secondary succession** takes place on areas that were previously occupied and where soil is present. Secondary succession occurs on such distributed areas as abandoned agricultural fields (figure 26.3), burned-over land, or land where vegetation has been cut and removed.

Primary Succession

An isolated sandy beach by the ocean often provides a firsthand look at primary succession. The vegetation changes as you walk into the dunes beyond the beach. Surface temperatures on the sand dunes during the summer are high, and moisture is in short supply. These conditions make it difficult for organisms to colonize the dunes. One of the most successful pioneering plants on the dunes is marram, or beach grass. Once this grass becomes established and stabilizes the dunes against movement by the wind, associated plants invade the area. These are followed by mat-forming shrubs, which in turn may be succeeded by trees—first pines and then oaks. As soil fertility and moisture availability increase, oaks may be replaced by maples, which are the dominant plants in a climax community.

Figure 26.4 The presence of decomposer fungi is confirmed by the development of mushrooms, which are the fruiting bodies of the fungus.

Secondary Succession

We can use a study of old-field succession in North Carolina as an example of secondary succession. When cornfields are abandoned after harvest in the fall, annual crabgrass, an opportunistic species, takes over. The seeds of another plant species, horseweed, also germinate and by early winter form clusters of growth in the disturbed soil. The following spring, horseweed gets a head start on crabgrass and crowds it out. But during that summer another plant, the white aster, germinates and begins to crowd out the horseweed. Dying horseweed opens up sites for the establishment of broomsedge, a bunchgrass. Eventually, broomsedge dominates the field. Because broomsedge grows in clumps, open ground still exists between individual plants. These moist, lightly shaded, and plant-free spots provide an ideal place for pine seeds to germinate. If a seed source is near, a pine stand develops and shades out the broomsedge. But since pine seedlings cannot grow in the shade of the adult pine tree, the pines eventually die and are replaced by climax species of oaks and hickories, whose seedlings can grow in the shade.

The Ecosystem

All of the organisms in a given area, together with the abiotic (nonliving) components of their environment, comprise an **ecosystem.** Ecosystems have both structure and function. Ecosystem structure is determined by the components that make up the system, while function is determined by the manner in which these components interact.

Ecosystem Structure

All ecosystems possess both biotic (living) and abiotic structural components. Biotic components include all of the communities of organisms in the system, while abiotic components include such things as soil, water, light, inorganic nutrients, and weather.

We can categorize the biotic components of an ecosystem on the basis of how they obtain energy and nutrients. The **producer** organisms in any ecosystem are chiefly green plants that use photosynthesis to convert radiant energy from the sun into the chemical energy of carbohydrates, fats, and proteins. In terrestrial ecosystems the producers are predominately herbaceous and woody plants, while in freshwater and marine ecosystems the dominant producers are various species of algae. From a nutritional point of view, producer organisms are **autotrophic;** that is, they can synthesize organic compounds using only inorganic raw materials and energy from an external source (usually sunlight).

Consumer organisms obtain energy and nutrients by ingesting producer organisms or other consumer organisms. **Primary consumers,** or **herbivores,** eat the producer organisms, while **secondary consumers,** or **carnivores,** eat primary consumers. Both types of consumers are **heterotrophic.** They must obtain organic nutrients by ingesting other organisms.

Decomposer organisms obtain their energy and nutrients by breaking down dead organic matter (figure 26.4). These heterotrophic decomposer organisms are mostly bacteria and fungi. The important end result of decomposer activity is that inorganic nutrients, originally bound up in the tissues of organisms, are converted to simple forms that are usable, once again, by producer organisms. We will examine an example of this when we consider the nitrogen cycle (page 567).

Figure 26.5 A hypothetical diagram showing energy relationships in a grazing food chain. We will explain the relationships for the herbivore feeding (trophic) level. In this example, 1,200 kcal (20 percent of the 6,000 kcal available at the producer level) are consumed. Herbivores are only 10 percent efficient in incorporating this energy into tissue; that is, only 120 kcal of the 1,200 consumed are incorporated into herbivore tissue. Although not indicated in the diagram, the rest of the 1,200 kcal is used to support the metabolism of the herbivores (and is lost as the heat of respiration) or passes unabsorbed through the digestive tracts of the herbivores. The diagram also indicates that only 36 kcal or 30 percent of the 120 kcal available at the herbivore trophic level is consumed by organisms in the next trophic level (carnivores). The rest of the 120 kcal, that is, 84 kcal, becomes available to decomposers. It should be clear after examining the carnivore level as well that only a small part of the energy available at any level in a food chain becomes incorporated into organisms in the next higher feeding level.

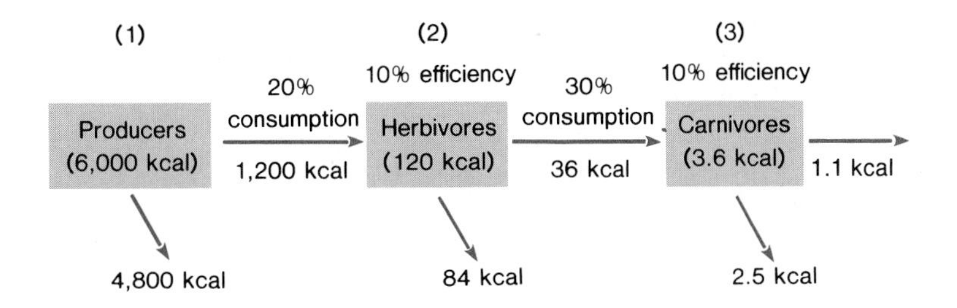

Energy Flow in Ecosystems

When a primary consumer eats a producer and, in turn, is eaten by a secondary consumer, we say that energy is flowing through the ecosystem. This energy flow is a one-way process, and as we shall shortly see, this means that ecosystems are unable to function unless there is a constant energy input from an external source.

Food Chains and Food Webs

One way to follow energy flow through an ecosystem is to identify the sequences of organisms through which the energy moves. Such a sequence of organisms is known as a **food chain.** A simple food chain involves a producer and a primary consumer (herbivore):

Sunlight → Plants → Meadow mouse

A longer chain would involve a secondary consumer (carnivore):

Sunlight → Plants → Meadow mouse → Weasel

A still longer food chain in the same ecosystem might be as follows:

Sunlight → Plants → Grasshopper → Meadowlark → Cooper's hawk (**tertiary consumer** or **top carnivore**)

Each step involves a transfer of energy from one feeding group, or link, to another.

Not all producer organisms are eaten by primary consumers and not all of the primary consumers are eaten by secondary consumers. In fact, most of the organisms at each feeding level die without being eaten and their bodies become part of the system's dead organic matter. Thus, a major portion of the energy incorporated at a given feeding level is not utilized by organisms of the next feeding level. An important characteristic of food chains is that there is less energy available at successively higher feeding levels.

Another significant portion of the energy that is consumed by organisms in the next higher feeding level is not digested and absorbed and thus passes through as feces. Of the portion that is absorbed, some is lost to the system as the heat of respiration. Heat is a by-product of the metabolic processes involved in the breakdown of food. Food energy that is absorbed but not lost during respiration is incorporated into the tissues of organisms.

On the average, in grazing food chains only about 10 percent of the energy ingested by the organisms at a particular feeding level is incorporated into the tissues of those organisms. This unavoidable inefficiency in energy transfer between feeding levels explains why food chains are relatively short—at the most four or five links—and why mice are more common than weasels, foxes, or hawks. This also explains why more people could be fed on grain than on meat from grain-consuming animals (figure 26.5).

A food chain represents just one part of energy flow through an ecosystem. But any ecosystem consists of numerous food chains, each linked to others to form complex **food webs** (figure 26.6).

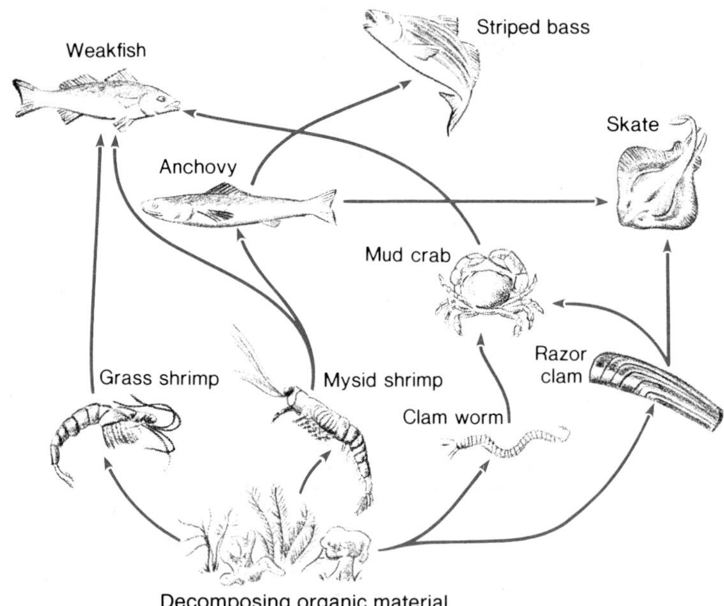

Figure 26.6 A food web, highly simplified, of a shallow coastal saltwater ecosystem. The shrimps, clam worm, and razor clam are primary consumers (herbivores). The mud crab and anchovy are secondary consumers (carnivores). The striped bass is a tertiary consumer (top carnivore). The weakfish and the skate function at both the secondary and tertiary consumer levels. The arrows show the direction of energy flow.

Figure 26.7 Diagrammatic representation of trophic levels in the ecosystem. Terms on the side indicate other names applied to these trophic levels.

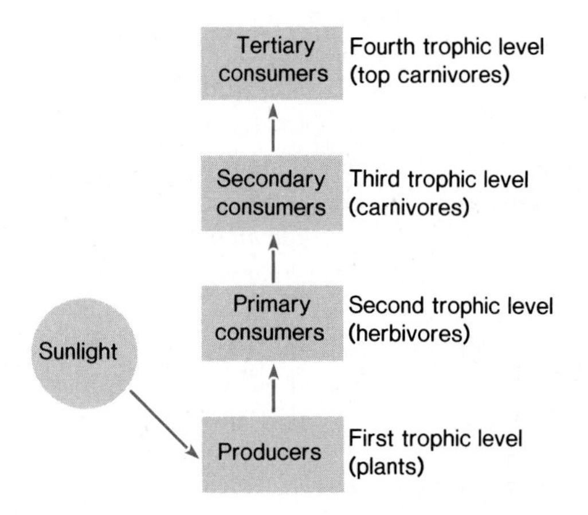

Trophic Levels and Pyramids

If we group all the organisms in each link of a food web according to their general source of nutrition, we can divide the food web into feeding or **trophic levels.**

Biologists have agreed to assign producer organisms to the first trophic level, primary consumers (herbivores) to the second trophic level, the secondary consumers (carnivores) to the third trophic level, and tertiary consumers (top carnivores) to the fourth trophic level (figure 26.7).

The trophic structure of an ecosystem can be summarized in the form of ecological pyramids (figure 26.8). The base of the pyramid represents the producer trophic level, the apex is the tertiary or some high level consumer, and other consumer trophic levels are in between.

There are three kinds of pyramids. One is based on the numbers of organisms at each trophic level and consequently is known as a **pyramid of numbers.** For most ecosystems, pyramids of numbers are right side up because numbers of organisms decrease at successively higher trophic levels. However, there are some ecological systems for which the pyramid of numbers is inverted. In a system involving a single tree and all of its insect predators, the part of the pyramid representing the producer trophic level (the tree) that is the base would be smaller than the part representing the consumer trophic level (the insects).

Figure 26.8 Ecological pyramids based on data for an experimental pond. (*a*) Pyramid of numbers (individuals/m^2). (*b*) Pyramid of biomass (dry g/m^2). (*c*) Pyramid of energy (dry mg/m^2/day).

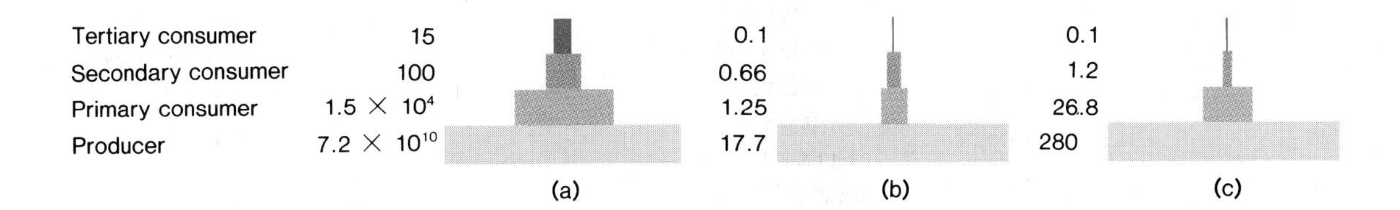

A second type of pyramid is the **pyramid of biomass.** Biomass is the weight of living material. Generally, the biomass of producers is much greater than the biomass of herbivores feeding on the producers, and the biomass of herbivores is much greater than the biomass of carnivores.

A third kind of pyramid is the **pyramid of energy.** The pyramid of energy is always upright because a given trophic level has a smaller energy content than does the trophic level immediately below it.

Biogeochemical Cycles

Energy, entering as sunlight and leaving as heat, is continually lost from an ecosystem. By contrast, inorganic nutrients—the chemical elements out of which organisms are made—continually cycle through the abiotic and biotic components of the ecosystem. Since the pathways by which inorganic nutrients circulate through ecosystems involve organisms and the "earth," they are known as **biogeochemical cycles.**

In general, inorganic nutrients begin moving through the various biotic components of an ecosystem by being incorporated into living tissue. For plants, sources of inorganic nutrients are the atmosphere and the soil. For animals, sources are plants and other animals. Recall that decomposer organisms return nutrients to the soil and atmosphere.

Some biogeochemical cycles are gaseous, while others are sedimentary. Elements whose main reservoirs exist in the atmosphere or ocean move through **gaseous cycles.** Four critical elements—oxygen, carbon, nitrogen, and hydrogen—are involved in gaseous cycles. These four elements constitute about 97 percent of living matter. The thirty-six or so other biologically important elements follow **sedimentary cycles** in which the main reservoir is the earth's crust. They initially become available as a result of rock weathering and are taken up by organisms in ionic form.

The Nitrogen Cycle

As an example of a gaseous cycle, let us examine the nitrogen cycle in some detail.

Nitrogen is an abundant element. Nitrogen gas (N_2) makes up about 78 percent of the atmosphere by volume. But vascular plants cannot incorporate N_2 into organic compounds. They absorb nitrogen from the soil mainly as nitrate (NO_3^-) or ammonium (NH_4^+) ions. Thus, these usable ions must be available in adequate quantities in the soil.

Let us trace the way nitrogen atoms cycle through ecosystems. Once NO_3^- or NH_4^+ is absorbed by plants, the plants—and the animals that eat the plants—can synthesize amino acids, proteins, and other nitrogen-containing compounds.

Some of these organic nitrogen compounds quickly return to the soil as fallen leaves or animal feces. Others return only when plants or animals die. Decay processes then break down organic compounds in several stages. In the last of these stages, which is called **ammonification,** soil microorganisms break down organic nitrogen compounds to release NH_4^+. Often, other bacteria oxidize NH_4^+ to produce NO_3^- in a process called **nitrification** (figure 26.9).

Thus, when decomposition is complete, nitrogen is once again available to green plants in the form of ammonium and nitrate ions. Not all of the nitrate produced is absorbed by green plants; some of it is converted to N_2 and N_2O by still other microorganisms in a process called **denitrification.** Nitrogen in the form of N_2 and N_2O is lost to the atmosphere.

Even though N_2 (nitrogen gas) is not directly usable by plants, there are some microorganisms that can convert N_2 from the atmosphere to NH_4^+. The NH_4^+ then can be oxidized to NO_3^- or be used directly by plants. The process by which this conversion takes place is known as **nitrogen fixation** and is critically important to ecosystems.

Figure 26.9 The nitrogen cycle.

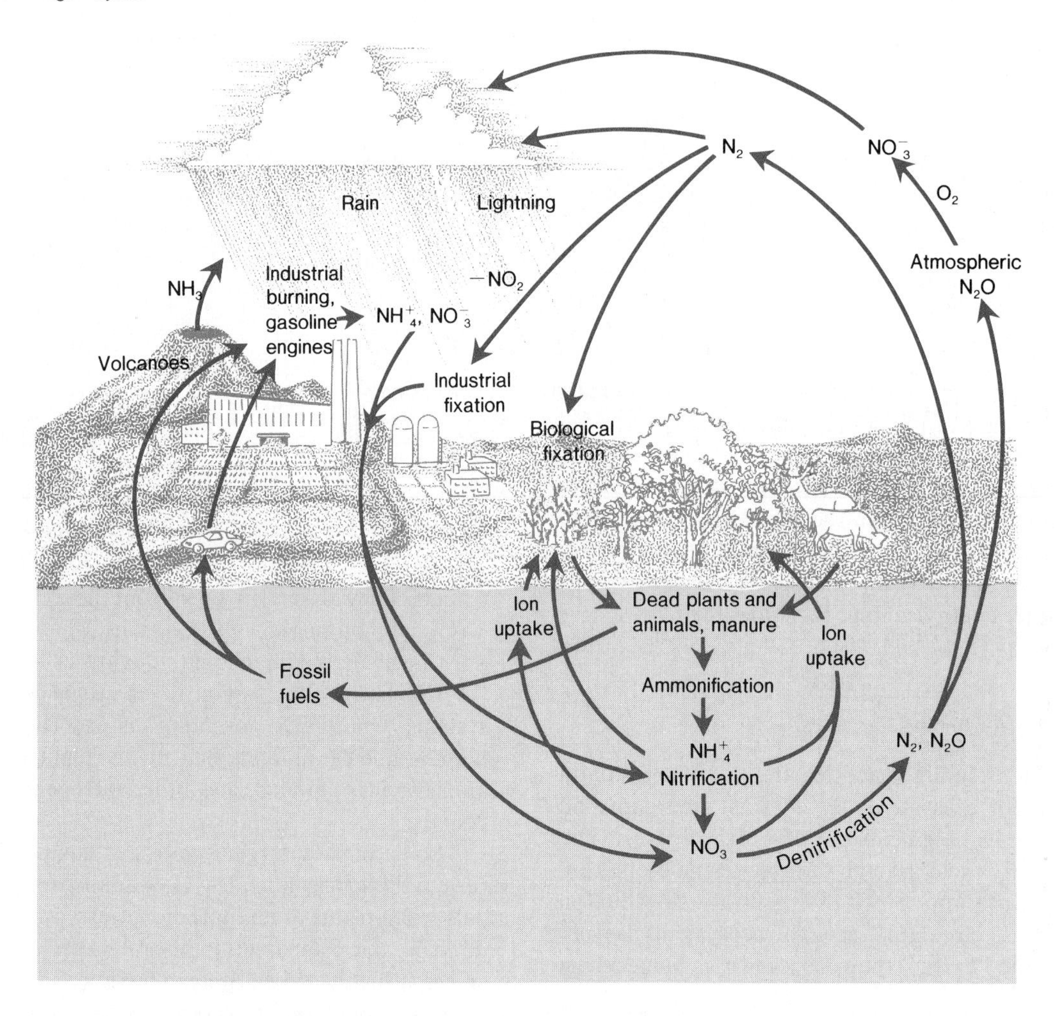

The Calcium Cycle

The calcium cycle is an example of a sedimentary cycle. It begins in the soil, where calcium is taken up by plants and deposited in leaves, twigs, stems, and trunks. Rain dripping from the foliage dissolves some of the calcium from the leaves and carries it back to the soil, where it is quickly taken up by the plants again. Herbivores obtain their supply of calcium from plants, and carnivorous animals obtain their calcium from herbivores and other animals that they consume. However, some animals have other means of obtaining calcium. For example, birds ingest calcium directly by picking up limestone particles from the soil and roadside. The death and decay of plants and animals return calcium to the soil where it can be taken up again by plants.

Major Ecosystem Types

Up to this point we have discussed ecosystems in a general way, paying particular attention to details of structure and function common to all ecosystems. Now we shall consider specific examples of the earth's ecosystems. The earth supports many types of ecosystems that collectively make up the **biosphere.** The biosphere in turn functions as one huge ecosystem.

Ecosystems are either water-based (**aquatic**) or land-based (**terrestrial**). Aquatic ecosystems may be fresh water or salt water (marine). Freshwater ecosystems include streams, rivers, lakes, swamps, marshes, and bogs. Marine ecosystems include salt marshes, estuaries, sandy and rocky ocean shores, shallow seas, and open oceans.

Figure 26.10 A stream, an example of a lotic ecosystem.

Figure 26.11 A lake, an example of a lentic ecosystem. Lakes that receive rich nutrient inputs from surrounding drainage areas are eutrophic, while lakes in nutrient poor areas are oligotrophic.

Freshwater Ecosystems

Freshwater ecosystems may be **lotic** or **lentic.** Lotic systems are characterized by flowing water and include brooks, streams, and rivers (figure 26.10). Lotic systems are inhabited by organisms well adapted to maintaining their position in flowing water. For example, fish in mountain streams, such as trout, have a streamlined shape. Algae grow tightly attached to the surface of rocks and may be encased in a slippery, gelatinous sheath.

The most important primary producers of lotic systems are algae, but in most streams the major energy source is organic matter carried in from surrounding terrestrial ecosystems.

Lakes and ponds are lentic systems and are characterized by still water (figure 26.11). Vertical stratification of temperature, light, and oxygen often occurs in lentic systems. For example, temperature strongly affects the structure of lakes and ponds in north temperate regions. During the summer, intense solar radiation is absorbed by surface waters so that a layer of warmer, lighter water "floats" on top of heavier, cooler water. The upper, well-lighted layer, where most of the phytoplankton (a collection of small photosynthesizing organisms) grows, is well aerated because of oxygen production by plants and mixing by the wind. The lower waters, however, may be deficient in oxygen because of bacterial decomposition. Because sediments and organic matter accumulate on the bottom, the deeper waters are relatively high in nutrients, while the upper waters become depleted of nutrients because of uptake by phytoplankton during the growing season.

Horizontal zonation is another characteristic of lentic systems. Horizontal zones in lakes and ponds include open water dominated by phytoplankton and a **littoral** zone, near the shore, that is dominated by floating and emergent vegetation rooted in the bottom. Characteristic bottom-dwelling (**benthic**) organisms occur in the "ooze" beneath each of these zones.

We can classify lakes and ponds by their nutrient status. **Oligotrophic** (nutrient-poor) lakes are characterized by low organic matter, low nutrient release from bottom sediments, and low productivity of phytoplankton. Such lakes are usually situated near nutrient-poor areas. On the other hand, **eutrophic** (nutrient-rich) lakes are characterized by high organic matter, high release of nutrients from bottom sediments, high productivity of phytoplankton, and often a well-developed littoral zone. Such lakes are usually situated where either naturally nutrient-rich water drains into them or where nutrients enter from agricultural areas or human settlements. Oligotrophic lakes can become eutrophic through large inputs of nutrients. This is usually the fate of oligotrophic lakes when they become polluted by sewage and other wastes.

Figure 26.12 A marsh is a wet grassland. Emergent vegetation includes species of sedges, cattails, and bulrushes. To be a suitable habitat for wildlife, the marsh should consist of patches of emergent vegetation interspersed with areas of open water.

Figure 26.13 A Florida swamp. A swamp is a wooded wetland.

Figure 26.14 A bog is a small body of water dominated by a wet mat of *Sphagnum* moss. Bogs often develop around lakes, but sometimes they develop on upland sites, in which case the moss holds water to create a perched water table.

Figure 26.15 A long stretch of sandy beach may appear to be a lifeless place, but its teeming life is buried beneath the sand and becomes active only at high tide. Also, during low tide, birds such as sandpipers, plovers, and gulls hunt the sandy beach for food located just beneath the surface.

Wetlands are another kind of lentic ecosystem and occur in areas where the water table is at or above ground level most of the year. In some cases, wetlands occur where lakes and pond basins have filled in with sediments. Wetlands include marshes, swamps, and bogs. **Marshes** are wet "grasslands" dominated by emergent vegetation, such as cattails, bulrushes, and sedges (figure 26.12). **Swamps** are wetlands dominated by woody vegetation, such as willow, swamp oak, and cypress (figure 26.13). **Bogs** have wet mats of vegetation, often dominated by *Sphagnum* moss (figure 26.14).

Marine Ecosystems

Marine ecosystems begin at the edges of oceans, where there is a transition between fresh water and salt water. Areas where fresh water meets the sea are called **estuaries.**

As you might imagine, the salinity (saltiness) of estuaries is intermediate between fresh water and salt water. Estuaries have high nutrient levels and serve as nursery grounds for a variety of organisms, including many kinds of fish.

On the other hand, because of incoming and outgoing tides, sandy beaches and rocky shores constitute rather severe environments for organisms. For almost twelve hours of each twenty-four-hour period, organisms of the seashore are exposed to the heat and drying effect of the sun. During the remaining time, they are under water. Many of the organisms on a sandy beach remain below the sand, becoming active and rising to the surface only during the period of high tide (figure 26.15). On rocky shores, organisms such as barnacles are tightly attached to the rocks to withstand the action of waves (figure 26.16). During low tides, many of these organisms close their shells to prevent drying out. At high tide they open their shells to feed.

Figure 26.16 The rocky seashore, especially along the northeastern coast of North America, shows strong zonation of life exposed at low tide. The dark zone consists mainly of blue-green algae. The zone below it contains barnacles. Below the barnacles is a zone dominated by rockweeds (*Fucus*).

Figure 26.17 A map showing the world's major biomes. The map is a generalized one, and indicated biome boundaries are only approximate.

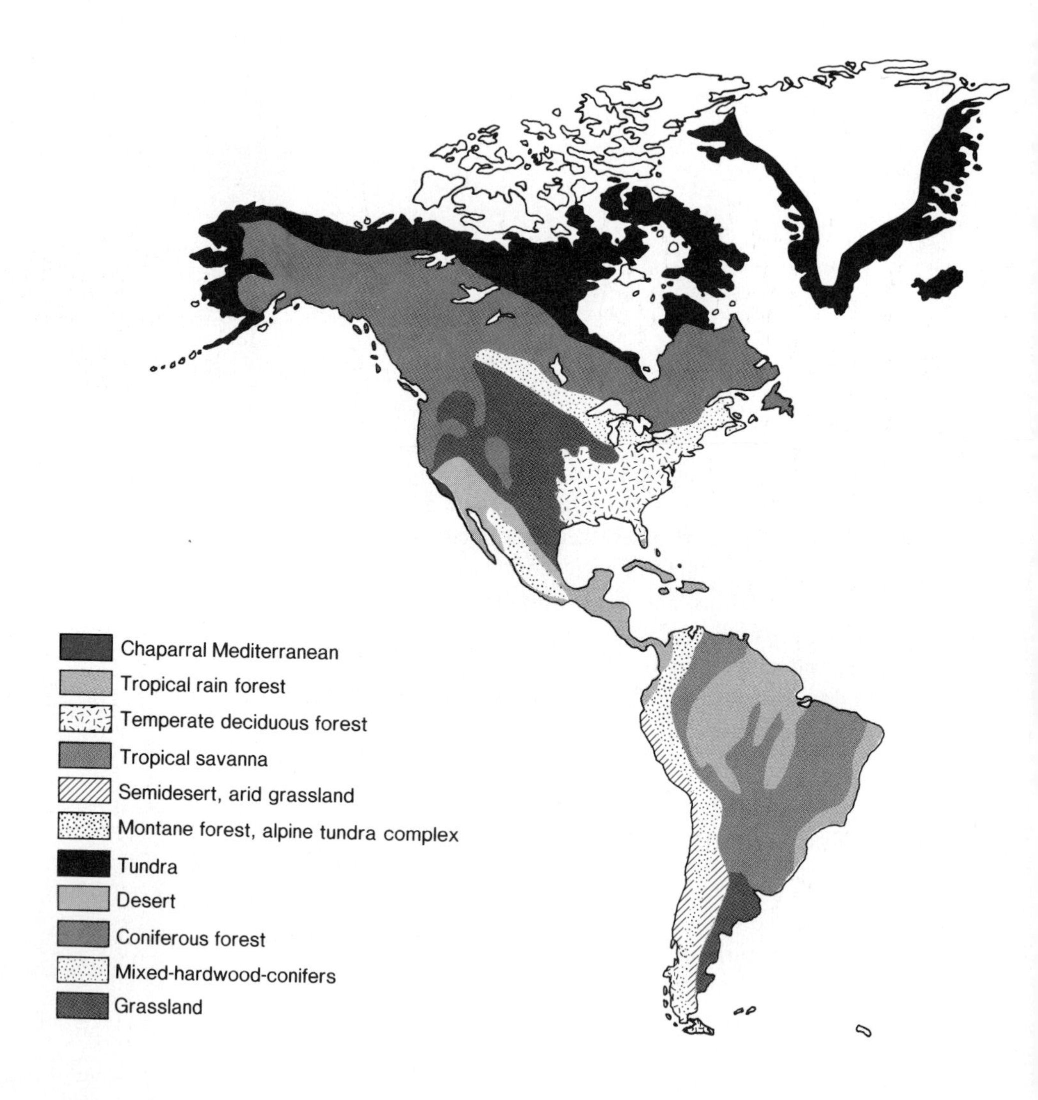

Oceans are relatively unproductive because there is little circulation between nutrient-containing bottom waters and light-absorbing upper waters. However, certain areas of shallow water over the continental shelf are highly productive. High productivity also is found in regions where cold and warm currents meet to produce upwellings that bring deep, nutrient-rich, bottom waters to the surface. This stimulates high phytoplankton production, which in turn supports dense populations of fish and fish-eating seabirds.

Terrestrial Ecosystems

We can categorize major terrestrial ecosystems on the basis of the climax community types they support. These community types are called **biomes** (figure 26.17).

Tundra

Lying largely north of latitude 60°N and encircling the top of the earth is a vast, treeless region called the **arctic tundra** (figure 26.18). The arctic tundra biome is dominated by mosses, lichens, grasses, and low-growing shrubs. The

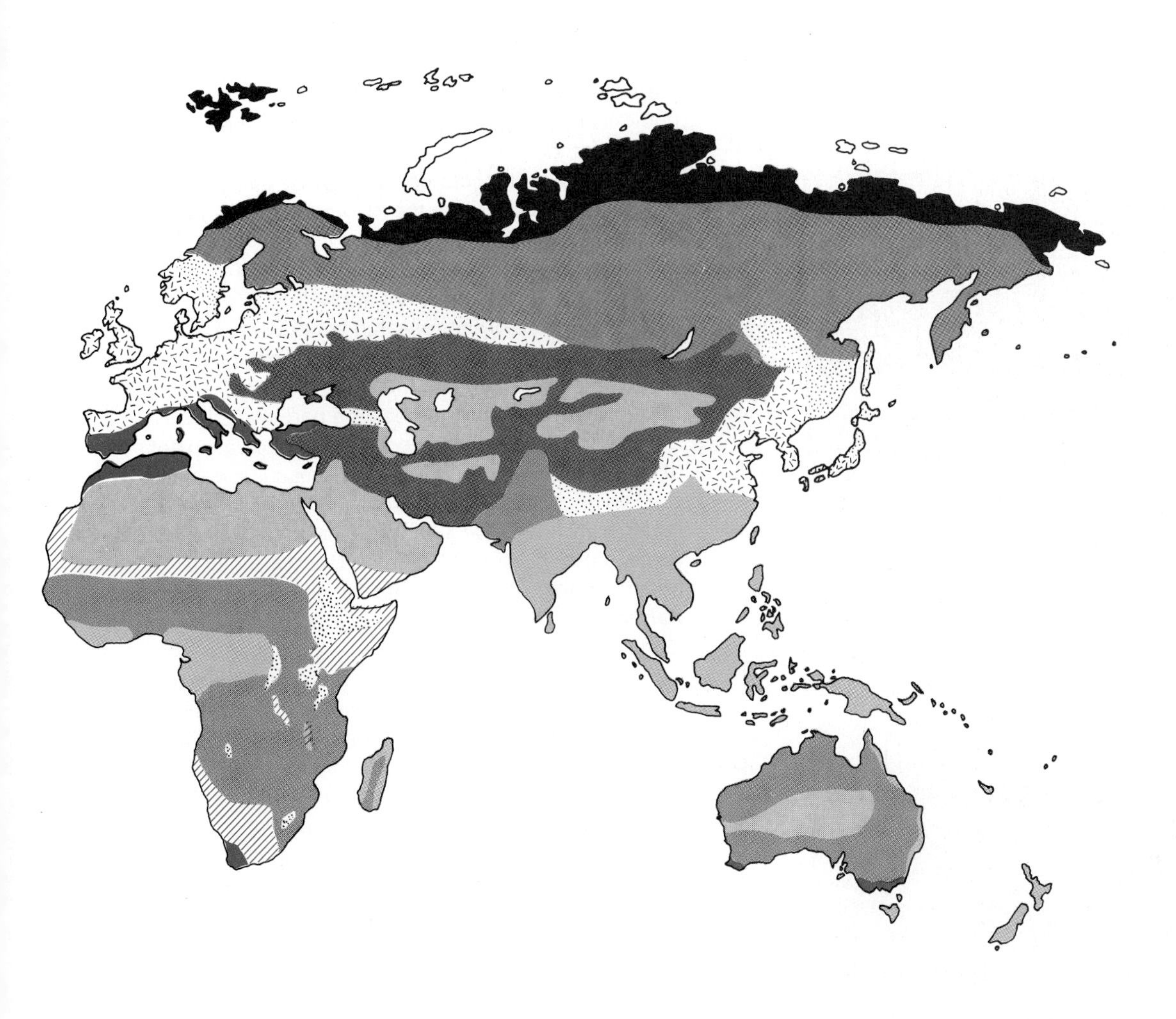

growing season (the period from the last frost of the spring to the first frost of fall) is short (sixty days or less). The winters are extremely cold, and the precipitation is usually less than 25 cm a year. However, because of poor drainage, low temperatures that reduce evaporation, and the presence of a permanently frozen soil (permafrost) a short distance below the surface, the land is wet and covered with numerous ponds, lakes, and bogs.

In much of the arctic tundra the most important herbivores are lemmings, small rodents related to meadow mice. In other parts of the tundra, musk ox and caribou are the major herbivores. The herbivores support a number of carnivores, including snowy owls, foxes, and weasels.

Above the timberline in the higher mountains of the world is the **alpine tundra** (figure 26.19). Unlike arctic tundra, alpine tundra either lacks a permafrost or has only certain parts that are permanently frozen.

Figure 26.18 The arctic tundra. The name tundra comes from the Finnish *tunturi,* meaning treeless plain. It is a land of bogs, sedge marshes, and lakes. Frost molds its landscape, and permafrost, a frozen layer of soil, impedes drainage and keeps the land constantly wet, even though precipitation is low.

Figure 26.19 The alpine tundra is a cold land of strong winds, snow, and widely fluctuating temperatures. It is a land of rock-strewn slopes, small bogs, alpine meadows, and shrubby thickets.

Box 26.1
Life Around Deep-Ocean Hot Springs

Bottom-dwelling animals are scarce on the seafloor beneath the open ocean. The few animals that do live in the perpetual darkness of the ocean depths must obtain nutrients either by ingesting material that sinks from areas near the surface, where there is light enough to permit photosynthetic production, or by eating other organisms that consume the material that falls from above.

However, there are some places on the deep ocean floor where these rules do not apply. In 1977 scientists aboard the deep-diving research submarine *Alvin* of the Woods Hole (Massachusetts) Oceanographic Institution discovered some areas of the ocean floor 2,500 to 2,600 m below the surface in the eastern part of the Pacific Ocean that contain surprisingly dense populations of relatively large animals. These areas occur in volcanic zones along rifts where the seafloor is spreading tectonically (see page 488). Water falls into volcanic hot spots in the rift, where it is heated, and it then emerges from hydrothermal vents. Water around the vents is much warmer (20° to 22°C) than the cold water (2°C) normally found on the ocean floor.

Large clams (many of which are up to 25 cm long) and mussels, crabs, polychaete worms, and fish have been discovered around hydrothermal vents along two spreading zones, the Galápagos Rift and the East Pacific Rise (box figure 26.1*A*). Some of the species found in these areas have not been observed elsewhere. Probably the most unusual animals in the hydrothermal vent communities are huge tube worms (*Riftia pachyptila*) that belong to the phylum Pogonophora, a relatively obscure group of marine worms. *Riftia* individuals live in tubes that they construct near vents, and they grow up to 3 m long and 10 cm in circumference (box figure 26.1*B*). One of the most remarkable characteristics of these worms is that they do not possess mouths or digestive tracts.

How do all the animals in these dense communities obtain nutrients? One suggestion is that the warm water rising from the vents sets up local convection circuits that sweep nutrients inward from large areas of the surrounding seafloor, thus concentrating nutrients around the vents. This may indeed be the case, but there is another intriguing factor in the nutrient equation of deep-ocean hydrothermal vent ecosystems.

The warm water in the vent areas contains high densities of bacteria. These bacteria are autotrophic; that is, they use carbon dioxide as a precursor in synthesis of reduced organic compounds. But they do not require light energy to accomplish this. In complete darkness, they utilize the hydrogen sulfide (H_2S), which is present in the vented water, in chemosynthesis. H_2S serves as a source of electrons for ATP-generating oxidation-reduction reactions and for the reduction reactions involved in CO_2 incorporation. Thus, a supply of H_2S coming out of the vents makes possible considerable primary production in an environment where photosynthesis is impossible.

These bacteria, therefore, are the primary producers in a food chain that is not based on photosynthesis. Filter-feeding animals such as clams and mussels can trap and digest the bacteria that grow in the water around them. Crabs and fish can live as scavengers, eating small animals or the debris from larger animals after their death. But how does the large, gutless worm *Riftia* manage nutritionally?

These worms may actively transport organic material from the seawater around them inward across body surfaces. But they also appear to utilize another nutritional strategy. There are symbiotic bacteria living in some tissues of *Riftia* that may produce nutrients utilized by the worms. However, more research will be needed to fully explain the nutrition of these large worms. There are a number of questions yet to be answered concerning these and other organisms that live around deep-ocean hydrothermal vents in the dark depths of the sea.

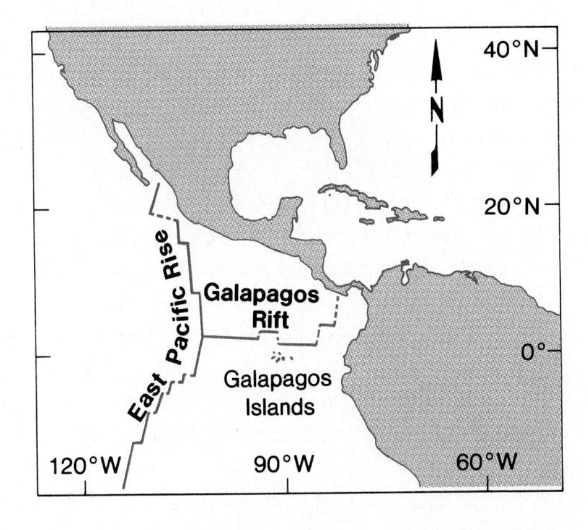

Box figure 26.1*A* Location of the spreading zones in the East Pacific along which deep-ocean hydrothermal vents occur. The "oases" of abundant life around the vents have been given such names as Dandelions, the Rose Garden, and the Garden of Eden.

Box figure 26.1*B* Giant tube worms, *Riftia pachyptila*, living near a deep-ocean hydrothermal vent. These large worms (up to 3 m long) have no digestive tract.

Figure 26.20 The coniferous forest, dominated by spruce and fir, forms a belt around the world below the tundra. It is also called the taiga.

Figure 26.21 A temperate deciduous forest in eastern North America.

Figure 26.22 The tropical rain forest forms a worldwide belt around the equator. The largest continuous rain forest is found in the Amazon basin, where its existence is threatened by land clearing. The tree species number in the thousands, climbing plants are common, and the interior is dark and moist.

Coniferous Forest or Taiga

South of the arctic tundra lies a worldwide belt of **coniferous forest (taiga)** (figure 26.20). The vegetation is dominated by spruce, firs, and pines. The climate is characterized by cool summers, cold winters, and a growing season of about 130 days. Precipitation ranges between 40 and 100 cm per year, with much of it coming as heavy snow. The soils are acid and infertile and have a thick litter of slowly decomposing needles. Much of the precipitation moves through the soil, carrying with it important nutrients.

Compared to temperate and tropical biomes, the coniferous forest has relatively few consumer species. However, the number of individuals within each consumer species tends to be large. Larger grazing herbivores include the snowshoe hare, which is preyed upon by the lynx, and moose and woodland caribou, which are prey of wolves.

Temperate Deciduous Forest

South of the coniferous forest lies the **temperate deciduous forest** (figure 26.21). Deciduous forest is found in areas of moderate climate with a well-defined winter and summer season and relatively high precipitation (75 to 150 cm/yr). The growing season ranges between 140 and 300 days. The soil is relatively fertile and mildly acid, and litter decomposes quite rapidly. The dominant trees are **deciduous;** that is, they lose their leaves in the fall.

The deciduous forest supports a high diversity of consumer organisms, the most numerous of which are insects. Insect populations seldom grow excessively large. The notable exception is the gypsy moth, which was accidentally introduced into the northeastern United States and feeds on a wide variety of hardwood trees, especially oaks.

A major large herbivore is the white-tailed deer. In areas of high population, the white-tailed deer can influence the structure and development of a forest by overutilizing certain species of forest tree seedlings and sprouts. Because of human activity, deer predators like the wolf and the mountain lion have disappeared from most areas.

Tropical Rain Forest

Tropical rain forest is located in the equatorial regions of Central America, central and northern South America, western Africa, and Indonesia (figure 26.22). Tropical rain forest occurs in areas where average temperatures are high (between 20° and 25°C) and precipitation is heavy (in excess of 200 cm per year). Temperatures fluctuate little, and the rainfall is evenly distributed throughout the year. The forest is dominated by broadleaf, nondeciduous trees, which are part of a very rich diversity of plant species.

The soil in tropical rain forests is nutrient-poor, but there is much plant growth because of high temperatures, a year-long growing season, and the rapid recycling of nutrients. Even though the soil is poorly suited for agriculture, the tropical rain forest is being converted to agricultural uses and may be destroyed by the turn of the century.

Tropical Savanna

Also associated with tropical regions is the **tropical savanna,** which is dominated by tall grasses and also supports scattered trees (figure 26.23). Tropical savanna occurs in regions where temperatures are high and where rainy and dry seasons are pronounced. African savanna supports a rich diversity of grazing animals. Tropical savanna has been highly disturbed by human activity, particularly agricultural development and overgrazing by domestic animals.

Figure 26.23 Tropical savannas are found over large areas in the interior of continents. The best known are the African savannas, characterized by scattered, flat-topped *Acacia* trees.

Figure 26.24 A preserved area of native prairie in South Dakota in autumn. Before the coming of European settlers, the interior of the North American continent supported one of the great grasslands of the world. The grassland varied from east to west across the plains: eastern tall-grass prairie gave way progressively to mixed prairie and short-grass plains. These plains once supported vast herds of bison. Today much of the native grassland around the world is under cultivation.

Figure 26.25 Desert in the southwestern United States. Deserts occur in two distinct belts about the earth, one near the Tropic of Cancer, the other near the Tropic of Capricorn. In North America there are two types of desert. One is the cool desert of central and northwestern North America, which lies in the rain shadow of the Sierra Nevada and Cascade mountains. This desert of the Great Basin is dominated by sagebrush. The other is the hot desert of the southwest in Arizona (pictured here), New Mexico, and southern California. This desert is dominated by cacti, creosote bush, paloverde, and ocotillo.

Grassland

Grassland is found on the plains areas of North America, Eurasia, and South America and supports a variety of grasses (figure 26.24). Rainfall is variable, and in many areas could support tree growth, but tree encroachment is halted by fires and periodic drought. Grasslands have very fertile soil that has been exploited worldwide for agriculture.

Grassland supports large populations of herbivores. In North America, the native grassland once supported millions of bison, and the African plains are dominated by a large assemblage of grazing animals, including many species of antelope, the wildebeest, zebra, and giraffe. These herbivores support a number of large predators, including lions and hunting dogs in Africa, and at one time the wolf in North America. In North America and much of Africa large native herbivores have been replaced by cattle, sheep, and goats, and in many places overgrazing has resulted in deterioration of the grassland.

Deserts

Desert forms a worldwide belt at about 30°N and 30°S latitude (figure 26.25). Desert is associated with the rain shadows of mountains, coasts next to cold ocean currents, and deep interiors of continents. It is characterized by rainfall of less than 25 cm per year, low humidity, a high evaporation rate, and high daytime temperatures and low nighttime temperatures. Desert plants are widely scattered shrubs with short, waxy, drought-resistant leaves. The soil is high in inorganic salts and low in organic matter.

Desert may be hot or cold. The North American hot desert is found in the southwestern United States and in Mexico. It is dominated by cactus and desert shrubs. Northern cold desert is dominated by sagebrush.

Even desert is subject to extensive human disturbance. Irrigation has made it possible to utilize desert for agriculture. However, water for irrigation often is pumped to the surface from deep wells. Often this water is removed faster than it can be replaced by natural processes. Irrigation also produces an accumulation of salts in surface soils as a result of the evaporation of irrigation water. Finally, there are grave questions concerning availability of water in the future for direct human use in areas such as the southwestern United States, where the population is growing very rapidly.

Environmental Problems

Human beings today face serious ecological problems. Many ecosystems are in trouble. Ecosystems generally can recover from disturbances, but there is a point beyond which an ecosystem loses its ability to return to normal. This happens when an ecosystem is subjected to a very large disturbance or to continuing disruptive influences that last for a long time.

Since ecosystems consist of interconnected parts, a disruption of one part of an ecosystem eventually affects the whole system. Despite our sometimes arrogant attitudes, we human beings definitely are an integral part of ecological systems and our activities very often have negative effects on our environment. Many of the environmental problems to which we contribute arise simply because there are so many of us.

World Population Growth

The total human population of the world has been growing rapidly for some time. However, the growth rate is smaller than it was a few years ago because birthrates have fallen dramatically in many developed countries. In fact, **replacement reproduction,** a situation in which the average couple has only two children, is spreading to some developing countries. But unfortunately, achieving replacement reproduction does not produce zero population growth immediately or even within a period of a few years.

This is because of the **age structure** of populations, especially in developing countries. We saw in chapter 25 that there will be growth in a population that has a large proportion of young individuals entering their reproductive period. This is inevitable, even if replacement reproduction is attained, and more than half of the world's people are under age 25. Thus, we can expect world population growth to continue well into the twenty-first century and we must consider some problems that will result from that growth.

World food production is not adequate to meet the needs of today's population. Can we find the means to increase production enough to supply a still larger world population or will massive starvation "solve" the problem?

There is an already great gap between people in developed countries and in developing countries in use of natural resources such as minerals and fossil fuels (coal, oil, and natural gas). What will happen as increasing world population rushes us toward depletion of many of those resources early in the next century?

Figure 26.26 A fish kill caused by oxygen depletion resulting from decay of algae produced in an algal bloom.

Pollution

Population growth also adds to problems of environmental pollution. **Pollution** is an undesirable change in characteristics of an ecosystem, and pollution can be very harmful to living things in the ecosystem, including humans.

Water Pollution

Toxic wastes from industries and sewage systems can poison living things in aquatic ecosystems directly. But release of toxic wastes can be detected and controlled. There is general support for enforcement of laws preventing direct water pollution by release of poisonous substances into streams, rivers, lakes, and oceans.

There are more subtle water pollution problems, however. Even very efficient sewage treatment plants release nutrients such as nitrates and phosphates into bodies of water. Nutrient molecules are also carried off farmland by rainwater runoff. This nutrient enrichment causes abnormally dense growth of photosynthetic organisms, especially algae in lakes. These algae "blooms" cause the water to become murky and often result in death of at least some organisms in the ecosystem because decay of dead algae uses up oxygen in the water (figure 26.26). Over time nutrient enrichment leads to rapid eutrophication (see chapter 25) and causes lakes to fill in much more rapidly than they would in the absence of the pollution.

Power plants often take water from rivers and lakes for cooling and release warmed water back into them. This **thermal pollution** also encourages excessive growth of organisms and produces some of the same results as nutrient pollution.

Air Pollution

Industries, automobiles and trucks, and home heating systems all release polluting substances into the air. Some of these combine with water vapor already there to make **smog.** Others are carried great distances and later fall with rain. Sulfur dioxide and nitrogen oxides combine with water vapor to make sulfuric and nitric acids. Thus, the rain that falls, often at great distances from the original source of the air pollution, has a low pH. This **acid rain** lowers the pH of natural waterways and has devastating effects on a wide range of living things.

As with water pollution, it is possible to recognize and regulate obvious causes of air pollution if we are willing to make the effort. But also as with water pollution, there is a more subtle form of air pollution.

Carbon dioxide produced through burning huge quantities of fossil fuels is accumulating in the atmosphere and may have a significant impact on world climate. The CO_2 in the atmosphere transmits shortwave radiation from the sun to the earth but absorbs the longer wave radiation that passes from the earth back toward outer space. Some of this absorbed heat energy is reradiated toward the earth. Thus, CO_2 makes a contribution to the **greenhouse effect** of the atmosphere, and an increase in CO_2 concentration could result in increased warming of the earth. This could, in turn, melt the polar caps, cause the oceans to rise, increase evaporation, and decrease oceanic circulation.

Box 26.2

Will We Have a Nuclear Winter?

By now everyone is aware of the devastating immediate effects resulting from explosion of nuclear weapons. Many schoolchildren can tell you that the average thermonuclear warhead has a yield of two megatons, the equivalent of two million tons of TNT. Our minds have become almost numb to the realization that one such weapon is more powerful than all of the explosive force of all weapons used in World War II. We are aware of the blast power and the intense, short-term lethal effects of radiation released by nuclear explosions. There are now enough nuclear warheads to destroy all military installations, most concentrations of industry, and *every* city with a population of more than 50,000 in both the U.S.S.R. and the United States.

For months after nuclear explosions, radioactive material would continue to fall from the upper atmosphere. But scientists from many countries including the United States and the U.S.S.R. now agree that dust and smoke raised by nuclear explosions of a moderately large war (with 5,000 megaton yield) could create yet another set of problems. Dust and smoke could, within a few days, reduce the amount of sunlight reaching the earth to just a few percent of the normal amount, far too little to support the photosynthesis conducted by primary producers in all ecosystems.

A chilling "nuclear winter" would soon follow, no matter what time of year the war occurred. Temperatures would fall to at least −20°C and remain there for several months. The combination of radiation, cold, and inhibited photosynthesis would probably destroy most ecosystems in the Northern Hemisphere and kill the majority of living things there. It could spread to the Southern Hemisphere as well if wind patterns were sufficiently disturbed.

When the dust cleared, chemical disturbance of the ozone layer would expose all remaining organisms not to the warming, benevolent rays of normal sunshine, but to harsh, unfiltered, burning ultraviolet radiation that could seriously burn human skin in minutes. Any extended exposure would greatly increase the incidence of skin cancer.

To most biologists, the potential effects of massive nuclear weapons exchanges on ecosystems are simply unimaginable. All normal patterns of ecosystem structure and function would disappear, leaving a bleak, unpredictable future for any organisms that might accidentally survive. Thus, the danger of nuclear war is not just another modern ecological problem, it is **THE** problem of our time, not just for human ecology, but for all life on our planet.

The Future

One view of the future is that natural resource and energy reserve depletion will soon bring human population growth to a screeching halt, no matter what we do about population control. Another view is that new technologies will expand human possibilities and allow for a much larger world population, though members of that population will need to abandon desires for solitude and access to undisturbed natural areas.

Faith in development of new technologies has faltered somewhat, however, as it has become apparent that "food from the sea," nuclear power, and even the famous "green revolution" in agriculture all have definite limitations. Thus, an acceptable human future depends on our will to act now to manage population size and to cut natural resource consumption (especially in the United States, the world's largest per capita resource consumer). Do we have the will to make hard, even sacrificial decisions to assure a decent future for generations to come or is the best that our grandchildren and great-grandchildren can hope for is a dreary life on a dismal planet?

Summary

All of the populations that inhabit a given area at a given time make up a community. Communities can be characterized in terms of structure, dominant species, and species diversity. Dominant species are those that exert more control over the character of the community than do other species. Species diversity relates to both the number of species in a community and the evenness with which individuals are apportioned among species.

Communities undergo predictable changes through time in a process called succession. Primary succession occurs on sites that have not previously supported life, while secondary succession occurs on sites where life has occurred and where there is a soil base. Succession reaches a climax when the vegetation dominating the site tends to replace itself and species composition tends to remain somewhat the same.

All of the organisms in a given area together with the abiotic components of their environment make up an ecosystem. The major living components of an ecosystem are the producers, consumers, and decomposers.

Energy and mineral nutrients move through an ecosystem. The flow of energy through an ecosystem is a one-way process. Nutrients, however, cycle through an ecosystem, and they can be used over and over.

The earth supports a number of recognizable ecosystem types. Some of these are aquatic (water-based) systems, while others are terrestrial (land-based) systems.

Aquatic ecosystems may be fresh water or salt water (marine). Lotic freshwater ecosystems are characterized by flowing water, while lentic freshwater systems are characterized by still water. Marine ecosystems include salt marshes, estuaries, rocky shores, sandy shores, and the open sea.

Major terrestrial ecosystems are categorized on the basis of the climax community types they support. These community types reflect differences in climate and soil and are called biomes. The major biomes are arctic tundra, coniferous forest (taiga), temperate deciduous forest, tropical rain forest, tropical savanna, grassland, and desert.

Worldwide population growth has created grave ecological problems for humans and many other species. Natural resources are being depleted, and water and air pollution are common.

Questions for Review

1. What is a community, and what are some of its characteristics?
2. What is succession? Describe the difference between primary succession and secondary succession.
3. Explain what we mean by a climax community.
4. What is an ecosystem? What are its basic components?
5. What is a food chain? a food web? a trophic level?
6. Distinguish among the three types of ecological pyramids.
7. What are the two types of biogeochemical cycles?
8. What are some characteristics of a lotic freshwater ecosystem? of a lentic freshwater ecosystem?
9. Distinguish between oligotrophic and eutrophic lakes.
10. What is an estuary?
11. Explain why environmental conditions are rather severe on rocky shores and sandy beaches.
12. Characterize each of the following as to location, general climate conditions, and climax communities: arctic tundra, taiga, temperate deciduous forest, tropical rain forest, tropical savanna, grassland, desert.

Questions for Analysis and Discussion

1. A pyramid of numbers may be inverted as when many insect herbivores feed on a single tree, but a pyramid of energy cannot be inverted. Why?
2. Could you draw a food web that includes human beings?
3. Explain why a lake isolated in the coniferous forest of northern Minnesota remains clear throughout the summer while a lake set in the agricultural area of southern Minnesota and lined with cottages becomes green and "soupy" by early August.

Suggested Readings

Books

Attenborough, D. 1984. *The living planet.* Boston: Little, Brown and Co.

Cole, G. A. 1979. *Textbook of limnology,* 2d ed. St. Louis: Mosby.

Krutch, J. W. 1960. *The desert year.* New York: Viking Press.

Odum, E. P. 1983. *Basic ecology.* Philadelphia: Saunders.

Whittaker, R. H. 1975. *Communities and ecosystems,* 2d ed. New York: Macmillan.

Articles

Brill, W. J. July 1977. A grazing system of the Serengeti. *Scientific American* (offprint 922).

Cooper, C. F. April 1961. The ecology of fire. *Scientific American* (offprint 1099).

Edmond, J. M., and Von Damm, K. April 1983. Hot springs on the ocean floor. *Scientific American.*

Ehrlich, P. R. 1984. North America after the war. *Natural History 93*(3):4–11.

Foin, T. C. 1984. Ecological energetics. *Carolina Biology Readers* no. 91. Burlington, N.C.: Carolina Biological Supply Co.

Richards, P. W. December 1973. The tropical rain forest. *Scientific American* (offprint 1286).

Tangley, L. 1984. After nuclear war—a nuclear winter. *BioScience* 34:6.

Woodwell, G. M. January 1978. The carbon dioxide question. *Scientific American* (offprint 1376).

The Diversity of Life

The most striking characteristic of the world of life is its diversity. Organisms of innumerable types possess a multitude of specializations that permit various living things to interact successfully with virtually every type of environment on earth. How do biologists make sense of this sometimes bewildering diversity?

Biologists divide living things into categories (taxonomic groups) on the basis of certain ranges of shared characteristics. No single classification scheme is, or probably ever will be, ideal. However, generally accepted taxonomic schemes are essential tools for biologists.

Taxonomic grouping aids in recognition and study of organisms. It also permits scientists of many lands to communicate clearly and directly with one another about the organisms they are studying. Furthermore, the systematic process of classifying organisms into taxonomic groups helps to clarify evolutionary relationships.

In chapters 27 through 29 we will sample the fantastic diversity of life on earth within the context of one of the most widely used general taxonomic schemes.

Facing page A sifaka, a tree-dwelling lemur from the island of Madagascar. Sifakas use their powerful hind legs to leap as much as 10 meters from tree to tree.

Viruses, Monera, and Protists

27

Chapter Outline

Chapter Concepts

1. All living organisms can be placed in one of five kingdoms based on their mode of nutrition, whether their cells are prokaryotic or eukaryotic, and their general organizational complexity.
2. Viruses are simply organized and can only reproduce within a living cell.
3. Bacteria and cyanobacteria are prokaryotic cells.
4. Bacteria are very flexible and diverse metabolically. This makes them important ecologically and industrially.
5. Eukaryotic cells may have arisen from endosymbiotic associations between amoeboid prokaryotes and bacteria.
6. Members of the protist kingdom are mainly unicellular or colonial eukaryotes.
7. Both autotrophs and heterotrophs are found among the protists.
8. Many protists are major primary producers and of great ecological importance. A number also cause serious diseases in humans, crops, and domesticated animals.

In this chapter and the two that follow, we will be surveying and sampling the diverse assemblage of living things that inhabit our earth. As we do so, we may well be reminded of William Shakespeare's famous lines, "What's in a name? That which we call a rose by any other name would smell as sweet."

Shakespeare may have had a point, but there is, nevertheless, a problem with the names of living things. For example, when an American scientist mentioned the English sparrow to a British colleague, the response was, "What's that?" In Britain the same sparrow is called the house sparrow or common sparrow. The problem can be even worse. The European white water lily has 245 different common names distributed among four languages!

Naming Living Things

Had the American scientist used the scientific name, *Passer domesticus,* his British colleague would have known immediately what bird he meant. The scientific name of each kind of organism is unique and universal because it is agreed upon and used by all scientists regardless of their native tongue. But it has not always been so.

Modern classification stems from the work of John Ray (1627–1705), an English naturalist. He outlined the species concept, that is, that there are small groups of organisms with similar characteristics that can be distinguished from other closely related organisms. This scheme was followed for the most part by Carolus Linnaeus (1707–1778), a Swedish physician and botanist. We consider Linnaeus to be the father of modern **taxonomy,** the part of biology devoted to describing and naming organisms and arranging them in a system of classification.

Linnaeus introduced the **binomial** ("two-name") system. Each organism is given a two-word name that is always in Latin or latinized. The first word is the name of the genus to which the organism belongs; the second word indicates the particular species. The initial letter of the genus name is capitalized, and the name is italicized in print or underlined when written or typed. For example, we are classified as *Homo sapiens.*

But individual species names alone will not help us to make sense of the world of life because there are literally millions of species of organisms.

Each species is placed in a succession of ever broader taxonomic categories. The major taxonomic ranks, from

Figure 27.1 This hypothetical example shows how a number of species (A–U) might be arranged taxonomically to form genera, families, and orders within a single class. The darkest colored area is a genus containing three species. The family containing this genus is composed of two genera and seven species; the order, of two families and three genera (the second family has only one genus).

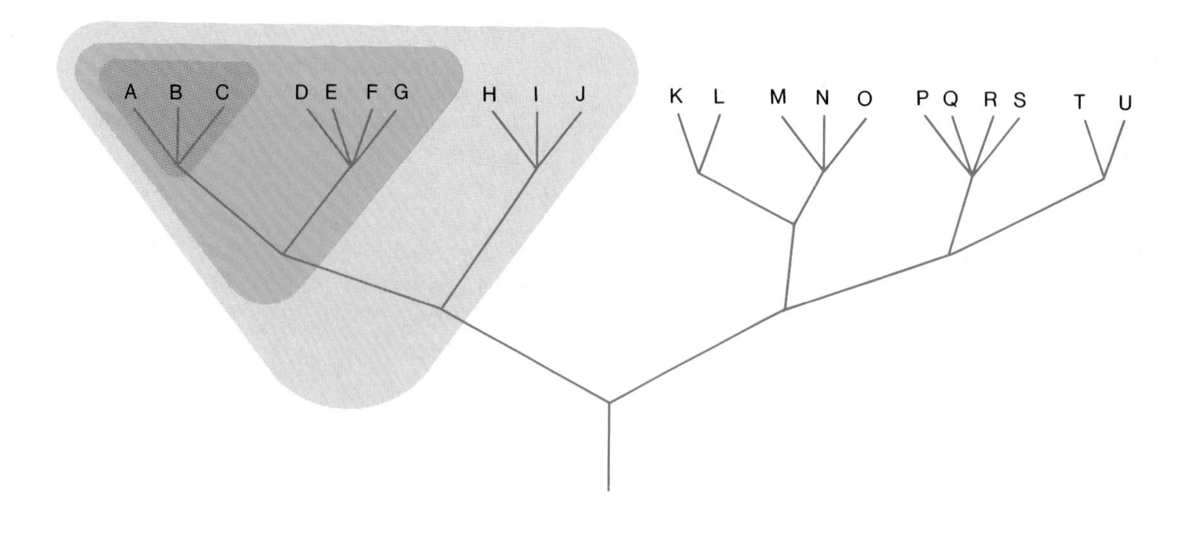

Table 27.1
Examples of Taxonomic Categories

Categories	Human	Domestic Dog
Kingdom	Animalia	Animalia
Phylum	Chordata	Chordata
Subphylum	Vertebrata	Vertebrata
Class	Mammalia	Mammalia
Order	Primates	Carnivora
Family	Hominidae	Canidae
Genus	*Homo*	*Canis*
Species	*sapiens*	*familiaris*

broader to most specific, are: kingdom, phylum or division, class, order, family, genus, and species. Usually, every rank in the hierarchy contains several groups in the level below it (for example, a particular family normally is composed of several genera, and each genus may have several species within it). You can see part of such a hierarchical classification system in figure 27.1. Table 27.1 illustrates the use of the system in classifying two kinds of organisms, human beings and domestic dogs.

The Kingdoms of Organisms

For centuries, the living world was divided into two kingdoms, plants and animals. This system seemed generally adequate. Trees, flowers, and mosses clearly are plants. Cougars, cod, caterpillars, and clams all move and eat other organisms (in one way or another), and clearly are animals. But there are problems in incorporating microorganisms into the scheme.

In recent years, it has been recognized that bacteria and cyanobacteria (blue-green algae) are radically different from other living things. Bacteria and cyanobacteria are **prokaryotes.** They are cells that lack a true membrane-bound nucleus, mitochondria, chloroplasts, and many other structures that are found in the cells of **eukaryotes** (see chapter 3 for more details). There are greater differences between prokaryotes and eukaryotes than between plants and animals, yet this distinction was not recognized in the two-kingdom system.

Furthermore, many microorganisms that are eukaryotic simply do not "fit" with either plants or animals. Some, for example, have both plant *and* animal characteristics.

Fungi also cause problems for the two-kingdom scheme.

The Five-Kingdom System

No system of classification is perfect and none will ever satisfy all biologists. However, the five-kingdom system developed primarily by R. H. Whittaker is fairly satisfactory. In Whittaker's scheme, living things are distributed into five kingdoms on the basis of level of organizational complexity and mode of nutrition (figure 27.2). The five kingdoms are Monera, Protista, Fungi, Animalia, and Plantae.

Kingdom Monera

All members of this kingdom are prokaryotic. All bacteria and cyanobacteria (blue-green algae) are included.

Kingdom Protista

This kingdom contains organisms that exist as single eukaryotic cells or as colonies of eukaryotic cells. The line dividing this kingdom from the other three kingdoms of organisms with eukaryotic cells is sometimes vague. For example, the chlorophyta (green algae) may be divided between the protist and plant kingdoms, or placed completely in either.

Kingdom Fungi

The fungi have walled cells that are arranged in networks of tubular structures. Their nutrition is absorptive.

Kingdom Animalia

Animals are multicellular with wall-less eukaryotic cells and are usually motile. Most have ingestive nutrition; that is, they eat other living things.

Kingdom Plantae

Plants are multicellular with walled eukaryotic cells and are usually nonmotile. They are photosynthetic and possess chloroplasts.

In this chapter and the chapters that follow, we will consider each of these kingdoms in more detail, and as we do, the distinctions among them should become clearer.

Figure 27.2 Whittaker's five-kingdom system. Major groups of organisms have been included. Dashed lines separate the protists from the three higher kingdoms because the boundaries are indistinct. For example, chlorophyta might be placed in either kingdom Plantae or Protista. The system is based on three levels of organization—prokaryotic; eukaryotic unicellular or colonial; and eukaryotic multicellular. The three higher kingdoms are distinguished from each other primarily on the basis of nutritional mode. The arrows indicate evolutionary relationships only in a general, simplified way.

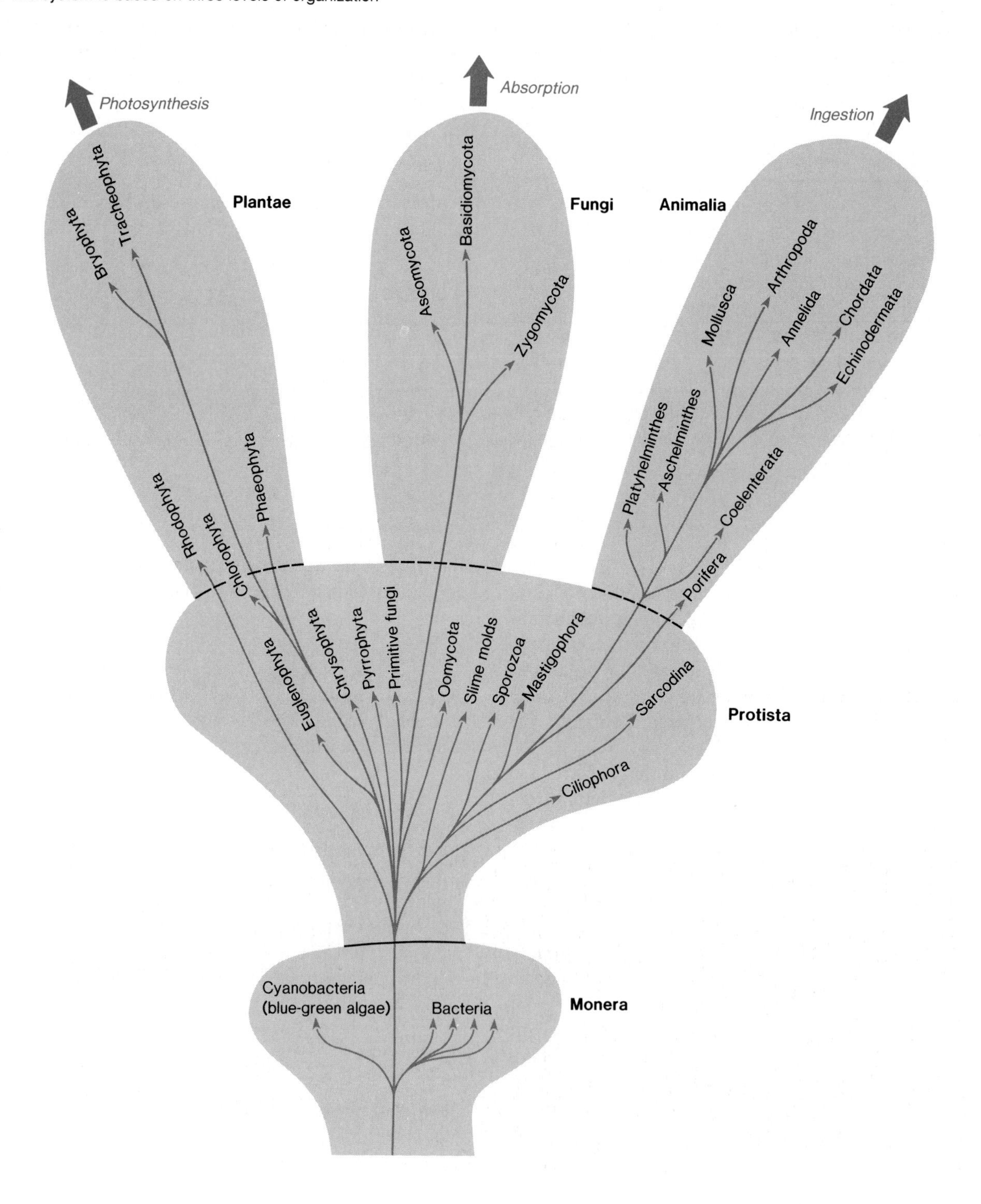

Viruses

Before we examine the kingdoms of living things, however, we will have a look at **viruses,** which exist at the border line of life. Strictly speaking, viruses are nonliving particles because they cannot reproduce themselves apart from the living cells that they infect. Most viruses consist simply of a nucleic acid core enclosed in a protein coat. But the interactions between viruses and living cells are interesting and complex aspects of biology.

Figure 27.3 Virus structure. (*a*) A simple virus with nucleic acid surrounded by a protein capsid. (*b*) A virus in which a lipid and glycoprotein envelope surrounds the capsid and nucleic acid.

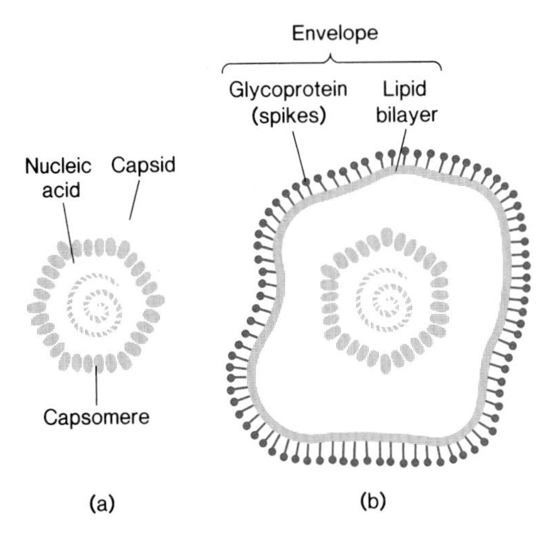

Discovery of Viruses

In the 1890s, Dimitri Ivanowsky, a Russian biologist, studied tobacco mosaic disease, a condition in which leaves of affected tobacco plants become wrinkled and mottled in appearance. Ivanowsky was able to transmit the disease to healthy plants by rubbing them with juice extracted from diseased plants. While attempting to isolate these "bacteria," he passed the infective extract through a finely meshed porcelain filter used to trap bacteria. To his surprise, the fluid that passed through the filter was still infective. Thus, the disease-producing agents of tobacco mosaic disease were smaller than any known bacteria. Any disease-producing agents that could pass through such a filter subsequently were known as filterable viruses and later, simply viruses.

The Structure of Viruses

The simplest virus particles are composed of a protein coat, the **capsid,** enclosing nucleic acid (figure 27.3). In some viruses, these basic elements are surrounded by an envelope that contains lipid and glycoprotein (molecules of protein combined with carbohydrate).

Careful examination of virus capsids with the electron microscope has shown that they are built from protein subunits, the **capsomeres,** which are arranged in different shapes in various kinds of viruses (figure 27.4).

Although living cells contain both DNA and RNA, viruses have either DNA or RNA, but not both. The nucleic

Figure 27.4 Electron micrographs of various kinds of virus particles. (*a*) Tobacco mosaic virus (magnification × 66,000). (*b*) Adenovirus showing the capsomeres that make up capsids (magnification × 81,200). (*c*) A bacteriophage, a virus that infects bacterial cells (magnification × 140,400). See figure 27.5 for an interpretation of its structure.

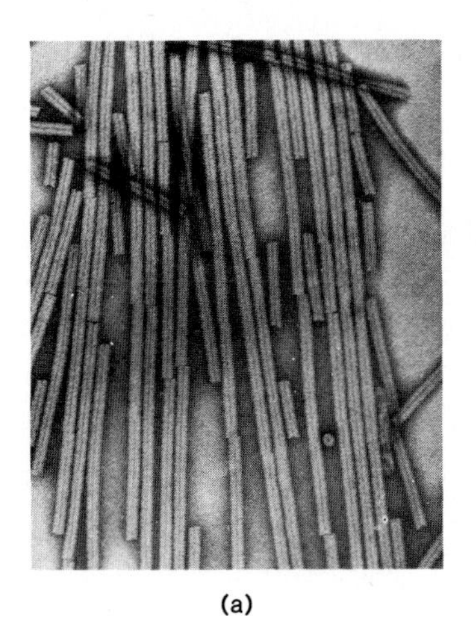

(a)

(b)

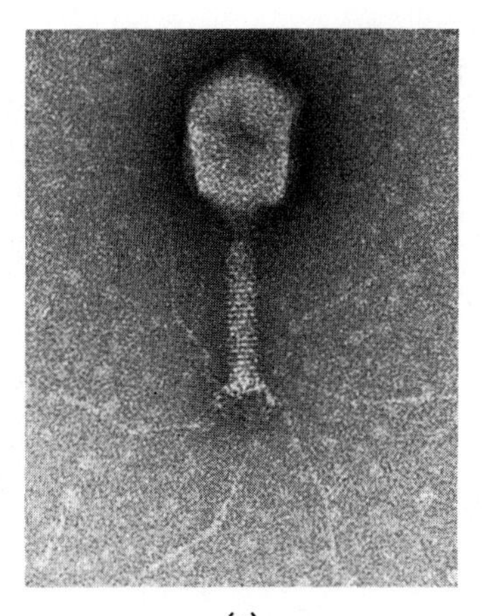

(c)

Figure 27.5 The structure of a bacteriophage. The virus attaches to a bacterial cell by its base plate and tail fibers. Then an enzyme makes a hole in the cell wall and the virus inserts its DNA into the bacterium. In (*a*) the tail sheath is relaxed; in (*b*) it has contracted as it injects its DNA into the bacterial cell.

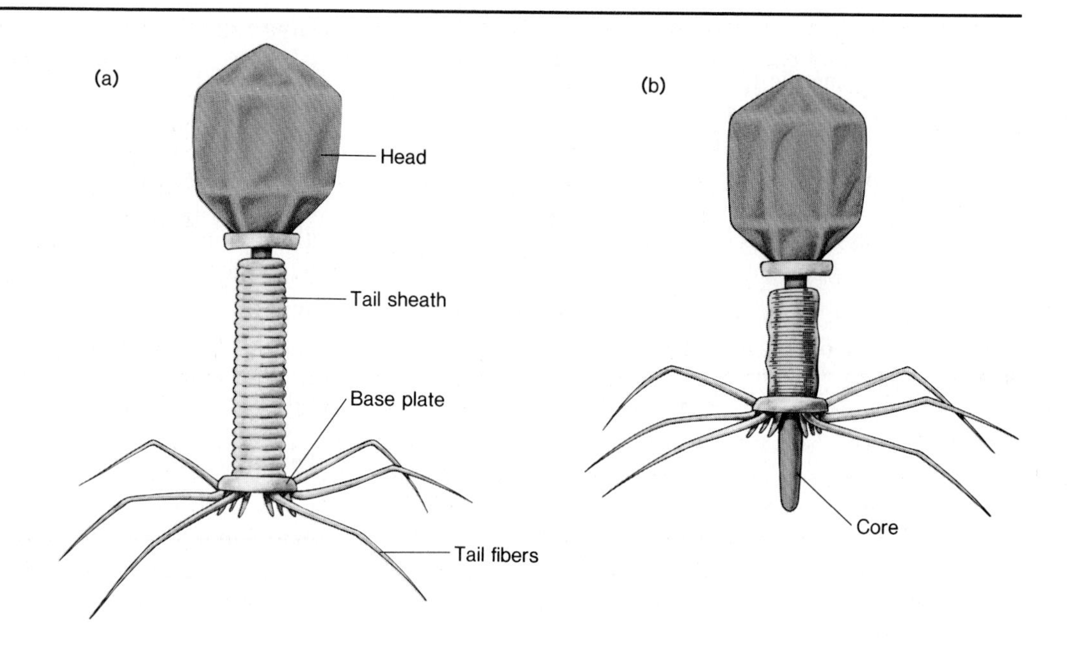

Figure 27.6 Infection of a bacterial cell (*Escherichia coli*) by a bacteriophage.

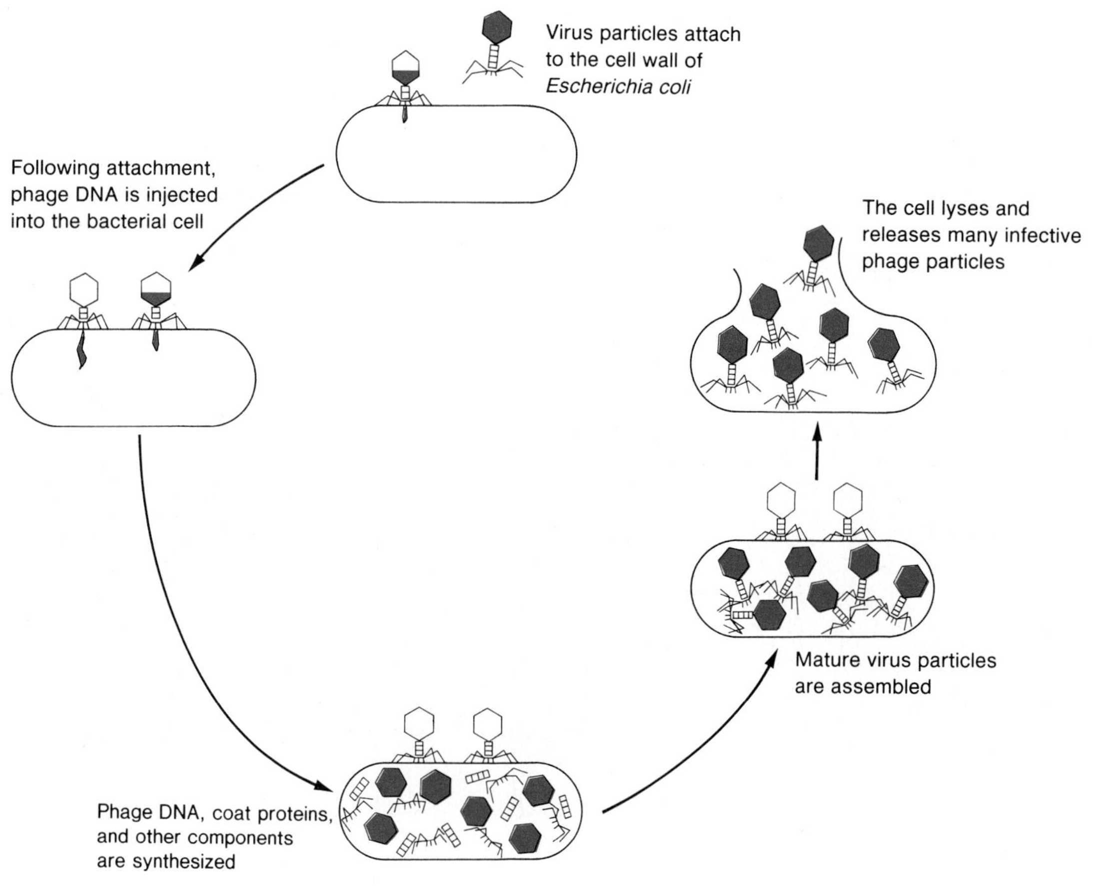

acids can range from very small amounts (fewer than five genes) up to several hundred genes. The nucleic acid carries genetic information, while the capsid protects the nucleic acid and aids in its transmission from cell to cell.

Viral Reproduction

Probably the best studied virus reproductive cycle is that of the large, complex viruses that infect the bacterium *Escherichia coli*. Since these viruses attack bacteria, they are referred to as **bacteriophages,** or **phages** for short (figure 27.5). We can use a bacteriophage to illustrate how a virus infects a living cell, takes control of it, and causes it to reproduce virus particles.

The bacteriophage attaches to a bacterium (figure 27.6). Then, a virus enzyme digests a portion of the cell wall. The tail sheath contracts, pushing the virus DNA into the bacterium. The virus genetic material takes control of the host cell within a few minutes and initiates the destruction of host DNA and inhibition of the synthesis of host cell proteins. Messenger RNA is synthesized, and it directs the production of virus proteins and any enzymes required for the manufacture of new virus particles. The host bacterium provides all the required energy and building blocks for virus synthesis. As soon as all virus components have been prepared, the virus particles are assembled. Finally, an enzyme synthesized under virus direction disrupts the cell wall and the mature virus particles—about 100 per bacterial cell infected by a bacteriophage—are released. Each of the new virus particles can now attack a neighboring *E. coli* cell.

The reproductive cycle just described is called a **lytic cycle** because the host cell is lysed (broken open) by the phage. In contrast, some bacteriophages do not immediately destroy their hosts but instead are reproduced along with the bacterium to generate more infected bacteria (figure 27.7).

Viral DNA is integrated into the bacterial DNA and becomes a **prophage.** The prophage is replicated along with the bacterial chromosome, and all the bacterial cell's offspring will carry the prophage. If bacterial cells carrying prophages

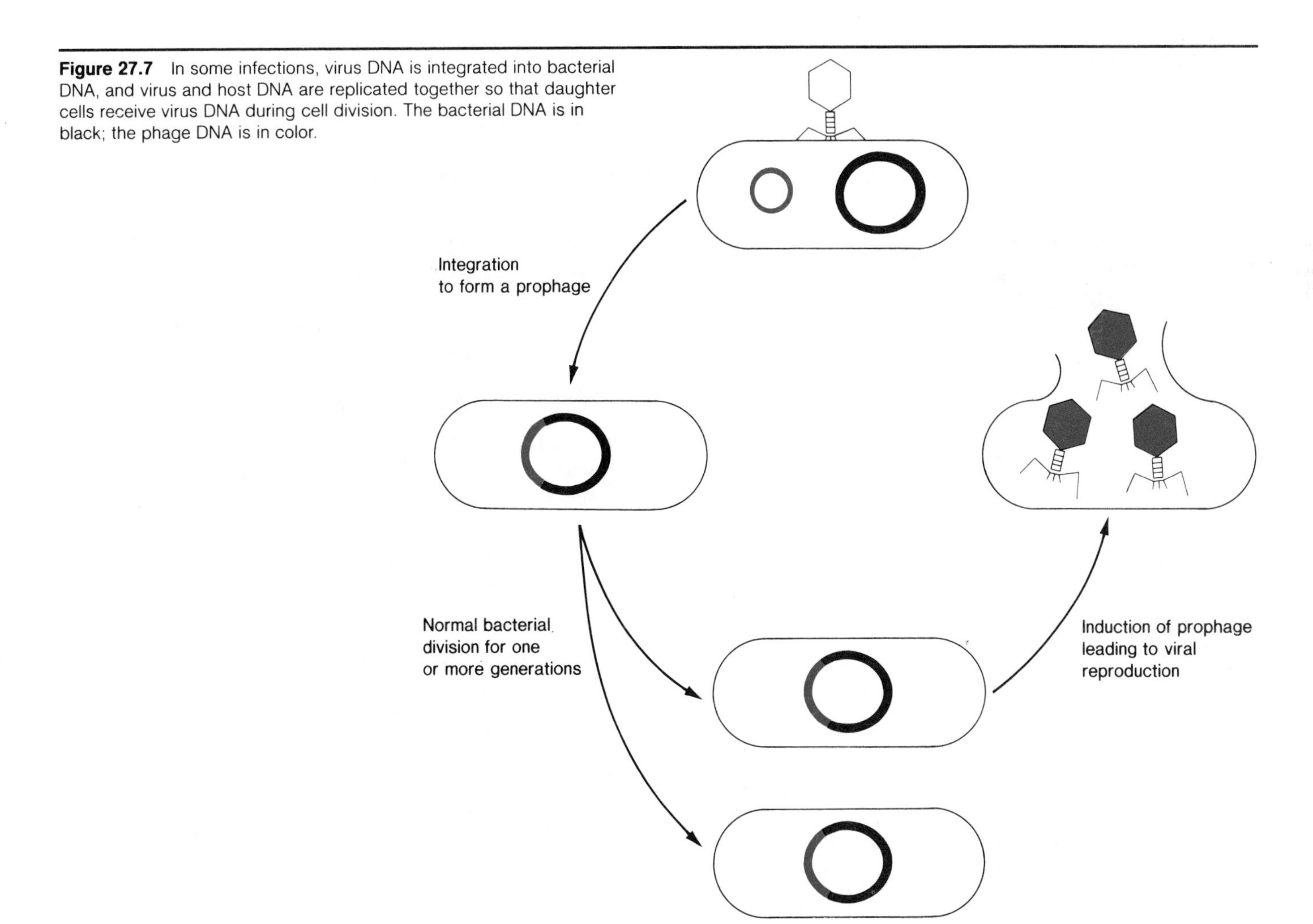

Figure 27.7 In some infections, virus DNA is integrated into bacterial DNA, and virus and host DNA are replicated together so that daughter cells receive virus DNA during cell division. The bacterial DNA is in black; the phage DNA is in color.

in their chromosomes are exposed to harsh environmental factors (such as ultraviolet light), **induction** occurs. The prophage leaves the bacterial chromosome, virus nucleic acids and proteins are synthesized, new virus particles are constructed, and the host cell breaks open. Such a life cycle is called a **lysogenic cycle.**

Viruses that infect animal cells take over host cells in much the same way that bacteriophages do. Animal virus DNA can also become incorporated into host cell DNA and be passed on to daughter cells at cell division.

Figure 27.8 This illustration compares *Escherichia coli,* several viruses, and the potato spindle-tuber viroid with respect to size and the amount of nucleic acid present. All dimensions are enlarged about 40,000 times. Bacteriophage f2 is one of the smallest known viruses (the capsid is about 20–25 nm in diameter).

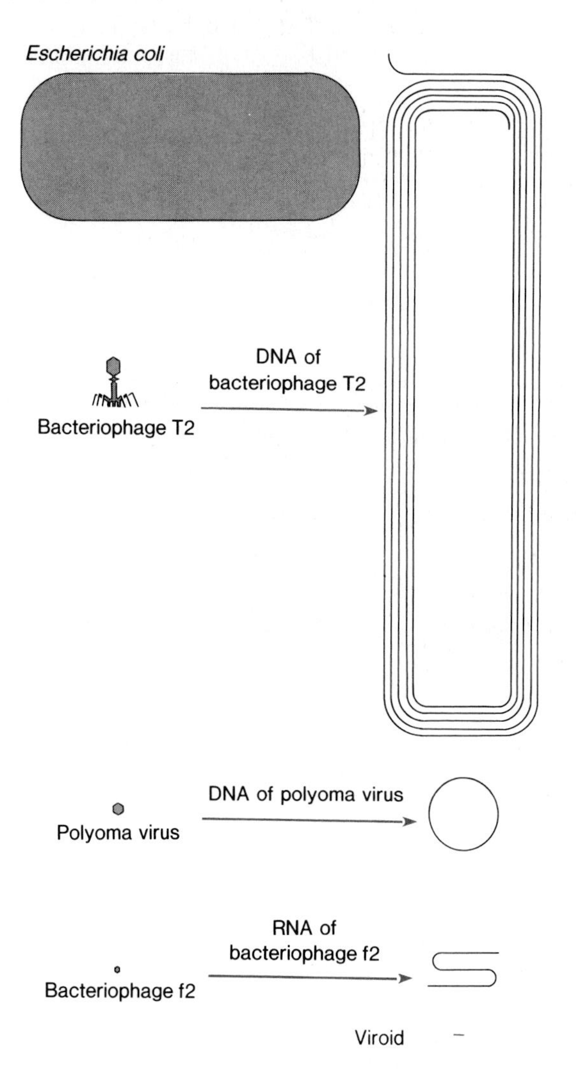

Host Resistance

Virus infections trigger immune responses (see chapter 21) in animals, but the immune response develops rather slowly. There is, however, another, quicker defense against virus infection. Infected animal cells can synthesize and release one or more types of small proteins called **interferons** (see page 456). Interferons bind to neighboring, uninfected cells and make them resistant to virus infection. Needless to say, there is great interest in the future manufacture of interferon for possible use in treatment of viral diseases.

Viral Diseases

Many human diseases result from viral infections. Some viruses attack the respiratory tract (influenza and rhinoviruses), while others grow in the skin (rubella, smallpox, measles), the liver (hepatitis), the central nervous system (rabies, polio), and the parotid salivary glands (mumps). In addition, viruses inflict extensive damage on other animals (swine influenza, rabies, hog cholera) and plants (mosaic diseases, leaf curls).

Viroids

A number of plant diseases are caused by a class of infectious agents called **viroids.** These are very short strands of RNA that can be transmitted between plants through mechanical means or carried by pollen. Viroids are smaller even than the smallest viruses (figure 27.8). They do not serve as messengers to direct protein synthesis, and it is not yet clear how they are reproduced or cause disease symptoms. Although the viroids so far discovered cause plant diseases, there is a possibility that similar agents are responsible for certain animal diseases as well.

Finally, there are some very mysterious particles named **prions** that also are suspected of causing infectious diseases. The puzzling thing about these tiny particles is that they do not seem to include any nucleic acid. At this point, we can only tell you to be on the lookout in the future for news about clues in the prion mystery.

The Kingdom Monera

The kingdom Monera includes two groups of organisms that are prokaryotic cells, the bacteria and the cyanobacteria (blue-green algae).

The Bacteria

Bacteria are all prokaryotes that range in size from spheres 0.2 or 0.3 μm in diameter up to rod-shaped bacteria hundreds of micrometers in length. The average cell is around 1 to 5 μm in size. A bacterium may be spherical (a **coccus**), rod-shaped (a **bacillus**), or helical (a **spirillum**) (figure 27.9). The rods and cocci can also associate to form clusters and linear filaments. For example, some cocci form chains of cells (streptococci) or grapelike clusters (staphylococci).

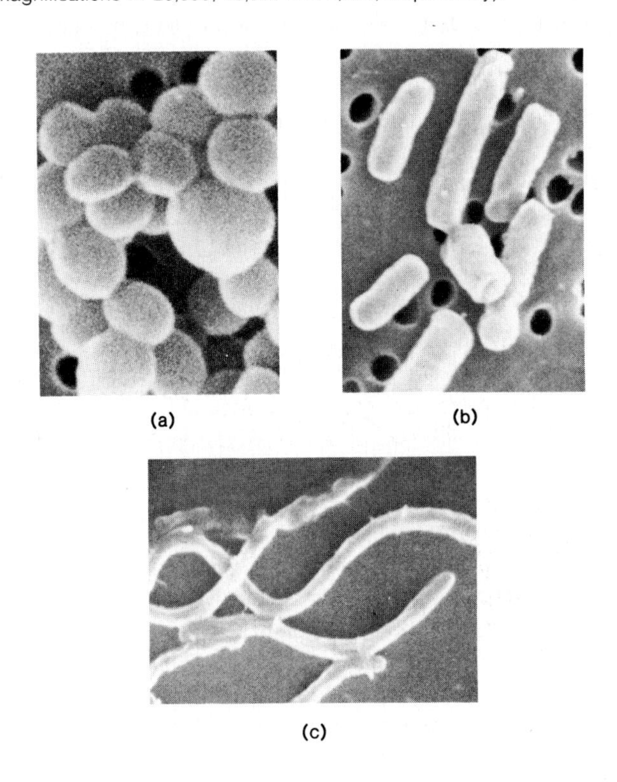

Figure 27.9 Scanning electron micrographs showing the three most common bacterial shapes—(*a*) cocci, (*b*) bacilli, and (*c*) spirilla (magnifications × 20,000, 13,000 and 5,000, respectively).

Bacterial Cell Structure

In distinct contrast with the variety in bacterial size, shape, and life-style, the internal structure of bacteria is moderately uniform (figure 27.10). Bacterial cytoplasm is bounded by a normal-looking **plasma membrane** (cell membrane). Most bacteria have a **cell wall** outside their plasma membrane. This wall gives the cell its shape and rigidity. It is chemically different from cell walls of eukaryotic cells.

Bacterial cytoplasm appears homogeneous under the electron microscope. **Ribosomes** are scattered in the cytoplasm. Bacterial ribosomes are smaller and lighter than eukaryotic cytoplasmic ribosomes.

Antibiotics like erythromycin, streptomycin, tetracycline, and chloramphenicol destroy disease-causing bacteria by acting on bacterial ribosomes. Thus, they inhibit bacterial

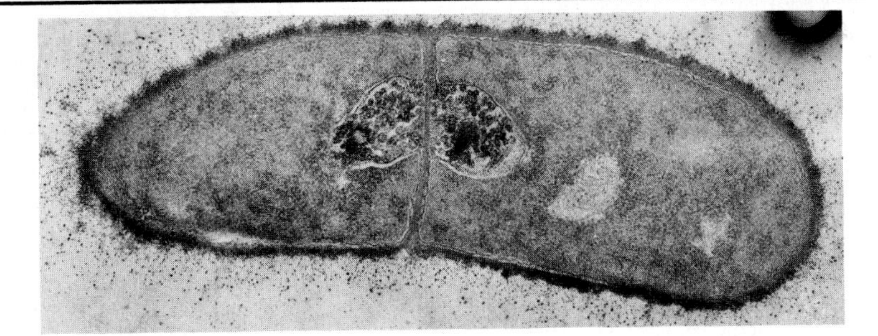

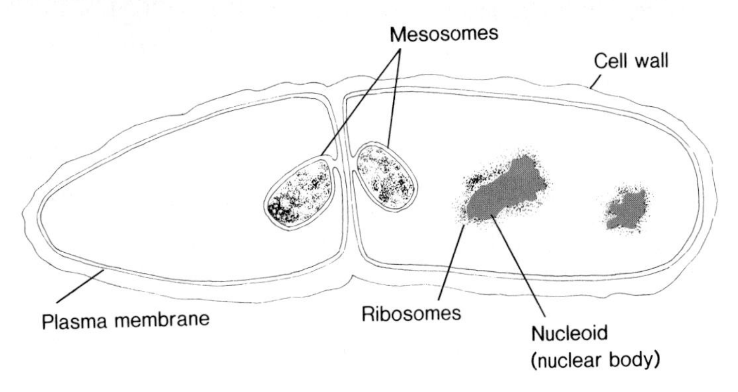

Figure 27.10 The internal structure of a dividing bacterial cell (*Bacillus megaterium*) as seen with the electron microscope (magnification × 61,500).

protein synthesis without inhibiting protein synthesis in the eukaryotic cells of the organisms being treated for a bacterial infection.

Bacteria do not have true nuclei as do eukaryotes. Their DNA is present as a single circular strand, about 1 mm long, associated with protein (but not histones). It is packed together in a discrete part of the bacterial cytoplasm, the **nucleoid.** Many bacteria also have large plasma membrane infoldings called **mesosomes.** Their function is still not clear.

Some bacteria secrete a shiny or gelatinous layer outside the cell wall. This **capsule** is composed of polysaccharides or polypeptides and protects the bacterium against phagocytosis by host white blood cells. For example, *Streptococcus pneumoniae* can cause pneumonia only when it possesses a capsule (see page 372).

Many bacteria can form **endospores.** The bacterial endospore has a thick wall of several layers (figure 27.11). Inside are DNA, ribosomes, and other cytoplasmic materials. Endospores are some of the most resistant living structures known. For example, many can survive several hours in boiling water. Endospores serve as a mechanism of survival for those bacteria that can form them. When a bacterium is subjected to unfavorable environmental conditions, it often responds by developing a dormant spore that may remain viable for years. When exposed to the proper stimuli, the spore germinates and produces a new bacterial cell.

Bacterial Nutrition and Metabolism

One of the most remarkable qualities of bacteria is their vast metabolic diversity. There probably is no natural organic molecule that cannot be degraded by at least one bacterial species. This makes bacteria extremely important components of the ecosystem. They play a critical part in the decomposition and recycling of organic matter, including many man-made materials.

Most bacteria require oxygen for growth; that is, they are **aerobic.** Other bacteria are **facultative anaerobes.** They do not require oxygen but grow better when it is present. *E. coli* is a well-known facultative bacterium. It grows adequately under anaerobic conditions in the human intestine but flourishes when incubated in the aerobic atmosphere of an incubator.

Anaerobic bacteria not only do not require oxygen for growth, but they are actually killed by it. Members of the genus *Clostridium*—the causative agents of botulism, tetanus, and other serious diseases—are **obligate anaerobes;** that is, they can grow only under anaerobic conditions. This explains why tetanus normally develops only from deep puncture wounds and not surface scratches.

Purple and green bacteria resemble plants in being **photoautotrophs,** using light as their energy source and producing reduced carbon compounds using carbon dioxide as a precursor. They lack the chloroplasts possessed by eukaryotes and have their photosynthetic membranes spread throughout the cytoplasm. Their distinctive colors show that their chlorophylls and other photosynthetic pigments differ from those of plants.

Bacteria and Diseases

When most people think of bacteria, they think of "germs" and disease. This is quite natural. Disease-causing (pathogenic) bacteria have had a great effect on the human race.

The **germ theory of disease** was developed during the nineteenth century. Louis Pasteur showed that the air is filled with microorganisms. The British surgeon Joseph Lister was very impressed by Pasteur's work. He worked to protect wounds from airborne microbes by sterilizing surgical instruments, dressings, and the operating room. His success provided indirect evidence in support of the idea that "germs" might cause disease.

Figure 27.11 Electron micrograph of a bacterial cell (*Clostridium botulinum*) containing an endospore. This organism causes the form of food poisoning known as botulism.

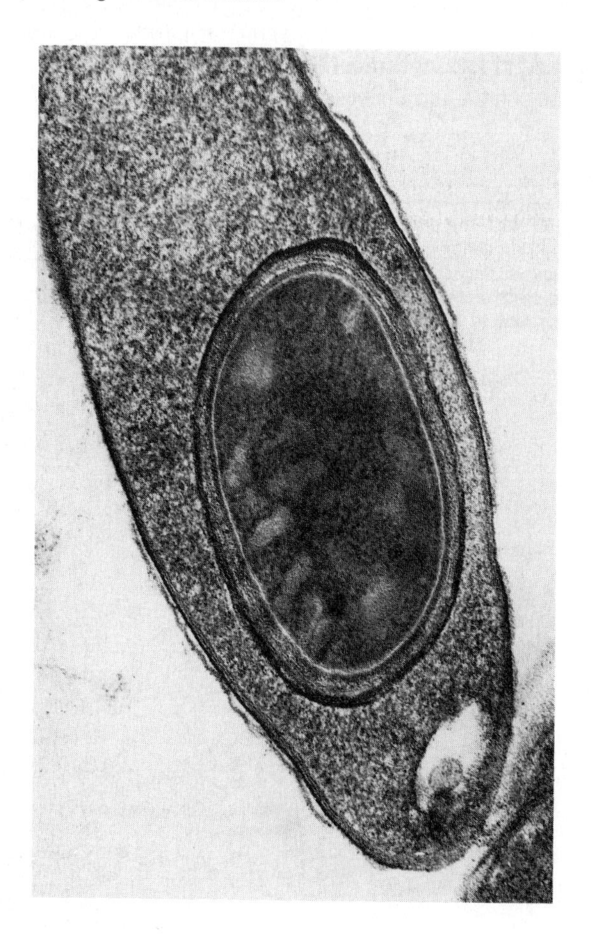

In 1876, the German physician Robert Koch published a detailed study of anthrax, a disease of cattle that could be transmitted to humans. He proved conclusively that anthrax was due to the spore-forming bacterium *Bacillus anthracis*. Koch's work was a milestone because it inspired others to search for other bacterial pathogens—the war against bacterial disease was under way. Bacterial pathogens for many of the most deadly diseases were discovered during the next thirty years: anthrax, gonorrhea, typhoid fever, tuberculosis, cholera, diphtheria, tetanus, pneumonia, meningitis, gas gangrene, plague, botulism, dysentery, whooping cough, and many more. These discoveries made possible the subsequent progress in disease control that has transformed human lives.

Beneficial Activities of Bacteria

Pathogenic bacteria have such widespread impact that we often consider all bacteria to be dangerous and harmful. In truth, beneficial bacteria vastly outnumber the pathogens and are so important that the ecosystem could not function without them. For example, bacteria play an indispensable role in the carbon cycle. They help degrade dead organic material, mostly soil humus, to CO_2. This makes the carbon available for photosynthetic incorporation by plants. Bacteria are also essential in the nitrogen cycle (see page 567), which supplies nitrogen to all living things.

Of course, bacteria are indispensable to many industries. Lactic acid bacteria are used by the dairy industry in the manufacture of yogurt, cottage cheese, and cheese. The distinctive flavors of cheddar, Swiss, Parmesan, and Limburger cheeses are the result of bacterial products. Vinegars are made by allowing acetic acid bacteria to oxidize the alcohol in wine, apple cider, or malt. With the exception of penicillin and ampicillin, most important antibiotics are produced by actinomycetes, a specialized group of bacteria.

The future of industrial microbiology looks even more promising. It is quite likely that methane for use as fuel can be manufactured from agricultural organic wastes by methane-producing bacteria. Several types of bacteria already are being grown for use as biological insecticides.

Cyanobacteria

The cyanobacteria (blue-green algae) are a group of prokaryotic organisms that contain membranes specialized for photosynthesis (figures 27.12 and 27.13). Like bacteria they lack a true nucleus and other membranous structures found in eukaryotic cells. Even their cell walls are similar to those of bacteria.

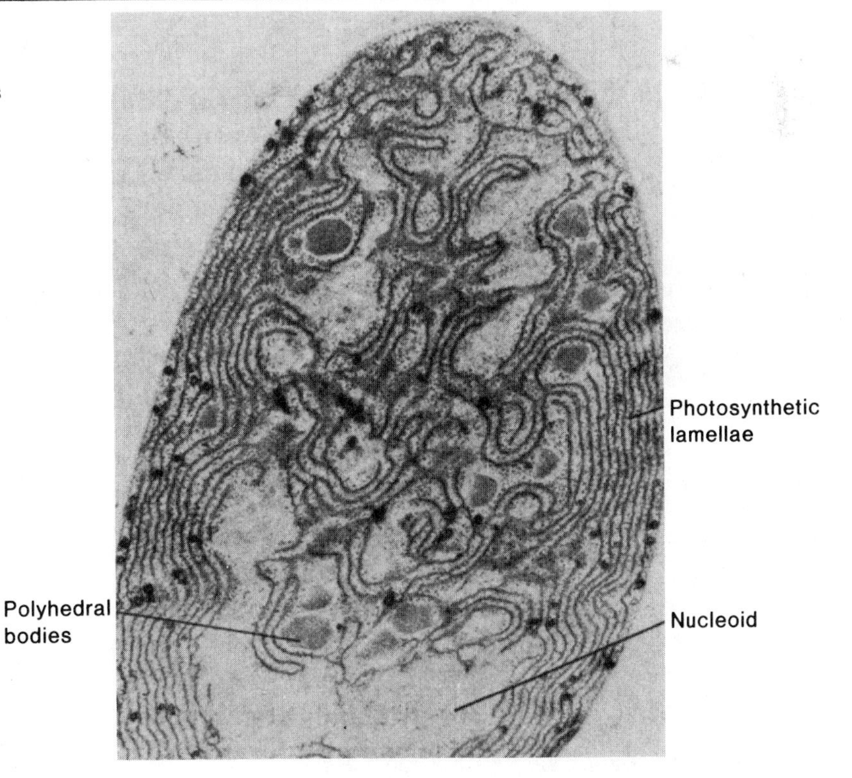

Figure 27.12 An electron micrograph of the cyanobacterium *Anabaena azollae*. Note the prokaryotic structure and the extensive photosynthetic membranes. Polyhedral bodies are structures that seem to be involved in photosynthetic functions.

Figure 27.13 Drawings of some common cyanobacteria (blue-green algae) that you might observe in water samples from lakes, ponds, or puddles. (*a*) *Lyngbya.* (*b*) *Gomphosphaeria.* (*c*) *Chamaesiphon.* (*d*) *Anabaena.* Akinetes are thick-walled resting structures that can survive periods of drying or cold and germinate under more favorable conditions. (*e*) *Spirulina.* (*f*) *Oscillatoria.* (*g*) *Aphanocapsa.* (*h*) *Gloeocapsa.* (*i*) *Nostoc.* (*j*) *Merismopedia.*

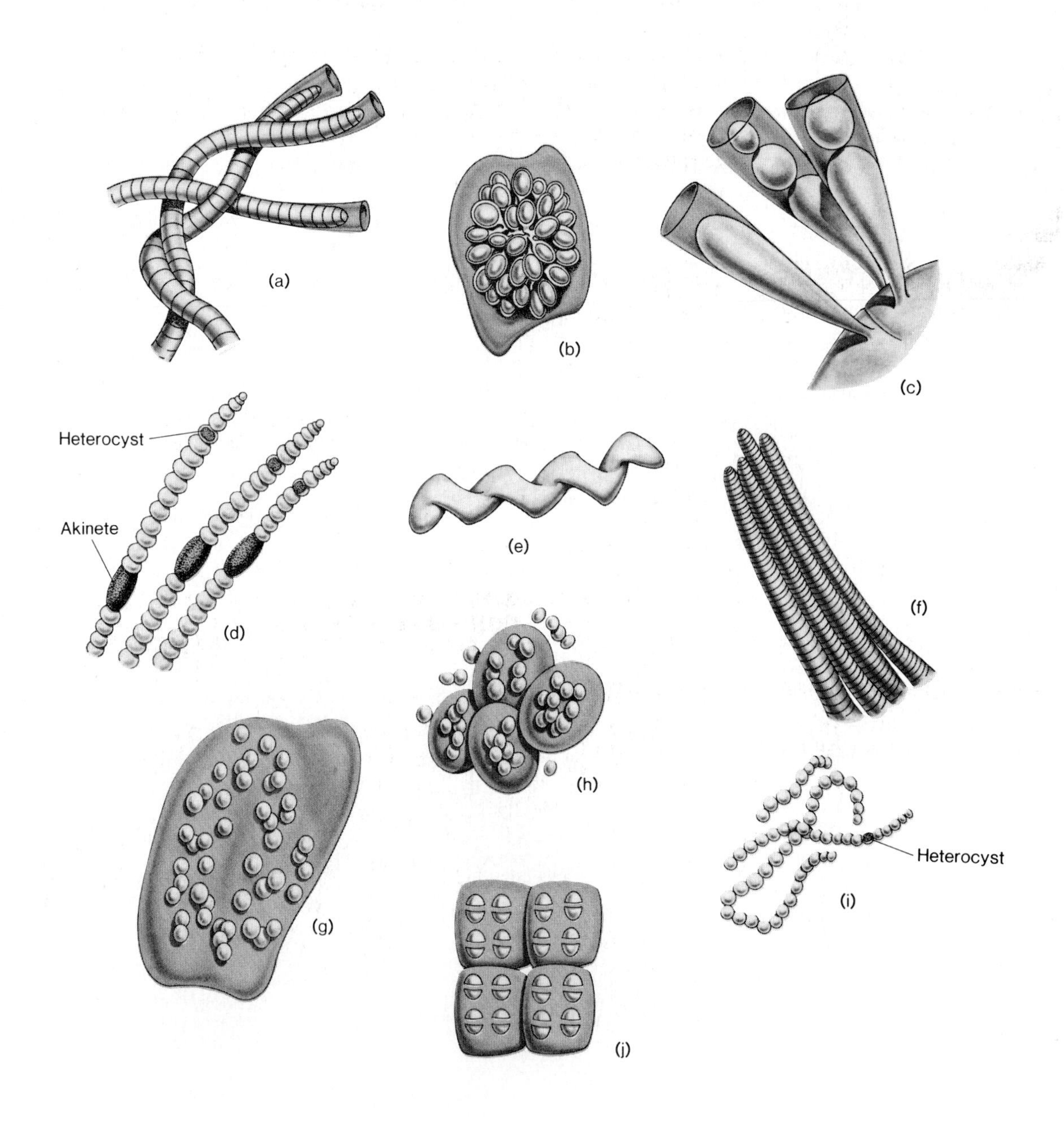

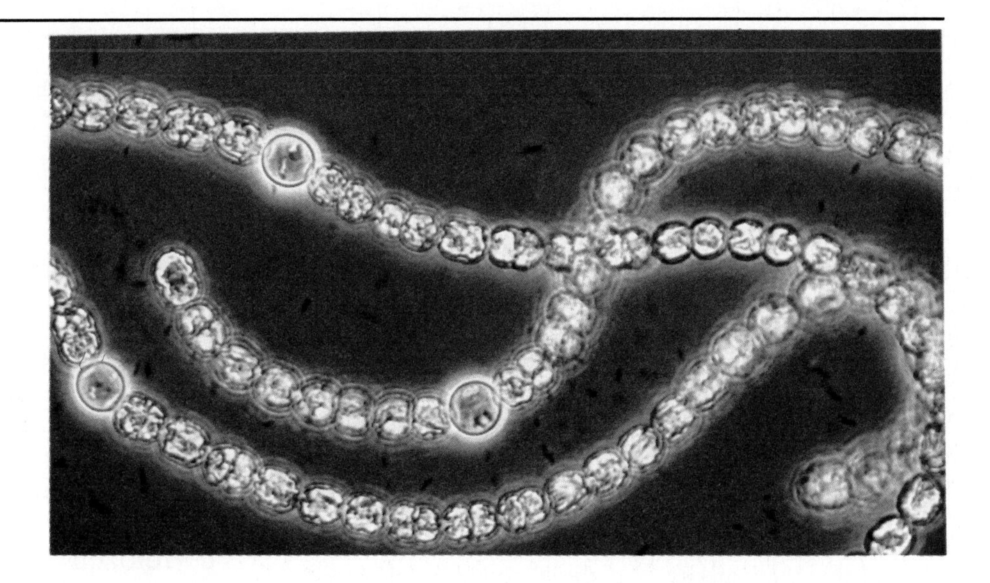

Figure 27.14 Filaments of the cyanobacterium *Anabaena* showing heterocysts, which form when N_2 is the only available nitrogen source.

Many cyanobacteria have blue pigments in addition to chlorophyll and thus are blue-green; others may be brown or red.

A number of cyanobacteria can fix nitrogen; that is, they can use nitrogen gas (N_2) from the atmosphere and incorporate nitrogen into forms that they and other photosynthetic organisms can use. Some cells specialize to become nitrogen fixation centers called **heterocysts** (figure 27.14) when there is no source of nitrogen other than nitrogen gas (N_2). In certain rice-growing regions of the world, nitrogen fertilizers are unnecessary because cyanobacteria abound on the surface water of paddy fields. Consequently, rice may be grown on the same land year after year without the addition of fertilizers.

Although exceedingly beneficial and ecologically important, cyanobacteria also may become a nuisance. They thrive in environments high in phosphates and nitrates. If care is not taken in the disposal of industrial, agricultural, and human wastes, phosphates and nitrates drain into lakes and ponds, resulting in a "bloom" of cyanobacteria. The surface of the water becomes cloudy. As a result, light available for photosynthesis by other aquatic algae is reduced. Furthermore, the toxic by-products of the cyanobacteria can kill fish.

Archaebacteria

A few groups of bacteria living in extreme environments possess a set of characteristics that suggests they may have an evolutionary history that is different from all other bacteria and cyanobacteria. These unusual bacteria are known as **archaebacteria.**

Archaebacteria differ from other prokaryotic organisms in several ways. The cell walls of archaebacteria differ chemically from the walls of all other bacteria. They also have a different form of ribosomal RNA and have protein synthesis mechanisms that differ from those of other bacteria.

Because of these fundamental differences, it has been suggested that archaebacteria should be placed in a separate, sixth kingdom of living things. For now, however, we will simply note these differences and leave archaebacteria in the kingdom Monera because they are prokaryotic cells just as the other bacteria and cyanobacteria are.

The Kingdom Protista

In many ways, this is the most diverse and fascinating of the five kingdoms because the protists include many different kinds of simple eukaryotic organisms. The protists that we will consider in this chapter are listed in table 27.2.

The Origin of the Eukaryotic Cell

All of the organisms that we will consider in the rest of this chapter and in chapters 28 and 29 are eukaryotic. Their cells have a true nucleus and contain an array of membranous organelles such as chloroplasts, mitochondria, and endoplasmic reticulum. The radical differences between prokaryotes and eukaryotes have provoked a great deal of research and speculation among biologists as to how eukaryotic cells might have arisen in the first place.

Probably the majority of biologists think the **endosymbiotic theory** suggested by Lynn Margulis and others to explain the evolution of eukaryotic cells is correct. **Symbiosis** is an intimate and long-term association between two organisms of different species. Symbiotic associations are widespread and important. Since symbiotic associations are prevalent today, such associations might have originated and been favored by natural selection early in the development of life. Thus, the mitochondria and chloroplasts of today may have been the endosymbionts (internal symbionts) of yesterday. This concept is the foundation of the endosymbiotic theory.

The fossil record indicates that prokaryotes may have arisen about 3,000 million years ago, while eukaryotes are thought to have first appeared about 1,500 million years ago. During the 1,500 million years between the appearance of prokaryotes and the origin of the eukaryotes, contact among different prokaryotes must have taken place frequently. The first step in the evolution of the eukaryotic cell is thought to have occurred when a large anaerobic amoeboid prokaryote ingested a smaller aerobic bacterium and stabilized its prey as an endosymbiont rather than digesting it (figure 27.15). This developed into the mitochondrion, which is the site of aerobic respiration in eukaryotic cells. Flagella might have arisen through the ingestion of prokaryotes similar to spirochetes. Ingestion of prokaryotes somewhat resembling present-day cyanobacteria (blue-green algae) could lead to the endosymbiotic development of chloroplasts in a fashion similar to formation of mitochondria from aerobic, nonphotosynthetic bacteria.

Regardless of the exact mechanism involved, the emergence of the eukaryotic cell led to a dramatic increase in the complexity and diversity of living things.

Table 27.2
The Kingdom Protista

Protozoa	
Mastigophora	Flagellated protozoa
Sarcodina	Pseudopodial protozoa
Sporozoa	Spore-forming protozoa
Ciliophora	Ciliated protozoa
Unicellular Algae	
Euglenophyta	*Euglena* and relatives
Pyrrophyta	Dinoflagellates
Chrysophyta	Diatoms and others
Funguslike Protists	
Water molds	
Slime molds	

The Protozoa

The protozoa are single-celled, heterotrophic organisms. They cannot manufacture fundamental organic compounds from inorganic molecules but must have a supply of organic molecules produced by other organisms. Although they are unicellular, the protozoa certainly are not uniformly simple organisms. Many are highly specialized cells in which organelles take the functional roles that organ systems play in complex multicellular organisms.

Protozoa are found in a wide variety of environments. The majority are free-living and inhabit freshwater or marine environments. There are also a number of parasitic protozoa. Some very widespread diseases such as malaria are caused by protozoa.

We can separate protozoa into four groups: Mastigophora, Sarcodina, Sporozoa, and Ciliophora (see table 27.2). These four groups are distinguished from one another on the basis of locomotor organelles and methods of reproduction.

Mastigophora

Mastigophorans have one or many flagella. The flagella have the characteristic 9 + 2 microtubular structure found in eukaryotic cells (see page 69). Mastigophorans (flagellates) are considered to be quite ancient. It is very likely that the other groups of protozoa arose from the flagellates. The majority of the mastigophorans are symbiotic and are of special interest because of their medical importance. Many are important human parasites. **Parasitism** is a symbiotic relationship in which the parasite derives benefit from the host at the host's expense. The parasite is dependent on the relationship, while the host might be far better off without the parasite. This is certainly true of the **trypanosomes,** for they cause severe human diseases, particularly in Africa (figure 27.16).

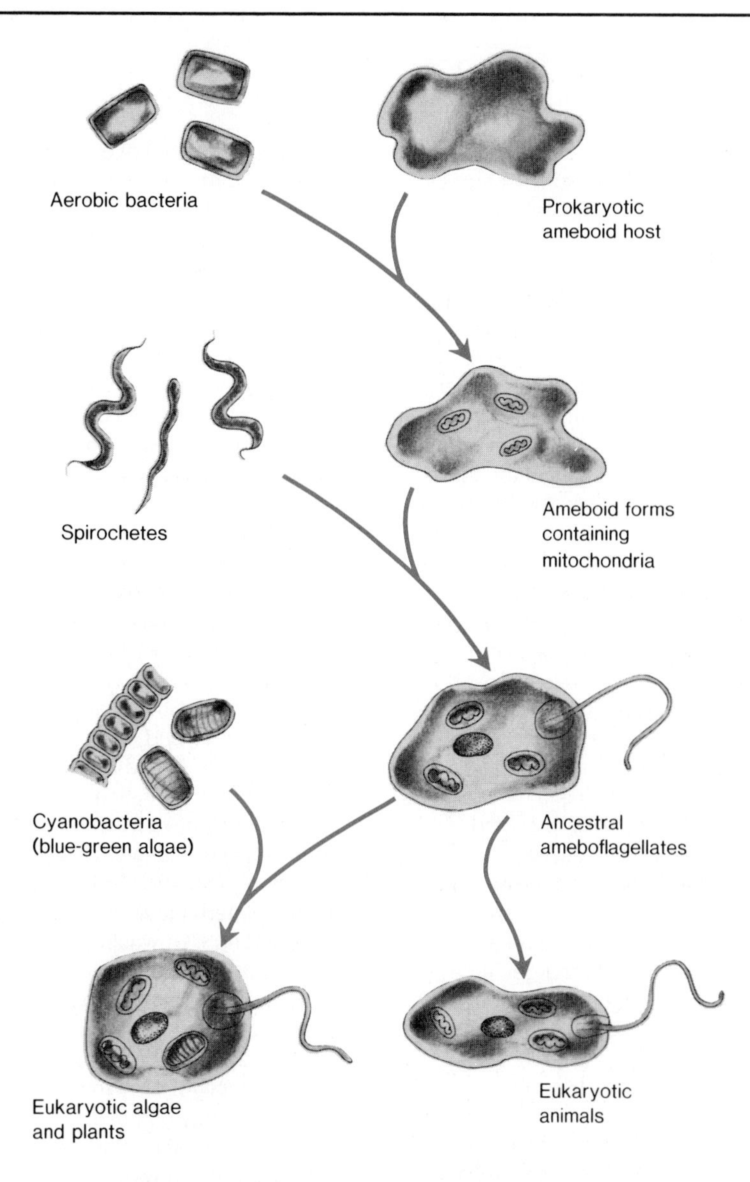

Figure 27.15 The endosymbiotic theory of how eukaryotic cells may have arisen.

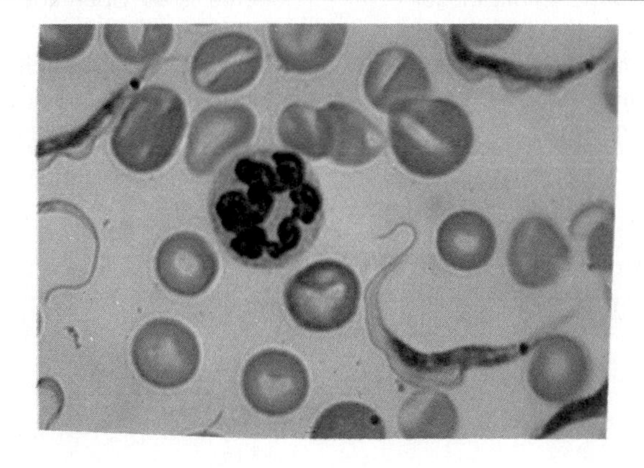

Figure 27.16 A stained blood smear from a patient suffering with African sleeping sickness showing trypanosomes among the blood cells. There is a white blood cell in the center of the picture (magnification × 1,000).

Box 27.1

The Origin of Life

When biologists survey the vast diversity of living things, they often wonder how life originated on earth. There is no way of knowing with certainty, but here are some ideas.

The earth probably formed by condensation of dust and gases approximately 4,500 to 5,000 million years ago. The lighter gases, such as hydrogen and helium, were lost as the planet formed because it was too small to hold them by gravity. It is thought that a primitive atmosphere was produced later by volcanic action, but there is considerable disagreement about the exact nature of this primitive atmosphere. Harold Urey proposed in the early 1950s that the atmosphere was strongly reducing due to the presence of hydrogen. He felt that it was composed of hydrogen, water, ammonia, and methane. More recent studies, however, favor the existence of a mildly reducing atmosphere consisting mostly of water, nitrogen gas, and carbon dioxide, with small amounts of hydrogen and carbon monoxide also present.

As early as the 1920s, the Soviet biochemist A. I. Oparin proposed that organic molecules could be produced from the inorganic constituents of the primitive atmosphere in the presence of an energy source, such as lightning or sunlight. In 1953, Stanley Miller provided support for Oparin's ideas through an ingenious experiment (box figure 27.1). Miller placed a mixture resembling the primitive atmosphere (methane, ammonia, hydrogen, and water) in a closed reaction vessel at 80°C. He then exposed it to electrical spark discharges for a week or more. At the end of this period, Miller discovered that a variety of amino acids and other organic acids had been produced.

A number of scientists have conducted similar experiments since the mid-1950s and have shown that a wide variety of biological molecules can be generated using several energy sources. Ultraviolet light or electrical discharges (like lightning, which undoubtedly acted on the earth's primitive atmosphere) can promote these syntheses.

The results are consistent with Oparin's original proposal that early oceans might have contained considerable numbers of organic molecules. Indeed, the early oceans might have resembled a very dilute broth or soup. The formation of this broth constitutes the first phase of what has been called **chemical evolution.**

The next phase would involve the joining of molecular building blocks to yield polymers, such as polypeptides, polynucleotides, and polysaccharides. Such molecules are relatively unstable but might have been synthesized under gentle conditions.

Macromolecules could have subsequently coalesced to form small droplets containing the polymers in an aqueous matrix. These **coacervate droplets,** as Oparin called them, could have begun to function as primitive cells. S. W. Fox has heated mixtures of amino acids at 130°C to 180°C to form amino acid polymers, which he calls **proteinoids.** When the hot solution is cooled, these proteinoid mixtures form small, cell-like structures, which Fox has named **microspheres.** Microspheres seem to possess a selectively permeable surface and show cell-like behavior. Fox has speculated that such microspheres might be able to incorporate or develop self-replicating polynucleotides and gradually evolve into protocells.

Eventually, in one way or another, biological evolution began with the formation of the first true cells. These must certainly have been heterotrophic forms, living at the expense of reduced organic molecules available in the "soup" that surrounded them. Eventually, as nutrients were depleted, the first autotrophs capable of incorporating CO_2 would have arisen. But these first autotrophs were not photosynthetic organisms. They were chemoautotrophs; that is, they derived energy for their synthetic activities not from light, but from chemical compounds available in their environment. This would have been followed by the development of the ability to trap light energy and use it in the process of photosynthesis. With the origin of photosynthetic cells that began to use water as an electron donor and produce oxygen, the primitive environment would have been altered. Spontaneous development of organic molecules would have ceased. As oxygen levels increased, cells capable of respiratory metabolism must have developed and flourished as a result of the cells' ability to make effective use of oxygen.

Having said all of this, however, we must caution you that this field is very speculative, since it obviously is impossible to observe or to describe precisely the original events. Many of the assumptions and theories about how the first life might have arisen are tentative and controversial.

(a)

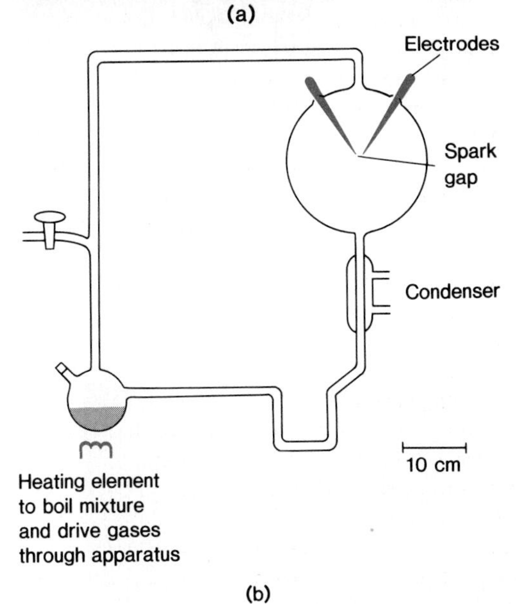

(b)

Box figure 27.1 The spark-discharge apparatus for synthesizing organic molecules from gases thought to be present in the primitive atmosphere. (*a*) Dr. Stanley Miller and his apparatus. (*b*) A diagram of the apparatus.

The majority of species of trypanosomes live in the blood of a variety of vertebrates, including humans. One of them, *Trypanosoma rhodesiense,* causes African sleeping sickness. A related flagellate, *T. brucei,* causes the fatal disease ngana in cattle.

Both *T. rhodesiense* and *T. brucei* are transmitted by the bite of the tsetse fly. The tsetse fly becomes infected by a diseased host when it takes a blood meal. The trypanosomes multiply and change into different life cycle forms within the fly's intestine. They then migrate to the fly's salivary glands, multiply further, and develop into infective forms. When the fly next feeds, the trypanosomes are transmitted in its saliva. Sleeping sickness and ngana make certain areas of Africa virtually uninhabitable. The only potentially effective control measure seems to be eradication of the tsetse fly.

Sarcodina

Sarcodines are capable of movement and feeding by extension of **pseudopodia** ("false feet"), which are cytoplasm-containing extensions of their plasma membranes. Pseudopodia come in a variety of shapes and forms (figure 27.17).

The majority of species of sarcodines are marine, but some species are freshwater amoebae living in the silt of streams, the ooze of ponds, on mosses, or in moist soil. Parasitic species are found in the digestive tracts and associated structures of vertebrates and invertebrates. A pathogenic species parasitizing humans, *Entamoeba histolytica,* causes amoebic dysentery.

Sarcodines feed by phagocytosis. They have a varied diet, ranging from algae and bacteria to protozoa of all kinds. They surround their prey with pseudopodia and engulf them.

One of the most common and popular sarcodines (at least in student labs) is *Amoeba proteus,* a large (up to 600 μm) freshwater amoeba usually found in slow-moving or still-water ponds, often on the underside of decaying leaves (figure 27.18).

Foraminiferans are important marine sarcodines that form tiny calcium carbonate shells (figure 27.19). The organisms extend pseudopodia through holes in their shells to feed. Upon death, these protozoa contribute to the mud on the ocean floor. They have been so populous through the ages that their shells have formed such massive deposits as the White Cliffs of Dover and the limestone used to build the Egyptian pyramids.

Figure 27.17 Three different types of pseudopodia. (*a*) *Amoeba proteus* with blunt pseudopodia. (*b*) A foraminiferan feeding with a net of fine pseudopodia. (*c*) *Actinophrys sol* with sharp pointed pseudopodia.

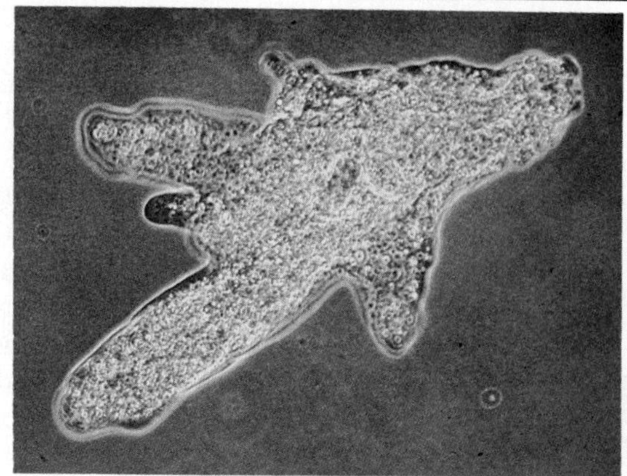

(a)

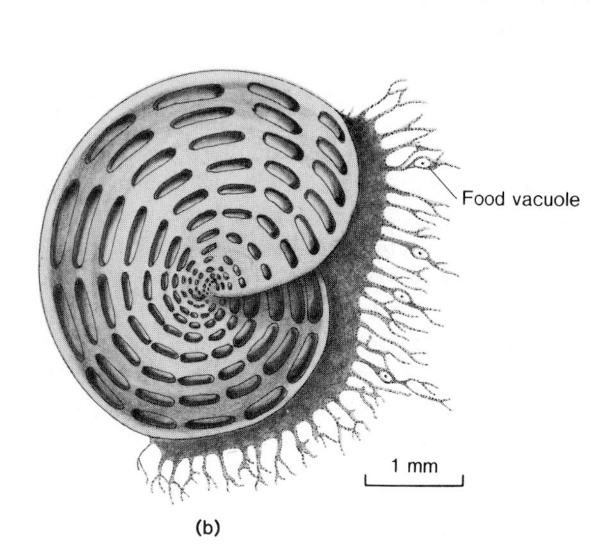

(b)

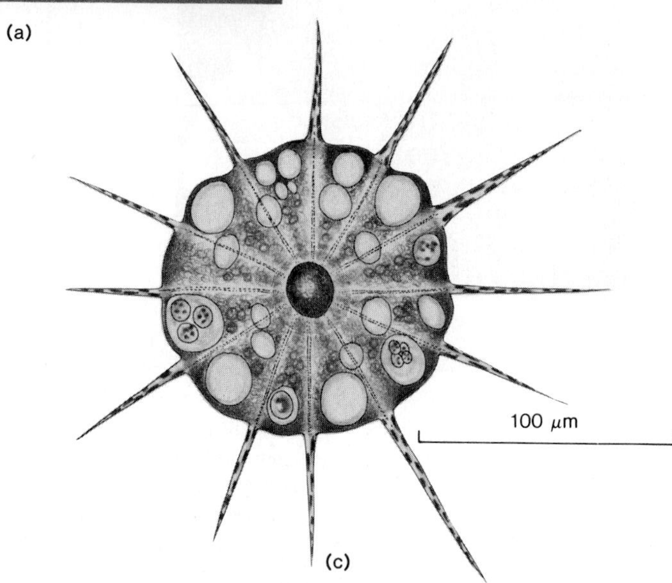

(c)

Figure 27.18 *Amoeba proteus*. (*a*) Sketch showing some structural details. (*b*) Scanning electron micrograph (magnification × 162).

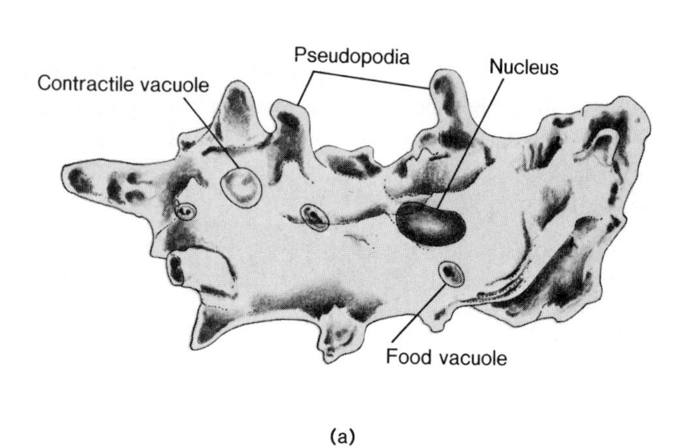

(a)

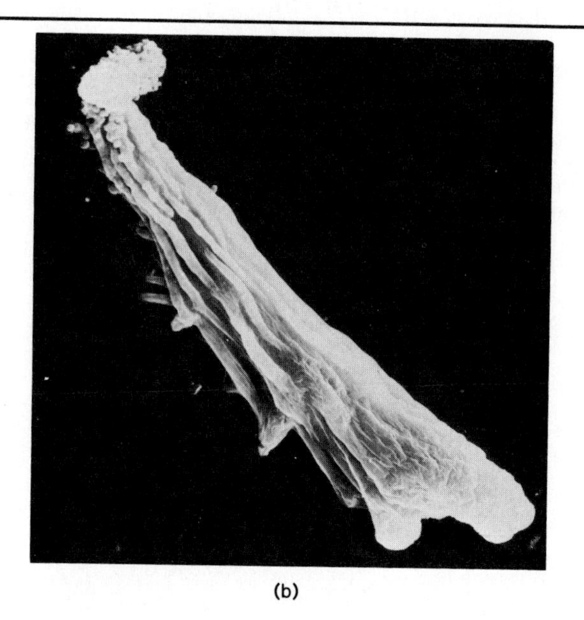

(b)

Figure 27.19 Scanning electron micrographs of two foraminiferan shells. (*a*) *Globigerina bulloides.* (*b*) *Elphidium crispum.*

Figure 27.20 Skeletons from a variety of radiolarians.

Radiolarians also contribute to the bottom ooze of the oceans. These marine forms float near the ocean surface. They have internal skeletons composed of silica or strontium sulfate (figure 27.20). Their skeletons are intricate, exquisite, and of almost infinite variety. These skeletons sink to the bottom after the death of the cells.

Sporozoa

Sporozoans are parasitic protozoa with complex life cycles that almost always involve the formation of infective spores at some point. Some sporozoans have been a scourge of humans and their domestic animals for longer than recorded history.

The most important human parasite among the sporozoa is *Plasmodium,* the causative agent of **malaria.** The life cycle of *Plasmodium vivax,* one of the species responsible for this devastating disease, is shown in figure 27.21. The infection is initiated when a female *Anopheles* mosquito feeds on a human host. The mosquito injects a small amount of saliva containing an anticoagulant, which prevents blood clotting. If the mosquito is infected, small haploid **sporozoites** may be injected too. Initially, sporozoites become established in liver cells of the host and undergo multiple divisions, their products being called **merozoites.** This phase within the host lasts

Figure 27.21 Life cycle of *Plasmodium vivax,* which causes a very common form of malaria in humans.

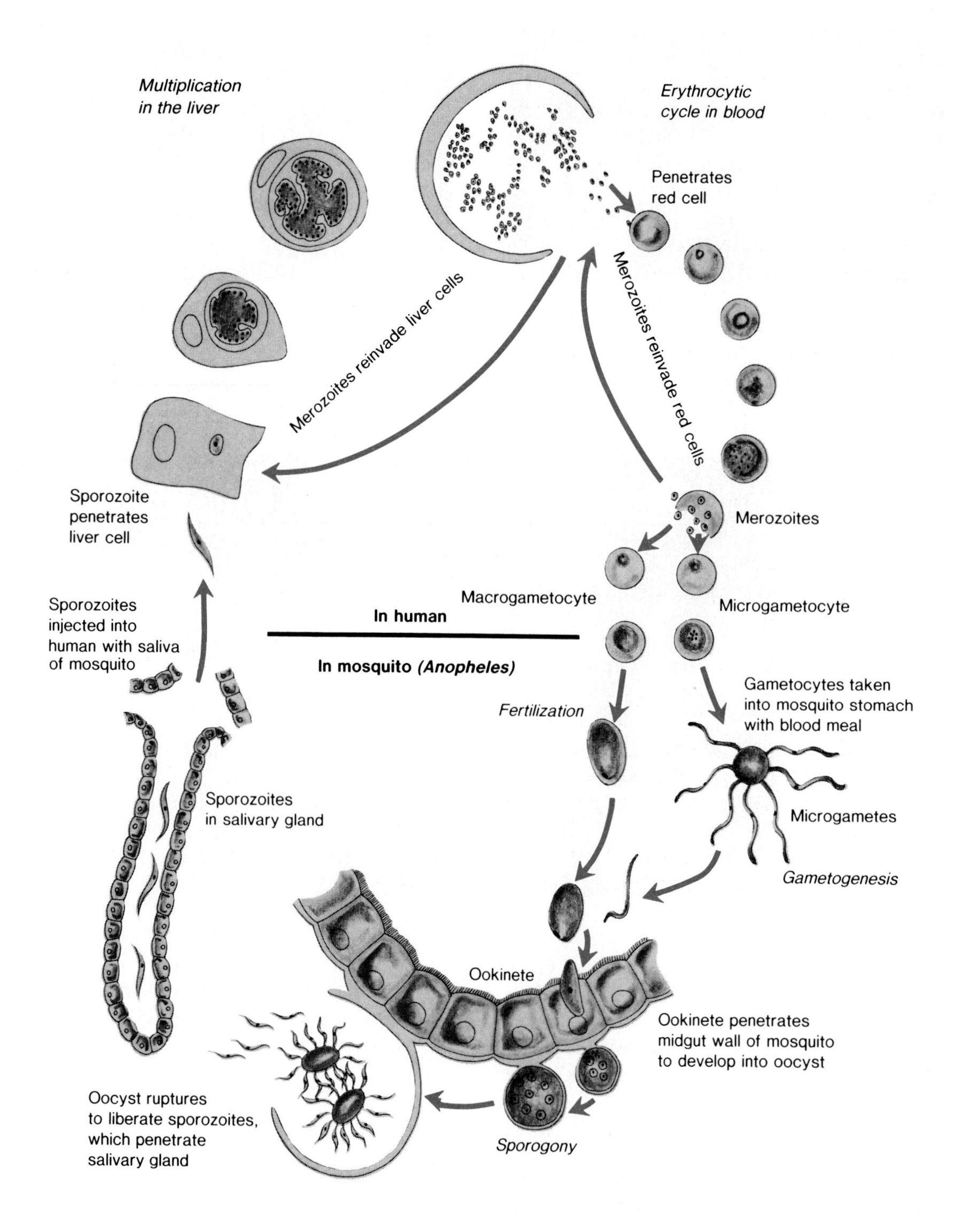

about ten days. Merozoites next invade erythrocytes (red blood cells) and repeat this division process. The erythrocytic phase is cyclic and repeats approximately every forty-eight hours in the most common form of malaria. Thus, every forty-eight hours a large number of the victim's erythrocytes simultaneously rupture and release merozoites that can invade uninfected erythrocytes. This sudden release of toxins and cell debris triggers an attack of the chills and fever characteristic of malaria. Occasionally, merozoites differentiate into **macrogametocytes** and **microgametocytes.** When taken up by a mosquito while feeding on the infected human host, the macrogametocytes and microgametocytes develop into female and male gametes. Fertilization results in a diploid zygote, the **ookinete.** The ookinete migrates to a position on the outside of the mosquito's gut wall and forms an **oocyst.** Once the zygote is established on the mosquito's gut wall, meiosis takes place, followed by the asexual formation of numerous sporozoites that migrate to the mosquito's salivary glands. The cycle is now complete, with only the bite of the infected mosquito needed to initiate a new infection in a human host.

Malaria is still a major killer despite modern medical advances. In fact, the number of people suffering from malaria seems to be increasing. Over a thousand million people live in areas where malaria occurs, and millions are infected each year. Generally, efforts at control have involved reduction of the mosquito population through massive spraying campaigns. The current resurgence of malaria is due to a number of causes—an increase in insecticide prices, development of insecticide-resistant strains of mosquitos, and parasite resistance to antimalarial drugs. Since William Trager at Rockefeller University managed to cultivate malarial parasites in tissue culture, the door has been opened to production of vaccines capable of protecting humans. Effective protection against malaria would drastically reduce the level of human suffering, but at the same time probably would result in a further increase in the rate of population growth in areas of the world already suffering problems due to overpopulation.

Ciliophora

Ciliophora are a complex group of ciliated protozoa. They are often simply called "ciliates." Many ciliates feed via a **cytostome** (cell mouth). In *Paramecium,* a ciliated protozoan often studied in biology labs, food is swept to the cytostome by cilia in the **oral groove,** and food vacuoles are formed (figure 27.22). Food is digested and nutrients absorbed as vacuoles circulate in the cell. Indigestible material is eliminated at the **cytoproct** (cell anus).

Figure 27.22 Structure of *Paramecium.* Contractile vacuoles pump out excess water, which enters the cell osmotically.

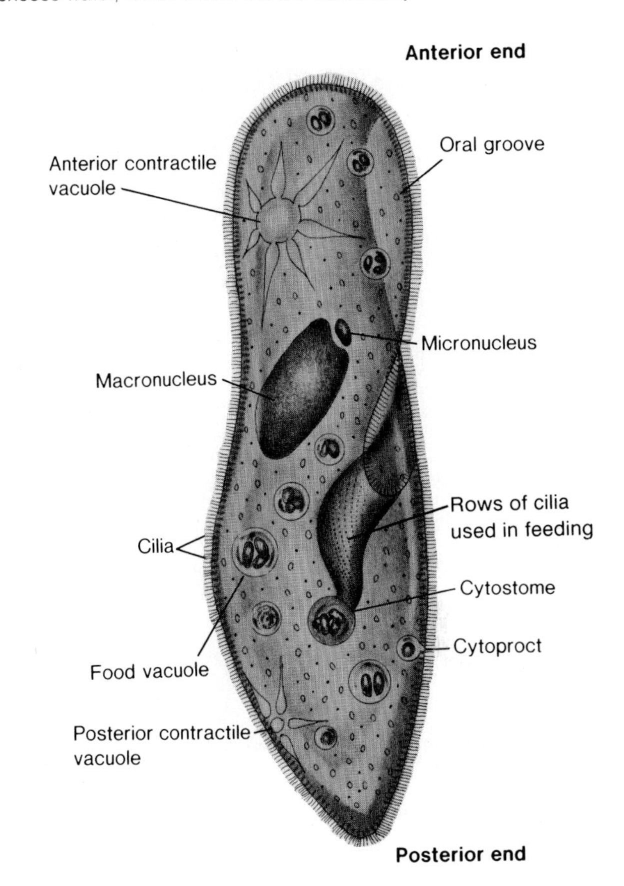

The ciliates are a diverse group. The barrel-shaped *Didinium* can gobble up a much larger *Paramecium* much like a snake swallowing a rabbit (figure 27.23). Some ciliates attach to surfaces. A colony of *Vorticella* looks much like a bouquet of beautiful flowers until they are disturbed and retract into a compact cluster (figure 27.24*a*). *Stentor coeruleus,* a favorite subject for research in regeneration, resembles a giant blue vase decorated with stripes when viewed under the microscope (figure 27.24*b*).

Figure 27.23 (*a*) *Didinium* feeding on *Paramecium*. A *Didinium* attacks a *Paramecium* twice its size (magnification × 520). (*b*) The *Paramecium* is folded in half, and the *Didinium* increases its length and engulfs it (magnification × 480).

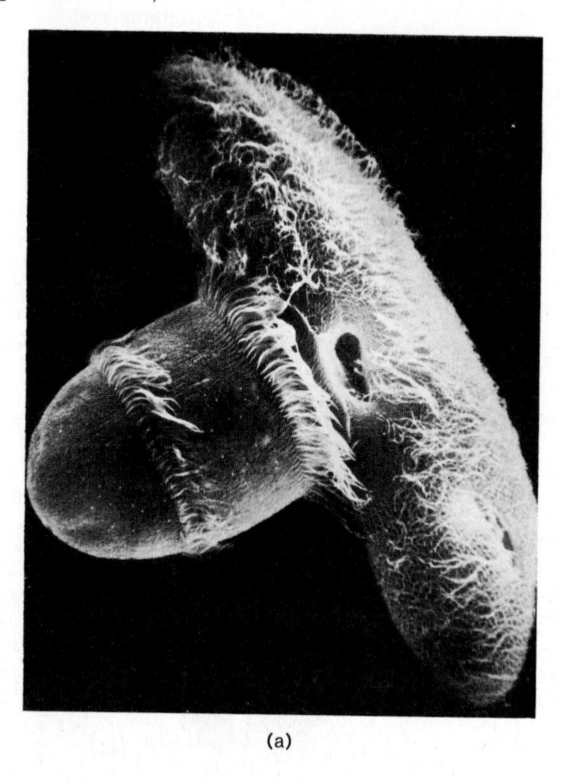

(a)

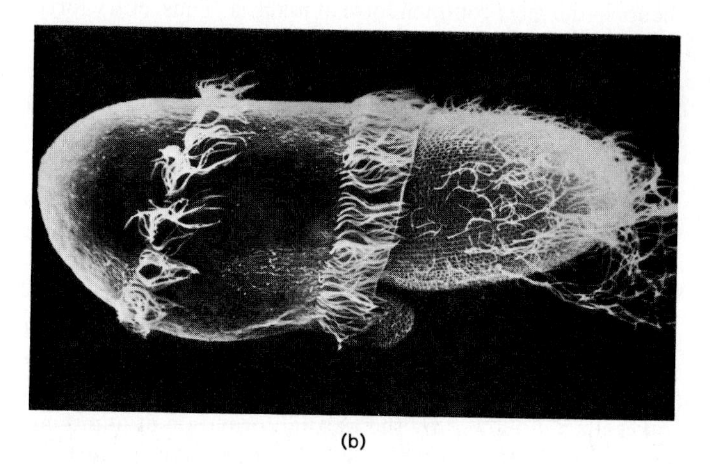

(b)

Figure 27.24 Ciliates that attach to surfaces. (*a*) *Vorticella* has a contractile stalk. (*b*) *Stentor*. This large ciliate may reach 2 mm in length. It sometimes attaches and sometimes moves about freely.

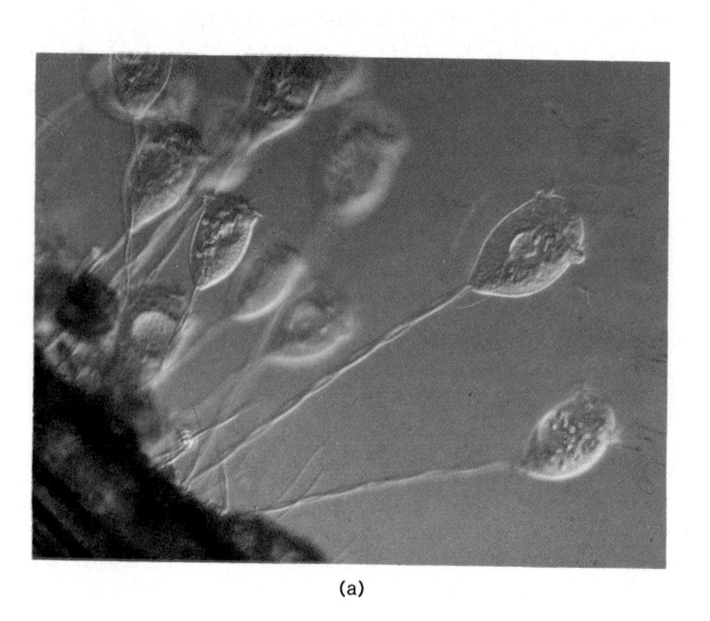

(a)

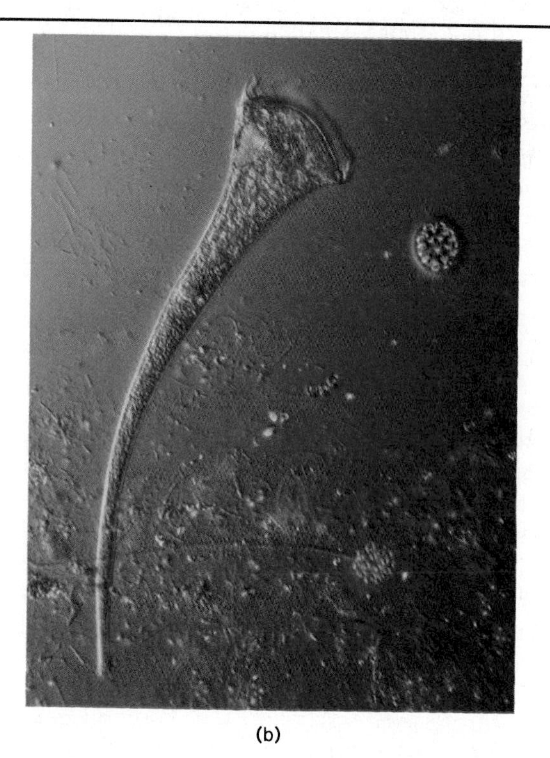

(b)

Figure 27.25 Dinoflagellates. (*a*) *Peridinium* has prominent surface plates. (*b*) The unarmored dinoflagellate *Gymnodinium*. Note the lack of surface plates. (*c*) A scanning electron micrograph of *Peridinium*. Note the plates and grooves.

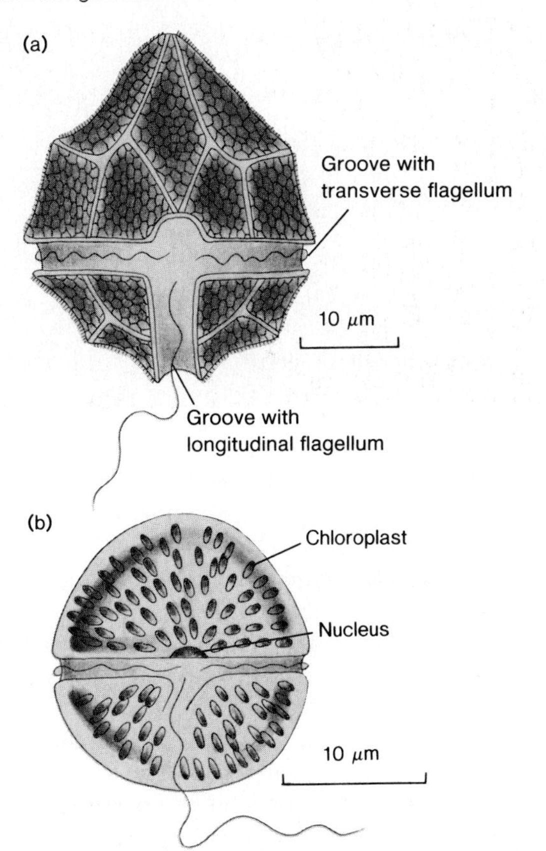

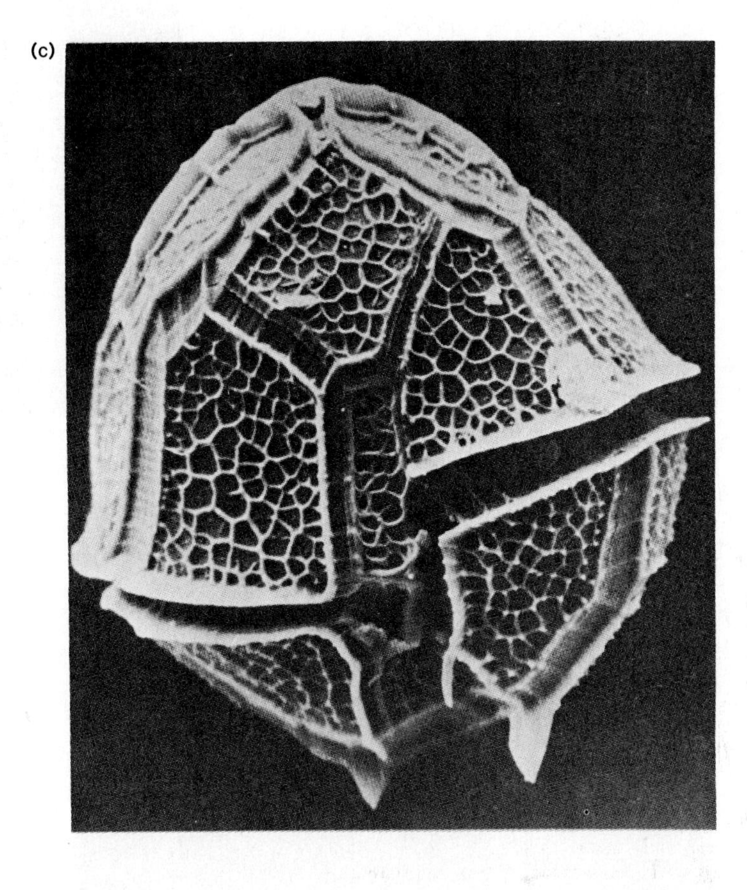

Unicellular Algae

There are several groups of unicellular algae. The **euglenoids** (Euglenophyta) are unicellular, flagellated photosynthesizing cells that are commonly found in fresh water. The genus *Euglena* is typical of this group. *Euglena* does not have a rigid cell wall, but rather has a flexible protein-containing wrapping outside its plasma membrane. *Euglena* can respond specifically to light from one direction and move toward it. Thus, many individuals in a *Euglena* culture will cluster on the side of a culture vessel toward a bright light source.

Dinoflagellates (Pyrrophyta) are extremely numerous photosynthetic marine organisms and are very important primary producers of organic nutrients in food chains in the ocean. Each dinoflagellate has two flagella situated in grooves on the cell surface (figure 27.25).

Dinoflagellates sometimes multiply extremely rapidly and enormous numbers can be produced in a short period of time. Certain pigmented species can discolor the water, leading to a condition called the **red tide.** These bursts of reproductive activity, or **blooms,** can have an economic impact on the region in which they occur. Toxic by-products can affect aquatic organisms, causing widespread fish kills. The same toxicity causes a condition known as paralytic shellfish poisoning (PSP). PSP is widespread throughout the world and occurs during the months of May through October.

Diatoms (Chrysophyta) are intriguing silica-clad organisms found in both freshwater and marine habitats (figure 27.26). Their glasslike remains accumulate as sediment on ocean floors. This **diatomaceous earth,** as it is called, is mined and used commercially in the making of insulation, as an abrasive in toothpaste and silver polish, and in water-filtering systems. From an ecological standpoint, diatoms are considered to be the most important primary producers in the marine ecosystem because of their enormous numbers. As such, they represent an important source of food for many aquatic animals.

Figure 27.26 A diatom. Diatoms are major producer organisms in aquatic food chains. Their silica-containing shells appear glassy.

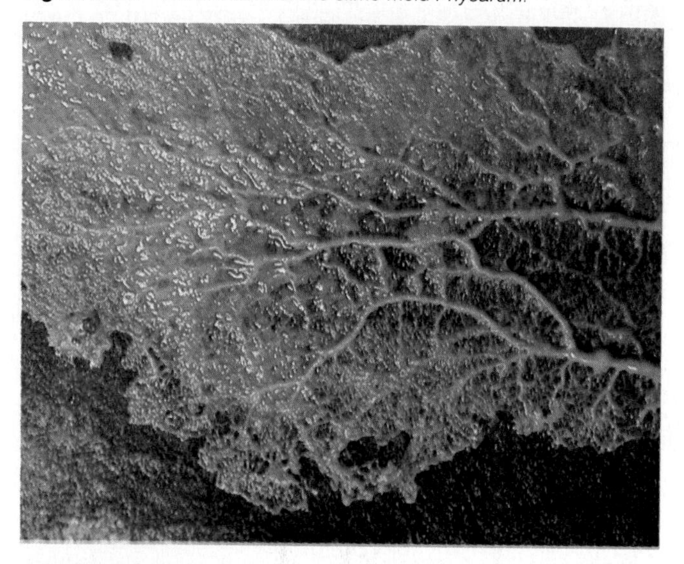

Figure 27.27 Plasmodium of the slime mold *Physarum.*

Funguslike Protists

A number of organisms that have cell walls and obtain nutrients by absorption (as fungi do) are nevertheless grouped with the protists because of their simple organization.

Many **water molds** (Oomycota) live in aquatic environments and form the cottony masses you may have seen on dead insects and fish. One of them, *Saprolegnia,* parasitizes fish, including aquarium fish, and can cover a fish completely and kill it.

A soil-dwelling parasitic member of this group has actually shaped history. *Phytophthora infestans* causes potato blight and was responsible for the 1840s potato famine in Ireland. The entire potato crop was destroyed in one week during the summer of 1846. The mass migration of the Irish to the United States during the last century was at least partially due to this organism.

The **slime molds** (Gymnomycota) are an unusual and varied collection of amoeboid organisms that possess both plant and animal characteristics. Plasmodial slime molds are found worldwide in moist, dark areas, such as under the bark of decaying trees or in layers of decomposing leaves. The active, vegetative form of the organism is a **plasmodium** that moves over surfaces, feeding on bacteria, protozoa, and other organisms (figure 27.27). As it grows, its nuclei undergo mitosis, but there is no division of the cytoplasm. Thus, the plasmodium contains a large mass of cytoplasm with many nuclei in it. Plasmodial contents are constantly mixed by cytoplasmic streaming. Plasmodia can become fairly large (30 cm or more across) and quite colorful. The sudden appearance of a large, bright yellow plasmodium on someone's lawn has more than once caused a panic in a neighborhood or community.

When the plasmodium has matured, it creeps out into a lighted area (light is often required for fruiting) and forms **fruiting bodies** (figure 27.28). Meiosis takes place as the spores develop within the **sporangium** (the part of the fruiting body that contains the spores) so that the mature spores are haploid. The spores can survive for years under unfavorable environmental conditions and then suddenly germinate in the presence of moisture to release **myxamoebae** or flagellated **swarm cells.** These haploid cells feed and reproduce. Eventually, they fuse to form a diploid zygote. The zygote then feeds, grows, and multiplies its nuclei through mitotic divisions. Finally, a mature plasmodium develops, and the life cycle has come full turn.

Slime mold fruiting bodies, though only a few millimeters tall, are among the most beautiful living things (figure 27.29).

Figure 27.28 The life cycle of a plasmodial slime mold. (Different parts are drawn at different magnifications.)

Figure 27.29 Plasmodial slime mold fruiting bodies. (*a*) *Hemitrichia*. (*b*) *Stemonitis*.

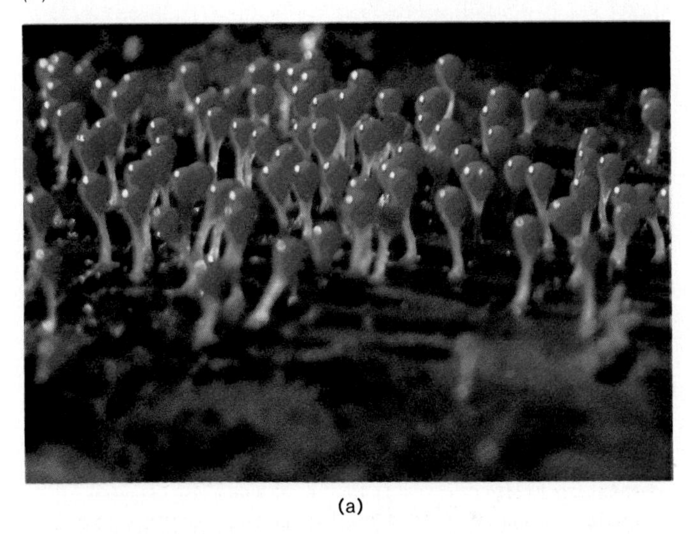

(a)

(b)

Summary

Taxonomy is concerned with naming organisms and arranging them in a system of classification. The genus name and the species name together identify a species. The major taxonomic ranks used, in decreasing order of breadth, are: kingdom, phylum, class, order, family, genus, and species.

We use the five-kingdom system, which uses cellular structure, general level of organizational complexity, and modes of nutrition (photosynthetic, absorptive, ingestive) to group organisms into one of five kingdoms: Monera, Protista, Fungi, Animalia, and Plantae.

Viruses are not considered to be living in the usual sense of the word. To reproduce, viruses enter host cells and cause them to replicate viral nucleic acids and synthesize virus capsids.

Bacteria and cyanobacteria are unicellular prokaryotes. Prokaryotic cells differ from eukaryotic cells in that the former do not possess a true nucleus and lack membrane-bound organelles, such as mitochondria and chloroplasts.

Bacteria and cyanobacteria are important in the ecosystem. They are critical components of the nitrogen and carbon cycles, since they are involved in nitrogen fixation and the decomposition of dead organic matter. Bacteria are useful industrially in the manufacture of food, antibiotics, and other important products. Some bacteria are pathogenic for humans and the plants and animals on which humans depend.

The protists are unicellular and colonial eukaryotes. The eukaryotic cell is thought to have arisen from the prokaryotes around 1,500 million years ago. The majority of biologists think that eukaryotes arose through endosymbiosis when anaerobic prokaryotes ingested aerobic bacteria and photosynthetic bacteria. These bacteria survived and eventually evolved into present-day mitochondria and chloroplasts.

The protists encompass three major varieties of organisms—protozoa, unicellular algae, and funguslike protists. The protozoa are single-celled or colonial heterotrophs that can be divided into four groups—Mastigophora, Sarcodina, Sporozoa, and Ciliophora. Protozoa are found in almost all environments, and a number of them are important parasites.

The unicellular algae consist of three major groups: the Euglenophyta, Pyrrophyta, and Chrysophyta. The dinoflagellates and diatoms are the most important primary producers in the oceans.

Funguslike protists include water molds and slime molds.

Questions for Review

1. Briefly describe the five-kingdom system and its advantages over the older two-kingdom system.
2. What is a virus? Write a description of the major features of virus structure.
3. What are viroids?
4. In what ways do bacteria differ from eukaryotic microorganisms?
5. Describe some ways in which bacteria benefit humans or the human ecosystem.
6. What are cyanobacteria, and how do they differ from other bacteria?
7. What are heterocysts, and what is their significance?
8. Briefly describe the endosymbiotic theory of the origin of eukaryotic cells.
9. What are protozoa? Describe some characteristics of the Mastigophora, Sarcodina, Sporozoa, and Ciliophora.
10. Briefly describe African sleeping sickness and malaria in terms of the nature of the organism causing the disease and the way in which the disease is transmitted.
11. With reference to slime molds, define the terms plasmodium and fruiting body.

Questions for Analysis and Discussion

1. Suggest some reasons why the discovery of virus diseases might have shaken the confidence of biologists in the germ theory of disease developed only a short time earlier.
2. Explain why you might expect to find few heterocysts in cyanobacteria filaments that are part of a "bloom" overgrowing a lake that receives drainage from heavily fertilized agricultural land.
3. It has been said that all flesh is grass. This may indeed be true in the terrestrial environment. Explain why it might be appropriate to say that in the marine environment all flesh is diatoms and dinoflagellates.

Suggested Readings

Books

Brock, T. D.; Smith, D. W.; and Madigan, M. J. 1984. *Biology of microorganisms,* 4th ed. Englewood Cliffs, N.J.: Prentice-Hall.

Margulis, L. 1981. *Symbiosis in cell evolution.* San Francisco: W. H. Freeman.

Stanier, R. Y.; Adelberg, E. A.; and Ingraham, J. L. 1976. *The microbial world,* 4th ed. Englewood Cliffs, N.J.: Prentice-Hall.

Articles

Demain, A. L., and Solomon, N. A. September 1981. Industrial microbiology. *Scientific American.*

Dickerson, R. E. September 1981. Chemical evolution and the origin of life. *Scientific American* (offprint 1401).

Diener, T. O. 1983. The viroid—a subviral particle. *American Scientist* 71:481.

Gould, S. J. 1976. The five kingdoms. *Natural History 85*(6):30.

Prusener, S. B. October 1984. Prions. *Scientific American.*

Sanders, F. K. 1981. Interferons: An example of communication. *Carolina Biology Readers* no. 88. Burlington, N.C.: Carolina Biological Supply Co.

Walsby, A. E. August 1977. The gas vacuoles of blue-green algae. *Scientific American* (offprint 1367).

Woese, C. R. June 1981. Archaebacteria. *Scientific American* (offprint 1516).

Fungi and Plants

28

Chapter Outline

Chapter Concepts

1. Fungi are multicellular heterotrophic eukaryotes that absorb nutrients after secreting extracellular digestive enzymes.
2. The plant kingdom includes the multicellular photosynthesizing organisms.
3. A plant life cycle has a haploid phase (the gametophyte generation) and a diploid phase (the sporophyte generation).
4. The relative prominence of the two life cycle phases differs among the major groups of plants.
5. Ecological distributions of various plant groups are determined by their body organizations and reproductive patterns.
6. Terrestrial plants have large, dominant sporophyte generations with well-developed vascular tissue.
7. Terrestrial plant reproduction is adapted to life on dry land.

Biology in the 1950s was ripe for change. Many biologists still favored the two-kingdom scheme of classification. This scheme grouped many organisms that we now call protists, all of the fungi, and the "higher plants" in one huge, diverse plant kingdom. All other organisms were placed in the animal kingdom.

Other biologists, however, were beginning to ask if there was not a better way. Surely fungi and complex plants should not be grouped together. Fungi are heterotrophic; they must obtain complex organic nutrients from the environment. Plants are photosynthesizing autotrophs; they can manufacture required organic molecules using light energy and simple inorganic molecules from the environment.

In the 1960s, new classifications such as the five-kingdom scheme we saw in chapter 27 were proposed. These new classifications strongly influenced our views of relationships among living things. In this chapter, we will examine and compare the fungi (kingdom Fungi) and the plants (kingdom Plantae).

The Fungi

Most fungi are nonmotile (nonmoving) organisms composed of tubes or filaments called **hyphae.** The mass of hyphae of a single organism forms a **mycelium.**

Like plant cells, fungal hyphae have rigid cell walls. But these cell walls usually are formed of chitin, a substance seldom found in plants. (Chitin is, however, a major component of the hard exoskeletons that cover the bodies of insects, spiders, crabs, and many other animals.)

Fungi are heterotrophic; they must absorb dissolved inorganic and organic food materials. Thus, they secrete digestive enzymes and then absorb the soluble digestion products. Fungi are particularly important as decomposers, aiding in decomposition of dead matter and the subsequent recycling of inorganic and organic molecules in an ecosystem.

Fungi reproduce both asexually and sexually. Asexual reproduction can be by simple fragmentation of a mycelium to produce two individuals. Most fungi also produce numerous spores, each of which can develop into a new individual.

We can separate the fungi into four divisions[1] on the basis of reproductive characteristics.

1. Botanists use the term "division" as an equivalent for the term "phylum" in the animal kingdom. The term "division" also designates the largest taxonomic categories among organisms in other kingdoms that used to be called plants in the old two-kingdom scheme. Thus, we speak of divisions of fungi and unicellular algae.

Molds

Members of the division Zygomycota are widespread and include some of the familiar molds you can often find growing on bread and fruit.

A common member of this division is *Rhizopus stolonifer,* the black bread mold, so-called because mature **sporangia** (structures producing spores) turn black. The life history of *R. stolonifer* begins when a spore germinates on bread (figure 28.1). Hyphae spread over the surface of the bread, growing with amazing rapidity. Once growth starts, the mold can completely cover a piece of bread in a very short time. Specialized hyphae called **rhizoids** extend down into the bread where they serve as anchors and at the same time secrete digestive enzymes and absorb nutrients that the enzymes digest. Additional hyphae rise into the air as sporangiophores, stalks that support sporangia. Mature sporangia eventually burst, releasing spores that are dispersed by air currents.

Sexual reproduction involves specialized structures called **gametangia** that fuse to form a diploid zygote. The zygote wall then thickens and blackens. The resulting **zygospore** can remain dormant for several months. Meiosis takes place upon germination. One or more haploid sporangia are immediately produced and release spores that germinate after dispersal. The adult fungus is haploid; only the zygospore is diploid.

Yeasts and Their Relatives

The Ascomycota is the largest and most varied of the fungal divisions and includes the single-celled yeasts, as well as multicellular forms with extensive mycelia. The Ascomycota are sometimes called sac fungi because one type of spore produced by these fungi develops inside a saclike structure called an **ascus.** These asci develop within a large, fruiting structure.

Certain of the Ascomycota (for example, morels and truffles) are edible and highly prized as food delicacies. Morels develop large, above-ground, fleshy fruiting structures (figure 28.2). Truffles, in contrast, form subterranean fruiting structures that must be dug from the ground.

Single-celled Ascomycota are called yeasts. Yeasts are of great economic importance. They usually are found in environments where there is a great deal of sugar, such as on flowers and fruit. Wild yeasts associated with grapes, for example, are important in wine making. The yeasts ferment sugars to ethanol and CO_2. Baker's yeast, *Saccharomyces cerevisiae,* is responsible for the rising of dough during baking as a result of CO_2 production.

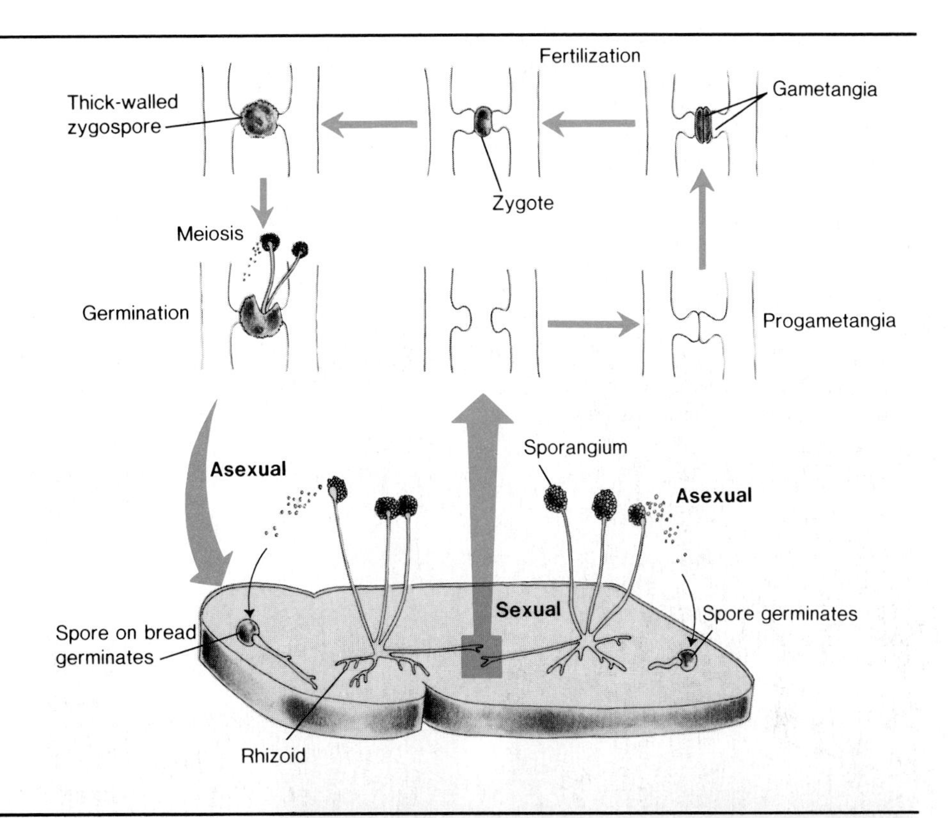

Figure 28.1 The life cycle of the black bread mold, *Rhizopus stolonifer*. Both asexual and sexual processes are shown.

Figure 28.2 An edible member of the Ascomycota. *Morchella esculenta* is a common morel that fruits in the spring.

Figure 28.3 Representative Basidiomycota. (*a*) *Pleurotus ostreatus*, the "oyster mushroom," is a common, gilled mushroom with a 2 to 30 cm cap that grows on tree trunks. (*b*) *Calvatia gigantea*, one of the largest puffballs, 20 to 50 cm across. (*c*) *Cantharellus cinnabarinus*, a funnel-shaped, red chanterelle mushroom with exposed gills.

(a)

(b)

(c)

Not all Ascomycota benefit humans. A number are plant parasites and are responsible for a variety of serious diseases, such as apple scab, powdery mildews, and Dutch elm disease.

Mushrooms, Rusts, and Smuts

Chances are that you are quite familiar with at least some of the Basidiomycota or club fungi. Basidiomycota fruiting bodies are often seen thrusting through the surface of a lawn or decorating the leafy floor of a forest. The division Basidiomycota is very large and includes mushrooms, puffballs, shelf or bracket fungi, rusts, and smuts. Many of the fruiting bodies of the Basidiomycota are quite beautiful (figure 28.3). They are distinguished by the formation of a swollen cell, the **basidium** on which **basidiospores** develop.

The life cycle of mushrooms is outlined in figure 28.4. A basidiospore, under suitable conditions, germinates and grows into a mycelium, which spreads in the soil.

A mushroom mycelium can sometimes grow outward for hundreds of years in an ever-expanding ring. The older mycelium in the center of the circle ages and dies even though the outer ring is still flourishing. This explains why mushrooms often develop in circles, which are sometimes called "fairy rings." (According to legend, mushrooms sprouted where the fairies had danced in a ring the night before.)

When the mycelium begins to form a fruiting body, a solid mass of hyphae called a **button** forms. This then pushes through the surface and develops into a mushroom. The mushroom cap supports a large number of platelike **gills.** The surfaces of these gills are coated with basidia, on which basidiospores develop. A large mushroom can produce and release many million basidiospores, which are then dispersed by the wind.

Many mushrooms are gourmet delicacies and have been prized since the time of the Roman Empire. The best-known edible mushroom is *Agaricus campestris,* one of the few gilled mushrooms that can be cultivated commercially (figure 28.5).

Figure 28.4 Life cycle of a mushroom.

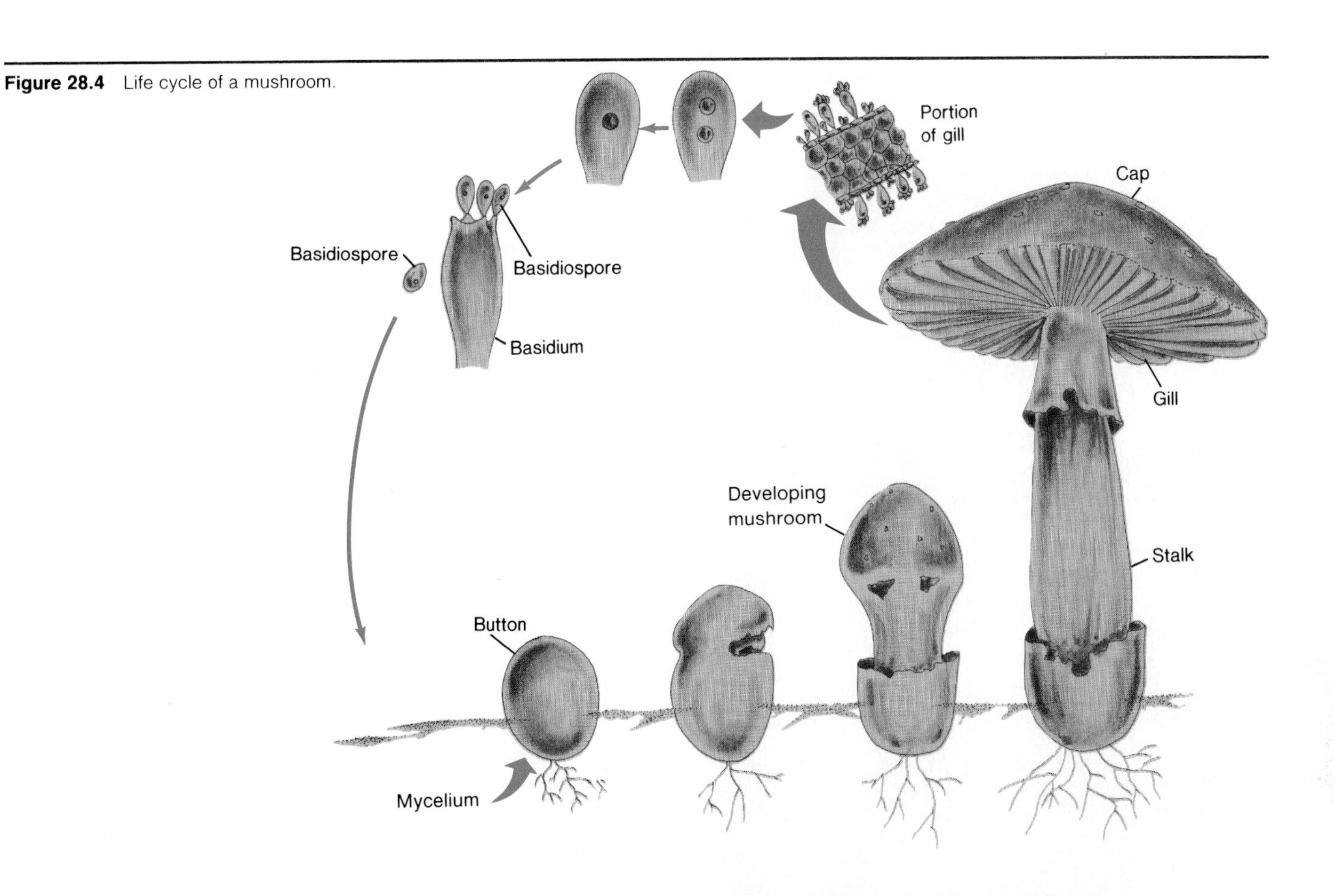

Figure 28.5 The meadow mushroom, *Agaricus campestris*, a common edible mushroom.

Figure 28.6 Typical lichens of the three major types. (*a*) A foliose lichen (leaflike) growing on a log. (*b*) A fruticose lichen (shrubby or hairlike). (*c*) Crustose (crustlike) lichens growing on a rock.

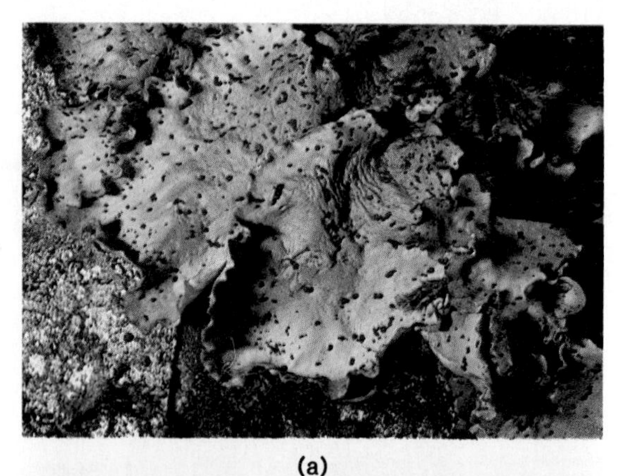

(a)

(b)

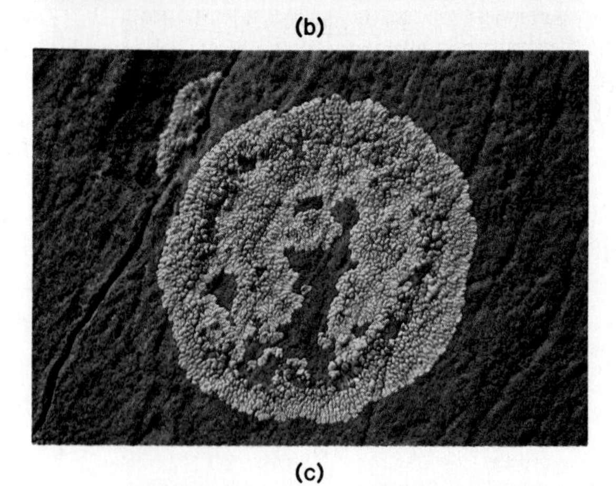

(c)

About 65,000 tons of mushrooms are produced annually in the United States. Unfortunately, some mushrooms (particularly members of the genus *Amanita*) are extremely poisonous.

Some Basidiomycota, the rusts and smuts, are serious plant pathogens and inflict extensive damage on agricultural crops. The wheat rust, *Puccinia graminis,* has been a severe problem for farmers. New rust-resistant strains of wheat must be bred continually to block or minimize outbreaks of new *Puccinia* strains.

Imperfect Fungi

When a fungus species lacks sexual reproduction or has not yet been observed reproducing sexually, it is placed in the Deuteromycota or Fungi Imperfecti division (only "perfect" fungi have sexual reproduction).

Many economically important fungi are found in this division. *Penicillium* molds are used in penicillin production and in the cheese industry, where they give cheeses like Roquefort and Gorgonzola their distinctive flavors.

Other imperfect fungi are important disease agents. *Candida albicans* grows on mucous membranes of the mouth and throat, causing the disease called "thrush." Skin can be affected by imperfect fungi that cause ringworm and the related "athlete's foot." Some species of *Aspergillus* produce **mycotoxins** when growing in stored food. These mycotoxins are quite poisonous and render the food unusable.

Lichens

Lichens are excellent examples of a symbiotic association. They are a combination of a green alga or cyanobacterium (blue-green alga) with a fungus, usually a member of the Ascomycota. The photosynthetic partner supplies the food, while the fungus protects the alga and absorbs water and minerals for both. There are around 20,000 known species of lichens, many with interesting shapes and beautiful coloring from fungal pigments (figure 28.6).

Lichens grow very slowly (a centimeter or less per year) but are remarkably resistant to environmental extremes. They are found on solid objects like rocks and trees in virtually all environments—from the desert to the arctic. Lichens produce acids while growing and thereby aid in breaking down rock in the initial stages of soil formation. They are eaten by many animals, including large mammals like reindeer. Despite their tolerance of harsh environmental conditions, lichens are very sensitive to air pollution and are being employed as pollution indicators in some places.

The Plants

The plant kingdom includes a diverse collection of multicellular photosynthesizing organisms. Most terrestrial (land-dwelling) members of this kingdom are vascular plants with highly specialized body parts: roots, stems, and leaves. Other plants are simpler. For example, algae are photosynthesizing organisms, living in aquatic environments, that have relatively unspecialized structure. Algae are difficult to fit neatly into the five-kingdom scheme. We included several groups of *unicellular* algae in the kingdom Protista in chapter 27. In the kingdom Plantae we can include the green algae, the brown algae, and the red algae, which, for the most part, are *multicellular* organisms (figure 28.7).

In the next few pages, we will compare structural and functional adaptations of various plant groups—from the simply organized green algae to the huge vascular plants with highly specialized body parts. We will begin by examining the three divisions of algae included in the plant kingdom.

Figure 28.7 The five-kingdom scheme of classification with the plant kingdom emphasized. Algae fall on the borderline between two kingdoms because some algae are classified as plants, and others are classified as protists.

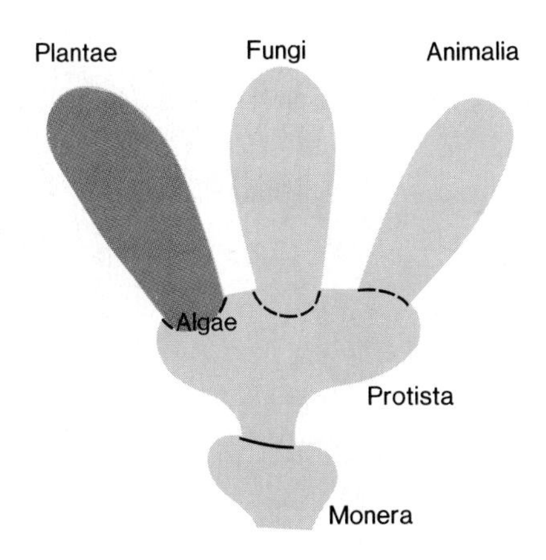

Green Algae

The division Chlorophyta is a diverse group of about 7,000 species of green algae. Some are unicellular; others are hollow balls of cells, filaments of end-to-end cells, or broad, flat sheets of cells. The majority of green algae live in fresh water, but some live in moist terrestrial environments, and many are marine.

Green algae possess the same types of chlorophylls found in all of the terrestrial plants. The photosynthetic protists do not. This is one of several reasons why green algae are included in the plant kingdom even though some of them are unicellular and thus do not fit the basic definition of plants as "multicellular photosynthesizing organisms."

Unicellular Green Algae

Some unicellular green algae are rather simply organized single cells. A good example is the flagellated unicellular alga *Chlamydomonas. Chlamydomonas* is found in pools, lakes, and even in damp soil. The ordinary **vegetative** (nonreproducing) *Chlamydomonas* cell has a cellulose cell wall and a pair of flagella. Each cell has a single, large, cup-shaped chloroplast. The chloroplast contains a conspicuous **pyrenoid,** which is the site of starch production. The **stigma,** or "**eyespot,**" is a modified portion of the chloroplast. It lies near the base of the flagella, as do two small **contractile vacuoles** that discharge rhythmically and expel excess water from the cell.

Chlamydomonas reproduces both by asexual and sexual processes (figure 28.8). Asexual reproduction may be either by mitotic cell division of one cell to produce two or by the production of a cluster of tiny **zoospores** inside a cell. The parent cell ruptures and releases the zoospores, each of which is a small copy of the parent cell. Subsequently, each zoospore grows to become an ordinary vegetative cell.

In sexual reproduction, vegetative cells either convert directly into gametes, or they divide mitotically to produce a number of gametes. Gametes of different **mating types** fuse in **syngamy** (fertilization). The different mating types look identical, but gametes of the same mating type will not fuse. Later, the zygote, which is produced by gamete fusion, secretes a thick wall and becomes dormant. The resistant, dormant cell, called a **zygospore,** can withstand adverse conditions, such as winter or a dried-up pond.

When conditions are suitable again, the zygospore germinates. Meiosis occurs, and four haploid, flagellated cells emerge through the ruptured wall. Each of these cells matures into an ordinary vegetative cell, and the sexual cycle is complete.

Although *Chlamydomonas* has mating types, the gametes are identical in appearance. Thus, they are classified as **isogametes** (iso = the same), and this pattern of reproduction is called **isogamy.**

The only diploid cell in the life history of *Chlamydomonas* is the zygote. All gametes, zoospores, and vegetative cells are haploid.

Figure 28.8 *Chlamydomonas.* A vegetative cell is shown in the center. In sexual reproduction, gametes must be of opposite mating types to fuse. The only diploid *Chlamydomonas* cell is the zygote, which develops a thick wall to become a resistant zygospore.

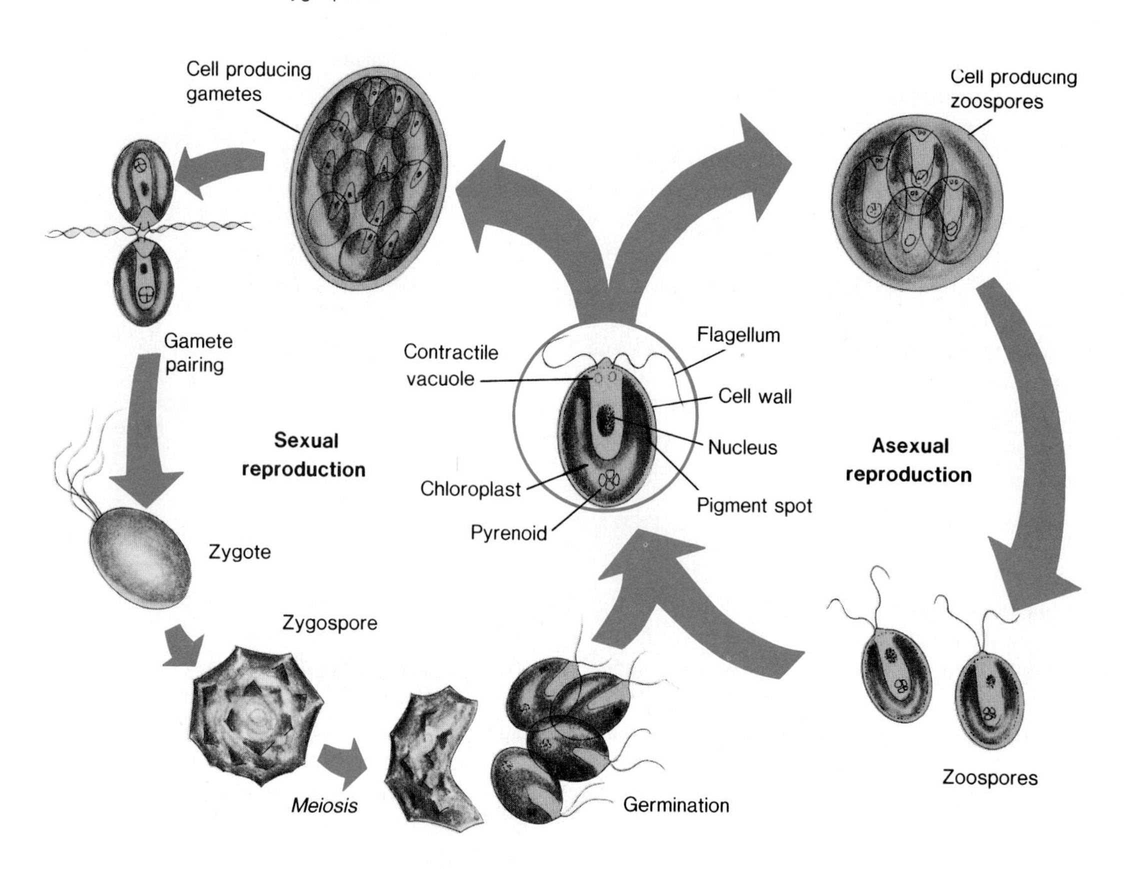

Colonial Green Algae

Some green algae occur in swimming colonies of cells. These **motile** (moving) colonies may be rather simple as in the algae *Gonium* and *Pandorina,* or large and complex as in *Volvox* (figure 28.9). *Volvox* colonies have thousands of cells whose flagella beat in a coordinated fashion. Colonies show some complex responses, such as movement toward moderate light, but away from very strong light. Daughter colonies develop inside a *Volvox* colony. They remain inside until the parent colony breaks apart and releases them.

Filamentous Green Algae

Some green algae form **filaments,** end-to-end chains of cells. Filaments form because cells remain attached to one another after cell divisions. *Spirogyra* is a common example of freshwater, filamentous green algae. *Spirogyra* is often an important part of "pond scum" when algal blooms cover the water surface.

Sexual reproduction in *Spirogyra* occurs by a special fusion process called **conjugation** (figure 28.10). Usually in the autumn, two filaments come to lie side by side. Bulges that form on the cells enlarge, meet, and fuse to form **conjugation tubes.** Once cells are paired in this way, one cell pulls away from its cell wall, rounds up, squeezes through the conjugation tube, and fuses with the other cell. Thus, individual cells in the filament function as gametes. This happens all along the paired filaments and results in the production of zygotes in one filament and a series of empty cell walls in the other filament. Each zygote forms a thick, resistant wall. Before germination, meiosis occurs, and one of the four haploid nuclei produced becomes the functional nucleus of the cell and grows out of the broken zygote wall to begin development of a new filament.

Just as in *Chlamydomonas,* all *Spirogyra* vegetative cells are haploid. The zygote is the only diploid cell present in *Spirogyra's* entire life history.

Figure 28.9 Colonial green algae. (*a*) *Gonium* is a flat colony of cells embedded in jelly. (*b*) *Pandorina* is an oval colony. (*c*) The *Volvox* colony is a large sphere with thousands of cells embedded in a jelly sphere and connected by cytoplasmic strands. Note the daughter colonies inside the *Volvox* colony.

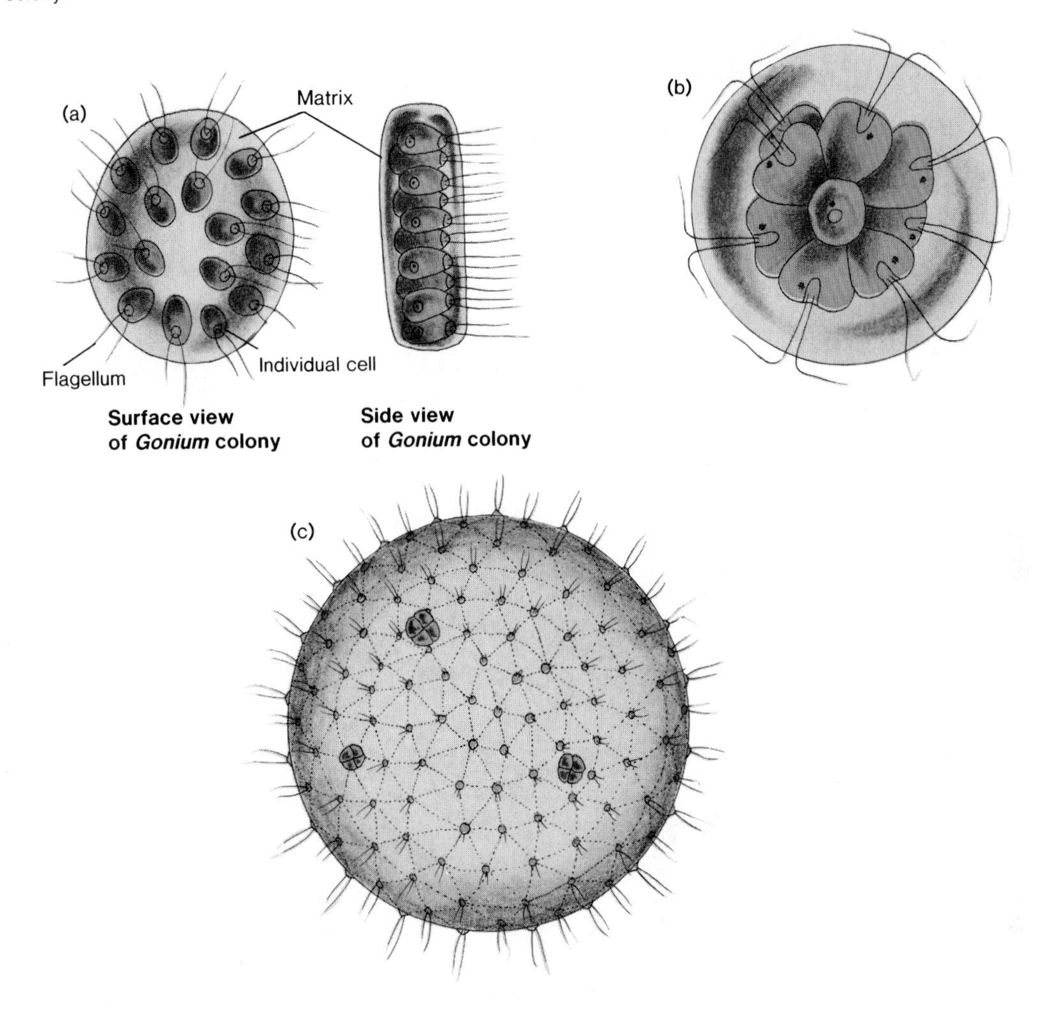

Oedogonium, another common freshwater alga, has a very different reproductive pattern (figure 28.11). *Oedogonium* has two distinctly different types of gametes. Small, motile sperm cells fuse with large, nonmotile egg cells. Sexual reproduction involving fusion of such unlike gametes is called **heterogamy,** because these are **heterogametes** (hetero = different). A vegetative cell can develop into an egg-forming cell (**oögonium**) or divide to produce several small sperm-forming cells (**antheridia;** singular: **antheridium**). An antheridium produces two swimming sperm cells.

Sperm are released and swim to eggs. As in other filamentous algae, the zygote secretes a thick wall and can withstand adverse conditions. The zygote eventually undergoes meiosis to produce four haploid zoospores. Each of them can initiate formation of a new *Oedogonium* filament.

Reproductive Patterns in Green Algae

In the filamentous algae, only one diploid cell is present in the entire life history. But in the green alga *Ulva* ("sea lettuce"), there is a different pattern.

Figure 28.10 *Spirogyra.* (*a*) Vegetative cell. (*b*) Stages of conjugation. (*c*) Photograph of the final stages of conjugation. Zygotes form in one filament of each pair, leaving empty cell walls in the other filament.

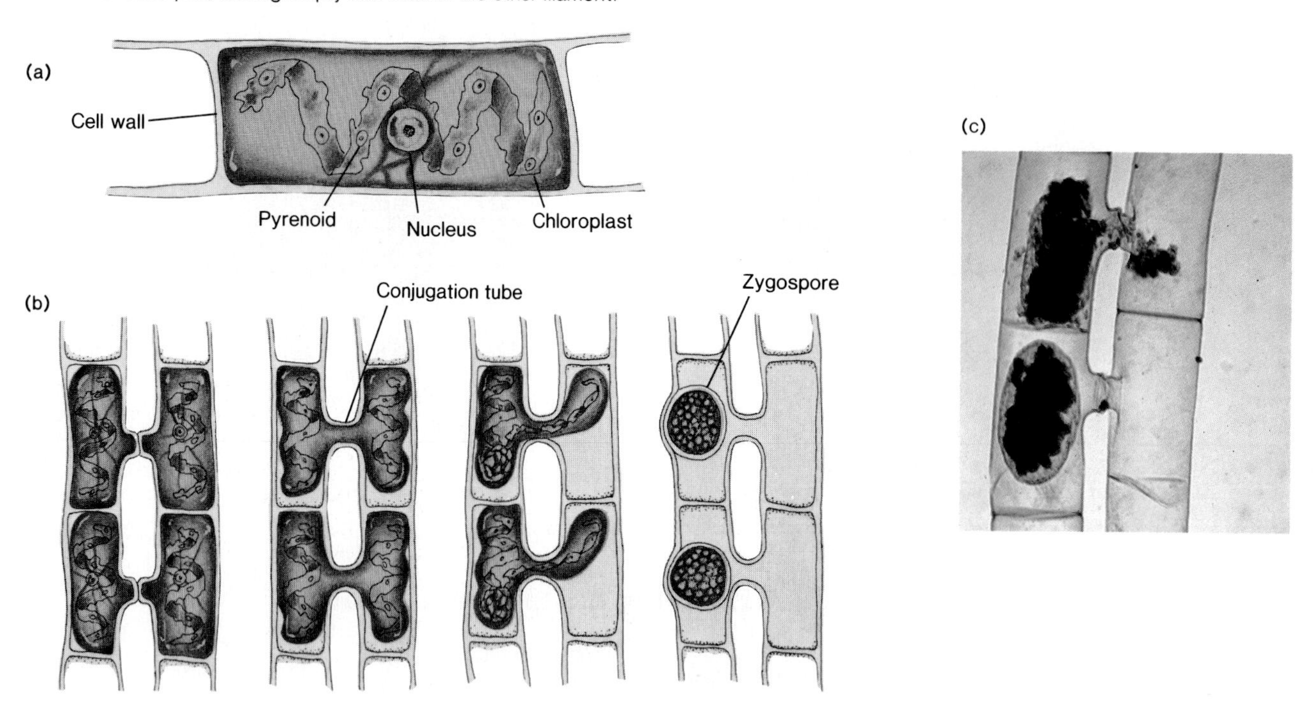

Figure 28.11 Reproductive processes in *Oedegonium.*

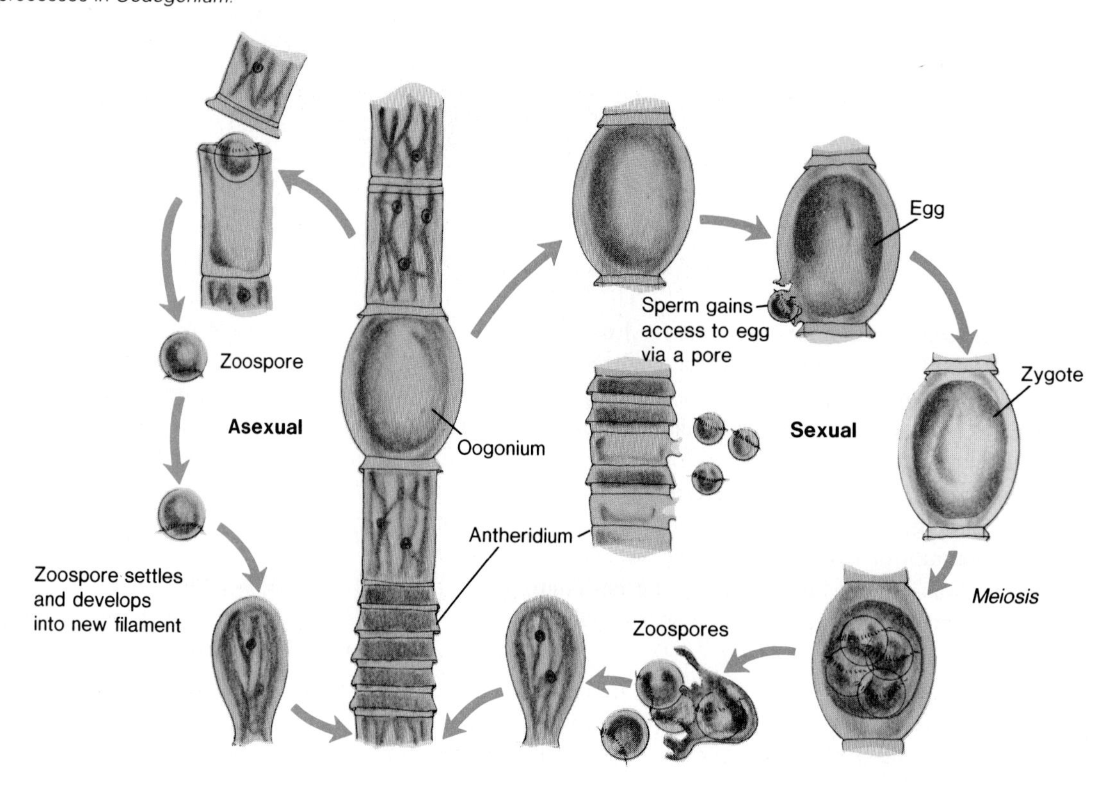

Figure 28.12 Plant reproductive patterns. (*a*) *Ulva* life history. There is clear alternation of generations between haploid (gametophyte) thalli and diploid (sporophyte) thalli. (*b*) The general pattern of alternation of generations in plant life histories.

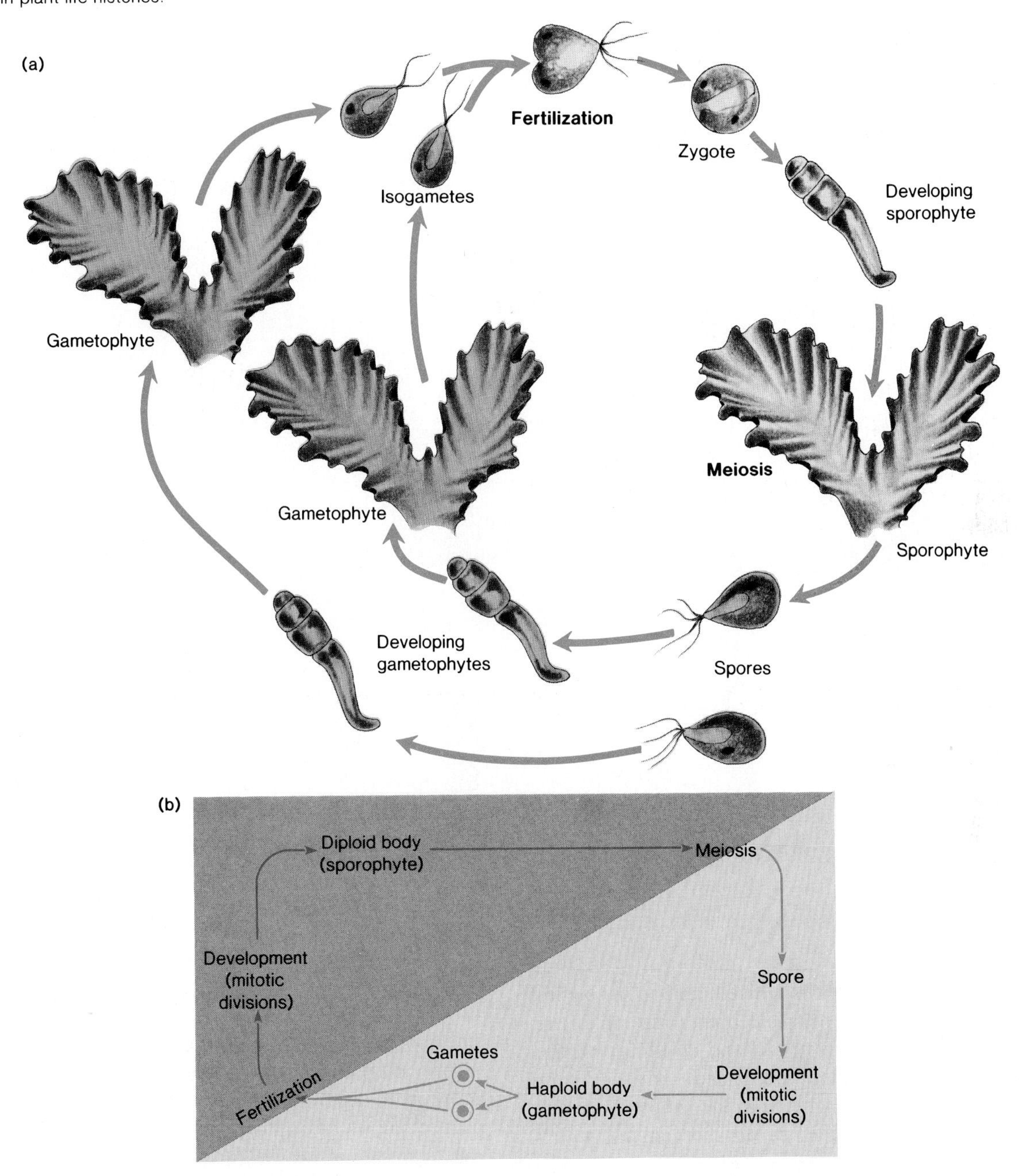

Ulva grows in a simple, flat, sheetlike form called a **thallus** (plural, **thalli**). There are actually two different kinds of thalli in *Ulva* (figure 28.12). One kind grows by mitosis from a spore and consists entirely of haploid cells. Gametes produced by these haploid bodies fuse to produce zygotes. The zygote divides by mitosis and grows into a thallus made up of diploid cells. This is a marked difference from the filamentous green algae, where the zygote is the only diploid cell and it divides meiotically to produce haploid spores. Eventually, specialized cells of *Ulva's* diploid body divide by meiosis to produce haploid zoospores that grow into haploid

Figure 28.13 Representative brown algae. These are not drawn to the same scale because *Macrocystis* (*a*) and *Nereocystis* (*b*) are giant kelp, which grow larger than *Laminaria* (*c*). *Fucus* (*d*) ("rockweed") is smaller. Brown algae are firmly anchored by holdfasts, and their broad flattened blades are connected to the holdfasts by stipes. Some have air bladders that keep the blades afloat.

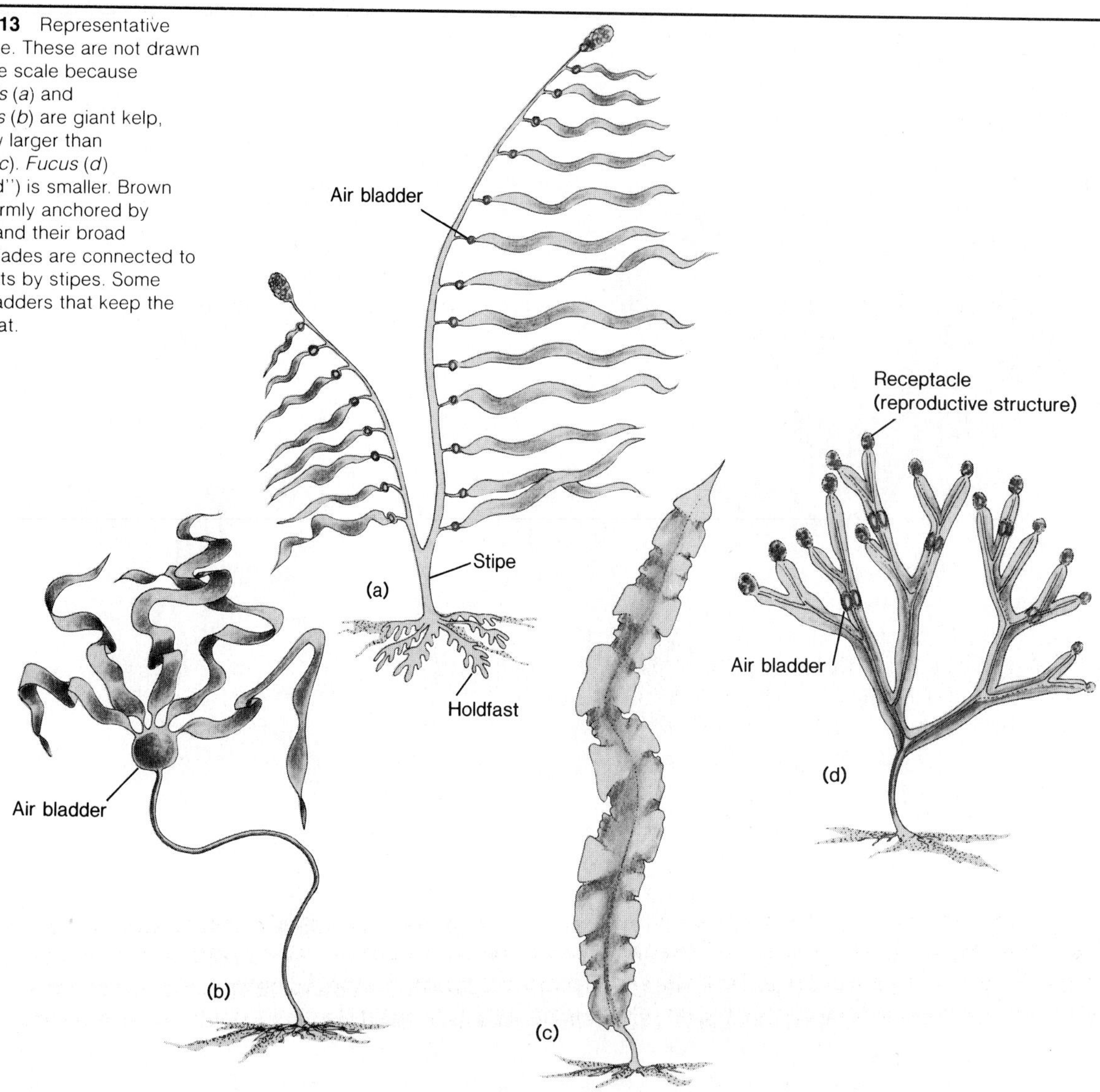

thalli. This life cycle pattern with multicellular haploid bodies alternating with multicellular diploid bodies resembles the life cycle pattern seen in terrestrial (land-dwelling) plants. Most plant life cycles include two phases: haploid plant bodies, called **gametophytes,** which produce and release gametes; and diploid plant bodies, called **sporophytes,** which produce spores by meiosis. These haploid and diploid bodies alternate in plants' life histories (**alternation of generations**), but the relative prominence of these two reproductive patterns is different in various groups of plants.

Brown Algae

The brown algae make up the division Phaeophyta (*phaeo* = brown). They contain the brown pigment **fucoxanthin** in addition to chlorophylls. The group includes many of the plants that we commonly call seaweeds. The approximately 1,000 species of brown algae are all multicellular, and the great majority of them are marine. Brown algae range from small plants with simple branched filaments to large plants with thalli that may be between 50 and 100 m long. Large brown algae, known as kelp, are common in the intertidal zone, and they spread over large areas of rocky shorelines in cooler regions of the world. In deeper water, giant kelp such as *Macrocystis* and *Nereocystis* often form spectacular underwater forests (figure 28.13).

Figure 28.14 Some red algae. Red algae are smaller and more delicate than the large brown algae. (*a*) *Antithamnium.* (*b*) *Batrachospermum.*

(a)

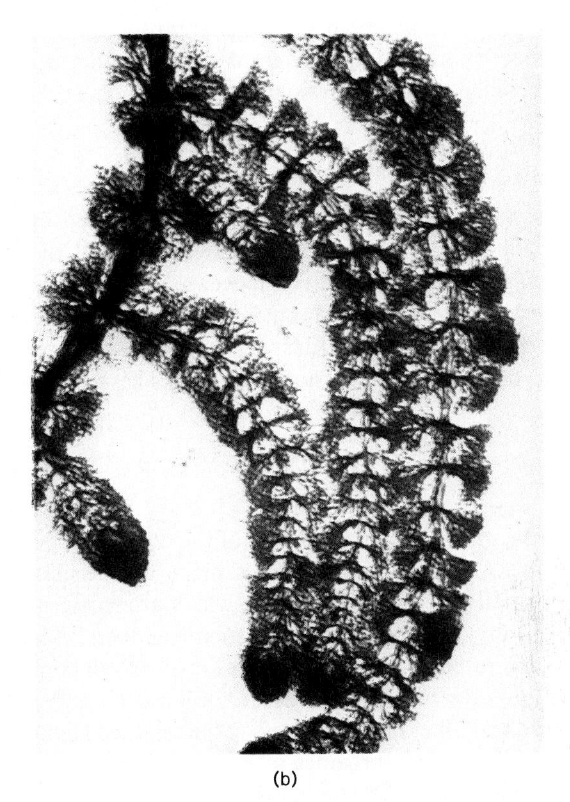

(b)

Brown algae provide food and habitat for many marine animals, and some kelps are used as human food in several parts of the world. Several kinds of kelp, such as *Macrocystis* and *Laminaria,* are economically important because a colloidal carbohydrate called algin can be extracted from them. Algin is added to ice cream, sherbet, cream cheese, and other products to give them a stable, smooth consistency. It also is used in the manufacture of adhesives and many other industrial products.

Red Algae

The division Rhodophyta, the red algae, includes approximately 4,000 species. In addition to chlorophylls, the red algae contain a red pigment called **phycoerythrin.** Some also contain a blue pigment called **phycocyanin.** Because of the various mixtures of pigments that they contain, "red" algae range from red to almost black. These pigments are involved in light absorption during photosynthesis, and they permit red algae to live in deep water. Red algae generally are much more common in deeper (up to 100 m) and warmer waters than the brown algae. A few red algae are up to a meter long, but most are smaller than that (figure 28.14).

Red algae are economically important because valuable substances can be extracted from the outer layer of their cell walls. The best known of these substances is **agar.** Agar is used in culture media for growing bacteria and fungi. It and other substances from red algae are used in many foods to produce a smooth consistency and to help retain moistness.

Terrestrial Plants

Terrestrial plants encounter opportunities and problems different from those of aquatic plants. For instance, the light supply for photosynthesis is much better on land. Also, carbon dioxide and oxygen are available in higher concentrations in air than in water, and gases diffuse more readily through air. However, the principal problem of terrestrial plants is the constant threat of losing excessive amounts of water to the environment. In addition, terrestrial plants must get mineral nutrients from the soil. Thus, terrestrial plants must have means of preventing excessive water loss and of obtaining an adequate supply of water and minerals from the soil.

Terrestrial plants face another important problem. Their spores, gametes, and developing zygotes are small and fragile. They can stand very little water loss or environmental stress of any kind. Different land plants have different solutions for the problems of spore and gamete transport in the dry terrestrial environment, which is so harsh for these delicate cells.

Bryophytes

Mosses, liverworts, and their relatives make up the division Bryophyta. They are all relatively small terrestrial plants that grow in moist places. The bryophytes have waxy cuticles that cover their external surfaces and control water loss, but they do not have vascular tissue to transport the soil's water and minerals through their bodies. Thus, their body sizes are limited.

Mosses

Mosses are the most familiar members of the division Bryophyta. About 14,500 species of mosses make up the class Musci. The small green bodies of moss, which grow densely in soft mats, are gametophyte generation members. The densely concentrated growth habit of mosses helps to compensate for the lack of transport tissue, because water becomes trapped and is held in spaces among the crowded bodies of the moss plants. This provides a water source close to aboveground body parts. **Rhizoids,** simple anchoring structures, absorb mineral nutrients from the soil.

Bryophytes depend on external surface water for sexual reproduction because they produce swimming sperm cells. In some cases, splashing raindrops or a heavy dew may provide enough water to make reproduction possible, but water is absolutely essential.

In bryophytes, the gametophyte generation is larger and more conspicuous than the sporophyte. Sex organs develop at the tips of gametophytes (figure 28.15). In some mosses, **antheridia** (sperm-producing structures) and **archegonia** (multicellular egg-producing structures) are borne on the same plant, but in others there are separate male and female plants.

Motile sperm swim from the antheridium to the archegonium. Fertilization occurs inside the archegonium, and the zygote develops into an embryo sporophyte in the archegonium's enclosed protected environment. Eventually, the sporophyte grows up out of the archegonium, but it remains attached by a foot that anchors it to the gametophyte plant. Moss sporophytes remain as dependent attachments on gametophytes and thus do not have independent lives of their own.

At the tip of a stalk (**seta**) extending up from the foot, a **sporangium** develops. In the sporangium, meiosis occurs, and haploid spores are produced. In some species of moss, a hoodlike covering of archegonium tissue, the **calyptra,** is carried along upward by the growing sporophyte. The calyptra comes off before spores are mature. Then a lid, the **operculum,** falls off the capsule, and mature spores escape.

When a spore lands in an appropriate site, it germinates. A single row of cells grows out and then branches. This algalike structure is called a **protonema.** After about three days of growth, "buds" appear at intervals along the protonema. Each of these sends down rhizoids and grows up into a gametophyte plant. This completes the moss life cycle.

Liverworts

About 9,500 species of liverworts (wort = plant or herb) make up the class Hepaticae. Most liverworts are small, flattened green plants that do not resemble mosses superficially, but they do have a reproductive cycle similar to mosses. Also like mosses, they have rhizoids and lack vascular tissue. The flat, lobed body of liverworts is called a thallus.

Marchantia is a very common liverwort that is often found on damp soil, such as sheltered areas around buildings where water runs off roofs. The *Marchantia* thallus is a gametophyte body. It branches dichotomously (in twos) as it grows. Rhizoids grow down into the soil from the bottom of the *Marchantia* thallus (figure 28.16). Pores in the upper epidermis permit CO_2 diffusion into air chambers occupied by chlorophyll-containing cells. Several layers of nonphotosynthetic cells lie below the air chambers.

Sexual reproduction in *Marchantia* is very similar to that of mosses. Male and female reproductive organs develop on upright stalks of separate gametophytes. Sperm swim to and enter the archegonium, where fertilization occurs. The zygote develops into an embryo within the archegonium. The mature sporophyte remains attached to and dependent on the gametophyte plant. Haploid spores, carried by air currents, land in favorable locations and grow into new gametophyte plants (figure 28.17).

In liverworts, as in mosses, sexual reproduction requires water for sperm to swim in, and the sporophyte (diploid generation) is a dependent attachment on the body of the female gametophyte.

Figure 28.15 Moss life history. The calyptra is a part of the female gametophyte that is carried along by the developing sporophyte.

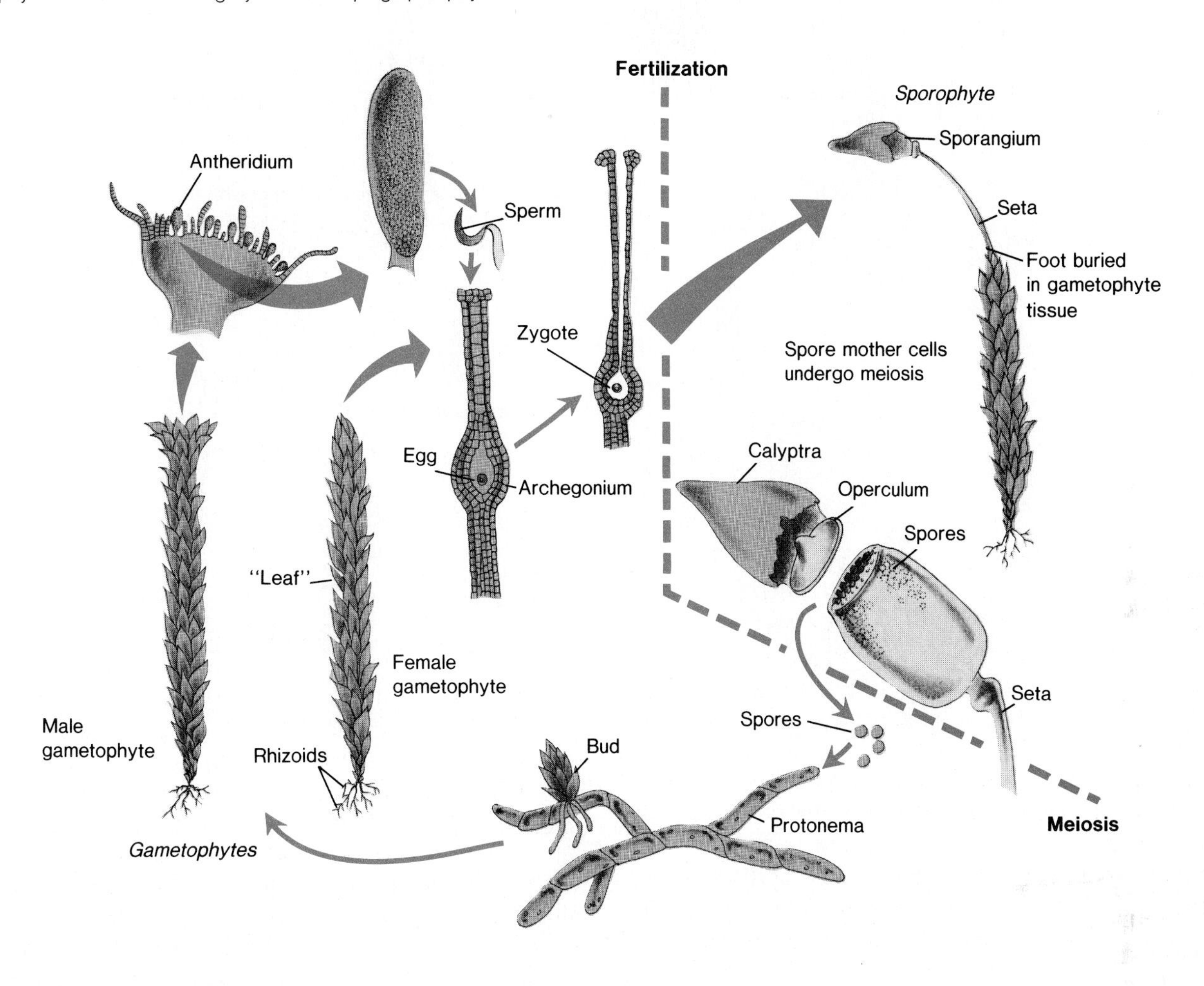

Figure 28.16 Sketch of a cross section of part of a thallus of *Marchantia*.

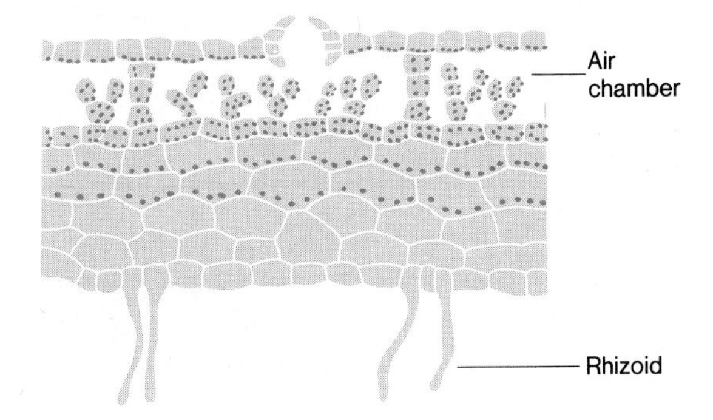

Figure 28.17 *Marchantia* life history.

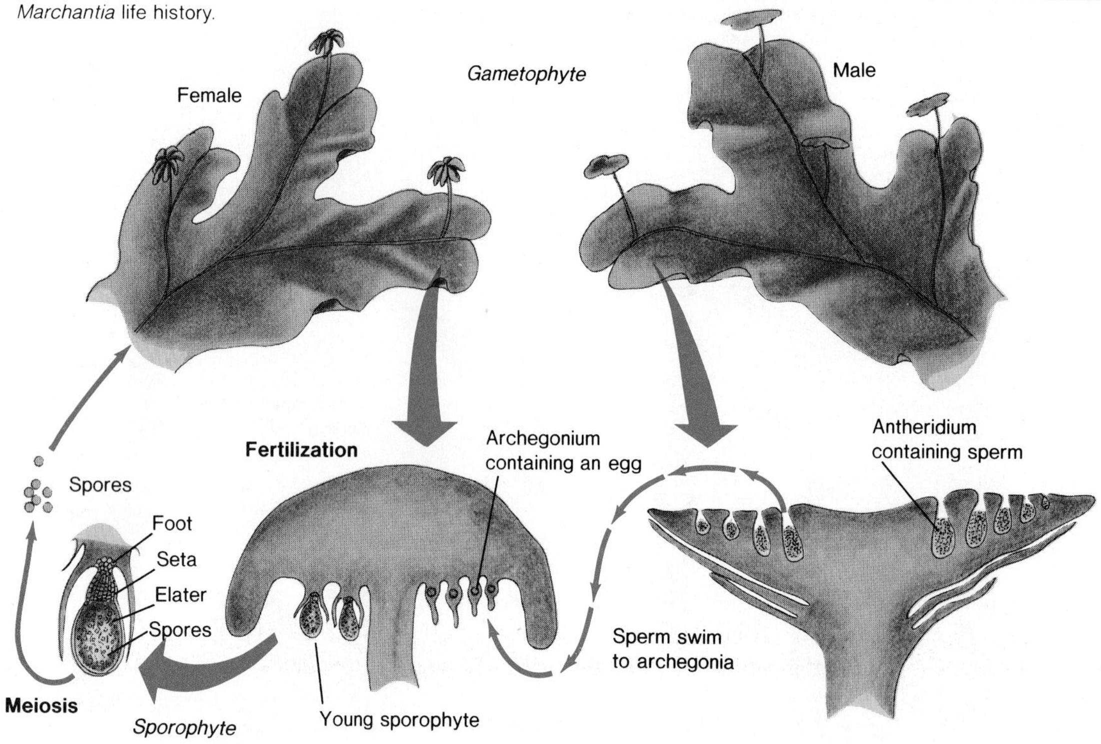

Vascular Plants

The division Tracheophyta is a large and diverse group that includes the most complex plants. The primary distinguishing characteristic of tracheophytes is the presence of vascular tissues. **Xylem** conducts water and minerals up from the soil, and **phloem** transports nutrients and other materials from one part of the plant body to another. Tracheophytes have specialized body parts—roots, stems, and leaves.

Tracheophyte reproduction involves a conspicuous alternation of generations, but the sporophyte clearly is the dominant generation. Sporophytes contain vascular tissues and specialized body organs. Gametophytes are small and relatively inconspicuous. Since the vascular tissues and specialized body parts of sporophyte generation plants are important adaptations for success as terrestrial plants, development of a dominant sporophyte generation has been an important trend in the evolution of tracheophytes.

Primitive Vascular Plants

We are all familiar with ferns, evergreens, and flowering plants. But besides these well-known vascular plants, a number are less familiar. They are, however, often encountered in nature and are fairly easy to recognize because their structure and growth patterns clearly set them aside from the more familiar plants.

The "whisk ferns" such as *Psilotum,* do not particularly resemble common ferns, but do look something like small green whisk brooms (figure 28.18). They may be members of a largely extinct group of very primitive vascular plants.

Modern club mosses, spike mosses (actually neither of them are mosses at all), and quillworts are small, inconspicuous plants (figure 28.19). But some extinct members of their group that lived millions of years ago were very large and treelike.

Probably the most common of these primitive vascular plants belongs to the genus *Equisetum* (figure 28.20). *Equisetum* is commonly called "horsetail" (because of the appearance of some stems) or "scouring rush" (because the plants have been used for scouring and cleaning metal). Silica deposits in stem cell walls make the stems very hard.

The Dominant Land Plants

The common plants that dominate the land environment belong to a large group collectively known as the subdivision Pteropsida. This group is composed of three classes of familiar plants: the **ferns** (class Filicineae); the **conifers** (class Gymnospermae); and the **flowering plants** (class Angiospermae). The sporophytes of all of these plants have well-developed roots and stems. But their conspicuous leaves

Figure 28.18 *Psilotum,* a whisk fern. The rhizome is a horizontal underground stem.

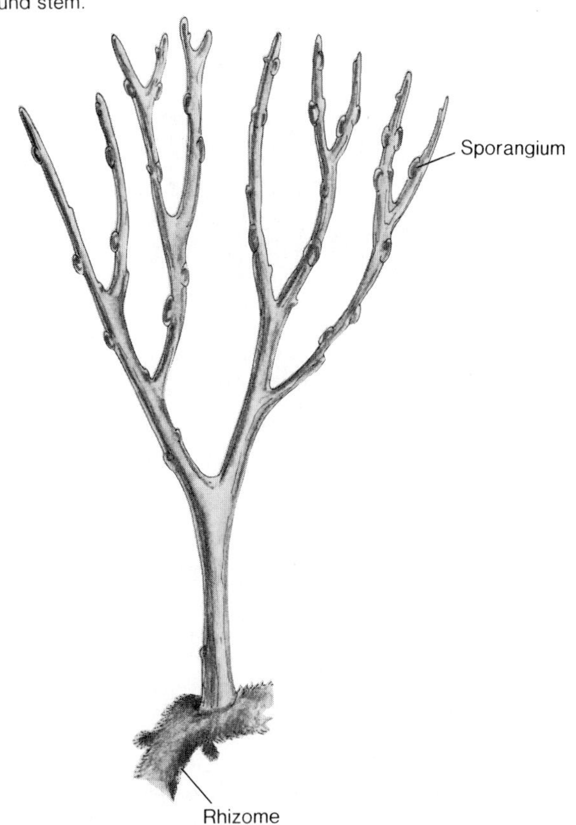

Figure 28.20 *Equisetum,* commonly called "horsetail" or "scouring rush."

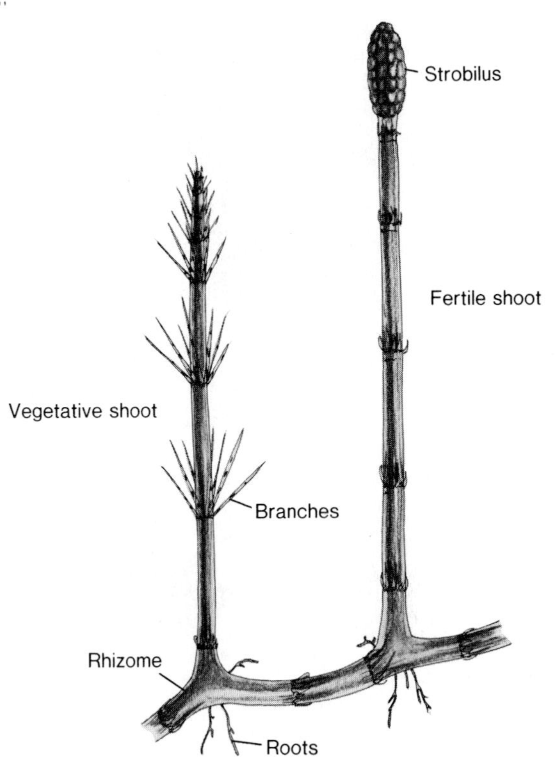

Figure 28.19 (*a*) Club moss (*Lycopodium*) is sometimes called "ground pine" because it is green year round. This evergreen condition has made club moss popular as a holiday decoration and has led to depletion of some populations formerly found in the eastern United States. A strobilus is a cluster of modified leaves that bears sporangia. (*b*) Spike moss. (*c*) Quillwort.

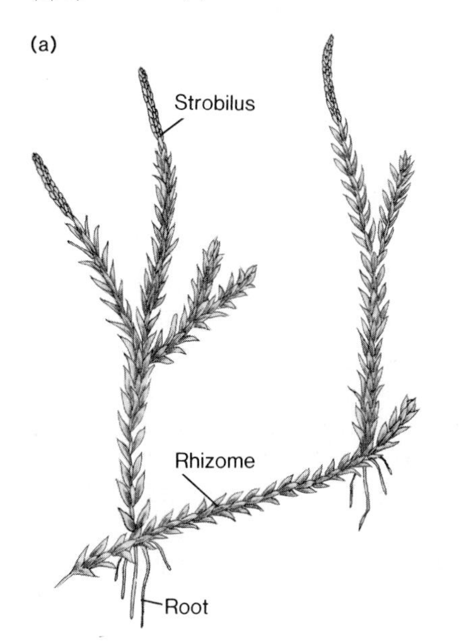

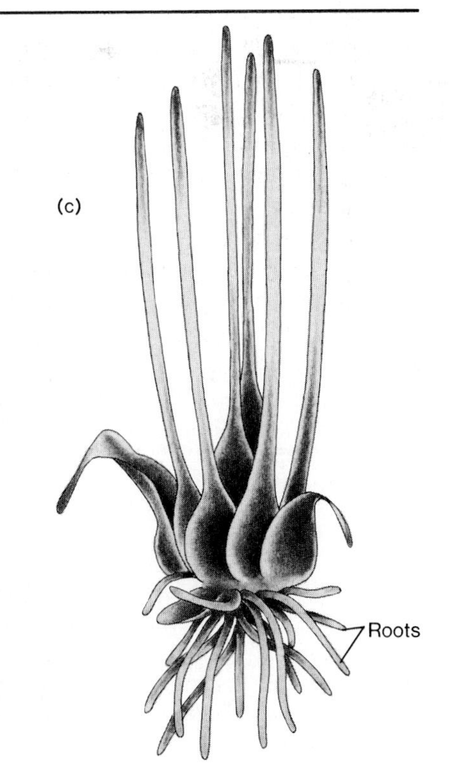

Figure 28.21 (a) Fern life history. (b) Fern frond and "fidd leheads."

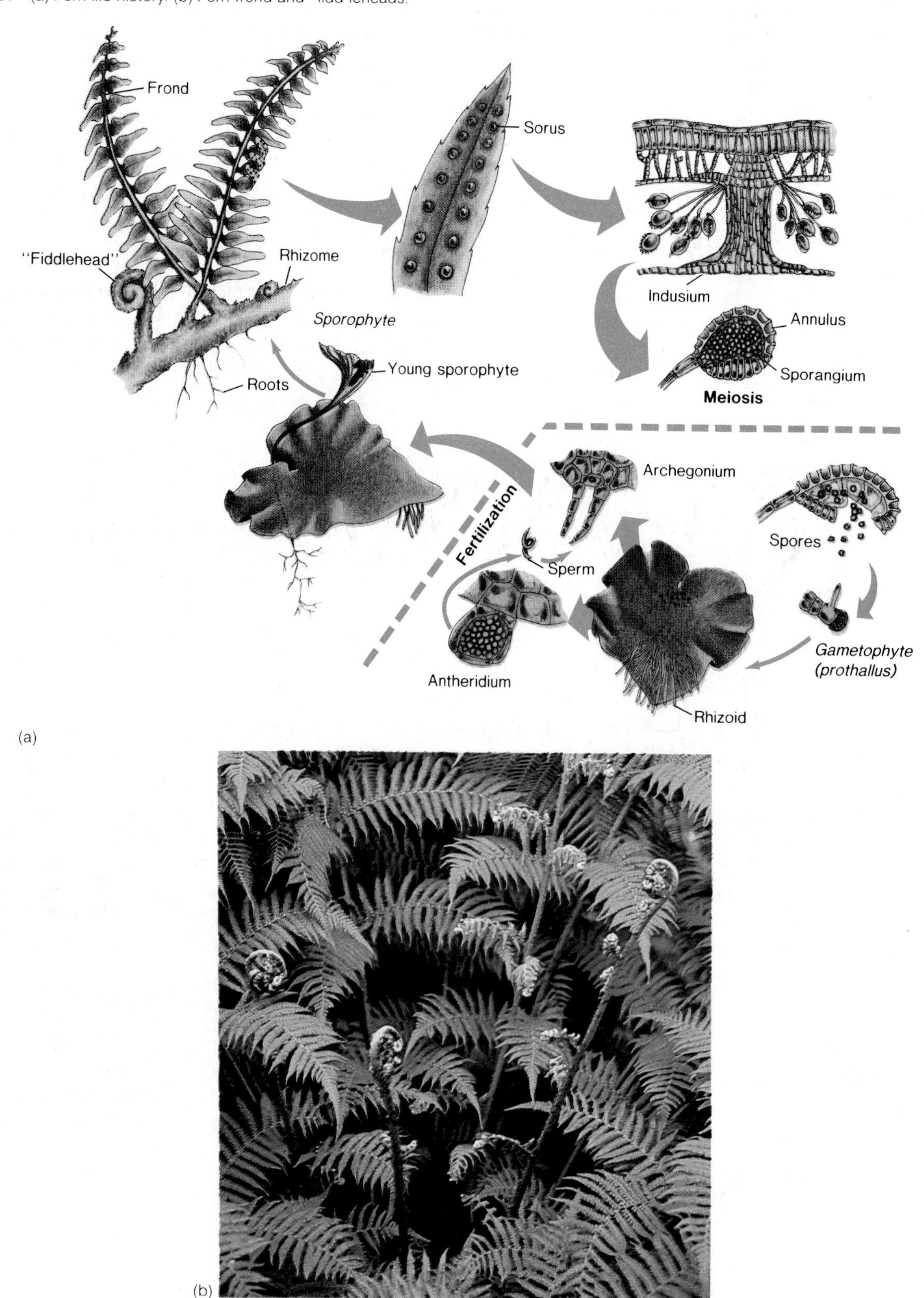

Figure 28.22 Comparison of gymnosperm (pine) and angiosperm (apple) seeds. Pine seeds (*a*) are "naked" on the surface of a cone scale. Apple seeds (*b*) are enclosed in a fruit.

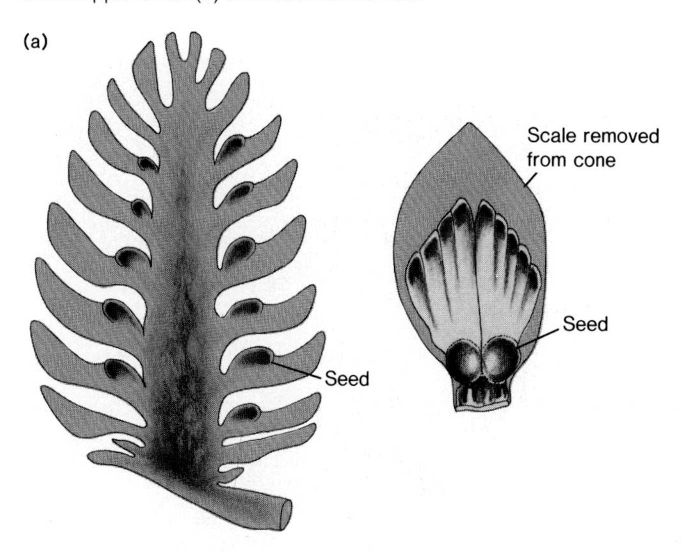

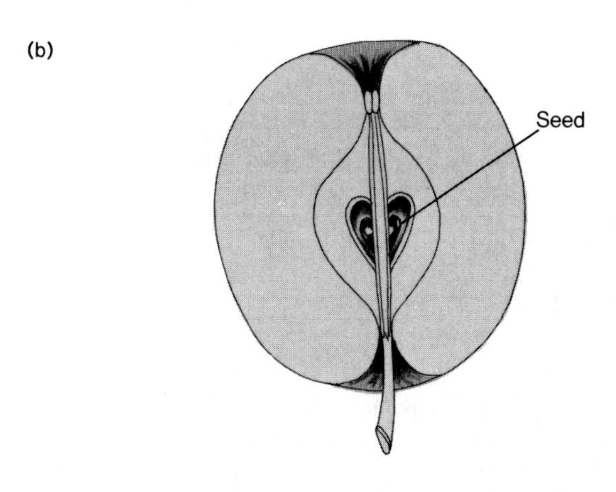

clearly distinguish them from other vascular plant subdivisions. Well-developed vascular and supporting tissues permit many of these plants to grow to very large sizes. The gametophytes of all pteropsids are small and inconspicuous.

Ferns

There are about 11,000 known species of ferns. They vary in size from tiny water ferns only a centimeter in diameter to the tree ferns of the tropics, which may be 25 m tall. In ferns that are common in the northern hemisphere, the familiar leafy fern plant is the sporophyte generation, and the entire aboveground growth consists of leaves, commonly called fronds. Both fronds and roots grow out of the rhizome, the underground stem. Young fronds grow in a curled-up form called "fiddle-heads" out of rhizomes. Then they unroll to produce mature fronds.

Sporangia develop in clusters called **sori** (singular: **sorus**) on the undersurface of fronds. A band of thickened cells known as the **annulus** functions in expelling mature spores from the sporangium. The annulus snaps in response to moisture changes and flings the spores out (figure 28.21).

Spores are carried by the wind, and a few end up in appropriate wet habitats. They germinate and develop into the heart-shaped **prothallus** (plural: **prothalli**), the gametophyte body. A prothallus is a thin plate of cells with rhizoids extending down from its lower surface. Antheridia form among the rhizoids. Archegonia develop close to the notch of the "heart," and each of them produces a single egg. Spiral-shaped sperm swim to the archegonium, where fertilization occurs. Thus, external water is required for sexual reproduction in ferns. The zygote begins its development inside the archegonium, but soon grows out of it. This young sporophyte grows and develops into a mature sporophyte, the familiar fern plant.

Seed Plants

The major groups of seed plants share two key characteristics. The first of these is the formation of **seeds,** protective structures that enclose the sporophyte embryo during a dormant stage. A seed includes the sporophyte embryo, a reserve of nutrients stored in nutritive tissue, and a tough, protective **seed coat.** Seeds are resistant to adverse conditions, such as dryness or temperature extremes. They provide for wide dispersal.

The second key characteristic of seed plants is **pollination,** a process by which male gametes are brought to eggs without a requirement for surface water.

Seed plants produce two different kinds of spores. Megaspores develop into female gametophytes, and microspores develop into **pollen** (immature male gametophytes). Grains of pollen are transferred by wind or by insects to the vicinity of the developing female gametophyte. Then a **pollen tube** grows out to carry the male gametes to the egg. Pollination and seed production have been critically important for the success of seed plants as terrestrial organisms.

The two groups of seed plants are the gymnosperms ("naked seeds") and the angiosperms ("enclosed seeds"). Gymnosperm seeds are produced exposed on the surface of the scales that make up cones, while angiosperm seeds usually are enclosed by a fruit produced from part of a flower (figure 28.22).

Figure 28.23 Gymnosperms. (*a*) The maidenhair tree (*Ginkgo biloba*). (*b*) *Welwitschia mirabilis*, a gymnosperm that grows in very harsh desert environments in Africa. The leaves of these plants are about 2 m long.

(a)

(b)

Gymnosperms

The gymnosperms include only about 700 living species, but some of them—especially conifers, such as pine, spruce, cedar, and fir—have huge numbers of individuals and cover large areas of the earth's surface. There are, however, other less familiar gymnosperms that are very different from the conifers (figure 28.23).

Cycads were very abundant during the age of dinosaurs, but many of the 100 or so living species that remain are near extinction. These tropical or subtropical plants are often mistaken for ferns or miniature palm trees.

Figure 28.24 Bristlecone pine (*Pinus aristata*). Some living bristlecone pines are known to be over 4,000 years old. What may have been the oldest living thing, a 4,900-year-old bristlecone pine, unfortunately was cut in 1965 to determine its age.

Ginkgo biloba, the maidenhair tree, is the only surviving species of a once widespread group. In fact, ginkgo trees probably do not now occur naturally anywhere in the world. All surviving ginkgos are cultivated, and the ginkgo is a popular ornamental tree in some areas of the United States.

One of the most peculiar gymnosperms is *Welwitschia,* which grows in the extremely dry deserts of southwestern Africa. Its short stem grows as a large, shallow cup, and leaves grow out from it. Leaves grow from their bases and continually replace tissue that is worn off at the tip by constant flapping in the wind. *Welwitschia* plants may live to be 100 years old, so they are obviously well adapted to their harsh environment.

The conifers are a widely distributed and economically important group of plants. They produce much of the wood used for building and paper, and they also produce many other valuable products, such as those extracted from **resins.** Resins contain waxy substances dissolved in the liquid solvent **turpentine.**

The oldest and largest trees in the world are conifers (figure 28.24). Bristlecone pines (*Pinus aristata*) in the Nevada mountains are known to be more than 4,500 years old, and a number of coastal redwood (*Sequoia sempervirens*) trees in California are 2,000 years old and over 90 m tall. Though not so tall as the coastal redwoods, the giant sequoias (*Sequoiadendron giganteum*) have the greatest mass of any living trees. Some of the famous "big trees" of California, which belong to this species, have trunks that are up to 10 m in diameter.

Figure 28.25 Pine tree life history.

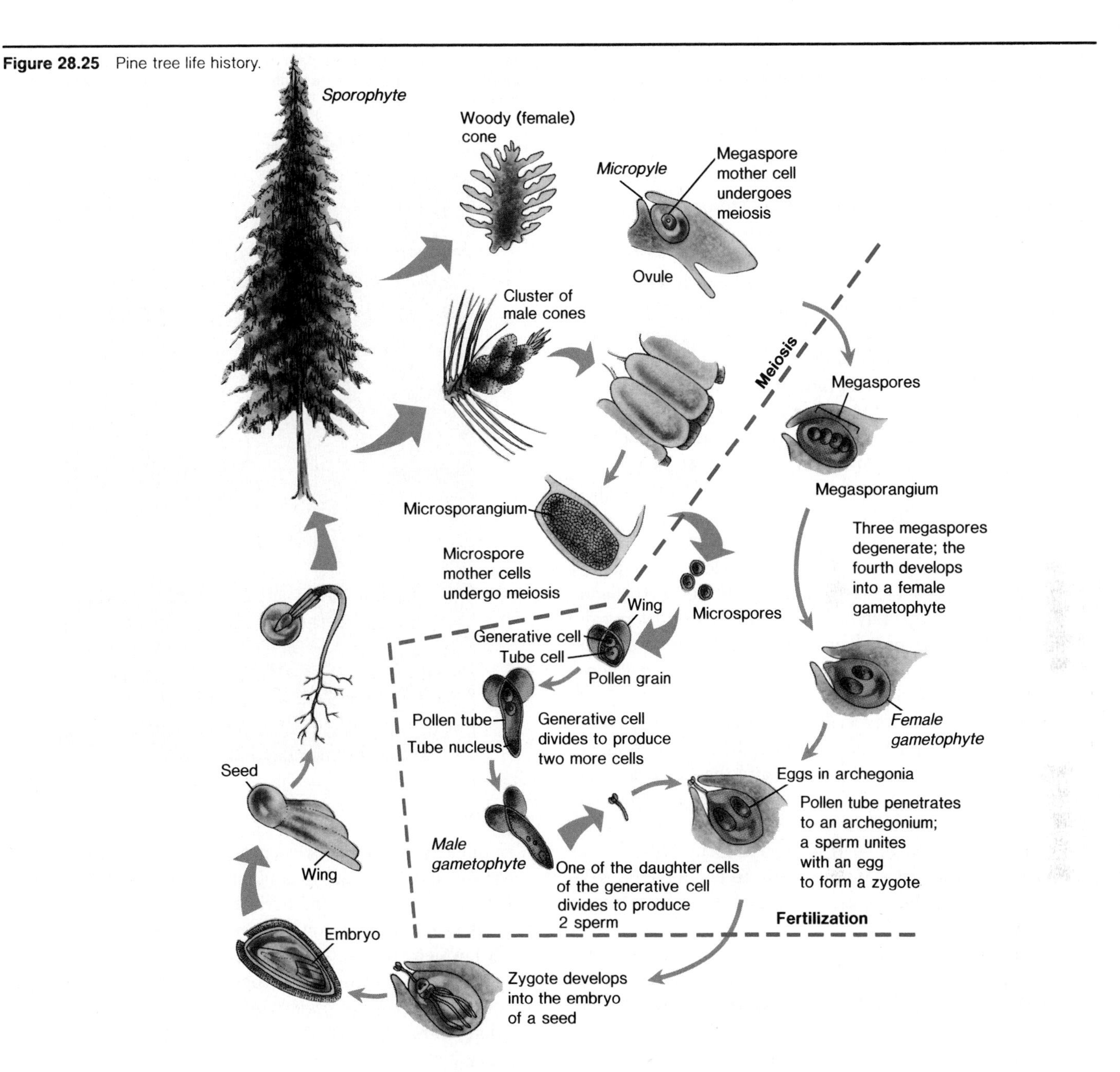

The pine life cycle is a good example of a conifer life history. Pines produce two types of spores that develop in separate types of cones. Cones are made up of **scales,** and sporangia are borne on the scales (figure 28.25).

Microspores are produced in **microsporangia** located on the scales of male cones that develop in clusters near the tips of branches. Male cones are usually not more than 1 or 2 cm long. Microspores develop into pollen grains that have a thickened, protective coat and a pair of flattened, winglike structures on their sides. Inside each pollen grain, mitotic cell division produces a **tube cell** and a **generative cell,** which are involved in further development after pollination. Pine trees release so much pollen that during the pollen season everything in the area around pine trees can be covered with a dusting of yellow, powdery, pine pollen.

Megaspores are produced inside ovules that occur in pairs on the upper surface of female cone scales. A single **megaspore mother cell** inside each ovule divides meiotically to produce a row of four megaspores. Three of the four degenerate, and the fourth develops slowly into a female gametophyte that contains several thousand cells. The mature female gametophyte produces several (two to six) archegonia at the end near the micropyle. A single large egg develops inside each archegonium. These events, from megaspore production to completion of female gametophyte development, require more than a year. During all of this time, the developing female gametophyte is enclosed in and dependent on the sporophyte plant body of the pine tree.

Pollination occurs during the first spring of a female cone's life, while the female gametophyte is still developing. The scales of the green cone separate, and pollen grains fall between the scales, where they become trapped in sticky material near the micropyles. After pollination, scales grow together, closing the cone again, and the sticky material shrinks back, drawing the pollen grains inward. The tube cell of the pollen grain develops a **pollen tube,** which slowly grows into the sporangium, and the generative cell enters the tube. The generative cell divides to produce two cells, one of which divides again to produce two sperm. A pollen grain with its pollen tube and two nonflagellated sperm is a fully developed pine male gametophyte.

The pollen tube reaches the archegonium and discharges sperm about fifteen months after pollination. One sperm unites with the egg to form a zygote. Clearly, pollination and fertilization are completely separate events in pine reproduction. Fertilization occurs in several archegonia in each ovule, but normally only one embryo completes its development as the seed is formed.

Seed development proceeds with accumulation of nutrients in the gametophyte tissue around the embryo. A layer of tissue around the ovule hardens to form the **seed coat,** and a thin membranous layer of the cone scale becomes the seed "wing." Finally, in the third season of their lives, pine cones, by now woody and hard, open to release their seeds. Seeds germinate under appropriate conditions to begin the growth of new pine trees, the sporophyte plants, and the cycle is completed (figure 28.25).

The reproductive pattern of conifers has several important advantages over reproduction in other plants that we have considered so far. These differences make the conifers better adapted for terrestrial life. Transfer of pollen grains and growth of the pollen tube eliminate the requirement of surface water for swimming sperm. Enclosure of the dependent female gametophyte inside a cone protects it during its development and shelters the developing zygote as well. Finally, the embryo is protected by the seed coat and provided with a store of nutrients that support development for the first period of its growth following germination. All of these factors increase chances for reproductive success.

Figure 28.26 Scanning electron micrograph of a duckweed (*Lemna paucicosta*). Other pond surface angiosperms are even smaller (magnification × 21).

Angiosperms

The angiosperms, or flowering plants, are an exceptionally large and successful group of plants. At least 250,000 species of trees, shrubs, vines, and herbs belong to this class. Angiosperms range in size from tiny pond surface plants (figure 28.26), which are only 0.5 mm in diameter, to very large trees. The oldest fossils definitely recognized as angiosperms come from the Cretaceous period, which began only about 130 million years ago. Yet during this relatively short evolutionary history, the angiosperms have diversified and multiplied tremendously. Angiosperms have well-developed vascular and supporting tissues that make them very well adapted for terrestrial life.

Reproduction in angiosperms is generally similar to gymnosperms in that the gametophyte generation members are small and inconspicuous. But angiosperm pollen and ovules are produced in flowers, rather than in cones. Details of angiosperm reproduction are discussed in chapter 19.

Angiosperms get their name from one of their most important characteristics. As we said earlier, angiosperm means "enclosed seed," and it describes the arrangement in which ovules are enclosed within an **ovary** that is situated in a flower.

Figure 28.27 Relative prominence of sporophyte (2N) and gametophyte (N) generations. Some algae (for example, *Ulva*) do have separate gametophyte and sporophyte thalli, but generally the gametophyte is more prominent in algae. The gametophyte is greatly reduced in vascular plants.

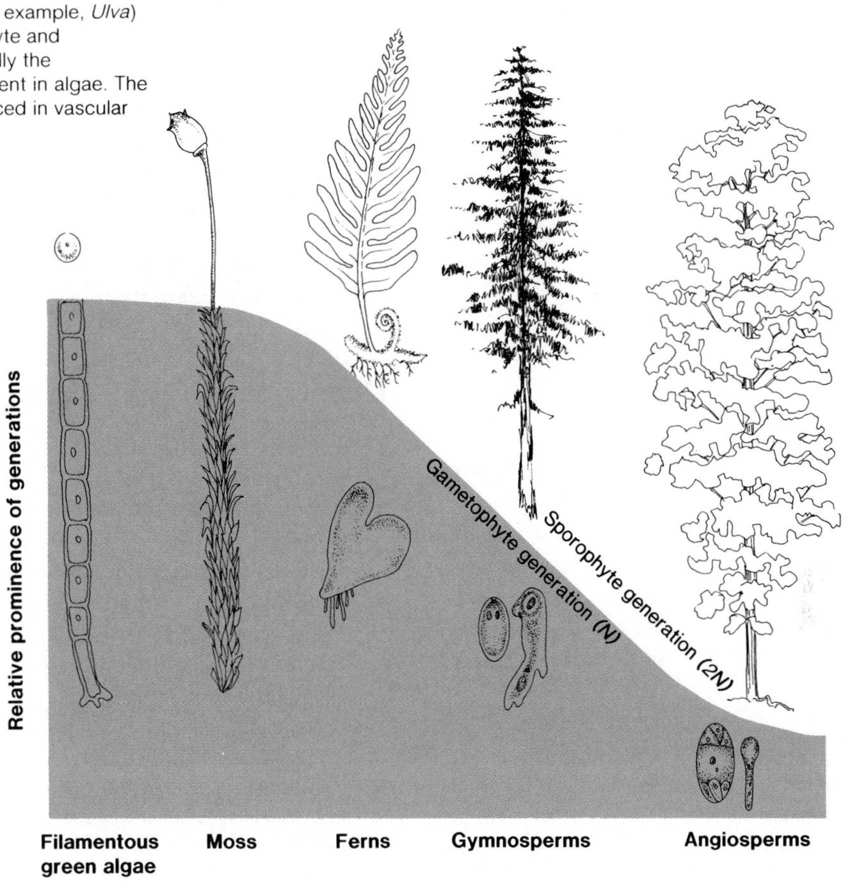

Relative prominence of generations
Gametophyte generation (N)
Sporophyte generation (2N)
Filamentous green algae
Moss
Ferns
Gymnosperms
Angiosperms

The ovary provides protection for the delicate developing female gametophytes inside the ovules and, later, for developing embryos and the seeds that enclose them.

Ovaries also produce fruits. A fruit is a covering that forms over seeds. It is derived from the ovary or from the ovary and other adjacent tissue. Fruits provide protection for seeds and a mechanism for their dispersal. Animals carry fruit and thereby transport seeds to locations often far from the plant that produced them.

The angiosperms are divided into two subclasses: the Dicotyledonae and the Monocotyledonae. Dicots and monocots differ in several important ways (see table 28.1). Common names for dicot families include many familiar plant groups, such as the buttercup, mustard, maple, cactus, carnation, pea, and rose families. The rose family, for example, includes roses, apples, plums, pears, cherries, peaches, strawberries, raspberries, and a number of other shrubs. Monocot families include lilies, palms, orchids, irises, and grasses. The grass family includes wheat, rice, corn (maize), and other agriculturally important plants.

Table 28.1
Differences between Monocots and Dicots

Monocots	Dicots
Embryo has one cotyledon (seed leaf).	Embryo has two cotyledons (seed leaves).
Leaves have parallel veins.	Leaves have nets of veins (branched and rebranched).
Leaf edges smooth.	Leaf edges usually lobed or indented.
Flower parts in threes or multiples of threes.	Flower parts in fours or fives or multiples of them.
Stem vascular bundles are scattered.	Stem vascular tissue is solid mass in center or ring of bundles between cortex and pith.
Cambium absent.	Cambium present.

Summary

Fungi are multicellular heterotrophic eukaryotes that secrete digestive enzymes and absorb the resulting soluble nutrients. As such, they act as important decomposers in the ecosystem. There are four major divisions of the fungi: the Zygomycota, Ascomycota, Basidiomycota, and Deuteromycota.

Fungi can cause diseases in both plants and animals. Many fungi produce toxic compounds and are poisonous if eaten. Others, however, are edible and a number are considered gourmet delicacies. Yeasts are employed in baking and the production of alcohol, while filamentous fungi are used to manufacture antibiotics, cheese, and other products.

Lichens are the result of an association between fungi and green algae or cyanobacteria. They serve as a food source for animals living in barren regions and initiate the colonization of rocky, sterile areas.

The roles of gametophyte and sporophyte generations in life cycles in various plant groups are related to the relative success of the groups as terrestrial organisms. Mosses and liverworts, with their dominant gametophytes and small, dependent sporophytes, have limited distributions determined by external water supply.

While ferns have well-developed sporophyte bodies with vascular tissues, they still require very wet conditions for the growth of their small, independent gametophytes and for fertilization with swimming sperm.

The entire lives of gymnosperms and angiosperms can be lived on dry land because the large, dominant sporophyte is well adapted to terrestrial life. Small dependent gametophytes and delicate spores, gametes, zygotes, and embryos are enclosed within protective coverings that are parts of the sporophyte plant.

Questions for Review

1. Define the term mycelium.
2. What are the basic characteristics of fungi?
3. Name the four major divisions of fungi and name one common representative of each.
4. What is a basidium?
5. What are lichens? How does each member of this symbiotic association contribute to the welfare of the whole?
6. Explain how classification of the green algae illustrates problems in assigning organisms to the categories of a classification scheme.
7. How does the life cycle of the alga *Ulva* illustrate the alternation of generations commonly seen in plants?
8. Explain the difference between sexual reproduction involving isogametes and that involving heterogametes.
9. It is often implied that "lower" plants (nonvascular plants) do not attain large size. Give an example that contradicts that implication.
10. What features of the division Bryophyta prevent its members from spreading widely in terrestrial environments?
11. Explain how the flowers and seeds of angiosperms function as adaptations for terrestrial life.

Questions for Analysis and Discussion

1. Lichens are often the very first living things to become established on barren rock surfaces. How can they live in environments that are unsuitable for virtually all other organisms?
2. Discuss the importance of vascular tissue as an adaptation for terrestrial life in plants.
3. Pollination and fertilization are not synonymous. Explain the distinction between them using pine reproduction as an example.

Suggested Readings

Books

Alexopoulos, C. J., and Mims, C. W. 1979. *Introductory mycology,* 3d ed. New York: Wiley.

Raven, P. H.; Evert, R. F.; and Curtis, H. 1981. *Biology of plants,* 3d ed. New York: Worth Publishers.

Saigo, R. H., and Saigo, B. W. 1983. *Botany: Principles and applications.* Englewood Cliffs, N.J.: Prentice Hall.

Stern, K. R. 1982. *Introductory plant biology,* 2d ed. Dubuque, Iowa: Wm. C. Brown Company Publishers.

Articles

Amadjian, V. 1982. The nature of lichens. *Natural History* 91(3):30.

Beadle, G. W. January 1980. The ancestry of corn. *Scientific American* (offprint 1458).

Meeuse, B. J. D. 1984. Pollination. *Carolina Biology Readers* no. 133. Burlington, N.C.: Carolina Biological Supply Co.

Mulcahy, D. L. 1981. Rise of the angiosperms. *Natural History* 90(9):30.

Swaminathan, M. S. January 1984. Rice. *Scientific American.*

Animals

29

Chapter Outline

Chapter Concepts

1. Animals are multicellular heterotrophic organisms.
2. Animals are classified according to body plan and level of organization.
3. Nutritional strategy is reflected in animal body organization.
4. Animals living in different environments face different sets of functional problems and possess different specific adaptations.
5. Throughout evolutionary history, there has been a series of dominant groups occupying each major habitat. Placental mammals and insects are the current dominant animals in the terrestrial environment.

Biologists classify more than one million species of organisms as **animals,** and within this huge array there is a very wide range of diversity. What common characteristics led biologists to establish a **kingdom Animalia** that includes organisms ranging from sponges to elephants?

All animals are multicellular and heterotrophic and most have ingestive nutrition (they eat other organisms). Animals have bodies that consist of aggregates of many cells, and they must take in organic compounds from their environments to obtain chemical energy. The ultimate source of organic compounds (reduced carbon compounds) is photosynthetic activity in autotrophic organisms. Animals can obtain nutrients by consuming autotrophic organisms directly or by consuming other animals that have eaten autotrophic organisms. For example, lions eat zebras that have eaten plants.

Most animals have internal **digestive cavities** lined with cells that function specifically in obtaining nutrients. Thus, animal bodies are hollow and have at least two body layers, an external covering layer **(ectoderm)** and a layer of cells lining a digestive cavity **(endoderm)**. All but the simplest animals also have a third middle body layer **(mesoderm)**.

The **vertebrate** group of animals (animals with backbones), which includes humans, is only one subphylum within the phylum Chordata. The great majority of the more than 1 million known animal species are **invertebrates** (animals without backbones). Our survey of the animal kingdom begins by considering eight "major" phyla of invertebrate animals as well as several interesting "minor" groups.

Phylum Porifera (The Sponges)

Sponges are very simply organized animals. They consist of loose aggregates of cells that have very little functional specialization. The majority of sponges are marine.

Adult sponges are **sessile;** that is, they remain attached to surfaces. They obtain nutrients from a stream of water that they continually move through their bodies. Water enters the central cavity through **porocytes** (pore cells) in the body wall and leaves through a single opening, the **osculum** (figure 29.1). This characteristic gives the phylum its name, as Porifera means "pore bearer." A number of special flagellated cells, called **choanocytes,** trap food particles from the water by phagocytosis.

Figure 29.1 Phylum Porifera. (*a*) Tube sponges, *Verongia*. (*b*) A simple sponge on the left and a diagram of the body organization of such a sponge shown in longitudinal section on the right. Flagella of choanocytes cause water movement through the sponge. Water passes through pores, which are enclosed in porocytes (pore cells), into the central cavity then out through the osculum. (*c*) Canal systems in sponges. In the simplest sponges, water flows in directly through pores. Other types of sponges have thicker bodies with networks of canals leading to internal pores.

(a)

(b)

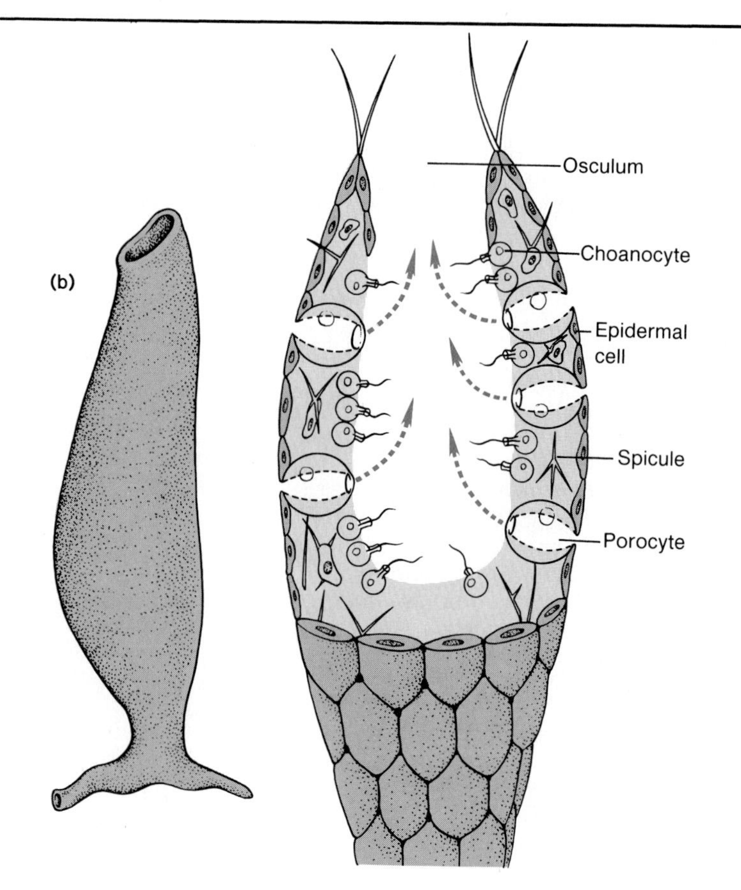

Some sponges are supported by **spicules,** which are crystalline structures made of calcium carbonate or silica (glass-like) compounds (figure 29.2). Other sponges have skeletons made of **spongin,** a fibrous protein. Natural bath sponges are prepared by beating spongin-containing sponges until all of the living cells are removed and just the skeleton remains.

Figure 29.2 Sponge spicules. (*a*) Photomicrograph of spicules. (*b*) Two steps in the process of spicule formation. Several cells cooperate in an organized fashion and secrete the material making up the spicule.

(a)

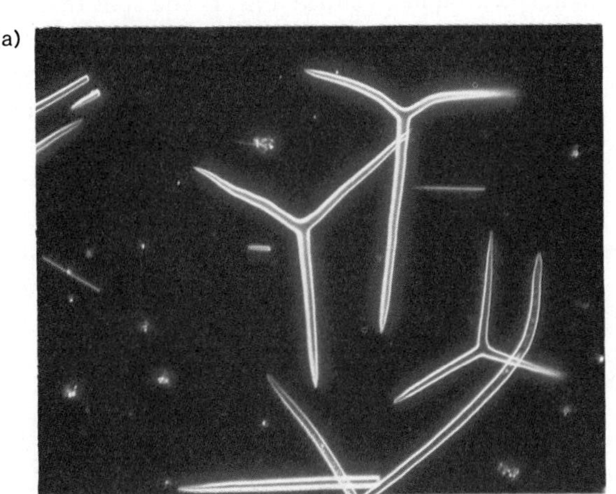

(b)

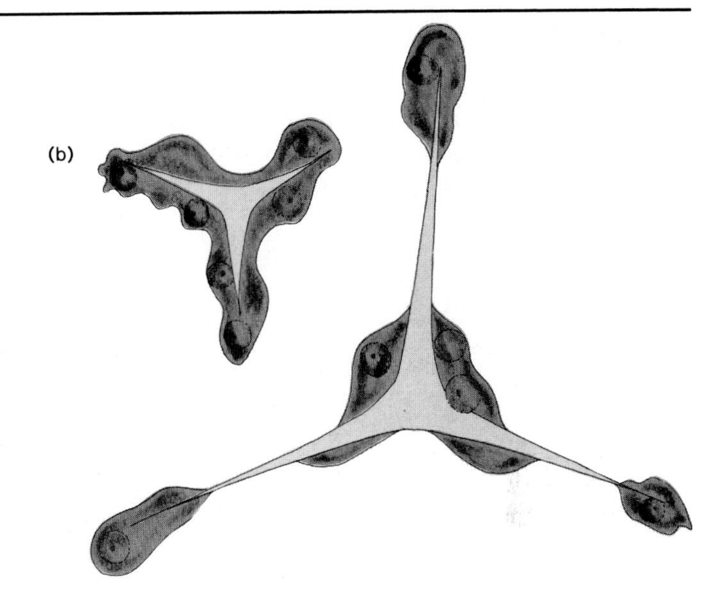

(c)

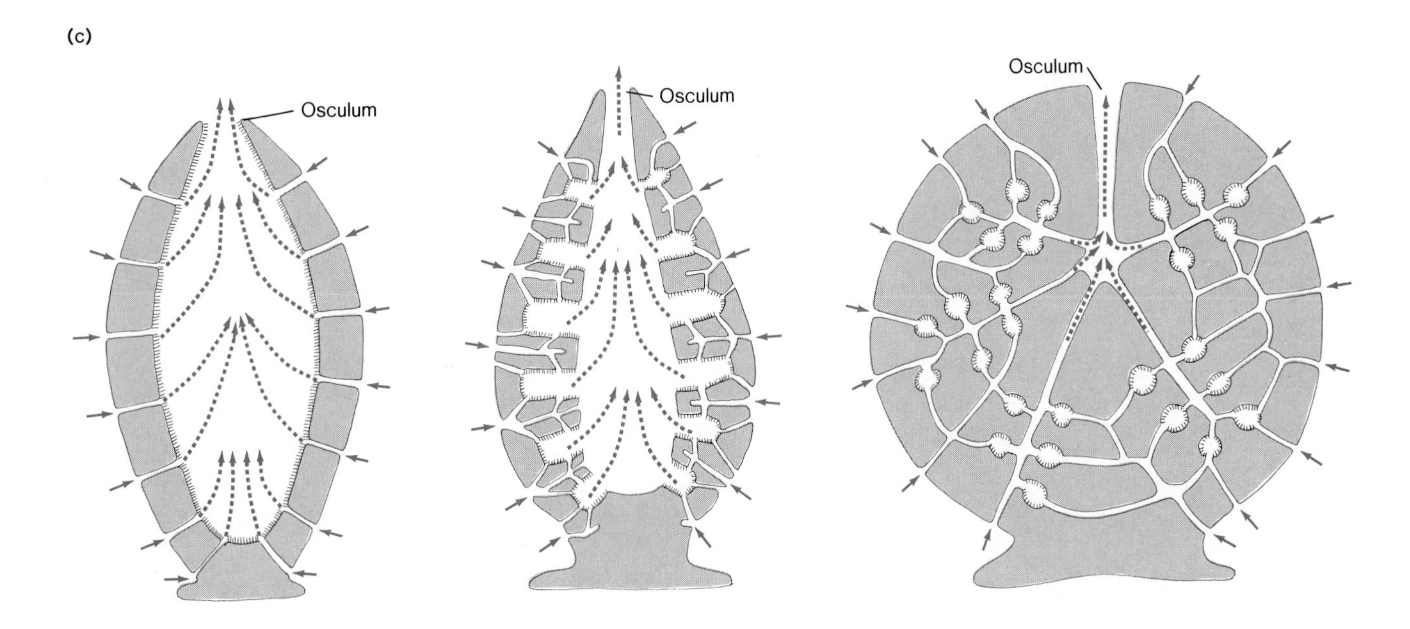

Sponges reproduce sexually. Eggs and sperm are released into the water, where fertilization occurs. The zygote develops into a ciliated larva that swims freely for a time before it settles, attaches, and develops into a small sponge.

Phylum Coelenterata

Coelenterates are a large group of simply organized aquatic animals. Some, such as the **hydras,** live in fresh water. But the great majority are marine, including **jellyfish, sea anemones, corals,** the **Portuguese man-of-war,** and others.

All coelenterates have the same basic body plan. Their bodies have well-developed ectodermal and endodermal layers, which are called the **epidermis** and **gastrodermis,** respectively. Between these layers is a jellylike layer, the **mesoglea.**

Coelenterates have a single opening, a mouth, leading to the **gastrovascular** cavity. It has both digestive ("gastro") and transport ("vascular") functions.

Coelenterates are **radially symmetrical.** Their bodies are arranged around a central axis. If you split them lengthwise along any plane that passes through the central axis, you will get two equal halves (figure 29.3). Most other animals are **bilaterally symmetrical;** they have definite anterior and posterior ends, and right and left halves. There is only one midline plane that separates them into equal halves.

Figure 29.3 Types of animal symmetry. (*a*) A radially symmetrical animal can be sliced on any of several or many planes along the main body axis to yield two equal, mirror-image halves. (*b*) This sketch of a horseshoe crab (an arthropod) shows that there is only one plane along which a bilaterally symmetrical animal can be sliced to yield two equal, mirror-image halves.

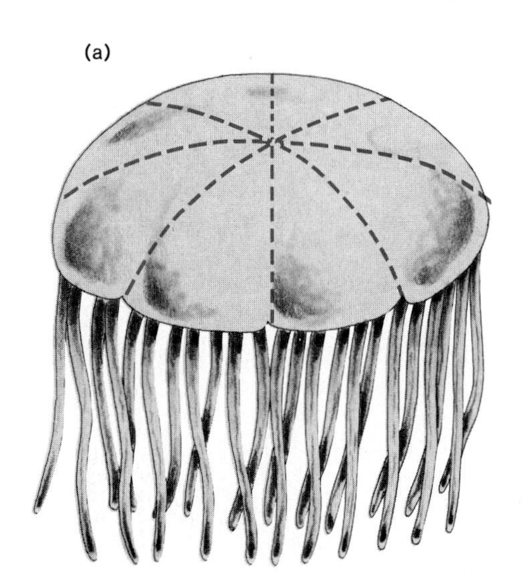

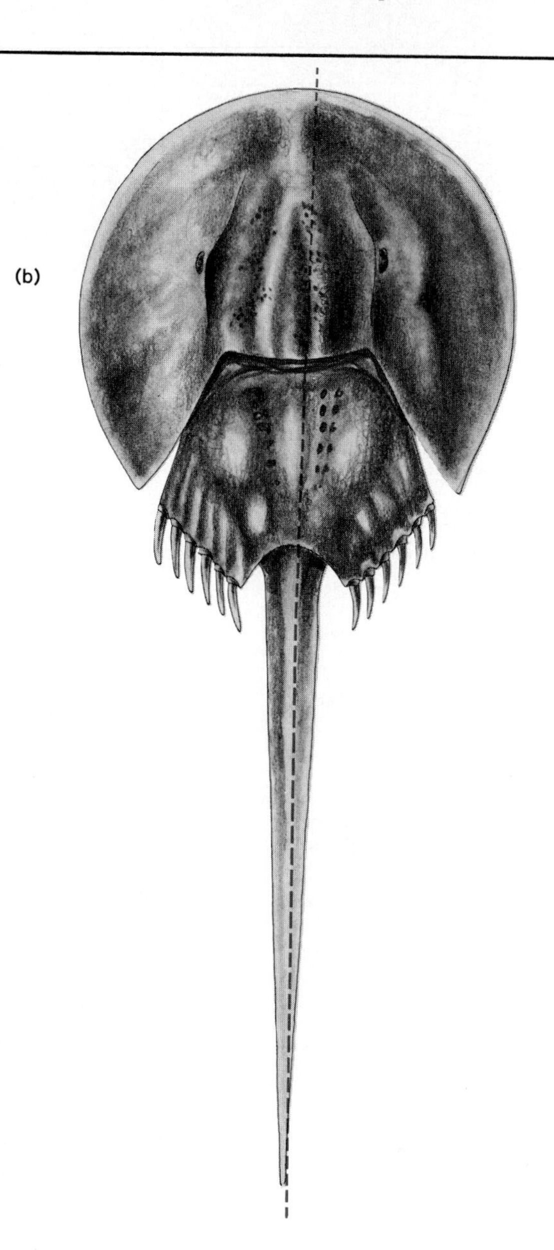

The basic coelenterate body plan has two expressions. The **medusa** has a rounded, somewhat flattened, jellyfish form and swims about with its mouth directed downward. A **polyp** generally is attached to a surface. Its mouth usually is directed upward (figure 29.4).

Coelenterates use their **tentacles** to maneuver food through their open mouths into their gastrovascular cavities. The epidermis of the tentacles has numerous cells containing stinging devices called **nematocysts,** which paralyze prey.

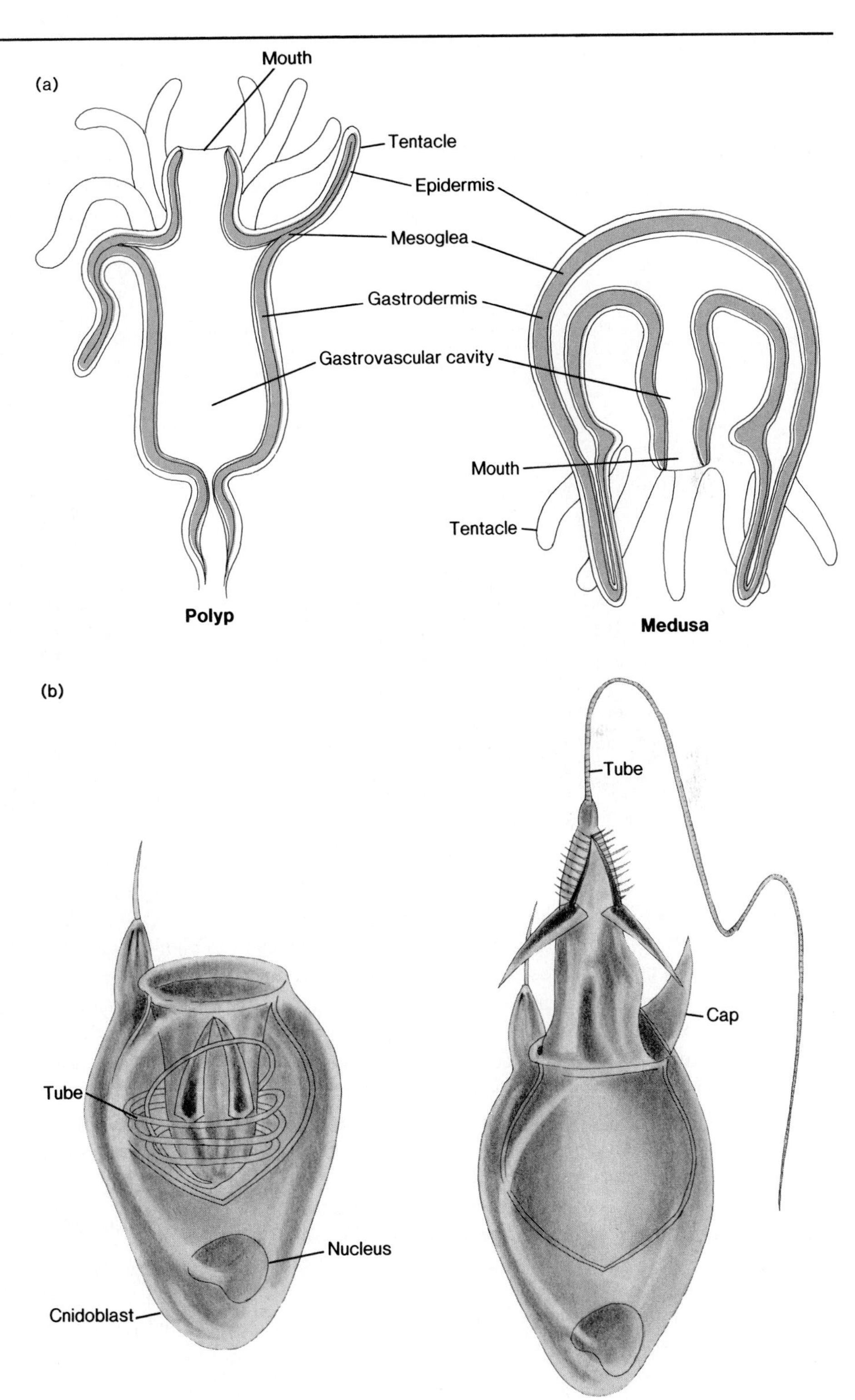

Figure 29.4 Coelenterates. (*a*) The two body forms of coelenterates. Polyps are attached to surfaces; medusae are free-swimming. (*b*) Enlarged, detailed drawings of nematocysts. A nematocyst is contained within a specialized cell, the **cnidoblast** (left). When a nematocyst is discharged, a cap opens and the tube shoots out (right).

Figure 29.5 Body organization of a hydra.

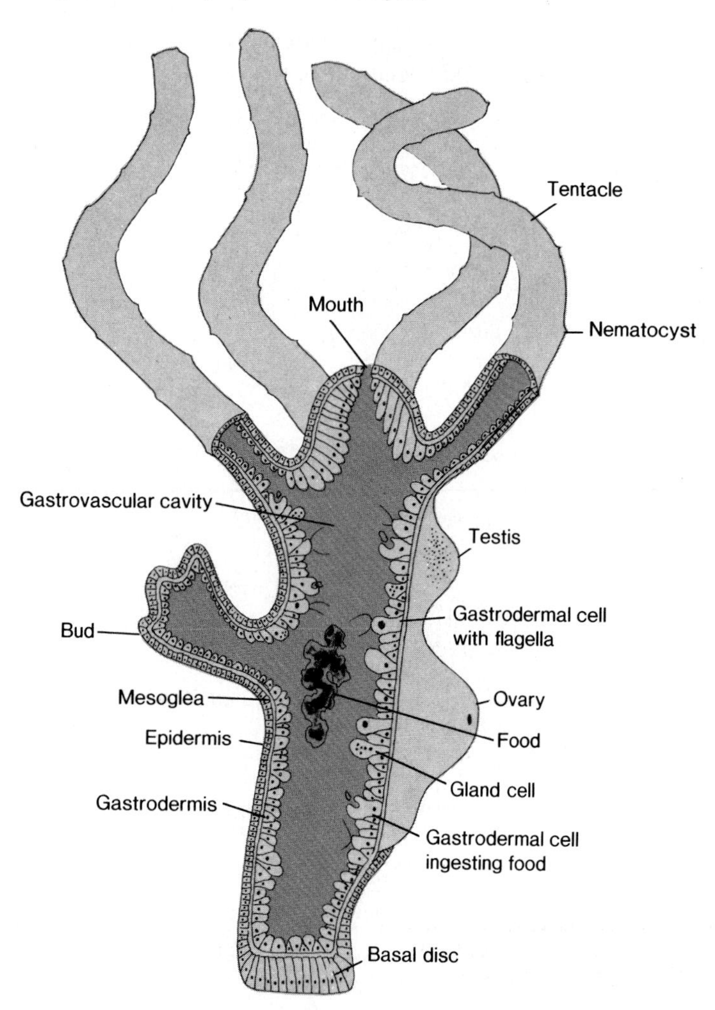

Figure 29.6 A hydra with two developing buds.

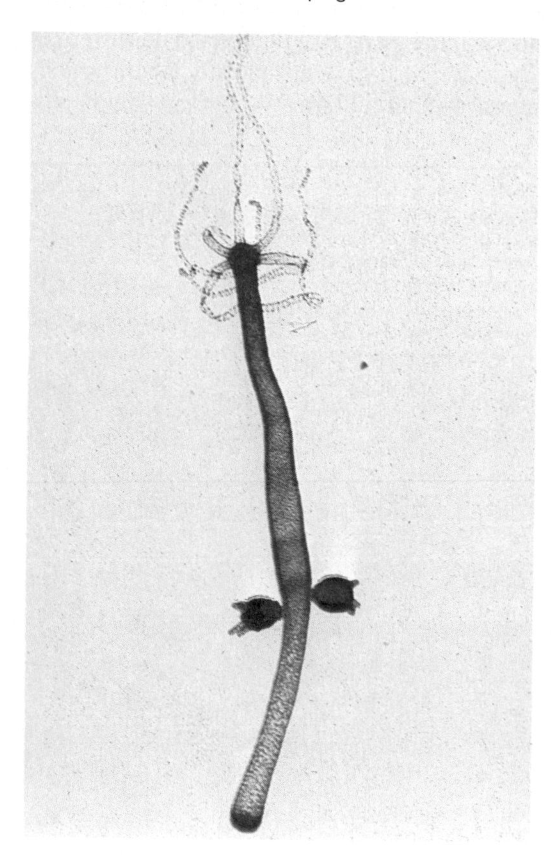

Class Hydrozoa

Freshwater hydras are the most familiar hydrozoan coelenterates (figure 29.5). A hydra captures food with its tentacles and pushes it into the gastrovascular cavity, where secreted enzymes begin digestion. Then gastrodermis cells phagocytize chunks of this partially digested food material. Thus, digestion in hydras (and other coelenterates as well) is a combination of **extracellular** digestion in the gastrovascular cavity and **intracellular** digestion inside gastrodermis cells.

Hydras can reproduce asexually and sexually. They reproduce asexually by forming **buds** (figure 29.6).

Sexual reproduction only occurs at certain times in response to environmental stimuli. Embryos produced by sexual reproduction become **dormant** (physiologically inactive) and are surrounded by a hard, protective shell. They emerge in spring to produce new polyps.

Obelia, a colonial, marine hydrozoan, has both polyp and medusa stages in its life cycle (figure 29.7). *Obelia* forms a colony of polyps, all of which are connected.

One of the most unusual hydrozoans is the Portuguese man-of-war *Physalia* (figure 29.8). *Physalia* is a colony of polyps. One polyp is specialized as a gas-filled float that provides buoyancy to keep the colony afloat. Other polyps are specialized for feeding or for reproduction. The Portuguese man-of-war also has fighting polyps armed with numerous nematocysts. Swimmers who accidently encounter Portuguese men-of-war can receive painful, sometimes serious, injuries from these fighting polyps.

Class Scyphozoa

Because the medusa stage is the dominant phase of scyphozoan life histories, the members of this class sometimes are called the "true jellyfish" (figure 29.9).

Figure 29.7 The life cycle of *Obelia,* a member of the class Hydrozoa. The polyp stage is the larger, more prominent stage of the life cycle. Polyps occur in a colony, and some are specialized for feeding while others are reproductive. Medusae are small, but they produce gametes.

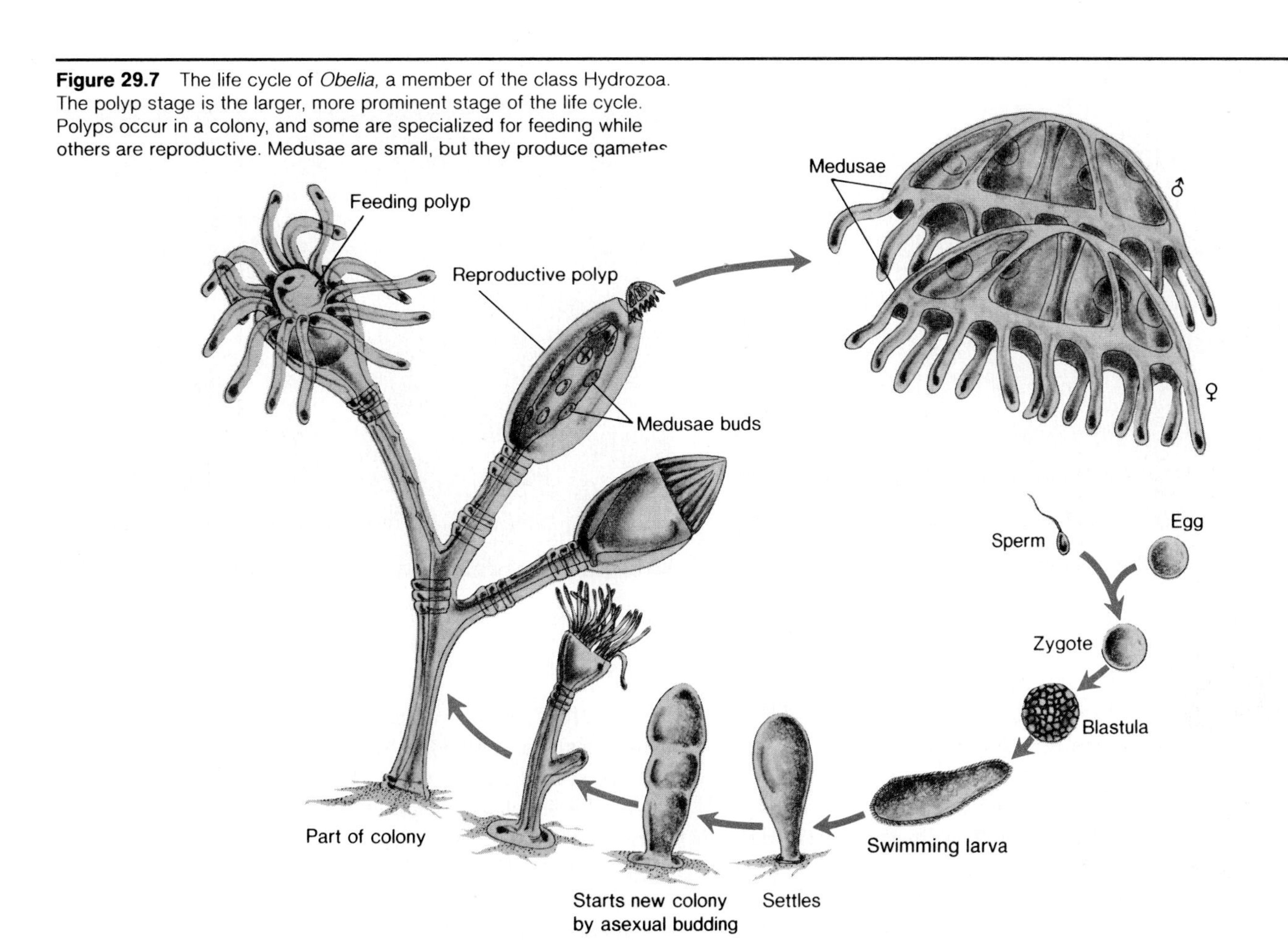

Figure 29.8 A Portuguese man-of-war (*Physalia*).

Figure 29.9 A scyphozoan, *Aurelia.* In scyphozoans, the medusa is the larger, more prominent stage of the life cycle and some of the jellyfish become quite large.

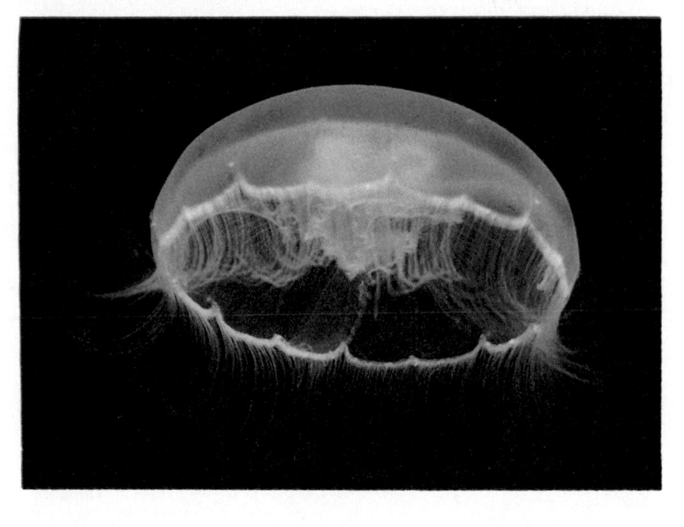

Figure 29.10 Class Anthozoa. (*a*) Sea anemones. (*b*) Coral in a reef off the coast of the Philippine Islands.

Class Anthozoa

Anthozoans include the sea anemones and the corals. These animals have no medusa stages. Some species have solitary polyps, and others are colonial.

Sea anemones are fleshy animals, some of which are large enough to capture and digest fish or crabs (figure 29.10*a*). Anemones often live along coasts in areas between the high and low tide water marks (the intertidal zone).

Corals live as solitary individuals in cold-water areas, but tropical corals form large, spreading colonies (figure 29.10*b*). Corals secrete hard, limy skeletons that remain in place even after the polyps die and degenerate. Each coral polyp is rather small and inconspicuous, but new generations of corals grow on the skeletal remains of past generations. In this way, massive deposits formed by corals build up over the years to make coral islands and coral reefs.

Phylum Platyhelminthes (Flatworms)

Flatworms are simply organized bilaterally symmetrical animals. They have elongate, flattened bodies with definite **organs** that function in reproduction and excretion.

There are three classes of flatworms. The **class Turbellaria** includes free-living animals such as the freshwater **planarians.** Members of the **class Trematoda (flukes)** and the **class Cestoda (tapeworms)** are all parasites.

Class Turbellaria

Planaria feed on dead organisms or on small living animals. A planarian has a **pharynx,** which is extended during feeding and retracted at other times. The pharynx is muscular and can produce such a strong sucking action that it tears off food chunks, which are then drawn into the gastrovascular cavity (figure 29.11). On their heads, planarians have pigmented **eyespots** that are sensitive to light intensity changes and **chemoreceptors** that detect potential food sources.

Planarians have tremendous powers of regeneration and actually can reproduce asexually by constricting their bodies and splitting in half. Each half regenerates the missing part and becomes a complete worm.

Class Trematoda (Flukes)

Flukes are parasites; they derive nourishment from a living **host** organism. A few flukes are **ectoparasites** (external parasites), but the majority are **endoparasites** (internal parasites). An adult fluke feeds on host tissues or body fluids by using its muscular pharynx to suck material into its digestive cavity.

Flukes are socially and economically important. They can infect domesticated animals and are responsible for some of the most widespread human parasitic diseases.

Much of the space inside a fluke's body is occupied by reproductive structures. Many species of flukes are **hermaphroditic** (each individual has both female and male reproductive organs). A fluke can produce huge numbers of eggs. For example, the Chinese liver fluke *Opisthorchis* (*Clonorchis*) *sinensis* produces literally hundreds of eggs every day during its lifetime in the human liver, where it can remain active for five, ten, or even twenty years (figure 29.12).

Figure 29.11 (*a*) A living planarian. Planarian worms glide along slimy mucous trails that they secrete as they go. They are propelled by ciliated cells on their body surfaces. (*b*) A planarian with its pharynx extended in feeding position. (*c*) Organization of the planarian digestive (gastrovascular) cavity. The mouth is the only opening into the cavity. Small branches of the cavity penetrate much of the body and distribute material to body tissues.

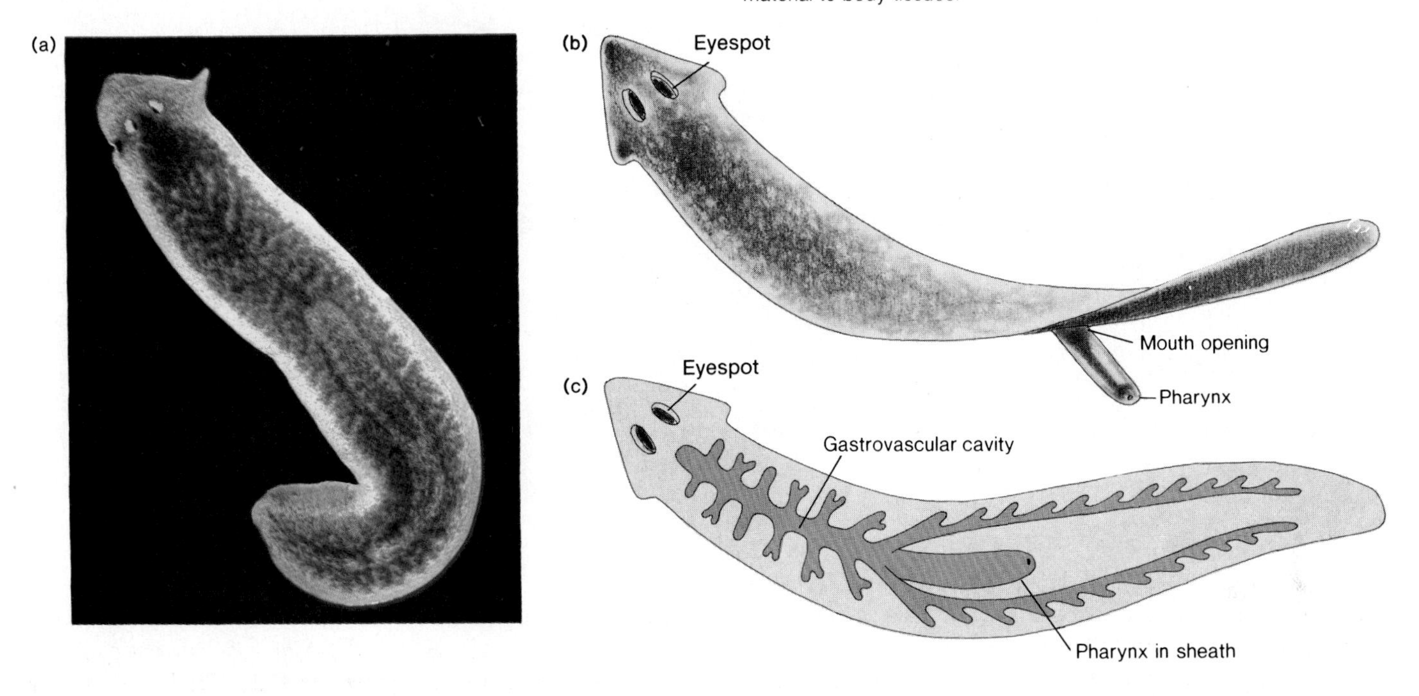

Figure 29.12 The human liver fluke, *Opisthorchis* (*Clonorchis*) *sinensis*. (*a*) An adult fluke. Each individual has both female and male reproductive structures. The fluke lives attached inside the liver and releases hundreds of eggs each day. (*b*) Simplified life history of the human liver fluke. If a snail eats a fluke egg, a miracidium larva hatches in the snail where it passes through several developmental stages and multiplies to produce several hundred cercaria larvae. The cercaria larvae break out and swim away. If they encounter a fish, they burrow into its muscle to form a cyst. If a human eats the fish raw or poorly cooked the cyst opens and the fluke journeys to the liver.

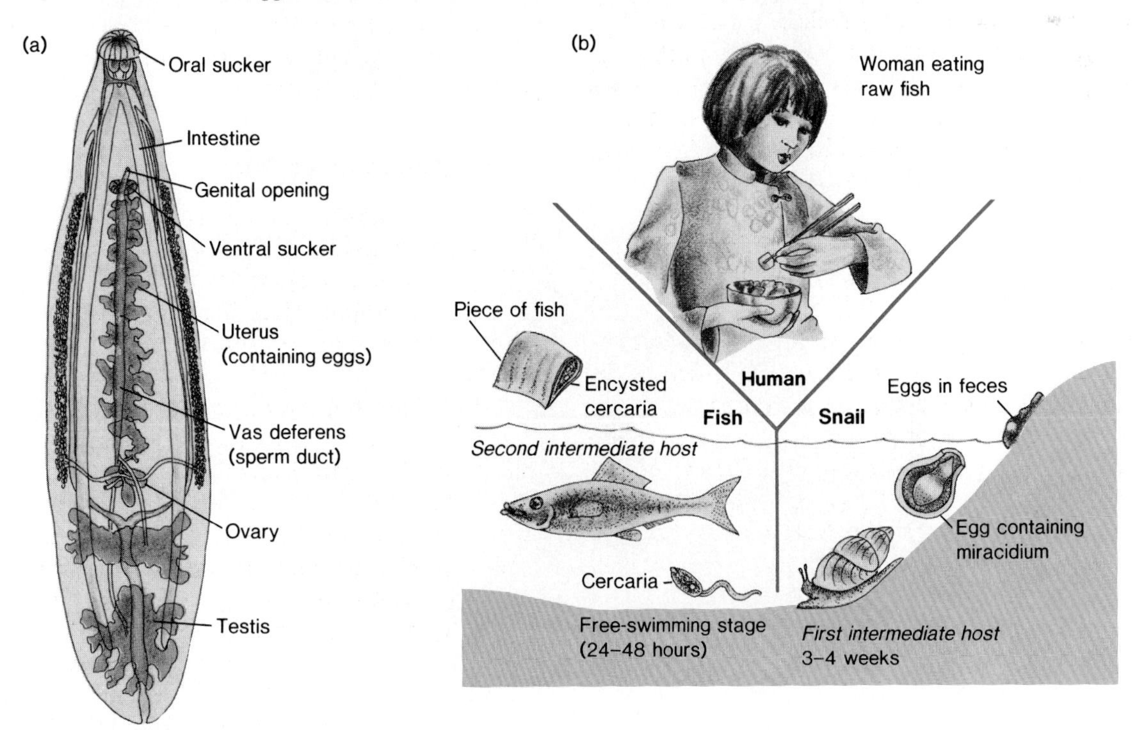

Figure 29.13 Distribution of the human schistosome fluke, *Schistosoma mansoni*.

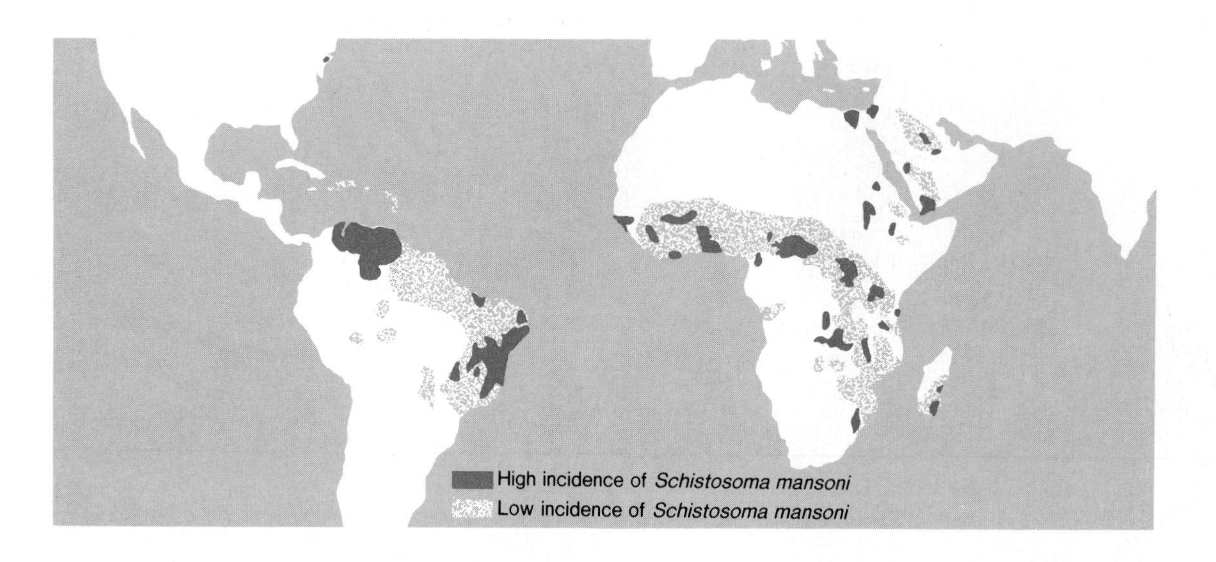

Figure 29.14 Scanning electron micrograph of a pair of *Schistosoma mansoni* adults copulating. The smaller female lies in a groove along the larger male's body.

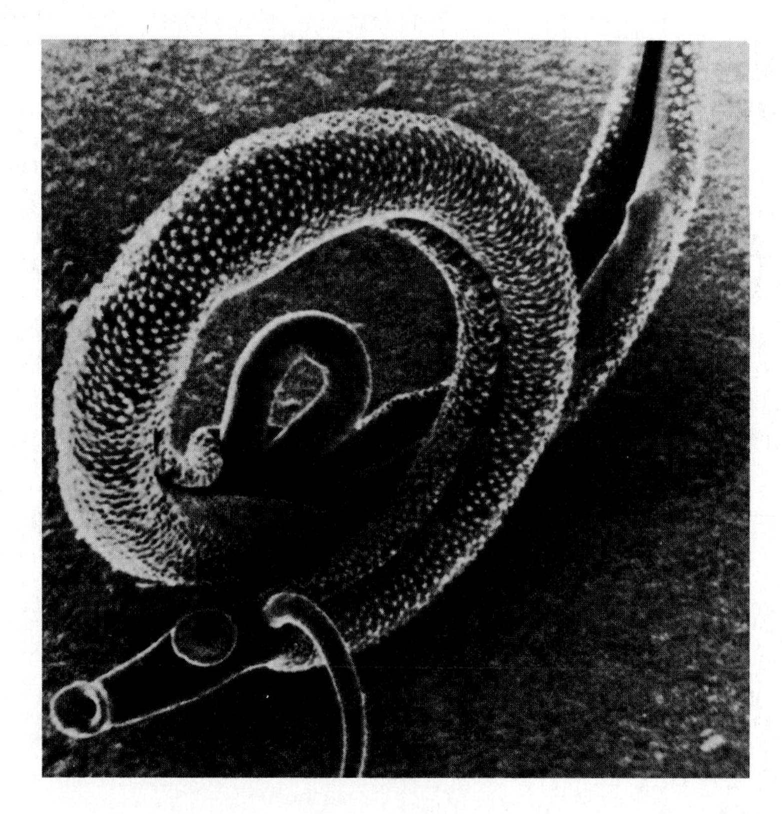

Figure 29.15 The life history of *Schistosoma mansoni*. All of the sketches are drawn to different scales. (1) Cercaria penetrate skin in water. (2) Adult worms live in blood vessels of intestines. (3) Copulating worms. (4) Eggshell containing developing miracidium. (5) Miracidium hatches in the water. It will enter a snail if it encounters one. (6) Inside snail, a mother sporocyst (7) forms. It produces many daughter sporocytes. Each one (8) produces many cercaria larvae. (9) Cercaria larvae break out of snail's body and enter the water.

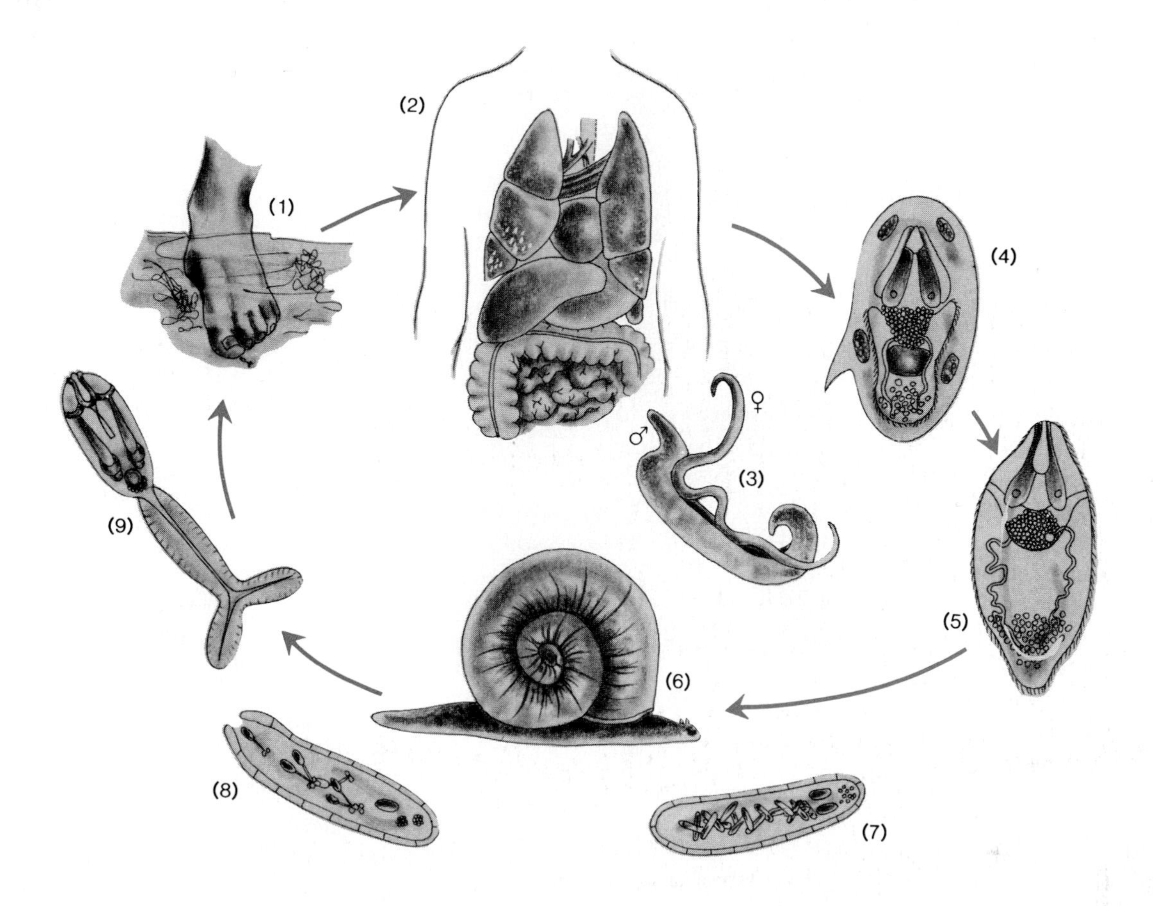

Most flukes have complex life cycles that involve several hosts. For example, humans or other mammals are **primary** (definitive) **hosts** of the liver flukes; that is, sexually reproducing adult flukes live in mammalian livers. Larval stages of flukes live in **intermediate hosts**—snails and then fish in the case of the liver flukes.

Human blood flukes of the genus *Schistosoma* are among the most important of human **helminth** (worm) parasites. It is estimated that nearly half of all of the people living in tropical and subtropical areas where schistosomes occur are infected by these flukes. *Schistosoma mansoni* is the most widely distributed of the schistosome flukes and causes major public health problems in many nations (figure 29.13).

Adult *Schistosoma mansoni* live in blood vessels of the human digestive tract. Schistosome flukes have separate sexes, and the males and females live together in more or less permanent copulatory contact (figure 29.14). Some eggs laid in the blood vessels penetrate the intestinal walls and leave the body with feces. The symptoms of **schistosomiasis** include dysentery, anemia, general weakness, and greatly reduced resistance to other infections.

Schistosome transmission depends on human feces reaching water, subsequent infection of the appropriate snail intermediate host, and contact between the cercaria larva of the fluke and human skin in the water (figures 29.15 and 29.16). Rice paddy farming, which is the heart of agriculture in many tropical countries, creates a nearly perfect environment for transmission of these flukes because human feces are used for fertilizer and people stand in the water tending the rice plants.

Figure 29.16 Scanning electron micrograph of a typical cercaria larva. This is the cercaria of one of the species of *Trichobilharzia,* which parasitizes ducks. This species cannot develop to adulthood in the human body, but the cercaria can penetrate the skin and cause a skin irritation called "swimmer's itch."

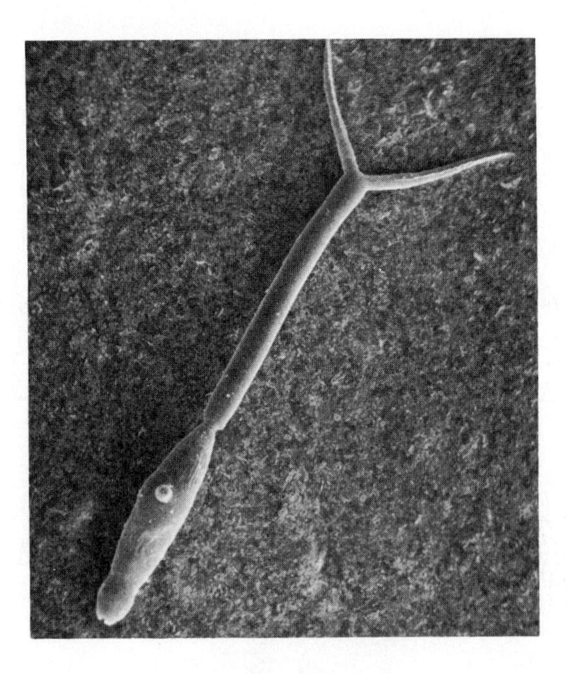

Class Cestoda (Tapeworms)

Tapeworms are very long, flattened worms that live as adults in the intestines of vertebrate animals, where they attach to the intestinal wall.

Tapeworms have a series of similar body subunits called **proglottids** (figure 29.17). The anterior tip of a tapeworm, the **scolex,** has hooks and suckers that permit firm anchoring in the host's intestinal wall. At the end of a narrow neck, new proglottids are produced continually. When a proglottid is mature, it contains a complete set of male and female reproductive organs, but no digestive structures. Tapeworms live by absorbing digested nutrients directly from the gut cavities of their hosts.

Each proglottid produces thousands of eggs and stores them in the uterus. Such "ripe" proglottids detach and pass out with the feces.

If an egg is swallowed by an appropriate animal, a larva hatches and enters a muscle, where it forms a hard-walled cyst. If a primary host animal eats the flesh raw or poorly cooked, the muscle cyst wall is digested. Then the **bladder** worm emerges and develops into a new tapeworm that attaches to the intestinal wall.

Beef tapeworm is rare in the United States, but pork tapeworm (*Taenia solium*) is common enough that great care should be taken to make certain that all pork is thoroughly cooked before it is eaten. Fish tapeworms also are relatively common in the United States and Canada.

Tapeworms generally grow to lengths of from 6 to 8 m in human intestines, and tapeworms approaching 20 m have been found. Tapeworms cause problems for their hosts because they absorb significant quantities of nutrients, excrete toxic wastes, and can sometimes interfere with passage of food through the gut.

Parasites and Hosts

Parasitic relationships are products of long evolutionary processes and very frequently involve a balance between host and parasite. Parasites can withstand the host's defenses effectively enough to become established and reproduce successfully. But a parasite is usually not so successful and

Figure 29.17 (*a*) The pork tapeworm, *Taenia solium*. (1) Tapeworm showing scolex, neck, and proglottids. (2) Details of scolex and neck. Compare the sketch and the photograph in (*b*). Suckers and hooks attach the worm to the intestinal wall. (3) A mature proglottid with a complete set of female and male reproductive structures. (4) "Ripe" proglottid that is nearly filled by an expanded uterus containing eggs. (*b*) Photomicrograph of the scolex, neck, and first few proglottids of a tapeworm.

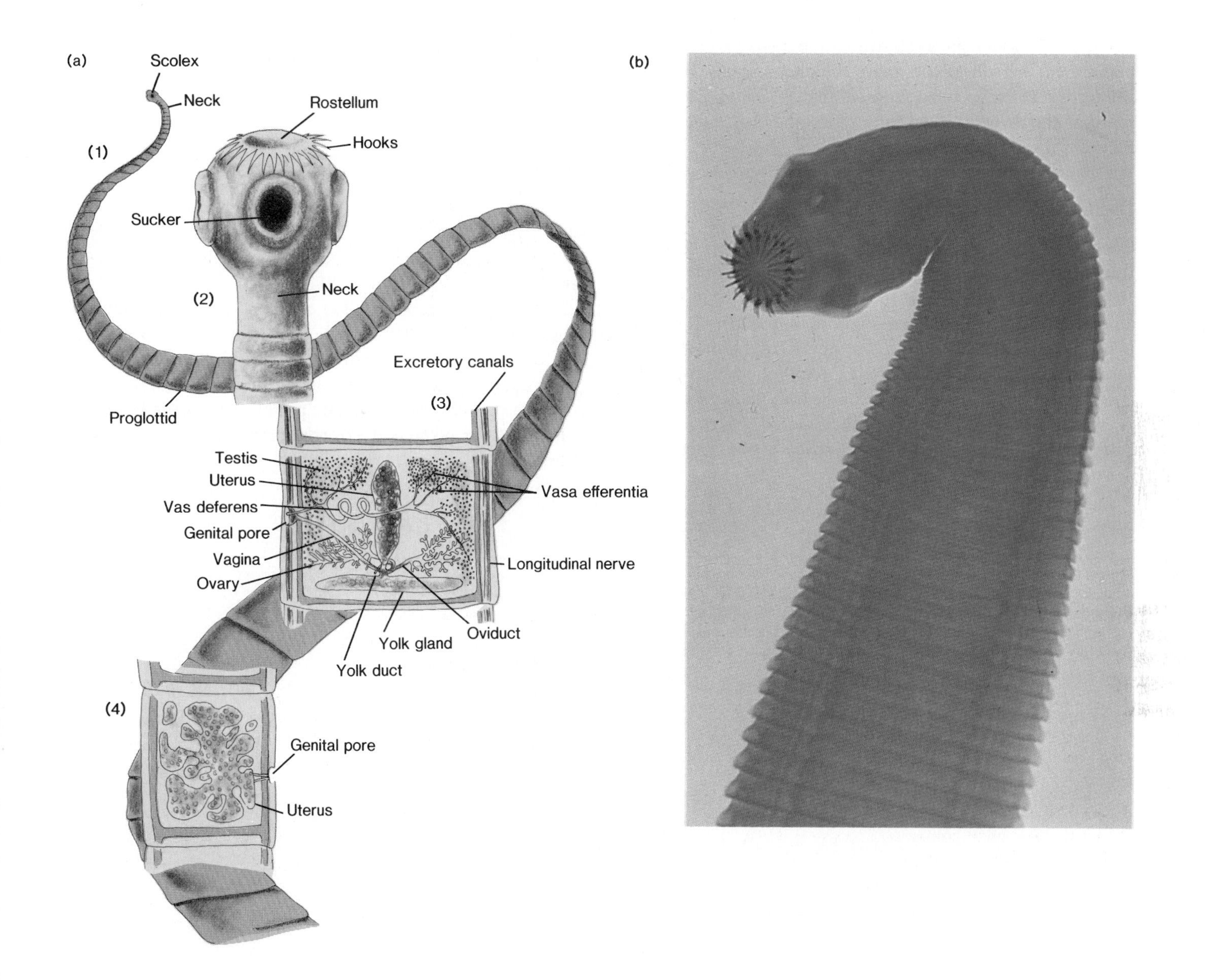

aggressive that the host is killed. Thus, a "good" parasite lives at the expense of its host, but does not destroy its own "livelihood" by killing the host.

While individual parasites usually do limited damage to a host, heavy parasite loads involving many parasites of one type or simultaneous parasitism by several types of parasites frequently do cause serious illness and death. Parasitic diseases interfere with progress in many of the world's underdeveloped areas, and they produce a tremendous burden of human misery.

Some Evolutionary Relationships of Animals

Many biologists think that primitive flagellated protists that formed colonies were the ancestors of all animals (figure 29.18).

We have examined sponges, coelenterates, and flatworms, all of which have relatively simple body organization plans. Next we will discuss more complex animals. But first we need to consider some features that we can use to make sense of the diversity seen in animals and the evolutionary relationships of the various groups.

Digestive Tracts and Body Cavities

Coelenterates and platyhelminthes have **incomplete digestive tracts** because a single opening must serve as both entrance (**mouth**) and exit (**anus**) of the digestive cavity. Most other animal groups have **complete digestive tracts.** In these animals, food is taken in through a mouth, digestion and absorption occur as food moves along through the digestive tract (**gut**), and undigested materials exit the body through an anus.

In the bodies of platyhelminthes, mesoderm, the middle body layer, is solid tissue filling the space between ectoderm and endoderm. In more complex animals, a body cavity separates internal organs from the outer body wall.

Body cavities are organized in two ways. Some animals, such as nematodes (roundworms), have a **pseudocoelom,** where mesoderm lies beneath the ectoderm and is separated from the endoderm of the digestive tract. Other animals have true **coeloms** and are called **coelomate** animals. A true coelom is a body cavity that is completely lined with mesoderm (figure 28.19).

Protostomes and Deuterostomes

All of the animals that have true coeloms fall into two separate groups, protostomes and deuterostomes. This division is based on differences in the way in which animals in the two groups develop (figure 29.20).

During development (see chapter 19 for more details), a set of rearrangements, called gastrulation, converts the hollow sphere of cells, known as a blastula, into the general form of an animal body. One side of the blastula sinks in and forms a pit. This indented part produces the endoderm (gut lining) of the body as it extends further into the interior of the blastula.

Figure 29.18 A family tree showing proposed evolutionary origins and relationships of all organisms. The broken line indicates that there are questions about sponges' evolutionary relationship to other animals. Some biologists think that sponges may have arisen separately from the rest of the animals because they are so different from all other animals.

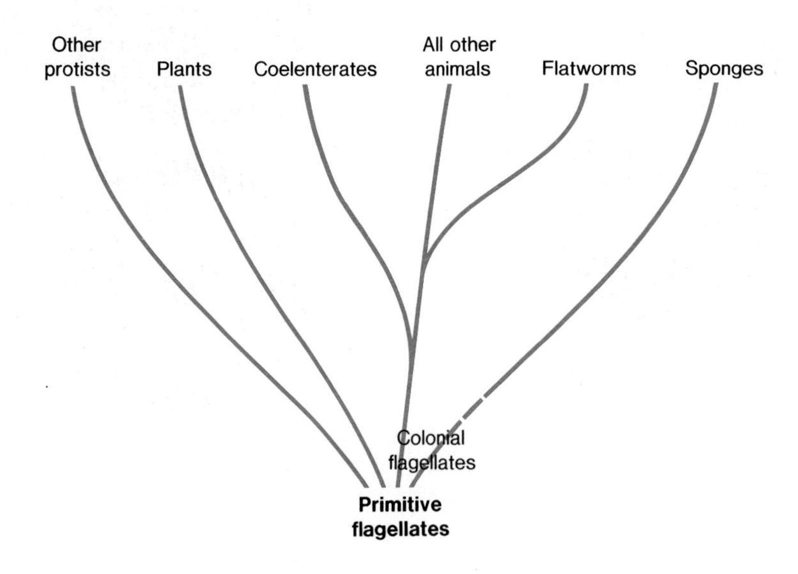

An opening called the **blastopore** is left at the surface by this process. In the embryos of some animals, the blastopore becomes the mouth, and a new opening forms the anus. These animals are called **protostomes** (proto = first, stome = mouth). Molluscs, annelids, and arthropods are protostomes. In the embryos of other animals, the blastopore becomes the anus, and a new opening forms the mouth. These animals are called **deuterostomes** (deutero = second, stome = mouth). The most familiar of the deuterostome phyla are the echinoderms and chordates.

Figure 29.19 Body organization. (*a*) In coelenterates (1), the ectoderm and mesoderm are separated by mesoglea, which contains scattered cells. In flatworms (2), a solidly packed mesoderm layer lies between the ectoderm and the endoderm. (*b*) Comparison of mesoderm organization. Flatworms (1) are acoelomate; that is, their mesoderm is packed solidly. Pseudocoelomate animals (2) have mesodermal tissues such as muscle inside their ectoderm, but not adjacent to their gut endoderm. Coelomate animals (3) also have mesodermal tissue covering their guts. True coeloms are body cavities completely lined by mesodermal tissue. Mesenteries hold organs in place within body cavities.

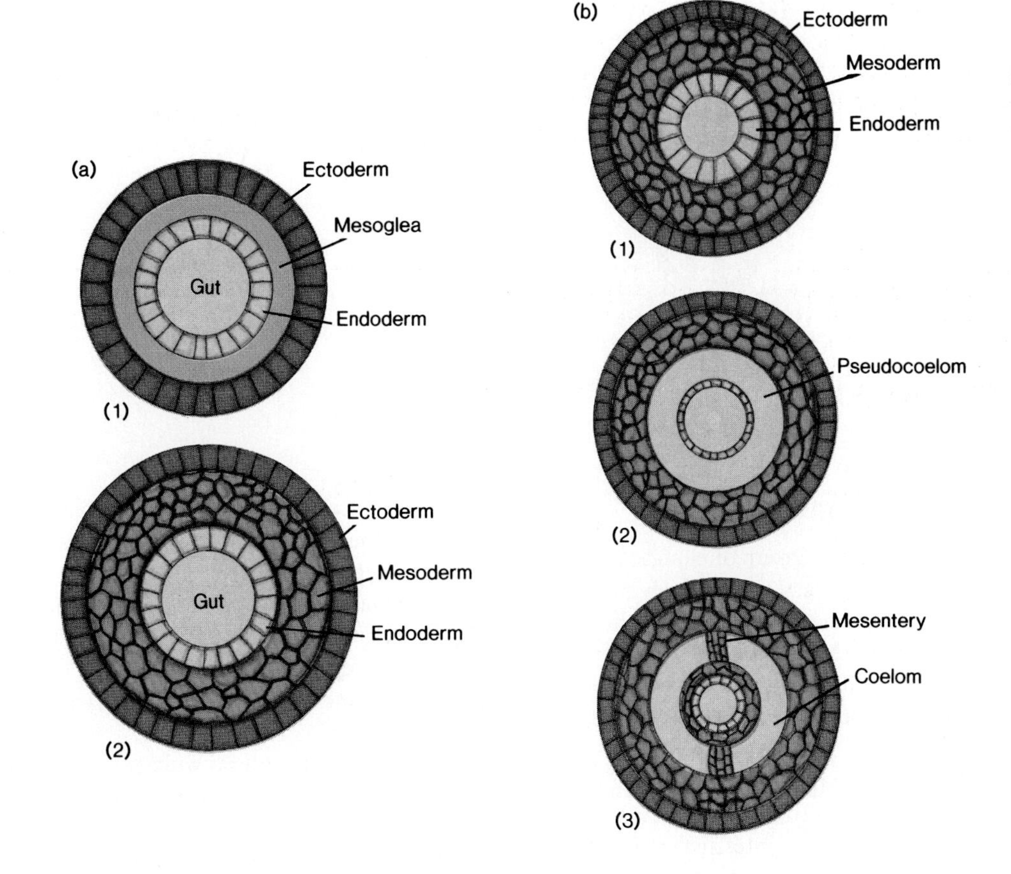

Figure 29.20 Protostomes and deuterostomes. (*a*) Differences in the developmental pattern of two major groups of animal phyla. In protostome embryos, the blastopore becomes the mouth. In deuterostome embryos, it becomes the anus. (*b*) "Family tree" of animals showing proposed evolutionary relationships.

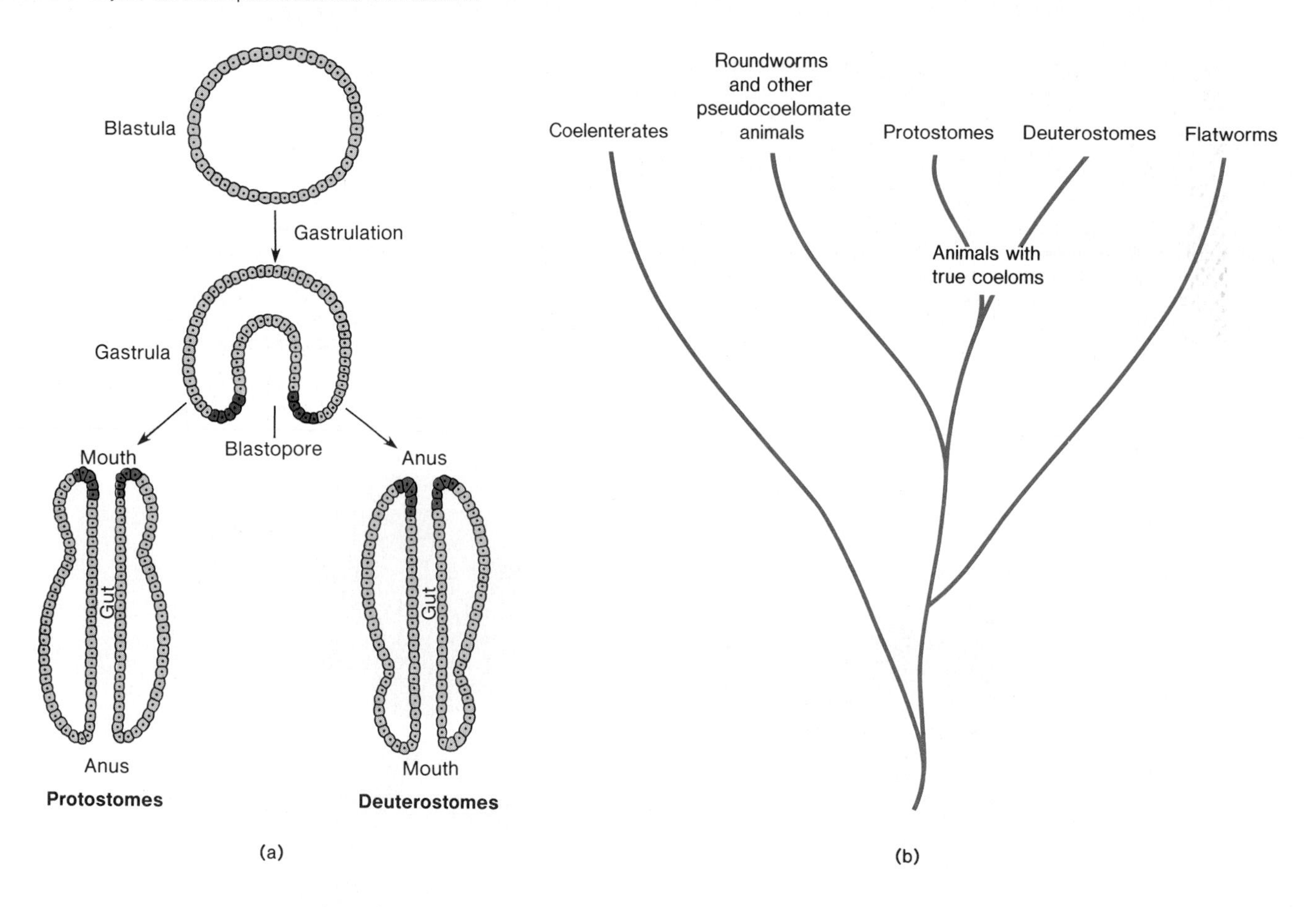

Phylum Nematoda (Roundworms)

The **nematodes (roundworms)** are a group of elongate, pseudocoelomate worms that all have very similar body plans.

Nematodes are among the most numerous of animals. They are everywhere: in soil, in the muddy bottoms of freshwater lakes and ocean shores, and also in some rather exotic environments. For example, some nematodes live in beer-soaked mats in German taverns and in vats of vinegar.

Virtually all animals have characteristic parasitic nematodes. But nematodes that parasitize vertebrates are the most familiar. The *Ascaris* worm, an intestinal parasite of vertebrates, is a relatively large roundworm (figure 29.21).

An *Ascaris* infection starts when a vertebrate animal such as a pig eats soil containing *Ascaris* eggs. The eggs hatch in the digestive tract, and the small worms penetrate the intestinal wall and enter blood vessels. They remain in the circulatory system for a time but eventually crawl into the lungs. They then work their way up the trachea and pass "over the top" into the esophagus. When they reach the gut, they mature and reproduce.

A female *Ascaris* can produce as many as 200,000 eggs per day. Considering this huge reproductive rate and the fact the *Ascaris* eggs remain viable in the soil for several years, it is easy to understand how soil contaminated with feces can be so infective.

There also are several widespread human nematode parasites. For example, intestinal **pinworms** infect almost all people (at least 90 percent) in every part of the world sometime during their lifetimes. **Trichinosis,** an infection resulting from eating poorly cooked meat that contains worm cysts, can cause serious illness and permanent muscle damage. **Filaria worms** are transmitted by mosquitoes in tropical areas. The worms enter the lymphatic system and cause large fluid accumulations and swellings in various body areas when they block lymphatic vessels (figure 29.22).

Figure 29.21 *Ascaris,* a parasitic roundworm. (*a*) Female and male *Ascaris* worms. Mature females are larger than mature males. Males have a curved posterior tip. (*b*) Internal anatomy of a female *Ascaris.* (*c*) Cross section of the body of a female *Ascaris.* Because it is extensively coiled, the reproductive tract is cut across several times. Note the pseudocoelomate arrangement. There is mesodermally derived muscle tissue inside the epidermis but no covering mesodermal tissue on the outside surface of the intestine.

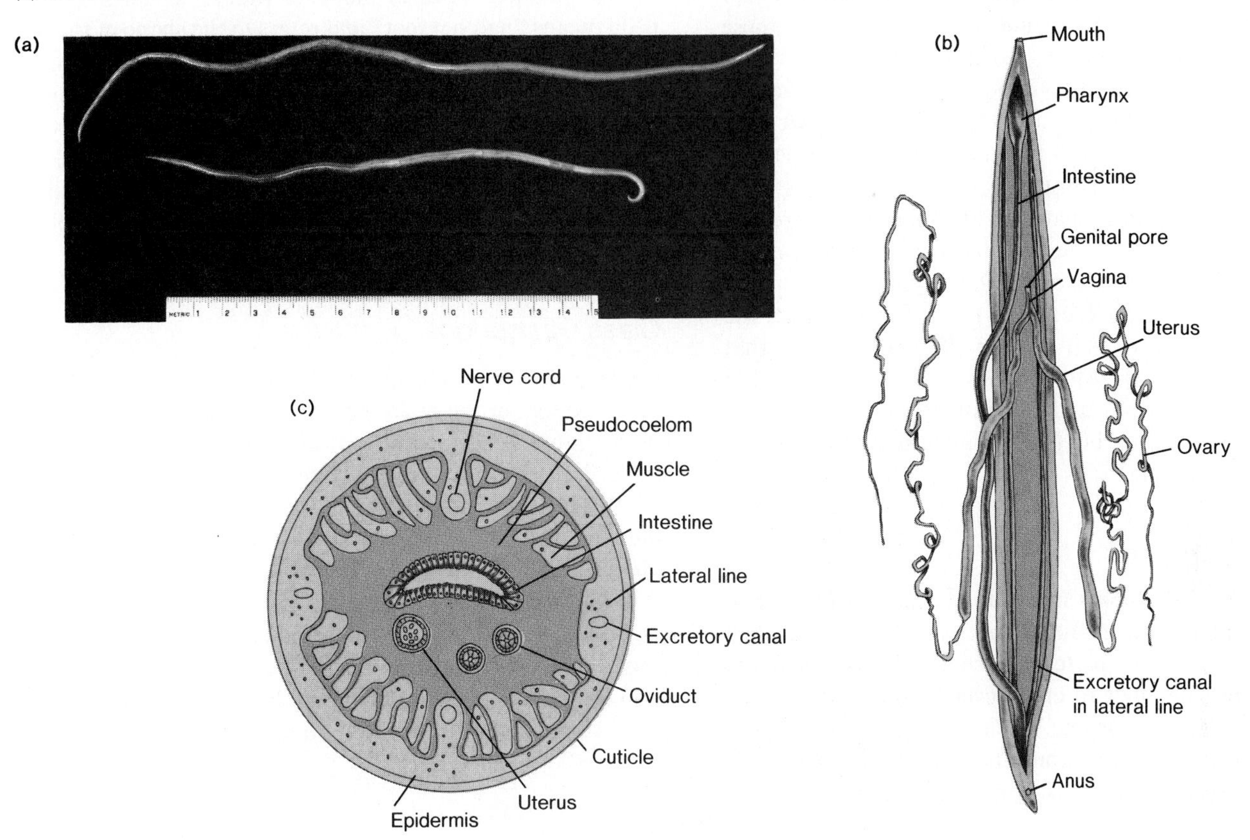

Figure 29.22 This man is suffering from **elephantiasis.** The condition results when filaria worms block lymphatic vessels. Fluid accumulates and causes extreme swellings of some body regions.

Phylum Mollusca

The **phylum Mollusca** includes a number of familiar animals, such as **snails, clams, oysters, squids,** and **octopuses.** There are at least 80,000 species of **molluscs,** and many of those species contain huge numbers of individuals.

Molluscs are bilaterally symmetrical, and they have true coeloms. All molluscs have a ventral muscular extension called the **foot** that functions in locomotion or is specialized in some other way—for example, as tentacles in squids.They have a soft **visceral mass** that contains digestive, excretory, and reproductive organs. The phylum name comes from the Latin *mollis,* meaning soft, and refers to this visceral mass. The visceral mass is enclosed by a heavy fold of tissue, the **mantle,** that also secretes a shell in the many species that have one (figure 29.23). Most molluscs have **gills** located in a **mantle cavity** that lies under the mantle. Molluscan gills are gas exchange organs, but in many molluscs, they also filter food out of the water.

Class Amphineura

Animals known as **chitons** make up the **class Amphineura.** They are flattened marine molluscs that have a dorsal shell arranged in a series of plates (figure 29.24*a*). Chitons use their flat, muscular foot to crawl over surfaces where they feed on algae. A hard, rasping, toothed strap, called a **radula,** can be protruded from the mouth to scrape algae loose from rocks and other surfaces.

Class Pelecypoda

The **class Pelecypoda** includes clams, oysters, and scallops—the two-shelled (**bivalve**) molluscs (figure 29.24*b*). Pelecypoda means "hatchet foot" and refers to the shape of the foot that is extended and used for digging by many bivalves.

Bivalves are sedentary and feed by moving a stream of water through their gills and filtering out food particles.

The two shells of these molluscs are hinged and are closed by powerful muscles. When a bivalve is disturbed, it pulls in and closes its shell firmly. Only a few predators are able to open tightly closed bivalve shells.

Class Gastropoda

Gastropoda means "belly foot" and describes the ventral, flattened foot of the snails, conchs, and other **gastropod** molluscs (figure 29.25). Most gastropods have coiled shells, and their visceral masses spiral up the inside of these shells.

Most gastropods browse by rasping off algae with their radulas, but some are predatory. The "oyster drill," *Urosalpinx,* perches on top of an oyster, bores through the shell, and eats the visceral mass.

Some gastropods, such as common terrestrial slugs, lack shells. The **nudibranchs,** which are often called "sea slugs," also lack shells. Certain of the nudibranchs are among the most colorful and beautiful of all animals.

Figure 29.23 The body organization of a hypothetical primitive mollusc showing relationships of the foot, mantle, shell, visceral mass, mantle cavity, and gills. This basic plan has been modified several different ways during the evolution of modern molluscs.

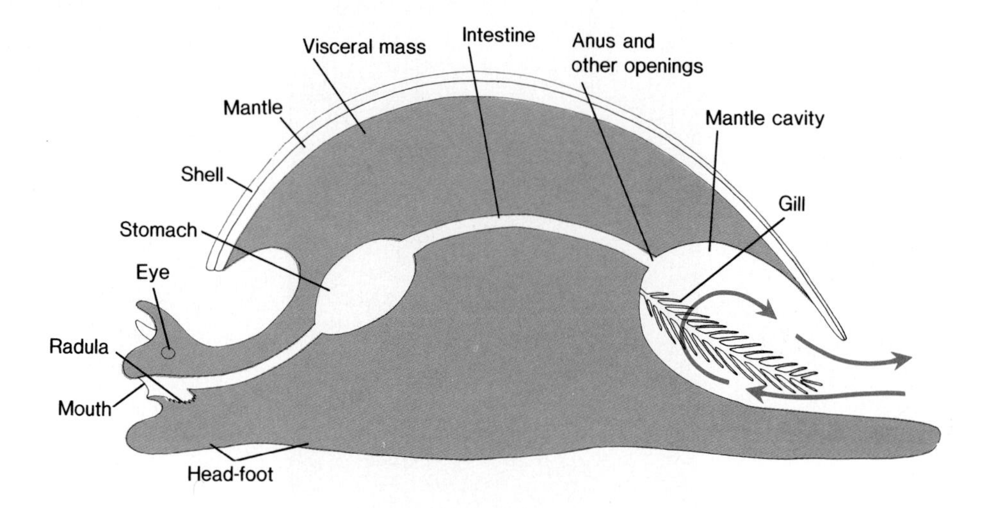

Figure 29.24 (*a*) Class Amphineura. A chiton. (*b*) Class Pelecypoda. The giant clam *Tridacna*. *Tridacna* shells can reach lengths of well over 1 m. Note the open siphons and the coral colonies growing on the clam's shells.

(a)

(b)

Figure 29.25 Class Gastropoda. (*a*) A snail. (*b*) A nudibranch (*Flabellina*).

(a)

(b)

Figure 29.26 Class Cephalopoda. (*a*) A nautilus shell cut open to show its chambers. The animal lives in the last chamber of its shell, but it carries the outgrown smaller chambers that it lived in earlier. (*b*) An octopus (*Octopus*) moving over the surface of coral in the Pacific Ocean.

(a)

(b)

Class Cephalopoda

The **cephalopod** ("head foot") molluscs are the most complex and specialized group in the phylum. This class includes octopuses, squids, and nautiluses (figure 29.26). Most of the cephalopods are fast-moving predatory animals with well-developed sense organs, including focusing camera-type eyes that are very similar to those of vertebrates. Cephalopods in general and octopuses in particular have well-developed brains and display complex behavior, including impressive learning ability.

Nautiluses are enclosed in shells, but the shells of squids are reduced and internal. Octopuses lack shells entirely. These cephalopods can squeeze their mantle cavities so that water is forced out, thus propelling them rapidly backwards by a sort of jet propulsion. Cephalopods also possess ink sacs from which they can squirt out a cloud of brown or black ink. This action often leaves a would-be predator completely confused.

Cephalopods use the ring of long tentacles around their mouth to capture prey, which they then tear up with a sharp **beak.** Experienced divers fear the beak of an octopus much more than its tentacles.

Although giant octopuses are a standard part of science fiction, octopuses seldom grow to be very large. Some giant squids, however, are enormous; they are by far the largest invertebrate animals. There are accurate records of a captured squid that was 18 m long from tentacle tip to tail and weighed at least two tons. Fairly reliable evidence also indicates that even larger squids may lurk in the ocean depths.

Phylum Annelida

The name **annelid** (from the Latin word *annelus,* meaning "little ring") is derived from the most obvious characteristic of this group of animals. Their bodies consist of a series of similar, but not identical, units or segments.

Annelids have true coeloms that are broken up into segmental compartments by partitions called **septa** (singular: **septum**). The digestive tract passes through the septa and runs the length of the body, but many body segments have their own pair of excretory tubules, the **nephridia.** Each segment also receives its own branches of major blood vessels, and the nervous system has an enlarged nerve center called a **ganglion** (plural: **ganglia**) in each segment.

Figure 29.27 Locomotion in a hypothetical, twenty-segment earthworm as an example of annelid movement. The worm's progress is shown at successive times in sketches (1) through (4). Contractions of circular or longitudinal muscles apply pressure against partitioned, fluid-filled body segments, altering their shapes in specific ways. Setae are bristles that anchor portions of the body that are expanded in circumference. The pattern of alternate anchoring and extending permits the worm to move forward. Such movement is especially effective, for example, as an earthworm burrows through the soil.

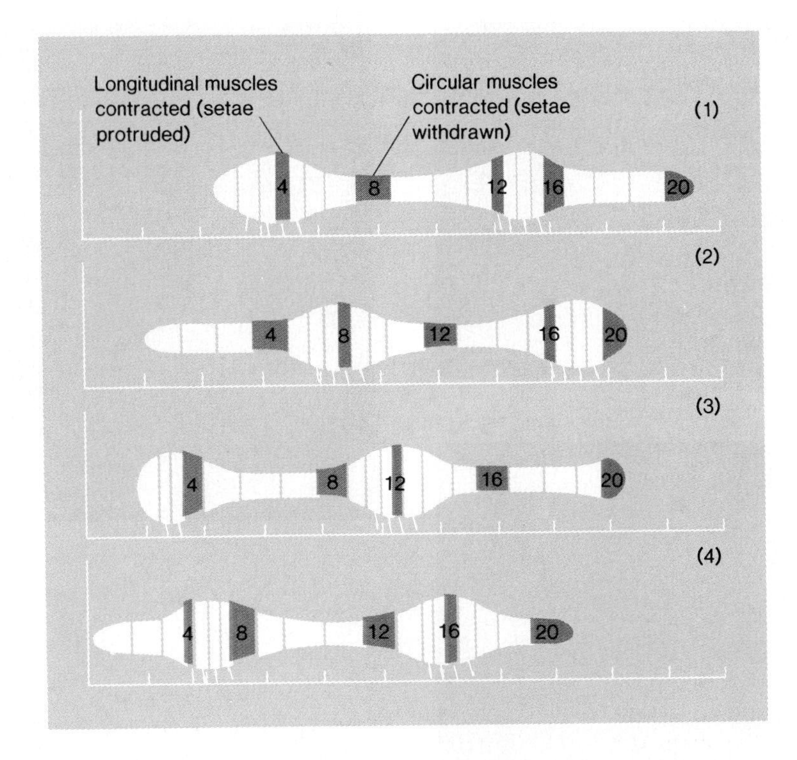

Movement in annelids depends on precisely coordinated contractions of two sets of muscles in the body wall. **Circular muscles** contract against the fluid contained inside segments and alter the shapes of the segments, causing them to narrow and lengthen. On the other hand, **longitudinal muscles** (muscles that run lengthwise) can contract to shorten and thicken the body as they force the fluid-filled segments to expand outward (figure 29.27).

Class Polychaeta

The **polychaetes** are marine annelids that have a pair of fleshy, lateral appendages called **parapodia** (singular: **parapodium**) on each of their body segments. Each parapodium has a number of stiff bristles, the **setae,** and the class name comes from this characteristic (polychaetae = "many setae"). Polychaetes have well-developed heads and anterior sense organs.

One group of polychaetes consists of active, free-swimming, crawling, or burrowing animals. Most of the polychaetes in this group are predators. The other group of polychaetes consists of worms that dwell permanently in tubes they have constructed (figure 29.28). Most of these are filter feeders.

All species of polychaetes have separate sexes, and gametes are released into the water, where the fertilization occurs. Special precise timing mechanisms that assure simultaneous gamete shedding by many individuals have evolved in many polychaete species. A rather spectacular example of this precision is provided by the Palolo worm, which lives in coral reefs in the South Pacific. About 99 percent of the worms shed their gametes as they swarm near the surface of the water during a single two-hour period on the night of the last quarter of the moon in November. A high percentage of fertilization is assured since the gametes of all animals are shed in the same place at the same time.

Figure 29.28 Polychaete annelids. (*a*) The sandworm (*Nereis*) is an example of an active polychaete that moves around seeking food. Note the characteristic fleshy parapodia on each of the body segments. (*b*) The fanworm (*Sabella*) is another sedentary polychaete. Its segmented body is enclosed in a tube. The worm feeds by sweeping food out of the water using its extended tentacles ("filaments"). (*c*) The tube of a tubeworm (*Chaetopterus*), which is an example of a tube-dwelling polychaete that feeds on material extracted from the stream of water that it moves through its tube.

(a)

(b)

(c)

Class Oligochaeta

The most familiar **oligochaetes** are the earthworms that live in moist soils throughout the world, but many oligochaetes also live in the muddy bottoms of freshwater streams, lakes, and ponds.

Oligochaetes lack parapodia and have fewer setae (oligochaetae = "few setae"). They also lack well-developed heads.

Earthworms are terrestrial, soil-dwelling animals. They feed by ingesting quantities of soil containing living and decaying organic matter. Much of what they eat passes through their bodies. They carry a good deal of this dirt up out of the soil where they leave small piles of dirt ("worm casts") on the soil surface.

Earthworms and other oligochaetes are hermaphroditic, but not self-fertilizing. Earthworms exchange sperm in copulation. Two worms line up facing opposite directions and exchange sperm (figure 29.29).

Figure 29.29 Reproduction of the earthworm. (*a*) Sketch of the internal anatomy of the anterior part of an earthworm's body with reproductive structures shown in color. Earthworms are hermaphroditic. Each worm has both male and female reproductive structures. The small sketch shows the location of the clitellum. (*b*) Copulation of earthworms. Worms face in opposite directions, and their clitellums secrete a mucous sheath that protects sperm as they are being exchanged. Sperm leave the vas deferens of one worm and enter seminal receptacles of the other worm (see *a*), where they are stored until fertilization. (*c*) Earthworm cocoons. A mucous sheath, secreted by the clitellum, slips forward as a worm backs up. Eggs from the ovaries and sperm from the seminal receptacles enter the sheath, where fertilization occurs. When the worm slips out of the sheath, it becomes a cocoon enclosing the developing zygotes.

(a)

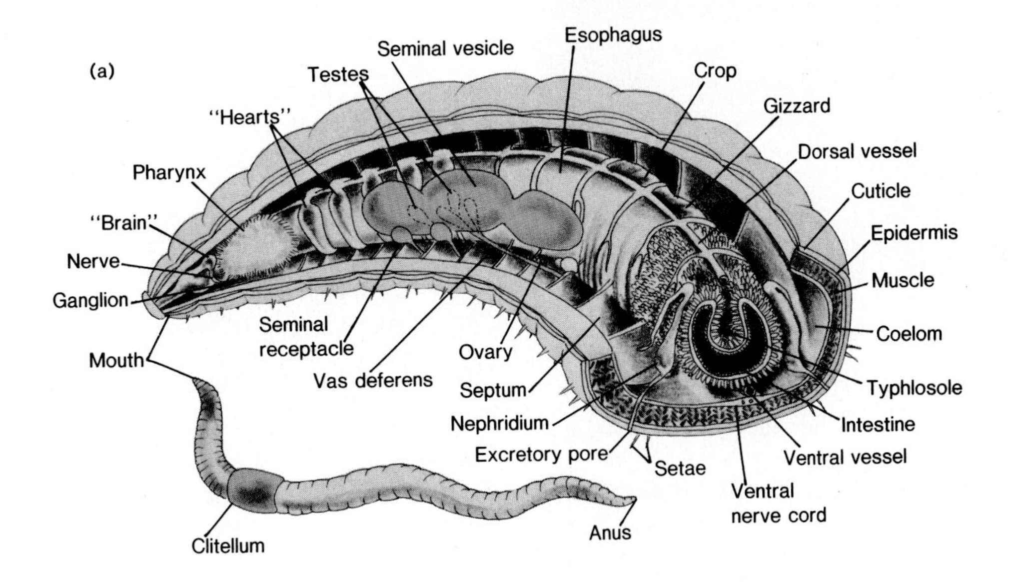

(b)

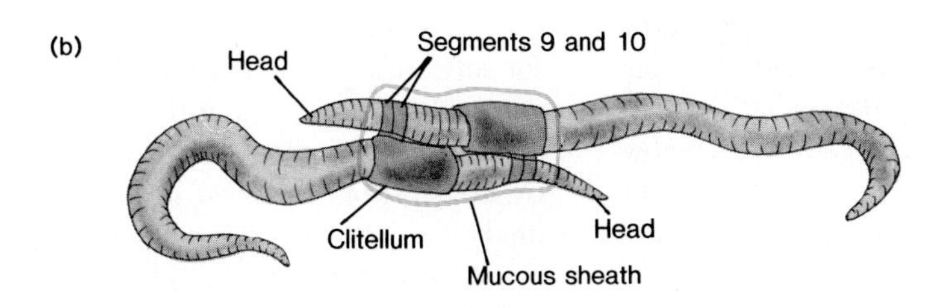

(c)

Figure 29.30 Leeches. (*a*) A blood-sucking leech on a human arm. (*b*) Locomotion in a leech. Colored regions are the same body areas in all diagrams. A leech moves by alternately attaching anterior and posterior suckers. Contraction of circular muscles makes the body stretch out and become long and thin. Contraction of longitudinal muscles shortens and thickens the body.

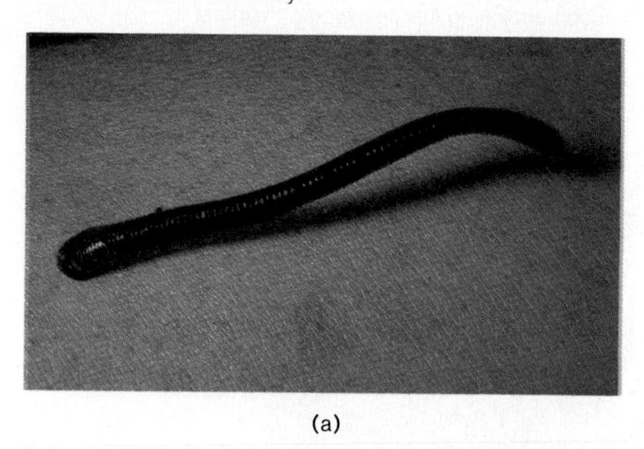
(a)

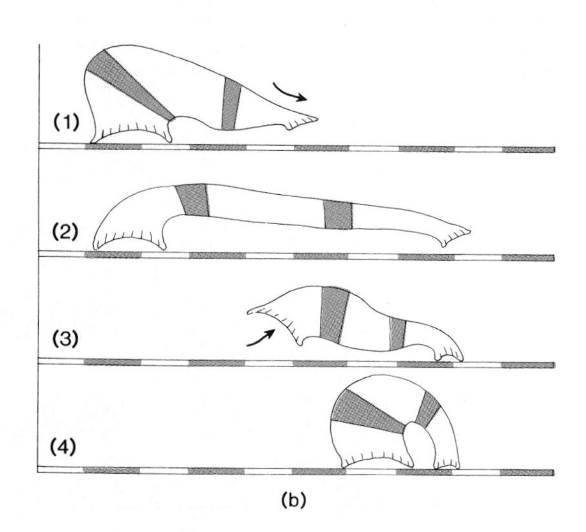

(b)

Class Hirudinea (Leeches)

This relatively small class of annelids has only about 300 species, most of which are aquatic.

A leech's first and last body segments are modified as suckers that are used for attachment and movement (figure 29.30). Most leeches are ectoparasites that attach by their suckers to the surface of other animals and feed by sucking their blood.

Leeches played an intriguing role in medical practice until about 100 years ago. Leeches were used to bleed people who were diagnosed as suffering from "bad blood." While the idea of bleeding an already weakened patient may now seem a rather bizarre treatment, the choice of leeches as an effective "instrument" for the bleeding process itself was excellent.

Phylum Arthropoda

The **phylum Arthropoda** includes many common and familiar animals: spiders and scorpions, crabs and lobsters, and insects, as well as many other less familiar creatures. Arthropod literally means "jointed foot," and possession of jointed appendages is an important characteristic of phylum members.

The phylum Arthropoda is by far the largest animal phylum: it contains about a million described and named species, as many species as there are in all other animal phyla *and* plant divisions combined. In addition, there probably are at least another several hundred thousand, as yet undescribed, species of arthropods.

The Arthropod Exoskeleton

In addition to their jointed legs, all arthropods have **exoskeletons** (exo = "outside") that are composed primarily of **chitin.** The exoskeleton is secreted by skin cells and remains attached to the skin, forming a tough, protective body armor. However, the exoskeleton permits movement because it is not a single hard shell, but rather a number of rigid plates connected by softer, flexible hinges at the joints (figure 29.31).

Hard, chitinous exoskeletons provide general protection for soft body parts, and they represent a crunchy deterrent to attacking predators. But the exoskeleton also is impermeable to water and thus helps to prevent excessive water loss by evaporation, the most urgent problem faced by land animals.

Arthropods' exoskeletons are nonexpandable. Thus, growth poses some special problems for arthropods that are solved by **molting,** a periodic shedding of the exoskeleton (figure 29.32).

Other Arthropod Characteristics

Arthropods probably evolved from annelidlike ancestors, and this ancestry is reflected in arthropods' segmented body arrangement. Ancient arthropods had segmented bodies with a pair of jointed appendages on each body segment. For example, the **trilobites,** a group of marine arthropods known only from fossil forms, had such a body plan (figure 29.33*a*).

In many modern arthropods, body segments tend to be combined into distinctive body regions. A familiar example of this union of body segments is the **head, thorax,** and **abdomen** arrangement in insect bodies (figure 29.33*b*). The arthropod head has specialized sense organs and contains large, prominent nerve cell masses.

The phylum Arthropoda has two major subgroups, the **subphyla Chelicerata** and **Mandibulata.**

Figure 29.31 The hinged exoskeleton of the arthropods. Hard, protective plates are connected by flexible hinges that permit movement. (*a*) One of the arthropods, a millipede, walking. (*b*) A millipede rolled up in its normal behavioral response to danger. Flexible hinges between exoskeleton plates permit this coiling.

(a)

(b)

Figure 29.32 Molting in arthropods. A male Jonah crab (*Cancer borealis*) standing over a recently molted female. Her shed exoskeleton is to the left. The new exoskeleton will increase in size and harden in a few days. Arthropods must molt as they grow because the arthropod exoskeleton is hard and nonexpandable.

Figure 29.33 Arthropod body plans. (*a*) "Primitive" segmented arthropod body plan with a pair of jointed appendages on each body segment. Fossil trilobites show this body arrangement. Two longitudinal folds divide the trilobite body into a median lobe and two lateral lobes. This three-part body arrangement suggested the name "trilobite." (*b*) Many modern arthropods have distinct combined body regions rather than numerous similar body segments. This ventral view of a carpenter ant shows the head-thorax-abdomen arrangement characteristic of the insect. Note that all three pairs of legs are attached to the thorax and that the abdomen has no appendages. This ant has a pair of large antennae on its head.

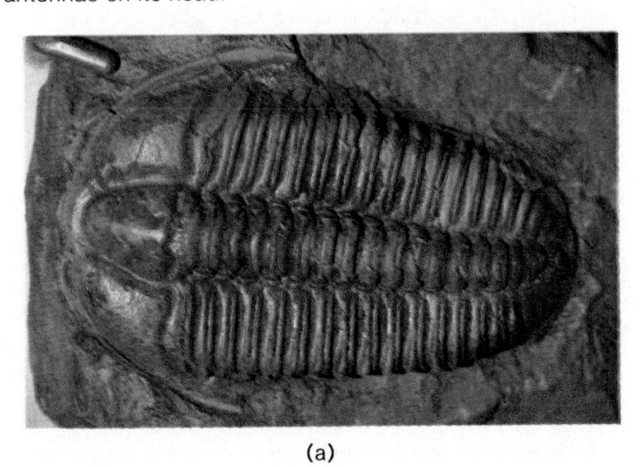

(a)

(b)

Chelicerate Arthropods

Chelicerates' bodies are divided into two major regions, a **cephalothorax** and an **abdomen.** The characteristic that gives the chelicerates their name is the specialization of their first (most anterior pair) appendages as mouthparts called **chelicerae,** which are either pincerlike or, as they are in spiders, fanglike. They are used for piercing prey animals and are often associated with poison glands. Chelicerate arthropods have four pairs of walking legs and do not have antennae.

There are several classes of chelicerate arthropods, but we will discuss only two of them here—class Xiphosura and class Arachnida.

Box 29.1

Peripatus

The Walking Worm

The "walking worm," *Peripatus,* belongs to the small **phylum Onychophora,** which includes only about seventy living species. These animals live in humid habitats mainly in tropical regions.

Onychophorans possess an interesting mixture of characteristics of two other phyla, the annelids and the arthropods (box figure 29.1*A*). Onychophorans have segmented bodies that are quite similar to those of annelids. They have a thin, flexible cuticle that resembles the annelid body covering and excretory structures that resemble those of annelids.

Onychophorans also have some arthropodlike characteristics. While onychophorans' legs are not jointed as arthropod legs are, they do have claws that resemble those of arthropods. Their respiratory system also resembles that of terrestrial arthropods.

As neat and clear-cut as taxonomic groupings of living things may sometimes appear, there always are organisms that seem to occupy intermediate positions astride the arbitrary boundary lines. The mixture of characteristics in the onychophorans illustrates the sometimes confusing complexity of the living world.

The evolutionary position of onychophorans is interesting because they appear to preserve some of the characteristics of the ancient wormlike animals from which modern arthropods arose (box figure 29.1*B*). Actually, living onychophorans very closely resemble fossil specimens from as long as 500 million years ago. Apparently, these "walking worms" have occupied a particular niche and remained relatively unchanged for millions of years, while ancient relatives of theirs were diversifying into the huge and evolutionarily successful group we know today as the phylum Arthropoda.

Box figure 29.1*A* The "walking worm," *Peripatus* (phylum Onychophora), showing its body segmentation, short, unjointed legs, and tentacles.

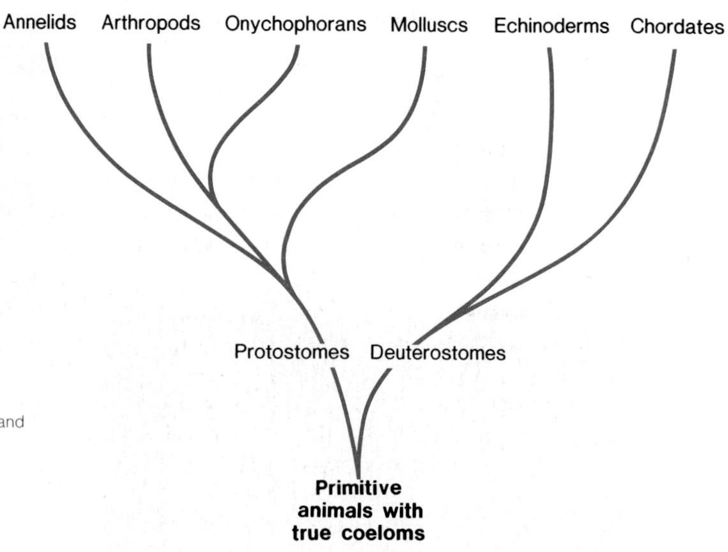

Box figure 29.1*B* Simplified "family tree" of the more complex animal phyla. Note the proposed evolutionary relationship of the onychophorans to the arthropods.

The **class Xiphosura** includes only four genera. One of them, the **horseshoe crab,** *Limulus polyphemus,* is common along the east coast of North America. *Limulus* can be found in large numbers on muddy shores, where they plow along through the mud, feeding on debris and organisms that they dredge up (figure 29.34).

The **class Arachnida** includes a number of familiar animals, such as spiders, scorpions, ticks, mites, and daddy longlegs (figure 29.35).

Spiders are the best-known arachnids. About 32,000 species of spiders have been described so far. Spiders can be very numerous in terrestrial habitats. For example, one study in Britain concluded that there were about $2\frac{1}{4}$ million spiders per acre (about $5\frac{1}{2}$ million per hectare) in some meadows.

Figure 29.34 The horseshoe crab (*Limulus polyphemus*). (*a*) Dorsal view of male (left) and female (right) crabs during the breeding season. Note the hard, protective carapace. (*b*) A ventral view of a *Limulus* showing appendages. Note the chelicerae. Pedipalps are the second appendages in chelicerate arthropods and are sensitive feeding devices. The operculum is a plate that covers and protects gills and reproductive structures.

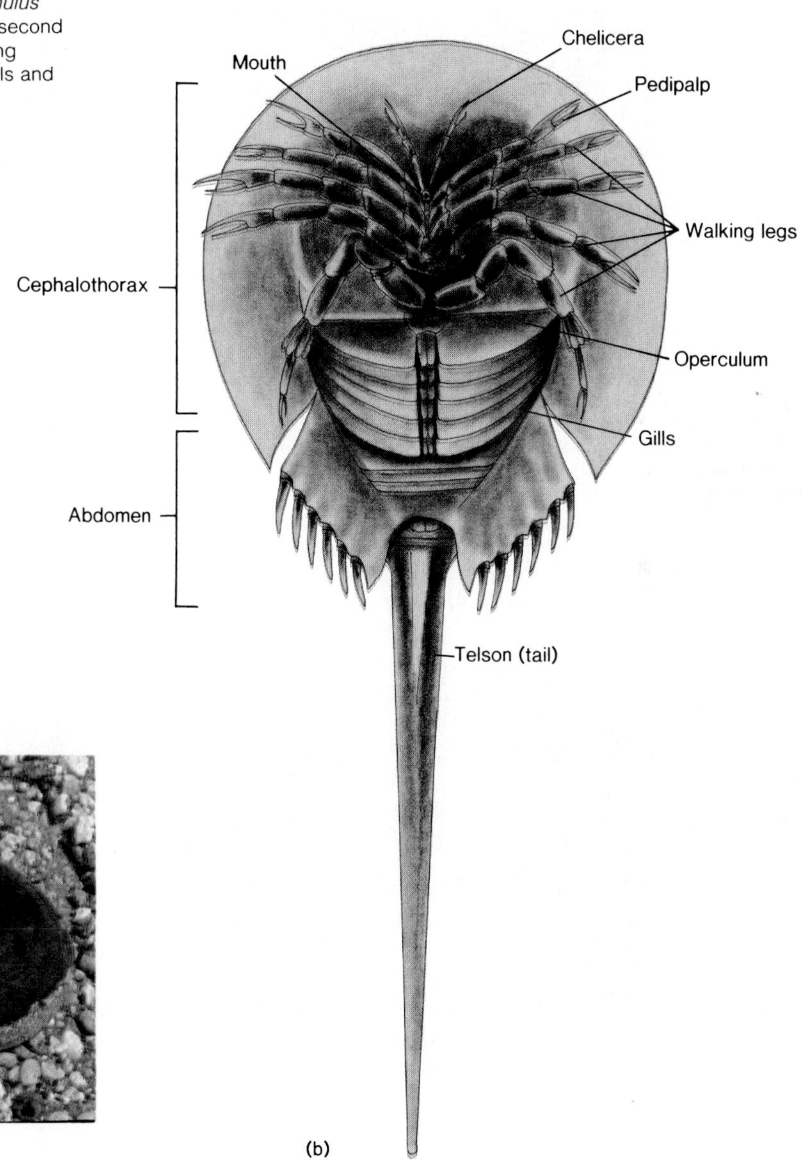

(a)

Figure 29.35 Scorpions and spiders, members of the class Arachnida. (*a*) A giant scorpion from Kenya. Note the large pincers, which are the second appendages of a scorpion. The scorpion's small chelicerae are not visible in this picture. The last body segment has a sharp stinging barb that injects venom with a stabbing motion. (*b*) A shamrock spider in its web.

(a)

(b)

Figure 29.36 Mites and ticks. (*a*) Scanning electron micrograph of a mite. Note the body hairs. Many mites live as scavengers on the surfaces of plant and animal hosts, but others, such as chiggers, suck blood from their hosts. Some mites are carriers (vectors) of disease-causing microorganisms, which they transmit from one host to another (magnification × 42). (*b*) The Rocky Mountain wood tick (*Dermacentor andersoni*). This tick carries the rickettsia (small bacterium) that causes Rocky Mountain spotted fever in humans. All ticks are parasites. Some ticks cause little direct damage themselves, but a number of ticks are vectors of serious animal and human diseases.

(*a*) Kessel, R. G., and Shih, C. Y. *Scanning Electron Microscopy in Biology.* © Springer-Verlag, 1976.

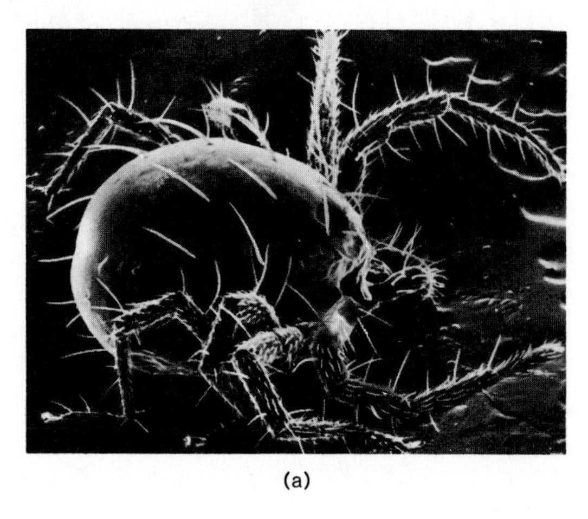

(a)

(b)

Figure 29.37 Head appendages of a hermit crab, an example of a mandibulate arthropod. Mandibulates have mandibles as their first mouthparts rather than chelicerae.

All spiders are carnivores that live by piercing their prey and sucking out their body fluids. Most spiders secrete poison that immobilizes prey animals.

Spiders have **spinnerets,** spinning organs on their abdomen, that secrete a liquid containing an elastic protein that hardens in the air to form a very fine silk thread. Many spiders spin elaborate **webs** of this silk, instinctively tracing out the same intricate geometric patterns that members of their species have spun for hundreds of generations.

Two groups of tiny arachnids, the **ticks** and **mites,** are specialized for life on a variety of plant and animal hosts, including humans (figure 29.36).

Class Crustacea

Because they are edible, some members of the **class Crustacea,** such as **lobsters, shrimps, crabs,** and **crayfish,** are quite familiar to most people. But **pill bugs, water fleas, brine shrimp,** and **barnacles** also are crustaceans.

Crustacea, along with the insects, millipedes, and centipedes, belong to the second subphylum of the phylum Arthropoda: subphylum Mandibulata. Mandibulate arthropods have one or two pairs of antennae, and they have **mandibles** rather than chelicerae. Mandibles usually are modified for biting or chewing, and they never are pincerlike. Most mandibulate arthropods have two additional pairs of mouthparts called **maxillae** (figure 29.37).

The large, familiar, bottom-dwelling crustaceans, such as crabs, lobsters, and crayfish, all belong to a group (order) known as **decapod** ("ten-foot") Crustacea. They are called decapods because they have five pairs of walking legs. Decapod crustaceans have rigid exoskeletons that are heavily impregnated with calcium carbonate (figure 29.38).

Despite a constant demand for the tasty decapods, some of the small, obscure crustaceans actually are more economically important than the ones that are food delicacies. These small crustaceans are extremely important links in aquatic food chains. For example, small shrimplike crustaceans known as **krill** provide food for many oceanic animals (figure 29.38). Krill occur in great swarms with more than 60,000 individuals per cubic meter. Such huge masses can provide food sources for even very large animals, despite the small size of the individual crustaceans. A blue whale can consume up to a ton of krill at a single feeding, and it may feed four times a day. Krill are now being considered as a possible direct source of human food.

Most crustaceans are free-swimming throughout their lives, but adults of one group, the **barnacles,** live attached to surfaces and strain food out of the surrounding water. Barnacles attach to rocks, other organisms (for example, snails, crabs, and even whales), wharf pilings, buoys, and ship bottoms.

Figure 29.38 Crustaceans small and large. The class Crustacea is a very diverse group of animals. (*a*) The water flea, *Daphnia*, a small freshwater crustacean, photographed by dark-field microscopy. The average *Daphnia* is 1 to 2 mm long. (*b*) *Homarus americanus*, the commercially important lobster that occurs along the northeast coast of the United States. (*c*) The King crab *Paralithodes*. This tasty crab occurs in the north Pacific and is sometimes called the Alaskan King crab. (*d*) Pill bugs (also called sow bugs or wood lice) are terrestrial crustaceans that live in moist places. (*e*) Krill. These animals grow to only about 6 cm in length, but are so numerous that they provide abundant food for large animals such as blue whales. (*f*) Barnacles live attached to surfaces and sweep food out of the water with their highly modified, feathery legs.

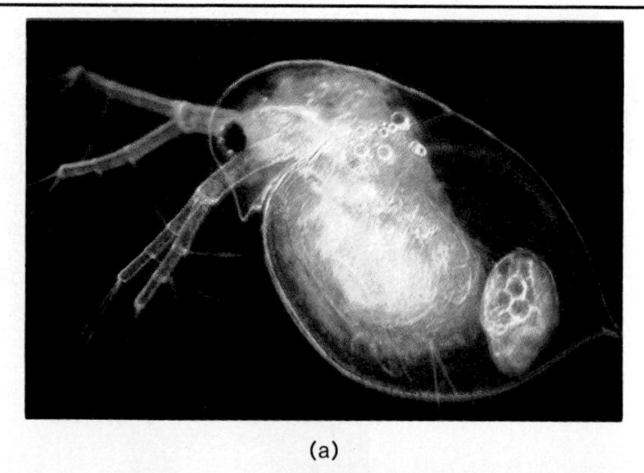

(a)

(b)

(c)

Centipedes and Millipedes

The **centipedes** ("hundred legs") and **millipedes** ("thousand legs") are both terrestrial and have an elongate, wormlike body with a head and a trunk consisting of many segments.

Centipedes (**class Chilopoda**) are active, fast-moving, carnivorous animals that use their first pair of legs, which are modified as poison claws, to kill their prey. Centipedes usually eat small insects and other small invertebrates, but larger centipedes have been known to feed on snakes, mice, and frogs (figure 29.39).

Millipedes (**class Diplopoda**) do not have nearly a thousand legs, but each body segment behind the first four or five does have two pairs of legs (see figure 29.31). These legs appear to move in waves as a millipede slowly moves along.

Millipedes are herbivores or scavengers that feed on decaying material and do not have poison claws. They are secretive animals that usually live beneath leaves, stones, or logs, and they avoid trouble by rolling up and feigning death.

Class Insecta

The **class Insecta** includes more species than any other class of organisms (figure 29.40). Insects are regarded as one of the most successful (possibly *the* most successful) groups of organisms that have ever lived. Ninety percent of the million or so species of arthropods are insects. Insects occupy almost every kind of terrestrial and freshwater habitat in the world, and they are variously specialized to utilize a tremendous variety of food sources.

(d)

(e)

(f)

Figure 29.39 A giant South American centipede (class Chilopoda) attacking a frog. Centipedes have paired antennae, poison claws, and flattened bodies with a single pair of walking legs attached to each trunk segment. Compare with the pictures of a millipede in figure 29.31.

Figure 29.40 Insect diversity. Representatives of some orders of insects drawn to different scales. (*a*) Silver fish (order Thysanura) (*b*) Dragonfly (order Odonata). (*c*) Camel cricket (order Orthoptera). (See also the grasshopper in figure 29.44.) (*d*) Termite soldier (order Isoptera). (*e*) Cinch bug (order Hemiptera). (*f*) Human body louse (order Anoplura). (*g*) Beetle (order Coleoptera). (See also dung beetles in figure 29.45.) (*h*) Royal walnut butterfly and its caterpillar (order Lepidoptera). (*i*) Gall gnat (order Diptera). (*j*) Buffalo treehopper (order Homoptera). (*k*) Flea (order Siphonaptera). (*l*) Wasp (order Hymenoptera).

Figure 29.41 Insect structure. (*a*) A grasshopper. Note that the three thorax segments each have a pair of walking legs. (*b*) Reproductive structures (in color) of a female grasshopper. Sperm are received in the seminal receptacle during mating. The female uses her ovipositor to insert eggs in the ground, where embryos diapause during winter months. (*c*) Insect (grasshopper) mouthparts in place (left) and spread out (right). (*d*) Scanning electron micrograph of the compound eye of the fruitfly *Drosophila*. Each unit, called an ommatidium, receives a separate image.

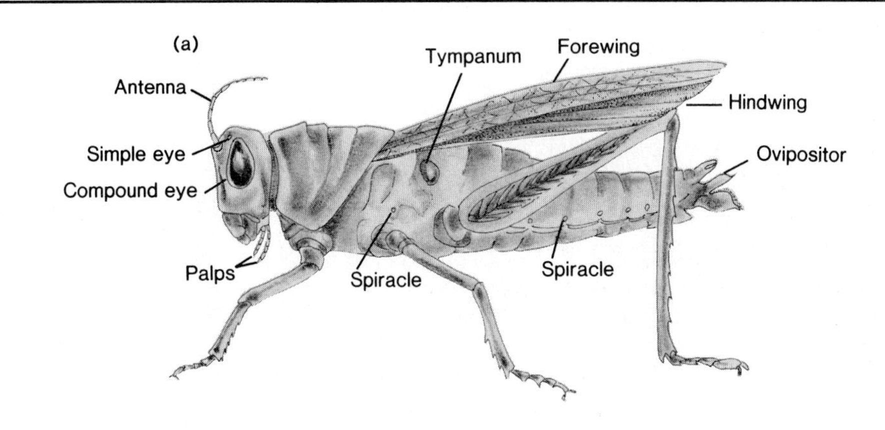

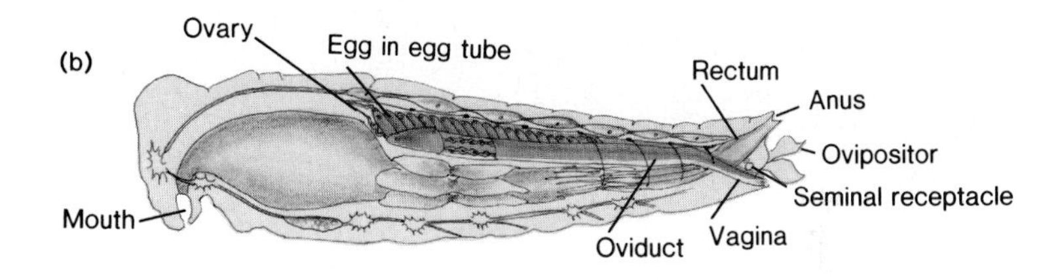

(d)

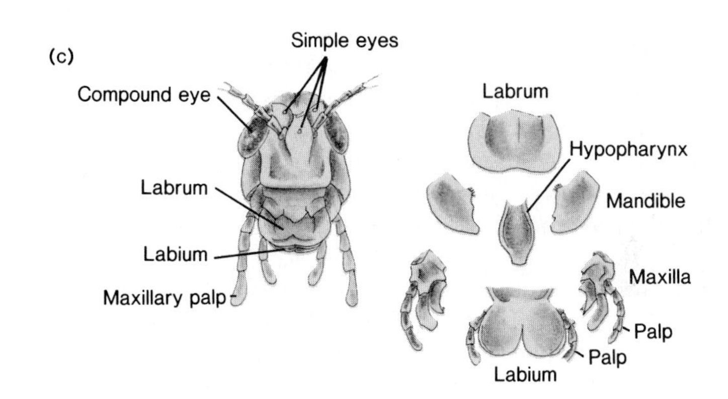

The Insect Body

A typical adult insect body is divided into a head, a thorax, and an abdomen. The head bears well-developed sense organs, including one pair of antennae, **compound eyes** (eyes with many separate focusing units), and sometimes simple eyes as well. Insect mouthparts include a pair of mandibles, a pair of maxillae, and a **labium,** which is a sort of lower lip that has evolved through the fusion of a second pair of maxillae (figure 29.41).

Insects have six legs. Many insects also have wings on the second and third thoracic segments and are excellent fliers (figure 29.42).

Many insect abdomens have prominent **spiracles,** the openings leading to the tracheal tubes that function in gas exchange.

Figure 29.42 Insect flight. A bee flying among wild aster flowers. The bee's wings appear blurred even in this high-speed, "stop-action" photo because they are beating so rapidly. Bee wings beat at frequencies up to 250 beats per second.

Figure 29.43 A parasitized tobacco hornworm with eggs attached.

Insect Reproduction

Insects' reproductive strategies are tremendously varied. For example, some insects lay eggs on plants that provide a ready food supply for their developing young. Some insects even use the bodies of other species of insects as "nurseries" for their young. They lay their eggs in or on the bodies of these unfortunate hosts, who eventually serve as a food source for the developing young after they hatch (figure 29.43).

Because it has a nonexpandable exoskeleton, an insect's development must involve a series of molts. But insect development involves much more than simple growth in size; there is also a marked change in body form that converts an immature individual into an adult. This change is called **metamorphosis.**

There are two basic patterns of molting and metamorphosis in insects. In **incomplete (gradual) metamorphosis,** the young insect (**nymph**) hatches in a form that generally resembles its parents (figure 29.44*a*). With each molt, the insect comes to resemble its parents more closely, and it emerges from a final molt as a fully developed adult.

The other major category of insect development involves much more drastic and abrupt changes in form. The familiar developmental stages of butterflies and moths—**egg, larva, pupa,** and **adult**—illustrate **complete (abrupt) metamorphosis** (figure 29.44*b*). A wormlike larva hatches from the egg, grows, molts several times, and then **pupates,** producing a hard case around itself.

In many species, the pupa may enter a period of suspended activity called **diapause,** in which it has lowered metabolism and increased resistance to environmental stresses, such as winter weather.

Dramatic developmental changes occur within the pupal case. Parts of the larval body are broken down and used as raw material for construction of new adult structures. Finally, the adult emerges as a reproductively mature individual.

Insects and Humans

Some insects play a vital role in the environment as pollinators of plants that depend entirely on insect pollination. Other insects are active in the important natural processes that break down dead plant and animal bodies.

Even dung is processed by insects. For example, in the African plains, busy little dung beetles do a spectacular job of attacking even football-size balls of elephant dung, which they can reduce to a 2 cm-thick mat overnight. Thousands

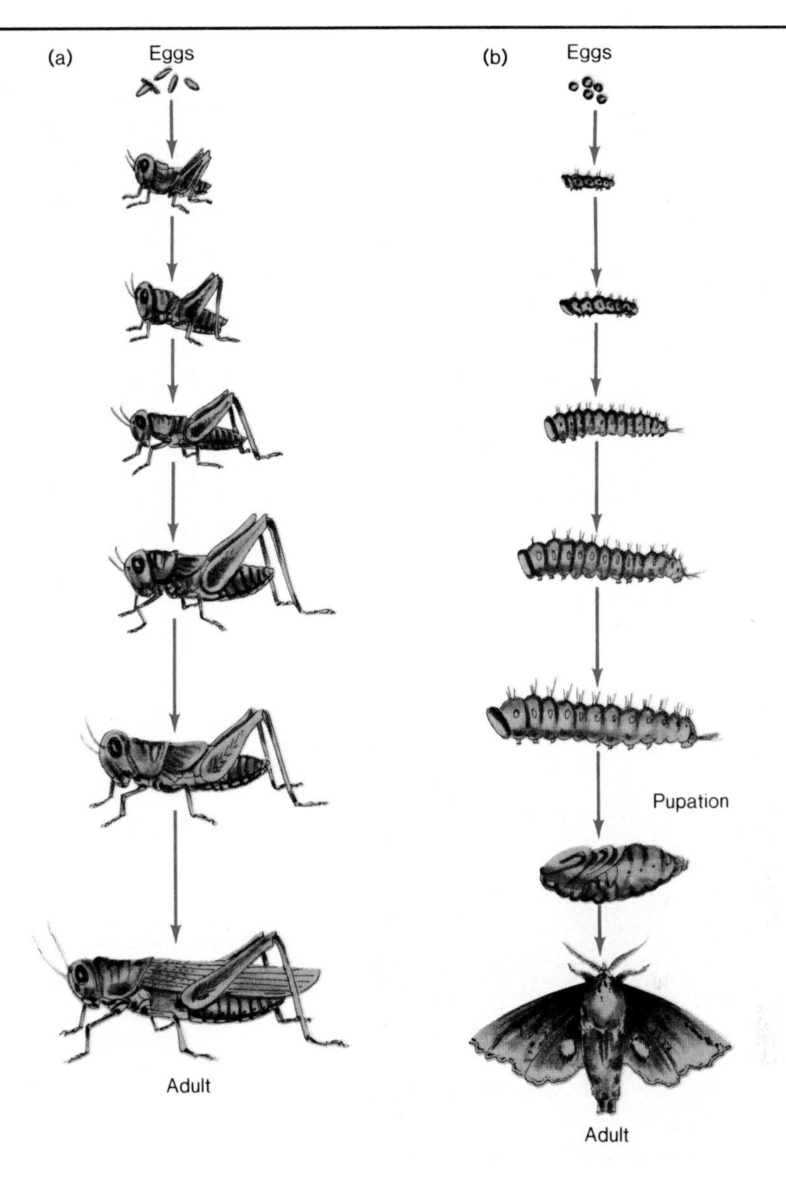

Figure 29.44 Insect development.
(*a*) Grasshopper development, an example of incomplete metamorphosis. The nymph that hatches generally resembles the adult. Following each molt, the emerging nymph more closely resembles the adult. (*b*) Moth development, an example of complete metamorphosis. The life history includes several wormlike larval stages, a pupa, and an adult.

of beetles congregate, cut out small balls of dung, and roll them away (at velocities up to 14 m per minute) to underground burrows, where adult beetles eat them or leave them as food for developing young (figure 29.45). All of this activity hurries the eventual return of vital resources contained in dung to forms that are usable by other organisms.

But even as dung beetles are doing their valuable work in African grazing areas, other insects are destroying more than half of all the crops in the fields of the same African countries.

And so it goes around the world. Insects everywhere eat food and fiber-producing plants in the field and attack agricultural products in storage. Other insects transmit diseases to humans and to animals and plants upon which we depend. This agricultural destruction and disease transmission has put human beings and insects in a direct conflict situation.

Our battle with insects has featured massive attempts to poison insects selectively with **insecticides.** Unfortunately,

Figure 29.45 A dung beetle cuts a small ball out of a mass of dung (*a*) and quickly rolls the ball away (*b*) toward an underground burrow.

(a)

some of these insecticides are not so selective, and poisons intended for harmful insects also kill beneficial insects, accumulate in the environment, enter food chains, and build up to toxic levels in other animals considered beneficial to humans. These poisons even threaten human health directly in some cases. Insects with resistance to low levels of these insecticides appear very quickly once insecticides are put into use.

In the future we may be able to fight the battle against insects through **biological control,** without use of chemical insecticides (table 29.1).

Table 29.1
Some Strategies for Biological Control of Insects

Control Involving Other Organisms

1. Introduce predators that selectively attack a particular harmful species.
2. Discover and disseminate viral or bacterial diseases of the insect in question.
3. Breed plants that produce insect-repelling substances without changing output of agricultural products.

Control Involving the Insect's Own Biology

1. Use sex attractants (pheromones) to lure large quantities of adults to traps.
2. For species in which females mate only once, release huge quantities of sterilized males that compete with normal wild males.
3. Apply hormones (for example, juvenile hormone) that disrupt development and prevent attainment of diapause stage by a critical time of year.

Phylum Echinodermata

The **phylum Echinodermata** consists of about 6,000 species, including **sea stars (starfish)**, **brittle stars, sea urchins, sea cucumbers,** and **sea lilies. Echinoderms** are radially symmetrical animals, most of whom have a five-part body organization. They are bottom dwellers that range from shoreline intertidal zones to very deep parts of the sea.

Echinoderms are the closest invertebrate relatives of the vertebrate animals and other members of the phylum to which vertebrates belong, the phylum Chordata. It seems likely that both echinoderms and chordates have descended from common or very similar bilaterally symmetrical ancestors.

Echinoderms have an **internal skeleton (endoskeleton)** that is covered by a separate skin. These skeletons consist of **calcareous ($CaCO_3$ containing)** plates, and in some echinoderms, especially sea urchins, these plates bear spines that stick out through the delicate skin. In fact, the name Echinodermata comes from Greek words meaning "spiny skin."

A unique characteristic of the echinoderms is their **water vascular system.** This system is a network of canals that contain watery fluid and extend throughout the body. Numerous **tube feet** that are attached to these canals function in body movement.

Echinoderms shed gametes into the water, where fertilization occurs. An echinoderm zygote develops into a free-swimming, bilaterally symmetrical larva that swims by means of cilia and feeds on plankton. These larvae eventually settle to the bottom as they undergo metamorphosis to become radially symmetrical adults (figure 29.46).

Figure 29.46 Starfish development as an example of echinoderm metamorphosis. (*a*) Advanced larval stage. (*b*) A young starfish during metamorphosis. (*c*) Young starfish.

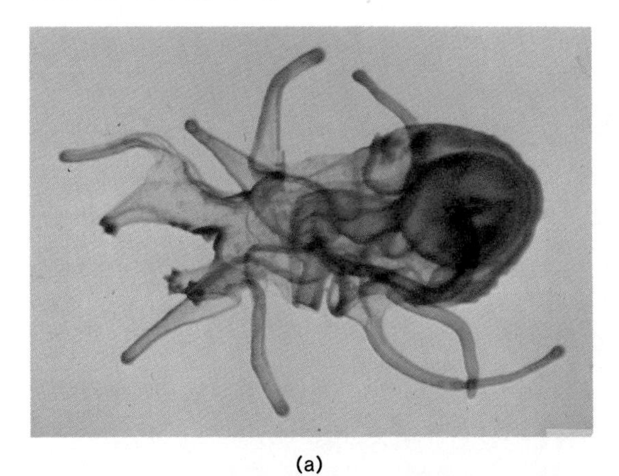

(a)

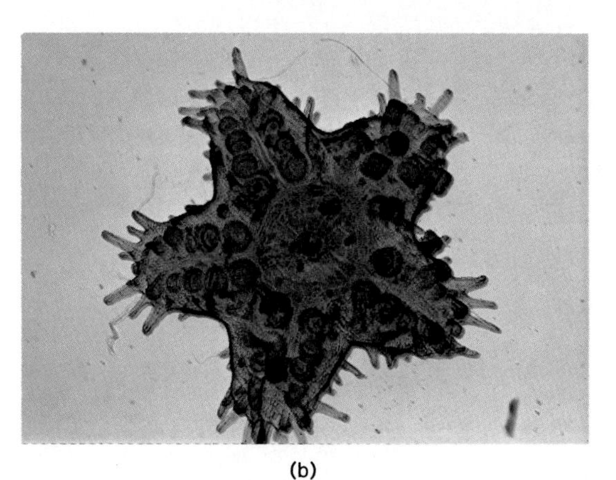

(b)

(c)

Class Asteroidea

The sea stars (starfish) are the most familiar of the echinoderms. The flattened starfish body has a central disc with five sturdy **arms (rays)** extending outward from it. Each arm has rows of tube feet running along a groove on its ventral side. Tube feet can be extended or shortened, and they can attach firmly to surfaces. Thus, a starfish moves by extending, attaching, and then contracting groups of tube feet (figure 29.47).

Starfish commonly feed on clams, oysters, and other bivalve molluscs (figure 29.48). A starfish's mouth is on the ventral side of its central disc, and its anus is on the dorsal side. To feed, a starfish positions itself over its intended victim, a clam, for example. It attaches some of its tube feet to the clam's shells, and these tube feet contract, pulling on the shells. The clam resists this pulling by forcefully contracting its shell-closing muscles. The starfish usually can win this tug-of-war, however, because it uses its tube feet in relays, attaching and contracting some while others rest. Eventually, the clam's muscles tire and weaken, and the shell begins to open ever so slightly.

A very small crack between the shells is enough for the starfish, which then pushes the large, lower part of its stomach out of its mouth and through the crack between the clam's shells. Because this everted stomach is inside out, exposed digestive surfaces can be spread over the soft body mass of the clam. The stomach secretes enzymes, and the digestion of the clam's body begins even while the clam is still working to close its shells. Later, partly digested food is taken into the starfish's body, where digestion continues.

Figure 29.47 Starfish anatomy. (*a*) Starfish with different arms dissected to show details of different systems. Canals of the water vascular system in each arm connect to a ring canal in the central disc. Water from outside the body can enter or leave the system via the madreporite. Pedicellariae are tiny pincers that work together to move objects off the top of the starfish's body. (*b*) A starfish with one of its arms lifted showing extended tube feet. Contractions of ampullae extend tube feet. The feet then attach and contract, pulling the starfish along.

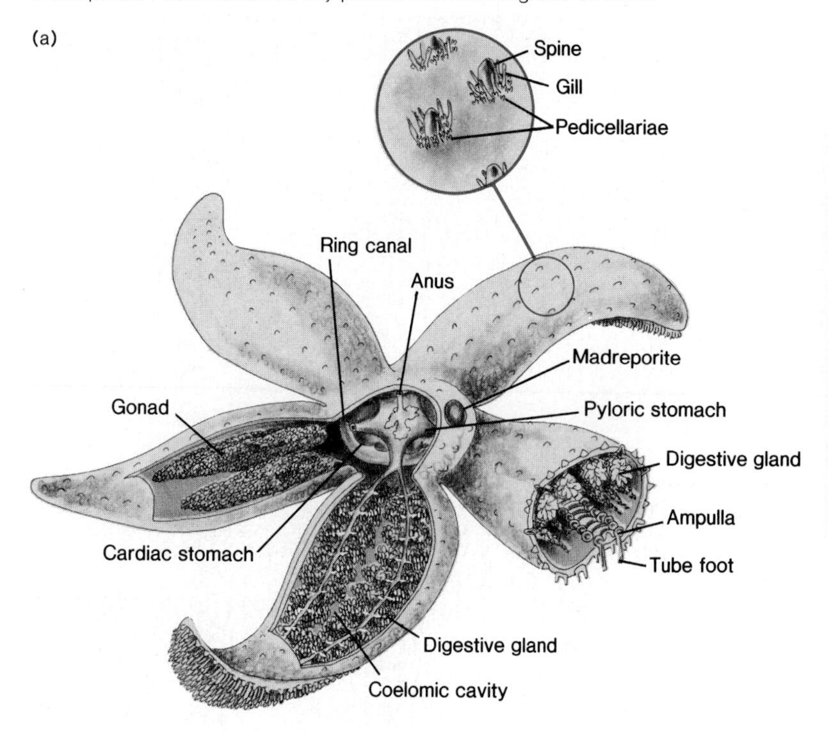

Figure 29.48 A starfish in feeding position wrapped around a clam (center).

Other Classes of Echinoderms

The **class Ophiuroidea** includes **brittle stars** and **basket stars** (figure 29.49). They have small central discs and long, slender, flexible arms.

Sea urchins and **sand dollars** belong to the **class Echinoidea** (figure 29.50). Sand dollars have skeletal plates that are fused into a single, flattened unit. Sea urchins also have fused skeletal plates, but they have more rounded bodies. Sea urchins have long, movable spines that protrude from their skin. Coordinated spine movement pushes urchins along and supplements the action of tube feet.

The **sea cucumbers, class Holothuroidea,** are very different from other echinoderms because they have leathery bodies and very reduced skeletons (figure 29.51*a*). A sea cucumber lies on its side and traps food particles in mucus on the surface of tentacles that are set in a ring around its mouth.

Figure 29.49 Members of the class Ophiuroidea. (*a*) Brittle stars. (*b*) A basket star. The five arms of a basket star branch and rebranch to produce a mass of coils.

(a)

(b)

Figure 29.50 Members of the class Echinoidea, sand dollars and sea urchins. (*a*) Skeleton of the arrowhead sand dollar *Encope*. (*b*) Some urchins, such as this *Diadema*, have long, sharp hollow spines containing an irritant. These spines can inflict painful puncture wounds if the urchin is handled or stepped on. (*c*) Skeleton of the sea urchin *Arbacia*. Note the definite five-part organization.

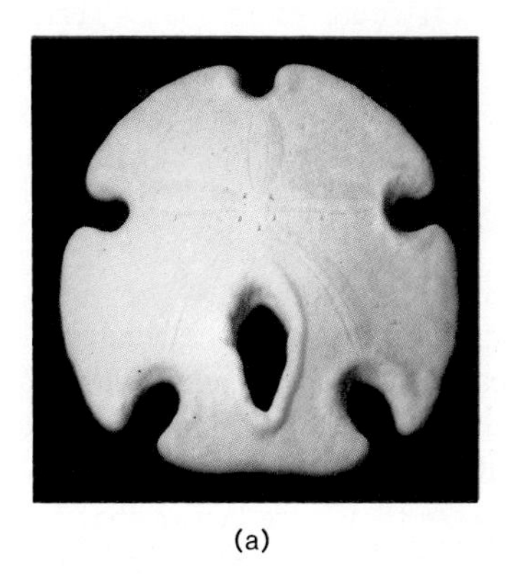

(a)

(b)

(c)

Sea cucumbers have great regenerative ability, which they sometimes put to unusual use. If bothered by a predator, a sea cucumber can eviscerate, that is, cast out large parts of its internal organs either by the mouth or anus, depending on the species. Often, this gives the sea cucumber an opportunity to move away and begin regeneration of the lost organs, while the startled predator is left to contemplate a writhing mass of body organs.

The final class of echinoderms are the oldest and most primitive in evolutionary terms. The **sea lilies,** members of the **class Crinoidea,** live attached to a substrate by a stalk (figure 29.51*b*).

Figure 29.51 Members of the class Holothuroidea. (*a*) A sea cucumber, *Cucumaria* (class Ophiuroidea). (*b*) A fossil sea lily (class Crinoidea). Some modern crinoids are quite similar to these very ancient echinoderms.

(a)

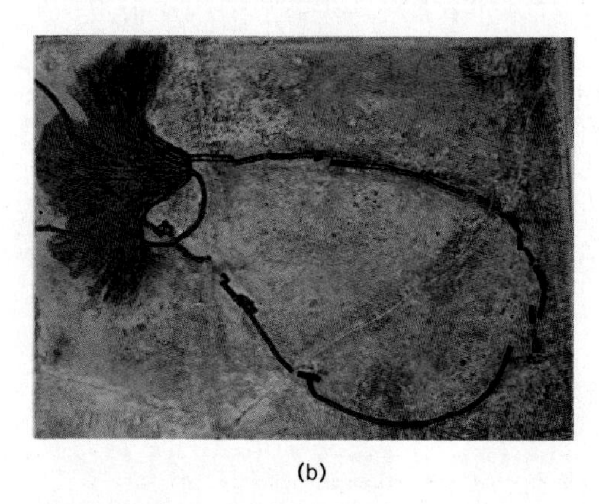

(b)

Figure 29.52 Distinguishing characteristics of chordates. Pharyngeal pouches do not open as gill slits in all chordates.

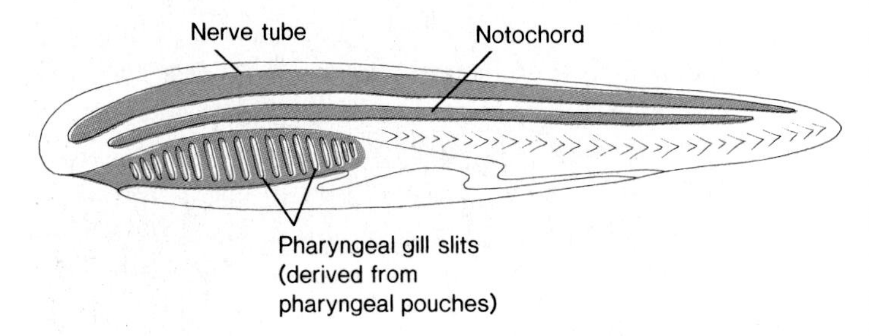

Phylum Chordata

There are three characteristics that distinguish animals in the **phylum Chordata** from all other animals.

First, every chordate, at least sometime during its life, has a rodlike **notochord,** a supporting structure that runs along its body just ventral to its nerve cord. This structural feature gives the phylum its name, chordata. Some nonvertebrate chordates have a notochord throughout life. However, in vertebrates, a notochord temporarily provides support for the developing embryo that is replaced during development by a **vertebral column.** The vertebral column is a chain of bones that provides support for the body axis and protection for the nerve tube.

Second, every chordate has a **dorsal hollow nerve tube,** an arrangement that contrasts sharply with the solid, usually ventral, nerve cords seen in the other animal phyla.

The third of the exclusively chordate characteristics is the possession at some time of **pharyngeal pouches** (figure 29.52). These are lateral expansions of the pharynx portion of the digestive tract. In many chordates, the pouches break through to the outside, producing **gill slits.** But the pouches are only temporary embryonic structures in many vertebrates, such as humans, and they do not break through to the outside.

Echinoderms, Hemichordates, and Chordates: Evolutionary Relatives

Although they somewhat resemble chordates, the **hemichordates** are classified as the separate **phylum Hemichordata,** a phylum that is distinct from, but closely related to, the phylum Chordata. Hemichordates and echinoderms also appear quite closely related because they have larvae that are virtually identical. This may indicate that echinoderms and hemichordates descended from a common deuterostome ancestor, probably the same group of primitive animals that gave rise to the chordates as well.

The most common hemichordates are the **acorn worms,** marine animals that live in burrows in coastal sand or mud (figure 29.53). At the anterior end of an acorn worm's body is a muscular **proboscis.** Its mouth leads to a pharynx with many gill slits in its wall. As water is drawn in through the mouth and forced out through the gill slits, food is trapped in mucus in the pharynx and passed on through the digestive tract.

Figure 29.53 Hemichordates and relatives. (*a*) Larva of a hemichordate (*Glossobalanus*). Evolutionary linkage of hemichordates and echinoderms depends on developmental patterns and similarities of larvae. (*b*) Larva of a sea cucumber (*Labidoplax*). (*c*) Model of a hemichordate (*Dolichoglossus*).

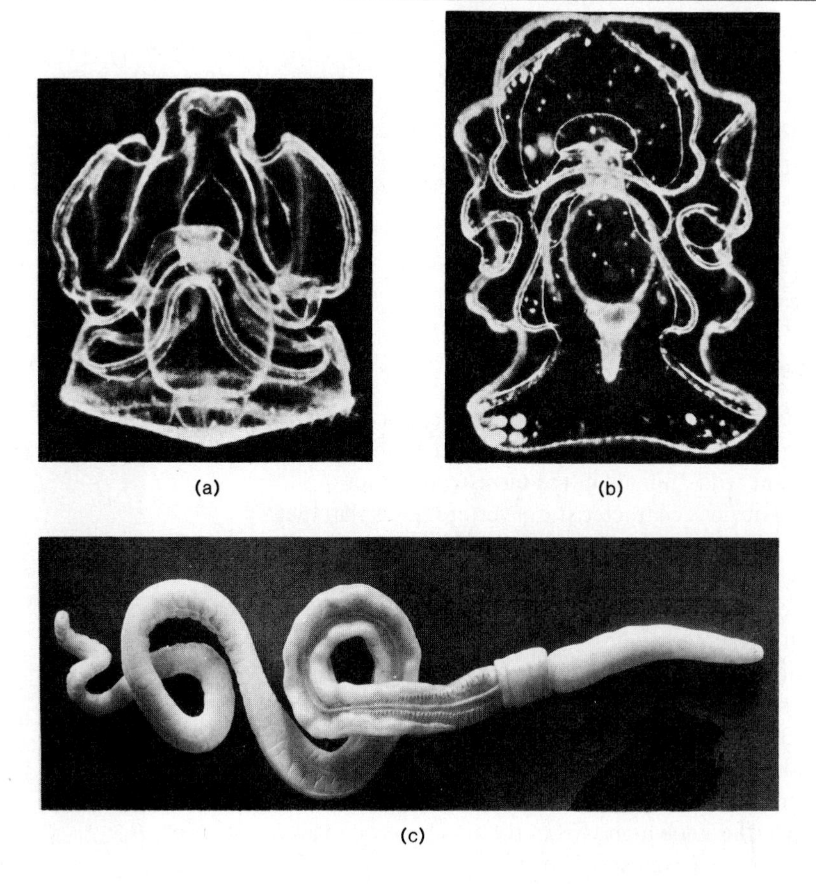

Nonvertebrate Chordates

Nonvertebrate chordates belong to two groups, the **subphylum Urochordata** and the **subphylum Cephalochordata.**

The **tunicates (sea squirts)** are the most common urochordates. Adult tunicates are sessile marine animals surrounded by a tough outer covering, the **tunic,** that is peculiar because it contains cellulose, a substance very rarely found in animal bodies (figure 29.54). Tunicates filter food from water pumped through hundreds of pharyngeal gill slits.

The common names for members of the subphylum Cephalochordata are **amphioxus** or **lancelet.** These small, fishlike, marine animals can swim freely but often remain stationary, partly buried in the mud. Amphioxus is a filter feeder with a pharyngeal gill apparatus, and it has a well-developed hollow nerve cord and notochord (figure 29.55). It is commonly thought to be quite similar to the ancient segmented animals from which vertebrates evolved.

Figure 29.54 A tunicate (sea squirt), *Ciona intestinalis.*

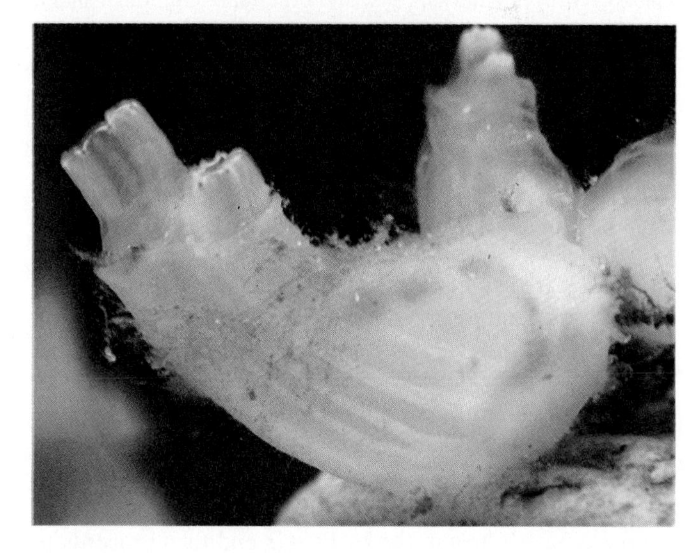

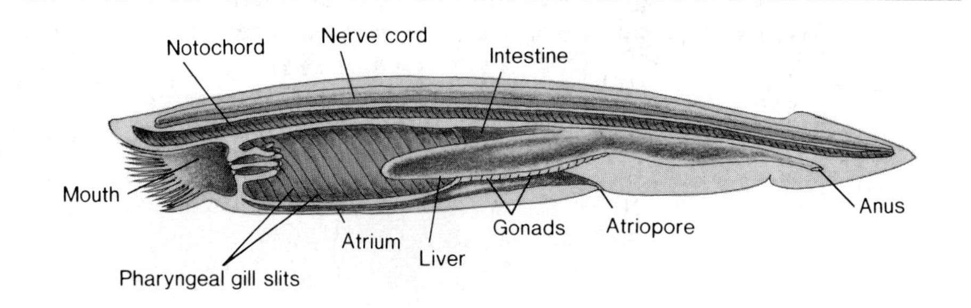

Figure 29.55 Amphioxus (lancelet), a member of the subphylum Cephalochordata. A longitudinal section showing internal anatomy. Water enters mouth, passes through pharyngeal gill slits into atrium, and then leaves body through atriopore.

Vertebrates

The **subphylum Vertebrata** needs little introduction. Humans are vertebrates, as are common domesticated animals and large dominant wild animals in the terrestrial habitat.

The most obvious characteristic of vertebrates is that they have a bony or cartilaginous endoskeleton that includes a vertebral column ("backbone"), made up of a series of **vertebrae.** Vertebrates have a brain, which is an obvious anterior enlargement of the dorsal, hollow nerve tube. The vertebrate brain is protected by a braincase, the **cranium.** Most vertebrates have several pairs of sense organs on their heads. Most also have a tail. A vertebrate digestive system characteristically includes a liver and a pancreas.

There are seven living classes of vertebrates and one class that is extinct and known only from the fossil record. Three living classes and the exclusively fossil class are fishes. These fish classes are the **Agnatha** (lampreys and other jawless fishes), **Placodermi** (extinct, primitive fishes with jaws), **Chondrichthyes** (sharks, rays, and other cartilaginous fishes), and **Osteichthyes** (bony fishes). The four-limbed (**tetrapod**) vertebrates make up the other four vertebrate classes: **Amphibia, Reptilia, Aves** (birds), and **Mammalia.**

Class Agnatha

Members of the class Agnatha ("without jaws") also are known collectively as the **cyclostomes** ("round mouths") because they are fish that have round, sucking mouths and lack hinged jaws. **Lampreys** and **hagfish** are living members of the class. They have skeletons constructed entirely of cartilage, but this condition may be a result of degenerative evolutionary changes, since the earliest fossil jawless fish had bony armor plates that covered their bodies.

A lamprey feeds by attaching its round, sucking mouth to the body of another fish (figure 29.56). It uses its horny tongue to rasp a hole in the host's skin and then proceeds to suck out blood and other body fluids.

Figure 29.56 The round, sucking mouth of a lamprey. Lampreys attach to other fish and feed on their blood and tissue fluids.

Class Placodermi

Although they have been extinct for millions of years, the **placoderms** represent an important milestone in vertebrate evolution because they were the first vertebrate animals with hinged jaws (figure 29.57). Jaws probably evolved from bars that supported gills in the anterior portion of the pharynx of their jawless ancestors.

Figure 29.57 Painting of an extinct placoderm. Placoderms were the first vertebrates to have hinged jaws.

Class Chondrichthyes

The class Chondrichthyes ("cartilage fishes") includes **sharks, skates,** and **rays** (figure 29.58). The chondrichthyes are descendants of ancient bony fishes, but all of them have exclusively cartilaginous skeletons. Their bodies are covered with small, toothlike scales called **denticles.** Thus, a shark's skin feels like sandpaper. The menacing teeth of sharks and their relatives are simply larger, specialized versions of these scales.

Rays and skates swim along the bottom and feed on animals, mostly invertebrates, that they dredge up. But most sharks are fast-swimming predators with beautifully streamlined bodies that slip easily through the water.

Sharks hold a terrifying fascination for many people, and they have a generally well-deserved reputation as ferocious and somewhat indiscriminate feeders. Tiger sharks, for example, have been known to bite off and swallow chunks of small boats. Great white sharks seem to bite first and then decide whether or not to swallow. This may explain why a number of human victims of great white shark attacks seem to have been "spit out."

However, some of the largest sharks, basking sharks and whale sharks, are not active hunters, but instead are filter feeders. The whale shark, which reaches a length of 16 m and thus is the largest known fish, filters huge amounts of water to obtain the great quantities of plankton that it eats.

Sharks have internal fertilization, and the eggs of some species develop inside the female reproductive tract. In those species, little sharks are born fully developed and swim away from their mothers immediately. An immediate escape is advantageous because some sharks do eat their own young just as they would eat any other small fish.

Figure 29.58 The class Chondrichthyes (cartilaginous fishes). (*a*) A white-tip reef shark. Note the streamlined body and the separate openings of all the gill slits. (*b*) Piece of shark skin showing scales (denticles). Note their toothlike shape. (*c*) A manta ray.

(a)

(b)

(c)

Class Osteichthyes

The class Osteichthyes includes most of the familiar fishes. It is a large, diverse class of bony fishes that may have as many as thirty or forty thousand species, although only about seventeen thousand species have been described so far (figure 29.59).

Members of the class Osteichthyes fall into two major groups. The **ray-fin fishes** have fins supported by spinelike rays. The other group includes **lungfishes** and **lobe-fin fishes.** Lobe-fin fishes have fins that are set in fleshy lobes that have bony, skeletal supports in them.

Most of the modern Osteichthyes are ray-fin fishes. Modern bony fish have a **swim bladder,** a gas-filled sac located in the dorsal part of the body cavity. By secreting gas into the swim bladder or absorbing gas from it, the fish can change its buoyancy. This allows the fish to hover effortlessly at different depths in the water. Modern bony fish have broad, flattened **scales** covering some or all of their body surfaces. Their gills are located in an enclosed **gill chamber** that is covered and protected by a hard bony flap, the **operculum.**

African, Australian, and South American lungfishes are rather obscure animals that live as their more numerous ancient ancestors did, in freshwater habitats that often become stagnant or even dry up entirely. The lungfishes regularly breathe air to supplement gas exchange in their gills.

From an evolutionary viewpoint, lobe-fin fishes are more significant than the lungfishes because some of the ancient lobe-fin fishes were the ancestors of all the land-dwelling vertebrates. The lobe-fin fishes could breathe air while they crawled clumsily on their stumpy fin lobes from pond to pond. This ability, marginal as it was, to survive on land opened a whole new realm of possibilities for lobe-fin fishes and their descendants. It is probable that the first amphibians evolved from some of the lobe-fin fishes (figure 29.60).

Until 1939, all of the lobe-fin fishes were believed to have been extinct for about 75 million years. But in that year, one was caught off the coast of South Africa. It was identified as a **coelacanth** and given the genus name *Latimeria* (figure 29.61). Since that time, other specimens have been caught and carefully examined. These "living fossils" probably are descendants of relatives of the lobe-fin fishes that gave rise to amphibians and, eventually, to other tetrapods.

Thus, while the lungfishes have remained in the same freshwater environments occupied by their ancient ancestors, the surviving lobe-fin fishes long ago entered the marine environment and "slipped off the continental shelf" into the depths of the ocean, where the coelacanth is found today.

Figure 29.59 The class Osteichthyes (bony fishes). (*a*) Sea perch. Note the flattened scales and the operculum that covers the gill chamber. (*b*) A spotted moray eel swimming among sea urchins in the Red Sea. (*c*) A sea horse. (*d*) A longlure frogfish with a "lure" near its mouth that attracts potential prey.

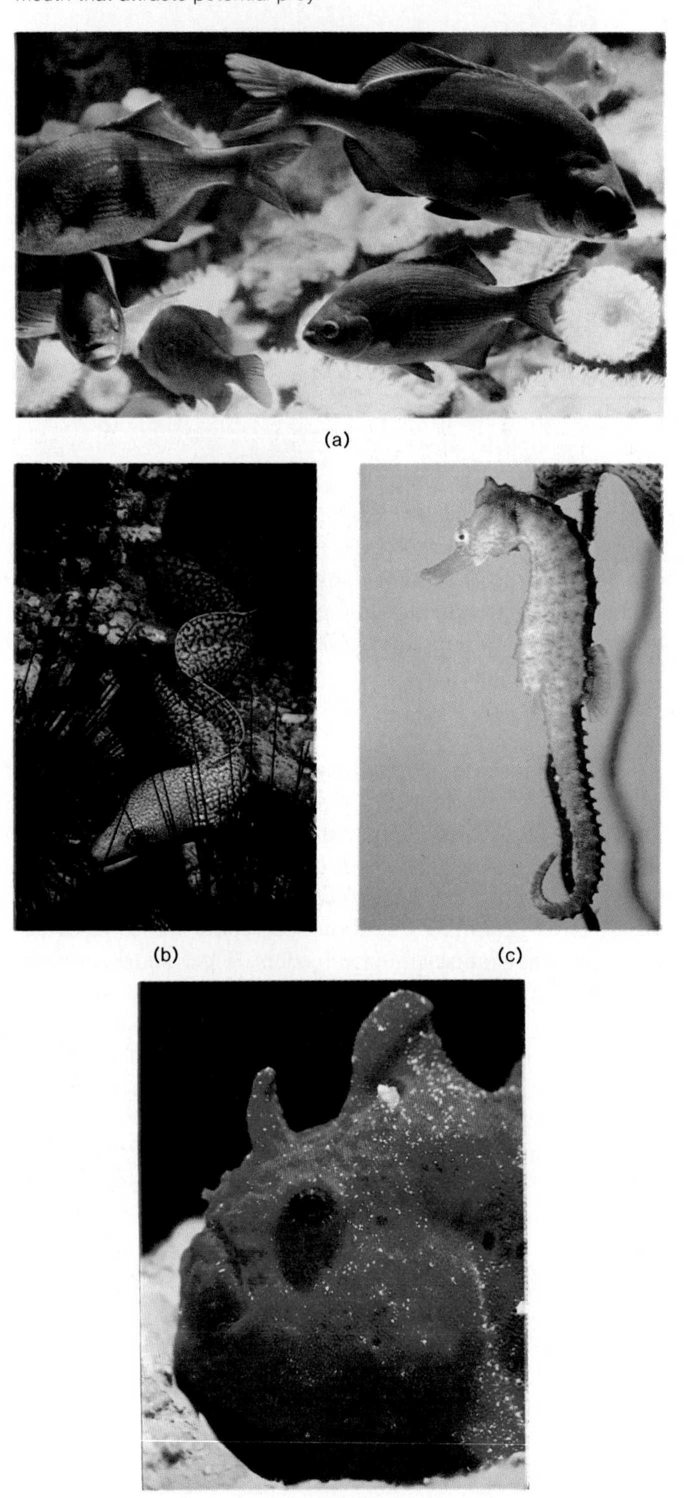

Figure 29.60 Lobe-fin fishes and early amphibians reconstructed from their fossils. (*a*) A lobe-fin fish (*Eusthenopteron*) that probably could move out of the water onto land. (*b*) An early amphibian (*Diplovertebron*). Notice that its body is very low to the ground and its legs stick out to the sides. It must have been an awkward walker, but its legs were an improvement over lobe-fins for locomotion on land.

Figure 29.61 Sketch of a coelacanth (*Latimeria*). This is the only living lobe-fin fish.

Figure 29.62 Amphibians. (a) A red-eyed tree frog, a member of the class Anura ("without a tail"). (b) A salamander (class Urodela). (c) A caecilian, a member of the class Apoda ("without limbs").

(a)

(b)

(c)

Class Amphibia

The three groups (orders) of modern amphibians are the **salamanders** and other amphibians with tails (**order Urodela**); the **frogs** and **toads** (**order Anura**); and a small group of limbless, burrowing amphibians, known as **caecilians,** that occur only in tropical environments (**order Apoda**) (figure 29.62).

The first amphibians were very similar to the lobe-fin fishes (see figure 29.60) but were better adapted for life on land for several reasons. Early amphibians had legs that ended in toes (usually five or less) and skeletons better suited to locomotion on land. Amphibians became the dominant animals for a period of time from about 350 million years ago until about 230 million years ago when reptiles replaced amphibians as the dominant land animals.

Amphibians are closely tied to water or, at least, very moist terrestrial habitats for several reasons. Amphibians have rather simple lungs and most depend on supplementary gas exchange through their skin. For skin to function efficiently as a gas exchange surface, it must be quite thin and must be kept moist. This means that amphibians can suffer excessive water loss in a dry environment.

However, the most important factor binding amphibians to water is their method of reproduction. Amphibians have external fertilization, with both eggs and sperm being shed into the water. Amphibian eggs are enclosed by a jelly coat that provides no protection against desiccation (drying) if the eggs are exposed to the air. Young amphibians hatch from their jelly as aquatic larvae (**tadpoles**) with gills. Tadpoles feed and grow in the water. Only after metamorphosis do amphibians emerge from the water as air-breathing adults (figure 29.63).

Amphibians' expansion into various terrestrial habitats has always been limited by their dependence on water. Modern amphibians still live on the border line between aquatic and terrestrial environments. Their class name, amphibia, means "double life" and accurately describes their life history, which is divided between two different habitats.

Class Reptilia

The reptiles arose from amphibian ancestors, but they were much better suited to life on land. They have internal fertilization, and they lay eggs that are protected by leathery **shells.** Enclosed within such a **cleidoic** ("boxlike") egg, a

Figure 29.63 The amphibian life history. (*a*) Eggs of the grass frog (*Rana temporaria*) surrounded by their jelly coats. (*b*) *Rana temporaria* tadpoles. Tadpoles are herbivores that feed on aquatic plants. (*c*) Young *Rana temporaria* just after metamorphosis.

(a)

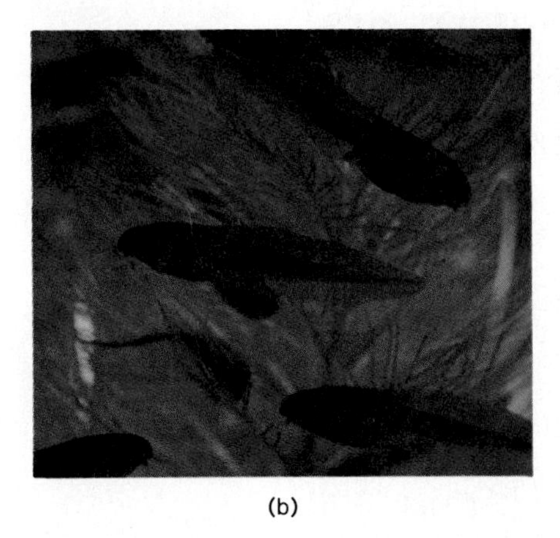

(b)

(c)

Figure 29.64 The cleidoic eggs of reptiles, an adaptation for reproduction in a terrestrial environment. (*a*) Baby king snake hatching out of its shell. Note that shells surrounding reptile eggs are leathery and flexible, not brittle like birds' eggs. (*b*) The arrangement of embryo and extraembryonic ("outside the embryo") membranes in a cleidoic egg. The yolk is drawn smaller than its normal proportions. The chorion encloses the whole system; the yolk sac absorbs nutrients; the allantois stores wastes, and blood vessels in its wall function in gas exchange; the amnion encloses amniotic fluid, which surrounds and protects the developing embryo.

(a)

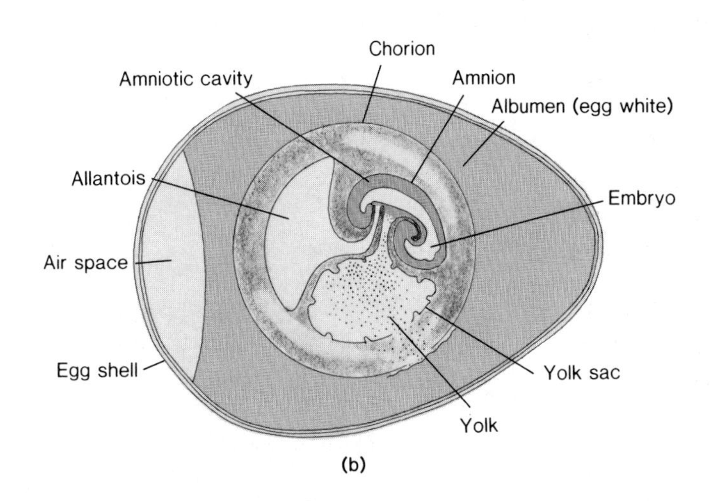

(b)

reptile embryo develops in a sheltered environment, where it is supplied with the water and nutrients required for its development (figure 29.64).

Reptiles have a thick, scaly skin that contains **keratin,** a protein that helps to make the skin impermeable to water. This is important in water conservation.

The four orders of modern reptiles are **turtles (order Chelonia)**, **crocodiles** and **alligators (order Crocodilia)**, **lizards** and **snakes (order Squamata)**, and the **tuatara (order Rhynchocephalia)** (figure 29.65). The tuatara is the only surviving member of its order and is found only on islands in New Zealand.

The earliest reptiles, called **stem reptiles,** gave rise to several lines of descent (figure 29.66). Some of those lines produced the modern reptiles; other lines led to the mammals and birds.

Possibly the most fascinating aspect of the history of reptiles, however, was the evolution of the great reptiles that dominated the earth for millions of years and then became extinct about 65 million years ago (figure 29.67). There were flying reptiles; large, swimming reptiles; and the **dinosaurs,** the reptiles that dominated the terrestrial environment.

Figure 29.65 The orders of reptiles. (*a*) A turtle (*Gopherus*) of the order Chelonia laying eggs in the Okefenokee swamp. (*b*) Tuatara (*Sphenedon*) of the order Rhynchocephalia. (*c*) Alligators (*Alligator*) of the order Crocodilia.

(a) (b) (c)

Figure 29.66 A family tree of the reptiles, including their relationships to other vertebrate classes.

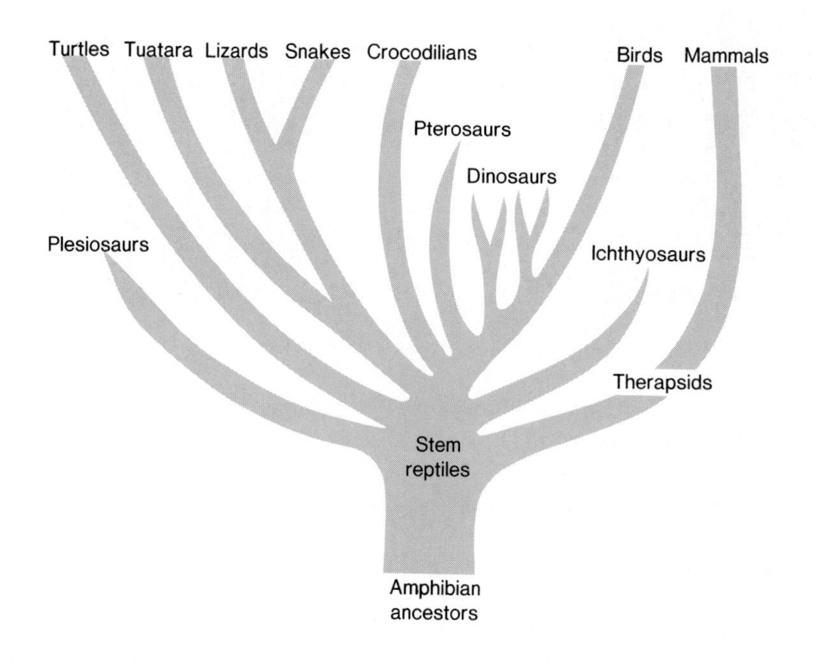

Figure 29.67 Fossil reptiles from the "age of reptiles." (*a*) Pterosaurs were flying reptiles that used broad, flat skin flaps for flying. Pterosaurs were not ancestors of birds, since birds descended from a separate group of reptiles. (*b*) Plesiosaurs (left) and Ichthyosaurs, swimming reptiles. (*c*) *Triceratops* (left), an armored herbivore, and *Tyrannosaurus* (right), a huge carnivore that was as much as 15 m long and 6 m tall when it stood upright.

(a)

(b)

(c)

The dinosaurs are well represented in the fossil record but still remain shrouded in mystery. One of the great mysteries of the dinosaurs concerns their extinction. One hypothesis is that climatic change brought about their demise because they could not stand cooler temperatures or because their food sources were reduced. Another hypothesis proposes that mammals evolved to a point where they preyed on dinosaur eggs to such an extent that the dinosaurs were no longer able to reproduce themselves. There is even a somewhat more radical suggestion that the dinosaurs died out, not gradually over millions of years, but quickly within a period of hundreds of years or even much less, following some catastrophic event (for example, an asteroid collision) that abruptly altered conditions worldwide.

In an evolutionary sense, the length of time that the dinosaurs dominated the earth is very impressive. Their reign lasted about 100 million years. Mammals existed during the dinosaurs' time, but they were small, relatively insignificant animals living where they could in the dinosaurs' world.

Mammals have dominated the terrestrial environment only for the last 65 million years (since the extinction of the dinosaurs). The life span of the human species is measured in only thousands of years, a moment in time compared to the length of the age of dinosaurs.

Another mystery concerns the basic nature of dinosaurs. They have long been described as stupid, sluggish, slow-moving animals that could function efficiently only in a uniformly warm environment that kept their body temperature high enough to permit a reasonable range of activity. But in the last few years, some biologists have challenged the view that the dinosaurs all were poikilotherms, animals that were unable to regulate their body temperatures metabolically. Dinosaurs lived successfully in northern latitudes, where very short winter days would have made it very difficult for them to heat up their giant bodies by solar radiation. Furthermore, present-day reptiles, which all are poikilotherms, have very sparsely vascularized bones. But studies of dinosaur bones reveal rich blood vessel networks such as those found in the bones of mammals, which are homeotherms ("warm-blooded animals").

It now appears that some reptiles, who were ancestors of the birds (but were not fliers themselves) had feathers or featherlike structures. Birds' feathers are involved in flying, but they also are critically important in body insulation. Thus, these nonflying, feathered reptiles may have been supplied with insulation that aided in maintaining a constant body temperature. This discovery of feathered, possibly homeothermic, dinosaurs has led some biologists to propose that the dinosaurs are not extinct after all, but that birds are just a highly specialized group of surviving dinosaurs.

Class Aves

Birds are the only modern animals that have **feathers.** Birds are descended from reptiles, and feathers have evolved as modified scales. Scales on the legs and feet of modern birds are reminders of their reptilian ancestry, as are the claws at the ends of their toes (figure 29.68).

Birds have specialized **contour feathers** that are attached to their wings in such a way that they overlap to produce a broad flat surface used in flight. Feathers also function in body temperature regulation. **Down feathers** provide excellent insulation against body heat loss (figure 29.69). This insulating effect is important because birds are homeothermic animals; that is, they maintain a constant, relatively high body temperature. This permits birds to be continuously active, even when there are rather drastic fluctuations in environmental temperature.

Figure 29.68 A representation of *Archeopteryx,* the oldest known fossil bird. *Archeopteryx* had several reptilian characteristics, including jaws with teeth and a long, jointed, reptilian tail. It was about the size of a crow and had rather frail wings for its size, but it was well feathered.

Some birds cannot fly (figure 29.70). Yet all birds share some fundamental features that clearly are flight adaptations. All birds have light horny **beaks** instead of heavy jaws with teeth. Birds' bones are hollow and thus very light. The general shape of the average bird's body also is a flight adaptation. A slender neck connects the head to a rounded, compact body. The weight of the digestive system, with a **crop** used in food storage and a thick muscular **gizzard** that grinds food, is well centered.

Birds have internal fertilization, and they lay cleidoic eggs that are enclosed in a brittle, calcium-containing shell. Generally, young birds hatch from their shells in fairly helpless condition. They require brooding to keep them warm as well as feeding by one or both parents.

Flight requires well-developed sense organs. Birds have particularly acute vision and excellent visual reflexes (figure 29.71).

Possibly the most remarkable aspect of the biology of birds, however, is the annual **migration** that so many of them accomplish. Some birds migrate thousands of kilometers over land and ocean, navigating successfully by day and night, through sunshine and cloudy weather, to reach very specific distant goals. Mysteries still remain regarding bird navigation, but birds almost certainly use celestial navigation and even variations in the earth's magnetic field intensity to find their way on these impressive journeys (see page 531).

Figure 29.69 Feathers. (*a*) A contour feather from a red-tailed hawk showing the many barbs that branch from the shaft of the feather. Note the downy branches near the base of the feather. (*b*) Scanning electron micrograph of barbs of a goose flight feather showing the little hooks that interlock the barbs so that each feather is firm. Wing feathers overlap to form a flat surface for flight (magnification × 420). (*c*) An embryonic down feather. Each of about a dozen barbs has many barbules branching from it. Thousands of down feathers together trap air and provide effective insulation (magnification × 5).

(a) (b) (c)

Figure 29.70 Flightless birds. (*a*) Adelie penguins. Their wings are modified as flippers used for swimming. (*b*) A brown kiwi from New Zealand. The kiwi's tiny wings are hidden under its plumage. The kiwi looks shaggy because all of its feathers resemble the juvenile down feathers of other birds. The kiwi uses its long bill to probe for earthworms.

(a)

(b)

Figure 29.71 Hunting birds in action demonstrate the acute visual reflexes of birds. Here brown pelicans are diving for fish off the Florida Keys.

Figure 29.72 The subclass Prototheria, the egg-laying mammals (monotremes). (*a*) A duck-billed platypus. (*b*) A spiny anteater (echidna).

(a)

(b)

Class Mammalia

Mammals are homeothermic vertebrate animals that have **mammary glands,** which produce milk, and body **hair,** which provides insulation against heat loss and thus aids in maintenance of a constant body temperature. Mammary glands permit female mammals to feed (nurse) their young regularly without having to find food every time that the young are hungry, as nesting birds must do. Parental care and protection helps to assure that a greater percentage of young mammals survive to reach adulthood than generally is the case in other animals.

The great majority of mammals are **viviparous.** Their embryos develop inside a specialized portion of the female reproductive tract, the **uterus.** A circulatory exchange organ called the **placenta** permits exchange of materials between the bloodstreams of the mother and the developing fetus. This viviparous arrangement shelters developing young from environmental changes and frees adults from having to tend a nest.

Mammals' limbs are oriented and connected in ways that permit them to run faster than other vertebrates can. Furthermore, mammals have well-developed sense organs and extensive development of brain centers, especially in the cerebral cortex, that are involved in flexible behavior patterns modifiable by experience and learning.

Living mammals are grouped into three subclasses. Two smaller subclasses consist of nonplacental mammals; the third, and largest, subclass contains the placental mammals and includes about 95 percent of living mammals.

Subclass Prototheria: Egg-laying Mammals

Only two kinds of egg-laying mammals, or **monotremes,** survive: the **duck-billed platypus** of Australia and the **spiny anteater,** which lives in Australia and New Guinea (figure 29.72). These animals lay their eggs, which resemble reptile eggs, in burrows in the ground, where they incubate the eggs and brood the young, much as birds do. Although the monotremes are egg-layers, they have mammary glands and nurse their young after they hatch. This curious combination puts the monotremes almost on the border line between reptiles and mammals.

Figure 29.73 Subclass Metatheria, the pouched mammals (marsupials). (*a*) Koala bears. The koala is a tree-climbing, herbivorous marsupial. (*b*) A red kangaroo, with its young (a "joey") in its pouch. A young kangaroo remains in the pouch even after it is no longer attached to a nipple. It ventures out occasionally and then returns to the pouch. Kangaroos are marsupials that occupy a grazing niche, which is occupied by ungulates on other continents.

(a)

(b)

Subclass Metatheria: Pouched Mammals

The young of pouched, or **marsupial,** animals begin their development inside the female's body, but they are born in a very immature condition. Newborn young enter a pouch, the **marsupium,** on their mother's abdomen. Inside the pouch, they attach to nipples of mammary glands. They continue their development in this sheltered environment and emerge later when they are much more mature.

Only a few marsupials, such as the American opossum, are found outside Australia (figure 29.73). Apparently, ancestral marsupials became isolated as virtually the only mammals in Australia. While placental mammals have become the dominant mammals elsewhere, in the absence of competition, Australian marsupials diversified to fill all of the specialized roles played by placental mammals on other continents.

Some of the marsupial mammals in Australia include the Koala bears, which are tree-climbing marsupial herbivores, and kangaroos, which fill the niche occupied by hoofed grazing mammals (**ungulates**) on other continents. The extinct Tasmanian "wolf" or "tiger" was a carnivorous marsupial mammal about the size of a collie dog.

Subclass Eutheria: Placental Mammals

In placental mammals, there is a long period of exchange, via the placenta, between the mother's bloodstream and the bloodstream of the fetus. During this exchange, the fetus is supplied with nutrients and oxygen and is rid of wastes and carbon dioxide. The pregnancy (**gestation**) periods of some placental mammals are long, and the young are rather mature at birth. Some mammals, such as horses, can stand and walk within minutes after they are born. Other placental mammals, however, are not nearly so mature at birth. For example, newborn kittens are blind and helpless, and a human infant is totally unable to meet any of its own basic needs for a long time after it is born.

Figure 29.74 Placental mammals. (*a*) A bat (order Chiroptera). (*b*) Two great blue whales (order Cetacea) photographed from the air. One of them is "spouting" (exhaling forcefully). Great blue whales are the largest animals that have ever lived. They can reach 30 m in length and weigh more than 110 metric tons. (*c*) A classic confrontation on the plains of east Africa. A lion (order Carnivora) with its prey, a wildebeest (order Artiodactyla). (*d*) Langur monkeys (order Primates).

(a) (b) (c) (d)

Placental mammals are a very diverse group of animals. Some major orders of mammals are listed in table 29.2. Most of them are animals that roam the terrestrial environment, but they are obvious exceptions (figure 29.74).

Bats (**order Chiroptera**) are flying mammals that feed at night, thereby avoiding direct competition with birds. Predatory bats are remarkably adapted for catching flying insects (see page 532).

Whales and their relatives (**order Cetacea**) are marine animals. Like all mammals, they have mammary glands and feed their young milk, but most whales have very little body hair. Their bodies are insulated by thick layers of fat under their skin. In recent years many kinds of whales have been hunted almost to extinction mainly for the oils extracted from their body fat.

As they migrate through the oceans, whales sing complex songs that apparently communicate messages great distances through the water to others of their species. The communication and behavior of whales and their relatives, such as dolphins, indicate a high level of intelligence.

In the terrestrial environment, the largest placental mammals are herbivores such as elephants (**order Proboscidea**) and ungulates (hoofed mammals). The ungulates fall into two groups, those with odd numbers of toes (**order Perissodactyla**), such as horses and rhinoceroses, and those with even numbers of toes (**order Artiodactyla**), such as cattle, bison, antelopes, and deer.

Table 29.2
Some Major Orders of Placental Mammals

Order	Description
Insectivora	Primitive, insect-eating mammals. Moles, shrews, and hedgehogs.
Chiroptera	Flying mammals with a broad skin flap extending from elongated fingers to the body and legs. Bats.
Primates	Omnivorous mammals, opposable thumb and fingers, eyes directed forward, well-developed cerebral cortex. Lemurs, monkeys, apes, and humans.
Edentata	Mammals with few or no teeth. Sloths, anteaters, and armadillos.
Rodentia	Mammals with sharp chisellike incisor teeth that grow continuously. Squirrels, beavers, rats, mice, moles, porcupines, hamsters, chinchillas, and guinea pigs.
Lagomorpha	Mammals with chewing teeth, tails reduced or absent, hindlimbs longer than forelimbs. Hares, rabbits, and pikas.
Proboscidea	Long muscular trunk (proboscis); thick, loose skin; incisors elongated as tusks. Elephants.
Cetacea	Marine mammals with fish-shaped bodies, finlike forelimbs, no hindlimbs, body insulated with a thick layer of fat (blubber). Whales, dolphins, and porpoises.
Perissodactyla	Herbivorous, odd-toed hoofed mammals. Horses, zebras, tapirs, and rhinoceroses.
Artiodactyla	Herbivorous, even-toed mammals. Cattle, sheep, pigs, giraffes, deer, antelopes, gazelles, hippopotamuses, camels, bison, and llamas.
Carnivora	Carnivorous mammals; sharp, pointed canine teeth and shearing molars. Cats, dogs, foxes, wolves, hyenas, bears, otters, minks, weasels, skunks, badgers, seals, walruses, and sea lions.

In natural ecosystems, the ungulates are prey to the largest members of the **order Carnivora,** which includes the great cats—lions, tigers, cheetahs, and leopards. In terms of physical characteristics, these large, fast, ferocious hunters may represent the pinnacle of evolution of placental mammals, but none of them is the single dominant species on earth today.

That distinction is reserved for the human species. Humans are members of the **order Primates.** We are not the fastest or strongest of animals. We do not have sense organs superior to those of all other animals. But we are dominant, despite our physical limitations, because human evolution has brought one key characteristic to the forefront: our brains are superior to those of other animals. We alone are able to contemplate our existence and the processes that have shaped the living world. Unfortunately, the power generated by our superior brains also has led to alterations of the natural world, many of which threaten our own welfare and that of many other species. It now seems that, for better or worse, the quality of future life on earth will be determined by human activities in the coming years.

Summary

This chapter examines and compares the major animal phyla and some evolutionary relationships among the groups.

Simply organized animals are restricted to relatively stable, aquatic environments or sheltered environments inside other organisms (internal parasites).

Terrestrial animals have more complex body organizations, including impermeable body surfaces that separate a stable internal body environment from the changeable external environment. Exchanges between inside and outside take place only in restricted body areas.

All animals are heterotrophs, but animals display a variety of nutritional specializations. Aquatic animals range from sessile filter feeders to active predators. Terrestrial animals include herbivores and the carnivores that feed on the herbivores. All of these animal nutritional strategies are associated with specific structural and functional specializations of the animals' bodies.

Reproductive patterns also are adapted to environmental conditions. This is particularly true of terrestrial animals. Well-adapted terrestrial animals have internal fertilization, which protects gametes from desiccation. They either lay eggs enclosed by protective shells (for example, insects, arachnids, reptiles, birds) or have reproductive adaptations that permit early development of their young to occur inside the female body (placental mammals).

Questions for Review

1. What are the three basic body layers of animals?
2. Define the term invertebrate.
3. What is extracellular digestion?
4. Explain why, in the United States, the danger of tapeworm infections is significant while the danger of fluke infections is very small.
5. Distinguish between incomplete and complete digestive tracts.
6. Describe some advantages and disadvantages of the arthropod exoskeleton as a body surface covering for terrestrial animals.
7. Name the two major animal phyla that are deuterostomes.
8. List and explain the three major distinguishing characteristics of the chordates.
9. What characteristics of amphibians limit their capacity to occupy large portions of the terrestrial environment?
10. What reproductive adaptations of reptiles, birds, and mammals contribute to their success as terrestrial animals?

Questions for Analysis and Discussion

1. Many animals that live attached to a substrate or move about very slowly are radially symmetrical. Animals that move about very actively are bilaterally symmetrical. How would you explain this difference?
2. What is the adaptive significance of the very large reproductive potentials of virtually all parasitic animals whose life histories involve several different host animals?
3. Homeothermic animals must obtain more food than poikilothermic animals of comparable size because homeotherms expend considerable energy for heat production. This would seem to be a disadvantage of being homeothermic. Discuss some advantages of being homeothermic.

Suggested Readings

Books

Attenborough, D. 1979. *Life on earth.* Boston: Little, Brown.

Attenborough, D. 1984. *The living planet.* Boston: Little, Brown.

Romer, A. S., and Parsons, R. S. 1977. *The vertebrate body,* 5th ed. Philadelphia: Saunders.

Villee, C. A.; Walker, W. F.; and Barnes, R. D. 1984. *General zoology,* 6th ed. Philadelphia: Saunders.

Articles

Alvarez, L. W.; Alvarez, W.; Asaro, F.; and Michel, H. V. 1980. Extraterrestrial cause for the Cretaceous-Tertiary extinction. *Science* 208:1095.

Calder, W. A., III. July 1978. The kiwi. *Scientific American* (offprint 1396).

Goreau, T. F.; Goreau, N. I.; and Goreau, T. J. August 1979. Corals and coral reefs. *Scientific American* (offprint 1434).

Langston, W., Jr. February 1981. Pterosaurs. *Scientific American* (offprint 1492).

McCosker, J. E. July/August 1981. Great white shark. *Science 81.*

McWhinnie, M. A., and Denys, C. J. 1980. The high importance of the lowly krill. *Natural History* 89 (3):66.

Marx, J. L. 1978. Warm-blooded dinosaurs: Evidence pro and con. *Science* 199:1424.

Roper, C. F. E., and Boss, K. J. April 1982. The giant squid. *Scientific American* (offprint 1515).

Würsig, B. March, 1979. Dolphins. *Scientific American* (offprint 1424).

Appendix 1
Classification of Organisms

This appendix summarizes the major taxonomic groups discussed in the survey of organisms in chapters 27 through 29 and lists some examples of members of many of the groups. The classification system we use here is only one of several systems currently in use by biologists and should not be considered the only possible classification of living things. There is, for example, considerable disagreement among biologists concerning the status of the slime molds and the assignment of various groups of algae to the protist and plant kingdoms.

Botanists use the term "division" for major groups of plants, while zoologists use the term "phylum" for the major groups of animals. We have used these terms in the same way. We also have used the term "division" in connection with groups of protists that formerly were included in the plant kingdom under the old two-kingdom classification system.

Kingdom Monera: prokaryotes—bacteria and cyanobacteria (blue-green algae)

Kingdom Protista: eukaryotes, unicellular or colonies without tissue differentiation

[Protozoa—Heterotrophic, Unicellular Protists]

Phylum Mastigophora: flagellated protozoa [*Trypanosoma*]

Phylum Sarcodina: pseudopodial protozoa [*Amoeba, Entamoeba,* foraminiferans, radiolarians]

Phylum Sporozoa: spore-forming protozoa [*Plasmodium*]

Phylum Ciliophora: ciliated protozoa [*Paramecium, Stentor, Vorticella, Didinium*]

[Unicellular Algae]

Division Euglenophyta: euglenoids [*Euglena*]

Division Pyrrophyta: dinoflagellates [*Ceratium, Peridinium, Gymnodinium*]

Division Chrysophyta: yellow-green and golden-brown algae [Diatoms]

[Funguslike Protists]

Division Oomycota: water molds, downy mildews [*Saprolegnia, Phytophthora*]

[Slime Molds]

Division Gymnomycota: plasmodial and cellular slime molds

Kingdom Fungi: eukaryotic heterotrophs (absorptive nutrition), mycelial organization

Division Zygomycota: zygospore-forming fungi [*Rhizopus,* many common fruit and bread molds]

Division Ascomycota: sac-fungi [*Saccharomyces* and other yeasts, morels, truffles, apple scabs, powdery mildews, Dutch elm disease, ergot disease of rye]

Division Basidiomycota: club fungi [Mushrooms, puffballs, shelf or bracket fungi, rusts, smuts]

Division Deuteromycota: "imperfect" fungi [*Penicillium, Aspergillus, Candida*]

Kingdom Plantae: multicellular eukaryotes with walled cells, photosynthetic

Division Chlorophyta: green algae [*Chlamydomonas, Volvox, Spirogyra, Oedogonium, Ulva*]

Division Phaeophyta: brown algae [*Fucus, Macrocystis, Nereocystis, Laminaria*]

Division Rhodophyta: red algae

Division Bryophyta: mosses and liverworts [*Marchantia*]

Division Tracheophyta: vascular plants
Subdivision Psilopsida: "whisk ferns" [*Psilotum*]
Subdivision Lycopsida: club mosses
Subdivision Sphenopsida: horsetails [*Equisetum*]
Subdivision Pteropsida: plants with complex conducting systems and large complex leaves
Class Filicineae: ferns
Class Gymnospermae: "naked seed" plants [Cycads, *Ginkgo, Welwitschia, Pinus, Sequoia*]
Class Angiospermae: "enclosed seed" plants
Subclass Monocotyledoneae: monocots [lilies, palms, orchids, grasses]
Subclass Dicotyledoneae: dicots [buttercups, maples, carnations, roses]

Kingdom Animalia: multicellular, eukaryotic heterotrophs

Phylum Porifera: sponges

Phylum Coelenterata
Class Hydrozoa: hydrozoans [hydras, *Obelia, Physalia*]
Class Scyphozoa: "true jelly fish" [*Aurelia*]
Class Anthozoa: sea anemones and corals

Phylum Platyhelminthes: flatworms
Class Turbellaria: planarians [*Dugesia*]
Class Trematoda: flukes [*Opisthorchis, Schistosoma*]
Class Cestoda: tapeworms [*Taenia*]

[Protostomes]

Phylum Nematoda: roundworms [*Ascaris*, pinworms, *Trichinella*, filaria worms]

Phylum Rotifera: "wheel animals"

Phylum Gastrotricha: gastrotrichs

Phylum Mollusca: molluscs
Class Amphineura: chitons
Class Pelecypoda: bivalves [clams, oysters, scallops]
Class Gastropoda: "belly-foot" molluscs [snails, slugs, nudibranchs]
Class Cephalopoda: "head-foot" molluscs [octopuses, squids, nautiluses]

Phylum Annelida: segmented worms
Class Polychaeta: marine annelids with parapodia [*Nereis, Chaetopterus*, Palolo worm]
Class Oligochaeta: terrestrial and freshwater annelids without parapodia [earthworms]
Class Hirudinea: leeches

Phylum Onychophora [*Peripatus*]

Phylum Arthropoda: "joint-footed" animals
Subphylum Chelicerata: first mouthparts are chelicerae; antennae are absent
Class Xiphosura: [*Limulus*]
Class Arachnida: [spiders, scorpions, ticks, mites, daddy longlegs]
Subphylum Mandibulata: possess mandibles and not chelicerae, antennae are present
Class Crustacea: [crabs, lobsters, crayfish, shrimp, barnacles]
Class Chilopoda: [centipedes]
Class Diplopoda: [millipedes]
Class Insecta: [silverfish, grasshoppers, termites, bugs, beetles, butterflies and moths, flies, bees, ants]

[Deuterostomes]

Phylum Echinodermata
Class Asteroidea: [sea stars]
Class Ophiuroidea: [brittle stars and basket stars]
Class Echinoidea: [sea urchins and sand dollars]
Class Holothuroidea: [sea cucumbers]

Phylum Hemichordata: acorn worms

Phylum Chordata: chordates
Subphylum Urochordata: [sea squirts]
Subphylum Cephalochordata: [amphioxus]
Subphylum Vertebrata:
Class Agnatha: cyclostomes [lampreys, hagfishes]
Class Placodermi: extinct, armored, jawed fishes
Class Chondrichthyes: "cartilage fishes" [sharks, skates, rays]
Class Osteichthyes: "bony fishes" [lobe-fin fishes, lungfishes, ray-fin fishes]
Class Amphibia: amphibians
Order Urodela: tailed amphibians [salamanders]
Order Anura: tail-less amphibians [frogs, toads]
Order Apoda: limbless amphibians
Class Reptilia: reptiles
Order Chelonia: [turtles]
Order Crocodilia: [crocodiles and alligators]
Order Squamata: [snakes and lizards]
Order Rhynchocephalia: [tuatara]
Class Aves: birds
Class Mammalia: mammals
Subclass Prototheria: egg-laying mammals [platypus, echidna]
Subclass Metatheria: pouched (marsupial) mammals [kangaroos, koala bears, opossums]
Subclass Eutheria: placental mammals [shrews, bats, monkeys, apes, humans, rats, rabbits, elephants, whales, horses, antelopes, cats, dogs, bears, walruses]

Appendix 2
Units of Measurement

Temperature Conversion Scale

The formula for conversion of Fahrenheit to Celsius:

$$°C = \frac{5}{9}(°F - 32)$$

The formula for conversion of Celsius to Fahrenheit:

$$°F = \frac{9}{5}°C + 32$$

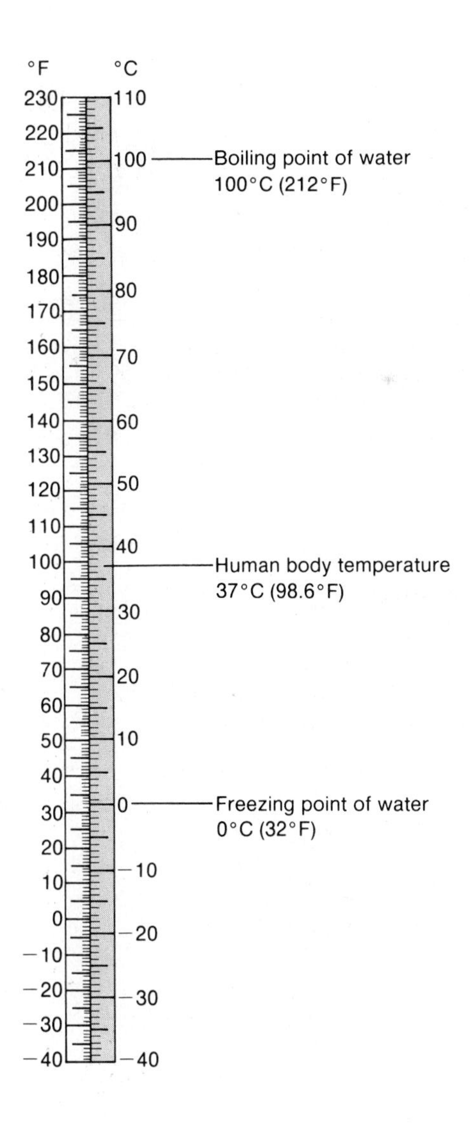

Standard Prefixes of the Metric System

kilo- (k)	1,000	10^3
hecto- (h)	100	10^2
deci- (d)	0.1	10^{-1}
centi- (c)	0.01	10^{-2}
milli- (m)	0.001	10^{-3}
micro- (μ)	0.000001	10^{-6}
nano- (n)	0.000000001	10^{-9}

Metric Conversions

	Metric Quantities	Metric to English conversion	English to metric conversion
Length			
	1 kilometer (km) = 1000 (10^3) meters 1 meter (m) = 100 centimeters 1 centimeter (cm) = 0.01 (10^{-2}) meter 1 millimeter (mm) = 0.001 (10^{-3}) meter 1 micrometer* (μm) = 0.000001 (10^{-6}) meter 1 nanometer (nm) = 0.000000001 (10^{-9}) meter *formerly called micron	1 km = 0.62 mile 1 m = 1.09 yards = 39.37 inches 1 cm = 0.394 inch 1 mm = 0.039 inch	1 mile = 1.609 km 1 yard = 0.914 m 1 foot = 0.305 m = 30.5 cm 1 inch = 2.54 cm
Area			
	1 square kilometer (km^2) = 100 hectares 1 hectare (ha) = 10,000 square meters 1 square meter (m^2) = 10,000 square centimeters 1 square centimeter (cm^2) = 100 square millimeters	1 km^2 = 0.3861 square mile 1 ha = 2.471 acres 1 m^2 = 1.1960 square yards = 10.764 square feet 1 cm^2 = 0.155 square inch	1 square mile = 2.590 km^2 1 acre = 0.4047 ha 1 square yard = 0.8361 m^2 1 square foot = 0.0929 m^2 1 square inch = 6.4516 cm^2
Mass			
	1 metric ton (t) = 1000 kilograms 1 metric ton (t) = 1,000,000 grams 1 kilogram (kg) = 1000 grams 1 gram (g) = 1000 milligrams 1 milligram (mg) = 0.001 gram 1 microgram (μg) = 0.000001 gram	1 t = 1.1025 ton (U.S.) 1 kg = 2.205 pounds 1 g = 0.0353 ounce	1 ton (U.S.) = 0.907 t 1 pound = 0.4536 kg 1 ounce = 28.35 g
Volume (solids)			
	1 cubic meter (m^3) = 1,000,000 cubic centimeters 1 cubic centimeter (cm^3) = 1000 cubic millimeters	1 m^3 = 1.3080 cubic yards = 35.315 cubic feet 1 cm^3 = 0.0610 cubic inch	1 cubic yard = 0.7646 m^3 1 cubic foot = 0.0283 m^3 1 cubic inch = 16.387 cm^3
Volume (liquids)			
	1 liter (l) = 1000 milliliters 1 milliliter (ml) = 0.001 liter 1 microliter (μl) = 0.000001 liter	1 l = 1.06 quarts (U.S.) 1 ml = 0.034 fluid ounce	1 quart (U.S.) = 0.94 l 1 pint (U.S.) = 0.47 l 1 fluid ounce = 29.57 ml
Time			
	1 second (sec) = 1000 milliseconds 1 millisecond (msec) = 0.001 second 1 microsecond (μsec) = 0.000001 second		

Glossary

a- [Gr.]
Not, without, lacking.

ab- [La.]
Off, away, away from.

abdomen [La.]
(1) In vertebrate animals, the posterior portion of the trunk containing visceral organs other than heart and lungs. (2) In invertebrates, especially arthropods, the posterior portion of the body.

abiotic [Gr. *bios:* life]
Nonliving.

absorption [La., to suck in, absorb]
(1) Movement of a substance into a cell or an organism, or through a surface within an organism. (2) Interception of radiant energy such as light waves.

acid [La.]
(1) A substance that can contribute a proton (hydrogen ion, H^+) in a reaction. (2) A substance that dissociates in solution to yield hydrogen ions (H^+), but not hydroxide ions (OH^+). (3) Sometimes used to describe a solution with a pH of less than 7.0.

actin
A protein that along with myosin accomplishes the physical shortening of muscle fibers. Actin also is found in many other kinds of cells that can shorten or contract.

action potential
An abrupt, localized change in the electrical potential difference across a cell membrane. A propagated series of action potentials result in transmission of a nerve impulse along a nerve cell membrane. Also occurs in muscle cells as part of the stimulus to contract.

active site
The specific portion of an enzyme molecule that binds reactants (substrates) during an enzyme-catalyzed reaction. Also called the catalytic site.

active transport
Movement of a substance across a membrane by a carrier molecule in a process that requires energy expenditure by the cell.

ad- [La.]
To, toward, next to, at.

adaptation
(1) An inherited characteristic that increases an organism's fitness for life in its environment. (2) Evolutionary changes that increase organisms' fitness to live and reproduce in their environment.

adenosine diphosphate (ADP)
The diphosphate of the nucleoside adenosine. ADP is a product of ATP hydrolysis. ADP is phosphorylated to produce ATP.

adenosine triphosphate (ATP)
The triphosphate of the nucleoside adenosine. Called the "cellular energy currency" because hydrolysis of ATP to ADP and phosphate makes energy available for energy-requiring processes in cells. Phosphorylation of ADP to produce ATP is a key process in cellular energy conversions such as photosynthesis and cell respiration.

adhesion [La., to stick to]
The force of attraction between unlike substances.

ADP
See adenosine diphospate.

adrenal
[La. *ad; renes:* kidney]
An endocrine gland of vertebrate animals located just anterior to the kidney.

adrenalin
See epinephrine.

aerobic [Gr. *aer:* air]
(1) In the presence of molecular oxygen (O_2). (2) Biological processes that require oxygen or that can occur in the presence of oxygen.

afferent [La. *ad; ferre:* to carry]
Something that leads or carries toward a given point, as an afferent blood vessel. Opposite of efferent.

alga, pl. **algae** [La., seaweed]
Any one of a diverse group of mainly aquatic photosynthesizing organisms that either are unicellular or are multicellular with little structural and functional differentiation. Various algae are classified in two kingdoms and a number of divisions.

alkaline
(1) Having a pH greater than 7.0. (2) Basic.

allantois
[Gr. *allantoeides:* sausage-shaped]
A saclike extraembryonic membrane, connected to the embryonic hindgut, that functions in nitrogenous waste shortage in some vertebrates. Allantoic blood vessels connect the embryo with important extraembryonic exchange areas.

alleles
Two or more alternative forms of a gene that can occur at a particular chromosomal locus.

allopatric
[Gr. *allos:* different; La. *patria:* homeland]
Describes populations that occupy physically separated geographic ranges.

altruism
Behavior that apparently demonstrates unselfish interest in the welfare of others.

amino acid
An organic compound with an amino group (NH_2) and a carboxyl group (COOH) bonded to the same carbon atom. Amino acids are called the "building blocks" of proteins because proteins are linear polymers of amino acids.

ammonia
NH_3, but usually present as an ammonium ion (NH_4^+) at the pH levels found inside animal bodies. Ammonia is a nitrogenous waste product of deamination reactions in animals.

amnion [Gr., lamb]
An extraembryonic membrane that forms as a fluid-filled sac surrounding and enclosing the embryo during development in reptiles, birds, and mammals.

amoeboid [Gr. *amoibē:* change]
Amoebalike; movement of a cell by cytoplasmic flow that produces protrusions called pseudopodia and results in cell shape changes.

amylase [La. *amylum:* starch]
Enzyme that catalyzes the hydrolysis of starch.

an- [Gr.]
Without, not.

anaerobic
[Gr. *an:* without; *aer:* air]
(1) In the absence of molecular oxygen (O_2).
(2) Biological processes that can occur without oxygen.

analogous
Structures of different organisms that are similar in function and general appearance, but differ in evolutionary origins and fundamental structural plans.

anatomy
(1) The gross structure of an organism. (2) The study of gross structure.

angio-, -angium [Gr.]
Vessel, container, case.

annual [La: *annus:* year]
Describes a plant whose entire life, from germination to seed production, is completed in one year. Annual plants die after reproduction or at the end of the growing season.

anterior
[La. *ante:* before, in front of]
The front end of an organism, or toward the front end.

antheridium
A sperm-producing structure in an alga or a plant.

anti- [Gr.]
Against, opposite.

antibiotic
A chemical substance produced by a microorganism that can kill or inhibit growth of other microorganisms.

antibody
A protein produced by the immune system in response to the presence of an antigen, to which the antibody binds specifically. This binding inactivates the antigen or leads to its destruction.

anticoagulant
A substance that inhibits blood clotting.

antigen
A foreign substance, usually a protein or polysaccharide, that stimulates production of specific antibody molecules that bind to the antigen. Cell surface antigens are particularly important in cellular recognition processes.

anus [La., ring]
The posterior opening of the digestive tract, through which feces are expelled.

aorta
(1) A major artery in a circulatory system.
(2) The main systemic artery.

apical
[La. *apic:* apex, summit, tip]
At or near the tip (apex), as of a plant shoot or root tip.

apo- [Gr.]
From, off, away from, different.

aquatic [La. *aqua:* water]
Water, of water, living in water.

aqueous [La. *aqua:* water]
Watery, dissolved in water, pertaining to water.

archegonium
[Gr. *archegonos:* first offspring or generation]
A multicellular, egg-producing organ in a plant.

archenteron
[Gr. *arche:* first; *enteron:* gut]
An early-developing cavity in an animal embryo that becomes the digestive tract.

arteriole
A small branch of an artery that supplies a capillary bed in a body tissue.

artery
A vessel that carries blood away from the heart toward body tissues.

artifact
(1) A product of human activity or intervention.
(2) In science, not an inherent property or part of the process or thing being observed.

-ase
Suffix added to the root name of a compound to identify an enzyme that catalyzes reactions involving the compound, *e.g.,* sucrase, which catalyzes reactions of sucrose.

asexual
A reproductive process that does not involve the union of gametes.

atom
[Gr. *atomos:* cannot be cut, indivisible]
The smallest unit that has the characteristic properties of a chemical element.

atomic number
The number of protons in a particular atomic nucleus.

atomic weight
The average weight of the atoms of an element relative to the weight of atoms of ^{12}carbon (^{12}C), which are arbitrarily assigned the atomic weight of 12.0.

ATP
See adenosine triphosphate.

auto- [Gr.]
Self, same.

autonomic nervous system
A division of the vertebrate nervous system including motor nerves and ganglia that innervate internal organs. Autonomic functions normally are not under voluntary control.

autosome [Gr. *soma:* body]
Any chromosome other than the sex chromosomes.

autotrophic
[Gr. *trophe:* food, nutrition]
Capable of synthesizing all required organic compounds from inorganic substances using an external energy source, most commonly sunlight. Contrast with heterotrophic.

auxin
[Gr. *auxein:* to grow or increase]
One of a class of plant hormones that promote cell elongation and have a variety of other growth-regulating effects.

axis [Gr., an axis, axle]
A line passing through a structure or an organism around which parts are symmetrically arranged.

axon [Gr. *axon:* axis, axle]
Usually the fiber of a nerve cell that carries impulses away from the cell body. Many axons can release a neurotransmitter that carries a message to another nerve cell.

bacillus, pl. **bacilli** [La., small rod]
A rod-shaped bacterium.

bacteriophage
[Gr. *phagein:* to eat]
A virus that infects a bacterial cell.

basal [La., base, foundation]
At or near the base or point of attachment, as of a plant shoot or an animal's limb.

base
(1) A substance that can accept a proton (hydrogen ion, H^+) in a reaction. (2) A substance that dissociates in solution to yield hydroxide ions (OH^-), but not hydrogen ions (H^+). (3) Sometimes used to describe a solution with a pH of more than 7.0.

benthic
[Gr. *benthos:* depths of the sea]
Pertaining to or living at the bottom of a body of water.

bi- [La.]
Two, twice, double.

biennial [La. *annus:* year]
A plant that lives through two growing seasons. Characteristically, biennial plants show only vegetative growth during the first season, then flower and set seed during the second. Biennials die after reproduction or at the end of the second growing season.

bilateral symmetry
Describes a body that has two essentially equal, mirror-image sides. A bilaterally symmetrical body can be cut along one (and only one) plane into similar right and left halves.

bile
The complex secretion of the vertebrate liver that is stored in the gall bladder and carried to the duodenum via the common bile duct. Bile contains sodium bicarbonate, bile salts, and bile pigments.

bio- [Gr. *bios:* life]
Life, living.

biological rhythm
Regular cyclic fluctuation in a physiological process or behavioral activity.

biomass
The total weight of all the living things, or of a selected group of living things, in a particular habitat.

biome
A major terrestrial ecosystem characterized by the climax community type that it supports.

biotic
(1) Living, pertaining to life. (2) In a complex system, the living components.

bivalve
A mollusc that has two shells (valves) hinged together; a member of the class Pelecypoda.

blade
A broad, relatively thin structure such as the expanded portion of a leaf.

blasto- [Gr. *blastos:* bud, sprout]
Embryo, embryonic, part of an embryo.

blastopore
The opening from the interior of the archenteron to the outside of an animal embryo.

blastula
An early stage in development of an animal embryo reached at the end of the period of most active cleavage divisions. In many embryos, the blastula consists of cells organized in a hollow sphere that surrounds a fluid-filled cavity, the blastocoel.

brackish
Describes water that is intermediate in saltiness between fresh water and seawater, such as water in estuaries.

buffer
A chemical system that resists pH change in a solution by binding H^+ ions in response to increases in the H^+ ion concentration, or releasing H^+ ions in response to decreases in the H^+ ion concentration.

caecum [La. *caecus:* blind]
A saclike pouch off a digestive tract; a blind diverticulum.

calcareous
Composed of calcium carbonate or containing quantities of calcium carbonate.

callus
A mass of undifferentiated tissue growing in response to a wound or in a tissue culture.

calorie [La. *calor:* heat]
The energy unit traditionally used by biologists. The amount of energy, in the form of heat, required to raise the temperature of one gram of water 1°C. Also called a gram calorie or small calorie. Often, the kilocalorie (kcal or Calorie, capitalized) is used in nutrition and metabolism studies. A kilocalorie equals 1,000 gram calories. The modern International System of Units recommends replacement of the calorie by the joule as a unit for energy measurement. *See* joule.

cambium [La. *cambialis:* change]
A layer of meristematic tissue in stems and roots of many vascular plants that gives rise to secondary xylem and phloem.

cAMP
See cyclic adenosine monophosphate.

capillaries [La. *capillus:* hair]
The smallest of blood vessels, capillaries form diffuse networks in body tissues. Exchanges between blood and tissue cells occur through capillary walls, which are only one cell thick. Capillaries receive blood from arteries and carry it to veins.

carbohydrate
An organic compound consisting of a chain of carbon atoms with hydrogen and oxygen attached, and having a carbon: hydrogen: oxygen ratio of about 1:2:1. Examples include sugars, starch, cellulose, and glycogen.

carcinogenic
Cancer-causing or cancer-promoting.

cardiac [Gr. *cardia:* heart]
Pertaining to the heart; near or part of the heart.

carnivore
[La. *carni:* flesh; *vorare:* to eat, devour]
An animal that feeds on animals. Contrast with herbivore and omnivore.

cartilage
A vertebrate skeletal tissue that provides support with a degree of flexibility because its intercellular material (matrix) is rubbery, as opposed to the matrix of bone, which is rigid.

catalyst
[Gr. *katalysis:* a dissolving]
A substance that accelerates the rate at which a chemical reaction proceeds toward equilibrium, but itself is not permanently altered by the reaction. Enzymes are catalysts.

caudal [La. *cauda:* tail]
Pertaining to the tail or posterior end of the body; at, near, or toward the posterior end.

cell cycle
The series of events during an entire cycle from the formation of a new cell until that cell has undergone mitosis.

cellulose
A polysaccharide that makes up fibers that form a major part of the rigid cell walls around plant cells. Cellulose is a polymer of β-glucose units.

central nervous system
(1) In vertebrates, the brain and spinal cord. (2) In general, a concentration of nerve cells that exerts a measure of control over the rest of the nervous system.

centri [La. *centrum:* center]
(1) The center. (2) A point.

centriole
A pair of cylindrical, microtubule-containing structures located together near the nucleus of animal cells and cells of some plants. Centrioles are duplicated before cell division, and a pair of centrioles moves to each of the poles of the developing spindle apparatus. Centrioles appear to be identical with the basal bodies of cilia and flagella.

centromere [Gr. *meros:* a part]
A small platelike structure located on a constricted area of each chromosome. Spindle microtubules attach to centromeres during mitosis and meiosis. Also called kinetochore.

cephalo- [Gr. *kephale:* head]
Pertaining to the head, or brain.

cephalothorax
One of the two major body regions (along with the abdomen) in arthropods that do not have the familiar head-thorax-abdomen arrangement seen in insects.

chemoreceptor
A specialized receptor cell that detects and responds to changes in the abundance of a particular chemical substance in its immediate environment.

chemosynthesis
Synthesis of organic compounds from inorganic precursors using energy obtained through oxidation of reduced substances from the environment.

chitin [Gr. *chiton:* a tunic]
A polymer of nitrogen-containing units that are synthesized using glucose as a precursor. Chitin is a major component of the hard exoskeleton of arthropods, such as insects, and many other invertebrates.

chloro- [Gr.]
Green.

chlorophyll [Gr. *phyllon:* leaf]
Any of several similar green pigments that absorb light energy in photosynthesis.

chloroplast
A membrane-bound organelle, found in some eukaryotic cells, that contains chlorophyll and is the site of photosynthesis.

chorion [Gr.]
A covering layer that encloses an embryo. In reptiles, birds, and mammals, the outer membrane that encloses the embryo and the other extraembryonic membranes; also contributes to placenta formation in mammals.

chrom-, -chrome [Gr.]
Color, colored, pigment.

chromatid
One of the two strands of a duplicated chromosome that are joined in the regions of their centromeres.

chromosome
A structure in the cell nucleus (nucleoid in prokaryotic cells) on which gene loci are located.

chyme
The thick, soupy mixture of food and gastric secretions that passes from the stomach into the duodenum.

cilium, pl. **cilia** [La., eyelash, small hair]
A hairlike locomotory structure on the surface of a cell. Characteristically, nine pairs of microtubules are arranged around two central microtubules in each cilium.

class
A taxonomic category between phylum or division and order. Usually, a phylum or division includes several classes, and a class includes several orders.

cleavage
The successive cell divisions that convert a zygote into a multicellular embryo.

cleidoic [Gr., boxlike]
An egg of a terrestrial organism, such as a bird or reptile, with a protective shell that encloses all nutrients and water required by the developing embryo as well as a provision for waste storage so that only gas exchange with the outside environment is required.

climax community
A final, relatively stable stage reached in an ecological succession.

cloaca [La., sewer]
A common exit chamber through which materials from digestive, excretory, and reproductive systems leave the body.

clone [Gr. *klon:* branch or twig]
(1) A group of genetically identical organisms descended by asexual reproduction from a single ancestor. (2) A line of cells descended from a single cell by mitotic cell division.

closed circulatory system
An animal circulatory arrangement in which blood is continuously enclosed within vessels and exchange of materials takes place through capillary walls.

co- [La.]
With, together.

coccus, pl. **cocci** [Gr., a berry]
A spherical bacterium.

codon
The basic unit of genetic coding, consisting of a three-nucleotide sequence in mRNA that specifies a particular amino acid.

coel-, -coel [Gr. *koilos:* hollow]
(1) Hollow. (2) A cavity or chamber.

coelom
A body cavity that is completely lined with mesoderm.

coenzyme
A nonprotein organic molecule that plays a necessary accessory role in an enzyme's catalytic action, often in the transfer of electrons, atoms, or molecules as part of the enzyme-catalyzed reaction.

collagen
A fibrous protein that is very abundant in animal bodies. It forms extracellular fibers that contribute to the flexibility and tensile strength of many tissues.

colon
The large intestine portion of the vertebrate digestive tract.

colony
An aggregated group of organisms living together in close association.

com-, con- [La.]
With, together.

common bile duct
The duct that delivers bile from the liver and gallbladder to the duodenum in vertebrate animals.

community
An ecological unit composed of all the populations of organisms living and interacting in a given area at a particular time.

complete digestive system
A tubular, flow-through digestive tract with an anterior mouth and a posterior exit (the anus) for digestive wastes.

condensation reaction
A reaction that joins two compounds while removing the components of a water molecule. Thus, each condensation reaction produces one water molecule.

conjugation
[La. *conjugatio:* a blending, joining]
A sexual process that occurs in some microorganisms in which genetic material passes from one cell to another through a tubelike cytoplasmic bridge between cells.

connective tissue
A connecting, supporting, or enclosing tissue in animals in which cells are distributed in a relatively extensive intercellular matrix. The matrix material may be rigid as in bone, flexible as in cartilage, or fluid as in blood.

consumer
An organism that obtains energy and nutrients by ingesting producer organisms or other consumer organisms in the ecosystem; that is, animals that eat plants or other animals.

contractile vacuole
An organelle that accumulates dilute fluid and then expels it through a tiny pore in the cell surface, thus ridding the cell of excess water gained from a hypotonic environment.

copulation [La. *copulare:* to join]
Physical joining of two animals during which sperm are transferred from male to female.

cortex [La., bark]
(1) The outer layer, such as the adrenal cortex or cerebral cortex. (2) In plants, the root or stem tissue between the epidermis and the central vascular tissues.

cotyledon [Gr., cup]
A "seed leaf," a leaflike portion of a plant embryo, which in some plants enlarges and functions as a storage site for nutrients that support early growth after seed germination.

countercurrent exchange
An exchange between two streams of fluid, flowing in opposite directions past each other, that allows for very efficient transfer from one to the other, as in gas exchange in fish gills, heat exchange in many vertebrate limb vessels, and salt exchange in vertebrate kidney tubules.

covalent bond
A chemical bond in which a pair of electrons are shared between two atoms.

crop
An anterior enlargement of a digestive tract that functions as a food storage organ.

crossing over
Exchange of segments between chromatids of synapsed chromosomes during meiosis.

crypt- [Gr. *kryptos:* hidden]
Hidden, concealed.

cutaneous [La. *cutis:* skin]
Pertaining to the skin.

cuticle [La. *cutis:* skin]
(1) A waxy layer on the outer surfaces of plant parts. (2) A relatively impermeable covering layer on the bodies of many invertebrate animals.

cyanobacterium [Gr. *cyano:* blue]
Also called blue-green algae. Very widespread, diverse, and abundant group of organisms that are prokaryotic, but resemble algae in general appearance and photosynthetic mechanisms.

cyclic adenosine monophosphate (cyclic AMP or **cAMP)**
Compound synthesized from ATP in a reaction catalyzed by adenylate cyclase that functions as a second messenger inside target cells responding to certain hormones.

cyst
[Gr. *kystis:* hollow place, bag]
(1) A capsule that encloses an organism during an inactive period of its life, usually providing protection against adverse environmental conditions. (2) A saclike abnormal growth.

-cyte, cyto-
[Gr. *kytos:* vessel, container]
Cell.

cytokinesis [Gr. *kinesis:* motion]
Division of the cytoplasm of a cell.

cytoplasm
All of the material in a cell except the nucleus.

de- [La.]
Away, from, off, down, out.

deamination
Removal of an amino group ($-NH_2$) from an organic compound, such as an amino acid.

decarboxylation
Removal of a carboxyl group ($-COOH$) from an organic compound. In cellular metabolism, results in release of carbon dioxide.

deciduous
[La. *decidere:* to fall off]
Plants that shed their leaves each year.

dendr-, dendro-
[Gr. *dendron:* tree]
Tree, treelike, branch, branching.

dendrite
Slender, branched extension that functions as a major receptor area of a nerve cell.

deoxyribonucleic acid (DNA)
A molecule that encodes genetic information in cells. DNA consists of two polynucleotide strands arranged in a double helix. The pentose sugar found in DNA nucleotides is deoxyribose.

derm [Gr., skin]
Skin, covering, layer.

dermis
The deeper layer of the skin of vertebrate animals, under the epidermis.

detritus
Dead plant and animal material; decaying organic matter.

deuterostome
One of a group of animals in which the embryonic blastopore produces the anus. Echinoderms and chordates are deuterostomes. Contrast with protostome.

di- [Gr.]
Two, double, separate, apart.

diapause
[Gr. *dia:* through; *pausis:* a stopping]
An extended period of reduced metabolic activity that is a normal part of many animal life cycles. Diapause occurs in many insects.

diaphragm
The dome-shaped muscular floor of the thoracic cavity of mammals that contracts and flattens during inhalation.

diatom [Gr. *diatomos:* cut in half]
A member of the division Chrysophyta, an extremely abundant group of aquatic algae that are the most important producers in marine ecosystems. A diatom is enclosed by a boxlike pair of silica-containing valves.

dicot
A member of one of the two subclasses of angiosperms, the flowering plants. Dicots possess two cotyledons ("seed leaves") and several other distinguishing features.

differentiation
The developmental process by which unspecialized cells and tissues become structurally and functionally specialized.

diffusion
The tendency for a net movement of particles from an area of higher initial concentration to an area of lower initial concentration due to spontaneous random movements of individual particles resulting from thermal agitation.

digestion
(1) Physical and chemical processes that convert ingested food materials into absorbable forms. (2) Enzyme-catalyzed hydrolysis of complex organic compounds that yields absorbable constituent molecules.

diploid (2N)
The condition in which cells have two of each type of chromosome; twice the number of chromosomes found in gametes that have the haploid (N) number of chromosomes.

disaccharide
A double sugar that yields two monosaccharides upon hydrolysis. Sucrose, maltose, and lactose are common disaccharides.

distal [La. *distare:* to stand apart]
Located away from a point of origin or attachment; farther from a central or reference point. Opposite of proximal.

diverticulum [La. *devertere:* to turn aside]
A sac with only one opening that branches off another structure such as a cavity, duct, or canal. A blind pouch.

division
A major taxonomic grouping applied to plants, autotrophic protists, and fungi that is equivalent to the term "phylum" applied to animals and heterotrophic protists. A division includes several to many classes that share certain basic features and are assumed to have common ancestry.

DNA
See deoxyribonucleic acid.

dormancy [La. *dormire:* to sleep]
A period of suspended activity and growth, usually associated with lowered metabolic rate and increased resistance to environmental stresses.

dorsal [La. *dorsum:* back]
Pertaining to the back, or located at or near the back. Opposite of ventral.

duodenum
[La. *duodecim:* twelve (related to its length, about 12 fingers' breadth)]
The first part of the vertebrate small intestine. Ducts carrying bile and pancreatic juice empty into it.

ecdysis
See molting.

ecosystem
[Gr. *oikos:* household or habitation]
An ecological unit composed of a community of organisms and the physical features of their environment.

ecto- [Gr.]
Outside, out, outer, external.

ectoderm
(1) The outer tissue layer of an animal embryo. (2) The outer primary body layer (germ layer) that gives rise to the epidermis and the nervous system in vertebrate animals.

-ectomy
[Gr. *ek:* out of; *tomein:* to cut]
Removal or excision of a structure, such as adrenalectomy (surgical removal of the adrenal gland).

effector
A part of an organism capable of responding to a stimulus, especially a tissue or organ, such as a muscle or gland capable of responding to a stimulus delivered by the nervous system.

efferent [La. *ex; ferre:* to carry]
Something that leads or carries away from a given point, such as an efferent blood vessel. Opposite of afferent.

egg
(1) A female gamete, usually nonmotile and larger than a male gamete. Eggs usually contain stored nutrients. (2) Sometimes used as a collective name for the egg proper together with accessory structures that surround the egg and the developing embryo, such as the albumen, shell membranes, and shell of birds' eggs.

electron
A negatively charged subatomic particle.

element
A subtance that cannot be broken down into simpler units by chemical reactions; it contains only one kind of atom.

embryo [Gr.]
(1) A plant or animal at an early stage of development. An embryo develops from a zygote. (2) In human development, the name applied during the first eight weeks after fertilization.

end-, endo- [Gr.]
Within, inner, inside.

endergonic [Gr. *ergon:* work]
Energy-requiring or energy-absorbing, as in an endergonic chemical reaction.

endocrine
[Gr. *krinein:* to separate]
Secreting internally. Applied to ductless glands that produce hormones and secrete them directly into the blood.

endocytosis
A process in which extracellular material is enclosed in a vesicle and taken into a cell. Phagocytosis and pinocytosis are forms of endocytosis.

endoderm
(1) The innermost layer of an animal embryo. (2) The inner primary body layer (germ layer) that gives rise to the linings of the digestive tract and its outgrowths, including the respiratory system, in vertebrate animals.

endodermis
A plant tissue that is the innermost layer of the cortex. In roots, endodermis cells' walls fit together tightly to form a barrier, the Casparian strip, between the cortex and the central vascular tissues.

endogenous
Originating inside a cell or body. Opposite of exogenous.

endoplasmic reticulum
[La. *reticulum:* network]
A system of branched, membranous tubules and sacs in the cytoplasm of eukaryotic cells. Sometimes has a rough surface due to a coating of ribosomes.

endoskeleton
An internal skeleton.

endosperm [Gr. *sperma:* seed]
A tissue found in seeds that functions in nutrient storage.

entropy
A measure of the disorder or randomness of a system.

enzyme [Gr. *zyme:* leaven]
A protein molecule that acts as a catalyst in a chemical reaction.

epi- [Gr.]
Upon, over, outer, outside.

epidermis
The outermost layer or layers of cells of a plant or animal body.

epiglottis
The flap of tissue that during swallowing covers the glottis (the opening from the pharynx into the larynx at the top of the trachea) and prevents food from entering the respiratory passage.

epinephrine
A hormone produced by the adrenal medulla that stimulates physiological changes that prepare the organism to respond to emergency and stressful situations. Also called adrenalin.

epithelium
A type of animal tissue that forms flattened covering and lining layers of external body surfaces and internal cavities.

erythrocyte [Gr. *erythros:* red]
A hemoglobin-containing red blood cell.

estivation [La. *estival:* summer]
A resting state with lowered metabolic rate in which some animals can remain inactive for a long period. Estivation usually occurs when the environment is too dry to support normal life activities. Sometimes called "summer sleep."

estuary
An area where fresh water meets the sea; thus, an area with salinity intermediate between fresh water and seawater.

etiolation
The result of growing a plant in darkness. An etiolated plant is pale in color and has a long, spindly stem, abnormal vascular tissue, and few or no leaves except cotyledons.

eu- [Gr.]
True, most typical, good, well.

eukaryotic cell
[Gr. *karyon:* nut or kernel]
A cell possessing a membrane-bounded nucleus and various membrane-bounded organelles in its cytoplasm, and having its DNA complexed with histones. None of these characteristics pertain to the prokaryotic cells of bacteria and cyanobacteria (blue-green algae).

evagination [La. *vagina:* sheath]
(1) Something that is folded or projected outward. (2) The process of becoming folded or projected outward.

eversible
[La. *evertere:* to turn out]
Can be turned inside out; usually something that is protruded at the same time.

evolution [La. *evolutio:* unrolling]
(1) Changes in the gene pool of a population with time. (2) Descent with modification.

ex- [La.]
Out, off, from, beyond.

excretion
The set of processes involved in removal of metabolic wastes, excess materials, and toxic substances.

exergonic [Gr. *ergon:* work]
Energy-releasing, as in an exergonic chemical reaction.

exhalation
Forcing air out of the lungs; breathing out.

exo- [Gr.]
Out, out of, outside, without.

exocytosis
A process in which an intracellular vesicle fuses with the plasma membrane so that the vesicle's contents are released to the cell's exterior.

exogenous
Originating or coming from a source outside a cell or body. Opposite of endogenous.

exoskeleton
An external skeleton that both supports and covers the body.

extra- [La.]
Outside, beyond, more, besides.

extracellular digestion
Digestion in which digestive enzymes are secreted into an extracellular space or a specialized digestive cavity where nutrients are hydrolyzed into absorbable forms.

extraembryonic
Pertains to structures outside the body of an embryo that enclose, protect, and supply the embryo.

extrinsic [La.]
From outside, external to, not a basic part of.

family
A taxonomic category between order and genus. Usually order includes several families, and a family includes several genera.

fauna
All of the animals present in a given area or during a particular period.

feces [La., dregs]
Undigested materials, various wastes, and intestinal bacteria discharged from the digestive tract.

feedback control
An arrangement in which a control mechanism is itself regulated by the function that it controls.

fertilization
(1) The fusion of two haploid gamete nuclei to form a diploid zygote nucleus. (2) More broadly, the complex set of interactions and responses of gametes, including nuclear fusion.

fetus [La., pregnant, fruitful]
(1) A developing vertebrate animal after it has completed embryonic development of major organs and systems and before birth or hatching. (2) A developing human from the ninth week after fertilization until birth.

filament [La. *filum:* a thread]
(1) A long, fibrous structure. (2) A chain of cells.

filter feeder
An animal that feeds on small organisms or particles strained out of the water around it or out of a stream of water that it moves through some part of its body.

fixation
Conversion of a substance into a biologically usable form or incorporation of a substance into a biologically usable compound, as in fixation of atmospheric nitrogen by microorganisms or fixation of carbon dioxide during photosynthesis.

flagellum [La., whip]
A hairlike locomotory structure on the surface of a cell. Longer than a cilium, but possesses the same microtubule arrangement with nine pairs around two central tubules.

flora
All of the plants present in a given area or during a particular period.

food chain
A sequence of organisms through which energy and materials move in an ecosystem. Includes producers and consumers.

free energy
Energy in a system that is available to do work.

fruiting body
A specialized structure produced by a fungus or slime mold on which spores are produced.

gamete [Gr., wife]
(1) A specialized haploid reproductive cell that fuses with another gamete in fertilization to produce a zygote. (2) In many kinds of organisms, an egg or sperm.

gametophyte [Gr. *phyton:* plant]
The haploid, gamete-producing phase (generation) in the typical plant life cycle with alternation of haploid and diploid phases (generations).

ganglion [Gr., knot, swelling]
(1) A mass or cluster of nerve cell bodies. (2) In vertebrate animals, a cluster of nerve cell bodies outside the central nervous system.

gastr-, gastro-
[Gr. *gaster:* stomach, belly]
(1) Pertaining to the stomach. (2) Ventral part or surface.

gastrovascular cavity
A blind, branched digestive cavity that also serves a circulatory (transport) function in animals that lack a circulatory system.

gastrulation
A set of processes that convert a single-layered embryo, the blastula, into a several-layered body with specialized areas.

-gen, -geny
[Gr., produce, bear, birth]
Producing, production.

gene
A hereditary unit located at a particular site (locus) on a chromosome. In Mendelian genetics, a gene determines the nature of a phenotypic character. Biochemically, a gene is a segment of DNA that encodes information for synthesis of a polypeptide or RNA molecule.

gene pool
The sum of all of the alleles of all genes present in a population.

genotype
The genetic constitution of an organism or a cell, as opposed to the phenotype, which is a set of observable characteristics.

genus, pl. **genera** [La., race]
A taxonomic category between family and species; a group of closely related species.

geo- [Gr., the earth]
Pertaining to the earth, especially to the earth's gravity.

geotaxis
A specific animal orientation with respect to gravity.

geotropism
A plant growth response oriented with respect to gravity.

germ cells
Gametes or the cells that will eventually give rise to gametes.

germination
The process by which a plant embryo in a seed or a spore of a microorganism or a plant ends dormancy and resumes normal metabolism and development.

gestation period
[La. *gest:* carried]
The time from fertilization to birth; the length of pregnancy.

gill
Projections of the body surface (or digestive tract) of aquatic animals that are specialized for gas exchange.

gizzard
A digestive organ with thick muscular walls that rub together and grind food.

gland [La. *glans:* an acorn, a gland]
A structure in an animal that is specialized for the production and release of one or several substances that act outside the gland.

glia
Nonneuron cells that are packed among neurons in the central nervous system. Glia are actually more numerous than neurons in the nervous system.

glottis
The opening from the pharynx into the larynx, which lies at the top of the trachea (windpipe).

glucose [Gr. *glykys:* sweet]
One of the six-carbon sugars ($C_6H_{12}O_6$) and a key substrate in cellular energy conversion processes.

glycogen
A polysaccharide that is the principal carbohydrate storage form in animals. Glycogen is a polymer made up of glucose units combined by condensation reactions.

Golgi apparatus
A membranous organelle composed of stacked, saclike cisternae that is involved in packaging and secretion of cell products.

gonad [Gr. *gone:* seed]
(1) A gamete-producing organ in an animal. (2) An ovary or testis.

granum, pl. **grana** [La., grain]
A stack of membranous bags or discs in a chloroplast.

guard cells
Specialized leaf epidermal cells that occur in pairs that surround each stoma and regulate its opening and closing.

habitat [La., to live in]
The kind of surroundings in which individuals of a given animal or plant species normally live.

haploid (N)
The condition in which cells have only one of each type of chromosome; one-half the diploid (2N) number of chromosomes.

helminth [Gr. *helminthos:* worm]
Worm, pertaining to worms, wormlike.

hem-, hema-, hemat-, hemo-
[Gr. *haima:* blood]
Blood, related to blood, filled with blood.

hemoglobin [La. *globus:* ball]
An iron-containing red pigment in blood that functions in oxygen transport.

hemorrhage
A large discharge of blood from a blood vessel, massive bleeding.

hepatic [Gr.]
Pertaining to the liver.

hept-, hepta- [Gr.]
Seven.

herbaceous
[La. *herba:* grass, herb]
A nonwoody plant.

herbivore
[La. *vorare:* to eat, devour]
An animal that feeds on plants. Contrast with carnivore and omnivore.

hermaphroditic
[Gr. *Hermes:* a god; *Aphrodite:* a goddess]
Describes the condition in which both female and male reproductive organs are present in the same individual.

hetero- [Gr. *heteros:* different]
Different, other.

heterogamy [Gr. *gamos:* union, marriage]
Reproduction involving fusion of two gametes that differ in size and structure.

heterotrophic
[Gr. *trophe:* food, nutrition]
Not capable of synthesizing required organic compounds from inorganic substances; requires organic nutrients produced by other organisms. Contrast with autotrophic.

heterozygous [Gr. *zygos:* yoked]
Having different alleles at the corresponding loci of homologous chromosomes. Opposite of homozygous.

hex-, hexa- [Gr.]
Six, as in hexose (a six-carbon sugar).

hist- [Gr. *histos:* web, sheet]
Pertaining to tissue.

homeo-, homo-
[Gr. *homos:* same]
Same, like, similar.

homeostasis
[Gr. *stasis:* standing, posture]
The dynamic steady state maintained within living things.

homeothermic [Gr. *therme:* heat]
Describes animals that maintain a stable body core temperature despite fluctuations in the temperature of their immediate environment.

hominid
A member of the family Hominidae, the family in which modern humans, fossil humans, and very closely related fossil primates are placed.

homologous
(1) Pairs of chromosomes that bear the same gene loci and synapse during prophase of the first meiotic division. (2) In evolution, fundamentally similar structures inherited from a common ancestor.

homozygous [Gr. *zygos:* yoked]
Having identical alleles at the corresponding loci of homologous chromosomes. Opposite of heterozygous.

hormone
[Gr. *hormaein:* to set in motion, to excite]
A chemical regulator substance that is produced in an organ or tissue and diffuses or is transported to target organs or tissues on which it has specific effects.

hybrid [La., mongrel, hybrid]
(1) Offspring of parents that differ in one or more heritable traits. (2) Offspring of parents belonging to two different species or varieties.

hydr-, hydro- [Gr. *hydro:* water]
(1) Water, fluid. (2) Hydrogen.

hydrolysis [Gr. *lysis:* loosening]
Splitting of a compound into parts in a reaction that involves addition of water, with the H^+ ion being incorporated in one fragment and the OH^- ion in the other.

hyper- [Gr.]
Over, above, excessive, more.

hypertonic [Gr. *tonos:* tension]
Describes a solution that tends to gain water osmotically from another solution that is separated from it by a selectively permeable membrane. Tonicity terms (hypertonic, hypotonic, isotonic) usually are used specifically with reference to the effects of solutions surrounding living cells.

hypha [Gr. *hyphe:* web]
One of the filaments that makes up a fungus mycelium.

hypo- [Gr.]
Under, below, deficient, less.

hypophysis
See pituitary.

hypothalamus
[Gr. *thalamos:* inner chamber or room]
A region in the floor of the forebrain that contains reflex centers that regulate the secretion of pituitary hormones and various activities of the autonomic nervous system. It is involved in regulation of body temperature and metabolic rate, water balance, circulation, hunger, thirst, sleep, and other functions and activities.

hypothesis
A supposition that is established by reasoning after consideration of available evidence and that can be tested by obtaining more data, often by experimentation.

hypotonic [Gr. *tonos:* tension]
Describes a solution that tends to lose water osmotically to another solution that is separated from it by a selectively permeable membrane. (*See* note under *hypertonic* regarding usage of tonicity terms.)

ichthy- [Gr. *ichthyos:* fish]
Pertaining to fish; sometimes fishlike.

immunity
Result of a complex set of responses through which an animal exposed to a foreign antigen, particularly an antigen associated with a pathogenic microorganism, develops capabilities to make effective responses to that antigen and thus is resistant to damage or disease that might be caused by subsequent exposure to it.

immunoglobulin
One of a class of globular proteins present in blood, tissue fluids, and various body secretions. Antibodies are immunoglobulins.

implantation
The entry of a mammalian embryo into the wall of the uterus.

incomplete digestive system
A gut that is a blind sac and that has only one opening that must serve as both entrance for food and exit for undigested residues.

inf-, infra- [La.]
Below, under.

inflammation
A tissue and blood vessel response to infection or injury in which fluid, plasma proteins, and phagocytic cells move out of blood vessels into interstitial spaces, making the inflamed tissue appear swollen.

ingestion
Taking in food from the outside environment; eating.

inhalation
Drawing air into the lungs; breathing in.

inorganic
Not organic; not a carbon compound (other than carbon dioxide).

in situ [La., in place]
In its original, natural, or proper place.

integument
[La. *integere:* to cover]
A covering, skin, outer surface.

inter- [La.]
Between, among.

interferon
A substance produced and released by cells infected by viruses that protects other cells from virus infection.

interneuron
A neuron that connects two other neurons, such as a sensory neuron and a motor neuron or two other interneurons. Complex brains contain huge numbers of interneurons arranged in very complex structural and functional networks.

internode [La. *nodus:* knot]
A length of stem between nodes (the attachment points of leaves or buds).

interstitial fluid
The extracellular fluid in the spaces around body cells.

intra- [La.]
Within, inside.

intracellular digestion
Digestion in which nutrients are taken into a cell by phagocytosis or pinocytosis and enclosed in a vesicle with which an enzyme-containing lysosome fuses. Digestion products are absorbed through the vesicle membrane.

intrinsic [La.]
Contained within, inherent in, a basic part of.

invagination [La. *vagina:* sheath]
(1) Something that is folded or projected inward. (2) The process of becoming folded or projected inward.

invertebrate
[La. *in:* without; *vertebra:* joint, vertebra]
Any animal not possessing a vertebral column. *See* vertebrate.

in vitro [La., in glass]
In a laboratory culture, in a laboratory vessel, not in place in an intact organism.

in vivo [La., in the living]
In the living thing, in an intact organism.

ion
An atom or molecule that has an electrical charge because of the loss or gain of one or more electrons.

irritability
Responsiveness to environmental stimuli. Irritability is especially well developed in nerve cells and specialized sensory receptor cells.

iso- [Gr.]
Equal.

isogamy
[Gr. *gamos:* union, marriage]
Reproduction involving fusion of two gametes that are alike in size and structure.

isotonic [Gr. *tonos:* tension]
Describes a solution that tends neither to gain nor to lose water osmotically to another solution that is separated from it by a selectively permeable membrane. (*See* note under *hypertonic* regarding usage of tonicity terms.)

isotope [Gr. *topos:* place]
Atom of an element that differs from another atom of that element in the number of neutrons in its nucleus. Different isotopes have the same atomic number but different mass numbers. Some isotopes are unstable (radioactive) and tend to decay by emission of radiation.

joule
The standard basic unit of energy measurement. 4.186 joules = 1 calorie, and 1 joule = 10^7 ergs.

juxta- [La.]
Near to, next to.

keratin [Gr. *keratos:* horn]
A tough, horny, water-insoluble protein produced by epidermal tissues of vertebrate animals. Especially abundant in nails, claws, hair, feathers, beaks, and hooves.

kilo- [Gr.]
A thousand.

lamella [La.]
A thin, platelike structure, thin sheet, layer.

larva [La., ghost]
An immature form of an animal that is very different in appearance from the mature adult.

lateral [La. *later:* the side]
Located away from a midline. Opposite of medial.

leuk-, leuko- [Gr. *leukos:* white]
White or clear.

leukocyte
A white blood cell.

lichen [Gr. *leichen:* a tree moss]
An intimate symbiotic association between an alga and a fungus.

ligament
A piece of tough connective tissue that links one bone to another in a joint.

lignin [La. *lignum:* wood]
A complex organic polymer produced by some plant cells that adds to the hardness of their walls. Wood is heavily lignified xylem tissue.

linkage
The tendency of certain genes to assort together because their loci are close together on the same chromosome. This results in linked inheritance of phenotypic traits determined by those genes.

lip-, lipo- [Gr. *lipos:* fat]
Fat, fatlike, fat-containing.

lipase
An enzyme that catalyzes the hydrolysis of fat molecules to fatty acids and glycerol.

lipid
One of a group of organic compounds that are insoluble in water but soluble in hydrophobic solvents. Includes fats, steroids, phospholipids, oils, and waxes.

locus, pl. **loci** [La., place]
A specific part of a chromosome that is the site of a particular gene. Any one of the alleles of that gene may occupy its locus.

-logy [Gr. *logos*]
Study, study of, discourse.

longitudinal
Along the length, lengthwise, as in a longitudinal section of a body or structure.

lumen [La., light, opening]
The cavity of space within a tube or sac.

-lysis, lyso- [Gr. *lysis:* loosening]
Loosening, disintegration.

lysosome [Gr. *soma:* a body]
A membrane-bound organelle containing hydrolytic enzymes.

macro- [Gr.]
Large, long.

macrophage
A large phagocytic cell in a body tissue outside the circulatory system. Macrophages develop from circulating monocytes that leave blood vessels and remain in a tissue.

marine [La. *mari:* sea]
Living in or pertaining to the sea or ocean.

marsupial [Gr. *marsypion:* little bag]
A pouched mammal; member of the subclass Metatheria.

matrix [La. *mater:* mother]
The material in which something is embedded or enclosed, or the homogeneous material in spaces among formed elements, as the extracellular matrix of bone or cartilage, or the matrix among membranous elements in a mitochondrion.

medial [La. *medi:* the middle]
Located at or toward a midline. Opposite of *lateral.*

medulla [La., marrow, pith]
The inner portion of a structure or organ, such as the adrenal medulla.

mega- [Gr.]
Large, great.

megaspore
A plant spore that will develop into female gametophyte (megametophyte).

meiosis [Gr., diminution]
A kind of nuclear division, usually as part of two successive cell divisions, that produces nuclei with half the number of chromosomes in the original nucleus.

membrane potential
An electrical charge difference between the inside and outside of a cell; thus, an electrical potential across the plasma membrane.

meristem [Gr. *meristos:* divisible]
A plant tissue made up of undifferentiated cells that function in production of new cells by mitosis.

meso- [Gr.]
Middle.

mesoderm
(1) The middle tissue layer of an animal embryo lying between ectoderm and endoderm. (2) The middle primary body layer (germ layer) that gives rise to skeletal, muscular, and circulatory systems as well as the lining and covering layers in the body cavities of vertebrate animals.

messenger RNA (mRNA)
Single-stranded RNA molecule that has a nucleotide sequence complementary to that of a segment of DNA and specifies amino acid sequence in polypeptides being assembled during translation. Thus, mRNA is a "messenger" from the genome to the cytoplasm.

meta- [Gr.]
(1) Change. (2) After, posterior.

metabolism
[Gr. *metabol:* change]
(1) The sum of chemical reactions within a cell, or the sum of all cellular activities in an organism. (2) Sometimes used in collective names for particular sets of biochemical reactions, as in amino acid metabolism, lipid metabolism, and nitrogen metabolism.

metamorphosis
[Gr. *morpho:* form, shape]
A change in body organization that transforms an immature animal (usually a larva) into an adult.

micro- [La.]
(1) Small. (2) In units of measurement, one millionth.

microbe [Gr. *bios:* life]
A microscopic organism, especially a bacterium.

microfilament
Minute protein threads present in many cells that function in cell shape changes, in the constriction that splits cells during cell division, and in movements of granules within cells.

microspore
A plant spore that will develop into a male gametophyte. In gymnosperms and angiosperms, microspores develop into pollen grains.

microtrabecular lattice
A complex meshwork in the cytoplasmic matrix of cells made up of microtubules, microfilaments, and other filamentous structures.

microtubule
Thin, hollow cylinders in the cytoplasm that are made up of spherical subunits. Microtubules function in chromosome movement during mitosis, and in the beating of cilia and flagella.

milli- [La.]
One thousandth.

mimicry
An organism presenting a showy appearance that resembles another organism, often one that is not attractive to predators.

mineral [La. *minera:* mine]
An inorganic material.

mitochondrion, pl. **mitochondria**
[Gr. *mitos:* thread; *chondros:* grain (or cartilage)]
A double-membraned organelle in which aerobic respiration occurs in eukaryotic cells. The site of Krebs cycle reactions and the electron transport system.

mitosis [Gr. *mitos:* thread]
Nuclear division that results in distribution of replicated chromosomes so that each of two daughter nuclei receives a set of chromosomes identical to that of the original nucleus.

molecule
A combination of like or different atoms bonded together; the smallest characteristic unit of a compound.

molting
Shedding and replacement of an outer body covering. Used especially in connection with the periodic shedding of the exoskeleton that is necessary for growth of arthropods.

mono- [Gr.]
One, single.

monocot
A member of one of the two subclasses of the angiosperms, the flowering plants. Monocots possess one cotyledon ("seed leaf") and several other distinguishing features.

monomer [Gr. *meris:* part]
A relatively small molecule that is linked to other molecules to form a polymer.

-morph, morpho- [Gr.]
Form, shape, structure.

morphogenesis
[Gr. *genesis:* origin]
The development of shape and form of an organism.

morphology
[Gr. *logos:* a word, discourse, study of]
(1) The structure of organisms or their parts.
(2) The study of structure and form.

motor neuron
A neuron that transmits nerve impulses from the central nervous system toward an effector, such as muscle.

muscle
[La. *musculus:* a little mouse, muscle]
A contractile tissue of animals that contains cells capable of exerting force by shortening when they are appropriately stimulated.

mutation [La. *muta:* change]
A stable, heritable change in the genetic material.

mutualism
A symbiotic relationship in which both organisms benefit from the association.

mycelium [Gr. *mykes:* fungus]
The mass of tubular filaments (hyphae) that constitute a fungus.

myo- [Gr. *mys:* muscle, mouse]
Muscle, having to do with muscle.

NAD
See nicotinamide adenine dinucleotide.

NADP
See nicotinamide adenine dinucleotide phosphate.

nano- [La. *nanus:* dwarf]
One thousand-millionth part (10^{-9}).

nasal [La., nose]
Pertaining to the nose or air passages through the nose.

necrotic [Gr. *necros:* death]
Having dead and degenerating cells.

-nema, -neme, nemato-
[Gr. *nema:* thread]
Threadlike, filamentous.

neo- [Gr.]
New, recent.

neonate [La. *nata:* birth, be born]
A newborn (or newly hatched) animal, especially a newborn mammal.

nephr-, nephro-
[Gr. *nephros:* kidney]
Kidney, pertaining to a kidney.

nephron
An individual functional unit of a vertebrate kidney. Each kidney contains many nephrons.

nerve [La. *nervus:* nerve, tendon]
A bundle of neuron fibers outside the central nervous system.

nerve impulse
A chain reaction series of action potentials conducted along the membrane of a neuron.

neur-, neuro- [Gr., nerve]
Having to do with nerve cells, nerves, or the nervous system.

neuron
A nerve cell.

neurosecretion
(1) A substance produced by a specialized nerve cell (neurosecretory cell) that is released in the blood and acts on target cells elsewhere in the body. (2) The process of releasing a neurosecretion.

neurotransmitter
A substance that is released by one neuron, crosses a synaptic cleft, binds with a receptor molecule, and causes a specific ion conductance change in the membrane of a second neuron.

neutron
A subatomic particle, found in the atomic nucleus, that is electrically uncharged and has approximately the same mass as a proton.

niche
An ecological description of the structural and functional role of a particular species in its natural community.

nicotinamide adenine dinucleotide (NAD)
An organic compound that functions as an electron acceptor or donor in various cellular oxidation-reduction reactions.

nicotinamide adenine dinucleotide phosphate (NADP)
Similar in function to nicotinamide adenine dinucleotide, but is structurally different because of presence of an additional phosphate group.

nitrogen fixation
The incorporation of nitrogen from the air into forms that can be used by organisms.

node [La. *nodus:* knot]
The point of attachment of a leaf or a bud on a plant's stem.

-nomy [Gr.]
The science of.

nucleic acid
One of several types of large organic molecules that are polymers of nucelotides and function in heredity and genetic expression in cells. The principal nucleic acids are deoxyribonucleic (DNA) and ribonucleic acid (RNA).

nucleoid
A region of a prokaryotic cell where the chromosome is located. Not membrane-bounded as in eukaryotic cells.

nucleolus [La., a little nut, a kernel]
A dense body, observable within the nucleus of a eukaryotic cell when it is not dividing, that is involved in synthesis of rRNA.

nucleosome
A basic structural unit of chromosomes in eukaryotic cells, consisting of an aggregate of eight histone molecules with a specific length of DNA wrapped in a helical coil around it.

nucleotide
An organic compound consisting of a nitrogen-containing base (either a purine or a pyrimidine), a five-carbon sugar, and phosphoric acid. Nucleic acids are polymers of nucleotides.

nucleus [La., a little nut, a kernel]
(1) A membrane-bounded spherical body that contains the chromosomes of a eukaryotic cell. (2) The central part of an atom. (3) A cluster of nerve cell bodies within the central nervous system.

nymph
[Gr. *nympha:* a bride, nymph]
A young insect that generally resembles its parents at hatching but lacks some adult characteristics, which it gains as it undergoes gradual metamorphosis through a series of molts.

oct-, octi-, octo- [La., Gr.]
Eight.

olfactory [La. *olfactere:* to smell]
Pertaining to the sense of smell.

olig-, oligo- [Gr.]
Few, small.

omnivore
[La. *omnis:* all; *vorare:* to eat, devour]
An animal that feeds on both plants and animals. Contrast with carnivore and herbivore.

ontogeny
[Gr. *onto:* being, existence; *genesis:* origin]
The developmental history of an individual organism.

oo- [Gr. *oion:* egg]
Egg or pertaining to an egg or its development.

oocyte
A cell that will undergo meiosis and give rise to an egg. Also, a prospective egg cell during meiosis.

oogamy
[Gr. *gamos:* union, marriage]
A type of sexual reproduction in which one of the gametes is a large, nonmotile egg. Oogamy is a form of heterogamy.

oogonium
An egg-producing cell in an alga.

open circulatory system
An animal circulatory arrangement in which blood leaves the confines of vessels and circulates in spaces among body organs and tissues before returning to vessels leading to the heart.

operculum [La., cover, lid]
A platelike structure that covers or encloses other structures, as the operculum covering the gills of a bony fish or the operculum covering the capsule of a moss sporophyte.

oral [La. *oris:* mouth]
Pertaining to the mouth.

order
A taxonomic category between class and family. Usually, a class includes several orders, and an order includes several families.

organ [Gr. *organon:* tool]
A body part specialized for a particular function or set of functions, usually consisting of several types of tissues.

organelle [La. *-ell:* small]
A distinctive intracellular structure specialized for a particular function.

organic
(1) A chemical compound that contains carbon. (2) Something produced by or derived from living things. (3) Pertaining generally to living things.

organism
A single cell or a multicellular aggregate that constitutes an individual living thing.

osmosis [Gr. *osmos:* thrust, push]
The net movement of a solvent such as water through a selectively permeable membrane.

osmotic pressure
The pressure that must be exerted to halt net water movement across a truly selectively permeable membrane that separates solutions having different solute concentrations. Usually measured with distilled water on one side of the membrane. Thus, a measure of the tendency of water to move osmotically.

osteo- [Gr. *osteon:* bone]
Pertaining to bone.

ov-, ovi- [La. *ovum:* egg]
Egg, or pertaining to an egg.

ovary
An egg-producing organ. Eggs are released from animal ovaries, but zygotes begin their development within plant ovaries that are converted into fruits in many flowering plants.

oviduct
A tube that transports eggs away from the ovary.

ovulation
Release of an egg from the ovary.

ovum, pl. **ova**
An egg cell, a female gamete.

oxidation
A chemical reaction that results in removal of one or more electrons from an atom, ion, or compound. Oxidation of one substance occurs simultaneously with reduction of another.

pale-, paleo- [Gr.]
Ancient, old.

paleontology
The study of fossils and the fossil record of life in past geologic times.

papilla [La., nipple]
A small protuberance; a nipplelike projection.

para- [Gr.]
Beside, near, beyond.

parasit- [Gr.]
(1) Near food. (2) To eat with another or at another's table. (3) A parasite.

parasitism
A symbiotic relationship in which one member benefits at the expense of the other.

parasympathetic nervous system
One of the two divisions of the autonomic (visceral motor) nervous system. Acts antagonistically to the sympathetic division in control of various effectors.

parenchyma
A plant tissue made up of loosely packed, relatively unspecialized cells.

parthenogenesis
[Gr. *parthenos:* virgin]
Development of an egg without fertilization.

pathogenic [Gr. *pathos:* suffering]
Disease-causing.

pelagic [Gr. *pelagos:* ocean]
Oceanic, pertaining to the open sea.

pent-, penta- [Gr.]
Five, as in pentaploid (possessing five sets of chromosomes).

peptide bond
A bond formed between two amino acids in a condensation reaction between the amino group of one and the carboxyl group of the other. Peptide bonds link amino acids in polypeptides.

perennial
[La. *per:* through; *annus:* year]
Describes a plant that lives throughout the year and grows during several to many growing seasons.

peri- [Gr.]
Around, surrounding.

peristalsis
[Gr. *stalsis:* constriction, contraction]
Rhythmic waves of contraction and relaxation that pass along the walls of hollow, tubular structures and function to move their contents along.

peritoneum
A very smooth membrane, of mesodermal origin, lining a wall of the coelomic cavity or covering the surface of an organ.

permeable
[La. *permeare:* to pass through]
Permitting a substance or substances to pass through. Usually pertains to properties of membranes in biology.

pH
Symbol for the negative logarithm of the hydrogen ion [H^+] concentration. pH values are measures of acidity in solutions and range from 0 to 14. pH 7 is neutral, less than 7 is acidic, and more than 7 is basic.

phage
See bacteriophage.

phagocytosis
[Gr. *phagein:* to eat]
Engulfment and intake of solid particles by cells.

pharyngeal
Pertaining to the pharynx.

pharynx
Portion of a digestive tract between the mouth cavity and the esophagus. Also associated with gas movement and exchange functions in vertebrate animals.

phenotype
[Gr. *phainein:* to show]
A set of observable characteristics of an organism or a cell, as opposed to the genotype, which is its genetic constitution.

pheromone
[Gr. *pherein:* to carry, bear]
A substance secreted and released into the environment by one organism that evokes behavioral, reproductive, or developmental responses in other individuals of the same species.

-phil, phili-, philo-
[Gr., love, loving]
Attraction, positive response.

phloem [Gr. *phlois:* bark]
A vascular tissue in plants that transports (translocates) dissolved material from place to place. Phloem may move material either up or down stems and roots.

-phob, -phobia [Gr., fear, dread]
Avoidance, negative response.

-phore [Gr. *pherein:* to carry]
Carrier.

phospholipid
One of a number of phosphate-containing lipids that are important constituents of cell membranes.

phosphorylation
Addition of one or more phosphate groups to a molecule.

photochemical reaction
A chemical reaction that involves absorption or release of light energy.

photoperiodism [Gr. *photos:* light]
Physiological responses of organisms to the lengths of light and dark periods in the twenty-four hour daily cycle, especially to changes in relative lengths of the two.

photosynthesis
Synthesis of organic compounds from inorganic compounds (commonly CO_2 and water) using light energy absorbed by chlorophyll.

phototropism
[Gr. *tropos:* a turning, change]
A movement or turning in response to a directional light source.

-phyll [Gr. *phyllon:* leaf]
Leaf, of a leaf, relating to leaves.

phylogeny [Gr. *phylon:* tribe]
The evolutionary history of a group of organisms.

phylum [Gr. *phylon:* tribe]
A major taxonomic grouping of organisms, usually including several to many classes of organisms that share certain basic characteristics and are assumed to have common ancestry.

physiology [Gr. *physis:* nature]
(1) Life functions and processes of cells, tissues, organs, and organisms. (2) The study of the functioning of one or more of these units.

-phyte, phyto- [Gr. *phyton:* plant]
Plant, pertaining to a plant or plants.

phytochrome
A plant pigment that exists in two different forms and is converted from one to the other in reversible reactions involving absorption of red or far-red light energy.

phytoplankton
[Gr. *planktos:* wandering]
Microscopic, free-floating, photosynthetic organisms that function as major producers in freshwater and marine ecosystems.

pigment [La. *pigmentum:* paint]
A colored substance that absorbs light energy of a particular set of wavelengths.

pinocytosis [Gr. *pinein:* to drink]
Engulfment and intake of small quantities of fluid by cells.

pistil [La. *pistillus:* pestle]
The portion of a flower that contains ovules, which are the sites of megaspore production and subsequent female gametophyte development. The pistil consists of stigma, style, and ovary.

pituitary
[La. *pituitarius:* secreting phlegm]
A small gland situated below, and attached to, the hypothalamus. The anterior lobe secretes several hormones that regulate functions of other endocrine glands. Other anterior lobe hormones and the hormones released in the posterior lobe regulate nonendocrine target cells. A functional intermediate lobe is present in some vertebrates.

placenta
[Gr. *plax:* flat object or surface]
An organ formed during mammalian development, made up of tissues of both mother and infant, within which materials are exchanged through a tissue barrier that separates elements of the two circulations.

plankton
[Gr. *planktos:* wandering]
Free-floating organisms, most of which are microscopic, found in freshwater and marine ecosystems. Includes both autotrophic (photosynthetic phytoplankton) and heterotrophic (zooplankton) organisms.

plasm-, plasmo-, -plasm
[Gr. *plasma:* something molded, modeled, or shaped]
Plasma; cytoplasm or part of cytoplasm; formed material.

plasma
The clear, fluid portion of vertebrate blood; that is, blood minus formed elements (cells and platelets).

plasma membrane
The outer membrane of a cell that forms the boundary between the cell and its environment.

plasmid
A relatively small, circular DNA molecule in a bacterial cell that replicates independently of the cell's chromosome and that also can be transferred to another bacterial cell.

plastid
A plant cell organelle that functions in synthesis (chloroplast) or nutrient storage (leucoplast).

pleura [Gr., the side, a rib]
The smooth membranes that line the thoracic cavity and cover lung surfaces.

pneumonia
A condition, caused by an infection, in which fluid and dead white blood cells accumulate in the alveoli of the lungs and interfere with gas exchange.

-pod, -podium
[Gr. *pod:* foot; *podion:* little foot]
A foot, leg, extension.

poikilothermic
Having an environmentally influenced, variable body core temperature; not capable of precise metabolic temperature regulation.

polarity
Having parts or ends that have contrasting properties. In molecules, having negative charge at one end, positive at the other. In organisms, having head and tail, or base and apex.

pollen [La., fine flour, dust]
Small male gametophytes, developed from microspores, that are released and carried by wind or animal pollinators.

pollination
Transfer of pollen to the stigma of a receptive flower. Pollination is not synonymous with fertilization, which occurs only after further development of the male gametophyte.

poly- [Gr.]
Many, much.

polymer [Gr. *meris:* part]
A large molecular chain of smaller molecular subunits (monomers) linked together.

polypeptide
A molecule composed of many amino acids linked together by peptide bonds.

polyploid
Having more than two complete sets of chromosomes per nucleus.

polysaccharide
A large carbohydrate molecule; that is, a polymer of single sugar monomers linked together by means of condensation reactions. Starch, glycogen, and cellulose are polysaccharides, being polymers of glucose.

polytene chromosome
A chromosome in which there has been repeated replication without mitosis, thus forming giant chromosomes.

population
A group of individuals of the same species that interbreed and occupy a given area at the same time.

portal system
[La. *porta:* gate, door]
A portion of the circulatory system that begins in a capillary bed and leads not to vessels that carry blood directly toward the heart, but rather to a second capillary bed.

posterior
[La. *post:* hinder, posterior]
The rear end of an organism, or toward the rear end.

precursor
Something that comes before or precedes. Chemically, a reactant that will be altered or incorporated into another form or compound.

predator
[La. *praedatio:* plundering]
A free-living organism that feeds on other organisms.

prey
An organism hunted for and eaten by a predator.

primate
A member of the order Primates, the order of mammals that includes monkeys, apes, and humans.

primitive [La. *primus:* first]
(1) Unspecialized, at an early stage of development or evolution. (2) Old, ancient, resembling an ancestral condition.

pro- [Gr.]
Before, in front of, forward.

proboscis [Gr. *boskein:* to feed]
A tubular extension of the head or snout of an animal, usually used in feeding, sometimes as a sucking tube.

producer
An organism that synthesizes organic compounds using only inorganic materials and energy from an external source, usually sunlight used in photosynthesis.

prohormone
A molecule that can be converted into an active hormone form.

prokaryotic cell
[Gr. *karyon:* nut or kernel]
A cell that lacks a membrane-enclosed nucleus and membrane-bounded organelles in its cytoplasm. Bacteria and cyanobacteria (blue-green algae) are prokaryotic cells.

protease
A general name for a proteolytic enzyme, an enzyme that catalyzes hydrolysis of peptide bonds in proteins. Thus, a protein-digesting enzyme.

protein
A large, complex organic molecule composed of one or several polypeptides.

proteolytic
Protein-hydrolyzing, as in proteolytic enzymes.

proto- [Gr.]
First, original, primary.

proton
A positively charged subatomic particle found in the atomic nucleus.

protonema [Gr. *nema:* a thread]
A simple, end-to-end chain of cells in an alga, or in the early development of a moss or fern gametophyte.

protostome
One of a group of animals in which the mouth develops from the blastopore of the embryo. Molluscs, annelids, and arthropods are protostomes. Compare with deuterostome.

proximal [La. *proxim:* nearest]
Located near a point of origin or attachment, nearer a central or reference point. Opposite of distal.

pseudo- [Gr., false]
False, substituting for, temporary.

pseudocoelom
A body cavity between mesoderm and endoderm; therefore, not completely lined with mesoderm as is a true coelom.

pseudopod, pseudopodium
A temporary membrane extension and cytoplasmic protrusion of an amoeboid cell, which functions in locomotion and phagocytosis.

pulmonary [La. *pulmonis:* lung]
Relating to the gas exchange organs, especially to lungs, as in pulmonary arteries and veins.

pupa, pl. **pupae** [La., doll]
A developmental stage in insects between larval and adult stages. Pupae are immobile (sometimes enclosed in cases) and undergo extensive body reorganization.

purine
A double-ringed nitrogenous base, such as adenine or guanine, that is a component of nucleotides and nucleic acids.

pyrimidine
A single-ringed nitrogenous base, such as cytosine, thymine, or uracil, that is a component of nucleotides and nucleic acids.

pyruvic acid
A three-carbon compound that is the product of the Embden-Meyerhof pathway of reactions.

quantum
A unit of electromagnetic energy. The energy of a quantum is inversely proportional to the wavelength of the radiation.

radi- [La.]
A spoke, ray, radius.

radial symmetry
Describes a body that is arranged around a central axis so that cutting it lengthwise along any plane that runs along that axis results in separation of two similar halves. Contrast with bilateral symmetry.

radiation
Energy transmitted as electromagnetic waves.

radioactive isotope
An unstable isotope that tends to decay by emitting radiation.

reduction
A chemical reaction that results in addition of one or more electrons to an atom, ion, or compound. Reduction of one substance occurs simultaneously with oxidation of another.

reflex [La. *reflexus:* bent back]
An automatic activation of an effector in response to a stimulus detected by a receptor. A neural reflex depends on a functional unit, including sensory and motor elements, and usually one or more interneurons that connect them.

refractory period
A period of time after a response or action during which a cell, tissue, organ, or organism does not show its characteristic response to a stimulus or cannot carry out a normal function.

regeneration
Replacement of a lost structure or body part.

renal [La. *renes:* kidneys]
Pertaining to the kidney.

replication
Of DNA, when the two strands of DNA molecule separate and a complementary strand is assembled for each original strand so that two DNA molecules identical to the original are produced.

reticulum [La.]
A fine network.

rhizoid [Gr. *rhiza:* root]
A colorless, hairlike extension of a plant or fungus that functions in nutrient absorption.

rhizome
[Gr. *rhizoma:* mass of roots]
An underground stem in some plants that gives rise to aboveground leaves. Some rhizomes serve as storage organs or in vegetative reproduction of the plants.

ribonucleic acid (RNA)
A class of nucleic acids that contain the pentose sugar ribose (not deoxyribose as in DNA) and the pyrimidine uracil (not found in DNA) and are involved in the several steps of cellular genetic expression.

ribosome
A minute granule assembled from two subunits composed of RNA and proteins that functions as a site of translation during protein synthesis.

RNA
See ribonucleic acid.

rudiment
[La. *rudis:* unformed, rough]
The first visible evidence of development of a structure or organ in an embryo.

rudimentary
[La. *rudis:* unformed, rough]
Incompletely developed; having a primitive form.

rumen
An enlarged, saclike area of the stomach of cattle and other ruminant mammals that houses symbiotic microorganisms that produce cellulase, an enzyme that catalyzes hydrolysis of cellulose in plants eaten by the ruminants.

salinity
[La. *salin:* salt pit, salt, salty]
Saltiness; measure of concentration of dissolved salts.

saprophyte
A heterotrophic bacterium or a fungus that absorbs nutrients directly from dead and decaying organic material.

sarcomere
[Gr. *sarx:* flesh; *meris:* part]
A basic contractile unit in a skeletal muscle fiber; the portion of the fiber between two Z lines.

savanna
A grassland with occasional trees or clumps of trees.

secretion
(1) Release of a cell product being "exported" from a cell. (2) Simultaneous secretion by many cells, as in glandular secretion.

section
(1) To cut apart or across. (2) A slice or a representation of a slice cut from a body or structure, such as a cross (transverse) section, which is a section cut at a right angle to the long axis of a body or structure.

sedentary
Relatively inactive, tending to stay or sit in one place.

seed
A plant embryo, together with a food reserve, enclosed in tough, protective seed coats that are derived from part of the ovule.

segmentation
Arrangement of an organism in a series of similar units or segments.

sensory neuron
A neuron that either responds directly to a stimulus or to a change in a specialized receptor cell and carries impulses toward the central nervous system.

septum, pl. **septa** [La., fence]
A dividing wall or membrane, a partition.

serum
[La., whey, watery part of a fluid]
Blood plasma minus the proteins involved in the clotting process.

sessile [La.]
Attached to a surface, sedentary, not free to move around.

sex chromosome
One of the chromosome pair that differs between females and males in animals that have a chromosomal sex determination mechanism. *See* X chromosome and Y chromosome.

sex-linked
Genes that are located on a sex chromosome, mostly on the X chromosome, but a very few on the Y chromosome.

sexual dimorphism
Describes the condition in which males and females of a species show clear and consistent differences in size or other general body characteristics.

shoot
Aboveground portions of a vascular plant; a stem with its branches and leaves.

sieve tube
A linear array of specialized cells (elements) running vertically through the phloem that functions in translocation of solutes.

sinus
[La., a fold, a hollow; curve, bend]
(1) A hollow space within a structure. (2) A large passage or channel in the circulatory system.

siphon
A tubular structure through which fluids are drawn in or expelled.

solute
A substance that is dissolved (uniformly dispersed) in another substance (the solvent).

solution
A homogeneous mixture in which one or more solutes are dissolved in a solvent. Water is the solvent in the great majority of biological solutions.

solvent
A medium in which a solute or several solutes are dissolved.

-soma, somat-, -some
[Gr. *soma:* body]
Body, entity, unit.

somatic cells
All of the cells of the body except the germ cells.

sorus, pl. **sori** [Gr., a heap]
A cluster of sporangia on a leaf, such as a fern leaf.

species
A population of morphologically similar organisms that can reproduce sexually among themselves but are reproductively isolated from other organisms. Among microorganisms, a group of organisms that share an extensive set of common characteristics.

sperm [Gr. *sperma:* seed]
A male gamete, usually motile and smaller than a female gamete.

sphincter [Gr., band]
A circular muscle that forms a ring around a tubular structure and can close it by contracting.

spindle
The spindle, or spindle apparatus, is a set of microtubules involved in chromosome movements during mitosis or meiosis.

spirillum, pl. **spirilla**
[La., little coil]
A helical or spiral-shaped bacterium.

sporangium
A structure within which spores are produced.

spore [Gr. *sporos:* seed]
(1) A reproductive cell that can develop directly into an organism. (2) A specialized bacterial structure that has a low metabolic rate, is protected by a thick capsule, can survive adverse conditions, and germinate under favorable conditions to produce a new cell.

sporophyll [Gr. *phyllon:* leaf]
A sporangium-bearing leaf. Flowers contain highly specialized and modified sporophylls.

sporophyte
The diploid phase (generation), which produces haploid spores by meiosis, in the typical plant life cycle with alternation of haploid and diploid phases (generations).

stamen [La., thread, fiber]
A portion of a flower consisting of a filament topped by an anther containing microsporangia within which are produced microspores that develop into pollen grains.

starch
A class of large, polymeric carbohydrate molecules that contain many glucose monomers and function as food-storage substances in plants.

stele
The vascular cylinder at the central core of a root or stem. Includes all tissues inside the endodermis.

stimulus
A change in the environment that is detected by an irritable cell, such as a sensory receptor, or a group of irritable cells.

stom-, -stome, stomo- [Gr.]
Mouth.

stoma, pl. **stomata**
A small opening in the epidermis of a leaf or other plant part that is bounded by guard cells that regulate its opening and closing.

stroma
[Gr., anything spread out, a mattress, bed]
The nonmembranous matrix or ground substance within chloroplasts that is the site of several important reactions.

structural gene
A gene that contains coded information for synthesis of a polypeptide chain with a particular amino acid sequence.

sub- [La.]
Under, below.

substrate
(1) Reactant in an enzyme-catalyzed reaction. (2) The base surface on which an organism lives.

succession
In ecology, a series of progressive changes in the plants and animals making up a community in a particular area.

sucrose
A common disaccharide (double sugar) consisting of a glucose molecule and a fructose molecule linked together. Sucrose is the major transport sugar in plants.

super-, supra- [La.]
Above, over.

sym-, syn- [Gr.]
With, together.

symbiosis [Gr. *bios:* life]
A long-term association between two organisms of different species living together in an intimate relationship.

sympathetic nervous system
One of the two divisions of the autonomic (visceral motor) nervous system. Acts antagonistically to the parasympathetic division in control of various effectors.

synapse
[Gr. *synapsis:* falling together, union]
A specialized junction between nerve cells at which communication of excitation occurs.

synapsis [Gr.]
The pairing of homologous chromosomes during prophase I of meiosis, a process that has no counterpart in mitosis.

syncytial
A multinucleate structure; several to many nuclei scattered in a mass of cytoplasm inside a single plasma membrane.

synergistic [Gr. *ergon:* work]
A factor, substance, or process acting together with and enhancing the effect of another.

syngamy
[Gr. *gamos:* union, marriage]
The union of gametes in sexual reproduction. Fertilization.

synthesis
(1) The formation of a more complex substance by combining simpler ones.
(2) Combining information and ideas to produce a meaningful and unified concept or set of concepts.

systemic circulation
The parts of the circulatory system supplying the remainder of the body apart from the circulation supplying the gas-exchange surfaces (the pulmonary circulation).

tadpole
(1) A tailed amphibian larva. (2) Sometimes applied to tailed larvae of some members of the subphylum Urochordata.

taiga [Russ.]
A worldwide belt of northern coniferous forest.

target cell
A cell that has appropriate receptors for a hormone and responds physiologically to the hormone.

taxis [Gr., arrangement, order]
A movement in which an organism assumes a specific orientation to a stimulus source, such as a phototaxis (a movement orientated to a light source).

taxonomy [Gr. *taxis:* arrangement]
The science concerned with naming organisms and arranging them in a system of classification based on their similarities and probable evolutionary relationships.

telo- [Gr.]
An end, complete, final.

tendon [La.]
A piece of tough connective tissue that attaches muscle to bone.

tentacle
[La. *tentare:* to handle, touch]
A slender, elongate, flexible process, extending from an animal body, that serves tactile and grasping functions. Tentacles usually occur around or near the mouth.

terrestrial [La.]
On land, land-dwelling, pertaining to the dry land environment.

testis, pl. **testes**
The sperm-producing organ. Also the source of male sex hormone in vertebrate animals.

tetra- [Gr. *tetrart:* the fourth]
Four, as in tetraploid (possessing four sets of chromosomes).

thalamus
[Gr. *thalamos:* chamber, inner room]
A major relay center derived from part of the embryonic forebrain that functions as an intermediary between the cerebrum and other portions of the nervous system.

thallus
[Gr. *thallos:* young shoot, twig]
A flattened, sheetlike plant body with relatively little tissue specialization and no differentiation of roots, stems, or leaves.

theory
A well-established scientific concept or set of concepts supported by a large body of data. *Not* a vague speculation or unsupported guess.

thorax [Gr., breastplate]
(1) In insects, the body region between the head and abdomen to which walking legs and wings are attached. (2) In vertebrates, the anterior portion of the trunk that contains lungs and heart.

thymus
A lymphoid organ that is involved in development and differentiation of immunologic capabilities in vertebrates.

thyroid [Gr. *thyreo:* shield]
An endocrine gland located at the base of the neck in vertebrate animals that produces several hormones, including thyroxin, an iodine-containing hormone involved in metabolic regulation.

tissue [La. *texere:* to weave]
A group of similar specialized cells, along with intercellular material that binds them together, that are organized as a structural and functional unit.

toxin [La.]
A substance, produced by an organism, that is very poisonous to another organism and will cause damage or death.

trachea, pl. **tracheae**
A tubular passage that conducts air inside the body, as the vertebrate windpipe or the tracheal tubes of insects.

tracheid
An elongate, spindle-shaped conducting cell in the xylem of vascular plants. Tracheids overlap at their tapered ends, which have pitted walls that allow fluid to pass from one cell to another.

trans- [La.]
Across, through, beyond.

transcription
[La. *scribere:* to write]
The process that results in the production of a strand of messenger RNA that is complementary to a segment of DNA.

transfer RNA (tRNA)
One of a group of small RNA molecules, each of which binds a specific amino acid and has a segment that is complementary to a messenger RNA codon. Thus, tRNA functions during polypeptide synthesis (translation) to position appropriate amino acids in the specified sequence.

transformation
(1) A genetic change in cells of a bacterial strain that results when the cells receive genetic information in the form of DNA fragments from another strain, including fragments released from dead cells. (2) The process by which normal body cells are converted into neoplastic (abnormally growing) cells.

translation
The complex interaction of mRNA, ribosomes, and tRNA that results in synthesis of a polypeptide having an amino acid sequence specified by the sequence of codons in mRNA.

translocation
[La. *locare:* to put or place]
(1) The transport of solutes from one part of the plant to another through the phloem. (2) In genetics, the breaking of a chromosome segment that becomes attached to a nonhomologous chromosome.

transpiration
[La. *spirare:* to breathe]
Evaporation of water from plants to the surrounding atmosphere, mainly through stomata.

transverse
[La. *transversere:* to cross]
On a plane that crosses the body from side to side and separates anterior and posterior parts, as a transverse (cross) section.

tri- [La.]
Three, as in triploid (possessing three sets of chromosomes).

trilobite
One of a group of extinct marine arthropods that had two longitudinal folds that appeared to divide the body into three lobes. Trilobites were very abundant during the Paleozoic era.

tropism [Gr. *tropos:* turn]
A turning response by a nonmotile organism to an external stimulus, primarily turning of a plant by a differential growth response.

trypsin
A general proteolytic enzyme that catalyzes hydrolysis of peptide bonds in many different kinds of proteins.

tumor [La. *tumere:* to swell]
An abnormal growth, especially one that produces a definite mass or nodule of tissue.

tundra [Russ.]
A treeless area characterized by short growing seasons, cold temperatures, and little rainfall, but, in the Arctic, wet conditions because of low evaporation, poor drainage, and the presence of permafrost.

turgor pressure
[La. *turgere:* to swell]
The pressure exerted by plant cells against their cell walls because the fluid in their environment usually is hypotonic.

ungulates [La. *ungula:* hoof]
Four-legged mammals with variously fused digits that are protected at their ends by a horny covering, a hoof.

-ura, uro- [Gr.]
Tail; pertaining to the posterior tip of the body.

urea
A water-soluble nitrogenous waste product of mammals and some other animals that functions in ammonia excretion, which is necessary as a result of amino acid deamination.

ureter
The duct that carries urine from the kidney to the urinary bladder (or cloaca) in vertebrate animals.

urethra
The duct that carries urine from the urinary bladder to the exterior in mammals.

uric acid
An insoluble nitrogenous waste product produced by reptiles, birds, and terrestrial arthropods.

urine
A fluid that contains waste substances produced and released by an animal.

uterus [La., womb]
A portion of a female reproductive tract enlarged for egg storage and, in some animals, embryo development. In mammals, the muscular chamber within which embryonic and fetal development take place.

vacuole [La. *vacuus:* empty]
A membrane-bounded, fluid-filled cellular organelle.

valve
(1) A cuplike flap or set of flaps that prevents backward flow of blood in a heart or blood vessel. (2) A hard shell or covering.

vas-, vasa-, vaso-
[La., vessel, duct]
(1) Blood vessel or pertaining to blood vessels. (2) Duct.

vascular tissue
[La. *vasculum:* small vessel]
Tissue made up of tubular vessels that function in fluid transport.

vegetative
(1) Plant parts not specialized for sexual reproduction. (2) Asexual, when applied to reproduction.

vein [La. *vena:* vein]
A vessel that carries blood away from body tissues toward the heart.

venous
Pertaining to veins.

ventilation
The process of renewing the gas-carrying medium over the surfaces where gas exchange occurs.

ventral [La. *venter:* stomach, belly]
Pertaining to the underside or belly, or located at or near the underside. Opposite of dorsal.

ventricle
[La. *venter:* stomach, belly]
(1) A cavity in an organ, such as one of the four ventricles of the vertebrate brain. (2) The large, muscular pumping chamber that pumps blood out from the heart.

venule
A small vessel that collects blood from a capillary bed and carries it toward a vein.

vertebrate
[La. *vertebra:* joint, vertebra]
A member of the subphylum Vertebrata of the phylum Chordata. Includes animals possessing a vertebral column, a series of skeletal elements that enclose and protect the posterior portion of the nervous system. Fish, amphibians, reptiles, birds, and mammals are vertebrates.

vesicle [La. *vesicula:* a small bladder]
A spherical fluid-filled sac.

vessel element
An individual vessel cell in the xylem of a vascular plant. During differentiation, a vessel element loses its nucleus and cytoplasm. End walls become perforated or disappear, leaving pipelike end-to-end chains of empty cell walls.

villus, pl. **villi** [La., shaggy hair]
A minute fingerlike projection that contains blood vessels and increases surface available for absorption or exchange of materials.

virion
A free virus particle. The reproductively inactive but infective form in which a virus exists outside a host cell.

viroid
An infectious agent consisting of only a short strand of nucleic acid, much smaller than the smallest virus.

virus [La., slimy liquid, poison]
An infectious submicroscopic particle composed of a nucleic acid core and a protein coat. Viruses reproduce only inside living cells.

viscera [La.]
The internal body organs of an animal.

vitamin [La. *vita:* life]
One of a set of relatively simple organic molecules that are required in small quantities for various biological processes and must be in an organism's diet because they cannot be synthesized in the organism's body.

viviparous
[La. *vivus:* alive; *parere:* to bring forth, produce]
Bearing young born at a relatively advanced stage after an extended period of development inside a sheltered environment in the female body (the uterus in mammals), where nutrient and other metabolic requirements are met through exchange of materials with the mother, as in the mammalian placenta.

woody
Plant tissue with lignin-containing (lignified) cell walls. The xylem of trees and shrubs is woody tissue.

X chromosome
One of the sex chromosomes. A pair of X chromosomes is present in the cells of one sex (*e.g.,* the female in *Drosophila* and in mammals, the male in birds). One X chromosome and one Y chromosome commonly are present in the other sex.

xylem [Gr. *xylon:* wood]
A vascular tissue that transports water and mineral solutes upward through the plant body.

Y chromosome
One of the sex chromosomes. One Y chromosome and one X chromosome are present in the cells of one sex (*e.g.,* the male in *Drosophila* and in mammals, the female in birds). Two X chromosomes are present in the other sex. Y chromosomes characteristically bear very few gene loci.

yolk
Stored nutrient material in an egg.

zoo- [Gr.]
(1) Animal, pertaining to animals.
(2) Animallike, motile.

zooplankton
A collective name for the small animals present in the plankton.

zoospore [Gr. *sporos:* seed]
A flagellated, motile plant spore.

zygote
[Gr. *zygotos:* yoked together]
The diploid (2N) cell formed by the union of two gametes, the product of fertilization.

Credits

Illustrations

Chapter 1

Figures 1.10, 1.11, and 1.12 From Johnson, Leland G., *Biology.* © 1983 Wm. C. Brown Publishers, Dubuque, Iowa. All Rights Reserved. Reprinted by permission.

Chapter 2

Figures 2.1, 2.2, 2.3, 2.4, 2.5, 2.6, 2.7, 2.8, 2.9, 2.11, 2.13, 2.14, 2.15, 2.16, 2.17, 2.19*a,* 2.20, 2.22, 2.24, 2.25*a,* 2.27, 2.28, 2.29, 2.30, 2.32*b,* Box figure 2.1*a,* Table 2.1, and Table 2.2 From Johnson, Leland G., *Biology.* © 1983 Wm. C. Brown Publishers, Dubuque, Iowa. All Rights Reserved. Reprinted by permission. Figure 2.12, 2.18 From Lehninger, A. L., *Principles of Biochemistry,* 2d ed., Worth Publishers, New York, 1982. Reprinted by permission. Figure 2.23 Reprinted with permission of Longman Group Limited from *Introduction to Molecular Biology* by G. H. Haggis et al., 1964. Figure 2.25*b* From Dickerson, R. E., and I. Geis, *The Structure and Action of Proteins.* © 1969 Benjamin/Cummings Publishing Co., Menlo Park, California. Reprinted by permission. Figure 2.32*a* from Linus Pauling and Robert B. Corey, "Specific Hydrogen-Bond Formation between Pyrimidines and Purines in Deoxyribonucleic Acids," in *Archives of Biochemistry and Biophysics, 65,* 164–81. (1956): Academic Press, Orlando, Florida. Reprinted by permission.

Chapter 3

Figure 3.2 From Dobell, C. E., *Antony Van Leeuwenhoek and his "Little Animals."* © 1932 Russell & Russell, Div. Atheneum Publications, New York. Figure 3.3 Reprinted with permission of Macmillan Publishing Company from *The Cell in Development and Heredity,* Third ed., by E. B. Wilson. Copyright 1925 by Macmillan Publishing Company, renewed 1953 by Anna N. K. Wilson. Figures 3.4, 3.5, 3.9, 3.12, 3.15, 3.17, 3.18 (bottom), 3.19 (bottom), 3.20, 3.22*b,* 3.25 (left), 3.26, 3.27, 3.28, 3.29, 3.30, 3.32, 3.33, 3.34, Box figure 3.1, Table 3.1, and Table 3.2 From Johnson, Leland G., *Biology.* © 1983 Wm. C. Brown Publishers, Dubuque, Iowa. All Rights Reserved. Reprinted by permission. Figure 3.6 From Thomas D. Brock, *Biology of Microorganisms,* 3d ed., © 1979, pp. 774, 775. Reprinted by permission of Prentice-Hall, Inc., Englewood Cliffs, NJ. Figure 3.7 From Singer, S. J., and G. L. Nicholson, "The Fluid Mosaic Model of the Structure of Cell Membranes." *Science* 175, pp. 720–731, 18 February 1972. Copyright 1972 by the American Association for the Advancement of Science. Figures 3.13*a* (bottom), 3.13*b* (bottom), and 3.14*b,* From Mader, Sylvia S., *Inquiry Into Life,* 3d ed., © 1976, 1979, 1982 Wm. C. Brown Publishers, Dubuque, Iowa. All Rights Reserved. Reprinted by permission.

Chapter 4

Figures 4.3, 4.4, 4.5, 4.6, 4.7, 4.8, 4.9, 4.11, 4.12, 4.13, 4.14, 4.15, 4.17, and 4.18 From Johnson, Leland G., *Biology.* © 1983 Wm. C. Brown Publishers, Dubuque, Iowa. All Rights Reserved. Reprinted by permission.

Chapter 5

Figures 5.2, 5.4*a,* 5.5 *a&c,* 5.6*a,* 5.7, 5.8, 5.9, 5.10, 5.11, 5.13, 5.14, 5.16, 5.17, 5.19 (bottom left and right), and 5.20 From Johnson, Leland G., *Biology.* © 1983 Wm. C. Brown Publishers, Dubuque, Iowa. All Rights Reserved. Reprinted by permission.

Chapter 6

Figures 6.2, 6.3, 6.4, 6.5, 6.6, 6.7, 6.8, 6.9, 6.10, 6.11, 6.12, 6.13, 6.14, 6.15, 6.16, 6.17, 6.18, 6.19, 6.20, 6.21, 6.22, 6.23, 6.24, 6.25, 6.26, Box figure 6.1, Table 6.1, and Table 6.2 From Johnson, Leland G., *Biology.* © 1983 Wm. C. Brown Publishers, Dubuque, Iowa. All Rights Reserved. Reprinted by permission.

Chapter 7

Figures 7.1, 7.2, 7.3, 7.4*a,* 7.5, 7.7, 7.8, 7.9, 7.10, 7.11, 7.14, 7.16, 7.18 (bottom), 7.19 (right), 7.20, 7.21, 7.22, 7.25, 7.26, and Box figure 7.1 From Johnson, Leland G., *Biology.* © 1983 Wm. C. Brown Publishers, Dubuque, Iowa. All Rights Reserved. Reprinted by permission. Figure 7.13 From *The Physiology of Plants,* edited and translated by A. J. Ewart, 1904. Oxford University Press, New York.

Chapter 8

Figures 8.1, 8.2, 8.3, 8.6, 8.7, 8.8, 8.9, 8.10, 8.12, 8.14, 8.17*a,* 8.18, 8.19, 8.20, 8.21*a,* 8.22, Table 8.1, Table 8.2, and Table 8.3 From Johnson, Leland G., *Biology.* © 1983 Wm. C. Brown Publishers, Dubuque, Iowa. All Rights Reserved. Reprinted by permission. Figures 8.15*a,* 8.16 From Mader, Sylvia S., *Inquiry Into Life,* 3d ed., © 1976, 1979, 1982 Wm. C. Brown Publishers, Dubuque, Iowa. All Rights Reserved. Reprinted by permission.

Chapter 9

Figures 9.1, 9.2, 9.3, 9.4, 9.5, 9.6, 9.7, 9.8*c*, 9.9 *a&b*, 9.10, 9.11, 9.12, 9.14, 9.15, 9.17, 9.18, 9.19, 9.21, 9.22, and 9.23 From Johnson, Leland G., *Biology*. © 1983 Wm. C. Brown Publishers, Dubuque, Iowa. All Rights Reserved. Reprinted by permission. Figure 9.8*a* Reprinted with permission of the American Society of Zoologists from "Fish Physiology" by D. J. Randall, in *American Zoologist, 8* (1968). Figure 9.16*a* Reproduced from *Biological Science* by William T. Keeton, Second Edition, Illustrated by Paula Di Santo Bensadoun, by permission of W. W. Norton & Company, Inc. Copyright © 1972 by W. W. Norton & Company, Inc. Box figure 9.2*b* From Juhl, John, *Essentials of Roentgen Interpretation*, 4th ed. © 1981 Lippincott/Harper and Row, Philadelphia, PA. Reprinted by permission.

Chapter 10

Figures 10.2, 10.3, 10.6, 10.7*b*, 10.9, 10.12, 10.14, 10.15, 10.20, 10.21, 10.22, 10.23, and Table 10.1 From Johnson, Leland G., *Biology*. © 1983 Wm. C. Brown Publishers, Dubuque, Iowa. All Rights Reserved. Reprinted by permission. Figures 10.5, 10.13, 10.16, 10.17, 10.18, and 10.19*a* From Mader, Sylvia S., *Inquiry Into Life*, 3d ed. © 1976, 1979, 1982 Wm. C. Brown Publishers, Dubuque, Iowa. All Rights Reserved. Reprinted by permission.

Chapter 11

Figures 11.2, 11.5, 11.6, 11.7, 11.8, 11.9, 11.10, 11.11*a*, 11.12, 11.15 *b&c*, 11.16, 11.17, 11.19, 11.20, 11.21, 11.22, Table 11.1, and Table 11.2 From Johnson, Leland G., *Biology*. © 1983 Wm. C. Brown Publishers, Dubuque, Iowa. All Rights Reserved. Reprinted by permission. Figure 11.3 After Wilson, et al., *Life on Earth*, Second Edition, 1978. Figure 11.4 Adapted from *General Endocrinology*, Sixth Edition, by C. Donnell Turner and Joseph T. Bagnara. Copyright © 1976 by W. B. Saunders Company. Reprinted by permission of CBS College Publishing. Figure 11.18 From R. E. Chance: *Recent Progress in Hormone Research 25* (1969): 274. Reprinted by permission of Academic Press, Inc., Orlando, Florida.

Chapter 12

Figures 12.2, 12.3, 12.4, 12.5, 12.9, 12.10, 12.11, 12.12, 12.14, 12.15, 12.16, 12.19, 12.20, 12.21, and Table 12.2 From Johnson, Leland G., *Biology*. © 1983 Wm. C. Brown Publishers, Dubuque, Iowa. All Rights Reserved. Reprinted by permission. Figures 12.17, 12.23 From Mader, Sylvia S., *Inquiry Into Life*, 3d ed. © 1976, 1979, 1982 Wm. C. Brown Publishers, Dubuque, Iowa. All Rights Reserved. Reprinted by permission. Figure 12.18 Courtesy Ward's Natural Science, Inc., Rochester, NY. Table 12.1 From Vander, A. J. et al., *Human Physiology* 3d ed. © 1980 McGraw-Hill Book Company, New York. Reprinted by permission.

Chapter 13

Figures 13.1, 13.2, 13.3, 13.4, 13.5, 13.6, 13.7, 13.8, 13.9, 13.10, 13.11, 13.12, 13.14, 13.16, 13.17, 13.18, and Table 13.1 From Johnson, Leland G., *Biology*. © 1983 Wm. C. Brown Publishers, Dubuque, Iowa. All Rights Reserved. Reprinted by permission.

Chapter 14

Figures 14.3, 14.4, 14.5, 14.7, 14.9, 14.10, 14.12, 14.15, 14.16, 14.17, 14.20, 14.21, 14.23, 14.25, 14.26, and Box figure 14.1*a* From Johnson, Leland G., *Biology*. © 1983 Wm. C. Brown Publishers, Dubuque, Iowa. All Rights Reserved. Reprinted by permission. Figure 14.2 Reprinted with permission from Wareing and Phillips, *The Control of Differentiation in Plants*. © 1970 Pergamon Press. Figure 14.24 From Palmer, John D., *An Introduction to Biological Rhythms*. Reprinted with permission from Academic Press and the author.

Chapter 15

Figures 15.1, 15.2, 15.6, 15.12, 15.15, 15.19, Box figure 15.1*a*, and Box figure 15.2*a* From Johnson, Leland G., *Biology*. © 1983 Wm. C. Brown Publishers, Dubuque, Iowa. All Rights Reserved. Reprinted by permission.

Chapter 16

Figures 16.3, 16.4, 16.5, 16.6, 16.7, 16.8, 16.9, 16.10, 16.11, 16.12, 16.13, 16.14, 16.15, 16.17, Box figure 16.1, Table 16.1, and Table 16.2 From Johnson, Leland G., *Biology*. © 1983 Wm. C. Brown Publishers, Dubuque, Iowa. All Rights Reserved. Reprinted by permission.

Chapter 17

Figures 17.2*b*, 17.3, 17.8, 17.9, 17.10, 17.11, 17.12, 17.14, 17.16, 17.17, 17.18, 17.19, 17.20, and Table 17.1 From Johnson, Leland G., *Biology*. © 1983 Wm. C. Brown Publishers, Dubuque, Iowa. All Rights Reserved. Reprinted by permission. Figure 17.4 From C. J. Avers, *Cell Biology*, 2d ed. Copyright © 1981 by PWS Publishers. Used by permission of Willard Grant Press. All Rights Reserved. Figure 17.21 From Gardner/Snustad, *Principles of Genetics*, 6th ed. © 1981 John Wiley & Sons, Inc., New York. Reprinted by permission.

Chapter 18

Figure 18.2 From R. Sage and F. J. Ryan, *Cell Heredity*. © 1981 John Wiley & Sons, Inc., New York. Reprinted by permission. Figure 18.3 From Kirk, David, *Biology Today*. © 1979 CRM Books, Division of Random House, New York. Reprinted by permission. Figures 18.4, 18.5, 18.6, 18.7, 18.8, 18.9, 18.10, 18.11, 18.14, 18.15, 18.16, 18.18, 18.19, Box figure 18.2, and Table 18.1 From Johnson, Leland G., *Biology*. © 1983 Wm. C. Brown Publishers, Dubuque, Iowa. All Rights Reserved. Reprinted by permission. Figures 18.12, 18.13 From Mader, Sylvia S., *Inquiry Into Life*, 3d ed. © 1976, 1979, 1982 Wm. C. Brown Publishers, Dubuque, Iowa. All Rights Reserved. Reprinted by permission.

Chapter 19

Figures 19.2, 19.4, 19.5, 19.6, 19.12, 19.16*b*, 19.17*c*, 19.18, 19.19, 19.20, 19.21, 19.22, 19.26*a*, and Box figure 19.1 From Johnson, Leland G., *Biology*. © 1983 Wm. C. Brown Publishers, Dubuque, Iowa. All Rights Reserved. Reprinted by permission. Figures 19.15 *a&b*, 19.16 *a&c* From Johnson, Leland G., and E. Peter Volpe, *Patterns and Experiments in Developmental Biology*. © 1973 Wm. C. Brown Publishers, Dubuque, Iowa. All Rights Reserved. Reprinted by permission. Figure 19.16*d* Adapted from *An Introduction to Embryology*, Fifth Edition, by B. I. Balinsky. Copyright © 1981 by CBS College Publishing. Reprinted by permission of CBS College Publishing. Figure 19.23 From Volpe, E. Peter, *Biology and Human Concerns*, 3d ed. © 1975, 1979, 1983 Wm. C. Brown Publishers, Dubuque, Iowa. All Rights Reserved. Reprinted by permission. Figure 19.24 From Steward, F. C., "Growth and Development of Cultured Plant Cells." *Science* 143, pp. 20–27, 3 January 1964. Copyright 1964 by the American Association for the Advancement of Science.

Chapter 20

Figures 20.2*a*, 20.3, 20.4, 20.5, 20.6, 20.7, 20.8, 20.9, 20.10, 20.11, 20.14, 20.16, 20.17, 20.18 *a,b&c*, 20.19, 20.24, Table 20.1, and Table 20.2 From Johnson, Leland G., *Biology*. © 1983 Wm. C. Brown Publishers, Dubuque, Iowa. All Rights Reserved. Reprinted by permission. Figure 20.2 *b&c* From Arey, L. B., *Developmental Anatomy*, 7th ed. Copyright 1974 W. B. Saunders Company, Philadelphia, PA. Reprinted by permission. Figure 20.18*d* From *Human Design* by William S. Beck. Copyright © 1971 by Harcourt Brace Jovanovich, Inc. Reproduced by permission of the publisher.

Chapter 21

Figures 21.1, 21.2, 21.3, 21.4, 21.5, 21.7, 21.8, 21.9, 21.10, 21.11, 21.12, 21.13, 21.15, 21.16, 21.19, 21.20, and Table 21.1 From Johnson, Leland G., *Biology*. © 1983 Wm. C. Brown Publishers, Dubuque, Iowa. All Rights Reserved. Reprinted by permission. Figure 21.14 From Mader, Sylvia S., *Inquiry Into Life*, 3d ed. © 1976, 1979, 1982 Wm. C. Brown Publishers, Dubuque, Iowa. All Rights Reserved. Reprinted by permission. Figure 21.23 From Burkitt, D., in *Cancer 16*, pp. 379–386, 1962. © 1962 J. B. Lippincott Co., Philadelphia, PA. Reprinted by permission. Figure 21.24 Reprinted by permission of the American Cancer Society, Inc. Figure 21.25 Reproduced, with permission, from Lamb, M. M., *Biology of Ageing*, Blackie, Glasgow and London. Figure 21.26 From Robert R. Kohn, *Principles of Mammalian Aging*, 2d ed. © 1978, p. 31. Reprinted by permission of Prentice-Hall, Inc., Englewood Cliffs, NJ. Table 21.2 From Volpe, E. Peter, *Biology and Human Concerns*, 3d ed. © 1975, 1979, 1983 Wm. C. Brown Publishers, Dubuque, Iowa. All Rights Reserved. Reprinted by permission.

Chapter 22

Figures 22.2, 22.4 *a&b*, 22.7, 22.12, Box figure 22.1, and Table 22.1 From Johnson, Leland G., *Biology.* © 1983 Wm. C. Brown Publishers, Dubuque, Iowa. All Rights Reserved. Reprinted by permission. Figure 22.5 From Lack, David, *Darwin's Finches.* © 1947 Cambridge University Press, New York. Reprinted by permission. Figure 22.6 Taken from Levi, W. M., *The Pigeon,* 1957, by permission of Levi Publishing Co., Sumter, South Carolina. Figure 22.9, 22.11 From Volpe, E. Peter, *Understanding Evolution,* 4th ed. © 1967, 1970, 1977, 1981 Wm. C. Brown Publishers, Dubuque, Iowa. All Rights Reserved. Reprinted by permission. Figure 22.10*a* Adapted from Villee, Claude and Vincent Dethier, *Biological Principles and Processes,* Second Edition. Copyright © 1976 by W. B. Saunders Company. Reprinted by permission of CBS College Publishing.

Chapter 23

Figures 23.1, 23.2 From Volpe, E. Peter, *Understanding Evolution,* 4th ed. © 1967, 1970, 1977, 1981 Wm. C. Brown Publishers, Dubuque, Iowa. All Rights Reserved. Reprinted by permission. Figures 23.4, 23.6, 23.10, 23.13, 23.17, 23.18, 23.19, 23.20, 23.23, 23.25, Box figure 23.1, Table 23.1, Table 23.2, Table 23.3, Table 23.4, Table 23.5, and Table 23.6 From Johnson, Leland G., *Biology.* © 1983 Wm. C. Brown Publishers, Dubuque, Iowa. All Rights Reserved. Reprinted by permission.

Chapter 24

Figures 24.3, 24.5, 24.8, 24.10, 24.11, 24.12, 24.14, 24.15, 24.16, and 24.17 From Johnson, Leland G., *Biology.* © 1983 Wm. C. Brown Publishers, Dubuque, Iowa. All Rights Reserved. Reprinted by permission. Figure 24.4 From Jennings, H. S., *Behavior of the Lower Organisms,* 1906. Columbia University Press, Macmillan Co., New York.

Chapter 25

Figures 25.1, 25.2, 25.6, 25.7, and 25.8 From Johnson, Leland G., *Biology.* © 1983 Wm. C. Brown Publishers, Dubuque, Iowa. All Rights Reserved. Reprinted by permission. Figure 25.3 From Kurt Schmidt-Nielson, *Animal Physiology,* 3d ed., © 1970, p. 51. Adapted by permission of Prentice-Hall, Inc., Englewood Cliffs, NJ. Figure 25.9 Courtesy Wisconson Department of Natural Resources. Figure 25.10 From Scheffer, V. B., "Rise and Fall of a Reindeer Herd." *Scientific Monthly,* 73, pp. 356–362, December 1951. Copyright © 1951 by the American Association for the Advancement of Science. Figure 25.11 After Fig. 4 (p. 52)/Fig. 2 (p. 51)/Fig. 3 (p. 52)/ Fig. 6 (p. 55) in *The Ecology of Man: An Ecosystem Approach,* 2d edition, by Robert Leo Smith. Copyright © 1972, 1976 by Robert Leo Smith. By permission of Harper & Row, Publishers, Inc. Figure 25.12 From Gause, G. F., *The Struggle for Existence.* © 1934 The Williams and Wilkins Company, Baltimore, MD. Reprinted by permission. Figure 25.13 From Wieland, N. K. and F. A. Bazzaz, "Physiological Ecology of Three Codominant Successional Annuals." *Ecology 56,* 681–688. © 1975 Ecological Society of America. Reprinted with permission. Figure 25.14 Reprinted from "Population Ecology of Some Warblers in Northeastern Coniferous Forests," by R. H. MacArthur. *Ecology 39,* 599–619, 1958.

Chapter 26

Figures 26.5, 26.6, 26.7, and Box figure 26.1*a* From Johnson, Leland G., *Biology.* © 1983 Wm. C. Brown Publishers, Dubuque, Iowa. All Rights Reserved. Reprinted by permission. Figure 26.8 Reprinted with permission of Macmillan Publishing Company from *Communities and Ecosystems,* Second Edition by Robert H. Whittaker. Copyright © 1975 by Robert H. Whittaker. Figure 26.9 from *Plant Physiology,* Second Edition by Frank B. Salisbury and Cleon W. Ross. © 1978 by Wadsworth Publishing Company, Belmont, California 94002.

Chapter 27

Figures 27.1, 27.5, 27.7, 27.8, 27.10 (bottom), 27.18*a*, 27.25 *a&b*, Box figure 27.1*b*, Table 27.1, and Table 27.2 From Johnson, Leland G., *Biology.* © 1983 Wm. C. Brown Publishers, Dubuque, Iowa. All Rights Reserved. Reprinted by permission. Figure 27.2 From R. H. Whittaker, "New Concepts of Kingdoms of Organisms." *Science 163,* 10 January 1969, pp. 150–160. Copyright 1969 by the American Association for the Advancement of Science. Figure 27.3 From Freeman (Ed.), *Burrows Textbook of Microbiology,* 21st ed. © 1979 W. B. Saunders Company, Philadelphia, PA. Reprinted with permission. Figure 27.6 From *Microbiology: Molecules, Microbes, and Man* by Eugene W. Nester, C. Evans Roberts, Brian J. McCarthy, and Nancy N. Pearsall. Copyright © 1975 by Holt, Rinehart and Winston, Inc. Reprinted by permission of CBS College Publishing. Figure 27.13 From Stern, Kingsley R., *Introductory Plant Biology,* 2d ed. © 1979, 1982 Wm. C. Brown Publishers, Dubuque, Iowa. All Rights Reserved. Reprinted by permission. Figure 27.15 From Lane, Theodore R. (coordinating editor): *Life: The Individual, The Species.* St. Louis, 1976, The C. V. Mosby Company; based on the work of L. Margulis. Figure 27.17 *b&c* Reprinted with permission of Macmillan Publishing Company from *Microbiology: An Introduction to Protista* by Jeanne S. Poindexter. Copyright © 1971 by Jeanne S. Poindexter. Figures 27.21, 27.22 Based on K. Vickerman and F. E. G. Cox, *The Protozoa.* Houghton Mifflin Company, 1967. Figure 27.24 Reprinted with permission of Macmillan Publishing Company from *The Invertebrates: Function and Form* by Irwin W. Sherman and Vilia G. Sherman. Copyright © 1970 by Macmillan Publishing Co., Inc. Figure 27.28 Reprinted with permission of Macmillan Publishing Company from *Algae and Fungi* by Constantine J. Alexopoulous and Harold C. Bold. Copyright © 1967 by Macmillan Publishing Co., Inc.

Chapter 28

Figure 28.1 From Stern, Kingsley R., *Introductory Plant Biology,* 2d ed. © 1979, 1982 Wm. C. Brown Publishers, Dubuque, Iowa. All Rights Reserved. Reprinted by permission. Figures 28.4, 28.11, 28.15, 28.17, 28.21, and 28.25 From Stern, Kingsley R., *Introductory Plant Biology.* © 1979 Wm. C. Brown Publishers, Dubuque, Iowa. All Rights Reserved. Reprinted by permission. Figures 28.7, 28.8, 28.9, 28.10, 28.12, 28.13, 28.18, 28.19, 28.20, 28.22, 28.27, and Table 28.1 From Johnson, Leland G., *Biology.* © 1983 Wm. C. Brown Publishers, Dubuque, Iowa. All Rights Reserved. Reprinted by permission. Figure 28.16 Adapted from *Botany,* Fifth Edition by Carl L. Wilson, Walter E. Loomis, and Taylor A. Steeves. Copyright © 1971 by Holt, Rinehart and Winston, Inc. Reprinted by permission of CBS College Publishing.

Chapter 29

Figures 29.1 *b&c*, 29.2*b*, 29.3, 29.4*a*, 29.5, 29.6, 29.11 *b&c*, 29.12, 29.18, 29.20, 29.21, 29.23, 29.29*a*, 29.34*b*, 29.40, 29.41 *a&b*, 29.44, 29.45, 29.47, 29.52, 29.55, 29.61, 29.64*b*, 29.66, Box figure 29.1*b*. Table 29.1, and Table 29.2 From Johnson, Leland G., *Biology.* © 1983 Wm. C. Brown Publishers, Dubuque, Iowa. All Rights Reserved. Reprinted by permission. Figures 29.4*b*, 29.17 Reprinted with permission of Macmillan Publishing Company from *The Invertebrates Function and Form* by Irwin W. Sherman and Vilia G. Sherman. Copyright © 1970 by Macmillan Publishing Co., Inc. Figure 29.7 Reproduced from *Biological Science* by William T. Keeton, Third Edition, Illustrated by Paula Di Santo Bensadoun, by permission of W. W. Norton & Company, Inc. Copyright © 1980, 1979, 1978, 1972, 1967 by W. W. Norton & Company, Inc. Figure 29.15 Courtesy Ward's Natural Science, Inc., Rochester, NY. Figure 29.19 From Mader, Sylvia S., *Inquiry Into Life,* 3d ed. © 1976, 1979, 1982 Wm. C. Brown Publishers, Dubuque, Iowa. All Rights Reserved. Reprinted by permission. Figures 29.27, 29.29*b* Reprinted with permission of Macmillan Publishing Company from *A Life of Invertebrates* by W. D. Russell-Hunter. Copyright © 1979 by W. D. Russell-Hunter. Figure 29.30*b* From Gray, J., H. W. Lissman, and R. J. Pumphrey, in *Journal of Experimental Biology 15,* (1938) 408. Reprinted by permission of Dr. R. J. Skaer, Company of Biologists, Ltd., Cambridge, England. Figure 29.41*c* From Storer, I. I., and R. L. Usinger, *General Zoology.* © 1957 McGraw-Hill Book Company, New York. Reprinted by permission.

Photos

Part Openers

Part One: © Richard J. Feldmann, National Institutes of Health; Part Two: © Juris Janavs; Part Three: © Manfred Kage/Peter Arnold, Inc.; Part Four: © Robert E. Waterman; Part Five: © Tom Pantages; Part Six: © Ted Schiffman/Peter Arnold, Inc.

Chapter 1

European Art: p. 2; Oriental Institute Museum, University of Chicago: 1.1; Historical Pictures Service, Inc.: 1.2; © Leonard Lee Rue/Tom Stack and Associates: 1.3; National Library of Medicine: 1.4, 1.5b, Box figure 1.1A a–c; Billings Microscope Collection, Armed Forces Institute of Pathology: 1.5a; Columbiana Collection/ © Columbia University, Low Memorial Library: 1.6; © Ralph Brinster/NATURE (16 December 1982): 1.7; © Robin Moyer/Black Star: 1.8; © George Wuerthner/Tom Stack and Associates: 1.9; © M. Abbey/Photo Researchers, Inc.: 1.13a; VU/© T. E. Adams: 1.13b; © Bruce Russell/BioMedia Associates: 1.14a, 1.16a; © Eric Gravé: 1.14b; © Rebecca Johnson: 1.15, 1.16f; © Doug Wechsler: 1.16b; © S. J. Krasemann/Photo Researchers, Inc.: 1.16c; © Carolina Biological Supply Company: 1.16d, 1.17b, 1.17f; © L. West/Bruce Coleman, Inc.: 1.16e; © M. P. L. Fogden/Bruce Coleman, Inc.: 1.16g; © D. Woodward/Tom Stack and Associates: 1.17a; © Ed Reschke/Peter Arnold, Inc.: 1.17c; © David Spier/Tom Stack and Associates: 1.17d; © Patrice/ Tom Stack and Associates: 1.17e; © John Lidington/ Photo Researchers, Inc.: 1.17g; © R. Andrew Odum/ Peter Arnold, Inc.: 1.17h; © Gregory G. Dimijian/ Photo Researchers, Inc.: 1.17i; © Ed Pacheco: 1.18

Chapter 2

© George Holton/Photo Researchers, Inc.: 2.9d; © John Bova/Photo Researchers, Inc.: 2.10; Courtesy of The Ealing Corporation: 2.18c; © J. Robert Waaland, University of Washington/Biological Photo Service: 2.19b; © AP/Wide World Photos: 2.25c; © Jerome Gross: 2.26; Courtesy of Department of Biophysics, King's College, London: 2.31b; from J. D. Watson, *The Double Helix.* Atheneum, NY. © 1968 by J. D. Watson/A. C. Barrington Brown, photographer: 2.32c; © David Spier/Tom Stack and Associates: Box figure 2.2 (left); © Bob Francis/Tom Stack and Associates: Box figure 2.2 (middle); © Ian F. Spellerberg: Box figure 2.2 (right)

Chapter 3

© Carnegie Institution of Washington, Department of Embryology, Davis Division: 3.1a; © Peter Karas: 3.1b; © G. F. Leedale/Biophoto Associates: 3.8, 3.14 (left), 3.25 (right); © Carolina Biological Supply: 3.10; © Paul Heidger: 3.11, 3.13a,b (top); © James Burbach: 3.16; © Charles Havel: 3.18 (top); © Herbert Israel, Cornell University: 3.19 (top); © Jean-Paul Revel: 3.21; © Elias Lazarides: 3.22a; © Jan Löfberg, Institute of Zoology, Upsala University, Sweden: 3.24a; © J. André and E. Favret-Fremiet: 3.24b; © Donald F. Lundgren: 3.28 (top); © S. J. Singer: 3.31a–c

Chapter 4

Courtesy of Boeing Company: 4.1; © Bob Coyle: 4.2; © Thomas Eisner: Box figure 4.1

Chapter 5

© Leonard Lee Rue III/Photo Researchers, Inc.: 5.1; © Don and Pat Valenti: 5.3; © John Troughton and Leslie Donaldson: 5.4b, 5.5b; Salisbury, F. and Ross, C. W.: *Plant Physiology,* 2nd edition. © 1978 Wadsworth Publishing Company, Inc. Reprinted by permission: 5.5d; © G. F. Leedale/Biophoto Associates: 5.6b (1–2); © Melvin Calvin: 5.15; © Nancy Vander Sluis: 5.19 (top)

Chapter 6

UPI/Bettmann: 6.1; © Charles Havel: 6.9 (right); © Brian Parker/Tom Stack and Associates: Box figure 6.2

Chapter 7

© John Troughton and Leslie Donaldson: 7.4b, 7.23; © Walter Hodge/Peter Arnold, Inc.: 7.6a,b; Kimball, J. W.: *Biology,* 2nd edition. © 1968 Addison-Wesley, Reading, MA: 7.7 (left); © Carolina Biological Supply: 7.8 (top), 7.17 (top); © G. F. Leedale/Biophoto Associates: 7.12; © E. J. Hewitt: 7.15a–c; © Charles Havel: 7.18 (top); © Al Bussewitz: 7.19 (right); Salisbury, F. and Ross, C. W.: *Plant Physiology,* 2nd edition. © 1978 Wadsworth Publishing Company, Inc. Reprinted by permission: 7.24; © Thomas Eisner: Box figure 7.1c; © Martin H. Zimmerman: Box figure 7.2a,b

Chapter 8

© Lester Bergman and Associates, Inc.: 8.4, 8.5a; © Niilo Hallman: 8.5b; © E. S. Ross: 8.11a; © Tom McHugh/Tom Stack and Associates: 8.11b; © Ira Block/Woodfin Camp: 8.11c; © Wallace Kirkland/Tom Stack and Associates: 8.11d; © Carolina Biological Supply: 8.13; © Omikron/Photo Researchers, Inc.: 8.17 (right); © Avery Gallup: Box figure 8.1A; © David Pramer: Box figure 8.1B (a,b)

Chapter 9

© G. M. Hughes: 9.8b; © Thomas Eisner: 9.9c; VU/© D. Philips: 9.13; Fulton/© British Heart Journal 18 (1956): 9.20; T. Kuwabara, from Bloom, W. and Fawcett, D. W.: *A Textbook of Histology,* 10th edition. © 1975 W. B. Saunders Company: 9.24; © D. W. Fawcett: 9.25a,b (top); © Martin Rotker/Taurus Photos: Box figure 9.1a–c; Juhl, J. H.: *Paul & Juhl's Essentials of Roentgen Interpretation,* 4th edition. © 1981 J. B. Lippincott Company: Box figure 9.2a; © American Heart Association: Box figure 9.2c (1–3)

Chapter 10

© John Crowe: 10.1a,b; © Carolina Biological Supply: 10.4; © Leo J. Le Beau, University of Illinois Hospital/Biological Photo Service: 10.7a; Reproduced with permission from "The Morphology of Human Blood Cells" by L. W. Diggs, D. Sturm and A. Bell. © 1984 Abbott Laboratories: 10.8; © Eila Kairinen and Emil Bernstein, Gillette Research Institute, Rockville, MD: 10.10; © R. B. Wilson: 10.19b; © Tom McHugh/Photo Researchers, Inc.: Box figure 10.1

Chapter 11

Bettmann Archives, Inc.: 11.1; © Carolina Biological Supply: 11.11b, 11.14; © Charles Havel: 11.13; From *General Endocrinology,* 6th edition, by C. D. Turner and J. T. Bagnara. © 1976 W. B. Saunders Company. Reprinted by permission of Holt, Rinehart and Winston: 11.15a; © Michael Borque/VALAN Photos: Box figure 11.1

Chapter 12

© Kim Taylor/Bruce Coleman, Inc.: 12.1; © David Denning/BioMedia Associates: 12.6a; © Randolph Perkins: 12.13a; © Manfred Kage/Peter Arnold: 12.22; © Field Museum of Natural History, Chicago: Box figure 12.1

Chapter 13

Carl May photographer, © Wm. C. Brown Publishers: 13.13a,b; © H. E. Huxley: 13.15

Chapter 14

© Keith Gunnar/Tom Stack and Associates: 14.1; © Frank B. Salisbury: 14.2a; © L. G. Johnson: 14.6a,b; Nitsch, J. P.: "Growth and Morphogenesis of the Strawberry as Related to Auxin," © *American Journal of Botany* 37 (3) March 1950: 14.8a–d; © Folke Skoog: 14.11; © Bernard Phinney/*Proceedings of the National Academy of Sciences,* 1956: 14.13; © Sylvan Wittwer: 14.14; from Kendrick, R. E. and Frankland, B.: *Phytochrome and Plant Growth* © Edward Arnold, Ltd.: 14.18; © Al Bussewitz: 14.19; Salisbury, F. and Ross, C. W.: *Plant Physiology,* 2nd edition. © 1978 Wadsworth Publishing Company, Inc.: 14.22; © E. J. Cable/Tom Stack and Associates: 14.26b

Chapter 15

© Marta Walters and Spencer Brown: 15.3, 15.4; © E. J. Dupraw: 15.5; © Carolina Biological Supply: 15.7, 15.8; © Ed Reschke: 15.9a,b; © Danial Mazia: 15.10; © W. P. Wergin and E. H. Newcomb, University of Wisconsin, Madison/Biological Photo Service: 15.11; © M. M. Rhoades: 15.13, 15.17, 15.18; © Christine Fastnaught: 15.14; © J. G. Gall: 15.16

Chapter 16

© Mary Eleanor Browning/Photo Researchers, Inc.: 16.1a; Courtesy of Eileen Hudak, American Belgian Tervuren Club: 16.1b; American Museum of Natural History: 16.2; Library of Congress Collection: 16.16

Chapter 17

© Jim Shaffer: 17.1; Olins, Donald and Ada: "The Structural Quantum in Chromosomes," *American Scientist* 66 (Nov. 1978): 17.2; Mittwoch, Ursula: *Sex Chromosomes.* © 1967 Academic Press: 17.5; From Wisniewski, L. P. and Hirschhorn, K: "A Guide to Human Chromosome Defects," 2nd edition, White Plains: The March of Dimes Birth Defects Foundation, BD: OAS 16(6), 1980: 17.6; © Digamber S. Borgoankar, Cytogenetics Laboratory, Wilmington Medical Center, Delaware: 17.7; © S. D. Sigamoni, from *American Journal of Human Genetics* 16 (4) 1964, p. 467: 17.13; © Bob Coyle: 17.15; © Murray L. Barr: Box figure 17.2

Chapter 18

Eli Lilly and Company: 18.1; Huntington Potter and David Droessler/LIFE Magazine, © 1980 *Time, Inc.*: 18.17a–c; © P. E. Polani: 18.20

Chapter 19

© Jeff Rotman/Peter Arnold, Inc.: 19.1a; © Robert A. Ross: 19.1b, 19.7; © E. S. Ross: 19.3a; © Bob Coyle: 19.3b; © William Byrd: 19.8a; © Gerald Schatten: 19.8b; © Mia Tegner, Scripps Institution of Oceanography: 19.8c,d; © L. G. Johnson: 19.9a,b, 19.10a–c, 19.11c; © Victor D. Vacquier: 19.11a,b; © Lynwood M. Chase/Photo Researchers, Inc.: 19.13; © Tony Huet: 19.17b; © Paul Roberts, Oregon State University/Biological Photo Service: 19.25; © Bo Lambert: 19.26b

Chapter 20

© Jason Birnholz: 20.1; © Landrum Shettles: 20.12, 20.13a, 20.13d, 20.15d, 20.20a–c; © Carnegie Institution of Washington, Department of Embryology, Davis Division: 20.13b, 20.15a–c, 20.21; Seitz, H. M., et al: "Cleavage of Human Ova in Vitro," *Fertility and Sterility* 22 (1971) p. 255: 20.13c; © Cleveland Health Museum: 20.22a–d; © Peter Karas: 20.23; Planned Parenthood of New York City: 20.25; © Biophoto Associates/Photo Researchers, Inc.: 20.26a; Atlanta Centers for Disease Control: 20.26b

Chapter 21

© R. Dourmashkin: 21.6; Goss, Richard J.: *Principles of Regeneration* © 1969 Academic Press: 21.17; Illingworth, Cynthia: *Journal of Pediatric Surgery* 9:853–858 (1974). Reprinted by permission of Grune & Stratton, Inc. and the author: 21.18a,b; © Beth Johnson: 21.21a,b; © R. W. Briggs: 21.22

Chapter 22

© John Moss/Black Star: 22.1; © George Kleiman/UniPhoto: 22.3a; © Alan Nelson/Tom Stack and Associates: 22.3b; © L. L. Rue III/Tom Stack and Associates: 22.3c; © Rudolf Freund: 22.4c; © E. S. Ross: 22.8; © American Museum of Natural History: 22.10b; © Don and Pat Valenti: 22.12a; © Ray Elliott, Jr.: 22.12b

Chapter 23

AP/Wide World Photos: 23.3; © J. A. Bishop and L. M. Cook: 23.5a,b; © E. S. Ross: 23.7a, 23.16; © Chesher/Photo Researchers, Inc.: 23.7b; © Carolina Biological Supply Co.: 23.8a, 23.24; © Thomas Eisner: 23.8b; © Hans Pfletschinger/Peter Arnold, Inc.: 23.9; Bob Anderson photo from "Animal Oddities," © Hubbard Scientific, Northbrook, IL 60062: 23.11; © W. H. Müller/Peter Arnold, Inc.: 23.12a; © Alan Nelson/Root Resources: 23.12b; © Tom McHugh/Photo Researchers, Inc.: 23.14a; © Doug Faulkner/Sally Faulkner Collection: 23.14b; © Annie Griffiths: 23.14c; © Russ Kinne/Photo Researchers, Inc.: 23.15; © National Geographic Society: 23.22a; David Brill/© National Geographic Society: 23.22b; © American Museum of Natural History: 23.26a,b; © Peabody Museum: 23.27

Chapter 24

© Alan Nelson/Tom Stack and Associates: 24.1; © Susan McCartney/Photo Researchers, Inc.: 24.2; © Sybille Kalas: 24.7; © E. S. Ross: 24.9; © Lee Miller: 24.13a–c; © J. Markam/Bruce Coleman, Inc.: 24.17; © A. Merciera/Photo Researchers, Inc.: 24.18; © Tom Walker/Photo Researchers, Inc.: Box figure 24.1

Chapter 25

© C. A. Morgan/Peter Arnold, Inc.: 25.4; © Lynne Rogers: 25.5; © L. W. Walker/Photo Researchers, Inc.: 25.15a; © L. L. Rue III: 25.15b, 25.16; © Ron West: 25.17; © Carolina Biological Supply Company: 25.18; © Tom Stack/Tom Stack and Associates: 25.19; © E. S. Ross: 25.20; © John Crowe: Box figure 25.1A,B

Chapter 26

© Jeff Rotman/Peter Arnold, Inc.: 26.1; © Tom Stack/Tom Stack and Associates: 26.2; © Ray Elliott, Jr.: 26.3; © R. L. Smith/EWPIC, Inc.: 26.4; © M. T. Cook/Tom Stack and Associates: 26.10; © Dave Spier/Tom Stack and Associates: 26.11; © Michael Quinton: 26.12; © Ron Church/Tom Stack and Associates: 26.13; © R. Hamilton Smith: 26.14; © Doug Faulkner/Sally Faulkner Collection: 26.15; © Jack Dermid: 26.16; © Rod Allin/Tom Stack and Associates: 26.18; © Rick McIntyre/Tom Stack and Associates: 26.19; © Brian Parker/Tom Stack and Associates: 26.20; © Anne Layman/Tom Stack and Associates: 26.21; © T. M. Smith/EWPIC, Inc.: 26.22; © E. S. Ross: 26.23; © G. W. Blankespoor: 26.24; © Stewart M. Green/Tom Stack and Associates: 26.25; © Steve Peterson/Tom Stack and Associates: 26.26; © Jack Donnelly, Woods Hole Oceanographic Institution: Box figure 26.1B

Chapter 27

© G. F. Leedale/Biophoto Associates: 27.4a–c; © Norman Hodgkin/Photo Network: 27.9a–c; © Donald F. Lundgren: 27.10 (top); © T. J. Beveridge, University of Guelph/Biological Photo Service: 27.11; © Norma J. Lang/*Journal of Phycology* 1: 127–134, 1965: 27.12; © J. Robert Waaland, University of Washington/Biological Photo Service: 27.14, 27.29a; © Carolina Biological Supply Company: 27.16; © Biophoto Associates/Photo Researchers, Inc.: 27.17a; © K. W. Jeon/Biological Photo Service: 27.18b; © J. W. Murray: 27.19a,b; © Gregory Antipa: 27.23a,b; © Bruce Russell/BioMedia Associates: 27.24a,b; © G. F. Leedale/Biophoto Associates: 27.27c; © Eric Grave: 27.26; © E. S. Ross: 27.27; © C. R. Wyttenbach, University of Kansas/Biological Photo Service: 27.29b; UPI/Bettmann: Box figure 27.1A(a)

Chapter 28

© 1982 Donald Breneman/Photographic Specialties: 28.2; © Fred and Marian Nickerson: 28.3a–c; © S. J. Krasemann/Peter Arnold, Inc.: 28.5; © Robert A. Ross: 28.6a–c; © Carolina Biological Supply Company: 28.10c, 28.14a,b; © John D. Cunningham: 28.21 (bottom); © L. G. Johnson: 28.23a; © E. S. Ross: 28.23b; USDA Forest Service: 28.24a; © C. Y. Shih: 28.26

Chapter 29

© Nancy Sefton/Photo Researchers, Inc.: 29.1a; © L. G. Johnson: 29.2a; © Charles Havel: 29.6, 29.21a, 29.38a; © Carolina Biological Supply Company: 29.8, 29.46b; © N. G. Daniel/Tom Stack and Associates: 29.9; © Robert Evans/Peter Arnold, Inc.: 29.10a; © Carl Roessler/Tom Stack and Associates: 29.10b, 29.25b, 29.58c; © Bruce Russell/BioMedia Associates: 29.11a; © Ward's Natural Science: 29.14; © Harvey Blankespoor: 29.16; © Biophoto Associates/Photo Researchers, Inc.: 29.17b; From Markell, E. K. and Voge, M.: *Medical Parasitology*, 3rd edition. © 1971 W. B. Saunders Co. Photo by John F. Kessel: 29.22; © Robert A. Ross: 29.24a; © Doug Faulkner/Sally Faulkner Collection: 29.24b, 29.32; © Dave Spier/Tom Stack and Associates: 29.25a; © Bud Higdon: 29.26a; © Michael DiSpezio: 29.26b, 29.28a–c, 29.29c, 29.31a,b, 29.33a, 29.36b, 29.38d, 29.46a, 29.47b, 29.48, 29.49b, 29.50a–c, 29.51a,b, 29.54, 29.58b, 29.59c, 29.62b; © E. S. Ross: 29.30a, 29.65a,c, 29.74a,c, Box figure 29.1A; © Russ Kinne/Photo Researchers, Inc.: 29.33b, 29.56, 29.74b; © L. L. Rue III/Photo Researchers, Inc.: 29.34a; © Tom McHugh/Photo Researchers, Inc.: 29.35a, 29.39, 29.59a, 29.62c, 29.72a,b, 29.73b; © Don and Pat Valenti: 29.35b; © Gary Milburn/Tom Stack and Associates: 29.37; © Zig Leszczynski/Animals, Animals: 29.38b; © Kent Dannen/Photo Researchers, Inc.: 29.38c; © Noel R. Kemp/Photo Researchers, Inc.: 29.38e; © Tom Stack/Tom Stack and Associates: 29.38f; © Manfred Kage/Peter Arnold, Inc.: 29.41d; © Treat Davidson/Photo Researchers, Inc.: 29.42; © Irene Vandermolen/Photo Researchers, Inc.: 29.43; © D. J. Rogers: 29.45b; © L. G. Johnson: 29.46c; © James Cribb: 29.49a; © American Museum of Natural History: 29.53c, 29.60a,b, 29.68; © Douglas P. Wilson: 29.53a,b; Field Museum of Natural History, Chicago: 29.57, 29.67a,b; (right), c; © Soames Summerhays/Photo Researchers, Inc.: 29.58a; © David Doubilet: 29.59b; VU/© R. Wallace: 29.59d; © George Porter/Photo Reseachers, Inc.: 29.62a; © Hans Pfletschinger/Peter Arnold, Inc.: 29.63a–c; © Cosmos Blank/Photo Researchers, Inc.: 29.64a; © John Cancalosi/Tom Stack and Associates: 29.65b; © American Museum of Natural History/Charles R. Knight, artist: 29.67b (left); © Bob Coyle: 29.69a; © Nancy Vander Sluis: 29.69b; © Ray L. Watterson: 29.69c; © L. Walker/Photo Researchers, Inc.: 29.70a; © Helen Williams/Photo Researchers, Inc.: 29.70b; © William R. Curtsinger/Photo Researchers, Inc.: 29.71; © Carl Purcell/Photo Researchers, Inc.: 29.73a; © Gunter Ziesler/Peter Arnold, Inc.: 29.74d

Index

Boldface page numbers indicate pages on which illustrations related to the subject appear. These pages may also contain text on the subject.